Riegel's

Handbook of Industrial Chemistry

Riegel's

Handbook of Industrial Chemistry

SEVENTH EDITION

Edited by

James A. Kent

College of Engineering
Michigan Technological University
Houghton, Michigan

VNR VAN NOSTRAND REINHOLD COMPANY
New York / Cincinnati / Toronto / London / Melbourne

Van Nostrand Reinhold Company Regional Offices:
New York Cincinnati Chicago Millbrae Dallas

Van Nostrand Reinhold Company International Offices:
London Toronto Melbourne

Library of Congress Catalog Card Number: 73-14798
ISBN: 0-442-24347-2

Manufactured in the United States of America

Published by Van Nostrand Reinhold Company
450 West 33rd Street, New York, N.Y. 10001

Published simultaneously in Canada by Van Nostrand Reinhold Ltd.

15 14 13 12 11 10 9 8 7 6 5 4 3 2 1

Library of Congress Cataloging in Publication Data

Riegel, Emil Raymond, 1882–1963
Riegel's handbook of industrial chemistry.

First–5th ed. published under title: Industrial chemistry; 6th ed. in 1962 under title: Riegel's industrial chemistry.
Includes bibliographical references.
1. Chemistry, Technical. I. Kent, James A., 1922– ed. II. Title. III. Title: Handbook of industrial chemistry.
TP145.R54 1974 660 73-14798
ISBN 0-442-24347-2

To my Wife
ANITA

Preface

The aim of this book is to present, in a single volume, an up-to-date account of the many facets of the chemical process industry. The originator of this book, Dr. E. R. Riegel, stated in the first edition which appeared in 1928 that, "Never before has this picture been more fascinating. Change is the order of the day." He cited this statement in a Foreword to the sixth edition (1962), saying that it "is even truer now than it was then. . . ." Indeed, the rapidity of development and change in many sectors of the chemical industry when viewed over the period of a decade is almost breathtaking. Thus, extensive revision was required for this edition and a great many of the chapters have been completely rewritten. Two new chapters, one an in-depth treatment of industrial wastewater technology and the other on air pollution, have been added. What we hope to have achieved is a greatly expanded, technologically up-to-date, and contemporary view of the chemical process industry.

The contributors have done an outstanding job and credit for the quality of the individual chapters goes to them. Errors of omission or duplication, and shortcomings in organization of the material are mine.

Grateful acknowledgment is made to the editors of technical magazines and publishing houses for permission to reproduce illustrations and other material and to the many industrial concerns which contributed drawings and photographs.

Comments and criticisms by readers will be welcome.

JAMES A. KENT,
Houghton, Michigan

Contents

1

Economic Aspects of the Chemical Industry

F. E. Bailey*

Within the formal departments of science at the traditional university, chemistry has grown to have a unique status because of its close correspondence with an industry, the chemical industry, and a branch of engineering–chemical engineering. There is no biology industry, though drugs, pharmaceuticals, and agriculture are clearly closely related applied disciplines. While there is no physics industry, there are power generation and electronics industries. But, matched with chemistry, there is an industry. This unusual correspondence probably came about because in chemistry one makes things–chemicals–and the science and the use of chemicals more or less grew up together during the past century.

Since there is a chemical industry and it serves a major part of any industrialized economy, providing in the end synthetic drugs, fertilizers, clothing, building materials, paints, elastomers, etc., there is also a subject, "chemical economics," and it is this subject, the economic aspects of the chemical industry, which is the concern of this chapter.

*Union Carbide Corp., South Charleston, W. Va.

DEFINITION OF THE CHEMICAL INDUSTRY

Early in this century, the chemical industry was considered to have two parts: the manufacture of inorganic chemicals and of organic chemicals. Today, the Standard Industrial Classification (SIC Index) of the United States Bureau of the Census defines "Chemicals and Allied Products" as comprising three general classes of products: (1) basic chemicals such as acids, alkalis, salts and organic chemicals; chemicals to be used in further manufacture such as synthetic fibers, plastics materials, dry colors and pigments; finished chemical products to be used for ultimate consumption as drugs, cosmetics, and soaps; or to be used as materials or supplies in other industries such as paints, fertilizers, explosives." An even broader description, often considered, is "chemical process industries," major segments of which include: chemical and allied products and petrochemicals; paper and pulp; petroleum refining; rubber and plastics; and stone, clay and glass products.

THE PLACE OF THE CHEMICAL INDUSTRY IN THE ECONOMY

The total value of manufacturers' sales and shipments in the United States in 1970 was about 670 billion dollars,[1] In comparison with this total, the value of sales and shipments of the chemical and allied products industry was about 55 billion dollars. For perspective, these chemical sales were about half that of the total value for food or machinery but about twice that of paper product shipments and sales. In Table 1.1, these chemical sales to selected markets are indicated for 1968. It is of interest to note that "Chemicals and allied products" is the industry's own best customer reflecting the sale of reactive chemical intermediates used in the manufacture of more complex chemical products.

To further gauge the place of chemicals in the economy of the United States, comparisons in growth can be made with the Gross National Product (GNP). The Gross National Product, an index of the size of the economy, is the sum for any one year of a nation's output in terms of expenditures for goods and services by consumers, government, business, and foreign interests; the total of personal consumption, government purchases, gross private domestic investment, and net export of goods and services. For 1970, the gross national product of the United States fell just short of one trillion dollars, about 980 billion dollars. The growth of the economy can be judged by the increase of GNP during the past two decades (Fig. 1.1). In terms of current dollars (reflecting both real growth and inflation) the United States' GNP doubled between 1950 and 1960 ($205 billion to $504 billion) and essentially doubled again by 1970. Many economists forecast a further doubling in the coming decade. These figures indicate average growth rates for the GNP of about 7.2 per cent per year of which about four per cent represents real growth in output of goods and services and the remainder inflation factors.

TABLE 1.1 Chemical Industry Sales to Selected Markets in the United States, 1968 (Billions of Dollars)

Market	*Sales*
Chemicals and allied products	12.56
Textile mill products (including apparel industry)	3.27
Construction	2.98
Rubber and plastics	2.79
Federal, state, and local government	2.70
Farming	2.18
Petroleum industry	1.14
Paper and allied products	0.85
Primary metals	0.82
Food, etc.	0.81
Motor vehicles and equipment	0.40
Other	~18.00

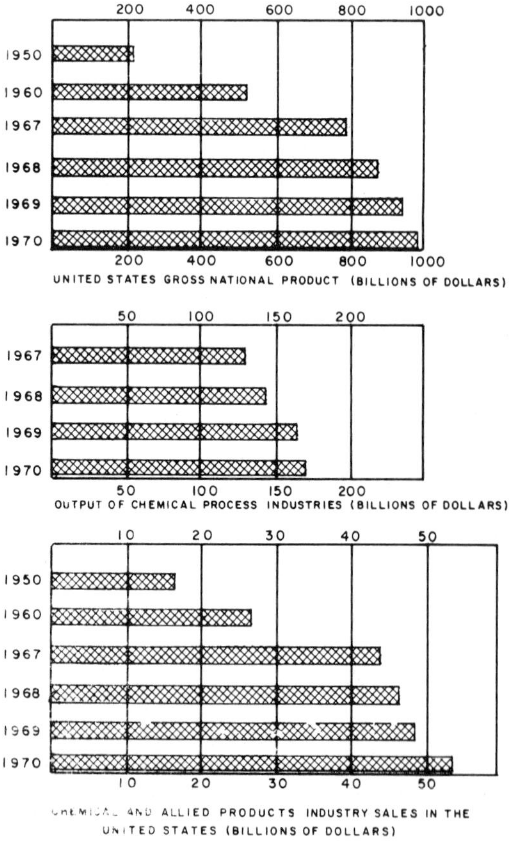

Fig. 1.1 Comparison of United States' GNP, CPI output, and chemical industry sales, 1950-1970 (est.).

In these terms, the chemical and allied products industry has kept pace with the general economy. Sales in 1950 amounted to about 15.4 billion dollars and 26.3 billion in 1960. In 1970, sales amounted to about 55 billion dollars—essentially a doubling of dollar value each decade, an industry annual growth rate of 7-8 per cent per year.

These average growth rates for an entire industry cover a broad spectrum of rates of growth for various product classes. The historic growth of chemicals has been characterized by rapid development of new products which in time achieve the status of high volume, bulk-shipment products having an established place in the economy and correspondingly slower over-all growth. Meanwhile, new materials are introduced which grow at a rapid rate. The net result is an industry growing with the economy, hence, the correspondence of chemical industry growth with the GNP.

If however, the spectrum is examined in more detail, Table 1.2, it is found that older products, originating with the early development of the industry itself, now grow at rates near that of the "real GNP" (sometimes given in terms of constant dollars, dollars deflated in value to a reference year e.g., the GNP in 1968 was about $831 billion in current dollars, but $600 billion in terms of 1958 dollars) of about 3-4 percent per year. In the intermediate range of 7-9 per cent growth rate is found the bulk of the petrochemical industry, comprising products introduced during the middle of this century. The highest rates of growth correspond to products for the most recently developed markets for chemicals and chemical intermediates.

TABLE 1.2 1958-1964 Growth Rates in Value Added in the United States for Selected Parts of the Chemical Process Industries

	Year
Organic, noncellulosic fibers	15.5%
Toilet preparations	12.1
Synthetic elastomers	10.5
Industrial gases	9.9
Basic organic chemicals	9.6
Sufface active agents	8.5
Agricultural chemicals	8.0
All chemical and allied products	7.5
Plastic materials and resins	7.5
Pharmaceutical (human and veterinary)	6.6
Inorganic chemicals	6.2
Chlorine	6.1
Paints	6.0
Soap and detergents	5.9
Printing ink	3.9
Cellulosic fibers	3.6
Wood chemicals	3.1

While growth in total dollar value of the chemicals produced has kept pace with the tremendous growth of the total United States' economy, the actual production of chemicals has grown even faster. In terms of physical volume, the tons produced annually, chemical process industries have grown faster than the industrial average, represented by the Federal Reserve Board index of industrial production (1957-59 = 100) of 173 in 1969, and the chemical and allied products industry has grown even faster, 239 in 1969 (Fig. 1.2). Again, however, these industry indices cover a broad spectrum of product classes, Table 1.3, ranging in production growth from the huge increases in plastics and synthetic fibers to the modest increases found for paints and fertilizers.

Clearly, if the amount in tons produced by the chemical industry has grown faster than average industrial production, while the total dollar sales have grown at a rate equivalent to the economy, it might be expected that on an average the unit price of chemicals has decreased. In Table 1.4, the wholesale price indices in the United States for a number of

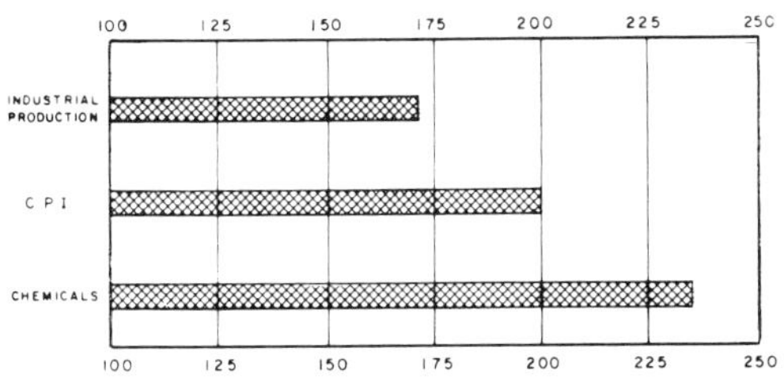

Fig. 1.2 United States Federal Reserve Board production indices, 1969, for industrial production, chemical process industries, and chemicals[1,2] (1957–59 = 100).

TABLE 1.3 Production Indices for Selected Parts of the Chemical Process Industries[4] (1957-59 = 100)

Year	All Chemicals	Basic Inorganics	Basic Organics	Plastics
1960	117	111	125	135
1965	173	185	173	299
1967	204	237	216	345
1968	239	263	240	450

Year	Synthetic Fibers	Synthetic Elastomers
1960	110	121
1965	237	153
1967	278	162
1968	365	180

Year	Soaps and Detergents	Paints	Fertilizers
1960	113	102	110
1965	134	119	134
1967	150	126	140
1968	157	129	139

industries are given for the years 1966 and 1969 relative to 1957-59. The average decline in price for chemical products is evident.

CHARACTERISTICS OF THE CHEMICAL INDUSTRY

A predominant characteristic of the chemical industry and its place in the economy is its tremendous growth in production. Older products, having reached a mature market position, tend to move with the general economy. Newer products, often supplanting other materials (e.g., plastic, synthetic fibers) have gained rapid acceptance and have led the economy in productive growth.

TABLE 1.4 Wholesale Price Indices—United States[3] (1957-59 = 100)

	1966	1969
Lumber and wood products	105.6	134
Textile products and apparel	102.1	107
Metal and metal products	108.3	117
Petroleum products, refined	99.5	102
Chemicals and allied products	97.5	98
Industrial chemicals	95.7	97
Plastics resins and materials	89.0	81
Agricultural chemicals	102.8	92

Investment Trends

The Chemical Industry tends to be a high investment business. Capital spending by the chemical and allied products industry in the United States has been a sizeable percentage of that spent for all manufacturing, some 10 per cent (Table 1.5). That spent for all chemical process industries is, of course, even larger. For perspective, annual expenditures for new plant and equipment in the United States for chemical and allied products industry in recent years has averaged about 25 per cent more than that for iron and steel and about two-thirds of that invested in the petroleum industry.

TABLE 1.5 Capital Spending by Industries in the United States[5] (Billions of Dollars)

	1959	1969
All manufacturing	12.0	31.7
Chemical process industries	3.2	11.0
Chemical and allied products	1.2	3.1

Much of this capital investment in the chemical industry is spent for facilities to produce major chemicals, Table 1.6, in truly enormous quantities. The growth in volume produced has been reflected in the size of plants being built to achieve the re-

TABLE 1.6 Production of Selected Chemicals and Plastics in the United States, 1970[1,3] (Billions of Pounds)

	Billion Pounds
Organic Chemicals	
Ethylene	17.0
Propylene (chemical usage)	8.8
Styrene (monomer)	4.9
Phenol (synthetic)	1.6
Inorganic Chemicals	
Chlorine	9.8
Ammonia	13.4
Sulfuric acid	29.0
Plastics	
Low density polyethylene	4.1
High density polyethylene	1.6
Polypropylene	1.1
Polystyrene (all types)	2.9
Poly(vinyl chloride)	3.4
Phenolics	1.2

quired economies of scale. That such economies are required is seen in the general decline in the wholesale price indices for the chemical industry as production volume and total sales have increased. (Economy of scale refers to the relative cost of building a larger plant. A rule of thumb is that the relative cost of building a smaller or a larger plant is the ratio of the productivities of the two plants being considered, raised to the 0.6 power. In other words, the unit cost of producing a chemical decreases markedly as the size of the plant producing it is increased, providing that the plant can be operated near capacity.)

Today, a typical base petrochemicals plant will consume the equivalent of 30,000 barrels per day of naphtha to produce about one billion pounds of ethylene a year, plus 2.5 billion pounds of coproducts. Vinyl chloride and styrene monomer plants most recently announced have been in the billion pounds-a-year range.

Along with these very large plants and associated enormous investment, most of the chemical industry is characterized by high investment versus low labor components in cost of manufacture. The National Industrial Conference Board statistics list the chemical industry as one of the highest in terms of capital investment per production worker. Once again, however, this index covers a spectrum of operations. The investment per worker in a base petrochemicals olefins plant may well exceed 100,000 dollars, while that for a profitable chemicals specialties manufacturer may be of the order of 10,000 dollars.

Commercial Development and Competition Factors

In an earlier period in the development of the chemical industry, chemical companies were generally production-oriented, exploiting a process to produce a chemical and selling it into rapidly expanding markets. Plant sizes and investments required to participate were small fractions of those of today. Raw materials were often purchased to produce chemical intermediates for sale.

As the industry has grown, there has been a strong tendency toward integration both forward and back. Petroleum producers have found opportunities based on their raw-materials position to move into chemical manufacturing as chemical companies, on the other hand, have moved to assure their access to low-cost raw materials. Similarly, producers of plastic materials have moved forward to produce fabricated products, such as films, fibers, bottles, and consumer items, while many fabricators have installed equipment to handle and formulate the plastic materials to provide a supply at the lowest possible cost.

With ever higher investment and increasing cross-industry competition, increasingly greater sophistication has been required of marketing analysis and selection of investments. The enormity of the investment required today to participate successfully does not generally permit multiple approaches for the private investor. Consequently, a high degree of market orientation now tends to predominate in the chemical industry.

Technological Orientation

The chemical industry is a high technology industry, albeit now more marketing oriented and competitive than in its early period of development. This orientation is seen in the number of scientists and engineers employed in Research and Development relative to other industries (Table 1.7). In general, the chemical industry is among the largest employers of scientists and engineers and invests a sizeable percentage of the total United States' business investment in R&D (Table. 1.8).

To illustrate, the Science and Technology Budget of the United States Government for the last three years has been about 17 billion dollars per year, of which about 70 per cent

TABLE 1.7 Scientists and Engineers in Research and Development in the United States, 1967[4]

Aircraft and missiles	98,700
Electrical equipment and communications	97,700
Chemical and allied products	42,100
Machinery	37,700
Motor vehicles and other transportation	24,400
Petroleum refining	10,600

TABLE 1.8 United States' Business Investment in Research and Development, 1969[2] (Billions of Dollars)

	Billion
All business	$19.2
Chemical process industries	2.7
Chemical companies	1.7

or 12 billion dollars has been for work carried out by private industry. R&D for the private sector of business, in which nearly all chemical-industry research is undertaken, has been about seven to eight billion dollars per year.[6]

Obsolescence and Dependence on Research

The high technology level that characterizes the chemical industry and which is reflected in heavy investments in R&D, generally concerns discovery and development of new products and improvement in the manufacture of known products. The first area is the more conspicuous: the new pharmaceutical for a specific disease; the narrow spectrum, transient pesticide; the new, super-performance, composite system used for the turbine blade in a jet engine or the spinnaker pole of the yacht "Intrepid" in the 1970 America's Cup race.[7] The second area makes viable the circumstances outlined earlier under which increasing investments can be made to produce larger quantities of materials, the wholesale price indices of which are declining. The development of a new, lower cost process for a commercial product can permit development of a profitable opportunity, or spell disaster for a company with existing investment in a now obsolete plant. Major reductions in manufacturing cost can be achieved, for example, by reducing the number of reaction steps required, changing to a lower cost raw material base, or eliminating coproducts or costly separations. Examples of these will make the economic consequences clear.

Acetic acid production in the United States has more than quadrupled in the last 20 years, yet the price has remained, in spite of inflation, about the same, roughly six cents per pound. From the 1930's acetic acid was produced by a three-step synthesis from ethylene: acid hydrolysis to ethanol, then catalytic dehydrogenation to acetaldehyde, then direct liquid-phase oxidation to give acetic acid and acetic anhydride as coproducts.

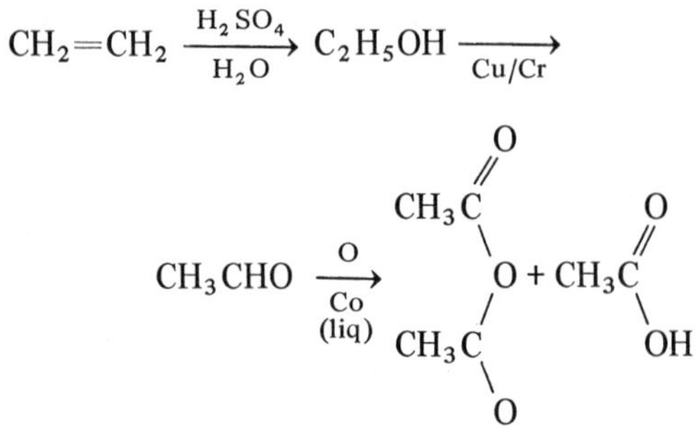

In the 40's, a major process change was introduced—direct oxidation of butane to acetic acid and coproducts (such as methyl ethyl ketone).

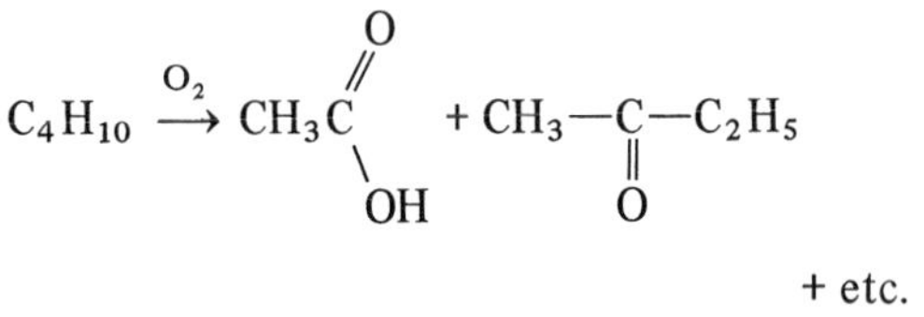

Fewer steps in synthesis were reflected in lower cost and investment. In 1969, another advance was announced, synthesis of acetic acid from methanol and carbon monoxide with essentially no coproducts.[8]

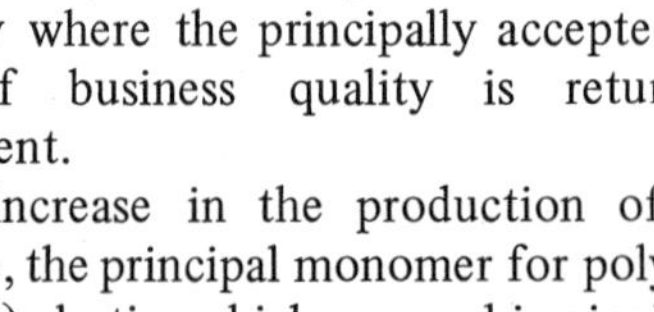

Lack of coproducts reduces costs and investment in distillation and other separations systems, very attractive process features in an industry where the principally accepted measure of business quality is return on investment.

The increase in the production of vinyl chloride, the principal monomer for poly(vinyl chloride) plastics which are used in vinyl flooring, phonograph records, shower curtains, raincoats, car seat upholstery, is more spectacular. Production in the United States has increased from 250 million pounds in 1950 to over one billion pounds in 1960, to about 3.6

billion pounds in 1970. Meanwhile, the price per pound has dropped from 14 cents per pound in 1950, to 10 cents in 1960, to five cents per pound or less in 1969.

During the early development period of vinyl resins in the 1930's, vinyl chloride was produced via catalytic addition of hydrogen chloride to acetylene.

$$CH\equiv CH + HCl \xrightarrow{HgCl} CH_2=CHCl$$

Later, a so-called "balanced" process was introduced in which, by addition of chlorine to ethylene, ethylene dichloride was produced. Ethylene dichloride could then be cracked to vinyl chloride and HCl, with the hydrogen chloride then cycled to produce vinyl chloride from acetylene.

$$CH_2=CH_2 + Cl_2 \longrightarrow CH_2Cl-CH_2Cl$$
$$\xrightarrow{\Delta} CH_2=CHCl + HCl$$

At this point, vinyl chloride was being produced from chlorine, acetylene, and ethylene. More recently, catalytic oxychlorination has been developed in which vinyl chloride is produced from ethylene and hydrogen chloride.[9]

$$CH_2=CH_2 + HCl \xrightarrow[Cu]{[O]} CH_2=CHCl + H_2O$$

The hydrogen chloride may be obtained via cracking of ethylene dichloride. The oxychlorination process has freed vinyl chloride from the economics of the more costly raw material, acetylene, a major cost improvement.

Propylene oxide is another basic chemical, used in manufacturing intermediates for urethane foams (used in cushioning and insulation) and brake and hydraulic fluids. In 1960, 309 million pounds of propylene oxide were produced in the United States and sold at about 14 cents per pound. In 1970, 1.2 billion pounds were produced with a selling price of eight cents per pound. The classical industrial synthesis has been the reaction of chlorine with propylene to produce the chlorohydrin followed by dehydrochlorination with caustic to produce the epoxide, propylene oxide, plus salt. In this case, both the chlorine and caustic used to effect this synthesis are discarded as a valueless salt by-product.

$$CH_3CH=CH_2 + Cl_2 + H_2O \longrightarrow$$
$$CH_3\underset{|}{\overset{OH}{C}}HCH_2Cl$$

$$CH_3\underset{|}{\overset{OH}{C}}HCH_2Cl \xrightarrow{Caustic} CH_3CH\overset{O}{-\!\!\!\diagup\diagdown\!\!\!-}CH_2 + Salt$$

Recently, a much more economic process has been commercialized. In one version a hydroperoxide is produced by catalytic air oxidation of a hydrocarbon such as ethylbenzene. Reaction of this hydroperoxide yields propylene oxide plus a coproduct. This peroxidation of propylene can be carried out with other agents, giving different coproducts such as t-butanol or benzoic acid.[10]

$$\text{Ph}-C_2H_5 \xrightarrow[V]{[O]} \text{Ph}-C_2H_4OOH$$

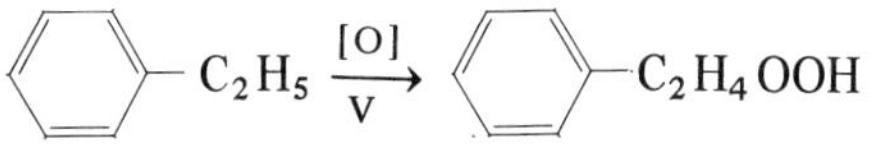

$$\text{Ph}-C_2H_4OOH + CH_3CH=CH_2 \longrightarrow$$
$$CH_3CH\overset{O}{-\!\!\!\diagup\diagdown\!\!\!-}CH_2 + \text{Ph}-C_2H_4OH$$

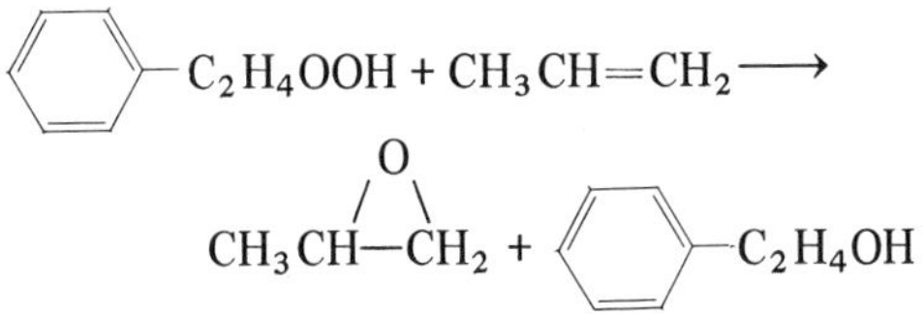

Again, when the economics are balanced, a significant cost reduction is achieved by eliminating the valueless coproduct, salt, which also presents distinct disposal problems, and developing a process which produces coproducts which can be used or sold as chemical intermediates.

If a company is in the business of making and selling products such as acetic acid, vinyl chloride, propylene oxide or other chemicals and has plans to stay in business and to expand its facilities to serve growing markets, it must have at least economically competitive processes. Today, this means competitive with new processes in the United States, Western Europe, Japan, the Soviet Union—for the chemical industry in the 1970's is a worldwide industry.

TABLE 1.9 1969 Overseas Sales of United States Chemical Producers[11] (Millions of Dollars)

Producer	*1969 Foreign Sales*	*1969 Total Sales*	*Foreign Sales as a Per Cent of Total Sales*
Du Pont	$847	$3,632	23
Union Carbide	768	2,933	26
Dow Chemical	643	1,797	35
Monsanto	451	1,939	23
American Cyanamid	206	1,087	19
Celanese	188	1,250	15

WORLD-WIDE CHEMICAL INDUSTRY

The major chemical producers in the United States have been developing their overseas business (Table 1.9). In companies such as Du Pont, Union Carbide, Monsanto, foreign sales amounted to about one-quarter of total sales. Dow Chemical derived an even higher proportion from foreign sales. The major chemical companies have truly begun to do business on a world-wide basis, for some time the familiar pattern in the petroleum industry. As United States' producers expand abroad, competitors based in the growing economies of Western Europe and Japan are developing plants competitive in size with those in the United States. They are competing not only in exports to developing nations, but also directly in the United States' domestic markets.

TABLE 1.10 United States Balance of Trade, 1969[12] (Millions of Dollars)

Total chemical exports	3,383
Total chemical imports	1,232
Chemical balance of trade	2,151
All export trade	37,314
All import trade	36,052
Official trade balance[a]	1,262
Chemical exports to	
Canada	510
Latin America	613
European Economic Community	828
European Free Trade Association	323
Japan	304
All other areas	805

[a]Including Department of Defense Shipments.

Chemical exports represent a very important part of favorable balance for the United States' balance of trade (Table 1.10). The United States' share of the world export market for both all manufacturers and chemicals has been from 20 to 25 per cent. In 1967, chemicals accounted for about 40 per cent of the favorable United States balance of trade which was in excess of four billion dollars. In 1970, United States chemical exports over imports were expected to approximate the nation's net balance of trade, about 2.5 billion dollars. Three-quarters of United States' chemical exports are shipped to the European Economic Community, Latin America, Canada, European Free Trade Association, and Japan, in order of decreasing value of dollar shipments.

While the United States is the world's leading chemical producer, and many of the largest chemical manufacturers are located principally in the United States, European-based chemical concerns are of comparable size in sales and resources (Table 1.11). The three giants of West German chemicals, Farben-

TABLE 1.11 World's Largest Chemical Companies[13, 14]

Company	*Country*	*Total Sales (Millions of Dollars) 1969*	*1967*
Du Pont	U.S.A.	3,655	3,102
I.C.I.	U.K.	3,250	2,692
Union Carbide	U.S.A.	2,933	2,546
Montecatini Edison	Italy	2,620	2,092
Farbenfabriken Bayer	W. Germany	2,550	1,459
Hoechst	W. Germany	2,550	1,650
B.A.S.F.	W. Germany	2,430	1,259
Monsanto	U.S.A.	1,939	1,632
Dow Chemical	U.S.A.	1,876	1,383
Rhone-Poulenc	France	1,840	1,063
Allied Chemical	U.S.A.	1,316	1,243
Celanese	U.S.A.	1,250	1,110
American Cyanamid	U.S.A.	1,067	937
Standard Oil (N.J.) (Chemical sales)	U.S.A.	1,004	

fabriken Bayer, Hoechst, and Badische Anilin & Soda Fabrik (BASF) have been notably aggressive in international trade. In 1967, the portion of business represented by export trade for these three was: Bayer, 63 per cent; Hoechst, 50 per cent, and BASF, 48 per cent. In addition, their foreign subsidiaries generated considerable sales: Bayer, 347 million dollars; Hoechst, 226 million dollars; and BASF, 471 million dollars. All three operate in the United States through joint ventures or subsidiaries. Bayer, with four plants, including Mobay in Pittsburgh and Chemagro in Kansas City, did about 112 million dollars in the United States in 1967. Hoechst's investments in the United States in 1969 were of the order of 200 million dollars.

In recent years, the Japanese chemical industry has grown to rank with the world's giants. The 1969 output of Japanese chemicals amounted to about 10 billion dollars. While in terms of total sales, the largest of the Japanese chemical producers do not seem, at first, to equal those of some of the largest in the United States or Europe, such cursory analysis is deceptive (Table 1.12).

The industry in Japan is carefully guided by the Ministry of International Trade and Industry. Capital demands are met by cooperation within an industry. For example, seven methanol producers agreed on joint investment for a new 800 million pound a year plant.[16] More than half of the olefins and ammonia plants are joint efforts, and vinyl chloride producers collaborated in bringing on-stream one billion pounds of new capacity in three collective plants.

TABLE 1.12 Major Japanese Chemical Producers[15]

Company	*Sales in 1969*
Sumitomo Chemical Company	520.7
Mitsubishi Chemical Industries	472.9
Ube Industries	344.9
Showa Denko	332.4
Mitsui Toatsu Chemicals	301.5
Mitsubishi Petrochemical	219.5

The Japanese chemical industry is also beginning to explore investment abroad. Mitsui Petrochemical has arranged to introduce its high density polyethylene technology in the United States in a joint venture with Hercules, ninth largest United States chemical producer in terms of chemical sales. Both Kanegafuchi Chemical and the Mitsui group have announced ventures under way in Belgium.

The trend of "foreign" expansion in the chemical industry on a world-wide basis by companies based in Japan, Europe, and in the United States can be expected to continue as more and more of the chemical giants become "multinational" in character.

REFERENCES

1. Fedor, W. S., *Chem. Eng. News*, **48**, No. 44, 102 (Oct. 19, 1970).
2. Piombino, A. J., *Chem. Week*, **106**, No. 2, 35 (Jan. 14, 1970).
3. Sixteenth Conference on Business Prospects, Oct. 16, 1969, Graduate School of Business, University of Pittsburgh.
4. Stanford Research Institute economic data.
5. *Chem. Week*, **106**, No. 18, 30 (May 6, 1970).
6. *New York Times* (Nov. 1, 1970).
7. *Chem. Eng. News*, **48**, No. 44, 38 (Oct. 19, 1970).
8. Belgian Patent 713,296 (to Monsanto).
9. British Patents 1,027,277 (to PPG) and 1,016,094 (to Toyo Soda).
10. French Patent 1,460,575 (to Halcon).
11. *Chem. Eng. News*, **48**, No. 17, 29 (Apr. 20, 1970).
12. U.S. Department of Commerce, Bureau of the Census.
13. *Chem. Ind.*, 1372 (Oct. 12, 1968).
14. *Chem. Eng. News*, **48**, No. 18, 21 (Apr. 27, 1970).
15. *Chem. Week*, **106**, No. 22, 45 (June 3, 1970).
16. *Chem. Eng. News*, **47**, No. 20, 40 (May 12, 1969).

2

Industrial Water Supply

Thomas J. Powers*

INDUSTRIAL WATER USES

The manufacturing industries need water for a great variety of purposes. Whenever locating a new plant or expanding an existing plant it is necessary to study the water demands and make the most economical use of the available water source. A flow diagram should be prepared which shows the relationship of water demand to process material flow. When properly evaluated and engineered, the flow diagram can point up many economies in water use.

Heat Transfer

The largest demand for water in the manufacturing industries is for heat transfer. Wherever heat is to be extracted, water is usually the transfer medium. Condensers, heat exchangers, coolers, refrigeration and air-conditioning equipment, and cooling towers fall under this classification. Large steam-power stations require 800 tons of water per ton of coal consumed to condense exhaust steam.[1] It is not uncommon for a steam-electric generating plant to pump 1,200 millions of gallons of water a day. It is stated that a cooling tower can reduce the water demand of a steam plant from 60 gals/kw-h to 1 gal/kw-h.[2] The cooling tower may have a 1 per cent loss from evaporation, a 0.2 per cent loss from entrainment and a blowdown of about 0.4 per cent. Thus on a 100 gpm recirculated system the need for make-up from supply might be 1.6 gpm. One large chemical complex on the Gulf of Mexico pumps 1,440 millions of gallons per day of sea water, most of which is for heat transfer.

Steam

A pound of *steam* requires a pound of water. In the case of electric steam generation, the amount of water required as make-up in the condensate is usually a very small demand. However, where large amounts of process steam or heating steam are used without condensate collection the water use can be very

*President, Operation Service and Supply Corp., Sarasota, Fla.

important. One large chemical complex uses over 5 million gallons per day of boiler feed water. A large utility furnishing steam to heat the business section of a city may require several millions of gallons per day for boiler feed.

Raw Material

Water is a *raw material* in the strict sense only when it becomes a part of the product. Thus water is a very important raw material in the beverage industry and the chemical industry. However, water finds many other applications in chemical processing. It is a *reactant* in the manufacture of acetylene from calcium carbide and in the manufacture of phosphoric acid.

The *solvent* uses of water are myriad. Good housekeeping and safety require that floors and equipment be kept clean. Since water is an almost universal solvent this use in common to all industry. Water is also the cheapest liquid available. Impurities in raw materials, intermediate products and final products are often removed by water washing. The solvent action of water on salt is the basis of mining for the production of salt, chlorine and caustic. The rinsing of pickled sheets and galvanized parts is an example of solvent use in the steel industry.

Miscellaneous Uses

Other common industrial water uses are for *material handling* such as sluicing sugar beets from stock pile to factory. Coal is slurried with water and pumped for many miles. Sulphur is mined by melting with hot water. Water handles paper pulp through processing to the paper machines. There are cases where *effluent dilution* requirements may exceed all other needs for water. Here the water requirements are based on restrictive standards on effluent quality.

The use of water for *kinetic energy* is best exemplified by the use of high pressure jets to clean metal parts or tube bundles and to debark logs.

Water is also used for *nuclear shielding* or as a moderator in nuclear energy reactors since water itself does not yield any important radioactivities when exposed to a neutron flux.[3]

Sanitary Uses

Every industry must furnish employees with a safe, palatable drinking water. This may be distributed in 5 gallon jugs, piped from a municipal supply, or be a part of the plant water system. Where potable water serves all sanitary fixtures the water needed in an industry is about 15 gallons per person per shift. It is important that the potable water system be engineered so that cross connections to other water systems are rigidly avoided. Potable water for sanitary and process use must meet the standards of the local Health Department. Most Health Departments accept the 1962 Drinking Water Standards established by the U.S. Public Health Service.

TABLE 2.1 Typical Industrial Water Demands[4]

Industry or Product	*Approximate Water Used*
Alcohol, industrial	600,000 gal per 1,000 bu grain
Gunpowder	200,000 gal per ton explosive
Oxygen, liquid	2,000 gal per 1,000 cu ft O_2
Soda ash	15,000 gal per ton 58% Na_2CO_3
Sulfuric acid (contact)	4,000 gal per ton 100% H_2SO_4
Beer	470 gal per bbl beer
Meat packing house	55,000 gal per 100 hog units
Milk receiving and bottling	450 gal per 100 gal milk
Sugar-beet	20,000 gal per ton sugar
Rayon	200,000 gal per ton yarn
Wool scouring and bleaching	40,000 gal per ton goods
Air conditioning	10,000 gal per person per season
Coke by-product	1,500 gal per ton coke
Tannery-leather	16,000 gal per ton hides
Oil refining	77,000 gal per 100 bbl crude
Steel plant	20,000 gal per ton of steel

SOURCES OF WATER FOR INDUSTRY

The primary source of all water supply is rain. The average rainfall on the United States is about 30 inches per year. Approximately 20 inches of this amount is lost by evaporation and transpiration. This leaves some 10 inches to replenish ground water and surface water.

Ground Water

Geologists estimate that ground water, in place, is equivalent to 10 years' precipitation, which is far more than all the country's surface water. It is found in saturated zones of sand, gravel and porous rock. Industry draws heavily on ground water supplies where these are available. Most state geologists and the U.S. Geological Survey have information about the location and yield of the various aquifers. Ground waters are recharged by rain percolating into the ground. Sometimes the recharge is from distant places. Water from a 150-foot-deep well at York, Nebraska, was estimated to have fallen fifty years previously.[2] Water pumped from wells in the Mattaponi formation just east of Williamsburg, Va. is estimated to have been rain which fell on the east side of the Blue Ridge Mountains thousands of years ago.

The extensive use of ground water has caused serious lowering of the ground water table in some areas, notably Long Island, New York, southern California, Texas and Arizona. This is the evident result of withdrawals in excess of recharge. The alluvial deposits near the Ohio and Mississippi rivers furnish tremendous volumes of water for industry in the St. Louis and Cincinnati areas. These aquifers are recharged from the rivers.

Surface Waters

Rivers and lakes supply about three-fourths of the total water pumped for industrial and municipal use. The Great Lakes are said to be the largest reservoirs of fresh surface water in the world. The rivers connecting the Great Lakes are also among the largest in the country. Lake Superior is probably one of the best and largest sources of cool fresh water. Information on the discharge of surface waters is available through the Water Supply Papers published by the U.S. Geological Survey–Surface Water Branch. The early industrial growth of the United States was due in part to the availability of water power and water transportation.

Industry must depend on surface waters for effluent dilution even though an adequate ground water supply is used. The existing quality of the water can restrict the amount of materials it can safely receive and the uses to which it can be put.

Water Quality

In the strict sense absolutely "pure" water is not found in nature. Since water is a solvent it dissolves gases even when it falls as rain. As soon as it reaches the ground it starts dissolving materials. The impurities dissolved depend on the chemical composition of the rocks and soils contacted. The most common impurity is calcium bicarbonate, formed by the CO_2 in water when it passes over limestone, $CaCO_3$. Water quality is defined by temperature, suspended solids, bacterial count, color, odor and dissolved impurities.

One of the important water-quality determinations is temperature since the total volume required is largely dependent upon temperature. Most deep wells will average close to 55°F while the range of maximum temperature of surface waters in the United States is from 65° to 85°F.

One of the early terms used to express water quality was *hardness*. This refers to the soap-consuming power of the water and can be actually measured by the formation of suds with a standard soap solution. Since soaps are the sodium salts of fatty acids such as stearic, palmitic and oleic, they will precipitate as the calcium, magnesium, iron or manganese salts if these ions are present in the water. Waters are broadly classified as "hard" or "soft" but this is arbitrary since in each section of the country we find a different interpretation of these relative terms. Hardness is the total parts per million (ppm) of the calcium as $CaCO_3$ plus the magnesium as $CaCO_3$ (ppm hardness as $CaCO_3$ = ppm Ca × 2.497 + ppm Mg × 4.116). Waters containing less than 60 ppm hardness are considered "soft" while waters with over 180 ppm can be considered "hard." Iron and manganese salts which may cause product discoloration will also cause hardness but are seldom found in waters in amounts significant to the total hardness.

Hardness is also subclassified as "temporary" or "permanent." "Temporary" hardness is the hardness which can be removed by boiling. The scale in the tea kettle is temporary hardness as $CaCO_3$ which precipitated when bound CO_2 was driven from $Ca(HCO_3)_2$. "Permanent" hardness is hardness which is noncarbonate and is usually $CaCl_2$, $CaSO_4$, $MgCl_2$, or $MgSO_4$.

The *alkalinity* of a water is its acid neutralizing capacity to pH 4.6. Thus carbonate hardness is a part of the total alkalinity. Alkalinity is also expressed in ppm as $CaCO_3$. Alkalinity can be subdivided by titration into caustic, carbonate and bicarbonate. The total alkalinity is determined by titration with acid to methyl orange end point pH 4.6. An identical sample is titrated with acid to the phenolphthalein end point, pH 8.3. This titration gives all the caustic and one-half of the carbonate alkalinity. The best sources of information on water quality are the state health departments and the U.S. Geological Survey Water Supply Papers. Many handbooks give comparative analyses of ground and surface waters.[4,5]

Re-use of Water

The limitations on water supply in certain areas and regulatory pressures to reduce thermal water pollution have forced industry to re-use water many times over. The cooling tower (Fig. 2.1) is a familiar structure where large heat transfer demands exist. The increasing demand for air conditioning has necessitated the installation of cooling towers of various sizes to permit transfer of building heat. The municipal water distribution and sewerage systems could not be expected to accommodate the water required for air conditioning on a once-through basis. As water is reused it dissolves more substances and through evaporation there is a build-up of dissolved solids. This quality change necessitates discarding or "purging" a portion of the recirculated stream to maintain a constant dissolved solids concentration. The increased dissolved solids and the oxygen saturation require treatment to keep the system operative.

The municipal water supply may be the cheapest water sources available if the user can recirculate heat exchange water. Normally a heat exchange and fire protection system can be obtained by an industry from an adequate surface water or well water for less than \$30 per million gallons. Due to the distribution system costs and maintenance of minimum pressures a city water is rarely available to industry for less than \$60 per million gallons. Treated, recirculated cooling tower water will cost about \$30 per million gallons. The economics of using the municipal supply will become apparent only after a careful study of all the cost factors involved.

Fig. 2.1 Forced-draft cooling tower incorporating biologic water treatment. (*Courtesy Sun Oil Co.*)

Sea Water: Brackish Water

Sea water is a raw material for the production of salt, bromine, and magnesium. By far the largest use of sea water is for heat exchange.

Research to make fresh water from sea water was sponsored by the U.S. Department of the Interior for many years, reaching a peak in 1966. The processes researched have been (1) solar distillation, (2) long tube distillation, (3) multistage flash distillation, (4) electrodialysis, (5) forced circulation vapor compression, (6) reverse osmosis, and (7) freezing.

Several ocean-island refineries have installed distillation plants to produce fresh water. The largest of these, 3.5 MGD, is on the island of Aruba. By 1966 there were over 200 known desalting plants operating around the world.

The combination of nuclear power and desalting has been under study since 1965. Distillation processes have produced fresh water from sea water at a cost of about $1.00 per 1000 gallons. Predictions are that by 1980 large plants will be able to produce fresh water from sea water at a cost of less than 50 cents per 1000 gallons. The cost of desalting is directly dependent on fuel cost and indirectly on size.

PROBLEMS CREATED BY INDUSTRIAL WATER USE

Corrosion

Since water is an almost universal solvent its use will create problems. Perhaps the most costly result of water use is the damage to equipment through *corrosion* (Fig. 2.2). Basically the corrosion of metal surfaces by water is the result of the solution of the metal. According to the electrolytic theory of corrosion, as iron atoms become positive ions and pass into solution, positive hydrogen ions are plated out.

At anode area	At cathode area
$Fe - 2e = Fe^{++}$	$2H^+ + 2e \longrightarrow 2H$

It is postulated that the hydrogen forms a molecular film on the metal surface and as long as this film is maintained there will be no further corrosion. If dissolved oxygen is present in the water it will react with the hydrogen and remove the protective film. Oxygen also reacts with the iron to form Fe_2O_3 or Fe_3O_4 at temperatures in excess of 100°C. These iron oxides will, in turn, form a protective coating on the metal surface if deposited. The erosive effect of high velocity removes these deposits. It can be seen then that corrosion is related to oxygen content of the water. The rate of corrosion is affected by (1) temperature, (2) dissolved solids, (3) hydrogen-ion concentration, (4) velocity. Corrosion can also be caused by iron bacteria and sulfate-reducing bacteria. These bacteria adhering to metal surfaces release acids which cause localized corrosion.

Fig. 2.2 Scaled and corroded open-box condenser tubes. (*Courtesy Dow Industrial Service.*)

Deposits

The maintenance of process efficiency is often related to water use. Heat transfer surfaces which become coated with deposits will have a reduced transfer capacity. Water flow is also restricted and hot spots in the system may develop, creating localized corrosion and further rapid deposition.

Deposits can result from corrosion, precipi-

Fig. 2.3 Calcium carbonate deposited in cast-iron water main. (*Courtesy Dow Industrial Service.*)

tation of dissolved solids, and bacterial growths. Deposits are classified as sludge, scale, biological deposits, and corrosion products. Sludge is soft, non-adhering material, whereas scale is tightly held to the metal surface and will not wash off. Biological deposits are bacterial slimes, fungi or algae which adhere to wetted surfaces and trap materials carried in suspension. Corrosion products which form deposits are not uniformly deposited and form "tubercules" on the metal surfaces. The most common deposit found in water systems is $CaCO_3$. The solubility of $CaCO_3$ at normal water temperatures is approximately 15 ppm; when CO_2 is driven off from $Ca(HCO_3)_2$ the resulting $CaCO_3$ will precipitate. Scales have been permitted to grow to the point of almost complete stoppage (Fig. 2.3). In such cases the operation must be shut down to permit chemical or mechanical cleaning (Fig. 2.4). Examination of the chemical analysis of a scale may show the presence of a wide variety of materials such as calcium carbonate, calcium sulfate, silicates, iron carbonates, magnesium hydrate and organic materials. A good analysis of the scale often points the way to corrective measures.

Erosion

High velocity and resulting turbulence in pipes and tubes can cause damage through erosion. Where suspensions exist the impact of these particles can wear away metal surfaces at an increased rate. The erosion of corrosion products also hastens the corrosive action. One of the best examples of erosion is the pumping of dredged materials where the solids separate and move along the bottom half of a pipe line. It is necessary to rotate this pipe quite often to avoid developing leaks along the bottom.

Biological Growths

Slime growths, which may be bacteria, fungi or algae, depending on the location, can affect the efficiency of heat transfer equipment. Biological growths can also contrib-

Fig. 2.4 Scale of ferric oxide and calcium carbonate in an industrial water line before and after chemical cleaning. (*Courtesy Dow Industrial Service.*)

ute to corrosion and cause odors in the water which might restrict usefulness. Flow rates through filters and ion-exchange beds can be seriously retarded by algae and flocculent bacteria.

Pollution

Since 1960 water pollution has become a matter of ever increasing concern to industry. The threat of an ecological crisis has, in many cases, changed water pollution from a technical problem to an emotional and political problem.

The common law doctrine of "reasonable" use has been replaced with water quality standards, and water pollution is defined as exceeding certain limits in the receiving water. Water pollution will be further defined as exceeding certain limits which will be established on factory effluents.

Man cannot use water without adding something to it. The more a water is used, the more impurities are added and the more chances for pollution exist.

Some products of industry which have widespread use and may cause environmental changes have been the subjects of many congressional hearings. The foam on water courses created by residual surfactant (ABS) in municipal waste waters caused such concern that the soap and detergent industry had, by the end of 1965, changed to the use of a biodegradable surfactant (LAS). (See Chapter 12).

In 1970 the Federal Government's Environmental Protection Agency had challenged the soap and detergent industry to find a substitute for phosphate builders in detergent formulations.

The inadvertent loss of elemental mercury and the use of mercurial fungicides became matters of national concern in 1970 after fish in Lake St. Clair were found to contain identifiable amounts of methylmercury.

The persistence of several insecticides and herbicides in the soil and water with subsequent accumulations of detectable quantities in fish and birds has caused further restrictions on the use of poisonous non-biodegradable materials.

The ever increasing sophistication of analytical techniques has opened new horizons and dimensions to studies of man's effect on his environment.

Industry can be held liable for the cost of dredging navigable waters where excessive deposition is caused by industrial wastes.[6] The Federal Government now requires permits for discharge to navigable waters and tributaries under the River and Harbor Act, March 3, 1899.

Of all the problems created by industrial water use, the pollution of the receiving water is the most difficult and expensive to completely control.

The trend in the control of water pollution has been toward the establishment of uniform industry effluent standards. This means that the dilution available and the "self purification" capacity of a receiving water are no longer usable in determining the degree of treatment necessary to avoid pollution.

WATER TREATMENT

Treatment of a water supply may be necessary to (1) make it potable; (2) prevent corrosion, scaling, or sliming; (3) permit economic boiler operations; and/or (4) prevent product damage.

Potable water is safe for drinking, practically colorless, and palatable. A water may be potable and yet not suitable for many industrial uses. Potable water must meet the standards of the state health departments, most of which accept the 1962 standards of the United States Public Health Service.

It is recommended that physical standards for public acceptance of water be:

1. Turbidity of less than 5 units.
2. Color should average less than 15 units.
3. Odor threshold should average less than 3.
4. The water should have no objectionable taste.

The simplest water system is a well system which requires no treatment to make the water potable. The well supply should be tested for chemical and bacterial quality and for yield. Pumping tests on wells can establish the well spacing and long-term yield. As a matter of safety most well supplies are disinfected.

Disinfection

This is usually the most important step in making a water potable. It can be accomplished in many ways but the simplest and most economic method has been through the use of chlorine or hypochlorite. Well waters normally require small amounts of chlorine in the range 0.5-1.0 ppm to disinfect since the chlorine demand of a water is dependent upon the amount of organic matter present. The choice between the use of cylinder chlorine and hypochlorite is largely one of economics. Where the demand exceeds 1.5 pounds per day it is usually economic to feed gaseous chlorine from 100-150 pound cylinders. The maximum amount of chlorine which can be removed from a 150-pound cylinder at 70°F is about 40 pounds per day and from a ton cylinder about 400 pounds per day. Demands in excess of these amounts are met by using multiple cylinders or by installing an evaporator. Completely automated chlorination systems are available through several equipment manufacturers.

Dry calcium or sodium hypochlorite [$Ca(ClO)_2$ or NaClO] can be purchased in 5-pound cans at 70 per cent available chlorine. One 5-pound can in 40 gallons of water makes approximately a 1 per cent chlorine water solution which may be fed to the water supply through proportioning pumps. These pumps are usually of the diaphragm type which can feed variable quantities of the solution and can be fully atomated.

The rapidity with which chlorine acts upon bacteria can be a function of pH. Complete kill in ten minutes might be achieved with 0.2 ppm at pH 6-8 while more than 1.0 ppm residual is required for equivalent kill at pH 10. Other agents for disinfection such as bromine, chlorine dioxide, ozone, ultraviolet light and silver salts have been used but have never been accepted widely.

Filtration

Surface waters for potable supplies usually must be filtered as well as disinfected. New York City does not filter its surface supply, but most of the other surface supplies such as those for Chicago, Detroit, Cleveland, New Orleans, Philadelphia and Los Angeles are filtered. The purpose of filtration is to remove suspended matter which might affect potability or might affect product purity. It is also used following sedimentation and coagulation to remove residual suspensoids. Filtration alone is not economic on waters containing over 30 ppm of suspended solids. Solids content above this figure should be coagulated and removed by settling.

The rapid sand filter (Fig. 2.5) will achieve filter rates of 2.0 gal per min per sq ft of horizontal filter area. Pressure filters with sand media achieve as high as 3.0 gal per sq ft per min. Backwash rates to remove filtered materials vary from 6 to 18 times the flow-through rate. Since backwash water must be filtered water it can be seen that with high suspended solids frequent backwash could result in no production. Normal filter practice limits backwash to 3 per cent of throughput. Some health departments have approved the diatomaceous earth filters which are operated at about 1.0 gal per sq ft per min.[7] Use of these filters provides a large amount of surface in a small area (Fig. 2.6). The precoat required is about $^1/_{16}$ inch or 0.1 lb per sq ft. The continuous feed of filter-aid is usually one to three parts per part of turbidity. The advantages of the diatomite filter are low backwash volume (0.25 per cent of throughput), small space requirement, and the ability to filter up to 60 ppm of turbidity. The disadvantages are high pumping costs, necessity for continuous filtering, and the cost of filter aid. The media used to support the precoat is usually a porous stone or a cloth backed by a metal screen.

Coagulation

The precipitation of non-settleable solids in water by chemical addition is termed *coagulation*. Most natural colloids in water carry a negative charge so the object is to add a cationic material which will neutralize or attract these particles to a matrix which will settle rapidly. The common coagulants used in water treatment are aluminum sulfate ($Al_2(SO_4)_3 \cdot 18H_2O$) and ferric sulfate ($Fe_2 \cdot$

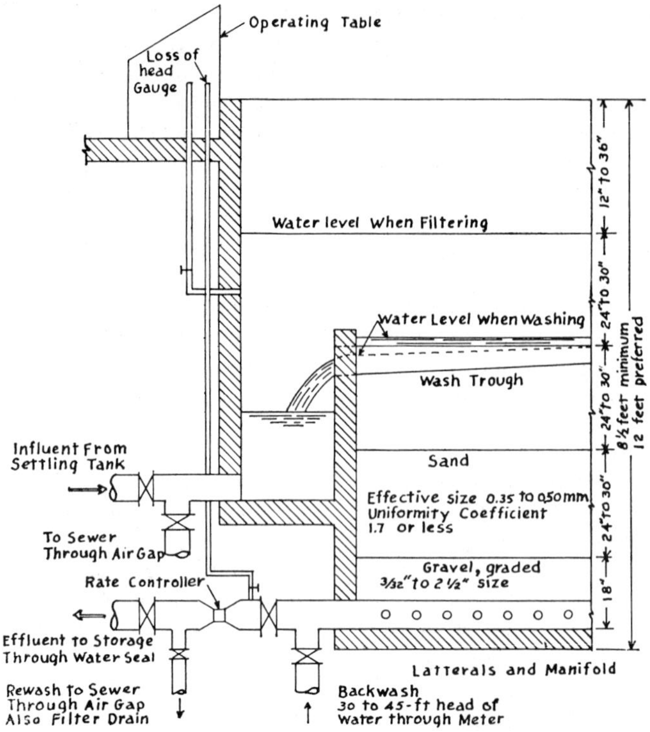

Fig. 2.5 Elements of the rapid sand filter. (*Redrawn from Public Works Magazine.*)

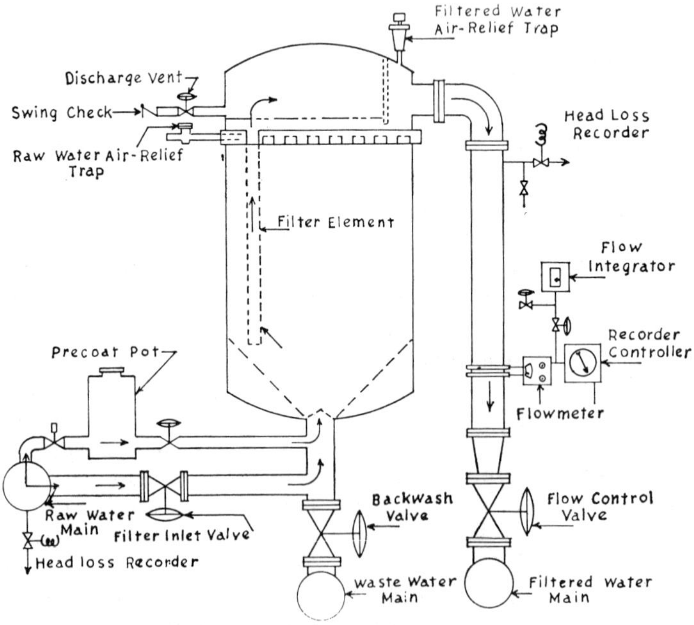

Fig. 2.6 Elements of a vertical tube diatomite filter.

$(SO_4)_3 \cdot 9H_2O$). If the system is sufficiently alkaline these trivalent ions form insoluble hydrates to which the colloids adhere. The reactions are:

$$Al_2(SO_4)_3 \cdot 18H_2O + 3Ca(HCO_3)_2 \longrightarrow 3CaSO_4 + 2AL(OH)_3\downarrow + 6CO_2 + 18H_2O$$

$$Fe_2(SO_4)_3 \cdot 9H_2O + 3Ca(HCO_3)_2 \longrightarrow 3CaSO_4 + 2Fe(OH)_3\downarrow + 6CO_2 + 9H_2O$$

When treating a water to remove suspensoids by coagulation care must be taken to leave 20-25 ppm alkalinity to prevent corrosion. In some cases it is necessary to add lime for pH control. Organic polymeric materials with long chain molecules such as the polyacrylamids can aid materially in the formation of rapidly settling *flocs*. The formation of flocs from coagulated materials is known as *flocculation*.

Other organic polymeric materials which have been made strongly cationic can be used in place of aluminum of iron salts. The one material then acts to neutralize the colloid charge as well as provide the attachment surface. In special applications it may be necessary to add anionic materials to aid coagulation. The organic polymers can also be made to carry a negative charge. The addition of both cationic and anionic materials is known as the "dual system" of coagulation.

Sedimentation

Prior to the filtration of water high in suspended solids or following coagulation for removal of non-settleable solids, it is necessary to use sedimentation tanks or basins where the suspended matter is allowed to settle out. The design of these tanks and basins is dependent upon the materials to be settled. Detention time may vary from two to twelve hours, or more.

Water Softening

In dye application plants, laundries, textile plants, and others, the water used in processing must, as a rule, be soft. If the only available water is hard, it must be treated to remove dissolved calcium and magnesium salts.

The earliest system of softening was the addition of lime to precipitate $CaCO_3$; thus

$$Ca(HCO_3)_2 + Ca(OH)_2 \longrightarrow 2CaCO_3\downarrow + 2H_2O$$

Where noncarbonate hardness exists it is necessary to add soda ash.

$$CaCl_2 + Na_2CO_3 \longrightarrow CaCO_3\downarrow + 2NaCl$$

The combined use of lime and soda ash is commonly called the lime-soda process of softening. Where magnesium is present it is necessary to provide excess caustic alkalinity in order to precipitate the magnesium as the hydrate since magnesium carbonate is soluble:

$$MgCl_2 + Na_2CO_3 \longrightarrow MgCO_3 + 2NaCl$$
$$MgCO_3 + Ca(OH)_2 \longrightarrow Mg(OH)_2\downarrow + CaCO3\downarrow$$

The lime-soda or caustic-soda softening processes are usually carried out in a sludge-blanket type of precipitator or settler where the sludge is accumulated, and freshly treated water is passed up through the preprecipitated material. This process results in a larger particle size and a more stabilized water. Almost all water-treating equipment companies manufacture equipment to accomplish sludge-blanket contact. The lime-soda process can theoretically achieve a hardness of 30 ppm.

Ion Exchange

The production of "zero" hardness water and demineralized water is generally accomplished through the use of ion exchange. This process can be defined as a reversible exchange of ions between a liquid and a solid during which there is no substantial change in the structure of the solid.[8] Although the process was recognized scientifically in 1850 it was not used until 1900. The original materials used were modified natural green sands (glauconite) which had exchange capacities of less than 5000 grains of $CaCO_3$ per cu ft. The name *zeolite* was applied to materials which had exchange properties. The science of manufacturing such materials from organic polymers has progressed rapidly since 1935 so that resins are available today which have exchange capacities of over 40,000 grains of $CaCO_3$ per cu ft.

The major use of ion-exchange resins at the present time is in the field of water treatment. The largest single use in water treatment is the softening of water using cation exchange resins in the sodium cycle. The reactions occurring in the common household water softener are:

Softening

$$Na_2R + Ca^{++} \longrightarrow CaR + 2Na^+$$

Regeneration

$$CaR + \underset{\text{(excess)}}{2Na^+} \longrightarrow Na_2R + Ca^{++}$$

A more complex procedure is to use the resin in the hydrogen cycle where the positive ions, calcium, magnesium and sodium, are replaced by hydrogen ions. The regenerant in this case is sulfuric or hydrochloric acid. The effluent from the hydrogen cycle is acidic through CO_2 from carbonate alkalinity and other anions. The CO_2 is removed by degasification. The acid water can then be mixed with a sodium cycle effluent to give a low alkalinity "zero" hardness water, or blended with unsoftened water to give the desired degree of alkalinity and hardness. Where completely deionized water is required the hydrogen cycle effluent, with or without degasification, is passed through an acid-absorbing strongly basic anion exchange resin. This system also effectively removes silica.

There have been many designs used to treat water by ion-exchange methods. Most water treatment equipment companies design, manufacture and assemble plants to deliver a specified water based on raw water quality. Completely deionized water systems must compete economically with evaporator systems. Compression distillation is said to produce distilled water at 65 cents per 1000 gal based on 10-cent fuel oil.[3]

Corrosion Prevention

Water treatment to avoid the problems accompanying water use is also accomplished by agents which will inhibit undesirable effects. For example, corrosion could be stopped if a protective molecular film would serve as a barrier to prevent oxygen from contacting the metal surface. Chromates, silicates, amines, polyphosphates, tannins and lignin have been used for this purpose in recirculated water systems. Controlled deposition of calcium carbonate through the addition of lime has also been used. Cathodic protection operates on the principle of maintaining a hydrogen film at the expense of sacrificial anodes or application of direct current.

The precipitation of calcium, magnesium and iron can be prevented by the addition of polyphosphates which form soluble complexes. Sequestering additives are generally too expensive to use on once-through systems unless most of the impurities have been removed by other treatment methods.

Odor Removal

Treatment of water to remove odors may involve aeration, chemical oxidation or adsorption. Chemical oxidation with chlorine in alkaline solution is sometimes called *breakpoint* chlorination. The amount of chlorine required for disinfection may be only 1.0 ppm while the oxidative chlorination may require as much as 10 ppm, depending on the amount of organic matter in the water. Chlorine dioxide (ClO_2), which is generated by acidifying or chlorinating sodium chlorite, has been very useful in the elimination of chlorophenol odors. Activated carbon is extensively used in large and small water works to adsorb odors from water. Granular activated carbon filters are used by breweries and bottling works to remove chlorine and odors which might affect product palatability. Powdered carbon which can be added as a slurry to water ahead of filtration is popular for odor reduction in municipal water-treatment plants.

Biological Growths

Control of biological growths is usually accomplished by intermittent chlorination of once-through water systems. The addition of chromates for corrosion control will aid in control of slimes and algae in recirculated systems. The Cooling Tower Institute recommends the use of non-oxidizing algaecides–fungicides to prevent aggravating chemical attack of wood.[9] It is also recommended that the pH of recirculated water be kept between 6.0–7.0 and that no more than 1.0 ppm of chlorine be used. There are several formulated biocides on the market to protect wood structures and clean recirculated water systems. It

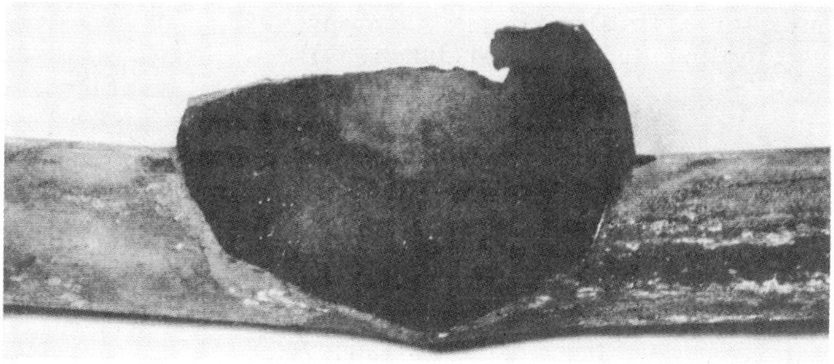

Fig. 2.7 Ruptured tube from a superheater. *(Courtesy Dow Industrial Service.)*

is often necessary to change treatment if growths develop resistance to the chemicals normally used. If chlorination becomes less effective a change to bromine many effectively kill the organism.

Boiler Water Treatment

Economic operation of a boiler may require special treatment of the boiler water. As water is evaporated the solids in the boiler water are concentrated. The loosely bound or entrained gases are released to the steam and can corrode boiler plates or tubes at a rapid rate. Deposits in the boiler can cause overheating of the tubes, resulting in bulging or splitting (Fig. 2.7). Complete clogging of boiler tubes can result. The higher the pressure at which a boiler is operated the purer the water make up must be. Concentration of solids in the boiler above 3,000 ppm might result in "wet" steam. Frequently blow-downs are necessary, with resultant loss of heat.

With low-pressure boilers the hot lime-soda process of water treatment plus internal treatment may suffice. The chemistry of the hot lime-soda process is the same as that of the cold process. However, the reactions proceed faster and since solubilities at near boiling are less, the hot process will achieve as low as 20 ppm hardness. Internal boiler treatment can reduce hardness to "zero" by use of phosphates which precipitate scale-forming materials as insoluble phosphate sludges which are removed with blow-down.

Adsorption and Absorption

Adsorption and absorption may be said to be a part of the coagulation and flocculation processes, removing colloidal materials which cause turbidity, color and odors. Activated carbon has been used to remove organics from water in water treatment for over 40 years. Until 1965 activated carbon was very seldom used in waste water treatment except as a threshold treatment following other treatment processes. The advanced waste treatment program of the Environmental Protection Agency has revived much interest in carbon adsorption and reactivation as a waste water treatment process.

REFERENCES

1. Perry, J. H., "Chemical Engineers Handbook," 3rd ed., 1724, New York, McGraw-Hill, 1960.
2. Lynch, R. G., "Our Growing Water Problems," National Wildlife Federation, 1959.
3. DePaul, D. J., "Corrosion and Wear Handbook for Water Cooled Reactors," New York. McGraw-Hill, 1957.
4. Excerpts from Manual on Industrial Water and Industrial Waste Water, 2nd ed., ASTM 1959.
5. Lange, N. A., "Handbook of Chemistry," 9th ed., 800–12, New York, McGraw-Hill 1956.
6. U.S. vs. Republic Steel Corporation *et al.*, 28LW4312 (1960).
7. Bell, G. R., Proc. 14th Annual Water Conference Engineers Society of Western Pennsylvania, 1953.

8. Dowex: Ion Exchange, The Dow Chemical Company, 1959.
9. Cooling Tower Institute Bull. WMS-104 (June 1959).

SELECTED REFERENCES

Nordell, E., "Water Treatment for Industrial and Other Uses," New York, Van Nostrand Reinhold, 1961.

Salvato, J. A., Jr., "Environmental Sanitation," New York, John Wiley & Sons, 1958.

McCabe and Eckenfelder, "Biological Treatment of Sewage and Industrial Waste," 2 vols., New York, Van Nostrand Reinhold, 1958.

Manual on Industrial Water and Industrial Waste Water, 2nd Ed. ASTM special technical publication No. 148 D, 1959.

Cooling Tower Institute Publications, Cooling Tower Institute, 1120 West 43rd, Houston, Texas.

"Saline Water Conversion," Advances in Chemistry Series, No. 27, American Chemical Society, Washington, 1960.

Projects of the Industrial Pollution Control Branch, U.S. Department of the Interior–FWQA, July, 1970.

Drinking Water Standards, Rev. Ed., U.S. Public Health Service, Rockville, Md., 1962.

Manual for Evaluating Public Drinking Water Supplies, U.S. Public Health Service, Environmental Control Administration, Cincinnati, O., 1969.

Water-Power, 330 West 42nd Street, New York (June 1966).

"Water in Industry," National Assoc. of Manufacturers, New York, Jan., 1965.

"Water: The Basic Chemical of the Chemical Process Industries," *Chem. Eng.,* New York, McGraw-Hill, 1966.

3

Coal Technology

Richard C. Corey*

INTRODUCTION

Coal technology comprises the engineering aspects of mining, cleaning and utilizing coal. Since energy is an indispensable handmaiden of industrial chemistry, and coal is the most abundant fossil-fuel source of energy in this country, it is useful to examine our energy patterns in the past, at present, and for the future, and coal's role in them.

The per capita consumption of energy to provide light, heat, and power is closely related to the standard of living. Also, the rate of growth of a nation's demand for energy is a reliable index of that nation's technological and economic growth. In the United States the demand for electricity, alone, is increasing more than five times faster than the population.[1]

Historically, this country's industrial growth has been the result of efficient and diversified utilization of its fossil fuels; coal, petroleum, and natural gas. The carbon and hydrogen in these fuels furnish both thermal energy for heat and power, and a wide variety of chemicals used either as intermediates in manufacturing processes, or as products for direct consumption. In the middle 60's, nuclear power plants began to meet part of the rapidly growing demand for electric energy.

The trend in energy consumption from fossil fuels and hydroelectric power in this country from 1920 to 1965 is shown, in British thermal units, in Fig. 3.1. Although total energy consumption rose rapidly, the percentage supplied by coal decreased owing to the growth of oil and gas in the fuel market. Projections to 1980 indicate an increase in coal consumption, but little change in coal's share of the energy market. Oil will probably continue to be the dominant fuel because of its extensive use in mass transportation. About 80 per cent of transportation energy is furnished by gasoline, diesel, and jet fuels.

The magnitude of the primary energy sources and related products is illustrated in Table 3.1, which gives the amounts consumed

*Senior Staff Scientist, Division of Coal, Energy, Bureau of Mines, U.S. Department of the Interior, Washington, D.C.

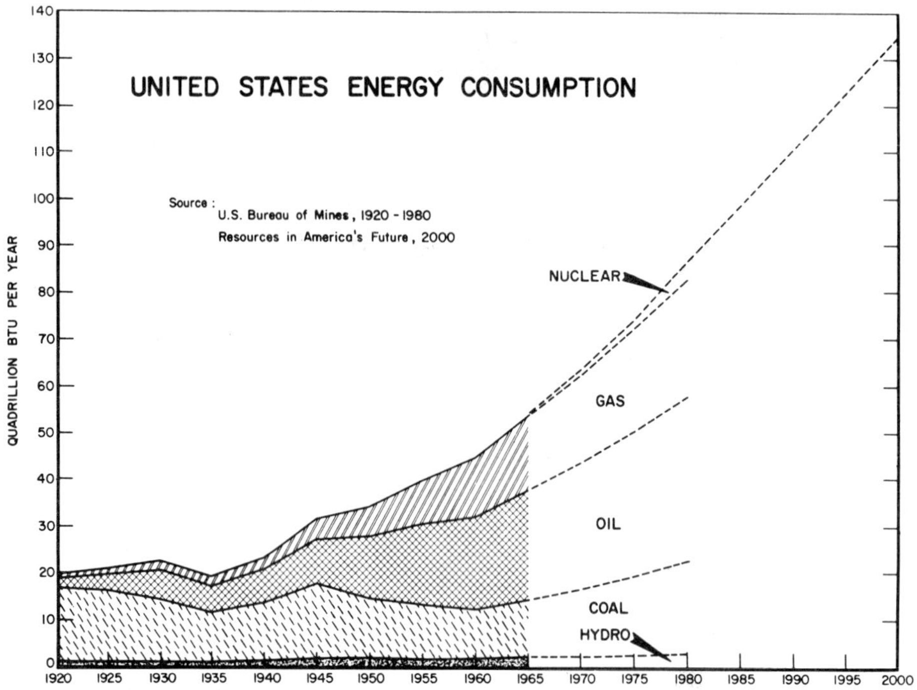

Fig. 3.1 United States energy consumption. (*Courtesy Bureau of Mines, U.S. Department of the Interior.*)

TABLE 3.1 Consumption of Primary Energy Resources and Related Products in the United States, 1968 (Source: U.S. Bureau of Mines)

Primary energy sources	
Bituminous coal and lignite, million short tons	499
Anthracite, million short tons	10
Crude petroleum, runs to stills, million barrels	3,767
Natural gas,[a] billion cubic feet	18,957
Hydroelectric power, utility, million kilowatt-hours	222,491
Nuclear power,[b] utility, million kilowatt-hours	12,528
Products	
All oils,[c] domestic product demand, million barrels	4,901
Coke, million short tons	64
Electricity, fossil-fuel-burning plants	
Utility,[b] million kilowatt-hours	1,094,424
Industrial,[b] million kilowatt-hours	106,586

[a]Residue gas—excludes extraction loss but includes transmission loss.
[b]Net generation.
[c]Includes natural gas liquids.
Note: See Appendix A for fuel and power conversion factors.

in 1968. Table 24.2 in Chapter 24 gives a breakdown of energy consumption in various areas of activity, viz., household, commercial, transportation, and industry in the United States in 1960.

Our fossil fuels are exhaustible; consequently the reserves concern us deeply, particularly when both the population and per capita use of energy are expanding rapidly. Table 3.2 gives the estimated recoverable reserves of these fuels, and the ratios of reserves to production in 1968. These ratios are useful only in terms of orders of magnitude because the estimates of reserves are subject to assumptions regarding how much can be recovered economically. However, it appears that coal will be the backbone of our energy system long after the reserves of petroleum and natural gas are depleted.

Other sources of energy include sunlight, geothermal power, wind, and tidal movements, but none of these are economical now for large-scale uses. However, if future national energy policies should require restrictions on the use of fossil fuels to conserve them, these sources of energy could become important factors in the energy mix.

TABLE 3.2 Estimated Fossil-Fuel Reserves in the United States

	Remaining Recoverable Reserves	*Ratio of Reserves to 1968 Production*
Coal, billion short tons[a]		
Bituminous	671	
Subbituminous	428	
Lignite	447	
Anthracite and semianthracite	13	
TOTAL, all ranks	1,560	235
Petroleum, crude, billion barrels[b]	30.7	9.3
Natural gas, trillion cubic feet[c]	287.3	14.8

[a]Averitt, Paul, "Coal Resources of the United States," *U.S. Geological Survey Bulletin* **1275** Jan. 1, 1967.
[b]*U.S. Bureau of Mines Minerals Yearbook*, Vol. 1–2, p. 817 (1968). (Reserves do not include oil discovered in 1968 on the north slope of Alaska, which is estimated to have a potential of 5 to 10 billion barrels.)
[c]*Ibid*, p. 727.

Converting coal to methane and oil would help to conserve the natural gas and petroleum reserves. Several processes for coal conversion are now in the experimental stage and will be described later.

COAL RESERVES, PRODUCTION AND CONSUMPTION

Coal reserves by states are given in Table 3.3. Illinois has the largest reserves of bituminous coal, Montana of subbituminous, and North Dakota of lignite. The leading coal producing states in 1968 are shown in Fig. 3.2. This pattern of production has been typical for many years. Domestic consumption of bituminous coal since 1945 is shown in Fig. 3.3; and it is seen that since 1955 the electric utilities have been the largest single consumer. The sharp decline in retail sales in large part reflects the change from coal to natural gas or fuel oil for home heating.

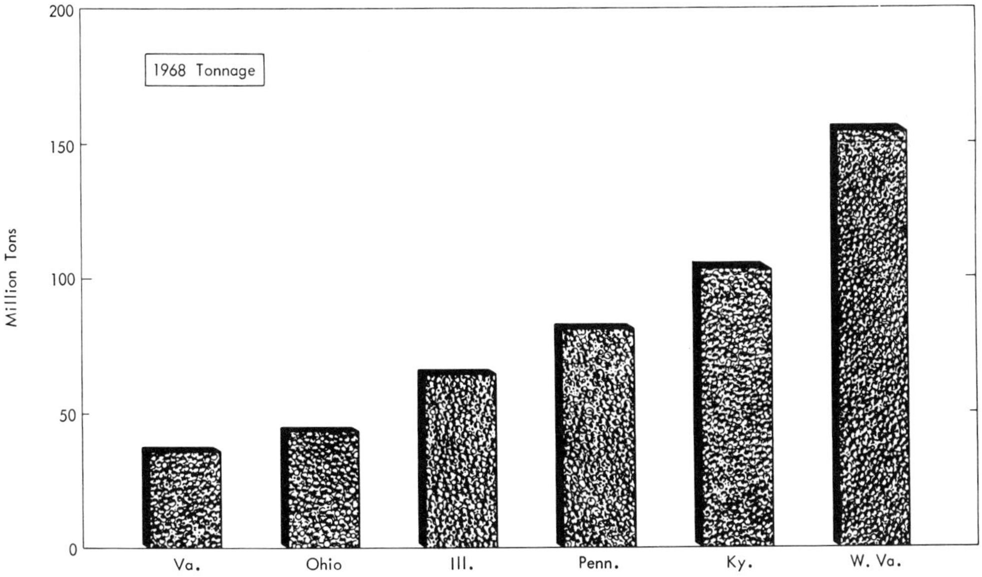

Fig. 3.2 Leading coal-producing states. (*Courtesy National Coal Assoc.*)

TABLE 3.3 Coal Reserves of the United States[a] by States (Millions of Tons)

	Overburden 0–3,000 ft thick								
	Resources Determined by Mapping and Exploration[b]								
	Bituminous Coal	*Subbituminous Coal*	*Lignite*	*Anthracite and Semi-anthracite*	*Total*	*Est. Addtl. Resources in Unmapped and Unexplored Areas[c]*	*Est. Total Remaining Resources in the Ground, 0–3,000 ft Overburden*	*Est. Resources in Deeper Structural Basins 3,000–6,000 ft Overburden[c]*	*Est. Total Remaining Resources in the Ground, 0–6,000 ft Overburden*
Alabama	13,518	0	20	0	13,538	20,000	33,538	6,000	39,538
Alaska	19,415	110,674	. . . .[d]	. . . .[e]	130,089	130,000	260,089	5,000	265,089
Arkansas	1,640	0	430	430	2,420	4,000	6,420	0	6,420
Colorado	62,389	18,248	0	78	80,715	146,000	226,715	145,000	371,715
Georgia	18	0	0	0	18	60	78	0	78
Illinois	139,756	0	0	0	139,756	100,000	239,756	0	239,756
Indiana	34,779	0	0	0	34,779	22,000	56,779	0	56,779
Iowa	6,519	0	0	0	6,519	14,000	20,519	0	20,519
Kansas	18,686	0	. . . .[f]	0	18,686	4,000	22,686	0	22,686
Kentucky	65,952	0	0	0	65,952	52,000	117,952	0	117,952
Maryland	1,172	0	0	0	1,172	400	1,572	0	1,572
Michigan	205	0	0	0	205	500	705	0	705
Missouri	23,359	0	0	0	23,359	0	23,359	0	23,359
Montana	2,299	131,877	87,525	0	221,701	157,000	378,701	0	378,701
New Mexico	10,760	50,715	0	4	61,479	27,000	88,479	21,000	109,479
North Carolina	110	0	0	0	110	20	130	5	135
North Dakota	0	0	350,680	0	350,680	180,000	530,680	0	530,680
Ohio	41,864	0	0	0	41,864	2,000	43,864	0	43,864
Oklahoma	3,299	0	. . . .[f]	0	3,299	20,000	23,299	10,000	33,299
Oregon	48	284	0	0	332	100	432	0	432
Pennsylvania	57,533	0	0	12,117	69,650	10,000	79,650	0	79,650

South Dakota	0	0	2,031	0	2,031	1,000	3,031	0	3,031
Tennessee	2,652	0	0	0	2,652	2,000	4,652	0	4,652
Texas	6,048	0	6,878	0	12,926	14,000	26,926	0	26,926
Utah	32,100	150	0	0	32,250	48,000	80,250	35,000	115,250
Virginia	9,710	0	0	335	10,045	3,000	13,045	100	13,145
Washington	1,867	4,194	117	5	6,183	30,000	36,183	15,000	51,183
West Virginia	102,034	0	0	0	102,034	0	102,034	0	102,034
Wyoming	12,699[g]	108,011	....[d]	0	120,710	325,000	445,710	100,000	545,710
Other States	618[g]	4,057[h]	46[i]	0	4,721	1,000	5,721	0	5,721
TOTAL	671,049	428,210	447,647	12,969	1,559,875	1,313,080	2,872,955	337,105	3,210,060

[a]Figures are for remaining resources in the ground, as of Jan. 1, 1967, about half of which may be considered recoverable. Includes beds of bituminous coal and anthracite 14 in. or more thick and beds of subbituminous coal and lignite 2½ ft or more thick. (Study by Paul Averitt, U.S. Geological Survey)

[b]Estimates from published reports of the U.S. Geological Survey and individual State Surveys reduced by production and losses in mining from date of estimate to Jan. 1, 1967. Losses assumed to be equal to production.

[c]Estimates by H. M. Beikman (Washington), H. L. Berryhill, Jr., (Virginia and Wyoming), R. A. Brant (Ohio and North Dakota), W. C. Culbertson (Alabama), K. J. Englund (Kentucky), B. R. Haley (Arkansas), E. R. Landis (Colorado and Iowa), E. T. Luther (Tennessee), R. S. Mason (Oregon), F. C. Peterson (Kaiparowits Plateau, Utah), J. A. Simon (Illinois), J. V. A. Trumbull (Oklahoma), C. E. Wier (Indiana), and Paul Averitt for the remaining states.

[d]Small resources of lignite included under subbituminous coal.

[e]Small resources of anthracite in the Bering River field believed to be too badly crushed and faulted to be economically recoverable.

[f]Small resources of lignite in beds generally less than 30 in. thick.

[g]Arizona, California, Idaho, Nebraska, and Nevada Bituminous coal in Black Mesa field, Arizona included under subbituminous coal.

[h]Arizona, California, Idaho.

[i]California, Idaho, Louisiana, Mississippi and Nevada.

Source: U.S. Geological Survey

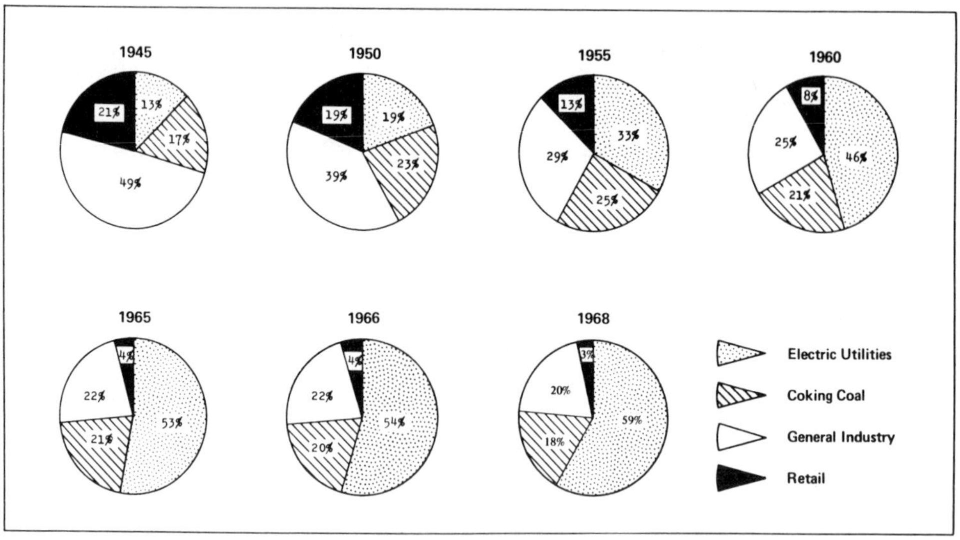

Fig. 3.3 Domestic consumption of bituminous coal. *(Courtesy National Coal Assoc.)*

ORIGIN AND CLASSIFICATION OF COAL

Coal originated from the remains of trees, bushes, ferns, mosses, vines, and other forms of plant life that flourished in huge swamps and bogs many millions of years ago, during prolonged periods of humid, tropical climate and abundant rainfall. The precursor of coal was peat, which was formed by bacterial and chemical action on heavy accumulations of plant debris comprising bark, leaves, seeds, spores, various secretions, and the remains of trunks and branches.

Exclusion of oxygen from the peat by overlying water retarded its rate of decay, and accumulations of more plant debris and sediments compressed and solidified it, resulting in the beginning of coalification. The extent to which coalification progressed, which determines the *rank* of the coal, depended largely on the temperatures existing during this process. When the prevailing temperature corresponded to a depth of only a few hundred feet below the surface, coalification was relatively slow and resulted in low-rank coal, such as lignite and subbituminous coal. At increased depths higher temperatures in the organic masses caused increases in rank, producing bituminous and anthracitic coal. Intrusions of igneous rocks sometimes accelerated these increases. The significant chemical changes that occur during coalification, from lignite to anthracite, are the decrease in hydrogen and oxygen content and the increase in carbon content. Some recognized authorities now believe that pressures on the coal deposits due to overburden affected the physical-structural development but had little influence on chemical coalification.

Coal is a heterogeneous substance, being grossly characterized by layers or bands having glossy or dull appearances. The glossy layers are composed mainly of *vitrinite*, which was formed from woody parts of plants: trunks, limbs, branches, twigs, and roots. The dull bands consist of finely divided material, formed from leaves, pollen, spores, seeds, and resins (collectively termed *exinite*), in addition to fine vitrinite and more coalified *micrinite* and *fusinite*. Besides these banded coals, there are two types of dull, nonbanded coals-*cannels*, which are rich in spores, and *bogheads*, which contain abundant remains of algae.

Table 3.4 gives a classification of coals in the United States according to rank. The main classes are grouped according to certain chemical and physical properties. It will be noted that the factors determining rank, or degree

TABLE 3.4 Classification of Coals by Rank[a]

Class	Group	Fixed Carbon Limits, per cent (Dry, Mineral-Matter-Free Basis): Equal or Greater Than	Fixed Carbon: Less Than	Volatile Matter Limits, per cent (Dry, Mineral-Matter-Free Basis): Greater Than	Volatile Matter: Equal or Less Than	Calorific Value Limits, Btu per pound (Moist,[b] Mineral-Matter-Free Basis): Equal or Greater Than	Calorific Value: Less Than	Agglomerating Character
I. Anthracitic	1. Meta-anthracite	98	-	-	2	-	-	-
	2. Anthracite	92	98	2	8	-	-	-
	3. Semianthracite	86	92	8	14	-	-	Nonagglomerating[c]
II. Bituminous	1. Low volatile bituminous coal	78	86	14	22	-	-	
	2. Medium volatile bituminous coal	69	78	22	31	-	-	Commonly agglomerating[e]
	3. High volatile *A* bituminous coal	-	69	31	-	14,000[d]	-	
	4. High volatile *B* bituminous coal	-	-	-	-	13,000[d]	14,000	
	5. High volatile *C* bituminous coal	-	-	-	-	11,500	13,000	
						10,500	11,500	Agglomerating
III. Subbituminous	1. Subbituminous *A* coal	-	-	-	-	10,500	11,500	Nonagglomerating
	2. Subbituminous *B* coal	-	-	-	-	9,500	10,500	-
	3. Subbituminous *C* coal	-	-	-	-	8,300	9,500	-
IV. Lignitic	1. Lignite *A*	-	-	-	-	6,300	8,300	-
	2. Lignite *B*	-	-	-	-	-	6,300	-

[a]This classification does not include a few coals, principally nonbanded varieties, which have unusual physical and chemical properties and which come within the limits of fixed carbon or calorific value of the high-volatile bituminous and subbituminous ranks. All of these coals either contain less than 48 per cent dry, mineral-matter-free fixed carbon or have more than 15,500 moist, mineral-matter-free British thermal units per pound.

[b]Moist refers to coal containing its natural inherent moisture but not including visible water on the surface of the coal.

[c]If agglomerating, classify in low-volatile group of the bituminous class.

[d]Coals having 69 per cent or more fixed carbon on the dry, mineral-matter-free basis shall be classified according to fixed carbon, regardless of calorific value.

[e]It is recognized that there may be nonagglomerating varieties in these groups of the bituminous class, and there are notable exceptions in high volatile *C* bituminous group.

of coalification, are moisture, volatile matter (material that is volatilized when the coal is heated in the absence of air at a certain temperature and for a certain length of time), fixed carbon (the residue after the loss of volatile matter), heating value, and caking and weathering properties.

An international classification of coals by type was recently developed by the Coal Committee, Economic Commission for Europe, to eliminate confusion in evaluating coals shipped in international trade. Figure 3.4 shows this system. It classifies high rank coals according to their volatile-matter content, calculated on a dry, ash-free (m.a.f.) basis. Since volatile matter is not an entirely suitable parameter for coals containing more than 33 percent volatile matter, the calorific value on a moist, ash-free basis is included as a parameter for such coals. The resulting nine classes of coal, based on volatile-matter content and calorific value, are grouped according to their caking properties when the coal is heated rapidly, employing either the free-swelling or the Roga test. These groups are further subgrouped according to coking properties, using either the Audibert-Arnu or the Gray-King test.

A three-figured code number is used to classify a coal. The first figure indicates the class of the coal, the second figure the group, and the third figure the subgroup. The details of its use and the test methods on which it is based have been described.[2]

GROUPS (determined by caking properties)			CODE NUMBERS										SUBGROUPS (determined by coking properties)		
GROUP NUMBER	ALTERNATIVE GROUP PARAMETERS: Free-swelling index (crucible-swelling number)	Roga index	The first figure of the code number indicates the class of the coal, determined by volatile-matter content up to 33% V. M. and by calorific parameter above 33% V. M. The second figure indicates the group of coal, determined by caking properties. The third figure indicates the subgroup, determined by coking properties.										SUBGROUP NUMBER	ALTERNATIVE SUBGROUP PARAMETERS: Dilatometer	Gray-King
3	>4	> 45					435	535	635				5	> 140	$> G_8$
						334	434	534	634				4	>50-140	G_5-G_8
						333	433	533	633	733			3	>0-50	G_1-G_4
						332 a / 332 b	432	532	632	732	832		2	⩽ 0	E-G
2	2½-4	> 20-45				323	423	523	623	723	823		3	>0-50	G_1-G_4
						322	422	522	622	722	822		2	⩽ 0	E-G
						321	421	521	621	721	821		1	Contraction only	B-D
1	1-2	> 5-20			212	312	412	512	612	712	812		2	⩽ 0	E-G
					211	311	411	511	611	711	811		1	Contraction only	B-D
0	0-½	0-5		100 (A, B)	200	300	400	500	600	700	800	900	0	Nonsoftening	A
CLASS NUMBER →			0	1	2	3	4	5	6	7	8	9			
CLASS PARAMETERS	Volatile matter (dry, ash-free) →		0-3	> 3-10 (A: >3-6.5; B: >6.5-10)	>10-14	>14-20	> 20-28	> 28-33	> 33	> 33	> 33	> 33			
	Calorific parameter a/ →		–	–	–	–	–	–	>13,950	> 12,960-13,950	>10,980-12,960	> 10,260-10,980			
CLASSES (Determined by volatile matter up to 33% V. M. and by calorific parameter above 33% V. M.)															

As an indication, the following classes have an approximate volatile-matter content of:
Class 6 33-41% volatile matter
7 33-44% "
8 35-50% "
9 42-50% "

Note: (i) Where the ash content of coal is too high to allow classification according to the present systems, it must be reduced by laboratory float-and-sink method (or any other appropriate means). The specific gravity selected for flotation should allow a maximum yield of coal with 5 to 10 percent of ash.
(ii) 332a . . . >14-16% V. M.
332b . . . >16-20% V. M.

a/ Gross calorific value on moist, ash-free basis (30 C., 96% relative humidity) B. t. u./ lb.

Fig. 3.4 International classification of coals by type. (*Courtesy Bureau of Mines, U.S. Department of the Interior.*)

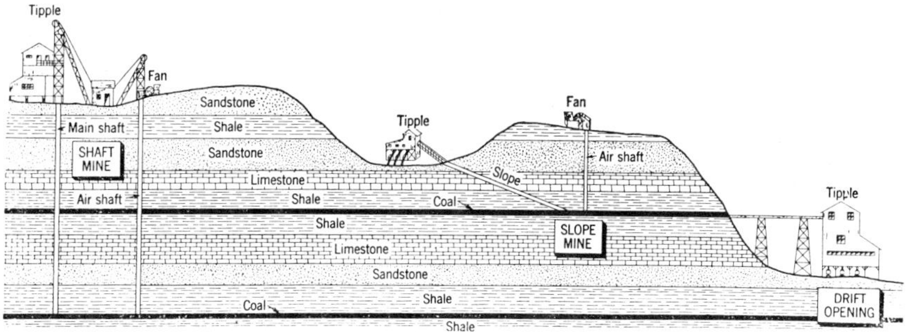

Fig. 3.5 Three types of entrances to underground mines—shaft, slope, and drift. (*Courtesy Bureau of Mines, U.S. Department of the Interior.*)

COAL MINING

There are two kinds of coal mines: *underground* and *open cut.* There are three types of underground mines. If the coal seam can be reached horizontally, say from the side of a hill on a level grade straight to the seam, it is called a *drift* mine. If the seam is at a perceptible angle to the entry, it is called a *slope* mine. If the seam must be reached by a vertical shaft from the surface, it is called a *shaft* mine. These types of underground mines are shown schematically in Fig. 3.5.

Underground Mining

Choice of coal fragmentation technique depends on factors such as seam height and uniformity, coal hardness, and economic factors such as capitalization and production rate. The most commonly used fragmentation techniques used in this country are *conventional* (blasting) and *continuous* (machine). However, the *longwall* machine, developed in Europe, is gaining favor here.

Conventional Mining. About 45 per cent of our underground mining; the basic steps are shown in Fig. 3.6. The advantages of this method of mining are:

1. Effective in coal beds having a high hardness, considerable bands of impurities, and varying dimensions.
2. Produces less fine coal.
3. Efficient where roof and floor planes undulate.

The comparative disadvantages are:

1. Requires numerous work cycles.
2. Requires more personnel.
3. Less production per man.
4. Face haulage is inefficient.
5. Inefficient where roof conditions are poor.

Continuous mining is performed by machines that bore, dig, or rip coal from the working face. This kind of underground min-

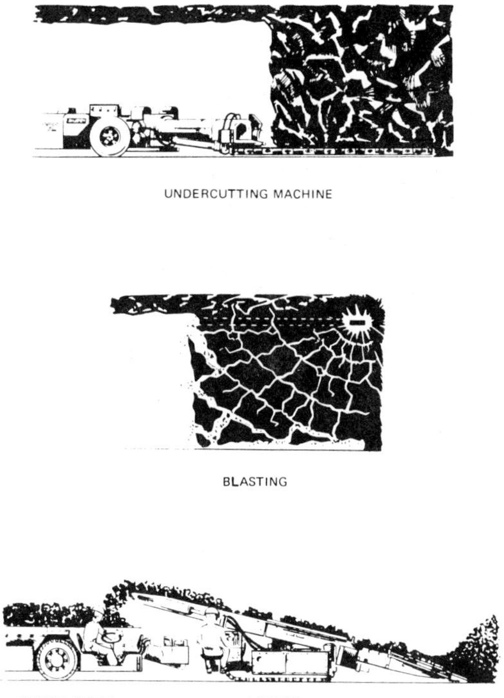

Fig. 3.6 Basic steps in conventional mining. (*Courtesy Bureau of Mines, U.S. Department of the Interior.*)

Fig. 3.7 Continuous mining machine. (*Courtesy Bureau of Mines, U.S. Department of the Interior.*)

ing is growing at the expense of conventional mining, and now accounts for about 50 per cent of our underground coal production. A continuous mining machine is shown in Fig. 3.7. It is a tracked vehicle operated by one man and capable of cutting up to 600 tons per hour. The comparative advantages of this kind of mining are:

1. Involves fewer work cycles, less equipment, and produces more coal per man than conventional mining.
2. Permits more concentrated mining, with fewer supervisory and ventilation problems.

The comparative disadvantages are:

1. Ineffective where coal hardness is high, bands of impurities are extensive, and floor and roof planes undulate.
2. Ineffective where seam thickness varies greatly.
3. Provides inefficient face haulage.

Longwall mining, first used here about 20 years ago, is currently most efficient for uniform coal seams 42 to 60 inches thick. A longwall mining machine is shown in Fig. 3.8. The comparative advantages of this kind of mining are:

1. High production rate.

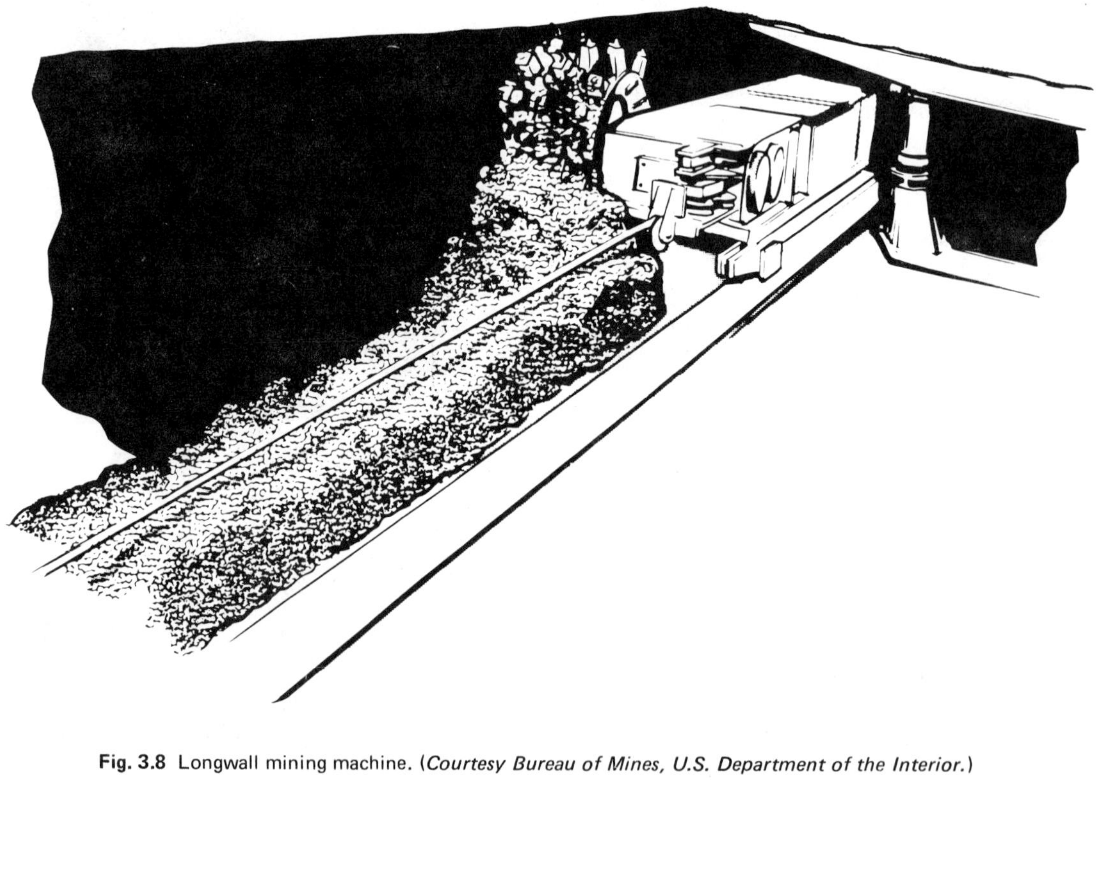

Fig. 3.8 Longwall mining machine. (*Courtesy Bureau of Mines, U.S. Department of the Interior.*)

2. Eliminates some permanent roof support costs.
3. Reduces ventilation, storage, and rock dusting costs.
4. Provides better ventilation and roof support.
5. Requires less supervision.
6. Safer where roof conditions are poor.

The comparative disadvantages are:

1. Requires large, level, straight blocks of coal free from obstructions.
2. Requires high capital investment for equipment.
3. Involves costly equipment moves.

Strip Mining

Open cut mines, commonly called *strip mines*, are used where the coal-to-overburden thickness ratio is not more than 1 to 20. The overlying earth and rock are stripped away to expose the coal, which is broken with explosives or dug without blasting.

Auger Mining

In some strip mines the overburden may become so thick that continued stripping is not economical. In such cases augers up to 60 inches in diameter are used to bore horizontal holes 100 to 200 feet into the coal seam. The coal falls from the augers into a conveyor and is elevated into a truck.

The number of coal mines by type is shown in Fig. 3.9 and the productivity of U.S. coal mines is shown in Fig. 3.10. The remarkable upward trend in productivity is due to increased mechanization, more efficient haulage procedures, and many auxiliary equipment improvements.

The U.S. Bureau of Mines publishes annually complete statistical data on coal production, consumption, distribution, and costs.[3]

COAL PREPARATION

Coal preparation is a collective term for the physical and mechanical processes applied to coal to make it suitable for a particular use. Some of these processes are breaking and crushing, screening, wet and dry concentrating, and dewatering. This section will discuss briefly only some wet methods of concentration to reduce the ash and sulfur content of coal. All aspects of coal preparation are described comprehensively in a recent book.[4]

Coal comes from the mine in a wide range of sizes and contains rock slate, pyrites, and

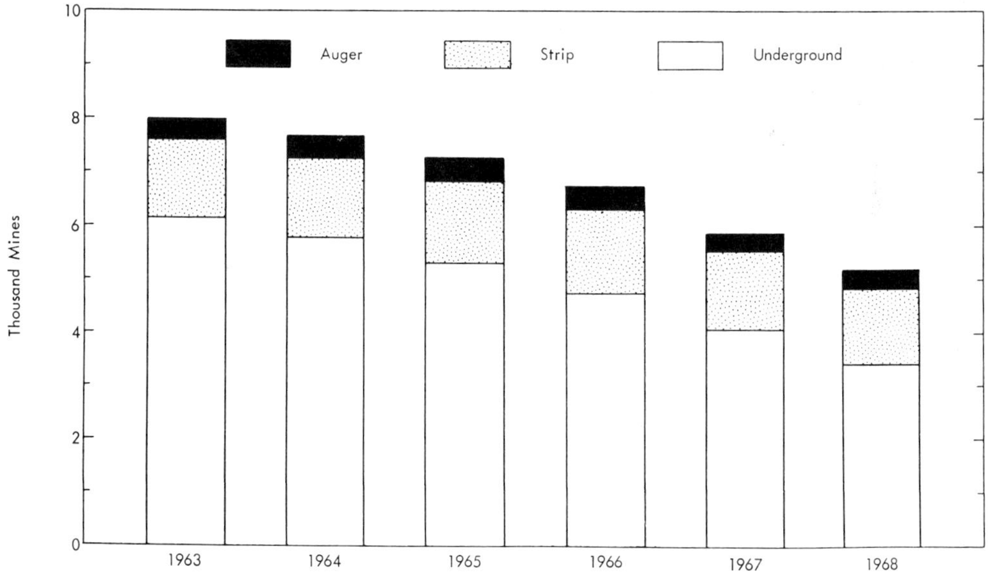

Fig. 3.9 Bituminous coal mines by type. (*Courtesy National Coal Assoc.*)

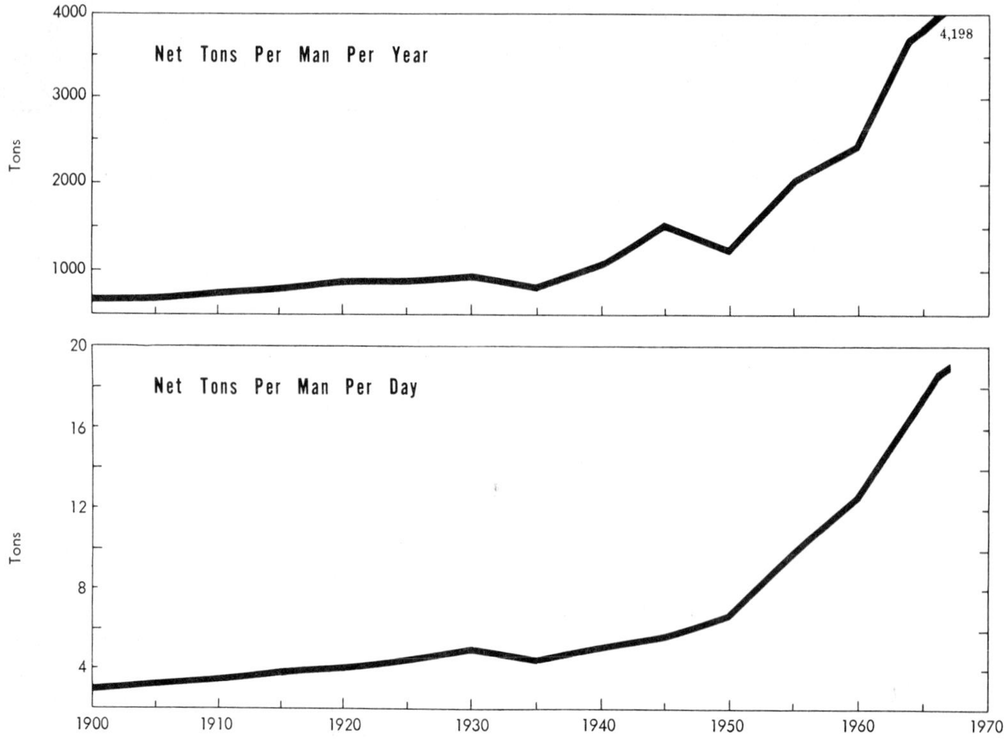

Fig. 3.10 Productivity of U.S. coal mines. (*Courtesy National Coal Assoc.*)

other impurities. Wet or dry concentration, hereafter called *cleaning*, is used to remove the impurities, thereby upgrading the coal for markets. Many users have rather rigid specifications concerning size and the ash and sulfur content. For example, coal for making metallurgical coke is specified as 80 per cent through a 1/8-in screen, 6 ± 1 per cent ash, and not more than 1.25 per cent sulfur.

The per cent of total coal production cleaned between 1940 and 1968 is shown in Table 3.5, which illustrates the importance of cleaning for the markets.

It is convenient to divide cleaning into two parts: *coarse-coal cleaning* and *fine-coal cleaning*, the dividing coal size being arbitrarily 3/8 in. Cleaning methods and devices are designed to separate coal from impurities by the differences in specific gravity, coal being less dense than its impurities. The exception is *flotation*, which depends on the differences in the surface characteristics of coal and its associated impurities.

Coarse-Coal Cleaning.

The most popular concentrating processes in this country are *jigging* and *dense-medium washing.*

Jigging. Separation of coal from its impurities by jigging consists of inducing pulsations in water in which the raw coal is suspended. An open-top rectangular box having a perforated bottom, or screen plate, is used. The pulsations are induced through the screen plate by reciprocating members, causing alternate expansion and compaction of the coal-water mixture, resulting in stratification of the coal according to the specific gravity of the particles. The lighter coal particles rise to the top, overflow the end of the jig, and are removed as clean product. The denser impurities settle on the screen plate and are withdrawn as *refuse*. Most jigs are of the Baum type, in which the pulsations are generated by air pressure. Plungers and diaphragms are also used as pulsers.

TABLE 3.5 Mechanically Cleaned Bituminous Coal (1000 Tons)

Year	*By Wet Methods*	*By Pneumatic Methods*	*Total Mechanically Cleaned*	*Per cent of Total Production Mechanically Cleaned*
1940	87,290	14,980	102,270	22.2
1945	130,470	17,416	147,886	25.6
1950	183,170	15,529	198,699	38.5
1955	252,420	20,295	272,715	58.7
1956	268,054	24,311	292,365	58.4
1957	279,259	24,768	304,027	61.7
1958	240,153	18,882	259,035	63.1
1959	251,538	18,249	269,787	65.5
1960	355,030	18,139	273,169	65.7
1961	247,020	17,691	264,711	65.7
1962	252,929	18,704	271,633	64.3
1963	269,527	19,935	289,462	63.1
1964	288,803	21,400	310,203	63.7
1965	306,872	25,384	332,256	64.9
1966	316,421	24,205	340,626	63.8
1967	328,135	21,268	349,402	63.2
1968	324,123	16,804	340,923	62.5

Dense-medium Washing. Coal is cleaned by immersing it in a fluid having a density between that of the coal and its impurities. Most dense-medium washers use a suspension of sand or magnetite in water to achieve the desired density. The modern highly automated dense-medium plants are capable of high throughput and sharp separation of the raw coal into a clean product and refuse.

Fine-Coal Cleaning

The principal fine-coal washers are the wet *concentrating table*, *dense-medium cyclone*, *hydrocyclone*, *launder*, *feldspar jig*, *hydrotator*, and *froth-flotation cell*. Of these, the concentrating table is the most popular, especially for cleaning bituminous coal. Cyclones are becoming increasingly important for washing both anthracite and bituminous coal, while launders and feldspar jigs are used to a limited extent for cleaning bituminous coal. The hydrotator is used extensively for anthracite.

Wet Concentrating Table. This is essentially a continuous mechanized form of the classical miner's pan. It consists of a transversely and longitudinally-tilted rectangular or rhombohedral table. A one-quarter size version of a commercial table is shown in Fig. 3.11. The diagonal ripples seen through the flowing coal-water mixture are caused by riffles in the deck. These riffles are directly responsible for the capacity of the table and represented a major improvement in tabling when it was first used to clean coal in 1890.

The upper side of the deck contains a feed-box at the corner near the head motion, and a wash-water distribution box. The deck is vibrated longitudinally with a low forward stroke and a rapid return. Normal decks are 8 feet wide and 16 feet long, and will clean up to 12 tons per hour per deck.

Dense-Medium Cyclone. Figure 3.12 shows the essential features of a dense-medium cyclone. The mixture of medium and raw coal enters tangentially near the top of the cylindrical section, producing free-vortex flow. The refuse moves downward along the wall and is discharged through the underflow orifice. The cleaned coal moves radially toward the cyclone axis, passes through the vortex

Fig. 3.11 A wet-concentrating table for cleaning fine coal. (*Courtesy Bureau of Mines, U.S. Department of the Interior.*)

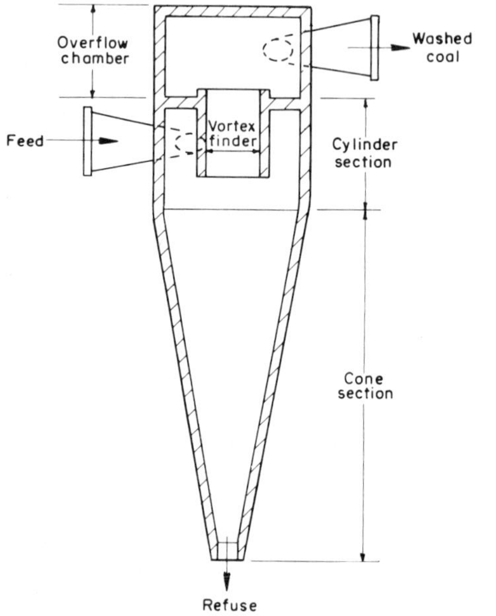

Fig. 3.12 Dense-medium cyclone for cleaning coal. (*Courtesy Bureau of Mines, U.S. Department of the Interior.*)

finder to the overflow chamber, and is discharged from a tangential outlet.

The dense medium used exclusively is pulverized magnetite suspended in water. The magnetite is recovered magnetically for re-use.

Cyclones are available in several sizes, 20-inch and 24-inch diameters being most common. A 20-inch unit will clean about 50 tons of raw coal per hour, and the 24-inch one about 75 tons per hour.

Hydrocyclone. The hydrocyclone, as the name implies, is a cyclone cleaning device that uses only water as a medium. Its design differs from that of the conventional dense-medium cyclone by having a greater cone angle and a longer vortex finder. Hydrocyclones currently are used to clean coal 0.025 inch and smaller, but can be used for coal as coarse as $^1/_4$ inch.

Other Cleaning Devices. Because of their limited use in the United States, the launder, feldspar jig, and hydrotator are not described in this section, and the reader is referred to an earlier reference[4] for complete information.

Froth Flotation. In the froth-flotation process fine coal slurry, to which a small quantity of flotation agents is added, is processed through a flotation cell. In the cell, finely disseminated air bubbles pass upward through the slurry. The hydrophobic coal particles attach themselves to the surface of the air bubbles and rise to the top to form a froth. The hydrophilic impurities sink in the cell and are removed as refuse.

There are three general classes of flotation agents; *frothers*, *collectors* or *promoters*, and *modifying agents.* The frother promotes stable froths. Examples are amyl and butyl alcohols and methylisobutyl carbinol (MIBC). The collector or promoter improves adherence of the coal particles to the air bubbles. Some frothers, e.g., MIBC, also function to a degree as collectors. Modifying agents have several functions, such as depressing flotation of unwanted material, altering the surface of the particles to aid the attachment of air bubbles, and providing the desired acidity or alkalinity of the flotation pulp.

The froth flotation process is used to clean minus 35-mesh coal.

COAL CLEANING AND AIR POLLUTION CONTROL

Until recently, coal cleaning was done mainly to reduce the amount of ash, and any concomitant reduction of sulfur was coincidental. However, owing to national efforts to reduce the amount of sulfur oxides discharged to the atmosphere, most of which comes from coal-fired electric utility plants, the removal of sulfur from steam coals is now an end in itself.

Sulfur occurs in coal in three forms: organic, pyrite (FeS_2), and sulfate. Sulfur as sulfate occurs up to only a few tenths of a per cent, the bulk occurring in the organic and pyritic forms. Since organic sulfur is part of the coal molecule, no method is now known for removing it economically. Consequently the amount of organic sulfur represents the lowest limit to which a coal can be mechanically freed of sulfur.

Since pyrite is nearly four times as dense as coal, it would seem possible to separate all of it from the coal substance. In practice, however, the amount of pyritic sulfur that can be removed varies considerably from one coal to another, depending on the way the pyrite is disseminated throughout the coal.

Where most of the pyrite can be removed, mechanical coal preparation is an economically-attractive process. And because pyritic sulfur comprises about one-half of the sulfur in coal, one of the more serious air pollution problems would be substantially reduced in such cases.

COAL COMPOSITION

Coal is composed chemically of carbon, hydrogen, oxygen, nitrogen, sulfur, and mineral matter (ash). The carbon, hydrogen, sulfur, nitrogen, and ash are determined directly, and the oxygen by difference, by means of an *ultimate analysis*, which has been standardized by the American Society for Testing and Materials (ASTM). The ultimate analysis is important for calculating material balances in thermal and chemical processes that use coal as a feed material (combustion, carbonization, gasification, hydrogenation, etc.). For certain uses, such as comfort and process heating, steam generation, and coking, it is usually sufficient to know the moisture, volatile matter, fixed carbon, ash, and sulfur contents, and the heating value* of the coal. The *proximate analysis* is used for this purpose. Both types of

---

*Heating value, Btu per lb, is the *high-heat value* (HHV), which is the heat of combustion at 20° C and constant volume when the fuel has burned to ash, CO_2 (gas), SO_2 (gas) and H_2O (liquid). The *low-heat value* (LHV) is calculated from the HHV by deducting 1030 Btu for each pound of water derived from a quantity of fuel, including both the water originally present as water and that formed by combustion. The HHV is most often used in the United States. The Dulong formula is quite accurate for calculating the heating value if the ultimate analysis is known, except for subbituminous and lignitic coals.

$$\text{Btu per pound} = 145.4C + 620\left(H - \frac{O}{8}\right) + 41S$$

C, H, O, and S are, respectively, the weight percentage of carbon, hydrogen, oxygen, and sulfur in the coal.

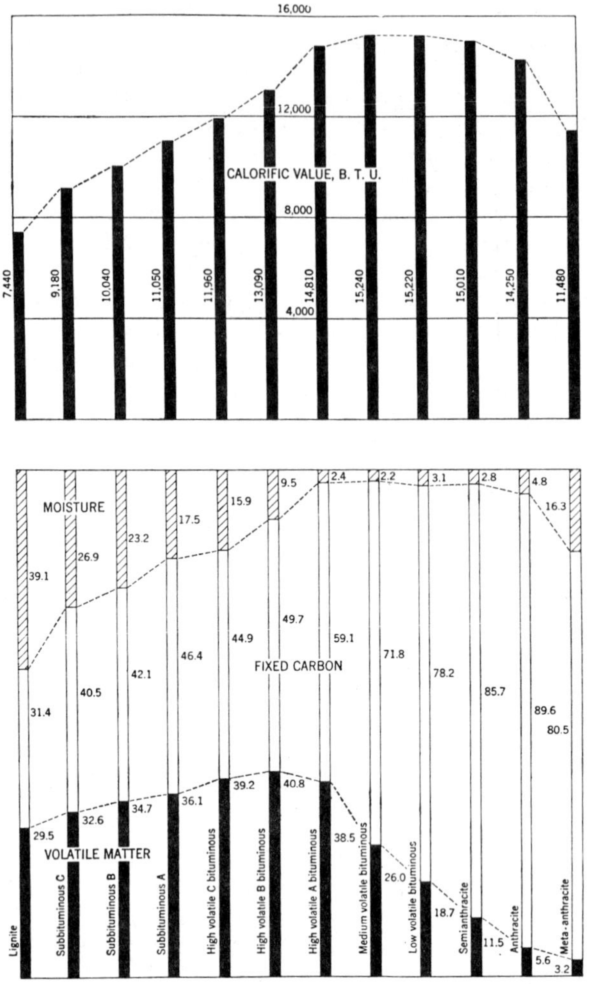

Fig. 3.13 Analysis of United States coals selected to represent the various ranks. *(Courtesy Bureau of Mines, U.S. Department of the Interior.)*

analyses are described in ASTM[5] and Bureau of Mines publications.[6]

Figure 3.13 shows the proximate analyses, on an ash-free basis, of coals selected to represent the various ranks. Analyses by state, country, and coal bed may be found in several Bureau of Mines publications; Bulletin 446 is a condensed, useful source of this information.

COAL UTILIZATION

The principal uses of coal may be classified broadly as follows:

1. *Combustion* to obtain thermal energy for power generation and process heating, including sintering and calcining.
2. *Carbonization* to convert coal to metallurgical and chemical coke, tars, chemicals, and industrial gases.
3. *Gasification* to obtain industrial gases for heating, chemical reduction, and hydrogenation and synthesis reactions.

Raw or processed coal, and fly ash from coal combustion, have several other uses, which will be described later.

Combustion

The greatest percentage of the coal consumed in this country is burned in boilers to generate steam for turboelectric plants in the electric utility industry. For example, of the 499 million tons consumed in 1968, the utilities used 298 million tons, or nearly 60 per cent of the total.

Remarkable advances have been made in the energy efficiency of electric utility plants because of improvements in the steam-generating equipment and turbines. The average heat rate of 1937 was 1.4 pounds of coal per kilowatt-hour, and 0.8 pound in 1968. Some plants recently placed in operation have heat rates of about 0.7 pound of coal per kilowatt-hour, which corresponds to an overall thermal efficiency of 39 per cent based on coal having an as-fired heating value of 12,500 Btu per pound.

Further improvements in thermal efficiency will be achieved with a new system of power generation that is now being developed. Known as MHD (magnetohydrodynamics), this system is expected to have a thermal efficiency between 50 and 55 per cent if coal is used as a fuel. The principle of MHD is to burn the fuel under conditions that will give a combustion gas temperature of about 4500°F, add an easily ionizable salt like potassium carbonate (seeding) so that the gas will have a relatively high concentration of free ions and electrons, and then pass this conducting gas through a magnetic field and draw off electric energy through electrodes placed in the path of the gas. By analogy, the hot conducting gas moving through a magnetic field behaves like the armature in a conventional turbogenerator. A coal-fired MHD generator would discharge less sulfur and nitrogen oxides to the atmosphere than a conventional plant of equivalent size because the seeding salt will absorb these oxides, which would be recovered during regeneration of the seed.

In the early days, boilers were rated by horsepower. One rated horsepower was defined as equivalent to 10 square feet of boiler heating surface. A developed horsepower was arbitrarily set as equal to the evaporation of 34.5 pounds of water per hour at 212°F. Later, an equivalent heat unit of 33,480 Btu per hour was adopted. From these definitions, R_b, per cent boiler rating, is obtained:

$$R_b = \frac{w(h - h_w) \times 100}{3348 \times S}$$

where w = steam flow, lb/hr
h = enthalpy of steam at boiler or superheater outlet, Btu/lb
h_w = enthalpy of water at boiler or economizer inlet, Btu/lb
S = boiler heating surface, sq ft

Per cent boiler rating is still used for small boilers and is a relative measure of "size" where furnace-wall cooling and superheating surfaces are absent or relatively small compared with boiler convection surface.

Utility and industrial boilers have a relatively large amount of heat-transfer surface other than boiler convection surface, and boiler horsepower and boiler rating are not suitable means of designating boiler output. Almost invariably output is now given in pounds per hour of steam flow at maximum continuous load, or in energy equivalent *megawatts.* Peak or overloads are given as percentages of full load.

Combustion Equipment. Industrial and electric utility boilers are fired with either stokers, pulverized-coal-fired burners, or cyclone burners, the choice depending on the kind of coal and the amount of steam needed. Table 3.6 gives the kinds of coal that can be burned with the various kinds of firing. Although good results will usually be obtained with the respective coal and firing equipment, this table should be used only as a rough guideline, and final equipment selection should be based on a sound engineering analysis.

(1) *Stoker Firing.* Table 3.7 gives the approximate range of capacity for each kind of stoker. The principles of stokers, showing the relative movement of fuel and air, are illustrated in Fig. 3.14. Both fuel and air have the same direction in retort stokers; this is called underfeed burning. The fuel moves across the air direction in chain- or traveling-grate stokers; this is called crossfeed burning. The spreader stoker approximates overfeed burn-

TABLE 3.6 Burning Equipment for Various Coals

	Stokers				
Fuel	*Under-feed*	*Traveling or chain grate*	*Spreader*	*Pulverized-coal burner*	*Cyclone burner*
Anthracite		x		x	
Bituminous:					
17–25% volatile	x		x	x	x
25–35% volatile				x	x
strongly coking	x	x	x	x	x
weakly coking		x	x	x	x
Lignite		x	x	x	x

ing, the incoming fuel moving toward the air. Except for certain types of coal gasifiers, in which lump coal moves downward toward a grate against air (or oxygen and steam) coming through the grate, no conventional combustion system operates purely in the overfeed mode.

(2) *Pulverized-Coal Firing and Cyclone Firing.* Electric utility and large industrial plants favor pulverized-coal-fired and cyclone-fired furnaces because of their inherent flexibility regarding the kind and quality of coal, their comparatively good availability, their quick response to load changes, and their extremely high steam-generating capacity. Recent units generate as high as 9 million pounds of steam per hour at 3500 psig and 1000°F.

The burner and furnace configurations for the main types of pulverized-coal firing (often called suspension firing), and cyclone firing, are shown in Fig. 3.15. There are some design variations in vertical, impact, and horizontal suspension firing, but these schematic drawings serve to illustrate the principles.

The first suspension-fired furnace in this country was designed like the one shown for vertical firing in Fig. 3.15. Pulverized coal (about 70 per cent through a 200-mesh screen) is transported to the burner with primary air, the amount of this air being about 20 per cent of that needed for complete combustion. The balance of the air, known as secondary air, is admitted through openings in the furnace wall. Because a large percentage of the total combustion air is withheld from the fuel stream until it projects well down into the

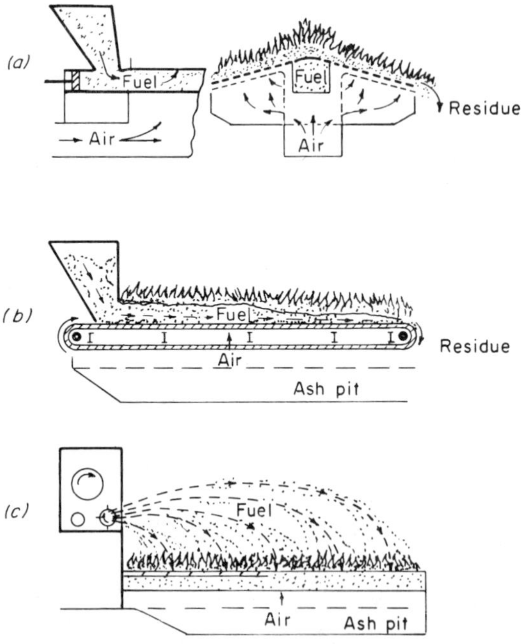

Fig. 3.14 Principles of mechanical stokers; (*a*) retort, (*b*) chain or traveling grate, and (*c*) spreader. (*Courtesy Bureau of Mines, U.S. Department of the Interior.*)

TABLE 3.7 Approximate Range of Capacity of Stokers

Type	*Steam, M lb/hr*	*Grate Heat Release, M Btu/hr sq ft (max.)*
Single retort	5–50	200
Multiple retort	40–300	300
Traveling or chain grate	10–300	300
Spreader	10–300	1,000

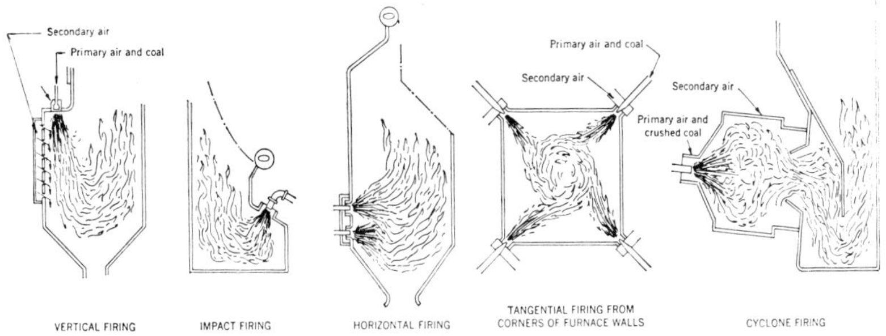

Fig. 3.15 Methods of firing pulverized and crushed coal. (*Courtesy Bureau of Mines, U.S. Department of the Interior.*)

furnace, ignition stability is good.[7] This type of firing is well suited for coals that are difficult to ignite, such as those with less than 15 per cent volatile matter. Although no longer used in central-station power plants, this design, with delayed admission of secondary air, may find favor again if low-volatile chars from various coal-conversion processes are burned for heat and power.

The other types of suspension firing use burners in which the primary air and coal, and the secondary air are mixed just before or immediately after entering the furnace. With tangential firing the burners are arranged in vertical banks at each corner of a square, or nearly square, furnace and directed toward an imaginary circle in the center of the furnace. This produces a vortex with its axis on the vertical center line. The burners consist of an arrangement of slots one above the other admitting through alternate slots, primary air-fuel mixture and secondary air. The burners can be tilted upward or downward 30 degrees from the horizontal plane, enabling the operator to control superheat and to permit selective utilization of furnace heat-absorbing surfaces. Basically, the turbulence needed for mixing the fuel and air is generated in the furnace instead of in the burners.

In cyclone firing, the coal is not pulverized as for suspension firing but is crushed to 4-mesh size and admitted tangentially with primary air to a water-cooled cylindrical chamber called a cyclone furnace. The finer particles burn in suspension, and the coarser ones are thrown by centrifugal force to the furnace wall. The wall, having a sticky coating of molten slag, retains the coal until it burns. The secondary air, which is admitted tangentially along the top of the furnace, completes the combustion of the coarse particles. Slag drains continuously into the main boiler furnace.

(3) *Fluidized-Bed Combustion.* Recently, interest has been growing in another mode for burning coal—the fluidized bed. Although still in the experimental stage, there are clearly seen potential advantages to this burning mode. a) The fuel-bed temperature is low, about 1800°F, which means less formation of nitrogen oxides, and retention of some of the sulfur in the ash of certain coals. (Adding dolomite to the fuel bed greatly improves sulfur retention.) Equally important is less volatilization of sodium and potassium in the coal, consequently, there are less deposits on and corrosion of furnace, superheater, and reheater tubes. b) Heat-transfer rates from the fluidized bed to immersed heat-transfer surfaces are relatively high, as much as 100 Btu per hr per sq ft per degree Fahrenheit.

It is believed that successful application of fluidized-bed boilers to electric-utility stations would reduce capital costs about 10 per cent compared with conventional plants of the same capacity. The principle of the fluidized-bed combustor is illustrated schematically in Fig. 3.16.

There is a commercial stoker that burns low-grade coals in a fluidized mode. It is called the *Ignifluid Stoker*, which was developed about 21 years ago. The Ignifluid combustion sys-

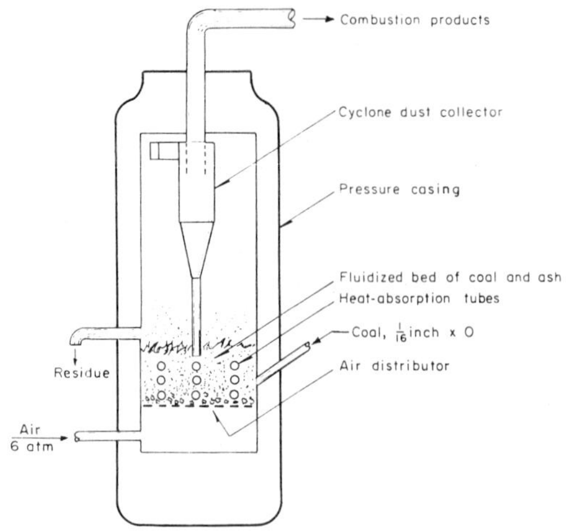

Fig. 3.16 Schematic of pressurized fluidized bed steam generator. *(Courtesy Bureau of Mines, U.S. Department of the Interior.)*

tem was first installed in France, and numerous industrial plants in Europe have since used it.

Basically, the Ignifluid system is a combustion chamber with a narrow chain grate inclined from the front to the rear of the chamber about 15 degrees. Naturally formed fuel banks between the front and sides of the chamber and the stoker surround the fluidized bed and protect the lower parts of the wall against excessive heat and adhering clinkers. The fuel banks have no part in the combustion because no air passes through them, but their surface is continually renewed by incandescent particles thrown up from the fluidized bed. Since some of the fuel is burned in suspension and the rest on the grate, this system differs from the fluidized-bed system described previously, in which the entire coal charge is fluidized. The Ignifluid system is not adaptable in its present form to steam-generating tubes within the bed, which would be necessary to achieve the high-steaming capacities of modern electric utility plants.

Boiler Types. There are various kinds of industrial and utility boilers, broadly classified as fire-tube and water-tube. In the former, the hot combustion gases pass through tubes, and heat is transferred to water outside the tubes. The most common and least expensive boiler of this type is the horizontal return tubular (HRT) boiler. However, because of the design and construction of fire-tube boilers, there is a definite limitation to their size and the pressure which they can tolerate. Water-tube boilers may be broadly classified as straight-tube and bent-tube types, the latter having several variations in design and being preferred for applications where higher capacities and steam pressures are required. In both types, heat is transferred by radiation or convection to the outside of the tubes, and water flows inside the tubes as the result of thermal circulation, or in the case of certain bent-tube boilers, as the result of forced circulation. A comparatively new version of the forced-circulation, bent-tube boiler for central-station power plants in the "once-through" type. The feedwater passes progressively through the heating, evaporation, and superheater sections; no drum is used for separating the steam from the boiler water as in other boilers, consequently, the ratio of water circulated in the boiler to steam generated is unity. The data in Table 3.8 approximate the range of steam capacities and pressures for the principal types of boilers.

Combustion Calculations. The amount of air or oxygen just sufficient to burn the carbon, net hydrogen, and sulfur to carbon dioxide is the theoretical or stoichiometric amount of air or oxygen. The general expression for combustion of a fuel is

$$C_mH_n + [(4m + n)/4]\,O_2 = mCO_2 + (n/2)H_2O$$

m and n being the number of atoms of carbon and net hydrogen, respectively, in the fuel. For example, one mole (an amount equal to the molecular weight of the fuel) of methane (CH_4) requires 2 moles of oxygen for complete combustion to 1 mole of carbon dioxide and 2 moles of water. If air is used, each mole of oxygen is accompanied by 3.76 moles of nitrogen.

The theoretical weight or volume of oxygen or air to burn a given weight of fuel is or primary interest in designing combustion equipment. The *volume of theoretical oxygen* to burn any fuel can be calculated from the ulti-

TABLE 3.8 Approximate Range of Capacities of Various Types of Industrial and Utility Boilers

Type	*Capacity, lb steam/hr*	*Maximum Design Pressure, psig*
Fire tube (HRT)	1,000–15,000	250
Water tube		
straight	15,000–150,000	2000
bent, 3-drum, low-head	1,000–35,000	400
2-drum, vertical	1,000–350,000	1000
electric utility	up to 9,000,000	3500

mate analysis of the fuel as follows:

$$359\left(\frac{C}{12}+\frac{H_2}{4}-\frac{O_2}{32}+\frac{S}{32}\right)$$
$$= \text{cu ft of oxygen/lb of fuel}$$

where C, H_2, O_2, and S are the fractional weights of these elements in 1 lb of fuel. The weight of oxygen in pounds is obtained by multiplying the cubic feet of oxygen by 0.0891.

The *volume of theoretical air* is obtained by using a coefficient of 1710 instead of 359 in the equation above.

More than the theoretical amount of air is always necessary in practice to achieve complete combustion. The *excess air* can be calculated by the equation:

$$A_{xs} = 100\ [(A/A_t - 1)]$$

in which A_{xs} is the percentage of excess air, A_t is the theoretical air in lb/lb, and A is the air actually used in lb/lb. If one wishes to know the percentage of excess air, and only the flue gas composition is known, it can be calculated from

$$A_{xs} = 100\ [O_2/(0.266N_2 - O_2)]$$

where O_2 and N_2 are the percentages by volume of these components in the dry flue gas. This equation is applicable only when the nitrogen in the fuel is negligible and there are no combustible gases, such as carbon monoxide or hydrogen, in the flue gas.

Sintering and Calcining

Sintering is a process of agglomerating finely sized fusible materials to obtain a product suitable in size for a particular use. Typical applications of sintering are the production of lightweight aggregate for use in concrete, and of sintered ore for use in blast furnaces. The Dwight Lloyd-type machine, which is widely used to sinter fine materials, is shown schematically in Fig. 3.17. The mixture of coal (or coke) and the material to be sintered, which may be pelletized ore or fly ash from power plants, is fed to the grate and ignited by a gas- or oil-fired hood. Air is drawn through the charge by induced draft, and ignition and burning occur from the top to the bottom of the bed. The quality of the sinter depends mainly on the composition of the mixture, the thickness of the bed, the speed of the grate, and the amount of air drawn through the bed. Usually, the windbox is compartmented to control the amount of air flowing through the bed at certain points.

Lightweight aggregate for concrete is also produced by burning pelletized refuse coal from coal-preparation plants on conventional traveling grates.

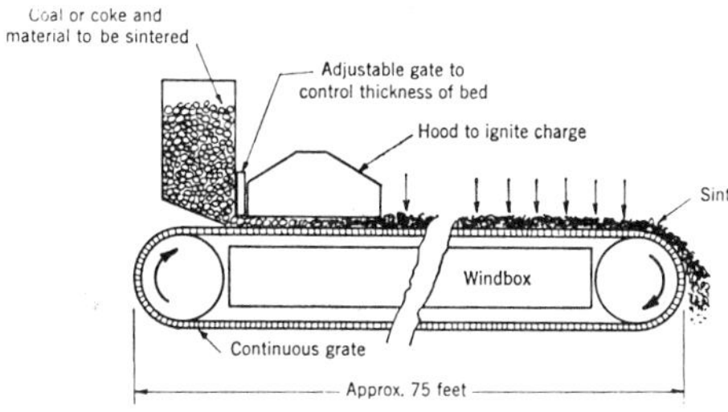

Fig. 3.17 Dwight Lloyd stoker for manufacturing sintered products. (*Courtesy Bureau of Mines, U.S. Department of the Interior.*)

A significant amount of coal is consumed annually for calcining to make cement, lime, gypsum, alumina, and magnesia, and for firing ceramic products. Calcining is usually done in rotary kilns, and ceramics are fired in tunnel kilns and muffles. Portland cement requires approximately 75 pounds of coal per barrel, lime 500 pounds of coal per ton, and gypsum 100 pounds of coal per ton.

Considerable attention is being given to the production of lightweight aggregate from pulverized fuel ash (PFA), also called fly ash. Several different processes have been devised, but they all depend on sintering.[8] The PFA, to which carbon in some form is added to bring the total amount up to 5 per cent, is usually pelleted with water to give pellets about the size of marbles. The pellets are sintered by heating them in a furnace or kiln to about 2200°F. The processes yield a hard, cellular, lightweight material having a bulk density of 50 pounds per cubic foot.

Coal Carbonization

This important part of coal technology is discussed in chapter 8. However, coal carbonization, or pyrolysis, will be discussed later in relation to some new experimental coal-conversion processes.

Coal Gasification

Gasification as used here means reacting coal with air, oxygen, steam, carbon dioxide, or a mixture of these, in a manner that yields a gaseous product suitable for use either as a source of energy, or as a raw material for the synthesis of industrial chemicals, liquid fuels, or other gaseous fuels. Some of the important synthesis products are synthetic pipeline gas (methane), hydrogen, ammonia, alcohols, olefins, and distillable oils. Still in the experimental stage is hydrogasification, which involves reacting coal with hydrogen at high pressure and temperature to obtain a methane-rich gas.

Currently in this country there is an urgent need to supplement the dwindling reserves of natural gas. Also, there is serious need for a sulfur-free gas from coal that can replace raw coal as a boiler fuel, thus abating atmospheric pollution by sulfur oxides. Processes based on coal gasification or hydrogasification are currently being investigated in the United States to satisfy these needs.

Coal gasification is discussed comprehensively by von Fredersdorff and Elliott,[9] consequently, the topic will be covered only briefly here.

There are two basic kinds of coal gasification. One method uses air and steam to react with the coal giving a gas with a low heating value, about 150 Btu per cubic foot. This is called *producer gas*. The other uses oxygen and steam, giving a nitrogen-free gas having a heating value up to 500 Btu per cubic foot depending on the equipment, coal, and operating conditions. This is called *synthesis gas*. The carbon in the coal reacts with the steam as follows:

$$C + H_2O \longrightarrow CO + H_2^*$$

The oxygen added with the coal and steam burns some of the coal to attain the high temperature needed for the reaction to occur at a reasonable rate.

Coal gasification can be conveniently divided into three categories:

1. Lump coal is supported on a grate or by other means, and the flow of gas and coal may be countercurrent or concurrent. The former is the most common process. This is called *fixed-bed gasification*.

2. Crushed or fine coal is fluidized by the gasifying medium, giving an expanded fuel bed that can best be visualized as boiling or jiggling. This is called *fluidized-bed gasification*.

3. Fine coal is suspended in the gasifying medium, the particles moving with the gas, either linearly or in a vortex pattern. This is called *suspension* or *entrainment gasification*.

These processes are shown schematically in Fig. 3.18.

Fixed-bed and fluidized-bed systems cannot be operated with strongly caking coals, which go through a plastic stage during heating and agglomerate into a mass that impedes the flow of gas. However, such coals can be pretreated to lessen their caking tendency, but this adds to the cost of the product gas. Or,

*Under certain conditions some methane is also formed during gasification.

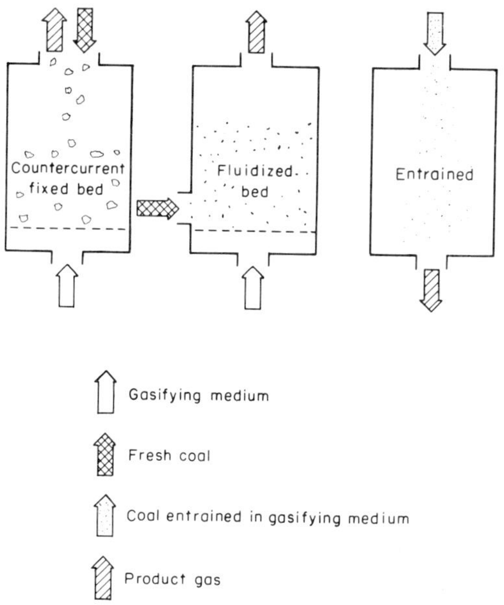

Fig. 3.18 Coal gasification systems. *(Courtesy Bureau of Mines, U.S. Department of the Interior.)*

under certain circumstances, agglomeration of caking coals in the fuel bed can be prevented by mechanical stirrers, thereby keeping the bed open to the flow of gases. Any rank of coal can be used in entrained gasification systems.

There exist, at least as theoretical schemes, over a hundred different forms of gasifiers and gas producers, but only a few of those that have been built and tested will be described here.

Fixed-Bed Gasifier Using Air and Steam at Atmospheric Pressure. The conventional gas producer operates at atmospheric pressure and gasifies coke, anthracite coal, and noncaking or weakly caking bituminous coal, using air and steam for the gasifying medium. The optimum fuel size is ½ to 2 in., and the ash-fusion temperature should exceed 2100°F. With coal, the heating value of the gas may be 170 Btu per cubic foot, and the gasification rate up to 50 pounds per hour per square foot of grate area. A well-known producer of this type, the *Wellman-Galusha*, is shown schematically in Fig. 3.19. It was used by the Bureau of Mines for an investigation to determine the technical feasibility of gasifying anthracite in a commercial-size unit.[10]

Fixed-Bed Gasifiers Using Oxygen and Steam at Atmospheric Pressure. These gasifiers can be divided into nonslagging and slagging, which mean, respectively, removing the ash from the gasifier dry or as a molten slag. Operating slagging simplifies gasifier design, reduces steam consumption, and permits operation with coals having low ash-fusion temperatures.

In addition to the Wellman-Galusha mentioned above, nonslagging operation with anthracite and coke has been reported for a *Kerpely* producer and a *Pintsch-Brassert* generator. Tests with anthracite showed gasification rates and heating value of the gas were nearly double with oxygen compared with air.

Slagging fixed-bed gasifiers operating at atmospheric pressure and having high specific throughputs, include the *Thyssen-Galocsy* and the *Koppers.*[14] Thyssen-Galocsy gasifiers were used in Hungary to make synthesis gas. This gasifier is essentially a small blast furnace with tuyeres at three levels. At the lowest row of tuyeres, some of the make gas is burned in a mixture of steam and oxygen to produce a highly-heated blast.

Koppers gasifiers were used in Switzerland to make synthesis gas for manufacturing methanol. Ten tuyeres are provided for the injection of an oxygen-carbon dioxide blast.

Fixed-bed Gasifiers Using Oxygen or Air at High Pressure. The Lurgi gasifier was developed in Germany, and before the war it was being used to furnish town gas having a heating value of 450 Btu per cu ft, by gasifying brown coal with oxygen and steam at 300 psi. In time, the Lurgi gasifier was able to handle weakly caking bituminous coal by the development of a rotating grate and a stirring arm, both attached to a water-cooled central drive shaft.

The largest user of nonslagging Lurgi gasifiers is South African Coal, Oil and Gas Corp., Ltd. (SASOL) in Sasolburg, Republic of South Africa. There, 13 gasifiers generate 205 million cubic feet per day of synthesis gas from bituminous coal to produce liquid and solid hydrocarbons and a wide variety of chemicals.

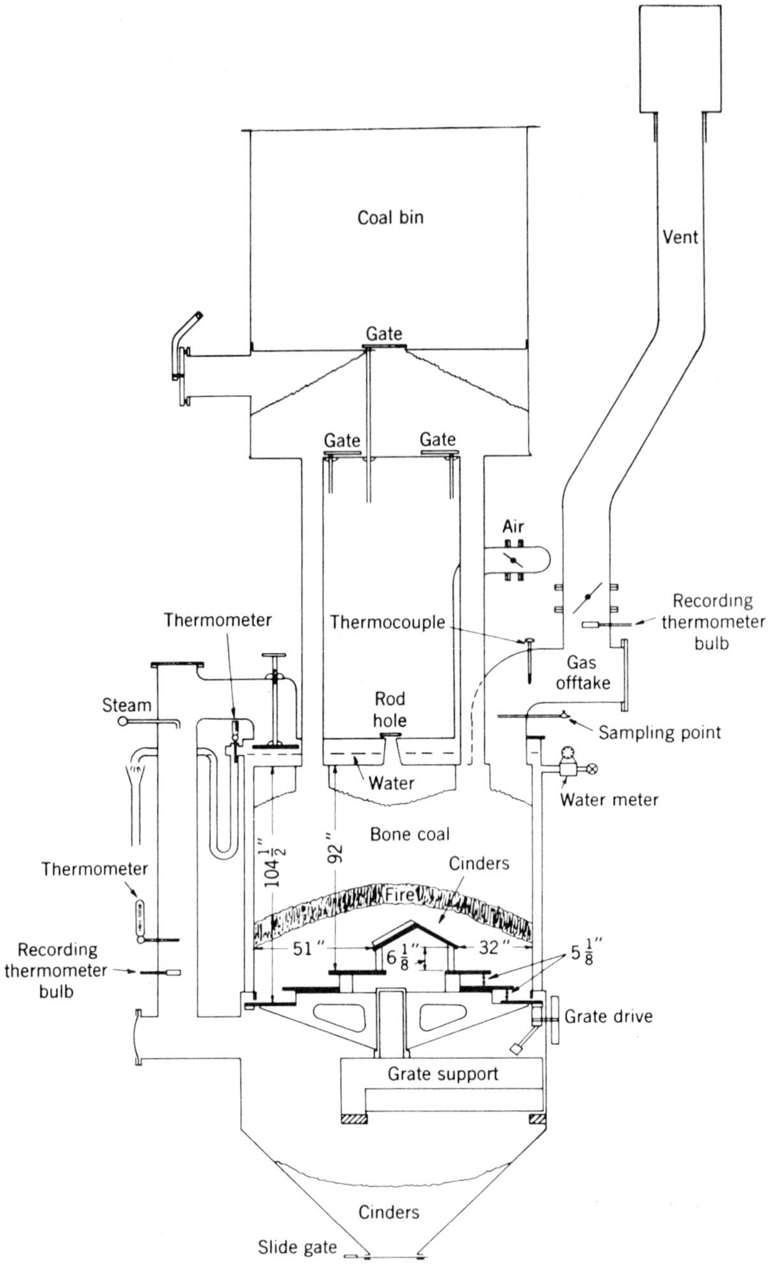

Fig. 3.19 Wellman-Galusha gas producer. (*Courtesy Bureau of Mines, U.S. Department of the Interior.*)

SASOL also supplies town gas to a public utility corporation for transmission and distribution.

The gasifiers have an internal diameter of 12 ft and are 26 ft high. A sectional view is shown in Fig. 3.20.

Currently, the Bureau of Mines is experimenting with a nonslagging fixed-bed pro-

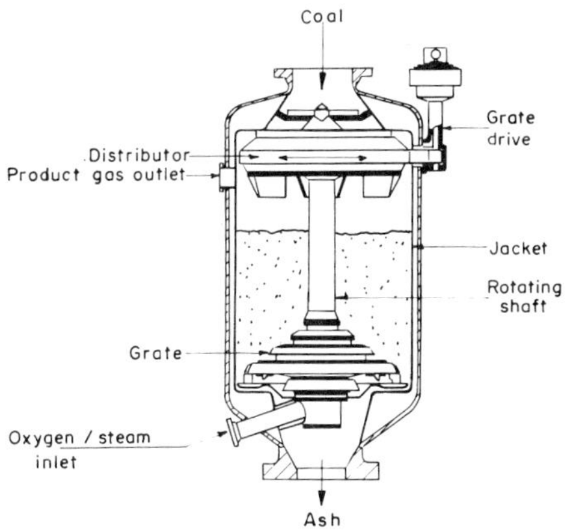

Fig. 3.20 Lurgi high-pressure gasifier. (*Courtesy Bureau of Mines, U.S. Department of the Interior.*)

ducer designed for 300 psig.[11] The aim is to generate producer gas under pressure from strongly caking coals, which cannot be done with conventional fixed-bed producers. The agitator or stirrer is an unique design, consisting of two rabble arms located two feet apart on a vertical shaft; both arms and the entire shaft are water cooled. A third rabble arm, uncooled and two feet above the middle arm, levels the fuel bed. The agitator rotates and moves up and down with an adjustable stroke. This compound motion both breaks up agglomerated coke masses, and fills in voids and blowholes. A schematic of the producer is shown in Fig. 3.21.

Slagging fixed-bed gasifiers for high pressure operation have not been developed to industrial size. The Lurgi Company constructed a special gasifier for slagging operation, but the work was discontinued prematurely. The Gas Council in England modified this same unit many ways to achieve continuous tapping of the slag, but used coke instead of coal.[12]

The Bureau of Mines designed, built, and tested a coal-gasification pilot plant to operate slagging with lignite.[13] Runs were made at 80 to 400 psig, yielding a gas with a heating value of about 350 Btu per cu ft. Although continuous slagging was achieved, no satisfactory means were found to protect the taphole from slag attack. However, these studies gave good indications of what the material requirements, gas composition, and gas and tar yields would be in an industrial-size slagging gasifier.

Fluidized-Bed Gasifiers. The *Winkler gasifier* was the first fluidized-bed gasifier on an industrial scale. It was developed in prewar Germany to make synthesis gas at atmospheric pressure from brown coal and bituminous coal, using oxygen and steam. It was also operated with air and steam to make producer gas. Typically, a Winkler gasifier is 13 to 15 feet in diameter and 65 ft high. Coal sized to $1/3$ in. by 0 is screw-fed into the side of the gasifier near the base, and the gasifying medium is fed through a grate or inverted cone in the bottom. Secondary oxygen or air is fed through multiple openings in the side, above the coal-feed point. The temperature in the fuel bed ranges from 1400 to 1800°F, depending on the reactivity of the coal. Although, coals having low reactivity and ash-fusion temperature are not suitable for the Winkler gasifier, it is well suited for high-ash coals.

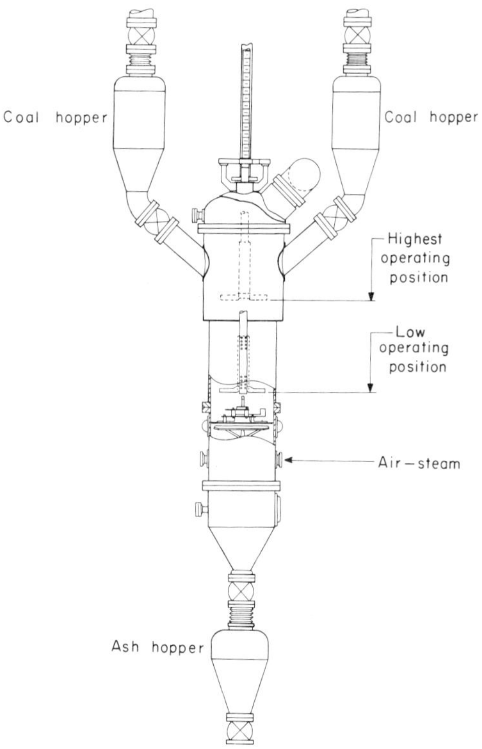

Fig. 3.21 Bureau of Mines fixed-bed producer. (*Courtesy Bureau of Mines, U.S. Department of the Interior.*)

Currently, the Bureau of Mines is developing a fluidized-bed gasifier that can handle strongly caking coals and make a gas containing up to 25 per cent methane (after purification). These characteristics are important for the coal-to-pipeline gas process that will be described later. It was mentioned earlier that strongly caking coals must be pretreated before they can be used in fixed-bed or fluidized-bed gasifiers. The Bureau of Mines gasifiers (Fig. 3.22) integrates three steps in the gasification process: pretreatment of freely falling coal particles with oxygen and steam in the upper section to destroy their caking properties; devolatilization in the expanded section; and gasification in the lower section. Ash is removed from the bottom. The unit operates at 600 psig with a fluid-bed temperature of about 1800°F.

Suspension Gasifiers. The suspension gasifier can handle any rank of pulverized coal;

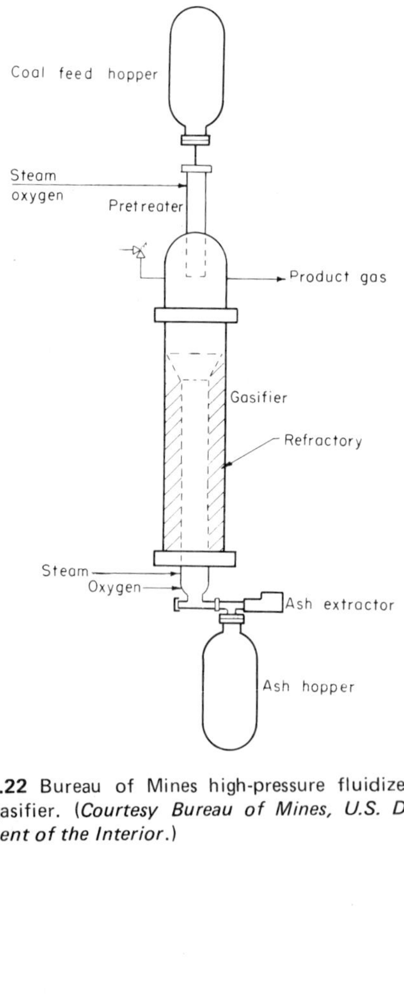

Fig. 3.22 Bureau of Mines high-pressure fluidized-bed gasifier. (*Courtesy Bureau of Mines, U.S. Department of the Interior.*)

produce a gas with no tar and very little methane, which is important for many synthesis processes; achieve high production rates per unit volume of reaction space; and operate either nonslagging or slagging. These advantages must be weighed against the need to recover the heat in the high-temperature gas, and the relatively low carbon conversion, 85 to 90 per cent, compared with 100 per cent in other gasification systems.

The *Koppers-Totzek gasifier* operates nonslagging at atmospheric pressure with oxygen and steam to make synthesis gas in commercial ammonia and chemical plants in Japan, Finland, France, and Belgium. The Bureau of Mines used the first large-scale unit in its synthetic liquid fuels demonstration plant, but the results were disappointing. Improved designs subsequently led to the commercial installations mentioned above.

The gasification chamber is essentially a horizontal cylinder with a central vertical gas outlet. Coal, oxygen, and steam are injected through the opposite ends of the chamber producing a high degree of turbulent mixing in the chamber.

In cooperation with the Bureau of Mines, the Babcock and Wilcox Co. (B&W) made pilot-plant studies on gasifiers for slagging operation at atmospheric pressure with coal, steam, and oxygen. These studies led to the design of a unit with a capacity of 17 tons of coal per hour and an output of 25 million cubic feet of carbon monoxide and hydrogen per day. Described as the *B&W-duPont gasifier*, it was to be installed at the Belle, West Virginia, Works of the Du Pont Co. Operating data on this unit have not been published, but information is available on an experimental unit of the same design with a coal capacity about one-tenth that of the commercial unit. Operating results with this unit are given in a table at the end of this section.

Basically, the B&W-duPont gasifier is a vertical refractory-lined vessel divided into a primary and a secondary zone. In the lower or primary zone the temperatures are high enough to melt the ash to a fluid slag that is tapped continuously from the bottom of the chamber. The upper or secondary zone is designed for lower gas temperatures and the accomodation of semiplastic coal ash. A heat-recovery section at the top, integral with the gasifier, generates steam from the heat in the product gases. Pulverized coal is conveyed by steam at 325°F. The coal, steam, and oxygen enter diametrically-opposed nozzles inclined 45 degrees downward in the primary zone.

The *Pandinco gasifier* operated nonslagging at atmospheric pressure with oxygen and steam to make synthesis gas from French coals. It is a refractory-lined tower 3 ft 8 in. in diameter and 38 ft high. Pulverized coal conveyed with oxygen is injected through a nozzle downward through the top of the chamber, and highly preheated oxygen and steam are fed through nozzles concentric with the coal nozzle.

The Bureau of Mines designed, constructed, and tested a series of experimental high-pressure gasifiers, which finally evolved to the design shown in Fig. 3.23.[15] These studies were aimed at developing a high-pressure gasifier that would produce a gas suitable for synthesizing liquid fuels from any rank of coal. Pulverized coal from a pressurized, fluidized

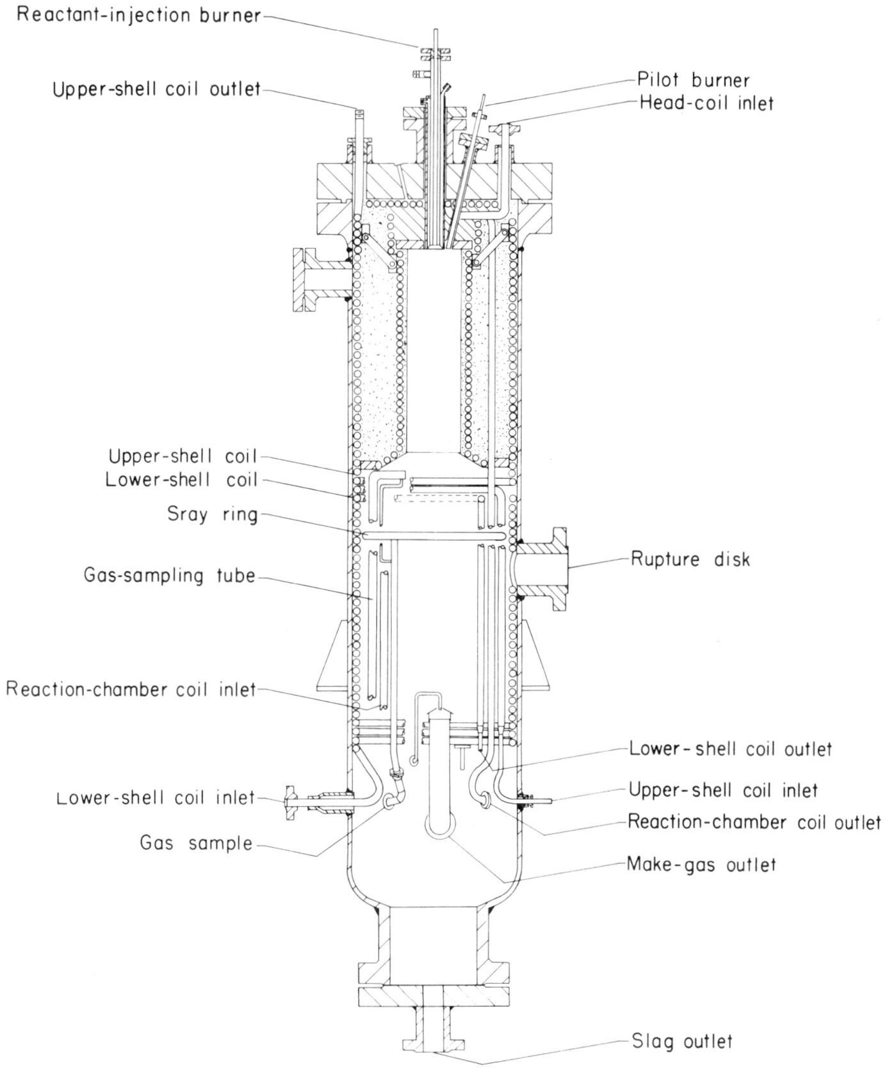

Fig. 3.23 Bureau of Mines high-pressure suspension-type gasifier. (*Courtesy Bureau of Mines, U.S. Department of the Interior.*)

feeder entered the central pipe of three concentric pipes comprising the axial reactant-injection burner. Oxygen and superheated steam entered the next concentric pipe. The outermost pipe was provided with a water-cooled jacket for insulation.

The Texaco gasifier is the only high-pressure, slagging, suspension gasifier that has ever been operated on an industrial scale. A unit operating at 400 to 500 psig in an Olin Matheson Chemical Corporation plant produced a gas for ammonia synthesis. The unique feature of this unit was the method for conveying the coal. Ground to pass a 40-mesh screen, the coal was mixed with water to make a slurry, which was preheated to 700–1000°F. Up to 75 per cent of the steam was removed in a cyclone separator, and the final feed to the gasifier contained 55–60 per cent coal by weight. No performance data have been reported.

A high-pressure, slagging, suspension gasifier is being developed by Bituminous Coal Research, Inc., aiming to obtain a high concentration of methane in the product gas.[16] It will be a two-stage gasifier with pulverized coal entering the upper stage. Here, it reacts with a rising stream of hot synthesis gas produced in the lower stage and is partly converted to methane and more synthesis gas. The residual char leaves the upper stage with the gas; the char is separated from the gas and returned to the lower stage of the gasifier. Here, the char is gasified completely with oxygen and steam, producing the synthesis gas needed in the upper stage and the heat needed for the endothermic reactions. The BCR gasifier is shown schematically in Fig. 3.24.

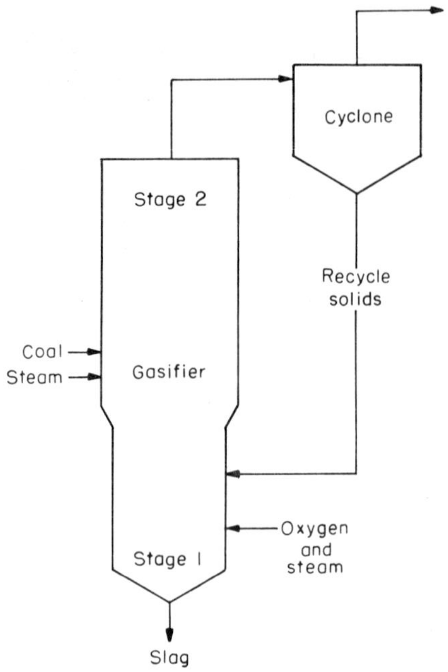

Fig. 3.24 Bituminous Coal Research high-pressure suspension-type gasifier. *(Courtesy Office of Coal Research, U.S. Department of the Interior.)*

Gasification Without Oxygen. Research and development are underway on coal gasification processes that do not use oxygen, which is a relatively expensive item. Such processes would reduce the cost of making synthetic pipeline gas, synthesis gas, or hydrogen. None of these gasification processes have yet come into industrial use.

The *Carbon Dioxide-Acceptor process*, developed by Consolidation Coal Co., makes use of the heat evolved when lime reacts with carbon dioxide and carbon with hydrogen. Since this process is designed specifically to make synthetic pipeline gas, more details will be given in a later section on this subject.

The *molten-salt process*, developed by the M.W. Kellog Co., involves injecting coal and steam continuously into molten sodium carbonate at about 1700°F. The heat needed to keep the salt molten is furnished by hot flue gas from the combustion of coal.

Comparison of Some Gasification Processes. Table 3.9 gives performance data for some oxygen-steam gasification systems. It is a composite modified version of tables compiled by Hoy and Wilkins,[14] and Shires[17] and includes some recent Bureau of Mines data.

SYNTHETIC PIPELINE GAS FROM COAL*

Basically, there are two preferred ways, alone or in various combinations, to make synthetic pipeline gas from coal.[19] These are *gasifi-*

*Synthetic pipeline gas must have a minimum heating value of 900 Btu per cubic foot and not more than 0.1 per cent by volume of carbon monoxide.

TABLE 3.9 Comparison of Some Industrial and Commercial Coal Gasifiers

	Industrial					*Experimental*		
Gasifier	*Wellman-Galusha*	*Lurgi*	*Winkler*	*B&W-duPont*	*Koppers-Totzek*	*Lurgi*[13]	*BuMines*[15]	*BuMines*
Type	fixed bed	fixed bed	fluid bed	suspension	suspension	fixed bed	suspension	fluid bed
Pressure, psig	1	300	1	1	1	400	300	600
Ash removed	dry	dry	dry	slag	dry	slag	slag	dry
Blast (including steam)	air	oxygen	oxygen	oxygen	oxygen	oxygen	oxygen	oxygen
Fuel size[a]	$1/2 \times 1$ in.	$1/4 \times 2$ in.	$1/3 \times 0$ in.	pulv.	pulv.	$3/4 \times 3/8$ in.	pulv.	pulv.
Fuel rate, lb/sq ft-h[b]	50	300	130	135		875	400	200
Gas heating value Btu/cu ft	140	285	270	270	270	350	280	350
Gas composition, per cent/vol.[c]								
CO_2	5	30	16	16	13	7	11	30
CO	27	16	44	40	51	57	52	23
H_2	14	43	36	40	34	29	32	29
CH_4	nil	9	1	1	nil	6	nil	18
N_2	53	–	–	–	–	–	–	–

[a]Pulv. usually means pulverized so that 70 to 80 per cent of the coal passes a 200-mesh screen (0.003 in.).

[b]Based on cross-sectional area.

[c]Rounded values for the major constituents only. Composition will vary somewhat for each gasifier according to kind of coal gasified and operating conditions, consequently, these values are useful only qualitatively.

cation-synthesis and *hydrogasification.* (Fig. 3.25.)

Gasification-Synthesis

Coal is gasified with oxygen and steam at high pressure, and the resulting synthesis gas is passed over a catalyst, preferably nickel, converting it to methane by the reaction:

$$3H_2 + CO \longrightarrow CH_4 + H_2O$$

The *Synthane* process, which is being developed by the Bureau of Mines, illustrates the conversion of coal to synthetic pipeline gas via the gasification-methanation route.[20] The unique feature of this process is the comparatively high concentration of methane in the synthesis gas, up to 25 per cent carbon dioxide-free basis. The advantages of this are: (1) because of the exothermicity of methane formation (70 Btu per cu ft of synthesis gas converted) less process oxygen is needed, and (2) a lower capital cost for purification, shifting, and methanation that if the feed gas contained no methane. For example, without methane in the purified feed gas the methanator must convert 400 cubic feet of synthesis gas to make 100 cubic feet of methane. With 25 per cent methane in the feed gas, the methanator would need to convert only 227 cu ft of synthesis gas—a volume reduction of 42.5 per cent.

The *Synthane* process is shown schematically in Fig. 3.26. The high-pressure (600 psig) fluidized-bed gasifier described earlier and shown in Fig. 3.22 is used for the experimental studies. The shift converter adjusts the hydrogen-carbon monoxide ratio to 3:1. The raw gas is purified with a hot potassium carbonate scrubber and activated carbon, so that the gas fed to the methanator contains no more carbon dioxide than 2 volume per cent and no more than 1 ppm of sulfur compounds. An alloy of nickel and aluminum, *Raney nickel*, is the methanation catalyst.

Gas from any of the high-pressure gasifiers described earlier, with or without oxygen, are feasible sources of gas for the front end of the process shown in Fig. 3.26, but the economics are favored by methane in the feed gas.

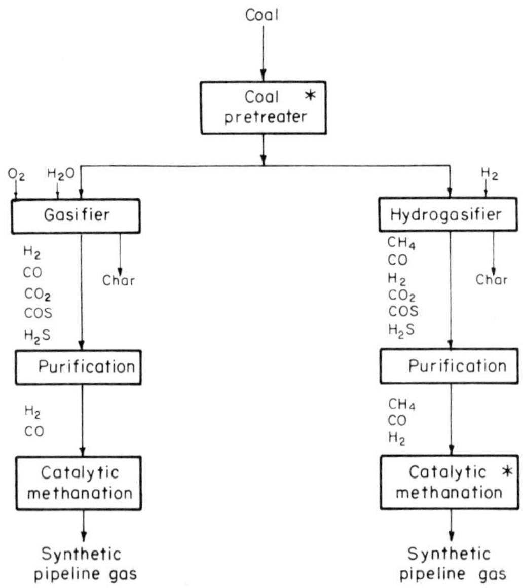

*Process step that may not be needed (see text).

Fig. 3.25 Processes for producing synthetic pipeline gas from coal. (*Courtesy Bureau of Mines, U.S. Department of the Interior.*)

Hydrogasification

In this process coal is reacted directly with hydrogen, or hydrogen and steam, at high temperature and pressure giving a methane-rich gas.

Char, which is a carbonaceous residue from incomplete conversion of the coal, is not necessarily a product from either process, but the economics of synthetic pipeline gas are favored by producing enough char to burn for steam and power, and to make hydrogen if it is needed for the process. Coal pretreatment would be needed only if agglomeration in the reactor were a problem. Methanation of the gas from a hydrogasifier would be needed only if it contained an excessive amount of carbon monoxide and hydrogen. Another step, *shift conversion*, would be needed if the hydrogen-carbon monoxide ratio of the gas going to the methanator were not 3:1.

Coal pyrolysis, that is, destructive distillation by heat in the absence of oxygen, also forms methane but the yield is too low for an industrial process for pipeline gas only. However, an experimental pyrolysis process for making several coproducts, of which methane is one, will be described later.

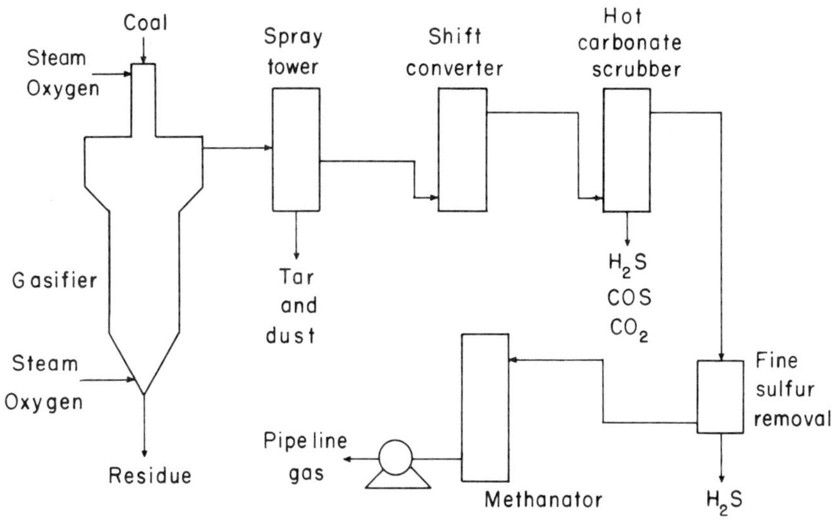

Fig. 3.26 Bureau of Mines Synthane process for synthetic pipeline gas from coal. (*Courtesy Bureau of Mines, U.S. Department of the Interior.*)

Both the Bureau of Mines and IGT are conducting research and development on the hydrogasification route to synthetic pipeline gas. The Bureau's Hydrane process is shown conceptually in Fig. 3.27. In the upper reactor, the hydrogasifier, fine coal falls in dilute phase concurrently with a mixture of methane and hydrogen. In the lower reactor the devolatized coal, or char, is fluidized with hot hydrogen. The gas from the fluidized bed, consisting mainly of methane and hydrogen, is sent to the top of the dilute-phase reactor. The raw product gas is mostly methane, and the residual carbon monoxide and hydrogen are methanated after gas purification, if that is necessary to attain the desired composition and heating value. In the IGT process, the hydrogasification is followed by purification and methanation steps. Char is sent to an electrothermal gasifier where some of it reacts with steam to make the process hydrogen. The rest of the char is burned to make steam and power. The electrothermal gasifier is based on passing an electric current through a conducting bed of char, thereby generating the heat needed for the steam-carbon reaction to make hydrogen.

The Institute of Gas Technology (IGT) also has developed a high-pressure hydrogasifier in which the conversion is done in two stages: (1) a low-temperature fluidized-bed stage (1300–1500°F) to obtain a high methane yield from the volatile matter in the coal; and (2) a high-temperature fluidized-bed stage (1700–1800°F) to generate hydrogen and more methane within the reactor.[18] The coal is first preheated externally if it is caking. Otherwise, it is fed directly to the gasifier as a coal-oil slurry; the oil is vaporized and recovered for recycle before hydrogasification of the coal occurs. Using a slurry makes it easier to feed coal at high pressure than with lock hoppers.

There are two major technical problems in hydrogasification: the high-pressure hydrogen increases the tendency of caking coals to agglomerate; and the exothermic heat of reaction must be controlled and utilized efficiently. Agglomeration can be prevented if the coal is first mildly oxidized or carbonized, or treated by a technique developed by the Bureau of Mines. This technique involves the free fall of coal particles (50 by 100 mesh) in a concurrent stream of hydrogen, or methane and hydrogen, through a reactor with a relatively large diameter. Because the coal particles comprise a *dilute* solid phase, they do not cohere.

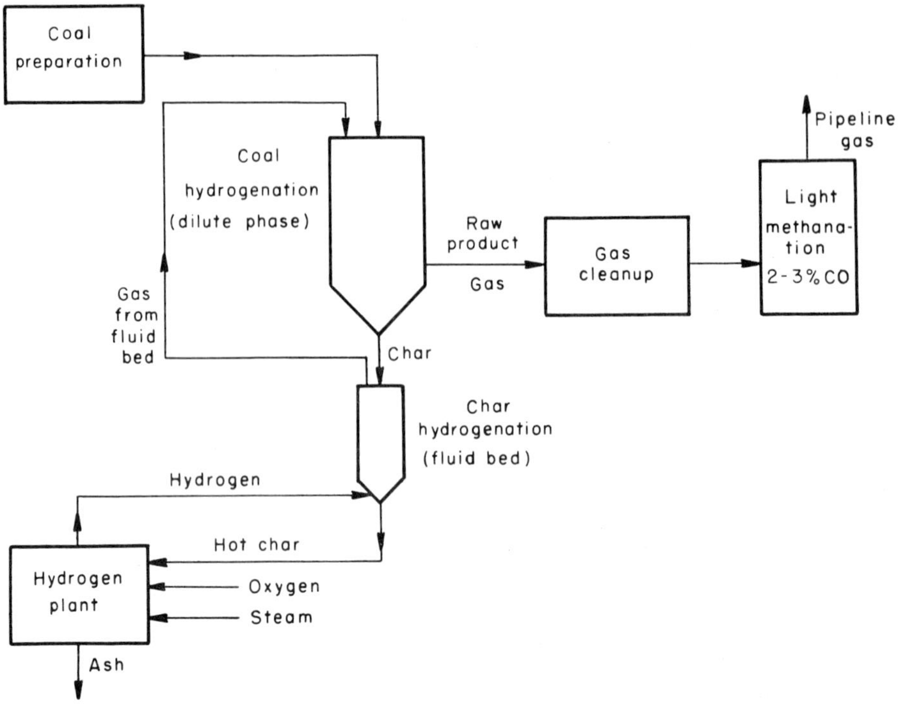

Fig. 3.27 Bureau of Mines hydrogasification process for synthetic pipeline gas from coal. (*Courtesy Bureau of Mines, U.S. Department of the Interior.*)

Non-Oxygen Process

The Carbon Dioxide-Acceptor process mentioned earlier is designed specifically to make synthetic pipeline gas or hydrogen from lignite. A report to the Office of Coal Research describes some of the features of a demonstration plant.[21]

A simplified flow sheet based on this report is shown in Fig. 3.28. Preheated lignite is lock-hoppered into the *fluidized-bed devolatilizer*. Other streams entering the devolatilizer are synthesis gas from the gasifier and calcined dolomite ($CaO \cdot MgO$) from the regenerator. The calcined dolomite is the core of the process because it reacts with carbon dioxide and liberates heat for devolatilization and gasification, thereby obviating the need for process oxygen.

Char from the devolatilizer is entrained with superheated steam (1200°F) and conveyed to the gasifier. Carbonated dolomite from both the devolatilizer and the gasifier is conveyed by

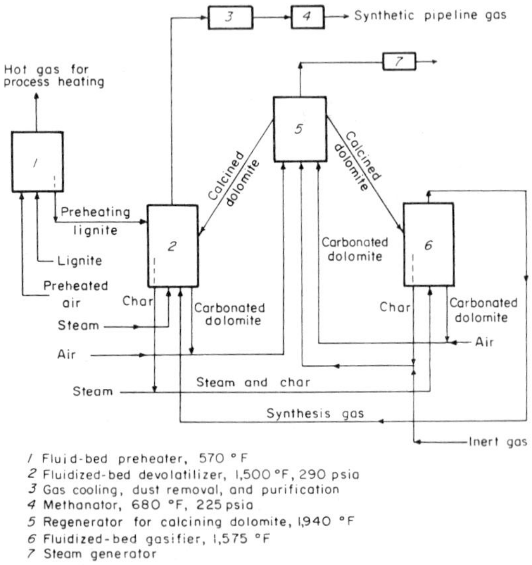

Fig. 3.28 Carbon dioxide-acceptor process for synthetic pipeline gas from lignite. (*Courtesy Bureau of Mines, U.S. Department of the Interior.*)

air to the regenerator. Char from the gasifier is conveyed to the regenerator with inert gas and burned with air to calcine the carbonated dolomite. The calcined dolomite then is recycled to the devolatizer and the gasifier. The enthalpy of the combustion products from the gasifier is sufficient to generate and superheat the steam needed for gasification and to drive air and gas compressors.

The product gas from the devolatizer consists mainly of methane, ethane, hydrogen, carbon monoxide, and carbon dioxide. After purification and methanation the synthetic pipeline gas has a heating value of about 950 Btu per cubic foot.

SYNTHETIC LIQUID FUELS AND CHEMICALS FROM COAL

The production of liquid fuels from indigenous coal was a national defense effort as early as the late twenties in countries that were short of petroleum reserves. The war effort in Germany in the thirties resulted in two successful processes. These were the *Bergius* process for converting coal directly to liquid fuels, and the *Fischer-Tropsch* process for synthesizing them catalytically from a mixture of hydrogen and carbon monoxide produced by coal gasification.

Despite considerable postwar research in the United States to improve these processes, neither one could produce gasoline at a price competitive with petroleum-based gasoline. However, the surging demands for energy in the 1970's and the need for both air pollution abatement and a secure supply of oil have revived research and development activity on catalytic synthesis, direct hydrogenation, and other ways to convert coal to oil. Projected costs today are more favorable than those of the 50's and 60's because they are approaching the costs for the manufacture of gasoline from petroleum. Also, the advent of modern petroleum-refining catalysts can contribute much to improving the efficiency of refining crude oils derived from coal.

Direct Hydrogenation

Hydrogenation of coal can convert it nearly completely to a mixture of gases, liquids, and solids that are raw materials for producing liquid fuels and numerous industrial chemicals. The pressure, temperature, catalyst, residence time, and rank of coal determine the nature and yield of the respective products.

The well-known coal hydrogenation process shown in Fig. 3.29 consists of two steps, viz., liquid-phase and vapor-phase hydrogenation. The light oils from the liquid-phase converter are comparatively rich in aromatic hydrocarbons and contain several different tar acids. The overhead stream from the light-oil still, usually boiling below 620°F, is fed to the vapor-phase converter where it is converted into quality gasoline. The gasoline is rich in aromatics, consequently benzene, toluene and xylene can be extracted readily with selective solvents. Additional aromatics can be obtained by hydroforming, platforming, or similar processes. The heavy products from the liquid-phase converter and the bottoms from the light-oil still are combined and recycled as pasting oil for the feed to the liquid-phase converter. Wu and Storch have reviewed comprehensively the history, economics and technology of coal hydrogenation in the conventional sense.[23]

In the early sixties research and development activities to improve the old direct hydrogenation processes, and particularly to find better catalysts, were intensified. According to Mills,[22] who has reviewed some of these activities, new hydrogenation catalytic systems can be grouped as follows:

1. Nascent–active hydrogen generated *in situ.*
2. Complexes of transition metals.
3. Halide catalysts (massive amounts required).
4. Organic hydrogen-donor solvents.
5. Alkali metals.
6. Reductive alkylation.

Mills states that these new catalytic systems result largely from work by the U.S. Bureau of Mines. Although promising, they have as yet not been developed into economical processes.

Coal solvents that are able to transfer hydrogen (hydrogen donors) to coal have been known for many years. A new process devel-

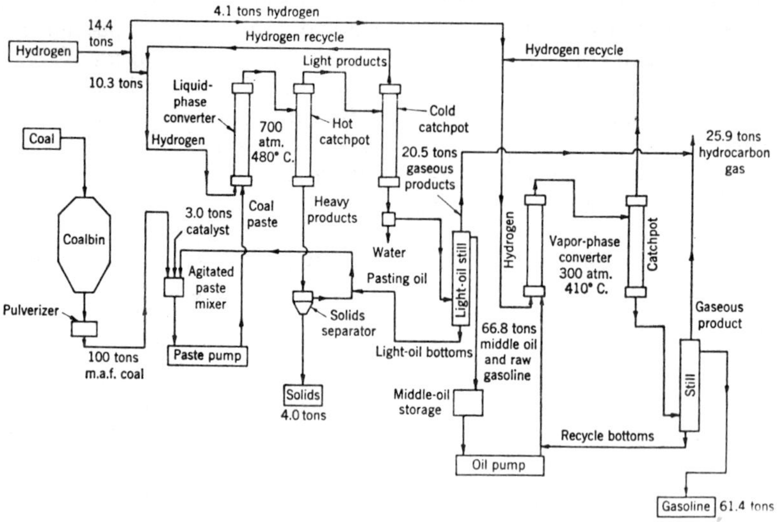

Fig. 3.29 Direct hydrogenation of coal to liquid and gaseous fuels and to chemicals. (*Courtesy Bureau of Mines, U.S. Department of the Interior.*)

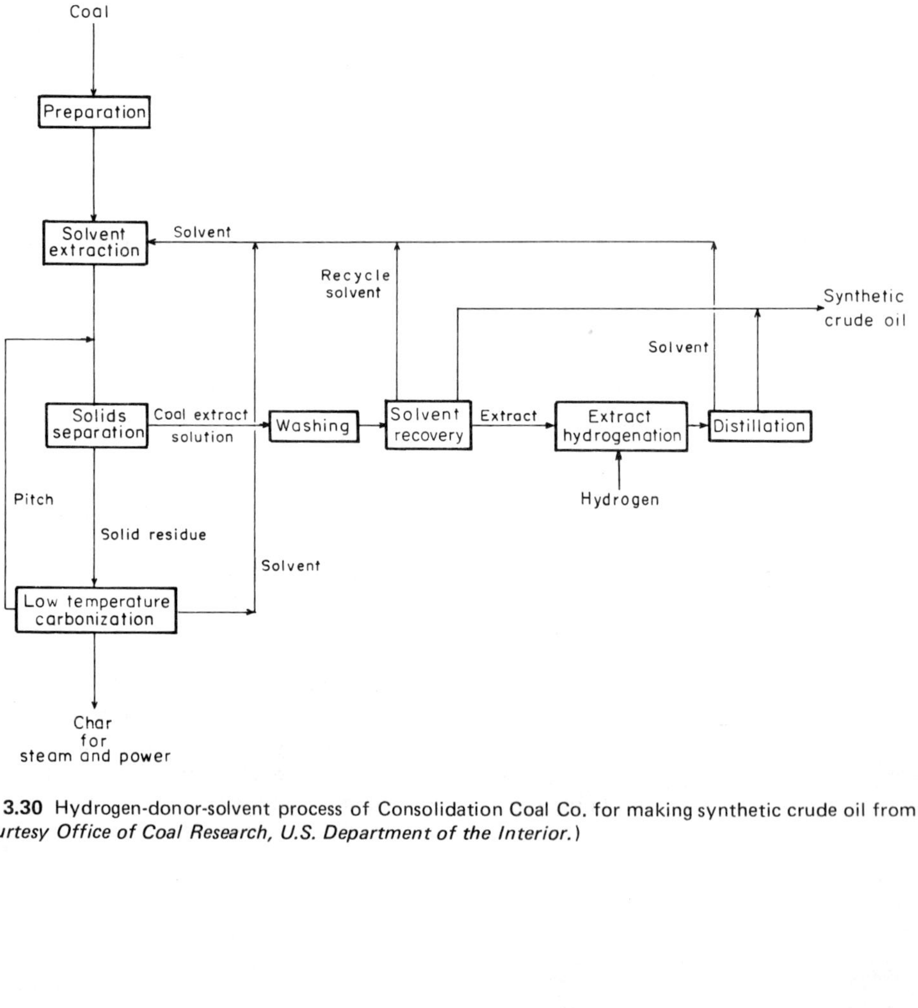

Fig. 3.30 Hydrogen-donor-solvent process of Consolidation Coal Co. for making synthetic crude oil from coal. (*Courtesy Office of Coal Research, U.S. Department of the Interior.*)

oped by the Consolidation Coal Co. employs a solvation step using a *hydrogen-donor solvent.* A flow sheet for this process is shown in Fig. 3.30. After the coal extract has been separated from mineral matter and unreacted coal it is washed to remove residual solids, stripped of some of the solvent, which is recycled, hydrogenated further, and finally distilled to a synthetic crude oil for refining to gasoline. The char is used to generate heat and power, or gasified to make hydrogen.

The *H-Coal* process of Hydrocarbon Research, Inc., is another direct hydrogenation process in an advanced state of development. Basically it is a coal-hydrogenation refinery using conventional petroleum refining operations to convert a crude liquid from a primary hydrogenation step to gasoline and salable distillates. A simplified flow diagram is shown in Fig. 3.31. Pulverized coal is slurried with heavy gas oil and charged continuously with hydrogen to a bed of cobalt-molybdenum

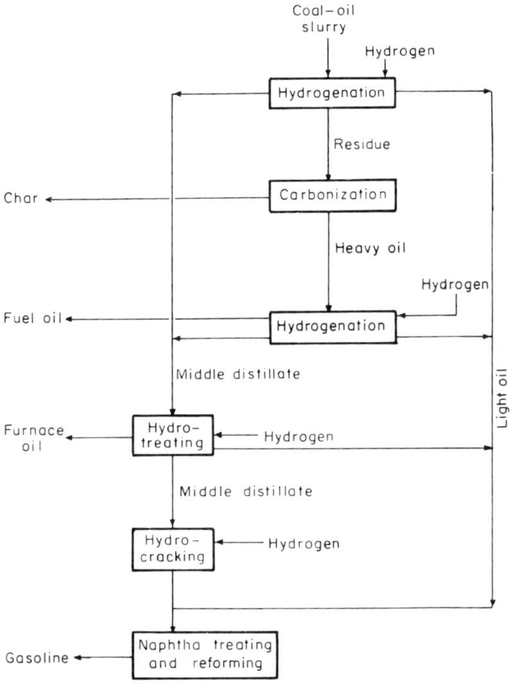

Fig. 3.31 H-coal process of Hydrocarbon Research, Inc. for producing gasoline from coal. (*Courtesy Office of Coal Research, U.S. Department of the Interior.*)

catalyst, forming gases and light, medium, and heavy oils. The upward passage of the solid, liquid, and gas feed through the bed keeps the catalyst in continuous random motion (ebullated), letting unconverted coal and mineral matter pass continuously from the reactor to a carbonizing section.

Catalytic Synthesis

The *Fischer-Tropsch* process, developed in Germany in the twenties, is the best known indirect way to synthesize liquid fuels and chemicals. Basically, it involves the catalytic hydrogenation of carbon monoxide. All the commercial Fischer-Tropsch plants in Germany used cobalt catalysts, producing mainly oxygenated chemicals, such as alcohols and ketones. The Bureau of Mines experimented with several versions of this process and different catalysts, with emphasis mostly on making liquid fuels. The flow sheet shown in Fig. 3.32 in the SASOL[24] plant in South Africa. This plant illustrates the wide variety of salable products that can be made from the Fischer-Tropsch process. The synthesis gas comes from high-pressure Lurgi gasifiers, which were described earlier.

Pyrolysis

The pyrolysis process uses neither hydrogen nor synthesis gas to convert coal, but depends on controlled thermal decomposition of coal in the absence of oxygen to produce synthetic crude oil, hydrogen or synthetic pipeline gas, and char. The process consists of heating coal to successively higher temperatures in a series of fluidized beds. In this manner, caking coals can be prevented from agglomerating in the bed. Still under development by the FMC Corp., this process is called *COED* (Char-Oil-Energy-Development).[25] Figure 3.33 is a simplified process flow sheet of the process development unit. Highly caking coals were found to require an additional pyrolysis stage.

In the four-stage system, oxygen-free flue gas heats the coal to 500–650°F in the first stage. The partially-devolatized coal, or char, is conveyed pneumatically to the second stage. Here, most of the remaining volatile matter is evolved at 740–1000°F and the char is sent to the third stage where the rest of the volatile

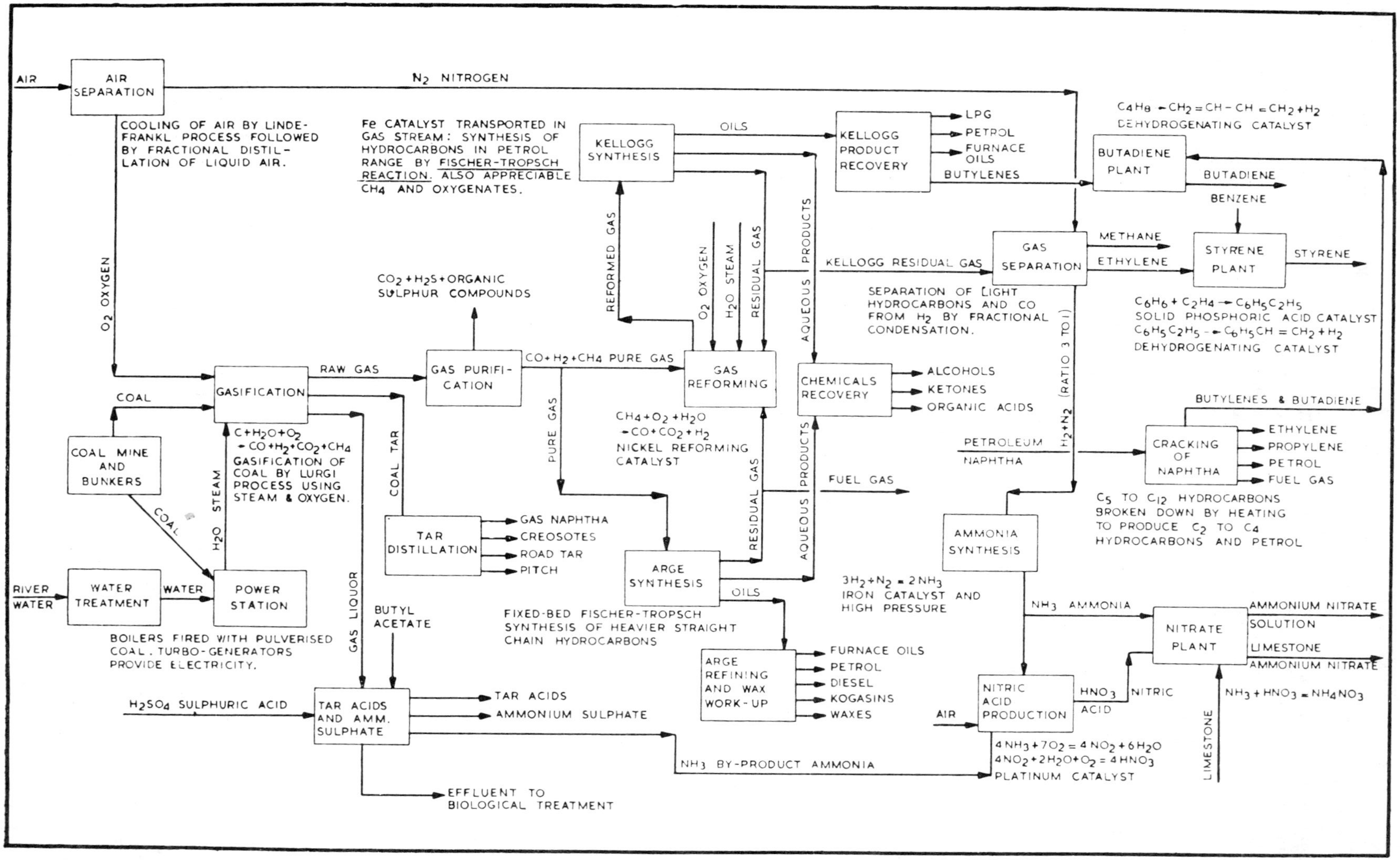

Fig. 3.32 SASOL plant for synthesis of gasoline, chemicals, and waxes from coal.[24]

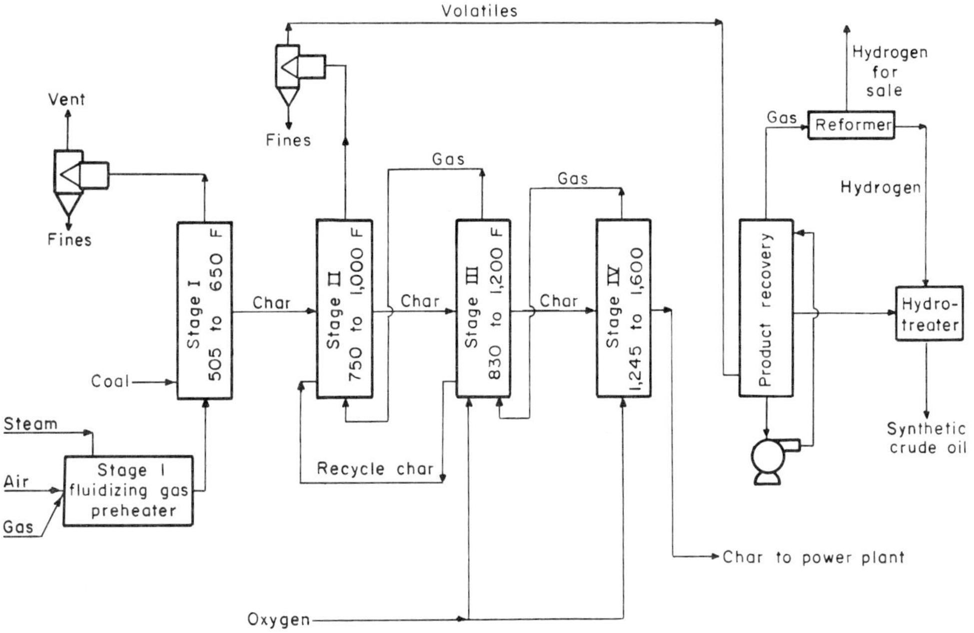

Fig. 3.33 Pyrolysis process of FMC Corp. for producing synthetic crude oil from coal. (*Courtesy Bureau of Mines, U.S. Department of the Interior.*)

matter is evolved at a still higher temperature. Oxygen diluted with nitrogen is used to fluidize the fourth-stage and burn the char from the third-stage; this provides the heat needed for all except the first stage. Part of the heat needed for the second stage is obtained by recycling a portion of the third-stage char. All the volatile products from the last three stages exit from the second pyrolyzer.

Solvation

The Pittsburg & Midway Coal Mining Co. is developing a process for making a low-ash, low-sulfur fuel from coal; light oil, coke, binder pitch, sulfur, phenol, and cresylic acid are by-products. The primary aim of this process is to produce a fuel that can be burned without polluting the atmosphere. The basic steps in the process are:

1. Pulverized coal is mixed with a solvent and heated to about 230°F.
2. The slurry then is heated to about 840°F and hydrogen is added, the pressure being held at 1000 psig.
3. The coal-solvent mixture is then sent to a dissolver and held there long enough to effect solvation of the coal.
4. Next the solvate is filtered from the mineral matter and sent to an evaporator where the solvent is recovered for recycling and the low-ash product is finally cooled to a solid.

Pilot plant experiments with Kentucky No. 11 coal yielded a solid having less than 0.5 per cent ash, and about 1 per cent sulfur, whereas the amounts in the original coal were 7.3 and 3.5 per cent, respectively.

NONFUEL USES OF COAL

Nonfuel uses of coal are those in which coal is used raw or converted to solid, liquid, or gaseous products for purposes other than generating heat and power. Some nonfuel uses of coal are industrially important now, some are being developed, some are in the experimental stage, and some have been experimented with and found to be technically feasible but unattractive economically. The latter situation could change, however, with changes in the national fossil-fuel-usage pattern and with a decline in coke-oven chemical production resulting from a declining blast furnace coke rate.*

The largest nonfuel products from coal are coke** and the chemical by-products from coke manufacture, which include crude tar, tar derivatives (tar acids and bases, naphthalene, and pitch), light oil and light-oil derivatives (benzene, toluene, xylene and solvent naphtha), and ammonia. See Chapter 8 for details on coal carbonization.

In addition to its extensive use in making iron and ferroalloys, substantial amounts of coke are used to make silicon carbide and calcium carbide; and smaller amounts are used as a reductant in making aluminum chloride, magnesium chloride, titanium tetrachloride and barium sulfide.

There are some nonfuel products based on the physical properties of raw or processed coal. These include activated carbon, graphite electrodes, and protective coatings. The Bureau of Mines obtained good quality carbon blacks in an experimental development involving free-fall or minus 325-mesh bituminious coal in the presence of a carrier gas (air, ammonia, argon, and nitrogen were used) at a temperature of 2280°F.[26] In the same equipment and using the same experimental procedures as for the carbon black studies, hydrogen cyanide was produced when the carrier gas was ammonia.[27]

The Bureau of Mines has shown that some coal and coal chars are effective absorbents for contaminants in secondary treated sewage.[28]

Earlier in this chapter processes were described for synthetic pipeline gas and gasoline that also yield nonfuel coproducts. Generally, salable coproducts from these fuel-oriented processes are essential for favorable process economics. Coal has enormous potential for many nonfuel products now derived from petroleum and natural gas, as well as for products containing features of coal's complex structure, whose the synthesis from petroleum or gas might be difficult.

The growth of coal as a source of nonfuel products, particularly novel chemicals and industrial chemicals, has been handicapped by the fact that it is a solid and contains mineral impurities. Both petroleum and natural gas can be converted to salable chemicals and chemical intermediates with little preliminary processing. Coal, on the other hand, usually needs extensive and costly processing to change it into a crude liquid or gas that can be converted to salable products. However, it is becoming increasingly evident that as coal-conversion and petroleum-refining technologies become more interdisciplinary, coal will have a greater part in the markets for nonfuel products.[29]

Chemicals can be obtained from coal by oxidation, solvent extraction, hydrolysis, hydrogenation, sulfonation, halogenation, and amination. These are processing methods that have been investigated for many years but have not become industrially important in this country.

New ways to convert coal to industrial chemicals are being investigated on a laboratory scale. These include treating coal with various kinds of plasmas (silent, glow, corona, arc discharges), various kinds of irradiation (sonic, ultrasonic, laser), novel chemical reactions (reductive alkylation, electrochemical reduction, decarboxylation of coal acids), and microbiological reactions. The latter effort is aimed toward producing synthetic food proteins from coal derivatives to supplement natural food proteins.

Some of these novel coal-conversion methods may emerge during the 70's as viable chemical industries. This will, of course, depend on demand, economics, and future policies on fossil-fuel usage.

*Coke rate is the weight of coke needed to make one ton of hot metal. Ten years ago it averaged about 1,600 pounds, currently it is about 1,250 pounds.

**Coke is considered as a nonfuel product from coal because it is used in blast furnaces and foundry cupolas to reduce ore to metal. However, it is also a fuel because it furnishes the thermal energy for the metallurgical reactions. About 63 million tons of coke were produced in 1968.

REFERENCES

1. Reid, W. T., "The Energy Explosion," *J. Inst. of Fuel*, **43**, 43–51 (Feb. 1970).
2. Ode, W. H., and Frederic, W. H., "The International Systems of Hard-Coal Classification and

Their Application to American Coals," *Bureau of Mines Report of Investigations*, 5435, 1958.
3. Minerals Yearbook, Bureau of Mines, U.S. Department of the Interior. Washington, D.C.
4. Leonard, J. W., and Mitchell, D. R., Eds., "Coal Preparation," American Inst. of Mining, Metallurgical, and Petroleum Engineers, New York, 1968.
5. American Society for Testing and Materials, Coal and Coke, Part 8 of 1970 Book of ASTM Standards, Philadelphia, Pa., 1970.
6. "Methods of Analyzing Coal and Coke," *U.S. Bur. Mines, Bull.*, No. 638, 1967.
7. "Combustion Engineering," Combustion Engineering, Inc., New York, 1966.
8. Barber, E. G., "The Utilization of Pulverized-Fuel Ash," *J. Inst. Fuel*, 4–9 (Jan. 1970).
9. von Fredersdorff, C. G., and Elliott, M. A., "Coal Gasification. Chemistry of Coal Utilization," Suppl. Vol., H. H. Lowry, Ed., pp. 892–1022, New York, John Wiley & Sons, 1963.
10. Eckerd, J. W., Clendenin, J. D., Sanner, W. S., and Morgan, R. E., "Gasification of Bone Anthracite," *U.S. Bur. Mines, Rept Invest.*, No. 5594, 1960.
11. Mills, G. A., "Progress in Gasification," U.S. Bureau of Mines, Third American Gas Association Synthetic Pipeline Gas Symposium, Nov. 17, 1970, Chicago, Ill.
12. Hebden, D., and Edge, R. F., "Experiments with a Slagging Pressure Gasifier," Research Communication GC 50, 1958.
13. Gronhovd, G. H., Harak, A. E., Fegley, M. M., and Severson, D. E., "Slagging Fixed Bed Gasification of North Dakota Lignite at Pressures to 400 Psig," *U.S. Bur. Mines, Rept Invest.*, No. 7408, 1970.
14. Hoy, H. R., and Wilkins, D. M., "Total Gasification of Coal," Review No. 174, *Brit. Coal Util. Res. Assoc., Monthly Bull.*, **22**, pp. 57–110 (Feb. and Mar. 1958).
15. Holden, J. H., Strimbeck, G. R., McGee, J. P., and Hirst, L. L., "Operation of Pressure-Gasification Pilot Plant Utilizing Coal and Oxygen, *U.S. Bur. Mines, Rept Invest.*, No. 5573, 1960.
16. Glenn, R. A., "Status of the BCR Two-Stage Super-Pressure Process," Third Synthetic Pipeline Gas Symposium, American Gas Assoc., Nov. 17, 1970, Chicago, Ill.
17. Shires, G. L., "Recent Development in Gas Producers," *Chem. Eng. Mining Rev.*, **50** (Aug. and Sept., 1958).
18. Lee, B. S., "Synthetic Pipeline Gas from Coal by the Hygas Process," Proceedings of the American Power Conference, Vol. 32, pp. 421–426, 1970.
19. Linden, H. R., "The Case for Synthetic Pipeline Gas from Coal and Oil Shale," Symposium Series 85, American Institute of Chemical Engineers, Vol. 64, pp. 57–72, 1968.
20. Forney, A. J., Gasior, S. J., Haynes, W. P., and Katell, S., "A Process to Make High-Btu Gas," *U.S. Bur. Mines, Tech. Progr. Rept*, No. 24, Apr. 1970.
21. "Pipeline Gas from Lignite Gasification," R & D Report No. 16, Interim Report No. 4, May 9, 1969, to Office of Coal Research, Department of the Interior, Contract No. 14-01-001-415.
22. Mills, G. A., "Conversion of Coal to Gasoline," *Ind. Eng. Chem.*, **61**, No. 7, 6–17 (Jul. 1969).
23. Wu, W., and Storch, H. H., "Hydrogenation of Coal and Tar," *U.S. Bur. Mines Bull.*, No. 633, 1968.
24. Rousseau, P. E., "The Production of Gas, Synthetic Oil and Chemicals from Low-Grade Coal in South Africa." Preprint 158, World Power Conference, Oct. 16, 1966, Tokyo, Japan.
25. Eddinger, R. T., Jones, J. F., and Blanc, F. E., "Development of the COED Process," *Chem. Eng. Progr.*, **64**, No. 10, 33–38 (Oct. 1968).
26. Johnson, G. E., Decker, W. A., Forney, A. J., and Field, J. H., "Carbon Black Produced by the Pyrolysis of Coal," *Prod. Res. Develop.*, **9**, 382–387 (Sept. 1970).
27. Johnson, G. E., Decker, W. A., Forney, A. J., and Field, J. H., "Hydrogen Cyanide from the Reaction of Coal with Ammonia," *U.S. Bur. Mines, Rept Invest.*, No. 6994, 1967.
28. Johnson, G. E., Kunka, L. M., Forney, A. J., and Field, J. H., "The Use of Coal and Modified Coals as Absorbents for Removing Contaminants from Waste Waters," *U.S. Bur. Mines, Rept Invest.*, No. 6884, 1966.
29. Berkowitz, N., "Objectives in Fuels Research," Paper No. ICFTC-NAFTC-1, North American Fuel Technology Conference, May 31, 1970, Ottawa, Canada.

4

Sulfuric Acid and Sulfur

James R. West* and Werner W. Duecker*

SULFURIC ACID

Sulfuric acid, a strong acid, is an oily, viscous, water-white, nonvolatile liquid. It absorbs water from the atmosphere. A drop of it on the skin causes a severe burn. It is made in large volume by the chemical industry. It is used as a solvent, a dehydrating agent, a reagent in chemical reactions or processes, an acid, a catalyst, an absorbent, etc. The concentrated acid is usually stored in steel tanks. The dilute acid may be stored in lead-lined or plastic tanks. It is used in very dilute concentrations and as strong fuming acid. It is often recovered and re-used. After use in some phases of the explosives, petroleum, and dye industries it is often recovered in a form unsuitable for re-use in that industry but suitable for use in another industry. It is a versatile, useful acid and has been called the "work horse" of the chemical industry.

*Texasgulf, Inc., New York, N.Y.
†4810 Boston Post Road, Pelham Manor, N.Y.

USES OF SULFURIC ACID

Sulfuric acid is one of the most widely used of all manufactured chemicals, and its rate of production has long been a reliable index of the total chemical production and the industrial activity of a nation. For 1969, in the United States, the average per capita consumption of sulfur was 103.8 pounds, of which most went into the manufacture of sulfuric acid. In the world the per capita consumption in 1969 was about 26 pounds.

The sulfuric acid-consuming industries in the United States are listed in Table 4.1. Of these, the largest consumer is the fertilizer industry which treats phosphate rock with sulfuric acid to produce superphosphate (a mixture of monocalcium phosphate and calcium sulfate) or crude ("wet process") phosphoric acid. There is hardly an article of commerce which has not come into contact with sulfuric acid at one time or another during its manufacture or in the manufacture of its components. The consumption of acid in the various industries is undergoing constant

TABLE 4.1 Sulfuric Acid-Consuming Industries in the United States (Thousands of Net Tons, 100% H_2SO_4)

	1957[(a)]		1970[(b)]		1975[(b)]	
Consuming Industries	*Tons*	*%*	*Tons*	*%*	*Tons*	*%*
Fertilizer Industries						
Phosphatic fertilizers	4,550	27	13,910	49	16,900	51
Ammonium sulfate or other	1,600	9	2,180	8	2,280	7
	6,150	36	16,090	57	19,180	58
Other Industries						
Chemicals	4,400	26	3,090	11	3,830	12
Petroleum refining	2,000	11	1,880	7	1,880	6
Iron and Steel	1,020	6	690	2	530	1
Other metals	270	1	710	2	1,300	4
Paints and pigments	1,380	8	1,490	5	1,160	3
Rayon and cellulose film	780	5	860	3	860	3
Miscellaneous	1,120	7	3,580	13	4,260	13
TOTAL	17,120	100	28,390	100	33,000	100

[a]Adapted from Table 1.3, "Distribution of Sulfuric Acid Consumed in U.S.," page 5, *The Manufacture of Sulfuric Acid*, Duecker and West, 1959. Based upon annual review issues of Chem. Eng.
[b]Unpublished data from The Sulphur Institute.

change. Progress demands that manufacturers strive to decrease the consumption of acid per unit product manufactured. Progress also is continually turning up new uses.

Kinds of Acid

Sulfuric acid is marketed in the United States as a large tonnage product. It is made in numerous grades and strengths, and shipments are made in both packaged containers and in bulk. It is produced in grades of exacting purity for use in storage batteries and for the rayon, dye, and pharmaceutical industries. It is produced, to less exacting purity specifications, for use in the steel, heavy chemicals, and fertilizer industries. Originally sulfuric acid was marketed in four grades known as: chamber acid, 50° Beaumé (Bé); tower acid, 60° Bé; oil of vitriol, 66° Bé; and fuming acid. At present it is marketed in the strengths listed in Tables 4.2 and 4.3.

The Manufacture of Sulfuric Acid

History. Sulfuric acid is formed in nature by the oxidation and chemical decomposition of naturally occurring sulfur and sulfur-containing compounds. It is formed by the weathering of coal brasses or iron disulfide, discarded on refuse dumps at coal mines. It is formed by bacteria in hot sulfur springs. It is formed in the atmosphere by the oxidation of sulfur dioxide emitted from the combustion of coal, oil, and other substances. It is formed by chemical decomposition resulting from geological changes.

In ancient times sulfuric acid was probably made by distilling niter (potassium nitrate) and green vitriol (ferrous sulfate heptahy-

TABLE 4.2 Standard Strengths of Sulfuric Acid[a]

Concentration Beaumé	*% H_2SO_4*	*Specific Gravity 60°F/60°F*	*Freezing Point °C*	*Freezing Point °F*
52°	65.13	1.5591	-40.0	-40
58°	74.36	1.6667	-44.0	-47
60°	77.67	1.7059	- 8.0	18
66°	93.19	1.8354	-32.0	-26
–	98.00	1.8438	3.0	37
–	100.00	1.8392	10.0	50

[a]Chemical Safety Data Sheet SD-20, "Properties and Essential Information for Safe Handling and Use of Sulfuric Acid," second revision, Washington, D.C., Manufacturing Chemists Association, 1963.

TABLE 4.3 Standard Strengths of Oleum[a]

Concentration % Free SO_3	% Equivalent H_2SO_4	*Specific Gravity* 100°F/60°F	*Freezing Point* °C	°F
20.0	104.50	1.8820	-9.0	15
30.0	106.75	1.9156	15.5	60
40.0	109.00	1.9473	33.0	94
65.0	114.63	1.9820	3.6	34
100.0 (Liquid SO_3)	122.50	1.8342	17.2	63

[a]Chemical Safety Data Sheet SD-20, "Properties and Essential Information for Safe Handling and Use of Sulfuric Acid," second revision, Washington, D.C., Manufacturing Chemists Association, 1963.

drate). Weathered iron pyrites were usually the source of the green vitriol. About 1740, the acid was made in England by burning sulfur in the presence of saltpeter (potassium nitrate) in a glass balloon flask. The vapors united with water to form acid which condensed on the walls of the flask. In 1746 the glass balloon flask was replaced by a large lead-lined box or chamber, giving rise to the name "chamber process." In 1827 Gay-Lussac, and in 1859 Glover, changed the circulation of gases in the plant by adding towers which are now known as Gay-Lussac and Glover towers. These permit the recovery from the exit gases of nitrogen oxides which are essential to the economic production of chamber acid. Today, most sulfuric acid made in the United States is produced by the contact process, based on scientific technology developed about 1900 and thereafter.

BASF built a successfully operating contact plant in the United States in 1898. General Chemical erected a pyrite-burning contact plant using the Herreshoff furnace in the United States in 1900.

Development of the Sulfuric Acid Industry in the United States

The manufacture of sulfuric acid has been a basic industry in the United States for many years. It has been made by two well-established methods, the chamber process and the contact process. Initially, the production of acid was concentrated on the Eastern seaboard. After the Civil War the industry spread to the west and into the South. Between 1899 and 1904 the number of acid manufacturers increased rapidly in Ohio and Illinois; and, just before the turn of the century, there was great activity in the South as a result of the discovery of phosphate deposits in South Carolina and Florida and the development of the phosphate fertilizer industry.

Over the years, the number of contact plants has increased while the number of chamber plants has decreased. During the mid 1940's the number of chamber plants was approximately equal to the number of contact plants. Today, the contact plants are in the majority and the trend is toward contact plants of larger and larger capacity. Contact plants producing 2000 tons of acid a day in a single train are not unusual and they may become larger in the future.

Consumers confronted with the task of disposing of waste or spent acid find it advantageous to arrange with an independent producer to exchange the waste acid for fresh acid. Methods have been developed which permit such producers to reprocess the waste acid and obtain a product of virgin quality. Also, they can operate centrally located plants of large size which can produce acid at a much lower cost than can be realized in the small plant needed by the average user. The end use of sulfuric acid, more than any other factor, determines the location of sulfuric acid plants. Data on the production of acid in the United States are listed in Table 4.4.

The Chamber Process for Making Sulfuric Acid

The chamber process for making sulfuric acid, at first glance, appears to be a rather simple

Fig. 4.1 Spent-acid regenerating plant built for Stauffer Chemical Co. by Chemical Construction Co. (Chemico). (*Courtesy Chemical Construction Co.*)

TABLE 4.4 Sulfuric Acid Production in the United States, 1919, 1957, 1969[a] (Net Tons, 100% H_2SO_4)

	Number of Establishments			*Sulfuric Acid Production*		
State	*1919*	*1957*	*1969*	*1919*	*1957*	*1969*
Alabama	13	12	7	98,138	314,669	338,319
California	10	9	12	257,330	843,636	1,432,267
Florida	5	14	16	–	1,738,945	7,574,730
Georgia	27	16	13	169,630	318,325	599,534
Illinois	11	15	14	337,502	1,241,474	1,658,828
Louisiana	5	8	9	51,067	727,144	2,296,816
Maryland	7	6	6	305,232	1,094,275	901,782
New Jersey	18	13	9	475,117 (New Jersey and New York)	1,469,591 (New Jersey and New York)	1,565,858 (New Jersey and New York)
New York	5	3	2			
Ohio	14	16	9	231,531	713,201	579,660
Pennsylvania	19	11	10	349,806	795,929	948,894
Texas	5	11	14	–	1,605,445	2,534,273
All Others	77	83	79	1,177,240	4,833,904	7,752,102
TOTAL	216	217	200	3,452,593	15,696,538	28,233,063
Chamber plants	192	79	30	2,686,117	1,988,303	513,134
Contact plants	31	144	170	766,476	13,708,235	27,719,929

[a]Data are from U.S. Government sources and refer to new acid production.

process requiring simple equipment. The reactions involved, however, are not simple and even today there is disagreement among experts as to just what does take place in the chambers. All agree, however, that the oxidation of sulfur dioxide to sulfuric acid in the chambers is not directly effected by oxygen but that intermediate compounds involving nitrogen oxides are formed and that the reaction is really a cyclic process involving the alternate formation and decomposition of the intermediate compounds. Many operators say that "operation of a chamber plant is more an art than a science."

In the chamber process chemical reactions take place between sulfur dioxide, oxygen, nitrogen oxides, and water vapor. A series of intermediate compounds are formed which decompose to yield sulfuric acid and nitrogen oxides. The over-all effect is that the sulfur dioxide is oxidized to sulfur trioxide, which combines with water vapor to form sulfuric acid ($2SO_2 + O_2 + 2H_2O \longrightarrow 2H_2SO_4$). The nitrogen dioxide acts as the oxidant and is reduced to nitric oxide, which must be continually reoxidized by oxygen in the air. When all the sulfur dioxide has been consumed, the nitrogen oxides appear as equal moles of nitric oxide and nitrogen dioxide, in which ratio they are absorbed in sulfuric acid in the Gay-Lussac tower as nitrosylsulfuric acid. The solution of nitrosylsulfuric acid (nitrose) from the Gay-Lussac tower is pumped to the denitration (Glover) tower where heat releases the nitrogen oxides for re-use in the cycle. In the Glover tower the denitrated sulfuric acid is concentrated to 60° Bé. Part of this acid is returned to the Gay-Lussac tower for recovery of the nitrogen oxides from the exit gases. The balance is available for use or sale.

A chamber plant, shown schematically in Fig. 4.2, consists of a sulfur burner, or ore roaster or other device for producing a gas containing sulfur dioxide, a Glover tower, several large lead chambers, one or more Gay-Lussac towers, and auxiliaries such as fans, pumps, and tanks.

If elemental sulfur is burned, a substantially clean gas containing 8 to 11 per cent sulfur

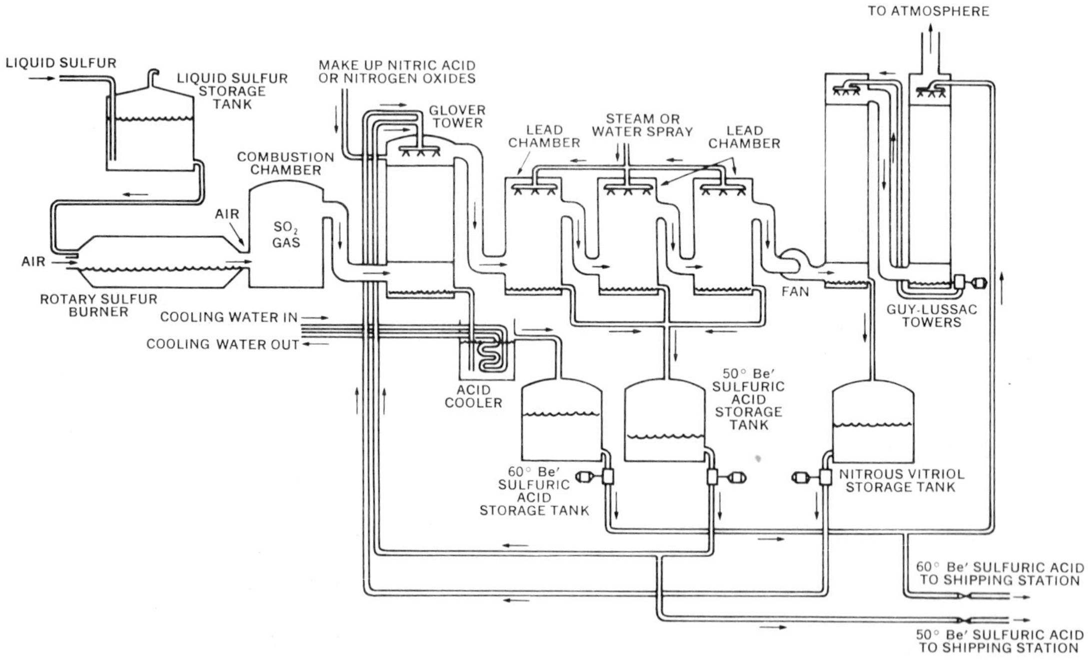

Fig. 4.2 Schematic flow diagram of chamber process for sulfuric acid manufacture. (*Courtesy Texasgulf Inc.*)

dioxide by volume is formed. If sulfide ores or other sulfur-bearing materials are burned, a gas containing about 7 per cent sulfur dioxide is produced. This gas is usually contaminated with varying amounts of dust, metallic fumes, water, and other gaseous impurities, which must be removed, at least in part. The sulfur dioxide gas of suitable purity is then conducted to the Glover tower of the chamber plant where it meets a countercurrent flow of sulfuric acid, 50° to 54° Bé, containing nitrosylsulfuric acid. The hot gas concentrates the acid to 60° Bé and decomposes the nitrosylsulfuric acid, releasing the oxides of nitrogen. The gas leaving the Glover tower and containing sulfur dioxide, nitrogen, oxides of nitrogen, water, and an excess of oxygen then enters the lead chambers.

The concentration of sulfur dioxide gas in the system must be controlled to allow an excess of oxygen throughout the system. Nitrogen oxides, obtained by adding nitric acid, by the decomposition of sodium nitrate reacted with sulfuric acid, or, in more recent plants, by burning ammonia, are added to the chambers. A fine spray of water is also added. The sulfuric acid produced condenses on the walls of the lead chambers. The unreacted gas flows to the Gay-Lussac towers. Here the nitrogen oxides in the gas are recovered by absorption in a counter-current flow of 60° Bé acid to form nitrosylsulfuric acid. This acid containing the nitrosylsulfuric acid is then pumped to the Glover tower and the cycle repeated. Up to 50 per cent of all the acid produced in the plant is formed in the Glover tower.

The acid produced is drawn from the pans in the bottom of the chambers or from the Glover tower. That produced in the first and intermediate chambers usually contains from 63.66 to 68.13 per cent sulfuric acid (52° to 54° Bé). That which is produced in the last chamber is weaker and contains about 59.32 per cent sulfuric acid (48° Bé). That which flows from the Glover tower contains from 75 to 85 per cent sulfuric acid.

Chamber acid may contain small amounts of impurities such as oxides of nitrogen, arsenic, and selenium and sulfates of iron, copper, zinc, mercury, lead, and antimony, depending in part on the kind of sulfur-bearing material used in making the sulfur dioxide gas charged to the plant. In some applications, as in the manufacture of fertilizers, these impurities are not harmful. In others they are harmful and must be removed.

The Contact Process

The basic features of the contact process for making sulfuric acid, as practiced today, were described in a patent issued in England in 1831. It disclosed that if sulfur dioxide, mixed with oxygen or air, is passed over heated platinum the sulfur dioxide is rapidly converted to sulfur trioxide, which can be dissolved in water to make sulfuric acid. The practical application of this disclosure, however, was delayed. An understanding of the complex reactions occurring in the gas phase over the catalyst required the development of that branch of physical chemistry known as chemical kinetics, and also the development of that branch of engineering known as chemical engineering. A demand for acid stronger than that which could be produced readily by the chamber process stimulated this development. The success of this process for making sulfuric acid led to the development of other catalytic processes for making many of the synthetic chemicals known today.

The heart of the contact sulfuric acid plant is the converter in which sulfur dioxide is converted catalytically to sulfur trioxide. Over the course of the years a variety of catalysts have been used, including platinum and the oxides of iron, chromium, copper, manganese, titanium, vanadium, and other metals. The first catalyst used was platinum. It proved to be extremely sensitive to poisons such as arsenic compounds present in small amounts in some sources of sulfur dioxide. The successful development of the contact process depended in part on the recognition of the existence of catalytic poisons, and in devising methods for their removal. Platinum and iron catalysts were the main catalysts used prior to World War I. At present, vanadium catalysts in various forms, combined with promoters, are generally used.

A number of different plant designs have been developed for the efficient production

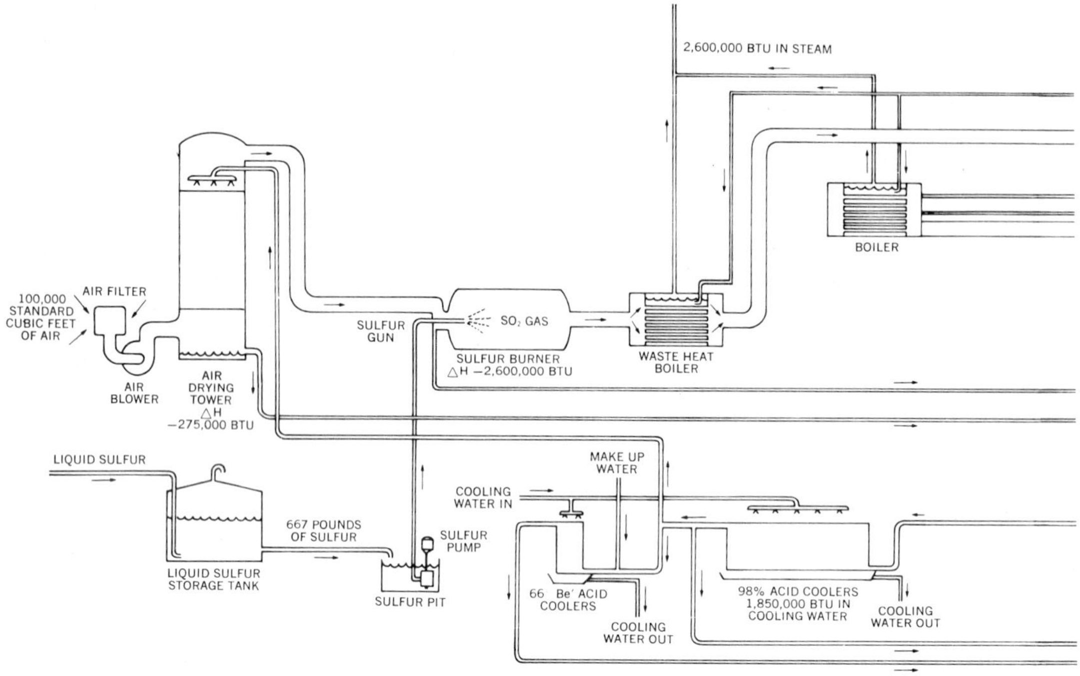

Fig. 4.3 Schematic flow diagram of contact process for

of sulfuric acid by the contact process. These, in the United States, are often referred to by the name of the builder or designer, e.g., a Chemical Construction (Chemico), a Leonard-Monsanto, or a Wellman-Lord plant. In Europe and other parts of the world Lurgi Gesellschaft für Chemie und Hüttenwesen mbH, Chemiebau Dr. A. Zieren GmbH & Co. KG, and Simon-Carves Chemical Engineering Ltd. are noted for designing and building contact sulfuric acid plants. Contact plants are also classified according to the material used in the production of the sulfur dioxide charged to the plant, e.g., sulfur, hydrogen sulfide, gypsum, iron pyrites, smelter gas, or spent and sludge acids.

The principal steps in a contact plant burning sulfur are shown in a simplified diagram in Fig. 4.3. These are: (1) the production of sulfur dioxide with dry air; (2) cooling the gas; (3) conversion or oxidation of the sulfur dioxide to sulfur trioxide; (4) cooling the sulfur trioxide gas; and (5) absorption of the sulfur trioxide in sulfuric acid.

In sulfur is the raw material, it is burned in a sulfur burner in the presence of dried air to yield a gas containing 8 to 11 per cent sulfur dioxide and 8 to 13 per cent oxygen. The gas produced by burning sulfur usually is clean, requires little further handling, and is cooled in a waste-heat boiler to about 760° to 840°F. If the raw material is a sulfide ore the equipment may include roasters, sintering machines, copper converters, reverberatory furnaces, flash smelters, etc. The gas from roasters usually contains from 7 to 14 per cent sulfur dioxide, and that from sintering machines seldom contains more than 8 per cent sulfur dioxide. Gases from metallurgical operations are contaminated with fumes, water and other vapors, and solid impurities which must be removed in cyclone dust collectors, electrostatic dust and mist precipitators, or scrubbing towers. The gases must then be dried. Cleaning metallurgical gases usually results in excessive cooling, so that they must be reheated before they enter the converter. The specific inlet temperature of the gas entering the converter is dependent upon the quantity and quality of the catalyst, and the composi-

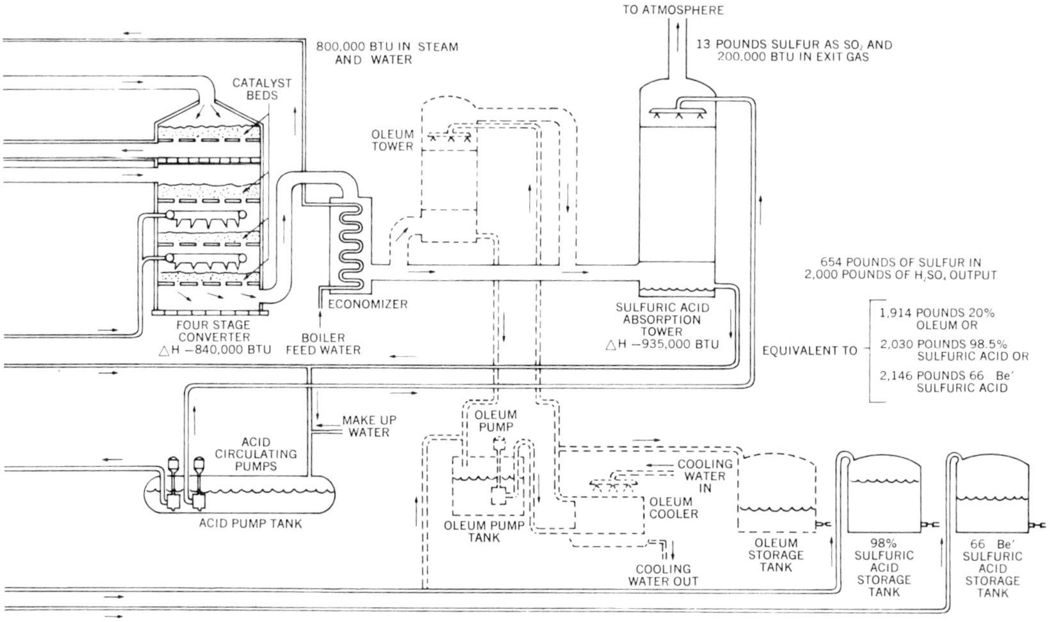

sulfuric acid manufacture. (*Courtesy Texasgulf Inc.*)

tion and flow rate of the sulfur dioxide gas, but it is usually somewhat in excess of 800°F.

The converter contains the catalyst, usually pentavalent vanadium in the form of individual pellets, similar in shape to aspirin tablets or, more commonly, as extruded cylinders. The catalyst is usually placed in horizontal trays or beds arranged so that the gas containing sulfur dioxide and an excess of oxygen passes through two, three, or four stages of catalyst. As the gas passes through the converter, approximately 95 to 98 per cent of the sulfur dioxide is converted to sulfur trioxide, with the evolution of considerable heat. Maximum conversion cannot be obtained if the temperature in any stage becomes too high. Therefore, gas coolers are employed between converter stages.

The concentration of sulfur trioxide leaving the converter at 800 to 850°F is approximately the same as that of the entering sulfur dioxide. The converter gas is cooled to 450 to 500°F in an economizer or tubular heat exchanger. The cooled gas then enters the absorption tower where the sulfur trioxide is absorbed with high efficiency in a circulating stream of 98 to 99 per cent sulfuric acid. The sulfur trioxide unites with the small excess of water in the acid to form more sulfuric acid.

A modern modification of the contact plant, which was developed to reduce the emission of sulfur dioxide in the exit gases, includes two stages of absorption. Because of chemical equilibria conditions it is impossible to obtain conversion efficiencies much higher than 98 per cent in plants using only one stage of absorption. The reversible reaction

$$SO_2 + \tfrac{1}{2}O_2 \rightleftharpoons SO_3$$

will, however, proceed further in the right-hand direction if the sulfur trioxide is removed. The modified process takes advantage of this fact.

The flow sheet of a double-absorption contact plant, depending on the raw material, is the same as that described in Fig. 4.3, plus the addition of equipment as follows:

The gases leaving the absorbing tower, instead of discharging directly to the atmosphere

Fig. 4.4 Sulfur-burning contact sulfuric acid plant (2 trains) of Texasgulf Inc. at Lee Creek, No. Car. (*Courtesy Texasgulf Inc.*)

as in a single absorption type plant, pass through a heat exchanger in which they are reheated to about 800°F before reentering the converter. They then pass through additional catalyst, are cooled, and flow through a second absorbing tower and then to the atmosphere. The overall conversion of SO_2 and SO_3 may exceed 99.5 per cent. The second conversion step may be made with one or two trays of catalyst contained in the same vessel used in the first conversion step. The heat exchanger may be substituted for the economizer in the first stage of the flow sheet and the economizer may be used to cool the gases in the last stage. Many variations are possible.

SULFUR

Uses

The 1970 consumption of sulfur in the free world is given in Table 4.5, together with estimates for 1975. It is seen that of the total sulfur consumed in 1970, nearly 86 per cent went to the manufacture of sulfuric acid. There are, however, many other important uses of sulfur. Nonacid consumption in the United States for 1970 and 1975 (estimated) is shown in Table 4.6.

Sources

For many years pyrite, iron disulfide, was the main sulfur-containing material used in the manufacture of sulfuric acid. At the turn of the century Herman Frasch developed his process for extracting sulfur from underground deposits by injecting hot water into the deposit, melting the sulfur, and recovering it in liquid form. The exceptional purity and quality of this sulfur appealed to the chemical industry, particularly to the manufacturers of sulfuric acid. It could be burned in relatively simple equipment to produce, at constant rates, a gas containing a high concentration of sulfur dioxide. The gas was clean and required little handling before use in either the chamber or contact sulfuric acid plant. As a consequence, Frasch's sulfur supplanted practically all other raw materials formerly used in the manufacture of sulfuric acid. In the United States, during the period of 1945 to 1970, approximately 70 to 80 per cent of all acid produced was made from brimstone, elemental sulfur. The remainder was made as a by-

TABLE 4.5 Free World Consumption of Sulfur[a] (Thousands of Long Tons)

	1970		*1975 (Estimated)*	
	Amount	*% of Total*	*Amount*	*% of Total*
Fertilizer industries				
Phosphatic fertilizers	10,464	36.9	14,766	40.8
Ammonium sulfate or other	2,909	10.3	3,180	8.8
	13,373	47.2	17,946	49.6
Other Industries				
Chemicals	2,367	8.4	3,066	8.5
Detergents	1,083	3.8	1,575	4.4
Petroleum refining	1,132	4.0	1,181	3.3
Iron and Steel	689	2.4	492	1.4
Other metals	664	2.3	1,083	3.0
Paints and pigments	1,437	5.1	1,619	4.5
Textiles and cellulose film	2,096	7.4	2,274	6.3
Pulp and paper	2,166	7.6	2,116	5.8
Rubber	143	0.5	172	0.5
Agriculture	325	1.1	345	1.0
Miscellaneous and new uses	2,875	10.2	4,223	11.7
TOTAL	28,350	100.0	36,092	100.0
Acid use	24,240	85.5	31,330	86.8
Nonacid use	4,110	14.5	4,762	13.2

[a]Unpublished data from The Sulphur Institute.

product of various metallurgical and smelting operations concerned with the production of copper, zinc, and lead.

Sulfur is a constituent of many materials consumed by industry, such as coal, natural gas, petroleum, gypsum, and, of course, the ores and minerals from which many metals are derived. The sulfur from sulfide ores has long been used in the manufacture of sulfuric acid largely because the sulfur dioxide evolved during the processing or conversion of the ore into metal had to be contained to prevent

TABLE 4.6 Nonacid Consumption of Sulfur in the United States[a]

	1970		*1975 (Estimated)*	
Use	*Amount (Long Tons)*	*% of Total*	*Amount (Long Tons)*	*% of Total*
Carbon disulfide	246,000	2.6	236,000	2.1
Pulp and paper	512,000	5.3	541,000	4.9
Rubber	39,000	0.4	39,000	0.4
Sulfur dioxide	49,000	0.5	59,000	0.5
Agriculture	79,000	0.8	89,000	0.8
Phosphorus pentasulfide	34,000	0.4	44,000	0.4
Other	197,000	2.0	295,000	2.6
TOTAL nonacid	1,156,000	12.0	1,303,000	11.7
Acid	8,466,000	88.0	9,839,000	88.3
TOTAL	9,622,000	100.0	11,142,000	100.0

[a]Unpublished data from The Sulphur Institute.

Fig. 4.5 Loading solid sulfur in rail cars from vat at Newgulf, Texas. (*Courtesy Texasgulf Inc.*)

excessive pollution of the atmosphere. Processes have been developed for making sulfuric acid from gypsum in countries deficient in other sources of sulfur. In these processes gypsum (calcium sulfate dihydrate) or anhydrite (anhydrous calcium sulfate) is heated in the presence of reducing gases to produce sulfur dioxide and lime or portland cement. In addition to the high fuel cost, the concomitant production of lime or portland cement has limited the attractiveness of gypsum as a raw material for the production of sulfuric acid. So, in spite of the availability of many competing forms of sulfur, elemental sulfur has remained the favorite for the manufacture of sulfuric acid.

Natural gas frequently contains hydrogen sulfide. To prepare such gas for domestic consumption, processes have been devised to remove the hydrogen sulfide and convert it to elemental sulfur. Large quantities of such sulfur are being recovered annually in the United States, Canada, France, and other countries and are now available for use in the manufacture of sulfuric acid and for nonacid uses.

The petroleum industry, for years, has been removing sulfur from the various products it manufactures, and is intensifying this effort in view of the prevailing significance accorded pollution of the atmosphere. Of all the sulfur dioxide escaping to the atmosphere 70 per cent is attributed to the combustion of coal, 20 per cent to the combustion of petroleum, and 10 per cent to metal smelting. The sulfur dioxide in the atmosphere, in a few days, is oxidized to a sulfate aerosol. Of the total sulfur in the atmosphere, about 65 per cent is in aerosol form and 23 per cent is present as sulfur dioxide.

It is difficult to predict what the future may hold as to new sources of sulfur. The combustion of coals containing sulfur represents a potential source of sulfur not hitherto used. At the present time not more than 50 per cent of the sulfur in coal can be removed by any simple treatment before the coal is burned. It may be more feasible to remove, and perhaps recover, the sulfur from the combustion gases. Combustion gases from some large coal-burning power plants may become a new source of sulfur and/or sulfuric acid if satisfactory scrubbing methods are developed. Alternatively, the coal may be gasified at the mine site and the sulfur removed from the gas there, perhaps as elemental sulfur. At the present time desulfurization of crude oils, or petroleum products containing sulfur, is practiced to some extent. Like sour natural gas, crude petroleum and petroleum products could become important sources of sulfur.

REFERENCES

1. Allied Chemical Corp., "Sulfuric Acid," 1960.
2. American Petroleum Institute, "API Toxicological Review: Sulfuric Acid," 2nd Ed., New York, 1963.
3. ASTM, Standard Methods for Analysis of Sulfuric Acid, Designation E-223-1968.
4. "Atmospheric Emissions from Sulfuric Acid Manufacturing Processes," Public Health Service Publication No. 999-AP-13, Cincinnati, U.S. Dept. of Health, Education, and Welfare, 1965.
5. *Chem. Eng.*, **70**, No. 8, 106 (Apr. 15, 1963).
6. *Chem. Eng. News*, **42**, No. 51, 42–43 (Dec. 21, 1964).
7. *Chem. Week*, **102**, No. 19, 55–56 (May 11, 1968).
8. *J. Air Poll. Control Assoc.*, **13**, No. 10, 499–502 (Oct. 1963).
9. *Oilweek*, **15**, No. 9, 23 (Apr. 13, 1964).
10. Bhattacharjya, S. K., *Chem. Age India*, **21**, No. 6, 585–590 (1970).
11. Boase, J. L., and Duckworth, R. A., *Chem. Process Eng.*, Part 1, 58–67 (Feb. 1963); Part 2, 132–137 (Mar. 1963).
12. Brink, J. A., Jr., Burggrabe, W. F., and Greenwell, L. E., *Chem. Eng. Progr.*, **64**, No. 11, 82–86 (Nov. 1968).
13. "Methods of Test for Sulfuric Acid," B.S. 3903, London, British Standards Institution, 1965.
14. Browder, T. J., "Latest Developments and Modern Technology in the Sulfuric Industry," Preprint 2a, American Institute of Chemical Engineers, Dec. 1970.
15. Burkhardt, D. B., *Chem. Eng. Progr.*, **64**, No. 11, 66–70 (Nov. 1968).
16. Chari, K. S., *Chem. Process. Eng.* (Bombay), **3**, No. 2, 35–36 (Feb. 1969).
17. Connor, J. M., *Chem. Eng. Progr.*, **64**, No. 11, 59–65 (Nov. 1968).
18. DeWolf, P., and Larison, E. L., "American Sulphuric Acid Practice," 1st Ed, New York, McGraw-Hill, 1921.
19. Donovan, J. R., and Stuber, P. J., *J. Metals*, **19**, No. 11, 45–50 (Nov. 1967).
20. Donovan, J. R. and Stuber, P. J., "The Technology and Economics of Interpass Absorption Sulfuric Acid Plants," presented Dec. 1968 at Annual Meeting, American Institute of Chemical Engineers.
21. Duecker, W. W., and West, J. R., Eds., "The Manufacture of Sulfuric Acid," New York, Van Nostrand Reinhold, 1959.
22. Du Pont, "Sulfuric Acid."
23. Fairlie, A. M., "Sulfuric Acid Manufacture," New York, Van Nostrand Reinhold, 1936.
24. Farkas, M. D., and Dukes, R. R., *Chem. Eng. Progr.*, **64**, No. 11, 54–58 (Nov. 1968).
25. Fasullo, O. T., "Sulfuric Acid Use and Handling," New York, McGraw-Hill, 1965.
26. Hensinger, C. E., *et al.*, *Chem. Eng.*, **75**, No. 12, 70–72 (Jun. 3, 1968).
27. Hiroshi, K., *Chem. Economy Eng. Rev.*, 29–31 (Dec. 1969).
28. Johnstone, H. F., and Moll, A. J., *Ind. Eng. Chem.*, **52**, No. 10, 861–863 (Oct. 1960).
29. Kappanna, A. N., and Chaudhuri, B. P., *Indian J. Appl. Chem.*, **26**, No. 4, 91–96 (1963).
30. Kennedy, J. G., *Metal Finishing J.*, 374–376 (Nov. 1967).
31. Kerr, J. R., *et al.*, "The Sulfur and Sulfuric Acid Industry of Eastern United States," *U.S. Bureau Mines*, Inform. Circ., No. 8255, 1965.
32. Kronseder, J. G., *Chem. Eng. Progr.*, **64**, No. 11, 71–74 (Nov. 1968).
33. Labine, R. A., *Chem. Eng.*, **67**, No. 1, 80–83 (1960).
34. Lunge, G., "The Manufacture of Sulphuric Acid and Alkali," Vol. 1 (3 parts), New York, Van Nostrand Reinhold, 1913.
35. Mandelik, B. G., and Pierson, C. U., *Chem. Eng. Progr.* **64**, No. 11, 75–81 (Nov. 1968).

36. Marshall, V. C., *Brit. Chem. Eng.*, **6**, No. 12, 841–850 (Dec. 1961).
37. Martin, G., and Foucar, J. L., "Sulphuric Acid and Sulphur Products," New York, Appleton-Century-Crofts, 1916.
38. Miles, F. D., "The Manufacture of Sulphuric Acid (Contract Process)," New York, Van Nostrand Reinhold, 1925.
39. Monsanto Co., "Monsanto Designed Sulfuric Acid Plants."
40. Moore, H. C., "Sulphuric Acid Tables; Hand Book," 2nd Ed., rev., Chicago, Hillson and Etten, 1926.
41. Olin Mathieson Chemical Corp., "Sulfuric Acid," 1958.
42. Parkes, J. W., "The Concentration of Sulphuric Acid," New York, Van Nostrand Reinhold, 1924.
43. Parrish, P., and Snelling, F. C., "Sulphuric Acid Concentration," New York, Van Nostrand Reinhold, 1925.
44. Rampacek, C., "Sulfuric Acid from Sulfur Dioxide by Autoxidation in Mechanical Cells," *U.S. Bur. Mines, Rept Invest.*, No. 6236, 1963.
45. Slin'ko, M. G., and Beskov, V. S., *Intern. Chem. Eng.*, **2**, No. 3, 388–393 (Jul. 1962).
46. Snowden, P. N., and Ryan, M. A., *J. Inst. Fuel*, **42**, No. 34, 188–189 (May 1969).
47. Sullivan, T. J., "Sulphuric Acid Handbook," 1st Ed., New York, McGraw-Hill, 1918.
48. Tennessee Corp., "Sulfuric Acid," 1962.
49. Texas Gulf Sulphur Co., "Sulphur Manual," Section VIII, Sulphuric Acid, 1965.
50. "Sulfuric Acid," U.S. Dept. of Commerce, Bureau of Census, Current Industrial Reports, Series M28A-13, Supplement I, issued annually.
51. U.S. Industrial Chemicals Co., "Sulfuric Acid User's Handbook;" New York, 1961.
52. Waeser, B., "Die Schwefelsäurefabrikation," Braunschweig, Friedr. Vieweg and Sohn, 1961.
53. Walitt, A., "A Process for the Manufacture of Sulfuric and Nitric Acids from Waste Flue Gases," presented Dec. 1970, at Second International Clean Air Congress of the International Union of Air Pollution Prevention Assoc.
54. Warren, I. H., *Australian J. Appl. Sci.*, **7**, 346–358 (1956).
55. Wells, A. E., and Fogg, D. E., "The Manufacture of Sulphuric Acid," *U.S. Bur. Mines, Bull.*, No. 184, 1920.
56. Wheelcock, T. D., and Boylan, D. R. *Chem. Eng. Prog.*, **64**, No. 11, 87–92 (Nov. 1968).
57. Wyld, W., "Raw Materials for the Manufacture of Sulphuric Acid and the Manufacture of Sulphur Dioxide," New York, Van Nostrand Reinhold, 1923.
58. Wyld, W., "The Manufacture of Sulphuric Acid (Chamber Process)," New York, Van Nostrand Reinhold, 1924.
59. Zwemer, J. H., and Dean, C. M., *Chem. Eng. Progr.*, **56**, No. 2, 39–41 (Feb. 1960).

5

Synthetic Nitrogen Products

Ralph V. Green*

NITROGEN FIXATION AND AMMONIA SYNTHESIS

As long ago as 1780 Cavendish caused atmospheric nitrogen and oxygen to combine by means of an electric spark. The first practical large-scale manufacture of a nitrogen compound from atmospheric nitrogen was that of Birkeland and Eyde at Nottoden, Norway, early in this century. In this process air is passed at a rapid rate through an arc spread out to form a flame. Bradley and Lovejoy used the arc method in 1902 at Niagara Falls, but the attampt failed because the arc flame area was too small and the gases were not removed from the reaction chamber fast enough. The Norwegian process benefited from the faults demonstrated in this installation.

The manufacture of synthetic ammonia, tried a little later, succeeded first in the Haber process, in which a mixture of nitrogen and hydrogen is passed at a moderately high temperature and under pressure over a contact catalyst, with a partial conversion of the elemental gases to ammonia. Several modifications for making ammonia from the elements have been so successful that this process is now more important than all other synthetic processes combined. Ammonia, some of its salts, and its derivatives are valuable fertilizers. Moreover, if nitric acid is called for, ammonia may be oxidized with atmospheric oxygen, aided by a contact catalyst, so that the synthetic ammonia process may also produce from atmospheric nitrogen, in an indirect way, what the arc process furnishes directly. In recent years the use of ammonia, ammonium nitrate, and urea as fertilizer materials has established a continuing need for nitrogen both as ammonia and as nitric acid.

An entirely different process for the fixation of atmospheric nitrogen is the calcium cyanamide process, which depends on the fact that metallic carbides, particularly calcium carbide, readily absorb nitrogen to form the solid cyanamide. This substance is a fertilizer. By further treatment it may be transformed

*Manufacturing Division, Biochemicals Dept., E. I. du Pont de Nemours & Company, Inc.

into cyanide; by another, into ammonia; but this ammonia is more costly than direct synthetic ammonia. The process was developed by Frank and Caro in Germany during the years 1895 to 1897, and has been introduced in many countries, including the United States and Canada, since that time.

Nitrogen can also be fixed by the high-temperature contacting of nitrogen with oxygen in the air. Experimental work on such a process has been undertaken by several investigators and results have been reported in the literature. The concentration of nitric oxides in the product gas has been too low to make the process commercially feasible. If this deficiency could be rectified, the process might become commercially attractive; however, there would still remain engineering problems associated with the design of the high temperature equipment.

This high-temperature nitrogen process operates in the range of 2,000°C, hence materials of construction are critical. The feed to the process is raised to the reaction temperature by means of a pebble preheater, which is heated by burning a fuel such as natural gas or fuel oil. Pebble attrition is a serious problem in the operation of the preheater.

Nitrogen may be passed over metals at suitable temperatures to form nitrides which, when treated with steam, will yield ammonia. The best known process embodying this principle is the Serpek process, introduced in France but not in America, for the manufacture of aluminum nitride.

Nitrogen Fixation

Recently new processes have been proposed for nitrogen fixation, but none of these have approached a commercially attractive stage.[1,2,3] One of these processes[1] is based upon reactions of nitrogen obtained from air in a nuclear reactor where the energy density corresponds to a temperature of 8000°K. This is high enough to break the nitrogen bonds and react the nitrogen and oxygen. Separation of the active particles for recycle and the NO_2 from the oxygen and nitrogen would be required. Also, radioactivity of the product and impurities must be overcome. One gram of fully enriched uranium is reported to produce two tons of nitric acid.

In another recent process[2] nitrogen is fixed by using a homogeneous molecular catalyst which allows a tremendous reduction in the normally used 2000 psig synthesis pressure. Molecular-nitrogen complexes are formed, in one case containing rhodium, and in another iridium. These react with an acid azide giving complexes in which the nitrogen shares two loosely bound electrons in the metal. The iridium complex contains two triphenyl phosphate groups, chloride, and a carbonyl group.

Research work in Russia and at the University of Michigan is based on the use of dicyclopentadienyltitanium dichloride in the presence of ethyl magnesium bromide.[2,3] This system is based on the fact that enzymes are able to fix nitrogen at room temperature and atmospheric pressure. The fixation of nitrogen occurs by allowing a mixture of diethyltitanium dichloride $[(C_2H_5)_2TiCl_2]$ and ethylmagnesium bromide (C_2H_5MgBr) in ether to react in a nitrogen atmosphere at room temperature and 150 atm pressure for 7 to 31 hours. It is reported that 0.8 to 0.9 mole of NH_3 is formed per mole of titanium dichloride.

A unique method for fixation of nitrogen is given in French Patent 1,391,595.[4] In this process, purified hydrogen and nitrogen are passed into an electrolyte solution of an alkali at a concentration of 5 to 15 per cent at 20 to 50 atm. A catalyst consisting of tablets of powdered nickel promoted with cobalt is used. The liquid electrolyte is removed from the reactor and the dissolved ammonia removed at lower pressure by distillation. The electrolyte is returned to the reactor where more ammonia is formed. A temperature of 50 to 80°C is used. Yields of 4.2 per cent from the gases per pass are reported. The process claims to reduce the plant investment and reduce compression costs.

The various processes with actual or potential value for the fixation of atmospheric nitrogen may be grouped together.

Arc Processes. Air is passed at a rapid rate through a broad or long electric arc to give nitrogen oxides. This was the earliest method for fixation of nitrogen and it is reported that the process is still used in Norway or Sweden. Other sources of high temperature are used such as preheating with natural gas and air, or by use of nuclear energy.

Direct Synthetic Processes. Nitrogen is combined with hydrogen to produce ammonia by contacting the two gases over a catalyst at elevated pressure and temperature. The ammonia can then be oxidized with air to produce nitric acid which will combine with additional ammonia to produce ammonium nitrate. The great expenditure of electric energy required by the arc process is avoided; however, energy requirements for compression of the hydrogen-nitrogen mixture are large. In addition, a large expenditure of energy is necessary to produce the hydrogen from natural gas and water, or from water by electrolysis.

Cyanamide Process. This process requires fairly pure nitrogen as well as calcium carbide, a product of the electric furnace ($CaC_2 + N_2 \xrightarrow{1000^\circ C} CaNCN + C$). Calcium cyanamide may be used as a fertilizer, or ammonia may be produced by steaming in autoclaves ($CaCN_2 + 3H_2O \longrightarrow CaCO_3 + 2NH_3$). However, the latter is rarely done.

Cyanide Process. Nitrogen is passed into a vessel containing an alkali and coal. The products are cyanides and ferrocyanides. This process has no commercial importance at present.

Nitride Process. Nitrogen reacts with certain metals to produce nitrides, such as aluminum nitride, which in turn, when treated with water, yield ammonia.

Fixation with Complexes. Nitrogen is reacted with acid azides in the presence of rhodium or iridium catalysts. Also, organic complex compounds have been shown to fix nitrogen in the presence of enzymes. These have no commercial importance.

Fixation of atmospheric nitrogen has been a difficult task because elemental nitrogen is comparatively unreactive. It combines with few other elements, and then usually only under drastic conditions (elevated temperatures).

Nitrogen Consumption

World consumption of nitrogen is increasing exponentially. During the period 1955–1965

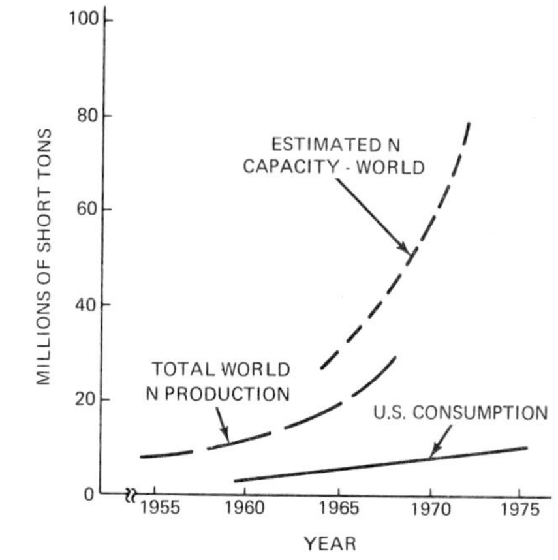

Fig. 5.1 Production and capacity for nitrogen fertilizers.[5]

consumption increased from 8 million tons to 19 million tons. It is anticipated that as the benefits of fertilization are fully realized, consumption will increase at an even more rapid rate as has happened in the United States following World War II. World production and estimated capacity for nitrogen fertilizers are shown in Fig. 5.1. Only a fraction of the amount that could be used is now consumed, or produced. In the period from 1965 to 1970 the capacity is estimated to have increased from 30 million to 60 million tons, and an additional 20 million tons were expected by 1972.

The rate of application in the United States is expected to be about 20 pounds per acre in the next ten years. However, a rate of 150 pounds per acre is reported in the Netherlands, 84 in Belgium, 46 in West Germany, and 15 in France.

Fertilizer consumption in various parts of the World is given in Table 5.1.

Total fertilizer consumption in the United States is greater than in any other area of the world. This emphasizes the importance that nitrogen plays in supplying food for the world. To provide adequate food for existing populations in many of these countries consumption of nitrogen would exceed that of the United States. For instance, India, with a population of about 450 million as compared to the 200 million in the United States, con-

TABLE 5.1 World Consumption of Fertilizer Nitrogen Compounds, 1967[a]

	1000 Tons
United States	6,000
Western Europe	5,500
U.S.S.R.	2,900
Eastern Europe	2,300
Communist Asia	2,150
Japan	940
India	915
Pacific Islands	430
Mexico	350
South America	310
United Arab Republic	275
Canada	275
Caribbean	215
Central Africa	186
Republic of South Africa	119

[a]National Fertilizer Development Center, TVA, p. 22.[5]

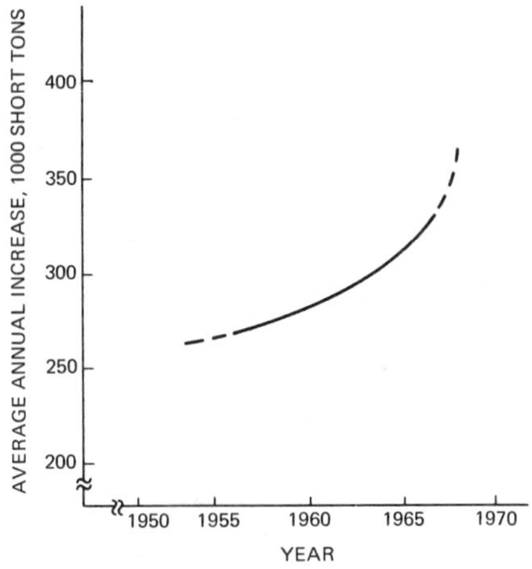

Fig. 5.2 Annual increase of U.S. nitrogen consumption.[6]

sumes 15 per cent as much fertilizer per year. Within the last decade the consumption in India has increased by a factor of 5.38.

Materials consumed as fertilizers in the United States are shown in Table 5.2.[6] Quantities and materials used, as indicated in this table, are for direct application. It is interesting to note that for the years reported, the largest gains in consumption have been made by anhydrous ammonia, nitrogen solutions, and urea. Consumption of all other materials such as calcium cyanamide, ammonium sulfate, and sodium nitrate have decreased. Urea is of interest since it contains the largest amount of nitrogen per pound for solid fertilizers. The rate of increase of nitrogen consumption was 293 thousand tons per year prior to 1965, and after 1965 the rate of increase has gone up to 328 thousand tons per year (see Fig. 5.2).

TABLE 5.2 Principal Nitrogen Materials Consumed as Fertilizers in the United States[a] (1000 Short Tons)

		Year	
Material	*1960*	*1965*	*1968*
Anhydrous ammonia	582	1282	2457
Aqua ammonia	85	164	163
Nitrogen solutions	195	596	803
Ammonium nitrate	416	547	786
Urea	65	194	243
Others	378	403	375
TOTAL	1721	3186	4827

[a]"Fertilizer Trends–1969," National Fertilizer Development Center, TVA, Muscle Shoals, Alabama.

Ammonia

Since 1963 there has been a revolution in the ammonia-manufacturing business. The advent of the large single-train plants has resulted in a large increase in production capacity, the shutdown of a number of smaller plants, and a reduction in the manufacturing cost. Since June 1967, at least twelve companies have shut down ammonia plants with a combined total capacity of about 6290 T/D. Table 5.3 gives a list of 28 current producers with ammonia plants having a daily capacity of over 600 T/D.[7] The combined capacity of these plants is 28,090 T/D. In January 1957, there were 41 producers with a total capacity of 4,975,000 tons per year or 13,600 T/D.[8]

Capacity for synthetic nitrogen is given in Figs. 5.3 and 5.4. Prior to World War II the capacity was relatively unchanged. During the war the need for explosives caused an increase in the production of nitric acid, and hence ammonia.

Available nitrogen capacity after the war

TABLE 5.3 Large U.S. Anhydrous Ammonia Producers[a]

Company	*Location*	*Capacity T/D*
Air Products	New Orleans, La.	600
Allied Chemical	Hopewell, Va.	1000
Allied Chemical	Geismar, La.	1000
American Cyanamid	Fortier, La.	1000
Amoco-Tuloma	Texas City, Texas	600
Apple River	East Dubuque, Ill.	600
Borden Chemical	Geismar, La.	1000
Chemical Distributors	El Centro, Calif.	600
Chevron Chemical	Pascagoola, Miss.	1500
Central Farmers	Donaldsonville, Pa.	1000
Collier Carbon and Chemical	Brea, Calif.	750
Collier Carbon and Chemical	Kenai, Alaska	1500
Commercial Solvents	Sterlington, La.	1000
Conoco	Blytheville, Ark.	1000
Du Pont	Beaumont, Texas	1000
Du Pont	Belle, W.Va.	1000
Farmers Chemical Assoc.	Tunis, N.C.	600
Farmland Industries	Ft. Dodge, Ia.	600
Farmland Industries	Dodge City, Kan.	600
First Nitrogen	Donaldsonville, La.	1000
Gulf Oil	Donaldsville, La.	1000
Hill Chemicals	Borger, Texas	1000
Mississippi Chemical	Yazoo City, Miss.	1000
Mobil Chemical	Beaumont, Texas	800
Monsanto	Luling, La.	600
Olin Mathieson	Lake Charles, La.	2 X 700
Phillips Petroleum	Beatrice, Neb.	600
Sinclair Petrochemical	Fort Madison, Ia.	1000
Terra Chemical	Sioux City, Ia.	600
Triad Chemicals	Donaldsonville, La.	1000
U.S. Steel	Clairton, Pa.	1140

[a]*Nitrogen*, No. 57 (Jan.-Feb., 1969).

was diverted to the manufacture of fertilizers. Accordingly, there was a rapid increase in fertilizer consumption. The advantages of fertilizer were therefore emphasized, and produc-

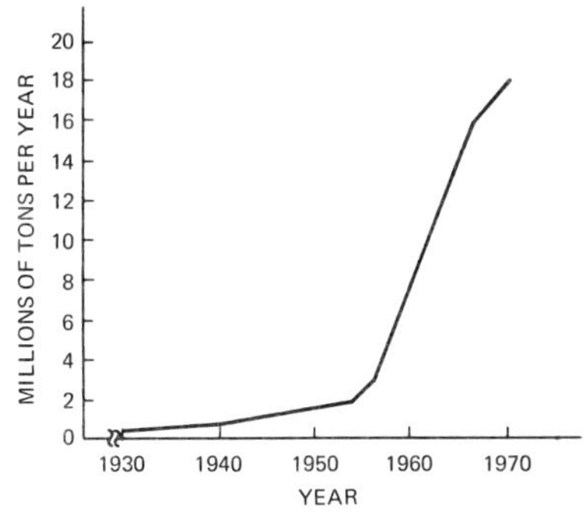

Fig. 5.3 Synthetic ammonia capacity in the U.S.[6]

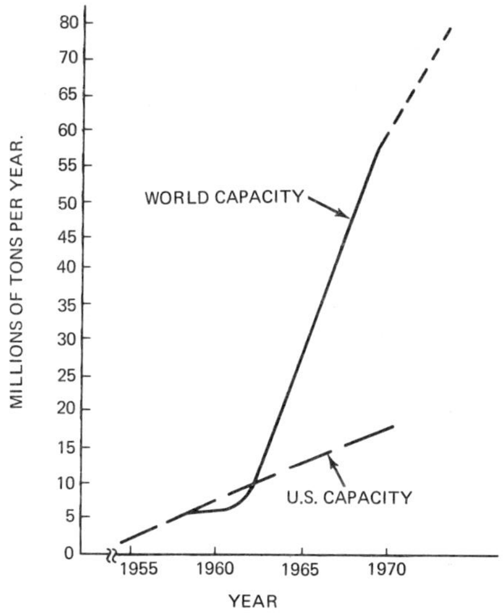

Fig. 5.4 U.S. and world synthetic ammonia capacity.[5]

tive capacity increased by leaps and bounds. From 1940 to 1950 the number of ammonia plants doubled, then from 1950 to 1960 the number more than doubled again. It was in the early sixties that the single-train concept was developed and many smaller plants were slated for shutdown. However, capacity tripled from about 1958 to 1968. A large increase in capacity is still needed throughout the world to meet the current and increasing food needs of an expanding population.

Early ammonia processes which have been used are indicated in Table 5.4. Other process modifications currently in use are:[9]

TVA	Topsoe
Chemico	Uhde
SBA	Lummus
OSAG–Linz, Austria	ICI
Kellogg	Braun

A promoted iron catalyst is used for all processes with the exception of Mont Cenis. Pressures employed vary from 100 to 900 atmospheres and the temperatures from 400° to 650° C. In general, gas recirculation is used to give good utilization of hydrogen and nitrogen since only 9 to 30 per cent conversion is obtained per pass over the catalyst. The synthesis gas source varies depending on the location. In some instances electrolytic hydrogen

TABLE 5.4 Ammonia Processes

Process	*Catalyst*	*Pressure (Atm)*	*Temperature (°C)*	*Gas Recycled*	*Conversion (%)*	*H_2 Source*
American	Promoted Iron	300	500	Yes	20–30	Natural gas or oil
Haber	Promoted Iron	200	550	Yes	8	Coke-steam
Mont Cenis	Iron cyanide complex	100	400	Yes	9–20	
Claude (France)	Promoted iron	900	650	No	40–85[a]	Coke-oven gas
Casale (Italy)	Promoted iron	750	500	Yes	15–18	
Fauser	Promoted iron	200	500	Yes	12–23	Electrolytic H_2

[a]Conversion of 85% by using two converters in series.

is used; in others, hydrogen from petroleum products or natural gas.

Processes[10,11] for producing ammonia from several raw materials are shown in Figs. 5.5, 5.6, 5.7, and 5.8. Figure 5.5 illustrates a process that employs coal or oil as a raw material. The process includes partial combustion of the coal or oil, followed by carbon and ash removal, carbon monoxide conversion, carbon dioxide and H_2S removal, low temperature scrubbing with liquid nitrogen, and compression for ammonia synthesis.

With natural gas as a raw material, the new plants depend on steam reforming to produce hydrogen. Air is bled into a secondary reformer to supply the nitrogen. The carbon monoxide is then reacted with steam to produce hydrogen and carbon dioxide; the latter is removed and the gas compressed. The residual carbon monoxide and carbon dioxide are removed by methanation ($CO + 3H_2 \longrightarrow CH_4 + H_2O$); the gas is then compressed and fed to the ammonia synthesis unit.

Partial combustion of natural gas with oxygen is widely used for producing hydrogen and carbon monoxide. Processes based on this initial step convert the carbon monoxide with steam to carbon dioxide and hydrogen directly after partial combustion. The carbon dioxide is removed; the gas is scrubbed with liquid nitrogen to remove residual carbon monoxide, argon, and methane, and then it is compressed and sent to ammonia synthesis.

With the advent of "platforming" of petroleum products to give unsaturated compounds for high test gasoline, large volumes of hydrogen off-gas have been produced.[12] The off-gas from the "platforming" operation will frequently contain over 90 per cent hydrogen. In such cases the processing requirements for

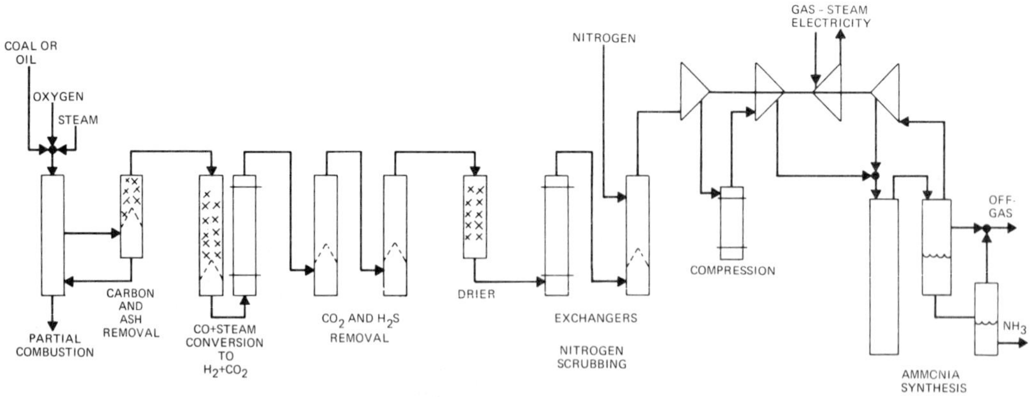

Fig. 5.5 Ammonia synthesis from coal or oil.

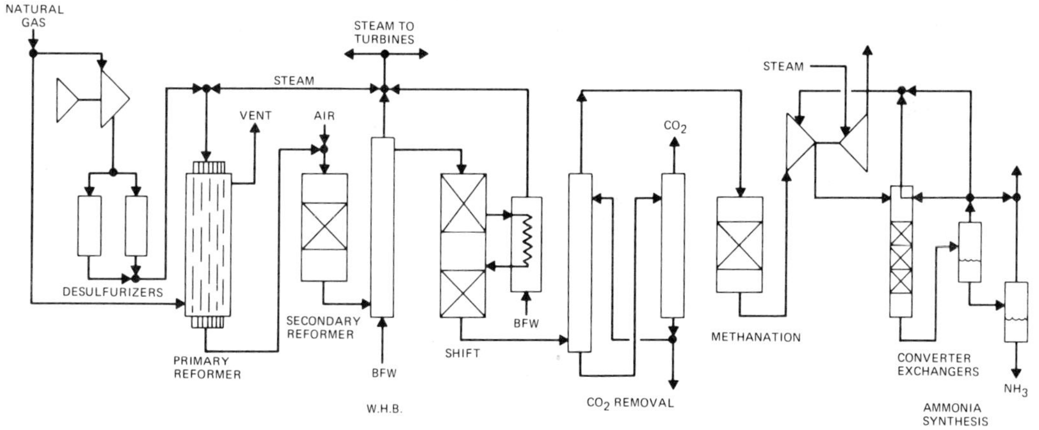

Fig. 5.6 Ammonia synthesis via natural gas reforming.

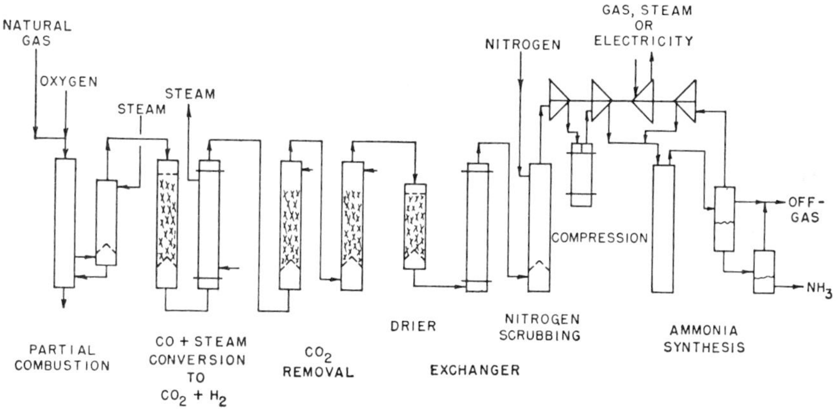

Fig. 5.7 Ammonia synthesis via partial combustion.

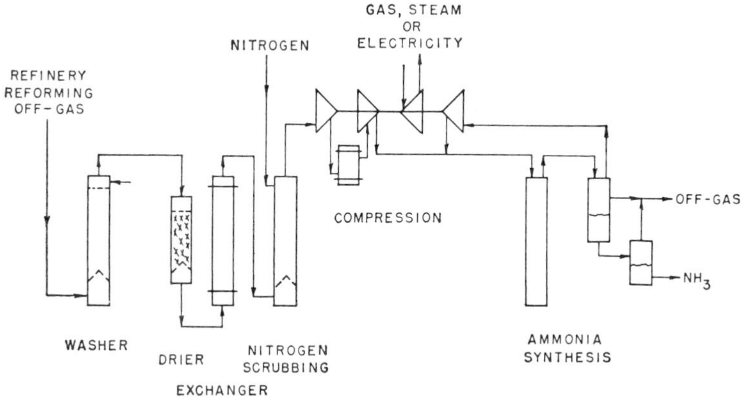

Fig. 5.8 Ammonia synthesis from refinery reforming off-gas.

Fig. 5.9 1000-T/D ammonia plant, Borger, Texas. (*Courtesy M. W. Kellogg Co., a Division of Pullman, Inc.*)

ammonia manufacture are greatly reduced. It is only necessary to wash out the higher hydrocarbons, and scrub the gas with liquid nitrogen to remove methane and carbon monoxide for the hydrogen-nitrogen mixture to be ready for ammonia synthesis. This process, or one based on H_2 from other cracking operations, is illustrated in Fig. 5.8. Currently this "platforming" hydrogen is generally used in a hydrocracking unit and many petroleum producers are short of hydrogen.

For an ammonia plant at a new location where by-product hydrogen is not available, selection of the most economical process will depend to a large extent on the raw material used. Production of hydrogen requires a large amount of energy, hence the energy reserves will dictate the type of process. The data[13,14] given in Table 5.5 indicate the relative availability of various energy sources.

At the 1969 production rate of 20.723 trillion cu ft (207 billion therms), it can be seen that the supply of natural gas will suffice for the immediate future. However, the type of raw material that will be consumed subsequently will no doubt depend on the process that is developed for utilization of the available energy.

The U.S. Bureau of Mines has worked on a process which uses nuclear fission as a source

TABLE 5.5 Energy Reserves[a]

	Billion Therms
Natural gas[b]	2,890
Natural gas liquids	250
Crude petroleum	1,760
Coal[b]	465,000
Recoverable coal	56,000
Oil shale	72,000
Nuclear	
Fissionable	5,600,000
Fusion	Very large
Solar (U.S.A.)	300,000/yr

[a]Data from I.G.T. Chicago News Letter, 1958.
[b]*Chem. & Eng. News* (Mar. 18, 1968).

of energy for producing hydrogen from coal. The Institute of Gas Technology in Chicago has studied the partial combustion of coal as a means of producing synthesis gas, as well as a high-Btu pipeline gas.

Natural gas is currently the preferred source of hydrogen and energy. Coal is a good source of energy, but a poor source of hydrogen. When coal is used, the hydrogen must be obtained from water. Nuclear fission will supply abundant energy; but the source of hydrogen is yet to be determined.

Feed Stock Purification

Since most of the ammonia plants today are based on the reforming of natural gas with steam over a nickel catalyst to produce hydrogen, it is imperative that the sulfur content of the feed stock be reduced to as low a level as possible. Natural gas for transportation by pipeline is usually desulfurized to minimize pipeline corrosion. Hence, by the time it reaches an ammonia producer, the only sulfur present is that used to odorize the gas. On the other hand, naphtha which is used as a raw material may require more drastic treatment to remove sulfur.

For natural gas purification, the usual treatment is to pass the gas over specially prepared activated carbon to remove the sulfur. The treatment takes place at ambient temperature and at the pressure level used in the reformer or natural gas line. The higher the pressure, the more effectively the carbon is used.

The activated carbon may be impregnated with iron or copper to enhance the removal of sulfur from the gas. About 100,000 to 200,000 cu ft of natural gas can be treated per cu ft of carbon. Then the carbon must be regenerated with high temperature steam at low pressure. The temperature of regeneration must exceed the boiling point of the adsorbed compound with the highest boiling point. Usually a temperature of 250 to 300°C is satisfactory. A two-bed system is normally used, one in operation while the other is being regenerated.

For desulfurization of naphtha a more complicated process is required. Hydrodesulfurization is quite often used. About 0.5 mole of H_2 is mixed with 1 mole of the vaporized naphtha or 250 SCF per barrel depending upon the sulfur and olefin content. The mixture is preheated to 600°F. It is then passed over a cobalt-molybdenum catalyst where the olefins are hydrogenated to hydrocarbons and the sulfur compounds are converted by the hydrogen to H_2S. The gas is then passed over a sulfur adsorbent such as iron or zinc oxide. It may or may not be necessary to condense the naphtha. This will depend upon the amount of hydrogen used and the need to remove it from the naphtha.

All olefins will be hydrogenated to hydrocarbons in this process. This is advantageous since olefins will readily crack to carbon in the next step (reforming) of the process. United States Patent 3,106,457[15] gives conditions under which olefins will crack to carbon. The minimum steam to carbon ratio reported to avoid carbon is given as $(Ri)_0 = (0.244p + 8.15 \times 10^{0.0052M})$ where $(Ri)_0$ is the minimum steam to carbon ratio (moles per mole organic carbon atom), p is the operating pressure in psia, and M is the molecular weight of the olefin.

Hydrogen Production

A list of processes for producing hydrogen is given in Table 5.6, along with the hydrogen to carbon monoxide ratio of the synthesis gas produced. When synthesis gas is made by the gasification of coke with steam and oxygen, the hydrogen to carbon monoxide ratio is 0.6; with the intermittent gasification of coke with air and steam the synthesis gas has a ratio of 0.9. By using excess steam with the reforming of methane, the hydrogen to carbon monoxide ratio can be increased to as high as 5. Thus

TABLE 5.6 Synthesis Gas H_2/CO Ratio

Method of Manufacture	*H_2/CO Ratio*
Oxygen-coke-steam	0.6
Air-coke-steam	0.9
Oxygen-coal-steam	1.0
Oxygen-fuel oil-steam	1.0
Propane-steam	1.33
Methane-oxygen	1.7
Methane-oxygen-steam	2.3
Methane-steam	3.0–5.0

by the selection of the proper process, the equipment requirements for conversion of carbon monoxide to hydrogen with steam can be reduced.

Very often the raw material available, such as natural gas, coal, or oil, will dictate the type of process employed. Table 5.7 gives raw material requirements for the various processes in the production of synthesis gas.

Selection of a process for hydrogen manufacture depends not only on the raw material and its cost, but also on the scale of operation, the purity of the synthesis gas produced, the pressure level of the natural gas, and other carbon monoxide and H_2 requirements.

Today hydrogen is manufactured by four principal processes: (1) steam reforming of natural gas; (2) partial combustion of natural gas or oil with pure oxygen; (3) recovery of hydrogen from petroleum refinery gases or other cracking operations; and (4) gasification of coal or coke with air or oxygen and steam. Small amounts of hydrogen are also manufactured by electrolysis.

Reforming. The reforming reaction is as follows: $CH_4 + H_2O \longrightarrow 3H_2 + CO$. The reaction is endothermic and requires the input of a large amount of heat. The heating-gas requirement is about 80 per cent of the process-gas requirement. The reaction is carried out at temperatures in the range of 800°C and at pressures ranging from atmospheric up to 500 psig.

TABLE 5.7 Synthesis Gas Raw Material Requirements

Method of Manufacture	*Requirements per 1,000 cu ft of ($CO + H_2$)*
Air-coke-steam	40 lb Coke
	80 lb Steam
	2,200 cu ft Air
Oxygen-coke-steam	30 lb Coke
	300 cu ft Oxygen
	34 lb Steam
Coal partial combustion	37 lb Coal
	350 cu ft Oxygen
	37 lb Steam
Natural gas partial combustion	360 cu ft Natural gas
	260 cu ft Oxygen
Natural gas reforming	285 cu ft Natural gas
	230 cu ft Natural gas for fuel
	60 lb Steam
Propane reforming	3 gal Propane
	1.67 gal Distillate for fuel
	380 lb Steam[a]
Naphtha reforming	14.7 lb naphtha
	5.8 lb Naphtha for fuel
	105 lb Steam
Fuel oil partial combustion	20.5 lb Oil
	320 cu ft Oxygen
Electrolysis	163 Kw-h
Steam-iron	290 lb Steam
	0.3 lb Ore
	206 cu ft Fuel gas
	65 lb Coke
Coke-oven gas	353 lb Coal[b]
"Platforming" off-gas	84 gal Feed stock

[a]Includes CO shift to CO_2 and H_2 with steam.
[b]Excludes $CH_4 + CO$ in coke oven gas which can be converted to H_2.

Other reactions which proceed at the same time as the reforming reaction are:

$$CO + H_2O \rightleftharpoons CO_2 + H_2$$
$$CH_4 \rightleftharpoons C + 2H_2$$
$$2CO \rightleftharpoons C + CO_2$$

The equilibrium composition dependends upon the steam to gas ratio entering the reactor, the temperature, the pressure, and the quantity of inerts in the reaction mixture. Hougen and Watson[16] give a general equation which simplifies to the following for reforming of CH_4 with steam:

$$K = \frac{(N_{CO})(N_{H_2})^3}{(N_{CH_4})(N_{H_2O})}(Kv)\left(\frac{\pi}{N}\right)^2$$

where K is the equilibrium constant; N_{CO}, N_{H_2}, N_{CH_4}, N_{H_2O} are the moles of each species present; Kv is the fugacity factor; π is the total pressure; and N is the total number of moles present. The equilibrium constant for the methane reforming reaction is as follows:

Temperature, °F	*Equilibrium Constant*
600	2.186×10^{-7}
800	2.659×10^{-4}
1,000	4.900×10^{-2}
1,200	2.679
1,400	0.6343×10^{2}
1,600	8.166×10^{2}
1,800	6.755×10^{3}

The final composition will also satisfy the carbon monoxide-shift reaction. The equilibrium constant for this reaction is:

Temperature, °F	*Equilibrium Constant*
600	31.4
800	9.03
1,000	3.75
1,200	1.97
1,400	1.20
1,600	0.819
1,800	0.604

To avoid carbon formation as indicated by the foregoing reactions, the steam to gas ratio must be maintained high enough to favor the reforming and shift reactions in preference to those forming carbon.

A catalyst is required for this reaction. Nickel oxide is normally used. It is prepared by mixing the oxide powder with alumina and calcium aluminate cement, forming the mixture into the desired shape (5/8 × 5/8 × 1/4-inch Raschig rings) and then calcining the catalyst at a temperature of 1,200 to 2,000°F.

A reaction furnace usually consists of a number of 2- to 8-inch centrifugally cast chrome-nickel tubes about 10 to 40 feet long, mounted vertically in a refractory-lined furnace. The process gas is fed downward over the catalyst and removed from the bottom of the tubes. The tubes are heated by burners located strategically in the sides of the furnace at the top or bottom. Close control of heat is essential to ensure uniform temperature distribution throughout the tubes.

During the change from atmospheric temperature to reaction temperature, expansion of the catalyst tubes is considerable, thus provision must be made for this growth. Usually, the tubes are suspended by springs, or counterweights are attached to the top of the tube. Feed materials are introduced through flexible tubing.

Steam and natural gas are mixed in the ratio of 3 to 5 prior to introduction into the furnace. These materials are preheated to avoid use of the catalyst bed as a preheating section. Usually the flue gases are sufficient to provide the preheat.

Steam to carbon ratios in the range of 4 to 8 are claimed to be advantageous; such ratios result in less fouling of waste-heat boilers and give better reformer-catalyst and tube life.[17]

The hydrogen to carbon monoxide ratio for the product gas is usually 4 to 5; however, this can be increased if additional steam is used in the reaction. This will reduce the demand on the carbon monoxide-shift converters which follow after the reformer in the production of hydrogen from steam and carbon monoxide. Steam is introduced after the reformer, before the gas is fed to the carbon monoxide-shift converter.

In the modern single-train ammonia plants the natural gas is reformed in two steps. In the first step, the reaction takes place in the primary reformer in tubes suspended in a refractory-lined furnace. The large endothermic heat is supplied by burning natural gas with air in the furnace. Heat flux in the tubes will be as high as 35,000 Btu/hr/sq ft. The methane leakage is about 10 per cent in the effluent dry gas, or about 60 to 65 per cent of the feed methane is converted to synthesis gas.

For ammonia manufacture it is necessary to introduce nitrogen into the system. This is done as indicated in Fig. 5.6 by mixing air with the primary reformer gas in a refractory-lined vessel which contains additional reforming catalyst. The temperature at the point of mixing will reach 2,300°F because of the fast reaction of the oxygen in the air. But, then the endothermic reforming reaction takes place and the gases cool to about 1,800°F. At this temperature the methane concentration has been reduced to about 0.3 per cent. The concentration is dependent upon the pressure, temperature, and the quantity of nitrogen and steam present.

The catalyst used in this reforming step does not need to be as active as that in the primary reformer. Hence, the usual nickel concentration is about 15 per cent compared with 25 per cent in the primary-reformer catalyst.

Mixing of the air and primary reformer gas is critical and specially designed burners are used.

The hot gases from this section of the process provide sufficient heat to generate the large amount of the steam necessary to supply the process and drive the compressors down stream in the plant.

Partial Combustion. By burning natural gas with a limited quantity of oxygen, a synthesis gas containing approximately 2 moles of hydrogen for each mole of carbon monoxide can be produced ($CH_4 + 1/2O_2 \longrightarrow CO + 2H_2$).[18] The reaction is exothermic and must be carried out in a refractory-lined or water-cooled vessel. High temperatures favor the formation of carbon monoxide and hydrogen.[19] Pressure will hinder the reaction; however, pressures up to 1,500 psi can be used and still produce a gas of low methane content. If steam is introduced into the reaction the hydrogen to carbon monoxide ratio can be increased to over 2. Because of the high temperatures involved, e.g., 1,400° C, special care must be observed in processing the gas after it leaves the reactor.

Preheating the natural gas and oxygen will reduce the requirements for oxygen and natural gas, since some of the natural gas must be burnt to carbon dioxide in order to raise the products to their proper temperature. The partial combustion reaction itself is not sufficiently exothermic to raise the products to the desired temperature.

The partial combustion of fuel oil or crude oil is practiced in many areas where natural gas is not available. When using crude oil, care must be taken to provide for removal of sulfur compounds and ash-containing materials in the crude oil. These have been known to damage the refractory and, of course, to introduce sulfur into the product gas. In all partial combustion reactions there is some residual carbon in the product gas. In the case of crude-oil or fuel-oil partial combustion, the carbon content is much greater than in the case of natural gas, and special design considerations are necessary to produce a satisfactory gas.

Nitrogen to produce the proper H_2/N_2 ratio is usually introduced later in the processing sequence.

Coal and Coke Gasification. An important process for the gasification of coal and coke is the Lurgi process, used in South Africa at the "coal to oil" plant and in Europe. This process requires the use of coke or a noncaking coal, and the product gas is usually high in methane. The reaction is carried out with a fixed bed of burning coal or coke into which are introduced steam and oxygen. The product gas will contain as impurities sulfur compounds, ungasified carbon, ash, and, compared to natural-gas partial combustion, large quantities of carbon dioxide. The gas may be further processed by converting the residual methane to carbon monoxide and hydrogen by passing the gas over a reformer catalyst with steam.

In the United States, coal has been gasified with steam and oxygen in processes developed by Babcock and Wilcox, E. I. du Pont de Nemours and Company, the Institute of Gas Technology in Chicago, and the U.S. Bureau of Mines. In all these processes pulverized coal is introduced into a refractory zone along with steam and oxygen. Product gas contains hydrogen and carbon monoxide and is relatively low in methane compared to that from the Lurgi process. Again, sulfur compounds in the coal appear in the product gas, and the carbon dioxide content runs much higher than in the gas made by the partial combustion of methane.

Nitrogen may be introduced for ammonia synthesis by using oxygen-enriched air for the gasification.

There are several processes for purifying synthesis gas made from coal, coke, or oil. The impurities usually present are H_2S, CS_2, COS, ungasified carbon, and residual ash. Other factors which influence the selection of a purification process are the hydrogen to carbon monoxide ratio, and the carbon dioxide content.

If the gas contains large amounts of sulfur, it is quite often the practice to remove this sulfur before the carbon monoxide-shift operation. There are a number of sulfur removal processes, namely, amine scrubbing, hot potassium carbonate scrubbing, and the Thylox process. Ash and carbon removal are usually carried out before the gas is treated for sulfur or carbon dioxide removal, before the shift operation.

If a gas does not contain large quantities of sulfur and ash, the carbon monoxide-shift reaction is carried out directly after the synthesis operation. In some instances it is also desirable to remove the carbon dioxide before the shift operation, since it has an adverse effect on the carbon monoxide-shift equilibrium. Carbon dioxide may be removed by scrubbing with water, monoethanolamine (MEA), or hot potassium carbonate. The preferred process depends to some extent on the local economics and the particular conditions under which the synthesis gas is produced.

Hydrogen from Petroleum Gases. Petroleum gases from platforming operations, hydroformers, and butadiene plants may be processed to recover H_2.[11] The gas from the platformer is particularly rich in H_2, containing as much as 90 to 95 per cent. This gas is usually purified by low temperature fractionation or washing with liquid nitrogen.

Electrolysis. In some areas hydrogen is obtained by electrolysis but normally that is not economical. Electric power must be available at a very low cost such as that from hydroelectric power.[113] In some cases, by-product hydrogen from the electrolysis of brine has been used to manufacture ammonia.

Carbon Monoxide Shift

Conversion of carbon monoxide to carbon dioxide and hydrogen with steam is necessary to make economical use of the raw synthesis gases produced by the foregoing processes. In this operation the following reaction is promoted: $CO + H_2O \rightleftharpoons CO_2 + H_2$. The reaction is exothermic; the equilibrium is not affected by pressure, but high temperatures are unfavorable for complete conversion.

The equilibrium constant for the reaction is as follows:

Temperature, °F	*Equilibrium Constant*
400	207
450	119
500	72.8
550	46.7
600	31.4
650	22.0
700	15.9

In recent years the shift operation has been carried out in two steps, one at high temperature 350 to 450°C, and one at low temperature 200 to 300°C. The high temperature shift reaction is carried out over an iron-chromium catalyst which is not particularly sensitive to sulfur, although high concentrations have been found to be detrimental. At the usual effluent temperature the dry gas will contain about 3 per cent carbon monoxide.

To operate the low temperature-shift reaction it is necessary to remove heat between the two catalyst beds. The low temperature-shift catalyst contains copper and zinc and is very sensitive to sulfur and halogens.[20] Therefore a guard catalyst is frequently used to protect the shift catalyst. High steam to gas ratios favor conversion of the carbon monoxide to hydrogen and also reduce the temperature rise, resulting in a more favorable equilibrium temperature. The normal carbon monoxide leakage is about 0.3 to 0.5 per cent at an exit temperature of 240°C.

The low temperature-shift catalyst can be purchased in a reduced state. If it is not, it must be reduced before palcing it in operation. The cupric oxide must be reduced to metallic copper.

Carbon Dioxide Removal

Following the shift conversion of CO with steam to CO_2 and H_2, the bulk of the CO_2 must be removed from the hydrogen and nitrogen. There are numerous processes for this operation and the selection of the most favorable one depends upon local circumstances.[21] Carbon dioxide can be removed by scrubbing the gas under pressure with water, hot potassium carbonate (with or without an additive), monoethanolamine, Sulfinol, propylene carbonate, and refrigerated acetone or methanol.

Water Scrubbing. This system, once used widely, has not remained competitive. However, it appears that such a system is currently being offered by Friedrich Uhde GMBH in Germany.[22] A large quantity of water is required to dissolve the carbon dioxide and special design is required to avoid excessive loss of nitrogen and hydrogen.

Hot Potassium Carbonate.[23] In this system an aqueous solution containing 40 per cent K_2CO_3 is circulated at 118° C in the absorber. The CO_2 picked up is released in the stripper at atmospheric pressure by heating the solution to 230° C. An inhibitor is generally needed to reduce corrosion. Stainless steel equipment is used in many places. This system is characterized by its low steam requirements and its compatibility with the catalysts.

The system may be modified by adding materials which enhance the solubility of CO_2 and hence result in more complete recovery. Alkanolamines and As_2O_3 are typical additives.

Monethanolamine (MEA). A 15 to 20 per cent solution of MEA in water is used to absorb the CO_2 under pressure. The solution is then regenerated by heating in a stripper in which the carbon dioxide is released. This system is characterized by good absorption properties at low pressure.

A corrosion inhibitor is generally used for the mild steel equipment. Stainless steel is used in critical areas, where hot carbon dioxide-laden solutions are present.

It is reported that antifoam agents, e.g., certain silicones, are used in some cases.

Sulfinol.[24] The sulfinol process is a recent development by the Shell Development Corp., Emoryville, Calif. The solution used consists of sulfolane ($C_4H_8SO_2$), a ring compound, diisopropanolamine and water. This process is characterized by its high CO_2 retention under pressure, and by the fact that steam requirements for regeneration are small. A side-stream regenerator is required to remove by-products which build up in the system. See U.S. Patent 3,347,621.[115]

Mild steel is suitable in a number of places; however, stainless is preferred where CO_2 concentrations are high.

The process must be carefully engineered to protect the downstream nickel catalysts from the sulfur which may carry over from the absorber. Likewise, the CO_2 that is produced must be protected from entrainment of the solvent and contamination with sulfur.

Propylene Carbonate.[25] This process was developed primarily for the removal of CO_2 from natural gas. A large plant is in operation in Terrel County, Texas, removing 50 per cent CO_2 from a natural gas stream. The absorbing solution consists of propylene carbonate with a low water content. The system is characterized by a small steam requirement for regeneration and by the fact that high CO_2 partial pressures are required for economical operation. The scrubbed gas will contain as much as 1 to 2 per cent CO_2 unless special precautions are used for regeneration and absorption.

The absorber operates at a relatively low temperature, made possible by the refrigerating effect of the released CO_2. Mild steel is suitable for construction in most areas of the system.

Refrigerated Acetone.[26] Refrigerated acetone has been studied for CO_2 removal. It is not widely used.

Refrigerated Methanol.[27] This system is in use at the South African coal to oil plant. The methanol is cooled to -50°C at which temperature the vapor pressure is less than 1 mm Hg.

Giammarco-Vetrocoke.[28] This process is based upon the absorptive power of an arsenic-activated potassium carbonate solution. Both H_2S and CO_2 are removed. The CO_2 level can be reduced to about 0.05 per cent, and CO_2 with a purity of 99 per cent can be produced.

$$CO_2 + K_2CO_3 + H_2O \rightleftharpoons 2KHCO_3$$

$$6CO_2 + 2K_3AsO_3 + 3H_2O \rightleftharpoons 6KHCO_3 + As_2O_3$$

The solution is reported to be noncorrosive to steel.

Final Purification

Before the synthesis gas ($3H_2$ and $1N_2$) is admitted to the ammonia synthesis step, essentially all of the carbon monoxide and carbon dioxide must be removed since they are temporary poisons for the ammonia catalyst. Also, the presence of carbon dioxide in the make-up gas causes difficulty in the synthesis loop because it forms carbamates and ammonium carbonate which are solids that can damage compressors and valves.

Methanation. This is the most popular process for final purification or removal of the last traces of carbon monoxide and carbon dioxide. This process is the reverse of the methane reforming process.

$$CO + 3H_2 \rightleftharpoons CH_4 + H_2O$$

$$CO_2 + 4H_2 \rightleftharpoons CH_4 + 2H_2O$$

The large excess of hydrogen present and the lower temperature favor the complete removal of the oxides of carbon. Usually, the remaining oxides are less than 10 ppm.

In this process, the synthesis gas which has been scrubbed to remove the bulk of the CO_2 is preheated to about 600° F and then passed over a nickel catalyst (NiO on Al_2O_3). Ruthenium is also effective as a catalyst. The space velocity used depends upon the pressure of the process. At 450 psig the space velocity (standard cu ft of gas per hour per cu ft of catalyst) is about 5,000 hr^{-1}. At lower pressures a space velocity of 2,000 hr^{-1} is used.

The catalyst life is normally one to three years. Sulfur and halogens are poisons and shorten the life.

Nitrogen Wash Operation. The last traces of carbon monoxide are removed by liquid nitrogen scrubbing in some plants; generally those using a partial oxidation process for generation of the synthesis gas. The nitrogen is obtained from the air-separation plant used for oxygen production. In the nitrogen-wash operation, the gas must be thoroughly dried and the carbon dioxide removed before the gas is cooled to the point where it can be scrubbed with liquid nitrogen. Otherwise, the moisture and carbon dioxide would freeze out in the equipment, causing a number of problems. Hence, the gas must be passed through silica gel driers and a caustic wash to remove the water and carbon dioxide. At this point the gas can safely be cooled to liquid-nitrogen temperature and scrubbed to remove the carbon monoxide, methane, and argon.

The overhead gases from the wash tower usually contain about 90 per cent hydrogen and 10 per cent nitrogen, thus requiring the addition of more nitrogen to give the required ratio for ammonia synthesis. The gas so produced is suitable for ammonia synthesis.

Copper Ammonium Carbonate Scrubbing. In some instances it is preferable to use copper scrubbing to remove the last traces of carbon monoxide in the ammonia synthesis gas. This process is usually carried out at pressures in the range of 1000 to 1500 psi, one of the intermediate pressures of the high pressure compressors. The copper liquor is circulated counter-current to the synthesis gas, picking up the carbon monoxide as a copper ammonium carbonate-carbon monoxide complex and producing a relatively carbon monoxide-free gas. Methane and argon, however, are not removed; these pass on with the hydrogen to the ammonia synthesis units.

The copper liquor is regenerated by reducing the pressure on the liquid, which allows the carbon monoxide complex to decompose. The carbon monoxide is separated from the copper liquor, and the liquid is compressed and recirculated to the scrubbing operation.

Selective Oxidation of Carbon Monoxide. This process has been studied by Halger C. Andersen and W. J. Green.[29] It is based on the fact that over a platinum catalyst at 250 to 320°F, carbon monoxide will be selectively oxidized to carbon dioxide. Excess oxygen is converted to water by hydrogen.

$$CO + 1/2O_2 \longrightarrow CO_2$$

$$H_2 + 1/2O_2 \longrightarrow H_2O$$

The disadvantage of the process is that a second scrubbing step is necessary to remove the CO_2 formed by oxidation of the carbon monoxide.

Cryogenic Purifier.[30] In the C. F. Braun process for ammonia manufacture excess air is added in the second reformer. The excess nitrogen thus introduced must be removed prior to the synthesis step to avoid excessive loss of hydrogen and excessive compression costs. At the same time that the excess nitrogen is condensed and removed, the remaining traces of carbon monoxide, the methane, and most of the argon are removed, leaving a gas comparable to that produced by the nitrogen-ash operation.

Compression

Within the last ten years, there has been a new approach to compression of the synthesis gas and recycle of the gases in the synthesis loop. With the advent of large-capacity plants (over 600 T/D) it has been possible to design centrifugal compressors for the services required. The pressure level has also dropped to about 2,000 psi to favor compressors which had been successfully used by industry. However, more and more machines are now being built for higher pressures, up to 5,000 psig. One of the advantages of the centrifugal compressors is that the need for lubricating oil is minimal, and this material does not cause difficulty in the synthesis loop.

The centrifugal compressors are driven by steam turbines. High-pressure, high-temperature steam is generated in the plant using the heat in the process gas leaving the secondary reformer and the flue gases from the primary reformer.

Design of the compressors is critical from the standpoint of efficiency and economy, and nonideality must be considered when working at high pressures. Compressibility factors for a mixture of 3 hydrogen and 1 nitrogen are available.[31]

For smaller plants, under 600 T/D, reciprocating compressors are still used. These often contain several services on a common drive shaft: air compression, natural gas compression, synthesis gas compression and recycle, and ammonia compression for refrigeration.

Ammonia Synthesis

Ammonia synthesis is carried out at pressures ranging from 2,000 to 10,000 psi. The preferred pressure depends largely on the quality of the synthesis gas and certain other conditions, such as production requirements per converter.

O. J. Quartulli, *et al.*,[32] published a detailed study of the factors to be considered in selecting the proper synthesis pressure. Reaction equilibrium, reaction rate, and condensation of ammonia are all favored by high pressure. Mechanical problems and cost considerations are also considered. In general they conclude that a pressure level of 3,000–3,200 psi is near the optimum although local economics could well change the level, up or down. Some of the factors considered were: catalyst volume, recycle flow, horsepower for make-up-gas compression, recycle-gas flow, refrigeration, equipment costs, and equipment reliability.

The ammonia reaction is $N_2 + 3H_2 \longrightarrow 2NH_3$, and it is exothermic. Equilibrium data for the reaction are given in Table 5.8. See also Reference (33) for a discussion of equilibrium values.

As can be seen from this table, lower temperatures favor higher equilibrium concentrations of ammonia. However, the rate of reaction at low temperature is so slow that the process is uneconomical at temperatures below about 400°C. Usually, a conversion of about 15 to 30 per cent per pass is obtained

TABLE 5.8 Percentage Ammonia at Equilibrium[a] (Ratio $H_2/N_2 = 3$)

Temperature (°C)	*Pressure (Atm Absolute)* 1	10	50	100	300	600	1000
200	15.30	50.66	74.38	81.54	89.94	95.37	98.29
300	2.18	14.73	39.41	52.04	70.96	84.21	92.55
350	0.90	7.41	25.23	37.35	59.12	75.62	87.46
400	0.44	3.85	15.27	25.12	47.00	65.20	79.82
500	–	1.21	5.56	10.61	26.44	42.15	57.47
600	–	0.49	2.25	4.52	13.77	23.10	31.43
700	–	–	1.05	2.18	7.28	–	12.87

[a]Compiled from publications of the Fixed Nitrogen Research Laboratory by Dr. Alfred T. Larson (JACS, 1924).

when the temperature is in the range of 400 to 600°C.

The heat of reaction for the synthesis of ammonia is given by the following equation:[33]

$$\Delta H\ (\text{cal/mole}) = -\left[0.54526 + \frac{840.609}{T} + \frac{459.734 \times 10^6}{T^3}\right]p - 5.34685T - (0.2525 \times 10^{-3})T^2 + (1.69167 \times 10^{-6})T^3 - 9157.09$$

where T is °K and p is in atmospheres.

Reaction Rate. The rate of reaction over the synthesis catalyst dependends upon the type of catalyst, rate constants for synthesis and decomposition, and the partial pressures of nitrogen, hydrogen, and ammonia. Temkin and Pyzhev[33] propose the following equation:

$$W = K_1 P_{N_2}\left(\frac{(P^3_{H_2})}{(P^2_{NH_3})}\right)^{\alpha} - K_2\left(\frac{(P^2_{NH_3})}{((P^3_{H_2})}\right)^{\beta}$$

where W is the reaction rate; K_1, K_2 are the rate constants for synthesis and decomposition, respectively; $P_{H_2}, P_{N_2}, P_{NH_3}$ are partial pressures, and α, β are constants. The constant α has been found to be in the range of 0.5 to 0.75 and β, 0.4 to 0.3; $\alpha + \beta = 1.0$. Tempkin, and others, established this equation on the basis that the rate determining steps are the chemisorption of nitrogen in a diatomic form and the combination of a chemisorbed nitrogen molecule with a hydrogen molecule. At high pressure and with some ammonia in the gas, the effect is to have only one rate-determining step.[34] Figure 5.10 gives some indication of the production rate as a function of pressure. Most systems are designed for 2,000 to 5,000 psi.

The gas composition for ammonia synthesis is also critical, relative to the cost of ammonia. High concentrations of inert materials such as argon or methane will build up in the circulating system and produce a low partial pressure of hydrogen and nitrogen resulting in low conversions to ammonia. In such cases it is necessary to purge these materials from the system and maintain the proper ratio of hydrogen to nitrogen. An excess of either one of these reactants will act as an inert material and lower the effective partial pressure of the

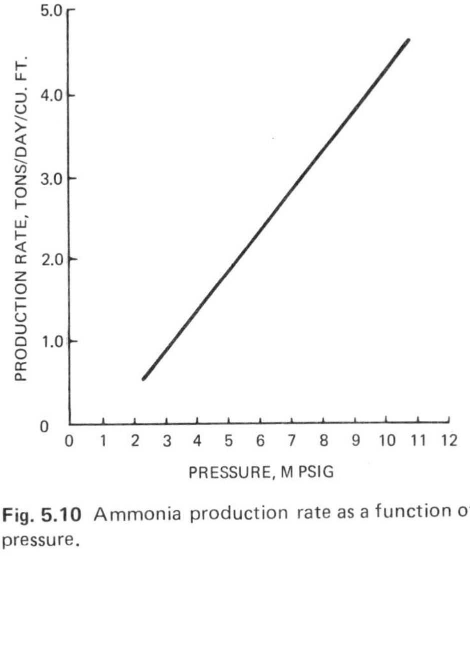

Fig. 5.10 Ammonia production rate as a function of pressure.

controlling component. Again, it would be necessary to purge excessively to restore the desired degree of conversion.

In design of the equipment, it is necessary to take into account the solubility of the various gases in the synthesis loop. The solubilities[35] of the gases in cc of gas at normal pressure and temperature per gram of liquid ammonia per atm partial pressure are:

Gas	*5°C*	*30°C*
Hydrogen	0.095	0.112
Nitrogen	0.117	0.126
Argon	0.151	0.161
Methane	0.270	0.392

It should be noted that all four gases are more soluble at the higher temperature.

The optimum concentration of inerts will depend on the temperature and pressure of the operation and the concentration of inerts in the make-up gas. The solubilities of argon, methane, hydrogen, and nitrogen in liquid ammonia determine the amount of these gases which is removed from the system with the liquid ammonia. The designer has some control over this purge. The conditions of synthesis can be selected to give optimum purge rates once solubility, make-up-gas composition, and degree of conversion are determined.

The production rate of ammonia is dependent upon the recycle rate for the gases in the loop, the degree of conversion attained in the catalyst (Table 5.8), and the efficiency with which the ammonia is removed. Table 5.9 gives some ranges for the saturated-vapor ammonia concentration in a $3H_2:1N_2$ gas mixture over liquid ammonia.

TABLE 5.9 Ammonia Concentration Over Liquid Ammonia

Pressure (Atm)	*Temperature (°C)*	*% NH_3 in Gas*
50	-20	5.7
	0	10.0
	20	19.0
	30	25.5
100	-20	3.25
	0	5.9
	20	11.0
	30	15
200	-20	2.05
	0	3.95
	20	7.5
	30	10.04
300	-20	1.62
	0	3.25
	20	6.4
	30	8.8
700	-20	1.19
	0	2.35
	20	4.90
	30	6.80

Converter Details. A modern converter arrangement designed by Chemico is shown in Fig. 5.11. It contains an internal heat exchanger and an internal electrical heater for start up. The bypass inlet is for temperature control purposes. As in most converters, the cold inlet gas passes between the catalyst basket and the shell to prevent overheating of the shell and hydrogen embrittlement.

In Fig. 5.12 a radial-flow converter is shown. This is a relatively new design by Topsoe. The main feature of this converter is the provision of a much greater cross section to the flow of gas, thereby reducing pressure drop. This low pressure drop makes possible the use of a smaller catalyst particles with some increase in activity per unit volume.

In addition to the care which must be exercised with respect to materials of construction, it is also necessary to design for heat removal. This is usually accomplished by internal heat exchangers in the converter. The incoming gas is heated by the exit gas which is cooled. A temperature rise of about 150° is used in the design of a converter.

One of the current circulating systems for ammonia synthesis is shown in Fig. 5.13. Control of the gas composition is important in the synthesis loop. Means of achieving this is revealed in U.S. Patent 2,894,821.[36] The hydrogen to nitrogen ratio must be maintained close to 3 and the concentration of inert materials, methane and argon, must be limited to levels which will permit a satisfactory partial pressure of the H_2 and N_2. In the process described in the patent, the circulating gas in the synthesis loop is analyzed for hydrogen and nitrogen and the inlet air flow to the secondary reformer is adjusted to maintain the ratio of hydrogen to nitrogen at 3.0.

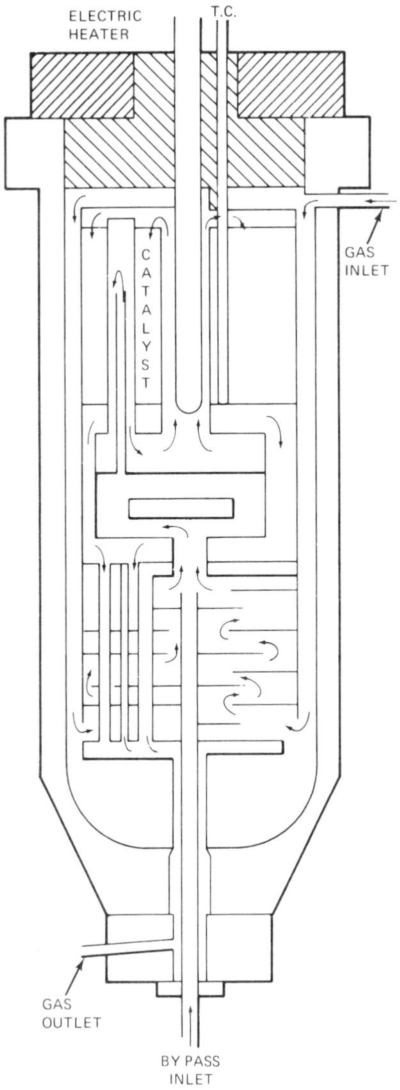

Fig. 5.11 Chemico converter. (*U.S. Patent 3,041,151.*)

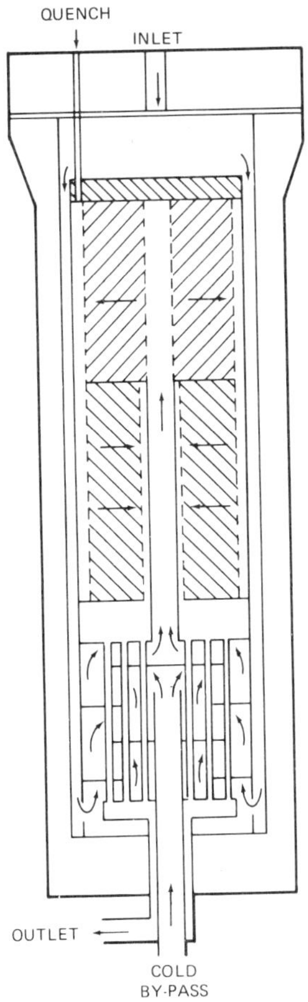

Fig. 5.12 Topsoe radial-flow converter. (*"Nitrogen," Sept. 1964, The British Sulfur Corp.*)

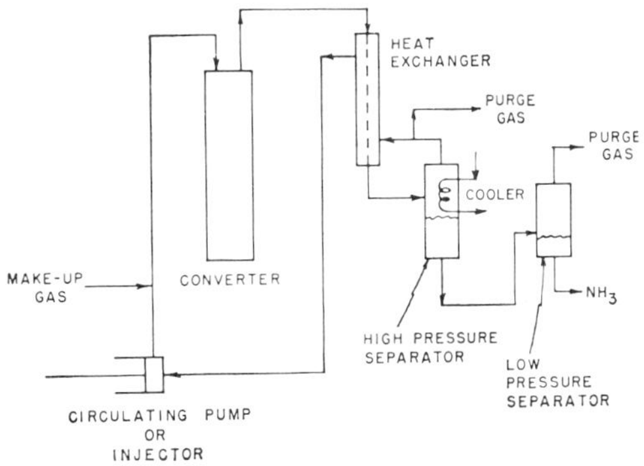

Fig. 5.13 Circulating system for ammonia synthesis.

Catalysts used for ammonia synthesis are basically reduced iron oxide (Fe_3O_4). The catalyst is prepared by melting relatively pure iron with an electric arc to produce the iron oxides. This material is then mixed with promoters, such as potassium, calcium, magnesium, or aluminum. The material is broken up, crushed, and sized for introduction to the converter. In the converter the oxide granules are reduced with hydrogen to their metallic constituents prior to the production of ammonia.

Although no known use is made of a synthesis system modified by an electrical potential[37] on the catalyst, the process is claimed to increase the conversion efficiency by 38.6 per cent. A static negative charge of about 6,500 volts was used on the catalyst.

The ammonia process can be combined with a urea process and some operating and investment efficiencies realized.[38] The synthesis gas generated by conventional means is mixed with liquid ammonia before the carbon dioxide is removed. This mixture of ammonia and carbon dioxide is subjected to urea-synthesis conditions and the off-gas from this operation used for ammonia synthesis. The carbon dioxide-removal step is thereby eliminated and the compression of carbon dioxide for urea is avoided.

Uses of Ammonia

The uses for ammonia are given in Table 5.10 and 5.11. By far the greatest consumption is in the manufacture of fertilizers or as direct application for fertilization purposes.

TABLE 5.10 Uses of Ammonia[a]

	1,000 Short Tons
Ammonium nitrate	2,900
Direct fertilizer application	3,300
Solid and liquid fertilizers	830
Urea	1,700
Ammonium sulfate	590
Nitric acid, industrial	380
Ammonium phosphates	2,000
Others	1,500
Total	13,200

[a]*Chem. Week* (September 11, 1965).[39]

TABLE 5.11 Industrial Uses of Ammonia[a]

Industry	*Usage*
Explosives	Nitrates, dynamite, azides
Plastics	Nitrocellulose, urea-formaldehyde, melamine
Metallurgy	Bright annealing of steel, dry reducing gas
Pulp and Paper	Ammonium bisulfite, melamine
Rubber	Aniline, acrylonitrile, polyurethanes, chemical blowing agents (for foam rubber)
Textiles	Nylons, acrylonitrile, terephthalates
Foods	Amino acids, sodium nitrate, sodium nitrite, and nitric oxides
Drugs	Vitamins, nitrofurans
Miscellaneous	Refrigerant, detergents, insecticides, nitroparaffins, hydrazine

[a]Based on data in *Chem. Eng.*, **62,** 11, 280–282 (1955).

Storage and Transport

Ammonia is usually stored in large 20,000 ton atmospheric-pressure tanks at a temperature of −28° F. With this system the ammonia vaporized by heat-leak must be compressed and condensed, and returned to the system. One unique storage system consists of an underground cavern mined from a strata of limestone.[40] Some of the low-pressure tanks are built with a double wall; this adds some protection from a leak and also provides a means for retaining insulation.

Normally, ammonia is transported by barge, ship, or tank cars to terminals located at strategic points in the consuming agricultural areas. Recently, some ammonia has been transported by interstate pipelines.

NITRIC ACID

Nitric acid has been known since the 13th century. Glauber devised its synthesis from strong sulfuric acid and sodium nitrate; however, it was Lavoisier who showed that nitric acid contained oxygen; and Cavendish who showed that it could be made from moist air by an electric spark.

In the oldest methods used, Chile saltpeter was reacted with concentrated sulfuric acid in heated cast iron retorts; the evolved nitric-acid vapors were condensed and collected in stoneware vessels.

Today nitric acid is made by oxidation of ammonia with air over a precious-metal cata-

TABLE 5.12 Nitric Acid Processes

	Temperature (°C)	*Pressure*	*Acid Strength (%)*
Low-pressure process	800	Atmospheric	50–52
Medium-pressure process (Montecatini)	850	40 psi	60
Medium pressure (Kuhlman)		40 psi	70
High pressure (Du Pont)	950	120 psi	60
Pintsch Bamag's "Hoko Process"	850	Atmospheric	98–99

lyst at atmospheric or higher pressures and at 800 to 950°C.[42] Nitric oxide is formed according to the reaction: $2NH_3 + 5/2O_2 = 2NO + 3H_2O$. The nitric oxide is then further oxidized by additional air to give $2NO_2$, which combines with water to give HNO_3 and nitric oxide. The nitric acid processes currently in use are listed in Table 5.12.

Pressures vary from atmospheric to 120 psig. The concentration of nitric acid that can be produced with conventional equipment is about 60 per cent. If higher strengths are needed, a special method of concentration is required.

Nitric acid can also be produced by the high-temperature combination of nitrogen with oxygen in the air, and by use of radiation.[43] One method which has received considerable attention is the so-called "Wisconsin" process. Although a 40-ton per day plant was built, it could not compete economically with the conventional ammonia-oxidation route.

Nitric acid is generally a light-amber liquid; however, the pure acid is colorless, strongly hygroscopic, and corrosive. It is a strong oxidizing acid. It boils at 78.2°C, freezes at -47°C, and forms a constant boiling mixture with water (68 per cent nitric acid by weight) which boils at approximately 122°C at atmospheric pressure. White fuming nitric acid usually contains 90 to 99 per cent by weight HNO_3, from 0 to 2 per cent by weight NO_2, and up to 10 per cent by weight water. Red fuming acid usually contains about 70 to 90 per cent by weight HNO_3, 2 to 25 per cent by weight NO_2, and up to 10 per cent by weight water.

Production.

In 1970 the production capacity for nitric acid in the United States reached approximately 10,000,000 tons, excluding that of U.S. Government ordnance plants. The estimated capacity in 1968 was 8,912,000 tons per year. Major producers with over 400,000 tons per year capacity are:

Company	*M Tons per Year*[44]
Du Pont	1,066
Allied Chemical	865
Monsanto	803
Hercules	674
Gulf	451

Processes[42]

Liquid ammonia is vaporized, mixed with preheated air, and introduced into the converter. The ammonia is oxidized to nitric oxide. ($4NH_3 + 5O_2 \longrightarrow 4NO + 6H_2O$).[45] The reaction is exothermic and liberates 216.7 kilocalories. The gases emerging from the converter pass through suitable heat exchangers and coolers where the following reaction takes place, $2NO + O_2 \longrightarrow 2NO_2$, accompanied by the evolution of 26.9 kilocalories.

The gases then pass to the absorber where the nitrogen dioxide combines with water in the absorber to produce nitric acid and nitric oxide ($3NO_2 + H_2O \longrightarrow 2HNO_3 + NO$). This reaction is exothermic and liberates 32.5 kilocalories. Sufficient air must be present in the absorber to oxidize the nitric oxide as it is liberated when the nitrogen dioxide and water combine.

In the first stage of the process where ammonia and air are mixed, it is important to hold the proper ratio of constituents. Usually a mixture of about 10 per cent ammonia and 90 per cent air is employed. At this point in the process it is important to avoid the decomposition of ammonia on the converter walls. Aluminum is superior to mild steel in this respect. At 350°C the rate of decomposition is 300 times faster on mild steel than on aluminum. Ammonia is decomposed 70 times faster on stainless steel than on aluminum.[45] By keeping the preheat temperature low, satisfactory yields can be obtained with mild steel. The temperature level used is below 200°C.

In the reaction which takes place over the catalyst bed, the rate-limiting step is the diffusion of ammonia to the catalyst surfaces.[45] With a high gas rate through the gauze, some of the ammonia will not reach the catalyst surfaces. It reacts instead with the nitric oxide product to give elemental nitrogen ($4NH_3 + 6NO \longrightarrow 5N_2 + 6H_2O$). The linear flow of gas across the catalyst is therefore important and will depend on the arrangement of the catalyst. Linear flow rates of 1 to 3 feet per second have given high efficiency. These rates are for gas at standard temperature and pressure. Without selection of proper flow rates or, conversely, the proper converter size for a given production, inefficient operation may result not only from too high a velocity with the aforementioned yield loss, but also from low velocities which would produce gas channels along the walls of the converter.

The catalyst generally used for ammonia oxidation is a platinum-rhodium alloy containing about 10 per cent rhodium. A gauze made of wire 0.003 inch in diameter with about 80 meshes to the inch is used. During the life of the catalyst it becomes polycrystalline in appearance. There is also some weight loss, the effect of which is more pronounced at the higher converter temperatures employed in the pressure process. With the atmospheric processes at lower temperature the platinum loss is about 45 mg per metric ton of acid produced. Recovery steps are required for operation at 120-pound pressure. This usually involves periodic cleaning of the converter and heat exchange equipment, and the use of a filter in the gas stream beyond these pieces of equipment.

There are a number of materials which will catalyze the oxidation of ammonia.[46] Platinum modified with other metals has been generally accepted. Recently, innovations have been introduced. Engelhard Industries has offered a Random Pack catalyst that uses 60 per cent less platinum and is offered along with a "getter" which will capture the volatilized platinum. Another new catalyst is offered by C & I Girdler, Inc. This is not a noble metal catalyst according to trade sources.[47] Yields are reported to be 93 to 94 per cent after two years, whereas the standard platinum catalyst yield drops from 98 to 91 per cent over a six-weeks period. I.C.I. in England[48] has also developed a new nonplatinum or nonprecious metal catalyst and the yield is as high as for other catalysts.

Absorption of nitric oxide takes place in a bubble-cap tower into which air is added to oxidize the nitric oxide to nitrogen dioxide. The nitrogen dioxide must be absorbed in water to liberate nitric acid and nitric oxide. There are two equilibria as follows: $N_2O_4 = 2NO_2 = 2NO + O_2$. The first of these two equilibria is established relatively quickly; the second more slowly. At 150°C the nitrogen tetroxide is almost completely decomposed into nitrogen dioxide. At 500°C the equilibria lies at 25 per cent nitrogen dioxide and 75 per cent nitric oxide. The formation of nitrogen dioxide from nitric oxide and oxygen is relatively slow at atmospheric pressure, but proceeds quite rapidly under higher pressure. It is interesting to note that this reaction has a negative temperature coefficient. The absorber is usually operated at the lowest temperature obtainable with available cooling water, in the temperature range of 10 to 40°C.

The absorption characteristics of nitrogen dioxide in water are of importance in the design of the absorber.[45] There are two factors which determine the strength of the nitric acid in the absorption tower, namely, total pressure on the system and gas composition. To increase by 10 per cent the strength of nitric acid in equilibrium with a gas of fixed composition it is necessary to raise the pressure about ten fold.

The gas composition also affects the con-

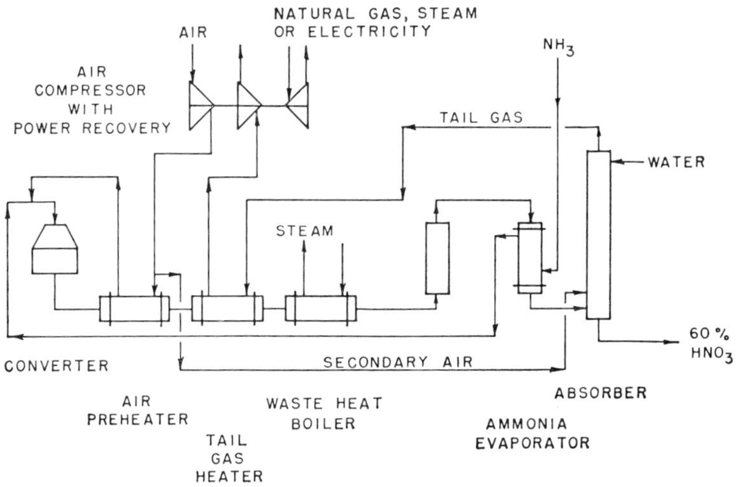

Fig. 5.14 Pressure process for nitric acid manufacture via ammonia oxidation. [*Adapted from Strelzoff, S., Chem. Eng., No. 5, 170 (1956).*]

centration of the nitric acid. The degree of oxidation of the nitric oxide to nitrogen dioxide should be as high as possible. Concentrations of nitric acid above 90 per cent are difficult to obtain.

There is some thought that the rate of diffusion of nitrogen dioxide to the liquid surfaces controls the rate of absorption. Some other data, however, indicate that the rate of absoprtion depends on the rate of the chemical reaction of the N_2O_4 with water. Thus diffusion and chemical rate may both be controlling factors in the rate of absorption. Absorption towers are currently designed by empirical methods. A novel tower design has been developed by Etablissements Kuhlman.[50]

The processes for nitric acid manufacture may be described as atmospheric, pressure, and medium pressure. These produce nitric acid in the concentration range of 50 to 60 per cent by weight. Figure 5.14 depicts the pressure process. Nitric acid of greater than 90 per cent strength can be produced by a process such as that shown in Fig. 5.15.

Figure 5.16 illustrates the process employed to concentrate nitric acid from 60 to 95 per cent strength by means of a dehydrating agent, in this case magnesium nitrate.

Atmospheric Plant. In an atmospheric plant ammonia efficiency and utilization of the precious metal catalyst is good. The temperature is about 800°C. The mixture fed to the converter contains about 9.5 to 11 per cent ammonia. Air is preheated by the exit gases from the converter; the gases are then fed through a heat exchanger which cools the reactants before they enter the oxidizer tower. Water is added to the tower in which the nitric acid solution is circulated. The product is drawn off through a cooler to an acid bleacher and storage. It contains approximately 50 to 52 per cent nitric acid.

Pressure Nitric Acid Plant. In the pressure nitric acid plant the equipment is much smaller, making the investment less. The ammonia content of the feed is about 10 per cent. The temperature is approximately 900°C. Ammonia and air are mixed ahead of the converter, having been preheated by the product gas. The tail gas from the absorber is preheated by the product gas and then sent to an expansion turbine which supplies power for air compression. Any of the additional power required for air compression is created by natural-gas-turbine, steam, or electrically driven power units.

To reduce catalyst loss in the pressure nitric acid plant, a catalyst filter is installed after the converter. At 820°C the platinum-catalyst requirement is about 0.03 troy ounce per ton of

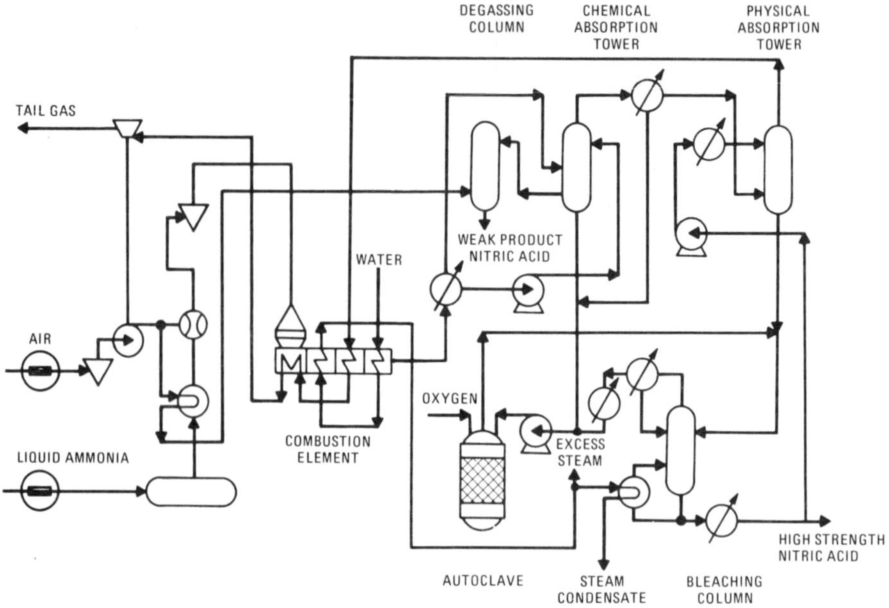

Fig. 5.15 Pintsch Bamag's "Hoko Process" for nitric acid. [*Hydrocarbon Processing,* **48,** *No. 11, 209 (Nov. 1969); copyright 1969 by Gulf Publishing Co.*]

nitric acid. About two-thirds of this is recovered. The ratio of platinum loss in the pressure plant to that in the atmospheric plant is 3 or 4 to 1.

Medium Pressure Plant. The medium pressure plant combines the advantages of both the high-pressure and low-pressure systems. In this process the converter operates at low pressure, the absorber and cooling system at high pressure. The disadvantages lie in the fact that the gases from the converter must be compressed before going to the absorption tower. For this reason the process has not received wide acceptance.

Table 5.13 gives the requirements per ton of nitric acid for the three processes.[42]

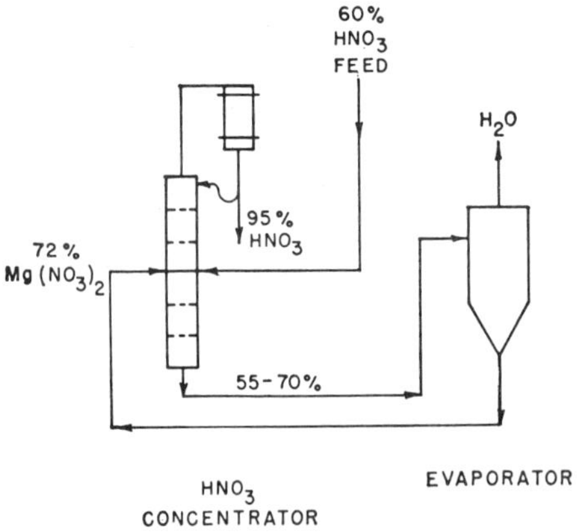

Fig. 5.16 Magnesium nitrate process for nitric acid concentration.[49]

High Strength Nitric Acid. Pintsch Bamag's "Hoko Process" for the manufacture of high strength nitric acid is shown in Fig. 5.15. In this process the concentration of nitric acid is based on the following equations:[49]

$$NH_3 + 5/4O_2 \longrightarrow NO + 3/2H_2O$$

$$NO + 1/2O_2 \longrightarrow NO_2$$

$$NO_2 + 1/4O_2 + 1/2H_2O \longrightarrow HNO_3$$

$$NO + 2HNO_3 \longrightarrow 3NO_2 + H_2O$$

After the oxidation step, the bulk of the water is removed by condensation without removing a great deal of the nitric acid present.

TABLE 5.13 Requirements per Ton of Nitric Acid

	NH_3 (Tons)	*Pt Cat. (Troy Oz)*	*Power[a] (Kw-h)*	*Steam Credit (Tons)*	*Water (1000 Gal)*	*Strength*	*Investment*
Atmospheric Plant	0.29–0.30	0.0025	85–90	1.0	27	45–52	Highest
Pressure	0.29–0.294	0.005–0.01	350	0.8–1.25	30	57–60	Lowest
Combination Pressure	0.187–0.29	0.0025	350	1.0	30	57–60	Intermediate

[a]In modern pressure plants, power requirements may be substantially less than shown in the foregoing table because of highly efficient recovery of power from the exhaust gases by means of an expansion turbine.

Atmospheric oxidation enhances this possibility. The remaining nitric oxide is oxidized to nitrogen dioxide, which is oxidized to nitric acid with high purity oxygen in the autoclave at about 60–80°C and 50 Kp/cm^2 gage.

Concentration of Nitric Acid. The 50 to 60 per cent nitric acid made by conventional processes cannot be concentrated by simple distillation because a constant boiling mixture is formed which contains approximately 68 per cent HNO_3 by weight; therefore, it is customary to use a dehydrating agent to aid in removal of the water. The most widely used method consists of mixing the 60-per cent nitric acid with strong (93 per cent) sulfuric acid, and then passing the mixture through a distillation system from which concentrated (95 to 98 per cent) nitric acid and denitrated, residual sulfuric acid containing approximately 70 per cent H_2SO_4 is obtained. The dilute, residual sulfuric acid may be reconcentrated for further use.

One process which employs magnesium nitrate as a dehydrating agent[51] is illustrated in Fig. 5.16. An aqueous solution containing 72 per cent magnesium nitrate is fed to the middle of a tower which also receives the 60-per cent nitric acid feed. Pure nitric acid of 95- to 98-per cent strength is taken overhead. A solution containing about 55 per cent magnesium nitrate is removed from the base. After reconcentration to 72 per cent, the magnesium nitrate is recirculated to the nitric acid tower.

A Pintsch Bamag process[52] concentrates nitric acid by using the difference in the composition of the nitric acid-water constant boiling mixtures at different pressures. A two column distillation system is employed.

Purification. Chemically pure nitric acid is required for some applications, and as a reagent for analytical measurements. The commercial grade may contain impurities, such as arsenic, sulfur, and phosphorous.

In one purification process[53] potassium or sodium permanganate is used to oxidize the impurities. Other oxidizing agents can also be used, such as persulfates or chlorates. Molybdenum trioxide is used to form a complex with the oxidized phosphorous impurities. The nitric acid is contacted with the oxidizing agent and the molybdenum compound at the boiling point of the acid, which is continuously removed by distillation. The oxidized impurities remain in the residue. The purified acid will contain less than 0.5 part per billion by weight of arsenic and phosphorous.

Stabilizers. Over a period of time concentrated nitric acids tend to decompose according to the equation

$$4HNO_3 \longrightarrow 4NO_2 + 2H_2O + O_2$$

As a result, pressures will build up in storage vessels. Nitric acid is also very corrosive, and some stabilizers or corrosion inhibitors are also used.

Corrosion of aluminum by red fuming nitric acid is reduced by adding 4 per cent by weight of hydrogen fluoride (52 per cent HF).[54]

Decomposition of the concentrated acid is reduced by such substances as quaternary ammonium compounds, organic sulfones, inorganic persulfates, and organic sulfonium compounds.[55]

Pollution Abatement. The tail gas from ammonia-oxidation processes will usually contain sufficient nitrogen oxides to result in serious atmospheric pollution (smog) or in serious corrosion of power-recovery equipment, if such is used, unless some means for removal of the oxide is employed. The oxides are usually reduced with natural gas, hydrogen, or a carbon monoxide-containing gas, or they can be absorbed by alakline liquors (soda, calcium, or magnesium) or ammonium sulfite-bisulfite.[55]

As an example of the reduction procedure, a gas containing the materials listed below was treated with a nickel catalyst on alumina at 90 psig, 1150°F (1600°F exit), and a space velocity of 20,000 cu ft of exit gas per cu ft of catalyst.

Gas	*%*
N_2	97.2
O_2	2.0
H_2O	0.6
$NO + NO_2$	0.2

The effluent gas contained about 75 ppm NO-NO_2 mixture; 0.1% CO; 0.3% H_2; and 1.5% CO_2.

With methane as the reducing gas, materials which are particularly effective as catalysts are rhodium, palladium, platinum, and ruthenium. Alumina or Torvex®, a ceramic honeycomb material, can be used as a base. These are usually used at about 0.5 per cent on Al_2O_3. With these materials the initial reaction temperature is about 700 to 1000°F depending upon the catalyst and oxygen concentration.

Dr. C. I. Harding and others at the University of Florida[56] have applied for a patent on an absorption-desorption process using zeolites. It is stated that this process will result in the recovery of the oxides and add up to 4 or 5 tons per day of capacity to a 300 T/D plant. Exit gas concentrations of 0.001 per cent for three hours and 0.02 per cent after three hours when feeding 2.4 scfm of tail gas (0.18 per cent nitric oxides) over 4.9 pounds of zeolite were obtained. The columns were regenerated with steam at 300-350°F, or by hot air.

Uses of Nitric Acid

The primary use of nitric acid is in the manufacture of ammonium nitrate for the fertilizer industry.[57] Approximately 65 per cent of the total production goes into fertilizers. Solution fertilizers, many of which contain nitrates, are growing in importance. Nitric acid is used in some cases for the acidulation of phosphate rock to produce mixed fertilizers.

About 25 per cent of the nitric acid produced is consumed in industrial explosives. Approximately 1500 thousand tons per year of nitric acid is used for explosives which, in turn, are used in coal mining, quarrying, metal and nonmetal mining, heavy construction, and petroleum exploration. Fertilizer grade ammonium nitrate may also be used as an explosive or blasting agent. It is mixed with oil and set off by priming with a high explosive.

Nitric acid has a number of other industrial applications. It is used in pickling stainless steel, in steel refining, and in the manufacture of dyes, plastics, and synthetic fibers. Most of the methods used for the recovery of uranium, such as ion exchange and solvent extraction, use nitric acid. Thus with the expansion of the nuclear program, nitric acid consumption may increase.

Nitric acid end-use pattern as of 1968 is as follows:[57]

Use	*1000 T/Yr*
Ammonium nitrate	
Fertilizers	3900
Explosives and other uses	500
Miscellaneous fertilizers	150
Dinitrotolunene (urethanes)	88
Nitrobenzene (rubber chemicals, urethanes, etc.)	87
High explosives and propellants	1050
Miscellaneous direct uses	325
Miscellaneous compounds	150
TOTAL	6250

AMMONIUM NITRATE

The capacity for ammonium nitrate[58] in North America has been increasing at the rate of approximately 200 thousand tons per year. In 1957 the production was approximately 2,561,000 tons; in 1970 the capacity was about 10,900,000 tons. A large part of this is consumed as fertilizer material; however, increasing amounts are being used for explosives; and some is being used for the manufacture of nitrous oxide.

The process for ammonium nitrate consists of neutralizing ammonia with nitric acid under controlled conditions. Water is evaporated, and the anhydrous, or nearly anhydrous, solution is prilled or dried to produce solid products[59] (see Fig. 5.17).

Precautions observed in ammonium nitrate manufacture:[60] operation is satisfactory below 250°F; use of type 304L stainless steel is possible when the pH may be less than 4.0 for a prolonged time; at a pH of over 4.0 for solids, and 6.0 for solutions, and temperatures less than 240°F aluminum is satisfactory. Above 250°F ammonium nitrate must be considered an explosive hazard.

Chemico[60] offers a vacuum process in which the ammonium nitrate solution is evaporated to 95 per cent concentration, rather than the usual 83 per cent. The heat liberated by the reaction (550 to 620 Btu/lb of ammonium nitrate, depending on acid strength) is used to concentrate the aqueous product, and mild steel, or stainless steel, is used in place of titanium.

There are four different crystalline forms of ammonium nitrate. One is stable at -16°C, another at -16 to 32°C, a third at 32 to 84°C, and the fourth at 84 to 169.6°C, the melting

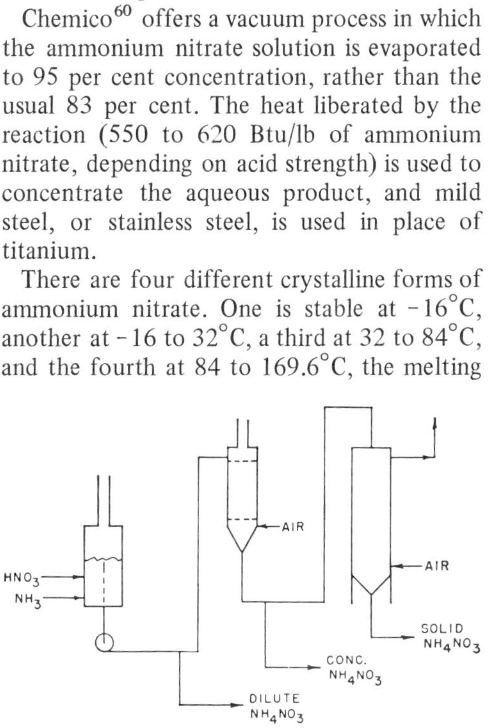

Fig. 5.17 Flow diagram for ammonium nitrate manufacture. [*Based on description in Chem. Eng. News,* **36,** *No. 34, 50 (Aug. 25, 1958).*]

point. When ammonium nitrate is dissolved in water, the negative heat of solution causes the temperature to drop. When heated, ammonium nitrate decomposes into N_2O and water.

It is important to mention the hazards involved in handling ammonium nitrate. Avoid using concentrated solutions in large amounts at high temperatures; in particular, avoid contamination with organic materials. The Texas City disaster was dramatic proof that organic material and ammonium nitrate decompose with explosive violence. Following a fire under conditions of confinement, an explosion occurred which took almost 600 lives, injured 3,500, and destroyed 33 million dollars worth of property. Ammonium nitrate solutions also explode with disastrous results.

To prill ammonium nitrate, a solution is evaporated in a vacuum or falling-film evaporator until it is almost free of water. The concentration of ammonium nitrate will vary from 95 to 99.95 per cent. The prilling tower itself is about 50 to 100 feet tall, depending on the process being used. The concentrated ammonium nitrate solution is fed through sprayformers which break up the stream into small droplets that solidify during free fall in the tower. In some instances limestone is mixed with the ammonium nitrate before prilling; in other cases the prills are coated with limestone dust. A mixture of limestone and ammonium nitrate is known as "Nitro chalk" in the fertilizer industry. Plants that use the 95-per cent ammonium nitrate feed to the prilling tower usually require a drying step before the material can be bagged.

Stafford, Samuels, and Croysdale[61] give a comprehensive review of ammonium nitrate-plant safety. They discuss how coating ammonium nitrate effects its explosibility. For instance, 2.5 per cent by weight of diatomaceous earth coated on ammonium nitrate reduces its detonation sensitivity, and 2.5 per cent of finely divided elemental sulfur increased the detonation sensitivity. At a 5-per cent concentration, urea decreases the thermal stability.

HEXAMETHYLENETETRAMINE

In the period 1957 to 1967 hexamethylenetetramine had a growth rate of 14.6 per cent per year.[62] The nonmilitary growth is expected to be about 5 per cent per year. Annual capacity is now (1971) estimated to be 145,000,000 pounds, with Wright Chemical, Tenneco, Hooker, and Borden being the major producers.

Hexamethylenetetramine or "Hexamine" is manufactured in the liquid phase by the following reaction:[65]

$$4NH_3 + 6HCHO \longrightarrow (CH_2)_6N_4 + 6H_2O$$

The heat of reaction is equal to 28.2 kilocalories per mole of tetramine.

The reaction mixture is controlled at a pH of 7 to 8 at a temperature of 30 to 50°C. Under acidic conditions, formic acid, carbon dioxide, and water are produced. At temperatures above 50°C decomposition of the tetramine occurs. Tetramine is produced either by continuous or batch operation. In batch operation, cycle times up to 6 to 8 hours have been used, whereas in continuous operation the holdup may be only 15 to 30 minutes.

Figure 5.18 shows the equipment which is used in the process. Thirty-seven per cent formaldehyde, usually containing less than 2 per cent methanol, is fed, together with ammonia, to a reactor which is cooled as a result of the solution being circulated through a water-cooled heat exchanger. The effluent from the reactor is fed to a vacuum evaporator. Here the product is concentrated, and crystals are formed. The slurry is centrifuged, and the crystals are washed and discharged to a drier. The mother liquor is recycled to the reaction system; however, a small bleed is necessary to avoid the build-up of impurities in the system.

The mechanism of the reaction is discussed by Baur and Ruetschi.[63] They indicate a one-step trimerization reaction of the third order in which the rate is proportional to the ammonia concentration and the square of the formaldehyde concentration. Boyd and Winkler[64] have indicated a more complex reaction mechanism. The exact mechanism may depend on whether ammonia or formaldehyde is in excess.

The reaction may be carried out at temperatures up to about 50°C. Above this temperature the rate of hexamine decomposition becomes excessive. The degree of conversion is essentially complete after 4 or 5 hours; only minor benefits are obtained by operating for 6 to 8 hours.

The vessels are usually made of stainless steel or aluminum. The heat exchanger for cooling the reaction mixtures is also made of stainless. Crystallization takes place in the evaporator, and it is important to note at this point that hexamine has an inverse solubility. At 25°C, 86.7 grams are dissolved in 100 cubic centimeters of water, whereas at 50°C the solubility is only 80 grams per 100 cubic

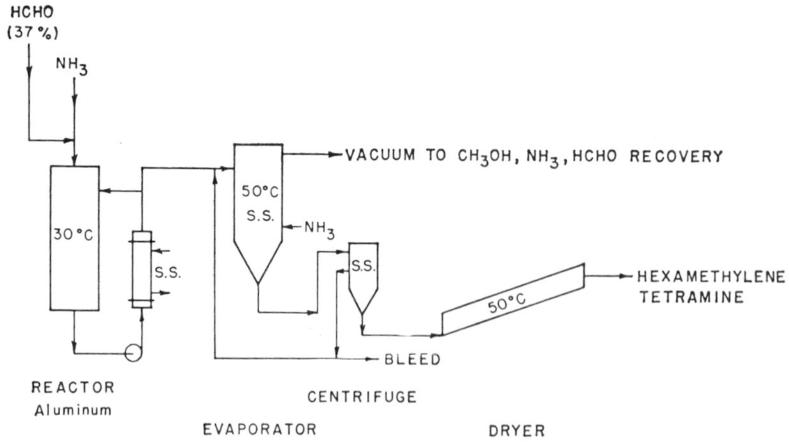

Fig. 5.18 Manufacture of hexamethylenetetramine. [*Adapted from Petrol. Refiner,* **37**, *No. 9, 351-353 (1958).*]

centimeters of water. The continuous addition of a small amount of ammonia during the evaporation step tends to reduce the decomposition of the tetramine during evaporation. If the formaldehyde feed contains methanol, it will be removed from the system along with the ammonia and water from the evaporator. The evaporators are usually made of stainless steel.

In the final drying step the temperature must be held to less than 50°C to avoid decomposition. The over-all yield is good, on the order of 95 to 96 per cent based on the formaldehyde.

Hexamine is used in the production of high explosives. The manufacture of "RDX" came into its own during World War II. By nitrating hexamine, a compound with high explosive properties is obtained. Hexamine is also used in thermosetting resin production as a curing agent. This is its principal use in a peacetime economy. It serves as a methylating agent in the curing of phenol-formaldehyde resins. The hexamine produces methylene groups without forming water, as is the case with paraformaldehyde. It is used as an accelerator in the rubber industry to prevent vulcanized rubber from blocking. Pharmaceutical applications exist for hexamine and some of its derivatives. It is also being studied as a fungicide in the citrus fruit industry,[65] as an inhibitor of corrosion caused by strong mineral acids, as a shrink-proofing agent in the textile industry, and as an agent to improve color fastness and to give better elasticity to cellulosic fibers.

HYDRAZINE

Manufacture

Hydrazine (N_2H_2) is becoming more important as an industrial chemical. Its manufacture involves the following reactions:[66]

$$NaOH + Cl_2 \longrightarrow NaOCl + HCl$$

$$NaOCl + 2NH_3 \longrightarrow N_2H_4 + NaCl + H_2O$$

$$NH_3 + NaOCl \longrightarrow NH_2Cl + NaOH$$

$$NH_2Cl + NaOH + NH_3 \longrightarrow N_2H_4 + NaCl + H_2O$$

Side reactions are as follows:

$$2NH_3 + 3NH_2Cl \xrightarrow{Cu^{++}} 3NH_4Cl + N_2$$

$$2NH_2Cl + N_2H_4 \longrightarrow 2NH_4Cl + N_2$$

Glue and gelatin are used to prevent these side reactions.

Hydrazine is a colorless liquid that boils at 113.5°C. It has a melting point of 1.4°C, and a specific gravity of 1.014 at 15°C. Hydrazine hydrate (H_2NNH_3OH) is obtained by fractional distillation from water. The hydrate boils at 119°C and is still liquid at -40° C. It has a specific gravity of 1.03 at 21°C. Hydrazine forms a sparingly soluble sulfate ($NH_2NH_2 \cdot H_2SO_4$) and the chloride. Hydrazine is a powerful reducing agent and is very poisonous, even through the skin.

The process, carried out in the liquid phase, is outlined in Fig. 5.19. Sodium hydroxide and chlorine are mixed in a cooler-reactor sys-

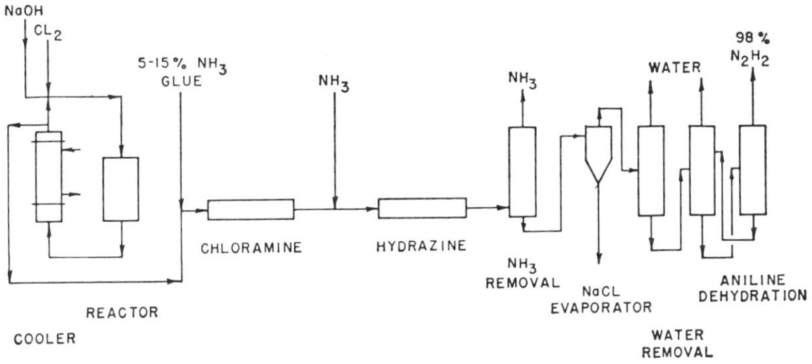

Fig. 5.19 Hydrazine manufacture. [*Adapted from Chilton, C. H., Chem. Eng.,* **65,** *No. 14, 123 (July 14, 1968).*]

tem to produce sodium hypochlorite. Glue is added to this solution as an inhibitor until the mix is viscous; a dilute solution of ammonia (5 to 15 per cent) is added until a mole ratio of 3 ammonia to 1 hypochlorite is obtained. This mixture forms chloramine, which, when reacted with anhydrous ammonia is a ratio of 20-30 to 1, produces hydrazine. The temperature reaches 130°C. The effluent from the hydrazine reactor is fed to an ammonia-removal still where excess ammonia is taken off overhead. The tails are fed to an evaporator where concentrated sodium chloride is removed. The vapors from the evaporator are dehydrated in three columns. In the first column the hydrazine is dehydrated until the water-hydrazine mixture approaches the hydrate NH_2NH_3OH (65 per cent NH_2NH_2). In the second column aniline is used as a dehydrating agent, the water being removed overhead with the aniline. A third column is required to remove the aniline from the hydrazine; 98 per cent hydrazine is produced. An over-all yield of about 70 per cent is obtained.[68]

Synthesis efficiency is increased by operating with a dilute system, but operating costs increase because of the low concentrations.

In some cases, anhydrous hydrazine is obtained by dehydrating with 50 per cent caustic. Pressures less than atmospheric are used to reduce decomposition; however, this may permit air leakage and result in decomposition of the hydrazine to nitric oxide, water, and ammonia.

There are other processes for the manufacture of hydrazine.[68] One of these is based on the nitration of urea to produce nitrourea, which in turn can be hydrogenated to the semicarbazide. When this is reacted with ammonia one mole of hydrazine is obtained along with one mole of urea.

In another process, hydrazine is obtained by reacting urea which a carbonyl-forming metal such as iron or nickel at 132 to 150°C.

Still another process is based on the reaction:

$$2NH_3 \rightleftharpoons N_2H_4 + H_2$$

In this process, ammonia is passed through an extended glow discharge. Highly turbulent flow and effective cooling are required. Allyl compounds are effective in increasing the yield of hydrazine.

Anhydrous hydrazine is also formed by the following reaction:

$$Cl_2 + 4NH_3 \longrightarrow N_2H_4 + 2NH_4Cl$$

It is postulated that the reaction occurs in stages.

$$Cl_2 + 2NH_3 \longrightarrow NH_2Cl + NH_4Cl$$

$$NH_2Cl + NH_4Cl + 2NH_3 \longrightarrow N_2H_4 + 2NH_4Cl$$

High temperature and pressure speed up the reaction. The larger the ratio of ammonia to chlorine (50:1 to 350:1), the better the yield.

Hydrazine is used as a scavenger for oxygen in boiler-feed water treatment and as a high-energy fuel. It is also used for making blowing agents for foaming rubbers and plastics, for the production maleic hydrazide to combat suckers on tobacco, and in the manufacture of other organic chemicals. Hydrazine is marketed as 64 per cent, 54.5 per cent, and 35 per cent solutions.

Stabilization

Hydrazine, hydrazine hydrate, or mixtures of these with water are decomposed by contact with metals, such as iron. It has also been shown that solutions containing 95 per cent N_2H_4 by weight are associated with its potential acidity.[68] This latter property is also associated with its corrosiveness. Organic compounds of the type $R^1SO_2NHR^2COOR^3$ (where R^1 is a hydrocarbon with 12 to 18 carbon atoms, R^2 is an aliphatic or aryl radical, and R^3 is a sodium, potassium, or ammonium core) are inhibitors. Substituted ammonium radicals, such as aniline, organic sulfides, sodium carbonate, and zinc oxide, at concentrations of 0.2 to 5 per cent by weight, will also inhibit decomposition.

UREA

Urea was first synthesized by Wohler in 1828 from ammonia and cyanic acid. It is a colorless crystalline material, soluble in water and in alcohol, but not in ether. Urea contains 46 per cent nitrogen, the most of any ordinarily

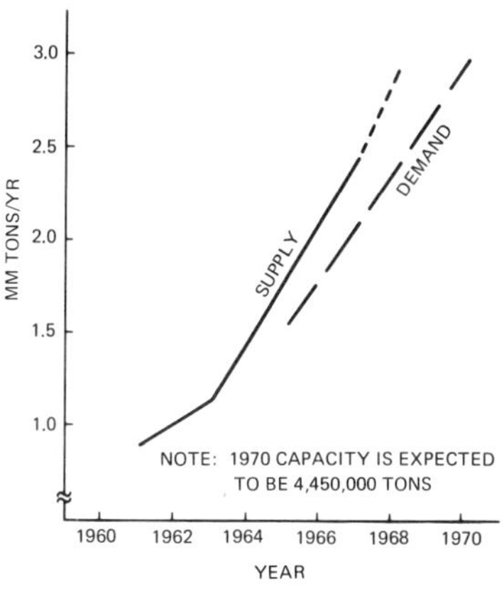

Fig. 5.20 Urea supply and demand.[70]

solid fertilizer material. Ammonium nitrate contains 34 per cent nitrogen. Since urea is converted into ammonia and then into nitrates in the soil, it makes a very concentrated form of nitrogen fertilizer. It is also used in resin and plastic manufacture and in the synthesis of organic materials.

The demand for urea increased three-fold during the ten years from 1960 to 1970 (see Fig. 5.20). In 1958 urea capacity was 623,000 tons; by 1970 the capacity increased to 4.45 million tons; demand in 1971 was about 3.1 million tons.[70]

A list of urea producers is given in Table 5.14.

Urea can be produced by the hydrolysis of cyanamide (melting point 44°C) according to the equation, $CNNH_2 + H_2O \longrightarrow CO(NH_2)_2$. At high pressure, 33 to 53 atm, urea can also be formed by heating ammonium carbonate. Equilibrium is obtained at 130 to 150°C with a 30 to 45 per cent yield. A process has been described by Raymond Fanz and Fred Applegath of Lion Oil Co. which produces urea from ammonia, carbon monoxide, and sulfur (dissolved in methanol).[72] The process operates at 100°C and 20 atm and produces H_2S along with the urea. The H_2S is oxidized to sulfur for recycle.

The common method of manufacture is to combine ammonia with carbon dioxide under pressure to form ammonium carbamate,

TABLE 5.14 Urea Producers, 1967[71]

	Capacity (Thousands of Tons/yr)
Agway, Inc.	58
Air Products and Chemical Co.	23
Allied Chemical	560
American Cyanamid	149
Arkla Chemical Corp.	65
Atlas Chemical Industries	59
Borden Chemical Co.	165
Central Farmers Fertilizer Co.	18
Collier Carbon & Chemical	400
Columbia Nitrogen Corp.	25
Cooperative Farm Chemical Association	50
Du Pont	205
Farmers Chemical Association	210
W. R. Grace	120
Gulf Oil Corp.	270
Hawkeye Chemical Co.	25
Hercules	98
Kaiser Agricultural Chemicals	60
Ketona Chemical Corp.	7
MisCoa (Mississippi Chemical Corp.)	92
Mobil Chemical.	49
Nipak, Inc.	268
Nitrin, Inc.	25
Olin	140
Phillips Pacific Chemical	40
Phillips Petroleum Co.	54
Premier Petrochemical Co.	70
Shell Chemical Co.	185
Sun Olin Chemical Co.	100
Terra Chemicals International	116
Triad Chemical	420
USS Agri-Chemicals, Inc.	18
Valley Nitrogen Producers, Inc.	182
Vistron Corp.	83
Wycon Chemical Co.	41
TOTAL	4450

which is then decomposed into urea and water. The unreacted carbon dioxide and ammonia are recovered and recycled to the synthesis operation

$$2NH_3 + CO_2 \rightleftharpoons NH_2COONH_4 \text{ (Exothermic)}$$

$$NH_2COONH_4 + \text{heat} \rightleftharpoons CO(NH_2)_2 + H_2O \text{ (Endothermic)}$$

A number of available processes are given in Table 5.15.

TABLE 5.15 Urea Processes[a]

	Du Pont[a]	*Pechiney*[a]	*Stamicarbon DSM*	*Allied Chemical C.P.I.*	*Chemico Thermo-Urea*[77]
Type of Operation	Carbamate recycled with NH_3 and water.	Carbamate recycled with oil.	CO_2 stripping at pressure to remove NH_3.	No carbamate recycled. NH_3 and CO_2 recycled via MEA system.	Gaseous NH_3, H_2O, and CO_2 compressed and recycled.
Temperature, °C	200	180	180	205	210
Pressure, atm	400	200	125	275	275
Converter material	Silver	Lead	Stainless steel	Zirconium	–
NH_3:CO_2;H_2O	5:1:0.73	2:1:0	2.5:1	4:1	–
CO_2 conversion, %					
In Autoclave	70	50	–	80–85	–
TOTAL	70	50	–	–	–
NH_3 conversion, %					
In Autoclave	24	50	–	–	–
TOTAL	24	50	–	–	–

[a]*Chem. Week*, **75**, 22, 90 (1954).

The main differences in the urea processes are in the methods used to handle the converter effluent, to decompose the carbamate and carbonate, to recover the urea, and to recover the unreacted ammonia and carbon dioxide for recycle with a minimum expenditure of energy and a maximum recovery of heat. In some processes, a liquid is used to recycle a solution of carbamate, carbon dioxide, ammonia, and water. In others, the amount of water recycled is minimized, and only carbon dioxide and ammonia are recycled. In the older plants, or once-through plants, the off-gases are used as feed to ammonium nitrate or ammonium sulfate plants.

Process requirements[74] per ton of urea are as follows:

Ammonia	0.58 tons
CO_2	0.76 tons
Steam	0.9 to 2.3 tons
Electricity	85 to 160 Kw-h
Water	15,000 to 36,000 gal

The *Du Pont* process represents a combination of a number of total recycle and partial recycle processes. Unconverted carbon dioxide and ammonia are recovered and recycled as a water solution.[73] The converter, lined with silver, operates at approximately 400 atm and 200°C. Conversion is approximately 70 per cent. In partial recycle processes only a portion of the unreacted ammonia is recycled, the rest is used to produce by-product ammonium sulfate or ammonium nitrate.

In the *Pechiney* process (Fig. 5.21) the reaction takes place in an oil medium later used to return the unconverted ammonia and carbon dioxide to the converter, which is lined with lead and operates at approximately 200 atm and 180°C. Conversion is approximately 50 per cent.

The *Stamicarbon* or *Dutch State Mines (DSM)* process is based upon the use of carbon dioxide to strip ammonia from the reactor effluent countercurrently. As the carbon dioxide removes the ammonia from the solu-

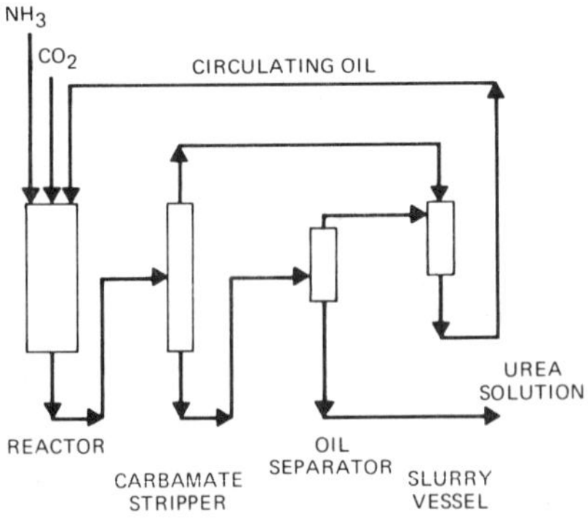

Fig. 5.21 The Pechiney process for urea. [*Adapted from Tonn, W. H., Chem. Eng.*, **62**, *No. 10, 188 (1955)*.]

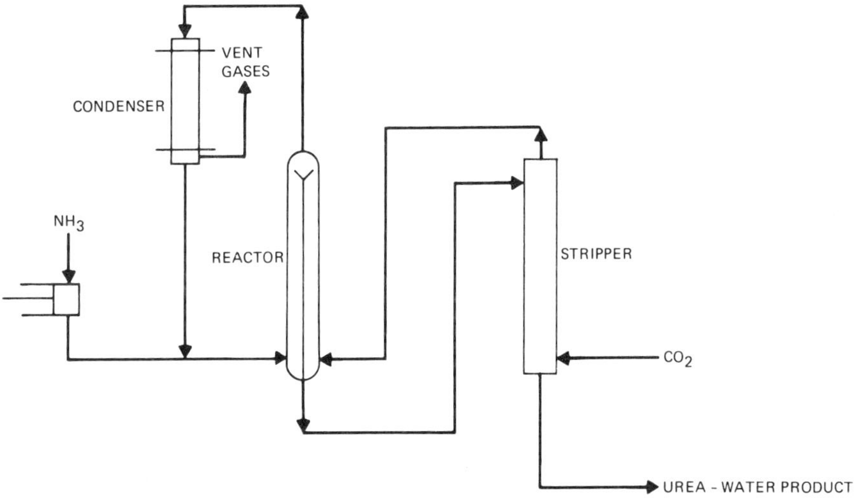

Fig. 5.22 DSM's urea process, carbon dioxide stripping.[76,77]

tion, the carbamate decomposes leaving a minimum amount in the effluent. The solution leaves the stripper at about 150-180°C. The stripped gases flow to the reactor along with ammonia, which is equivalent to the amount of carbon dioxide added for stripping. Most of the off-gas from the reactor is condensed, and inert gases are bled from the system before the condensate is returned to the base of the reactor. This process is shown in Fig. 5.22.[76,77]

The *Allied Chemical–C.P.I.* process, which was developed by T. O. Wentworth (U.S. Patent 3,107,149),[78] is based on recycle of NH_3 which is separated from the reactor effluent gases by scrubbing with monoethanolamine. The carbon dioxide is absorbed in the amine, and the gaseous ammonia is recycled to the reactor. The reactor effluent is first decomposed; then the urea solution is removed and concentrated. A specially designed reactor lined with zirconium is needed. Carbon dioxide conversion is relatively high–80 to 85 per cent per pass.[77]

The *Chemico "Thermo-Urea"* process[80,69]

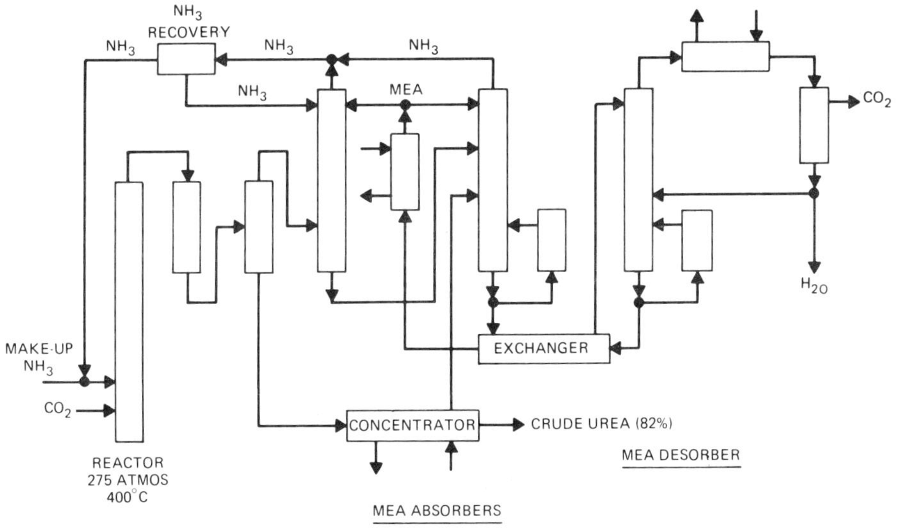

Fig. 5.23 Allied Chemical–C.P.I. process.[77,78,79]

employs centrifugal compressors to recycle the carbon dioxide, ammonia, and water vapor to the reactor. The reactor effluent is passed into a separator from which the hot gases pass to the compressor, and the liquid effluent enters a carbamate decomposer where the remaining gases are removed and sent to the compressor. The liquid effluent from the first decomposer enters a low-pressure decomposer which removes the last traces of carbon dioxide, ammonia, and inert gas. The ammonia and carbon dioxide are absorbed in water for recycle to the compressor after vaporization; and the inert gas is removed from the top of the absorber. Aqueous urea is removed from the low pressure decomposer (see Fig. 5.24). An improved process is described in U.S. Patent 3,301,897.[81,77]

The largest commercial urea plant, 1500 tons per day, is a single-train plant using the *Mitsui Toatsu* total recycle process shown in Fig. 5.25.

Investment per daily ton of urea capacity dropped from about $34,000 in 1960 to $8,000 in 1969 for a 1000 ton a day plant.[87]

To avoid a yield loss and minimize product contamination, it is important to minimize biuret formation. The concentration of biuret in urea used for foliar feeding must be low to avoid mottling of the leaves. Consequently, a foliar grade is produced containing less than 0.035 per cent biuret. Its formation is as follows:

$$CO(NH_2)_2 \rightleftharpoons HNCO + NH_3$$

$$CO(NH_2)_2 + HNCO \rightleftharpoons NH_2CONHCONH_2$$

At 140°C and ammonia pressures of 760 and 70 mm Hg, the velocities of formation were reported to be 0.9×10^{-4} mole/sec. and 3.4×10^{-4} mole/sec., respectively. The rate of conversion is increased by basic compounds, and decreased by acidic compounds.[82]

A low-biuret product can be produced by recycling the biuret-containing effluents from the purification zone back to the converter (see U.S. Patent 3,232,984).[83]

The urea-water solution produced by the various processes must be further processed to obtain the final products, in the form of crystals, flakes, prills, or as granular material. By far the largest amount of urea is produced as a fertilizer in the form of prills, which are produced by removing water from the solution, and dropping the molten urea down through spray formers in a tower countercurrent to a stream of air. The droplets of molten urea, which solidify before hitting the base of the tower, are cooled, and then bagged or sold in bulk.

Crystallization of urea from the urea-water solution produces a relatively pure product.

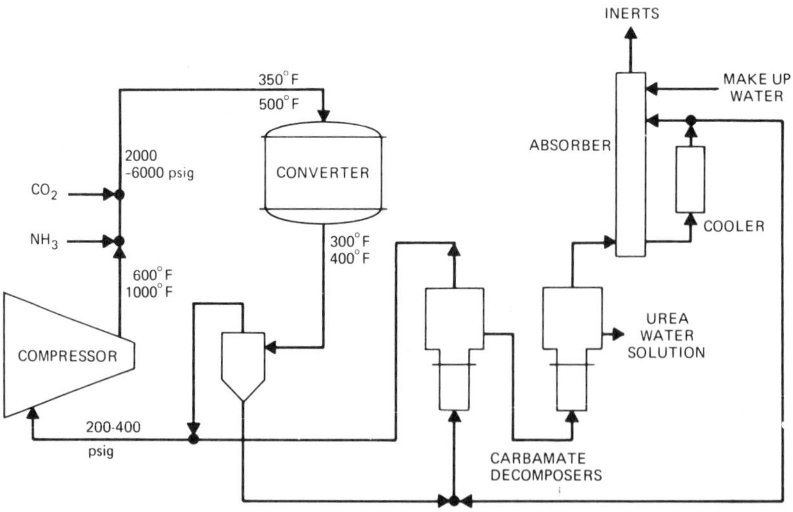

Fig. 5.24 Chemico thermo-urea process.[80]

Fig. 5.25 1000-T/D urea plant. (*Courtesy Chemical Construction Corp., Chemico.*)

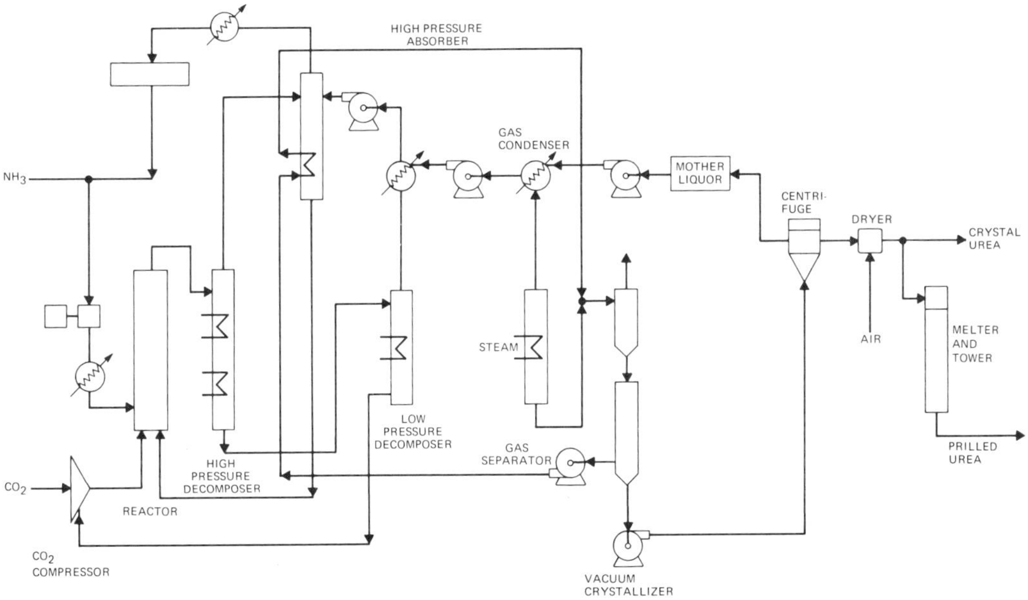

Fig. 5.26 Mitsui Toatsu total recycle process for urea. [*Adapted from Hydrocarbon Processing,* **49,** *No. 6 (June 1970); Yoshimura, S., "Optimize New Urea Process"; copyright 1970 by Gulf Publishing Co.*]

The biuret content is lower than that in the standard prilled product, since the biuret in the urea-water solution is removed with the mother liquor in the crystallization step. Prills with low biuret content can be produced by starting with crystallized urea (see U.S. Patent 3,025,571).[84]

Granules are produced by introducing into an agitated mass of dry particles a concentrated solution of urea at a temperature a little higher than its set point. Agitation is continued until granules are obtained, then they are dried at a temperature of no more than 120°C. The usual ratio of urea solution to dry particles is 0.4:1 to 0.7:1.

The caking properties of urea make the product difficult to handle.[77] The techniques used to reduce caking are:

1. Coating.
2. Mixing with additives.
3. Thermal treatment.

Additives such as ammonium sulfate, acetylene-diurea, magnesium carbonate, and substituted ureas, e.g., *p*-toluidine, are used.

Coatings which are reported to be effective are sugar, formaldehyde, and acetaldehyde.

In the thermal treatment, the prills are passed through a zone of temperature higher than the melting point of urea.[77]

Anhydrous area can be produced directly, according to J. S. Mackay,[85] by synthesizing the urea at a temperature above its melting point and simultaneously keeping the partial pressure of the water below the condensation point. A pressure above 1000 psig and a temperature above 200°C are used. The preferred conditions are 2000 psig and 225°C. Nitrogen, but preferably ammonia, can be used to reduce the partial pressure of the water vapor.

$$10NH_3 + CO_2 \longrightarrow (NH_2)_2CO + H_2O + 8NH_3$$

Uses of Urea

Industrially, when combined with formaldehyde, urea gives resins that can be machined. Urea is used as an ingredient in softeners for cellulose, cellophane, and wood. It is used to retard end checking of boards during drying. Added to glue, gelatin, or starch, it reduces the viscosity and permits the use of higher concentrations of the active material. Urea is used in the petroleum industry to separate straight-chain hydrocarbons by forming crystalline complexes between the urea and a hydrocarbon with a long straight carbon chain. Urea is used in the preparation of barbituric acid, caffein, ethyl urea, hydrazine, melamine, guanidine, and sulfamic acid. It is also used in the manufacture of medicinals, in some cosmetic applications and deodorants, and in skin creams to improve the texture. With hydrogen peroxide, urea forms a crystalline additive $CO(NH_2)_2 \cdot H_2O_2$ which is employed as a disinfectant and oxidizing agent.

Urea-formaldehyde, urea-furfural-formaldehyde, and urea-resorcinol mixtures form thermosetting resins which are used in molding applications such as radio and TV cabinets, in the treatment of paper and fabrics, and as adhesives for bonding wood. The resins are nearly colorless but opaque, and may be pigmented as desired. They improve crush and crease resistance in textiles, and impart wet strength to paper and paper coatings. A small amount of acid, or acid generating catalysts, may be added to make the resin set.

The principal outlet for urea is in the fertilizer field, not only as a solid fertilizer but also as a liquid in the production of granular materials.

In 1958, 44,714 tons of urea nitrogen were consumed as direct application materials; by 1968 this use had increased to 243,359 tons, an increase of 542 per cent.[6] Total production of urea as a primary solution increased from 530,098 tons in 1958 to 2,358,579 tons in 1968.[6] In 1970, total capacity was about 4,500,000 tons. When combined with superphosphate, urea produces a fertilizer known as "Phosphazote." A salt with calcium nitrate is also recommended as a fertilizer. Urea is combined with formaldehyde to produce a slowly soluble compound which will supply nitrogen to the soil during the entire growing season.

Urea has found acceptance as a protein substitute for ruminant animals (cattle, goats, and sheep). In this application it can efficiently replace about one-third the natural protein in ruminant feeds, but care in formulating must be exercised to optimize the concentration. Urea is toxic in large doses. It is reported in a recent patent[86] that urea-hydrocarbon solids adducts are more palatable than urea alone, making the product more readily accepted by the animals. The hydrocarbons used are alkanes, such as *n*-hexane to *n*-tetradecane and olefins, such as *n*-hexene and *n*-octadecene. The following table shows the advantage of an urea adduct over urea.

Feed Additive	*Nitrogen Retained (Grams per Day)*
Urea	1.73
n-Tetradecane-urea adduct	2.16
1-Tetradecene-urea adduct	2.22

HYDROGEN CYANIDE

Hydrogen cyanide is a relatively new commercial chemical. It is increasing in importance with the rapid growth of methyl methacrylate, acrylonitrile, and other chemicals.[75] Ordinarily hydrogen cyanide is a gas, and it is a very toxic material. It has a melting point of -14°C and can be liquefied by cooling to 26°C. The liquid is colorless. It is soluble in cold water, alcohol, and ether. It forms explosive mixtures with air. The lower limit of flammability is 5.6 per cent, and the upper limit is 40 per cent.

As of October 1, 1967,[88] the estimated annual capacity for HCN in the United States was 185,000 tons, excluding some internal commercial consumptions, such as in the manufacture of adiponitrile and ferrocyanide pigments. Over the years 1956 to 1966, the growth rate was 3.5 per cent per year. Future growth will partly depend on the use in chelates for detergent manufacture.

Sodium cyanide can be manufactured in several ways. One of the earliest methods involves the action of sulfuric acid on sodium cyanide, a process which is convenient for small quantities. Sodium cyanide may also be produced by reacting sodium carbonate with carbon and nitrogen, or by reacting sodamide with charcoal, or by passing ammonia and carbon monoxide through molten sodium to form a mixture sodium cyanide and carbon dioxide.

Hydrogen cyanide may be obtained from hydrogen, nitrogen, and carbon by passing the two gases through an electric arc formed by two carbon electrodes. The conversion is low even at 2,000°C, hence the reaction is not of commercial importance.

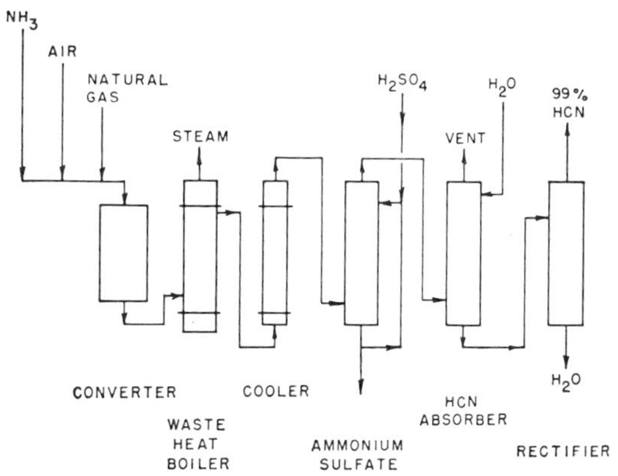

Fig. 5.27 Hydrogen cyanide synthesis. [*Adapted from Chem. Eng.,* **62,** *No. 9, 289 (1955).*]

Formamide has been dehydrated to produce hydrogen cyanide and water. Today, however, most hydrogen cyanide is produced by the reaction of ammonia with methane and air[89] (Fig. 5.27).[90] Since the reaction will proceed only if sufficient heat is supplied, the air is used to burn the methane and produce the heat required. The reaction of methane with ammonia is as follows:

$$CH_4 + NH_3 \longrightarrow HCN + 3H_2$$

The formation of hydrogen cyanide requires the addition of heat–60 kilocalories per mole. In some instances the heat is supplied to the reaction zone through small-diameter converter tubes, in which case the converter gases are not diluted with large volumes of nitrogen from the air.

The reaction of methane, air, and ammonia is believed to proceed in accordance with the following equations:[89]

$$NH_3 + O_2 \longrightarrow NH_3O_2 \longrightarrow HNO + H_2O \quad (1)$$

$$HNO + CH_4 \longrightarrow HNCH_2 + H_2O \longrightarrow HCN + H_2 + H_2O \quad (2)$$

The net result of these two reactions is:

$$NH_3 + O_2 + CH_4 \longrightarrow HCN + H_2 + 2H_2O \quad (3)$$

Equation (2) evolves 66 kilocalories per mole. The other reaction is also exothermic and the net effect is 144 kilocalories liberated per mole of hydrogen cyanide produced. Reactions (2) and (3) occur at very high rates below 850°C. Temperatures above 1,100°C are necessary to promote reaction (1) at a significant rate. The feed composition for the conveter, which is selected to meet the demands of equation (3), is as follows:

$$NH_3 + CH_4 + 3/2O_2 \longrightarrow HCN + 3H_2O + 144 \text{ kcal per mole}$$

Some side reactions occur and produce carbon monoxide and hydrogen, which are in turn partially consumed by the oxygen to give carbon dioxide and water. The effluent gas normally contains about 6 per cent hydrogen cyanide, 1.6 per cent ammonia, 7.5 per cent hydrogen, 57 per cent nitrogen, 23 per cent water, and smaller amounts of carbon monoxide, carbon dioxide, methane, and oxygen.

A catalyst is universally employed. A platinum catalyst may be used. However, it is usually alloyed with 10 to 20 per cent rhodium and deposited on a ceramic material to give it mechanical strength, or it may be used as a gauze.

Under conditions of hydrogen cyanide synthesis, platinum alone will volatilize, and in some cases recrystallize. Both these phenomena will result in the loss of catalyst. When alloyed with rhodium, the catalyst will give satisfactory performance for as long as 4,000 hours.

It is important that the ingredients be properly metered to the system and that foreign materials be removed before the feed enters the catalyst chamber. Oil filters for the NH_3 and dust filters for the air and methane are sometimes used. Phosphorous compounds will deactivate the catalyst and must be removed from the feed materials. Short-term exposure will not permanently injure the catalyst, but long-term exposure will result in permanent loss of catalyst activity. Hydrocarbon impurities will crack to carbon and hydrogen, and the carbon will deposit on the catalyst resulting in temporary loss of activity. Unsaturated materials and hydrocarbons with more than three carbon atoms are particularly detrimental. Iron rust from pipelines must also be avoided, since it is a catalyst poison. Feed materials are usually filtered separately, since acidic components in the air react with ammonia to produce solids which might plug the filter screen or the catalyst bed.

The manner in which the feed materials are introduced into the reactor zone is important. Extreme precautions may be necessary to distribute the reactants over the catalyst bed. The catalyst shape may be other than flat to ensure proper distribution of the gases.

Gas fed to the converter usually has the following composition:

Ammonia	11 to 12%
Methane	12 to 13%
Air	74 to 78%

In many cases the converters are provided with rupture discs to relieve the pressures which may result from explosions caused by improper control of feed materials.

Ordinarily the converter is started by heating a portion of the gauze electrically, or with a torch, to reach reaction temperature. The reaction then spreads throughout the gauze as a result of the overall exothermic reaction.

Approximately 60 to 67 per cent ammonia and 53 per cent methane are converted in a single pass. Higher conversions may be obtained at pressures of 30 to 40 psig. The choice of pressure is influenced more by the recovery system than by the advantages to be gained in the synthesis step.

It is important that the reaction mixture be cooled quickly to avoid cracking the hydrogen cyanide. Basic materials, such as ammonia promote polymerization. Accordingly, the gases are introduced into a waste-heat boiler where about 6 pounds of steam are produced per pound of hydrogen cyanide. From the waste-heat boiler the gases are passed through a cooler and then to an ammonia absorption system. At this point in the process several innovations have been developed. The most common method for removing ammonia is to use sulfuric acid as the absorbent. However, a polyhydroxy-boric acid complex, which is also used, has one advantage—the ammonia may be recovered for recycle. The bottoms from the scrubbing tower contain an aqueous ammonium sulfate, which is processed further to produce solid ammonium sulfate. The overhead gases which are free from ammonia are then absorbed in water prior to purification. The water-hydrogen cyanide mixture is rectified in a final scrubber to produce 99 per cent hydrogen cyanide.

Materials selection is important in the construction of the equipment. Aluminum has good corrosion resistance to the reaction products but it lacks strength above 600°C, and it promotes polymerization of hydrogen cyanide in the presence of ammonia and moisture.

In the *Degussa* process[91] the methane and ammonia are reacted in the presence of a special platinum-containing catalyst at 1200–1300°C. The reaction is endothermic and heat is supplied through tubes mounted in a furnace. The product gases contain 22.8 per cent HCN; 2.7 per cent NH_3; 2.4 per cent CH_4; 0.9 per cent N_2; and 71.2 per cent H_2. The HCN can be removed by the means mentioned previously, in the discussion on the removal of the ammonia with dilute sulfuric acid. The advantage of the process is that the tail gas contains 95.6 per cent hydrogen by volume. Yield is 84 per cent based on ammonia, and 90 per cent based on methane.

U.S. Patent 2,776,872[92] describes a process in which methane is reacted with nitrogen from the air at 2200°C to produce hydrogen cyanide.

$$2CH_4 + N_2 \longrightarrow 2HCN + 3H_2$$

In a more recent patent, U.S. 3,501,267,[93] a process is proposed which is based on the reaction of ammoniated coal with ammonia at 1000 to 1300°C to produce HCN. The yields obtained were: 1.56 cu ft of HCN per pound of coal, and 0.57 cu ft of HCN per cu ft of ammonia.

The synthesis of HCN from its elements, and from ammonia and carbon is covered in a paper by T. K. Sherwood, E. R. Gilleland, and S. W. Ing, Jr.[94]

Hydrogen cyanide is used principally in the manufacture of methyl methacrylate, acrylonitrile, sodium cyanide, adiponitrile, chelates, acrylic acid, ferrocyanide pigments, and in the production of tertiary alkylamines. These materials are obtained by adding HCN to tertiary olefins, followed by hydrolysis of the formamides produced.

ACRYLONITRILE

Acrylonitrile is a colorless liquid with a boiling point of 77.3°C, a melting point of −82°C and a density of 0.806 at 20°C. It is a volatile liquid that is only slightly soluble in water, the freezing point is −83.4°C. It forms constant boiling mixtures with water, methanol, and benzene. And it forms explosive mixtures with air.

Acrylonitrile is toxic due to release of the cyanide radical. Hence, the symptoms may be delayed. Poisoning may result from inhalation, oral ingestion, or skin penetration. Emergency kits are available for treatment of exposure.

In 1967 the demand for acrylonitrile was 670,000,000 pounds; but, by 1972 the demand was expected to have grown to 1,400,000,000 pounds.[96] The capacity in 1971 was estimated at 1,775,000,000[97] pounds with the distribution being as follows:

	Millions of Pounds
American Cyanamid	170
Du Pont	630
Goodrich	45
Monsanto	540
Vistron (Sohio)	270
Union Carbide	90
TOTAL	1,775

The demand is closely tied to the textile-fibers market since most of the acrylonitrile is used in the production of acrylic and modacrylic fibers.

Acrylonitrile can be manufactured by reacting hydrogen cyanide with acetylene according to the following reaction:

$$HCN + HC{\equiv}CH \longrightarrow HCH{=}CHC{\equiv}N$$

In practice, about 10 moles of acetylene are used per mole of hydrogen cyanide.[95]

Another process for the manufacture of acrylonitrile employs ethylene oxide. In this process, hydrogen cyanide is reacted with ethylene oxide to produce the cyanohydrin which, when dehydrated, yields acrylonitrile. The production of acetylene and ethylene oxide are described in Chapters 17 and 25, respectively, and will not be discussed here. It should be mentioned, however, that from the standpoint of both safety and yield, care must be taken in handling these materials to avoid loss. Acetylene must be purified before it is introduced into the reaction mixture. The actual combination of hydrogen cyanide and acetylene takes place in a liquid phase at 80° C with an aqueous catalyst containing cuprous chloride. All the hydrogen cyanide reacts, and about 10 per cent of the acetylene. Other products are also formed, such as acetaldehyde, vinyl chloride, methyl vinyl ketone, chloroprene, and condensation polymers. The reactor is usually rubber lined.

Reactor gases containing acrylonitrile, water, by-products, acetylene, and hydrogen cyanide are scrubbed with water to remove the acrylonitrile and unreacted hydrogen cyanide. The solution is then sent to a water stripper where the acrylonitrile is removed overhead. The low boilers are removed in a topping column. The bottoms from the topper are sent to the acrylonitrile dryer which removes the water by azeotropic distillation. Finally, acrylonitrile is recovered in a rectification column, the acrylonitrile being removed overhead, the high boilers from the base of the column.

The *Sohio* Process (Vistron Corp.) for acrylonitrile is based on the reaction of propylene, air, and ammonia in a fluid-bed catalytic reactor at 5-30 psig and 750-790°F[98]

$$CH_2{=}CHCH_3 + NH_3 + 3/2O_2 \longrightarrow CH_2{=}CHCN + 3H_2O$$

By-products are HCN, CH_3CN, water, light ends, and high boiling impurities. These are removed by distillation. A molybdenum and bismuth catalyst was used originally, but a new catalyst is reported to give less by-products. This process is shown in Fig. 5.28. Yields of 0.73 lb of acrylonitrile per lb of propylene are obtained. Hydrogen cyanide produced is 0.15 to 0.20 lb per lb acrylonitrile, and acetonitrile is also produced at the rate of 0.10 lb per lb of acrylonitrile.

A similar process (*Distillers-Ugine*) based on the same raw materials uses a fixed-bed reactor.[98] Yields are about the same.

One fixed-bed catalyst disclosed is a platinum, or 90:10 platinum and rhodium, metallic gauze or screen.[99] With this catalyst the olefin to oxygen ratio in the feed is between 2.5:3 and 3.5:3.0. Sufficient oxygen or air must be added to make the process thermally self-sufficient. The olefin to ammonia ratio preferred is between 1.1:1.0 and 1.5:1.0. For this catalyst the preferred temperature is 800 to 900°C. Pressures from 50 mm of Hg up to

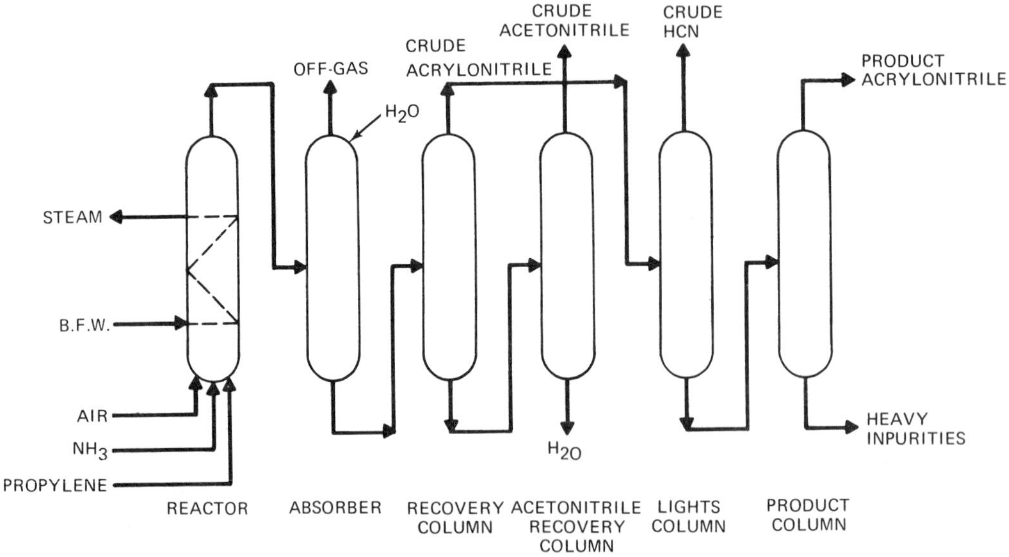

Fig. 5.28 Sohio process for acrylonitrile synthesis. [*Hydrocarbon Processing,* **46**, *No. 11, 141 (Nov. 1967); copyright 1967 by Gulf Publishing Co.*]

5 atm and space velocities of 5000 to 10,000 volumes per hour may be used. Fertilizer grade ammonia and 85 to 90 per cent propylene are satisfactory raw materials.

Acrylonitrile may also be made from propylene and nitric oxide, from acrolein and ammonia, and from propionitrile.[100]

Acrylonitrile is used in the production of acrylic and modacrylic fibers, plastics and resins, and nitrile rubbers or elastomers. There is an increasing demand for acrylonitrile for use in synthetic fibers which are shrink resistant to both dry and wet cleaning and can be used in window curtains, tents, sails, nets, awnings and auto fabrics. Small quantities of other monomers are sometimes added for special properties. Monomers that can be used include methyl acrylate, styrene, vinyl acetate, vinylidenes, and isobutylene.

Polyacrylonitrile is a very hard resin. Acrylonitrile formulated with other materials produces hardness, solvent resistance, and greater resistance to high temperatures. Such materials are suitable for moldings, tubings, sink and table tops, wall panels, artificial leathers, shoe soles, luggage, etc.

Copolymerication of acrylonitrile with butadiene and butadiene-styrene produces elastomers which have good oil resistance, hardness, toughness, and tensile strength.

The end use distribution is as follows:[97]

Outlet	*Per Cent*
Acrylic and modacrylic fibers	60
Acrylonitrile-butadiene-styrene and styrene-acrylonitrile resins	12
Nitrile rubbers	8
Others (including exports)	20

Cyanoethylation is receiving considerable attention, since cyanoethylated cotton is more receptive to dyes, is more resistant to rot, mildew, and bacteria, and has a higher abrasion and stretch resistance than untreated cotton.

MELAMINE

Melamine or cyanuramide, [$C_3N_3(NH_2)_3$] a monoclinic compound, decomposes at its melting point of 360°C. Its molecular weight is 126.13.

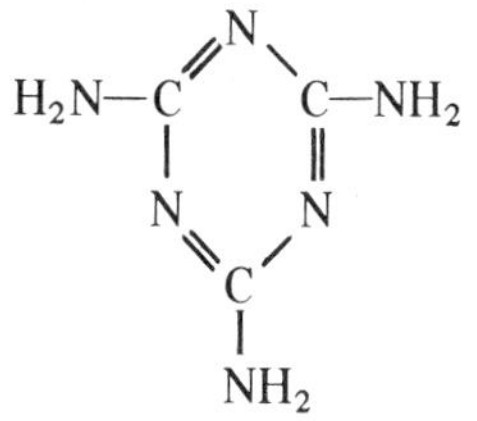

The crystalline product has a bulk density of 20.6 lb/cu ft, a nitrogen content of 66.6 per cent, and forms a solution in formalin with a color of 15 APHA.

In 1966 the demand was 96 million pounds and in 1971 it was expected to be 123 million pounds.[101] Estimated capacity in 1967 was 135 million pounds with Allied, Cyanamide, Fisher, and Reichhold the major producers.

Melamine is produced by heating urea and passing the resulting mixture of isocyanic acid and ammonia over a suitable catalyst, quenching the products with water or aqueous mother liquor, and working them up by filtration, centrifuging, or crystallization.[49]

The reaction takes place in two stages:

$$CO(NH_2)_2 \longrightarrow HNCO + NH_3 \quad (1)$$

This reaction is endothermic requiring 800 kilocalories per kg of urea or 21,800 kilocalories per lb mole of urea.

$$6HNCO \longrightarrow C_3N_3(NH_2)_3 + 3CO_2 \quad (2)$$

This reaction is exothermic, liberating 1100 kcal per kg of melamine or 63,000 kcal/lb mole of melamine. The over-all reaction is endothermic.

The net yield is 84 per cent of theoretical; but if the mother liquor can be recycled to a urea plant a better yield is possible.

The *O.S.W.* process is shown in Fig. 5.29.[49] Urea, along with recycle ammonia, is fed into a fluidized-bed reactor containing inert solids, where the temperature is quickly raised to 350°C within a few seconds. The resulting mixture of isocyanic acid and ammonia is fed into a reactor containing a solid catalyst, such as silica gel, alumina, or boron phosphate. Products from the isocyanic acid reactor are quenched with water, the resulting slurry is

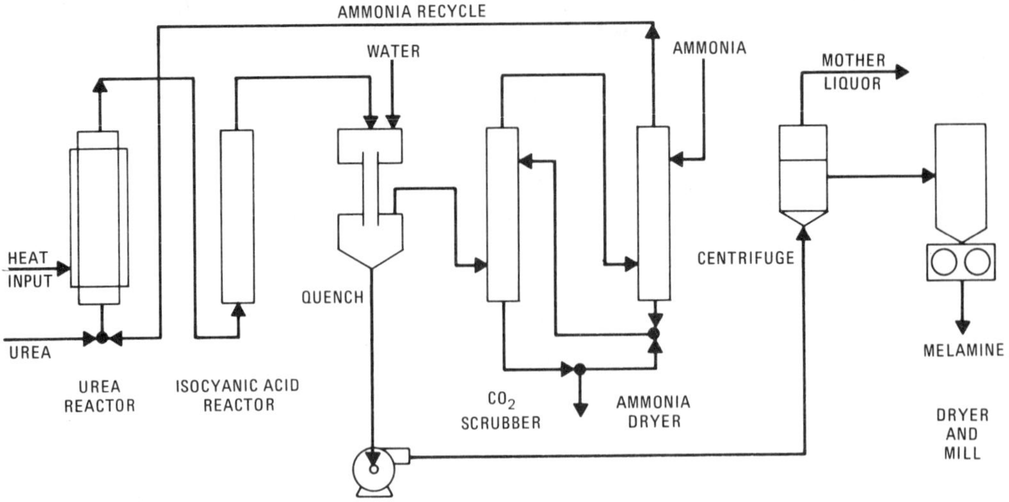

Fig. 5.29 Melamine synthesis—O.S.W.[49]

centrifuged, and the mother liquor is sent to recovery or a urea plant.

Off-gases from the quench system contain ammonia, carbon dioxide, and water vapor. The CO_2 is removed as ammonium carbonate, and the ammonia is returned to the urea reactor.

The design of the urea reactor is critical. The wall temperature must not be too high or corrosion is a problem.[102] This can be avoided by using stationary layers of catalyst along the inner wall, held in place by corrosion-resistant flights secured to the periphery at suitable internals. By this method the reactor wall can be made of steel containing 0.16 per cent carbon, 0.35 per cent silicon, 0.40 to 2.00 per cent manganese, not more than 0.05 per cent phosphorous, and not more than 0.05 per cent sulfur.

The formation of melamine from isocyanic acid takes place at 400 to 430°C. The heat of reaction is absorbed as sensible heat by raising the entering vapors from 320 to 430°C. Melam ($C_6H_9N_{11}$) and melem ($C_6H_6N_{10}$), condensation products of melamine, are by-products that form by splitting off ammonia when the temperature is too high. Melamine will also hydrolyze if allowed to remain too long with water at high temperature. The products are ammeline [$(CH)_3(NH_2)_2OH$] and ammelide [$(CN)_3(NH_2)(OH)_2$].

The product is milled to produce a fine particle size, 5-10 microns.

Melamine is used almost exclusively for resin manufacture; in laminates, 38 per cent; in molding compounds, 33 per cent; in textile treating agents, 10 per cent; in coatings 8 per cent; in paper treating agents, 6 per cent; and in adhesives, 5 per cent.[101]

AMINES

Ethanolamines are produced by the reaction of ethylene oxide and ammonia at elevated temperature and pressure.[103]

$$CH_2CH_2O + NH_3 \longrightarrow NH_2CH_2CH_2OH$$

$$CH_2CH_2O + NH_2CH_2CH_2OH \longrightarrow NH(CH_2CH_2OH)_2$$

$$NH(CH_2CH_2OH)_2 + CH_2CH_2O \longrightarrow N(CH_2CH_2OH)_3$$

The raw materials are fed at controlled rates to give the proper ratio of products. The reaction is highly exothermic and it is necessary to remove the heat to obtain good yields. Variation of the ratio of ethylene oxide to ammonia will vary the yield of the three amines. An over-all yeild of 95 per cent is possible.

The crude product is worked up by distillation. In the first column ammonia is removed, and in the second water is removed and recycled to the reactor. In the next three columns, monoethanolamine, diethanolamine, and triethanolamine are produced.

Average utility consumptions are:[103]

Item	*Per 100 Lbs of Product*
Steam	1,200 lb
Cooling water	8,650 gal
Electricity	3.3 kw-h
Heat	36,300 Btu

Demand for ethanolamines will increase from 218,154,000 pounds in 1967 to an estimated 320 million pounds by 1972.[104] Most of this demand is for use in detergents, 40 per cent, gas treating processes, 20 per cent; and morpholine synthesis, 7 per cent. Other uses consume 19 per cent, and 14 per cent is exported. Olin and Shell Chemical are the major producers.

Methylamines are produced by reacting ammonia and methanol.[105] The reaction produces the three amines: monomethylamine, dimethylamine, and trimethylamine. The ratio produced depends upon the ratio of ammonia and methanol used in the reaction. U.S. Patent 2,153,405[106] describes a process for trimethylamine production which uses a ratio of 1 to 2 moles of methanol per mole equivalent of ammonia, monomethylamine and dimethylamine.

$$CH_3OH + NH_3 \longrightarrow NH_2CH_3 + H_2O$$

$$CH_3OH + NH_2CH_3 \longrightarrow NH(CH_3)_2 + H_2O$$

$$CH_3OH + NH(CH_3)_2 \longrightarrow N(CH_3)_3 + H_2O$$

The mixed feed materials are preheated and passed into a converter operating at 300 to 350°C and 30 to 75 psig, where they come in contact with a catalyst, such as alumina. The product from the reactor is cooled, and the unreacted ammonia and methanol, water, and amines are separated by distillation at a pressure which may be between 100 and 300 psig.

Demand for methylamines was 103 million pounds in 1966 and was expected to be 155 million pounds by 1971.[107] The uses for the three amines are:

Monomethylamine	
1-Naphthyl methyl carbamate	45%
Surfactants	33%
Others	22%
Dimethylamine	
Dimethylformamide and acetamide	50%
Lauryl dimethylamine oxide	16%
Rubber chemicals	12%
Pesticides	10%
Rocket propellant and others	12%
Trimethylamine	
Choline chloride	90%
Others	10%

Ethylenediamine is produced from ethylene glycol or ethylene dichloride.[100] The reaction from ethylene glycol is via ammonia over a hydrogenation catalyst, such as nickel or copper supported on alumina, titania, or similar metal oxide.

$$CH_2OHCH_2OH + 2NH_3 \longrightarrow CH_2NH_2CH_2NH_2 + 2H_2O$$

According to U.S. Patent 3,137,730[108] the preferred ratio of ammonia to glycol is 20 to 30 moles per mole, with water present to the extent of at least 30 per cent by weight of the glycol. Hydrogen is also required to activate the catalyst, about 0.2 per cent by weight of the glycol. The liquid phase reaction temperature is 220 to 270°C, and the pressure 3,000 to 6,000 psig.

Ethylenediamine is produced from ammonia and ethylenedichloride by the following reaction:

$$ClCH_2CH_2Cl + 2NH_3 \longrightarrow CH_2NH_2CH_2NH_2 + 2HCl$$

The reaction is carried out in a fractionating tower operating at about 300 psig and temperatures at the bottom of 100 to 200°C and at the top of 30 to 40°C. No catalyst is required. The ethylenediamine is freed from the crude product by adding NaOH which frees the ammonia from the ammonium chloride and the amines from their hydrochlorides.

Ethylenediamine demand was expected to reach 60 million pounds in 1971.[109] The principal producers are Union Carbide, Dow, and Jefferson. It is used primarily for the manufacture of fungicides, resins, and sequestrants.

Hexamethylenediamine is produced by the hydrogenation of adiponitrile.[100]

$$CN(CH_2)_4CN + 4H_2 \longrightarrow NH_2(CH_2)_6NH_2$$

The process is carried out in the presence of ammonia at a temperature of 120 to 170°C and a pressure of 4,000 to 6,000 psig over a fixed-bed hydrogenation catalyst of cobalt or nickel on kieselguhr, alumina, or silica gel.

The product from the reaction must be distilled to obtain the desired purity for production of polyamides. One distillation scheme is given in U.S. Patent 3,017,331[110] (Jan. 16, 1962), which C. R. Campbell assigned to Chemstrand Corp. The system consists of four columns: the first for removal of foreshots, the second for removal of intermediates, the third for recovery of products from the intermediates and removal of the purge materials, and the fourth for removal of the refined product. Operating pressure is 50 to 400 mm Hg absolute. By-products are 1,2-diaminocyclohexane, pentamethylenediamine, and epsilon-aminocapronitrile (the half-hydrogenated product).

The nitrile for hexamethylenediamine may be obtained from adipic acid and ammonia using a phosphoric acid catalyst. An alternative method for production is via acrylonitrile, which on heating gives dicyanocyclobutane. This latter product will produce adiponitrile when reacted with hydrogen in a stirred autoclave at 250-300°C and 100 atm in a solvent, such as dioxane, and in the presence of a nickel catalyst. The reaction time required is reported to be about three hours.[100]

Hexamethylenediamine is produced in large quantities for the manufacture of nylon. Production figures are not generally reported.

ANILINE

Aniline or aminobenzene, $C_6H_5 \cdot NH_2$, is a colorless oil with a boiling point of 184.4°C and a freezing point of -6.1°C. It is slightly soluble in cold water and infinitely soluble in alcohol and ether. It is highly toxic.

Aniline is produced from nitrobenzene[100] by hydrogenation of the nitro group to an amine group. A catalyst is used; metals of Groups I, V, VI, and VII, as well as iron, copper, tungsten, molybdenum, nickel, and cobalt are effective.

The reaction can be carried out in the vapor phase using a fluidized bed with a molar ratio of 3 hydrogen to 1 nitrobenzene, at a temperature of 250 to 300°C and a pressure of 10 to 25 atm. A linear flow rate of 0.1 to 5.0 ft/sec is used. Reduction of nitrobenzene can also be carried out in the liquid phase with H_2S.

Aniline can also be produced from chlorobenzene by ammonolysis.

$$C_6H_5Cl + 2NH_3 \longrightarrow C_6H_5NH_2 + NH_4Cl$$

Three moles of ammonia are usually used at an operating temperature of 180 to 220°C and a pressure greater than the vapor pressure of the reactants. This reaction is carried out in the liquid phase with copper compounds as the catalyst.

The demand for aniline in 1968 was 263 million pounds and by 1973 it was expected to be 320 million pounds.[111]

Aniline is used in the production of rubber chemicals, isocyanates, and dyestuffs.

ISOCYANATES

Toluene diisocyanate (TDI) has recently become industrially important for the production of polyurethane foams, resins, and rubbers. The material is made according to the following reactions:[100,49]

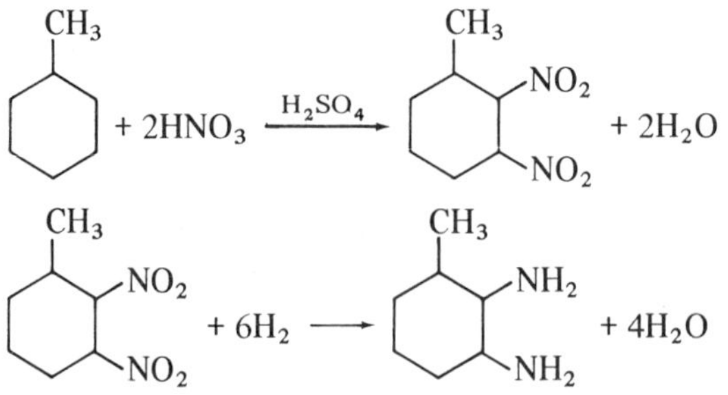

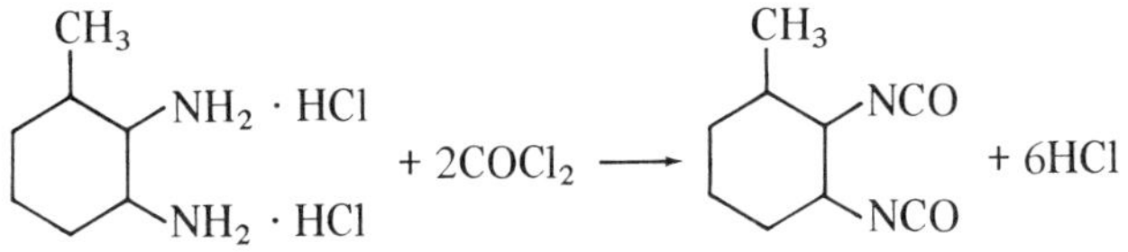

This latter reaction is characterized by the reaction of a primary amine with phosgene.

$$RNH_2 + COCl_2 \longrightarrow RNCO + 2HCl$$

In this reaction the phosgene is introduced at such a rate that at least 25 per cent excess phosgene is present.

It has been shown[49] that polymer formation usually results in poor yields. Free amine and TDI react readily to form a urea derivative which polymerizes. By dissolving the toluenediamine in *o*-dichlorobenzene and adding dry HCl to form the hydrochlorides before adding the phosgene, the yield can be improved.

The consumption of raw materials is as follows:

	Per Ton TDI
Toluene	0.593 ton
HNO_3 (98%)	0.838 ton
H_2SO_4	0.680 ton
Cl_2	0.858 ton
H_2	30,800 cu ft
CO	24,800 cu ft

A mixture of 2,4- and 2,6-TDI is obtained. By varying conditions and recovery techniques, the proportion of each can be varied to meet the demand.

Capacity for toluene diisocyanate (80/20) production in the United States is stated to be about 400 million pounds and was expected to go to 600 million pounds by 1973.

The uses for TDI are:

Use	*%*
Flexible urethane foam	50
Rigid foam	23
Coatings	7
Elastomers	5
Export	8
Miscellaneous	7

OTHER NITROGEN COMPOUNDS

Additional nitrogen compounds receiving commercial attention are the nitroparaffins,[112,12] dimethylformamide, dimethylacetamide, and nitrilotriacetic acid.

The *nitroparaffins* are produced by Commercial Solvents Corp. by nitration of propane at 400°C and 100 psi. Seventy-five per cent nitric acid is sprayed into the reactor to give a high hydrocarbon to nitric acid ratio. A yield of 40 per cent per pass based on nitric acid is obtained. By-products such as, aldehydes, ketones, acids, alcohols, carbon dioxide, and water are obtained.

Physical properties of the nitroparaffins are given in Table 5.16.

Nitroalcohols are produced from the nitroparaffins by reaction with formaldehyde. These alcohols are used in the synthesis of aminoalcohols which are emulsifying agents for waxes, paints, and cleaners. The nitro-

TABLE 5.16 Physical Properties of the Commercially Available Nitroparaffins[a]

	Abbreviation	*Freezing Point (°C)*	*Boiling Point (°C, 760 mm)*	*Specific Gravity (20/20 °C)*	*Refractive Index (20/D)*	*Flash Point[b] (°F)*
Nitromethane	NM	-29	101.2	1.139	1.3818	112
Nitroethane	NE	-90	114.0	1.052	1.3916	106
1-Nitropropane	1-NP	-108	131.6	1.003	1.4015	120
2-Nitropropane	2-NP	-93	120.3	0.992	1.3941	103

[a] *Pet. Ref.*, **35**, 8, 151 (1956).
[b] Tag open cup.

alcohols can also be converted to chloronitro compounds which have insecticidal and fumigant properties.

Dimethylformamide, $HCON(CH_3)_2$, is a colorless liquid with a boiling point of 153°C and a freezing point of -61°C. It is miscible with water and common organic solvents. Because of these properties it is a good reaction medium. It is used as a solvent for vinyl resins in lacquers, films and printing inks, for polyurethanes, pigments, dyes and inorganic salts. It can be made from dimethylamine and methyl formate.

Dimethylacetamide, $CH_3CON(CH_3)_2$, is a colorless liquid with a boiling point of 165.5°C and a freezing point of -20°C. It is miscible with water and common organic solvents. Dimethylacetamide is used as a solvent for polyacrylonitrile, vinyl resins, cellulose derivatives, styrenes, linear polyesters, and paint removers. It is also used as a reaction medium. It can be made from acetic acid and dimethylamine.

Nitrilotriacetic acid (NTA) had passing importance as a substitute for phosphates in detergents, shampoos, soaps, and bleaching agents. It is used as the sodium salt ($NTANa_3$). Nitrilotriacetic acid is a chelating agent which eliminates the undesirable effects of polyvalent cations such as calcium (Ca^{++}). This compound is important because it is biologically destroyed, whereas other agents which reduce the undesirable effects of polyvalent cations are not. However, potential health hazards resulting from its use led to its withdrawal after only a short period (see Chapter 13).

The nitration of benzene and other aromatic compounds is also becoming more important. These materials are used in the manufacture of aniline, trinitrotoluene (TNT), and dyes. The reactions are controlled by varying the catalyst (sulfuric acid) concentration, the excess nitric acid present, the temperature, and the pressure. In toluene the orientation of the nitro group is predominately in the ortho-para direction. In nitrating alkylbenzenes the size of the alkyl group influences the position of nitro group. Dinitration is more difficult than mononitration.

REFERENCES

1. *Chem. Eng. News*, **37**, No. 32, 46–47 (Aug. 10, 1959).
2. *Chem. Week*, **100**, No. 3, 64 (Jan. 21, 1967).
3. *Akad. Nauk S.S.S.R, Seriya Khimicheskaya Izvestiya* No. 9, 1728–1729, (1964).
4. French Patent 1,391,595, *Chem. Eng.* **65**, 58 (Jul. 28, 1958).
5. "Estimated World Fertilizer Production as Related to Future Needs–1967 to 1972–1980," TVA, National Fertilizer Development Center.
6. "Fertilizer Trends–1969," TVA, National Fertilizer Development Center, Nursch Schoals, Ala.
7. *Nitrogen*, No. 57 (Jan. and Feb. 1969).
8. *Chem. Week* **80**, 66 (Jan. 12, 1957).
9. *Chem. Process Eng.*, 473–483 (Sept. 1965).
10. Duff, B. S., *Chem. Eng. Progr.*, **51**, No. 1, 12-J (Jan. 1955).
11. *Petrol. Process.* (Sept. 1956).
12. Updegraff, N. C. and Mayland, B. J., *Petrol. Refiner*, **33** No. 12, 156–159 (Dec. 1954).
13. *News Letter*, Institute of Gas Technology; **10**, No. 3 (1958).
14. *Chem. Eng. News* **46**, No. B. (Mar. 18, 1968).
15. Lockerbie, T. E. *et al.*, U.S. Patent 3,106,457.
16. Hougen, O. A., *et al.*, "Chemical Process Principles," Part II, p. 1017, New York, John Wiley & Sons, 1959.
17. U.S. Patent 3,081,268.
18. Synthetic Fuels Symposium, A.C.S., Sept. 12–15, 1955.
19. "Selected Values of Properties of Hydrocarbons," Am. Petrol. Inst. Research Project 44, National Bureau of Standards.
20. Lombard, J. F., *Hydrocarbon Process.*, **48**, No. 8, 111 (Aug. 1969).
21. Swaim, C. D., *Hydrocarbon Process.*, **49**, No. 3, 127–30 (Mar. 1970).
22. "Water Scrubbing for CO_2 Removal," Fredrich Uhds-Gmbh.

23. *Nitrogen*, No. 59, 42 (May–Jun. 1969).
24. U.S. Patent 3,352,631.
25. *Hydrocarbon Process.*, **43**, No. 4, 113 (Apr. 1964).
26. U.S. Patent 2,880,591.
27. U.S. Patent 2,863,527.
28. *Chem. Eng.*, **67**, No. 19, 166–9 (Sept. 19, 1960).
29. "Removing Carbon Monoxide from Ammonia Synthesis Gas," *Ind. Eng. Chem.*, **53**, No. 8, 645 (Aug. 1961.
30. Grotz, B. J. *Hydrocarbon Process.*, **46**, No. 4, 197 (1967).
31. Perry, J. H., "Chemical Engineers Handbook," New York, McGraw-Hill Book Co., 1963.
32. *Nitrogen*, No. 58, 25 (Mar.-Apr. 1969).
33. Nielsen, Anders, "An Investigation of Promoted Iron Catalysts for the Synthesis of Ammonia," 3rd Ed., Copenhagen, Jul. Gjellerup Forlag, 1956.
34. *Catalyst Reviews*, 1–25 (Apr. 9, 1970).
35. "Ulmanns Encyklopadie de technischen Chemie," München, Berlin, Urban and Schwarzenberg, 1953.
36. U.S. Patent 2,894,821.
37. U.S. Patent 3,344,052.
38. U.S. Patent 3,310,376.
39. *Chem. Week*, **97**, No. 11 (Sept. 11, 1965).
40. "Mined Underground Storage," C.E.P., Vol. 11, Safety in Air and Ammonia Plants.
41. *Chem. Week*, **97**, No. 11 (Sept. 11, 1965).
42. Strelzoff, S., *Chem. Eng.* **63**, 170–174 (May 1956).
43. *Chem. Eng. News*, **36**, No. 38, 62–64. (Sept. 22, 1958).
44. *Chem. Week*, **103**, No. 15, 51–6 (Oct. 12, 1968).
45. Spratt, D. A., *Proc. Fertilizer Soc. of England* (Mar. 25, 1958).
46. Scott, W. W., *Ind. Eng. Chem.* **16**, 74 (1924).
47. *Chem. Eng.*, **77**, No. 14, 24 (Jun. 29, 1970).
48. *Chem. Eng. News*, **48**, No. 22, 25 (May 25, 1970).
49. *Hydrocarb. Process.*, **48**, No. 11, 209 (Nov. 1969).
50. *Chem. Eng.* **65**, 58 (Jul. 28, 1958).
51. *Chem. Trade J.*, 75 (Jul. 11, 1958).
52. *Chem. Eng.*, **67**, No. 8, 94 (1960).
53. U.S. Patent 3,202,481.
54. U.S. Patent 2,760,845.
55. Powell, R. "Nitric Acid Technology–Recent Developments," Noyes Development Corp., 1969.
56. *Chem. Eng. News*, **44**, No. 27, 13 (Jul. 4, 1966).
57. *Chem. Week*, **103**, No. 15, 51 (Oct. 12, 1968).
58. "World Nitrogen Plants 1968–1973," Stanford Research Institute, May 1969.
59. *Chem. Eng. News* **36**, No. 34, 50 (Aug. 25, 1958).
60. *Chem. Week*, **101**, No. 2, 75 (Jul. 8, 1967).
61. Stafford, J. D., Samuels, W. E., and Croysdale, L. G., "Ammonium Nitrate Plant Safety," American Institute of Chemical Engineers, Sept. 27–29, 1965.
62. Chemical Profile, April 1, 1969, Schnell Publishing Co.
63. *Helv. Chim. Acta.* **24**, 754 (1951).
64. *Can. J. Res.*, **25**, Sec. B.
65. *Petrol. Refiner*, **37**, No. 9, 351–353 (1958).
66. *Chem. Eng.*, **65**, No. 14, 120–123 (July 14, 1958).
67. *Chem. Trade J.*, p. 260 (Aug. 1, 1958).
68. Powell, R., "Hydrazine Manufacturing Processes," Park Ridge, N.J., Noyes Development Corp., 1968.
69. *Chem. Eng. News*, **43**, No. 36 (Sept. 6, 1965).
70. Chemical Profile "Urea," Schnell Publishing Co.
71. *Chem. Eng. News*, p. 17 (Jul. 27, 1970).
72. *Chem. Eng.*, **67**, No. 25, 71 (1960).
73. *Chem. Week*, **75**, No. 22, 90 (1954).

74. *Chem. Eng.*, **73**, No. 25, 76 (Dec. 5, 1966).
75. Tonn, W. H., *Chem. Eng.*, **62**, No. 10, 188 (1955).
76. *Nitrogen*, No. 38 (Nov. 1965).
77. "Urea Process Technology," p. 139, Park Ridge, N.J., Noyes Development Corp.
78. U.S. Patent 3,107,149.
79. U.S. Patent 3,236,888.
80. U.S. Patent 3,200,148.
81. U.S. Patent 3,301,897.
82. *Nitrogen*, No. 15, 37 (Jan. 1962).
83. U.S. Patent 3,232,984.
84. U.S. Patent 3,025,571.
85. U.S. Patent 2,527,315.
86. U.S. Patent 3,502,478.
87. *Chem. Week*, **105**, No. 3 (Jul. 19, 1969).
88. Chemical Profile, "Hydrogen Cyanide," Schnell Publishing Co., Oct. 1, 1967.
89. *Refining Engineer*, C-22-C-27 (Feb. 1959).
90. *Chem. Eng.*, **62**, No. 9, 289 (1955).
91. *Hydrocarbon Process.*, **46**, No. 11, 189 (Nov. 1967).
92. U.S. Patent 2,776,872.
93. U.S. Patent 3,501,267.
94. *Ind. Eng. Chem.*, **52**, No. 7 (Jul. 1960).
95. *Chem. Eng.*, **62**, 288–291 (Sept. 1955).
96. Chemical Profile, Schnell Publishing Co., 1969.
97. *Chem. Week*, **106**, No. 20, 75 (May 20, 1970).
98. *Hydrocarbon Process.*, **46**, No. 11, 141 (Nov. 1967).
99. Thomas, C. C., U.S. Patent 3,499,025 (to Sun Oil).
100. Sittig, M., "Amines, Nitriles and Isocyanates, Processes and Products," Park Ridge, N.J., Noyes Development Corp., 1969.
101. Chemical Profile, "Melamine," Schnell Publishing Co., July 1, 1967.
102. Schwartzmann, M., U.S. Patent 3,498,982 (to B.A.S.F.).
103. *Hydrocarbon Process.*, **48**, No. 11, 175 (Nov. 1969).
104. Chemical Profiles, Schnell Publishing Co., Apr. 1, 1968.
105. *Hydrocarbon Process.*, **44**, No. 11, 241 (Nov. 1965).
106. U.S. Patent 2,153,405.
107. Chemical Profile, Schnell Publishing Co., Apr. 1, 1967.
108. U.S. Patent 3,137,730.
109. Chemical Profile, Schnell Publishing Co., July 1, 1967.
110. Campbell, C. R., U.S. Patent 3,017,331 (to Chemstrand).
111. Chemical Profile, Schnell Publishing Co., Apr. 1, 1969.
112. Hatch, L. F., *Petrol. Refiner*, **35**, No. 8, 147–152 (Aug. 1956).
113. "Abundant Nuclear Energy," U.S. Atomic Energy Commission, CFSTI, May 1969.
114. *Nitrogen*, No. 31 (Sept. 1964).
115. U.S. Patent 3,347,621.

6

Miscellaneous Heavy Chemicals

Richard W. Clough*

SODIUM CHLORIDE

Sodium chloride (common salt, NaCl) occurs in nature in almost unlimited quantities. It is the direct source of such sodium compounds as soda ash, caustic soda, and sodium sulfate; and indirectly, through soda ash, it furnishes the sodium for sodium phosphate and many other salts. Moreover, it is the source of chlorine and of hydrochloric acid. It is these uses which come to the chemist's mind when he thinks of salt; but even without these, salt has an imposing list of uses, which places it among the important substances in the economic world. It serves to preserve meat, fish, and hides; it is a necessary article of the diet, and as such appears on every table; it is used in dairies to give temperatures below the ice point. Large quantities are used for ice control on streets in the northern states. Salt enables the soap maker to separate the soap from the glycerin and lye, and the dye manufacturer to precipitate his products. In addition, salt is an important agent for regeneration of water-softening resins.

Salt is used in the form of rock salt, in nearly all the chemical industries,** in the northern part of the United States at any rate. It is cut from a solid-salt deposit by means of a shaft. The shaft mines number 18 in all, distributed as follows: five in Kansas, six in Louisiana, one in Michigan, two in New York, two in Texas, and two in Ohio. In addition, a new salt mine is under construction near Ithaca, New York. One of the operating New York mines is at Retsof, Livingston County; the other is at Myers, Tompkins County. The Michigan mine is near Detroit. The depths of the shafts are 1063 ft (Retsof), 1950 ft (Myers) and 1050 ft (Detroit). At Retsof, the rock is lifted to the mouth of the shaft, crushed and screened to size, without any other operation for purification. The color is a light gray-white, and it is essentially pure (98.5 per cent NaCl).

*Allied Chemical Corporation, Solvay, N.Y.

**Except soda ash and chlorine, for which the brine is better adapted.

TABLE 6.1. United States Salt Production in Short Tons, 1968[a]

Manufactured (evaporated) salt	5,522,000
In brine	23,291,000
Rock salt	12,461,000
Total	41,274,000

[a]"Mineral Industry Surveys," 1968.

The shaft of the Avery Salt Company at Avery, La., is 518 ft deep. The quality of the salt runs 99.4 per cent NaCl. There is one place at Avery Island where salt comes within 10 or 12 ft of the surface.

Salt is more frequently obtained by means of water sent down one pipe, and after becoming saturated, brought up by another pipe concentric with the first (partly by hydrostatic pressure), and then pumped to a refining plant. Such artificial brines permit a cheaper operating cost, and are well adapted to soda ash manufacture and, particularly, to the making of white table salt. Artificial brines are obtained in New York State at Watkins Glen, Silver Springs, Myers, and Tully near Syracuse. Artificial brines are made in Michigan, where natural brines also occur but are of less importance; one field is near Detroit (Wyandotte), and a second one is in the center of the State, near Midland. Kansas has many artificial brine wells. Ohio produces both natural and artificial brines. A number of other states produce artificial brines.

In the dry climate of the Western states, salt is found as an outcrop at the surface;* in some of these states this salt is utilized to some extent.

In southern California, as in Spain and southern France and various other locations, sea water is concentrated in wide basins, by solar evaporation, until the salt deposits; by running off the mother liquors at that point, the bittern magnesium salts are removed. An teresting application of the same method is used at Salt Lake, Utah.

From the brines, whether artificial or natural, a grade of salt suitable for table and dairy

*From the streets of the town of Salina, Sevier County, Utah, outcrops of salt may be seen at the side of the hills; its color is red, and it may be fed to cattle.

TABLE 6.2 Production of Salt by Method of Manufacture, U.S., 1968[a]

	Short Tons	*Value per Ton*
Evaporated salt		
Bulk		
Open pans or grainers	322,000	$26.45
Vacuum pans	2,943,000	23.53
Solar	1,900,000	6.74
Pressed blocks	357,000	25.90
Rock		
Bulk	12,376,000	6.27
Pressed blocks	85,000	27.31
Salt in brine	23,291,000	3.98
TOTAL	41,274,000	$ 6.60

[a]"Mineral Industry Surveys," 1968.

use is made by solar evaporation (in sunny climates), by open-pan evaporation, or by evaporation in vacuum pans, with one pan (single effect) or several pans (double or triple effect); in the latter the steam raised from the salt solution in one pan becomes the heating steam for the next pan. The difficulty in the crystallization of salt solutions is due to the fact that, as salt is about as soluble in cold as in hot water, merely cooling a hot concentrated solution does not form crystals. The water must actually be removed from the hot solution, until the salt drops out for lack of solvent.

The precipitated salt is dried in rotary or fluid bed driers, constructed of Monel metal or 316 stainless steel.

TABLE 6.3 Salt Sold or Used by Producers, U.S., 1968[a]

	Short Tons	*Per Cent of Total*
Louisiana	10,908,000	26.4
Texas	8,534,000	20.7
Ohio	5,713,000	13.8
New York	5,218,000	12.6
Michigan	4,893,000	11.8
California	1,901,000	4.6
West Virginia	1,308,000	3.2
Kansas	1,128,000	2.7
Utah	405,000	1.0
Oklahoma	7,000	0.2
Other States[b]	1,259,000	3.0

[a]Mineral Industry Surveys" (1968).
[b]Ala., Colo., Hawaii, Kan. (brine only); Nev., N.D., Va., N.M.

There were 100 plants, belonging to 55 companies operating in 1968.

The several states in which salt is produced in any form are shown in Table 6.3.

The total world production in 1968 was estimated at over 124,000,000 short tons, of which the United States produced 33 per cent. West Germany produced 7,980,000 short tons of salt in 1969. Other large producers are India, Canada, France, and Italy.

SODA ASH, THE COMMERCIAL SODIUM CARBONATE

Soda ash is made commercially by three different processes:

1. The ammonia-soda process.
2. Recovery from naturally occurring trona ore.
3. Recovery from naturally occurring alkaline brines.

The original Leblanc process for making synthetic soda ash is now obsolete because it could not compete economically with the ammonia-soda process. In the United States, the ammonia-soda process and recovery from trona ore account for 63 and 32 per cent, respectively, of soda ash production capacity.

Solvay Process

The ammonia-soda process is usually called the *Solvay* process, because the successful manufacture, in 1865, is due to Ernest Solvay, a Belgian, who developed and perfected the process. The process is based on the fact that when ammonium bicarbonate is added to a saturated solution of common salt, the ammonium salt dissolves and sodium bicarbonate separates as a solid; if filtered off and calcined, it is changed to soda ash, or sodium carbonate, Na_2CO_3.

$$NH_4HCO_3 + NaCl \rightleftharpoons NaHCO_3 + NH_4Cl$$

$$2NaHCO_3 + \text{heat} \longrightarrow Na_2CO_3 + H_2O + CO_2$$

Application of this simple principle proved difficult, for ammonia is comparatively expensive, and unless it is almost completely recovered, the process cannot sustain itself economically. It was the nearly complete recovery of the ammonia which enabled Solvay to defeat the well-established Leblanc soda process.

In practice, a saturated salt solution is treated with ammonia gas, and this solution is then saturated with carbon dioxide; the resulting suspension of sodium bicarbonate in an ammonium chloride solution is filtered, and the sodium bicarbonate is calcined. The ammonium chloride filtrate is treated with lime and steam to recover the ammonia which is reabsorbed in fresh incoming brine (Fig. 6.1).

The carbon dioxide is obtained by burning limestone, at the same time furnishing the lime necessary for treating the ammonium chloride solution. It is clear that this process requires a large amount of fuel: to burn the limestone, to calcine the sodium bicarbonate, and to raise steam for the ammonium chloride still. For the reaction proper, no fuel is required. In fact, large volumes of cooling water are required to remove the heat generated by the absorption and reaction of ammonia and carbon dioxide.

The ammonia-soda process has one imperfection, in that the chlorine which the common salt furnishes is not recovered, except to a small extent; it is discarded as a solution of calcium chloride.

Plant Location. As a rule, soda ash plants are located near the source of common salt. Thus the Syracuse, New York, plant formerly drew natural brines from the plant site property; and now that a greater supply is needed, artificial brine from Tully is piped to Syracuse, about 20 miles away; the brines flow by gravity. At Wyandotte, Michigan, huge deposits of salt are available. The great Dombasle plant in Lorraine, France, was located there because of the almost limitless supply of salt in the region. Rock salt may be shipped in by rail, which, however, involves extra expense for freight; and rock salt costs more than the salt in brine.

Brine Preparation. The brine must be purified to remove calcium and magnesium salts present in low concentrations. Otherwise, they will be precipitated when the brine is subsequently ammoniated, producing objec-

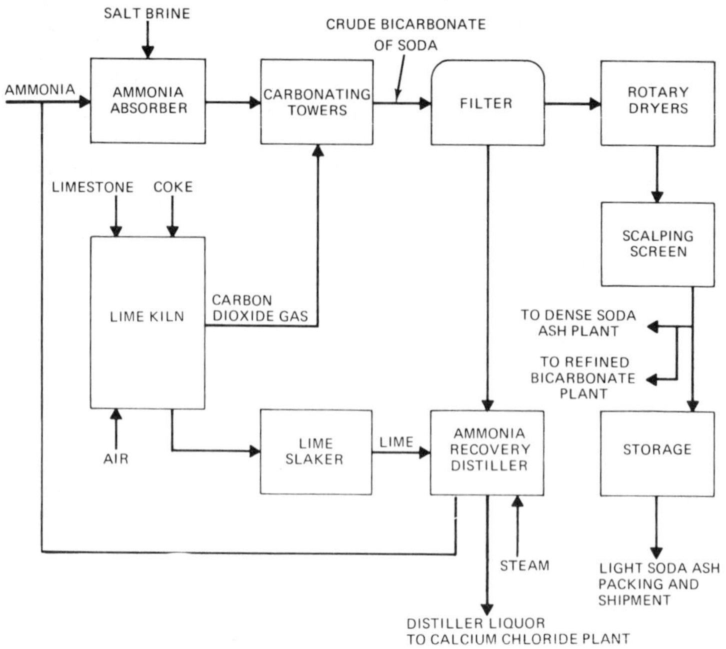

Fig. 6.1 Simplified diagrammatic flow sheet for the Solvay ammonia soda process.

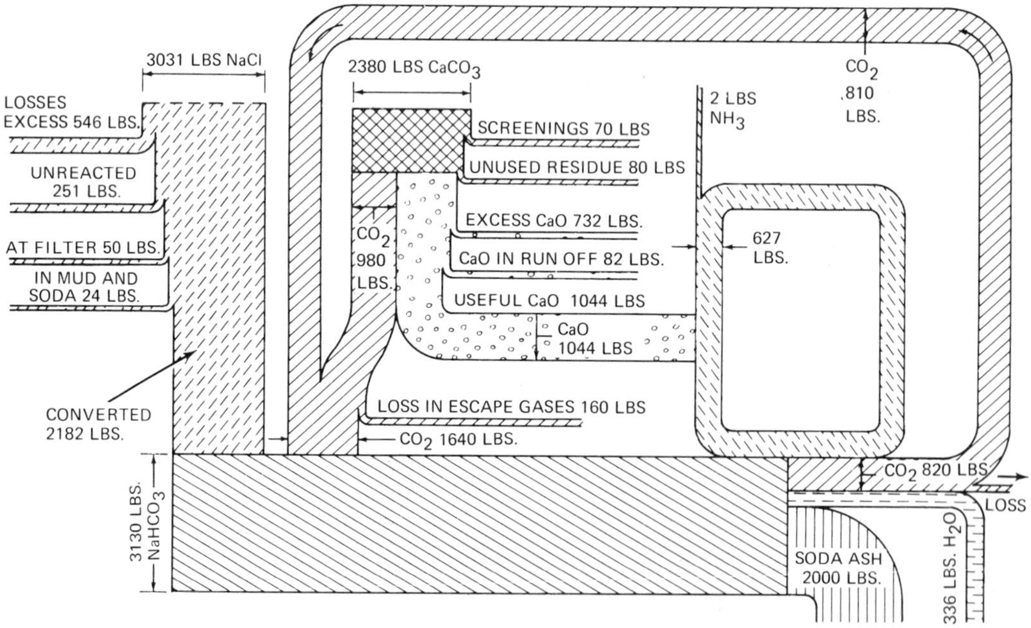

Fig. 6.2 Diagrammatic sketch of the amounts of raw materials, circulated chemicals, products, and discards in the Solvay process for soda ash. The widths of the columns and frames are in proportion to the weight of the materials they represent. (*After Kirchner.*)

tionable deposits on the equipment surfaces and contaminating the product with insoluble material. Soda ash is used to precipitate the calcium as calcium carbonate; and lime or caustic soda is used to precipitate the magnesium as magnesium hydroxide. The reactions are as follows:

$$CaSO_4 + Na_2CO_3 \longrightarrow CaCO_3 + Na_2SO_4$$

$$CaCl_2 + Na_2CO_3 \longrightarrow CaCO_3 + 2NaCl$$

$$Ca(OH)_2 + Na_2CO_3 \rightleftharpoons CaCO_3 + 2NaOH$$

$$MgCl_2 + 2NaOH \longrightarrow Mg(OH)_2 + 2NaCl$$

A slight excess of the precipitating reagents is used to assure complete precipitation. The brine is mixed with the reagents, followed by moderate agitation to flocculate the precipitates, which are then allowed to settle. The clear, purified brine is fed to the ammonia absorbers.

Lime and Carbon Dioxide Generation. Lime for recovering the ammonia from ammonium chloride is produced by burning the best available limestone with foundry coke, or anthracite coal, in vertical shaft kilns. This type of equipment is preferred for producing the maximum yield of active lime and the maximum concentration of carbon dioxide in the kiln gas. The dry lime, drawn from the vertical kiln, is cooled by the entering air, which is thereby preheated. Likewise, for maximum fuel economy, the exiting gas preheats the entering limestone and fuel, while cooling the gas. Gas composition should exceed 40 per cent CO_2 with only fractional percentages of CO and O_2 present.

The primary reactions in the kiln are:

$$C + O_2 + 4N_2 \longrightarrow CO_2 + 4N_2 + \text{heat}$$

$$CaCO_3 + \text{heat} \rightleftharpoons CaO + CO_2$$

$$CO_2 + C \rightleftharpoons 2CO$$

$$MgCO_3 + \text{heat} \longrightarrow MgO + CO_2$$

The dry lime and hot water are fed to a rotating cylindrical slaker to produce milk of lime ($CaO + H_2O \longrightarrow Ca(OH)_2$). The milk of lime is pumped to the distillers to react with ammonium chloride for ammonia recovery.

Ammonia Absorption. The Solvay process recycles large quantities of ammonia; therefore, it is necessary to minimize loss in scrubbing the various ammonia-containing gas streams. Thus, the incoming purified brine is used to wash the ammonia-bearing air which is pulled through the cake of the vacuum filters. This is accomplished in a packed absorption unit. The brine then passes through a second packed section where it absorbs the ammonia brought with the gases leaving the carbonating towers.

Then the partially ammoniated brine exiting the tower washer falls into the main ammonia absorber, which is also packed. The brine is recirculated through water-cooled heat exchangers to remove the heat of absorption of the ammonia. Ammonia, and the carbon dioxide and hydrogen sulfide released in the distiller are absorbed in this section of the column. In addition, a small amount of make-up ammonia is added. Reactions taking place in the absorption train are:

$$NH_3(g) + H_2O \rightleftharpoons NH_4OH$$

$$CO_2(g) + H_2O \rightleftharpoons H_2CO_3$$

$$2NH_4OH(aq) + H_2CO_3 \rightleftharpoons (NH_4)_2CO_3(aq) + 2H_2O$$

$$2NH_4OH(aq) + H_2S \rightleftharpoons (NH_4)_2S(aq) + 2H_2O$$

Analysis of the cooled ammoniated brine is approximately as follows:

Temperature	38°C
NH_3	90 g/l
CO_2	40 g/l
NaCl	260 g/l
H_2S	0.1 g/l

The total heat removed from the absorption of ammonia amounts to 1.25 million Btu/net ton soda ash produced.

Carbonation and Crystallization. Next, the ammoniated brine produced in the absorption system must be carbonated to a point just short of crystallization. The brine is given a final carbonation and cooled to produce crystalline crude sodium bicarbonate. These operations are performed separately in order to optimize the utilization of the heat-transfer surface, and to control the crystal quality of the product.

The equipment used in the carbonation step consists of groups of five identical towers having alternate rings and discs in the upper section to assure mixing of falling liquor and rising gas stream without becoming plugged with crystallizing solid phase. The lower section is made up of a series of heat-exchange-tube bundles alternating with rings and discs.

The ammoniated brine is passed downward through one of the group of five columns which has become fouled with sodium bicarbonate after four days' operation as a crystallizing unit. The 40 per cent CO_2 gas from the kilns is pumped into the bottom of the column to provide agitation and heat to dissolve the crystalline scale and bring the liquor to a composition just short of crystallization. The liquor is adjusted to the desired temperature by passage through a heat exchanger in preparation for feeding to the crystallizing towers. The liquor is then fed into the top of each of the other four columns in the group.

A mixed gas of 60-75 per cent CO_2 derived from mixing 40 per cent CO_2 from the kilns and 90 per cent gas from the calcination of bicarbonate, is fed to the bottom of these crystallizing units. Absorption of CO_2 in the highly alkaline ammoniated brine results in crystallization of crude sodium bicarbonate. Due to the heat evolved in the absorption and neutralization of the carbonic acid gas, and from crystallization of the sodium bicarbonate, the liquor in the column rises from 38°C to a maximum of about 62-64°C. In normal operation temperature of the discharge slurry is maintained at about 27°C by automatic adjustment of the water flow through the cooling tubes. A tower such as that shown in Figure 6.3 has a capacity of 50 tons of finished soda per day.

The reactions involved in carbonation and crystallization are:

$$CO_2(g) + H_2O(l) \rightleftharpoons CO_2(aq) \rightleftharpoons H_2CO$$

$$H_2CO_3 + 2NH_4OH \rightleftharpoons (NH_4)_2CO_3 + 2H_2O$$

$$(NH_4)_2CO_3 + H_2CO_3 \rightleftharpoons 2NH_4HCO_3$$

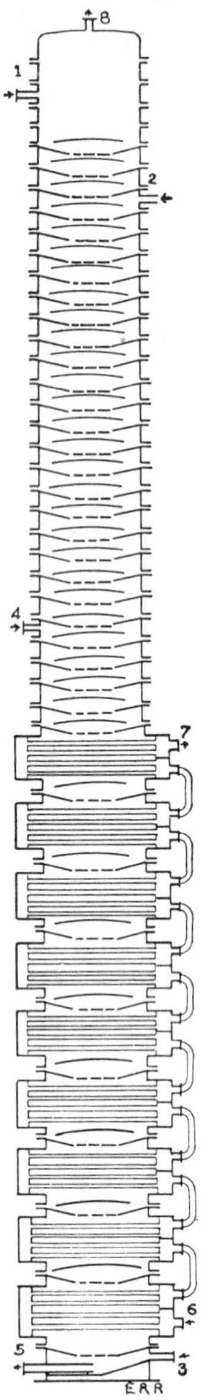

Fig. 6.3 A carbonating tower in the ammonia soda process (Solvay); it is 69 feet high and 6 feet in diameter: (1) entry for ammoniated brine, used when the tower is being cleaned; (2) entry for the ammoniated brine for the regular bicarbonate precipitation; (3) and (4) carbon dioxide entries; (5) outlet for bicarbonate slurries; (6) cooling water inlet; (7) cooling water outlet; (8) escape for uncondensed gases. *(Modeled after Kirschner.)*

$$NaCl(aq) + NH_4HCO_3 \rightleftharpoons NaHCO_3(aq) + NH_4Cl$$

$$NaHCO_3(aq) \rightleftharpoons NaHCO_3(s)$$

The heat removed from the carbonator liquor is about 0.26 million Btu/net ton soda ash while that removed from the crystallizing units is 1.25 million Btu/net ton soda ash.

The presence of sulfide in the liquid feed to the column serves to maintain a protective iron sulfide film on the cast iron equipment, which minimizes contamination of the product crystals with iron.

Filtration. Slurry drawn from the crystallization columns flows to the basins of the rotary vacuum filters. Agitators in the basins prevent settling of the slurry. The filter consists of a rotating drum with filter cloth covering the slotted surface of the drum. The slurry is picked up from the basin, in which the drum is only partially immersed, using an axially applied vacuum.

The crystals are washed with water from a weir across the face of the filter drum to remove the chloride-rich mother liquor. Air drawn through the cake, which is compressed by compaction rolls turning on the surface of the drum, displaces much of the residual water and liquid phase. The liquid phase flows to the distillation unit for recovery of ammonia. The cake of crude bicarbonate of soda containing approximately 3 per cent NH_3HCO_3 and 12–15 per cent free water is removed with a cutoff knife and falls to the calciner-feed floor.

Distillation. To recover ammonia, the filter liquor is pumped to the ammonia-still preheater. At this point the sulfide solution required for corrosion protection is usually added. The preheater not only heats the feed to the still, but cools the gas from the still and serves to strip part of the CO_2 from the filter liquor. Constructionwise, the preheater is made up of a series of heat-exchange-tube bundles stacked one above the other. Liquor passes through the tube bundles in series and is heated by rising vapors in the shell. At the exit of each heater is a separator from which CO_2, evolved as a result of heating, is separated from the liquid before it enters the next lower exchanger. The gas, being essentially CO_2 with some ammonia, a trace of H_2S, and saturated with water vapor is vented to the gas cooler at the dryers, which will be described later.

The liquor exiting the separator of the bottom preheater box empties into the next lower section of the distillation unit. This is a packed column and serves to strip off the residual CO_2 from the distiller feed before it comes into contact with lime. Any CO_2 remaining in the solution consumes lime for no useful purpose. The stripping is done by the rising steam and ammonia vapor. The reactions taking place in the preheater and stripper are represented by the equations:

$$NH_4HCO_3 + H_2O \rightleftharpoons NH_4OH + H_2CO_3$$

$$H_2CO_3 \rightleftharpoons CO_2 + H_2O$$

$$NH_4HS + H_2O \rightleftharpoons NH_4OH + H_2S$$

The hot liquor from the stripper is then conveyed to a separate agitated vessel known as the pre-limer, into which is also added the milk of lime produced at the lime kiln section. The calcium hydroxide, or milk of lime, reacts with the ammonium chloride of the filter liquor to produce free ammonia and calcium chloride.

$$Ca(OH)_2 + 2NH_4Cl \longrightarrow 2NH_4OH + CaCl_2$$

$$NH_4OH \rightleftharpoons NH_3 + H_2O$$

Liquor from the pre-limer containing calcium chloride, sodium chloride, ammonium hydroxide, and some excess calcium hydroxide, flows to the top of the last, or distillation, section of the distiller. Steam is injected at the bottom of the unit. The steam strips ammonia from this solution to a residual level of about 0.001 per cent. The rising vapors, as previously stated, rise through the packed section above, where they strip CO_2 from the downflowing liquor. The distillation section is constructed with bubble cap-type plates.

Discharge liquor from the distiller is allowed to flash in a separate chamber, or heat trap, before being discarded. The vapor recovered from the flash is used in a separate vacuum distillation unit to recover ammonia from various liquors which do not contain ammonium chloride and therefore do not require any lime. These liquors are primarily the condensates resulting from cooling the vapors exiting from the lime stills and dryers.

To avoid unwanted dilution, the vapor phase exiting from the preheater is further cooled to 58°C in water-cooled heat exchangers to condense water from the gas before absorption of the ammonia in the brine. Excessive cooling must be avoided or the ammonium carbonate will condense and plug the coolers and gas mains leading to the absorber.

Gas Compression. Before proceeding to a discussion of the drying operation reference should be made to the gas compressors which pump the kiln gas and mixed gas to the carbonation and crystallization columns. These are usually steam driven reciprocating compressors. Some of the compressors pump washed kiln gas directly to the carbonating columns, while the others operate on a balanced suction system to draw gas from the dryers and kilns simultaneously to supply the crystallizing columns. Steam exhausted from the drivers on the compressors is used in the ammonia stills described above.

Drying. Crude bicarbonate of soda is calcined by indirect heating to permit recovery of decomposition gases at high concentration and efficiency. The solid product from the filtration operation is blended with hot recycle product from the calciners and fed continuously to the dryers. The dry recycle product absorbs and evaporates some of the free water of the filter cake to prevent scaling of the hot dryer surface. Gases evolved, which are water vapor, carbon dioxide, and ammonia, are drawn by suction of the compressors through dry and wet separators to remove entrained dust. They are then cooled by contact with water-cooled tubes. The CO_2 rich gas which is vented from the distillation preheating system is blended with dryer gas in this unit. Much of the water vapor is condensed, and ammonia is absorbed here. The gas is then contacted directly with fresh water for cooling and recovery of residual ammonia before recycling to the carbonation system.

Two types of dryers or calciners are used to convert the crude bicarbonate to soda ash. The steam-tube dryer is an axially rotating cylinder with concentric banks of tubes into which high pressure steam is admitted. The indirectly fired dryer is a rotating steel cylinder mounted in a brick furnace fired by gas, oil, or coal with the combustion gases circulating around the dryer. As the process gases released inside either type of dryer must be recovered, the dryers are constructed with rotating seals and gas-tight feed and discharge mechanisms. The decomposition of the crude bicarbonate may be represented by the following equations:

$$2NaHCO_3(s) \longrightarrow Na_2CO_3(s) + CO_2(g) + H_2O(g)$$

$$NH_4HCO_3(s) \longrightarrow NH_3(g) + H_2O(g) + CO_2(g)$$

$$H_2O(l) \longrightarrow H_2O(g)$$

The theoretical heat requirement for the above reactions amounts to about 2.0 million Btu/net ton soda ash. Product from the dryers is cooled for shipment, or converted to dense ash or other products.

An analysis of the composition of good commercial light soda ash made by the ammonia-soda process follows:

Na_2CO_3	99.70 per cent
NaCl	0.12 per cent
H_2O	0.12 per cent
Fe	15 ppm
Ca and Mg	75 ppm

Waste Disposal. The discharge liquor from the distillation heat traps containing calcium chloride, unconverted sodium chloride, a slight excess of lime, and the impurities introduced with the milk of lime may be used as a source of merchant calcium chloride and sodium chloride. The major portion, however, is pumped to waste ponds or settling basins where the solids are allowed to settle and accumulate and the supernatant liquor is decanted and run off into water courses having natural flow sufficient to provide the necessary dilution.

Soda Ash from Other Sources

Production of soda ash from naturally occurring trona ore in the Green River area of Wyoming is rapidly expanding due to favorable production costs compared with the ammonia-soda process. Two basic processes are employed. The original process, started in 1953, involves dry mining of trona ore from a depth of about 1500 feet, crushing and dissolving the ore at the surface. After purification of the solution by settling and filtration, sodium sesquicarbonate crystals are formed by evaporative crystallization. The crystals are separated from the slurry with centrifuges and calcined to soda ash in steam-tube dryers according to the reaction:

$$2Na_2CO_3 \cdot NaHCO_3 \cdot 2H_2O(s) \longrightarrow 3Na_2CO_3(s) + CO_2(g) + 5H_2O(g).$$

In a more recent process, the trona ore is also dry mined and crushed at the surface, but the sesquicarbonate in the crushed ore is immediately calcined to soda ash in direct gas-fired rotary units. The calcined ore is then dissolved, purified by settling and filtration, and the solution fed to evaporative crystallizers where sodium carbonate monohydrate is precipitated. The crystals are centrifuged and converted to anhydrous dense soda ash in steam-tube dryers according to the reaction:

$$Na_2CO_3 \cdot H_2O(s) \longrightarrow Na_2CO_3(s) + H_2O(g)$$

Relatively small quantities of soda ash are produced from natural alkaline brines at Searles Lake, California, by a complex fractional crystallization process which also produces other sodium and potassium salts.

The production of soda ash in the United States is given in Table 6.4; and the distribution of soda ash for 1969 is given in Table 6.5.

TABLE 6.4 Production of Soda Ash in the United States, 1969[a]

	Short Tons
Synthetic soda ash	4,502,839
Natural soda ash	2,494,690

[a]"Current Industrial Reports, Inorganic Chemicals," U.S. Dept. of Commerce, Bureau of the Census, 1970.

TABLE 6.5 Distribution of Soda Ash in the United States, 1969[a]

Consuming Industries	*Per Cent*
Glass	47
Chemical manufacturing	23
Pulp and paper	9
Alkaline cleaners	5
Municipal water treatment	3
Exports	4
Miscellaneous	9

[a]"Chemical Profile," Schnell Publishing Co., Inc., Oct. 1, 1969; (by permission).

The 1970 price of soda ash in carload lots in the East was $33 per ton in bulk and $43 per ton in 100-lb paper bags. Prices in the West were $2 per ton lower.

SODA ASH-RELATED PRODUCTS

Most manufacturers convert a large portion of the light ash produced by the ammonia-soda process to dense soda ash because this is the form preferred by the major consumer, the glass industry. Dense ash has about twice the bulk density of light ash, 65 lbs/cu ft versus 35 lbs/cu ft, and is much coarser in granulation, approximating that of the sand used in the glass batch, thus giving uniform mixing. Dense ash is manufactured by hydrating light soda ash, then calcining the monohydrate back to anyhydrous sodium carbonate.

Some light ash manufacturers convert a portion of their production to refined sodium bicarbonate, or baking soda. This is accomplished by carbonating a soda ash solution to crystallize the bicarbonate according to the reaction:

$$Na_2CO_3 + H_2O + CO_2 \longrightarrow 2NaHCO_3$$

The bicarbonate crystals are centrifuged and dried at a low temperature to avoid decomposition by reversal of the above reaction. The resulting food product is much purer than the soda ash from which it was formed with respect to salt, ammonia, and iron content. Annually, about 150,000 tons of refined sodium bicarbonate are manufactured in the United States.

The waste from the distillers in the ammonia-soda process is a source of salable calcium chloride and salt. Clarified waste is fed to multiple-effect evaporators. During concentration, sodium chloride precipitates out. It is filtered off, washed, and dried for sale. The remaining calcium chloride solution is either sold as liquor, or concentrated further and sold in flake form. Major uses are for road ice and dust control, and for refrigeration brines. Estimated production of calcium chloride from soda ash waste and natural brines in 1970 was 800,000 tons (as 100 per cent $CaCl_2$).

SODIUM SULFATE

Sodium sulfate is generally sold in three different grades. Anhydrous sodium sulfate is the technical grade which generally exceeds 99 per cent Na_2SO_4 in purity. Glauber's salt, $Na_2SO_4 \cdot 10H_2O$, was important in the past in textile production, but it is now produced in very small quantities. In 1961, the last year for which production figures for Glauber's salt were reported, it accounted for less than 6 per cent of the total sodium sulfate produced. The largest-volume grade of sodium sulfate is known as salt cake. Salt cake is the trade name which in the past referred to the co-product of the reaction between sulfuric acid and sodium chloride in the Mannheim furnace, the other product being hydrochloric acid. Due to impurities, salt cake contains considerably less than 99 per cent Na_2SO_4. In 1967, anhydrous sodium sulfate accounted for 30 per cent of the total Na_2SO_4 production and salt cake accounted for most of the balance.

Production of sodium sulfate by manufacturing process is shown in Table 6.6.

TABLE 6.6 Sodium Sulfate, Production by Process, 1968[a]

Process	*Short Tons*	*Per Cent*
Mannheim, Hargreaves	165,016	11.1
Natural	660,227	44.5
Viscose rayon	403,733	27.2
Sodium dichromate, phenol, boric acid, formic acid, and other	253,706	17.2

[a]"Current Industrial Reports, Inorganic Chemicals and Gases," U.S. Department of Commerce, Bureau of the Census, 1968.

Salt Cake

Salt cake is produced in the Mannheim process according to the following reactions:

$$NaCl + H_2SO_4 \longrightarrow NaHSO_4 + HCl$$

$$NaHSO_4 + NaCl \longrightarrow Na_2SO_4 + HCl$$

The mechanical furnace may be fed with salt and sulfuric acid directly, and the mass plowed and heated until the reaction to sulfate is complete. Alternatively, niter cake ($NaHSO_4$), obtained from the manufacture of nitric acid by treatment of sodium nitrate with sulfuric acid, may be mixed with the salt and fed to the furnace. The formation of salt cake goes to near completion at 650°C.

The Mannheim furnace consists of a circular muffle of cast iron, 12 feet in diameter, with dish-shaped bottom and top; the inner height at the circumference is 20 inches, in the center 40 inches. A shaft penetrates it from below and carries four arms, each of which carries two cast-iron plows. The shaft is rotated slowly, 1 revolution in 2 min, by the gear and pinion indicated in Fig. 6.4. The mixed salt and niter cake is fed in from the top near the center, and is moved to the circumference by the plows. Plow 8 is wider than the others, and discharges the burned cake (now salt cake) through the opening of the chute. Here the cake accumulates to some extent and is removed in small trucks to the storage bins. The discharge cake is yellow and turns white on cooling.

Some of the details of construction of the muffle will be plain from the illustration; the bottom and top are single castings. The sides consist of twelve curved castings which, when assembled, provide three doors; one of these is over the discharge opening and chute. Each plow differs in length of shank; and each is slightly turned so that the cake is swept outward. The heat is furnished by a small fireplace. The fire gases enter over the muffle, heating the top; then travel to the underside by a passage in the brickwork and heat the lower side; from here they pass out to the chimney. The temperature is registered by a platinum resistance pyrometer placed in a protecting cast-iron tube reaching into the center of the muffle from above.

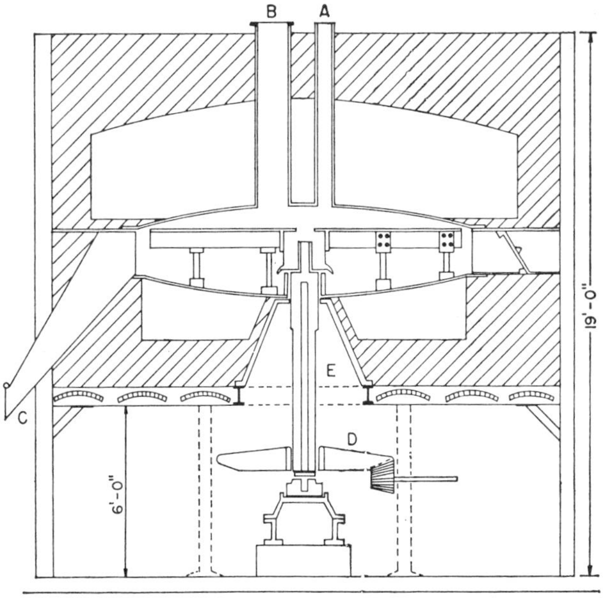

Fig. 6.4 Mannheim mechanical salt-cake furnace, with hydrogen chloride as by-product. The charge enters at *A* and is discharged at *C*; the gases pass out at *B*. The eight plows are rotated from below by gear *D*; the shaft may be water-cooled at *E*.

A well-burned salt cake should contain 2 per cent or less of sodium chloride, and 1.5 per cent or less of H_2SO_4. In the 12-foot diameter muffle, six tons of good cake may be produced per day.

In 1968 three U.S. plants manufactured salt cake by the Mannheim process. A single U.S. plant used the *Hargreaves* process:

$$4NaCl + 2SO_2 + 2H_2O + O_2 \longrightarrow 2Na_2SO_4 + 4HCl$$

This is carried out by passing hot sulfur dioxide and air over salt lumps in vertical steel chambers with false bottoms. Several chambers are run in series with countercurrent gas flow. The complete cycle for reaction, cleaning, and repacking is three days. A large amount of manual labor is required.

The production of sodium sulfate from natural brine is steadily increasing.[1] At Searles Lake, California, two major producers recover sodium sulfate from the lake brine. In one process, the brine is evaporated until sodium chloride and a double salt of sodium carbonate and sodium sulfate crystallize out. The mixed salts are separated from the liquor and the double salt separated from the sodium chloride in hydraulic classifiers. The double salt is treated in several leaching and washing steps and finally redissolved in water. Lithium salts are removed and the solution is cooled to crystallize Glauber's salt. Anhydrous sodium sulfate is recovered by mixing the Glauber's salt with sodium chloride. This causes Glauber's salt to dissolve and allows the anhydrous sodium sulfate to precipitate because of its lowered solubility in the presence of sodium chloride.

In another process practiced at Searles Lake, the brine is first carbonated and chilled to remove sodium carbonate and borax. Further chilling causes crystallization of Glauber's salt and some remaining borax. The coarse crystals of Glauber's salt are separated from the fine borax crystals in a hydraulic classifier. They are filtered off, washed, dried, and fed to a melter where hot sodium sulfate solution is added. Here, Glauber's salt melts in its own water of crystallization. The concentrated solution is evaporated to crystallize anhydrous sodium sulfate.

In Texas, natural brines are processed by chilling to precipitate Glauber's salt, which is filtered, washed, and melted. Then the water of crystallization is evaporated with submerged combustion equipment to give anhydrous sodium sulfate.

A large amount of sodium sulfate is manufactured as a by-product to other materials, chief among which is *viscose rayon*. In the viscose process, sodium sulfate is formed by the reaction between sulfuric acid and cellulose xanthate plus free caustic soda in the spin bath. Spin-bath liquor is evaporated to crystallize Glauber's salt, which is centrifuged off and melted; and the melt is concentrated to precipitate anhydrous sodium sulfate for sale.

In the manufacture of *sodium dichromate*, by-product sodium sulfate crystallizes in the anhydrous state when sulfuric acid is used to convert sodium chromate to dichromate in a boiling hot solution, according to the reaction:

$$2Na_2CrO_4 + H_2SO_4 + H_2O \longrightarrow Na_2Cr_2O_7 \cdot 2H_2O + Na_2SO_4$$

The production of many organic and inorganic chemicals involves the use of sulfuric acid to react with a sodium salt or to neutralize free caustic soda. These processes yield sodium sulfate which may be recovered as a salable by-product.

The uses of sodium sulfate are indicated in Table 6.7.

TABLE 6.7 End Use Distribution for Sodium Sulfate[a]

	Per Cent
Kraft paper pulp	74
Detergents	16
Glass and other	10

[a]"Chemical Profile," Schnell Publishing Co., Inc., July 1, 1968 (by permission).

In 1970, the price of salt cake in bulk shipments was quoted at $28.00 per ton in the East. Technical grade sodium sulfate in bulk was quoted at $34.00 per ton in the East. The prices of both grades were $9.50 per ton lower in the West.

GLAUBER SALT

Glauber salt ($Na_2SO_4 \cdot 10H_2O$) is a purified salt cake; the Glauber plant is usually adjacent to the salt-cake storage bin.

Salt cake is dissolved in hot water, preferably in a circular wooden tank with stirrer. Steam is passed in during the dissolution to make up for cooling to the air. The solution is made as strong as possible (32° Bé hot), lime is added to neutralize the sulfuric acid which is invariably present, and to precipitate iron hydroxide and alumina. The liquor is allowed to settle, and the clear portion is run into the crystallizers. The muddy bottom is filter-pressed, and the filtrate sent to the crystallizers.

The latter are usually wooden forms lined with lead, 15 ft long, 6 ft wide and 2 ft deep. On standing overnight, crystals form; the mother liquor is then run off by removing a wooden plug from the outlet in the bottom of the crystallizer, and the crystals are shoveled into low trucks on wheels; these are pushed to one of several openings in the floor, through which the crystals are dumped into the storage bin and shipping room below. The mother liquor is collected in a low tank and pumped into the dissolver, where, after being heated it replaces part of the water.

A typical analysis of Glauber salt is:

	Per Cent
$Na_2SO_4 \cdot 10H_2O$	97.52
NaCl	0.21
Moisture	2.23
$Fe_2(SO_4)_3$	0.01
$CaSO_4$	0.022
Free acid	0.008

The manufacture of Glauber salt is interesting not only because of the high purity obtained, but also because, by making the first hot solution strong enough, no concentration step of any kind is needed. This principle is followed in the chemical industry whenever possible.

HYDROCHLORIC ACID

Hydrochloric acid is made (1) from common salt, in salt-cake furnaces and by the Hargreaves process; (2) by burning electrolytically produced chlorine in excess hydrogen; and (3) as a by-product from the chlorination of hydrocarbons such as pentane and benzene ($C_6H_6 + Cl_2 \longrightarrow C_6H_5Cl + HCl$). Table 6.8 gives the production of hydrochloric acid by method of manufacture.

TABLE 6.8 Production of Hydrochloric Acid, 1969[a]

Source	*Short Tons (100% HCl)*	*Per Cent*
Salt	134,674	7.3
Chlorine and hydrogen	123,852	6.7
By-product and other	1,590,055	86.0

[a]"Current Industrial Reports, Inorganic Chemicals," U.S. Dept. of Commerce, Bureau of the Census, 1970.

Hydrogen chloride is a gas; on cooling to room temperature, it does not condense to a liquid, as does nitric acid, but must be dissolved in water. The ordinary commercial strength is 20°Bé, at 60°F (15.5°C), containing 32.46 per cent HCl. The system for absorption is essentially the same, whichever method is used for the production of hydrogen chloride. The basic differences lie in the method employed to purify the HCl vapor in preparation for absorption.

The hydrogen chloride generated in salt cake furnaces must first have sulfuric acid mist and sodium sulfate particles removed. This is accomplished by first passing the gas through a series of S-shaped coolers made of Karbate.®* These units have external water cooling. The cooled gas is then passed upward through a packed tower to remove suspended particles.

After leaving the chlorinator, by-product HCl from the chlorination of hydrocarbons is scrubbed in packed towers which absorb the hydrocarbon vapors. The scrubbing medium is usually the hydrocarbon which is circulated in the scrubbing towers before being sent to the chlorinators. A refrigerated cooling system is frequently employed to maintain a low temperature in the circulating scrubbing liquor, thus maximizing absorption efficiency.

Synthetic acid generated by burning chlorine in hydrogen requires no purification prior to absorption.

The absorption system usually consists of two or three towers in series. Because of the very high solubility of HCl in water and the rapidity with which the reaction with water occurs, the absorption rate of HCl is controlled by its concentration in the gas phase. With low concentrations of inerts in the gas fed to the first absorber, a very simple absorption device may be used. This is frequently a falling-film type of absorber in which the strong HCl vapor and weak acid solution (about 10 per cent HCl) from the second tower meet at the top and pass downward on the inside of Karbate tubes. Cooling water is passed through a shell surrounding the tubes. The product, 20° Bé acid (32 per cent), is discharged to storage at the bottom together with unabsorbed vapors which pass to the bottom of the second tower.

The second, or tails, tower is a ceramic-packed unit which provides adequate contact surface for the absorption of HCl vapor from the less concentrated gas stream. The tower shell is usually Karbate. Fresh water or weak acid is added to the top of the tower. The degree of absorption of HCl from the vapor depends on the concentration of inerts in the gas stream. With low inerts, HCl absorption is complete. With gas generated by a process giving relatively high dilution with inerts, such as by-product HCl from an organic chlorination using low-test "tail" gas from the chlorine plant, a third tower is required to absorb the last traces of HCl before discharging the gas to the atmosphere. This is also a ceramic-packed tower fed at the top with water. The weak acid discharged from the last tower becomes feed for the second tower.

The heat generated in the absorption of HCl in water is considerable; it amounts to about

*Trade-mark of National Carbon Company for impervious carbon and graphite products.

700 Btu per pound of HCl absorbed. In the absorption process described above, all of the heat is removed by water cooling on the first, or "absorber-cooler," tower.

Acid made by burning chlorine in hydrogen, with absorption in towers with iron-free packing, is water-white and essentially chemically pure. It more than meets the requirements for the chemists' **C.P.** acid. The price of **C.P.** acid in carboys was quoted at 17¼¢ per lb ($345/ton) in 1970, compared with $35/ton for 20° Bé commercial acid in tank cars.

Anhydrous hydrogen chloride is manufactured by passing hot gasses from the hydrogen-chlorine burner through anhydrous calcium chloride, which absorbs any water vapor present. The dry hydrogen chloride is then compressed and loaded into steel cylinders, special tank trucks, or tank cars. Anhydrous hydrogen chloride was quoted at 39¢ per lb in 50 lb cylinders and $110/ton in tank cars in 1970.

The uses of hydrochloric acid include steel pickling (cleaning); the manufacture of dyes, phenol, and plastics; glucose and food processing; oil well acidizing; and a number of miscellaneous purposes requiring a strong and easily neutralized inorganic acid.

SODIUM SILICATE

Sodium silicate is made by fusing sand and soda ash in a furnace at about 1300°C, according to the reaction:

$$Na_2CO_3 + nSiO_2 \longrightarrow Na_2O \cdot nSiO_2 + CO_2$$

Sodium silicate is commonly called *water glass*, because, when solid, it actually is a glass, and because, unlike lime-soda glass (ordinary window glass) it is soluble in water. The melting is performed in large tank furnaces similar to window-glass furnaces. The materials are introduced in batches, at intervals; the product may be drawn off continuously or periodically. A mixture of sodium sulfate and coal may replace part of the soda ash.

As the melt leaves the furnace, a stream of cold water shatters it into fragments; these are dissolved by means of superheated steam in tall, rather narrow steel cylinders with false bottoms,[2] and the resulting liquor is clarified.[3] Sodium silicates are sold in solutions which vary from the most viscous, 69° Bé, to thinner ones, reaching finally 22° Bé solutions, suitable for paints. The dry material in the form of a powder is also on the market. This is made by forcing the thick liquor through a very fine opening into a chamber which is swept by a rapid current of cold air, which carries off the moisture.[4] Because sodium silicate is hygroscopic, powdered sodium sulfate is sometimes incorporated with the solid silicate[5] to prevent caking. In composition, the sodium silicates may vary from $2Na_2O \cdot SiO_2$ (sodium orthosilicate) to $Na_2O \cdot 4SiO_2$. In sodium sesquisilicate the ratio of Na_2O to SiO_2 is 1.5 to 1. The compound with the composition $Na_2O \cdot SiO_2$ is called sodium metasilicate. Silicates that have a ratio of Na_2O to SiO_2 of 1 to 1.6 up to 4 are known as colloidal silicates. A large number of the colloidal silicates are sold as water solutions. Each of the commercial ratios results in a liquid with special properties, rendering it the proper selection for a specific purpose.

The uses of sodium silicate are surprisingly numerous as shown in Table 6.9.

TABLE 6.9 Uses of Sodium Silicate, 1970[a]

	Per Cent
Catalysts and gels	35
Soaps and detergents	15
Boxboard adhesives	13
Pigments	12
Water, paper, and ore treatment	6
Other	19

[a] "Chemical Profiles," Schnell Publishing Co., Inc., Jan. 1, 1970 (by permission).

In 1968 production was 813,000 tons (on the anhydrous basis).[6] Sodium silicate as 40° Bé solution, bulk lots, was $30 a ton in 1970.

BROMINE AND BROMIDES

Bromine is obtained from sea water, natural brines, and oil field brines. In 1969, the production of bromine and bromine compounds by the nation's primary producers was 386.3 million pounds, valued at $93 million. The U.S. bromine-producing industry consisted of eight companies operating 12 plants in four states. Bromine was extracted from natural brines at five plants in Michigan and at one plant in California, from oil-field brines at five plants in Arkansas, and from seawater at one plant in Texas. Michigan was the major producing State, followed by Arkansas, Texas, and California.[7]

The natural brines of Michigan contain about 1300 ppm bromine. The brine is preheated to about 90°C and fed to a packed chlorinator tower. The partially chlorinated brine is next fed to the top of a "steaming-out" tower, where steam and the balance of the necessary chlorine are injected at the bottom. Bromine is liberated according to the reaction:

$$CaBr_2 + Cl_2 \longrightarrow Br_2 + CaCl_2$$

and stripped out of the brine by the steam. The debrominated brine may be processed for recovery of calcium chloride.

The vapors from the top of the steaming-out tower, containing bromine, a little chlorine, and steam are condensed and flow to a decanter. The mixture separates into a lower layer of crude bromine and an upper layer of bromine-water solution which is returned to the steaming-out tower. The crude bromine is purified by distillation which removes most of the residual chlorine. Purity of the bromine is about 99.8 per cent.

Bromine in sea water occurs in concentrations of 60 to 70 ppm. Because of the large volumes of sea water which must be processed (about 4 million gallons per ton of bromine produced), air is a more economical blowing-out agent than steam. The sea water is acidified to a pH of about 3.5 to minimize chlorine consumption. After chlorination to liberate bromine by the reaction

$$2NaBr + Cl_2 \longrightarrow Br_2 + 2NaCl$$

the sea water is fed to the top of an air-blow tower where it is contacted countercurrently by a large volume of air which is introduced by a blower at the base of the tower. The free bromine is stripped out by the air, and the debrominated sea water is returned to the sea.

The bromine-containing air stream is scrubbed with sulfurous acid in a packed tower where the following reaction occurs:

$$Br_2 + H_2SO_3 + H_2O \longrightarrow 2HBr + H_2SO_4$$

The acid solution is circulated in the tower to build up the concentration of hydrobromic acid. A portion of the concentrated solution is continuously withdrawn and fed to a reactor to which chlorine is added to free the bromine by the reaction

$$2HBr + Cl_2 \longrightarrow Br_2 + 2HCl$$

Steam is added to the reactor to strip out the free bromine. The bromine and water vapors are condensed and collected in a decanter. The lower layer of crude liquid bromine is drawn off and purified by distillation. The upper layer of water which is saturated with bromine is returned to the reactor. The sulfuric-hydrochloric acid solution from the bottom of the reactor is sent to the beginning of the process to supply part of the acid required to treat the incoming sea water.

Alkali bromides are made by treating iron borings in water with bromine to give iron bromide according to the reactions:

$$Fe + Br_2 \longrightarrow FeBr_2$$

$$3FeBr_2 + Br_2 \longrightarrow Fe_3Br_8$$

The iron bromide is then reacted with an alkali carbonate to give the desired alkali bromide, e.g.,

$$Fe_3Br_8 + 4Na_2CO_3 + nH_2O \longrightarrow 8NaBr + 4CO_2 + Fe_3O_4 \cdot nH_2O$$

The iron hydroxides are filtered off, and the filtrate evaporated to crystallize the product bromide.

Table 6.10 gives the uses of bromine.

TABLE 6.10 Uses of Bromine, 1970[a]

	Per Cent
Ethylene dibromide	73
Elemental bromine and bromine compounds	22
Methyl bromide	5

[a]"Chemical Profile," Schnell Publishing Co., Inc., Oct. 1, 1970 (by permission).

Ethylene dibromide is a constituent of gasoline antiknock fluids. This compound prevents lead oxide deposits which would otherwise result from tetraethyllead. The future of bromine production will be significantly affected by current activity toward reduction of lead gasoline additives for air-pollution control reasons.

Ethylene dibromide and methyl bromide are used as soil and seed fumigants. Methyl bromide is also used as an insecticide, rodenticide, and methylating agent. The alkali bromides are used in photography and as sedatives in the pharmaceutical industry. Bromine is employed as a laboratory reagent, as a disinfectant, and in the preparation of dyes. The price of bromine was quoted at 27¢ per lb in drums and 17¢ per lb in tank cars in 1970.

SODIUM SULFIDES

Sodium sulfide (Na_2S) and sodium hydrosulfide (NaHS) will be discussed together, since their manufacturing processes are related and they are similar in their industrial uses. Production of these two compounds is given in Table 6.11.

TABLE 6.11 Production of Sodium Sulfides, 1968[a]

Product	*Short Tons*
Sodium sulfide (100% Na_2S)	116,502
Sodium hydrosulfide (100% NaHS)	35,307

[a]"Current Industrial Report, Inorganic Chemicals and Gases," U.S. Dept. of Commerce, Bureau of Census, 1968.

Sodium sulfide was originally made on a commercial scale by the reduction of salt cake with coal in reverberatory furnaces. This method has been replaced by newer processes which give the advantage of continuous operation, less corrosive conditions, and a purer product.

Chief among these processes is the saturation of caustic soda with hydrogen sulfide, which is manufactured by combining hydrogen and sulfur catalytically, or is obtained from by-product sources.

$$H_2S + NaOH \longrightarrow NaHS + H_2O$$

The sodium hydrosulfide solution is filtered to remove the sulfides of heavy metals introduced with the caustic. These include iron, nickel, copper, manganese, and mercury. The clear filtrate may be sold as a 44–46 per cent sodium hydrosulfide solution, or it may be evaporated in stainless steel equipment to crystallize a solid hydrate containing 70–72 per cent NaHS which is sold as a flake product.

Sodium hydrosulfide is converted to sodium sulfide by reaction with caustic soda.

$$NaHS + NaOH \longrightarrow Na_2S + H_2O$$

By using a sodium hydrosulfide solution of the proper concentration with flake caustic soda, a hydrated product of 60–62 per cent sodium sulfide is obtained which is sold as flake or fused solid in drums. When the raw materials comprise catalytically produced hydrogen sulfide and mercury-cell caustic soda, this process results in a very high quality sodium sulfide, suitable for use in dyes, rayon, photography, and leather unhairing.

Sodium sulfide and sulfhydrate of somewhat lower quality can be obtained by utilizing hydrogen sulfide produced as a by-product in the manufacture of carbon disulfide from methane or another low molecular weight hydrocarbon, e.g.,

$$CH_4 + 4S \longrightarrow CS_2 + 2H_2S$$

The gas mixture from this catalytic reaction is cooled and scrubbed with caustic soda solution to produce liquid carbon disulfide and a sodium hydrosulfide solution. This solution is filtered and concentrated for sale, or converted to sodium sulfide by reaction with another mole of NaOH.

The sulfides produced by this process are somewhat contaminated by the mercaptans produced in the reaction of the hydrocarbon with sulfur.

$$CH_4 + S \longrightarrow CH_3SH \text{ (methyl mercaptan)}$$

Since the mercaptans are extremely odorous, the uses of the product sulfide are limited to applications where the odor is not objectionable.

TABLE 6.12 Estimated Uses of Sodium Sulfide[a]

Use	*Per Cent*
Leather unhairing	20
Colors and dyes	18
Chemicals, miscellaneous	13
Rayon and film	13
Metals and minerals	12
Pulp and paper	5
Export	19

[a]Kirk & Othmer, "Encyclopedia of Chemical Technology," Vol. 18, 2nd Ed., 1969 (by permission).

Sodium sulfide is produced as a by-product in the manufacture of barium carbonate. Barite is roasted with coal and the kiln product leached with water to give barium sulfide.

$$BaSO_4 + 4C \longrightarrow BaS + 4CO$$

Treatment of the barium sulfide solution with soda ash produces a precipitate of barium carbonate and a sodium sulfide solution:

$$BaS + Na_2CO_3 \longrightarrow BaCO_3 + Na_2S$$

The sulfide solution is filtered and concentrated for sale.

Sodium sulfide and hydrosulfide are used extensively to remove hair from leather hides before tanning. They are used in the manufacture of sulfur dyes, kraft pulp, rayon and cellophane, and miscellaneous chemicals. They also find application in the mineral industry for the flotation of copper ore, and for the separation of copper and nickel sulfides in Canadian-ore concentrates.

In 1970 carload prices for sodium sulfide in drums were \$137/ton for flake and \$127/ton for fused product. Sodium hydrosulfide sold for \$158/ton for flake in carloads of drums and \$146/ton (100 per cent basis) for tank cars of 45 per cent liquid.

TABLE 6.13 Uses of Sodium Hydrosulfide[a]

Use	*Per Cent*
Paper pulping	49
Dyestuffs processing	15
Rayon and cellophane	13
Leather dipilatory	9
Export and other	14

[a]"Chemical Profile," Schnell Publishing Co., Inc., July 1970 (by permission).

SODIUM THIOSULFATE

Sodium thiosulfate, $Na_2S_2O_3 \cdot 5H_2O$ or *hypo*, is the agent employed in photography for dissolving the unreduced silver salts, and in textile mills as an *antichlor.* It is made in three ways: the first is independent of any other process, and requires soda ash and sulfur; the second makes use of sulfide liquors and is dependent upon a sulfide plant; and the third consists of recovery of thiosulfate from waste liquor produced in sulfur dye manufacture.

In the *soda ash-sulfur* process soda ash is dissolved in hot water and the solution (26° Bé) is pumped to a small storage tank at the top of the first of two absorption towers. These are frequently made of lead, and they are packed with wooden grids. The nearly spent sulfur dioxide gas from the second tower (second with respect to the liquor) enters at the base and meets the descending soda ash solution which absorbs all the remaining sulfur dioxide. The partly gassed liquor from the first tower is elevated to the top of the second tower where it meets the sulfur dioxide gas fresh from the burners; the soda ash is completely converted to sodium bisulfite, which runs out through a seal at the base of the tower and is collected in lead-lined receiving tanks.

$$Na_2CO_3 + H_2O + 2SO_2 \longrightarrow 2NaHSO_3 + CO_2$$

Figure 6.5 shows the disposition of the apparatus. The nitrogen, oxygen, and carbon dioxide pass out of the first tower to a small stack.

Sulfur dioxide may be made by burning sulfur in iron pans cooled from below by air, and set in brick work; or, a special patented burner may be installed, such as the Glens Falls, New York, rotary burner, the Chemico spray burner, and the Vesuvius burner with shelves. Formerly the busulfite liquor, with 22 per cent SO_2 content, was an important product, but it has been largely replaced by anhydrous bisulfite of soda, $Na_2S_2O_5$, a powder containing 67 per cent SO_2. Soda ash changes the bisulfite liquor into neutral sulfite, which is heated with powdered sulfur in an agitated stainless steel digestion tank to

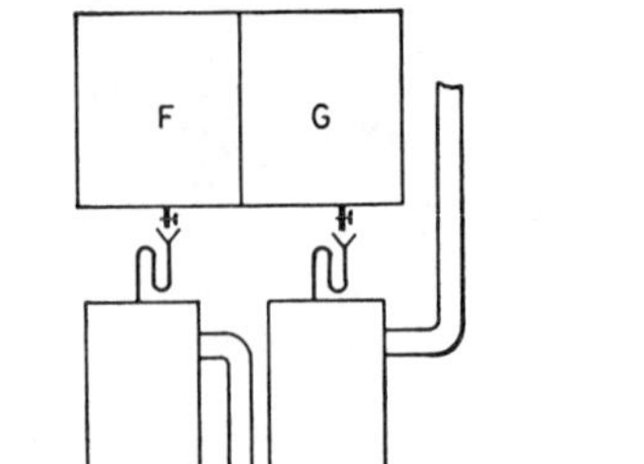

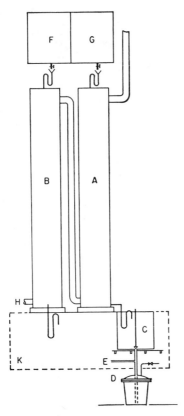

Fig. 6.5 A simple plant showing the working principles for making sodium bisulfite solution. Soda ash solution in tank *G* is fed to the first lead tower, *A*, where it meets the weak gas from the top of the second lead tower, *B*. The liquor from *A* collects in *C*, is run into the lead-lined cast-iron pot *D*, used as a blowcase, and is forced through *E* to tank *F*, from which it is fed to tower *B*, meeting fresh gas entering at *H*. The finished liquor collects in the storage tank *K*. The towers are packed with wooden grids.

give a solution of sodium thiosulfate:

$$2NaHSO_3 + Na_2CO_3 \longrightarrow 2Na_2SO_3 + H_2O + CO_2$$

$$Na_2SO_3 + S \longrightarrow Na_2S_2O_3$$

The thiosulfate solution is filtered and evaporated from 36° Bé to 51° Bé. It is then cooled to produce sodium thiosulfate crystals, ($Na_2S_2O_3 \cdot 5H_2O$), which are centrifuged, washed, dried, and packed for sale.

Sodium thiosulfate crystals effloresce so readily in air that they must be packed at once in air-tight containers.

In the *sulfide liquor* process, a solution of sodium sulfide and soda ash is prepared containing about 10 per cent Na_2S and 7 per cent Na_2CO_3. This solution is treated with sulfur dioxide to give thiosulfate according to the reaction:

$$2Na_2S + Na_2CO_3 + 4SO_2 \longrightarrow 3Na_2S_2O_3 + CO_2$$

An excess of sulfide must be avoided, as it may lead to a loss of useful materials due to the generation of H_2S.

$$Na_2S + SO_2 + H_2O \longrightarrow Na_2SO_3 + H_2S$$

Some of this hydrogen sulfide will be changed into sulfur by sulfur dioxide, but it is manifestly cheaper to supply sulfur as such if any is needed. Should the carbonate run too high and the sulfur too low, sulfur may be added to change the sulfite formed (the bisulfite is avoided*) into thiosulfate.

The sulfide-carbonate liquor may be treated with sulfur dioxide gas in towers. Instead of a single passage of the liquor through a rather tall tower, many passages through shorter tower may be substituted; this is called a *circulating tower.* The liquor is run into a large tank and pumped at a rapid rate to the top of the tower, which has wooden grids. As the liquor cascades through the tower, fresh surfaces are exposed and the absorbing liquid readily exhausts the gas. The gassed liquor at the base of the tower runs back into the original tank. The tower may be 20 ft high, 4 ft in diameter, and made of steel lined with lead. Two or more towers may be used, the gas passing from one to the other to insure its exhaustion.

After sufficient gassing, the liquor is filtered and concentrated by evaporation with steam. The crystallization and further steps are the same as in the soda ash-sulfur process.

By-product sodium thiosulfate is obtained in the manufacture of sulfur black and other sulfur dyes. Organic nitro compounds are boiled with a solution of sodium polysulfide

*Acidity due to $NaHSO_3$ is detected and measured by titration with caustic in presence of phenolphthalein; Na_2SO_3 is neutral to phenolphthalein, but alkaline to methyl orange; it is measured, with any carbonate present, by titration with standard acid, in the presence of methyl orange; $NaHSO_3$ is neutral to methyl orange.

TABLE 6.14 Estimated Uses of Sodium Thiosulfate, 1966[a]

Use	*Per Cent*
Photography	85
Leather tanning	5
Thioglycolates, paper and textiles, pharmaceuticals, and miscellaneous chemical uses	10

[a]Kirk & Othmer, "Encyclopedia of Chemical Technology," Vol. 18, 2nd ed., 1969 (By permission).

which is almost completely converted to the thiosulfate. The dyes precipitate and are filtered off. The filtrate is purified with activated carbon, and the sodium thiosulfate is recovered by concentration and crystallization.

Thiosulfates are nearly universally used in fixing baths in photography because of their ability to dissolve the unreduced silver bromide on developed film as soluble thiosulfato-silver complex compounds. The reaction may be shown as follows.

$$AgBr + 3Na_2S_2O_3 \longrightarrow Na_5[Ag(S_2O_3)_3] + NaBr$$

The end use distribution of commercial sodium thiosulfate in 1966 is shown in Table 6.14.

Two grades of sodium thiosulfate were marketed in 1970, the pentahydrate or "crystal" grade quoted at $7.05 per 100-lb bag in carloads, and the anhydrous, or technical grade, quoted at $9.15 per 100-lb bag in carloads. The anhydrous grade offers the advantage of lower cost per pound of $Na_2S_2O_3$, plus freight savings. However, the material is hygroscopic.

The production of sodium thiosulfate as $Na_2S_2O_3 \cdot 5H_2O$ was 30,545 short tons in 1968.[6]

SODIUM BISULFITE, ANHYDROUS

The sodium bisulfite of commerce, generally listed as "sodium bisulfite, anhydrous," is in reality the anhydride of two molecules of sodium bisulfite, $NaHSO_3$. This anhydride is sodium metabisulfite or sodium pyrosulfite, $Na_2S_2O_5$. Most commercial products contain at least 98 per cent $Na_2S_2O_5$ and approximately 0.5 per cent sodium sulfate, with the remaining 1.5 per cent being largely sodium sulfite, Na_2SO_3. Low moisture content enhances the storage characteristics of the material. The commercial methods for the manufacture of sodium bisulfite, a white crystalline powder, differ in details but generally follow the same principle. The mother liquor from previous batches, saturated with sodium bisulfite, is reinforced with soda ash, which converts the dissolved bisulfite to sodium sulfite and forms a suspension of soda ash. Sulfur dioxide gas from a sulfur or pyrite burner is fed into the soda ash suspension and forms the metabisulfite.

In some processes the absorbers are stainless steel vessels through which the suspension is continually circulated. The sulfur dioxide gas is added through multiple sparger pipes below the liquid level in the absorber. The gas enters first into the second-stage absorber, is drawn next into the first-stage absorber, and is finally withdrawn by a fan and discharged into the air. Countercurrent absorption columns, packed with rings or saddles, can also be used to absorb the sulfur dioxide.

Formation of metabisulfite crystals from the saturated solution is promoted by a moderate lowering of the temperature. The crystals are removed by centrifuging, and the mother liquor is returned to the absorption system. Rapid drying of the crystalline product is important in order to avoid excessive oxidation to sulfate, or decomposition with loss of sulfur dioxide.

Processes described in patents indicate that the wet powder can be dried by passing it through a shelf drier having six shelves. Rotating arms move the material from shelf to shelf, dropping it through alternate circumferential and central openings. The shelves are steam heated. Flash driers, which give especially rapid drying, can also be used. The function of the drier is merely to remove adhering moisture. The anhydrous bisulfite (or metabisulfite) is produced as such in the reaction during the absorption of sulfur dioxide.

$$Na_2CO_3 + 2SO_2 \longrightarrow Na_2S_2O_5 + CO_2$$

By using a building with several stories, the material may be moved by gravity, except for feeding the soda ash to the suspension tank, and lifting the soda ash suspension to the absorbers. These operate in two stages. The first-stage absorbers, on the top floor, receive the fresh soda ash suspension and the sulfur dioxide gas which has already passed through the second-stage absorbers. The latter, on the floor below, receive the suspension from the first-stage absorbers by gravity, and the fresh sulfur dioxide from the sulfur or pyrite burners. Here the soda ash is completely transformed to anhydrous bisulfite, and this suspension is fed to the centrifugals on the floor below. The drier is on the ground floor on a level with a railroad car floor.

Anhydrous sodium bisulfite is used in the manufacture of dyes and in their application to fibers, in the leather industry, in the photographic industry as an antichlor, in the manufacture of sodium hydrosulfite, and for many other purposes.

The neutral sulfite, Na_2SO_3, can be made by the action of soda ash on sodium bisulfite. It is sold both as the anhydrous form and as crystals, $Na_2SO_3 \cdot 7H_2O$.

The annual United States production of sodium bisulfite, anhydrous, may be estimated at about 600 million pounds. The published price of 100 lb bags (August 1970) is carload quantities was 5.85¢ per lb, f.o.b. producing point. The product is shipped in 100-lb net, polyethylene-lined bags and 400-lb net, fiber drums.

SODIUM HYPOSULFITE (HYDROSULFITE)

Sodium hyposulfite, anhydrous $Na_2S_2O_4$, called "hydrosulfite" in the trade, is a convenient form of the powerful reducer which is used principally for the reduction of vat dyes. Before its advent in the trade, it had to be prepared as a solution in the textile mills and used at once. Formerly it had to be imported, chiefly from Germany. There are now five manufacturers in the United States.

Two methods of manufacture employed in the United States use zinc dust. In one process sodium bisulfite is reduced with zinc dust in the presence of sulfurous acid, according to the reaction

$$2NaHSO_3 + Zn + H_2SO_3 \longrightarrow Na_2S_2O_4 + ZnSO_3 + 2H_2O$$

Milk of lime is added to neutralize any remaining acid and the zinc and calcium sulfite precipitates are removed by filtration. Sodium chloride is added to salt out the hydrate, $Na_2S_2O_4 \cdot 2H_2O$. The suspension is heated to 60°C and held there until the crystals have been transformed into anhydrous sodium hydrosulfite, $Na_2S_2O_4$ (transition point is 52°C). The crystals are dried by extraction of the water with alcohol followed by vacuum drying. The crystals are stable only when completely dry.

A second method of manufacture consists of treating an aqueous suspension of zinc dust with sulfur dioxide at 80°C.

$$Zn + 2\,SO_2 \longrightarrow ZnS_2O_4$$

The resulting solution is treated with soda ash.

$$ZnS_2O_4 + Na_2CO_3 \longrightarrow Na_2S_2O_4 + ZnCO_3$$

A basic zinc carbonate is precipitated, while sodium hydrosulfite remains in the alkaline solution. The suspension is filtered, and the dry crystalline product is obtained by salting out, alcohol extraction, and vacuum drying as in the previously described process.

A third process, operated in the Netherlands, is used to manufacture sodium hydrosulfite by the reduction of a sodium bisulfite solution at a pH of 5 to 7 with sodium amalgam produced in electrolytic cells.[8] The principal reaction is the amalgam reactor is

$$4NaHSO_3 + 2NaHg \longrightarrow Na_2S_2O_4 + 2Na_2SO_3 + 2H_2O + Hg$$

TABLE 6.15 Uses of Sodium Hydrosulfite, 1970[a]

	Per Cent
Vat dyeing of cotton	60
Vat dyeing of other fibers and blends	20
Miscellaneous bleaching, reducing, and stripping	14
Exports	6

[a]"Chemical Profile," Schnell Publishing Co., Inc., Oct. 1, 1970 (by permission).

In a second reactor in circuit with the amalgam reactor, the sodium sulfite coproduced with the hydrosulfite is converted back to bisulfite by treatment with sulfur dioxide.

$$Na_2SO_3 + SO_2 + H_2O \longrightarrow 2NaHSO_3$$

The production of sodium hydrosulfite in 1969 was 39,630 short tons.[9]

The published price (September 1970) of sodium hydrosulfite in carloads of drums was 27.5¢ per lb, delivered.

CAUSTIC SODA AND CHLORINE

Caustic soda, NaOH, a white solid which is extremely soluble in water, has been made in the past by two commercial processes: (1) causticizing soda ash with lime; and (2) by the electrolysis of common salt, NaCl, in water solution, with the simultaneous production of chlorine and hydrogen. Both processes date back to the 19th century; lime-soda caustic was first marketed about 1850 and electrolytic caustic in 1892. Lime-soda caustic dominated the market until 1940 when the increasing demand for chlorine caused the coproduced eledtrolytic caustic to exceed lime-soda caustic production for the first time. The trend of a relatively greater growth rate for chlorine than caustic has continued and, in 1968, forced the shutdown of the last merchant lime-soda caustic plant.

Productions of caustic soda and chlorine in 1969 are given in Table 6.16.

TABLE 6.16 Production of Chlorine and Caustic Soda, 1969[a]

	Short Tons
Chlorine gas, 100%	9,422,013
Sodium hydroxide (caustic soda), 100%, liquid	9,618,666

[a]"Current Industrial Reports, Inorganic Chemicals," U.S. Dept. of Commerce, Bureau of Census, 1970.

Electrolysis of Brine

Many cells have been devised in which the decomposition of salt in water solution may be performed. The commercially successful cells are of two basic types: diaphragm cells, and mercury cells. The first percolating diaphragm cell was the LeSueur cell, from which evolved the Townsend, Allen-Moore, Nelson, Gibbs, Vorce, Hooker, Dow Bi-polar, and other cells. The original mercury cell was the Castner, the invention of an American.

For all cells it is customary to purify the salt solution and to use it as concentrated as possible, that is, almost saturated, about 25 per cent NaCl. The manufacture of caustic soda in diaphragm cells, in a few words, involves the preparation and purification of the saturated salt solution; electrolysis of this solution in the cell, wherein half the salt is transformed into caustic; concentration of this mixed solution with separation of the salt as the concentration rises; and finally evaporation to anhydrous caustic. For the mercury cells, the procedure is about the same, except that the caustic liquor issuing from the cell is free from salt. This is an important advantage if the liquor is to be used as produced, since it requires no concentration. The liquor issuing from the diaphragm cell, with as much salt as caustic, could not be used as such for most purposes. In the manufacture of solid caustic from diaphragm-cell liquor, a final product containing 2 per cent salt is produced. This amount is not objectionable in many of the important applications of caustic.

The decomposition efficiency of cells varies between 50 and 60 per cent. Decomposition efficiency is the ratio of the salt decomposed to the total salt; thus if 6 parts of salt are decomposed into caustic, and 4 parts of salt remain unchanged in the liquor (both determined by titration), the decomposition efficiency is 60 per cent. The decomposition voltage for commercial cells lies between 3.5 and 5 with 4.5 perhaps an average. The theoretical decomposition voltage for NaCl in aqueous solution is 2.3.* The voltage efficiency is the theoretical decomposition voltage divided by the actual decomposition voltage, here, $(2.3 \div 4.5) \times 100 = 51\%$. The actual

*This value is based on the Gibbs-Helmholtz equation, and takes into account the heats of formation of the substances represented in the equation $NaCl + H_2O \longrightarrow NaOH + 1/2Cl_2 + 1/2H_2$.

voltage applied will vary with the current density desired, that is, the number of amperes per square unit of surface of electrode; the higher the current density, the higher the voltage necessary. A part of the difference in energy represented by the voltage figures appears as heat; the temperature of the cell, for instance, maintains itself at about 80° to 90°C. In general, the rough estimate is made that the potential drop for each cell is 5 volts.

The current efficiency, or better, the cathodic current efficiency, varies from 90 to 96 per cent in the best-designed cells, under even running conditions; it may drop to 75 per cent or even less, for a variety of causes. The cathodic current efficiency is the weight of caustic formed divided by the theoretical weight of caustic which the amount of current per hour (ampere hour, independent of voltage) should have formed (1.491 g), multiplied by 100.

The energy efficiency is the product of the voltage efficiency and the current efficiency, here, $(0.96 \times 0.51) \times 100 = 49\%$.

The amount of current required to liberate a gram-equivalent at each pole is the faraday, which is equal to 96,500 coulombs (1 coulomb = 1 ampere-second). One ampere-hour equals 3600 coulombs, so that the amount of material liberated by 1 ampere-hour is:

$$\frac{3600}{96{,}500} \times 35.45 = 1.322 \text{ g chlorine}$$

$$\frac{3600}{96{,}500} \times 40.00 = 1.492 \text{ g caustic}$$

The Electrolytic Cell

The two cells in Fig. 6.6 show the essential parts of an electrolytic cell. The current, which is always direct current, is said to enter at the anode and to leave at the cathode.* A brine contains sodium ions (Na^+) and chloride ions (Cl^-), as well as ions resulting from the ionization of water, i.e., the hydrogen ion (H^+) and the hydroxyl ion (OH^-). After the current is applied, the chloride ions give up their negative charge (one electron) and be-

*It would be more satisfactory to say that electrons enter at the cathode and leave at the anode, but the general custom will be observed.

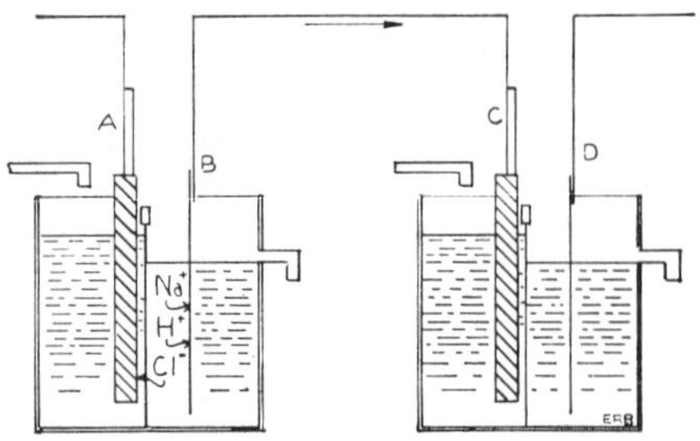

Fig. 6.6 Cells for the electrolysis of solutions of sodium chloride in water. The chloride ion, Cl^- forms chlorine gas at the anodes, marked *A* and *C*; the hydrogen ion, H^+ forms hydrogen gas at the cathode, marked *B* and *D*. A diaphragm divides the anode brine compartment from the cathode compartment.

come yellow chlorine gas, Cl_2. The hydrogen ions acquire an electron each at the cathode, and form the hydrogen molecule (hydrogen gas). There are many more Na^+ ions than H^+ ions present at any time, but the single potential for hydrogen is lower than that for sodium.** As soon as the hydrogen ions present have deposited, more are formed by ionization of the water molecules, a process which takes place at a rate approaching the speed of light. The Na^+ ions thus remain undisturbed during the action of the current of the brine. The OH^- ions, constantly formed by the ionization of the water, accumulate at a rate which equals that of chlorine-ion discharge, so that the sodium ions are matched in number by the newly formed OH^- ions. The reaction may be written as follows:

$$Na^+ + Cl^- + H^+ + OH^- \longrightarrow Na^+ + \tfrac{1}{2}Cl_2 + \tfrac{1}{2}H_2 + OH^-$$

That is the process which goes on in the diaphragm cell. In the mercury cell, the deposition of hydrogen ions at the mercury cathode does not occur, because the mercury cannot accept them; here, the Na^+ ion functions, by acquiring an electron and dissolving in the mercury as sodium metal. In a second

**The deposition potential for H^+ from a solution with pH 7 is about 0.414 volt; even with a hydrogen overvoltage on iron of 0.8 v, the total is only 1.213 v. The single potential for sodium is 2.7146 v in a normal solution of sodium ions, somewhat less in the 4.3 normal solution here. The hydrogen ions therefore deposit first.

chamber, the previously formed amalgam is denuded of its sodium, which reacts with water and again forms sodium ions and liberates hydrogen, which in turn leaves an equivalent number of OH^- ions to form the caustic soda in solution.

$$2Na + 2H_2O \longrightarrow 2NaOH + H_2$$

Returning to the diaphragm cells, it should be noted that they are connected in series so that the current leaving at the cathode of one cell enters at the anode of the next. The path of the current lies through the metallic bar to the anode, through the liquid to the opposite cathode, through a metallic connection to the next anode, and so on to the last cathode, which is connected to the terminal at the generator. As it passes through the metal, the current does not cause any change (i.e., chemical change); but as it goes through the brine, the decomposition takes place. In the diaphragm cell, caustic generally accumulates near the cathode; to prevent its diffusion toward the anode, a wall may be placed in the cell, forming two compartments. The wall allows slow passage of the solution and free passage to the sodium ions; but by keeping the level in the anode chamber higher than in the cathode chamber, the hydrostatic flow is toward the cathode, nullifying the diffusion tendency of the NaOH toward the anode. The permeable wall is usually asbestos fibers supported on an iron screen, and is called the *diaphragm.*

Purification of the Salt Solution

The salt may be in the form of a natural brine, an artificial brine, or rock salt. The purification of the latter includes the same steps as those required for the brines. The salt is conveyed from hopper cars into an underground hopper which feeds an inclined belt elevator, or bucket elevator, which raises the salt to the dissolving tank. This is kept filled by pumping in warm water at the base and allowing it to overflow at the top; during its passage through the salt the water becomes saturated. Instead of a wooden tank, a concrete tower may be used. The brine is collected in the treating tank, where sodium carbonate and sodium hydroxide are added in amounts just sufficient to precipitate the calcium and magnesium salts. After settling to remove the coarser suspended particles, the brine is passed through sand or diatomaceous earth filters. The filtered salt solution, containing about 25 per cent NaCl, is elevated to a tank above the cells and fed through a constant-level boot to a pipe having side branch for each cell. The methods for regulating the amount fed to the cell vary considerably; one of the simpler schemes is to insert a horizontal plate with a small orifice in the vertical branch leading to the anode compartment, with a glass sleeve immediately below the orifice-plate to permit observation of the flow. The size of the orifice determines the rate of flow.

Diaphragm Cells

All diaphragm cells embody the principle of enclosing the anode compartment with a diaphragm. However, the modifications in the construction of the later cells have been of the greatest importance; furthermore, they differ from each other radically. One fundamental difference in cell design is the presence or absence of liquor in the cathode compartment. The Hooker cell may be taken as typical of the diaphragm cells which have liquor in the cathode compartment.

The Hooker "S" Cell. The Hooker "S" cell (Fig. 6.8) is essentially square in cross section. It consists of a concrete top piece, a concrete bottom piece (set on short legs), and a central steel frame, which carries the iron-wire mesh forming the multifingered cathode.[10] The latter is also the support for the asbestos diaphragm, an unbroken mat prepared by plunging the steel cathode into a suspension of asbestos fibers in cell liquor and applying suction to the cathode chamber. The anode is part of the bottom section. It is made of impregnated graphite slabs set in a bitumen-covered lead base plate which receives the current; the slabs reach up from below between the fingers of the cathode, so that every vertical face of the cathode and diaphragm is close to a similar graphite surface. A single broad-faced connection at the cathode of one cell carries the current the short distance to the single stout rod feeding the anode of the next cell. The construction will be clearer

Fig. 6.7 Electrolytic cell room before the installation of cells is complete. Graphite anodes are exposed here. Later, the cathode assembly and concrete cell tops are placed over the anodes. (*Courtesy Hooker Chemical Corp.*)

after examination of Fig. 6.9. The cell, closely packed as it is with anode and cathode branches, has a 4-inch central passage for anolyte circulation.

The brine is preheated to 60°C (140°F) and enters the cell in the dome (see sketch); the stream of brine breaks into drops, so that there is no electrical conductivity from the cell to pipes and heaters outside the cell. Its content of salt (NaCl) is 322 g per liter, essentially a saturated solution. The cell works at a temperature of 90°C (194°F). The chlorine-gas outlet is in the dome, a stoneware pipe connected to a larger stoneware header. The hydrogen leaves the upper level of the cathode chamber through an iron pipe, insulated by a rubber sleeve coupling. The caustic dribbles out from the lower part of cathode chamber through an adjustable outlet, so that the level of caustic in the chamber can be controlled.

Fig. 6.8 View of Hooker "S" cell with the middle section, the cathode section, in place. (*Courtesy Hooker Chemical Corp.*)

The voltage drop per cell when the cell is new is 3.4 v; as time passes, the anode surfaces become imperfect, and the voltage gradually rises to a final 4.4 v, when the anode section is renewed. The current density* at the anode for the 7000-amp cell is 0.446 amp per sq in, and at the cathode, 0.377 amp per sq in. The level of the brine may be observed by means of a sight glass; as the cell ages, it rises, producing a higher head, in order to overcome plugging in the diaphragm and maintain a given flow of cathode liquor. The diaphragm is replaced on an average of 4 times per anode run. The life of the anode section is about 365 days. In the 10,500-amp cell, the loss of graphite averages about 5½ lb per ton of

*The current density means the current is amperes per unit surface. An increase in current density requires an increase in voltage.

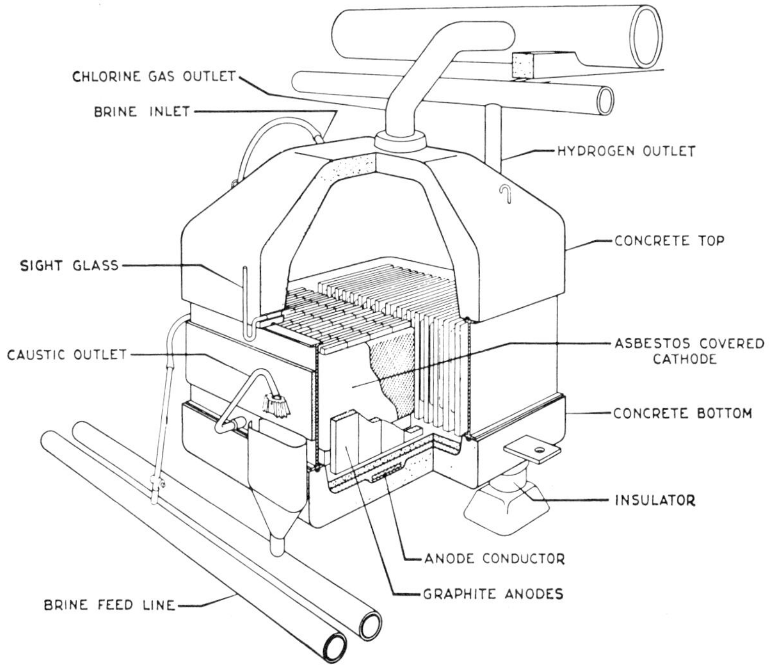

Fig. 6.9 Hooker-type "S-3B" cell. (*Courtesy Hooker Chemical Corp.*).

chlorine produced. The construction of the cell is such that the upper section and the cathode section can be removed vertically, a feature which saves floor space. As spare sections are always in readiness, a change can be made in a few hours. The energy efficiency is 65 per cent; the ampere efficiency is of the order of 97 per cent.

At the higher temperature of operation,† chlorine is less soluble in the anolyte, and a smaller amount of chlorate forms.

The caustic liquor produced contains on an average 11.3 per cent NaOH, 15 per cent NaCl and 0.1 per cent $NaClO_3$. The gas in the chlorine header contains on a dry basis, 97.5 per cent Cl_2, 1.4 per cent CO_2, 0.8 per cent O_2, 0.2 per cent H_2 and 0.1 per cent N_2. The hydrogen gas, except for moisture, is essentially pure; it is under a slight positive pressure. The chlorine line is under a slight negative pressure.

The 10,500-amp cell produces 790 lb of caustic (NaOH), 685 lb of chlorine and 3560 cu ft of hydrogen per day.

Nelson Cell. In the Nelson cell, as in the Allen-Moore, Gibbs, Vorce, and several other

Fig. 6.10 Several of Hooker's latest design, type "S-3B" electrolytic cells in four rows which have a rated daily capacity to produce 110 tons of caustic soda, 100 tons of chlorine, and 1,000,000 cu ft of hydrogen. (*Courtesy Hooker Chemicals Limited, Canada.*)

†The higher the temperature, the less the resistance; however higher than 90°C is avoided because of the greater action of the products on the cell walls.

cells, the cathode compartment contains no liquor; the caustic solution, with residual salt, runs down the diaphragm and collects at the base. In the Nelson cell, steam is sent into the cathode space to maintain the temperature near 65°C (149°F). The form of the cell is again a narrow rectangular box set up in the same manner as the Hooker cell; the Allen-Moore has a similar outside appearance. There is only one row of suspended graphite anodes, and the usual size unit receives 1000 amp; the current density is 50 amp per sq ft, half of that in the Hooker cell. The voltage is lower, averaging 3.7 v. Among other advantages, this cell produces a very pure chlorine; for this reason it was chosen for the installation at Edgewood arsenal during the first World War,[11] 3500 Nelson cells furnished 100 tons of chlorine per day. The anodic ampere efficiency (for chlorine) was 90 per cent, and the caustic liquor was maintained 10 to 12 per cent NaOH and 14 to 16 per cent NaCl.

Allen-Moore Cell. The Allen-Moore cell[12] usually has 1700 ampere units, working with a voltage of 3.6 v; the cathodic current efficiency is maintained at 95 per cent over extended periods. The caustic liquor contains 8 to 10 per cent NaOH and about 12 per cent NaCl. The cell is constructed of concrete and cast-iron sides, and in general, has the shape of a narrow rectangular box. The basic principle of the unsubmerged cathode, that is, of using an empty cathode compartment, was first proposed by the designers of the Allen-Moore cell.[13]

The cell described by L. D. Vorce[14] is cylindrical, and is said to furnish more caustic per square foot of floor space than any other cell. Another cylindrical cell used in Canada, and by the United Alkali Company of Great Britain, is the Gibbs cell,[15] patented in 1907.[16] The units are of 1000-amp capacity, requiring 3.6 v; the floor space occupied is small.

Concentration of the Caustic Liquor

The liquor flowing from the cathode compartment of the diaphragm cells just discussed contains both caustic and salt. The solution contains 10 to 12 per cent NaOH and 14 to 16 per cent NaCl. It is concentrated in a triple-effect evaporator, for example, and each boiling pan may have its own separator in which the salt is collected as fast as it separates from solution, so that the heating surface in the pan may always be swept by liquor. When cell liquor has been concentrated to 50 per cent NaOH, only 1 per cent of salt remains in the solution. The salt in suspension is removed by settling and cooling, followed by a final filtration. The solid salt which is removed is redissolved in the evaporator condensate to make cell-feed brine. In some plants, the 50 per cent NaOH liquor is concentrated further to a 73 per cent concentration in a single vacuum pan heated with high-pressure steam. The liquid caustic, which is sold as such, offers several advantages: it may be handled in pipes; no drum containers are necessary; less labor and less fuel are required in its production in comparison with solid caustic; and it is in the form in which the customer applies it. In order to produce anhydrous caustic, the 50 or 73 per cent is boiled down in cast-iron pots heated by an oil or gas fire, until all the water is evaporated. The liquid anhydrous caustic, with perhaps two per cent NaCl, is removed from the pots by a centrifugal pump. The liquid caustic is pumped to thin steel drums on low trucks; after cooling, the mass is solid. The anhydrous caustic is also produced in flake and granular form by using a rotary drum flaker and Hummer screen to obtain desired size. It is packaged in 400 and 450 lb drums. A recent innovation has been the manufacture of caustic soda beads in prilling towers. This product offers the advantage of being free flowing and containing no dust. Some anhydrous caustic soda manufacturers have achieved success in packaging the products in 100-lb polyethylene-lined multiwall bags. The use of bags is more economical since they cost less than the drums and are more easily handled.

In some recent installations for the manufacture of anhydrous caustic soda, the liquor is evaporated to the anhydrous state continuously with rising film evaporators. The tubes in such an evaporator are made of "Inconel" and the heat-transfer agent is either molten salt or "Dowtherm" (diphenyl and diphenyl

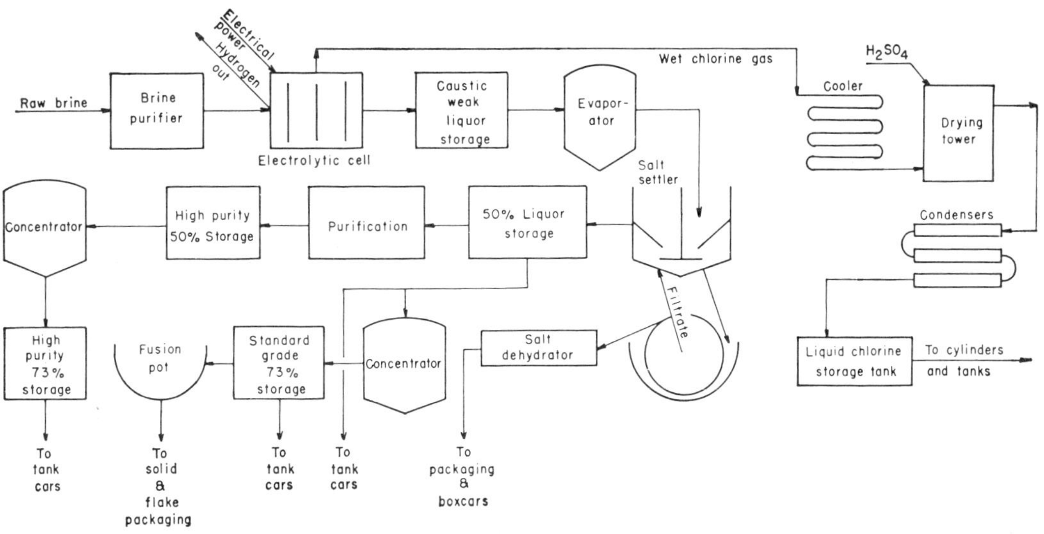

Fig. 6.11 A more complete flow diagram for making caustic soda and chlorine. (*Courtesy Columbia-Southern Chemical Corp.*)

oxide). Product quality, cost, and safety favor the continuous process for the manufacture of anhydrous caustic soda over the old batch-pot method.

For use in making viscose rayon, a caustic soda with a low chloride content, such as 0.15 per cent NaCl on the 100 per cent basis, is required. The excess NaCl in the caustic liquor from diaphragm cells may be reduced in one of several ways:

1. Hydrates such as $NaOH \cdot 3.5H_2O$ and $NaOH \cdot 2H_2O$ may be crystallized in jacketed crystallizers with agitators; the crystals are filtered off, and the liquor carries away the bulk of the sodium chloride.

2. A triple salt may be formed by adding sodium sulfate, which binds the NaCl as $NaOH{\cdot}NaCl{\cdot}Na_2SO_4$, this is insoluble in 35 per cent caustic solution, and is removed by filtration.[17]

3. Liquid ammonia may be used for extraction of the NaCl and the small amount of sodium chlorate present.[18]

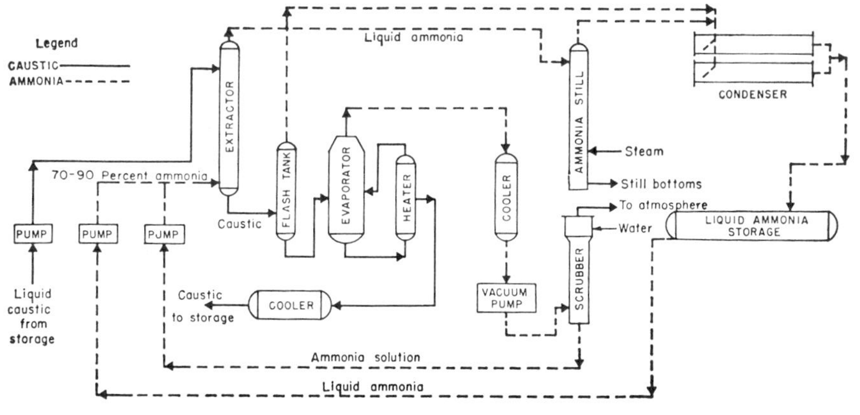

Fig. 6.12 Purification process for diaphragm caustic liquor using anhydrous ammonia as extracting medium. (*Courtesy Columbia Southern Chemical Corp.*)

The Mercury Cell

In the mercury cell working on brine, the cathode consists of mercury, and the anode of graphite. There is no diaphragm. The salt in solution is decomposed, the sodium ion accepts an electron and forms sodium metal, which dissolves in the mercury.

The chlorine is liberated, as before, in the form of a yellow gas. These events take place in one of the chambers of the cell, the electrolyzing chamber. The amalgam is then transferred to the denuding chamber containing water, where the sodium metal reacts to form hydrogen gas and a solution of caustic $2Na + 2H_2O \longrightarrow 2NaOH + H_2$.

In the United States, the mercury cells now installed are the Sorensen, I.C.I.-Wyandotte, Solvay et Cie, De Nora, Hoechst-Uhde, Olin Mathieson, Dow, and Krebs-BASF. Mercury cells have been in favor with many European countries for many years and consequently many developments and improvements in mercury-cell technology have resulted. Today, some cells, depending on cathodic and anodic surface, are known to operate as high as 300,000 amp.[19]

A mercury cell of the Olin Mathieson type is shown in Fig. 6.13. It has a long, narrow electrolyzing chamber; the amalgam passes to a smaller, vertical, cylindrical decomposer where pure water is fed in to make caustic solutions containing up to 50 per cent very pure NaOH. A pump functions between the electrolyzer and decomposer.[20]

Mercury and diaphragm cells have been compared from time to time, with mercury cells being the most favored. However, the advantages offered by mercury cells are far from clear and are often negated by disadvantages. Their superiority in producing low-chloride-content caustic soda is unchallenged.

Typical mercury-cell caustic contains 0.002–0.001 per cent sodium chloride, compared with about 1 per cent in the product from diaphragm cells. The latter may be reduced to 0.16 per cent when purified. Although the rayon trade must have low chloride material, most users do not require it.

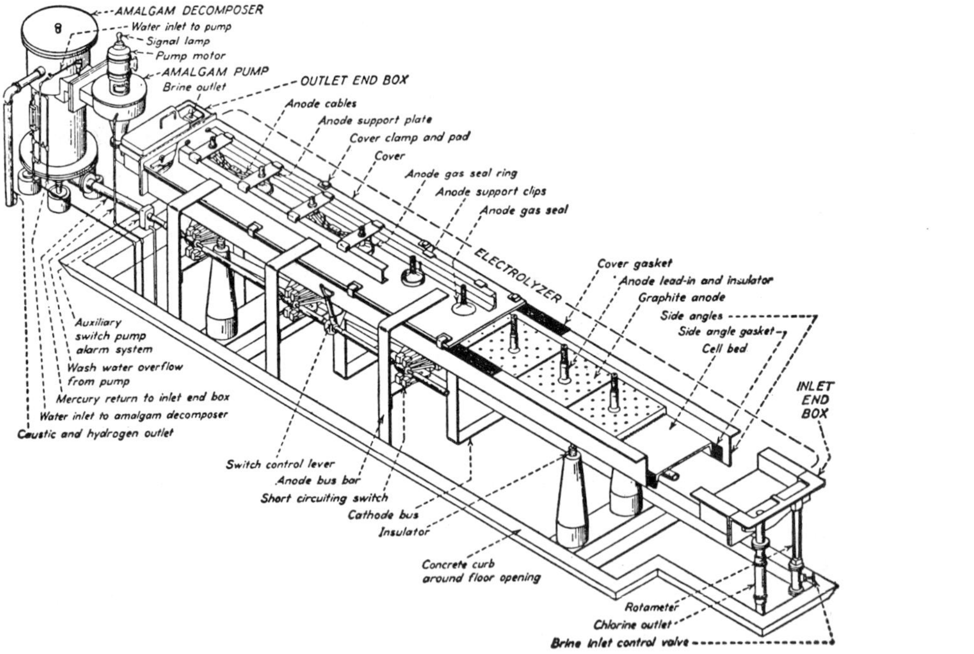

Fig. 6.13 Stationary mercury cell with electrolyzing chamber separate from decomposing chamber. (*Courtesy Mathieson Chemical Corp.*) [*Reproduced from Chem. Eng.,* **54** *(1947) by permission.*]

Mercury cells require solid salt to resaturate the recirculated spent cell brine, in contrast with diaphragm cells which are normally operated on purified natural brine. Thus the relative cost of the two salt sources is an important factor in cell selection. Dry salt generally costs four to eight times as much as salt in natural brine. The predominance of the diaphragm cell in the United States is largely due to the availability of well brines.

Mercury cells require 15 to 20 per cent more electric power per ton of cell product than diaphragm cells, but this is more than offset by the steam required to concentrate to 50 per cent the caustic soda liquor from the diaphragm cells.

Regardless of inroads made by mercury cells, diaphragm cells maintain an important position as indicated by Table 6.17.

TABLE 6.17 U.S. Chlorine Production Routes, 1970[a]

Process	*Capacity (Short Tons per Day)*	*Per Cent*
Diaphragm cells	19,567	69.2
Mercury cells	7,804	27.6
Fused salt and nonelectrolytic	905	3.2
TOTAL	28,276	100.0

[a]"North American Chlor-Alkali Industry Plants and Production Data Book," The Chlorine Institute, Inc., July, 1970.

The estimated distribution of chlorine is given in Table 6.18. The end use distribution of caustic soda is given in Table 6.19.

TABLE 6.18 Chlorine End Use Pattern, 1970[a]

	Per Cent
Chlorinated hydrocarbons	59
Paper	18
Other chemicals	11
Miscellaneous	12
TOTAL	100

[a]"Chemical Profile," Schnell Publishing Co., Inc., April 1, 1969 (by permission).

TABLE 6.19 Caustic Soda End Use Pattern, 1970[a]

	Per Cent
Chemicals and metals processing	56
Cellulosics (paper, rayon, etc.)	16
Petroleum, textiles, soaps, food, and other merchant uses	15
Exports	8
Miscellaneous	5
TOTAL	100

[a]"Chemical Profile," Schnell Publishing Co., Inc., August 3, 1970 (by permission).

In September 1970, the quoted price for chlorine was $75 per net ton, f.o.b. works, in single unit tank cars. Quotations for caustic soda in carloads, f.o.b. works, 100 per cent NaOH basis, were as follows:

	$/Net Ton
50% Liquor	66
50% Liquor (rayon grade)	68
73% Liquor	68
73% Liquor (rayon grade)	70
Solid, 700 lb drums	116
Flake, 400 lb drums	120
Granular, 450 lb drums	120
Beads, 500 lb drums	120

Hydrogen Disposal

Hydrogen from the cathode chamber is the by-product of the electrolysis of brine. It may be used to make extremely pure synthetic hydrogen chloride by burning chlorine in an excess of hydrogen. This may be dryed and sold as anhydrous hydrogen chloride or converted to C.P. hydrochloric acid by absorption in water. Cell hydrogen is also used for the catalytic hydrogenation of organic compounds. Lacking a profitable productive outlet for the hydrogen, it may be burned as fuel in chemical processing operations or in boilers for steam generation.

Other Processes for the Production of Chlorine

Estimated U.S. production of chlorine by process is shown in Table 6.20.

Chlorine and caustic potash are produced by electrolyzing a potassium chloride brine in equipment which is essentially the same as

TABLE 6.20 U.S. Sources of Chlorine, 1967[a]

Source	*Annual Production (Thousand Short Tons)*	*Per Cent*
Coproduct of caustic soda	7,266	95.0
Coproduct of caustic potash	110	1.4
Coproduct of sodium	254	3.3
Coproduct of potassium nitrate	20	0.3
TOTAL chlorine production	7,650	100.0

[a]"World Chlor-Alkali Survey," A. D. Little, Inc., 1968.

that used for caustic soda production. The mercury cell decomposers are operated to produce a 45 per cent caustic potash solution for the market. Diaphragm cell-discharge liquor at 10 to 15 per cent KOH is concentrated to 45 per cent for sale using multiple effect evaporators as in the concentration of diaphragm cell-caustic soda. Caustic potash solutions may be further concentrated in batch, nickel-lined pots or in continuous, Inconel-tubed rising-film dehydrators to 88–92 per cent KOH concentration for sale in solid or flake form.

Sodium metal is manufactured by the electrolysis of fused sodium chloride using the Downs cell. The cell operates on a eutectic mixture of sodium and calcium chlorides at a temperature of 600°C. The cell produces hot concentrated dry chlorine, which in most plants is cooled and liquefied for sale.

Potassium nitrate is made at one plant by reacting potassium chloride with nitric acid to give coproduct chlorine.[21] The reactions are:

1. $3KCl + 4HNO_3 \longrightarrow 3KNO_3 + NOCl + Cl_2 + 2H_2O$

2. $NOCl + 2HNO_3 \longrightarrow 3NO_2 + \frac{1}{2}Cl_2 + H_2O$

3. $2NO_2 + H_2O + \frac{1}{2}O_2 \longrightarrow 2HNO_3$

Over-all

$$2KCl + 2HNO_3 + \tfrac{1}{2}O_2 \longrightarrow 2KNO_3 + Cl_2 + H_2O$$

Liquid Chlorine

Production of Liquid Chlorine. Gaseous chlorine is easily changed to an amber liquid; this circumstance has extended its usefulness considerably over that of the gas. At 0°C (32°F), a pressure slightly over 39 lb will cause liquefaction; at -20°C (-4°F), slightly over 12 lb is needed; at -33.5°C (-28.5°F), only a little over atmospheric pressure is required.

The chlorine gas leaving the anode or electrolyzer portion of the cell is warm and moist. In such a condition, the wet gas can be handled only in pipes and equipment impervious to its acidic attack. Leakage of hydrogen, or production of hydrogen in the electrolyzer portion of the cell is carefully guarded against because the two gases, chlorine and hydrogen, form explosive mixtures over a very wide range of concentrations. The chlorine gas passes along pipes of stoneware, rubber- or saran-lined steel, or polyvinyl chloride (PVC) to a cooling tower where the moisture content of the raw gas is reduced by cooling. Cold water is used as the coolant, passing countercurrent to the warm moist gas from the cells. The gas is then dried by passing it countercurrent to a stream of strong sulfuric acid. Once the gas is cooled and dried it can be handled in steel pipes or equipment. The gas next reaches a Nash Hytor compressor whose multibladed impeller works in the sulfuric acid which is held in an elliptical casting. For each revolution, there are two compressions and two expansions of the working fluid caused by the shape of the casing; the alternating compression and expansion move the gas. Pressures up to 60 lb are developed with these compressors. The chlorine gas may also be compressed with reciprocating or centrifugal compressors, the latter being common in the larger plants. The compressed gas is passed through steel pipes to a liquefying unit where it is subjected to cold surface temperatures which cause it to liquefy at the pressure existing in the system. Many systems exist for liquefying chlorine, the most common today being shell and tube heat exchangers using "Freon" and "Genetron" refrigerants. The gaseous chlorine liquefies at a rapid rate, and is piped to a steel storage tank. Any of the less compressible gases present as impurities, such as hydrogen, carbon dioxide and air, are vented to bleach chambers or special absorbers.

Shipping Liquid Chlorine. Liquid chlorine is shipped in 100 and 150 lb steel cylinders and in 1-ton steel containers 15 of which are placed on a specially equipped railroad flatcar. In addition to these containers which are usually purchased by consumers of small amounts of chlorine, liquid chlorine is shipped in single-unit tank cars with capacities of 16, 30, 55, 85, and 90 net tons. Moreover, liquid chlorine is also shipped in barges along inland waterways. Such barges have capacities up to 640 net tons of liquid chlorine. A barge is equipped with four or six tanks to hold the chlorine. A recent innovation is the shipment of liquid chlorine by tank truck. This truck is of special design and has a carrying capacity of 16 net tons of liquid chlorine. Many chlorine-producing installations also pipe liquid or gaseous chlorine to other plant units for various chlorination operations. Pumps of special design are used in many installations to move liquid chlorine from one part of a plant to another.

The 1-ton unit is very convenient for making the bleach at a mill, such as a pulp mill, because once the required quantity of milk of lime has been prepared, all the chlorine in the cylinder may be allowed to pass in without danger of wasting any. Calcium hypochlorite, $Ca(OCl)_2$, and calcium chloride, $CaCl_2$, are formed. All containers have one or two internal pipes, so that either the gas or the liquid may be drawn. If the gas is drawn, heat must be supplied to make up for the heat of vaporization, or else the process is very slow. It is more convenient and simpler to use the liquid, mixing it directly with the milk of lime in a small mixing chamber;[22] in that way the vaporization process uses part of the heat of reaction of the chlorine on the lime. This is important in keeping down the temperature of the mixture to prevent chlorate formation.

Bleaches

Bleaching Powder. Bleaching powder is the product of the interaction of chlorine gas and hydrated lime.

$$Ca(OH)_2 + Cl_2 \longrightarrow CaOCl_2 + H_2O$$

The resulting chloride of lime, when dissolved in water, gives equal moles of calcium chloride, which is useless as far as bleaching is concerned, and calcium hypochlorite, which retains the total bleaching power of the original material.

$$2CaOCl_2 \text{ (dissolved)} \longrightarrow CaCl_2 + Ca(OCl)_2$$

Chloride of lime or "bleach" must not be confused with calcium hypochlorite.

The chlorine absorption is performed in a rotating steel cylinder equipped with inner flights which lift and shower the solids in the path of the gas. The countercurrent principle is applied, the fresh gas meeting the richest bleach; the lean gas, the new hydrated lime. The bleach prepared in this manner has been displaced to a considerable degree (1) by liquid chlorine, in mills and factories; and (2) by calcium hypochlorite, $Ca(OCl)_2$, a stable material of high test, whereas ordinary bleach spoils after several months, and has a low test (35 per cent available chlorine). Available chlorine means chlorine evolved on addition of acid; 35 per cent available chlorine means the material has the effectiveness of 35 parts of liquid chlorine. Pure calcium hypochlorite, by the laboratory test applied to all bleaching agents, rates 99.2 per cent available chlorine.*

High Test Hypochlorite (H.T.H.). Calcium hypochlorite, $Ca(OCl)_2$, essentially free from any other material, is stable in the crystal form, and efforts to produce it in bulk have been earnest and successful.

One method for its manufacture is the chlorination of a lime slurry, followed by salting

---

*The test for available chlorine in bleaching powder consists of acidifying in the presence of potassium iodide. The iodide liberated is titrated. It may be liberated by Cl, but also by oxygen.

Calcium hypochlorite contains 99.2 per cent available chlorine, yet only 49.6 per cent total chlorine. The available chlorine for calcium hypochlorite is shown by the reactions:

$$Ca(OCl)_2 + H_2SO_4 \longrightarrow CaSO_4 + 2HOCl$$

$$2HOCl + 4HI \longrightarrow 2H_2O + 4I + 2HCl, \text{ and } 4I \text{ are equivalent to } 4Cl$$

Bleaching is an oxidation reaction; hence oxygen as well as chlorine is a bleaching agent, a fact already well known in an empirical way.

out of the calcium hypochlorite by means of common salt (NaCl). No organic solvent requiring recovery is needed. The product may be made essentially 100 per cent $Ca(OCl)_2$, but an additional operation is then required, so that the material actually marketed is 75 per cent $Ca(OCl)_2$. Therefore it is twice as strong as ordinary bleach; also it does not spoil on standing, it is not hygroscopic, and when it is made up with water the solution is practically clear.

The latest and most successful method for making high test hypochlorite (H.T.H.) is by the formation of the triple salt $Ca(OCl)_2 \cdot NaOCl \cdot NaCl \cdot 12H_2O$ and its subsequent reaction with calcium chloride. The triple salt is made as follows: 40 parts NaOH, 37 parts $Ca(OH)_2$ and 100 parts water are chlorinated at a temperature below 16°C (60.8°F), such as 10°C (50°F). This reaction is:

$$4NaOH + Ca(OH)_2 + 3Cl_2 + 9H_2O \longrightarrow Ca(OCl)_2 \cdot NaOCl \cdot NaCl \cdot 12H_2O + 2NaCl \quad (1)$$

The comparatively large hexagonal crystals—the triple salt—separate and are centrifuged. In the meantime a special calcium chloride mixture is prepared by chlorinating a milk of lime in these proportions: 74 parts $Ca(OH)_2$, 213 parts water, and 71 parts chlorine; the temperature is held at 25°C (77°F). The reaction is:

$$2Ca(OH)_2 + 2Cl_2 \longrightarrow Ca(OCl)_2 + CaCl_2 + 2H_2O \quad (2)$$

Next, this solution is cooled to 10°C (50°F), the centrifuged crystals are added in the proportion required by the reaction:

$$2NaOCl + CaCl_2 \longrightarrow 2NaCl + Ca(OCl)_2 \quad (3)$$

The suspension is agitated with paddles. On warming to 16°C (60.8°F), reaction (3) takes place and the whole sets to a rigid mass. The triple sale $Ca(OCl)_2 \cdot NaOCl \cdot NaCl \cdot 12H_2O$ becomes $\frac{3}{2}$ $Ca(OCl)_2 \cdot 2H_2O$, the dihydrate, and 2NaCl and water. After vacuum drying, a material testing 65 to 70 per cent $Ca(OCl)_2$ results. Such calcium hypochlorite as accompanied the calcium chloride is just that much more product. The NaCl content from (3) is not removed.

Sodium Hypochlorite. Sodium hypochlorite is prepared by reacting liquid or gaseous chlorine with a caustic soda solution accompanied by cooling.

$$Cl_2 + 2NaOH \longrightarrow NaOCl + NaCl + H_2O$$

Sodium hypochlorite is employed as a disinfectant and deodorant in treating water supplies and sewage effluent, in swimming pools, dairies, and in the household. It is used as a bleach in commercial and home laundries.

Soda bleach solutions are prepared in various strengths up to 16.5 per cent NaOCl, but household bleach is limited to about 5.5 per cent NaOCl, because less excess caustic soda is required to maintain stability. A 5.5 per cent sodium hypochlorite solution contains 5.25 per cent available chlorine.

Sodium Chlorite and Chlorine Dioxide. Sodium chlorite, $NaClO_2$, is a powerful but stable oxidizing agent. The 80 per cent commercial material has about 125 per cent available chlorine. It is manufactured from chlorine dioxide which is obtained from the reduction of sodium chlorate with a reducing agent in a strong-acid medium. The chlorine dioxide is reacted with lime, caustic soda, and carbon as follows:

$$4ClO_2 + 4NaOH + Ca(OH)_2 + C \longrightarrow 4NaClO_2 + 3H_2O + CaCO_3$$

The calcium carbonate is filtered off and the $NaClO_2$ solution is evaporated and dried. Sodium chlorite is used chiefly as a bleach in the paper and textile industries. It may be used to generate chlorine dioxide by the reaction:

$$NaClO_2 + \tfrac{1}{2}Cl_2 \longrightarrow ClO_2 + NaCl$$

Chlorine dioxide, which has 2.63 times the oxidizing power of chlorine, is used in water purification and pulp bleaching. In the latter application, however, it is normally obtained directly by reducing sodium chlorate in strong acid.

REFERENCES

1. Weisman, W. I., "Sodium Sulfate from Brine," *Chem. Eng. Progr.*, **60**, No. 11, 47 (1964).
2. U.S. Patent 1,385,595.
3. U.S. Patent 1,132,640.
4. German Patent 249,222.
5. U.S. Patent 1,139,741.
6. "Current Industrial Reports, Inorganic Chemicals and Gases," U.S. Dept. of Commerce, Bureau of the Census, 1968.
7. "Mineral Industry Surveys," U.S. Dept. of Interior, Bureau of Mines, 1969.
8. Sconce, J., "Chlorine," ACS Monograph No. 154, 180 (1962).
9. "Current Industrial Reports, Inorganic Chemicals," U.S. Dept. of Commerce, Bureau of the Census, 1970.
10. U.S. Patents 1,862,244 and 1,865,152.
11. Green, S. M., *Chem. Met. Eng.*, **21**, 17 (1919).
12. Mitchell, F. H., *Chem. Met. Eng.*, **21**, 370 (1919).
13. *Trans. Am. Inst. Chem. Eng.*, **13**, I, 11 (1920).
14. *Trans. Am. Inst. Chem. Eng.*, **13**, I, 47 (1920).
15. *Ind. Eng. Chem.*, **16**, 1056–7 (1924).
16. British Patent 28,147.
17. German Patent 522,676; and U.S. Patents 1,888,886 and 1,998,471.
18. *Chem. Met. Eng.*, **51**, 119 (Aug. 1944)
19. Sommers, H. A., "The Design of Large Mercury Cells," ECT, **5**, No. 3–4, 108–23 (1967).
20. Gardiner, W. C., *Chem. Eng.*, **54**, 108 (1947).
21. Spealman, M. L., "New Route to Chlorine and Saltpeter," *Chem. Eng.*, **72**, 23, 198 (1965).
22. U.S. Patent 1,481,106.

SELECTED REFERENCES

Hou, T. P., "The Manufacture of Soda," 2nd Ed., New York, Van Nostrand Reinhold, 1942.

Kaufman, D. W., "Sodium Chloride," New York, Van Nostrand Reinhold, 1960.

Brighton, T. B., "Salt-making on the Great Salt Lake," *J. Chem. Educ.*, **9**, 407 (1932).

Robertson, G. R., "California Desert Soda," *Ind. Eng. Chem.*, **23**, 478 (1931).

Maude, A. H., "Synthetic Hydrogen Chloride," *Chem. Eng. Progr.*, **44**, 179 (1948).

Oldershaw, C. F., Simenson, L., Brown, T., and Radcliffe, F., "Absorption and Purification of Hydrogen Chloride from Chlorination of Hydrocarbons," *Chem. Eng. Progr.*, **43**, 371 (1947).

Stuart, K. E., Lyster, T. L. B., and Murray, R. L., "The Story of the Hooker Cell," *Chem. Met. Eng.*, **45**, 354–8 (1938).

MacMullin, R. B., "Diaphragm vs. Amalgam Cells (for chlorine-caustic production)," *Chem. Ind.*, **61**, 41 (1947).

Gardiner, W. C., "New Mercury Cell Makes Its Bow," *Chem. Eng.*, **54**, 108 (1947).

Maude, A. H., "Synthetic Hydrogen Chloride," *Chem. Eng. Progr.*, **44**, 179 (1948).

Twiehaus, H. C., and Ehlers, N. J., "Caustic Purification by Liquid-Liquid Extraction," *Chem. Ind.*, **63**, 230 (1948).

Sommers, H. A., "The Chlor-Alkali Industry," *Chem. Eng. Progr.*, **61**, No. 3, 94–109 (1965).

Sconce, J. S., "Chlorine, its Manufacture, Propertier, and Uses," ACS Monograph No. 154, Van Nostrand Reinhold, 1962.

7

Industrial Fermentation Processes

Samuel C. Beesch* and Fred W. Tanner, Jr.**

INTRODUCTION

From time immemorial, one of the processes which depends on the existence and growth of microorganisms has been utilized, namely, that of alcoholic fermentation of grape juice and fruit juices. Its exact nature remained a mystery until Louis Pasteur discovered, isolated, and classified the several kinds of organisms. He succeeded in demonstrating that fermentation and bacterial disturbances in general are not due to spontaneously generated plant organisms, but to organisms which already exist elsewhere and which are carried in by air currents, on the skin of the fruit, or in other ways. In the century since Pasteur's recognition of the significant role of microorganisms, knowledge of their capabilities has gradually evolved into a flourishing branch of the chemical industry.

*Biochemical consultant, Route 2, Box 151, Rhinebeck, N.Y.
**Biochemical consultant, 2714 Norfolk Rd., Orlando, Fla.

Industrial fermentation processes may be best described as that portion of biological science which deals with the possible utilization of microorganisms in processes in which their activity becomes of industrial significance.

Bacteria, yeasts, molds, and actinomycetes are utilized to produce a variety of foods and industrial and medicinal chemicals, many of which cannot be obtained from other sources. The industry is not static; it faces intense competition from the synthetic chemist, and in fact has gradually changed its products from the rather simple carbon compounds to relatively large molecules, with complex stereochemical and structural characteristics. Ethanol, butanol, and acetone, and some other products produced by fermentation prior to World War II, are now supplied more economically by the petroleum industry. Their place in the fermentation picture is now occupied by many important medicinals including vitamins, antibiotics, enzymes, and steroids unknown a decade or two ago. At the same time, several useful organic acids of fermentation origin have become established. Fermentation citric acid has displaced the acid made from cull fruit and new acids such as glutamic, itaconic, kojic, and gluconic are finding industrial applications.

Some of the newest applications of microorganisms are in the manufacture of enzymes for use in detergent products, and in the synthesis of single-cell protein (SCP), particularly from hydrocarbons, by a fermentation process. It is hoped that this process will result in making a cheap source of protein available for underdeveloped countries, for use in animal feeds, and possibly in human nutrition.

A treatment of industrial fermentation processes involves, of necessity, the use of terms which are in general not familiar to the chemist and chemical engineer. The following discussion should help to define some of the terms used throughout this chapter.

The Microorganism

Microorganisms are found in both plant and animal kingdoms. For the fermentation industry useful members are confined to lower forms of plant life (chorophyll-free) and comprise bacteria, fungi, yeasts, and actinomyces.

All are named by a binomial system according to rules of botanical nomenclature. The first word designates the major group, the second, the particular type or subgroup.

Bacteria are unicellular organisms which multiply by fission. Each cell divides every 20–30 minutes under optimum conditions. Cells may be nearly spherical (coccus) or rod-shaped (Bacterium, Bacillus). Some are characterized by the formation of chains (Streptococcus), but usually the cells grow as separate entities. The average cell length varies from 2-6 microns long, the width being half or less the length (1 micron = .001 mm). Bacteria are classified according to their morphological and physiological characteristics. Among the various reactions are motility, spore formation, and ability to use various sugars for growth. A comprehensive compilation of the characteristics of bacteria and their botanical classification is periodically issued in "Bergey's Manual for Determinative Bacteriology."

Fungi comprise two large groups of microorganisms in the family Eumycetes. These microorganisms normally grow as a filamentous structure known as mycelium (or hyphae). Those with nonseptate hyphae are in the subfamily Phycomycetes; those with septate mycelium are in the subfamily Mycomycetes. Though the diameters of hyphae are usually only a fraction of a micron, their lengths may extend to centimeters. Within these groups, they are further subdivided on the basis of modes of spore formation. However, many do not form true spores and are placed in the *Fungi Imperfecti.*

Yeasts are generally distinguished from fungi by their ability to grow as individual cells, either spherical or ellipsoidal, usually

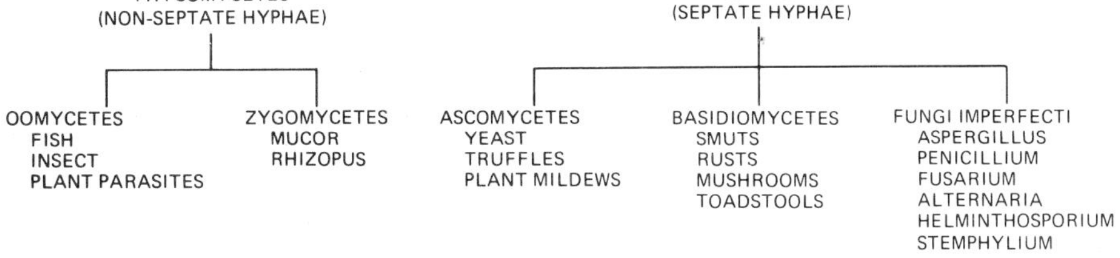

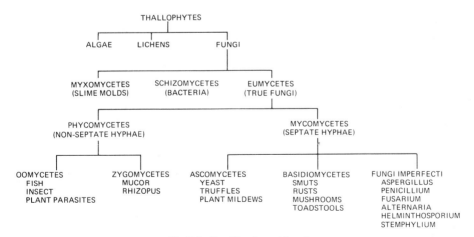

Fig. 7.1 Classification of fungi.

1-5 microns in width and from 1-9 microns in length. They possess a thick membrane or cell wall. Yeasts usually multiply by budding. The true yeasts form ascospores, although non-sporulating yeasts are also found among the *Fungi Imperfecti.*

Standing midway between the true bacteria and the more complex fungi are the *Actinomycetes*, a group which is extremely difficult to classify. They are characterized by a delicate mycelium, septate hyphae, and generally show branching. The mycelium frequently fragments into rod- and coccal-like forms.

The general classification of fungi is shown in Fig. 7.1.

Microbial Nutrition

The growth and multiplication of microorganisms, like higher plants and animals, depend on proper nutrition. The fundamental food elements necessary for higher forms of life are also required by microorganisms. They include carbohydrate, protein or nitrogen sources, trace metals, and vitamins. Some microorganisms possess amazing synthesizing abilities; when supplied with relatively simple substrates, such as glucose, inorganic nitrogen, and metal salts, they produce all of the required amino acids and proteins of their cellular structures, the host of complex enzyme systems, and growth (as measured by reproduction) is rapid and satisfactory. Some microbial strains, however, are unable to synthesize one or more essential components of their metabolic systems and these must be supplied in the substrate. Studies of these metabolic deficiencies have led to the discovery of several members of the vitamin B complex and have permitted elucidation of the roles of many biochemical intermediates. One can generalize that those organisms capable of growing rapidly on very simple substrates are endowed with the most prolific synthesizing systems and, conversely, the nutritionally fastidious cell is less versatile.

Carbohydrates supply energy for the cell's metabolic system, and serve as a source of carbon for its cellular needs. In general, glucose is the fundamental sugar. Disaccharides such as sucrose and maltose, are usually (but not always) hydrolyzed by appropriate enzymes to monosaccharides. Thus maltose is hydrolyzed to 2 moles of glucose. Higher sugars are frequently suitable substrates. Many microorganisms can utilize the various complex carbohydrates, such as starch, while some organisms utilize pentoses.

Carbohydrates frequently represent a substantial portion of the substrate-cost so that cheap sources are attractive. Blackstrap molasses, which usually contains about 50 per cent fermentable carbohydrate, is the most abundant low-cost source. Invert or hi-test molasses may contain more than 70 per cent fermentable carbohydrate. Beet molasses is also used. Glucose and xylose are found in sulfite liquor now used for feed-yeast manufacture in some European countries and the United States.

Hydrocarbons. In recent years petroleum hydrocarbons have attracted considerable interest as the least expensive carbon source, replacing carbohydrates. Some hydrocarbon fermentations probably will become firmly established in the near future.

One of the first applications suggested was the production of microbial proteins for animal feed and human food use, to help alleviate the protein deficiencies in many parts of the world. Microbial cells such as, bacteria and yeasts, contain more than 50 per cent protein, which the cells can synthesize from very simple forms of carbon, nitrogen, and sulfur. Several large single-cell-protein plants have been announced, and they are expected to produce several hundred tons of single-cell protein annually.

Some fermentation products now made from carbohydrates may soon be made from hydrocarbons. And carbohydrates may be replaced with petrochemicals, such as cheap acetic acid. This has already been done in certain fermentations which produce amino acids in Japan.

Nitrogen sources range from simple inorganic ions (NH_4^+, NO_3^-) and urea to proteolytically digested proteins of plant and animal origin. Although growth is satisfactory with the simple nitrogen sources, it is not unusual to obtain improved yields of fermentation products from the more complex materials. Corn steep liquor (a by-product of starch manufacture by the corn wet-milling industry) is almost universally used in penicillin fermentations. Soybean meal is commonly used in *Streptomyces* fermentations for antibiotic production. Corn meal may supply both carbohydrates and nitrogen. The diversity of usable materials depends on the ingenuity of the research worker, availability of the material, and its relative cost. By-products of the processing of agricultural commodities are commonly used.

Metals are frequently required in trace amounts. These usually enter cellular metabolism as components of coenzymes. Natural materials often contain sufficient trace metals to satisfy the requirements. In fermentations producing organic acids, $CaCO_3$ is frequently employed for maintaining the pH in the most productive range by forming calcium salts of the sugar acids. Calcium ions probably serve also in a more fundamental nutritional capacity. Calcium salts of organic acids have limited water solubility and their precipitation in the fermentation medium may increase the problems of product recovery. In such instances, more soluble salts are formed by intermittent neutralizations with NaOH or NH_4OH solutions. These compounds may be introduced automatically if the fermentors are equipped with the proper type of pH control equipment.

Accessory growth factors are required by some organisms not capable of growing on simple inorganic substrates. These are usually resolved into known members of the vitamin B complex (e.g., thiamin, riboflavin, and pyridoxine), perhaps some specific amino acids, and even "unidentified growth factors." Materials such as corn steep liquor, dried yeast, and oil-seed meals furnish good sources of these growth requirements.

Intermediary metabolism of microorganisms is a fascinating study of the mechanisms involved in the release of energy from food. The general principles established for microorganisms are applicable to all forms of life. The classical glucose dissimulation pathways depicted by the Embden-Meyerhoff scheme, coupled with the equally classical Krebs "citric acid cycle," explain the accumulation of such fermentation products as ethanol, butanol, and acetone, and acids such as acetic, fumaric, and citric. Gluconic acid results from an oxidation of glucose by the enzyme "glucose oxidase."

Intermediary metabolism involves a long series of one-step reactions, each mediated by a specific biological catalyst (enzyme). The products of one reaction serve as substrates for the succeeding reaction. If a particular enzyme is missing or inoperable, intermediary products accumulate, and it is this particular feature which explains the production of most fermentation products.

On the other hand, useful complex molecules such as antibiotics result from unexplained syntheses accomplished by the respective microorganisms. The biosynthesis of penicillin has been shown to involve cystine and 1-valine.

Fermentation as a Unit Process

Rather than discuss the methodology and problems of each commercial fermentation, this space can better be devoted to a more general discussion of the principles of fermentation as a unit process. The various steps are similar to those employed in the more common chemical synthesis processes, except for the fact that a biological system is involved. Raw materials for the various fermentation processes are similar, the particular product being dependent on selection of the proper organism. The environment surrounding the organism during the fermentation markedly influences product yields. Finally, the fermentation products must be isolated from the aqueous medium.

The products of the fermentation industry vary widely. Some have a structure very similar to the principal raw material (substrate), while others are quite complex and have no structural relationship to the substrate. Furthermore, fermentation yields may represent nearly stoichiometric conversions of substrate to product, or the product may occur in the broth in minute quantities, measured in parts per million.

Five basic prerequisites of a good fermentation have been suggested:

1. A microorganism must form a useful product. The microorganism must be readily propagated, and must maintain uniform bio-

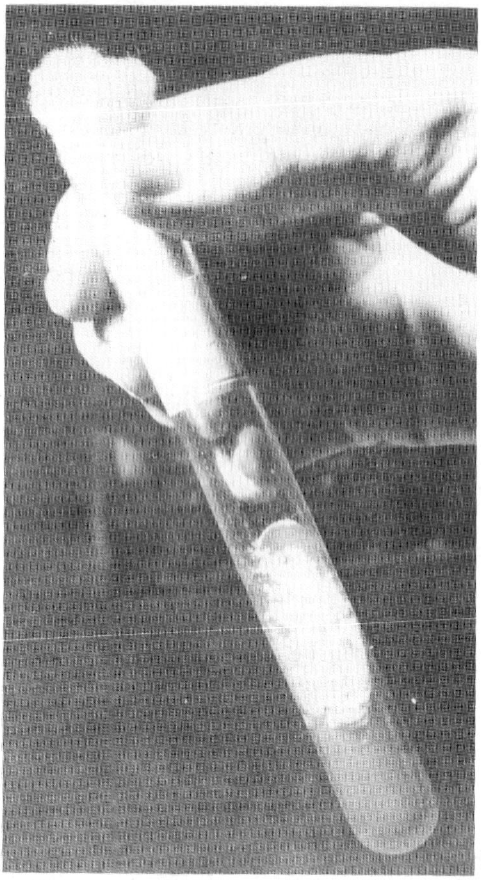

Fig. 7.2 An agar slant containing growth of a pure culture of a selected microorganism. (*Courtesy Chas. Pfizer Inc.*)

logical and biochemical characteristics to give consistent yields.

2. Economically attractive raw materials in good supply and predictably uniform in composition must be available.
3. Acceptable yields must be based on economics.
4. A rapid fermentation time is desirable.
5. The product should be easy to recover and purify.

The microorganism is the heart of the whole synthetic process. It must produce the desired product, preferably with a minimum of by-products. Microbiologists have examined the fermentation products of thousands of cultures isolated from natural habitats, usually soil. From these have been selected strains producing the desired products.

In the laboratory it is generally most convenient to propagate the microorganism on a solid medium. These comprise water-soluble sugars or plant or animal tissue extracts to which a nonnutritive polysaccharide material called agar (manufactured from certain seaweeds) is added. Agar liquefies at 45-50°C and solidifies when cooled to 40°C. The hot medium is placed in test tubes, plugged with cotton, and sterilized by heating under steam pressure (15 psi or 121°C) for 15-30 minutes. During cooling, the tube is tilted to form a "slant" with considerable surface for growth of a microorganism. The same material is also used for culturing microorganisms in flat-bottomed, round, Petri dishes. Viable cultures are maintained on the slants, and stored in refrigerators after good growth is attained.

Preservation of the stock culture is imperative, and should permit long storage with a minimum of effort on the part of laboratory personnel. For instance, viable cells or spores can be maintained in a sterile dry state by lyophilization. Cell or spore suspensions are frozen and the water removed by sublimation, after which the glass tubes are sealed with an oxygen flame. Under these conditions the cells are dormant, but are readily regenerated by transfer to a suitable liquid or solid culture medium. Some spores are easily maintained for long periods of time suspended in sterile dry sand or soil.

Culture improvement is always desirable, either to obtain greater yields or faster fermentations. Since there is a natural tendency for some variation to arise, improvements are sometimes attained through single cell isolations from the mass population of a culture. However, it is recognized that fermentation processes are gene controlled and substantial yield improvements are frequently attained by destruction of certain genes. Gene destruction is accomplished by ultraviolet or X-ray irradiation of cell or spore suspensions, or culturing in the presence of mutagenic substances, such as the nitrogen mustards, morpholine, and camphor.

All these problems are within the scope of the microbiologist responsible for at least the laboratory phases of the process.

Fig. 7.3 Shake flasks containing a *streptomyces* culture. One flask has sterile medium, the other contains medium with a well-grown-out microorganism. (*Courtesy Chas. Pfizer Inc.*)

The Medium. It is axiomatic that the medium shall be as cheap as possible, consistent with yields and the product. Raw materials should, if possible, be available year round, be stable, and not require unusual storage conditions. Since the formulated liquid medium must be sterilized, most conveniently by steam under pressure, heat-labile culture medium ingredients create special handling problems.

Yields. As in chemical synthesis, yields or concentration of product in the final fermentation liquor can vary inversely with the market value. Some products (organic acids) may be produced in broth concentrations of 150 grams per liter (perhaps 90 per cent stoichiometric yield from glucose) whereas others, such as antibiotics and some vitamins, are present in concentrations of hundreds of parts per million (1 ppm = 1 microgram/ml). At the low extreme is vitamin B_{12}, which occurs in broths in yields ranging from less than 1 ppm to a few ppm.

Fermentation Rate. The length of time devoted to fermentation is dependent on the productive capacity of the microorganism. Alcohol fermentations by yeasts are generally complete in 50-60 hours; other fermentations may be shorter or longer. The longer the fermentation cycle, the greater the risk of contamination with an undesirable microorganism. Antibiotic fermentations have the advantage of producing a substance inhibitory to at least some segments of the microbial population. Fermentations which operate at low pH's are also less susceptible to contamination.

Product Stability and Isolation

Chemical compounds vary considerably in stability, and fermentation chemicals cover the whole range. Ease of recovery is also dependent on solubility factors. Some are precipitated as calcium or barium salts (organic acids), some are solvent extracted from the medium (penicillin), others may be adsorbed on ion-exchange resins (streptomycin). Distillation is used for the volatile solvents (ethanol, butanol, acetone).

FERMENTATION

The main fermentation operation methodology depends on the physiological properties of the particular microorganism. Some bacteria are anaerobes, that is, they require no oxygen; in fact the butanol-acetone-producing bacteria *Clostridium acetobutylicum* grows poorly in the presence of oxygen. Yeasts produce ethanol under anaerobic conditions, but in the presence of adequate oxygen the alcohol is further metabolized to water and CO_2 with a resultant increase in yeast crop.

Anaerobic Fermentations

The anaerobic fermentation is a comparatively simple operation. The sterile medium contained in a sterile fermentor is inoculated with the appropriate organism. The organism multiplies rapidly and the "fermentation" is established. In both the important anaerobic fermentations, ethanol by the yeast *Saccharomyces cerevisiae* and butanol-acetone by the bacterium *Clostridium acetobutylicum*, substantial quantities of CO_2, and CO_2 and H_2, respectively, are produced. These must be suitably vented from the fermentor and may be recovered as by-products. In the butanol plants, the gases from one fermentor may be used in the head space of a new fermentation to exclude atmospheric oxygen.

Alcohol Fermentation. Beverage and industrial ethanol (95 per cent ethanol) are produced by yeast fermentations, though the latter is rapidly being displaced by ethanol produced from petroleum. Industrial alcohol is itself a raw material for chemicals; it is also a solvent. It is not subject to the federal alcohol tax. In order to prevent the diversion of industrial alcohol to potable uses, it is "denatured" by the addition of some material which cannot be separated by any physical or chemical process and which renders the alcohol so treated unfit for use as a beverage. Many different formulas are authorized by the government, so that the industrial user may select the particular formula which will have least effect on his particular process. Under the supervision of federal inspectors, chemical processes which require pure industrial ethyl alcohol may be operated.

The fermentation may be conducted on nearly any carbohydrate-rich substrate. Molasses, which is the final mother syrup in the crystallization of sucrose or table sugar, is preferred because of cost and ease of handling. Blackstrap molasses is a viscous liquid containing 50 per cent fermentable carbohydrate. It is diluted to 10–14 per cent sugar, ammonium salts are added to relieve a nutritional deficiency, and it is adjusted to about pH 5–6. The medium is pasteurized, cooled, and inoculated. Grain fermentations require additional pretreatment since yeast cannot metabolize starch. The grain (usually corn) is ground, heated in an aqueous slurry to gelatinize or solubilize the starch, cooled to about 145°F, and treated with malt (germinated, dried barley) which is rich in the starch hydrolyzing enzyme diastase, principally beta-amylase. More recently, a fungal amylase produced by *Aspergillus niger* has found favor in some plants. The enzyme system converts the starch to the disaccharide maltose, which is readily fermented by the yeast. The final "beer" contains 5.5–8 per cent alcohol by volume, which is recovered by distillation, and other products including fusel oil, higher alcohols (of which isoamyl alcohol is the principal one), glycerin, and a small amount of organic acids.

The efficiency of alcohol fermentations may be compared to the classical Gay-Lussac equation:

$$C_6H_{12}O_6 \longrightarrow 2C_2H_5OH + 2CO_2\uparrow \qquad \Delta H;\ 31{,}200 \text{ cal}$$

which predicts that 0.51 pound of ethanol shall be produced from each pound of glucose. Note that considerable heat is evolved during the fermentation; this is dissipated by appropriate cooling coils in the fermentor. A flow diagram for the manufacture of industrial alcohol is given in Fig. 7.4.

Butanol-Acetone. The history of the butanol-acetone fermentation process has been fully compiled by several authors (see Prescott and Dunn, and Underkofler and Hickey) and

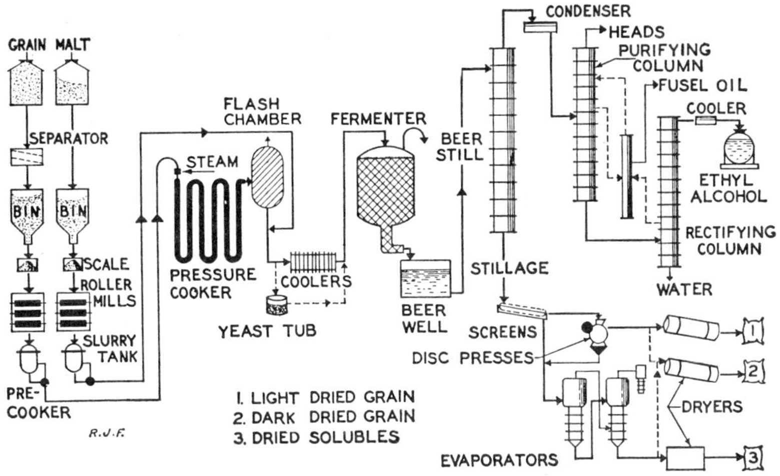

Fig. 7.4 Flow diagram for the manufacture of industrial alcohol (ethyl alcohol 95 per cent) by fermentation of grain. (*Based on a sketch in Manufacturing Process Guide for Am. Inst. Chem. Eng. Tour, 1947, Joseph E. Seagram and Sons, Louisville, Ky.*)

need not be repeated here. It also represents the first commercial fermentation process requiring aseptic techniques.

The fermentation was an outgrowth of the need for acetone for the manufacture of the explosive cordite during World War I. Butanol was an unwanted by-product. However, with the development of fast-drying automobile lacquers in the 1920's, large quantities of butanol esters were needed as solvents, and butanol became the principal product. This was indeed fortunate since the fermentation yields roughly 2 parts butanol to 1 part acetone.

The first strains of *Clostridium acetobutylicum* discovered were starch-fermenting types; thus grains such as corn were the principal media ingredients. By the 1930's some butanol, and especially acetone, were being produced by chemical synthesis, and the fermentation industry faced economic difficulties. This problem was finally solved by the discovery of molasses-fermenting strains, and the industry thrived until the end of World War II. Petrochemical processes have since dominated.

In the grain fermentation process, 8–10 per cent corn mashes were fermented (corn contains 70–72 per cent starch, dry basis). Fermentation yields were on the order of 29–32 grams mixed solvents per 100 grams starch used, with solvent ratios of approximately 60-30-10 (butanol-acetone-ethanol, respectively). The organisms possessed good diastatic activity, thus malting was not required. The cooked sterile corn meal suspension was aseptically transferred to sterile fermentors, inoculated, and incubated for about 65 hours at 37°C, after which the solvents were recovered by distillation. The aqueous residue (slop or stillage) was concentrated in multiple effect evaporators and drum dried for use in animal and poultry feeds.

With the advent of molasses-fermenting strains of *Clostridium*, more rapid fermentations were attained (40–48 hr) and the solvents produced contained as much as 65–75 per cent butanol, principally at the expense of ethanol.

After distillation of the solvents, valuable by-product feeds are obtained by drying the stillage. With proper control of the soluble iron level in the medium (1–3 ppm), the grain-fermenting clostridia produce significant amounts of riboflavin (vitamin B_2) which increase the value of the feed products. Lesser amounts of riboflavin are found in molasses stillage. Fermentation sources of riboflavin are more fully discussed elsewhere in this chapter.

The butanol-acetone fermentations are conducted in large scale equipment; fermentation of 50,000 to 500,000 gallons capacity is com-

Fig. 7.5 800-gallon bacteria—propagation tanks for butyl alcohol fermentation; one tank furnishes the right amount of culture for one 50,000-gallon fermentation tank. (*Courtesy Commercial Solvents Corp., Terre Haute, Ind.*)

monly used. A photograph of an installation for propagating bacteria for butanol fermentation is shown in Fig. 7.5.

Aerobic Fermentations

With the development of aerobic fermentations, the industry was able to produce a wide array of new products, most of which are not in serious competition with chemical synthesis. New and improved techniques and better equipment were required; new production problems were encountered and solved. This is partially explained by the fact that more fastidious conditions must be maintained.

Aerobic fermentations, as the name implies, require the use of substantial amounts of oxygen (from air). Aerobic microorganisms fail to proliferate in anaerobic systems and the limited oxygen solubility of air in water (7 ppm at 37°C) prevents the initial absorption of sufficient dissolved oxygen to satisfy the continued requirements of the organisms. In the laboratory, aerobic or at least semi-aerobic conditions in the medium can be attained by using a thin layer of liquid medium in an Erlenmeyer or Fernbach flask. The microorganism usually grows on the surface as air diffuses into the medium. Absorption may be further increased by shaker culture methods wherein the flasks are incubated on appropriate reciprocal or rotary platforms.

In large scale fermentations these conditions are obtained in a number of ways. The first is simply a scale-up of the shallow liquid method, or the pan method. A sterilized medium is placed in shallow pans and inoculated with the appropriate organism, which establishes itself as a skin over the surface of the medium. The pans are stacked in cabinets in such a way that sterile air may be forced over their surface to provide an oxygen-rich atmosphere and to sweep away CO_2 and other gases which might be produced by the fermentation. The obvious disadvantages are the considerable manual labor and equipment involved.

A modification of this method has been used for the microbial production of certain useful enzymes from molds such as *Aspergillus oryzae.* A moist, solid medium, essentially moist bran, is placed on the trays and inoculated. The bran may be stirred or mixed occasionally, but the general methodology remains. The desired enzymes are extracted from the final fermented solids, and the extract is processed to a stable liquid enzyme concentrate or dry powder.

In vinegar generators the bacterium *Aceto-*

bacter aceti grows as a heavy slime on the surface of wood chips in a packed tower. The dilute alcohol solution trickles over the surface, countercurrent to a stream of air. By recycling, a complete oxidation is attained. Edible acetic acid or vinegar, by law, must be of biological origin.

By far the most practical method is the so-called "deep tank" or "submerged aerobic" fermentation and it is universally used except for a few special situations. In this method, continuous aeration of the medium, with adequate agitation, permits luxurious growth of the microorganism throughout the medium. Fermentation rates are more rapid by this method than by others, less manpower is required, and large scale equipment may be employed. These advantages, however, have accompanying problems. Sterile air in large volumes is required since it is customary to inject into the medium $^{1}/_{4}$ to 1 volume of air per volume of medium per minute. The air must be finely dispersed so that a large surface area is available for the transfer of oxygen to the surrounding medium. Fermentors are frequently equipped at the bottom with coils or crossed pipes in which a large number of small holes have been drilled (spargers). Mechanical stirrers may be provided above the air spargers. The air droplets rise through the height of the medium and out the exhaust vents of the tanks. Coalescence of the droplets is undesirable, but little can be done to prevent it.

Sterile air is commonly produced by passing compressed air through sterile filters packed with glass wool or carbon. In such filters microorganisms are removed from the air, presumably by electrostatic and inertial impaction forces. The bulk density, fiber or particle diameter, and fiber distribution all contribute to filtration efficiency. Unfortunately, 99.9 per cent efficiency is not sufficient since a few cells, once passed through the filter and into the fermentation medium, can rapidly multiply and create a serious problem.

Mechanical agitators introduce additional problems. The packing gland where the agitator shaft enters the vessel must be more than liquid proof—it must be sterilized with the vessel and must remain sterile throughout the fermentation cycle. Some plants find it advantageous to equip such packing glands with a "steam seal," a means of maintaining a sterilized environment for the gland. This, however, accelerates thermal decomposition of the packing material. With small equipment, agitators are usually top entry; in large fermentors they may be side entry. Some recent European plants have installed bottom entry agitators. Sometimes aeration alone provides sufficient agitation.

Microbial contamination problems are more prevalent in aerobic fermentations than in anaerobic. Most microorganisms are aerobes, with all degrees of nutritional requirements. Thus there is seldom a lack of organisms capable of thriving on the production medium. Likewise, it is not uncommon to encounter contaminants which multiply as rapidly as the desired culture; competition always produces an undesirable effect on the wanted organism.

Introducing large volumes of air into a fermenting system frequently induces foaming. Antifoam agents such as sterile vegetable or animal fats (soybean or lard oil), or synthetic defoamers such as some of the liquid silicones, are commonly used. The fats are frequently metabolized and so must be added intermittently throughout the fermentation cycle. A satisfactory defoamer must be nontoxic to the microorganism, stable to heat for sterilizing purposes, and must not interfere with the recovery methods.

FERMENTATION EQUIPMENT

As in other branches of the chemical industry, a properly equipped laboratory must be provided to support the production department. Its services include the usual analytical work to indicate the progress of the process. However, a fermentation laboratory performs several other necessary services, chiefly that of maintaining the valuable stock cultures and preparation of inoculum for the production plant. Microbiological processes must be scaled up stepwise from a test-tube scale. The laboratory, under aseptic conditions, prepares a pure culture of actively growing organism. This is done in a properly designed container from which the organism can be transferred to a small inoculum tank (1-10 per cent by volume used as inoculum in each step). The inoculum tank contains a previously sterilized and cooled medium, suitable for the microorganism. Depending on the size of the final fermentation volume, the microorganism is scaled up through one or more successive inoculum stages.

Besides the usual glassware common to laboratories, a fermentation laboratory is equipped with a steam autoclave, dry-heat ovens for sterilizing glassware, incubators, and culture shakers. Sterile rooms are usually employed for transferring the growing organism from stock cultures to inoculum propagation vessels.

Production Equipment

The production plant is equipped with a battery of fermentation tanks, and auxiliary tanks used for propagating the inoculum. There is no standardization of equipment, since each manufacturer adopts equipment based on past experience. Usually, the inoculum fermentors are scaled in the range 5-10 per cent of the volume of the production fermentors. It is not unusual to have more than one size inoculum tank; the smallest may receive an inoculum from the laboratory and, after suitable growth has been attained, this in turn may be used to propagate a larger inoculum batch. Thus the scale up is step-wise, and each production fermentor can be traced back to the culture laboratory.

The size and design of fermentors probably vary a good deal throughout the industry. Fermentors of 500,000-gallons capacity have been used for ethanol and butanol-acetone fermentations; many companies have reported using 20,000-80,000-gallon fermentors for antibiotic fermentations. Fermentor construction may be of carbon steel, stainless steel, stainless-steel clad, or carbon steel with a relatively stable organic polymer coating. Others have been glass-lined or constructed of nickel alloys (Inconel) or aluminum. Probably stainless steel is the material of choice since it tends to minimize corrosion problems and the deleterious metal toxicities with respect to the microorganism. Virtually all are pressure vessels, capable of withstanding 20 psi steam pressure for sterilization purposes.

The fermentor requires certain auxiliary

Fig. 7.6 A pilot-plant scale fermentor. (*Courtesy Chas. Pfizer Inc.*)

equipment. After sterilization it must be cooled to the optimum temperature of the microorganism before being used. This is usually in the range 25-37°C, but may be as high at 55°C in the case of certain thermophilic (heat-loving) organisms. In actual operation the smaller-size tanks may be used for batch sterilization of the medium and, after cooling, operated as the fermentor. With large tanks, the sterile fermentor receives, through sterile pipelines, a previously sterilized and cooled medium. Because considerable carbohydrate is metabolized, fermentations are exothermic and heat must be removed during the more vigorous phases. For instance, in some antibiotic fermentations as much as 20-150 Btu per hour per gallon is given off. Cooling is accomplished with internal coils, jackets, or "film cooling" (running water over the outside of tank wall), or external heat exchangers.

Fig. 7.7 Some commercial-sized fermentors. (*Courtesy Chas. Pfizer Inc.*)

Media Sterilization

Sterilization implies the destruction of all forms of life in the medium. Heat is the most effective and economic method. Vegetative cells, especially of bacteria, are less heat resistant than spores. Generally, 10-30 minutes at 120°C renders any medium sterile, but one must correct for slow heat penetration rates under certain conditions. The rate of attaining sterility also depends on the microbial population; quantitative death curves are conveniently expressed as logarithmic functions.

In commercial practice media are sterilized by vigorous heating in batch or continuous systems. Two types of batch processes are found. In one, the fermentor itself is used as the batch "cooker." The liquid medium is raised to the boiling point by steam, either injected (ingredient concentrations adjusted for condensate dilution) or in coils, with suitable vents open to displace air; then, with vents closed, the temperature is raised to 121° C (15 psi) for 1 hour. In the second method, a separate batch "cooker" is used, and the sterile medium is transferred through sterile lines to a sterile fermentor.

In continuous sterilization, a heat exchanger may be used, but usually direct injection steam raises the temperature instantaneously, the medium then passing through an insulated pipe system ("retention tubes") followed by a cooling section (heat exchanger) to adjust it to the desired fermentation temperature. This is probably the preferred method; it is flexible with regard to time and temperature, fairly accurately controlled (important for heat-labile media), and economic in its use of steam and cooling water.

The Fermentation

For the main fermentation operation three basic processes have been proposed:

1. Batch fermentation.
2. Continuous fermentation (single vessel system).
3. Continuous fermentation (multiple-vessel system).

In commercial practice batch operations predominate. Most of the theoretical advantages of the continuous fermentations are discounted by the extreme difficulties inherent in maintaining a pure culture system when large volumes of fermenting broths are handled. Continuous fermentations also require a microbial culture which does not deteriorate as successive generations of cells are produced, a requirement that cannot be fulfilled readily.

However, beer, yeast from petroleum, baker's yeast, and vinegar have been made on a large scale by continuous culture, and undoubtedly other substances will be made as the problems are solved. One fermentation that might be particularly suitable for continuous culture (industrial alcohol production) has failed to compete with purely synthetic production from a price standpoint.

Fermentation time is that required for attaining the most economic product concentration. Rates of product formation vary during the cycle. At first, a "lag phase" is encountered during which the inoculum is multiplying to attain a maximum cell count. Near the end of this phase, the product begins to accumulate, and a phase of product-formation is then established. A third phase, one of declining productivity, eventually arises as a result of various factors, such as substrate exhaustion, accumulation of substances toxic to the organism, and culture deterioration.

Factors Influencing Yields

1. For peak efficiency, the environment of the microorganism must be optimum. The substrate must be selected for its nutritional value. Many microorganisms produce organic acids during carbohydrate metabolism, even though these acids are not the prime desired product. Each microorganism has a limited pH range for maximum yields. Various buffers are employed to control pH; $CaCO_3$ is undoubtedly the most widely used. Intermittent addition of mineral acids or bases, such as lime, NaOH, NH_3, and NH_4OH are not uncommon. It is possible for medium ingredients to be balanced and to control their utilization so that the remaining ions stabilize the pH.

2. Temperature control within 1-2°F is desirable. High temperatures are particularly destructive to the culture; low temperatures usually only depress the rate. Thus cooling capacity must be adequate for the short period of maximum heat of combustion (metabolism). Lesser amounts of heat arise from mechanical stirrers.

3. Aerobic fermentations require adequate aeration with efficient mixing to facilitate maintenance of desired dissolved oxygen levels. Actual measurements of dissolved oxygen in a dynamic system such as a fermenting substrate are difficult. However, these measurements and their proper interpretation could have a marked influence on equipment design and fermentation technique.

4. Temperature and the length of time of the sterilization of the media may influence yields. Such treatment may produce toxic substances in the medium or destroy labile, unidentified nutrient factors.

5. Size of inoculum frequently influences ultimate yields. No ready explanation can be offered, but through experience it is well established that maximum yields in some fermentations, for example, riboflavin production by *Eremothecium ashbyi* and *Ashbya gossypii*, are attained with young, small inocula (1 per cent). In others, 10-20 per cent inocula are preferred.

ANTIBIOTICS

The science of microbiology was established 110 years ago by the French chemist, Louis Pasteur. Among his many contributions should be mentioned the discovery of the existence of microscopic forms of life. Yeasts were established as the cause of the alcohol fermentation of wine. He later established pathogenic bacteria as the cause of many infectious diseases and even recognized viruses as causative agents of other infections.

Shortly before World War II another significant role of microorganisms began to emerge. Alexander Fleming, an English bacteriologist, is credited with first suggesting that the product of one microorganism might be used to inhibit the growth of another. Fleming observed that a chance contaminant (a Penicillium mold) clearly prevented the growth of a pathogenic *Staphylococcus* he was culturing in a Petri dish. Fleming succeeded in establishing some of the simple properties of the mold product, penicillin, and published his results in 1929. Nearly a decade later, another group of English biochemists undertook to further examine the phenomenon in the course of a broad research program for better chemotherapeutic agents. Florey, Chain, Heatley, and Abraham, of Oxford University, succeeded in isolating a small quantity of penicillin concentrate and by 1941 unequivocally demonstrated its potential usefulness in *Staphylococcus* septicemia. British government officials recognized its possible usefulness in military medicine, but could not further develop the discovery because of the serious war conditions then prevailing. The research was brought to the attention of government and industry in the United States, with the result that an international government-industry research program was established to produce the remarkable chemotherapeutic agent. Success of this program established a new class of powerful therapeutic agents, the antibiotics, which have revolutionized medical practice. Thus the product of one microorganism is used to combat an infection caused by another. To supply the huge amounts of antibiotics needed in modern medicine, the fermentation industry, too, has undergone a virtual revolution.

Since the early 1940's an intensive search for new and useful antibiotics has been in progress throughout the world. Researchers have succeeded in at least partially describing some 1200 antibiotic substances produced from microorganisms, over sixty of which have been produced on a commercial scale.

Penicillin

The original mold observed and preserved by Fleming was a strain of *Penicillium notatum*, a common laboratory contaminant. Later, cultures of *Penicillium chrysogenum* were found to be better producers of penicillin, and the present industrial strains have been derived from this species. The original strains produced the antibiotic only by surface fermentation methods and in very low yields, a few ppm. Gradually, improved media and the eventual discovery of strains productive under submerged aerobic fermentation conditions led to dramatic yield increases which made commercial production a reality. Subsequent improvements, principally in culture selection and mutation to more productive strains, further improved yields until today broths often contain 25,000 units/ml (1667 units = 1 mg potassium penicillin G). With improved production have come dramatic price reductions. For instance, a 100,000 unit vial of penicillin had a wholesale price in 1943 of $20.00. In 1952, the same vial had a wholesale value of as little as $0.13. Table 7.1 indicates the production of penicillin in 1968 and its value, which is now over 30 million dollars per year.

The original *P. chrysogenum* strains produced large amounts of unwanted yellow pigments which were difficult to remove from the recovered penicillin. Today, nonpigmented mutants, a strain known in the industry as Wisconsin 49-133 (or progeny therefrom), are universally employed. Fermentation is effected by the submerged aerobic process outlined earlier in this chapter. The desired culture is propagaged from a laboratory stock in small flasks and transferred to plant inoculum tanks. After 24 hours these are used to inoculate larger fermentors which contain a typical production medium composed of:

	Grams per 1000 Ml
Corn steep liquor	30
Lactose	30
Glucose	5.0
$NaNO_3$	3.0
$MgSO_4$	0.25
$ZnSO_4$	0.044
Phenyl acetamide (precursor)	0.05
$CaCO_3$	3.0

The medium is usually sterilized batch-wise or by means of a retention tube, cooled to 24°C, and inoculated. The time of fermentation may vary from 60–200 hours. Sterile air is blown through the tank, usually at a rate of one volume per volume per minute.

When penicillin concentration reaches its peak potency, as determined by microbiological or chemical assays, the broth is clarified by means of rotary vacuum filters. The penicillin, being acidic, is extracted from the aqueous

TABLE 7.1 United States Production, Sales and Dollar Value on Some Antibiotic Products, 1968[a]

Antibiotic Product	*Unit or Quantity*	*Production*	*Sales*	*Dollar Value*
Bacitracin total	B.U.[c]	6,274	5,844	$4,963,000
For medicinal use	B.U.[c]	371	237	894,000
For other uses	B.U.[c]	5,903	5,607	4,069,000
Neomycin for all uses	kg[d]	141,312	34,254	1,451,000
Penicillin, total	B.U.[c]	2,473,189	930,133	33,427,000
Penicillin G potassium for medical use	B.U.[c]	1,130,990	NA	NA
Penicillin G procaine for all uses	B.U.[c]	825,082	579,210	9,981,000
Semisynthetic penicillin for medical use, total	B.U.[c]	262,984	NA	NA
Ampicillin	B.U.[c]	194,138	NA	NA
Dicloxacillin, sodium	B.U.[c]	14,101	NA	NA
All other		54,745	NA	NA
All other penicillin for all uses	B.U.[c]	254,130	350,923	23,446,000
Tetracyclines for all uses	kg[d]	1,273,484	388,810	19,913,000
Antibiotics, total	lbs	10,262,000	4,383,000	93,589,000

[a]Source: U.S. Tariff Commission, Washington, D.C.
[b]NA = Not available
[c]B.U. = Billions of U.S.P. units.
[d]kg = Kilograms of Antibiotic Base.

phase into a solvent, such as methyl isobutyl ketone or amyl acetate, at a pH of 2.5 by means of a continuous countercurrent extractor, such as a Podbielniak. The penicillin extract is then re-extracted with an aqueous alkaline solution or a buffer at a pH of 6.5–7.0. A 90 per cent recovery is made at this step. The aqueous solution is chilled, acidified, and extracted again with a solvent, such as ether or chloroform. The solvent extract is

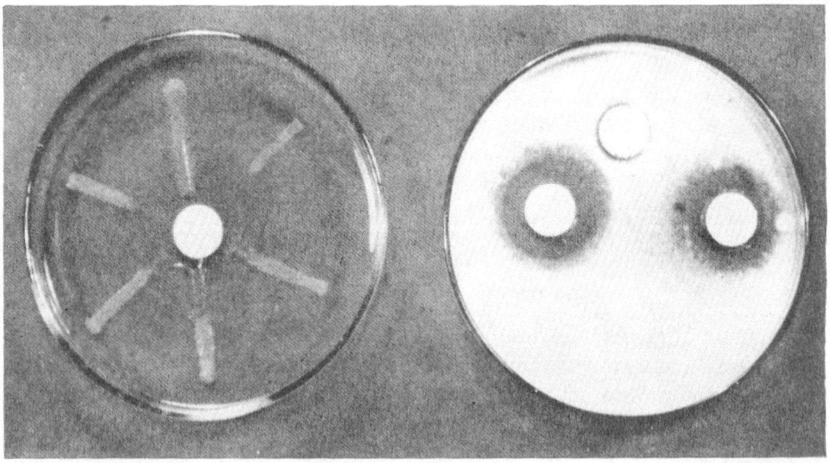

Fig. 7.8 Two methods of qualitative determination of antibiotic activity. The test broth is absorbed on filter-paper discs in the left-hand Petri dish. The disc is placed on a nutrient agar and six different test microorganisms are streaked against the disc. Those test organisms sensitive to the antibiotic have an inhibition zone near the disc. Right-hand plates and discs are placed on agar medium preseeded with test organism. Clear zone denotes inhibition or presence of antibiotic (*Courtesy Chas. Pfizer Inc.*)

Fig. 7.9 Commercial broths are usually filtered on rotary precoat filters. Antibiotics are then recovered from the clear filtrates. (*Courtesy Chas. Pfizer Inc.*)

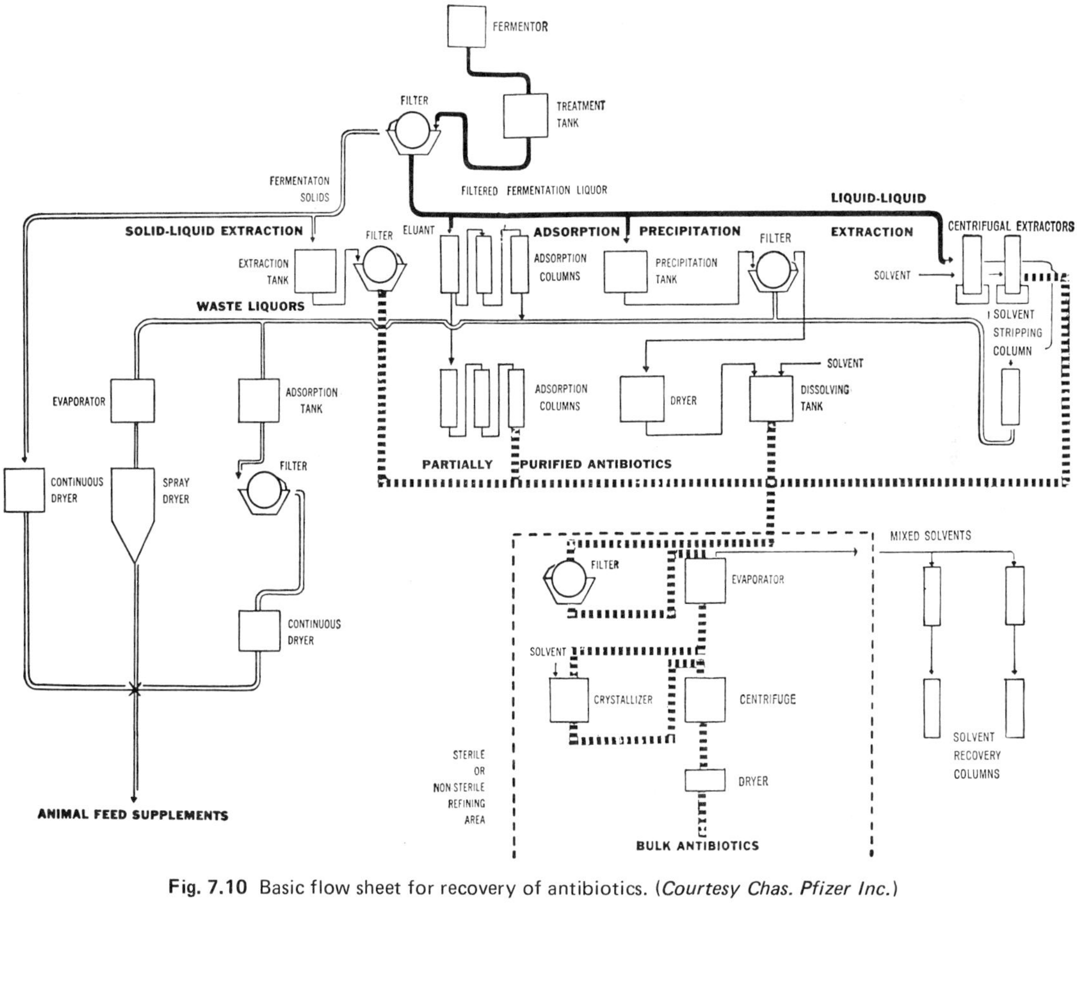

Fig. 7.10 Basic flow sheet for recovery of antibiotics. (*Courtesy Chas. Pfizer Inc.*)

then re-extracted into water at a pH of 6.5–7.0 by titration with a solution of base. The base used depends on which salt of penicillin is desired. The popular forms are sodium or potassium salts. A typical flow sheet for antibiotic recovery is shown in Fig. 7.10.

Table 7.2 gives the structural formulae of the "natural penicillins" (except penicillin V), comprising several closely related structures with aliphatic and aromatic substitutions to the common nucleus. The early impure product contained mixtures of these types. For several reasons penicillin G became the preferred type and the crystalline product of commerce. Phenylacetic acid or its derivatives are used as precursors in the fermentation medium to enhance penicillin G biosynthesis and suppress the production of the less desirable types.

The fact that proper selection of precursors could lead to new variations in the penicillin side chain offered the first source of synthetic penicillins. Penicillin V, derived from a phenoxyacetic acid precursor attracted clinical use because of greater acid tolerance, which made it more useful in oral administration.

The widespread use of penicillin eventually led to a clinical problem of penicillin-resistant staphylococci and streptococci. Resistance for the most part involved the penicillin-destroying enzyme, penicillinase which attacked the beta-lactam structure of the 6-aminopenicellanic acid nucleus (6-APA).

In 1959 Batchelor and co-workers in the Beecham Research Laboratories in England discovered that the penicillin nucleus, 6-APA, accumulated during fermentation when side chain precursors were omitted. This 6-APA could be used for the chemical synthesis of entirely new types of penicillin by coupling with new side-chains. Shortly thereafter several sources of penicillin amidase were found which would cleave the phenylacetyl side chain from penicillin G, thus producing a more economical source of 6-APA. A vast number of "synthetic penicillins" have been generated and a few have achieved clinical importance. Several objectives were sought: (1) to broaden the inherent utility of penicillin to include gram-negative pathogens not inhibited by the natural penicillins; (2) to improve its stability and absorption; (3) to increase its re-

TABLE 7.2 Structural Formula of Penicillins

Antibiotic— Penicillin	Microorganism— *Penicillium chrysogenum*	Activity Against— $Gram^+$ bacteria

Formula—

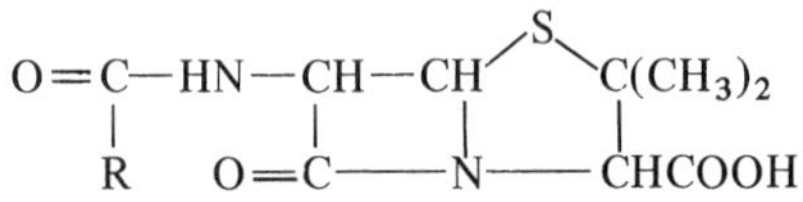

Type of Penicillin	Side Chain R Substitutions
(G) Benzyl	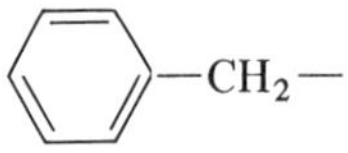
(X) *p*-Hydroxybenzyl	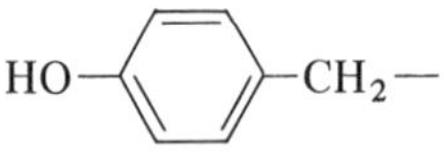
(F) 2-Pentenyl	$CH_3—CH_2—CH{=}CH—CH_2—$
(Dihydro F) *n*-Pentyl	$CH_3—CH_2—CH_2—CH_2—CH_2—$
(K) *n*-Heptyl	$CH_3—CH_2—CH_2—CH_2—CH_2—CH_2—CH_2—$
(V) Phenoxy	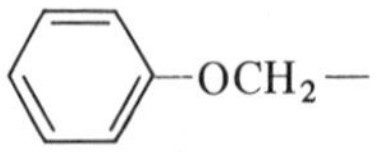

TABLE 7.3 Semisynthetic Penicillins

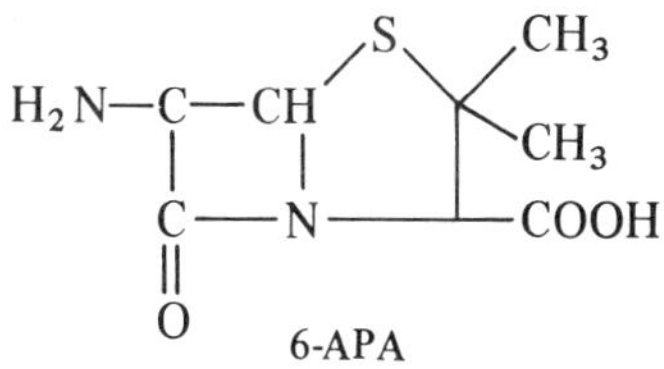

6-APA

Name	*R Group (Subst'd at NH_2)*	*Activity and Absorption*
Phenethicillin	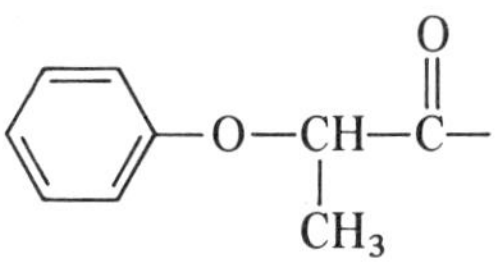	Active against some penicillin-resistant staphylococci, orally absorbed.
Methicillin	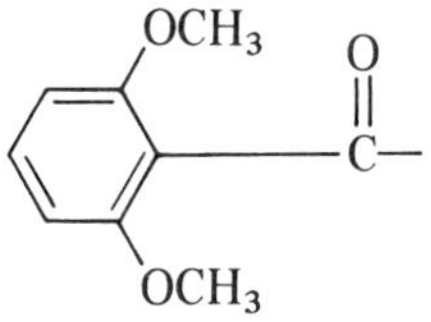	Effective against resistant staphylococci, but less active in general; not absorbed orally.
Oxacillin	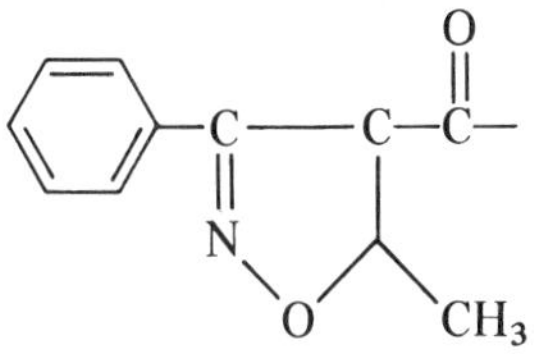	Effective against resistant staphylococci; orally absorbed.
Ampicillin	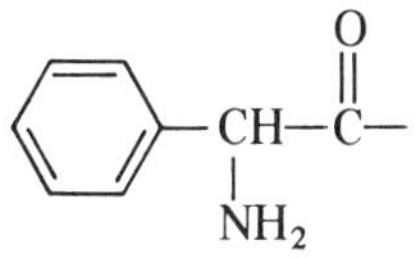	Not active against penicillin-resistant staphylococci; but active against selected gram-negative infections; orally absorbed.
Carbencillin	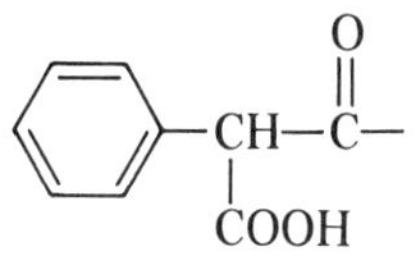	Excellent activity in selected gram-negative infections.

sistance to penicillinase-producing pathogens; (4) to decrease allergenicity; and (5) to improve other factors pertinent to clinical use. These broad objectives have been achieved with varying degrees of success.

Table 7.3 shows the structures of some of the semisynthetic penicillins which have become important chemotherapeutics.

Cephalosporins and 7-ACA

In 1948 Prof. Guiseppe Brotzu isolated a *Cephalosporium* culture from sea water near the sewer discharge of Cagliarri, Sardinia. This culture produced a broth inhibiting both gram-positive and gram-negative bacteria. The team of Florey, Abraham, and Newton at Oxford University isolated a compound identified as cephalosporin N. During the same period a group in the Michigan Department of Health isolated synnematin B from another strain of *Cephalosporium*. Synnematin B and cephalosporin N proved to be identical. Structure studies eventually proved these antibiotics to be a new type of penicillin, alpha-aminoadipyl-6-APA (also called penicillin N), a naturally produced penicillin with gram-negative activity.

In the course of studies on the Brotzu strain of *Cephalosporium*, Abraham and Newton detected small quantities of a second antibiotic, cephalosporin C. Painstaking work proved it to be chemically similar to penicillin N, but not a penicillin. It had pronounced gram-negative activity, was more stable to acid, and was not destroyed by penicillinase. It possessed the same alpha-aminoadipyl side chain as the new penicillin, but the nucleus was 7-aminocephalosporanic acid (7-ACA). 7-ACA contains a 6-membered 1,3-dihydrothiazine ring instead of the 5-membered thiazole ring in 6-APA.

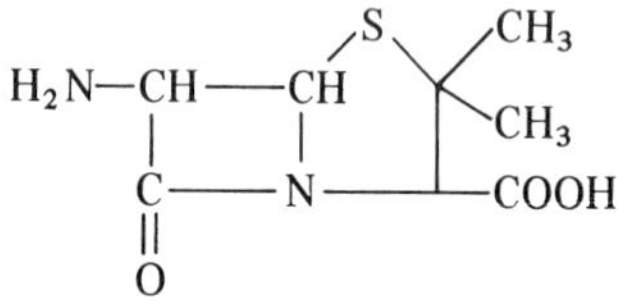

6-APA

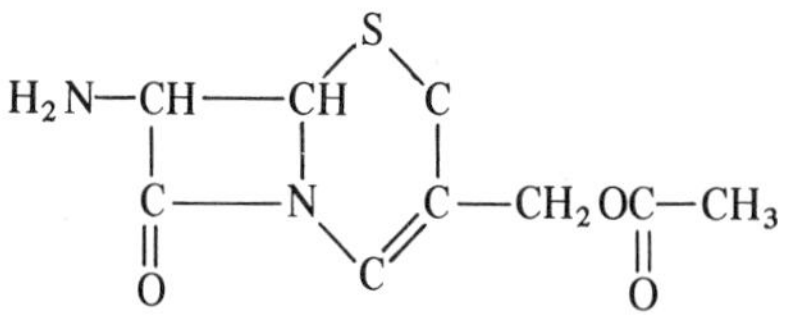

7ACA

As in the penicillin studies, the possibility of further improving the chemotherapeutic properties of cephalosporin C was apparent if the 7-ACA nucleus could be obtained. Enzymatic cleavage of the side chain failed, and the use of precursors to generate new side chains also failed; but successful chemical methods have been found. Several semisynthetic cephalosporins have been produced and are used clinically.

Very recently the chemical transformation of phenoxypenicillin (V) to the cephalosporin analog has been accomplished, a transformation which converts 6-APA to 7-ACA. This brilliant achievement may change the economics of all penicillin-cephalosporin chemistry.

Streptomycin

The second major antibiotic introduced to the medical profession was streptomycin, discovered by Waksman and associates at Rutgers University in 1944. It was particularly useful because it was inhibitory to a large group of pathogenic bacteria unaffected by penicillin, namely, gram-negative bacteria and *M. tuberculosis.* Streptomycin was also significant as the first useful antibiotic from the group of microorganisms known as the *Actinomycetes*. This group has since proved to be a prolific source of antibiotics.

Streptomycin is produced by strains of *Streptomyces griseus* when cultivated by submerged aerobic conditions at 27–29°C on media comprising starch or sugar, soybean meal, and certain inorganic salts. Yields of 15,000 mcg/ml have been reported. Streptomycin has found wide use as a chemotherapeutic agent.

Chemically, streptomycin is an organic base, not extractable with water-immiscible solvents. Commercially, it is recovered from the broth filtrates by ion-exchange methods, eluted as a mineral acid salt, and further processed to pure white powder, usually as the sulfate.

By catalytic hydrogenation, streptomycin is converted to dihydrostreptomycin. This compound has properties similar to streptomycin; but its quantitative toxicity response is somewhat modified.

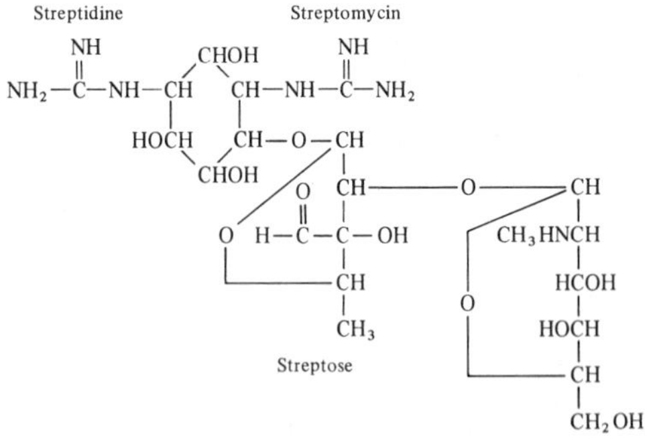

Streptomycin has also been used to treat plant diseases. It is absorbed and stored by the plant leaf. Since it is not removed by rainfall, the antibiotic within the leaf acts as a reservoir of protective agent against infection. It is particularly effective against blights.

Chloramphenicol

Following the discovery of streptomycin, attention was centered on the *Actinomycetes* as possible sources of new antibiotics. This resulted in the discovery of many more antibiotics, some of which have attained major importance. The first of these newer drugs, chloramphenicol (trade marked "Chloromycetin"), was discovered in 1947 by a group of researchers at Parke Davis and Company.

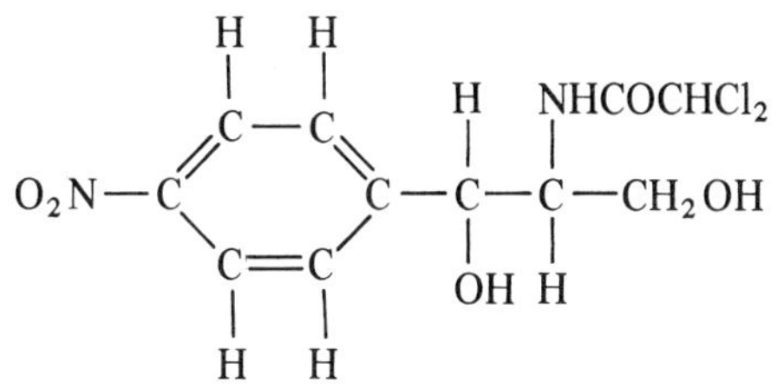

Chloramphenicol

The antibiotic was first produced by an aerobic fermentation using *Streptomyces venezuelae*. This antibiotic, in spite of its stereospecific structure, can also be produced by chemical synthesis, and the drug is now produced in this manner.

This important antibiotic attracted immediate interest because it was the first "broad-spectrum" antibiotic, effective against many diseases caused by gram-positive and gram-negative bacteria as well as against infections caused by Rickettsiae.

The Tetracycline Group

In 1948 another broad-spectrum antibiotic, chlortetracycline ("Aureomycin"), was announced from the Lederle Laboratories, Division of American Cyanamid Company. This antibiotic is produced by *Streptomyces aureofaciens* when grown under submerged aerobic conditions on media composed of sugar, corn steep liquor, and mineral salts. The crystalline compound has a golden yellow color, which suggested the trade name.

The following year a second related antibiotic, oxytetracycline ("Terramycin"), a product of *Streptomyces rimosus*, was announced by Pfizer Inc. It also is a yellow substance, chemically and biologically similar to chlortetracycline. Independent research by both companies eventually led to the disclosure of the structure of these two important chemotherapeutic agents; this has been regarded as one of the brilliant achievements of modern organic chemistry.

Both compounds may be regarded as derivatives of a nucleus known as tetracycline. Thus chlortetracycline contains a chlorine atom in the 7-position (R^2 below), whereas oxytetracycline contains a hydroxyl group in the 5-position (R^1 below). The parent compound was synthesized by catalytic hydrogenation of chlortetracycline, which displaced the chlorine atom with a hydrogen atom.

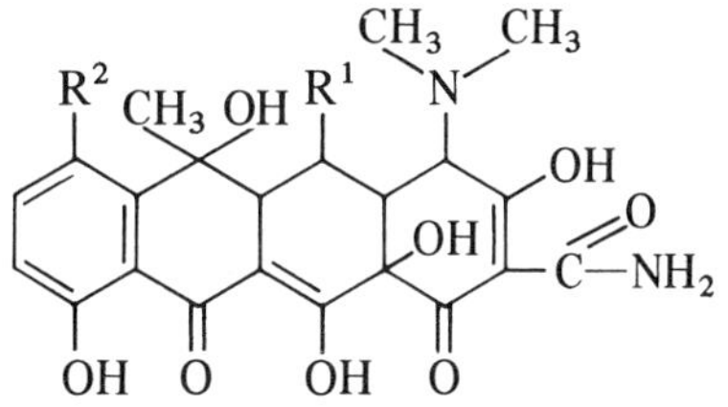

R^1 = OH; R^2 = H Oxytetracycline
R^1 = H; R^2 = Cl Chlortetracycline
R^1 = H; R^2 = H Tetracycline

Tetracycline, together with chlortetracycline, can also be produced by *S. aureofaciens* fermentations under special conditions, i.e., chloride starvation or special strains of the organism which fail to halogenate efficiently.

Tetracycline possesses the many chemotherapeutic qualities of chlortetracycline and oxytetracycline, and is currently the largest selling broadspectrum antibiotic.

Mutation of the organisms have led to other tetracycline analogs, of which 6-demethyltetracycline has clinical use. Chemical modifications of oxytetracycline has generated two other useful members of the family, known as Rondomycin® and Vibramycin®.

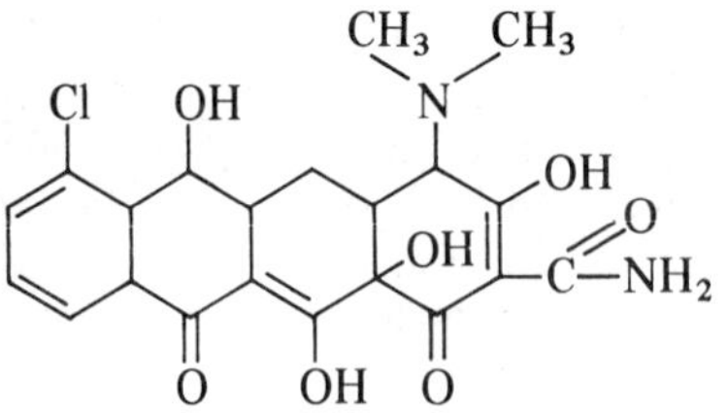

6-Demethylchlortetracycline
Declomycin® (American Cyanamid)

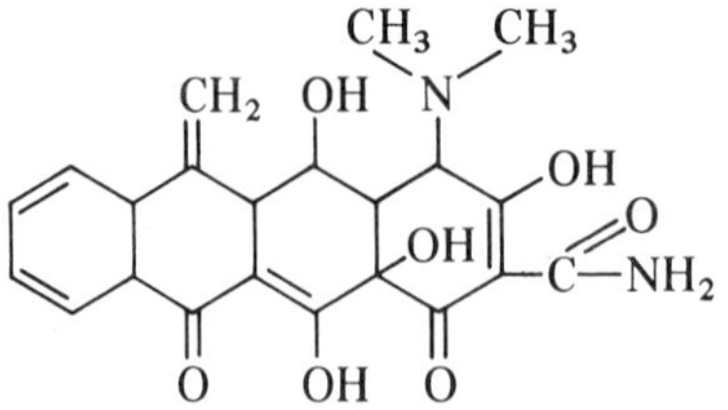

6-Methylene-6-deoxy-6-demethyl-5-hydroxytetracycline
Rondomycin® (Pfizer Inc.)
Methacycline

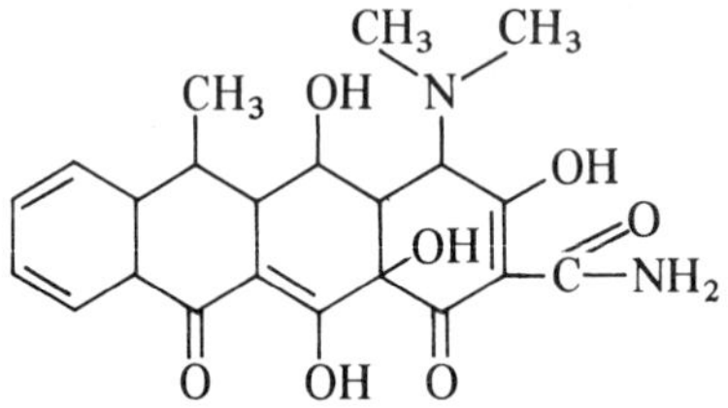

6-Deoxy-5-hydroxytetracycline
Vibramycin® (Pfizer Inc.)
Doxycycline

Erythromycin

From the broths of a fermentation carried out by *Streptomyces erythreus*, workers of Eli Lilly and Company obtained a new antibiotic known as erythromycin (Erythrocin, Ilotycin) in 1952. It was produced by submerged aerobic fermentation of soybean meal, corn steep liquor medium.

This antibiotic is the most useful in combating infections caused by strains resistant to penicillin. It is effective against infections caused by gram-positive bacteria and Rickettsiae.

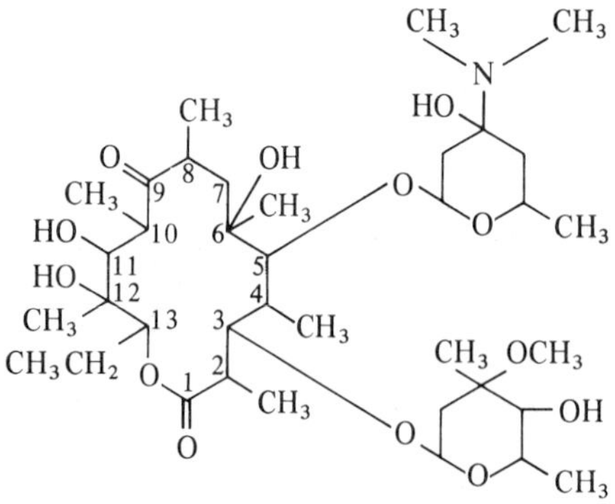

The structure of erythromycin has recently been determined. This is one of several antibiotics in a group, for which Woodward coined the term "Macrolide." Essentially, the compound comprises a sugar, cladinose, an aminosugar, desosamine, and a 14-membered lactone.

Other Antibiotics

Table 7.4 lists some antibiotics produced on a commercial scale.

The importance of new antibiotics for use in clinical medicine is stressed in most discussions of antibiotics, but their importance in other areas should not be forgotten. Antibiotics such as streptomycin and tetracycline are being used against bacterial plant pathogens, while cycloheximide and blastomycin S are being used against fungi. Antibiotics are also being used in livestock production where they improve marketable weight, and increase food utilization. Those which stimulate animal growth are bacitracin for poultry and swine; chlorotetracycline for poultry, swine, calves, cattle, and sheep; erythromycin for chickens; nystatin, oleandomycin, and procaine penicillin for poultry and swine; streptomycin for poultry; oxytetracycline for poultry, swine, cattle, and calves; and tylosin for poultry and swine. Combinations of antibiotics and their mixtures with sulpha drugs are also being used. Another potential use of antibiotics (antimycin A) in agriculture is for the elimination of fish in fresh-water ponds.

Further research on new agents to treat both human and animal diseases is certainly in order, particularly to treat those diseases which are not successfully controlled at present.

AMINO ACIDS

In recent years, amino acids have become important fermentation-industry products, largely through the ingenuity of Japanese industry. It is only natural that Japan is a major producer of amino acids, since many enzyme-modified proteins are so widely used in its dietary culture. The first amino acid to be sold for food flavoring was Japanese-produced monosodium glutamate, made from acid-hydrolyzed soy and wheat proteins.

Amino acids are the monomers of proteins. Some 20 amino acids are found in most proteins, in varying ratios. Cereal proteins are deficient in some "essential" amino acids, and the combined ingestion of animal and fish proteins with cereals provides for nutritional balance. Lysine, methionine, tryptophan, threonine, and phenylalanine levels in cereal proteins are lower than desired for the best human and animal nutrition.

Inexpensive DL-methionine is chemically synthesized in large quantities for the animal-feed industry. Its use permits lesser amounts of expensive animal protein in poultry rations. Lysine is sometimes used when fish-meal costs are too high. Lysine has been recommended as a means of vastly improving the protein value of wheat flour. Some of the smaller amino acids have utility as flavoring agents in food. Beta-alanine is in demand for the manufacture of the vitamin pantothenic acid, again used in large amounts in animal feeds and human vitamin products.

With new methods of synthesis, amino acid prices have been reduced drastically; and the essential amino acids not supplied in adequate amounts in cheap proteins can now be added. Vitamin fortification of foods was adopted in much the same fashion during World War II. Synthetic amino acids will probably play an important role in solving the protein shortages so prevalent in underdeveloped countries.

Natural amino acids are of the L-configuration, and almost without exception, the D-forms are not metabolized by animals. Whereas chemical syntheses produce DL-mixtures which must be racemized, fermentation methods produce only the natural L-forms, a particular advantage.

To develop amino acid-synthesizing fermentations some very fascinating biochemistry has unraveled the intermediary biochemical metabolism of these compounds, including the interconvertability of one amino acid into another.

All amino acid fermentations involve the use of microbial mutants which are blocked somewhere in the normal disposition pathway of the amino acid, thus permitting it or its immediate precursor to accumulate in abnormal quantities in the broth. All are aerobic fermentations. Usually ion-exchange processes are used for isolation of the amino acids from broths, and the eluted amino acids are crystallized from water

Lysine

The first amino acid to be produced commercially by a fermentation method was lysine, using a process developed by Pfizer Inc. *Escherichia coli* synthesizes its own lysine requirements by converting carbohydrate and ammonia to α,ϵ-diaminopimelic acid (DAP). Decarboxylation results in lysine. The industrial fermentation employed an *E. coli* mutant which accumulated substantial diaminopimelic acid in the culture medium. This mutant was devoid of DAP-decarboxylase. After maximum DAP yields were attained, a second organism, such as another *E. coli* strain or *Aerobacter aerogenes* was used as a source of

TABLE 7.4 Some Antibiotics Produced on a Commercial Scale[a]

Antibiotic	*Microbial Source*	*Antibiotic Spectrum*	*Chemical Type*	*Route of Administration*	*Some Commercial Sources*
Amphomycin	*Streptomyces canus*	Gram+ bacteria	Polypeptide	Topical	Bristol Laboratories
Amphotericin B	*Streptomyces nodosus*	Yeast; fungi	Polyene	Oral and parenteral	E. R. Squibb & Sons
Aminosidine	*Streptomyces species*	Gram− bacteria	Carbohydrate derived	Parenteral	Farmitalia
Bacitracin	*Bacillus subtilis*	Gram+ bacteria	Polypeptide	Topical; animal feed	Bioferm Corp.; Commercial Solvents; Kayakukosei-bussitsu; S. B. Penick; Chas. Pfizer & Co.
Blasticidin S	*Streptomyces griseo-chromogenes*	Fungi		Agricultural use (rice diseases)	Kaken Kagaku
Candicidin B	*Streptomyces griseus*	Yeast; fungi	Polyene	Topical (spermatocide)	S. B. Penick
Cephaloridine	Chemical derivative of 7-aminocephalos-sporanic acid	Gram+ and Gram− bacteria		Parenteral	Glaxo Laboratories Ltd.; Eli Lilly Co.
Cephalothin	Chemical derivative of 7-aminocephalos-sporanic acid	Gram+ and gram− bacteria		Parenteral	Eli Lilly Co.; Glaxo Laboratories Ltd.
Chloramphenicol	Chemical synthesis (formerly prepared from *Streptomyces venezuelae* fermenta-tion)	Gram+ and gram− bacteria; rickettsia		Oral and parenteral	Parke-Davis & Co.; Sankyo
Colistin	*Bacillus colistinus*	Gram− bacteria	Polypeptide	Topical and parenteral	Kayakukoseibussitsu; Bayu Seiyaku
Cycloheximide	*Streptomyces griseus*	Fungi		Agricultural fungicide	Upjohn Co.
Cycloserine	*Streptomyces orchidaceus*	Gram+ and TB bacteria	Amino acid derived	Parenteral (TB infections)	Commercial Solvents; Shionogi Seiyaku; Sumitomo Kagaku
Dactinomycin (actinomycin D)	*Streptomyces antibioticus*	Gram+ bacteria; antitumor	Polypeptide	Wilm's disease (systemic)	Merck Sharp and Dohme
Erythromycin	*Streptomyces erythreus*	Gram+ bacteria	Macrolide	Oral and parenteral	Eli Lilly & Co.; Abbott Laboratories; Roussel

Fusidic acid	*Fusidium coccineum*	Gram+ bacteria	Steroid	Oral and parenteral	Leo Pharmaceutical Products
Gentamicin	*Micromonospora purpurea*	Gram− bacteria	Carbohydrate derived	Parenteral	Schering Corp.
Gramicidin	*Bacillus brevis*	Gram+ bacteria	Polypeptide	Topical	S. B. Penick
Griseofulvin	*Penicillium griseofulvum*	Fungi		Oral and topical	Glaxo Laboratories Ltd.; ICI
Hygromycin B	*Streptomyces hygroscopicus*	Gram+ and Gram− bacteria; helminths	Carbohydrate derived	Agricultural purposes (anthelmintic)	Eli Lilly Co.
Kanamycin	*Streptomyces kanamyceticus*	Gram+, gram−, and TB bacteria	Carbohydrate derived	Parenteral	Bristol Laboratories; Banyu Seiyaku; Meiji Seika
Leucomycin	*Streptomyces kitasoensis*	Gram+ bacteria		Oral and parenteral	Toyo Jyozo
Lincomycin	*Streptomyces lincolnensis*	Gram+ bacteria		Oral and parenteral	Upjohn Co.
Neomycins	*Streptomyces fradiae*	Gram+, gram−, and TB bacteria	Carbohydrate derived	Topical	Chas. Pfizer & Co.; S. B. Penick; E. R. Squibb & Sons; Takeda Yakuhin; Upjohn Co.; Boots
Novobiocin	*Streptomyces niveus*	Gram+ bacteria		Parenteral and oral	Merck Sharp & Dohme; Upjohn Co.; Lepetit
Nystatin	*Streptomyces noursei*	Fungi; yeast	Polyene	Oral and topical	E. R. Squibb & Sons
Oleandomycin	*Streptomyces antibioticus*	Gram+ bacteria	Macrolide	Oral and parenteral	Chas. Pfizer & Co.
Paromomycin	*Streptomyces rimosus*	Gram+, Gram−, and TB bacteria; protozoa	Carbohydrate derived	Oral	Farmitalia; Parke-Davis & Co.
Penicillins					
Pencillin G	*Penicillin chrysogenum*	Gram+ bacteria	Amino acid derived	Oral (buffered) parenteral	Abbott Laboratories; Banyu Seiyaku; Glaxo Laboratories Ltd.; ICI; Eli Lilly & Co.; Meiji Seika; Nihon Kayaku; Chas. Pfizer & Co.; E. R. Squibb & Sons; Takeda Yakuhin; Toyo Jyozo; Wyeth
Pencillin V	*Penicillium chrysogenum*	Gram+ bacteria	Amino acid derived	Oral	Abbott Laboratories; Eli Lilly & Co.

Table 7.4 (contd.)

Antibiotic	*Microbial Source*	*Antibiotic Spectrum*	*Chemical Type*	*Route of Administration*	*Some Commercial Sources*
Penicillin O	Penicillium	Gram+ bacteria	Amino acid	Parenteral	Upjohn Co.; Meiji Seika; Toyo Jyozo; Wyeth
Cloxacillin	Chemical derivative of 6-aminopenicillanic acid	Gram+ bacteria		Oral and parenteral	Beecham Laboratories; Bristol Laboratories
Dicloxacillin	Chemical derivative of 6-aminopenicillanic acid	Gram+ bacteria		Oral and parenteral	Beecham Laboratories
Methicillin	Chemical derivative of 6-aminopenicillanic acid	Gram+ bacteria		Parenteral	Bristol Laboratories; Beecham Laboratories; E. R. Squibb & Sons
Nafcillin	Chemical derivative of 6-aminopenicillanic acid	Gram+ bacteria		Oral and parenteral	Wyeth
Oxacillin	Chemical derivative of 6-aminopenicillanic acid	Gram+ bacteria		Oral and parenteral	Bristol Laboratories; E. R. Squibb & Sons; Banyu Seiyaku
Phenethicillin	Chemical derivative of 6-aminopenicillanic acid	Gram+ bacteria		Oral and parenteral	Bristol Laboratories; Meiji Seika; Chas. Pfizer & Co.; E. R. Squibb & Sons; Pfizer Taito
Ampicillin	Chemical derivative of 6-aminopenicillanic acid	Gram+ and Gram− bacteria		Oral and parenteral	Beecham Laboratories; Bristol Laboratories
Polymyxin B	*Aerobacillus polymyxa*	Gram− bacteria	Polypeptide	Topical	Burroughs-Wellcome; Chas. Pfizer & Co.
Pristinamycin	*Streptomyces sp.*	Gram+ bacteria	Polypeptide	Oral	Rhone-Poulenc
Rifomycin SV	*Streptomyces mediterranei*	Gram+ and TB bacteria		Oral and parenteral	Lepetit
Spiramycin	*Streptomyces ambofaciens*	Gram+ and Gram− bacteria; rickettsia	Macrolide	Oral	Kyowa Hakko; Rhone-Poulenc
Staphylomyan	*Streptomyces virginiae*	Gram+ bacteria	Peptide	Oral	Recherche Industie Therapie

Streptomycin	*Streptomyces griseus*	Gram+, Gram–, and TB bacteria	Carbohydrate derived	Parenteral; agricultural uses	Eli Lilly & Co.; Merck Sharp & Dohme; Chas. Pfizer & Co.; E. R. Squibb & Sons; Kyowa Hakko; Meiji Seika
Dihydro-streptomycin	Chemical derivative of streptomycin	Gram+, Gram–, and TB bacteria		Parenteral	Eli Lilly & Co.; Merck Sharp & Dohme; Chas. Pfizer & Co.; E. R. Squibb & Sons
Tetracyclines					
Chlortetracycline	*Streptomyces aureofaciens*	Gram+ and Gram– bacteria;	Anthracycline	Oral and parenteral	Lederle Laboratories; Nopco
6-Dimethyl-7-chlortetracycline	*Streptomyces aureofaciens*	Gram+ and Gram– bacteria; rickettsiae	Anthracycline	Oral and parenteral	Lederle Laboratories
5-Hydroxy-tetracycline	*Streptomyces rimosus*	Gram+ and Gram– bacteria; rickettsiae	Anthracycline	Oral and parenteral	Chas. Pfizer & Co.
6-Methylene-6-deoxy-5-hydroxytetra-cycline	Chemical derivative of 5-hydroxytetracycline	Gram+ and Gram– bacteria; rickettsiae	Anthracycline	Oral and parenteral	Chas. Pfizer & Co.
Tetracycline	*Streptomyces aureofaciens;* chemical derivative of chlortetracycline	Gram+ and Gram– bacteria; rickettsiae	Anthracycline	Oral and parenteral	Rachelle Laboratories; Bristol Laboratories; Lederle Laboratories; E. R. Squibb & Sons; Lepetit Spa; Chas. Pfizer
Thiostrepton	*Streptomyces azureus*	Gram+ bacteria	Polypeptide	Topical	E. R. Squibb & Sons
Trichomycin	*Streptomyces hachijoensis*	Fungi; yeast	Polyene	Topical	Fijisawa Yakuhin
Tylosin	*Streptomyces fradiae*	Gram+ bacteria	Macrolide	Agricultural uses	Eli Lilly & Co.
Tyrothricin	*Bacillus brevis*	Gram+ and Gram– bacteria	Polypeptide	Topical	S. B. Penick
Vancomycin	*Streptomyces orientalis*	Gram+ and TB bacteria		Parenteral	Eli Lilly & Co.
Variotin	*Paecilomyces varioti*	Fungi; yeast		Topical	Nihon Kayaku
Viomycin	*Streptomyces floridae*	Gram+, Gram–, and TB bacteria	Amino acid derived	Parenteral	CIBA; Parke-Davis & Co.; Chas. Pfizer & Co.

[a]Perlman, D., "Process Biochemistry," Morgan-Grampian Ltd., London, April, 1968 (by permission).

DAP-decarboxylase, resulting in the production of lysine.

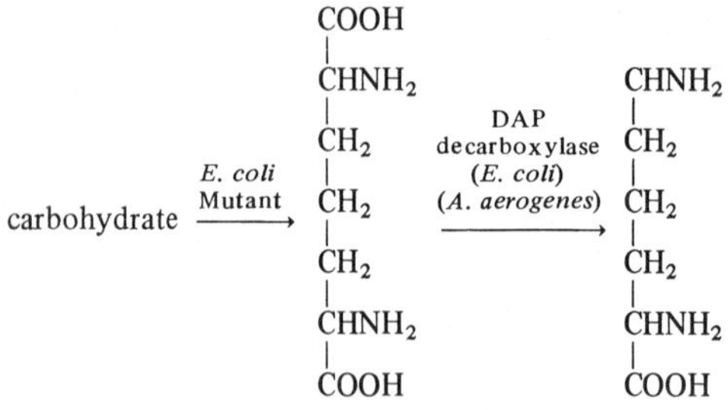

Diaminopimelic Acid (DAP) L-Lysine

Shortly thereafter a single-fermentation method was devised by Dr. S. Kinoshita of Kyowa Hakko Kogyo Co., Ltd. in Japan. By mutation he secured an *Arthrobacter* strain which accumulated lysine directly.

Glutamic Acid

Marketed as monosodium glutamate, a good flavor adjunct, glutamic acid is the most widely used amino acid. Formerly made from proteins, it is now made by chemical synthesis and several different fermentation processes. In Japan alone, annual production is greater than 100,000 tons per year. The total worldwide production is approximately 200,000 tons.

Again, a mutant of *Arthrobacter* was found to produce large amounts of this amino acid from glucose. Lately, compounds such as *n*-paraffins or acetic acid have supplanted carbohydrate substrates.

Other Amino Acids

Fermentation methods for several other amino acids are known, but have not yet reached commercial development on a large scale. Additional cost-reductions are required. Among these can be included tryptophan, phenylalanine, threonine, valine, and tyrosine. Both tyrosine and phenylalanine are receiving new impetus because of the clinical use of L-dopa (L-3,4-dihydroxyphenylalanine) for the oral treatment of Parkinsonism, a disease syndrome reflecting lowered amounts of dopamine in the brain. L-dopa provides a precursor of dopamine capable of being transmitted across the blood-brain barrier.

Other Uses

Stimulated by the need for amino acids in foods and feeds, mass production has lowered prices to levels where amino acids can be seriously considered as chemical raw materials. Some exciting new synthetic fibers have been developed which may have important textile uses. The unique chemistry of amino acids will probably yield other products of utility to mankind.

VITAMINS

Another important area of microbial biosynthesis is the production of vitamins.

Riboflavin

Riboflavin was first produced microbiologically by species of the genus *Clostridium*, the anerobic bacteria used for the microbial production of acetone and butyl alcohol. Riboflavin was purely a by-product and was found in the dried stillage residues in amounts ranging from 40–70 mcg per gram of dried residue. Further research developed improvements, adaptable only to the fermentation of cereal, grains, and milk products, by *Clostridium acetobutylicum* to yield residues containing as high as 7000 mcg per gram. This was effected principally by reducing the iron content of the medium to 1–3 ppm, and fermenting in stainless steel, aluminum, or other iron-free tanks.

Later investigations disclosed that riboflavin could be produced by species of a yeast, *Candida flareri* and *C. guilliermondi*, when grown under aerobic conditions in a medium containing a fermentable sugar, an assimilable source of nitrogen, biotin, and less than 10 mcg of iron per 100 ml of medium. Yields as high as 200 mcg per ml were obtained.

Other studies on a fungus, *Eremothecium ashbyii*, and a closely related organism known as *Ashbya gossypii* resulted in the production of much larger amounts of riboflavin. An aerobic process was used in which the iron content was not critical. Riboflavin can be produced in large amounts by the fermentation industry using either the *Eremothecium* or *Ashbya* strains. Yields as high as 10–15 thousand mcg per ml are possible.

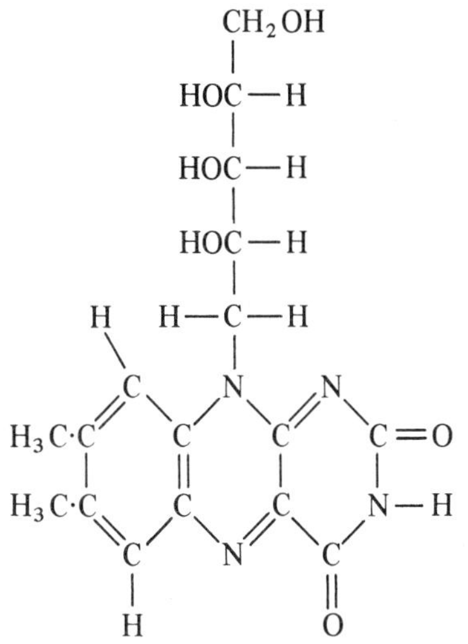

Riboflavin: Vitamin B_2

Concurrently with the development of a fermentation process, a synthetic means of producing riboflavin was discovered, and this process predominates today.

Vitamin B_{12}

The microbiological production of vitamin B_{12}, or cyanocobalamin, arose from an interesting sequence of events. For many years liver extract was used to check cases of pernicious anemia. Investigators at Merck and Company discovered that crystalline extracts made from liver contained the highly active compound responsible for the therapeutic action. Identity with the antianemia factor in liver was established and the compound was called vitamin B_{12}. Later it was found that spent liquors from streptomycin and other antibiotic fermentations contained appreciable quantities of vitamin B_{12}. Further investigations resulted in separate fermentations for the production of vitamin B_{12}, using selected strains of *Actinomycetes* and bacteria. Subsequent investigations disclosed that dried sewage residues from the activated-sludge process also contained vitamin B_{12}. Today vitamin B_{12} is obtained from various antibiotic-fermentation broths or separate fermentations using selected strains of *Streptomyces*, *Propionobacterium*, or *Pseudomonas* cultures.

The isolation of vitamin B_{12} from fermentation media, where it is normally present in parts per million, is a brilliant achievement on the part of the chemist and the chemical engineer. Vitamin B_{12} is an exceedingly active compound biologically. It is currently (1970) selling for \$8.00 per gram of **U.S.P.** material. This is a considerable reduction from 10 years ago, when it was selling for \$139.00 per gram.

ORGANIC ACIDS

The microbiological production of organic acids represents one of the earlier areas of fermentation, necessary for the accumulation of information which made possible the large scale production of antibiotics and other microbial products of more recent date.

Citric Acid

Citric acid is one of the most important organic acids used in foods, beverages, and pharmaceuticals. During the past few years it has also become important as an organic intermediate.

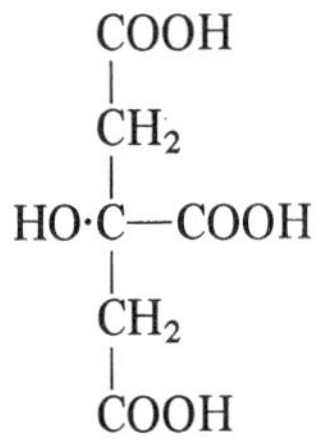

Citric Acid

Citric acid was first isolated from lemon juice in 1784 by Scheele. In 1917 Currie, of the U.S. Department of Agriculture, found citric acid could be produced microbiologically by using *Aspergillus niger* grown on a sugar-mineral salts solution. Since then, many other microorganisms have been shown to produce citric acid; however, *A. niger* has always given the best result in industrial production. Citric acid may be fermented either by using shallow pans or by employing a submerged or deep fermentation process with aeration.

Sucrose in the form of cane or beet molasses is the principal source of sugar. A 12-20 per cent sugar solution is normally used along with mineral supplements. The duration of shallow pan fermentation is 7-10 days at

26–28°C. Submerged fermentation periods are shorter but yields are less. On shallow pans, yields on the sugar used may be 90–95 per cent, while the submerged process normally runs 75–85 per cent. The citric acid is recovered as the calcium salt and treated with sulfuric acid to precipitate calcium sulfate, which is removed. Citric acid crystallizes upon concentrating the resulting solution. Some oxalic acid is recovered as a by-product of the citric fermentation process.

Recently, it has been shown that certain strains of *Candida* (a yeast) can produce citric and isocitric acid from *n*-paraffins or carbohydrates. The impact of this research development may affect the future production method.

Current production of citric acid in the U.S. is approximately 130,000,000 lbs as it continues to be the leading food acidulant.

Gluconic Acid

Gluconic acid is produced by the oxidation of the aldehyde grouping of glucose.

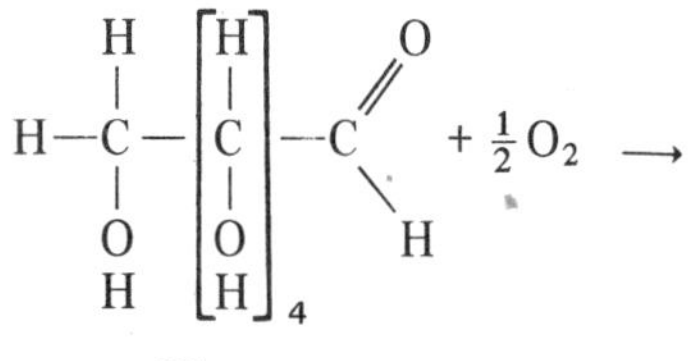

Glucose

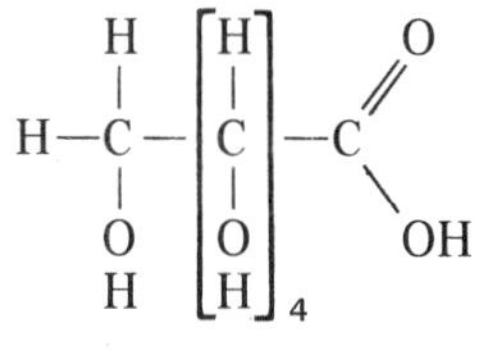

Gluconic Acid

Gluconic acid may be prepared from glucose by oxidation with a hypochlorite solution, by electrolysis of a solution of sugar containing a measured amount of bromide, or by fermentation of glucose by molds or bacteria. The latter method is now preferred, from an economic standpoint. The most important microorganisms used are *Aspergillus niger*, and *Acetobacter suboxydans* grown on a glucose-salt solution in deep tank fermentation. Yields as high as 90 per cent on the sugar consumed have been reported. Gluconic acid is marketed in the form of several crystalline metal salts, 50 per cent aqueous acid, and the delta-lactone. Calcium gluconate is frequently used as a nutritional calcium source because of its solubility. The sequestering properties of sodium gluconate, particularly for Ca^{++} and heavy metal ions in strong caustic solution, make it useful in cleaning operations.

Acetic Acid

Acetic acid in the form of vinegar (by law, 5 per cent acetic acid) is a widely used food adjunct. Vinegar is produced by the oxidation of ethanol by bacteria of the *Acetobacter* genus. In the food industry many vinegar types are classified on the basis of the source of alcohol. Most vinegar is made from apple cider; a yeast converts the sugar to ethanol, and the acetification is accomplished by *Acetobacter aceti* strains.

$$C_6H_{12}O_6 + \text{yeast} \longrightarrow 2C_2H_5OH + 2CO_2\uparrow$$

$$C_2H_5OH + O_2 + \textit{Acetobacter aceti} \longrightarrow CH_3COOH + H_2O$$

Ethanol may also be converted into acetic acid by catalytic oxidation at high temperatures, but synthetic acid cannot be used in foods.

Itaconic Acid

Itaconic acid is an unsaturated dibasic acid which may be used for the preparation of resins or surface active agents, or in the manufacture of synthetic organic chemical compounds. Its esters may be polymerized.

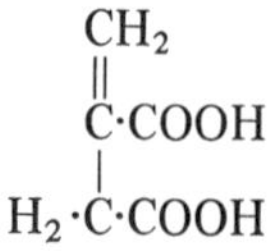

Itaconic Acid

Itaconic acid may be produced by either a shallow-pan or a deep-tank fermentation process by growing *Aspergillus terreus* or *A. itaconicus* on lactose-, glucose-, or molasses-salt media. Fermentation of solutions of 20–25 per cent glucose gives yields equivalent to 50–70 per cent based on the sugar consumed.

Kojic Acid

Kojic acid was first discovered in Japan in 1907 by Saito; it was a by-product of the fermentation of steamed rice by *Aspergillus oryzae*. Various other investigators have found that numerous species of *Aspergillus* and some *Acetobacter* bacterial strains produce kojic acid. In 1955 it was first produced on a commercial scale by a fermentation process. Kojic acid is an acid weaker than carbonic. It is reactive at every position and forms a number of products.

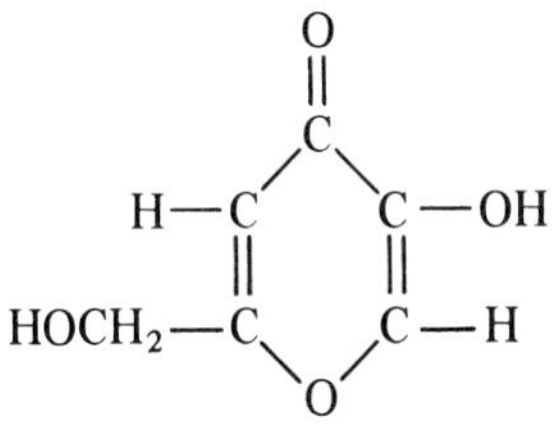

Kojic Acid

One of its major uses has been for the manufacture of maltol and ethyl maltol, widely used in foods as flavor-enhancing agents. Chemically the $—CH_2OH$ group is oxidized to —COOH (comenic acid) which is removed by pyrolysis (pyromeconic acid). The 1,4-pyrone nucleus is reactive at the 5-position with formaldehyde or acetaldehyde, and the reduction of the respective aldehydes gives maltol or ethyl maltol.

OTHER KETOGENIC FERMENTATIONS

Sorbose

Sorbose fermentation, the first bacterial ketogenic fermentation discovered, is one of the simplest. L-Sorbose is produced from the polyhydric alcohol sorbitol by the action of several species of bacteria of the genus *Acetobacter*. Sorbitol is made by the catalytic hydrogenation of glucose. The most commonly used microorganism is *Acetobacter suboxydans*. Since this organism is very sensitive to nickel ions, it is important that the medium and fermentor be free of nickel. The medium normally consists of 100-200 grams per liter sorbitol, 2.5 grams per liter corn steep liquor, and antifoam such as soybean oil. The medium is sterilized and cooled to 30-35° C, where about 2.5 per cent inoculum is added. The tank is aerated and sometimes stirred. Yields of 80-90 per cent of the sugar used are commonly obtained in 20-30 hours.

The only commercial use of L-sorbose is in the manufacture of ascorbic acid (vitamin C). The chemical steps in the conversion of sorbose to ascorbic acid involve the preparation of the diacetone derivative which is then oxidized; the acetone groups are removed and the resultant 2-keto-L-gluconic acid is isomerized to the enediol with ring closure. This concept of using combined microbiological and chemical conversions has recently been applied with commercial success to the preparation of new steroid drugs.

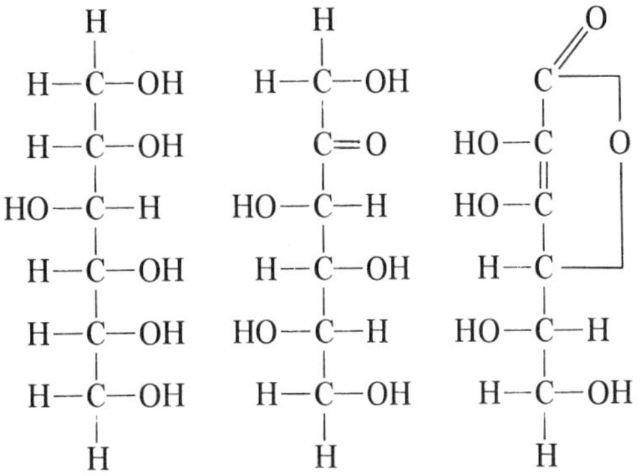

Sorbitol   Sorbose   L-Ascorbic Acid (Vitamin C)

2-Ketogluconic Acid

2-Ketogluconic acid may be produced by a bacterial fermentation involving various strains of *Acetobacter* and *Pseudomonas*. Selected strains of *Pseudomonas fluorescens* have been reported as giving the highest yields (up to 70 per cent) when glucose or gluconate is used in the medium in highly aerated processes. D-Gluconic acid is an intermediate in the oxidation of glucose to 2-ketogluconic acid. 2-Ketogluconic acid is structurally related to both gluconic acid and glucosone, and may be derived from both by oxidation. The 2-ketogluconic acid is recovered as the calcium salt. The principal use of 2-ketogluconic acid is as an intermediate in the preparation of D-araboascorbic acid (isoascorbic acid), now known to the trade as erythorbic acid.

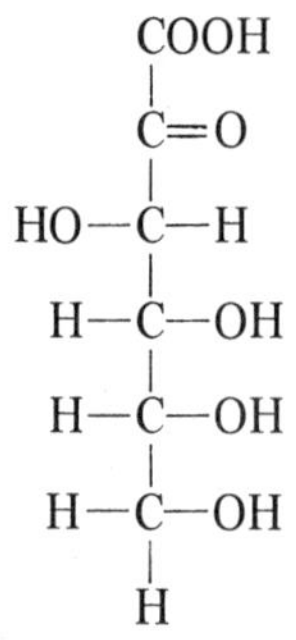

D-2-Ketogluconic Acid

STEROID TRANSFORMATIONS

As one peruses biochemical literature dealing with the activities of microorganisms, it becomes apparent that many reactions are effected which cannot easily be duplicated by classical chemical methods. Some of the very useful ones have established a vigorous branch of the pharmaceutical industry. Chemical and biochemical reactions have been joined to convert relatively inexpensive raw materials into powerful chemotherapeutics which have brought dramatic changes in the treatment of severe chronic inflammatory diseases. Bacteria, filamentous fungi, streptomycetes, and even protozoa, produce useful enzymes, but the filamentous fungi are the most versatile.

With the discovery that cortisone might be useful in the treatment of arthritis, and the final elucidation of its structure by the brilliant work of Kendall, production of the chemotherapeutic attracted attention. Purely chemical routes of synthesis from bile acids gradually developed. Another route, probably more widely employed, utilized biochemical reactions in part of the process.

Microorganisms are known to perform a variety of transformations of the steroid nucleus, principally hydroxylation, dehydrogenation, hydrogenation, epoxidation, perhaps direct ketone formation, and even cleavage of the side chain at carbon atom 17.

Steroid transformations are conducted somewhat differently from the usual fermentations. The selected microorganism is propagated in the fermentor until a good cellular mass is attained (usually aerobically). Then an appropriate steroid substrate is added as a solution or well dispersed suspension in a nontoxic organic solvent. The enzymatic reaction proceeds until the maximum product and minimum substrate are attained. Paper chromatographic methods are used almost exclusively for following the progress of the reaction.

Further information on steroid transformation may be obtained by consulting the reference by Charney and Herzog.

OTHER MICROBIAL TRANSFORMATIONS

There are other transformations, particularly hydroxylations, that microorganisms can do more efficiently than chemists. Recent examples are the conversion of acetanilide to the 2-hydroxy derivative by the higher fungus *Amanita muscaria* and to the 4-hydroxy derivative by a species of *Streptomyces*, and the hydroxylation of N-(3-chloro-4-methylphenyl)-N, N′-diethylenediamine and of Lucanthone by *Aspergillus sclerotiorum*. The latter gives hycanthone, which is used as a schistosomacide. The conversion of phenylalanine to tyrosine and tyrosine to L-3,4-dihydroxyphenylalanine (L-dopa), currently showing promise in the treatment of Parkinson's disease, also fall under the same heading.

The last conversion is of interest in that the normal first step in the microbiological decomposition of tyrosine is deamination. Therefore the amino group must be chemically protected. This can be done by preparing an N-formyl or similar derivative. It is also necessary to add ascorbic acid to stop melanin-formation. The ability of microorganisms to effect a specific step or steps in a reaction sequence may be of considerable use in the next decade in the manufacture of pharmaceutically-active compounds.

ENZYMES

All fermentation processes are the result of the enzyme activity of microorganisms. In fact, life itself, whether plant or animal, involves a complex myriad of enzyme actions. Enzymes are specific proteins which have biocatalytic activity. Each enzyme catalyses a specific chemical reaction.

The practical application of enzymes dates back many centuries, long before the nature of enzymes was understood. The coagulation of milk solids, observed by the ancients when they used calf stomachs as storage vessels, is now recognized as due to the special proteases, renin and pepsin, exclusively used in cheese making. Flax is still retted by submersion of the stalks in water; a microbial flora especially productive of pectinase develops. The mucilages are dissolved leaving cellulose fibers used for textiles. Germinated barley was used in ancient brewing. A whole industry has evolved, producing germinated barley malt for modern brewing and the alcohol fermentation industries.

The natives of Africa, South and Central America, and the Pacific Islands developed special treatments with food which are now known to contain proteases, like papain from papaya fruit, bromelin from pineapples, and ficin from wild figs.

The oriental cultures, especially in Japan and China, developed many unique foods which involved enzymatic modification of plant proteins. Of these, perhaps soy sauce is the most widely known in the Western world.

Even breadmaking involves many enzyme reactions. Wheat flour contains varying amounts of amylases and proteases. It is not uncommon in the baking industry to adjust these enzyme levels to secure the type of bread desired.

At the turn of the century the famous Japanese biochemist Dr. J. Takamine, recognizing that microorganisms were the enzyme sources so desired in oriental foods, pioneered the production of enzymes from microbial sources. The production of microbial enzymes is now a significant segment of the fermentation industry.

Bacterial α-amylase (*Bacillus subtilis*) is widely used to modify the viscosity of the starch used in coating printing papers. The same enzyme removes the starch sizes applied to cotton thread before weaving. Fungal amylases are used in the production of approximately 3 billion pounds of corn syrups in the U.S. They are widely used in food manufacture. Pectinases provide clarity to fruit juices and wines. In jelly manufacture, natural pectins are removed from fruit juices so that specific amounts of pectin can be added to ensure uniform products.

Crystalline glucose is a major product of the corn-wet milling industry. Besides its many uses in foods, it is a key chemical raw material for important products, such as ascorbic acid (vitamin C), erythorbic acid, gluconic acid, and penicillin. Formerly made by a laborious, controlled acid hydrolysis of starch, it is now made enzymatically using microbial amyloglucosidase (glucosidase). Starch is first gelatinized in hot water, then thinned by mild acid or, preferably, bacterial alpha-amylase, after which the amyloglucosidase enzyme is applied. Yields are higher, and glucose crystallization is improved over the older acid-hydrolysis method. In 1970 the U.S. production of glucose was over a billion pounds.

Another new enzyme, glucose isomerase, is commercially used by the starch industry. This enzyme will isomerize glucose to fructose, and the starch industry is producing a form of invert sugar by this method.

Certain alkaline-active proteases and amylases have been added to household detergents to improve their clothes-washing power.

Medically related uses of enzymes are gradually receiving attention. In some countries mixtures of enzymes are commonly used as digestive aids. Streptokinase and streptodornase are used in wound debridement. L-Asparaginase has shown some utility in the remission of lymphatic leukemia. Dextranase has been studied as a means of removing dental plaque and in reducing the occurrence of cavities.

Because many chemical reactions which are difficult by conventional chemical means can be easily effected by enzymes, the future for industrial applications of enzyme processes is believed to be bright.

Historically, enzymes have been used in aqueous solutions. However, recent technology has shown that they can be bound chemically or ionically to insoluble supports and be used in this form to modify soluble substrates, then recovered and used again. This is a very interesting concept which holds much future promise.

MICROBIAL POLYSACCHARIDES

Microbial polysaccharides can be roughly classified into two groups: homo- and heteropolysaccharides. In the first group are those that are produced from sucrose by a variety of bacteria. The best known example of such polymers is dextran, which is of particular interest because of its use as a blood-plasma extender. It is ideally suited for this purpose as it can be stored for prolonged periods without deterioration. Another use of dextran, which has been patented, covers its addition to syrups and candies to increase moisture retention and to inhibit crystallization of the sugar. Purified dextransucrase can be added to sucrose solutions to produce syrups of high viscosity which contain all the fructose originally present in the sucrose. Dextran has been proposed as a coating for a variety of foods to prevent their drying out while still allowing the escape of gases. It has also been used in paper sizing, and in textile coatings and sizes, and in water-based enamel paints. Another use has been in the preparation of molecular sieves identified by the trade name "Sephadex."

The Northern Regional Research Laboratories of the U.S.D.A., Peoria, Illinois has been instrumental in the development of microbial polysaccharides.

Dextran is produced by fermentation using *Leuconostoc mestenteroides* N.R.R.L. B-512 grown on a nutrient medium containing sucrose, minerals, vitamins, and water. It also can be produced by enzymatic synthesis consisting of the production of dextransucrase; removal of the bacterial cells from the production medium; synthesis of dextran in a reaction mixture containing sucrose, the dextransucrase, and a primer under carefully controlled conditions; and, finally, a fractionation and purification of the dextran.

Following their work on dextran the N.R.R.L. investigated other sources of microbial polysaccharides. One such product was a phosphomannan produced by a diploid strain of *Hansenula holstii* N.R.R.L. 2448. This polysaccharide is composed of mannose and mannose-6-phosphate. Production is relatively simple, using glucose, corn steep liquor, and mineral salts.

Another interesting microbial polymer is polysaccharide N.R.R.L. B-1973. This polymer is produced by a newly discovered species of *Arthrobacter viscosus.* The polysaccharide produced from glucose is composed of nearly equal ratios of glucose, galactose, and mannuronic acid. The best known polysaccharide polymer found at the Northern Division is polysaccharide B-1459 produced from *Xanthomonas campestris*. It is produced on a medium containing glucose and distillers' solubles. The polymer contains D-glucose, D-mannose and D-glucoronic acid in a ratio of 3:3:1.

Polymer B-1459 has proved its value in a variety of applications. It is an excellent fluid for oil well-drilling muds, and oil field flooding. It is very effective in fire-fighting fluids needed for timber and brush fires.

For further information on microbial polysaccharides, the review by Smiley should be consulted.

THE BREWING INDUSTRY

Beer is made by boiling, in the presence of hops, certain extractable materials contained in malted barley and other grains, and then fermenting this brew by the addition of brewers' yeast. The methods employed by various breweries in the preparation of beers and ales differ in important details, yet are sufficiently similar to be represented by the practice in a typical brewery in upper New York State.

The capacity of the plant is 600,000 barrels per year (1 barrel = 31 gallons). The raw materials are rice, corn grits (both oil-free), wheat flakes, soybean (optional), malt, hops, pure yeast and some others in smaller quantities. Besides these, an abundance of water is required for making the beverage itself and for cooling. The corn grits, which must be free from germ, and the rice, for example, are taken in quantities sufficient to furnish 30 to 40 per cent of the required starch, while enough malt is added to furnish the remainder. After passing cleaners and crushing rolls, the rice and the corn, made up with water, and mixed with some malt, are heated in a pressure cooker with an agitator under 5 pounds of pressure for 45-60 minutes. The action is primarily the liquefaction of the starch. The main portion of the malt, made up with water, is placed in the mash tub, which is just like the cooker, and thoroughly mixed at a temperature between 100 and 120°F, then it is allowed to rest. During the rest period the proteolytic enzymes begin to degrade the malt and other proteins into soluble forms. After this period the contents of the cooker are added; these raise the temperature to 149-158°F, at which point the material is held for conversion. The temperature is then raised to 167°F, high enough to arrest the action of the conversion. The purpose of mashing first is to extract from the malt its starch and ferments, including the diastase, which changes the starch to sugars during the conversion period. The material next passes to the Lauter tub the bottom of which has slots, through which the clear liquor, or wort, passes while the arms of the agitator turn slowly. "Sparge water" in spray form is introduced to exhaust the material on the screen. The wort is run into the boiling kettle, a jacketed copper kettle with capacity, in this plant, of 450 barrels. During the period of boiling, which lasts 2½-3 hours, the hops are added at fixed intervals. While in the kettle, the enzymes are destroyed, the hop "bitter" is extracted, undesirable proteins are coagulated, other proteins are precipitated by the tannin in the hops, and the whole is sterilized by the heat. Also, the material is concentrated, about 10 per cent. Next, it is strained to remove all solids, especially the spent hops, and the clear liquor, after settling for a while in the hot wort tank, is run over coolers, such as the Baudelot. From there it goes to fermentors, which may be of wood, glass-lined steel, or aluminum, and the yeast is added (e.g., ¾ pound liquid yeast per barrel). The fermentation lasts 7-10 days, after which the beer is run through a cooler into glass-lined storage tanks which in this plant have a capacity of 60,000 barrels. Here some 5-15 per cent of beer in high state of fermentation is added to the fully fermented beer and al-

TABLE 7.5 Typical Analyses of Brewery Products

	Lager Beer	*Ale*
Appearance on arrival	Clear	Clear
Color	5.75° Lovibond	6.0° Lovibond
Odor	Pure	Mildly aromatic
Taste	Pure, somewhat full-bodied, good hop flavor	Pure, vinous, good hop flavor
Foam-keeping capacity	Good	Good
Specific gravity	1.01598	1.01399
Saccharometer indication	4.07° Plato	3.57° Plato
Alcohol, by weight	3.80%	4.20%
Alcohol, by volume	4.89%	5.40%
Extract	5.79%	5.49%
Sugar	1.63%	1.54%
Dextrin	2.52%	2.23%
Unconverted starch	None	None
Acidity as lactic	0.16%	0.16%
Protein	0.40%	0.42%
Heavy metals	None	None
pH value	4.60	4.30
Carbon dioxide, 2 bottles	0.48%, 0.48%	0.45%, 0.45%
Original gravity	13.1° Plato	13.6° Plato
Real degree of fermentation	55.8%	59.6%
Apparent degree of fermentation	68.9%	73.8%
Chilling test, clarity (absolute turbidity) on 2 bottles chilled 24 hours in ice pack	a.–0.0029. b.–0.0030	a.–0.0028. b.–0.0028

lowed to work out; as it ferments it provides carbonation, since the tank is scaled to a pressure of 6-8 pounds. The beer absorbs the carbon dioxide. After a period of a week or more the beer is filtered through a glass filter under pressure; it is recarbonated sufficiently while passing through the pump, at a temperature of 32°F, to bring the carbon dioxide content to 0.48 per cent in the final beer, and the product is bottled or barreled. The carbonation may be performed entirely by pumping sterile carbon dioxide into the cooled beer from the fermentors. The bottled beer is pasteurized at 140°F (60°C) after capping.

Obviously, brewing would seem to be a process which should be operated continuously; but it has required many years of development in many breweries before the introduction of continuous operation on a large scale. A brewery was first operated continuously on a production scale in New Zealand. And more recently, Watney Mann Ltd., one of the largest of the British brewing combines, announced a continuous fermentation capacity of 20,000 barrels per week.

WASTE DISPOSAL

A single family, living on sufficient land, can responsibly treat its own wastes by underground septic tanks and leach fields. Anaerobic digestion processes convert family wastes to innocuous, soluble substances that can be further consumed by soil microorganisms and in an earlier time small communities could safely discharge their wastes to nearby rivers and lakes where natural aeration provided oxygen for aerobic digestion and destruction of waste material. Now, due to increasing population levels, nature has been overwhelmed, and man has had to create artificial facilities to stimulate the aerobic process of nature. These facilities permit more rapid treatment of concentrated wastes and insure that discharges to a river or lake will be adequately stabilized. Aeration or oxygen treatment stimulates the microbial activity which minimizes odors and reduces dissolved solids; thus the biochemical oxygen demand of the discharge is reduced to levels which do not interfere with higher forms of life. Biochemical oxygen demands (B.O.D.) are measured in terms of mg of oxygen per liter required by microorganisms to consume the biodegradable organics in waste water under aerobic conditions. Sanitary engineers also use a similar term, chemical oxygen demand (C.O.D.), which measures the oxygen in mg per liter to oxidize both organic and oxidizable inorganic compounds.

Our technologically advanced society has learned how to make many new compositions of matter which are not readily biodegradable. These include many polymers and plastics, insecticides and herbicides, detergents, and pharmaceuticals. Thus the problem of disposal has become quite complex.

Anaerobic digestion is widely used to stabilize concentrated organic solids which are removed from settling tanks, biological filters, and activated sludge plants. The waste is mixed with large quantities of microbes, and oxygen is excluded. Highly specialized bacteria which grow under these conditions convert about 80-90 per cent of the degradable organics into carbon dioxide and methane gas.

Unfortunately, all such facilities are expensive to build and operate, and they have not been constructed in sufficient numbers. Also, certain industrial effluents upset the useful and natural microbial flora. In the future, these abnormal effluents will require special facilities where microorganisms adapted to the substrates can perform their services either by complete treatment or by producing partially treated effluents which can then be accommodated by municipal systems.

Nature has always used combinations of aerobic and anaerobic digestion of organic materials. Foliage and dead trees of forests are digested to humus which adds to the soil. Organic gardeners are familiar with compost methods of creating humus and fertilizer by anaerobic digestion. Some municipalities are now developing large compost operations to dispose of garbage, and fill methods use similar technics, creating new soil from organic wastes.

New developments in waste-disposal technology will be of increasing importance as new antipollution efforts are undertaken. Many of the processes are modifications of fermentation processes, but use mixed flora instead of pure cultures. Many of the processes of the sewage-treatment works may be extended in the future to produce potable water.

For further information on waste-water treatment the papers by Marks, and by Abson and Clark give a rather complete and interesting story.

SELECTED REFERENCES

Books

1. Prescott, S. C. and Dunn, C. G., "Industry Microbiology," 3rd Ed., New York, McGraw-Hill, 1959.
2. "Industrial Fermentation," Vols I and II, L. A. Underkofler and R. J. Hickey, Eds., New York, Chemical Publishing, 1954.
3. "Advances in Applied Microbiology," Vols I–XII Perlman, D., Ed., New York, Academic Press, 1959–1970.
4. Solomons, G. L., "Materials and Methods in Fermentation," New York, Academic Press, 1969.
5. "Biochemical and Biological Engineering Science," Vol. I–II, N. Blakebrough, Ed., New York, Academic Press, 1967–1968.
6. Aiba, S., Humphrey, A. E. and Millis, N., "Biochemical Engineering," New York, Academic Press, 1965.
7. Hahn, P. A., "Chemicals from Fermentation," New York, Doubleday, 1968.
8. "Microbial Technology," H. Peppler, Ed., New York, Van Nostrand Reinhold, 1967.
9. "Progress in Industrial Microbiology" Edited by D. J. D. Hockenhull, Vol. I through VIII, London, Heywood, 1959–1968.
10. "Fermentation Advances," D. Perlman, Ed., New York, Academic Press, 1969.
11. "Methods in Microbiology," Vol. I, II, IIIA, and IIIB, J. R. Norris and D. W. Ribbons, Eds., New York, Academic Press, 1969–1970.
12. "Antimicrobial Agents and Chemotherapy," G. L. Hobby, Ed., Baltimore, Williams & Wilkins, 1969.
13. Charney, W. L., and Herzog, H. L., "A Handbook of Microbial Transformations of Steroids," New York, Academic Press, 1967.
14. Miller, M. W., "The Pfizer Handbook of Microbial Metabolites," New York, McGraw-Hill, 1961.
15. Vanek, Z., and Hostalek, Z., "Biogenesis of Antibiotic Substances," New York, Academic Press, 1965.

Papers

1. Perlman, D., "Fermentation Industry–Evolution," *Process Biochemistry*, **4**, No. 6, 29–32 (June, 1969).
2. Curtis, N. S., "Brewing Advances," *Process Biochemistry*, **3**, No. 4, 17–21 (April, 1968).
3. Berdy, J., and Magyam, K., "Antibiotics–A Review," *Process Biochemistry*, **3**, No. 10, 45–50 (Oct. 1968).
4. Perlman, D., "Fermentation Industry," *Chemi. Week*, p. 83 (Dec. 16, 1967).
5. Perlman, D., "Are New Antibiotics Needed?" *Process Biochemistry*, **3**, No. 4, 54–60 (Apr. 1968).
6. Marks, R. H., "Waste-Water Treatment," *Power* (June 1967).
7. Jones, R. G., "Antibiotics of the Penicillin and Cephalosporin Family," *Am. Scientist* **58** (Jul.-Aug., 1970).
8. Smiley, K. L., "Microbial Polysaccharides," *Food Technol.* (Sept. 1966).
9. Miall, L. M. "Fermentation–The Last Ten Years and the Next Ten Years," *R.I.C. Rev.*, **3**, No. 2, 135 (Oct. 1970).
10. Price, K. E., "Structure–Activity Relationships of Semisynthetic Penicillin," *Advan. Appl. Microbiol.*, **11** (1969).
11. Steel, R., and Miller, R. L., "Fermenter Design," *Advanc. Appl. Microbiol.*, **12** (1970).
12. Kulhanek, M., "Fermentation Processes Employed in Vitamin C Synthesis," *Appl. Microbiol.*, **12**, (1970).

13. Abson, J. W., and Clark, E. I., "Trends in Sewage Treatment," *Process Biochemistry*, **6**, No. 1, 15-21 (Jan. 1971).
14. Morihiro, E., "The Amino Acid Industry," *Chem. Economy Eng. Rev.* (Japan) (Sept. 1970).
15. Appleweig, N., "Steroids," *Chem. Week*, **104**, No. 20, 57 (May 17, 1969).
16. Dunnill, P., and Lilly, M. D., "Large Scale Isolation of Enzymes," *Process Biochemistry*, **2**, No. 7, 13-19 (July 1967).
17. Burbidge, E. and Collier, B., "Production of Bacterial Amylases," *Process Biochemistry*, **3**, No. 11, 53-59 (Nov., 1968).
18. Wang, D. I. C., "Biochemical Engineering," *Chem. Eng.*, **76**, No. 27, 108 (Dec. 15, 1969).
19. Sassiver, M. L., and Lewis, A., "Structure-Activity Relationships Among Semisynthetic Cephalosporins," *Advan. Appl. Microbiol.*, **13** (1970).
20. Perlman, D., "Microbial Transformation of Antibiotics," *Process Biochemistry*, **6**, No. 7, 13-15 (July 1971).
21. Keay, L., "Microbial Proteases," *Process Biochemistry*, **6**, No. 8, 17-23 (Aug. 1971).

8

Coal Carbonization and Recovery of Coal Chemicals

Michael Perch* and Richard E. Muder*

INTRODUCTION

The carbonization of coal is the pyrolysis or heat treatment of bituminous coal in the absence of air which results in the production of a residue, called *coke*, and vapors, which when condensed and separated produce crude coal chemicals commonly referred to as *tar*, *light oil*, *ammonia liquor*, and *coke-oven gas*. These crude fractions are further refined, in varying degree from plant-to-plant, to produce commodity chemicals, such as benzene, toluene, xylene, ammonium sulfate, pyridine, naphthalene, anthracene, phenanthrene, creosote, road tars, and industrial pitches, to name a few.

During the early years of this century a large part of the chemical industry used these products of coal carbonization as starting materials. For many years the coke plant was the principal, and, in some cases, the only source of these important chemicals. However, as the years went by, and particularly since World War II, the demand for these materials has increased enormously, and the demand has been met by products from other sources. In the United States there has been a curious turnaround. The first by-product-coke plant was built primarily as a source of ammonia needed in the Solvay process, with the coke as a useful by-product. Today, coke plants are built solely for the purpose of providing fuel for blast furnaces and foundry cupolas. In some quarters, it is considered unfortunate that the production of a desirable solid must be accompanied by the evolution of various gases and liquids. The usefulness and economic value of these by-products depend on the location and the accounting practices of the coke plant. The gas is generally welcomed as a useful fuel. However, in order to be distributed to the points of use it must be cooled, and this drop in temperature results in condensation of various compounds. One of these, coal tar, under most circumstances is still a rich source of chemicals and bituminous materials, at worst it is a useful fuel. Another, ammonia liquor, the aqueous condensate, con-

*Koppers Co., Inc., Pittsburgh, Pa.

tains a large number of organic and inorganic substances in solution and remains a source of recoverable ammonia, but it is viewed more and more as a nuisance since to dispose of it without causing water or air pollution is difficult. The removal of certain other materials, such as benzene and hydrogen sulfide, requires more than simple cooling of the gas and, therefore, is somewhat elective.

The total amount of coal carbonized in 1971 amounted to about 83 million tons which produced about 52 million tons of blast-furnace coke, 3 million tons of foundry coke, and about 3 million tons of coke for miscellaneous uses, such as sintering ore fines and residential heating. Coal chemicals produced during this carbonization amounted to about 750 million gallons of crude tar, 650 thousand tons of ammonia, 200 million gallons of crude light oil, and 860 million cubic feet of coke-oven gas. The total value of the chemicals produced in the United States alone is of the order of one-quarter billion dollars.

For many years a large amount of coke was produced for home heating and for water-gas manufacture in many large cities. With the installation of large pipe lines from the natural gas fields in the southwest, the use of coke for domestic heating and the manufacture of gas dwindled to small amounts. At the present time, the coal-carbonization industry is highly dependent on the steel industry and is, therefore, influenced greatly by its vagaries. Steel industry strikes not only halt the production of steel but curtail the operations of the coke ovens and the production of coal chemicals. Coal mine strikes have the same effect. Other factors also affect this source of chemicals. With the recent substantial advances in blast-furnace technology the coke rate (the pounds of coke required per ton of hot metal produced) decreased from an industry average of about 1900 in 1950 to about 1260 in 1971, most of the decrease occurring in the last ten years. However, this 35 per cent drop in coke rate and the equivalent reduction in the production of coal chemicals was more than offset by the introduction and rapid growth of the basic-oxygen furnace process. This process has been rapidly displacing the open-hearth furnace method of steel making; since the basic-oxygen furnace cannot use more than about 30 per cent scrap in its burden, more hot metal is required from the blast furnace and, thus, more coke. This is well illustrated by the actual production and consumption figures given in Table 8.1. The coke rate is expected to decrease even further in the future.

In spite of the significant reduction in coke rate and substantial improvement in the economy of coke ovens in recent years, there is a considerable amount of developmental activity on processes for the direct reduction of iron ore. These would bypass the blast furnace and eliminate the need for coke. This threat to the carbonization industry, however, is not expected to be significant in the near future.

Another threat to the industry stems from the environmental pollution problem. However, rapid strides are being made to solve the severe environmental problems associated with coke plants. Stricter enforcement of air and water pollution controls could result in the retirement or replacement of some of the older batteries, as well as the remaining beehive or nonrecovery ovens. The new batteries of ovens being constructed are increasingly being equipped with devices to reduce emissions.

In spite of the decreased coke rates and the above threats to the carbonization industry, the prognostication is that the industry will hold its own for many years. The economics of coke manufacture have been enhanced considerably by the trend toward larger coke ovens and faster coking rates. The projected total coke consumption in 1980 is expected to be about 71 million tons, which is about 14 million more than in 1971. It is anticipated that continued reductions in the average coke rate in the blast furnace will be counteracted by increased hot metal production.

METHODS OF CARBONIZATION

Practically all coke is made in chemical-recovery, slot-type ovens with only about one per cent being produced in beehive or non-recovery ovens. Pollution regulations may soon lead to the complete demise of the latter type of oven. Horizontal and vertical retorts have also been used to carbonize coal, chiefly

TABLE 8.1 Actual Hot Metal Production and Coke Consumption in the United States (Millions of Net Tons)

Year	*Hot Metal Production*[a]	*Coke Rate*[b] *lbs/ton*	*Coke Consumed*[c] *Blast Furnace*	*Total*	*Total*[c] *Coal Carbonized*
1955	77.8	1760	68.5	76.1	107.4
1956	76.0	1750	65.3	73.3	106.0
1957	79.3	1730	67.6	74.4	108.0
1958	57.8	1640	46.6	52.7	76.6
1959	60.8	1610	48.2	54.7	79.2
1960	67.3	1540	51.0	56.9	81.0
1961	65.3	1435	46.8	52.1	73.9
1962	66.3	1395	46.2	51.8	74.3
1963	72.4	1350	48.9	55.0	77.6
1964	86.2	1325	57.1	62.6	88.8
1965	88.9	1330	59.1	65.4	94.8
1966	92.2	1295	59.6	66.0	95.9
1967	87.6	1285	56.2	61.6	92.3
1968	89.3	1260	56.2	62.4	90.8
1969	99.5	1260	60.2	66.2	92.8
1970	91.9	1265	58.2	63.2	96.1
1971	81.7	1260	51.5	56.7	83.2

[a]AISI Annual Statistical Reports, "Total Blast Furnace Production of Pig Iron and Ferroalloys."
[b]AISI Annual Statistical Reports, "Coke Consumption by Use," U.S. Bureau of Mines, Coke and Coal Chemicals Monthly, Disposal of Coke.
[c]U.S. Bureau of Mines Minerals Yearbooks, "Coke and Coal Chemicals (incl. beehives)."

in Europe, when gas was desired as the principal product; coke and tar were the by-products. With the availability of natural gas the use of these retorts has also dwindled into insignificance. In the United Kingdom, plants have recently been installed for the production of smokeless fuel for household heating.

High-temperature carbonization in coke ovens is the predominant process used for making metallurgical coke. This is because of its high productivity and the suitability of the coke and the coal chemicals produced by this process. High-temperature carbonization uses a temperature in the range of 900 to 1150°C. Low-temperature carbonization, at temperatures of 500 to 750°C, has been practiced to only a small extent in the United States. It is used to a greater extent in some foreign countries, principally to produce a smokeless household fuel. The coal chemicals are useful, but the tar produced is less desirable than that obtained from high-temperature carbonization because of its smaller content aromatic hydrocarbons, even though yields are higher.

High-temperature carbonization is carried on in the United States in about 65 plants containing over 13,000 chemical-recovery ovens. These consist of Koppers, Koppers-Becker (see Fig. 8.1), Wilputte, Semet-Solvay, Otto, and Simon-Carves types. In general, the oven chambers are of 17 to 20 inches in width, though in a few special cases they are as narrow as 14 inches, 10 to 20 feet high, and 37 to 50 feet long. A number of these chambers are arranged side-by-side in a battery which may consist of 15 to 100, or more ovens. Flue chambers are interposed between the walls of each two adjacent oven chambers and gas is burned in the flues to provide heat for the carbonization. An oven may hold 12 to 35 tons of coal depending on its dimensions. Depending on the oven width and flue temperatures, the coking time of each charge may range between 15 and 20 hours for the pro-

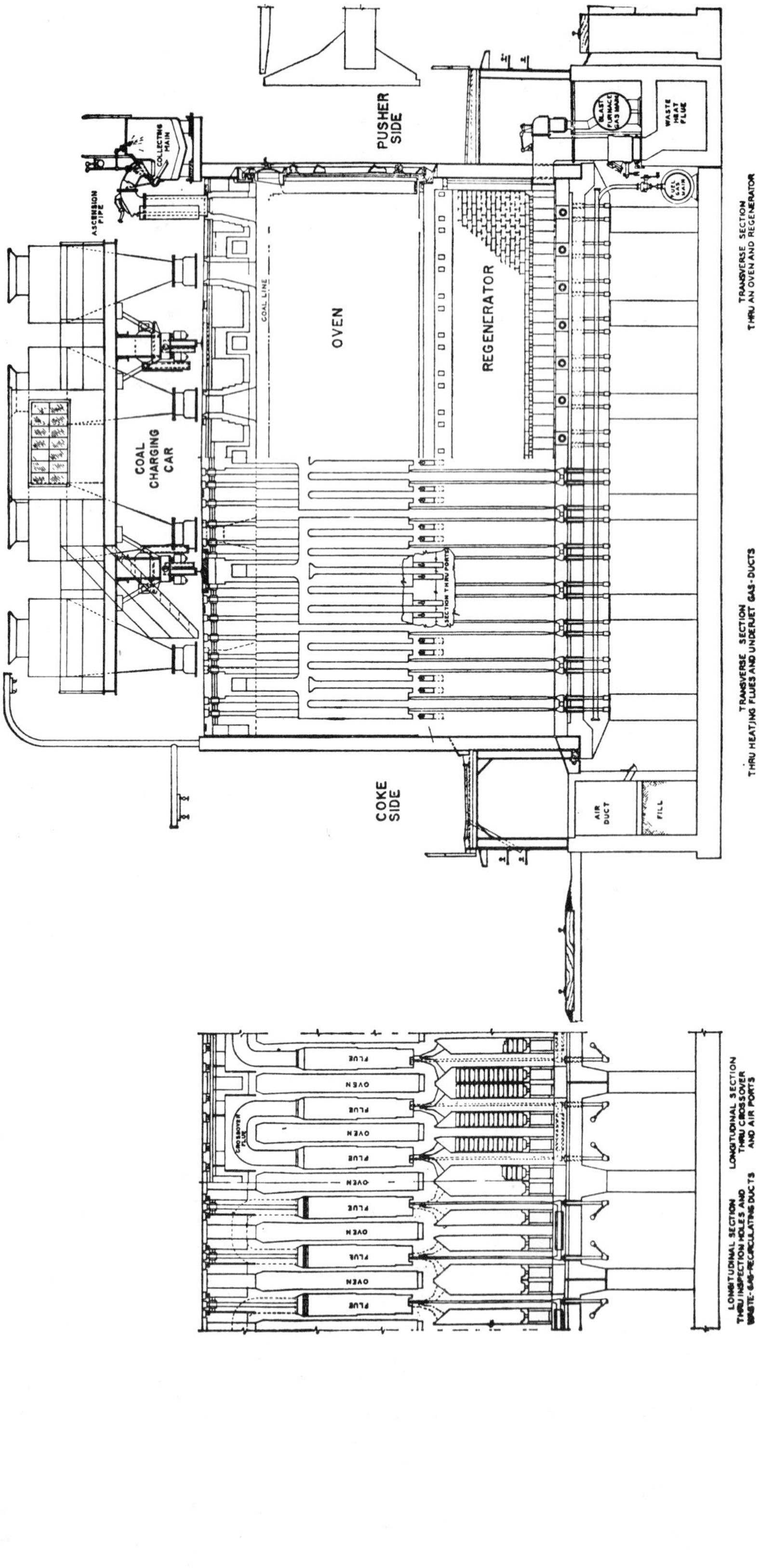

Fig. 8.1 Koppers-Becker coke oven. *(Courtesy Koppers Co., Inc.)*

duction of blast-furnace coke, and be 30 hours or more for foundry coke. For many years a coking rate of 1.0 to 1.1 inches of oven width per hour was common, but now, with walls containing high-density refractories, coking rates up to 1.25 inches per hour are being obtained.

Coal Selection and Preparation

In order to produce suitable metallurgical coke, it is necessary to select the coals for carbonization carefully. Only bituminous coals of low ash content, low sulfur content, and suitable coking properties are used. Generally two or more of these coals are blended. High-volatile coals containing 32 to 38 per cent volatile matter are blended with low-volatile coals containing 15 to 20 per cent volatile matter. At times, a medium-volatile coal is included. In a typical case, 80 per cent of a high-volatile coal containing 35 per cent volatile matter is blended with 20 per cent of a low-volatile coal containing 17 per cent volatile matter. One plant has been blending five coals regularly, using three high- and two low-volatile coals. The amount of low-volatile coal may range from 15 to 40 per cent, but the amount in the mixture is limited by the amount of pressure developed by the blend during carbonization. Excessively high carbonization pressure can cause movement and damage to the walls. Because of the large investment in the ovens, extreme care must be exercised to safeguard them from this potential damage. Many coke plants have movable-wall test ovens in which proposed mixtures containing new coals are tested, or the current mixture is continually monitored, so that the safety of the mixture is ascertained before and during use. Generally, the maximum pressure tolerated when the mixture is tested under standard conditions in the test oven is about $1\frac{1}{2}$ lbs per square in.

The individual coals are usually pre-crushed to a maximum size of about one-inch and stored in individual bins. These coals are withdrawn as needed from the bottoms of the bins and proportioned on a conveyor belt feeding a hammermill or impact-type pulverizer, which crushes and mixes the blend. Generally 70 to 80 per cent of the coal in the pulverized mixture is smaller than $\frac{1}{8}$ inch. In an alternate form of preparation, the individual coals are pulverized to their final size, and proportioned onto a belt, and finally blended by a paddle wheel or other type of mixer.

Bulk-density control of the coal in the ovens is practiced in nearly all plants, and this is done by adding oil or water as required, depending on the moisture content of the coal. The presence of oil on the surface of the coal particles lubricates them so that they consolidate to a higher and more uniform bulk density in the oven chamber, thus counteracting the fluffing action of moisture. A more uniform density contributes to uniform heating of the charge and improved coke quality.

Carbonization

The blended coal for carbonization is stored in a bin atop one end of the battery or between two batteries of ovens, and is loaded out into a *larry car* for charging into the ovens. When an oven is ready for charging, the larry car, filled with a measured amount of coal, is transported on rails running on top of the battery to the oven being charged. The car is spotted directly over opened charging holes (usually 3 or 4) in the top of the oven, and the coal is charged by mechanical feeders or by gravity. A levelling bar levels out the top surface of the coal charge and also provides a gas space through which the carbonization gases are conducted.

As soon as the coal is dropped into the oven, the coal layer adjacent to the heated walls becomes soft, begins to decompose, and then sets to a solid semicoke. As time passes the heat coming from the direction of the wall penetrates deeper into the charge. In a short time layers of newly formed coke appear at the walls, layers of plastic coal appear adjacent to the coke, and unconverted coal lies between the layers of plastic coal. As coking proceeds further, the coke layers become thicker and the plastic layers move toward the center of the oven where they finally meet, coalesce, and solidify; when the temperature at the center approaches the wall temperature the coking is considered complete. A great deal of the decomposition of the coal occurs within the plastic layers; most of the evolved vapors pass through shrinkage-cracks in the coke to the heated walls, hence upward along the wall to the gas space at the top and finally out of the oven. While the primary decomposition products evolved from the plastic zones are of high molecular weight, during passage through the hot coke and along the heated wall they undergo secondary decomposition into lower molecular weight products. If the temperature is sufficiently high and residence time long enough, some of this decomposition results in the deposition of carbon on the coke, the walls, and the roof. Those factors, as well as the character of the coal blends used, influence the chemical composition of the decomposition product and the yield of the various chemicals. The yield of coke, its size, and strength are the dominant economic factors in the coking process; the yield and quality of chemical products are of secondary importance.

The coke ovens are charged and pushed on a predetermined schedule. When an oven is ready for discharge it is taken "off-the-main," the doors of both ends of the oven are removed, and the coke cake is pushed out with the aid of the pusher machine ram into a quenching car. The coke is transported to a quenching station where it is sprayed with water and cooled, then it is deposited on an inclined refractory wharf from which it is transported by belt conveyor to a screening station for sizing and subsequent use.

Outline of Recovery Methods

The following discussion will first describe the methods for gas treatment and recovery of coal-chemicals that have prevailed for many years and are still in use at most coke plants. Later, alternative procedures will be described. These are currently becoming more popular as the economic incentives to recover by-products lessen. However, if shortages of natural gas and petroleum continue to develop, recovery of carbonization chemicals may become more profitable.

The hot gas resulting from carbonization leaves the coke oven through an opening in the roof and passes through a standpipe into the collecting main (see Fig. 8.2). The standpipe consists of a vertical section rising from the oven and a gooseneck, which diverts the gas downward into the main and which contains a water-sealed valve for shutting off the oven from the main when coking is completed and the oven is ready to be pushed. It also contains a spray of flushing liquor that cools the gas chiefly by the evaporation of water. The collecting main, which runs the length of the battery, is a large pipe, either circular or troughlike in cross section, that also contains cooling sprays. Usually, only one collecting main is provided; this is on the pusher side of the battery. Sometimes, a second main is provided on the other side to minimize emissions to the atmosphere while coal is being charged into the ovens. The flushing liquor that is pumped through the sprays is simply a recirculated aqueous condensate from the gas. Since the rate of recirculation of the liquor is high, its temperature is raised only a few degrees. The gas which leaves the coke oven at about 1200 to 1500°F, leaves the collecting main at about 185°F. As a result of this cooling, most of the tar is condensed. The tar, along with the unevaporated flushing liquor, is conveyed to decanters, which are large tanks that allow time for the gravity separation of the tar and the liquor. The former is pumped to storage tanks to await further processing; the latter is recirculated to the battery.

In the ovens and the collecting mains, the gas is under a slight positive pressure regulated by butterfly valves in the off-take main connected to the suction main that leads to the primary coolers. After the butterfly valves, the gas is maintained under a slight suction by the exhausters which pull it through the primary coolers and then pump it through the rest of the coal-chemical plant under a slight positive pressure.

In the primary coolers the gas is cooled to 75–95°F. Cooling may be indirect by contact with water-cooled tubes, or direct by contact with sprays of cooled ammonia liquor. In either case, more tar is condensed. Since this tar has somewhat different properties from the collecting-main tar, it is sometimes processed separately, but most often the two are combined. This second cooling of the gas also results in the condensation of water and an increase in the volume of liquor, an increase that must be dealt with by means that will be discussed later.

The exhausters that keep the gas moving

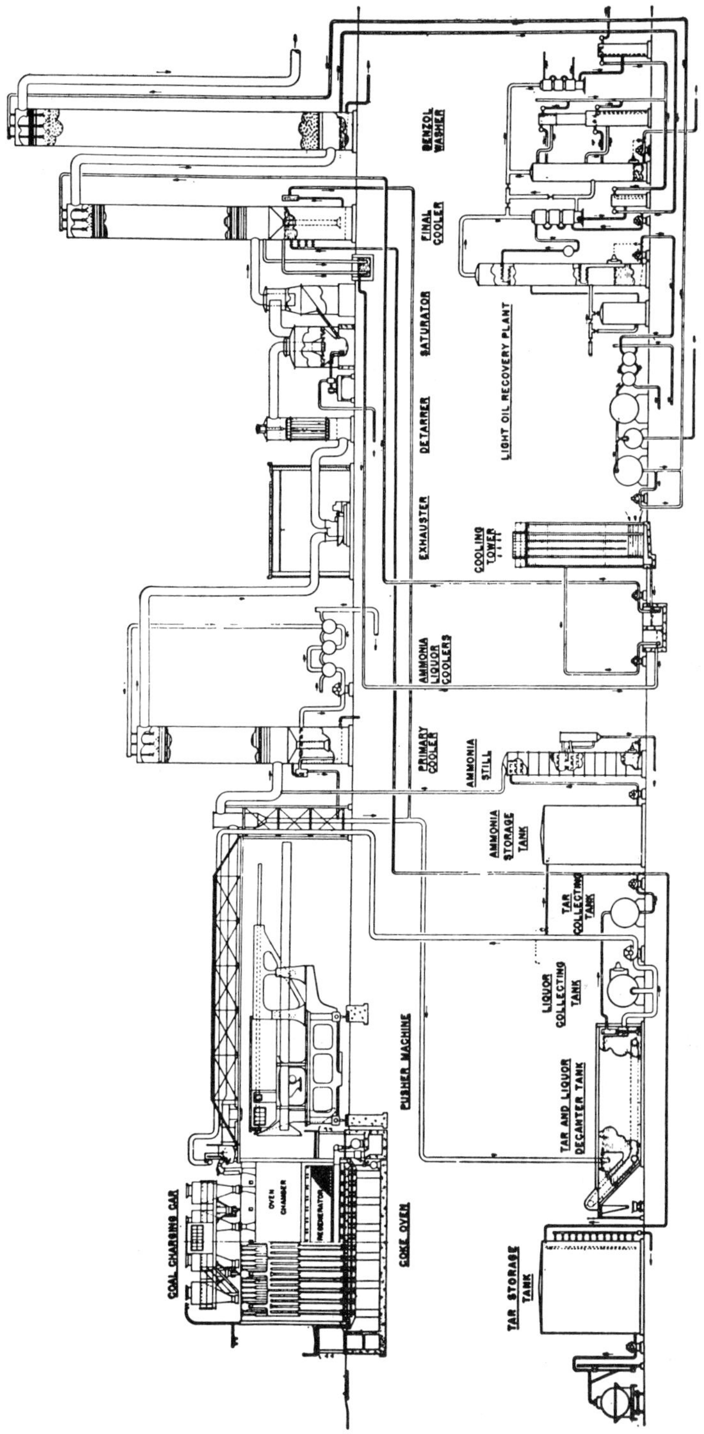

Fig. 8.2 Coke plant sectional flow diagram. *(Courtesy Koppers Co., Inc.)*

throughout the whole plant are generally of the turboblower or high-speed centrifugal type, driven by steam turbines. The amount of compression imposed varies widely depending upon the complexity of the coal-chemical recovery system and on what is done with the gas subsequent to this. But commonly the gas enters the exhauster with a suction of about 10 inches of water column and leaves under a pressure of about 60 inches of water. The gas temperature is raised about 20°F by this compression. A great deal of tar fog that is present is removed by the high-speed impellers of the exhausters. Removal of all but traces of this tar fog is next accomplished by passage through electrostatic tar pricipitators.

Removal of ammonia is usually the next step in gas treatment. Before the gas enters the saturator or ammonia absorber, vapors from the ammonia still (to be described later) are added to the gas stream. In the older type of saturator, the gas bubbles through a bath consisting of a saturated solution of ammonium sulfate containing about 5 per cent free sulfuric acid. In the more modern absorber, the gas is washed by a series of sprays. In either case, crystals of ammonium sulfate are formed. A slurry of these crystals is fed to a high-speed centrifuge for drying. The product is then sold for use as a fertilizer.

For good light-oil recovery the gas must be further cooled to about 75°F. This is generally accomplished by direct contact with water in the final coolers. This results in the condensation of naphthalene. If the water from the cooler is run to a sump, this naphthalene can be skimmed off since the crystals float due to the presence of adhering gas bubbles. Alternatively, the water can pass through a bath of tar in the base of the cooler and the naphthalene will be dissolved in the tar. In another method, a petroleum-oil fraction is used as the cooling medium and for dissolving the naphthalene from the oil stream. The naphthalene is recovered by distillation. The cooling medium, whether water or oil, is usually recirculated in a closed system, from which the heat is removed either in indirect coolers or in cooling towers.

Next, light oil, consisting mainly of benzene, toluene, and xylene, is recovered by countercurrent contact with a petroleum wash oil in the light-oil scrubbers, which are either packed towers or a series of spray chambers. The enriched wash oil is then stripped with steam. The vaporized light oil is condensed, and the stripped wash oil is cooled and recirculated to the scrubbers. The light oil is generally washed with sulfuric acid to remove unsaturated materials and most of the thiophenic compounds present, and then it is fractionally distilled into salable benzene, toluene, and xylene. Other methods of refining, which also remove paraffinic hydrocarbons, include cracking in the presence of hydrogen, and catalytic hydrogenation followed by solvent extraction.

The gas is generally given no further treatment and is used as fuel in a variety of ways, such as underfiring coke ovens, heating soaking pits and reheating furnaces, and raising steam in boiler houses. For some purposes, it is desirable that the hydrogen sulfide be removed from the gas. A large number of processes are available for accomplishing this, the most commonly used being the vacuum-carbonate process in which the H_2S is absorbed by countercurrent scrubbing with a solution containing sodium carbonate and sodium bicarbonate. Hydrogen sulfide is expelled from the solution by boiling under vacuum and can be converted into elemental sulfur in a Claus kiln, or into sulfuric acid in a contact-acid plant.

It was mentioned previously that the vapors from the ammonia still are put back in the gas before contact with sulfuric acid. This is the semidirect process of ammonia recovery. In the indirect process, the gas does not come into contact with sulfuric acid. Rather, it is washed with water in a series of scrubbers and the resulting liquor is mixed with the liquor from the collecting mains and distilled. In liquors from either process, that portion of the ammonia that exists as a salt of a weak acid, such as carbonic or hydrosulfuric, is called "free" and is liberated by simple boiling. However, the "fixed" ammonia that exists as a chloride or thiocyanate requires liberation by the addition of lime before it can be recovered by distillation. Most often, the vapors are brought into contact with sulfuric acid. However, the resulting ammonium sulfate contains only 25 per cent of useful plant nutrient and sometimes cannot be sold at a price that covers the cost of the acid necessary for its manufacture. An alternative is the use of phosphoric acid in the absorbers and the manufacture of mono- or diammonium phosphate. A small amount of ammonia produced in coke-oven plants is recovered by absorbing the vapors from the ammonia still in water and forming an aqueous solution containing 15 to 25 per cent of NH_3. Pure anhydrous ammonia can be recovered by scrubbing the gas with phosphoric acid, stripping out the ammonia with steam, and then fractionally distilling the resultant mixture of steam and ammonia vapors. When no satisfactory market exists for the relatively small quantities of ammonia recoverable at a coke plant, it is possible to burn the vapors from the ammonia still, and release the products of combustion to the atmosphere in the form of nitrogen and water vapor.

No matter what form of ammonia recovery is practiced, carbonization of coal results in the formation of an aqueous liquor that must be disposed of. In the past, the release of this liquor to a stream was objected to chiefly because the liquor contained phenols which were converted to chlorophenols during the usual methods of water purification; these products, when present in concentrations of even a few parts per billion, imparted an obnoxious taste to the water. The phenols can be readily removed from the liquors by extraction with solvents, usually aromatic hydrocarbons, and recovered as sodium phenolates by contact with sodium hydroxide. However, some present effluent standards prevent the release of even dephenolized liquors because of other oxygen-consuming constituents. Treatment of the liquor by bacterial oxidation, using the activated sludge process, results in an effluent of low oxygen demand.

An alternative method of recovering coal chemicals uses a cryogenic system in which the gas is cooled to very low temperatures in a stepwise fashion, and various condensates are removed until only hydrogen remains in a gaseous state. This method is normally used only when it is desirable to separate the hydrogen for ammonia or methanol synthesis, or for hydrogenation of a hydrocarbon fraction.

Yields and Uses of Coal-Carbonization Products

Both the quality and the quantity of the products of coal carbonization vary depending on the properties of the coal used, carbonization conditions, and the type of recovery facilities used. However, a typical plant is one that produces from each ton of coal the various products listed in Table 8.2, in the amounts given.

Of the total coke produced in chemical-recovery ovens, about 90 per cent is used in blast furnaces. Of the remainder, the largest amount is used in foundry cupolas for melting iron for castings. The principal use of coke breeze, which is small coke less than $^3/_4$-inch size, is as a fuel in sintering iron ore. Most of the remaining coke breeze is used as a boiler fuel.

Coke-oven gas is a mixture of a large number of substances, but two of them, hydrogen and methane, make up about 75 to 85 per cent by volume of the total. The heating value of the gas is about half that of natural gas. In the past, a large quantity of coke-oven gas was distributed through city gas mains, but this use now accounts for only 2 per cent of the total consumption of coke-oven gas. Most of the gas is burned in the steel plants; 36 per cent for heating coke ovens, 10 per cent for raising steam, and 52 per cent for other uses.

About 16 per cent of the coal tar produced is burned without further treatment, mainly in open-hearth furnaces. About 38 per cent is distilled, the distillate being sold as a source of valuable chemicals, with the residue being burned. About 46 per cent is sold to tar processors who use the residues for preparing road tars, roofing pitches, pipeline enamels, and carbon electrodes. From the distillates such materials as naphthalene, tar acids, tar bases, and other chemicals are recovered, and these are important starting materials in a wide variety of chemical syntheses. The heavier distillate is creosote, whose principal use is as a wood preservative.

About 60 per cent of the light oil produced is refined at coke plants, with the principal

TABLE 8.2 Typical Yields of Coke and Chemicals from High-Temperature Coal Carbonization

	Per Cent	*Lbs Per Ton*
Coke (incl. coke breeze)	75	1500
Coal Tar	4	80 (8 gal)
Light oil		1
Naphthalene oil		8
Heavy creosote oil		9
Anthracene oil		14
Soft pitch		16
Medium pitch		14
Hard pitch		18
Light Oil	1	22 (3 gal)
Benzene		15
Toluene		3
Xylene		1.5
Other		2.5
Liquor	6	120
Ammonium sulfate		20
Gas	14	280 (10,500 cu ft)
Hydrogen sulfide		6
Carbon dioxide		18
Nitrogen		8
Hydrogen		32
Carbon monoxide		45
Methane		130
Ethane		11
Ethylene		20
Propylene		3
Light oil		3
Other (butylene, HCN, etc.)		4

products being benzene, toluene, and xylene, which have many uses as chemicals and solvents. The remainder of the light oil is sold to other processors, usually petroleum refiners who handle it along with their own materials.

Practically all of the ammonia recovered is used as a fertilizer. About ninety per cent is removed as ammonia sulfate. Small quantities of diammonium phosphate and concentrated ammonia liquor are also produced. One large plant recovers the ammonia in coke-oven gas in the pure anhydrous state, and also synthesizes large quantities of ammonia from the hydrogen in the gas.

COAL TAR PROCESSING

Practically all the tar in the United States, and in most other countries, is produced by high-temperature carbonization in slot-type coke ovens by the steel industry. Although no low-temperature tar is produced in the United States, small quantities are available in some countries. The tar obtained from horizontal and vertical retorts in England and Europe has dwindled to a less significant amount as a result of the influx of natural gas and the subsequent shutting down of retorts in gas-producing plants, but the amount is still sizable. Owing to an increase in the production of smokeless-fuel briquettes in the United Kingdom the amount of low-temperature tar has been increasing recently.

Much of the coal tar produced in the U.S. is burned in the steel industry open-hearth furnaces because of its availability, low sulfur content, and high heating value. About 16 per cent of the total tar produced is consumed in this way. Some tar is also injected into the blast furnace.

Fig. 8.3 Continuous tar distillation unit. (*Courtesy U.S. Steel News.*)

The remainder of the tar is processed by fractional distillation either by the steel producer or by an independent tar distiller. Continuous fractionation (Fig. 8.3) is the preferred technique because of high-volume throughput and low fuel cost. Also, operating conditions can be closely controlled. However, batch stills are very flexible and tar can be easily processed in them to produce pitch products of any desired melting point. Because of this, batch stills are frequently used for small volume specialties, even in plants having high-capacity continuous units.

The design of continuous plants varies considerably; but the following operation may be considered typical. The essential equipment items are tube heaters, fractionating columns, heat exchangers, pumps, tanks, and instruments, which are arranged in such a way that up to eight different fractions, ranging from naphthalene to high-melting point pitches, are produced.

Crude tar is pumped from heated storage tanks through a series of heat exchangers which preheat it to 300°F by recovering heat from product streams. The tar then enters a dehydrator where water and solvent-naphtha components are flashed off.

The dehydrated tar, after further heating by a tube heater, enters a dual or multiple-column fractionation system containing from 12 to 75 trays generally operated under vacuum. The tar fractions are heated to progressively higher temperatures between each column so that tar fractions of increasingly higher boiling point will be removed in each succeeding column. Vapors from each column are condensed and processed separately. In some installations, a flash tank operating under vacuum is used after the first column. The boiling point of the distillate removed from a column depends upon the feed, the softening point of the desired residue, the temperature to which the feed is heated, and the vacuum applied. Vapors taken overhead, and sometimes as side streams, are condensed and processed separately.

Chemical oil containing light oil, solvent naphtha, naphthalene, tar acids, and tar bases is generally removed in the first column, sometimes with a side stream of light creosote. In some plants the chemical oil is split into several fractions by using additional columns, from which progressively higher-boiling fractions are removed. Creosote fractions and various grades of pitch are produced in the second, or succeeding column, or in a flash pot.

Tar acids are removed from the chemical oil by extraction with caustic, which forms sodium cresylate. After concentrating the solution and removing oil impurities, the acids are released by neutralizing the alkalinity with CO_2 and/or sulfuric acid. After neutralization and separation with water, the tar acids are refined by fractionation. Fairly pure compounds, such as phenol, o-cresol, m- and p-cresol, xylenols, and other homologues are produced.

After removal of tar acids from the refined chemical oil, naphthalene may be recovered by crystallization, distillation, or a combination of both. If the naphthalene is refined by distillation, it is fed to a two-column unit in which light oils, water, and solvent naphtha are removed in one column. The naphthalene is recovered in a relatively pure form in the overhead stream from the second column, which may have one or more side streams. If the naphthalene is recovered by fractional distillation, a naphthalene-oil cut is used. This can be obtained by fractional distillation during the original separation or by redistilling the chemical oil to provide an enriched naphthalene fraction. This fraction is passed through one or more jacketed crystallizers or pans wheie crude crystals of naphthalene form on cooling. These crystals are centrifuged to remove adhering oil and washed with water.

About 16 per cent of the coal tar produced is burned without further treatment, mainly in open-hearth furnaces. About 38 per cent is distilled; the distillate is sold as a source of valuable chemicals and the residue is burned. About 46 per cent is sold to tar processors who use the residues for preparing road tars, roofing pitches, pipeline enamels, and carbon electrodes. From the distillates such materials as naphthalene, tar acids, tar bases, and other chemicals are recovered, and these are important starting materials in a wide variety of chemical syntheses. The heavier distillate is creosote, whose principal use is as a wood preservative.

About 60 per cent of the light oil produced is refined at coke plants; the principal products are benzene, toluene, and xylene, which have many uses as chemicals and solvents. The remainder of the light oil is sold to other processors, usually petroleum refiners who handle it along with their own materials.

Practically all of the ammonia recovered is used as a fertilizer. About 90 per cent is removed as ammonium sulfate. Small quantities of diammonium phosphate and concentrated ammonia liquor are also produced. One large plant recovers the ammonia in coke-oven gas in the pure anhydrous state, and also synthesizes large quantities of ammonia from the hydrogen in the gas.

Tar bases are recovered from the solvent naphtha, side-stream products, and the naphthalene-still residue by washing with sulfuric acid. The tar bases, removed by decanting, are neutralized with ammonia, sodium carbonate, or caustic solution, separated by decanting, and then rectified into relatively pure fractions containing compounds, such as pyridine, picolines, collidines, lutidines, and quinoline.

The distillate and residue obtained at the higher distillation temperatures are creosote and pitch, respectively. Together, these represent the largest portion of the tar. If the plant is connected with a steel mill, topped tar, pitch, or creosote may be consumed as fuel. However, most of the residues produced in a tar distiller's plant go into road and refined tars, roofing pitch, and other pitch products. The various creosote fractions produced are blended to meet different customer specifications.

The composition of high-temperature tar produced in slot-type ovens is given in Table 8.3.

Over one-half of the tar distilled is pitch, for which the major market is in the production of electrodes used in the refining of aluminum and other electrolytic processes. Pitch is also used as roofing pitch, as a binder for graphite and industrial carbons, in fiber-pipe impregnation, for refractory impregnation, and as core pitch for foundry cores. The next largest constituent of tar is creosote, a commodity that is widely used in the preservation of wooden railroad ties and telephone poles. Naphthalene is the next largest fraction. Its

Table 8.3 Composition of High-Temperature Coke-Oven Tar[a]

	Per cent by Weight
Liquor	1.6–5.8
Benzol	0.1–0.3
Toluol	0.1–0.4
Xylol	0.1–0.5
Total tar acids (phenols, cresols, xylenols)	2.0–3.9
Total tar bases (pyridine, picolines, quinolines)	1.4–2.0
Naphtha (coumarone, indene)	0.4–2.0
Crude naphthalene	7.7–11.7
Methylnaphthalene oil	2.1–2.9
Biphenyl oil	0.9–1.5
Acenaphthene oil	1.4–2.8
Fluorene oil (fluorene, diphenyl oxide)	1.9–3.6
Anthracene-heavy oil (anthracene, phenanthrene, carbazole)	9.6–12.3
Pitch	60.2–64.2
Distillation losses	0.9–2.8

[a]Ranges of composition of five typical tars from "The Coal Tar Data Book." The Coal Tar Research Association, 2nd Ed., Section Al, 2–4, 1965.

Table 8.4 Production of Tar and Products from Tar, in 1970[a]

Product	*Production*[b]	*Unit Value*[c]
Tar	760.9	
Crude light oil	244.1	0.12
Intermediate light oil	5.1	0.09
Light-oil distillate		
Benzene, total	1133.5	0.22
Benzene, coke-oven operators	93.5	0.21
Benzene, petroleum operators	1040.0	0.22
Toluene, total	829.6	0.18
Toluene, coke-oven operators	17.0	0.18
Toluene, petroleum operators	812.6	0.18
Xylene, total	537.6	0.17
Xylene, coke-oven operators	4.5	0.20
Xylene, petroleum operators	533.1	0.13
Solvent naphtha, coke-oven operators	3.7	–
Naphthalene, crude, coke-oven operators	428.1[d]	–
Tar acid, crude, coke-oven operators	19.0	0.17
Creosote oils, coke-oven operators	128.9	–
Road tar	53.0	0.14
Tar for other uses	9.7	0.24
Soft and medium pitch	859.0[e]	32.65[e]
Hard pitch	899.0[e]	38.19[e]

[a]U.S. Tariff Commission, "Synthetic Organic Chemicals, U.S. Production and Sales, 1970," TC Publication 479.
[b]Millions of gallons, unless otherwise stated.
[c]Dollars per gallon.
[d]Millions of pounds–value per pound.
[e]Thousands of tons–value per ton.

major use is as a starting material for the production of phthalic anhydride. Naphthalene is also the source material for manufacturing alpha-naphthol, tetralin, and beta-naphthol, compounds which are used primarily in the production of insecticides, rubber chemical, and dye intermediates. Solvent naphtha is the raw material used for the production of coumarone-indene resin. Tar acids are used in the manufacture of resins, plasticizers, solvents, insecticides, pharmaceuticals, and in various other organic syntheses. Tar bases are used as solvents in rubbers and paints, and as basic materials for organic chemical synthesis.

The production of tar and products from tar in 1970 is given in Table 8.4. The total production of tar in 1970 was about 761 million gallons. The production of benzene, toluene, and xylene in the coke oven now amounts to a comparatively small fraction of the total produced; most of these commodities are supplied by petroleum refineries. But the coke oven is still the principal source of the other bulk chemicals and commodities listed. The output of steel is expected to grow in the future. And the production of coke and coal chemicals is expected to hold its own, thus assuring supplies of tars and coal chemicals.

SUGGESTED READING

1. Denig, F., Industrial Coal Carbonization in *Chemistry of Coal Utilization*, H. H. Lowry, Vol. I, Chap. 21, pp. 774–833, New York, John Wiley & Sons, Inc., 1945.
2. Hill, W. H., Recovery of Ammonia, Cyanogen, Pyridine and other Nitrogenous Compounds from Industrial Gases in *Chemistry of Coal Utilization*, H. H. Lowry, Vol. II, Chap. 27, pp. 1008–1135, New York, John Wiley & Sons, Inc., 1945.
3. Rhodes, E. O., The Chemical Nature of Coal Tar in *Chemistry of Coal Utilization*. H. H. Lowry, Vol. II, Chap. 31, pp. 1287–1370, New York, John Wiley & Sons, Inc., 1945.

4. Beimann W. *et al.*, High-Temperature Carbonization in *Chemistry of Coal Utilization*, H. H. Lowry, Supplementary Vol., Chap. 11, pp. 461–493, New York, John Wiley & Sons, Inc., 1963.
5. Muder, R., Light-Oil and Other Products of Coal Carbonization in *Chemistry of Coal Utilization*, H. H. Lowry, Supplementary Vol., Chap. 15, pp. 629–674, New York, John Wiley & Sons, Inc., 1963.
6. Wilson, P. J., and Clendenin, J. D., Low-Temperature Carbonization in *Chemistry of Coal Utilization*, H. H. Lowry, Supplementary Vol., Chap. 10, pp. 395–460, New York, John Wiley & Sons, Inc., 1963.
7. Wilson, P. J., and Wells, J. H., *Coal, Coke and Coal Chemicals*, New York, McGraw-Hill Book Co., Inc., 1950.
8. Russell, C. C., Carbonization in *Encyclopedia of Chemical Technology*, Kirk-Othmer, Vol. 4, pp. 400–423, New York, John Wiley & Sons, Inc., 1964.
9. McNeil, D., Tar and Pitch in *Encyclopedia of Chemical Technology*, Kirk-Othmer, Vol. 19, pp. 653–682, New York, John Wiley & Sons, Inc., 1969.
10. Hoiberg, A. J., *Bituminous Materials: Asphalts, Tars, and Pitches*, Vol. III, Coal Tars and Pitches, New York, Interscience Publishers, 1966.
11. Abraham, Herbert, *Asphalts and Allied Substances*, Sixth Ed., Vol. II, Industrial Raw Materials, New York, Nostrand Reinhold, 1961.
12. McNeil, Donald, *Coal Carbonization Products*, New York, Pergamon, 1966.
13. *Metallurgical Coke and Coal Chemicals*, in *Making, Shaping and Treating of Steel*, H. E. McGannon, Editor, Eighth Ed., United States Steel Corp., Chap. 4, pp. 98–147 (1964).

9

Rubber

Dr. R. L. Bebb*

INTRODUCTION

The strategic position of rubber in our civilization is evident from the effect of wars on the supply and from the way in which countries have reacted to find substitutes when the supply of natural rubber was threatened.

During World War I, Germany produced 2500 tons of methyl rubber from 2,3-dimethylbutadiene, which they synthesized with considerable difficulty. This was part of a program to make them independent of curtailed supplies of natural rubber. Methyl rubber was better suited for making hard rubber, and its production was stopped as natural rubber became available at the end of the war.

In Germany, research in the synthetic rubber field continued and production of the numbered "Bunas" was started. This name was developed from *Bu*tadiene *Na*trium, since butadiene was polymerized by sodium in the German process. Three types, buna 32, 85, and 115, were made in this series and used for special applications. Buna 32 was a softener used in rubber products as well as in hard rubbers. It was not used alone. Buna 85 and 115 saw some use in conventional rubber applications.

Emulsion polymerizations entered the picture in Germany in 1927, first with the polymerization of butadiene, and then with its copolymerization with styrene and with acrylonitrile. By the beginning of World War II these bunas (identified by letters–S, SS, N) had achieved commercial importance, and a number of types were produced as rubber substitutes when natural rubber supplies were cut off.[37]

The first synthetic rubbers to be commercially available in the United States were "Thiokol" (1930) and "Neoprene" (1931), or "Duprene" as it was first called. Both of these are still being produced commercially because they have special properties that are not matched by natural rubber; and therefore neither is competing for the natural rubber market.

Prior to the beginning of World War II a number of U.S. laboratories had investigated the German products and had initiated pro-

*Firestone Tire & Rubber Co.

TABLE 9.1 Production Capacities of U.S. Synthetic Rubber Plants, (1969)[a]

			Long Tons
American Rubber & Chemical	Louisville, Ky.	BR	75,000
American Synthetic Rubber	Louisville, Ky.	SBR	125,000
Ameripol Inc. (Orig. Goodrich-Gulf)	Port Neches, Tex.	SBR	192,800
	Orange, Tex.	BR	54,000
	Orange, Tex.	IR	44,600
	Institute, W.Va.	SBR	(Operation ceased, 1966)
	Institute, W.Va.	IR	13,500 LT (transferred to Orange, 1968)
Armstrong Rubber Co.	Borger, Tex.	BR and other stereo polymers	11,250
Ashland Chemical Co.	Baytown, Tex.	SBR	60,000
Columbian Carbon Co.	Lake Charles, La.	Butyl	37,500
Copolymer Rubber & Chemical	Baton Rouge, La.	SBR	125,000
	Baton Rouge, La.	EPDM	Pilot unit
	Addis, La.	EPDM	24,750[4] (40,000, 1971)
Dow Chemical	Baton Rouge, La.	NBR	5,000
	Midland, Mich.; Freeport, Tex.; Pittsburg, Cal.; Allyns Point, Conn.; Dalton, Ga.	SBR latex	>20,000 (est.)
E. I. du Pont de Nemours	Louisville, Ky.	CR	122,000
	Montague, Mich.	CR	30,000
	LaPlace, La.	CR	35,000 (end '69)
	Beaumont, Tex.	EPDM	45,000 ('71)
Enjay Chemical Co.	Baytown, Tex.	Butyl	80,000
	Baton Rouge, La.	EPM/EPDM	22,500 (40,000 est. by '71)
	Baton Rouge, La.	Butyl and chlorobutyl	46,000
Firestone Tire & Rubber	Akron, O.	SBR	47,500
	Akron, O.	NBR	5,000
	Lake Charles, La.	SBR	235,000 (1970)
	Lake Charles, La.	Solution SBR	80,000 (1970)
	Orange, Tex.	BR	70,000
GAF Corp.	Chattanooga, Tenn.	SBR latex	>20,000
General Tire & Rubber	Odessa, Tex.	SBR	70,000
	Mogadore, O.	SBR latex	20,000
	Borger, Tex.	BR and other stereo polymers	22,500
B. F. Goodrich Chemical	Akron, O.	NBR	14,000
	Louisville, Ky.	NBR	28,000
	Baton Rouge, La.	EPDM	25,000 (end '71)
B. F. Goodrich Industrial Products	Shelton, Conn.	SBR and BR latex	15,000
Goodyear Tire & Rubber	Houston, Tex.	SBR and nitrile	285,000 (1970) 10,000
	Akron, O.	SBR latex	40,500
	Akron, O.	SBR and NBR latex	17,250

			Long Tons
	Beaumont, Tex.	IR	60,000
	Beaumont, Tex.	BR	100,000 (1970)
W. R. Grace	Owensboro, Ky. Acton, Mass.	SBR latex } NBR latex }	15,000
Hooker Chemical Corp.	Hicksville, N.Y.	SBR latex	<5,000 (est., not official)
Petrotex Chemical Corp.	Houston, Tex.	CR	20,000
Phillips Petroleum	Bolger, Tex.	BR	57,500
	Bolger, Tex.	Solution SBR	23,750
	Bolger, Tex.	SBR	66,500
Shell Chemical Co.	Marietta, O.	IR	30,000
	Marietta, O.	IR latex	7,000
	Torrance, Cal.	SBR	97,000
Sinclair-Koppers	Monaca, Pa.	SBR latex	20,000
Southwest Latex Corp.	Bayport, Tex.	SBR latex	5,000
Standard Brands Chemical	Cheswold, Del.	SBR/NBR latex	26,000
Texas-US Chemical	Port Neches, Tex.	SBR and emulsion BR	148,000
Uniroyal	Naugatuck, Conn.	SBR and latex	26,000
	Painesville, O.	NBR and NBR latex	26,500
	Geismar, La.	EPDM	55,000 (1971)
Wica Chemicals	Charlotte, N.C.	SBR latex	11,500

SBR–Styrene-butabiene rubber
BR–Butadiene rubber
NBR–Acrylonitrile-butadiene rubber
EPM–Ethylene-propylene rubber
EPDM–Ethylene-propylene terpolymer
IR–Isoprene rubber
CR–Polychloroprene rubber
[a]Source: *Rubber & Plastics Age* (Nov. 1969).

grams of their own to develop practical synthetic rubbers both for tire applications and for specialty uses. By the date of Pearl Harbor, December 7, 1941, Firestone had over 100,000 tires tested on the road. Other companies also had active programs.

The number and size of the synthetic rubber plants in the United States alone (Table 9.1) and the spectrum of properties offered to the manufacturers of rubber goods has led to a usage pattern (Table 9.2) heavily weighted in favor of the synthetic rubbers. While similar rubbers are being produced throughout the world, outside the U.S. a higher proportion of natural rubbers is being used. But the synthetic rubbers are gaining in relative volume (Table 9.3).

The United States synthetic-rubber industry was a very effective stabilizer for the price of natural rubber. After the start of the Korean

TABLE 9.2 Rubber Consumption in U.S. by Major Fields, 1969[a] (MT)

Tire and Tire Products		
Natural		435,670
S-type synthetic	907,569	
Butyl rubber	70,282	
Acrylonitrile copolymers	160	
Stereoregular polymers	329,809	
Others	5,680	
TOTAL synthetic		1,313,500
Nontire Products		
Natural		172,202
S-type synthetic	423,522	
Butyl rubber	25,655	
Acrylonitrile copolymers	68,864	
Stereoregular polymers	70,272	
Others	154,729	
TOTAL synthetic		743,042
TOTAL consumption		2,664,414

[a]Rubber Statistical Bulletin, International Rubber Study Group, Vol. **25**, No. 2 (1970).

TABLE 9.3 World Production and Consumption, 1969[a]

	Production MT	*Consumption MT*
Natural rubber (and latex)	2,900,000	2,945,000
Synthetic rubber	4,572,500	4,452,500

	Percentage Consumption Synthetic Rubber
U.S.	77.19
Total, Western Europe	60,49
Total, World	65.93

[a]Rubber Statistical Bulletin, International Rubber Study Group Vol. **25**, No. 2 (1970).

War the price of natural rubber rose from 15 to 73.4 cents in February, 1951. Synthetic rubber replaced an increasing amount of the natural rubber then in use and the price of natural rubber was forced down. As natural rubber came down to 20 cents in 1954, some consumers changed from synthetic rubber to natural product, but, with each rise in the price of natural rubber, synthetic rubber became more established, and a swing back to natural rubber was resisted. There has been a fairly free interchange of the two products depending on the price of each. Synthetic rubber now has the advantage of having many suppliers who are located in the United States and produce a uniform product at competitive prices (Fig. 9.1).

As the United States entered into World War II, many companies entered into research, development, and production contracts with the government, and for the period from December 1941 through April 1955 conducted, under government sponsorship, closely coordinated programs in which open exchange of background and experience in the production of synthetic rubber was encouraged. Without close cooperation, especially at the beginning of the program, it would have been impossible to build up the synthetic rubber industry as fast as it was. That it did succeed is evident from the production of synthetic rubber and the price stability of both natural and synthetic rubber.

The emergency program under which the U.S. synthetic-rubber industry was set up provided that the plants should ultimately be sold. After advertising the sales and considering the bids, the government sold the plants to private companies who, in turn, were required to maintain the plants so that synthetic rubber of the GR-S type could be produced immediately in the event of an emergency. Production capacity for both monomers and rubber has increased considerably under private operation.

The number of plants producing various synthetic rubbers now includes many producers who were not active in the government program, and the types of synthetic rubber have been considerably expanded especially by the new stereospecific solution rubbers which will be discussed later in this chapter.

The following are some important dates in the commercial history of synthetic rubber:

1914–18	Methyl rubber produced in Germany.
1930	Thiokol, an organic polysulfide rubber resistant to oils and solvents, introduced to oils and solvents, introduced in the U.S.
1931	Neoprene (originally DuPrene), a polymer of 2-chlorobutadiene-1,3, production started.
1933	Buna S, a butadiene-styrene copolymer, made in emulsion in Germany.
1936	Perbunans, or Buna N, specialty oil-resistant rubbers from butadiene and acrylonitrile, manufactured in Germany.
1939	Mercaptan-persulfate emulsion recipes patented. French Patent 843,903 (July 12, 1939) served as the basis for the butadiene-styrene "Mutual" hot-polymerization recipe later adopted as standard for hot GR-S production in the U.S.
1940	Butyl rubber, a copolymer of isobutylene and isoprene characterized by a very low permeability to air and especially suited to innertubes, production started.
1942	GR-S (hot rubber), a copolymer of butadiene and styrene, produced in the U.S.
1944	Silicone elastomers, characterized by retention of elastomeric properties over a wide range of temperatures, introduced.

1946 Polyurethane rubbers, prepared by diisocyanate-coupling of dihydroxy compounds, introduced in Germany.

1947 Cold rubber, a copolymer of butadiene and styrene emulsion polymerized at 41°F, produced in the U.S.

1954–55 Synthetic natural rubber, prepared by the polymerization of isoprene, announced.

1955 Government-owned synthetic rubber plants purchased by private industry; expansion of production capacity begun.

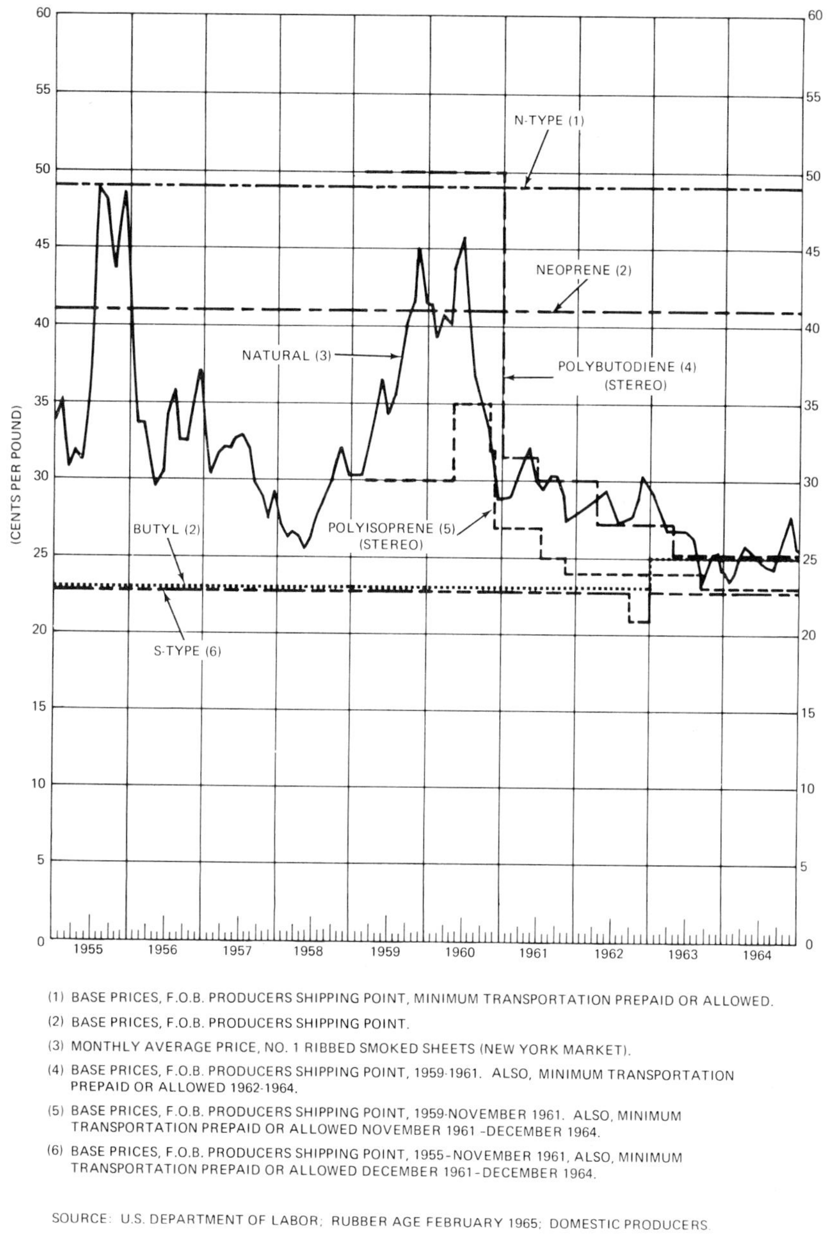
N-TYPE (1)
NEOPRENE (2)
NATURAL (3)
POLYBUTODIENE (4) (STEREO)
POLYISOPRENE (5) (STEREO)
BUTYL (2)
S-TYPE (6)
(CENTS PER POUND)
1955
1956
1957
1958
1959
1960
1961
1962
1963
1964
(1) BASE PRICES, F.O.B. PRODUCERS SHIPPING POINT, MINIMUM TRANSPORTATION PREPAID OR ALLOWED.
(2) BASE PRICES, F.O.B. PRODUCERS SHIPPING POINT.
(3) MONTHLY AVERAGE PRICE, NO. 1 RIBBED SMOKED SHEETS (NEW YORK MARKET).
(4) BASE PRICES, F.O.B. PRODUCERS SHIPPING POINT, 1959-1961. ALSO, MINIMUM TRANSPORTATION PREPAID OR ALLOWED 1962-1964.
(5) BASE PRICES, F.O.B. PRODUCERS SHIPPING POINT, 1959-NOVEMBER 1961. ALSO, MINIMUM TRANSPORTATION PREPAID OR ALLOWED NOVEMBER 1961 -DECEMBER 1964.
(6) BASE PRICES, F.O.B. PRODUCERS SHIPPING POINT, 1955-NOVEMBER 1961, ALSO, MINIMUM TRANSPORTATION PREPAID OR ALLOWED DECEMBER 1961-DECEMBER 1964.
SOURCE: U.S. DEPARTMENT OF LABOR; RUBBER AGE FEBRUARY 1965; DOMESTIC PRODUCERS

Fig. 9.1 Natural and synthetic rubber prices, 1955–1964. [*U.S. Department of Labor; Rubber Age (Feb. 1965); Domestic Producers.*]

This chapter will include a description of both emulsion and nonaqeous routes for manufacturing synthetic rubbers, the production of natural rubber and natural rubber latex, the preparations of monomers used in the synthetic rubber industry, and a short review of the compounding and curing of rubbers.

SYNTHETIC RUBBERS BY THE EMULSION PROCESS

Butadiene-Styrene Copolymers

The highly simplified flow sheet shown in Fig. 9.2 represents the route used for making the synthetic rubber which is produced in a larger volume than any other elastomer in the world. Butadiene ($CH_2{=}CH{-}CH{=}CH_2$) and styrene ($C_6H_5{-}CH{=}CH_2$) are mixed with an appropriate emulsifying solution, catalyst, and modifying agent (mercaptan), and heated under agitation in a pressure vessel until the desired conversion of the monomers (butadiene and styrene) is achieved. The polymerization is then stopped with a suitable chemical, the latex is stripped of unreacted monomer, antioxidant is added, and the rubber is isolated from the latex by coagulation with salt (NaCl), salt-acid, or aluminum sulfate solution.

Polybutadiene rubber can be made in the same way by omitting the styrene; oil-resistant rubbers result from substituting acrylonitrile for the styrene.

The formula adopted March 26, 1942 by the Technical Advisory Committee of Rubber Reserve for the preparation of "hot" rubber has remained the basis for its commercial production, with only minor modifications:[18,22]

	Parts Per 100 Monomer
Butadiene	75.0
Styrene	25.0
Water	180.0
Soap	5.0
"Lorol" mercaptan (n-$C_{12}H_{25}SH$)	0.50
Potassium persulfate	0.30
Polymerization temperature	50°C
Time	12 hrs
Conversion	75%

The object of the polymerization is to form rubbery polymers of butadiene and styrene of a desired molecular weight. Polymerization is initiated by potassium persulfate, and the molecular weight is regulated by a mercaptan. The role of the mercaptan in this polymerization is particularly important since its type and concentration regulates the molecular weight of the polymer and determines its processing characteristics. Chemical reactions occurring in hot (50°C) emulsion polymerization are given below:

(1) $K_2S_2O_8 + C_{12}H_{25}SH \longrightarrow C_{12}H_{25}S^{\bullet}$ — Generation of free radicals

Potassium persulfate; Dodecyl mercaptan; Mercaptyl free radical

(2) $C_{12}H_{25}S^{\bullet} + CH_2{=}CHCH{=}CH_2 \longrightarrow C_{12}H_{25}SCH_2CH{=}CHCH_2^{\bullet}$ — Initiation

Butadiene; Monomer radical

(3) $C_{12}H_{25}SCH_2CH{=}CHCH_2^{\bullet} + n(CH_2{=}CHCH{=}CH_2) \longrightarrow C_{12}H_{25}S(CH_2CH{=}CHCH_2)_{n+1}^{\bullet}$ — Propagation

Butadiene; Homopolymer radical

and/or

$m(C_6H_5CH{=}CH_2) \longrightarrow C_{12}H_{25}S(CH_2CH{=}CHCH_2)(CH_2\underset{\underset{C_6H_5}{|}}{CH})_m^{\bullet}$

Styrene; Copolymer radical

(4) $C_{12}H_{25}S(CH_2CH{=}CHCH_2)_{n+1}^{\bullet} + C_{12}H_{25}SH \longrightarrow C_{12}H_{25}S(CH_2CH{=}CHCH_2)_{n+1}H + C_{12}H_{25}S^{\bullet}$ — Termination and chain transfer

Polymer

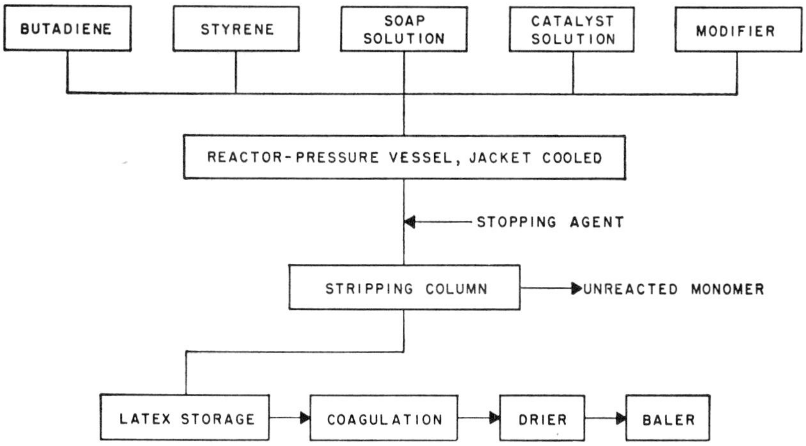

Fig. 9.2 Simplified flow sheet for preparation of butadiene-styrene synthetic rubber.

A review of the mechanism of emulsion polymerization may be found in References 23 and 66.

The polymerization reactors are designed to withstand pressures over 100 psig. The temperature of the reactants is adjusted to the desired level by controlling the temperature in the reactor jacket or in the internal coils.

The reactors vary in size from 3700 gallons, in the early plants, to over 5000 gallons. They are either constructed of stainless steel or of iron clad with stainless steel; they are jacketed and agitated. Different types of agitators have been used. Their selection depends considerably on the type of mercaptan used, since the efficiency of the mercaptan action is affected by the type and rate of agitation. In a hot polymerization the pressure within the reactor may reach 70 psig, dropping to about 40 psig at approximately 70 per cent conversion of the monomer.

The reactors are usually located in a chain and may be equipped for either batch or continuous operation. In the batch route butadiene, styrene, soap solution, catalyst, and modifier are charged directly into the reactor. In the continuous operation the various ingredients are metered into the bottom of the first reactor; the low conversion latex passes out the top, and into the bottom of the second reactor. This process is repeated throughout the chain, and the extent of polymerization is monitored by measurement of the total solids.

During polymerization the hydrocarbons are emulsified in the soap solution; a portion is dissolved in the soap micelles where polymerization is initiated. As polymerization proceeds the soap micelle loses importance and the balance of the reaction takes place in the polymer-monomer particles.[27] Butadiene and styrene do not enter the polymer at the same rate. In a 75/25 charge the initial polymer contains 17.2 per cent styrene, whereas at 80 per cent conversion the polymer contains 21.2 per cent styrene and at total conversion, 25.0 per cent.

At the desired conversion, polymerization is stopped by the addition of a material such as hydroquinone which destroys the catalyst and arrests further polymerization. At this stage the rubber is contained in a stable milky suspension known as latex. Before being used as such, or coagulated to isolate the rubber, the latex must be stripped of unreacted monomers. This is done in two steps; butadiene is removed with a compressor and styrene by vacuum steam distillation.

Before the latex is coagulated, an antioxidant must be added to prevent deterioration of the rubber during drying, storage, and processing. One material frequently used is phenyl-beta-naphthylamine, about 1.25 per cent being added, as a suspension, on the dry rubber. The latex is then coagulated with an acidified brine solution, or by means of an aluminum sulfate solution. One practice involves creaming the latex with concentrated brine and adding it to an agitated vessel along with dilute acid until the particle size of the crumb reaches the desired value. The fatty acids are released from the soap on the surface of the copolymer particles and large porous crumb aggregates are formed. The crumbs pass over a vibrating screen or an Oliver filter to remove the serum. The rubber is reslurried, washed, and filtered. It then passes over a continuous belt dryer where it is dried at the lowest possible temperature in the shortest time. The dried rubber is pressed into 75-pound bales for shipment.

The finished butadiene-styrene rubber contains fatty acids, rosin acids, antioxidants, moisture, and some inorganic materials, mainly sodium chloride.

Since termination of the government program, many types of hot rubber have been introduced to the trade for different uses. These represented variations in butadiene-styrene ratios, changes in plasticity, changes in conversion, and changes in antioxidant and methods of coagulation. Important objectives have been improvements in properties, processing characteristics, and color.

Cold Rubber

Throughout the government program there was a persistent belief that a better rubber could be made if lower polymerization temperatures were used. Efforts to activate the hot recipe were not particularly successful, and the polymerization time was long at temperatures lower than 50°C (122°F). In the early part of the program a number of revised recipes were proposed, but these did not reduce the polymerization temperature substantially below 30°C. There was no major improvement in the quality of this rubber over that of a normal hot control.

The successful development of ways to accelerate polymerization at low temperatures centered around the "Redox" system which was developed independently in Germany, Great Britain, and the United States. The Redox system involves the presence of an oxidizing agent, usually a peroxide or hydroperoxide, a reducing agent, and a soluble salt of a metal capable of existing in several states of oxidation.

Chemical reactions in cold (5°C) emulsion polymerization are:

(1) $\underset{\text{Diisopropylbenzene hydroperoxide}}{ROOH} + \underset{\text{Ferrous ion}}{Fe^{++}} \longrightarrow \underset{\text{Peroxyl free radical}}{RO^{\bullet}} + \underset{\text{Ferric ion}}{Fe^{+++}}$ — Generation of free radicals

(2) $RO^{\bullet} + \underset{\text{Butadiene}}{CH_2{=}CHCH{=}CH_2} \longrightarrow \underset{\text{Monomer radical}}{ROCH_2CH{=}CHCH_2^{\bullet}}$ — Initiation

(3) $ROCH_2CH{=}CHCH_2^{\bullet} + \underset{\text{Butadiene}}{n(CH_2{=}CHCH{=}CH_2)} \longrightarrow \underset{\text{Homopolymer radical}}{RO(CH_2CH{=}CHCH_2)_{n+1}^{\bullet}}$ — Propagation

and/or

$\underset{\text{Styrene}}{m(C_6H_5CH{=}CH_2)} \longrightarrow \underset{\text{Copolymer radical}}{RO(CH_2CH{=}CHCH_2)(CH_2CHC_6H_5)_m^{\bullet}}$

(4) $RO(CH_2CH{=}CHCH_2)_{n+1}^{\bullet} + RSH \longrightarrow \underset{\text{Polymer}}{RO(CH_2CH{=}CHCH_2)_{n+1}H}$ — Termination

At the end of the war it was learned that the Germans had developed a system using benzoyl peroxide as the oxidizing agent and sugar as the reducing agent. Good rubbers were produced from these systems in the United States, but polymerization characteristics were considered unsatisfactory. The German process was therefore modified in the United States by the introduction of hydroperoxides and by the replacement of sugar with amines. Since synthetic rubber plants are now privately owned, polymerization recipes are no longer published. It is felt, however that the "cold" recipe still remains a variation of the SFS type developed toward the end of the government program. This was such a dependable recipe in 1955, that it is most likely that it continues in use with, at most, small modifications of the ingredients.

	Parts per 100 Monomer
Butadiene	71.5
Styrene	28.5
Water	200.0
Mixed *tert.*-mercaptans	0.125–0.15
Potassium fatty acid soap	4.7
"Daxad-11"	0.1
KCl	0.5
$FeSO_4\ 7H_2O$	0.004
Sodium formaldehyde sulfoxylate (SFS)	0.0228
Ethylene diamine tetraacetic acid (Sequestrene AA)	0.0246
NaOH	0.0024
Diisopropylbenzene hydroperoxide	0.03–0.10
Sodium dimethyl dithiocarbamate (SDD) Stopping agent	0.10

With the introduction of cold polymerization, it was found that the physical properties of synthetic polymers, and the wear characteristics of tires made from them, were superior to those of hot rubbers (Fig. 9.3). As a consequence, the synthetic rubber program moved increasingly toward cold rubbers. Variations of the polymerization recipe resulted in smoother polymerization rates, better reaction times, and more reproducible performance.

The production equipment used for cold polymerization is similar to that used for hot-rubber manufacture except provision is made for cooling the reactors with a refrigerant in jackets or in coils within the reactor.

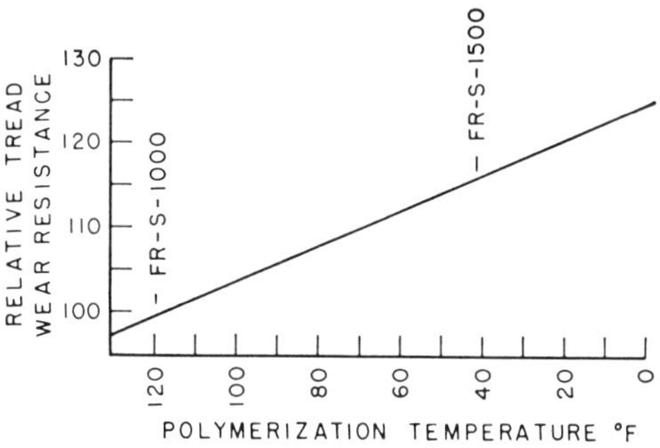

Fig. 9.3 Effect of polymerization temperature on tread wear resistance. [*Sjothun, Rubber Age,* **74** *(1953)*]

Butadiene-Acrylonitrile Copolymers

Butadiene-acrylonitrile copolymers are produced in much the same way as butadiene-styrene rubber. Because this rubber was not made under government operation there are many trade names for its various types. Each supplier offers a range of plasticities and acrylonitrile contents, thereby achieving considerable variation in oil resistance. Some are stabilized with discoloring antioxidants; others contain light-stable materials for use in special applications.

$$nCH_2{=}\underset{\substack{|\\ CN}}{CH} + mCH_2{=}CH{-}CH{=}CH_2 \longrightarrow$$

Acrylonitrile Butadiene

$$(-CH_2-CH{=}CH-CH_2-CH_2-\underset{\substack{|\\ CN}}{CH}-)_{n+m}$$

Butadiene-acrylonitrile polymer

The outstanding characteristics of the acrylonitrile rubbers are their resistance to oils, fats and solvents in general, and their performance at high and low temperatures (-70 to 300°F).

Since these rubbers are unsaturated, as are the butadiene-styrene rubbers, they may be used for hard rubber by incorporating sufficient sulfur in the classical compounding routes. Latices of butadiene-acrylonitrile copolymers may be used for making foam rubber products as well as special-purpose papers.

Neoprene[11,19,66]

The polymerization of chloroprene to neoprene has been described by Walker and Mochel[50], and resembles the preparation of butadiene-styrene rubbers.

	Parts per 100 Monomer	
Chloroprene	100	
N-Wood rosin	4.0	Dissolved in the monomer
Sulfur	0.6	Dissolved in the monomer
Water	150	
Sodium hydroxide	0.8	
Sodium salt of naphthalene sulfonic acid–formaldehyde condensation product	0.7	
Potassium persulfate	0.2–1.0	

Change in the specific gravity of the contents of the reactor during polymerization is large enough for the conversion to be followed by measuring the specific gravity. At the desired conversion, tetramethylthiuram disulfide is introduced to stop polymerization; the latex is ready for removal of unreacted chloroprene and coagulation. The stripping is done by vacuum steam distillation.

One method for isolating neoprene from the latex involves freezing.[72] Acetic acid is added to the alkaline latex just short of coagulation. The sensitized latex is then passed over the surface of a large brine-cooled drum where a sheet is frozen and rubber is coagulated. The rubber is removed from the drum, washed with water, passed through squeeze rolls, and dried in air at 120°C. The dried film is gathered into ropes and cut into short lengths for bagging and use. The finished rubber has a specific gravity of 1.23.

There are a number of different types of neoprene polymers available. These are sulfur-modified polymers (GN, GNR, GRT) which can be cured by the addition of metal oxides alone. The unmodified types (W and WRT) require accelerators as well as the metal oxides. For vulcanization the neoprene rubbers are usually compounded with light-calcined magnesia and zinc oxide; cross-linking occurs at the 1,2-addition sections. The cured strength may reach 4000 psi with an elongation of 900 per cent, and a 600 per cent modulus value near 1000 psi.

Neoprene is both stable to oxidation and flame resistant; it swells only moderately in oils and chemicals, and has good retention of properties in the swollen state. It has been used in wire insulation, cable jackets, gaskets for aliphatic liquids, belts for power transmission, and conveyor belt covers, especially where oil and heat are encountered. Since it is available as a latex, there are many applications where its unusual properties are an advantage.

The differences between the various synthetic rubbers have been summarized in Table 9.4.

Latex [5,31,69]

A latex is a stable dispersion of a polymeric substance in an aqueous medium. The dispersed component of the two-member system thus consists of essentially spherical particles of varying size suspended in a continuous phase containing suitable emulsifiers, or stabilizers.

Natural latex came into commercial use in the early 1920's, and by the beginning of World War II had become an important raw material. It was therefore logical that as the natural latex supply became limited, synthetic rubber latex should be substituted as far as possible.

In the commercial manufacture of synthetic rubbers where water and a dispersing system are charged into the reactor, the process also forms a latex from which the rubber is isolated. A disadvantage of latices made in early polymerization recipes was the very small particle size of the dispersed phase. As these latices made by "hot" recipes were concentrated, the viscosity increased so rapidly that 40 per cent solids was the upper limit. These were sold at 30–40 per cent total solids.

The low-temperature "cold" polymerization recipes with proper selection of dispersing agents led to 62-65-per cent solids latex (2105) which became the generally accepted high-solids type.

The total-solids percentage to which a latex can be concentrated is limited by the *packing effect* of the particles themselves; thus a latex of very small particle size is limited to 40 per cent solids. Some improvement can be made by adding electrolytes, such as trisodium phosphate, to the charge; but a greater improvement results from increasing the particle-diameter so that an optimum of 74 per cent of the rubber occurs as large particles. Several methods have been used commercially for increasing this diameter by destabilizing the particles: (1) careful additions of a hydrocarbon; (2) freezing the latex under rigid pH control; (3) passing the latex through a tight colloid mill (pressure process); and (4) addition of a polyether glycol (chemical process). The agglomeration may be effected in the presence of other latices such as polystyrene.

Latices have been prepared to a more limited extent by dissolving rubbers in suitable solvents, dispersing the resulting solutions, and subsequently recovering the solvent. The "solution" rubbers described later are well

TABLE 9.4 Relative Properties of Natural and Synthetic Rubber[a]

Property	*Natural Rubber*	*Grade R GR-S or Buna S*	*Butyl*	*Grade SA Thiokol (GR-P)*	*Grade SB Nitrile (Buna N)*	*Grade SC Neoprene (GR-M)*	*Grade T Silicone*
Composition	–	Butadiene-styrene	Isobutylene-isoprene	Organic polysulfide	Butadiene-acrylonitrile	Chloroprene	Polysiloxane polymer
Tensile strength (psi)							
Pure gum	Over 3000	Below 1000	Over 1500	–	Below 1000	Over 3000	Below 1500
Black load stocks	Over 3000	Over 2000	Over 2000	–	Over 2000	Over 3000	–
Hardness range (shore durom A)	30–90	40–90	40–75	35–80	40–95	40–95	40–85
Specific gravity (base material)	0.93	0.94	0.92	1.34	1.00	1.23	–
Tear resistance	Good	Fair	Good	Poor	Fair	Good	Poor
Abrasion resistance	Excellent	Good to Excellent	Good	Poor	Good	Excellent	Poor
Solvent resistance							
Aliphatic hydrocarbons	Poor	Poor	Poor	Excellent	Excellent	Good	Poor
Aromatic hydrocarbons	Poor	Poor	Poor	Good	Good	Fair	Poor
Acid resistance							
Dilute	Fair to good	Fair to good	Excellent	Fair	Good	Excellent	Excellent
Concentrated	Fair to good	Fair to good	Excellent	Fair	Good	Good	Fair
Oxygenated solvents (ketones, etc.)	Good	Good	Good	Good	Poor	Poor	Fair
Permeability to gasses	Fair	Fair	Very low	Low	Fair	Low	Fair
Oil and gasoline resistance	Poor	Poor	Poor	Excellent	Excellent	Good	Fair
Animal and vegetable-oil resistance	Poor to good	Poor to good	Excellent	Excellent	Excellent	Good	Fair
Oxidation	Good	Good	Excellent	Good	Good	Excellent	Excellent
Sunlight aging	Poor	Poor	Very good	Good	Poor	Very good	Excellent
Heat aging	Good	Very good	Excellent	Fair	Excellent	Excellent	Outstanding
Resistance to swelling in lubricating oil	Poor	Poor	Poor	Excellent	Very good	Good	Fair
Resistance to water absorption	Very good	Good to very good	Very good	Fair	Fair to good	Good	Good
Resistance to lacquer solvents	Poor	Poor	Poor	Good	Fair	Poor	Poor
Flame resistance	Poor	Poor	Poor	Poor	Poor	Good	Fair
Cold resistance	Excellent	Excellent	Good	Fair	Good	Good	Excellent
Heat resistance	Good	Excellent	Excellent	Poor	Excellent	Excellent	Excellent
Ozone resistance	Fair	Fair	Excellent	Excellent	Fair	Excellent	Excellent
Rebound							
Cold	Excellent	Good	Bad	Fair	Good	Very good	Excellent
Hot	Excellent	Good	Very good	Fair	Good	Very good	Excellent
Dielectric strength	Excellent	Excellent	Excellent	Fair	Poor	Good	Good
Electrical Insulation	Good to excellent	Good to excellent	Good to excellent	Fair to good	Poor	Fair to good	Excellent
Compression set	Good	Good	Fair	Poor	Good	Fair to good	Fair
Vulcanizing properties	Excellent	Excellent	Good	Fair	Excellent	Excellent	–
Adhesion to metals	Excellent	Excellent	Good	Poor	Excellent	Excellent	–
Adhesion to fabric	Excellent	Good	Good	Fair	Good	Excellent	–

[a]Krause, R. A. *Machine Design*, April, 19, p. 129 (1956) (by Permission).

suited to this technique, and latices of high *cis*-polyisoprene have been offered commercially. This process permits the latex to be tailored for particular applications by changing the type and level of the stabilizing system. Since one or more additional steps are involved in manufacturing these latices, they may be more expensive to produce than the latices produced by emulsion polymerization.

Synthetic latices are widely used in foam rubbers, paper-coating processes, saturation of wet and dry web in paper manufacture, carpet backings, adhesives, tire-cord treatments, and "latex" paints.

POLYMER STRUCTURE

Polymerization to macromolecules can follow two routes: *condensation polymerization* or *addition polymerization*. In condensation polymerization, monomeric molecules react with the elimination of a simple molecule such as water, ammonia, hydrogen chloride, or sodium chloride. This condensation may proceed until the desired molecular weight is achieved. Or a preliminary condensation product may be reacted with other types of monomers accompanied by the elimination of similar or other by-products.[20] In this process the molecular weight increases from the very beginning; and the molecular weight distribution can be calculated from the amount of conversion. The final molecular weight is limited by the availability of monomer and the degree to which complete conversion can be achieved. The organic polysulfide rubbers "Thiokols" and the silicone rubbers are examples of this type of polymerization.

Addition polymerization involves the reaction of monomer units with no elimination of by-product. Thus a homopolymer has the same composition as the initial monomer. As developed earlier in this chapter, *free radical initiation* was used to prepare the first synthetic rubbers. These are produced largely by an emulsion process, but the free radical mechanism can occur in solution systems.

Ionic addition polymerization was used at an early stage in the preparation of synthetic rubbers by the Germans and the Russians. The problem of heat removal made it difficult to reach high conversions and to adequately control this process. Early in the Rubber Reserve Program, alfin-catalyzed polymerization systems were developed, but these were used in solution to overcome the heat problem.

Friedel-Crafts-type catalysts played an important role when aluminum trichloride was used to prepare butyl rubber by the copolymerization of isobutylene and isoprene in an inert diluent such as methyl chloride.

The characterization of a synthetic rubber is much more complex than the characterization of a simple organic compound, and a highly specialized science has developed for this field.[40] It has become customary to separate *microstructure* from *macrostructure* in discussing a polymer.

Microstructure refers to the way in which monomer units are ordered along a polymer chain and includes the geometric formation in which they are assembled.

Macrostructure, on the other hand, covers the molecular weight distribution of the individual polymer molecules, the average molecular weight, and characteristics of the macromolecule itself.

Microstructure

In diene polymers, the isoprene units found in natural rubbers, and the butadiene and isoprene units found in polybutadiene and polyisoprene synthetic rubbers, are arranged in various combinations of *cis*-1,4-and *trans*-1,4-additions as well as the 1,2 structure, and in the case of isoprene, additions between the 3,4-positions. Natural rubber of the Hevea type is essentially all *cis* structure; the hard, more resilient plastic counterparts (balata and gutta percha) are largely *trans* polyisoprene structures.

(1)

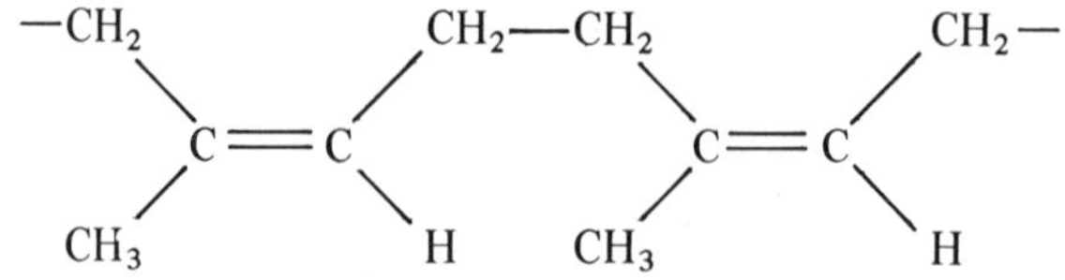

cis, head-to-tail 1,4-addition

(2) 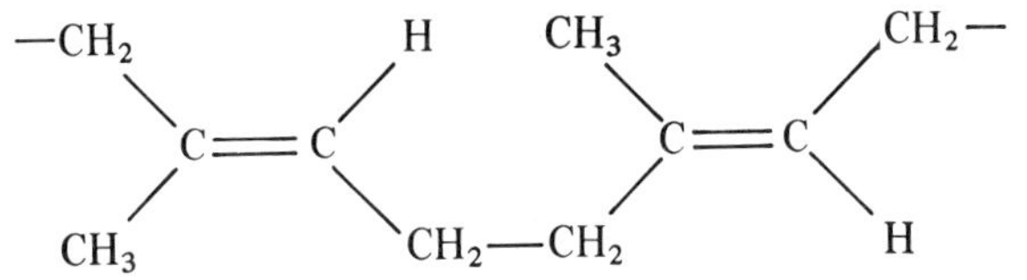 *trans*, head-to-tail 1,4-addition

(3) 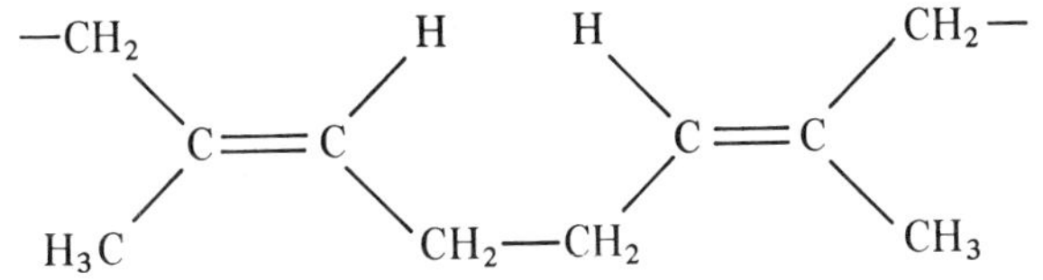 *trans*, head-to-head 1,4-addition

(4) 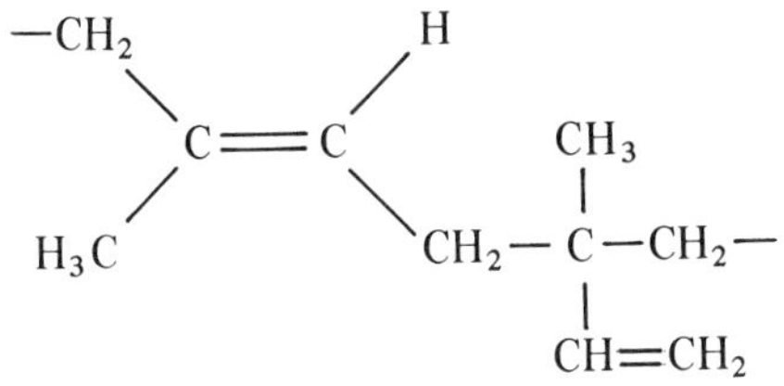 *trans*, 1,4; 1,2-addition

Since about 1955 considerable work has been published on stereospecific catalysts capable of producing polymers with larger amounts of *cis*-addition. In order to define more completely the complexity of these polymers, Professor Natta of Milan, Italy has coined the words *atactic*, *isotactic*, and *syndiotactic.* A polymer with an ordered arrangement like that of (1) is an isotactic polymer; a polymer with a random arrangement of any sort is an atactic polymer. With the recently developed anionic polymerization catalysts it is possible to produce polymers with a regular spiral arrangement of atoms along the chain. These may be divided further into spirals toward the right and toward the left. This class of compounds represents an isotactic arrangement. If, however, the polymer chain is made up of alternating right and left spirals, the polymer has a *syndiotactic* configuration. Examples of the latter type are currently found in polypropylene, polyethylene, and polystyrene.

In polymers based on diene monomers, the structure is determined by: (1) the nature of the monomer or mixture of monomers selected; (2) the conversion–in the case of a single monomer feed, chain branching or crosslinking may occur at the higher conversion, while in copolymerization the two monomers may enter at different rates; (3) the mechanism of polymerization, whether free radical or ionic; (4) the nature of the catalyst involved; and (5) the temperature of polymerization.

When two or more monomers are charged into a system, copolymers may result. Depending on the relative reactivities of the monomers, the second, or other additional monomer, may enter the polymer chain at a different rate than that of the first monomer. As a consequence examination of the microstructure may involve a study of the tendency of the second or third monomer to enter the polymer chain at a given statistical rate or to form blocks of varying sizes. The greater the size of the blocks in the final rubber, the more the rubber is affected by the characteristics of the homopolymer of the second or third monomer. Thus, as the polystyrene blocks increase in size in butadiene-styrene copolymers, the elastomers begin to show the hardness of polystyrene at room temperature and its tendency to soften on the mill. Infrared, NMR, and chemical analysis techniques are being developed for characterizing the size of these blocks and their length. Block copolymers of butadiene and styrene are becoming commercially important with Shell's introduction of the "Kratons"[61] for special nontire applications.

An interesting comparison of the *cis-trans*-1,2- and 3,4-contents of a series of polymers

TABLE 9.5 Microstructures of Synthetic Rubbers

	Percent			
	Cis-1,4	Trans-1,4	1,2	3,4
Emulsion rubbers				
SBR "hot"[a]	17.6	65.5	16.9	–
SBR "cold"[a]	12.3	71.8	15.9	–
Solution rubbers				
Polybutadiene				
Butyllithium catalyst	35.0	55.0	10.0	–
Ziegler cobalt system	98.0	–	2.0	–
Ziegler nickel system	95.0	4.0	1.0	–
Alfin catalyst	5.0	75.0	20.0	–
Polyisoprene				
Butyllithium catalyst	94.0	–	–	6
Ziegler aluminum system	99.0	–	–	1

[a]Structure calculated on butadiene content only.

shows the type of control offered by the various polymerization techniques (Table 9.5).

Macrostructure

The molecular weight of a polymer represents an average expression of the various chain lengths in a particular sample. There are now four ways of expressing the average molecular weight.[7]

1. $M_{\bar{v}}$, viscosity average molecular weight based on viscometric techniques.
2. $M_{\bar{n}}$, number average molecular weight based on osmometer determinations.
3. $M_{\bar{w}}$, weight average molecular weight based on light scattering.
4. $M_{\bar{k}}$, kinetic molecular weight based on polymer yield divided by the moles of initiator used in the charge.

For a further description of these see Reference 21.

Molecular weight distribution has classically been determined by fractional precipitation. It may also be done using an ultracentrifuge, by precipitation chromatography, and more recently, by the increasingly important technique of gel permeation chromatography (GPC)[42].

Other properties that belong in the macrostructure category include the degree of branching, the rheology of a polymer, the glass transition temperature (Tg) and Young's modulus[21].

SYNTHETIC RUBBERS BY THE NONAQUEOUS PROCESSES

Nonaqueous processes for making synthetic rubbers have a commercial history dating back to the German numbered Buna rubbers, the Russian polybutadienes, polyisobutylene (Vistanex) in Germany, butyl rubber in the United States, Thiokol (U.S.), and silicone rubbers (U.S.). These processes cover the spectrum extending from those methods which produce an insoluble polymer which separates during polymerization (butyl rubber) to the numbered Bunas which remain in solution throughout.

The solution method of making synthetic rubber has been the source of many new and interesting products; it promises to become more important in the synthetic rubber field than the emulsion process previously described.

Butyl Rubber [66]

Butyl rubber is a copolymer of isobutylene with about 3 per cent isoprene. Its polymerization differs considerably from that of butadiene-styrene copolymers. In this case the monomers are polymerized in a solvent containing the dissolved catalyst at polymerization temperatures near $-150°F$. The original catalyst, aluminum chloride, is added as a dilute solution in methyl chloride. As polymerization proceeds, the polymer precipitates and the resultant slurry overflows into an agitated hot-

$$n\underset{\underset{CH_3}{|}}{\overset{\overset{CH_3}{|}}{C}}=CH_2 + m CH_2=\overset{\overset{CH_3}{|}}{C}-CH=CH_2 \rightarrow$$

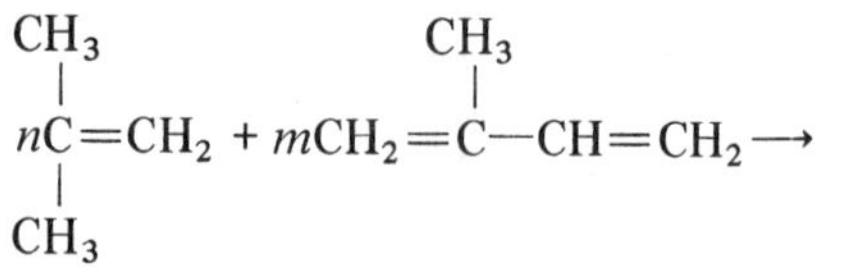

Isobutylene   Isoprene
(boiling point 34°C)

$$(-CH_2-\overset{\overset{CH_3}{|}}{C}=CH-CH_2-\underset{\underset{CH_3}{|}}{\overset{\overset{CH_3}{\|}}{C}}-CH_2-)_{n+m}$$

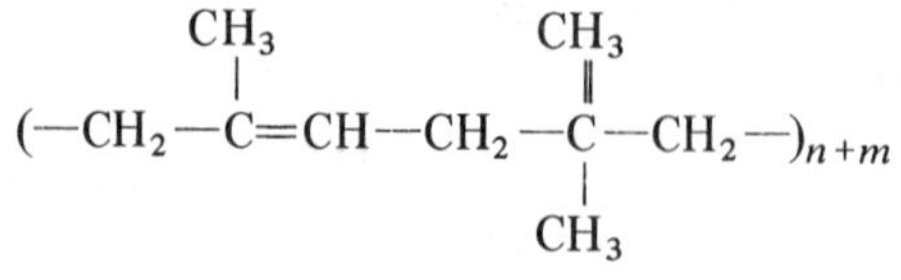

Butyl rubber

water tank. Unreacted monomer flashes off and is recovered by drying and condensation. As in the preparation of butadiene-styrene polymers, an antioxidant is added to prevent deterioration during drying and storage. The crumb is passed through a tunnel dryer at 200 to 350°F to remove most of the water; then it is fed into an extruder and onto rubber mills. It is removed continuously to a suitable packaging machine.

Butyl rubber resembles natural rubber in appearance; it is unsaturated and can be vulcanized, but it cannot be made into hard rubber because of extremely low (3 per cent) unsaturation. Butyl rubber does not pass through a latex stage as do butadiene-styrene copolymers. Butyl rubber dispersions have been prepared indirectly by dissolving the rubber in a solvent, dispersing the solution, and subsequently removing the solvent to leave a dispersion of butyl rubber in water. Since butyl rubber is the least permeable to gases of all rubbers, much of it is consumed in the manufacture of inner tubes for automobile tires. It is also of considerable value in making air bags for curing tires where repeated exposure to high temperatures would greatly shorten the life of normal rubbers.

A somewhat similar polymer known as "Vistanex," a polymer of isobutylene without the addition of isoprene, was developed in Germany. Its range of plasticity varied from a soft sticky gum to tough elastic material, depending on the degree of polymerization. Since it could not be vulcanized, it was mixed with other rubbers for tire treads, for electrical insulation of high-voltage wires, for tapes, and for many other special purposes. The seal-strength of wax wrappings for frozen foods was found to be improved by blending Vistanex with the coating wax.

Silicone Rubbers

Silicone rubbers differ from those previously described in that their chains consist of alternate atoms of silicon and oxygen, with no carbon atoms. Although work on the preparation of the silicone polymers was started in the Corning Research Laboratories, the General Electric Company initiated a program at about the same time, and in 1945 both companies announced the development of silicone rubber.

Most of the silicone rubbers are derived from dimethyldichlorosilane, but variations include a partial substitution of other groups, such as the phenyl for the methyl radical, and the preparation of polymers containing vinyl or allyl radicals.

Dimethyldichlorosilane (boiling point 70°C) is prepared by passing methyl chloride over powdered silicon with copper catalysts at 275 to 375°F. The general reaction is,

$$2CH_3Cl + Si \longrightarrow (CH_3)_2SiCl_2$$

The conversion of dimethyldichlorosilane into the polymer results from the addition of water and subsequent hydrolysis in the presence of small proportions of iron chloride, sulfuric acid, or sodium hydroxide. These catalysts must be washed out of the polymer. Low polymeric materials are removed by distillation.

$$R_2SiCl_2 + 2H_2O \longrightarrow R_2Si(OH)_2 + 2HCl$$

$$nR_2Si(OH)_2 \longrightarrow HO(SiR_2O)_nH + (n - 1)H_2O$$

Silicone rubbers are useful over a remarkably wide range of temperatures extending from -130 to 550°F. Although their tensile strengths are low, other properties compensate. One advantage is the fact that the rubbers are white and can be used for the preparation of light-colored stocks. They are not attacked by ozone and have good electrical properties. Swelling in oils is relatively low, and they are affected by very few chemicals.

Silicone rubbers are generally vulcanized by the addition of a peroxide such as di-*tert*-butyl peroxide or dicumyl peroxide. Oxides of certain metals, such as lead and zinc, accelerate vulcanization; silica, titania, ferric oxide, and alumina are good fillers.[39,55]

High *Cis*-Polyisoprene (Synthetic Natural Rubber)

In preparing butadiene-styrene polymers during World War II, it was recognized that the products were not true duplicates of natural rubbers. Although Hevea is composed of isoprene units, an emulsion polyisoprene does not duplicate the high gum tensile strength of natural rubber nor many of its other properties. Furthermore, emulsion diene polymers are composed primarily of *trans*-diene units whereas natural rubber is largely *cis*-1,4-polyisoprene. In the latter case, the regular arrangement of molecules permits natural rubber to crystallize during extension,[16,53] butadiene-styrene polymers show no comparable tendency to crystallize.

In November 1955 the Firestone Tire and Rubber Company announced the successful synthesis of a *cis*-1,4-polyisoprene supported by tire-test results. This polymer, known as "Coral" rubber,[62] was prepared by the polymerization of isoprene catalyzed by powdered lithium. The catalyst was made by melting the metal in petroleum jelly and agitating the product in a special high speed stirrer to produce a fine dispersion of lithium. A stirrer speed of 18,000 revolutions per minute for about twenty minutes at 200°C was used to produce a 35 per cent dispersion of lithium having a mean diameter of 20 microns.

A reactor was charged with 100 parts of pure dry isoprene and 0.1 part lithium. After an induction period at 30 to 40°C the charge thickened, and it became solid toward the end of the polymerization. The catalyst was decomposed with isopropyl alcohol, and the rubber was stabilized with a suitable antioxidant. The polymer was washed, then dried at 50°C.

Lithium was reported to be specific in its effect; other alkali metals gave mixtures of *cis*- and *trans*-polymers.

Tests have been reported[2,3] in which 100 per cent Coral rubber was used in truck tires. The heat build-up during service at 50 miles per hour was identical to that of natural rubber. Carcass stocks gave satisfactory performance and the polymer was considered a satisfactory substitute for natural rubber.

In November 1955 B. F. Goodrich Company reviewed for the American Chemical Society the structure and properties of *cis*-polyisoprene polymerized according to information supplied by Dr. Karl Ziegler. From the infrared absorption spectra, x-ray diffraction patterns, and second-order transition temperatures, the synthetic polymer was considered substantially the same as natural rubber.

Polyisoprenes require antioxidants as do the emulsion butadiene-styrene copolymers. Since fatty acid is not present in the final polymer, it is necessary to compensate by adjustments in the compounding recipe. Physical tests in the laboratory for pure gum, and body and tread stocks indicate that the polymer and Hevea are virtually identical.[30]

According to a Belgian patent[24] issued to Goodrich-Gulf Chemicals Company, *cis*-1,4-polyisoprene and mixtures of *cis*- with *trans*-1,4-polyisoprene may be prepared by controlling the catalyst consisting of a reaction product of an alkylaluminum with titanium tetrachloride in a hydrocarbon medium. A *cis*-polymer is obtained if the mole ratio of Ti to Al is 1; if the mole ratio is 1.5 to 3, a *trans*-polymer results.

In 1955 the Goodyear Tire and Rubber and Rubber Company disclosed the preparation of *cis*-1,4-polyisoprene[13] in which triethylaluminum was used as a catalyst together with a cocatalyst. The infrared spectrum was comparable to that of Hevea rubber.

A description of the pilot plant for the preparation of *cis*-polyisoprene has been given by the Goodyear Tire and Rubber Company.[67]

The heart of the system consists of two 500-gallon stainless steel autoclaves set up for interchangeable operation. The turbine-type agitators are driven by 10-horsepower motors; they can move material which shows a viscosity as high as 100,000 centipoises at a solids content of 15 per cent.

Extremely pure isoprene and a hydrocarbon solvent are used in this process. Special precautions are required, since oxygen and certain unsaturated compounds are severe catalyst poisons. Both isoprene and the solvent are distilled, mixed, and passed through a silica gel or alumina dryer. They are pumped through the dehydrating bed into the reactor where the catalyst is added. The temperature is controlled with water or brine in the reactor jacket at a polymerization temperature of 50°C. In this system, lower temperatures give higher molecular weight polymers.

At about 7-per cent solids the contents of the reactor become viscous and temperature-control is difficult. The final solids content of the cement is reported as being about 25 per cent. At this stage the cement is extremely viscous.

At the end of polymerization the cement is pumped to a tank where the catalysts can be deactivated and the necessary antioxidant is added. The cement is then heated and transferred to an extruder dryer. In this equipment the solvent is vaporized for recovery and reuse. The product, as extruded, contains less than 1 per cent volatile material; the rubber is packaged into 50-pound bales.

The catalyst for the polymerization consists of two parts. Triisobutylaluminum, which is spontaneously flammable, must be handled carefully. It should be mixed with the correct proportion of a second component in storage cylinders and be pressured into the reactor.

As in the case of the manufacture of butadiene, gas chromatography has been used to trace the purity of the monomer.

Solution Polybutadiene

Whereas *cis*-polyisoprene described previously was developed in an effort to synthesize natural rubber, polybutadiene took a place in the synthetic rubber field because of the unique character it imparted to mixtures with natural rubber, emulsion SBR, polyisoprene, and solution butadiene-styrene copolymers. Furthermore certain polybutadienes have achieved a considerable position in the production of high-impact polystyrene.

Compounds containing polybutadiene show high rebound and low heat build-up, characteristics which make this product valuable for trucks and bus-tire manufacture[12], and equally valuable in the production of passenger-car tires.

The catalysts for producing polybutadiene fall into two categories—organolithium compounds[21] and Ziegler-type systems.[17]

High 1,4 Polybutadiene

While the polymerization of butadiene catalyzed by *alkyllithium* compounds in ethers had been known since 1934, it was polymerization in the absence of ethers that gave high 1,4-configuration in polybutadiene and polyisoprene. Polyisoprene was announced in 1955 as a natural rubber replacement. Of all the alkali metals, only lithium produced the desired high *cis* content and properties.

The polymerizations are run in very pure, dry hexane or heptane in the presence of an organoalkali catalyst, butyllithium being used most frequently. Pure butadiene may be added in a batchwise operation or in a continuous process involving one or more stirred pressure vessels. Heat is removed from the jackets. In general, the polybutadienes are similar regardless of the alkyl- or aryllithium selected.

The linear polymeryllithium remains active at the end of the polymerization unless the charge is overheated. These are known as "living polymers," for they can continue polymerizing if more monomer is added.

If a second monomer is charged to a living-polymer solution, the block polymer that results may have unusual properties that have special applications.

The molecular weight distribution from a single polybutadiene preparation is very narrow and the solutions are quite viscous.

High Cis-Polybutadiene

The *Ziegler-Natta catalysts*[38,46,73] were reaction products of alkylaluminum compounds with $TiCl_4$ or $TiCl_3$ at the start. They were extended to salts of other transition metals, with Ti, V, Mo, Co and Ni being the most important. Commercial practice seems directed toward the reaction products of $TiCl_4$ and TiI_4 with alkylaluminum compounds; alkylaluminum halides with cobalt compounds; and aluminum trihalides with cobalt compounds.

Condensation Rubbers

The urethane rubbers constitute examples of *condensation* polymers, wherein monomers combine with the elimination of simple molecules, such as water or ammonia. Occasionally, as in this case, nothing is eliminated. Here isocyanates react with dihydroxy alcohols:

$$\mathrm{RNCO} + \mathrm{HO{-}R'{-}OH} \longrightarrow \mathrm{RNHCOO{-}R'{-}OCONHR}$$

There are many possibilities for varying the nature of the polymer by selecting different diisocyanates as well as different dialcohols, such as polyether glycols, branched glycols, and polyester glycols for the starting materials.[59] Either millable of castable polymers may then be mixed with suitable crosslinking agents and cured to usable rubbers.[44]

Urethane rubbers have high tensile strengths and resilience; they are generally resistant to oils, ozone, and oxygen.

Solution Butadiene-Styrene Rubbers

Sodium catalysts for the copolymerization of butadiene and styrene were developed in the mid-1940's by A. A. Morton and termed "Alfin" catalysts. These catalysts were reaction products of sodium, isopropyl *al*cohol and ole*fins* (Alfin).[43] The molecular weight was very high and the dilute-solution viscosity of 8–10 made processing impractical.

Recently Greenberg and Hansley[25] discovered that 1,4-dihydronaphthalene is a very effective chain-length regulator, which can control the molecular weight to 300,000 and the Mooney viscosity to 40.

Alkyllithium catalysts are excellent for the manufacture of butadiene-styrene copolymers. By charging a mixture of the monomers into a hexane solution of the catalyst, one can make a copolymer which is adequate for some uses. Figures 9.4 and 9.5 show that styrene has much less tendency to enter the copolymerization in the early stages than does

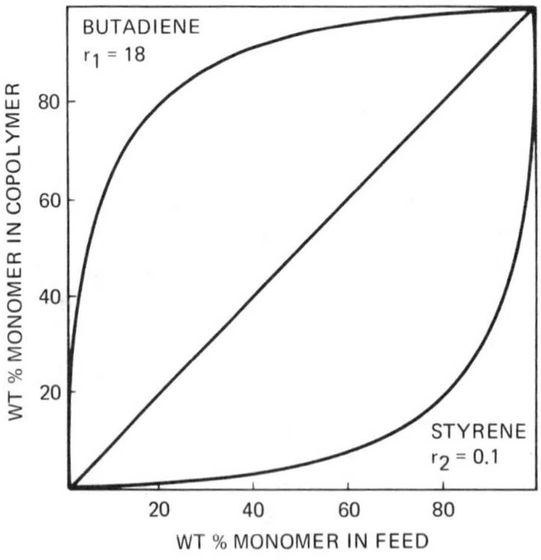

Fig. 9.4 Copolymerization of budadiene and styrene in benzene with butyllithium. (*"Polymer Chemistry of Synthetic Elastomers," J. P. Kennedy and E. G. Tornquist, eds., Interscience, 1969; by permission of John Wiley and Sons, Inc.*)

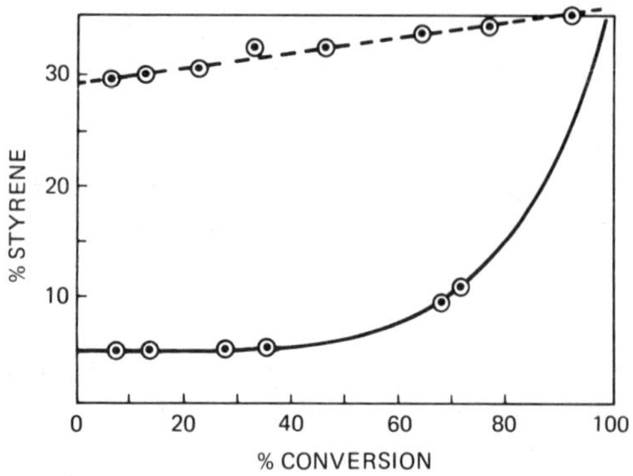

Fig. 9.5 Butadiene/styrene 65/35 copolymerization at 90° F; —— butyllithium initiator, no modifier; ---- butyllithium/TMEDA, 1/2.[48]

butadiene. As the butadiene supply is depleted in the neighborhood of 80 per cent conversion, styrene starts entering very rapidly.

Solution butadiene-styrene copolymers may be prepared by careful regulation of the monomer mixture present at any specific time, or by the addition of randomizing agents, such as alkali alkoxides or tetramethylethylenediamine (see Fig. 9.5). These polymers are being marketed under the tradename Duradene® or Stereon® by The Firestone Tire and Rubber Co., and Solprene® by the Phillips Petroleum Co.

Recently attention has been turning to the use of "living" systems in making block polymers. This can be done by:

1. Sequential addition of monomers:
   $A^- + B \longrightarrow AB^-$
   $AB^- + A \longrightarrow ABA$
2. Dilithium initiators:
   $^-BB^- + A \longrightarrow ABBA$
3. Coupling of diblock polymers:
   $A^- + B \longrightarrow AB^-$
   $2AB^- + RCl_2 \longrightarrow AB-R-BA$
4. Initial charge sequential addition
   a) $A + B \longrightarrow AB^-$
   $AB^- + A + B \longrightarrow ABAB$
   b) $B^- + A + B \longrightarrow BAB$

Polyolefin Rubbers

The transition metal-catalyst systems have been extended from low-pressure polymerization of ethylene to mixtures of ethylene and propylene, and interesting rubbers have resulted[45]. These have required specific catalyst systems, usually involving mixtures of alkyl aluminum compounds with derivatives of vanadium (VCl_4, VCl_3, $VOCl_3$, V triacetate) or titanium ($TiCl_4$, $TiCl_3$), and again, hydrocarbon solvents.

Ethylene-propylene rubbers (EPR) have been offered commercially in the range 55-60 per cent ethylene content. These contain no unsaturation and must, therefore, be cured with peroxide or radiation, or grafted with unsaturated diacids.

Therefore, terpolymers (EPDM or EPT) have been made containing 2.5-4 per cent of a nonconjugated diene (dicyclopentadiene, 1,5-hexadiene, ethylenenorbornene, methylenenorbornene). These can be vulcanized either with sulfur systems, with peroxides, or by radiation.

The EPR and EPDM vulcanizates have good low-temperature properties and excellent resistance to oxygen and ozone. They were first used in mechanical rubber goods but are expanding into the tire field, especially in sidewalls, as polymer variations and compounding techniques are making them more adaptable.

NATURAL RUBBER

Rubber is found in varying amounts in many plants throughout the world. Although other sources have been used in time of war, only two are commercially important at present. The first is the *Hevea brasiliensis*, a tree native to Brazil but now grown on plantations throughout the tropics, and the *Kok Saghys*, a dandelion grown as a biennial in Russia.

Rubber occurs in plants as a milky latex. The dry product is obtained by a process known as coagulation, during which the latex is destabilized by the addition of acids or salts.

Hevea brasiliensis was found growing wild in the tropics of Brazil and was taken into the Far East by the British. Vast natural rubber plantations have grown from this beginning (see Table 9.6). Stock from carefully selected trees giving high yields were grafted onto ordinary seedlings to produce a family of descendants from a single tree known as a "clone," and production was improved from an average of 250 to 500 pounds per acre to more than 2000 pounds per acre.

TABLE 9.6 Production of Natural Rubber, 1969[a]

	Metric Tons
Asia and Oceania	
Malaysia	1,279,227
Indonesia	790,430
Thailand	281,843
Ceylon	150,834
Vietnam	26,151
Cambodia	51,836
India	79,951
Other	26,250 (est.)
Africa	182,000 (est.)
Latin America	
Brazil	23,950
Other	7,000
TOTAL	2,900,000 (est.)

[a]Taken from International Rubber Study Group Rubber Statistical Bulletin **25**, No. 2 (Nov. 1970).

As new seedlings develop in the plantation nursery they are arranged in regular plantings around a collecting house. After the trees are six years old a program is established for tapping them and collecting the latex. In Hevea brasiliensis the latex occurs in tiny ducts or tubes found under the bark and just outside the green cambium or growing layer. Each morning a diagonal cut just deep enough to produce the latex is made with a special knife. The liquid is collected in a small cup at the bottom of the diagonal cut. When the cut is first made a small amount of preservative is placed in the cup to prevent coagulation. During the several hours before the latex stops flowing a tree will yield about 100 cubic centimeters of normal latex (30 to 40 per cent solids). This is strained in the collecting station to remove dirt and bark, and treated with more preservative. Then it is transferred to a central factory where it is centrifuged or coagulated.

Because latex is very sensitive to bacterial action, an adequate preservative must be added to protect it from the time it leaves the tree until it is used. Dilute ammonia is commonly used despite its volatility. It does not kill bacteria but reduces their growth. Some plantations use a small amount (0.15-0.30 per cent) of formaldehyde to sterilize the latex, following it up with ammonia before shipment. Still other producers use Santobrite (sodium pentachlorophenate) at approximately 0.3 per cent based on the latex along with 0.1 per cent ammonia to produce a latex with especially good keeping qualities.[68]

Natural Latex

When latex is brought to the central processing station on a plantation it has about 30-40 per cent solids. Almost no latex is sold at that level, but is concentrated to 62–68 per cent total solids. About 90 per cent is concentrated by centrifuging. Most of the balance is concentrated by the creaming method, with a little being concentrated by evaporation. Electrodecantation has been used for concentration, but only to a limited extent.

The *centrifuge* method depends on the difference in specific gravities between the rubber and the serum; rubber has a specific gravity of 0.91, and the serum, 1.02. In this case a special centrifuge separates a cream from a low solids skim containing about 10 per cent solids. The cream fraction is further stabilized by the addition of preservatives and adjusted to the correct solids level for shipment. Special concentrates for applications requiring low water absorption and high dielectric properties have also been developed by diluting and recentrifuging the cream.

In the *creaming* process a small quantity of a gum, such as ammonium alginate, gum tragacanth, or Irish moss is added to produce a reversible agglomeration of the rubber particles. With an increased size and slower Brownian movement, the particles cream and a high solids fraction can be removed from the top. Creamed latices, like centrifuged latices, lose the major water-soluble impurities with removal of the serum. These latices are, therefore, of particular value where a low level of impurities is desirable.

Concentration by *evaporation* requires the addition of stabilizers, alkalis, and soap to the latex. Concentration is effected in a rotating drum in which a smaller rolling drum furnishes additional evaporation and agitates the latex. This route differs from the two described previously in that all ingredients in the original latex, plus any additives introduced for stability, remain in the finished product. The nonrubber ingredients in the evaporated latex may amount to 6 to 7 per cent. Since latex concentrated by this method can reach as much as 75-per cent solids it is of use in special applications.

In the *electrodecantation* method[60] latex is added to a rectangular tank with an electrode at each end and many grooves, about 1 centimeter apart, in which sheets of cellophane are placed. When an electric current is applied, particles build up on the cellophane and float to the top as a cream. Fresh latex is added continuously at about the mid-point of the tank to displace the cream from the top of the cell. Latex of 60- to 62-per cent concentration has been produced by this process.

Dry Rubber

The latex is stabilized with more preservative, such as sodium sulfite, if necessary, diluted to about 15 per cent, and coagulated by the addition of dilute formic or acetic acid. In this process the fine particles of rubber agglomerate to large masses. These are transferred to washing and dewatering mills. In the manufacture of "pale crepe" rubber, the material is washed thoroughly before it is dried in a hot oven. In the preparation of "smoked sheets," the freshly coagulated rubber is not washed but dewatered in mills with even-speed rolls, and the wet sheets are dried in wood smoke. The slow smoking process produces a brown rubber which resists deterioration by mold and bacteria. In addition to pale crepe and smoked sheets, plantations produce a number of different grades of rubber including bark, earth scrap, and factory salvage.

The following table, based on an analysis of 35 samples of smoked sheets and 102 samples of pale crepe, gives some idea of the impurities present in both types of rubber.[15]

	Smoked Sheet (Average) %	*Pale Crepe (Average)* %
Moisture	0.61	0.42
Acetone extract	2.89	2.88
Protein (N × 6.25)	2.82	2.82
Ash	0.38	0.30
Rubber hydrocarbon (by difference)	93.30	93.58

The acetone-extract fraction contains fatty acids, sterols, and esters. The fatty acids have an important effect on vulcanization; the sterols and esters are believed[56] to contain the natural antioxidants which protect the rubber during processing and storage. The protein fraction has an important effect on the vulcanization rate of the natural rubber. If the coagulum deteriorates during drying or if putrefaction has occurred prior to coagulation, cure rates will vary from lot to lot. The ash content in natural rubber is generally not important unless it is found to contain copper and manganese. These particular elements catalyze oxidation of the rubber and subsequent deterioration. Since these elements are concentrated in the bark of the tree it is important to remove the bark from the latex as soon as possible.

Grades of natural rubber have been established by the Rubber Manufacturers Association[57,68] in an effort to classify the various rubbers produced throughout the world.

RUBBER TECHNOLOGY

The commercial application of either raw natural or raw synthetic rubber is very limited. Since raw polymers are plastic and soluble their uses are restricted to adhesives and sealants, for example, friction tape and electrical tape.

When rubber is mixed with sulfur and heated, vulcanization occurs. The rubber changes from a plastic material to a strong elastic substance which is tack-free, abrasion resistant, and no longer readily soluble in common solvents.

For Hevea and the butadiene-based polymers, sulfur is the normal vulcanizing, or curing agent; however, a material with a high sulfur content such as an organic polysulfide (tetramethylthiuram disulfide, alkylphenol disulfides, and aliphatic polysulfides) may be substituted. Another class of curing agent is found among the organic peroxides, such as di-*tert*-butyl and dicumyl peroxides.

Some synthetic rubbers, such as neoprene and copolymers containing methacrylic acid, possess functional groups through which cross-linking may occur without a sulfur cure. In the case of neoprene, zinc oxide and magnesium oxide are the normal curing agents; with butyl rubber, alkylphenolformaldehyde resins may serve the purpose. Otherwise the cure depends on reaction with sulfur or sulfur-containing agents.

The type and concentration of the various ingredients to be added in the development of a rubber stock depend on the properties desired in the finished product. The purpose of the stock will determine the preferred kind of rubber and compounding recipe.

The most frequently used curing system is based on sulfur. Although it may be used alone, vulcanization may be shortened by adding accelerators. These are usually selected from the aldehyde amines, guanidines, thiazoles, or the ultra-accelerators ordinarily derived from dithiocarbamic acid. The usual concentration of these materials is in the range of 0.1 to 1.5 parts per 100 parts of rubber.

In the development of a usable compounding recipe, activators, such as zinc oxide, and fatty acids such as stearic acid to solubilize the zinc, are often added. The concentration of these materials ranges from 1 to 5 per cent and 0.5 to 4 per cent, respectively. The compounded rubber ordinarily contains an antioxidant to improve aging. This is often a secondary aromatic amine or a substituted phenol depending on whether the rubber is to be used in a dark- or light-colored application, respectively.

Because "pure gum" products containing the ingredients listed above are not suitable for applications such as tires, reinforcing agents must be added to the stock. Selected for economy and for development of optimum performance, these are normally carbon blacks, inorganic fillers, or reinforcing resins. The many different types of carbon black available for the manufacture of black stocks vary in particle size and in their effect on the properties of rubber stocks. Proper choice of a black or reinforcing pigment is possible only after careful study of available materials.

For a detailed review of compounding ingredients and vulcanization see Reference 1.

Stock Preparation

The operations involved in preparing stocks of Hevea or butadiene-styrene rubbers may involve some of the following: breakdown in an oven, a plasticator, or on open-roll mills (in the case of natural rubber); batch mixing in a mill or a Banbury; warm-up on a mill or Banbury if the stock has been allowed to cool during storage; calendering for preparation of sheet skim coating, or skimming onto fabric to produce stocks or plies of the desired dimensions; extrusion, for treads, onto wires, over hose carcasses; building into green tires, shoes, and hose.

Natural rubber differs from the synthetic butadiene-styrene types in its greater tendency to soften during milling. For this reason the synthetic rubbers are manufactured to a low plasticity suitable for subsequent processing in the factory. The natural product is prepared by mill mastication at as low a temperature as is practical, normally around 200°F but at least below 270°F; in a plasticator the rubber may be heated as far as 350°F for breakdown.

The rubber mills used in preparing stocks consist of two parallel steel rolls which vary in size from 6- to 10-inch models in the laboratory to 84- and 120-inch mills in the factory. The selection of surfaces, roll speed, and ratio of speed of the two rolls depends on the particular types of rubber being handled.

The Banbury mixer is an enclosed machine containing two water-cooled rotors operating in a water-cooled chamber. The Banbury must be of such a size that the batch will completely fill the mixing chamber. Advances in the design of the Banbury mixer now permit a mixing time of 8 minutes or less.

After the Banburys, the stocks are fed to a three- or four-roll calendar used for three types of work: frictioning, skim coating, and sheeting (Fig. 9.6). Fabric is frictioned when it is passed over Roll C with Roll B running at a higher speed, the rubber compound being wiped into the fabric. Skim coating is a similar operation but the speed of roll B is the same as that of roll C, so that a film of rubber is pressed into the fabric. For sheeting, roll C is lowered and a cotton liner is used; the sheet of rubber formed between rolls A and B is led into the liner in which it is wrapped. Fabric used for manufacturing tire plies is usually dipped in a bath containing either a latex-casein compound or a latex-resocinol-formaldehyde mixture to improve adhesion between

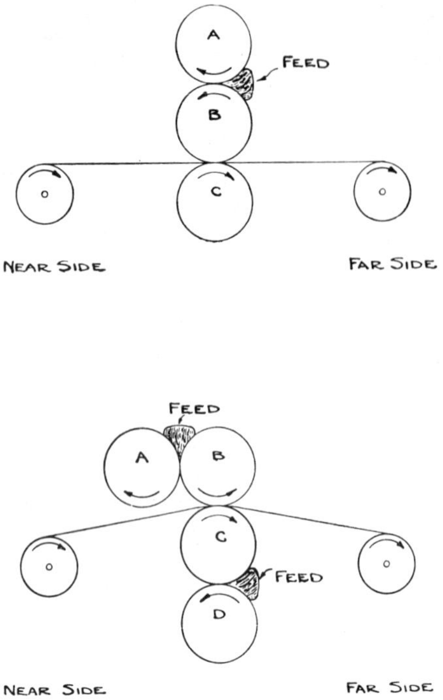

Fig. 9.6 Diagram of a three- and a four-roll calendar.

the rubber and the cords. It is then dried and fed into the calendar.

Automobile Tires and Tubes

One of the most important applications of all rubbers is tires. Because of the increased demand for speed and endurance in modern vehicles, it has been necessary to design rubber compounds and build tires with these factors in mind. Tires must be designed to withstand heavy loads and high speeds for long periods; the body of the tires must be able to withstand severe shocks and must not show excessive heat build-up.

For years tires have been built on a flat drum. Successive plies of fabric, which has been friction-coated and cut on a bias, are applied on the drum in an alternating fashion. Circles of wire known as "beads" which will reinforce the tire where it touches the rim have been prepared previously from a number of wires embedded in rubber. The beads are attached at the end of the fabric on the drum. The beads may be further protected by "chafer strips." The center of the tire is protected by "breaker strips" topped by a slab of highly abrasion-resistant stock comprising the tread. The last part added to the tire is a sidewall, compounded for white or black and for maximum resistance to ozone deterioration and curb chafing.

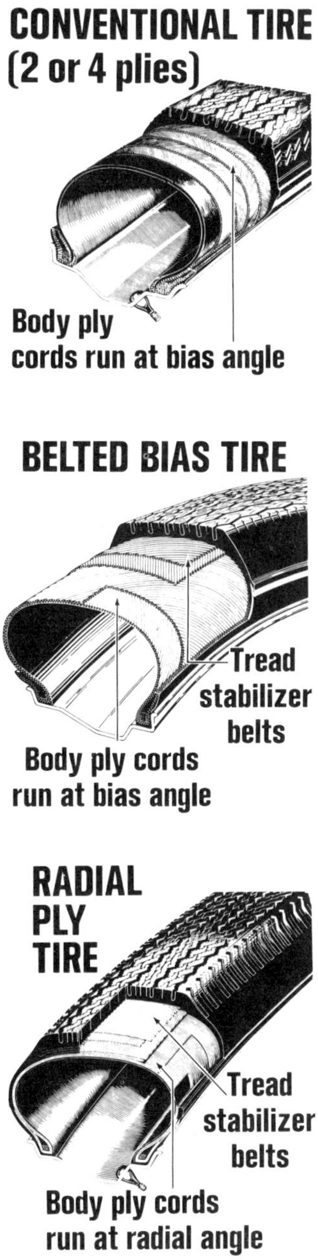

Fig. 9.7 Variations of tire construction. (*Courtesy Firestone Tire and Rubber Co.*)

Recent developments in tire construction have modified the "bias" design described above by adding a belt of tire cord in rubber laid almost circumferentially on top of the bias plies. This is the bias-belted tire design shown in Fig. 9.7. Most recently a radial design has appeared. Here, the first plies are laid around the tire and the belt is again applied circumferentially.

The rayon tire cord of a few years ago has received considerable competition from nylon and polyester cords, especially for bias plies and for the radial under plies. Rayon, glass, and steel cords have taken active roles in the belt plies.

Because of the high temperatures at which tires must operate on passenger cars, trucks, and buses, considerable attention has been given to polymers which generate less heat as well as to replacements for cotton fabric which tends to lose strength and fail at higher temperatures. Rayon and nylon cords are finding application for high speed operation.

Tires from the flat drum are shaped around an air bag in a mold for vulcanization or "cur-

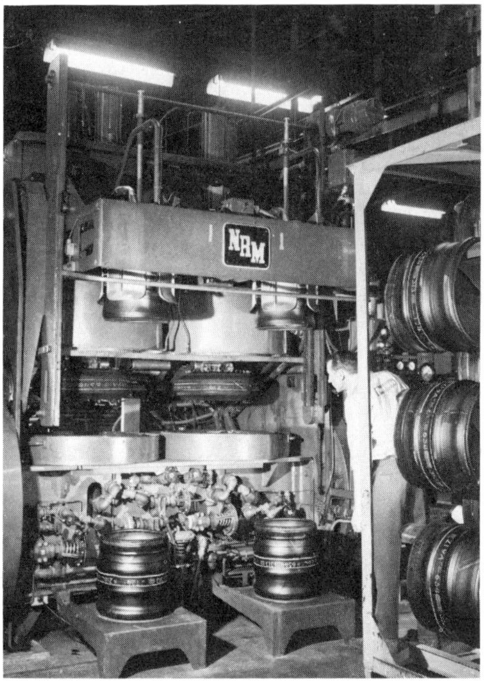

Fig. 9.8 Apparatus for shaping and curing tires. (*Courtesy Firestone Tire and Rubber Co.*)

ing" (Fig. 9.8). In modern molds, the green tire is cured by means of steam in the jacket of the mold; steam and air are introduced into the air bag to expand the tire into the mold. The mold is heated for a specified time at a definite temperature to produce an optimum state of cure throughout the tire. Then, the heat is shut off, the internal pressure is released, the mold is opened, and the tire is removed for final inspection.

Mechanical Rubber Goods

This term is applied to the large class of rubber products including hose, belts, gaskets, and molded, extruded, and compounded items used in the automotive, industrial, household, and appliance fields. These may be made from either natural or synthetic rubber, or mixtures, depending on whether the stocks are expected to show oil resistance or some particular characteristic, such as resistance to aging or abrasion-resistance. Since price is often important in the manufacture of mechanical rubber goods, these stocks may include more filler than those used for tires.

For applications in the automotive field, mechanical rubber goods must withstand oil and heat, and show a high degree of resilience. Therefore, they are frequently manufactured from neoprene or butadiene-acrylonitrile polymers.

Hard rubber may be made from natural or synthetic rubber by increasing the amount of sulfur to 30 or 35 parts per 100 of polymer. Hard rubber stocks may contain extenders such as hard rubber dust, coal dust, carbon black, or inorganic pigments.

Self Sealing Fuel Tanks

During World War II, fuel tanks made from synthetic rubber were of considerable value to fighter planes. They were constructed of multiple layers of rubber and fabric with a butadiene-acrylonitrile lining designed to resist the material which would be contained in the tank. The liner was surrounded by alternating layers of sealant and fabric to achieve the desired thickness. When the resistant liner was punctured, the unvulcanized rubber stock outside swelled in contact with the fuel. For successful performance the hole must seal within an extremely short time.

MONOMER MANUFACTURE

Butadiene

Butadiene ($CH_2=CH-CH=CH_2$) has been identified as the product from thermal cracking of a wide variety of organic substances. Its commercial production involves one of the following routes:

1. Thermal cracking of C_4 or higher hydrocarbons.
2. Catalytic dehydrogenation of butenes.
3. Catalytic dehydrogenation of *n*-butane.
4. Dehydrochlorination of chlorinated butanes.
5. Catalytic treatment of ethyl alcohol and ethanol-acetaldehyde mixtures.
6. Aldol condensation of acetaldehyde followed by conversion of aldol to butadiene.
7. Reaction of acetylene and formaldehyde to form 2-butyne-1,3-diol followed by hydrogenation and dehydration.

In 1939 only one commercial producer of butadiene was listed; by 1941 there were five manufacturers, using different routes, including recovery from refinery streams.

At the initiation of the government program in 1941 it was necessary that a large volume of butadiene be immediately available. "Quickie" plants were started to recover butadiene from refinery streams as a fast temporary solution. A review of the long range routes showed that the alcohol-based plants could be put into production rapidly but that the petroleum routes would ultimately be cheaper.

The following table compares production of butadiene from petroleum and alcohol. The alcohol route was important at the beginning of World War II and again during the Korean War, but it lost favor during the time when maximum production was less urgent.

Thermal Cracking. Butadiene may be recovered from an existing refinery stream, or a plant may be designed for cracking a gasoline fraction.[19] The feed must be carefully selected for optimum production; the preferred material is reported to be cyclohexane. The butadiene yield is about 3.5 weight per cent of the naphtha cracked. This process requires a cyclic operation in a lined, baffled furnace. Economy

TABLE 9.7 Production of Butadiene[a]

	Thousands of Short Tons		
	Petroleum	*Alcohols*	*Total*
1940	0.2	0.0	0.2
1942	12.1	0.0	12.1
1943	49.3	129.7	179.0
1944	244.5	361.7	606.2
1945	389.9	233.4	623.3
1950	302.4	2.7	305.1
1951	482.9	128.3	611.2
1952	406.6	146.6	553.2
1953	526.9	49.2	576.1
1954	404.4	0.0	404.4
1955	703.2	2.5	705.7
1956	738.4	16.6	755.0

[a]Synthetic Organic Chemicals–U. S. Production and Sales, U. S. Tariff Commission and Schedule 1, March 31, 1952, Synthetic Rubber Div., Reconstruction Finance Corporation.

of production depends on the recovery of non-butadiene products, essential if this route is to compete with other processes; therefore extraction is tied in with a refinery operation.

The recovery of by-product butadiene from refinery streams continues to be an important route; it will become increasingly important as synthetic-rubber production expands. The volume will depend on trends in cracking technology, which is continually changing with the requirements of the refining industry, and improved technology leading to higher yields of more valuable components.

It has been possible to produce the very pure butadiene required for solution polymerization where such impurities as acetylenes are detrimental. This has been achieved through selective hydrogenation or extractive distillation with such solvents as acetonitrile, N-methylpyrrolidone, or dimethylformamide.

Catalytic Dehydrogenation of n-Butylenes; This process depends on the fractionation of refinery gases to produce a cut of normal C_4 olefins. Extractive distillation is used for this purpose.[10,26] Furfural or acetone is added to the butane-butene stream to alter the relative volatilities of the components so that separation can be effected more easily. The butanes are taken overhead; butylene, dissolved in the solvent, is removed at the bottom and separated by distillation. Isobutylene must be removed for efficient production of butadiene since this material is inert in the cracking step and becomes concentrated in the recycle. Removal can be achieved by sulfuric acid oligomerization to produce diisobutylene or higher oligomers which may be used for high-octane gasoline.

Catalyst 1707, used initially in all plants for cracking butylene, contains a complex mixture of magnesium, copper, iron, and potassium compounds. The butylene feed is diluted with steam to prevent the formation of carbon on the catalyst surface.[36] Catalyst 105, used later for this dehydrogenation, contains iron, chromium, and potassium.[52]

The reactors are designed to supply heat at a temperature of 1150 to 1250°F. The feed must be exposed at a low partial pressure for a short contact time; and facilities must be arranged for regenerating the deactivated catalysts. The original reactors[58] have been revised for operation with some of the more recent catalysts.

Another catalyst is Dow Type B, calcium nickel phosphate stabilized with 2 per cent chromium oxides.[9,52] Butadiene selectivities of 90 mole per cent at 35 per cent conversion of the butene compare with 70 per cent for catalyst 105 at the same conversion level. The Dow B catalyst requires frequent regeneration with an air-steam mixture and needs a steam to hydrocarbon feed of about 20 to 1. This particular route produces as much as 4 per cent acetone based on the butadiene.

Butane dehydration.[65] This route combines two operations in a one-step production of butadiene from butane. During the government operation, Phillips used a two-step process, dehydrogenating the butane in fixed beds contained in heated tubes. An 80 per cent yield of butenes resulted from a 30- to 40-per cent conversion per pass using 1 to 2 psig and 1100°F inlet temperature. The reactors operated one hour on stream and one hour on regeneration using a flue gas mixture containing 2 to 3 per cent oyxgen. The butenes were dehydrogenated by the same method previously described.

A direct route for producing butadiene from butane (*Houdry* process) is based on a regenerative operation[70] in which a chromia-alumina catalyst is employed at low temperatures and pressures. The system is operated on a cyclic basis. First, the feed stream is cracked for a short period (10 minutes), then the catalyst chamber is evacuated, and given an air blow to bring the catalyst back to operating temperatures, and finally the chamber evacuated to complete the cycle.

Lebedev Process.[8,63] Ethanol is fed into a converter containing metal oxides capable of effecting simultaneous dehydrogenation and dehydration. Recycling of nonbutadiene components gives yields of 60 per cent of theoretical. This process uses an Al_2O_3-ZnO or MgO-CoO catalyst. No United States plant has used this process.

Ostromislenski Process. An ethanol-acetaldehyde mixture is fed over copper-containing catalysts or silica gel containing tantalum oxide.[14,34,35,51,64] Part of the ethanol is converted to acetaldehyde by dehydrogenation over a copper catalyst at temperatures above 200°C and atmospheric pressure. Ethanol and acetaldehyde are then mixed at about 3 to 1 mole ratios. The catalyst gradually becomes covered with carbon and must be burned off periodically to restore its activity.

The products of this cracking reaction are simpler than those derived from the petroleum cracking processes. They consist primarily of ethylene, propylene, butadiene, unchanged starting materials, and some oxygenated by-products. The butadiene fraction containing 96 per cent butadiene is refined in extractive distillation by treatment with β,β'-dichloroethyl ether ("Chlorex"). The resulting purity is at least 98.5 per cent.

Butadiene, regardless of the method of preparation, is sensitive to polymerization. Unless it is to be used immediately, an inhibitor such as tertiary butylcatechol must be added to avoid spontaneous polymerization. The inhibitor may be removed by either distillation or a caustic wash.

Isoprene

Isoprene (2-methyl-1,3-butadiene) is a logical starting point for making synthetic rubber since this is the basic unit found in natural rubber. It was isolated from natural rubber by Williams in 1860, and was polymerized to a rubber in 1912 (German Patent 254,672).

Isoprene, however, is much more difficult to produce than butadiene, and processes for making it frequently require chemical feed stocks. For many years, isoprene found little place in the synthetic rubber picture beyond its use in butyl rubber, where several per cent isoprene supplied the unsaturated sites used for sulfur cure. Stereopolymerization systems have now increased the demand for isoprene for "synthetic natural rubber," and commercial plants have been built using a number of routes from different starting materials.

Recovery from C_5 Streams. This was the earliest route to this monomer, and it still remains the first route for a new producer. Isoprene recovery is a highly competitive situation, and the economics depend on the availability of suitable feed stock in commercial volumes, possibly from adjacent refineries. It is costly to transport the non-isoprene components to and from the isoprene extraction plant. The purification process resembles that used for butadiene but with a greater limitation on the extractive solvent (acetonitrile, dimethylformamide, or N-methylpyrrolidone). If the purity requirements are very stringent, selective hydrogenation may be included to lower the level of acetylenes.

Dehydrogenation of Isoamylenes. This is effected in the same manner as the catalytic dehydrogenation of *n*-butylenes, which was described in the section on manufacture of butadiene. The economics of this route depend on having access to a favorably priced feed stream. The value of this feed-stock varies with changes in the production of gasoline. Isoamylenes can be used for alkylation or for direct blending in gasoline itself. Its availability may be further limited by the current trend toward low-lead gasolines.

Dimerization of Propylene. The following equations illustrate this process used by Goodyear and developed by that company and Scientific Design:

$$CH_3{-}CH{=}CH_2 \longrightarrow CH_2{=}C(CH_3){-}CH_2{-}CH_2{-}CH_3 \quad (1)$$

$$CH_2{=}C(CH_3){-}CH_2{-}CH_2{-}CH_3 \longrightarrow CH_3{-}C(CH_3){=}CH{-}CH_2{-}CH_3 \quad (2)$$

$$CH_3{-}C(CH_3){=}CH{-}CH_2{-}CH_3 \longrightarrow CH_2{=}C(CH_3){-}CH{=}CH_2 + CH_4 \quad (3)$$

In step (1) high purity, polymerization grade propylene is dimerized in the presence of a suitable catalyst, for example, tri-*n*-propyl-aluminum. The process is a continuous one using high pressures (300 psig) at 150–250°C. Considerable heat is evolved and provision must be made for removing it.

The resulting 2-methyl-1-pentene must then be separated from the catalyst and unreacted propylene, before it can be isomerized to 2-methyl-2-pentene. The latter compound gives the best yields of isoprene and the least amount of coproduct dienes. An acidic catalyst is used for the isomerization, step (2), and essentially quantitative conversion is obtained. One example is a fixed-bed silica-alumina catalyst at 150–300°C at space velocities varying from 0.5 to 15.0 hr^{-1}.[49]

Step (3) is a conventional cracking step in the presence of HBr or an unspecified catalyst.[49] The reaction temperature is 650–800°C at residence times of 0.05 to 0.3 sec.

The economics of this process depend on the cost of the required high-purity propylene feed and the ability to dispose of by-products at an attractive price. It has been reported that up to 250 pounds of propylene are used per 100 lbs of finished product, but further experimental work may change this pattern in commercial practice.

Isobutylene and Formaldehyde. The reaction product of the compounds has been explored by many workers, including some in France and Russia:

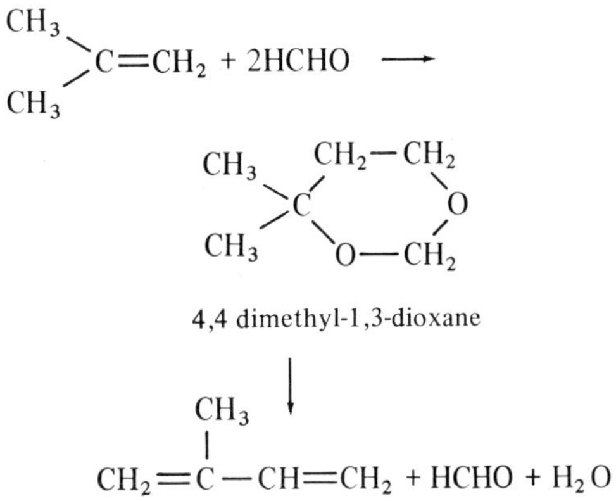

Depending on the available feed stocks, a crude isobutylene stream may be used, and formaldehyde may be manufactured from methanol.

Aqueous formaldehyde and isobutylene are reacted at low pressures at about 95°C. Unreacted components are flashed off and the dimethyldioxane is recovered by distillation from a first bottoms stream known as Residols I. The catalytic decomposition of dimethyldioxane occurs at atmospheric pressure below 400°C. Both fixed and moving catalyst beds have been used. In the final purification a second bottoms stream (Residols II) is separated.

The economics of this process again depend on the prices and availability of feed stocks, as well as the value of the recovered Residol streams. While these can be burned, they are being considered as a source of chemicals, thus ultimately reducing the isoprene cost.

Acetone and Acetylene. The reaction products of these compounds is developed by SNAM in Italy.[38]

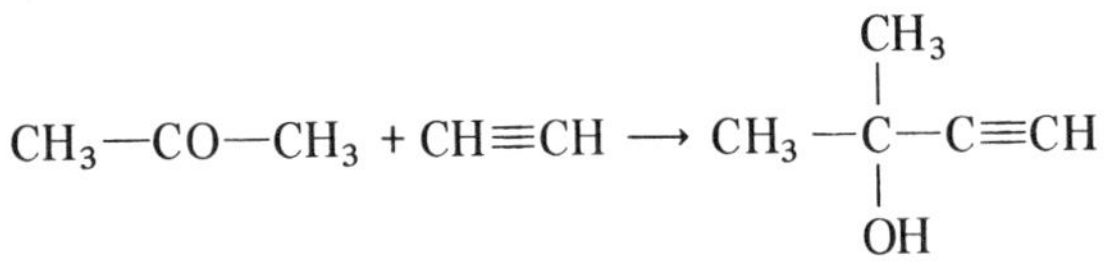
(1)

2-methyl-3-butyn-2-ol

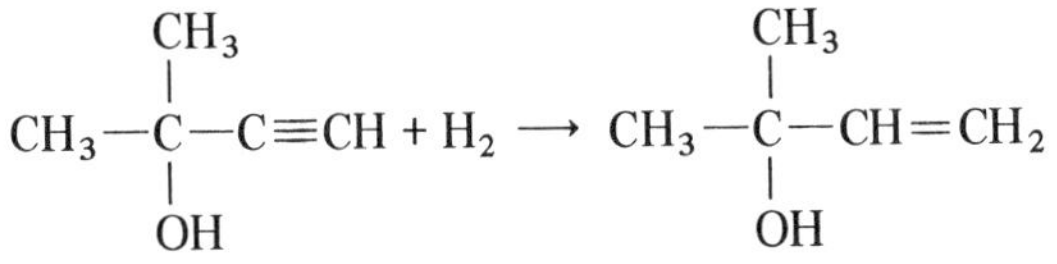
(2)

2-methyl-3-butene-2-ol

$$CH_3-\overset{\displaystyle CH_3}{\underset{\displaystyle OH}{\overset{|}{\underset{|}{C}}}}-CH=CH_2 \longrightarrow CH_2=\overset{\displaystyle CH_3}{\overset{|}{C}}-CH=CH_2 + H_2O \qquad (3)$$

In step (1) acetone and an excess of acetylene react in liquid ammonia over a suitable catalyst (aqueous KOH). The yield at 10–40°C and 285 psi is high. The catalyst is neutralized at the end of the reaction and the unreacted acetylene is recycled.

Two distillation steps are used. First, the unreacted acetone is separated and then the methylbutynol; the catalyst and heavier components are left in the bottoms. Partial hydrogenation is effected at 30–80°C and 70–140 psi hydrogen over a supported palladium catalyst. This is separated from the product stream by a centrifuge, and re-used. The methylbutenol is fed, as an azeotrope with water, over high purity aluminas at atmospheric pressure and 250–300°C. After a water wash, the product is separated by distillation. The economics of this route obviously depend on having a source of low-cost acetylene.

Dehydrogenation of Isopentane. This has been explored in the Houdry reactors in a pilot plant[17] but so far no commercial plant has used this route. Isoprene alone may be produced by feeding an isopentane stream, or isoprene and butadiene may be produced jointly by feeding butane and isopentane in the same stream. One by-product of this operation is piperylene.

Chloroprene[19]

The manufacture of chloroprene ($CH_2=CCl-CH=CH_2$) starts with the conversion of acetylene to monovinylacetylene[47] by polymerizing the acetylene in the presence of an aqueous catalyst solution of cuprous chloride and ammonium chloride. The off-gas is cooled to -70°C to condense the vinylacetylene ($CH\equiv C-CH=CH_2$, boiling point 5°C) while the excess acetylene is recycled. The condensate is distilled in a column, pure vinylacetylene comprising the overhead while the bottoms contain divinylacetylene. Monovinylacetylene is then converted to chloroprene by reaction with hydrogen chloride in the presence of a cuprous chloride solution.[11]

Styrene[6]

The most important route for preparing styrene involves the preparation of ethyl benzene followed by its dehydrogenation to styrene. The reactions are as follows:

$$(1)\quad \underset{\text{Benzene}}{C_6H_6} + \underset{\text{Ethylene}}{CH_2=CH_2} \xrightarrow{AlCl_3} \underset{\text{Ethylbenzene}}{C_6H_5CH_2-CH_3}$$

$$(2)\quad C_6H_5CH_2-CH_3 \longrightarrow \underset{\text{Styrene}}{C_6H_5CH=CH_2} + H_2$$

Ethyl Benzene. The reactor, a vertical vessel lined with acid-resisting bricks, contains the reaction mixture to a depth of about 35 feet. Ethylene, added at the bottom of the column, causes circulation of the liquid. The reactor is cooled by means of an overhead condenser and a flow of cold water on the outside of the shell to maintain the temperature of 95°C. Pressure within the reactor is 5 pounds per square inch or less. Aluminum chloride, and some ethyl chloride, are added continuously throughout the process to make up for catalyst depletion. The ethylene-benzene ratio is controlled at 0.58 moles ethylene per mole of benzene.

The tendency for polyethylated derivatives to be produced during the process is suppressed by the recirculation of polyethylbenzene. Since moisture in the feed would increase the aluminum chloride requirements, special precautions are taken to make sure that the feed is anhydrous. Both fresh and recovered benzene are dried by azeotropic distillation.

The reactor is operated in a continuous manner; crude ethylbenzene is withdrawn from the top, cooled, and then decanted from a lower layer consisting of an aluminum complex-hydrocarbon mixture that can be recovered to conserve catalyst. The crude ethylbenzene is washed with casutic soda and fed to a stripping column to separate ethylbenzene and benzene from the higher polyethylbenzenes. The overhead stream consisting of ethylbenzene and benzene is again distilled, washed in caustic, and dried over a bed of caustic. The ethylbenzene purity at this stage is 99.5 per cent.

The dehydrogenation of ethylbenzene to styrene is an endothermic reaction requiring the presence of superheated steam which acts as an inert gas diluent. A solid catalyst is used. Since reaction (2) runs, from left to right, with an increase in volume, a decrease in partial pressure would favor higher yields; consequently, a steam-ethylbenzene (2.6 lb to 1 lb) mix is used to permit cracking under reduced partial pressure. This reduces the partial pressure of the reactants to 0.1 atm and shifts the equilibrium so that a theoretical conversion of 70 to 80 per cent is possible. The superheated steam functions in two ways: it supplies the heat of reaction; and it keeps the catalyst clean by reacting with any deposited carbon. The catalyst increases the reaction rate for this temperature to a satisfactory level with a minimum of outside reaction.

The reactor consists of an insulated brick chamber containing the catalyst in granular form. Catalyst 1707 (see the butadiene section) was found to work satisfactorily for this process and was used extensively during the war period. About 90 per cent of the steam is raised to 383°C by heat exchange with product vapors, then to 710°C in a superheating furnace. The remainder of the steam is mixed with the ethylbenzene charge at 160°C and raised to 520°C by heat exchange with product vapors (ahead of the pure steam). The two streams meet in concentric inlet tubes, mix thoroughly, and enter the catalyst chamber. The base of the reactor is held at 630°C. The outgoing vapors are cooled by a heat exchanger, a spray-type desuperheater, and finally a condenser. The condensate contains steam, ethylbenzene, styrene, benzene, toluene, and tar. Conversion with fresh catalyst is approximately 37 per cent. Vent gas contains hydrogen, carbon monoxide, carbon dioxide, methane, and other gases. The life of the catalyst operating uninterruptedly is about a year.

In the styrene-finishing step, pure styrene is isolated from the crude condensate, containing approximately 37 per cent styrene, by distillation at reduced pressure in the presence of elemental sulfur as a polymerization inhibitor. The small differential in boiling points between ethylbenzene (136.2°C) and styrene (145°C) necessitates a 70 plate tower which may, if desired, be divided into two columns operating in series. The bottoms, consisting of the styrene, may pass to two batch-finishing stills with packed towers, operating under vacuum, where styrene is the overhead stream and tar and sulfur are the raffinate. The styrene, 99.7 per cent pure, is cooled and stabilized by the addition of 10 to 15 parts per million of *tert*-butylcatechol inhibitor.

The over-all yield, benzene to styrene, is 88 to 92 per cent as is the ethylene to styrene yield.

Acrylonitrile

There is at present only one major route for making acrylonitrile.

Propylene-Ammonia. This route uses ammoxidation.

$$2CH_2{=}CH{-}CH_3 + 3O_2 + 2NH_3 \longrightarrow 2CH_2{=}CHCN + 6H_2O$$

Standard Oil Co. of Ohio built a plant in 1960 using a fluid-bed reactor into which propylene is fed at <500°C and 2-3 atm pressure. The literature[32] shows a bismuth phosphomolybdate or silica gel catalyst giving a 33 per cent yield based on propylene. Many other companies have worked on this route, but Standard Oil remains a major source of the technology. The process is so efficient that it is now used by all but one producer in the U.S.

Acetonitrile is a by-product and amounts to 10-15 per cent of the acrylonitrile produced. This by-product is valuable as an extractive solvent for purifying both butadiene and isoprene.

Other Routes for Acrylonitrile are:

1. Acetylene and hydrogen cyanide can be reacted in liquid phase or in vapor phase requiring only a catalyst for the reaction.

$$CH \equiv CH + HCN \longrightarrow CH_2 = CHCN$$

At one time this was an important commercial process for making acrylonitrile, but the price of acetylene has made it a relatively expensive and, therefore, little used route.

2. Ethylene cyanhydrin, produced from ethylene oxide, was used by American Cyanamid in the first acrylonitrile production.

$$\underset{\diagdown O \diagup}{CH_2 - CH_2} + HCN \longrightarrow \underset{\substack{| \\ OH}}{CH_2} - CH_2 - CN \longrightarrow$$

Cyanhydrin

$$CH_2 = CHCN + H_2O$$

The ethylene oxide route is no longer of commercial interest in the U.S.

REFERENCES

1. Alliger, G. and Sjothun, I. J., "Vulcanization of Elastomers," New York, Van Nostrand Reinhold 1964.
2. Alliger, G., Willis, J. M., Smith, W. A., and Allen, J. J., *Mech. Eng.* 1098–1102 (1956).
3. Alliger, G., Willis, J. M., Smith, W. A., and Allen, J. J., *Rubber World*, **134**, 549–59 (1956).
4. Brit. Pat. 1,130,770 (1968; to Asahi).
5. Blackley, D. C., "High Polymer Latices," Vols. I and II, New York, Palmerton, 1966.
6. Boundy, R. H., Boyer, R. F., and Stoesser, S. M., "Styrene," New York, Van Nostrand Reinhold, 1952.
7. Bovey, F. A., "Polymer Conformation and Configuration," New York, Academic Press, 1964.
8. British Intelligence Objectives Subcommittee, Report No. 1060, 1947.
9. Britton, E. C., Dietzler, A. J., and Noddings, C. R., *Ind. Eng. Chem.*, **43**, 2871–4 (1951).
10. Buell, C. K., and Boatright, R. G., *Ind. Eng. Chem.*, **39**, 695–705 (1947).
11. Carothers, W. H., Williams, I., Collins, A. M., and Kirby, J. E., *J. Am. Chem. Soc.*, **53**, 4203–25 (1931).
12. *Chem. Eng. News*, **37**, 23 (1959).
13. *Chem. Eng. News*, **33**, 4518 (1955).
14. Corson, B. B., Jones, H. E., Welling, C. E., Hinckley, J. A., and Stahly, E. E., *Ind. Eng. Chem.*, **42**, 359–73 (1950).
15. Davis, C. C., Ed., "Chemistry and Technology of Rubber," New York, Van Nostrand Reinhold, 1937.
16. D'Ianni, J. D., *Eng. Chem.*, **40**, 253–6 (1948).
17. DiGiacomo, A. A., Maerker, T. B., and School, J. W., *Chem. Eng. Progr.* **57**, 35 (1961).
18. Dunbrook, R. F., *India Rubber World*, **117**, 203–7 (1947).
19. Du Pont, "The Neoprenes."
20. Flory, P. J., *Chem. Rev.*, **39**, 137–97 (1946).
21. Forman, L. E., Chap. 6 in Polymer Chemistry of Synthetic Elastomers, J. P. Kennedy and E. G. M. Tornquist, Eds., Part II, 491–596, John Wiley & Sons, 1969.
22. Fryling, C. F., Private Communication (March 26, 1942).
23. Gardon, J. L., "Mechanism of Emulsion Polymerization," *Rubber Chem. Technol.* **43**, 74–94 (1970).
24. Belgian Patent 543,292 (1956; to Goodrich-Gulf).
25. Greenberg and Hansley, U. S. Patents, 3,067,187 (1962) and 3,223,691 (1965).
26. Happel, J., Cornell, P. W., Eastman, D. B., Fowle, M. J., Porter, C. A., and Schutte, A. H., *Trans. Am. Inst. Chem. Eng.*, **42**, 189–214, 1001–7 (1946).
27. Harkins, W. D., *J. Am. Chem. Soc.*, **69**, 1428–44 (1947).
28. Heilman, H. H., *Petrol. World*, **44**, No. 3, 51–5 (1947).
29. Holden, G., and Milkovich, R., U.S. Pat. 3,265,765 (1966; to Shell Oil).
30. Horne, S. E., Jr., Kiehl, J. P., Shipman, J. J., Folt, V. L., Gibbs, C. F., Willson, E. A., Newton, E. B., and Reinhart, M. A., *Ind. Eng. Chem.*, **48**, 784–91 (1956).
31. Howland, L. H., and Brown, R. A., *Rubber Chem. Technol.* **34**, 1501 (1961).
32. Idol, J. D., U.S. Pat. 2,904,580 (1959; to Standard Oil of Ohio).
33. International Rubber Study Group, Rubber Statistical Bulletin 24, No. 2 Nov. 1970.
34. Jones, H. E., Stahly, E. E., and Corson, B. B., *J. Am. Chem. Soc.*, **71**, 1822–8 (1949).

35. Kampmeyer, P. M., and Stahly, E. E., *Ind. Eng. Chem.*, **41**, 550–5 (1949).
36. Kearby, K. K., *Ind. Eng. Chem*, **42**, 295–30 (1950).
37. Logemann, H. and Pampus, G., *Kautschuk Gummi–Kunststoffe*, **23** 479–486 (1970).
38. Malde, M. de, DiCio, A., and Mauri, M. M., *Hydrocarbon Process. Petrol. Refiner* **43**, 149 (1964); *Chem. Eng.* **71**, No. 20, 78 (1964); and *Chem. Eng.* 78 (Sept. 28, 1969).
39. McGregor, R. R., "Silicones and their Uses," New York, McGraw-Hill, 1954.
40. Miller, M. L., "The Structure of Polymers," New York, Van Nostrand Reinhold, 1966.
41. Mitchell, J. M., Enbree, W. H., and MacFarlane, R. B., *Ind. Eng. Chem.*, **48**, 345–8 (1956).
42. Moore, J. E., *J. Polymer Sci.* **A2**, 835 (1964); Gamble, L. W., Westerman, L. and Knipp, E. A., *Rubber Chem. Technol.*, **38**, No. 4, 823 (1965); Cazes, J., *J. Chem. Educ.*, **43**, No. 7, A-567 and No. 8, A-625 (1966).
43. Morton, A. A., "Alfin Catalysts" in "Encyclopedia of Polymer Science and Technology," Vol. I, 629–238, New York, John Wiley & Sons, 1964.
44. Myer, D. A., Chapter 10 in Reference 1.
45. Natta, G., Crespi, G., Valvassori, A. and Sartori, J., "Polyolefin Elastomers," *Rubber Chem. Technol.*, **36**, 1583 (1963).
46. Natta, G., and Porri, L., in "Polymer Chemistry of Synthetic Elastomers," J. P. Kennedy and E. G. M. Tornquist, Eds., Part II, Chapter 7, New York, John Wiley & Sons, 1969.
47. Nieuwland, J. A., Calcott, W. S., Downing, F. B., and Carter, A. S., *J. Am. Chem. Soc.* **53**, 4197–4202 (1931).
48. Oberster, A. E., and Bebb, R. L., *Angew. Makromol. Chem.*, **16-17**, in press (1971).
49. Osterhof, H. J., *Rev. Gen. Caoutchouc Plastiques*, **42**, 529 (1965).
50. "Proceedings Second Rubber Technical Conference," 69–78 (1948).
51. Quattlebaum, W. M., Toussaint, W. J., Jr., and Dunn, J. T., *J. Am. Chem. Soc.*, **69**, 593–9 (1947).
52. Reilly, P. M., *Chem. Can.*, **5**, No. 3, 25–9 (1953).
53. Richardson, W. S., and Sacher, A., *J. Polymer Sci.*, **10**, 353–70 (1953).
54. Riegel, E. R., "Industrial Chemistry," New York, Van Nostrand Reinhold, 1942.
55. Rochow, E. G., "Introduction to the Chemistry of Silicones," 2nd Ed., New York, John Wiley & Sons, 1951.
56. *Rubber Chem. Technol*, **7**, 633 (1934).
57. Rubber Manufacturers Assoc., Inc., New York, "Type Description and Packing Specifications for Natural Rubber," Revised, Dec. 1954.
58. Russell, R. P., Murphree, E. V., and Asbury, W. C., *Trans. Am. Inst. Chem. Engrs.*, **42**, 1–14 (1946).
59. Saunders, J. H., and Frisch, K. C., "High Polymers," Vol. XVI, New York, John Wiley & Sons, 1962.
60. British Patent 459,972 (Jan. 19, 1937) ("Semperit" Oesterreichisch-Amerikanische Gummiwerke A. B., to Metallgesellschaft A.-G.).
61. Kraton, S. *Mod. Plastics*, 77–79 (June 1966); and *Rubber J.*, 20–33 (Dec. 1968).
62. Stavely, F. W., *et al*, *Ind. Eng. Chem.*, **48**, 778–83 (1956).
63. Talalay, A., and Talalay, L., *Rubber Chem. Tech.*, **15**, 403–29 (1942).
64. Toussaint, W. J., Dunn, J. T., and Jackson, D. R., *Ind. Eng. Chem.*, **39**, 120 (1947).
65. Watson, C. C., Newton, F., McCausland, J. W., McGrew, E. H., and Kassel, L. S., *Trans. Am. Inst. Chem. Eng.*, **40**, 309–15 (1944).
66. Whitby, G. S., Davis, C. C., and Dunbrook, R. F., eds., "Synthetic Rubber," New York, John Wiley & Sons, 1954.
67. Winchester, C. T., *Ind. Eng. Chem.*, **51**, 19 (1959).
68. Winspear, G. G., "Vanderbilt Latex Handbook," R. T. Vanderbilt Co., 1954.
69. Winspear, George G. and Watermann, R. R. in "Introduction to Rubber Technology," M. Morton, Ed., New York, Van Nostrand Reinhold, 1959.
70. Womeldorph, D. E., Stevenson, D. H., and Friedman, L., *Am. Petrol. Inst.* (May 14, 1958).
71. Woods, L. A., Chapter 10 in Reference 66.
72. Youker, M. A., *Chem. Eng. Progr.*, **43**, No. 8, 391 (1947).
73. Ziegler, K., Dersch, F., and Wollthau, H. *Ann. Chem.*, **511**, 13 (1934); Ziegler, K. and Jakob, L. *Ann. Chem.*, **511**, 45 (1934); Ziegler, K., Jakob, L., Wollthau, H., and Wenz, A., *Ann. Chem.*, **511**, 64 (1934).

10

Synthetic Plastics

Robert W. Jones* and K. T. Chandy**

INTRODUCTION

The word "plastic" was originally used as an adjective to denote a degree of mobility or formability. In the 1909 edition of "Webster's International Dictionary," the noun was not listed. Shortly thereafter, with the introduction of "Bakelite" by Dr. Baekeland, the word was often used as a noun, most frequently referring to "Bakelite," "Celluloid," and casein plastics.

The American Society for Testing Materials (D-833-69) has defined a plastic as "a material that contains as an essential ingredient an organic substance of large molecular weight, is solid in its finished state, and, at some stage in its manufacture or in its processing into finished articles, can be shaped by flow." According to this definition, synthetic fibers, all rubbers, and even bread doughs are plastics, but glass is not. Those who insist that glass is a plastic omit "organic substance" as part of the definition. In this chapter, synthetic fibers, regenerated cellulose, e.g., rayon, (Chapter 11), rubber (Chapter 9), glass, those materials used exclusively in surface coatings (Chapter 22), and, of course, bread dough will not be considered.†

The word "resin" is an old one derived from the Latin "resina" and the Greek "rhetine." Originally it referred to the natural exudates (or their fossil remains) of vegetable origin. The ancient Egyptians used such materials to help preserve their mummies.[80] Frankincense and myrrh, the Wise Men's gifts to the infant Jesus, are both natural resins. There are many such natural products: accroide, congo, rosin, copal, dammar, sandarac, elemi, kauri, manila, mastic, batu, pontianak, and shellac. Today they are used principally in surface coatings or as binders and adhesives. When identified as "synthetic" (the adjective is frequently omitted for brevity), the current meaning of

*Monsanto Co., Hydrocarbons and Polymers Division, Springfield, Mass.
**C. G. Systems, Madras, India.

†Except in connection with production statistics where the United States Department of Commerce chooses to include as plastics all synthetic resins used as surface coatings.

"resin" in the plastics industry is "that base substance of high molecular weight" before it has been mixed with colorants, fillers, plasticizers,* lubricants and/or stabilizers to make a finished commercial plastic molding powder.† In the surface-coatings industry, similarly, resin refers to the base binding material of high molecular weight before it has been formulated into a paint, varnish, or enamel.

The commercial history of synthetic plastics begins with the development of a practical molding process for cellulose nitrate (Celluloid), in 1869, by John Wesley Hyatt and his brother Isaiah, who were seeking a substitute for ivory. In about 1899, Adolf Spitteler treated casein with formaldehyde to make commercial plastic materials. In 1909, Leo Baekeland developed the first practical process for making moldings from phenol-formaldehyde resins. The organic and physical chemistry of these inventions remained art for a long time awaiting two concepts. One, championed by Hermann Staudinger and a few others in the 1920's, is that plastics owe their most significant properties to the extremely large size of their "giant" molecules. The other, first clearly implied by the experiments of Wallace Hume Carothers in 1929, is that the reaction-mechanisms and thermodynamic equilibria normally associated with reactive organic groups were almost the same for groups attached to large molecules.

With these concepts as guideposts, many thousands of polymers have since been made, or rediscovered. Less than fifty basic types have attained commercial success.

VIEWPOINT OF THE CONSUMER

The ultimate judgment about the utility of plastics is made by the consuming public; and judging by the response to the use of plastics in a variety of areas, the consumer is beginning to realize that, properly applied, plastics are superior to the materials formerly used and not just a "cheap substitute," unfortunately still a common image. The superior stain, abrasion, and mar resistance of laminated melamine (e.g., Formica) table and counter tops is well know. The superiority of plastic laminates (with fiber glass) for boats is also well known. (However, note the use of the term "fiber glass boat hulls" rather than "plastic boat hulls" due to the early poor image of the word "plastic.") Automobile, furniture, airplane, and bus upholstery of vinyl sheeting is superior to previous materials in many respects; replacement slip covers for automobiles are nearly obsolete due to the superior wear resistance of vinyl sheeting. The use of plastics in football helmets, in golf club shafting, in fishing rods, and in shampoo bottles is beginning to destroy the myth that all plastics are brittle. Developers of the highly touted synthetic leathers (e.g., Corfam®) were careful in avoiding the use of the word "plastic" although that is what the materials are. The industry must constantly guard against misapplication which leads to the image of a "cheap substitute."

The typical consumer is unaware of the extensive use of plastics where their superior price-performance dictates their use, e.g., insulation for nearly all wiring, drain and vent plumbing, printed circuits, automobile instrument panels and front grills—now chrome plated, in windshields to make safety glass, agricultural piping, greenhouses, and mulching.

Frequently, plastics will give a necessary balance of properties which cannot be matched by other materials, e.g., color, light weight, warmth of touch, low thermal and electrical conductivity, and resistance to biological and environmental degradation. Ablating shields for space vehicles and coatings for the leading edges of supersonic aircraft wings are two recent exotic examples of such applications.

Some of the major areas of application include: packaging, housings (e.g., for appliances, tools, TV's, radios, and telephones), furniture, automobiles, electrical and thermal insulation, toys, building products, boats and

*A plasticizer is a substance ordinarily (though not necessarily) of lower molecular weight that makes the plastic more flexible (lower elastic modulus); hopefully it should improve impact strength without markedly increasing creep or lowering ultimate strength.

†The term "powder" is a misnomer. Most commercial molding powder is in the form of granules or pellets which provide high bulk density, good flow, no dusting, and other desirable handling characteristics.

TABLE 10.1 Basic Commercial Plastics

Descriptive Name	*Chemical Structure of Monomers*	*Outstanding and/or Typical Properties or Uses*[a]	*Thermosetting, Thermoplastic, or Casting*	*Commercial Methods of Manufacture*
ADDITION POLYMERS				
Polyethylene	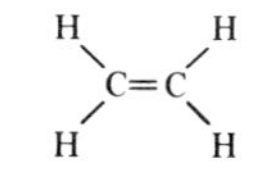	Excellent electrical properties. Good impact strength. Excellent colorability but translucent in thick sections. Good chemical resistance. Used in film, moldings, wire insulation, pipe, paper coating, and flexible bottles. Available in flexible and semirigid forms only unless heavily filled. Film widely used in packaging.	Thermoplastic	(a) High pressure; dense gas with free radical catalysts. (b) Medium pressure, solvent-nonsolvent; transition-metal oxide catalysts. (c) Low- and medium-pressure, solvent-nonsolvent, Ziegler catalysts.
Polypropylene	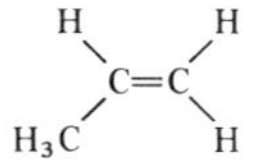	Lowest density of any plastic. Fair to good impact. Fair rigidity and dimensional stability. Excellent colorability. Translucent in thick sections. Good chemical resistance Properties vary widely with degree of crystallinity.	Thermoplastic	Solvent-nonsolvent; Ziegler catalysts.
Polyvinyl chloride and copolymers with vinyl acetate and vinylidine chloride ("Vinylite")	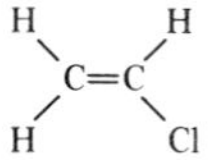	Good electrical properties and flame resistance with proper plasticizers. Rigid as polymerized–made flexible with plasticizers. Good impact strength especially in polyblends with elastomers. Good chemical resistance. Requires heat stabilizers. Many diverse uses.	Thermoplastic	(a) Suspension. (b) Solvent-nonsolvent. (c) Emulsion, used primarily for surface coatings, textile and paper treating, and "Plastisols."
Polyvinylidene chloride, always as a copolymer with vinyl chloride	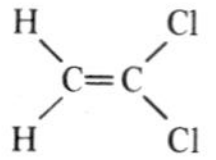	Very low moisture vapor transmission. Good chemical resistance. Self-extinguishing flame resistance. Used in pipe linings, film.	Thermoplastic	(a) Suspension. (b) Emulsion.
Polystyrene; also copolymerized with α-methyl styrene and vinyl toluenes to improve heat resistance	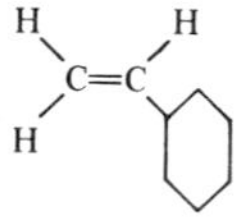	Excellent color, transparency, rigidity, dimensional stability, electrical properties, molding speeds. Used principally in moldings; also in phonograph records and rigid foams.	Thermoplastic	(a) Mass (polymer soluble in monomer). (b) Suspension. (c) Emulsion, used principally for surface coatings and polishes.

Polystyrene-based graft and/or polyblend polymers with styrene/butadiene copolymer synthetic rubbers	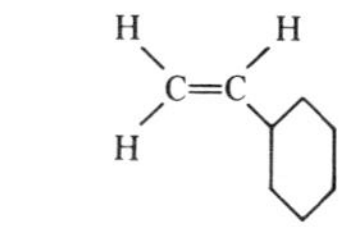 $H_2C{=}CH{-}CH{=}CH_2$	Excellent rigidity, dimensional stability, molding speeds. Good impact strength. Good colorability in opaques. Used extensively in moldings and for extruded sheet which is then vacuum formed. Packaging is large market.	Thermoplastic	(a) Mass. (b) Emulsion. (c) Suspension.
Styrene/acrylonitrile copolymer grafted or polyblended to butadiene/acrylonitrile rubber (ABS)	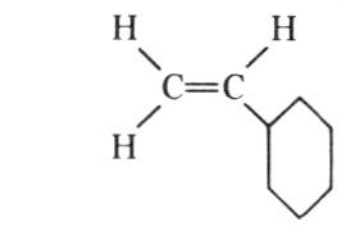 $H_2C{=}CH{-}CH{=}CH_2$ $CH_2{=}CHCN$	Outstanding impact. Good chemical resistance. Fair weatherability. Fair colorability in opaques. Used in pipe, moldings, extruded sheet.	Thermoplastic	(a) Emulsion. (b) Suspension. (c) Mass.
Styrene/acrylonitrile copolymer	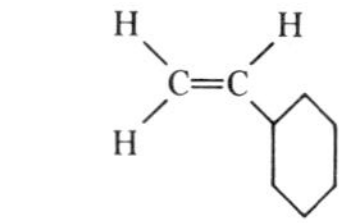 $CH_2{=}CHCN$	Better chemical resistance, crazing resistance, weatherability, impact strength than polystyrene. Good colorability including transparents. Used in moldings, film.	Thermoplastic	(a) Solution. (b) Mass to low conversion followed by devolatilization.
Polymethylmethacrylate ("Plexiglass")	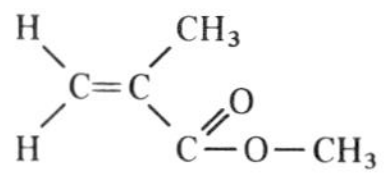	Excellent color, transparency, rigidity, dimensional stability, outdoor stability. Good impact strength. Used in moldings and as heavy sheeting. High impact graft copolymers available.	Thermoplastic and casting	(a) Suspension. (b) Casting between glass for transparent sheeting.
Polyesters (Combination of condensation and addition polymers.)	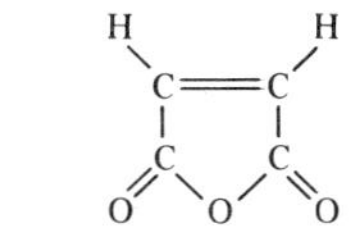 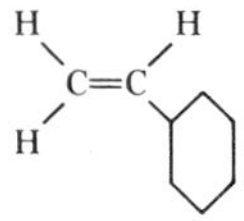 HOROH	Used with glass cloth or fibers to achieve very high tensile and flexural strengths approaching those of steel. Used to make laminates in aircraft, boats, furniture. Some potting of electrical components. Some compression and low-pressure molding.	Thermosetting and casting	See text.

Descriptive Name	Chemical Structure of Monomers	Outstanding and/or Typical Properties or Uses[a]	Thermosetting, Thermoplastic, or Casting	Commercial Methods of Manufacture
Polyvinyl butyral (Chemically modified addition polymer)	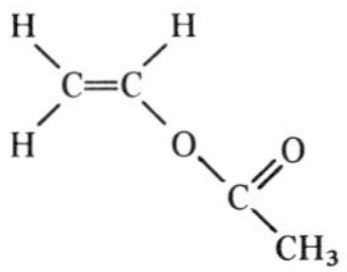 $CH_3CH_2CH_2\overset{H}{C}=O$	Primarily used for safety glass interlayer. Has excellent adhesion to glass. Excellent ultraviolet light resistance in absence of air.	Thermoplastic	See text.
Polychlorotrifluoroethylene ("Kel-F")	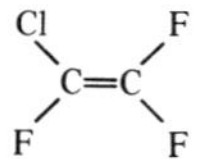	Excellent resistance to chemicals. Used as gasketing, packing, and corrosion resistant coatings.	Thermoplastic	(a) Emulsion. (b) Solvent-nonsolvent.
Polytetrafluoroethylene ("Teflon")	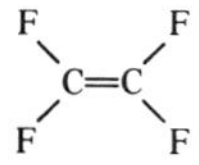	Excellent resistance to chemicals. Used as gasketing, packing, and corrosion resistant coatings. Attacked only by alkali metals and hot strong bases.	Fuse slowly under high pressure and 450–500°F	Emulsion.
CONDENSATION POLYMERS				
Phenol-formaldehyde ("Bakelite")	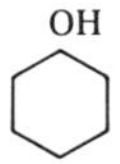 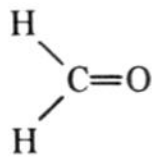	Good electrical properties with proper formulation. High heat resistance. Good to excellent impact and tensile strengths. Very slow to essentially no burning rate depending on filler. Acid resistant. Used for moldings, particularly for industrial and electrical applications where its poor colorability is not important. Used extensively for decorative laminate backing.	Thermosetting and casting	See text.
Nylon 66	$HO\overset{\overset{O}{\|}}{C}(CH_2)_4\overset{\overset{O}{\|}}{C}OH$ $NH_2(CH_2)_6NH_2$	Excellent wear resistance, tensile and impact strengths. Low coefficient of friction. High heat distortion temperature. Self-extinguishing to fire. Used for gears, rolls, and other moving parts.	Thermoplastic	See Chapter 11.

ϵ-Caprolactam (nylon 6)	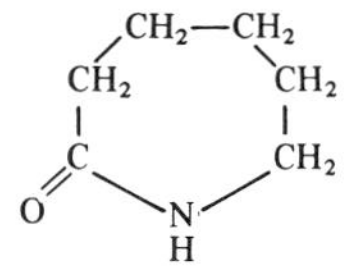	Properties very similar to nylon 66.	Thermo-plastic	See Chapter 11.
Melamine-formaldehyde	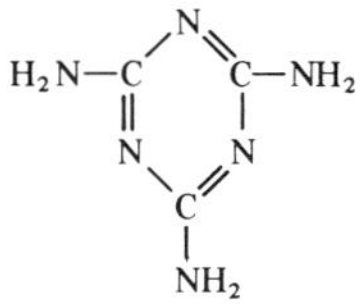 $CH_2{=}O$	Excellent hardness abrasion resistance, flame resistance, heat resistance. Good colorability in opaques, stain resistance, electrical properties. Used in laminates. Used with alkyd surface coatings and for adhesives.	Thermo-setting	See text.
Urea-formaldehyde	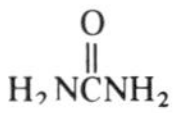 $CH_2{=}O$	Similar to melamine but poorer alkali resistance, crazing resistance, heat resistance, dimensional stability. Also poorer wear and hardness. Used in adhesives.	Thermo-setting	See text.
Epoxy	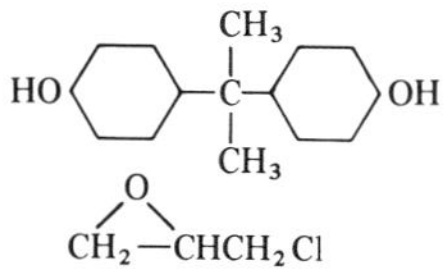 amines, organic anhydrides (see text for other monomers)	Outstanding adhesion to metals. Good chemical resistance. Fair to excellent impact in flexible formulations. Glass laminates have excellent tensile and flexural strengths plus good chemical and outdoor resistance if properly formulated. Widely used in surface coatings and adhesives.	Thermo-setting and casting	See text.
Ethylene glycol-terephthalic acid ("Mylar")	$HOCH_2CH_2OH$ 	Outstanding toughness and tear strength in films. Also widely used in fibers. (See Chapter 11.)	Thermo-plastic	

Descriptive Name	*Chemical Structure of Monomers*	*Outstanding and/or Typical Properties or Uses*[a]	*Thermosetting, Thermoplastic, or Casting*	*Commercial Methods of Manufacture*
Silicones	$(CH_3)_2SiCl_2$ 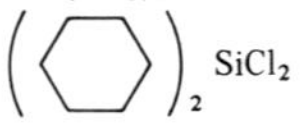 (see text)	Outstanding heat resistance. Unusual solubilities give anti-foam and water-repellent properties. Used for insulating varnish on electric motors, for high temperature rubbers, greases, and lubricants, for high temperature glass laminating, to increase adhesion of polyesters to glass fibers, and a multitude of other specialty uses most of which are not "plastic" applications in the strict sense.	Thermo-setting in plastic applications	See text.
Polyurethanes	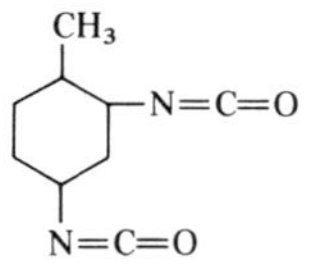 HOC(=O)RC(=O)OH (see text) HOROH	Best properties as rubbers with excellent wear resistance. Widely used as flexible foams; also for rigid foams and surface coating applications.	Thermo-setting	See text.
Polyoxymethylene	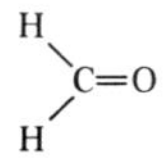	Excellent abrasion resistance. Tough. Excellent colorability in opaques only. Good chemical resistance. Excellent dimensional stability.	Thermo-plastics	Ionic solvent-nonsolvent (probably a fast chain reaction mechanism like vinyl polymers).
Polyoxetanes ("Penton")	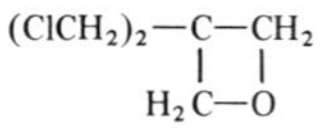 (see text)	Best chemical resistance of easily molded rigid polymers. Self-extinguishing to fire.	Thermo-plastic	Ionic solvent-nonsolvent.
Polycarbonates	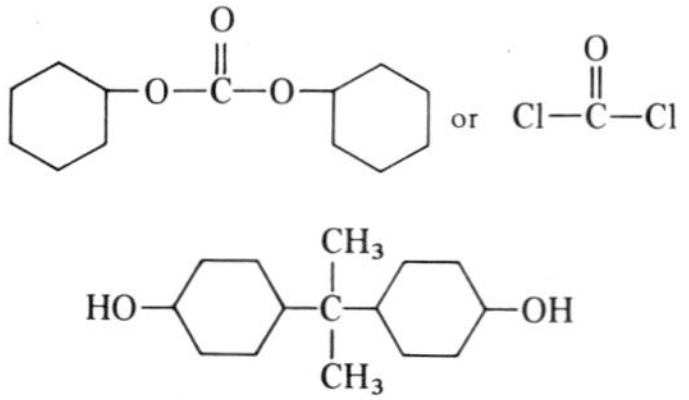	Excellent impact, dimensional stability at high temperatures, colorability. Good electrical properties. Self-extinguishing.	Thermo-plastic	See text.

Polyphenylene oxides	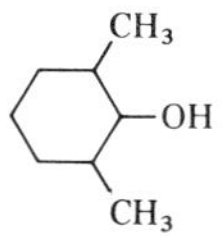	Very high heat distortion resistance for a thermoplastic; nondripping and self-extinguishing in flammability tests.	Thermo-plastic	See text.
Polysulfone	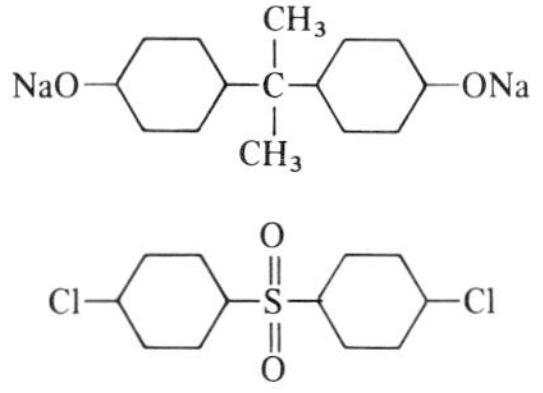	Very high heat distortion resistance for a thermoplastic; nondripping and self-extinguishing in flammability tests.	Thermo-plastic	See text.
CELLULOSIC	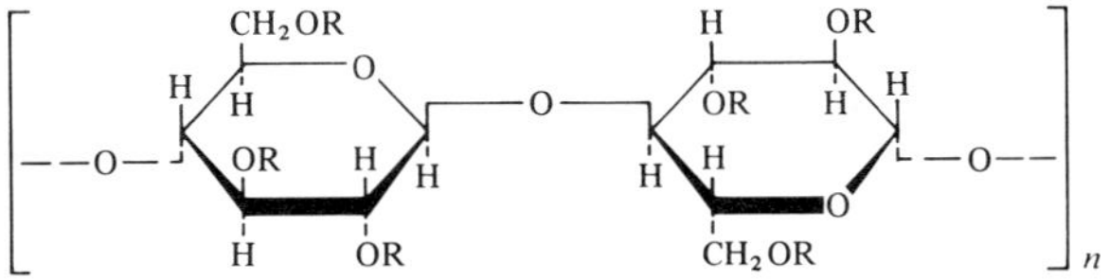			
Cellulose acetate	$n = 50$–100 $R = —C(=O)—CH_3$ (2.2–2.3 R's per ring)	Good impact strength. Good transparency. Excellent colorability. Fair outdoor stability. Slow to self-extinguishing burning rate. Requires plasticizers to form. Used for photographic film and moldings.	Thermo-plastic	See text.
Cellulose nitrate	$n = 250$ $R = NO_2$ (1.9–20 R's per ring)	Outstanding impact strength. Uses as plastic declining.	See text	See text.
Cellulose acetate butyrate	$R = —C(=O)—CH_3$ (1 per ring) $R' = —C(=O)CH_2CH_2CH_3$ (1.7 per ring)	Similar to cellulose acetate with better dimensional stability and weatherability. Used for pipe, telephone housings, moldings.	Thermo-plastic	See text.
Ethyl cellulose	$n = 250$ $R = —CH_2CH_3$	Outstanding impact of all cellulosics, otherwise similar.	Thermo-plastic	See text.

[a]Within each class of polymer the properties can be widely varied depending on the plasticizer, copolymer type and the ratio of filler and stabilizer in the basic formulation, particularly in condensation polymers. The properties given are only the outstanding and characteristic ones.

recreational equipment, and piping. New and novel uses include: synthetic printing paper—principally in Japan; artificial organs and tissues, and prosthetic devices in medicine; unbreakable glazing and "safety-glass" laminates; soft drink and beer bottles; synthetic leather for shoes; chrome-plated plastic parts for superior wear and corrosion resistance.

VIEWPOINT OF THE INDUSTRIAL DESIGNER

In plastics, the industrial designer has found a versatile raw material. Plastics have a tremendous advantage over natural products in that they can be tailor-made for a specific use. Plastics may be classified in many ways. The industrial designer uses conventional physical properties to characterize plastics, e.g., density, tensile strength, impact strength, modulus of elasticity, and creep rate. These, however, are not enough. Price, fabricating costs, electrical characteristics, thermal conductivity, colorability, other appearance factors, and chemical resistance frequently determine the selection of a particular plastic in preference to another plastic or other materials.

For information about specific properties of various plastics, consult Table 10.1 and Selected References PP-1 through PP-5.

Testing plastics is a highly specialized art, which has been developed in the U.S. primarily through the efforts of the members of the American Society for Testing Materials (ASTM). Plastics are usually excellent electrical insulators; but their electrical properties must be measured over a wide range of electrical and climatic conditions. Few plastics obey Hooke's Law; many types creep even at very low temperatures and stresses. Specialized stress-strain testing machinery has been developed which permits extensive variations in temperature and rate of loading. Since coloring versatility and simplicity are prime factors in many plastics markets, tests for color reproducibility and stability are often essential. (See Selected References T-1 through T-5 for testing procedures for plastics.)

VIEWPOINT OF THE FABRICATOR

Not only must the fabricator keep in mind the industrial designer and consumer; he must also consider the types of equipment required and the behavior of the plastics in relation to this equipment. Plastics are divided into thermosetting and thermoplastic materials. Thermoplastics are materials which become fluid upon heating above a certain temperature frequently called the "heat distortion temperature." Upon cooling they set to an elastic solid. The process can be repeated many times since no primary chemical bonds are normally made or broken. Thermosetting materials, on the other hand, may normally be heated to the fluid state only once. The chemical structure of a thermosetting plastic is altered by heat, and crosslinked products are formed which cannot be resoftened. Thermosetting plastics are not normally used in injection molding machines and extruders since these machines contain spaces where hot molten plastic may remain for indefinite lengths of time, thus causing varying levels of cure and possibly set-up in the machine. Chemically, the molecules of a thermosetting resin crosslink* irreversibly through primary chemical bonds to form one large molecule. It may seem strange to call a washing machine agitator a single molecule, but this is the modern view of highly crosslinked structures.†

Compression and transfer molding are the two main methods used to produce molded parts from thermosetting plastics[64]; however, injection molding is under development and may become important in the future.[71,141] In compression or transfer molding the thermosetting molding compound is heated to about 150°C so as to soften it sufficiently so that it will flow into the mold cavity. The soft plastic is held under a pressure often greater than 140 kg per cm^2 long enough for the material to polymerize or crosslink, resulting is a hard, rigid product. A simple compression molding machine is driven by a vertical hydraulic ram press as shown in Fig. 10.1.

---

*A term more inclusive than thermosetting resin is crosslinked resin, applied equally well to resins such as polyurethane, some types of which form crosslinked structure at room temperature.

†There are undoubtedly many individual molecules of varying sizes trapped in the over-all network, but it is probable that any macroscopic region of the agitator is connected through primary chemical bonds to all other regions of the agitator.

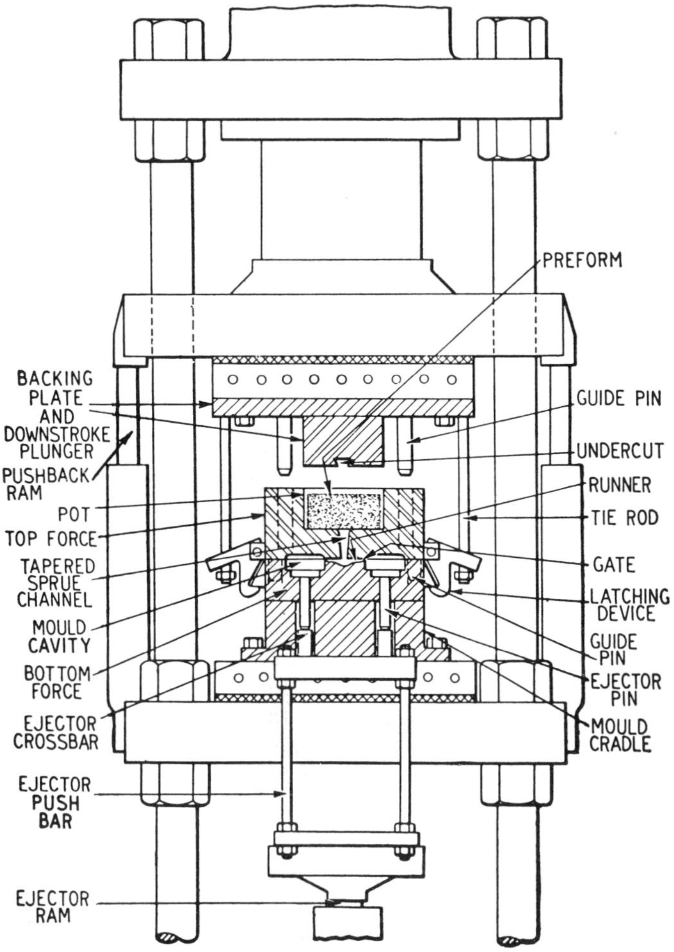

Fig. 10.1 Transfer mold for thermosetting plastics which permits heating of preform pellet during curing time of piece in mold cavity. (*Redfarn, C. A., "A Guide to Plastics," Cliffe and Sons, Ltd., London, 1958; by permission.*)

Thermoplastics may be molded in a compression molding machine or in the faster, more versatile, and economical injection molding machine (Fig. 10.2). The concept of maintaining a stable thermoplastic material in a fluid state and squirting it under pressure into a cooler mold was first developed in the middle 1920's by Dr. Arthur Eichengum and the German firm of Eckert and Ziegler.[123] It was not until the late 1930's that successful automatic machines were used in the United States. Once a thermally stable material became commercially available, development of the injection molding machine followed logically. The first such commercial plastic was cellulose acetate, introduced in 1933. Injection molding of a thermoplastic is a physical process and is based on the ability of the thermoplastic to reversibly soften with heat and harden on cooling. No chemical reaction takes place as in the case of thermosetting resins. Heating and cooling may be repeated many times if desired. Polymer beads or pellets are heated in a cylinder or, in more modern machines, with the aid of an extruder screw until molten. The melt is pushed out of the cylinder through a nozzle by either a piston or the extruder screw, now acting as a pis-

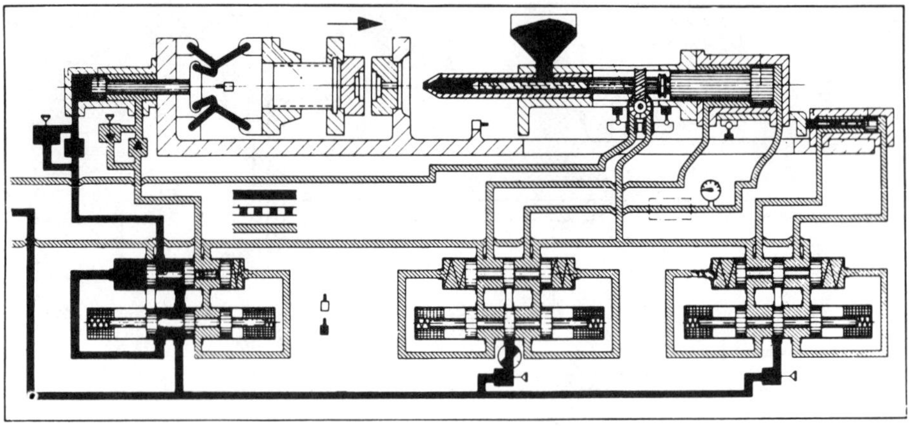

A

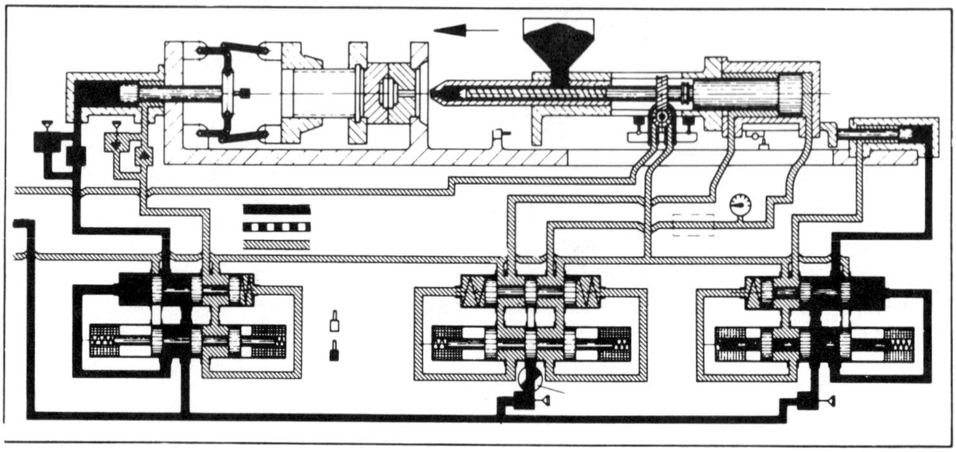

B

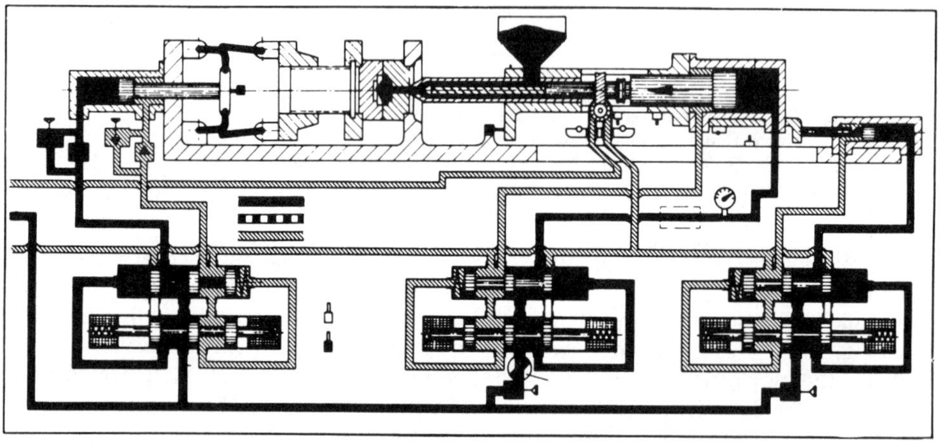

C

Fig. 10.2 Operational details of in-line reciprocating screw injection molder. *A.* When the mold starts to close, the screw has just finished charging the front end of the cylinder. *B.* With the mold closed, the heating cylinder then moves toward the sprue bushing and is ready to commence the injection cycle. *C.* The hydraulic cylinder then forces the screw forward under the injection head and fills the cavity.

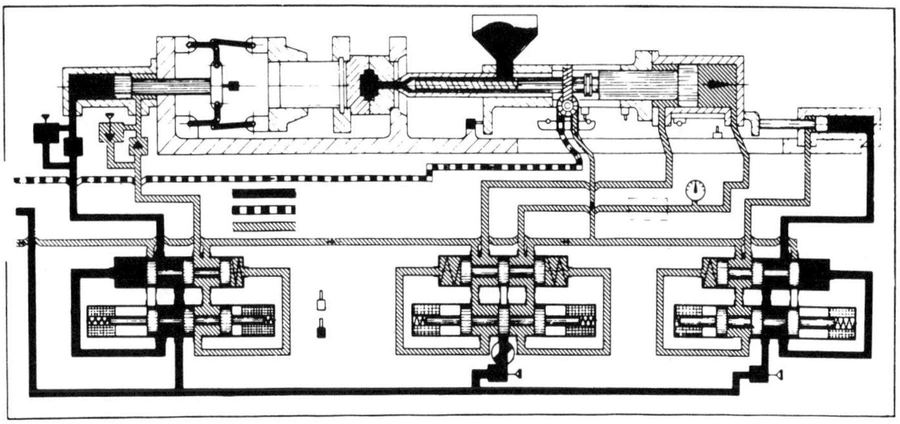

D

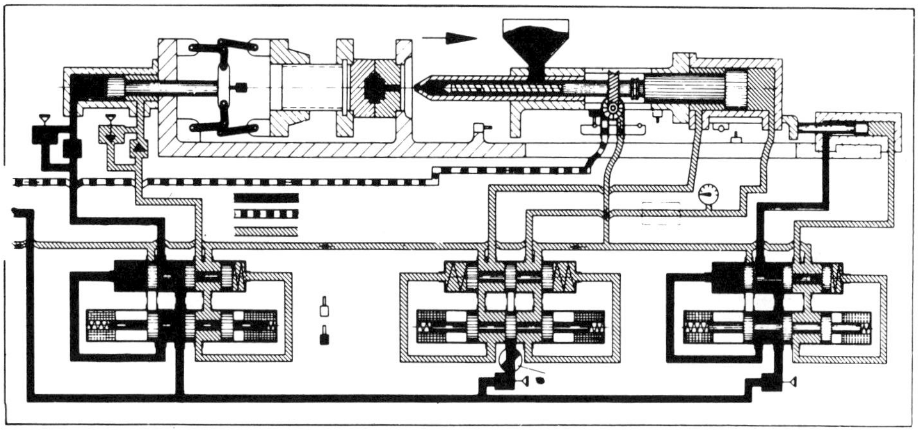

E

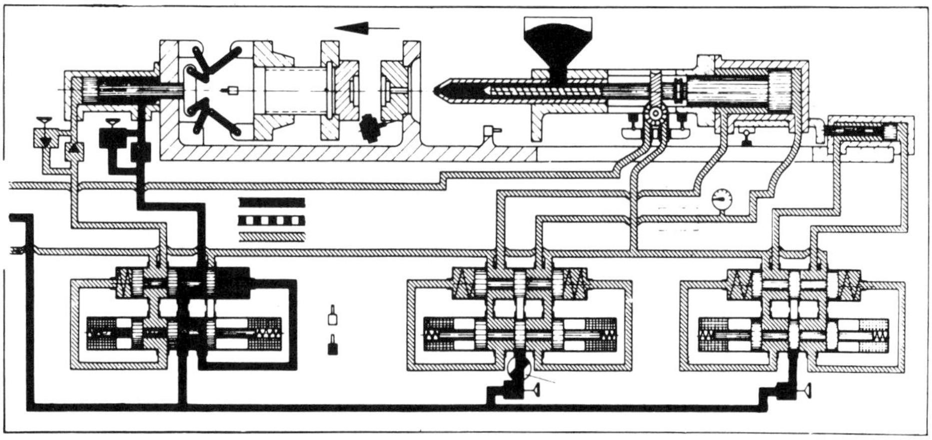

F

Fig. 10.2 (*continued*). *D.* Cooling and charging cycle: the mold is filled, and, after holding time, the screw starts to rotate and charge the front end of the cylinder. *E.* The screw is still charging the front end of the cylinder, and the cylinder carriage starts to move away from the sprue bushing. *F.* The finished cycle: mold opens and molded part is ejected with the sprue bushing. (*Du Bois, J. H., and John, F. W., "Plastics," Reinhold, 1967; by permission Van Nostrand Reinhold Co.*)

ton, into a relatively cool, closed mold. Here it solidifies to form a part shaped like the reverse image of the mold. A schematic of a screw injection molding machine is shown in Fig. 10.2. If sheets, rods, tubes or profiles of various lengths are desired, thermoplastics can be conveniently extruded (Fig. 10.3). Rigid sheets of rubber-modified polystyrene and styrene acrylonitrile copolymers are frequently vacuum formed into complex shapes, e.g., containers, refrigerator liners, boats (Fig. 10.4). Many thermoplastics are calendered into films or sheets (Fig. 10.5).

Blow molding (Fig. 10.6) is a very popular technique for fabricating hollow containers (bottles) from a wide variety of polymers, e.g., rigid and flexible polyvinylchloride, polyethylene, polypropylene, rubber-modified polystyrene and rubber-modified styrene-acrylonitrile copolymers (ABS).

Most polyethylene film is made by a blow extrusion process (in Fig. 10.7) where air is maintained in a "balloon" formed at the exit of a circular extrusion die. The top (cool exit end) of the balloon is closed via a set of pinch rolls. Film, particularly polyethylene, up to 40 feet wide is routinely manufactured in tremendous volume by such a process.

There are a few methods of fabricating plastics in which either thermoplastic or thermosetting base resins may be used. Laminates are made by pressing—under heat and pressure—layers of paper, cloth, glass fiber, or glass cloth which have been impregnated with a liquid resin. Thermosetting resins are the only types used commercially since they make a harder, more solvent resistant, and more rigid laminate. Kitchen counter and table tops are one of the better known applications of such materials.

Vinyl plastisol is a unique plastic consisting of a highly fluid dispersion of finely divided polyvinylchloride resin in a plasticizer. Mild heat causes the resin and plasticizer to fuse into a homogeneous solid. Vinyl plastisols—and also other thermoplastics, such as polyethylene, polypropylene, rubber-modified polystyrene—are frequently formed into large hollow shapes by a technique known as rotational molding or casting. The plastic is normally added to the mold while it is rotated about at least two axes to evenly distribute the plastic. The mold is heated either prior to or during the addition of the plastic. After the plastic has fused, the mold is cooled and opened, and the part is removed.

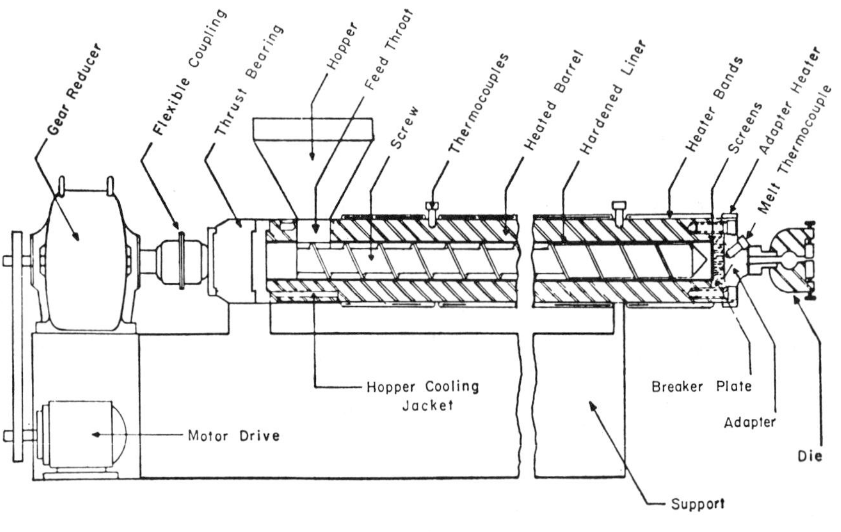

Fig. 10.3 Elements of an extruder. (*Bernhardt, E. C. "Processing of Thermoplastic Materials," copyright by The Society of Plastics Engineers, Inc., 1959. Van Nostrand Reinhold, New York.*)

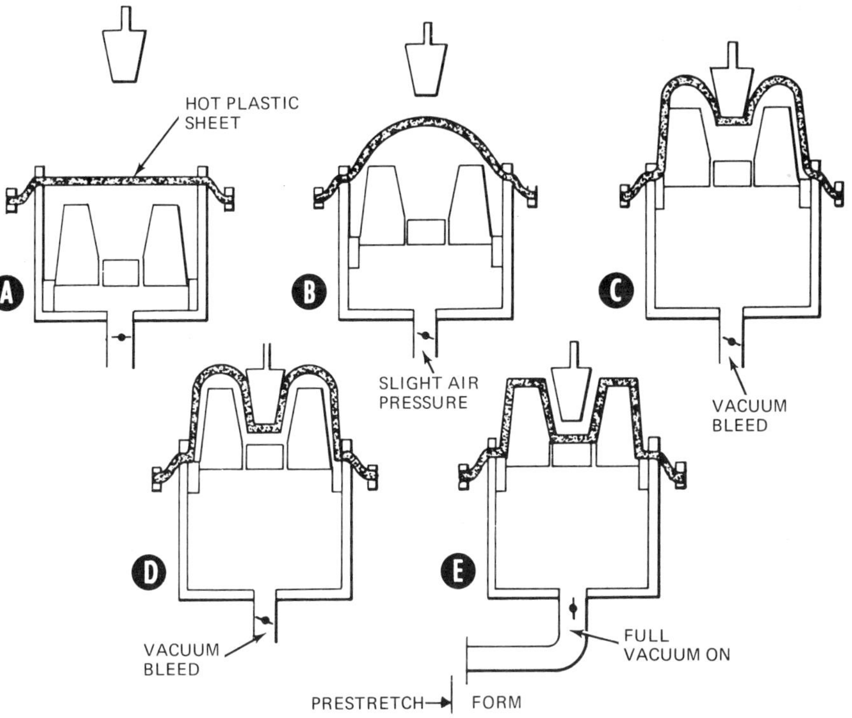

Fig. 10.4 Vacuum-forming using combination or air-slip-plug assist. (*Basdekis, C. H., "ABS Plastics," Reinhold, 1964; by permission Van Nostrand Reinhold*)

Plastic foams made from a wide variety of polymers have a wide range of applications and are made by a variety of methods depending upon the polymer and the application. Polystyrene beads containing some pentane blowing agent are prefoamed with high pressure steam and then formed into a wide variety of shapes from drinking cups to fish-net floats and insulated paneling. Rigid urethane foams provide the insulating material for most home refrigerators. They are formed in place by filling the enclosed space between the inner and outer shells (both of which may be vacuum formed plastic sheets) with the urethane monomers plus a small amount of water to generate the "pneumatogen" carbon diox-

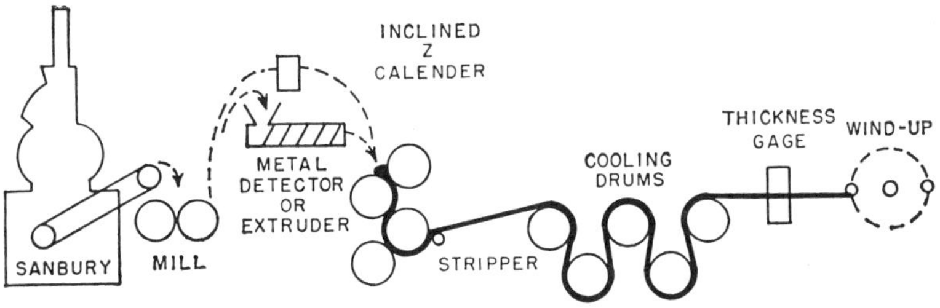

Fig. 10.5 Parts of calender layout for plastic film production. (*Bernhardt, E. C., "Processing of Thermoplastic Materials," Van Nostrand Reinhold, copyright by The Society of Plastics Engineers, Inc., 1959; by permission.*)

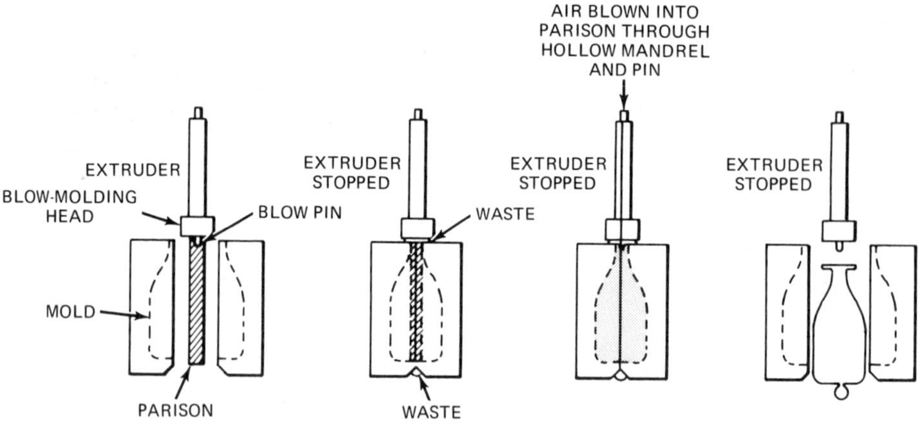

Fig. 10.6 Simple intermittent extrusion blow molding of hollow articles. *A.* Complete extrusion of parison. *B.* Close mold. *C.* Blow and cool. *D.* Eject. (*"Encyclopedia of Polymer Science and Technology," Vol. 9, H. F. Mark and W. G. Gaylord, Eds., Interscience, 1968; by permission of John Wiley and Sons, Inc.*)

ide. The above materials are rapidly mixed in an injection nozzle or gun to prevent premature foaming and/or setting. Other techniques use "blowing agents" which will decompose into gases—usually nitrogen—upon heating. Commercial blowing agents include azobisformamide (ABFA), azobisisobutyronitrile (AIBN), N,N′-dimethyl-N,N′-dinitrosoterephthalamide (DNTA), 4,4′-oxybis(benzenesulfonylhydrazide) (DBSA), and, more recently, several sulfonyl semicarbazides. Recent methods have been devised to permit injection molding of foamed articles, e.g., furniture pieces resembling wood from rubber-modified polystyrene.

VIEWPOINT OF THE PHYSICAL CHEMIST

To the physical chemist the most significant characteristic of plastics is high molecular weight. Man has used giant molecules—or high polymers—to feed, clothe, and house himself for centuries. Wood is a high polymer, so are cotton, silk, wool, and meat. The realization that plastics are high polymers wherein the atoms are joined by primary chemical bonds, and vice versa, is a surprisingly recent concept championed by pioneers like Hermann Staudinger and Theodor Svedberg, both of whom received the Nobel Prize for their work characterizing "makromolekules."* To the physical chemist the separation of polymers into plastics, resins for surface coatings, synthetic fibers, rubbers (or more recently, elastromers) destroys some of the unity in the subject of high polymers. To a physical chemist, a rubber is a polymer with certain rheological properties which can be fairly well explained by a crosslinked network present in a model of a polymer above its second-order ("melting") transition point. Similarly, the physical chemist can explain and correlate many of the properties which make certain polymers desirable for fibers, films, moldings, surface coatings, and other uses. He can measure, with great precision in some instances, the size and size-distribution of the molecules, the number and size of branches in the chain, the randomness of the copolymers, the degree and type of stereospecificity, the arrangement and patterns of the crystalline regions (Fig. 10.8) and the complex morphology of rubber-reinforced rigid "polyblends" (Fig. 10.9). He has achieved considerable success in relating these measurements to gross physical properties, e.g., tensile strength, impact strength, elonga-

*For a particularly interesting history of the development of the modern concept of the structure of organic polymers, see P. J. Flory, "Principles of Polymer Chemistry," pages 3–27.

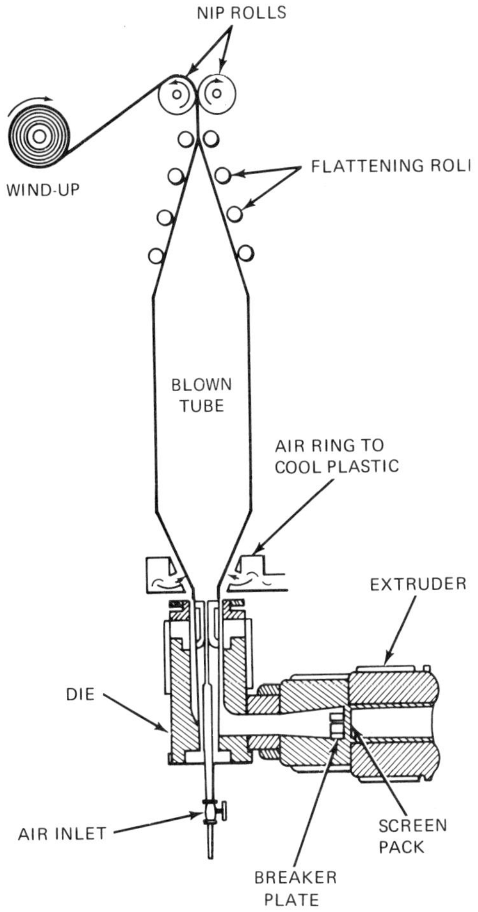

Fig. 10.7 Formation of sheet by extrusion through a circular die (blown-bubble extrusion). Ingredients are resins, stabilizers, pigments, and other additives. (*"Encyclopedia of Polymer Science and Technology," Vol. 6, H. G. Mark and W. G. Gaylord, Eds., Interscience Publishers, 1968; by permission John Wiley and Sons, Inc.*)

Fig. 10.8 Spherulites of crystalline polystyrene growing in a melt. Electron micrograph. (*Courtesy G. C. Claver, Monsanto Co.*)

tion, and solubility. However, although the physical chemist has explained a great deal, much remains to be done in this field; it is still necessary to synthesize and test a new polymer before its properties can be known.

VIEWPOINT OF THE ORGANIC CHEMIST

The organic chemist, usually interested in describing chemical reactions and the resulting products, ignored much of polymer chemistry for a long time. The traditional organic chemist of the nineteenth century regarded the uncrystallizable "tars" that formed from certain reactions as unsuccessful experiments, since the objective was the preparation of crystals whose physical chemistry was better understood. The progeny of the polymeric "gunks" which the nineteenth and early twentieth century chemist threw away in disgust, today, constitute the largest tonnage market for synthetic organic materials.*

Now with better methods of characterizing

*To make this statement precisely true, fuels must be excepted and the colorants, plasticizers, lubricants, stabilizers, and solvents utilized by the plastics, rubber, synthetic fibers, and synthetic surface coatings industries must be included. However, world usage of purified or chemically modified cellulose, about 9 billion tons annually, need not be included as synthetic high polymers if, likewise, it is not called "synthetic." It has also been estimated that one third of the American chemists and chemical engineers are now employed in industries connected with polymeric materials.

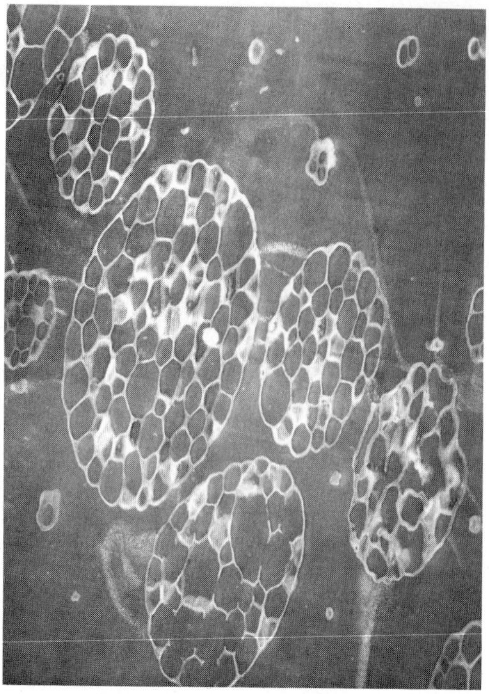

Fig. 10.9 Complex "honeycomb" structure of rubber-modified ("impact") polystyrene. Rubber is white, polystyrene is dark. Electron photomicrographs of thin (ca. 0.1 micron) sections stained with OsO_4 per Kato.[70a] *(Courtesy G. C. Claver, Monsanto Co.)*

the molecular structure of polymers, the modern organic chemist has developed well understood syntheses for many polymers and logical mechanisms which explain vast bodies of data. The inventor of nylon, Wallace Hume Carothers,[19] beginning in 1929, showed that the reaction mechanisms and the thermodynamic equilibria ordinarily associated with reactive organic groups were not altered merely because these groups were attached to large rather than small molecules. This vital concept removed the mystery from polymer synthesis and funneled the vast knowledge of organic chemistry into the polymer field, resulting, in the early 1930's, in the dramatic development of new polymers.

Carothers defined two types of polymers: (1) *Addition polymers* are those like polyethylene in which the monomer units, on polymerizing, do not completely rupture a chemical bond but, in the usual case, utilize the π bonding electrons in a double bond to form a new bond between the monomers. Most addition polymers are obtained from monomers containing carbon-carbon double bonds, e.g., vinyl monomers ($H_2C{=}CHR$), vinylidene monomers ($H_2C{=}CR_2$), diene monomers ($H_2C{=}CR{-}CH{=}CH_2$), but some commercial addition polymers are also made from $CH_2{=}O$, and ethylenic monomers ($CF_2{=}CF_2$); (2) *Condensation polymers* are those wherein chemical bonds between atoms completely rupture during polymerization, causing formation of low molecular weight fragments, frequently water. By using pure materials and forcing the reactions to completion through removal of the low molecular weight fragment, many varieties of high polymers can be made from multifunctional organic molecules, e.g., synthetic fibers (Chapter 11) prepared from organic acids and amines (nylon) and from organic acids and alcohols (polyesters such as "Dacron" and "Mylar"), as well as alkyd paints (Chapter 22), from organic aldehydes and phenols or amines. Catalysts vary, depending on the nature of the functional organic group. Usually Lewis acids or bases are required.

Some reactions do not fit neatly into either the addition- or condensation-polymer classification, but are, nonetheless, usually considered condensation polymers, e.g., reactions between a diisocyanate and a glycol to form a polyurethane,

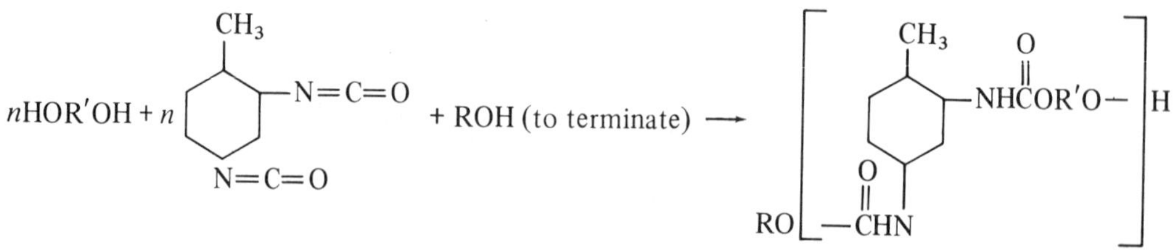

or ring-opening reactions such as the formation of nylon 6 from caprolactam,

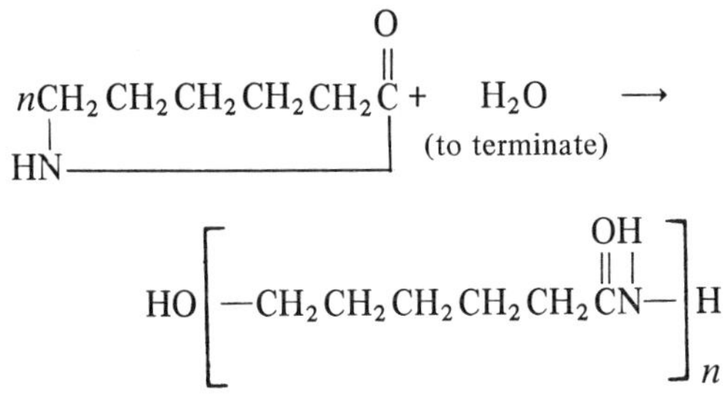

In addition to the large number of vinyl and vinylidene monomers which form addition polymers, and the even larger number of difunctional and multifunctional organic chemicals which form condensation polymers, the combination of various addition monomers into addition copolymers, and mixtures of condensation reactants form literally millions of combinations. Also, by adding monomers to previously formed polymers, e.g., adding styrene to a synthetic rubber, graft polymers* with unusual properties can be formed; this is done commercially to produce vinyl chloride copolymers, styrene-acrylonitrile copolymers (ABS), polystyrenes, and polymethylmethacrylates possessing high impact strength. Also, variations in the molecular weight, molecular-weight distribution, amount of chain branching and crosslinking, molding or forming conditions, types and amounts of plasticizers, stabilizers, colorants, and fillers further increases the possibilities for various material specifications to astronomical levels. Table 10.1 gives a very brief summary of the principal plastics and the monomers from which they are made.

Addition polymerization may be initiated by free radicals or ions. At the present time free-radical polymerization is commercially the most important. However, the discovery that polymer properties can be greatly influenced by their degree of stereoregularity is currently creating great interest in ionic polymerization. Free radicals can be generated by the action of ultraviolet light, X-rays, and high-energy particles. They are commercially produced for industrial polymerizations by heat, peroxides, oxygen, and diazo compounds. The polymerization propagates itself by a chain-reaction mechanism, i.e., when a free radical and a monomer molecule react a new free radical is formed. The reacting monomer unit captures the odd electron and becomes an initiating free radical itself. By such a mechanism, polymer molecules with thousands of monomeric units are generated in a few seconds or less; the monomers are normally linked together in a head-to-tail fashion.

The chain reaction may be stopped by the combination of two growing chains or by an initiating radical colliding with the end of a growing chain rather than with a monomer molecule (this is unlikely), in which both ends of the polymer chains will contain initiator fragments. Chain transfer agents used to control molecular weight, e.g., mercaptans and halogenated aliphatic hydrocarbons, may also terminate a growing chain by splitting into two free-radical fragments, one terminating the growing chain and the other behaving as a free-radical initiator. Disproportionation, i.e., the rearrangement of the polymeric free radical into a more stable configuration, usually with the formation of a double bond, is also

---

*Graft polymers are a special type of copolymer and resemble block polymers, e.g.,

—ABAABBBABAAA— Random copolymer

—AAAAAABBBBBBB— Block polymer

Graft polymer

possible. In addition, chain terminators or inhibitors, such as quinones, oxygen, sulfur, and amines, may stop a growing chain by forming stable adducts with the initiator and/or the polymeric free radical. Early chemists were plagued with the question of what stops the chain; this is still a difficult problem to solve quantitatively for any particular polymerization scheme.

Ionic catalysis of vinyl and diene monomers is a very old art. Matthews[84] in England and Harris[56] in Germany discovered simultaneously, in 1910, that metallic sodium would polymerize butadiene. Yet little commercialization was achieved until very recently. Then around 1952 Karl Ziegler, who had been working in the field of metal-organic compounds for nearly 30 years, discovered that aluminum trialkyls complexed with titanium tetrachloride would polymerize ethylene to high molecular weight at low pressure.[31,146,147,148] At about the same time John Hogan and Robert Banks[62] of Phillips Petroleum discovered that certain hexavalent chromium oxides on silica or aluminum gel would polymerize ethylene to commercial polymers. Shortly thereafter G. Natta and coworkers in Italy discovered that a modified "Ziegler catalyst" would cause "stereospecific" propagation* of the growing chain for α-olefins. Other soluble ionic catalysts, particularly lithium alkyls and cobaltous alkyls, have been found to influence chain configuration, particularly in dienes.

Frequently, polymer molecules possessing a regular structure crystallize readily, producing unusual properties, i.e., they are higher melting, more rigid, stronger, and less soluble than their noncrystalline stereo-irregular counterparts.

This new field has excited great commercial and scientific interest. To date *cis*-polypropylene, *cis*-polyisoprene, and *cis*-polybutadiene have been produced for sale, but crystalline polymers of styrene, methyl methacrylate, 3-methyl-1-butene, and several methylpentenes and methylhexenes have all been produced in the laboratory.

In contrast to the general nature of catalysis in addition polymerization, condensation polymerization uses the acid or base catalyst appropriate for the organic groups involved. Some pairs or groups require no catalyst at all. No chain reaction mechanism occurs, and the molecules gradually increase in size as the polymerization proceeds. If trifunctional or higher functional molecules are present, crosslinking will occur, leading first to gelation and eventually to complete hardening.

VIEWPOINT OF THE CHEMICAL ENGINEER

The plastics industry provides a fascinating field for the chemical engineer. He can investigate polymerization kinetics, reactor design, process control, the rheology of polymer melts, slurries, solutions or latices, the unique problems of mixing or compounding various polymer blends, and the problems of fabricating the polymer into the final product. In addition, the entire range of transport phenomena (mass, heat, and momentum transfer) are involved in the engineering of polymer plants. The objective of the chemical engineer is the commercial development of a process which will produce the highest quality polymer for a particular use at the lowest possible cost.

Polymerization mechanisms and kinetics are central to the design of the optimum process for the optimum product. For example, the process by which unsaturated monomers are converted to polymers of high molecular weight by addition polymerization is a chain reaction. In free-radical polymerization in the presence of an initiator, the first step is the decomposition of the initiator I (e.g., benzoyl peroxide) to yield two free radicals ($I \xrightarrow{kd} 2R^{\bullet}$). (The $^{\bullet}$ symbolizes the odd electron of the free radical.) Then a chain is initiated by the addition of a monomer molecule M to the primary radical $R^{\bullet}$ ($R^{\bullet} + M \xrightarrow{ki} M_1^{\bullet}$). The polymer molecule grows with successive addition of monomer molecules:

$$M_1^{\bullet} + M \xrightarrow{kp} M_2^{\bullet}$$

$$M_2^{\bullet} + M \xrightarrow{kp} M_3^{\bullet}$$

*Natta's modified Ziegler catalyst forbids *trans* addition of an incoming α-olefin molecule; hence all *cis*-polymer is produced.[19,91]

or in general:

$$M_x^\bullet + M_s^\bullet \xrightarrow{kp} M_{x+i}$$

Normally, the propagation reaction above proceeds extremely rapidly. Thousands of monomer units add in fractions of a second. Finally the growth of the polymer molecule may be terminated by its combination with another free radical, for example, by coupling

$$M_x^\bullet + M_y^\bullet \xrightarrow{k_t} M_{x+y}$$

Alternatively, the chain may be interupted by chain transfer agents, which terminate the "chain" but initiate a new chain simultaneously. Such agents permit control of the molecular weight to nearly any degree. Control of molecular-weight distribution is also possible, but very narrow distributions can only be obtained by using "living," growing anionic chains. The above simplified kinetic scheme for addition (vinyl) polymers illustrates the complexity of most polymerizations; condensation polymerizations, "living" ionic chain-growth reactions, and polymerizations of olefins using the heterogeneous solid catalysts exhibit different but equally or more complex kinetics.

The chemical engineer in the plastic industry has a few unique problems which he does not have to the same degree in other branches of his industry. These problems stem from the fact that polymers cannot be purified economically, i.e., the separating operations of distillation, extraction, adsorption, etc., are not commercially practical methods for isolating good polymeric molecules from undesirable ones. Ordinarily, all polymeric molecules produced by a particular polymerization scheme will end up in the final product sent to the customer. This means that any changes in the polymerization process will be reflected in the properties of the final product. The process engineer in the plastics industry must be completely aware of polymerization mechanisms, kinetics, and quality evaluation schemes. In addition, he must appreciate that tests for "quality" cannot be simple or definitive; variations in the process due to scale-up or modifications made for economic gain may alter "quality" irrevocably in subtle ways not always measurable by simple routine control tests.

VIEWPOINT OF THE ECONOMIST

The plastics industry began a period of rapid growth in the late 1930's and is continuing to expand. Total U.S. production of all thermoplastic and thermosetting resins increased from 6.3 billion pounds in 1960 to 11.4 billion pounds in 1965, and in 1968 to 16 billion pounds valued at \$3.9 billion, which accounts for 14.8 per cent of the basic chemical industry's sales of \$26.4 billion (see Fig. 10.10 and Table 10.2).

Thermoplastics, which have been growing at about 14.6 per cent per year (U.S.) for the 1960–68 period, have the lion's share of sales volume at 12.3 billion pounds in 1968. Thermosetting resins have grown only at 6.3 per cent per year during the same period. Growth rates are expected to decline to an annual rate of 11 per cent for thermoplastics and 5 per cent for thermosets during the 1969–73 pe-

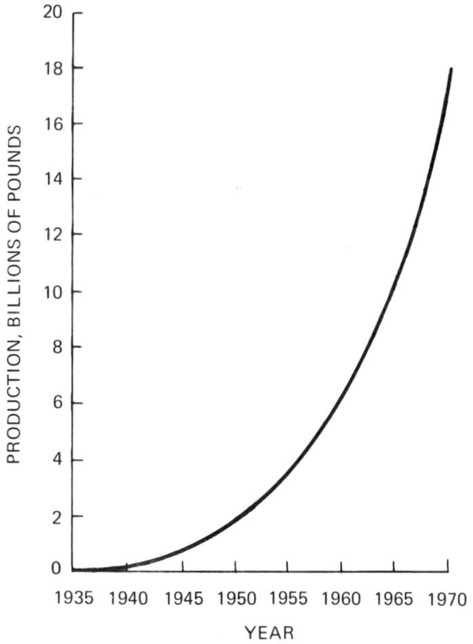

Fig. 10.10 Production trend of plastic materials. *(From U.S. Department of Commerce data, including Std. Ind. Class 2821 and cellulosics.)*

TABLE 10.2 Production in 1971 of Synthetic Plastics and Resin Materials in the United States[1]

Material	*Production, millions of lb*	*Approx. Avg. $ Value/ Pound, 1971*	*Approx. Avg. $ Value/ Pound, 1957*
Polyethylenes	6,396	0.15	0.32
Polyvinylchlorides	3,433	0.13	0.29
Polystyrenes	2,443	0.14	0.30
Polypropylenes	1,257	0.20	(2)
ABS and P(S/AN)	768	0.35	0.55
Phenolics	1,142	0.20	0.27
Polyesters	637	0.30	0.40
Polyacrylates	410	0.45	0.60
Cellulosics	195	0.48	0.52
Urea and melamines	683	0.20	0.30
Polycarbonates	50	0.75	(2)
Nylons (for non-textile use)	110	0.80	1.25
Polysulfones	n.a.	1.00	(2)
Polyacetals	62	0.60	(2)
Epoxies	175	0.50	0.70
Polyphenylene oxides	50 (est.)	0.70	(2)
Fluorocarbons	n.a.	4.00	(2)

(1) Predominantly based on U.S. Tariff Commission reports.
(2) Either not in large scale production in 1957 or price uncertain.

riod. The above rates should be compared with the increase in the "gross national product" of about 6 per cent per year in the same time period.

Production costs for both types of polymers are determined by

	Per Cent
Monomer costs	45–65
Processing or conversion costs	25–40
Utility costs	5–10

Polymer selling prices are based not only on production costs, but also on the capital requirements, the price-performance characteristics of the polymers compared to alternative materials, and the current supply versus demand situation. The selling price/production cost ratio ranged from 1.2 for low density polyethylene (11¢/lb) to 2.1 for polypropylene (20.5¢/lb). As sales volume increases, this ratio decreases as larger markets, economics of scale, improved productivity and increased competition result in lower selling prices. The Census Bureau price index for all plastics (1957-59 = 100) fell from 101 in 1958 to less than 89 in 1968.[96]

MANUFACTURING PROCESSES

The manufacturing processes for polymers is nearly as diverse as the number of polymers, but it is possible to classify many of them as follows:

A. Addition polymerization processes.
   1. Mass
   2. Emulsion
   3. Suspension
   4. Solvent
   5. Solvent-nonsolvent

B. Condensation polymerization processes.

C. Ring opening and other "addition" polymerization processes in which the polymer grows in a manner other than by a chain reaction.

Addition Polymerization

Addition polymerization has evolved along many paths which try to solve the major engineering problem of removing the large heat of polymerization, e.g., 12–26 kg-cal/mole of vinyl monomer.*[109] For polyethylene this amounts to more than 800 calories per gram or a temperature rise of over 1000°C under adiabatic reaction conditions. Such temperature changes are intolerable, causing complete breakdown. For example, carbonization of ethylene occurs under high pressure at about 300-350°C. The thermal conductivity of such organic materials is very low—less than in most common insulating materials—and thus magnifies the already difficult heat transfer problem. Unusual methods are required to solve this problem which is peculiar to the polymer industry.

Commercial reaction temperatures range from -90°C for isobutylene, to 180-270°C for high pressure polyethylene. For polymers that dissolve in their monomers, extreme viscosities are reached at low conversions, e.g., 100,000-200,000 centipoise is the viscosity of 30-per cent converted styrene syrups at the polymerization temperature of 85-90°C. For polymers which will precipitate from their monomers (polyvinylchloride, polyacrylonitrile), slurries or pastes of low mobility result at 25-35 per cent conversion.

Mass Polymerization. The term mass or bulk polymerization, as used here, refers only to monomers which dissolve their polymers (styrene, methyl methacrylate) and not those from which the polymer precipitates (vinyl chloride, vinylidene chloride, acrylonitrile). Mass polymerization is the method frequently used in the laboratory to study a new monomer and its copolymers; no extra variables are introduced and the heat removal problem is trivial provided thin (10-25 mm) glass tubing is used. However, scaling-up such a process in a practical manner presents tremendous problems. Only polystyrene and some of its copolymers are commercially manufactured today by a pure mass polymerization process.**

In the middle 1930's, Badische Anilin in Germany developed a continuous process for manufacturing polystyrene which may still be used in at least Germany and Italy.[15,24] The process consists of polymerizing styrene monomer to 30-33 per cent in stirred jacketed kettles at about 90°C and then feeding this viscous syrup to the top of a vertical jacketed tower, about 30 inches in diameter by 19 feet long, containing a few temperature regulating coils. This tower is operated at atmospheric pressure and is not quite full; thus temperatures are prevented from rising much above the boiling point of styrene monomer (145°C) until high conversions are attained, at which point temperatures rise to about 200°C. Molten polystyrene is pumped from the bottom of the tower at a rate of about 900 lb/hr by means of screw extruders. The molten strands from the extruder are cooled and then cut into transparent granules or pellets.

In the United States, the degree of temperature control attained in the German tower process has not been adequate to achieve optimum quality and versatility. Equipment has been devised[2,3,4,85,111,126] which will control the reaction temperatures more effectively to higher conversions than in the German tower process. Most of these processes use slowly revolving screws or paddles which gently agitate the viscous polymerizing syrup past tubes filled with circulating oil, thereby aiding heat transfer. A small amount of ethylbenzene may also be added to aid viscosity reduction and heat transfer. To reach very low monomer contents, devolatilization is carried out under high vacuum (29 in. of Hg) using vented twin-screw extruders or similar devices[3,5,22] (see Fig. 10.11).

Methyl methacrylate is mass polymerized to make sheets using plate glass as the mold material.[113] Also, certain allyl resins containing some trifunctional molecules for crosslinking are mass polymerized to encase or "pot" electrical components. The volume manufactured by these latter processes is very small.

---

*The difference in energy between a "standard" carbon-carbon double bond and two single bonds is about 16 kg-cal. Variations from this level are due to substituents.

**Methyl methacrylate and "allyl" casting-resin syrups are cast and polymerized in thin sections by a pure mass process, but these processes, though practiced commercially on a small scale for speciality items, will not produce molding powders economically. Methyl methacrylate molding powder is made via a suspension process.

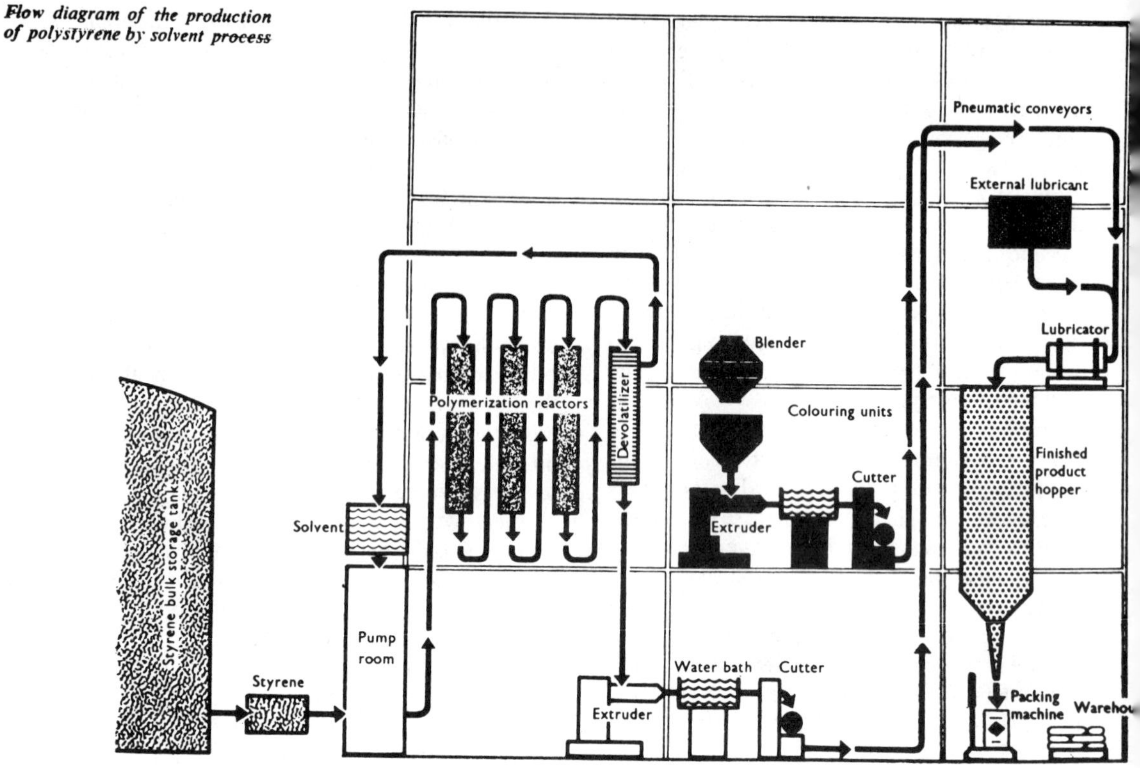

Fig. 10.11 Styrene polymerization, continuous mass process of Dystrene Ltd. [*Reproduced with permission from British Plastics,* **30,** *26 (Jan. 1957)*]

Emulsion Polymerization. Emulsion polymerization is a technique by which an addition polymer or copolymer is produced in a two-phase system. Its application requires the emulsification of the monomer in a medium, usually water, through the use of emulsifiers, such as soaps, alkyl sulfates, and alkyl sulfonates. These are supplied in addition to the other ingredients that go into most polymerizations, such as the initiator and chain transfer agents.* The use of this simple, additional reagent, the emulsifier, leads to several important characteristics which make the emulsion process unique.[127]

Three of the most important of these are:

1. The ability to form a polymer of high molecular weight at a very high rate of polymerization.
2. The maintenance of low viscosity throughout the reaction.
3. The relative ease of heat transfer.

The product of an emulsion-polymerization reactor is a latex, which is simply a stable colloidal suspension of polymer particles in a continuous phase, generally water. The small polymer spheres are much smaller than the initial monomer droplets and range in size between 0.01 and 1.0 micron in diameter (Fig. 10.12).

Emulsion polymerization was first attempted with natural gums and other hydrophilic protective colloids in an effort to duplicate natural rubber. Though patented in 1913,

---

*Chain transfer agents lower molecular weight with slight or no change in the over-all polymerization rate, i.e.,

$$M_n^{\cdot} + RH \longrightarrow M_nH + R^{\cdot}$$

where

$M_n^{\cdot}$ is a growing polymer chain; RH is a chain transfer agent, e.g., mercaptan, halogenated aliphatic hydrocarbons, hydrocarbons with active hydrogen (cumene); $R^{\cdot}$ is a new radical ready to begin generation of a new polymer chain.

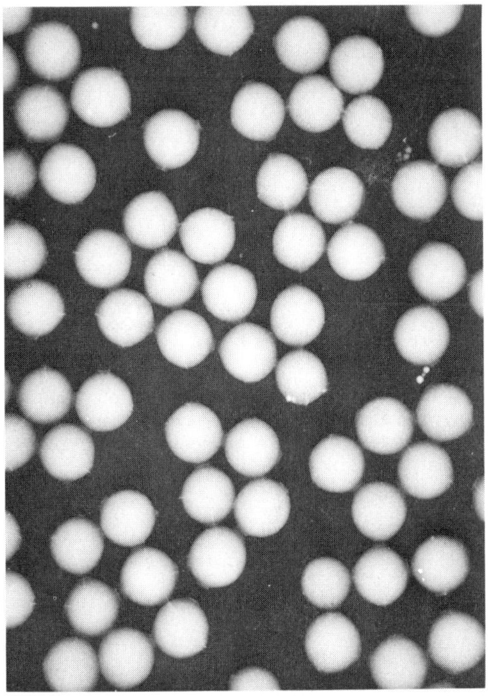

Fig. 10.12 Electron microgram of polyvinyl chloride latex particles with unusually narrow distribution of particle size. *(Courtesy G. C. Claver, Monsanto Co.)*

these methods were neither practical nor reproducible. Then in 1927, Dinsmore,[30] and Luther and Heuck[79] in Germany tried to make synthetic rubber from normal soaps and surface active agents. Later on Luther and Heuck used water-soluble initiators. Much of the development and study of emulsion technology was spurred by the need, just prior to and during World War II, for synthetic rubber, first in Germany and then in the United States where the late W. D. Harkins[53,54,55] is credited with establishing the fundamental postulates of the mechanism of emulsion polymerization.

Emulsifiers from micelles when they are added to water above a certain concentration, which is called "the critical micelle concentration." This is a characteristic of the particular emulsifier. Micelles are aggregates of emulsifier molecules whose hydrophobic ends are clustered together at the center and whose hydrophilic ends extend into the water phase. The micelles, which have the ionic character of the emulsifier from which they are formed, are either spherical or lamellar[72] with dimensions on the order of 25-50 Å, depending on the emulsifier used, and number around 10^{17}-10^{18} micelles per cm^3 of water. The micelles "solubilize" or absorb monomer, swelling slightly in the process. Because of their size they are able to account for only a small fraction of the total monomer. The remainder of the monomer is present as emulsified monomer droplets 10-30 thousand angstroms in size, stabilized like the micelles and numbering about 10^{10}-10^{11} particles per cm^3 of water.[43] If a dissociative free-radical initiator is added to the aqueous phase, the free radicals generated by thermal decomposition of the initiator diffuse through the water to the micelles where they initiate a polymer chain. The micelle now becomes a polymer particle. Harkins established that the micelles are the source of polymer particles and that these particles are then the locus of polymerization. Termination of a growing polymer chain occurs when a new free radical enters the particle and terminates the chain. Harkins showed that:[54]

1. There are more polymer particles produced than there are emulsified monomer droplets initially.
2. The sizes of the polymer particles are several orders of magnitude smaller than the emulsified monomer droplets.
3. The emulsified monomer droplets, if tested for polymer formation during the reaction, show only insignificant amounts of polymer.

Harkins postulated that the emulsified monomer droplets must serve as a reservoir of monomer supplying the growing particles. A pictorial representation of the emulsion polymerization system very early in the reaction is shown in Fig. 10.13.[127] A typical emulsion polymerization recipe would be as follows (in parts by weight):

Styrene	100
Sodium oleate	6
Potasium persulfate	0.3
Water	150

The reaction temperature is 70°C.

The kinetics of emulsion polymerization is considerably more complex than that of mass polymerization. The initiator and emulsifier concentrations, the concentration of monomer present, the number of particles formed and their size, and the effect of temperature on the decomposition of the initiator and the propagation reaction of the polymer chain must all be considered. Extremely important work providing a quantitative understanding of the emulsion-polymerization mechanism has been carried out by Smith and Ewart,[124] and more recently, by Gordon.[44] Sundberg[127] has provided an excellent review of the current quantitative understanding of emulsion polymerization and has developed a set of mathematical models based on Harkins' postulates which provide a means of predicting the physical characteristics of the latex as well as the chemical characteristics of the polymer.

Modern emulsion polymerization solves the major problem of heat transfer admirably. Rapid polymerization rates may be obtained at precisely defined low temperatures while high molecular weights are maintained, particularly when "redox" catalyst systems are used.[16] Also, the use of water-soluble catalysts frequently promotes stability of the latex particles which are much smaller than those obtained by dispersion of a polymer through intensive mixing. (Fig. 10.13) Continuous polymerization is possible in tubular reactors, in towers, or in a series of agitated reactors.

In spite of its many advantages, however, emulsion polymerization has not been used extensively to make solid plastics.* Since it is commercially impractical to remove the emulsifying agents completely, emulsion-produced polymers are poorer in haze, color, heat stability, and electrical properties than products made by other processes. Also, some water-soluble initiators such as potassium persulfate make comparatively unstable chain-end groups for at least vinyl chloride and vinylidene chloride polymers. In addition, the cost of obtaining a dry, pelleted polymer from a latex is considerably higher than for mass or suspension polymer. First, the emulsion must be coagulated. This may be accomplished in many ways, such as adding acid or other electrolytes, vigorous mixing, ultrasonics, freezing, forcing the latex through jets. The coagulated crumb will ordinarily contain most of the water originally present in the latex, e.g., from 40–65 per cent water. "Synerizing," i.e., heating near the softening point, the coagulated crumb can squeeze out some of the water inclusions.[120] This water is not on the surface but must slowly diffuse out through the polymer. Drying is therefore slow and expensive. After drying, the crumb (or sometimes powder) must be densified into pellets. Drum drying[24] and spray drying (commercially used for polyvinyl chloride plastisols) achieve coagulation and drying in a single step, but both these methods are comparatively expensive.

Emulsion polymerization equipment is almost always glass-lined because plating or scaling of polymer on metals is frequently very

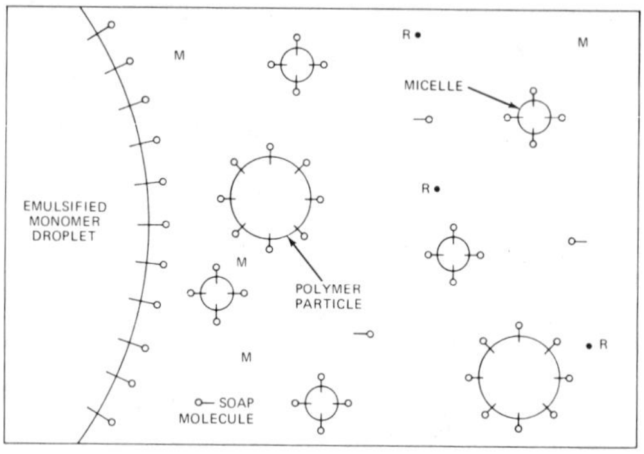

Fig. 10.13 Schematic representation of emulsion polymerization: M = solubilized monomer; R$^{\cdot}$ = catalyst, monomer, or low molecular weight polymer radical. (*Courtesy D. C. Sundberg.*[127])

*Latices commercially processed to plastics include: polyvinyl chloride plastisol resin (produced by spray drying the latex); "Kralastic," "Cycolac," and "Lustran" (styrene-acrylonitrile graft polyblends with synthetic rubbers);[10] "Teflon" (polytetrafluoroethylene); and possibly, "Kel-F" (polytrifluorochloroethylene). Many other latices are produced in large volume to be sold as such for use in formulations for surface coatings, such as textile finishes, adhesives, paper binders and coatings, waxes and polishes.

It is important to distinguish between plastics, rubbers, resins for surface coatings, and synthetic fibers; the latter three industries utilize emulsion technology extensively.

severe. In the laboratory, tumbling 6-ounce soft-drink bottles may be used for pressures up to 150 psig.*

Suspension Polymerization. For a long time suspension and emulsion polymerization were considered synonymous. Then Mark and Hohenstein[81] pointed out the characteristics which today serve to separate the two processes: emulsion polymerization utilizes comparatively low molecular weight substances of high surface activity as emulsifiers, i.e., substances which markedly lower the interfacial tension; suspension polymerization, on the other hand, requires finely divided solids or a protective colloid,† which, though they preferentially locate themselves at the interface, do not greatly lower the surface tension. In the emulsion process, initiation begins in the continuous water phase; in suspension polymerization the catalyst is soluble in the monomer phase and initiation begins there. The most apparent physical effect resulting from these essential differences is the much larger size of the polymer particle produced by suspension polymerization, e.g., the diameters of suspended particles are usually between 50 and 2,000 microns (about 0.002-0.08 inch), while those of most emulsion particles are between 0.01 and 1.0 micron (Fig. 10.14). The large suspension particles are easily separated from the water phase by filtration or centrifugation, whereas the smaller emulsion particles are separated with greater difficulty. Also, the importance of agitation differs in the two systems. In emulsion polymerization a certain minimum level of agitation is required to break up the pure monomer phase, but after this minimum level is attained agitation may vary greatly up to the point where it causes coagulation of the emulsion particles; nor does agitation ordinarily affect particle size or particle-size distribution as it does in suspension polymerization.

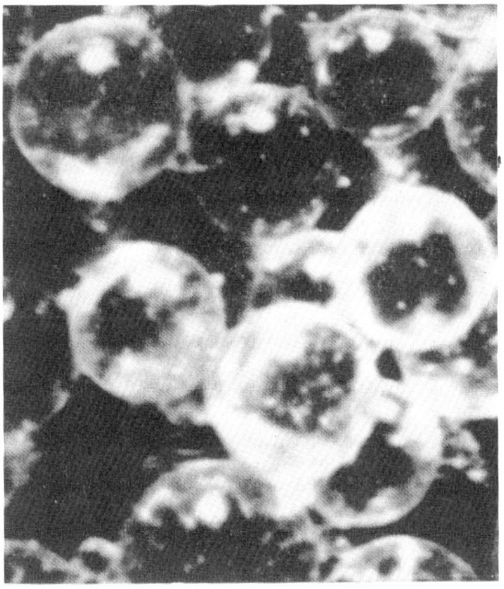

Fig. 10.14 Polystyrene beads. *(Courtesy the Monsanto Co.)*

Many types of finely divided solids, frequently combined with protective colloids or very small quantities of low-molecular weight surface active agents have been used as suspension agents. These include commercial tricalcium phosphate (hydroxy apatite),[48] specially precipitated barium sulfate, calcium oxalate, bentonite clay, and aluminum hydroxide. Water-soluble high polymers are also effective, e.g., polyvinyl alcohol, carboxymethylcellulose, hydroxyethylcellulose, polyacrylic and polymethacrylic acid, and copolymers of acrylic and methacrylic acid and their esters.

The use of an oil-soluble catalyst gives reaction kinetics and mechanisms essentially identical to those of mass polymerization rather than of emulsion polymerization. Therefore, reaction rates are slower in suspension polymerizations than in emulsion polymerizations aiming for the same molecular weight product.

*Permission should be obtained from the appropriate company for such use, and the bottles should be kept segregated.

†Protective colloids are water-soluble substances of great molecular weight. They usually contain sufficient amounts of hydrophobic groups to cause them to locate preferentially at the oil-water interface; however, because of their great molecular weight the interfacial tension is not markedly lowered. Water-soluble polymers that have been utilized for suspension polymerization include gelatin, natural water-soluble gums (usually polysaccharides), polyvinyl pyrrolidone, polyvinyl alcohol, sulfonated polystyrene, polyacrylic acid and its salts, and polymethacrylic acid and its salts.

Suspension polymerization is of great importance in the plastics industry. It not only solves the heat-transfer problem but also permits simple and complete separation of the polymer from water and suspending agents. Haze, water absorption, color, and electrical properties of polymer made by a good suspension process will essentially equal those of a polymer made by a laboratory or commercial mass process.

A typical suspension process would be carried out as follows:

Into a glass-lined, stirred reaction vessel are charged 200 parts freshly distilled water and 0.1 part of a water soluble interpolymer of acrylic acid and 2-ethylhexyl acetate. The atmosphere in the reaction vessel is swept free of oxygen with nitrogen and then 100 parts of styrene and 0.1 part benzoyl perioxide are charged to the reaction vessel. The reaction mixture is heated and stirred for 6 hours at 90°C and then for 8 hours at 130°C. The polymer is obtained at close to 100% conversion, in the form of small spherical beads of an average diameter of about 3 mm, suspended in the water as a slurry.[94]

The slurry is transferred to a slurry hold tank, the polymer is separated by centrifugation, washed, and dried. (Fig. 10.15). From the dryer the beads are mixed with colorants, stabilizers, plasticizers (if polyvinyl chloride), etc; melted in a Banbury mixer or an extruder; extruded, cooled, and finally chopped into pellets for bagging as a molding powder.

Plastic molding powders that are commercially made via a suspension process include:

1. Polymethylmethacrylate.[63]
2. Polystyrene and rubber-modified polystyrene.[129]
3. Polyvinylidene chloride-vinylchloride copolymer.
4. Polyvinyl chloride and copolymers.
5. Polyvinyl acetate.
6. Poly(styrene-acrylonitrile) copolymer.
7. Rubber-modified styrene-acrylonitrile copolymer, i.e., ABS.[10]

As far as is known, commercial suspension processes are all batch, but several designs for continuous suspension polymerization have been patented.[11,119] A semicontinuous suspension polymerization process is employed by Wacker-Chemie[135] and Tenneco for the production of PVC paste resin. The reactors

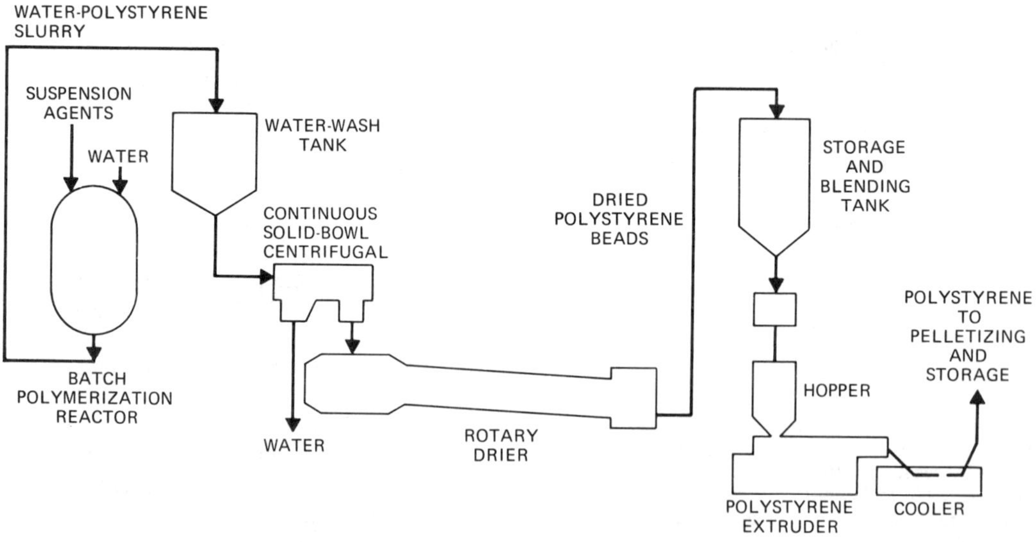

Fig. 10.15 Flow sheet for polystyrene by suspension process. (*Reprinted by special permission from Chem. Eng.,* **65**, *100 (Dec. 1958); copyright 1958 by McGraw-Hill, Inc.*)

may be glass-lined or of stainless steel and up to 16,000 gallons in volume. The difficulties of commercially obtaining a "clean" bead with no contamination or agglomeration are treated by Tromsdorff.[133] In vinyl chloride polymerization a further difficulty is that of obtaining a porous bead that will soak up plasticizer rapidly and uniformly in subsequent processing (Fig. 10.16). The commercial art of suspension polymerication is complicated and many details have not been disclosed. The complete suspending-agent recipe is the most important part of the process, although agitation also plays a vital role.

Solution Polymerization. Addition-type polymerization carried out in a solvent which will dissolve the monomer, the polymer, and the polymerization catalyst, is termed solution polymerization. Polymerization of pure monomers that dissolve their polymer is solution polymerization provided the polymerization is halted at a low enough conversion to maintain a reasonable level of fluidity. The eventual high-viscosity levels of complete bulk polymerization are avoided in solution polymerization, and heat transfer is greatly facilitated. By proper adjustment of the reaction temperature and pressure, reflux of monomer or solvent may remove heat. (Because of the high molecular weight of polymers, boiling elevation is insignificant until high-per cent conversions by weight are reached.)

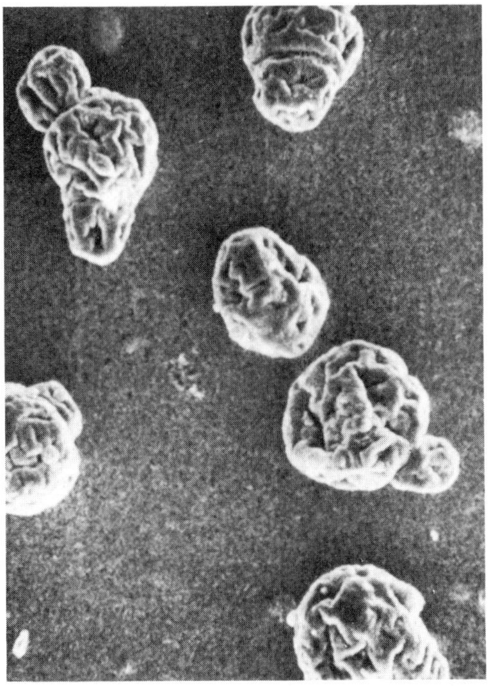

Fig. 10.16 Polyvinyl chloride suspension particles showing porosity. Taken with a scanning (surface reflecting) electron microscope. (*Courtesy G. C. Claver, Monsanto Co.*)

Solution polymerization will generally slow down the reaction rate and lower the molecular weight, particularly when free radical initiators are used. Depending on the polymer, such effects may be commercially desirable or undesirable. In some instances the rate at which monomers will graft onto preformed polymer molecules can be lessened; this leads to lower levels of chain branching and crosslinking.

Solution polymerization, however, has been used sparingly in the plastics industry, primarily because most of the polymers marketed to date have been successfully polymerized by the more economical mass and suspension processes. The cost of solvent removal and recovery may be appreciable. Machinery that can successfully and economically devolatilize a solvent from a polymer is not yet standardized. Although vacuum drum driers will do the job they are extremely costly for the rates attainable. Multiple-vented single or twin screw extruders have been used commercially though a vacuum in excess of 29 in Hg is usually required.[3] Another commercially used technique involves strands or sheets falling through a vacuum chamber.[5] Production rates are slow because of the slow diffusion rates in polymers.

A solution process for making styrene-acrylonitrile copolymers and other copolymers of styrene[51,52,68] which is probably being used for the commercial production, has been disclosed. (Styrene-alpha-methyl styrene copolymers are probably also being made commercially by a similar process, but here the solvent is merely the monomer.) For styrene-acrylonitrile copolymers, ethylbenzene or toluene at 30 per cent concentration by weight are the preferred solvents; they ensure a favorable balance between the reaction rate and chain transfer, i.e., reaction rates and molecular rates are both lowered to commercially desirable levels. The process is a continuous one with the reactant (70/30 styrene/acrylonitrile) and solvent being fed to a recycling loop of a 20 per cent by weight polymer solution maintained at 150°C by a heat exchanger. The reaction rate under these conditions is 27 per cent/hour. A form of twin-screw extruder called the "Plastruder"[50] recovers the solvent and the unreacted monomer for recycle to the reactor.

Solution polymerizations of the addition type using ionic catalysis have been studied in the laboratory without notable commercial success as yet, except in the rubber industry where nearly all plant expansions are currently using this route (see Chapter 9). Cationic catalysts, e.g., $AlCl_3$, HF, are used extensively to polymerize olefins to a very low degree in the petroleum industry (Chapfer 14). Also, many tons of low-molecular weight petroleum resins and asphaltic polymers are produced using cationic catalysts, e.g., concentrated sulfuric acid, aluminum halides, ferric chloride, boron halides.[37] To describe all these resins and their manufacture is beyond the scope of this chapter.

Solvent-Nonsolvent Polymerization (Precipitation Polymerization). Some polymers are insoluble in their monomers, e.g., polymers of vinyl chloride, vinylidene chloride, acrylonitrile, chlorotrifluoroethylene, ethylene (at least at lower temperatures and pressures). If polymerization is initiated in the pure monomers a polymeric precipitate forms. Sometimes other solvents are added to increase or decrease solubility of the monomer in the solid polymer. Surprisingly, with free radical catalysts, molecular weights are frequently much higher in such systems than in nearly equivalent homogeneous systems. The reason for this molecular weight effect is probably similar to the explanation given for the higher molecular weight of emulsion polymers, i.e., initiation occurs primarily in the continuous phase so that each precipitated particle is likely to have only one growing radical, to which fresh monomer can readily diffuse while other polymeric radicals cannot. Termination by combination of two polymeric radicals, therefore, is infrequent. And because of the relative immobility of the growing chains, even if two precipitated particles should fuse, the polymeric radicals will probably not combine.

Chain termination mechanisms have not been explored in detail for any particular system, but chain transfer to monomer, to the polymeric chain, to the solvent, or to a chain transfer agent is a possibility as is disproportionation or termination by catalysts, inhibitors, or retarders. If chain transfer to polymer takes place as it does with polyvinyl chloride, branched structures develop.

Solvent-nonsolvent polymerization was the method selected for preparing the vinyl chloride-vinyl acetate copolymers ("Vinylites") first introduced commercially in 1935. The monomer itself is the principal nonsolvent. The initial patents[33,105] have expired, but none of the many other manufacturers of vinyl chloride polymers and copolymers in the United States use a similar process. Except for a very few patents, little has been disclosed. The material is polymerized at moderate temperatures, 30-70°C, and moderate pressures of 80-140 psig using reactive free radical catalysts (aliphatic acid peroxides). At about 25-30 per cent conversion, the polymer adsorbs most of the monomer and heat transfer is substantially lowered. The unreacted vinyl chloride monomer, which is a gas at normal temperature and pressure, must be recovered by flash drying or other methods. A recent French process[9,130] uses reflux in a ribbon-agitated batch reactor of special design to achieve sufficient heat transfer to permit conversions of about 80-85 per cent.

Recently, polyoxymethylene (polymerized formaldehyde) has been developed commercially using basic ionic catalysts, such as quarternary ammonium bases, trialkyl amines of high molecular weight, and trialkyl phosphines.[1,115] Thoroughly dried hydrocarbon solvents which do not dissolve the polymer are used, e.g., hexane, cyclohexane, and toluene. Pure formaldehyde completely free of water and acids must be used to obtain a polymer of high molecular weight with reasonably narrow molecular weight distribution and good thermal stability. After polymerization is completed the polymer is stabilized by acetylating the end groups at high temperatures, 700°F, with acetic anhydride in the presence of pyridine or an alkali acetate as a catalyst.[35] Additional stability can be incorporated by copolymerizing ethylene oxide with the formaldehyde.

A continuous low-temperature (32°F) solvent-nonsolvent process has been described[89,104,142] for polytrifluorochloroethylene ("Kel-F"). The reactive, free-radical catalyst, bis-trichloroacetyl peroxide, dissolved in trichlorofluoromethane, is added continuously to a stirred, jacketed, stainless steel reactor along with the monomer. The polymer is separated from the slurry containing monomer and trichlorofluoromethane by means of a continuous filter or centrifuge, and it is then dried further in a vacuum drier to remove all the remaining monomer. The unreacted monomer and diluent are recycled to the reactor with a portion being sent to a repurification distillation column.

Low-pressure polymerization of ethylene and propylene with heterogeneous catalysts could be described as a solvent-nonsolvent type of polymerization; however, the insoluble solid catalysts used here make the reaction kinetics and purification methods substantially different. These processes will be described later in a separate section.

Condensation Polymerization

Generalizations concerning processes and equipment for the various commercial condensation polymerizations would be meaningless. These subjects will be treated in the following sections where each type of condensation polymer is discussed separately.

Phenol-Formaldehyde Resins. It is not surprising that the first completely synthetic plastic was made from the common and highly reactive multifunctional chemicals, phenol and formaldehyde. As early as 1870, Baeyer, in Germany, observed the reactions between phenol and formaldehyde and characterized the phenol alcohols which were formed. Around the turn of the century Delair, Smith, and Leback attempted to utilize the resinous reaction products as shellac substitutes. Beginning in the early 1900's Dr. Leo Baekeland discovered that useful moldings could be made if the final stages of the reaction were carried out under heat and pressure, preferably in the presence of a suitable filler, e.g., sawdust, wood flour, and cotton cloth. Since that time phenolic resins have become the "work-horse" of the plastics industry. They are used in electrical components, insulating varnishes, industrial laminates, binders, etc.

The chemistry of this complex reaction is, even today, not completely understood. When crosslinked structures of large size are formed, isolation and characterization of these structures is very difficult. Much investigation has been done, however, and a great deal is known—so much, in fact, that the brief description given below suffers from oversimplification.

The first series of reactions is the formation of phenol alcohols under the influence of either acids or bases.

(1)

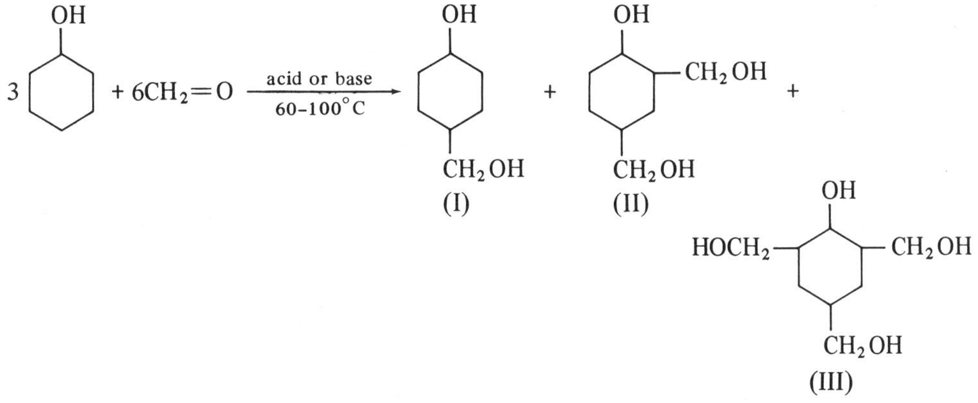

Next, these phenol alcohols condense in a complex manner. Using acid catalysts the following reactions are favored.

(2) 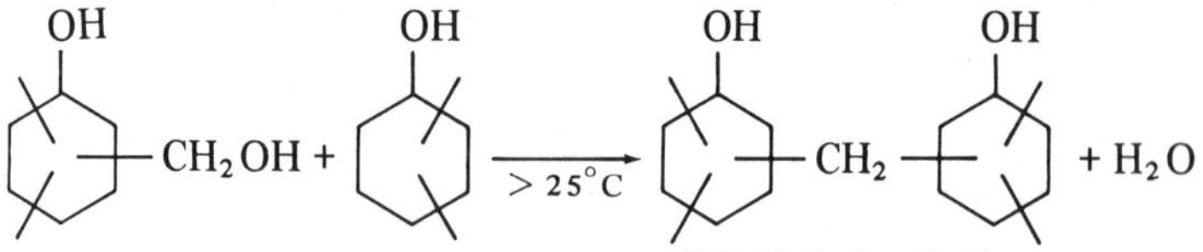

(IV) Methylene bridge

(3) 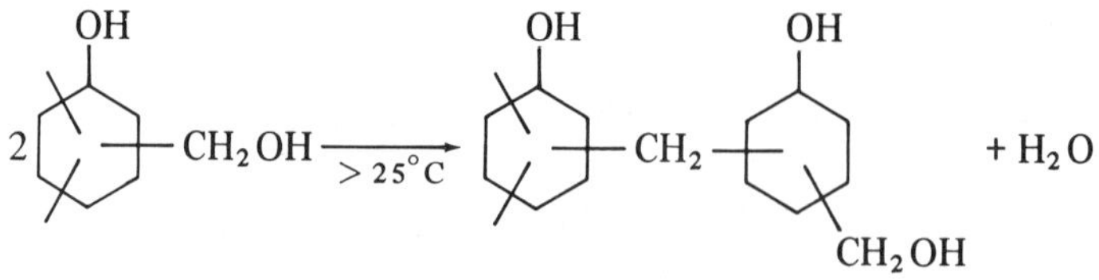

(V) Methylene bridge

(4) 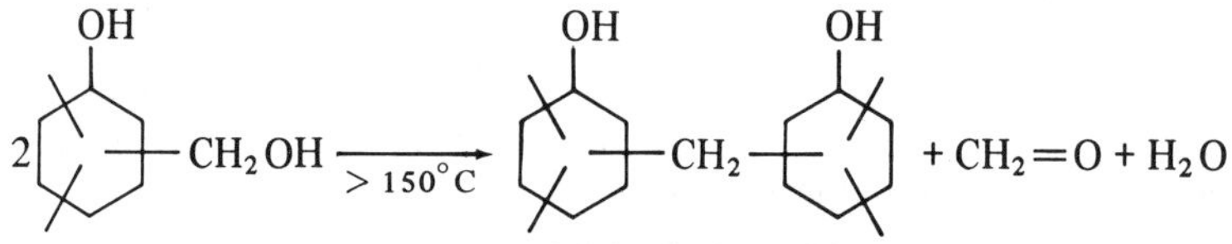

(VI) Methylene bridge

(5) 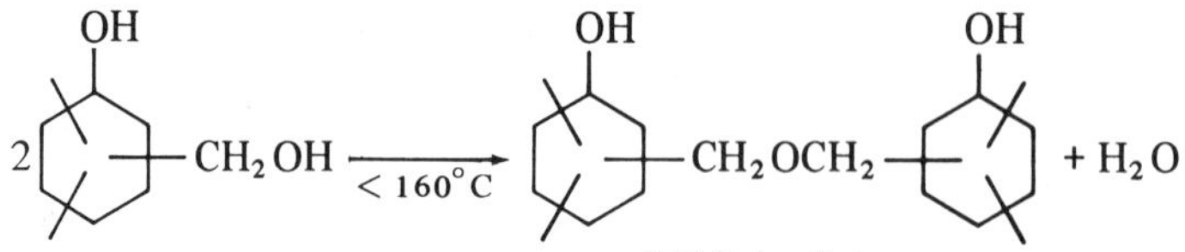

(VI) Ether linkage

Reactions (2) and (3) with the benzene nucleus proceed quickly with acid catalysts even at temperatures below the point where phenol alcohols form rapidly. Hence, under acid-catalyzed conditions unreacted phenol alcohols cannot ordinarily be isolated, but are intermediates. Reaction (5), though catalyzed by weakly acidic conditions, is slower than Reactions (2) and (3) and occurs, therefore, at neutral pH's and/or when all of the *ortho* and *para* positions on the phenol nuclei are occupied by either alcoholic groups or unreactive ones. At high temperatures (around 160°C) the ether linkage becomes unstable, losing formaldehyde and reverting to a methylene bridge (Reaction (4)).

With basic catalysts Reaction (5) is slow at the high pH's used commercially. The condensation reactions (2), (3), and (4) also take place, but at a much slower rate than under acid catalysis. These reactions are catalyzed by a surprisingly low concentration of strong base; and further amounts of base do not increase their rate. Since the formation of phenol alcohols (Reaction (1)) is more strongly catalyzed by bases, pure or partially condensed alcohols can be isolated.

Commercial equipment for the above reactions usually includes jacketed anchor-agitated steel or stainless steel kettles equipped with vacuum reflux condensers, although scraped-wall heat exchangers might be used for a continuous process.

Two major types of commercial molding powders are based on either one- or two-stage resins. One-stage resins are made with basic catalysts, e.g., 1 to 2 parts calcium hydroxide to 100 parts phenol, and use a formaldehyde phenol mole ratio between 1.1 and 1.5. Formalin or paraformaldehyde* may be used to supply the formaldehyde. Phenol alcohols may be formed in a few minutes at 100°C or in a few hours at about 70°C, depending on the heat-exchange capacity of the reaction equipment, the catalyst concentration, and the type of resin desired. After most of the formaldehyde has combined with the phenol, the water is removed and condensation is continued under 27–29 inches of Hg vacuum at approximately 75°C. When the desired level of condensation has been reached the entire contents of the kettle must be cooled uniformly and rapidly to ensure a uniform degree of polymerization. One commercial method involves dumping the entire contents onto a large steel cooling floor through which cold water flows. After cooling, the resin is ground to a fine powder and blended about half and half with wood flour or other fillter and colorants plus small quantities of mold release agents and/or cure accelerators (magnesium oxide, calcium oxide). This blend is then densified and fused on hot mill rolls, cooled, and ground into molding powder (Fig. 10.17).

*In the past formalin (a 37 per cent solution of formaldehyde in water) was almost always used, but with the advent of cheap solid paraformaldehyde (a solid polymer of formaldehyde) from natural gas, the latter is frequently used to increase capacity and save on subsequent dehydrating costs.

One-stage resins, though still manufactured, are not as important for molding powders as two-stage resins (novalaks) which usually have superior flow characteristics. The polymerization equipment used for two-stage resins is essentially the same as that used for one-stage resins although acid resistant kettles are preferred.

The first stage of a two-stage resin may be prepared as follows:

One hundred parts of phenol and 0.5 part of concentrated H_2SO_4 are charged into an anchor-agitated stainless-steel reaction vessel equipped with a reflux condenser. The charge is heated at 100°C and 69 parts (0.8 mole CH_2O/mole of phenol) of 37 per cent forma-

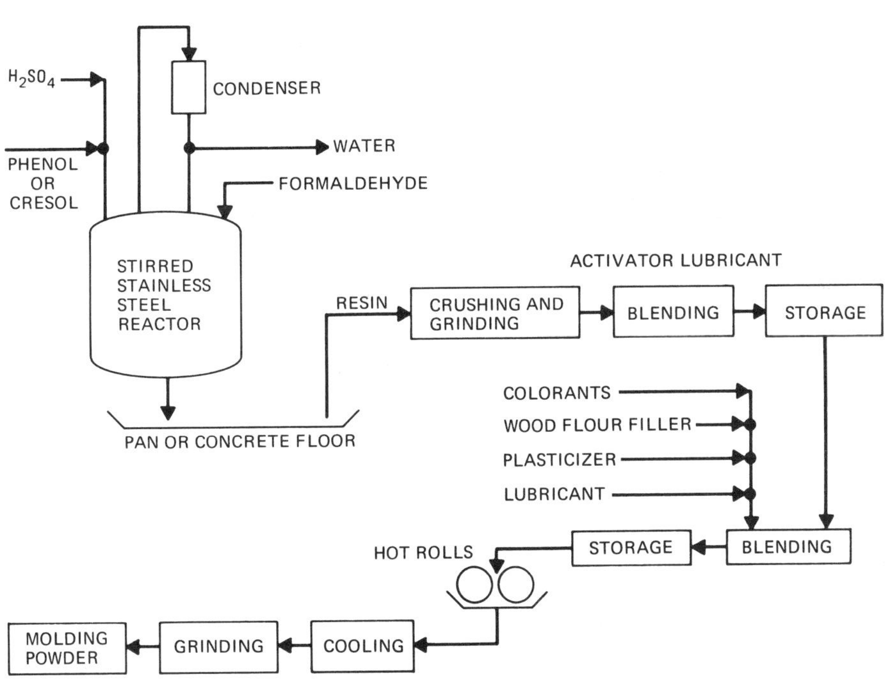

Fig. 10.17 Flow diagram for phenol-formaldehyde molding powder.

lin added at a rate compatible with the heat-exchange capacity of the reflux condenser. After all the formalin has been added, the charge is refluxed for an additional 30 minutes, then dehydration is begun by switching the condensate flow from reflux to a distillate receiver. Vacuum is applied as the boiling point begins to rise until a vacuum of 28 inches of mercury and temperature of 215°F are reached, corresponding to approximately 4 per cent free phenol in the final resin. The charge is neutralized with lime and dumped to a cooling floor. The solid resin is then ground into a fine powder.

The large heat of reaction*,[46] and the rapid rate of condensation under acid conditions makes the above procedure desirable for safety reasons. The resulting resin is thermoplastic and is termed a novalak. To make a thermosetting molding powder, hexamethylenetetramine† crystals (about 10 to 15 parts/100 parts of novalak resin) are blended with wood flour, colorants, mold release agents, and cure accelerators using a fifty/fifty ratio of total resin including "hexa" to total fillers. After blending, the resin is fused on hot mill rolls, cooled, and ground into a molding powder in a manner similar to the method used for one-stage resins.

To explain the final cure of one- and two-stage resins, a highly reactive quinone methide is postulated as an intermediate which combines to form complex condensed rings of red or brown color. These quinone methides may also explain the reaction of phenolic resins with the double bonds in rosin, drying oils, rubber, and other addition-type monomers. Addition polymerization with rosin and drying oils has been used extensively to increase the oil solubility of phenolic-resin varnishes.

To explain the reaction of hexa, the following principal reactions are postulated:

(6)

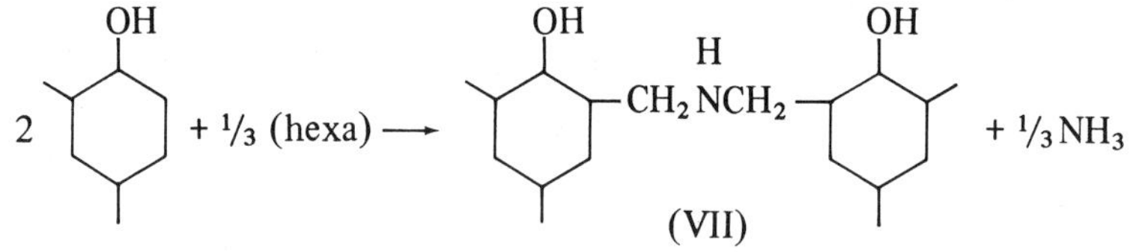

(VII)

(7)

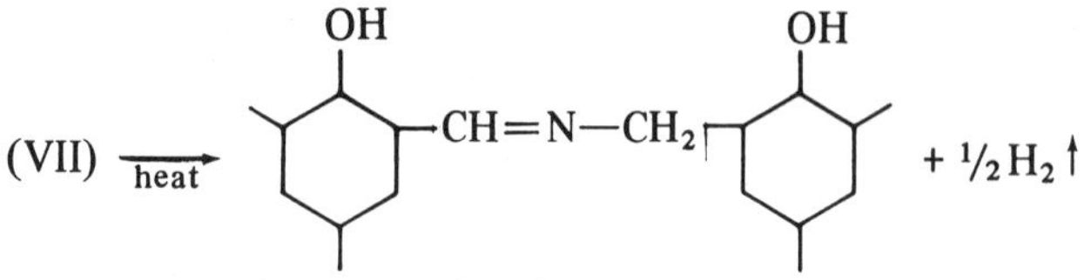

(VIII) yellow

(8)

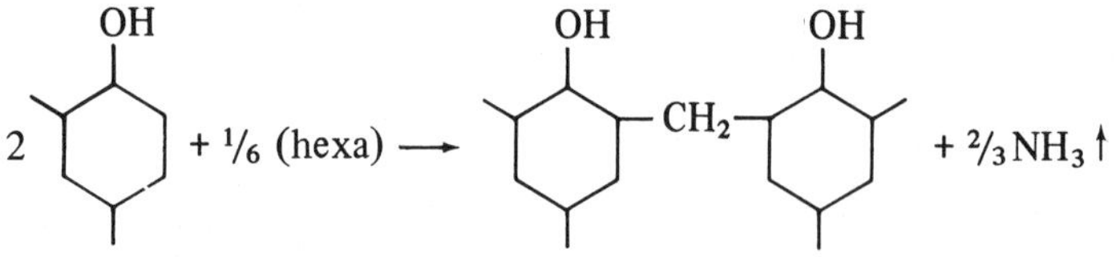

(IV)

*For the complete condensation of formaldehyde to form a methylene-bridge structure plus liquid water, about 21 kcal/mole of formaldehyde are released. This can be split into 4.1 kcal/mole for the formation of the phenol alcohol and 16.9 kcal/mole for the formation of the methylene bridges [reaction (2)].

†Commonly called "hexa," it is the nearly instantaneous reaction product of ammonia and formaldehyde,

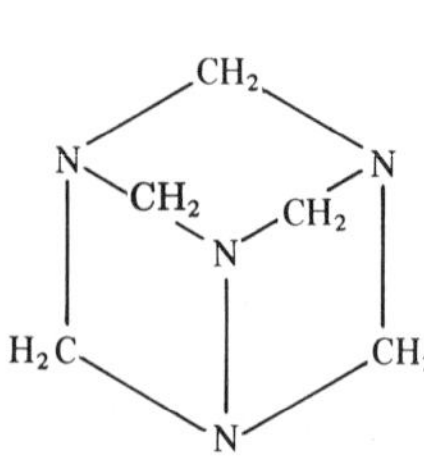

Many more reactions during curing have been proposed, and undoubtedly many of them do occur to some extent. Several authors[20,45,82,87,110,145] have given excellent summaries of experimental evidence for the above and other possible reactions.

Aminoplast Resins. Urea was known to react with formaldehyde as early as 1884, but commercial molding powers were not developed until 1926. As in the case of phenolic resins, commercial development preceded an understanding of the basic chemistry involved. Again, this is primarily due to the difficulty of characterizing complicated crosslinked structures. The following equations oversimplify a very complex picture:[13,29,66,134,136]

$$(1)\quad CH_2{=}O + H_2N\overset{\overset{O}{\|}}{C}NH_2 \rightleftarrows$$

$$H_2N\overset{\overset{O}{\|}}{C}\overset{H}{N}CH_2OH \quad \text{(I)}$$

$$\Big\updownarrow + CH_2{=}O$$

$$HOCH_2\overset{H}{N}\overset{\overset{O}{\|}}{C}\overset{H}{N}CH_2OH \quad \text{(II)}$$

$$\Big\updownarrow + CH_2{=}O$$

$$(HOCH_2)_2N\overset{\overset{O}{\|}}{C}\overset{H}{N}CH_2OH \quad \text{(III)}$$

Reaction (1) is catalyzed by base or acid and is an equilibrium type. Under commercial conditions very little trimethylolurea is formed. If acid catalysts are used the methylolureas cannot be isolated since further condensation occurs as follows:

$$(2)\quad \left[-\overset{\overset{O}{\|}}{C}\overset{R'}{N}CH_2OH + H_2N\overset{\overset{O}{\|}}{C}\overset{R'}{N}-\right] \longrightarrow \left[-\overset{\overset{O}{\|}}{C}\overset{R'}{N}CH_2\overset{H}{N}\overset{\overset{O}{\|}}{C}\overset{R'}{N}-\right] + H_2O$$

$$\text{(IV)}$$

$$(3)\quad \left[-\overset{\overset{O}{\|}}{C}\overset{R'}{N}CH_2OH + H\overset{R}{N}\overset{\overset{O}{\|}}{C}\overset{R'}{N}-\right] \longrightarrow \left[-\overset{\overset{O}{\|}}{C}\overset{R'}{N}CH_2\overset{R}{N}\overset{\overset{O}{\|}}{C}\overset{R'}{N}-\right] + H_2O$$

$$(4)\quad 2\left[-\overset{\overset{O}{\|}}{C}\overset{R'}{N}CH_2OH\right] \longrightarrow \left[-\overset{\overset{O}{\|}}{C}\overset{R'}{N}CH_2OCH_2\overset{R'}{N}\overset{\overset{O}{\|}}{C}-\right] + H_2O$$

$$\text{(V)}$$

$$\text{where } R = -CH_2OH \text{ or } -CH_2\overset{R'}{N}\underset{\underset{O}{\|}}{C}\overset{R'}{N}R$$

$$R' = R \text{ or } H$$

With equal concentrations of possible reaction sites, Reaction (2) with an unsubstituted amide hydrogen to form methylene bridges (IV) and Reaction (4) to form ether linkages (V) are favored over Reaction (3) with a substituted amide hydrogen to form methylene bridges. Under basic conditions Reactions (3) and (4) are not favored and Reaction (2) will proceed very slowly.

At the higher temperatures and acid conditions which exist during molding, the ether linkages are unstable, breaking to form methylene bridges and formaldehyde. For this reason low formaldehyde/urea ratios of about 1.5 are used in molding power or laminating resins to minimize ether formation and subsequent shrinkage and cracking during molding or after long use.

Melamine was an expensive laboratory curiosity until about 1939. Low cost commercial manufacture was then stimulated when the outstanding properties of melamine resins were discovered (Table 10.1).

Melamine has the following structure. The possibility of many resonating configurations account for its excellent thermal stability.

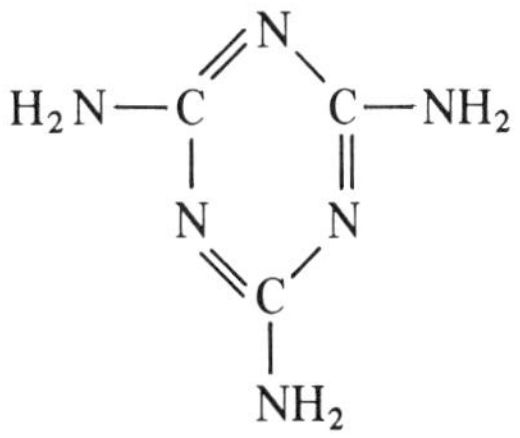

The chemistry of the condensation of melamine with formaldehyde is analogous to that of urea, with the following exceptions:

1. Six reactive hydrogen atoms exist, three

of which are the highly reactive unsubstituted amine* type.

2. The reactivity of these hydrogens is greater than that for urea, and methylene bridges are readily formed under basic conditions.

3. Since the melamine molecule contains three highly reactive sites it can more readily form crosslinked structures than can urea with only two highly reactive sites, e.g., formaldehyde/melamine ratios for molding and laminating resin are usually 2.5 compared to 1.5 for urea.

4. The methylolmelamines are much more resistant to dissociation into formaldehyde and melamine. The hexamethylol compound is readily formed, and the di- and trimethylol compounds do not dissociate rapidly.

The equipment and processing steps for the preparation of melamine and urea resins are very similar to those described earlier for phenolic resins. Corrosion resistant equipment is used to retain the excellent color inherent in these resins.[97] Although water may be removed by simple vacuum distillation as it is for phenolic resins, spray drying is frequently the method employed. If the laminates or molding powders are made at the resin manufacturing site, impregnating the laminating paper or the molding powder filler with the wet resin is the most economical method; followed by drying at atmospheric pressure in tunnel or tray driers. The dried impregnated sheets or molding powders are then completely cured under heat and pressure as are phenolic molding powders. The spray-dried, low molecular weight resins used for laminates are redissolved in water and alcohol to make an impregnating solution. Fillers are ordinarily highly refined α-cellulose and cotton fibers which retain the excellent color inherent in the base resins. Paper with a naturally high wet strength is used for laminates.

The adhesive, textile, paper-treating, and surface-coating industries comprise a slightly larger market for aminoplast resins than do the molding powder and laminating markets. To ensure compatibility with surface coatings, butyl ethers are usually formed with butanol, using higher formaldehyde ratios to increase ether formation. Other nitrogen-containing chemicals used in aminoplast formulations include dicyandiamide, substituted melamines, cyclic ethylene urea, and thiourea.

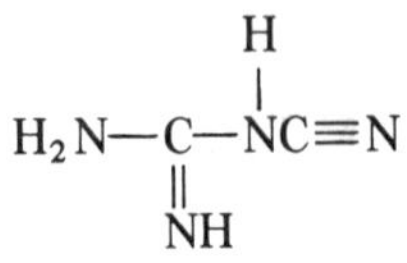

Dicyandiamide

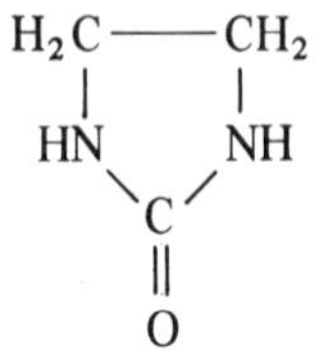

Cyclic ethylene urea

The manufacture of melamine is discussed in Chapter 5, Synthetic Nitrogen Products.

Polycarbonates. Condensation polymers need not be thermosetting. Polycarbonates, which are condensation polymers, are linear thermoplastic polyesters of carbonic acid with aliphatic or aromatic hydroxy compounds. They may be represented by the general structure.[14,24,116,117,132]

$$\mathrm{H}\!\left[\mathrm{ORO}\overset{\overset{\mathrm{O}}{\|}}{\mathrm{C}}\right]_n\mathrm{OROH}$$

where R is normally the hydrocarbon radical in bisphenol A

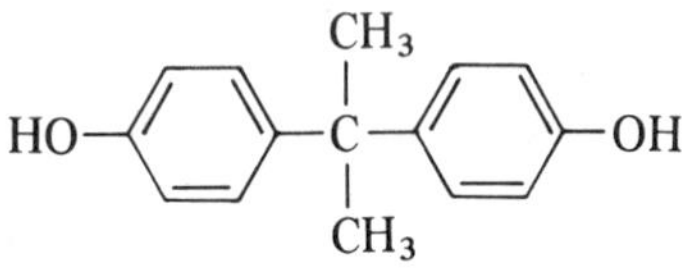

Though Einhorn[36] first reported preparing linear polyesters of carbonic acid in 1898, polycarbonates are relatively new polymers industrially, with the first commercially important patents being issued in 1956[118] and 1959.[40] The most readily available monomer is bisphenol-A and the polycarbonate that derives from this monomer has been found to have the best balance of properties. Hence, when full scale commercial production began in Germany in 1959 and in the U.S. in 1960, it was bisphenol-A polycarbonate that was produced. The polymer found quick applications in cast-film form as electrical foil and as

*The $-NH_2$ groups in melamine behave even more like amide groups than do the $-NH_2$ groups in urea.

a base for photographic film, and in pellet form as an injection molding and extrusion compound for use in parts for electrical and electronic components.[24]

Polycarbonates can be broadly divided into aliphatic polycarbonates, aliphatic-aromatic polycarbonates, and aromatic polycarbonates. Bisphenol-A polycarbonate is an aromatic polycarbonate. As free carbonic acid, H_2CO_3, cannot be used directly but only as a derivative, a classical esterification reaction where the acid is heated with an aliphatic or aromatic alcohol is not possible. Aliphatic polycarbonates can be prepared by the reaction of aliphatic dihydroxy compounds with phosgene, bis-chloroformates, or bis-chlorocarbonates by transesterification of esters of aliphatic dihydroxy compounds, or by polymerization of cyclic carbonates of aliphatic dihydroxy compounds. They have no commercial significance due to their low melting points (below 120°C), their high solubility, their hydrophilic nature, and their low thermal stability.

Aliphatic-aromatic polycarbonates are prepared by polycondensation of the bisalkyl or bisaryl carbonate of p-xylene glycol in the presence of titanium catalysts, such as tetrabutyl titanate. These have not found any practical application either.

Aromatic polycarbonates, of which bisphenol-A polycarbonates are the most important, are the polycarbonates which have found wide commercial application. They are prepared by phosgenation of aromatic dihydroxy compounds in the presence of pyridine, by interfacial polycondensation between aromatic dihydroxy compounds in aqueous alkaline solutions with phosgene or bischlorocarbonic acid esters of aromatic dihydroxy compounds in the presence of inert solvents, or by transesterification.

The transesterification reaction of bisphenol-A and diphenyl carbonate is shown in the following equation:

$$2(C_6H_5O)_2CO + HOC_6H_4C(CH_3)_2C_6H_4OH \longrightarrow C_6H_5OCO\left[C_6H_4C(CH_3)_2C_6H_4OCO\right] - C_6H_5 + 2C_6H_5OH$$

The reaction is run in a well stirred reactor between 180 and 300°C and 1 to 30 mm Hg vacuum. The polymer molecular weight is limited by the high melt viscosity which is characteristic of polycarbonates due to the relative inflexibility of the polymer chains (e.g., 5×10^5 poise at 240°C for a 30,000 osmotic molecular weight).

A plant for the preparation of bisphenol-A polycarbonate by phosgenation in the presence of pyridine has been in operation since 1960.[7] The reaction is:

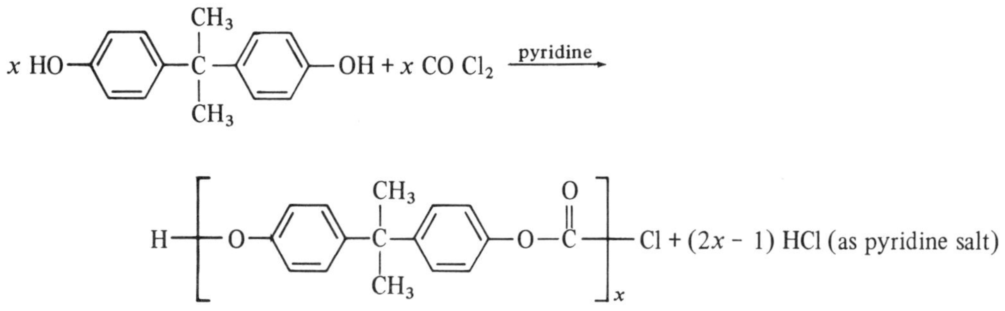

A simplified-process flow sheet is shown in Fig. 10.18. A solution of bisphenol-A in an inert solvent, such as methylene chloride, and pyridine is phosgenated. The polymer solution is washed with dilute hydrochloric acid to convert the excess pyridine to the hydrochloride which dissolves in the aqueous phase together with the pyridine hydrochloride formed during the reaction. After phase separation, the polymer is precipitated out of the organic solvent by the addition of an aliphatic hydrocarbon. The polycarbonate produced is

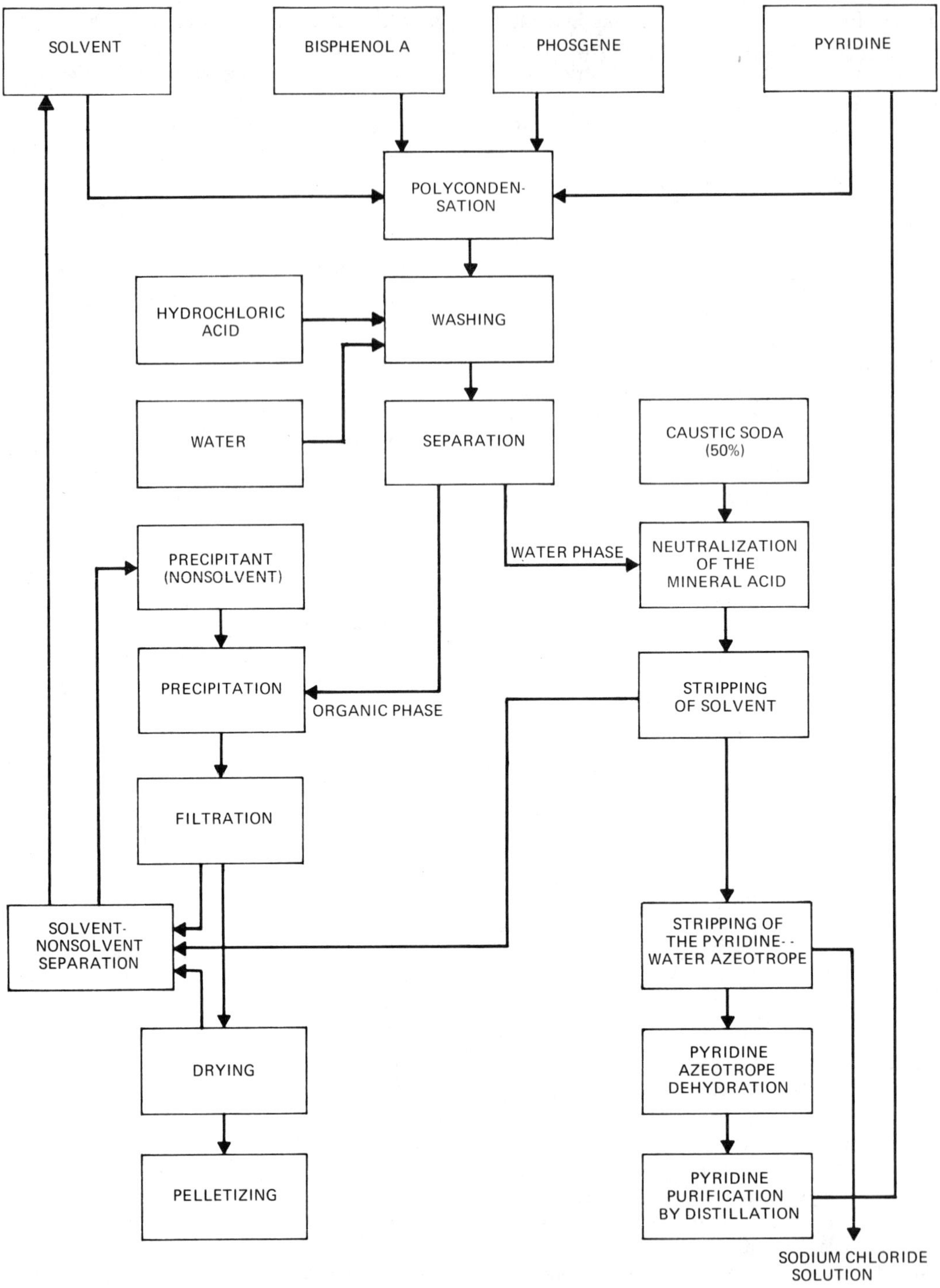

Fig. 10.18 Preparation of polycarbonate from bisphenol A by reaction with phosgene in the presence of pyridine. (*"Encyclopedia of Polymer Science and Technology." Vol. 10, H. G. Mark and N. G. Gaylord, eds., Interscience Publishers, 1968; by permission of John Wiley and Sons, Inc.*)

a white powder which is filtered off, dried, extruded, and pelletized.

The advantage of the process is that polycondensation is carried out at low temperatures in a homogeneous liquid phase. Its disadvantages are that it uses pyridine, which is expensive and causes odor problems, and the solvent and precipitant have to be separated.

Epoxy Resins. Epoxy resins, commercially introduced at the end of World War II, reached a capacity of 35 million pounds in 1957 and around 220 million pounds in 1970. The initial application was in surface coatings, but there was rapid growth in other areas, such as in potting compositions for electrical components, in laminates, in castings, particularly for metal-forming tools, and due to their unique property of exceptional adhesion to metals, as metal bonders. Thus, epoxies rather than metal rivets are used to apply the outer skin of aluminum on high-speed aircraft. Recently the automotive industry has begun using a one-component epoxy resin-based solder to replace the conventional lead solders. Epoxy-based adhesives are replacing brazing methods for joining aluminum to copper in low-pressure refrigerator coils. It is reported that out of one million units there have been only thirty failures. As a result of this success, epoxy adhesives are being tested in high-pressure refrigeration coils, dehumidifiers, and room air-conditioners. The next few years will see expanded use of epoxy resins and hardeners in reinforced plastics, especially in filament winding. With the development of specially designed electrical grade resins, the electrical industry is increasing its use of epoxies. Architects and builders are seeing the possibilities in epoxy resins for industrial and decorative building applications. The resins played a part in the Lunar Landing Module-test program and were used as adhesives and fillers in conjunction with tapes and honeycombs in the construction of the Apollo Command Module.[57]

The reactive "epoxy" group $-OCH_2\overset{O}{\overbrace{CH-CH_2}}$* was first used in resin chemistry by the Germans in the late 1930's. In the United States Pierre Castan, a Swiss chemist, obtained the first patent.[21] Later, S. O. Greenlee of Devoe and Raynolds, the large paint manufacturer, greatly extended and developed the use of epoxies.[47] Still later, chemists from Shell Development, interested in extending markets for propylene derivatives such as epichlorohydrin, $ClCH_2\overset{O}{\overbrace{CH-CH_2}}$, became active in this field.[139]

Most epoxies are based on the glycidyl ether resulting from the condensation of bisphenol A and epichlorohydrin, and may be represented by the formulas:

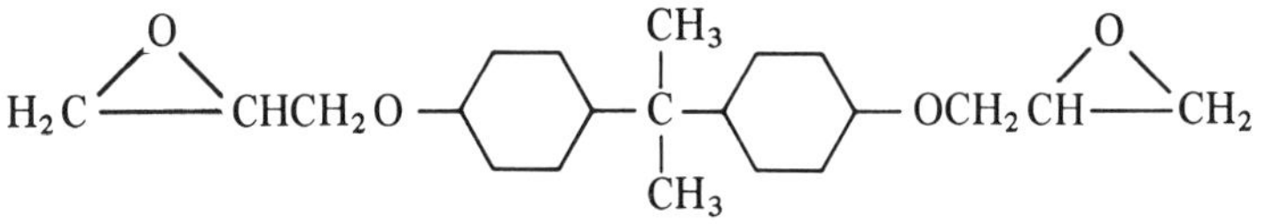

(I)

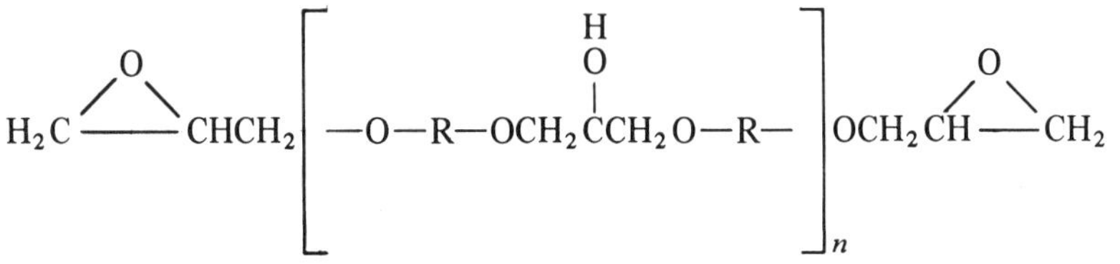

(II)

*Glycidyl ether is a more correct term since epoxy groups without oxygen on the *a* carbon atom do not have the same reaction characteristics as glycidyl ether groups. However, in resin chemistry the term epoxy is synonymous with glycidyl ether and will be so used here.

where, for most commercial resins, and $n = 1$ to 8, and

R = –⟨⟩–C(CH$_3$)(CH$_3$)–⟨⟩–

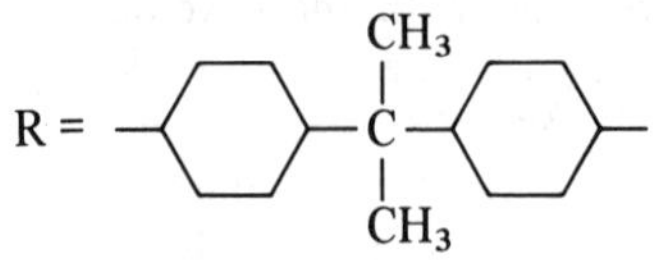

The reaction scheme by which these bis-glycidyl ethers are produced is discussed in greater detail by Skeist,[122] Lee and Neville,[77] and Coderre;[26] a brief summary follows:

(1) HO–⟨⟩–C(CH$_3$)(CH$_3$)–⟨⟩–OH + 2NaOH ⟶
    (III)

Na$^⊕$ $^⊖$O–⟨⟩–C(CH$_3$)(CH$_3$)–⟨⟩–O$^⊖$Na$^⊕$ + 2H$_2$O
    (IIIa)

(2) Na$^⊕$ $^⊖$O–⟨⟩–C(CH$_3$)(CH$_3$)–⟨⟩–O$^⊖$Na$^⊕$ + 2CH$_2$–CHCH$_2$Cl + 2H$_2$O ⟶
    (IIIa)⁻   \\O/

ClCH$_2$C(O$^⊖$H$^⊕$)(H)CH$_2$O–⟨⟩–C(CH$_3$)(CH$_3$)–⟨⟩–OCH$_2$C(O$^⊖$H$^⊕$)(H)CH$_2$Cl + 2Na$^⊕$OH$^⊖$
    (IV)

↓

CH$_2$–CHCH$_2$O–⟨⟩–C(CH$_3$)(CH$_3$)–⟨⟩–OCH$_2$CH–CH$_2$ + 2Cl$^⊖$ + 2Na$^⊕$ + 2H$_2$O
\\O/   \\O/
    (I)

(3) Na$^⊕$ + Cl$^⊖$ ⟶ NaCl ↓

(4) 2(IIIa) + (I) + 2H$_2$O ⟶

Na$^⊕$ $^⊖$OR–OCH$_2$CH(OH)CH$_2$O–R–OCH$_2$CH(OH)CH$_2$ORO$^⊖$Na$^⊕$ + 2NaOH
    (V)

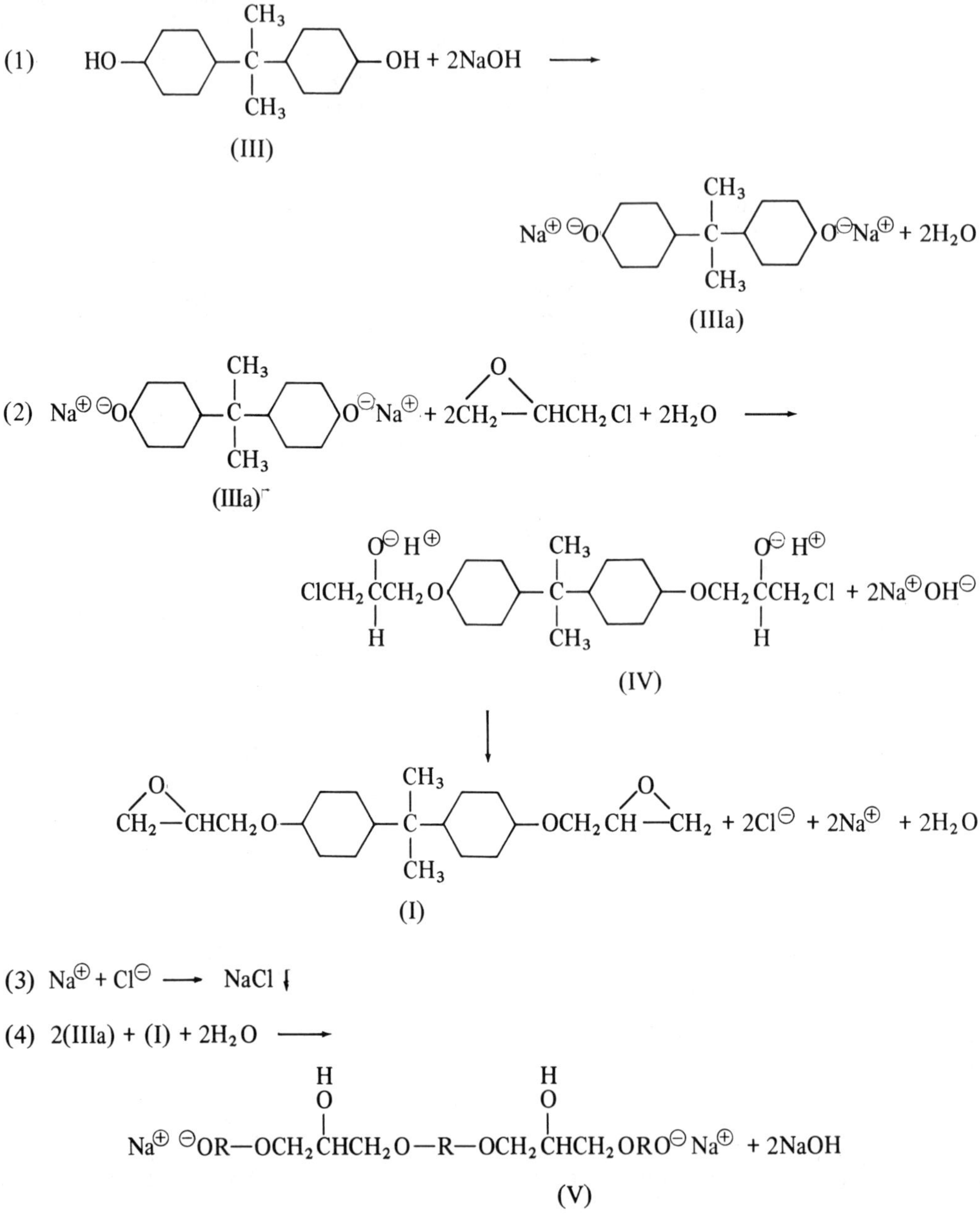

where R = 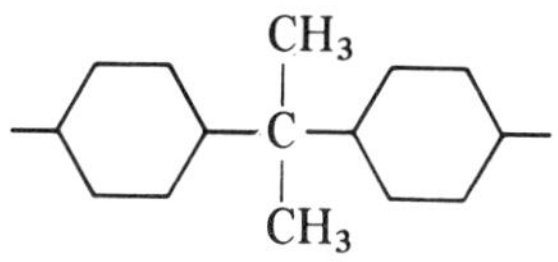

(5) (V) + $\overbrace{CH_2-CH}^{O}CH_2Cl \rightarrow$ higher molecular weight analogs of (IV) and (I).

Reaction (1) in the above requires a small amount of water (about 1 per cent) to permit formation of the ionized species (IIIa). Reaction (2) to form the chlorohydrin (IV) is slow compared to the formation of the epoxy from the chlorohydrin and to the formation of salt [Reaction (3)]. If an excess of base or water is present the epichlorohydrin and the epoxy group in the resin (I) will undergo reaction to the glycol. The rate of addition of base must be adjusted to the rate at which Reaction (2) occurs if optimum basicity is to be maintained. Also, to make the monomeric bisepoxy (I), an excess of epichlorohydrin will favor Reaction (2) over Reaction (4). The above principles are embodied in the following preparation of a nearly pure monomeric bisepoxy resin (I).

A reaction vessel fitted with a heater, stirrer, thermometer, and distilling head having a separator providing return to the reactor of the lower layer was charged with a solution containing a mole ratio of epichlorohydrin to bisphenol of 10:1. The solution was heated to about 100°C and maintained at this temperature during addition of 1.90 moles of sodium hydroxide per mole of the bisphenol, the caustic being introduced as a 40 per cent aqueous solution. Water and epichlorohydrin distilled from the reaction mixture was condensed in the head, and only the epichlorohydrin layer returned to the reaction mixture. The rate of addition of caustic and rate of distilling kept the temperature at about 100°C so that the reaction mixture contained about 1.5 per cent water, the addition taking approximately 2 hours. Upon completing the caustic addition the bulk of the unreacted epichlorohydrin was distilled from the reaction mixture followed by application of vacuum to a pressure of 1 mm Hg and 160°C to remove residual epichlorohydrin. The residue consisting of ether product and salt was cooled and to it was added an equal weight based on the ether of methyl isobutyl ketone along with three times this weight of water. The mixture was agitated at about 25°C and then allowed to separate into two phases. The brine phase containing about 9.5 per cent salt was removed and discarded. The organic phase with ether product containing about 1.0 per cent chlorine was then contacted with an equal weight of 5 per cent aqueous sodium hydroxide solution, and the mixture was agitated for an hour at about 80°C. This quantity of excess caustic amounted to about 8.9 times that needed to react with the organically bound chlorine [Product IV in the preceeding reaction scheme] in the ether product. The mixture was next cooled to about 50°C and the aqueous phase separated. The organic phase was then agitated with about half an equal weight of 2 per cent aqueous solution of sodium dihydrogen phosphate at about 25°C to neutralize any residual sodium hydroxide. After separation of phases the methyl isobutyl ketone was distilled from the organic phase, first up to a temperature of 160°C under atmospheric pressure, then down to a pressure of about 1 mm Hg at the same temperature. The resulting diglycidyl ether of bisphenol [Product I above] was a pale yellow liquid which analyzed as containing 0.25 per cent chlorine and 0.521 epoxy equivalents per 100 grams, and had a molecular weight of 355. The product had high reactivity with added bisphenol, giving 100 per cent reaction when heated at 190°C for 6 hours with an added 35.6 per cent bisphenol.[98]

An alternate method of removing salt from the bisepoxy monomer involves filtering it, then washing the filter cake with isopropanol to recover the adhering resin, and finally distilling the isopropanol from the filter cake wash.

If higher molecular weight solid epoxy ethers are desired, as is the case in surface coatings, lower mole ratios of epichlorohydrin to bisphenol A, e.g., 1.2 to 2.0, are used.

Also, less caustic catalyst is required per mole of bisphenol A since this is partially consumed by the condensation reaction (4) which does not destroy the caustic catalyst through formation of NaCl. For these higher molecular weight resins, salt is ordinarily removed by stirring the resins with water that is above the melting point of the resin, then decanting off the salt layer. Washing is repeated with hot water until the resin reaches a neutral pH.

Equipment for preparing base epoxy resin is similar to that used for phenolics and aminoplasts, i.e., batch operations are performed in an anchor-agitated kettle which has a vacuum reflux condenser. The high temperatures (400°F) required for alkyds are not necessary, but corrosion-resistant material is usually used to maintain a good level of color.

Besides bisphenol A, other polyhydroxyl molecules are used for commercial base resins, e.g., glycerol and novalac (phenol/formaldehyde) resins; however, at present bisphenol A is by far the principal polyol. Other epoxy materials which have been used in specific formulations include butadiene dioxide, the dioxide of vinyl cyclohexene, styrene oxide, and an epoxidized Diels-Alder condensation product (see below) of butadiene and crotonaldehyde.

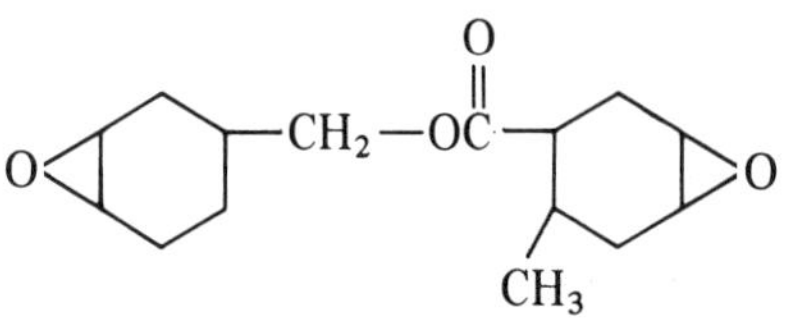

Condensation product

Epoxy groups may be introduced into olefins by means of the newly perfected peracetic acid-epoxidation process,[66] valuable in manufacturing epoxy plasticizers from unsaturated natural oils (e.g., soybean oil). The epoxy group acts as a heat stabilizer for vinyl chloride resins and also promotes resin compatibility:

The base bisepoxy resins are very stable, requiring catalysts and/or coreactants, e.g., amines, organic acid anhydrides, boron trifluoridetertiary amine complexes, to effect cure.

In the curing reactions no volatile products* are evolved; thus shrinkage strains, which cause problems in other thermosetting condensations where volatile products are evolved during the final curing operations, are minimized. The lack of volatiles during cure and the inherent stability of most of the bonds in the cured epoxy, plus the high softening point imparted by the aromatic groups account for many of the desirable properties of epoxy materials. The ester bond formed with acid anhydride catalysts is the weakest with respect to caustic or water hydrolysis while the heat distortion temperature of aliphatic amine-hardened resins is frequently lower than acid-hardened resins, not because of bond rupture but because of the flexibility of the aliphatic crosslinkages.

The secondary amine first formed when using amine coreactants is even more reactive than the original primary amine, giving an autocatalytic effect to the reaction rate even if temperatures are held constant. In addition, the reaction produces a large amount of heat, on the order of 20–25 kg-cal/mole of condensing epoxide group. For liquid resins with a large number of epoxy groups per unit weight, adiabatic temperature rises of 350°F or more are easily possible in either thick laminates or potting molds. Many amines, particularly those aliphatic amines partially reacted with ethylene oxide or bisphenol A to decrease the dermatitis frequently caused by amines, will react with epoxy groups even at room temperature, giving handling times (i.e., "pot lives") sometimes as short as 15 minutes. To solve the short-pot life problem, automatic two-fluid metering devices have been developed. Aliphatic amine hardeners used commercially include ethylene diamine, diethylenetriamine, triethylene tetramine, their partially hydroxyethylated or cyano-

*Although no volatile products result, the catalysts themselves are sometimes volatile. At the high temperatures reached in the exotherm with amine coreactants, many of the amines are quite volatile, causing toxicity and odor as well as foaming. Also, BF_3 is a toxic gas and poses similar problems.

ethylated counterparts, and their glycidyl ether derivatives. Commercially important aromatic amines are *m*-phenylenediamine, 4,4′-methylenedianiline, 4,4′-diaminodiphenyl sulfone. The aromatic groups increase the softening temperature of the cured resin.

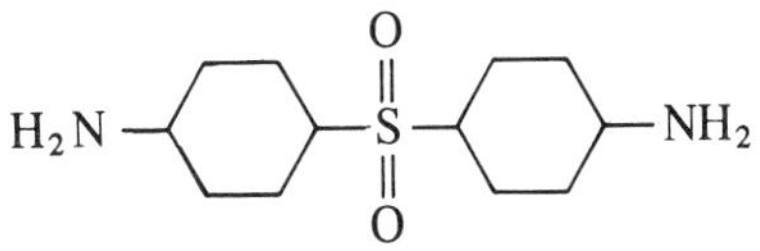

4,4′-Diaminodiphenyl sulfone

The polymerization catalyzed by tertiary amines is believed to be a true chain reaction type of addition polymerization, but the length of the chain is probably very low. The reaction is sensitive, and the presence of alcohols such as the secondary alcohol in the epoxy-base resins containing more than one bisphenol A molecule, or water, though accelerating the reaction, will cause chain transfer: the molecular weight is thereby drastically lowered, and the properties of the cured resin are significantly altered. One commercial tertiary amine, 2,4,6-tris(dimethylaminomethyl)-phenol, is commonly used, however. In addition, this polymerization reaction undoubtedly occurs with the tertiary amines that are formed by epoxide condensation with primary and secondary amine-curing agents, thus accounting for greater than stoichiometric yields from many amines, particularly piperidine.

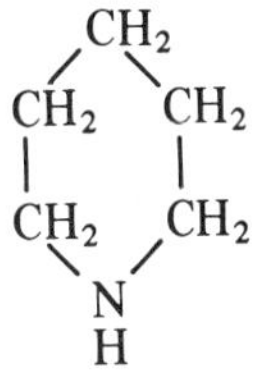

Piperidine

The amine hardeners discussed so far give resins with high softening temperatures but poor impact resistance. To improve impact strength and flexibility, multifunctional amines having long flexible aliphatic chains between the amino groups are used, e.g., hexamethylenediamine and mixed diamines derived from fatty acids. Similar products are the so-called polyamides ("Versamids"), the reaction products of aliphatic polyamines, e.g., diethylenetetramine and the diacid obtained via a Diels-Alder dimerization of unsaturated fatty acids (e.g., linoleic isomers). Another means of imparting flexibility to the cured resin is to include a Thiokol "liquid polymer" which has a structure similar to $HS(CH_2CH_2OCH_2CH_2SS)_nCH_2CH_2OCH_2$-$CH_2SH$. In the presence of amine hardeners, the mercaptan groups will react with epoxy groups incorporating the flexible "rubber" chain in the molecule.

Acid anhydrides react with the epoxy group to form ester linkages. These reactions are only mildly exothermic and do not cause the "exotherm" problems associated with amine hardeners. Also, the acid conditions catalyze formation of ethers between the secondary alcohol in (V) and other epoxy groups; therefore less than stoichiometric quantities of acids are frequently used to achieve cure. Ether-formation, it is believed, does not occur under the basic conditions existing in amine curing. Pot lives of acid anhydride-catalyzed formulations are very long at room temperatures, but high-temperature treatment or baking ovens are necessary to effect cure (300–400°F *vs.* 150–250°F for amine catalysts). Commercial acid anhydrides are classified as solids, liquids, or chlorinated derivatives. The solids include phthalic anhydride, hexahydrophthalic anhydride, and pyromellitic dianhydride.

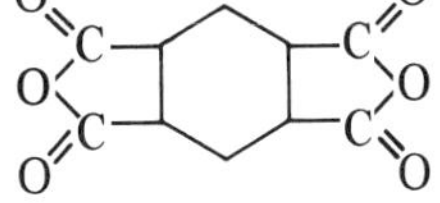

Pyromellitic dianhydride

The liquids include dodecenylsuccinic anhydride and a methylated maleic adduct of phthalic anhydride ("Methyl Nadic Anhydride"). Flame resistance is given by chlorinated anhydrides, such as dichloromaleic anhydride and hexachloroendomethylenetetrahydrophthalic anhydride.

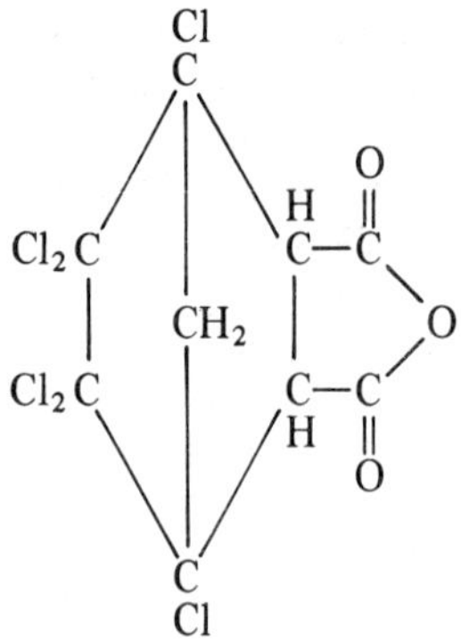

Hexachloroendomethylenetetrahydrophthalic anhydride

The diversity of chemicals used during curing and the complexity of the curing reactions explain why those who use epoxies (in contrast to many other plastics) frequently have an independent chemical laboratory directed by a competent industrial chemist or chemical engineer.

Polyesters. In the plastics industry the term *polyesters* has a far narrower connotation than is implied chemically. In the plastics field a polyester is ordinarily the base resin consisting of a liquid unsaturated polyester plus a vinyl-type monomer; this liquid mixture is capable of reacting to form infusible crosslinked solids under the influence of catalysts and/or heat. Such polyesters (in the strict chemical sence) as ethylene glycol-terephthalic acid polycondensates for textiles (Chapter 11) and alkyds for paints (Chapter 22) are not polyesters in the plastics industry.

One of the largest uses of polyesters is in fibers for textiles and automobile tires. However, this use will not be discussed in this chapter. In the "plastics" industry unsaturated polyesters, also called polyester resins, are used primarily in rigid laminates, moldings, or castings and are almost always reinforced with glass cloth or glass fibers. They were developed during World War II and were first used very successfully in self-sealing gas tanks containing a rubber liner. When pierced by a bullet, metal tanks splay or "flower" and prevent the rubber tank liner from swelling shut and closing the hole. Polyester laminates do not splay and therefore allow the rubber liner to close the hole. When reinforced with glass, their high flexural strength is outstanding, approaching that of metals. The major markets are the automotive and marine industries. Parts now being molded include fender extensions, rocker panels, and roofs. In housing, the use of polyesters in modular bathrooms has become significant. Approximately 770 million pounds of polyester resins are expected to be produced in the United States in 1970.

Maleic anhydride is the principal unsaturated dibasic acid used in the polyester, although fumaric acid is also used in limited amounts. Uniquely among ordinary unsaturated monomers, the double bond in maleic anhydride, its acid, its esters, and similar α,β-carboxyl-substituted olefins will not undergo homogeneous polymerization even at high temperatures but will copolymerize rapidly with a wide variety of vinyl monomers at even faster rates than these monomers will homopolymerize. It is because of this peculiar property of maleic anhydride that polyesters are prepared by first making (at the necessarily high temperatures) a linear low-viscosity gel-free polyester containing several double bonds per molecule and then mixing this unsaturated polyester with a vinyl type monomer, usually styrene, which, under the influence of catalysts and/or heat, will crosslink the polyester molecules to form a rigid infusible polymer.

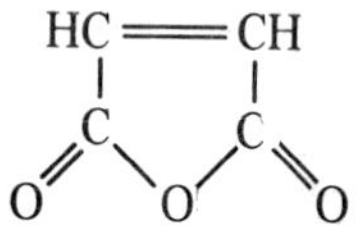

Maleic anhydride

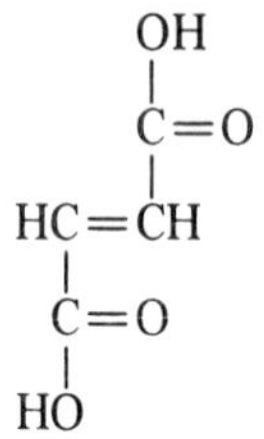

Fumaric acid

The difunctional alcohol ethylene glycol is frequently used for the coreactant, but it is supplemented with propylene glycol, diethylene glycol, or dipropylene glycol to decrease the tendency for the liquid resin to crystallize and to increase the flexibility of the cured resin.

To promote compatibility with the styrene monomer used to crosslink the polyester, phthalic anhydride is incorporated into the polyester backbone in mole ratios with maleic anhydride of from 1:1.5 to 1:1. Besides giving compatibility with styrene, phthalic anhydride imparts some flexibility to the cured resin and lowers the cost. Adipic acid, because of its long flexible aliphatic carbon chain, is used to promote a high degree of flexibility.

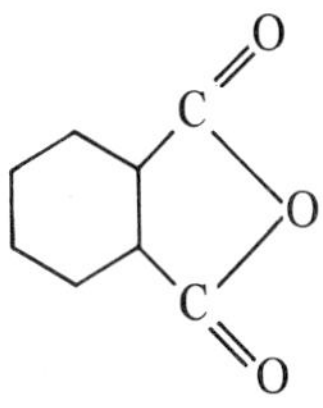

Phthalic anhydride

$$HO\overset{\overset{O}{\|}}{C}(CH_2)_4\overset{\overset{O}{\|}}{C}OH$$

Adipic acid

Styrene is by far the most common crosslinking agent and is usually added to the base polyester by the manufacturer of the polyester. To ensure that most of the maleic unsaturated groups are reacted and optimum strength is achieved, an excess of styrene must be used (30–40 per cent is the average) because the styrene monomer will polymerize not only with an active maleic radical but also with an active styrene radical; maleic radicals, however, will react only with styrene monomer. Low temperatures favor the maleic-styrene reaction while high temperatures favor the styrene-styrene reaction, accounting for the advantages frequently found in initial low temperature cures.

Besides styrene, many other vinyl monomers may be used to develop special properties, e.g., triallyl cyanurate to promote heat resistance, dially phthalate to reduce volatility during cure, acrylic acid ester to promote flexibility via internal plasticization and to impart improved weatherability.

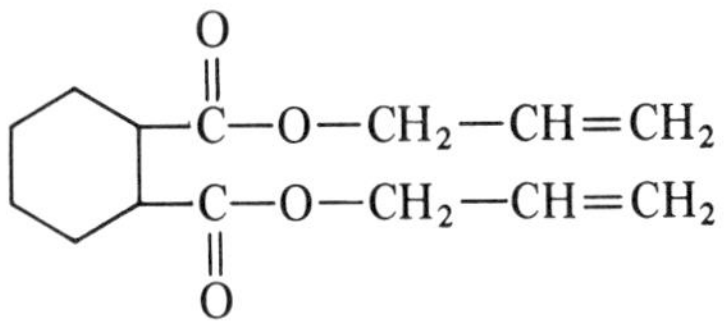

Diallylphthalate

$$CH_2{=}\overset{H}{C}{-}\overset{\overset{O}{\diagup\!\!\diagup}}{C}{-}OR$$

Acrylic acid ester

In addition to the difunctional acids and alcohols, monofunctional acids and alcohols may be added in small amounts to limit polyester molecular weight. Allyl alcohol has been used to give additional unsaturation for subsequent crosslinking.

The properties of polyesters depend on the nature of the reactants. High levels of crosslinking and high percentages of aromatic rings promote hardness, rigidity, strength, and brittleness. Lower levels of crosslinking (less maleic anhydride) and long flexible aliphatic chains impart flexibility and some impact strength. Larger polyester molecules promote some strength although this effect has a definite upper limit which is reached at a moderate molecular weight. The above brief picture illustrates how complicated and varied the formulation of polyesters can be. As an example[12] the preparation of a particular polyester might be carried out in a jacketed stainless steel reaction vessel equipped with an agitator, and a reflux condenser followed by a total condenser. The charge is 5 moles of maleic anhydride, 3 moles of phthalic anhydride, 4 moles of ethylene glycol, and 4 moles of diethylene glycol. The temperature is rasied to 375°C until an acid number* of 60–65 is obtained, while maintaining an inert atmosphere. The inert purge gas (CO_2 or N_2) is introduced through a submerged sparger. In addition to excluding oxygen, which discolors the resin, the purge gas aids in removing the

*Acid number is the number of milligrams of potassium hydroxide necessary to neutralize the free acid in one gram of sample. Since each free carboxyl group represents, on the average, one molecule, the number average molecular weight of the resin, Mn, is given by the following:[39]

$$Mn = \frac{1000 \times 56}{\text{Acid Number}}$$

water formed during the condensation reaction. The temperature is then raised and held at 440°C until the acid number is 45-50 or until a Gardner-Holt* viscosity of N to Q is reached when 100 parts of polymer are combined with 50 parts of styrene. When the reaction is completed (about 5 hours total time) the reaction mass is cooled to about 150°C and mixed with 6.5 moles of styrene containing *p-t*-butylcatechol inhibitor (0.02 per cent of final mixture); the temperature during mixing is maintained at about 40°C. The mixture is cooled to 20°C and pumped to drums or tank cars.

Many catalysts, most of which are free radical generators, have been proposed to cure the resins. Benzoyl peroxide, mixed with 50 per cent dibutyl phthalate to facilitate mixing with the resin, is most frequently use for hot cures (those above 50°C) at levels of 1 to 2 per cent. Although methyl ethyl ketone peroxide, as a 60 per cent solution in dimethyl phthalate, is sometimes used for lower-temperature cures, particularly when combined with paint "driers" such as cobalt napthenate. Sometimes combinations of these lower-temperature catalysts with a much higher-temperature catalyst, e.g., di-*t*-butyl peroxide (activity beginning at about 115°C) or dicumyl peroxide (activity beginning at about 105°C), may be used for composite cures leading to improved properties.

$$(CH_3)_3—C—O—O—C—(CH_3)_3$$

Di-*t*-butyl peroxide

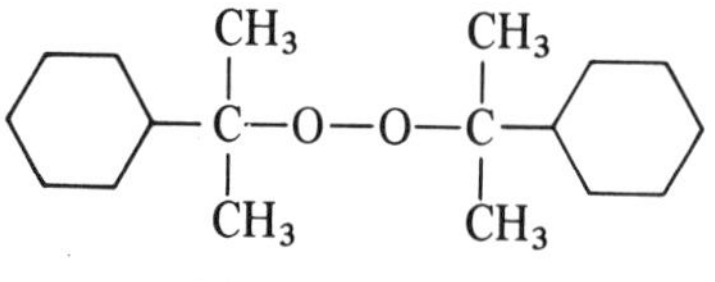

Dicumyl peroxide

*Gardner-Holt viscosities are widely used in the surface-coating industry and are obtained by comparing the time it takes for a standard bubble of air to rise in a sample tube with the time it takes for a similar bubble to rise in a set of standard tubes identified by letters. N to Q Gardner-Holt corresponds roughly to 3.40 to 4.35 posie at 77°F.

Promoters (amines) are sometimes used to lower the temperatures at which the catalysts become effective. Many other catalyst systems are used, for example, the ozonides in styrenated polymers, and organic azo compounds and azines as crosslinking catalysts.

In addition to the action of free radicals, some cure is possible by further condensing the polyester itself, using acids or bases which will catalyze esterification, e.g., calcium hydroxide, barium hydroxide, *p*-toluenesulfonic acid.

The properties, e.g., filament size, type of weave, etc., of the laminating material (almost always glass fibers or glass cloth) is as important to the final properties of the cured laminate or casting as the resin.[76,125] To ensure a good bond between resin and glass, the starch lubricants, necessary during weaving, must be burned off; adhesive coatings such as special vinyl silicones are also used.

Polyphenylene Oxide Resins. In 1959, Hay and co-workers[58,59] reported that certain 2,6-di-substituted phenols could be oxidatively polymerized using the cuprous ion as a catalyst to give aromatic polyethers

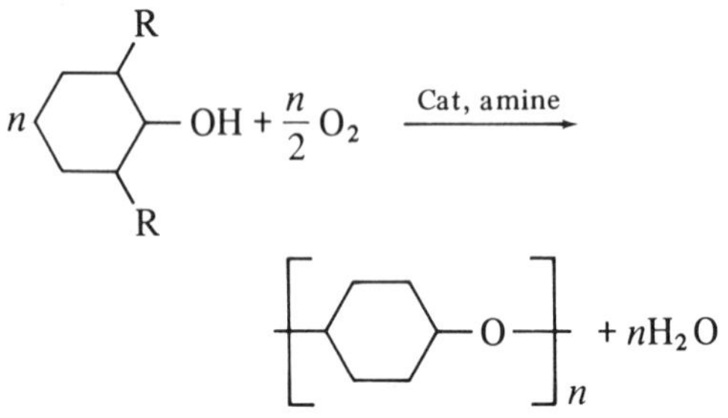

In the commercial PPO, R is a methyl group. The product has an exceptionally high softening point, about 210°C, higher than that of most other commercially available thermoplastics. (See polysulfones in the following pages.) It is hard, tough, modestly transparent and self-extinguishing. It is somewhat difficult to mold, although opaque polyblends are available with better flow characteristics.

A typical laboratory polymerization process has been described:

Charge 200 ml of nitrobenzene, 70 ml of pyridine, 1 gm of cuprous chloride to a vigorously agitated glass flask. Oxygen is bubbled in at 300 ml/min followed by 15 g of 2,6-dimethylphenol. The reaction is continued for about 30 minutes maintaining the temperature at about 30°C. After dilution of the polymeric solution with 100 ml of chloroform, the solution is precipitated in 1.1 liters of methanol containing 3 ml of concentrated hydrochloric acid. Filtering, methanol washing and/or redissolving in chloroform followed by reprecipitation in methanol are used to separate the pyridine and catalyst. The yield was about 91%.

The details of commercial practice are not known.[59]

Polysulfone Resins. In the early 1960's chemists at Union Carbide developed a new type of polyer called "polysulfones" containing both ether linkages and sulfoxide linkages between phenylene groups. They are prepared by reacting bisphenol A with di-*p*-chlorophenyl sulfoxide in the presence of exactly two moles of sodium hydroxide per mole of bisphenol A. A dipolar aprotic solvent, completely free of water, e.g., dimethyl sulfoxide, is required.

$$n\,\text{NaO}-\!\!\bigcirc\!\!-\underset{\underset{\text{CH}_3}{|}}{\overset{\overset{\text{CH}_3}{|}}{\text{C}}}-\!\!\bigcirc\!\!-\text{ONa} + n\,\text{Cl}-\!\!\bigcirc\!\!-\underset{\underset{\text{O}}{\|}}{\overset{\overset{\text{O}}{\|}}{\text{S}}}-\!\!\bigcirc\!\!-\text{Cl}$$

$$\xrightarrow[\text{DMS solvent}]{130-160°\text{C}} \left[-\text{O}-\!\!\bigcirc\!\!-\underset{\underset{\text{CH}_3}{|}}{\overset{\overset{\text{CH}_3}{|}}{\text{C}}}-\!\!\bigcirc\!\!-\text{O}-\!\!\bigcirc\!\!-\underset{\underset{\text{O}}{\|}}{\overset{\overset{\text{O}}{\|}}{\text{S}}}-\!\!\bigcirc\!\!-\right]_n + 2n\,\text{NaCl}$$

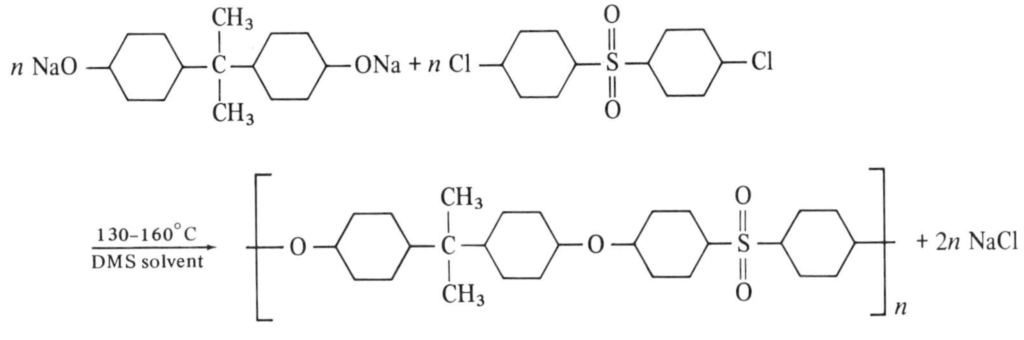

Molecular-weight control is achieved by the addition of a monofunctional chain terminator, e.g., sodium phenate or methyl chloride.

The structure of these materials is similar to the pure polyphenylene oxides (PPO); hence, their properties are similar. The color and transparency is somewhat inferior to PPO, but the heat resistance is claimed to be slightly superior.[67]

Polychloromethylether (a polyoxetane). Hercules Chemical Co. has commercialized a poly(3,3-bis(chloromethyl)oxetane) under the trade name of "Penton."

$$\text{H}-\!\!\left[-\text{O}-\text{CH}_2-\underset{\underset{\text{CH}_2}{|}}{\overset{\overset{\text{CH}_2\text{Cl}}{|}}{\text{C}}}-\text{CH}_2-\right]-\text{OH}$$

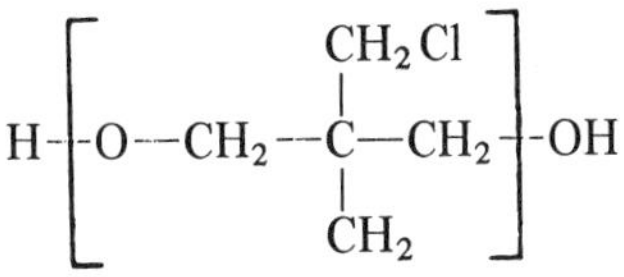

The oxetane monomer is synthesized from pentaerythritol by esterification with acetic acid (or anhydride) followed by chlorination with dry hydrochloric acid at 200°C in the presence of zinc chloride catalyst. Treatment with sodium hydroxide cyclizes the trichloroacetate to the oxetane monomer.

$$(\text{HOCH}_2)_4\text{C} + 4\text{CH}_3\text{COOH} \longrightarrow$$

$$\text{C}(\text{CH}_2\text{O}\overset{\overset{\text{O}}{\|}}{\text{C}}\text{CH}_3)_4 + 4\text{H}_2\text{O}$$

$$\text{C}(\text{CH}_2\text{O}\overset{\overset{\text{O}}{\|}}{\text{C}}\text{CH}_3)_4 + 3\text{HCl} \xrightarrow[\text{ZnCl}_2]{200°\text{C}}$$

$$(\text{ClCH}_2)_3\text{C}(\text{CH}_2\text{O}\overset{\overset{\text{O}}{\|}}{\text{C}}\text{CH}_3) + 3\text{CH}_3\text{COOH}$$

$$(\text{ClCH}_2)_3\text{C}(\text{CH}_2\text{O}\overset{\overset{\text{O}}{\|}}{\text{C}}\text{CH}_3) + 2\text{NaOH} \xrightarrow[25°\text{C}]{\text{H}_2\text{O}}$$

$$\text{ClCH}_2-\underset{\underset{\text{CH}_2-\text{O}}{|}}{\overset{\overset{\text{CH}_2\text{Cl}}{|}}{\text{C}}}-\text{CH}_2 + \text{NaCl} + \text{NaO}\overset{\overset{\text{O}}{\|}}{\text{C}}\text{CH}_3 + \text{H}_2\text{O}$$

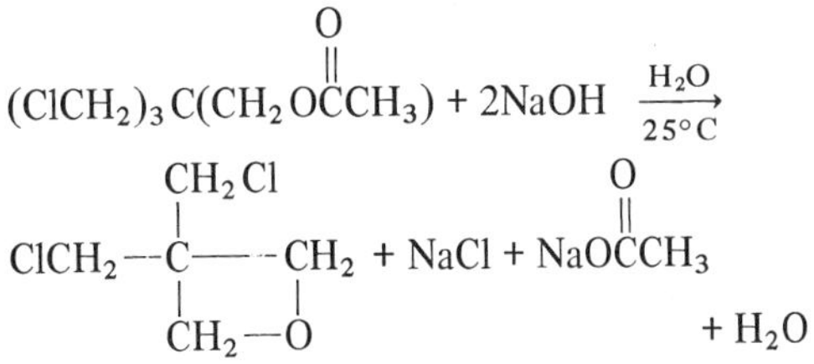

To polymerize the oxetane monomer, a rapid continuous mass process has been developed using a special trialkylaluminum catalyst (plus promoters); the heat of reaction raises the melt to the exit temperature of about 200°C. To control temperature, viscosity, and molecular weight, catalyst levels are kept very low and conversions are about 80 per cent.[23]

Other oxetanes can be polymerized, but this bischloromethyl derivation is the only one commercialized to date.

The principal uses are for fabricating corrosion-resistant equipment, particularly where the Penton is bonded to metals. Such composite fabrication is simpler for Penton than for the fluorinated polymers, Teflon or Kel-F.

Polyolefin Polymerization

Polyolefin polymerization processes are too unique to be fitted into the previous classifications.

Polyethylene has had a spectacular rate of growth. From early exploratory investigations by English chemists in 1933-35 (8 grams of polymer were finally made in December 1935) and initial limited production during World War II for specialty electronics work, production has increased so rapidly that, with more than 5.5 billion pounds of produced for 1970 in the United States alone, this is perhaps the most important polymer in industry today. The current annual capacity is more than the cummulative total of all plastics produced in all the years prior to 1940. The copolymers of ethylene with other olefins are versatile plastics or elastomers with excellent resistance to oxidation. The copolymers with polar monomers often show improved physical properties over polyethylene itself and are finding increasing use. The chlorinated, chlorosulfonated, and oxidized derivatives of the ethylene polymers have been found to be particularly useful.[18]

Polyethylene–High–Pressure Processes. In 1933 a group of chemists at the British company Imperial Chemical Industries, using new high-pressure laboratory equipment designed by Professor Michels at Amsterdam University, explored the effect of high pressure on polymerization and other reaction. When benzaldehyde was subjected to a pressure of 1400 atmospheres at 170°C, using ethylene to produce the pressure, a waxy polymer was produced on the walls of the vessel. This was found to be a polymer of ethylene rather than of benzaldehyde. Repeating the experiment with pure ethylene resulted in an explosive reaction which discouraged further experiments with ethylene until 1935. Repeating the experiments with "pure" ethylene in December 1935, 8 grams of polymer were produced in an 80-cc autoclave because of an almost miraculous combination of events. A small leak developed at 180°C and the ethylene used to repressurize the autoclave must have contained a very small but precisely proper quantity of oxygen. It took several months to explain the success of the above experiment.*

There are three types of processes for the manufacturer of polyethylene that are of commercial importance. These are:

1. High-pressure polymerization by free-radical catalysis.
2. Medium-pressure polymerization with catalysts of transition metal oxides, such as molybdenum oxide or chromium oxide.
3. Low- and medium-pressure polymerization using Ziegler catalysts.

The high pressure process, which was chronologically first, produces polymers with more branches on the chain, a lower level of crystallinity, and a lower density. Commercial free-radical processes use pressures between 1000 and 3000 atm at temperatures up to 350°C. Polymerization is initiated by molecular oxygen or other free radical generators.[37a] Oxygen has an effective half-life of one minute at 160°C. Other free-radical generators used as initiators include caprylyl or lauroyl peroxides with one-minute half-lives at about 115°C and cumene peroxide with a one-minute half-life at about 255°C. The polymerization reaction is a typical free-radical polymerization involving free-radical initiation, polymer propagation, and termination

*This account of the early history of the development of polyethylene has been condensed from J. C. Swallow's[128] entertaining account of the exploratory work which led to the discovery of polyethylene.

by combination. Several important side reactions involving chain transfer also occur, due to the highly reactive nature of the polyethylene free radical and the high temperatures involved. Chain-transfer agents such as alkanes, olefins, ketones, aldehydes, and hydrogen are used to control molecular weight. A double "back-biting" mechanism is probably responsible for the rearrangement which forms vinylidene unsaturation, which is the principal molecular weight-determining reaction in the absence of a chain transfer agent.[25]

A flow diagram of a commercial high-pressure process for the manufacture of polyethylene is shown in Fig. 10.19 and highlights the basic operations. The reactor may be either tubular or a stirred tank. If a tubular reactor is used the ethylene-containing initiator is passed through a preheater to raise its temperature to the 100-200°C-level where polymerization begins. Some of the heat of polymerization is removed by a cooling medium circumlating around the tubes, while the balance of the thermal energy generated raises the reaction temperature to 300°C or higher. When the stirred-tank reactor is used the cold ethylene is added directly to the hot polymerization mixture which is maintained at a fairly constant temperature around 150-300°C.[1a,46,102,121,143,144] The reaction medium is actually a dense gas, since the polymerization temperature exceeds the critical temperature of ethylene (99°C), and the process may be comparable to a bulk process.[37a] A suspension of ethylene in water may also be used,[75] and benzene may be added.

Normal conversions per pass for the tubular process are reported to lie in the range of 15-25 per cent, while somewhat higher conversions may be obtained in the stirred-tank processes utilizing the cold feed to absorb some of the heat of reaction. The following example from a patent uses laboratory apparatus in which flow surges were purposely

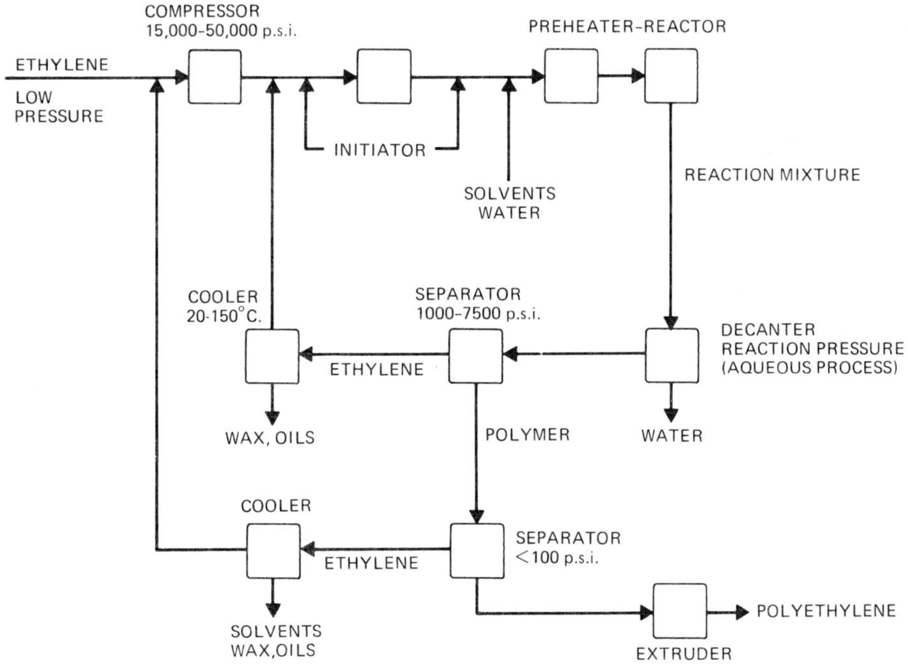

Fig. 10.19 Generalized flow diagram for high-pressure polyethylene processes. (*"Crystalline Olefin Polymers," part I, R. A. V. Raff and K. W. Doak, Eds., Interscience Publishers, 1965; by permission of John Wiley and Sons, Inc.*)

generated, presumably to agitate the two-phase* system believed to be present:

In a small continuous experimental polymerization apparatus, a series of runs was made in which ethylene containing free oxygen was subjected to polymerization to make polyethylene. The let-down valve was controlled manually. In the first ten runs irregular and severe pressure drops occurred, temperature control was poor, most of the runs were terminated by explosions, and the products were severely contaminated with carbon black. In the eleventh run the pressure was deliberately dropped at timed intervals. Procedure was as follows: A standard operating pressure of 25,000 pounds per square inch was used. The let-down valve was rapidly opened by hand and pressure allowed to drop approximately 1500 pounds per square inch after which the valve was closed. The pressure was then allowed to build up until it reached the standard operating pressure of 25,000 pounds per square inch; then a period of 5 seconds was allowed to elapse during which the pressure was maintained at that value by automatic control of the pump. At the end of the 5 seconds the valve was again opened and the cycle repeated. Process conditions were: oxygen content 100 parts per million, dwell time 11 minutes, temperature approximately 240°C. As a result of operating in this manner, process control was good, polyethylene was produced in a conversion of 25 weight per cent, and no carbon black was made, nor did any explosion occur.[108]

The stirred-autoclave process may resemble the following example from a patent:

A silver-lined steel reaction vessel is charged with 10 parts of water and 1 part of benzene. The vessel is closed, an internal mechanical stirrer put in motion, and ethylene injected at a pressure such that at the reaction temperature of 200 to 220°C, to which the vessel is heated, the pressure is approximately 1000 atmospheres.

As soon as the reaction commences, from the bottom of the reaction vessel a liquid phase comprising a mixture of polymer, water, and benzene is continuously withdrawn and passed into a separating vessel maintained at a lower pressure and temperature than that of the reaction vessel.

From the top of the separating vessel, unreacted ethylene is conducted, together with additional benzene and water, to the reaction vessel for further reaction. The water in the liquid phase remaining in the separating vessel, together with benzene in admixture therewith may be recirculated to the reaction zone after removal therefrom of its polymer content.

The ethylene employed has an oxygen content of approximately 30 parts per million. The water and benzene ratios, based upon the ethylene introduced, are maintained at 4 parts of water and 0.4 parts of benzene per part of ethylene introduced.[75]

To separate this large quantity of ethylene from the polymer, the pressure may be lowered directly to atmospheric, but this highly irreversible process is very wasteful of compressor energy—a major production-cost item. By putting in at least one separator operating at moderate pressures (100–300 atm), considerable savings in recompression costs are possible. If the pressures are too high, considerable low-molecular weight polyethylene may remain dissolved in the recycle ethylene, causing recompression problems. Polyethylene, still molten, may be removed from the final atmospheric-pressure-separation pot by means of gear pumps, screw pumps, or extruders. Further blending of the base resin with colorants, lubricants, stabilizers, etc. may then be made in Banburys or extruders, followed by chopping or cutting to pelleted molding powder.

Polyethylene—Lower-Pressure Processes. There were three separate and independent developments of polymerization methods for ethylene at lower pressures. Around 1953 Karl Ziegler[146,147,148] in Germany discovered the unusual synergistic effect of titanium tetrachloride in the presence of aluminum tri- or dialkyls. At about the same time John Hogan and Robert Banks at Phillips Petroleum[62] discovered that chromium salts deposited on conventional cracking catalysts and

*Some data have been presented[65,73] which show that low-molecular weight polyethylenes will dissolve in high-density ethylene gas above its critical temperature. Hence, in a tubular reactor two phases probably are present: one, a higher-molecular weight molten polyethylene containing large amounts of dissolved ethylene, and the other, "gaseous" ethylene, above its critical temperature, in which some lower-molecular weight polyethylenes are dissolved.

oxidized with air at high temperatures to produce hexavalent chromium oxide would polymerize ethylene to high molecular weight. Even earlier, Standard Oil of Indiana found several catalyst systems, notably nickel oxide on charcoal partially reduced with hydrogen, and molybdena-alumina activated by hydrogenation. The Standard Oil process is not in use in the United States, but it is reported to be in operation in Japan. Polyethylene produced by this process has a density range of 0.90-0.96 gm/cm^3. The polymerization may be carried out in a fixed catalyst bed or with the catalyst slurried or suspended in the reactant mixture. A solvent may be used. Polymerization temperatures range from 200-300°C at pressures of 500-1500 psig.[131]

The Phillips Petroleum process uses a "hexavalent" chromium oxide deposited on a microspheroidal commercial cracking catalyst support, e.g., a 90 SiO_2/10 Al_2O_3 silica gel catalyst with large pore size and small surface area. The catalyst may be made by slurrying the support with an equal weight of a 0.78-molar-$Cr(NO_3)_3$ and 0.78-molar-$Sr(NO_3)_2$ water solution, filtering off the excess solution, drying the catalyst, and finally activating the catalyst by heating it at 950°F while passing thoroughly dry air through the catalyst at a space velocity of 300 reciprocal hours.[62]

The flow sheet in Fig. 10.20 illustrates the over-all polymerization process. A catalyst slurry in cyclohexane solvent is fed along with the ethylene into a stirred moderate-pressure (200-300 psig) 4000-gallon steel autoclave. The preferred proportions of the reactants have not been disclosed, but hourly rates to a 4000-gallon continuous reactor may be 2500-pounds of ethylene, 25 pounds of catalyst, and 22,500 pounds of cyclohexane. Reaction temperatures are probably in the range of 275-300°F. Higher pressures corresponding to

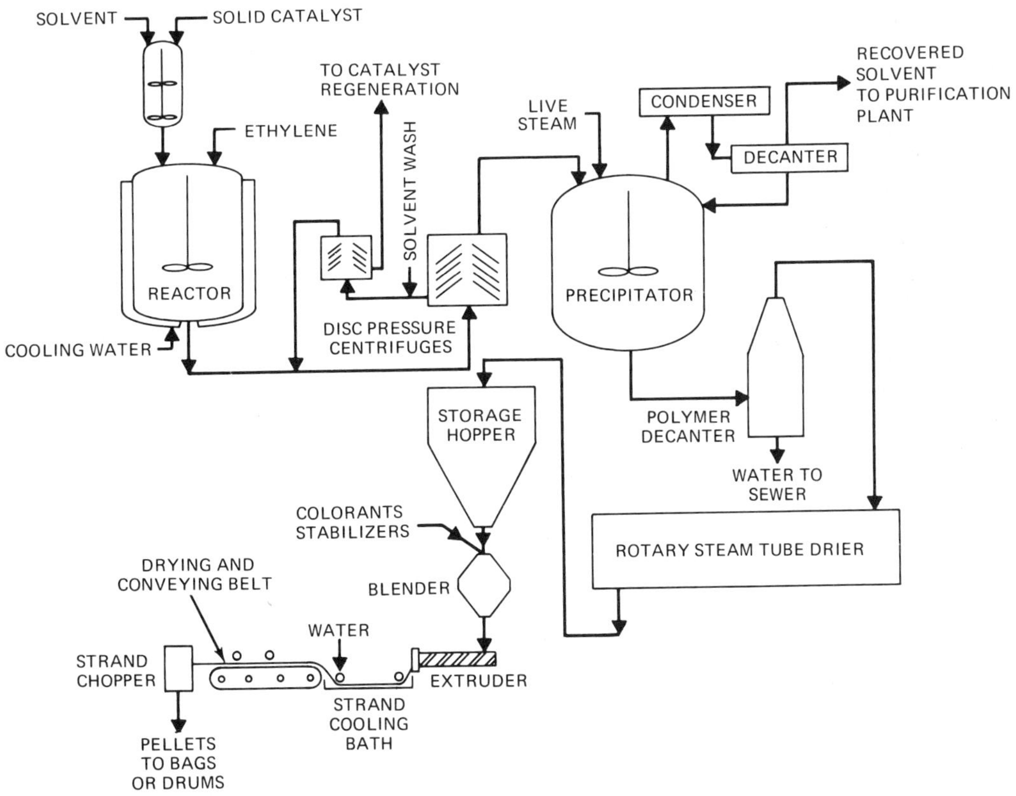

Fig. 10.20 Phillips low-pressure polyethylene process.

higher ethylene concentrations produce higher molecular weights, while lower temperatures produce lower-molecular weight products. From the reactors the solution of polymer in cyclohexane, catalyst, and the very small amount of unreacted ethylene are fed into specially developed continuous pressure centrifuges[138] of the sloped-internal-disc type which remove the catalyst as a slurry through fine discharge nozzles. The catalyst is washed with hot cyclohexane to recover the polyethylene and fed to a second centrifuge which sends the washed catalyst to the catalyst regeneration plant and recycles the wash liquor to the main centrifuge. From the main centrifuge the polymer solution is fed to a water precipitator which precipitates the polymer and steam distills the cyclohexane solvent and any unreacted ethylene. From the precipitators the polymer is separated from the bulk of the water by flotation and then dried in a rotary steam-tube drier.[140] After drying, the polymer dust is blended with colorants, stabilizers, etc. and densified into pellets by extruders.

Ziegler catalysts are the reaction products of trialkylaluminum and $TiCl_4$. The precise chemical nature of these unstable complexes has not been determined. It is known that the aluminum alkyl first reduces the $TiCl_4$ to the "β" crystalline form of $TiCl_3$. The alkylaluminum chloride formed by this reaction complexes with the $TiCl_3$ as does additional "unoxidized" trialkylaluminum. Aluminum/titanium mole ratios are ordinarily around 2.0. A simplified explanation of the polymerization mechanism is that coordinate Al-Ti linkages on particular sites in the $TiCl_3$ crystal activate the corresponding $\overset{\oplus}{Al}—\overset{\ominus}{C}$ bonds. Ethylene monomer then feeds into this activated $\overset{\oplus}{Al}—\overset{\ominus}{C}$ bond making a new alkylaluminum with a higher molecular weight. Reaction is completed when an active hydrogen-containing molecule, e.g., an alcohol, severs the bond and terminates the polymer.

$$\leftarrow \underset{R''}{\overset{R''}{Al}}(CH_2—CH_2)_nR + 3R'OH \longrightarrow$$

$$\uparrow \quad AL(OR')_3 + 2R''H + H(CH_2—CH_2)_nR$$

low energy coordinate linkage to $TiCl_3$, R'' = alkyl or Cl.

To make high-molecular weight polymer, such active hydrogen compounds as acids, water, and alcohol must be rigorously excluded. An inert aliphatic hydrocarbon (e.g., hexane, refined kerosene, isooctane, or cyclohexane) is the usual reaction medium.

Maximum polymer content of the slurry ranges from 20 to 30 per cent depending on the solvent, temperature, and reaction system used. Catalyst efficiency depends primarily on 1) the purity of the solvent and the reactants, and 2) the temperature and other factors which affect molecular weight. Efficiencies as high as 12,000 pounds of polymer per pound mole of trialkylaluminum have been reported,[92] thus, keeping the cost of the expensive trialkylaluminum—$TiCl_4$ catalyst within reason. The density of the polyethylene produced is in the range of 0.940–0.965 gm/cm^3.

Separation of the transition metal catalyst from the polymer is essential to good heat stability and is achieved by a complicated procedure. First, excess solvent is wrung out with a centrifuge. The polymer cake is then successively extracted as a slurry in a higher alcohol (C_4 or higher) to remove the catalysts. Finally, the alcohol is removed by steam distillation; and drying follows. The dried fluffy powder must then be densified, stabilized against heat and light, colored, lubricated, etc. to make a molding powder. These last steps are ordinarily performed simultaneously in an extruder or Banbury.

Polypropylene. Polypropylene,[74] which was commercially produced for the first time in 1957, exceeded a sales volume of one billion pounds in 1970. This thermoplastic possesses excellent film optics, high resiliency in molded parts, high-tensile properties and excellent resistance to stress cracking. Like most crystalline polyolefins it has excellent chemical resistance. It is, however, susceptible to oxidative and ultraviolet attack unless stabilized, and has poor low-temperature properties.

The oldest and largest-volume polypropylene is the unmodified homopolymer which is available in a wide range of molecular weights. With proper additives, polypropylene homopolymer is widely used in molded items, film, and fiber applications. The characteristics of polypropylene can be significantly changed through the use of fillers, such as talc or asbestos, which are added to improve scuff resistance, dimensional stability, and heat distortion.

For applications where the homopolymer does not possess adequate impact strength, blends with polypropylene copolymers are used. These may be random or block copolymers.[137] Ethylene is the most common comonomer.

Although several catalyst systems are applicable to the manufacture of crystalline polypropylene, those based on titanium halides are the most extensively used and studied. For the formation of isotactic polypropylene the catalyst must be heterogeneous, i.e., the catalyst must be a solid on which the linear head-to-tail polymer chain can grow. The form δ of violet titanium trichloride and $Al(C_2H_5)_2Cl$ is one type used.[28,61,70] The purity requirements for the propylene monomer and the diluents used in the polymerization are stringent because of the extreme reactivity of the Ziegler-Natta catalyst. Polar impurities, particularly water, destroy catalyst activity. Similarly, oxygen, carbon monoxide, carbon dioxide, hydrogen sulphide, etc. must be rigorously excluded for optimum process operation and efficiency. The polymerization may be carried out in either a batch or continuous process in a hydrocarbon diluent, e.g., *n*-heptane. Reaction conditions most often cited are 40–80°C temperatures and 1–25 atm. pressures. Stirred tank reactors are generally used, and the catalyst is added to the vessel as a slurry in the polymerization diluent. (Hydrogen is an effective chain transfer agent to control molecular weight.) The purified propylene monomer is added to the reactor, and the polymer which is formed precipitates out as a finely divided solid enveloping the catalyst particles. Slurry concentrations are not allowed to exceed 25–40 per cent solids. Residence time varies from minutes to hours depending on catalyst activity and reaction conditions. The slurry is transferred to a stripping vessel where the unreacted monomer is flashed off. The catalyst is solubilized by addition of an alcohol, diketone, or alkylene oxide. Diluent and solubilized catalyst are removed by centrifuging,

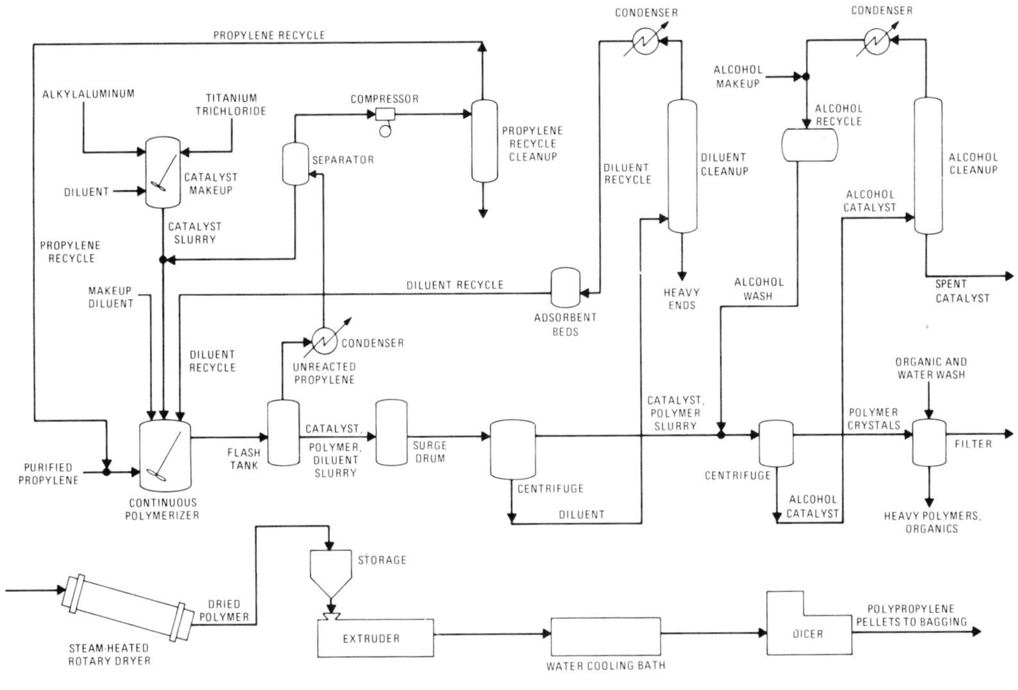

Fig. 10.21 A continuous process for polypropylene manufacture which is similar to the process for "Ziegler" polyethylene. (*"Encyclopedia of Polymer Science and Technology," Vol. 2, H. G. Mark and N. G. Gaylord, Eds., Interscience Publishers, 1968; by permission of John Wiley and Sons, Inc.*)

filtration, or aqueous extraction, and the polymer is dried (see Fig. 10.21).[27,31,101,107]

CELLULOSE DERIVATIVES

Cellulose is the most common naturally occurring high polymer of organic chemical nature. The cell walls of nearly all plants consist of cellulose.* Throughout the world, over a trillion pounds of cellulose are converted annually to paper products, regenerated cellulose (rayon and cellophane), and cellulose-based plastics, with less than 0.1 per cent being utilized in the plastics industry.

Although cellulose is a linear molecule, heating it below its charring temperature will not soften it sufficiently to permit plastic flow. The cause for this behavior is believed to be the strong hydrogen bonds between neighboring molecules forming highly stable crystalline structures with regular lattice spacings. To solubilize or fluidize the molecule, the crystalline structure is broken up by substituting various groups at the hydroxyl groups of the cellulose. The xanthate group, $-\mathrm{O}\overset{\mathrm{S}}{\overset{\|}{\mathrm{C}}}\mathrm{SNa}$, and the cuprammonium complex are used commercially to temporarily dissolve cellulose in caustic solutions for eventual regeneration to rayon or cellophane. More permanent substitution is achieved with organic acid groups, ether groups, and inorganic acid groups. Cellulose acetate, propionate, acetate-butyrate, and nitrate are commercial plastics, as is the ethyl ether of cellulose.

Preparation of *nitrocellulose* uses both nitric acid and sulfuric acid which aids nitration (1) by forming cellulose sulfate which causes the fibers to swell rapidly, and (2) by complexing with the water formed, thus forcing nitration to a high level. Since only about 1.9–2.0 of the three hydroxyls available in a "glucose" ring are substituted for plastic applications, more water is used in the nitrating acid than would be used in the preparation of guncotton. A suitable nitrating acid for plastics consists of 61 per cent H_2SO_4, 21 per cent HNO_3 and 18 per cent H_2O. After nitration at low temperatures (75–85°F) to minimize acid cleavage of the "glucose" rings which lowers the molecular weight and impairs impact strength, the spent acid is centrifuged off; the centrifuged fibers are washed thoroughly and boiled in water which is periodically drained off and replaced. This boiling, which takes from 10 to 60 hours, removes the sulfate groups that cause product instability. After boiling, the fibers are bleached with 1 per cent chlorine and treated with sodium sulfite. To remove the water the wet fibers are pressed tightly together and alcohol is percolated through the pressed cake. The alcohol-wet cake (30 per cent alcohol by weight) is then charged to a powerful mixer together with 26 per cent camphor to plasticize the polymer.

The alcohol, as well as the camphor, fluid-

*Native cellulose may contain some rings other than the "glucose" (or more properly, anhydro-β-D-glucopyranose) ring, but such portions of the "cellulose" chain are considered anomalous and "pure" cellulose usually means the following:

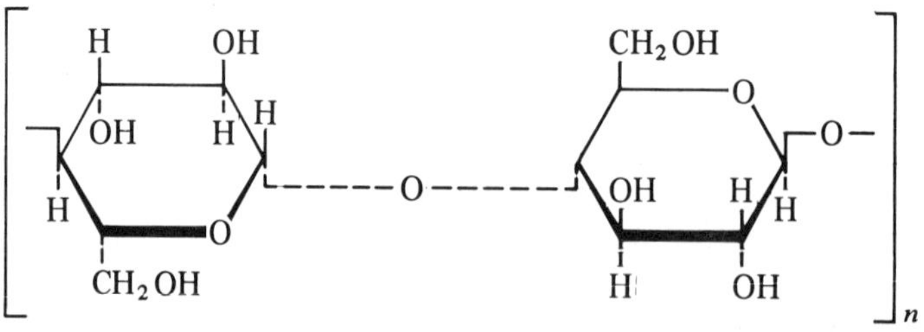

where the end groups in the naturally occurring product, though somewhat uncertain, are believed to consist equally of reducing end groups and nonreducing end groups, in cellulose which has undergone hydrolytic chain cleavage to lower the molecular weight, these end groups are sometimes acidic in nature, presumably because of oxidation.

izes the polymer, permitting extrusion or other forming operations to be carried out in comparative safety. After forming, the alcohol is removed by very slow aging or drying to minimize distortion. During all the above processes prior to the final aging, it is essential that the nitrated cellulose be kept wet with water or alcohol since serious fires can easily be started by slight frictional heating. Because of the necessity for final aging and because of great sensitivity to heat degradation, nitrocellulose is not a thermoplastic in the ordinary sense; however, neither is it a thermosetting plastic.

To make cellulose acetate (or propionate or butyrate), the purified alpha-cellulose pulp [usually obtained by the sulfite process from wood (Chapter 15)] is shredded and mixed with acetic acid to wet and swell the fibers. This mixture is then fed to the acetylation reactor where sulfuric acid catalyst (1 per cent cellulose), methylene chloride solvent (or more acetic acid as solvent), and acetic anhydride are added. Acetylation is done at moderate temperatures, e.g., 90–120°F, to minimize acid cleavage. Following acetylation, wet acetic acid is added to remove the sulfate ester groups and to deesterify the cellulose to the desired level. During the acetylation process the fibers dissolve in the solvents. This requires precipitation in water and subsequent milling to make a flake and to squeeze out most of the acid and water. The flake is then washed with water in a countercurrent band filter; centrifuging, drying, and blending follow (Fig. 10.22). To make a molding powder the flake is rapidly cold-mixed with a plasticizer but not homogeneously blended. The mix is then fed to a set of hot mill rolls where colorants are added; the plasticizer is then homogeneously mixed. The mill-rolled sheet is diced to molding powder pellets. Alternatively, the flake and the plasticizer containing dispersed colorants are fed continuously to a heated extruder which homogenizes the mixture and forms strands which are subsequently cooled and chopped into molding powder. To make the mixed acetate-butyrate* or acetate-propionate, butyric acid or propionic acid is added to the acetylating mixture.

Ethyl cellulose for plastic purposes usually contains 2.3–2.5 ether groups per "glucose" ring. It is made by reacting cellulose under alkaline conditions in the presence of water and ethyl chloride. By-products are ethyl alcohol and ethyl ether formed via hydrolysis of the ethyl chloride.

Following the reaction, the reaction mass is mixed and simultaneously precipitated and atomized to fine granules; a special two-fluid spray nozzle is used in which the "gaseous" fluid stream is steam.

The polymer spray is contained in a precipitating tank where the volatile benzene, ethyl chloride, ethyl ether, and ethyl alcohol are steam distilled off. After precipitation the polymer is centrifuged, washed with hot

*Usually about 1.7 butyryl and 1.0 acetyl groups are substituted per "glucose" ring.

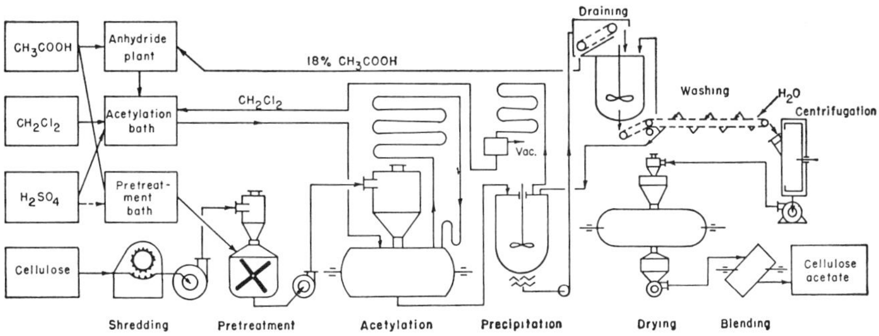

Fig. 10.22 Manufacture of cellulose acetate by the methylene dichloride method. (*After Ranby, B. G. and Rydholm, S. A., "Cellulose and Cellulose Derivatives," in "Polymer Processes," C. E. Schildknecht, Ed., Interscience Publishers, 1956; by permission John Wiley and Sons, Inc.*)

water, and bleached with dilute sodium hypochlorite (pH 10.5–11.3 at 175°F); another centrifuging and washing follow. Stabilization (deashing) is achieved with a dilute acetic acid rinse (pH 4–4.2) and a final centrifuging and water rinsing; this is followed by drying in a countercurrent rotary steam tube or a concurrent rotary hot air drier. The dried crumb is then mixed with colorants, stabilizers, and plasticizers, and extruded to strands which are chopped into molding-powder pellets.

The above is but a brief description of the cellulosic plastics; Ott and Spurlin[94] and others[17,60,88,95,103] give more complete accounts.

OTHER PLASTICS

Besides the plastics already discussed, there are several others that are used either in small amounts as plastics or extensively for applications other than as plastics as narrowly defined in this chapter.

Polyvinyl butyral is used for the interlayer of safety glass. Although for this purpose it is the best plastic on the market, it has few other uses. Its formula may be represented by:

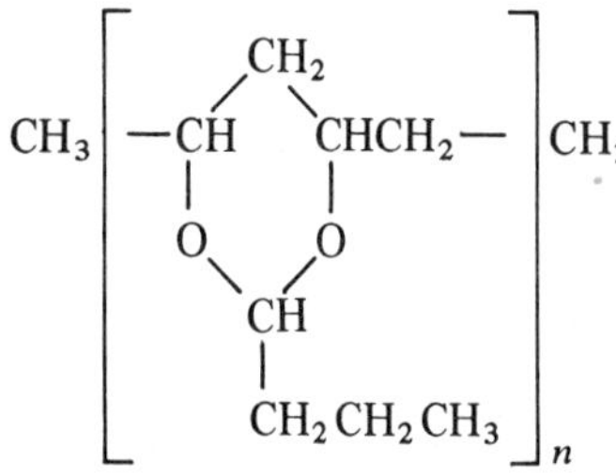

although some acetyl groups and alcohol groups are also present. It is made by polymerizing vinyl acetate $(CH_2{=}CHO\overset{\overset{O}{\|}}{C}CH_3)$ either in suspension or solution with free-radical catalysts, hydrolysis of the resulting polymer with sulfuric acid in an alcoholic solution to yield polyvinyl alcohol, and then acetal formation with butyraldehyde using a mineral acid catalyst.[6]

Polyvinyl acetate is used extensively for coatings, textile and paper finishes, and adhesives. *Polyvinyl alcohol* also has a large specialty market. *Polyvinyl formal* is made by a similar process for use as an insulating varnish.

Silicone plastics are very well known and much publicized. They may be represented by the generic formula

$$R'\left[\begin{matrix}R\\ SiO\\ R\end{matrix}\right]_n R'$$

where R is usually $—CH_3$ or 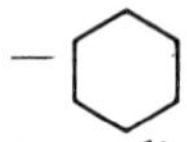. They are produced from the corresponding diorganosilicon chlorides (R_2SiCl_2) plus water, and acid or base condensation catalysts.[78,86] A very small amount of the triorgano-substituted silicon chloride will terminate the chain and control molecular weight. Branching, and eventual crosslinking may be achieved by using a monoorganosilicon chloride ($RSiCl_3$) or by adding oxidizing agents (e.g., benzoyl peroxide) to oxidize the methyl groups as in the vulcanization of silicone rubbers. Allyl groups have been utilized to permit vinyl-type crosslinking to occur with free radical-generating catalysts.

Depending on the conditions, oils, heat-hardenable resins, or rubbers may be formed. Silicones are used commercially for potting compounds, surface-coating enamels (particularly in electrical applications): stable rubbers, lubricating oils and greases which have extraordinarily low variations in viscosity with temperature changes. Specialty uses include mold release agents, antifoam agents, water repellents, polyester-glass bonding agents, and more recently, due to their inert character, surgical uses, espeically in cosmetic surgery. Their use in special heat-stable moldings, castings, and laminates is minimal.

Polyurethanes were manufactured in Germany during World War II when the Bayer laboratories developed their use in rigid foams, adhesives, and coatings. Polyurethanes have developed rapidly in the United States. Since their discovery in 1937, the application of polyurethanes has grown until in 1968, 500 million pounds of the polymer were produced in the United States alone. Following World War II, in the late 1940's, Du Pont and Monsanto Company began supplying 2,4-tolylene diisocyanate in pilot plant quantities.[100] Polyurethanes are those polymers which contain a significant number of urethane groups, regardless of the composition of the rest of the molecule, and are now largely used in flexible (rubbery) foams and surface coatings, though they are used to some extent in rigid foams and laminates. Two methods are used for producing foam:[32,112] (1) The "one-shot" process involves simultaneous mixing of all reactants, catalyst, and additives. The reactions begin at once; foaming is completed in a couple of minutes, but several hours may be required to complete the cure. (2) In the "prepolymer" method the isocyanate and a diol are reacted to produce a prepolymer. This product is foamed later by reaction with water. The molecular weight is increased during foaming. Typical reactions for the second method are:

(1) $2R(C{=}N{=}O)_2 + HOR'OH \longrightarrow$

$$O{=}C{=}N{-}R{-}\overset{H}{N}{-}\overset{O}{\overset{\|}{C}}{-}O{-}R'{-}O{-}\overset{O}{\overset{\|}{C}}{-}\overset{H}{N}{-}R{-}N{=}C{=}O \quad \text{“Prepolymer”}$$

(2) $$n(\text{Prepolymer}) + nH_2O \longrightarrow \left[{-}\overset{H}{N}\overset{O}{\overset{\|}{C}}\overset{H}{N}{-}R{-}\overset{H}{N}\overset{O}{\overset{\|}{C}}O{-}R'{-}O\overset{O}{\overset{\|}{C}}\overset{H}{N}{-}R{-}\right]_n + nCO_2$$

A number of new *heat-resistant polymers* were developed during the 1960's. Heat resistance, or thermal stability, is a measure of the ability of a material to maintain mechanical properties such as strength, toughness, and elasticity at a given temperature.[90] The "use temperature" of many polymers is limited by changes in their physical character at high temperatures, that is they soften or melt, rather than by changes in their chemical character. A typical linear polymer such as polystyrene has a maximum use-temperature of about 80–100°C. The melting point of a polymer may be increased by the introduction of polar substituents, such as halogen atoms and nitrile groups–polytetrafluoroethylene, for example, has a maximum use-temperature of about 290–310°C; or hydrogen-bonding groups such as amides.

In addition to physical factors, there are chemical factors which determine the thermal stability of a polymer. Polymers with high heat resistance must have bonds of high dissociation energy as well as structural features which do not allow degradation by low-energy processes. Another chemical factor affecting heat stability is the formation of crosslinked structures under the influence of heat with resultant changes in the physical properties of the polymer. Further, for a thermally stable polymer to be useful it must also be chemically inert to oxygen, moisture, acids, bases, or other substances to which it might be exposed.

The synthesis of polymers containing aromatic units in the polymer-chain backbone was prompted by the known high bond energies, low degree of reactivity, and the rigidity of aromatic structures. These are the most successful of the new heat-resistant polymers. One of the first developed by Marvel[41,83] is *poly*-p-*phenylene* which is an insoluble, infusible material with good thermal stability. *Polythiazoles*, among the first fully aromatic condensation polymers to be made, are composed of alternating phenyl and thiazole rings. Though they are quite stable thermally, they cannot be produced at a sufficiently high molecular weight to form films. *Polyimides* are extremely heat stable, in addition to being flame-, radiation-, and oxidation-resistant; they also possess excellent electrical characteristics. For example, they have a thermal stability of over a year at 275°C.[90] Due to this unique combination of properties polyimides, e.g., Skybond (Monsanto Co.) or Kaplon (DuPont), are used in many areas. For example, polyimides reinforced with glass or boron, or glass-based polyimide honeycombs are used as structural components in aircraft. Polyimide-graphite composites show promise in the construction of fan blades in jet-engine aircraft because of their high heat resistance and the weight savings over titanium blades. Polyimide adhesives are used for metal surfaces, such as titanium and aluminum. They are also used in the production of printed circuit boards with high-temperature operating capabilities.

Other polymers with superior thermal stability are the *polybenzimidazoles* and derivatives, *polytetrazopyrenes*, *polyquinoxalines*, and the various *ladder* and *spiro polymers*. Spiro polymers are noncrosslinked structures in which two molecular strands are joined to each other. These polymers could be soluble and fusible, and would be expected to have better thermal properties than the simple stranded type as at least two bonds must be broken for a reduction in molecular weight to take place. Ladder polymers are still in the developmental stage and techniques to make structures with higher molecular weights and no breaks in the ladder are being investigated.[8] Recently Marvel and co-workers have developed a pyrrolone-type ladder polymer which is thermally stable above 500°C.

Nylon is used extensively as a plastic as well as a fiber. Over 50 million pounds were used in plastic applications in 1970. Gears, bearings, and other unlubricated machine parts are particularly appropriate applications. Its manufacture is discussed in Chapter 11.

PLASTICIZERS

Besides the large industry engaged in manufacturing base resins, a sizable synthetic organic chemical industry is based on plasticizers for these resins. Over 1.5 billion pounds of plasticizers were made in 1957. These materials impart flexibility and formability to many polymers, particularly those based on vinyl chloride and cellulose. Chemically they are frequently aliphatic esters (usually C_4 or higher) of dibasic acids, e.g., phthalic, adipic, azelaic, and sebacic acids. Epoxidized unsaturated natural esters (soybean oil) and similar unsaturated esters are increasing in importance. Phosphate esters, polyester resins, and aromatic hydrocarbons of high molecular weight are also common.

The list of commercial polymeric materials is still growing despite the fact that each new product must compete with the properties and prices of existing polymers. The perfect polymer for all plastic applications is an impossibility. A plastic that is resistant to high temperatures yet is heat-sealable and readily molded at low temperatures, that has excellent outdoor stability but is transparent, that has excellent hardness and abrasion resistance yet is strong and tough, that may be either rigid or flexible as desired, that is an excellent electrical insulator but also dissipates static charge at a high rate, and that is resistant to all chemicals yet can form solution coatings—such a plastic exists only in salesmen's dreams and chemist's nightmares.

REFERENCES

1. Akin, R. B., "Acetal Resins," New York, Van Nostrand Reinhold Co., 1962.

1a. Albright, L. F., *Chem. Eng.* **73,** 113 (Dec. 19, 1966).

2. Allen, I., *et al.*, U.S. Patent 2,496,653 (Feb. 1952; to Union Carbide).

3. Allen, I., *et al.*, U.S. Patent 2,614,910 (Oct. 1952; to Union Carbide).

4. Amos, J. L., *et al.*, U.S. Patents 2,494,924; 2,530,409; 2,714,101; 2,941,985; 3,058,965 (to Dow Chemical).

5. Amos, J. L., *et al.*, U.S. Patent 2,849,430 (to Dow Chemical).

6. *Chem. Eng.*, **61,** 346 (Feb. 1954).

6a. *Brit. Plastics*, **30,** 26 (Jan. 1957).

7. *Chem. Eng.*, **38,** 124 (Nov. 14, 1960).

8. *Chem. Eng. News*, **48,** No. 34, 36 (1970).

9. Baeyaert, A. E. M., U.S. Patent 2,715,117 (Aug. 1955); U.S. Patent 2,856,272 (Oct. 1958; to Saint-Gobain).

10. Basdekis, C. H., "ABS Plastics," New York, Van Nostrand Reinhold, 1964.

11. German Patent 1,125,175 (1962; to B.A.S.F.).

12. Bjorksten Research Laboratories, Inc., "Polyesters and Their Applications," New York, Van Nostrand Reinhold, 1956.
13. Blais, J. F., "Amino Resins," New York, Van Nostrand Reinhold, 1959.
14. Bottenbruch, L., in "Encyclopedia of Polymer Science and Technology, Vol. X, p. 710, New York, John Wiley & Sons, 1969.
15. Boundy, R. H., and Boyer, R. F., "Styrene," New York, Van Nostrand Reinhold, 1952.
16. Bovey, F. A., Kolthaff, I. M., Modulin, A. I., and Mechon, E. J., "Emulsion Polymerization," pp. 71–93, New York, John Wiley & Sons, 1955.
17. Bracken, W. O., "Cellulose Plastics," in "Encyclopedia of Chemical Technology," R. E. Kirk and D. F. Othmer, Vol. III, pp. 391–411, New York, John Wiley & Sons, 1949.
18. Canterino, P. J., in "Encyclopedia of Polymer Science and Technology, Vol. VI, p. 275, New York, John Wiley & Sons, 1967.
19. Carothers, W. H., "Collected Papers," H. Mark and G. S. Whitby, Eds., New York, John Wiley & Sons, 1940.
20. Carswell, T. S., "Phenoplast S," New York, John Wiley & Sons, 1947.
21. Castan, Pierre, *et al.*, U.S. Patents 2,324,483 (Jul. 1943); 2,444,333; 2,458,796; 2,637,715 (to Ciba).
22. Charlesworth, R. K., Murdock, S. A., Shaw, K. G., U.S. Patent 3,201,365 (to Dow Chemical).
23. Chopey, N. P., *Chem. Eng.*, **68**, No. 2, 112 (1961).
24. Christopher, W. F., and Fox, D. W., "Polycarbonates," New York, Van Nostrand Reinhold, 1962.
25. Clegg, P. L., in "Encyclopedia of Chemical Technology," R. E. Kirk and D. F. Othmer, Eds., Vol. XIV, p. 217, New York, John Wiley & Sons, 1967.
26. Coderre, R. A., "Epoxy Resins," in "Encyclopedia of Chemical Technology," R. E. Kirk and D. F. Othmer, Eds., Vol. I (Supplement), pp. 312–329, New York, John Wiley & Sons, 1957.
27. Compostella, M., "The Manufacture and Commercial Applications of Stereoregular Polymers," in "The Stereochemistry of Macromolecules," A. D. Ketley, Ed., Vol. A, Chapter 6, pp. 309–387, New York, Marcel Dekker, 1967.
28. Cossee, P., "The Mechanism of Ziegler-Natta Polymerization. II. Quantum-Chemical and Crystal-Chemical Aspects," in "The Stereochemistry of Macromolecules," A. D. Ketley, Ed., Vol. I, Chapter 3, pp. 145–175, New York, Marcel Dekker, 1967.
29. DeJong, J. I., *Rec. Trav. Chim.*, **71**, 643 (1952); **72**, 88, 653, 1027 (1953); **73**, 139 (1953).
30. Dinsmore, R. P., U.S. Patent 1,732,795 (Oct. 22, 1929; filed Sept. 13, 1927; to Goodyear).
31. Doak, K. W., and Schrage, A., "Commercial Polymerization and Copolymerization Processes," in "Crystalline Olefin Polymers," R. A. V. Raff and K. W. Doak, Eds., Part I, Chapter 8, New York, John Wiley & Sons, 1964.
32. Dombrow, B. A., "Polyurethanes," Van Nostrand Reinhold, 1965.
33. Douglas, S. D., U.S. Patent 2,075,429 (to Union Carbide).
34. Dunlop, R. D., and Reese, F. E., *Ind. Eng. Chem.*, **40**, 654 (1948).
35. British Patent 770,717 (April 1955; to Du Pont).
36. Einhorn, A., *Ann. Chem.* **300**, 135 (1898).
37. Ellis, C., "The Chemistry of Synthetic Resins," pp. 123–141, 201, 231, New York, Van Nostrand Reinhold, 1935.
37a. Ehrlich, P., and Mortimer, G. A., *Advan. Polymer Sci.*, **7**, 386 (1970).
38. Fawcett, E. W., Gibson, R. D., Perrin, W., Paton, J. G., and Williams, E. G., British Patent 471,590 (Sept. 1937; to I.C.I.); U.S. Patent 2,153,553 (April 1939; U.S. Patent 2,188,465 (June 1940).
39. Flory, P. J., "Principles of Polymer Chemistry," Ithaca, Cornell University Press, 1953.
40. Fox, D. W., Australian Patent 221,192 (1959; to General Electric).
41. Frey, D. A., Hasegawa, M., and Marvel, C. S., *J. Polymer Sci.*, A, **1**, 2067 (1963).
42. Gaylord, N. G., and Mark, A., "Linear and Stereospecific Addition Polymers," New York, John Wiley & Sons, Inc., 1959.
43. Gemens, H., *Advan. Polymer Sci.*, **1**, 2349 (1959).
44. Gordon, J. L., *J. Polymer Sci.*, A-1, **6**, 623, 643, 665, 687 (1968).
45. Gould, D. F., "Phenolic Resins," Van Nostrand Reinhold Co., 1959.
46. Greenwalt, C. H., U.S. Patent 2,388,138 (Oct. 1945; to Du Pont).

47. Greenlee, S. O., *et al.*, U.S. Patents 2,456,408 (Dec. 1948); 2,503,726; 2,510,885-6; 2,511,913; 2,512,996; 2,521,911-12; 2,528,359-60; 2,538,072; 2,558,949; 2,581,464; 2,582,985; 2,589,245; 2,592,560; 2,615,007-8; 2,694,694; 2,698,315; 2,712,000; 2,717,885 (to Devoe and Raynolds).
48. Grim, J. M., U.S. Patent 2,715,118 (Aug. 1955; to Koppers).
49. Hanford, W. E., and Sargent, D. E., "Reactions of Organic Gases Under Pressure," in "Organic Chemistry," H. Gilman *et al.*, Eds., Vol. IV, pp. 1024–1042, New York, John Wiley & Sons, 1953.
50. Hanson, A. W., Heston, A. L., and Buecken, H. E., U.S. Patent 2,519,834 (1950; to Dow Chemical).
51. Hanson, A. W., U.S. Patents 2,488,198 (1949; to Dow Chemical), and 2,769,804 (Nov. 1956; to Dow Chemical).
52. Hanson, A. W., and Zimmerman, R. L., *Chem. Eng. News*, **77** (1957).
53. Harkins, W. D., "Physical Chemistry of Surface Films," p. 332, New York, Van Nostrand Reinhold, 1952.
54. *Ibid*, p. 339.
55. Harkins, W. D., *J. Chem. Phys.*, **13**, 47, 381 (1945); and *J. Am. Chem. Soc.*, **69**, 1928 (1947).
56. Harries, C., *Ann. Chem.*, **383**, 213 (1911).
57. Hartman, S. J., *Plastics World*, **27**, No. 8, 42 (1969).
58. Hay, A. S., Blanchard, H. S., Endres, E. F., and Estance, J. W., *J. Am. Chem. Soc.*, **81**, 6335 (1959).
59. Hay, A. S., *et al.*, in "Encyclopedia of Polymer Science and Technology," Vol. X, p. 94, New York, John Wiley & Sons, 1969.
60. Hill, R. O., "Cellulose, Esters, Organic," in Encyclopedia of Polymer Science and Technology, Vol. III, p. 307, New York, John Wiley & Sons, 1965.
61. Hoeg, D. F., "The Mechanism of Ziegler-Natta Catalysis. I. Experimental Foundations," in "The Stereochemistry of Macromolecules," A. D. Ketley, Ed., Vol. I, Chap. 2, pp. 47–144, New York, Marcel Dekker, 1967.
62. Hogan, J. P., and Banks, R. L., U.S. Patent 2,825,721 (March 1958; to Phillips Petroleum).
63. Horn, M. B., "Acrylic Resins," Van Nostrand Reinhold, 1960.
64. Hull, J. L., in "Encyclopedia of Polymer Science and Technology," Vol. IX, pp. 1–47, New York, John Wiley & Sons.
65. Hunter, E., and Richards, R. B., U.S. Patent 2,457,238 (Dec. 1948; to I.C.I.).
66. Ilicoto, *Chimi. Ind.* (Milan), **34**, 688 (1952).
66a. Jezl, J. L., Honeycutt, E. M., in "Encyclopedia of Polymer Science and Technology," Vol. XI, p. 603, New York, John Wiley & Sons, 1970.
67. Johnson, R. N., in "Encyclopedia of Polymer Science and Technology," Vol. XI, pp. 447–463, New York, John Wiley & Sons, 1969.
68. Jones, C., Harris, B., and Ingley, F. L., U.S. Patent 2,739,142 (March, 1956; to Dow Chemical).
69. Jones, T. T., *J. Soc. Chem. Ind.*, **69**, No. 9, 264 (1946).
70. Jordan, D. O., "Ziegler-Natta Polymerization: Catalysts, Monomers, and Polymerization Procedures," in "The Stereochemistry of Macromolecules," A. D. Ketley, Ed., Vol. I, Chapter 1, pp. 1–45, New York, Marcel Dekker, 1967.
70a. Kato, K., *Polymer Eng. Sci.*, **7**, 38 (1967).
71. Kestler, J., *Mod. Plastics*, pp. 58–59, Aug. 1969.
72. Klevins, H. B., *Chem. Revs.*, **47**, 1 (1950).
73. Krase, N. E., U.S. Patent 2,388,160 (Oct. 1945; to Du Pont).
74. Kresser, T. O. J., "Polypropylene," New York, Van Nostrand Reinhold, 1960.
75. Larson, A. T., U.S. Patent 2,405,962 (Aug. 1946; to Du Pont).
76. Lawrence, J. R., "Polyester Resins," New York, Van Nostrand Reinhold, 1960.
77. Lee, H., and Neville, K., "Epoxy Resins," New York, McGraw-Hill, 1957.
78. Lichtenwalner, H. K., Sprung, M. N., "Silicones," in Encyclopedia of Polymer Science and Technology, Vol. XII, p. 464, New York, John Wiley & Sons, 1970.
79. Luther, M., and Heuck, C., U.S. Patent 1,860,681 (filed July 17, 1928; issued June 21, 1932; filed in Germany Jan. 8, 1927; to I. G. Farben).

80. Mantell, C. L., Kopf, C. W., Curtis, J. L., and Rodgers, E. M., "The Technology of Natural Resins," New York, John Wiley & Sons, 1942.
81. Mark, H., and Hohenstein, W. P., *J. Polymer Sci.*, **1**, 127 (1946).
82. Martin, R. W., "The Chemistry of Phenolic Resins," New York, John Wiley & Sons, 1956.
83. Marvel, C. S. and Hartzell, G. E., Jr., *J. Am. Chem. Soc.*, **81**, 488 (1959).
84. Matthews, F. E., and Strange, E. H., German Patent 249,868.
85. McDonald, D. L., Colomer, Coultor, K. E., and McCurdy, J. L., U.S. Patent 2,727,884 (to Dow Chemical).
86. Meals, R., in "Encyclopedia of Chemical Technology," R. E. Kirk and D. F. Othmer, Eds., 2nd Ed., Vol. XVIII, p. 221, New York, John Wiley and Sons, 1969; (Meals, R. N., and Lewis, F. M. "Silicones" New York, Van Nostrand Reinhold, 1959.)
87. Megson, J. I. L., "Phenolic Resin Chemistry," New York, Academic Press, 1958.
88. Miles, F. D., "Cellulose Nitrate," New York, John Wiley & Sons, 1955.
89. Miller, W. T., U.S. Patent 2,579,437 (Dec. 1951; to M. W. Kellogg).
90. Mulvaney, J. G., "Heat-Resistant Polymers," Encyclopedia of Polymer Science, Vol. VII, p. 478, New York, John Wiley & Sons, 1967.
91. Natta, G. and Danusso, F., Eds., "Stereoregular Polymers and Stereospecific Polymerizations," 2 Volumes, Elmsford, N.Y., Pergamon Press, 1967.
92. Orzechowski, A., *J. Polymer Sci.*, **30**, 62 (1959).
93. Ott, E., Spurlin, H. M., and Grafflin, M. W., Eds., "Cellulose and Cellulose Derivatives," Parts I, II, III, New York, John Wiley & Sons, 1954.
94. Ott, J. B., U.S. Patents 2,862,912 and 3,051,682 (to Monsanto).
95. Paist, W. D., "Cellulosics," New York, Van Nostrand Reinhold, 1958.
96. Parker, G. R., *Chem. Eng. News*, **47**, No. 36, 64A (1969).
97. Perry, E., and Reese, F. E., *Ind. Eng. Chem.*, **40**, 2039 (1948).
98. Pezzoglia, P.,U.S. Patent 2,841,595 (July 1958; to Shell Development).
99. Phillips, B., and MacPeek, D. L., "Peracetic Acid," in "Encyclopedia of Chemical Technology," Kirk, R. E., and Othmer, D. F., Eds., Vol. I (Supplement), pp. 622–642, New York, John Wiley and Sons, 1957.
100. Pigott, K. A., in "Encyclopedia of Polymer Science and Technology, Vol. XI, p. 506, New York, John Wiley and Sons, 1969.
101. Platzer, N., *Ind. Eng. Chem.*, **62**, No. 1, 6 (Jan. 1970).
102. Raff, R. A. V., and Allison, J. B., "Polyethylene," New York, John Wiley and Sons, 1958.
103. Ranby, B. G., and Rydholm, S. A., "Cellulose and Cellulose Derivatives," in "Polymer Processes," Schildknecht, C. E., Ed., pp. 351–428, New York, John Wiley and Sons, 1956.
104. Rearick, J. S., U.S. Patent 2,600,804 (Jun. 1952; to M. W. Kellogg).
105. Reid, E. W., U.S. Patent 2,064,565 (to Union Carbide).
106. Renfrew, A., Morgan, A., "Polythene," New York, John Wiley & Sons, 1960.
107. Repka, Jr., B. C., in "Encyclopedia of Chemical Technology," Vol. XIV, p. 282, New York, John Wiley & Sons, 1967.
108. Richard, W. R., Stewart, R. K., and Calfee, J. D.,U.S. Patent 2,852,501 (Sept. 1958; to Monsanto).
109. Roberts, D. E., *J. Res. Nat. Bu. St.*, **44**, 221 (1950).
110. Robitschek, P., and Lewin, A., "Phenolic Resins," London, Iliffe and Sons, 1950.
111. Ruffing, N. R., U.S. Patent 3,243,481 (to Dow Chemical).
112. Saunders, J. H., Frisch, K. C., "Polyurethanes. Part I: Chemistry; Part II: Technology," New York, John Wiley & Sons, 1962, 1964.
113. Schildknecht, C. E., "Vinyl and Related Polymers," pp. 197–203, New York, John Wiley and Sons, 1952.
114. Tromsdorff, E., "Polymerizations in Suspension," in "Polymer Processes," C. E. Schildknecht, Ed., Chap. III, New York, John Wiley & Sons, 1956.
115. Schneider, A. K., McDonald, R. N., Cairns, T. L., *et al.*, U.S. Patent 2,768,994 (Oct. 1956); 2,775,570 (Dec. 1956); 2,795,571 (June 1957); 2,828,286 (March 1958); 2,828,287 (March 1958; to Du Pont).
116. Schnell, H., *Angew. Chem.*, **68**, 633 (1956).
117. Schnell, H., "Chemistry and Physics of Polycarbonates," New York, John Wiley & Sons, 1964.

118. Schnell, H., Bollenbruch, L., and Krim, H., Belgian Patent 532,543 (1956; to Fabenfabriken Bayer).
119. Shanta, P. L., U.S. Patent 2,694,700 (1954).
120. Simon, R. H. M., and Oster, B., U.S. Patent 3,345,430 (Oct. 1967; to Monsanto).
121. Sittig, M., "Polyolefin Resin Processes," Houston, Tex., Gulf Publishing, 1961.
122. Skeist, I., and Somerville, G. R., "Epoxy Resins," New York, Van Nostrand Reinhold, 1958.
123. Sloane, D. J., "Injection Molding Machinery," in "Machinery and Equipment for Rubber and Plastics," Seaman and Merril, Eds., Vol. I, New York, Rubber World, 1952.
124. Smith, W. V., and Ewart, R. H., *J. Chem. Phys.*, **16**, 592 (1948).
125. Sonneborn, R. H., "Fiberglass Reinforced Plastics," New York, Van Nostrand Reinhold, 1954.
126. Stober, K. E., et al., U.S. Patent 2,530,409 (May 1948; to Dow Chemical).
127. Sundberg, D. C., Ph.D. Thesis, Univ. of Delaware, 1970.
128. Swallow, J. C., "The History of Polythene," Renfrew and Morgan, Eds., Chap. I, London, Iliffe and Sons, 1957.
129. Teach, W. C., Kiessling, G. C., "Polystyrene," Reinhold, 1960.
130. Thomas, J., *Hydrocarbon Process.*, **47**, No. 11, 192, (Nov. 1968).
131. Thomasson, R. L., *et al.*, *Petrol. Refiner*, **35**, 191 (Dec. 1956).
132. Thompson, R. J., and Goldblum, K. B., *Mod. Plastics*, **35**, 131 (1958).
133. Tromsdorf, E., "Polymerizations in Suspension," in "Polymer Processes," Schildknecht, C. E., Ed., Chapter 3, New York, John Wiley and Sons, 1956.
134. Vale, C. P., "Aminoplastics," London, Cleaver-Hume Press, 1950.
135. U.S. Patent 2,981,722 (1961; to Wacker-Chemie).
136. Walker, J. F., "Formaldehyde," pp. 281-317, New York, Van Nostrand Reinhold, 1953.
137. Walton, R. J., "1969-1970 Modern Plastics Encyclopedia," Vol. 46, No. 10A, 185, 1969.
138. Weedman, J. A., Payne, W. E., and Johnson, O. W., *Chem. Eng. Progr.*, **55**, 49 (1959).
139. Werner, E. G., Wiles, Q. T., *et al.*, U.S. Patents 2,467,171 (April 1949); 2,528,932-3-4; 2,541,027; 2,575,558; 2,640,037; 2,642,412; 2,682,515; 2,735,329; 2,716,099; 2,752,269; 2,840,511; 2,841,595 (to Shell Development).
140. Weyermuller, G., *Chem. Process.*, **21**, 87 (April 1958).
141. Wright, V., *Modern Plastics*, pp. 54-58 (Oct. 1969).
142. Wrightson, J. M., U.S. Patent 2,600,821 (June 1952; to M. W. Kellogg Co.).
143. Young, H. S., U.S. Patent 2,394,960 (Feb. 1946; to Du Pont).
144. Young, Peterson, Brooks, Brubaker, Hanford, *et al.*, U.S. Patents 2,388,178; 2,388,225; 2,394,960; 2,395,327; 2,395,381; 2,396,677; 2,396,785; 2,396,920; 2,402,137; 2,405,962; 2,409,996; 2,414,311; 2,436,256; 2,449,489; 2,462,678; 2,462,680; 2,467,234; 2,511,480; 2,703,794; 2,762,791 (to Du Pont).
145. Zavistsas, A. A., Beaulieu, R. D., Leblanc, J. R., *J. Polymer Sci.*, A-1, **6** No. 9, 2541 (1968).
146. Ziegler, K., and Gellert, H. G., U.S. Patents 2,699,457 (1955); 2,695,327.
147. Ziegler, K., Holzkamp, E., Breil, H., and Martin, H., *Angew. Chem.*, **67**, 541 (1955).
148. Ziegler, K., Martin, H., *Makromol. Chem.*, **18-19**, 186 (1956).

SELECTED REFERENCES*

Bibliography

B-1 Kaback, S. M., "Literature of Polymers," p. 273-94, in "Encyclopedia of Polymer Science and Technology," H. F. Mark, and N. G. Gaylord, Eds., Vol. VIII, pp. 273-294, New York, John Wiley and Sons, 1968.
Listing of periodicals specializing in polymers, and books through 1967.
B-2 Yescombo, E. R., "Sources of Information on the Rubber, Plastics and Allied Industries," Elmsford, N.Y., Pergamon Press, 1968.

---

*Limited to books in English.

Chemistry, General

C-1 Flory, P. J., "Principles of Polymer Chemistry," Ithaca, N.Y., Cornell University Press, 1953.
A classic, but somewhat difficult for undergraduates.
C-2 Kaufman, M., "Giant Molecules," New York, Doubleday, 1968.
Very elementary, high school level.
C-3 Lenz, R. W., "Organic Chemistry of Synthetic High Polymers," New York, John Wiley and Sons, 1967.
Excellent graduate text.
C-4 Margerison, D., East, G. C., "An Introduction to Polymer Chemistry," Elmsford, N.Y., Pergamon Press, 1966. (An introductory undergraduate text.)
C-5 Stille, J. K., "Introduction to Polymer Chemistry," New York, John Wiley and Sons, 1962.
Undergraduate text.

Commercial Preparation

CP-1 Schildknecht, C. E., Ed., "Polymer Processes," New York, John Wiley & Sons, 1956.
CP-2 Smith, W. M., "Manufacture of Plastics," New York, Van Nostrand Reinhold, 1964. (See also E-2.)

Dictionary

D-1 Whittington, L. R., "Dictionary of Plastics," Stamford, Conn., Technomic, 1968.
A SPE-sponsored publication.

Encyclopedia and Handbooks

E-1 Brandrup, J., and Immergut, E. H., Eds., "Polymer Handbook," New York, John Wiley and Sons, 1966.
E-2 Mark, H. F., and Gaylord, N. G., Eds., "Encyclopedia of Polymer Science and Technology," New York, John Wiley and Sons, 1968.

Processing and Fabrication

P-1 Benjamin, B. S., "Structural Design with Plastics," New York, Van Nostrand Reinhold, 1969.
P-2 Bernhardt, E. C., Ed., "Processing of Thermoplastic Materials," New York, Van Nostrand Reinhold, 1959.
P-3 Butler, J., Barker, H., and Ritchie, P. D., "Compression and Transfer Molding of Plastics," New York, John Wiley & Sons, 1960.
P-4 DuBois, J. H. and John, F. W., "Plastics," New York, Van Nostrand Reinhold, 1967.
P-5 Griff, A. L., "Plastics Extrusion Technology," New York, Van Nostrand Reinhold, 1968.
P-6 Kobayaski, A., "Machining of Plastics," New York, McGraw-Hill, 1967.
P-7 McKelvey, J. M., "Polymer Processing," New York, John Wiley & Sons, 1962.
P-8 Ovorkiewski, R. M., "Thermoplastics: Effects of Processing," New York, Van Nostrand Reinhold, 1969.
P-9 Person, J. R. A., "Mechanical Principles of Polymer Melt Processing," Elmsford, N.Y., Pergamon Press, 1966.
P-10 Plastics Technology Staff, "Processing Handbook," New York, Bill Publications, 1969.
P-11 Schenkel, G., "Plastics Extrusion Technology and Theory," London, Iliffe, 1966.
P-12 Simonds, H. R., ed., "The Encyclopedia of Plastic Equipment," New York, Van Nostrand Reinhold, 1964.
P-13 Tadmor, Z., and Klein, I., "Engineering Principles of Plasticating Extrusion," New York, Van Nostrand Reinhold, 1970.
P-14 Welch, C., Ed., "Plastics International—an Industrial Guide and Catalog of Plant, Materials, and Processes," London, Temple Press, 1961.
(See also E-2.)

Properties, Practical

PP-1 ASTM Standards, Part 26, "Plastics-Specifications; Methods of Testing Pipe, Film, Reinforced and Cellular Plastics," reissued annually, Philadelphia, American Society for Testing and Materials.
PP-2 Baer, E., Ed., "Engineering Design for Plastics," New York, Van Nostrand Reinhold, 1964.
PP-3 Brydson, J. A., "Plastic Materials," New York, Van Nostrand Reinhold, 1966.
PP-4 Lever, A. E., Ed., "The Plastics Manual," London, Scientific Press, 1966.
PP-5 Modern Plastics Encyclopedia, New York, McGraw-Hill, Annually.
(See also E-2, P-1, P-4, P-10, P-14.)

Properties, Theoretical vs. Structure

PT-1 Andrews, E. H., "Fracture in Polymers," New York, American Elsevier, 1968.
PT-2 Bueche, F., "Physical Properties of Polymers," New York, John Wiley & Sons, 1962.
PT-3 Meares, P., "Polymers: Structure and Bulk Properties," New York, Van Nostrand Reinhold, 1965.
PT-4 Nielson, L. E., "Mechanical Properties of Polymers," New York, Van Nostrand Reinhold, 1962.
PT-5 Tobolsky, A. V., "Properties and Structure of Polymers," New York, John Wiley & Sons, 1968.
(See also R-1 and R-2.)

Rheology

R-1 Eirich, F. R., Ed., "Rheology—Theory and Application," 5 vol., New York, Academic Press, 1956, 1958, 1960, 1967, 1969.
R-2 Ferry, J. D., "Viscometric Properties of Polymers," New York, John Wiley and Sons, 1961.
R-3 Middleman, S., "The Flow of High Polymers," New York, John Wiley and Sons, 1968.
R-4 Pearson, J. R. A., "Mechanical Principles of Polymer Melt Processing," Elmsford, N.Y., Pergamon Press, 1966.
R-5 Severs, E. T., "Rheology of Polymers," New York, Van Nostrand Reinhold, 1962. Introductory text.
(See also PT-2, PT-3, PT-5.)

Structure, Characterization of

S-1 McIntyre, D., Ed., "Characterization of Macromolecular Structure," Publication 1573, Washington, D.C., National Academy of Sciences, 1968.
S-2 Miller, M. L., "The Structure of Polymers," New York, Van Nostrand Reinhold, 1966.
S-3 Morawetz, H., "Macromolecules in Solution," New York, John Wiley and Sons, 1965.
S-4 Tanford, C., "Physical Chemistry of Macromolecules," New York, John Wiley and Sons, 1961.
(See also C-1.)

Testing of Polymers

T-1 ASTM Standards, Part 27, "Plastics—General Methods of Testing, Nomenclature," reissued annually, Philadelphia, American Society for Testing and Materials.
T-2 Level, A. E., and Rhys, J. A., "The Properties and Testing of Plastic Material," 3rd Ed., Temple Press, 1968.
T-3 Schmitz, J. V. and Brown, W. E., "Testing of Polymers," 4 volumes, New York, John Wiley and Sons, 1965, 1966, 1967, 1969.
T-4 Schmitz, J. V., "Bibliography of Polymer Testing, Processing, and Applications," Plastics Institute of America and Society for Plastics Engineers, 1965.
(See also PP-1.)

11

Man-Made Textile Fibers

Robert W. Work*

HISTORY

The first conversion of naturally occurring fibers into threads strong enough to be looped into snares, knotted to form nets, or woven into fabrics is lost in prehistory. Unlike stone weapons, such threads, cords, and fabrics, being organic in nature, have in most part disappeared although in some dry caves traces remain. There is ample evidence to indicate that spindles used to assist in the twisting of fibers together had been developed long before the dawn of recorded history. In that spinning process, fibers such as wool were drawn out of a loose mass, perhaps held in a distaff, and made parallel by human fingers. (A servant girl so spins in Giotto's, "The Annunciation to Anne," c. 1306 A.D., Arena Chapel, Padua, Italy.)[1] A rod (spindle), hooked to the lengthening thread, was rotated so that the fibers while so held were twisted together to form additional thread. The finished length was then wound by hand around the spindle, which in becoming the core on which the finished product was accumulated, served the dual role of twisting and storing, and in so doing, established a principle still in use today. (Even now, a "spindle" is 14,400 yards of course linen thread.) Thus, the formation of any thread-like structure became known as spinning, and it followed that a spider spins a web, a silkworm spins a cocoon and man-made fibers are spun by extrusion, although no rotation is involved.

It is not surprising that words from this ancient craft still carry specialized meanings within the textile industry and have entered everyday literary parlance, quite often with very different meanings. Explanations are in order for some of those that will be used in the pages that follow. For example, as already indicated, "spinning" describes either the twisting of a bundle of essentially parallel short pieces of wool, cotton, or precut man-made fibers into thread or the extrusion of continuous long lengths of man-made fibers. In the former case the short lengths are known

*Director of Textile Research, North Carolina State University, Raleigh, N.C.

as "staple" fibers and the resulting product is a "spun yarn," the long lengths are called "continuous filament yarn," or merely "filament yarn." Neither is called a "thread," for in the textile industry that term is reserved for sewing thread and rubber or metallic threads. Although to the layman "yarn" connotes a material used in hand knitting, the term will be used in the textile sense hereinafter.

Before discussing man-made fibers it is necessary to define some of the terms used.* The "denier" of a fiber or a yarn defines its linear density, i.e., the weight in grams of a 9000-meter length of the material at standard conditions of 70°F and 65 per cent relative humidity. Although denier is actually a measure of linear density, in the textile industry the word connotes the size of the filament or yarn. Fibers usually range from 1 to 15 denier, yarns from 15 to 1650. Single fibers, usually 15 denier or larger, used singly, are termed "monofils." The cross-sectional area of fibers of identical deniers will be inversely related to their densities which range from 0.92 for polypropylene to 2.54 for glass. Since by definition denier is measured at standard conditions, it describes the amount of "bone-dry" materials plus the moisture regain, which ranges from zero for glass and polyethylene to 13 per cent for rayon. It should be mentioned that some years ago scientific organizations throughout the world accepted the word "tex," this being the weight of one kilometer of the material, as a more useful term than denier, but general acceptance is apparently still far away. Furthermore, the sizes of cotton, wool, and worsted yarns and yarns containing man-made fibers but produced by the traditional cotton, wool, or worsted systems are still expressed in the inverse-count system used by each for centuries past.

The "breaking tenacity" or more commonly, "tenacity," is the breaking strength of a fiber or a yarn expressed in force per unit denier, i.e., in grams per denier, calculated from the denier of the original unstretched specimen. "Breaking length" expresses the theoretical length of yarn which would break under its own weight and is used mostly in Europe. "Elongation" means "breaking elongation" and is expressed in units of length calculated as a percentage of the original specimen length.

Typical force-elongation curves of some man-made and natural staple fibers, and textile-type man-made fibers are shown in Fig. 11.1 and 11.2.

An examination of the history of man-made fibers, as they have come to be described in recent years, reveals the existence of two major periods. The start of each was slow, there having been many problems requiring study and research. As these difficulties were solved, however, production surged ahead. As might be expected, the rate of change was more rapid in the later period than in the earlier.

At first man attempted to duplicate the fibers found in nature. Materials were developed which differed from the natural fibers but which were based, nevertheless, on such natural polymers as cellulose and proteins. The former was the basis of the two cellulosics that became successful: rayon and cellulose

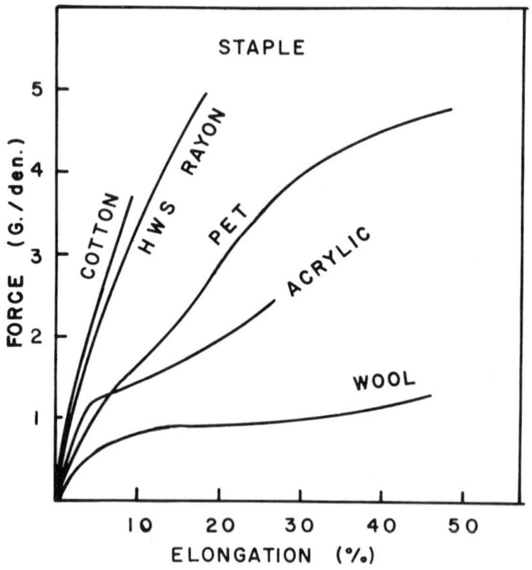

Fig. 11.1 Force-elongation relationships of natural and man-made staple fibers at standard conditions of 75°F and 65% relative humidity.

*Each year the ASTM publishes in its "Book of Standards" the most recent and accepted definitions and test methods used in the textile and fiber industries.

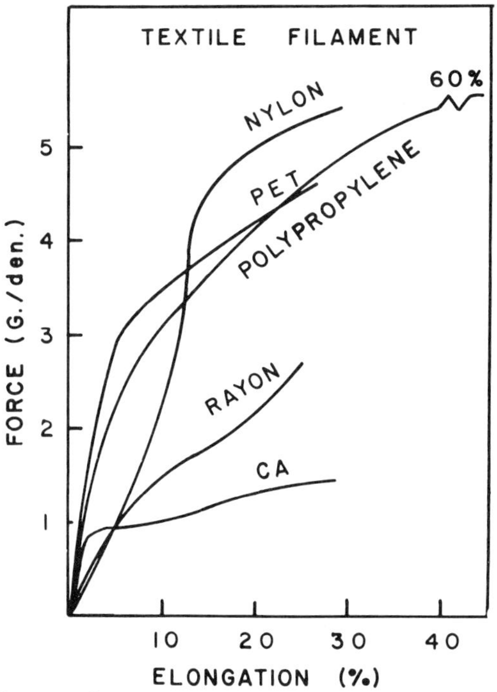

Fig. 11.2 Force-elongation relationships of man-made textile continuous filament yarns at standard conditions of 75°F and 65% relative humidity.

acetate. The second period, initiated by a technological breakthrough, was marked by the work of W. H. Carothers.[2] Fibers were described as being composed of high-molecular weight linear polymers, and the first of these, nylon 66, was synthesized and produced in large volumes. It was quickly followed by nylon 6, polyester, and acrylic fibers.

Much earlier, Hooke in 1664, remarked that not only should it be possible to make a silk-like fiber, but also that such a fiber would be of value in the market place. This observation may be regarded as the beginning of the idea of man-made fibers. Yet almost two hundred years elapsed before this concept was realized by Andemars, who drew fibers from a solution of cellulose nitrate containing some rubber; and only after considerable development and the passage of a generation did Chardonnet's patent (1885) open the door for commercialization of man-made fibers. During the last years of the nineteenth century and the beginning of the twentieth, progress was so rapid that production of the fiber known then as artificial silk, and since named rayon, increased from several thousands of pounds in 1891 to over 2 million pounds in 1910. Although in the United States, the production of rayon had risen to about 150 million pounds a year by the time the revised first edition of this book was published (1933), its future was still so unpredictable that the following statement was made: "the artificial silk industry may therefore be regarded as supplementing worm silk production rather than rivaling it." That same edition also mentioned that worm-silk importation into the United States in 1931 totaled almost 90 million pounds. It is with this background in mind that the development and production of man-made textile fibers should be examined.

VOLUME OF PRODUCTION*

Figure 11.3 compares population growth with the production of man-made fibers and the mill consumption of natural fibers in the United States. In this connection it should be mentioned that the per capita consumption of all fibers, starting at a level in the 1920's of about 30 pounds, rose to approximately 40 pounds following World War II, and topped 59 pounds during 1972. But clearly overshadowing the increases resulting from population growth and a higher standard of living are the volumes produced of first the cellulosic fibers and then the noncellulosic, or completely synthetic fibers. As may be seen from Fig. 11.3, the rise of the noncellulosics had great impact on the cellulosics. The combination of the two has reduced wool to a quite secondary position while at the same time deposing "King Cotton."

Perhaps the most dramatic changes have taken place in the use of the materials required in the manufacture of tire cords. Originally made from cotton, rayon took a commanding position during World War II. But as late as 1951 cotton comprised about 40 per

*Developed from data contained in various issues of the monthly "Textile Organon" published by the Textile Economics Bureau, Inc., New York, N.Y.

cent of the total output of approximately half a billion pounds, and nylon was at a negligible level of 4 million pounds. By 1960, cotton had all but disappeared; nylon represented about 37 per cent of the total (on a weight basis), even though only about 0.8 pounds of nylon is needed to replace 1.0 pounds of rayon. Whereas rayon for several years had dominated the so-called original-equipment tire market and nylon had held a corresponding position for replacement tires, more recently glass and polyester have made heavy inroads into both, especially in belted constructions. Indeed, in the United States in 1972, the situation in a dynamically changing field showed rayon down to 14 per cent, nylon off to 42 per cent, polyester up to 32 per cent, glass up to 7 per cent, and steel at 5 per cent all on a weight basis. Early in 1972, DuPont announced fibers B and B′, and Monsanto announced a series X-500 as new tire yarns possessed of extremely high moduli and previously unattainable breaking strengths. Fibers B and B′ are reputed to be

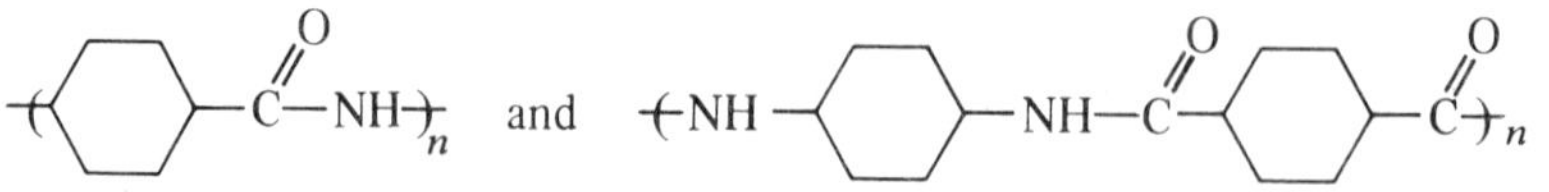

and the series X-500 is represented by

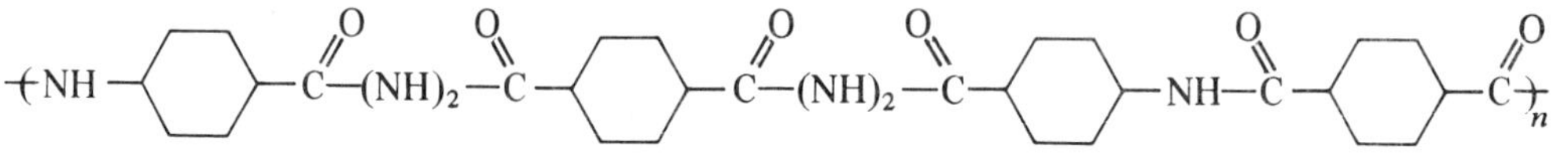

Apparently all are prepared by the reaction between an acid chloride and the active hydrogen of an amine or hydrazide to form ordered phenylene ring systems. Thus, at long last, an analogue of the classical lecture demonstration of the interfacial polymerization of adipyl chloride and hexamethylene diamine to form nylon 66 is converted to industrial chemistry.

The production of man-made fibers throughout the world has developed in a manner that rather parallels the situation in the United States, as may be seen by reference to Fig. 11.4. There are some expected differences and obviously, the data for the world usage are strongly influenced by the large components attributable to the United States. The output of the cellulosics has leveled off, but the expansion of the noncellulosics has continued unabated. The use, or at least the recorded use, of the natural fibers, cotton and wool, rose rapidly in the 1950's, as the world economy rebounded at the conclusion of World War II. Since then (1960–1970) a modest increase has continued, being essentially parallel with the growth of world population. But, when compared with population trends, it appears that the great demand area has been for the man-

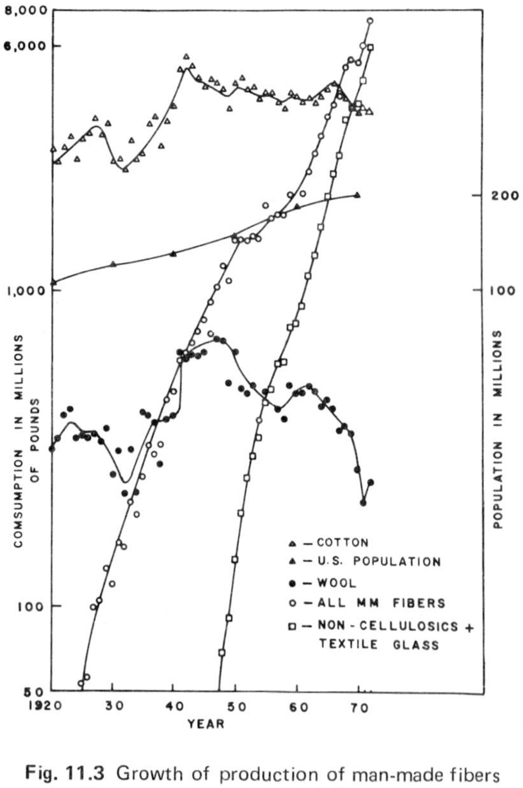

Fig. 11.3 Growth of production of man-made fibers and mill consumption of natural fibers, and population in the U.S., 1920–1972.

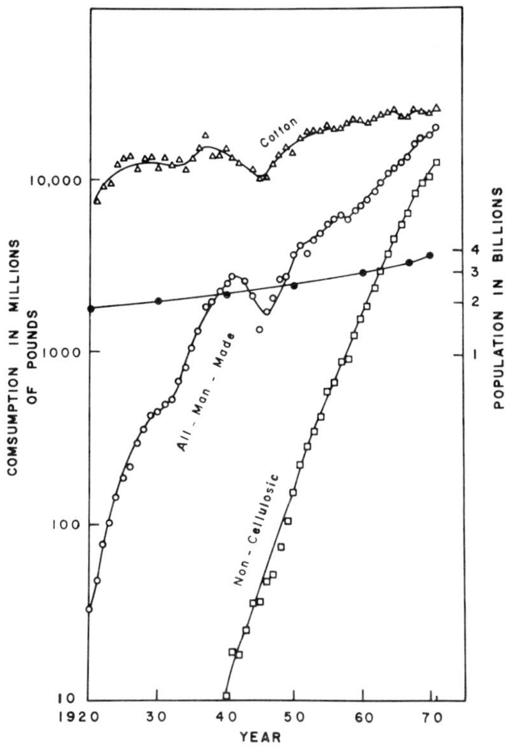

Fig. 11.4 Production of fibers in the world, 1910–1971.

made textile fibers. Much has resulted from an improved standard of living and the absence of major wars.

A detailed economic examination of the prices of fibers and the changes that have taken place during the last half century would show two rather vivid occurences. The first of these is the rapid decrease in the prices of the newer fibers as they became established, followed by a leveling out and stabilization. The second is the relative stability of prices of the man-made fibers on short-term and even long-term bases as compared with fluctuations in the prices for the natural fibers where governmentally imposed stability has not been in effect. Data are not presented in this text, but suffice it to say that in the first half of the twentieth century there was a truism in the textile industry to the effect that the man who made or lost money for the company was the one who was responsible for buying cotton and wool "futures." But it should also be emphasized that list prices of man-made fibers are ceiling prices and do not reflect the short-term discounts, allowances, and special arrangements that are given in a free market place when the demand for any man-made fiber gets soft.

A presentation of complete information about consumption of raw materials, chemical reactions, reagents and catalysts used, and efficiencies of operation in the manufacture of man-made fibers would undoubtedly contribute to a better understanding of the industrial chemistry involved. There are several factors which have prevented this, however. In the first half of the twentieth century the historical belief in the efficacy of trade secrets still permeated the chemical industry. Retired laboratory employees still exchange anecdotes about the use in the rayon industry of thermometers calibrated in arbitrary units and of tire manufacturers whose rubber compounds might contain three portions of the same chemical taken from three differently coded bins. Even with the increased mobility of technical and scientific personnel during and following World War II, the idea still pervaded that if nothing other than patents were allowed to become public knowledge, so much the better. But the situation has changed considerably since about 1960, as can be noted from the availability of information contained in the list of suggested reading that follows this chapter. Nevertheless, secrecy tends to be maintained despite the fact that key employees move from company to company and the chemical engineering knowledge available in chemical concerns producing large volumes of fibers permits an almost complete appraisal of a competitor's activities.

In general, in the early period of production of a fiber, the cost of the original raw material may have had very little bearing on the selling price of the final fiber. A most important factor is the action of the producer's competitors and the conditions of the market and the demand that can be developed. But the complexity of the processes involved in conversion determines the base cost of the fiber at the point of manufacture. As a process becomes older, research reduces this complexity; with simplification, there may be rapid drops in plant cost. These are reflected in the selling price, with little or no variation in the cost of raw materials having occurred. A comparison of rayon and nylon will serve as illustration. "Chemical cellulose" for rayon was priced at 9 to 10 cents per pound when rayon staple was selling at about 31 to 35 cents per pound. On the other hand, benzene at 4 1/2 cents per pound became adipic acid at 32 cents, and finally, nylon continuous filament yarn at well over a dollar per pound. The great difference in the ratio of raw material cost to selling price of fibers lies in the intermediate manufacturing operations. The great importance of man-made fibers in the chemical industry and in the over-all economy of the United States becomes apparent when the volume of production of these materials is considered and compared with the market value of even the least expensive of the raw materials used by them.

RAYON

Chemical Manufacture

Rayon, the first of the man-made fibers produced in large volume, depends on the natural polymer cellulose for its raw material. In the early days the main source of raw material was cotton linters, the short fibers left on the cotton seed after the lint was removed for direct use in textile yarns. This was natural since linters, a relatively pure source of cellulose, were readily available and inexpensive, and the noncellulosic portions could be removed without excessively drastic treatment. The combination of improved technologies for purifying cellulose derived from wood, and the shortage of cotton linters during World War II when they were used for cellulose nitrate, resulted in the virtual elimination of cotton linters as a raw material for rayon.

The manufacture of rayon starts with the cellulose raw material, i.e., wood pulp. (A flow diagram for the complete process is shown in Fig. 11.5.) This is received by the rayon manufacturer in sheet form and in appearance is not unlike a thick unbacked blotter. In the manufacture of this pulp, impurities are removed, special care being given to traces of metallic elements, such as manganese and iron, which affect the manufacturing process and the quality of the final product. Since the production of the wood pulp from wood involves drastic chemical action at elevated temperatures, it is not surprising that degradation occurs which substantially reduces the originally very high molecular weight of the cellulose. Some of this breakdown is essential, but it also results in a wide distribution of molecular weights. That portion which is not soluble in 17 to 18

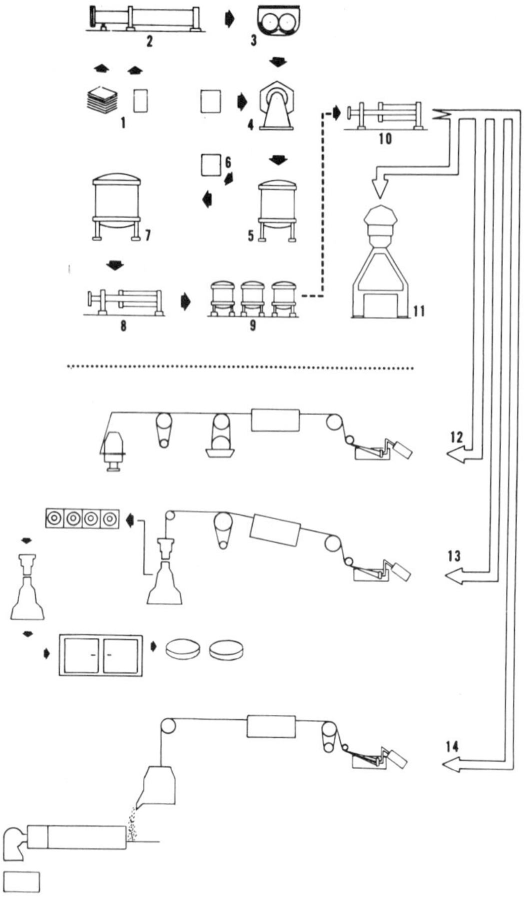

Fig. 11.5 Flow diagram for manufacture of viscose rayon: (1) cellulose sheets and caustic soda; (2) steeping press; (3) shredder; (4) xanthating churn; (5) dissolver; (6) caustic supply; (7) ripener; (8) filtration; (9) deaeration; (10) filtration; (11) continuous process; (12) tire cord; (13) pot spinning; (14) staple spinning.

per cent aqueous caustic is known as alpha cellulose. The lower-molecular weight beta and gamma fractions are soluble and are lost in the first step of the manufacture of rayon. The composition of the pulp is therefore aimed at high alpha content. A typical economic trade-off is involved. The pulp producers can secure an alpha content up to 96 per cent or even higher by means of a cold caustic extraction or, on the other hand, the rayon manufacturer can use a less expensive, lower alpha-content pulp and expect to secure a lower yield. It goes without saying that the sellers have numerous grades available to meet the specific process needs and end-product requirements of each of the buyers.

In the manufacture of rayon it is the usual practice to begin "blending" at the first step, which involves steeping the pulp. Further blending proceeds throughout succeeding steps. The warehouse supply of pulp consists of numerous shipments from the pulp source or, in some cases, sources. In making up the batches for the conventional process, a few sheets are taken from each of several shipments, the ratio being so predetermined that all units of the stock are used, and every exhausted shipment is replaced by a new one. This serves two purposes. It prevents a slight variation in a single pulp lot from unduly affecting any given volume of production, and it provides a moving average so that changes in time are reduced to a minimum.

The cellulose sheets are loaded vertically, but loosely, into a combination steeping bath and press (Fig. 11.6), which is slowly filled with a solution of 17 to 20 per cent caustic, and in which they remain for about 1 hour. In the steeping, the alpha cellulose is converted into alkali or "soda" cellulose; at the same time and, as already mentioned, the caustic solution removes the beta and gamma celluloses (also called hemicelluloses) which are too low in molecular weight to be used in

Fig. 11.6 Steeping of cellulose in the manufacture of viscose rayon. (*Courtesy FMC Corp., American Viscose Division.*)

the final rayon fiber. The exact chemical composition of the soda cellulose is not known, but there is evidence that one mole of NaOH is associated with two anhydro-glucose units in the polymer chain.

$$(C_6H_{10}O_5)_n \text{ (cellulose)} + 18\% \text{ aqueous NaOH} \longrightarrow [(C_6H_{10}O_5)_2 \cdot NaOH]_n \text{ (swollen, insoluble, soda cellulose I)} + \text{soluble soda celluloses from } \beta \text{ and } \gamma \text{ celluloses}$$

The excess caustic solution is drained off for re-use. Additional amounts are removed by forcing the sheets between press plates with a hydraulic ram. Even after this operation the sheets remain in a highly swollen condition and retain from 2.7 to 3.0 parts of the alkali solution. The spent steeping solution squeezed out of the pulp is pumped to dialyzers which separate and recover the caustic from the organic material.

As might be expected, attempts have been made to convert this batch operation to a continuous process. In order to do so the pulp in roll form can be fed through the caustic bath with the excess solution being removed by vacuum or other means. Another approach is to shred the sheet so that upon immersion of the free fibers in caustic a slurry is formed. This provides a rapid and complete contact between the reactants and a correspondingly accelerated formation of the soda cellulose. But, as is the case with every continuous process, there is the added burden of securing a flow of raw materials of identical properties and the need for exact control of processing conditions.

shredder where the already soft sheets are torn into small pieces or "crumbs"; cooling is provided to prevent thermal degradation. Shredding is controlled to produce crumbs that are open and fluffy, and that will allow air to penetrate the mass readily; this is essential in aging.

Soda cellulose is aged by holding it at a constant temperature in perforated containers so that air can contact all the material. The oxygen in the air produces uniform aging accompanied by a reduction in molecular weight and an increase in the number of carboxyl groups present. The target of aging is an average molecular weight high enough to produce satisfactory strength in the final rayon but low enough that the viscosity of the spinning solution will not be excessively high at the desired concentration and not require reduction of the concentration to an uneconomically low point. The aging proceeds for periods of up to two or three days, although the tendency is to speed up the operation by using higher temperatures and traces of metal ions to catalyze the reaction. A combination of experience and constant quality-control testing guarantees that the material will reach the correct point for conversion to cellulose xanthate.

Cellulose xanthate, or more exactly, sodium cellulose xanthate, is obtained by mixing the aged soda cellulose with carbon disulfide in a vapor-tight xanthating churn. Based upon weight of cellulose, the amount of carbon disulfide used will be in the range of 30 per cent for regular rayon to 50-60 per cent for improved and modified varieties.

$$[(C_6H_{10}O_5)_2 \cdot NaOH]_n + CS_2 \text{ (30–60\% based on weight of cellulose in soda cellulose)} \longrightarrow (C_6H_{10}O_5)_n[C_6H_7O_2(OH)_x(O{-}\underset{\underset{S}{\|}}{C}{-}S \cdot Na^+)_{3-x}]_m \text{ or,}$$

$$\text{for simplicity, } (\text{cellulose}{-}O{-}\underset{\underset{S}{\|}}{C}{-}SNa)$$

Following removal of the solution, the sheets of soda cellulose in the batch process are discharged into a shredder. Blending is again accomplished by mixing the charges from two or more steeping presses in a single

The xanthate is soluble in a dilute solution of sodium hydroxide—a characteristic discovered by Cross and Bevan in 1892—and this property makes the spinning of rayon possible. It is a yellow solid; when dissolved in a dilute solution of alkali it becomes a viscous, honey-colored liquid, hence the word "viscose." At this stage the viscous solution may contain about 7.25 per cent cellulose as xanthate in about 6.5 per cent solution of sodium hydroxide, although concentrations of both of these vary, depending on what end products are desired. The solution is ready for mixing with other batches to promote uniformity, to be followed by filtration, ripening, deaeration, and spinning. Filtration takes place in plate and frame filter presses, usually in several stages so that filter dressings of decreasing pore size may be used to secure a balance of through-put and step-wise particle and gel removal.

Such an operation is a straightforward one for "bright" rayon, but only in the days of "artificial silk" did the shiny fiber alone satisfy the market. After a few years a dull appearing fiber was also demanded. At first, fine droplets of oil in the filaments were used to produce dullness until it was discovered that titanium dioxide pigment having a particle size smaller than one micron in diameter was even more satisfactory. It has since become the universal delustrant for all man-made fibers. With the use of pigments of any type, problems of dispersion and agglomerate-formation must be faced. The usual practice has been to add this pigment when mixing the cellulose xanthate into the dilute solution of caustic. However, it must be realized that not only bright and dull, but also semidull fiber is needed, and the demand for each of these three kinds of rayon varies from time to time. It should be mentioned that the producer may add a few parts per million of a tracer element to the solution so that the product can be identified in the event of later complaints.

When the "dope dyeing" of rayon (the use of colored pigments in the fiber) became important, it was no longer possible in all cases to make the final spinning solution during the process of dissolving the xanthate. ("Dope dyeing" will be discussed in greater detail under another heading; it is sufficient to point out here that a large volume of this type of rayon is manufactured and is of considerable importance.) Cleaning or flushing equipment, made necessary by change-overs, involves additional costs. Thus, development of a process which would inject colored pigments as close to the spinning as is possible is desirable. Moreover, in recent years, chemicals other than pigments have had to be added in order to obtain a rayon different from the simple material which was the standard product in the early days of the industry. In order to produce rayon of different physical properties, the rate of precipitation and regeneration of the cellulose is modified by the addition of various chemicals to the spinning bath.

From the standpoint of chemical processing it is obvious that pigments and other chemicals may be added when the sodium cellulose xanthate is dissolved in dilute caustic solution, or at any time up to the moment before it enters the spinnerette prior to being extruded. To keep the operations as flexible as possible the additives should be injected at the last possible moment so that when a change-over is desired there will be a minimum amount of equipment to be cleaned. On the other hand, the farther along in the operation that additives are placed into the stream the greater the problem of obtaining uniformity in an extremely viscous medium, and the greater the difficulty in maintaining exact control of proportions before the viscous solution is passed forward and spun. Furthermore, all insoluble additives must be of extremely small particle size, and all injected slurries must be freed of agglomerates by prefiltration; if not, the viscous solution containing the additives must be filtered. Each manufacturer of viscose rayon develops his own particular conditions for making additions, depending on a multitude of factors, not the least of which is the existing investment in equipment. All manufacturers must face the universal necessity of filtering the solution, with or without pigments or other additives, so that all impurities and agglomerates which might block the holes in the spinnerette are removed.

Although it was known in the years following the discovery by Cross and Bevan that a viscose type of solution could be used in the preparation of regenerated cellulose, the conversion of this solution into useful fibers was not possible until the discovery that the solution required aging until "ripe." Ripening is the first part of the actual chemical decomposition of cellulose xanthate which, if allowed to proceed unhampered, would result in gelation of the viscose solution.

$$\text{Cellulose}-\text{O}-\underset{\underset{\text{S}}{\|}}{\text{C}}-\text{SNa} \overset{H_2O}{\rightleftarrows} \text{Cellulose}-\text{O}\underset{\underset{\text{S}}{\|}}{\text{C}}-\text{SH} + \text{NaOH} \overset{H_2O}{\rightleftarrows}$$

$$\text{Cellulose} + \text{HO}\underset{\underset{\text{S}}{\|}}{\text{C}}-\text{SH} \longrightarrow \text{Cellulose} + CS_2 + H_2O$$

Experience has taught the manufacturer the correct time and conditions for his aging operation, but the requirement of aging itself demands that the entire process be so planned that the viscose solution will arrive at the spinnerette possessing, as nearly as possible, the optimum degree of ripeness so as to produce fibers having the desired characteristics. This degree of ripeness is determined by an empirical test made periodically. It is a measurement of the resistance of the solution to precipitation of the soda cellulose when a salt solution is titrated into it. Thus it is known as the "salt index" or "Hottenroth number" after its originator.

An additional step in the over-all ripening operation involves the removal by vacuum of residual, dissolved and mechanically-held air. More recent practice is to use a vacuum on a moving thin film rather than on the bulk solution in a tank.

It should be mentioned that so inevitable is decomposition of cellulose xanthate and consequent gelation of the contents of pipes and tanks, that all viscose rayon plants must be prepared to pump in-process viscose solutions to a waste receiver, purging the entire system with dilute caustic solution in the event of a long delay in spinning.

Wet Spinning

Spinning a viscose solution into rayon fibers (wet spinning) is the oldest of the three common ways of making man-made fibers. In this method, the polymer is dissolved in an appropriate solvent, and this solution is forced through fine holes in the face of the spinnerette which is submerged in a bath of such composition that the polymer precipitates. The pressure necessary for this extrusion is supplied by a gear pump which also acts as a metering device; the solution is moved through a final or "candle" filter before it emerges from the holes of the spinnerette. There is immediate contact between these tiny streams and the liquid or "wet" bath. As the bath solution makes contact with the material extruded from the holes, chemical or physical changes take place. These changes, whether of lesser or greater complexity, convert the solution of high-molecular weight linear polymer first to a gel structure and then to a fiber. It is an interesting fact that the spinning of viscose rayon, with all of the ramifications made possible by variations in the composition of the solution and the precipitating bath, as well as in the operating conditions, presents the chemist and chemical engineer with both the oldest and the most complex wet-spinning process.

The formation of rayon fibers from viscose solution is far from being simple, either from the physical or chemical standpoint. The spinning bath usually contains 1 to 5 per cent zinc sulfate and 7 to 10 per cent sulfuric acid, as well as an surface-active agent, without which minute deposits will form around the holes in the spinnerette. Sodium sulfate (15 to 22 per cent) is present, formed by the reactions, and as sulfuric acid is depleted and sodium sulfate concentration builds up, an appropriate replenishment of the acid is required. There is a coagulation of the organic material as the sulfuric acid in the spinning bath neutralizes the sodium hydroxide in the viscose solution; at the same time, chemical decomposition of the sodium cellulose xanthate takes place to regenerate the cellulose. If zinc ions are present, which is the usual situation in the production of the improved types of rayon, an interchange takes place so that the zinc cellulose

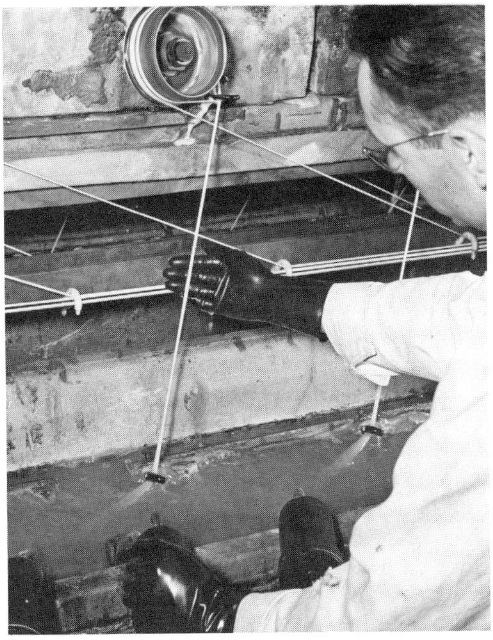

Fig. 11.7 Spinning of viscose rayon. *(Courtesy FMC Corp., American Viscose Division.)*

xanthate which precipitates less quickly, becomes an intermediate. Chemical additives are usually present to repress hydrogen-ion action. The gel-like structure, the first state through which the material passes, is not capable of supporting itself outside the spinning bath. As it travels through the bath, however, it quickly becomes transformed into a fiber that can be drawn from the spinning bath and that can support itself in subsequent operation (Fig. 11.7). The reactions between the bath and the fiber which is forming are paramount in determining the characteristics of the final product; it is for this reason that additives (previously mentioned), as well as zinc ions, may be used to control both the rate of coagulation and regeneration. In this manner the arrangement of the cellulose molecules may be controlled to produce the structural order desired.

(Rapid reaction)

$$2\ \text{Cellulose}-O-\underset{\underset{S}{\|}}{C}-SNa + H_2SO_4 \longrightarrow \text{Cellulose} + Na_2SO_4 + CS_2$$

(Slow reaction)

$$2\ \text{Cellulose}-O-\underset{\underset{S}{\|}}{C}-SNa + ZnSO_4 \longrightarrow (\text{Cellulose}\ O-\underset{\underset{S}{\|}}{C}-S)_2\,Zn + Na_2SO_4$$

$$(\text{Cellulose}\ O-\underset{\underset{S}{\|}}{C}-S)_2\,Zn + H_2SO_4 \longrightarrow \text{Cellulose} + ZnSO_4 + CS_2$$

Because of hydraulic drag, stretching occurs in the bath and also in a separate step after the yarn leaves the bath. In both cases the linear molecules of cellulose are oriented from random positions to positions more parallel to the yarn axis. If a rayon tire cord is to be the final product the fibers must be severely stretched to produce a very high orientation of the molecules, this being the basis of the tire cord's high strength and ability to withstand stretching without which growth of the tire body would occur. For regular textile fibers such high strengths are not desired, and the spinning and stretching operations are controlled to produce rayon of lower strength and greater stretchability under stress.

In order to stretch the yarn uniformly during the manufacturing process, two sets of paired rollers or "godets" are employed, each of the two sets operating at different rotational speeds. The yarn is passed around the first set of godets several times to prevent slippage and is supplied to the stretching area at a constant speed. A second set of godets moves it forward at a more rapid rate, also without slippage. Stretching may range from a few to 100 or more per cent. Spinning speeds are on the order of 100 meters per minute, but these may vary with both the size of the yarn and the process used.

Spinning conditions, composition of the spinning bath, and additives to the viscose solution determine the physical characteristics of the rayon, its breaking strength and elongation, modulus, ability to resist swelling in water, and its characteristics in the wet state as compared with those of the dry material. Not only must the chemical composition of the spinning bath be carefully controlled; the temperature is regulated at a selected point, somewhere in the range of 40° to 60°C, to ensure those precipitation and regeneration conditions essential to the manufacture of any particular viscose rayon having the properties needed for a selected end use.

After the precipitating bath has reacted chemically with the viscose solution to regenerate the cellulose, and raw rayon fiber has been formed, the subsequent steps must be controlled so that differences in treatment are minimized; otherwise such sensitive properties as "dye acceptance" will be affected and the appearance of the final product will vary. It is not the purpose of this description of processes to include information on dyeing and finishing of fabrics used for garments and in households. But it is well to emphasize an old adage at this point, as true today as ever. In essence, fibers that cannot be supplied to the public in a full range of colors and shades are fated to fit only into limited specialty or industrial uses. The manufacturers of rayon learned early how to meet these requirements.

Minute traces of suspended sulfur resulting from the chemical decomposition of cellulose xanthate must be removed by washing with a solution of sodium sulphide. It is expedient to bleach the newly formed fibers with hypochlorite to improve their whiteness; an "antichlor" follows. The chemicals present originally and those used to purify the fibers must be removed by washing. As a final step a small amount of lubricant must be placed on the filaments to reduce friction in subsequent textile operations.

There are several different processes used for the steps involved in spinning and purifying continuous filament rayon. One of the most common involves the formation of packages of yarn, each weighing several pounds, for sepa-

Fig. 11.8 A modern integrated plant, designed for the continuous flow of materials from raw chemicals to the shipment of final textile yarns. (*Courtesy Celanese Fibers Co.*)

rate treatment. After it has been passed upward out of the spinning bath and stretched to the desired degree, the yarn is fed downward vertically into a rapidly rotating can-like container called a spinning pot or "Topham" box (after the man who invented it in 1900). It is thrown outward to the wall of the pot by centrifugal force and gradually builds up like a cake with excess water being removed by the same centrifugal force. This cake is firm, although it must be handled with care, and is sufficiently permeable to aqueous solutions to permit purification.

In another method of package-spinning, the yarn is wound onto a mandrel from the side at a uniform peripheral speed. With this process the yarn may be purified and dried in the package thus formed. In any of these systems the spinning and stretching, as well as subsequent steps, may involve separate baths.

The continuous process for spinning and purifying textile-grade rayon yarn merits particular mention from the standpoint of industrial chemistry since it is rather an axiom that a continuous process is to be preferred over a batch or discontinuous operation. This method employs "advancing rolls' or godets which make it possible for the yarn to dwell for a sufficient length of time on each pair, thus allowing the several chemical operations to take place in a relatively small area. Their operation depends on the geometry existing when the shaft of a pair of adjacent cylindrical rolls are oriented slightly askew. Yarn led onto the end of one of these and then around the pair will progress toward the other end of

Fig. 11.9 A continuous process reel. (*Courtesy IRC Fibers Co., a subsidiary of American Cyanamid Co.*)

the set with every pass, the rate of traversing, and therefore the number of wraps, being determined by the degree of skewness. The same result may be obtained by using a pair of advancing reels, these being circularly arranged interdigitating sets of fingers, the axes of which are also slightly askew. As these rotate the yarn carried on them forms a helix and each element of the yarn moves through this helix. Dwell time is controlled by the degree of skewness of the axes.

The production of rayon to be converted to staple fiber is also amenable to line operation. Here the spinnerette has many thousands of holes and a correspondingly large number of filaments are formed in the precipitation bath. The resulting tow is then stretched to the desired degree and immediately cut in the wet and unpurified condition. The mass of short lengths can be belt conveyed through the usual chemical treatments after which it is washed and dried. It is fluffed to prevent matting and packaged for shipment in large cases.

Nitro, Cuprammonium, and Cellulose Acetate Processes for Rayon

Nitrocellulose. Although nitrocellulose was spun and regenerated into cellulose in the early days of the rayon industry, it is a good many years since this process has been used commercially, and today it is only of historical interest.

Cuprammonium Cellulose. Cellulose forms a soluble complex with copper salts and ammonia. Thus, when cellulose is added to an ammonical solution of copper sulfate which also contains sodium hydroxide, it dissolves to form a viscous blue solution and in this form it is known as cuprammonium cellulose. When used in determining the viscosity of cellulose as a measure of its molecular weight, it has been called "Schweizer solution." The principles on which the chemical and spinning steps of this process are based are the same as those for the viscose process. Cellulose is dissolved, in this case in a solution containing ammonia, copper sulfate, and sodium hydroxide. Unlike the vicose solution, the cuprammonium solution need not be aged and will not precipitate spontaneously on standing except after long periods. It is, however, sensitive to light and oxygen. It is spun into water and given an acid wash to remove the last traces of ammonia and copper ions. Although this rayon was never manufactured in volume even approaching that produced by the viscose method, it produced finer filaments than the former and continues to have a specialized market.

Cellulose Acetate. The conversion of cellulose to cellulose acetate is still another method of making cellulose soluble, and therefore, spinnable. (The manufacture of cellulose acetate will be described in a later section.) A solution of the acetate in acetone may be wet spun under conditions that cause it to coagulate and precipitate in typical fiber form and then it may be saponified back to cellulose in batch operation or continuous process. On the other hand, cellulose acetate fibers, dry spun in a manner that will be described in a later section, can be stretched in the presence of steam, which plastifies them, followed by saponification in a caustic solution containing a relatively large amount of sodium acetate to prevent disorientation from taking place.

Regenerated cellulose yarn of extremely high strength but low breaking elongation was produced in small volume by these methods under the trade name Fortisan.®* but its production was discontinued in 1970.

Textile Operations

After the filament rayon fiber has been spun and chemically purified, much of it passes through what are known as "textile operations" before it is ready to be knitted or woven. Since these steps of twisting and packaging or beaming are common to the manufacture of all man-made fibers, it is advisable to review briefly the background and processes.

Rayon, the first man-made fiber, not only had to compete in an established field, but also had to break into a conservative industry. Silk was the only continuous filament yarn, and products made from it were expensive and had high prestige. This offered a tempting market for rayon. Thus, the new product entered as a competitor to silk and, as already noted, became known as "artificial silk." Under the circumstances it was necessary for rayon to adapt itself to the then existing processing operations and technologies.

When making yarn it was customary to twist several of the silk filaments together in order to secure a yarn of desired size and strength and prevent the breaking of a single filament from pushing back to form a fuzz ball. Since rayon was weaker than silk and the individual filaments were smaller, it required the same amount or even more twisting. In pot spinning, the twists were two to three per inch at the start; in package spinning, none. Experience proved that both the degree of twisting and the size of the package desired by the weaver or knitter varied according to the final application.

These operations could have been carried out in the same plant where the yarn was spun, but the existence of silk "throwsters" (from the Anglo-Saxon "*thrāwan*," to twist or revolve) made this unnecessary. As the rayon industry developed, the amount of yarn twisted in the producing plant or sent forward to throwsters was the result of many factors. Over the years the trend has been to use less twist and to place, instead, several thousand parallel ends directly on a "beam," to form a package weighing as much as 300 to 400 pounds. Such a beam is shipped directly to a weaving or knitting mill. The advent of stronger rayons, as well as other strong fibers, and the diminishing market for crepe fabrics which required highly twisted yarns, accelerated the trend away from twisting.

In all twisting and packaging operations the yarn makes contact with guide surfaces and tensioning devices, often at very high speeds. To reduce friction it is necessary to add a lubricant as a protective coating for the filaments. This is generally true of all man-made fibers, and it is customary to apply the lubricant or "spinning finish" or "spinning lubricant" as early in the manufacturing process as possible. In the case of those materials which develop static charges in passing over surfaces this lubricant must also provide antistatic characteristics.

---

*Registered trade-mark, Celanese Corporation.

Spun Yarn. After rayon became established in the textile industry where it could be used as a silk-like fiber, and its selling price was greatly reduced, other markets were developed. The cotton, wool, worsted, and linen systems of converting short discontinuous fibers to yarns were well established and their products were universally accepted. Here again it was necessary to make rayon fit in with the requirements of available and historically acceptable operations. The first of these was that it be cut into lengths the same as those found in cotton and wool. The viscose rayon process was and is eminently suited to the production of tows containing thousands of filaments. The pressure required to force the solution through the holes is so low that neither heavy metal nor reinforcement of the surface is necessary to prevent bulging and large spinerettes containing several thousand holes can be used. Furthermore, the spinning bath succeeds in making contact with all the filaments uniformly. As a result, the spinning of viscose rayon tow is very similar in principle to the production of the smaller continuout filament yarns.

Since both cotton and wool possess distortions from a straight rod-like structure, machinery for their processing was designed to operate best with such crimped fibers. It was necessary for rayon staple to possess similar lengths and crimpiness in order to be adapted to existing equipment. The crimp, that is, several deformations from straightness per inch, may be produced in rayon either "chemically" or mechanically. In the former case the precipitation-stretching step in spinning is carried out so that the skin and core of the individual filaments are radially nonuniform and constantly changing over very short lengths along the filaments. Since the skin and core differ in sensitivity to moisture, a situation occurs which is not unlike the thermal effect on a bimetallic strip, with consequent distortion of the filaments. Mechanical crimp is produced by feeding the tow between two wheels, which in turn force it into a chamber already tightly filled with tow. As soon as the tow leaves the nip of the feeding wheels it is forced against the compacted material ahead of it, and the straight filaments collapse immediately. As the mass of material is pressed forward it becomes tightly compacted in the distorted condition and remains that way until it escapes through a pressure loaded door at the opposite end of the chamber.

CELLULOSE ACETATE

Historical

Cellulose acetate was known as a chemical compound long before its potential use as a plastic or fiber-forming material was recognized. The presence of hydroxyl groups had made it possible to prepare cellulose esters from various organic acids, since cellulose consists of a long molecular chain of beta-anhydro-glucose units, each of which carries three hydroxyl groups—one primary, the other two secondary. The formula for cellulose (already noted) is $[C_6H_7O_2(OH)_3]_n$; when this is fully esterified a triester results. It was learned quite early that whereas cellulose triacetate is soluble only in chlorinated solvents, a product obtained by partial hydrolysis of the triester to a "secondary" ester (having about 2.35 to 2.40 acetyl groups per anhydro-glucose unit) was easily soluble in acetone containing a small amount of water. Many other cellulose esters have been prepared, but only the acetate has been commercialized successfully as a man-made fiber. Propionates and butyrates, and mixed esters of one or both with acetate, have applications as plastics.

Cellulose nitrate, a well-known material, was used to "dope" the fabric wings of fighter planes in World War I to render them taut and impermeable to air. With the advent of tracer bullets these aircraft were dubbed "flaming coffins," and the need for expanded production of the less flammable secondary cellulose acetate became imperative. Thus at the end of the war the brothers Camille and Henri Dreyfus found themselves in possession of a large factory and a product having no apparent usefulness. Necessity being the mother of invention, they not only developed a method for spinning cellulose acetate into a fiber, they also sponsored the research which resulted in new dyestuffs to color it and thus make it a broadly salable textile product. Their story is often told because their contributions to fiber science seem to have concluded an era in the minds of many scientists. They were among the last inventors who built and controlled industries based upon their inventions and reaped hugh financial rewards therefrom.

Manufacture of Secondary Cellulose Acetate

Cellulose acetate was originally made from purified cotton linters but this raw material has been entirely replaced by wood pulp. Other raw materials used in the manufacture of cellulose acetate are acetic acid, acetic anhydride, and sulfuric acid. For many years acetic acid and acetic anhydride have been produced from natural gas by the petrochemical industry.

The manufacture (See Fig. 11.10) of cellulose acetate is a batch operation; the "charge" of cellulose is of the order of 800 to 1500 pounds. It is pretreated with about one-third its weight of acetic acid and a very necessary amount of moisture, about 6 per cent of its weight. If it is too dry at the time of use, more water must be added to the acetic acid. The pretreatment swells the cellulose and makes it "accessible" to the esterifying mixture.

Although there has been much discussion of the chemistry of cellulose acetylation, it is now generally agreed that the sulfuric acid is not a "catalyst" in the normal sense of the word, but rather that it reacts with the cellulose to form a sulfo ester. The acetic anhydride is the reactant which provides the acetate groups for esterification. The acetylation mixture consists of the output from the acetic anhydride recovery unit, being about 60 per cent acetic acid and 40 per cent acetic anhydride, in an amount 5 to 10 per cent above the stoichiometric requirement, to which has been added 10 to 14 per cent sulfuric acid based on the weight of cellulose used. The reaction is exothermic and requires that the heat be dissipated.

In preparing for acetylation, the liquid reactants are cooled to a point where the acetic acid crystallizes, the heat of crystalliza-

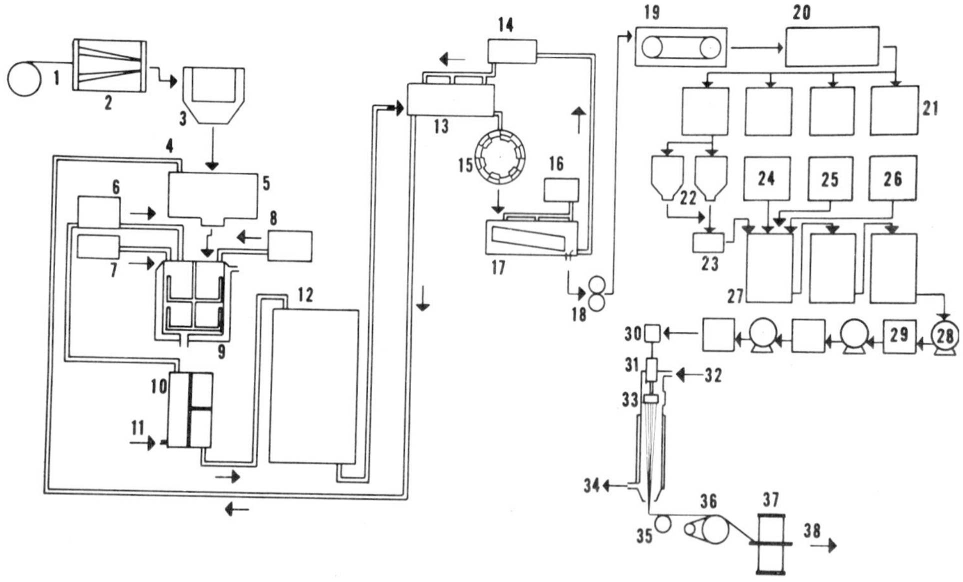

Fig. 11.10 Flow diagram for manufacture of cellulose acetate yarn: (1) wood pulp; (2) attrition mill; (3) cyclone; (4) 35% acetic acid; (5) pretreater; (6) magnesium acetate solution; (7) precooled acetylation mix; (8) sulfuric acid; (9) acetylator; (10) ripener; (11) steam; (12) blender; (13) precipitator; (14) dilute acetic acid; (15) hammer mill; (16) water; (17) rotary screen washer; (18) squeeze rolls; (19) drying oven; (20) blender; (21) storage bins; (22) silos; (23) weight bins; (24) acetone; (25) wood pulp, (26) pigment; (27) mixers; (28) hold tanks; (29) filter press; (30) pump; (31) filter; (32) air; (33) jet; (34) acetone recovery; (35) oiling wheel; (36) feed roll; (37) bobbin; (38) inspection.

Fig. 11.11 Process vessel for acetylation of cellulose. (*Courtesy Celanese Fibers Co.*)

tion being removed by an appropriate cooling system. When a temperature of about 0°C is reached, the slush of acetic acid crystals in the acetic anhydride-sulfuric acid mixture is pumped to the acetylizer, a brine-cooled mixer of heavy construction. This may be either a unit equipped with sigma blades on horizontal axes or a tank carrying a vertically mounted stirrer. The pretreated cellulose is dropped in from the pretreating unit located above. The reaction is highly exothermic, and at the start large amounts of heat are produced. As the temperature of the reaction mixture rises to the melting point of the acetic acid (16.6°C) its large heat of fusion (45.91 cal/gram) prevents a dangerous rise in temperature which would degrade the molecular weight of the cellulose chain. As the reaction proceeds, brine in the jacket of the acetylizer provides additional cooling.

$$\text{Cellulose} + (CH_3CO)_2O + H_2SO_4 \text{ (10–15\% based on weight of cellulose)} \xrightarrow[\text{Anhydrous}]{CH_3COOH} [C_6H_7O_2(OSO_3H)_{0.2}(CH_3COO)_{2.8}]_n$$

The reaction product is a cellulose sulfoacetate of uncertain chemical composition insofar as the substitution of primary and secondary hydroxyls are concerned, but it appears that complete substitution takes place. This compound is soluble in the acetylation mixture; as it is formed and dissolved, new surfaces of the cellulose are presented to the reagents. One variation of this procedure uses methylene chloride, rather than an excess of acetic acid in the reaction mixture. This chemical is used both to dissipate the heat by refluxing (boiling point, 41.2°C) and to dissolve the cellulose ester as it is formed. As the reaction proceeds, the temperature is allowed to rise. Since the cellulose is a natural product obtained from many sources it varies slightly in composition and the end of the reaction cannot be predicted exactly; the disappearance of fibers as determined by microscopic examination is therefore the usual means of following its progress.

During the acetylation operation a certain amount of chain fission is allowed to take place in the cellulose molecule. This is to ensure that the viscosity of the cellulose acetate spinning solution will be as low as possible for ease of handling but high enough to produce fibers with the preferred physical characteristics. The temperature of the reaction controls the rates of both acetylation and molecular-weight degradation. By carefully regulating the temperature, acetylation can be carried to completion in less time than is needed to reduce the molecular weight to the desired value. The final adjustment to this target can then be easily made by continued stirring at a constant temperature.

The next step in the manufacture of cellulose acetate is "ripening," whose object is to convert the triester, that is, the "primary" cellulose acetate, to a "secondary" acetate having an average of about 2.35 to 2.40 acetyl and no sulfo groups per anhydro-glucose unit. While the cellulose sulfo-acetate is still in the acetylizer, sufficient water is added to react with the excess anhydride and start the hydrolysis of the ester. Usually the water is used as a solution of sodium or magnesium acetate which increases the pH and promotes hydrolysis. In the early years of the industry, ripening was allowed to proceed slowly at room temperature for a period of over 24 hours. More recently the trend is to raise the temperature to about 70 to 80°C, by direct injection of steam, to speed the reaction. Although this introduces additional water, the total amount is far below the point where insolubility occurs; with constant stirring, hydrolysis, first of the sulfo and then of the acetyl groups, is relatively homogeneous. Hydrolysis is continued until the desired acetyl content is obtained. When this value is reached, an aqueous solution of magnesium or sodium acetate is added to cool the batch and stop the hydrolysis. It is then ready for precipitation. For example,

$$[C_6H_7O_2(OSO_3H)_{0.2}(CH_3COO)_{2.8}]_n + (CH_3COO)_2Mg \xrightarrow[\text{conc. } CH_3COOH]{\text{aqueous}} [C_6H_7O_2(OH)_{0.65}(CH_3COO)_{2.35}]_n + MgSO_4$$

The regeneration of sulfuric acid and the presence of magnesium ions in a concentrated solution of acetic acid results in the formation of magnesium sulfate which subsequently may be crystallized, with cooling, into long needle-like shapes. These crystals may be used as a "filter aid" if it is desirable to filter the solution of secondary cellulose acetate in the diluted acetic acid solution to remove any unacetylated fibers and other impurities.

The solution is carried to the verge of precipitation by adding dilute acetic acid. It is then flooded with more dilute acetic acid and mixed vigorously, so that the cellulose acetate comes out as a "flake" rather than a gelatinous mass or a fine powder. The flake is then washed by standard countercurrent methods to remove the last traces of acid, and dried in a suitable dryer.

Manufacture of Cellulose Triacetate

When completely acetylated cellulose, rather than the secondary ester, is the desired product, the reaction must be carried out with perchloric acid rather than sulfuric acid as the catalyst. In the presence of 1 per cent perchloric acid, a mixture of acetic acid and acetic anhydride converts a previously "pretreated" cellulose to triacetate without changing the morphology of the fibers. If methylene chloride rather than an excess of acetic acid is present in the acetylation mixture, a solution is obtained. However, it is not imperative to secure an ester with no hydroxyl groups whatsoever in order to obtain a fiber which will behave in essentially the same way as the triester. It is possible to use about 1 per cent sulfuric acid instead of perchloric acid. When the sulfoacetate obtained from such a reaction is hydrolyzed with the objective of removing only the sulfo-ester groups, the resulting product has about 2.94 acetyl groups per anhydroglucose unit instead of the theoretical 3.00 of the triacetate. The preparation, hydrolysis, precipitation, and washing of "triacetate" are in all other respects similar to the corresponding steps in the manufacture of the more common secondary acetate.

Acid Recovery. In the manufacture of every pound of cellulose acetate about 4 pounds of acetic acid are produced in 30 to 35 per cent aqueous solution. Needless to say, every attempt is made to move the more dilute solutions countercurrent to the main product so as to raise the concentration of the material to be recovered. For example, as already men-

Fig. 11.12 Recovery of acetic acid. *(Courtesy of Celanese Fibers Co.)*

tioned, when the acid dope is diluted and precipitation is produced by flooding, previously obtained dilute acid rather than water is used.

The accumulated acid contains a small amount of suspended fines and some dissolved cellulose esters of low acetyl value and molecular weight. To remove the suspended material the acid is passed slowly through settling tanks. Then it is mixed with organic solvents, for example, a mixture of ethyl acetate and benzene, which concentrates the acid in an organic layer which is decanted. Distillation separates the acid from the solvent and concentrates the acid to a glacial grade needed for conversion to the anhydride.

To produce the acetic anhydride, the acid is dehydrated to ketene and reacted with acetic acid using a phosphate catalyst at 500°C or higher in a tubular furnace.

$$CH_3COOH \xrightarrow[\text{catalyst}]{\text{heat}} H_2O + CH_2{=}C{=}O$$

$$CH_2{=}C{=}O + CH_3COOH \longrightarrow (CH_3CO)_2O$$

Both the dehydration and the reaction with acetic acid may be carried out in separate steps; or they may be allowed to take place with the products quenched so that the water produced does not react with the anhydride. In the latter operation the mixture of unreacted acid, water, and anhydride is fed directly to a still which yields dilute acetic acid overhead and an anhydride-acetic acid mixture at the bottom. Conditions are controlled in such a way that the raffinate is about 40 per cent anhydride and 60 per cent acetic acid. As already mentioned, this is the desired ratio for the reaction mixture used for acetylation of cellulose.

Blending of Flake. As in the manufacture of viscose, the products of batch operations are blended to promote uniformity in the manufacture of cellulose acetate, since it has not been found possible to manufacture a consistently uniform product. Although a blend of different celluloses is selected in the beginning, the pretreatment, acetylation, and ripening are batch operations with little or no mixing. Before precipitation, a holding tank provides an opportunity for mixing; then precipitation, washing, and drying—all continuous—promote uniformity. The dried cellulose acetate flake moves to holding bins for analysis; the moisture content, acetyl value, and viscosity being especially important. The results of the analyses determine what further blending is necessary to obtain a uniform product; rather elaborate facilities are used to select, weigh, and mix the flake. After blending and mixing portions of selected batches, the lot is air-conveyed to large storage bins or "silos" which are filled from the center of the top and emptied from the center of the bottom, thus bringing about further mixing.

Spinning Cellulose Acetate

Although cellulose acetate was originally dissolved in acetone containing a small amount of water in large rotating drum-like mixers, this has long since been replaced by a continuous operation. The acetone is metered into a vertical tank equipped with a stirrer, and the cellulose acetate flake and filter aid are weighed in an automatic hopper; all operations are controlled by proportioning methods common to the chemical industry. The ratio of materials is about 25 per cent cellulose acetate, 4 per cent water, less than 1 per cent ground wood pulp as a filter aid, and the remainder acetone. The mixture moves forward through two or three stages at the rate at which it is used, the hold-time being determined by experience. After dissolution is completed, filtration is carried out in batteries of plate and frame filter presses in three or even four stages. The filter medium of the first stages may be any of several highly open-type pulps; the latter are often the blotter-type cellulose sheets so that the passage of the "dope" is through presses of decreasing porosity. The wood-pulp filter aid is removed early in the process and by building up in the presses, presents new filter surfaces.

Much of the cellulose acetate is delustered by the addition of titanium dioxide pigment, as with viscose rayon. Similarly, from the standpoint of thorough dispersing and mixing, it is desirable to add the pigment as early as possible in the dissolving and mixing operation. But this creates cleaning problems if the equipment is to be used alternately for bright and dull dope. Generally speaking the ratio of demand for these two products is fairly constant and it is preferable to use separate systems to produce them. Each manufacturer must make compromises according to his particular needs.

Between each (and after the last) filtration, the dope goes to storage tanks which serve to remove bubbles; in this case vacuum is not necessary. From the final storage tank it is pumped into a header located at the top of each spinning machine; then it is directed to a series of metering gear pumps, one for each spinnerette, which in some sectors of the acetate industry is traditionally called a "jet." Since the holes in the cellulose acetate-spinning spinnerette are smaller (0.03 to 0.05 mm) than those in the corresponding viscose devices, great care must be taken with the final filtration. An additional filter for the removal of any small particles that may have passed through the large filters is placed in the fixture, sometimes called the "candle," to which the spinnerette assembly is fastened. A final filter is placed in the spinnerette-assembly unit over the top of the spinnerette itself.

The operating unit for spinning cellulose acetate illustrates one of the three common spinning methods, that is "dry" spinning. Having come after the development of viscose rayon manufacture, which had been named "wet" spinning, the absence of water and its replacement by heated air was logic enough for the use of the contrasting word. The dope is heated (in some cases above the boiling point of acetone, 56.5°C) to lower its viscosity and thus reduce the pressure required to extrude it, and to supply some of the heat needed for evaporating the acetone solvent. Heat may be produced in a thermostatically controlled preheater or supplied from the heated cabinet itself.

The spinnerette is stainless steel, and because the filaments must be heated and prevented from sticking together, and space must be allowed for the escape of acetone vapor, the holes must be kept farther apart than those of the spinnerettes used for wet spinnerettes used for wet spinning. As the hot solution of cellulose acetate in acetone emerges downward into the spinning cabinet, an instantaneous loss of acetone takes place from the surface of the filaments which tend to form a solid skin over the still liquid or plastic interior. A current of air, either in the direction the filaments are moving, or counter-current, heats the filaments, and as the acetone is diffused from the center through the more solid skin, each filament collapses to form the indented cross-sectional shape typical of cellulose acetate. The heated air removes the vaporized acetone. Each manufacturer uses a preferred updraft, down-draft, or mixed-draft operation, as his needs dictate.

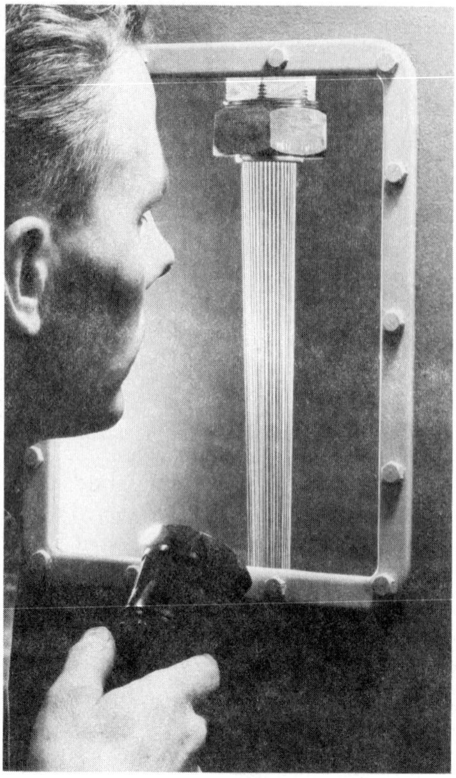

Fig. 11.13 Dry spinning of cellulose acetate. *(Courtesy Tennessee Eastman Co.)*

The cabinet through which the yarn passes vertically downward must be long enough to allow sufficient acetone to diffuse outward and evaporate from the surfaces of the filaments so that the latter will not stick to the first surface contacted nor fuse to each other. The temperature of the air in the cabinet, the rate of flow, the length of the cabinet, the size and number of filaments, and the rate of travel are all interrelated in the spinning process. Since it is desirable to increase spinning speeds to the limit of the equipment, the tendency has been to construct longer spinning cabinets as each new plant is built. Present spinning speeds are of the order of 600 or more meters per minute, measured as the yarn emerges from the cabinet.

Other dry-spinning operations follow essentially the same pattern. For example, the dry spinning of cellulose triacetate is identical to the process used for the older and more familiar secondary acetate except that the acetone solvent is replaced by a chlorinated hydrocarbon such as methylene chloride. The solubility is improved by the addition of a small amount of methanol (5 to 15 per cent).

The acetate or triacetate yarn may emerge from the cabinet through an opening on the side near the bottom after passing over a deflecting pin or it may be brought out of the bottom directly below the spinnerette. In either case, it makes contact with an applicator which provides the lubricant required to reduce both friction and static formation in subsequent operations. With its surface lubricated, the yarn passes around a "feed" roll which determines the rate of withdrawal from the spinning cabinet, and then to any of several desired packaging devices.

Unlike the packaging of rayon yarn, cellulose acetates are either "ring" spun or wound into

Fig. 11.14 Beaming cellulose acetate yarn from a reel holding about 800 packages of yarn. (*Courtesy Tennessee Eastman Co.*)

a package called a "disc," "zero twist," or "cam wound." In the ring-spun package, the yarn carries a slight twist of less than one turn per inch, but it requires a relatively expensive bobbin. Since the trend is toward less twisting, much acetate yarn is "beamed" in the producers plant after little or no twisting, the heavy beams being shipped directly to knitters or weavers.

Solvent Recovery. The air containing the acetone vapor is drawn out of the spinning cabinet and passed through beds of activated carbon which sorb the organic solvent. The acetone is recovered by steaming and then separating it from the water by distillation. Efficiency of recovery is above 95 per cent, and about 3 pounds of acetone are recovered per pound of cellulose acetate yarn. The recovery of methylene chloride and methanol from the manufacture of cellulose triacetate follows the same general procedure but the combination of chlorinated hydrocarbon, steam, and activated charcoal tend to produce traces of corrosive products which must be guarded against.

Dope-Dyed Yarn. As with viscose rayon, colored pigments or dyestuffs may be added to the spinning solution so that the yarn will be colored as it is produced, thus eliminating the necessity of dyeing the final fabric. Although methods of dope dyeing may differ in detail depending on the fiber and the process used, certain essentials are the same throughout. As mentioned earlier, even when using titanium dioxide, a compromise must be made on the basis of two competing needs. Complete mixing, uniformity, and filtration require that the addition be made early in the operation; minimal cleaning problems during change-overs require just the opposite. This problem is much more difficult to solve when colored pigments are involved since the demands for colors are determined by ever-changing style trends. A manufacturer may plan a long run of a given color, arranging the changes from shade to shade so that the differences in successive runs will be negligible. This might, for example, be through a series of blues ranging from a pastel to navy; but such is not always possible. There exists two solutions to the problem. If a manufacturer must produce a multitude of colors in relatively small amounts, it is desirable to premix individual batches of spinning dope. Each batch should be pretested on a small scale, even sampled and submitted to the customer, to ensure that the desired color will be acceptable when it is produced. Facilities must be provided to allow each batch of colored dope to be cut into the system very close to the spinning operation in order to minimize pipe cleaning. Permanent piping must be flushed with solvent or the new batch of colored dope; some of the equipment may be disassembled for mechanical cleaning after each change of color.

Another method of producing spun-dyed yarn involves using a group of "master" dopes of such color versatility that when they are injected by appropriate proportioning pumps into a mixer located near the spinning operation they will produce the final desired color. The advantages of such an operation are obvious; the disadvantage lies in the textile industry's demand for an infinite number of colors. No small group of known pigments will produce final colors of every desired variety.

PROTEINS

As previously mentioned, the use of naturally existing polymers to produce fibers has had a long history. In the case of cellulose the results were fabulously successful. An initial investment of $930,000 produced net profits of $354,000,000 in 24 years for one rayon company.[3] On the other hand, another family of natural polymers—proteins—has thus far resulted in failure or at best very limited production.

These regenerated proteins are obtained from milk (casein), soya beans, corn, and peanuts. More or less complex chemical separation and purification processes are required to isolate them from the parent materials. They may be dissolved in aqueous solutions of caustic, and wet spun to form fibers which usually require further chemical treatment as, for example, with formaldehyde. This reduces the tendency to swell or dissolve in subsequent wet-processing operations or final end uses. These fibers are characterized by a wool-like feel, low strength, and ease of dyeing. Nevertheless, for economic and other reasons they have not been able to compete successfully with either wool (after which they were modeled) or other man-made fibers.

NYLON

Historical

Nylon was the first direct product of the technological breakthrough achieved by W. H. Carothers of E. I. du Pont de Nemours & Co. Until he began his classic research on high polymers, the manufacture of man-made fibers was based almost completely on natural linear polymers. Such materials included rayon, cellulose acetate, and the proteins. His research showed that chemicals of low molecular weight could be reacted to form polymers of high molecular weight. By selecting reactants which produce molecules having great length in comparison with their cross section, that is, linear molecules, fiber-forming polymers are obtained. With this discovery the man-made fiber industry entered a new and dramatic era.

Manufacture

Nylon 66. The word "nylon" was established as a generic name for polyamides, one class of the new high-molecular weight linear polymers. The first of these, and the one still produced in the largest volume, is nylon 66 or polyhexamethylene adipamide. Numbers are used with the word nylon to indicate the number of carbon atoms in the constituents, in this case hexamethylenediamine and adipic acid.

To emphasize the fact that it does not depend on a naturally occurring polymer as a source of raw material, nylon has often been called a "truly synthetic fiber." To start the synthesis, benzene, as a by-product from the coking of coal, may be hydrogenated to cyclohexane,

$$C_6H_6 + 3H_2 \xrightarrow{\text{catalyst}} C_6H_{12}$$

or the cyclohexane may be obtained from petroleum. The next step is oxidation to a cyclohexanol-cyclohexanone mixture by means of air.

$$x\,C_6H_{12} + O_2\ (\text{air}) \xrightarrow{\text{catalyst}} y\,C_6H_{11}OH + z\,C_6H_{10}O$$

In turn, this mixture is oxidized by nitric acid to adipic acid.

$$C_6H_{11}OH + C_6H_{10}O + HNO_3 \xrightarrow{\text{catalyst}} (CH_2)_4(COOH)_2$$

Adipic acid so obtained is both a reactant for the production of nylon and the raw-material source for hexamethylenediamine, the other reactant. The adipic acid is first converted to adiponitrile by ammonolysis and then to hexamethylenediamine by hydrogenation.

$$(CH_2)_4(COOH)_2 + 2NH_3 \xrightarrow{\text{catalyst}} (CH_2)_4(CN)_2 + 4H_2O$$

$$(CH_2)_4(CN)_2 + 4H_2 \xrightarrow{\text{catalyst}} (CH_2)_6(NH_2)_2$$

Another approach is through the series of compounds furfural, furane, cyclotetramethylene oxide, 1,4-dichlorobutane, and adiponitrile. The furfural is obtained from oat hulls and corn cobs.

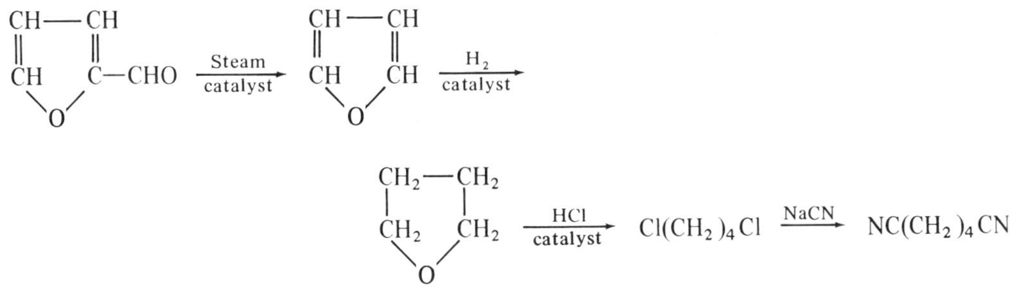

Or 1,4-butadiene, obtained from petroleum, may be used as a starting raw material to make the adiponitrile via 1,4-dichloro-2-butene and 1,4-dicyano-2-butene.

$$CH_2{=}CHCH{=}CH_2 \longrightarrow$$

$$ClCH_2CH{=}CHCH_2Cl \xrightarrow[\text{catalyst}]{HCN}$$

$$NCCH_2CH{=}CHCH_2CN \xrightarrow[\text{catalyst}]{H_2}$$

$$NC(CH_2)_4CN$$

When hexamethylenediamine and adipic acid are mixed in solution in a one to one molar ratio, the so-called "nylon salt" hexamethylenediammoniumadipate, the direct progenitor of the polymer, is precipitated. After purification, this nylon salt is polymerized to obtain a material of the desired molecular weight. It is heated to about 280°C under vacuum while being stirred in an autoclave for 2 to 3 hours; a shorter holding period follows; and the process is finished off at 300°C. The molecular weight must be raised to a level high enough to provide a fiber-forming material, yet no higher. If it is too high, the corresponding viscosity in the subsequent spinning operation will require extremely high temperatures and pressures to make it flow. Accordingly, a small amount of acetic acid is added to terminate the long-chain molecules by reaction with the end amino groups.

The polymerized product is an extremely insoluble material and must be melt spun (see below). Therefore, should a delustered fiber be desired it is necessary to add the titanium dioxide pigment to the polymerization batch before that reaction occurs. For ease of handling, the batch of nylon polymer may be extruded from the autoclave to form a thin ribbon which is easily broken down into chips after rapid cooling. But, whenever possible, the liquid polymer is pumped directly to the spinning operation.

Nylon 6. Nylon 6 is made from caprolactam and is known as Perlon® in Germany where it was originally developed by Dr. Paul Schlack.[4] Its production has reached a very large volume in the United States in recent years.

Like nylon 66, nylon 6 uses benzene as raw material, which converted through previously mentioned steps to cyclohexanone. This compound is in turn converted to the corresponding oxime by reaction with hydroxylamine, and cyclohexanone oxime is made into caprolactam by the Beckmann rearrangement.

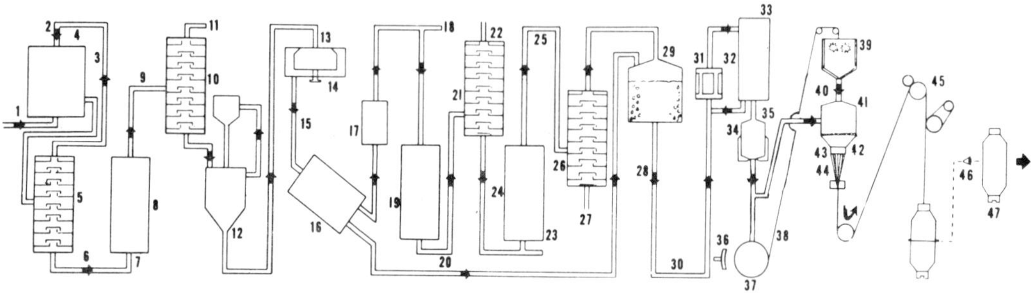

Fig. 11.15 Flow diagram for the manufacture of nylon 66 yarn: (1) air; (2) cyclohexane from petroleum; (3) reactor; (4) recycle cyclohexane; (5) still; (6) cyclohexanol-cyclohexanone; (7) nitric acid; (8) converter; (9) adipic acid solution; (10) still; (11) impurities; (12) crystallizer; (13) centrifuge; (14) impurities; (15) adipic acid crystals; (16) drier; (17) vaporizer; (18) ammonia; (19) converter; (20) crude adiponitrile; (21) still; (22) impurities; (23) hydrogen; (24) converter; (25) crude diamine; (26) still; (27) impurities; (28) nylon salt solution; (29) reactor; (30) stabilizer; (31) calandria; (32) evaporator; (33) excess water; (34) autoclave; (35) delustrant; (36) water sprays; (37) casting wheel; (38) polymer ribbon; (39) grinder; (40) polymer flake; (41) spinning machine; (42) heating cells; (43) spinnerette; (44) air; (45) drawtwisting; (46) inspection; (47) nylon bobbin. (Note: Wherever the demand for liquid polymer at a spinnerette is large, as for example in the spinning of tire yarn, it is pumped directly from the autoclave.)

Fig. 11.16 Aerial view of a nylon filament yarn plant: shipping docks, textile spinning area, and plant office buildings (left center), chemical intermediates (right center); and the technical center (left foreground). (*Courtesy Monsanto Co.*)

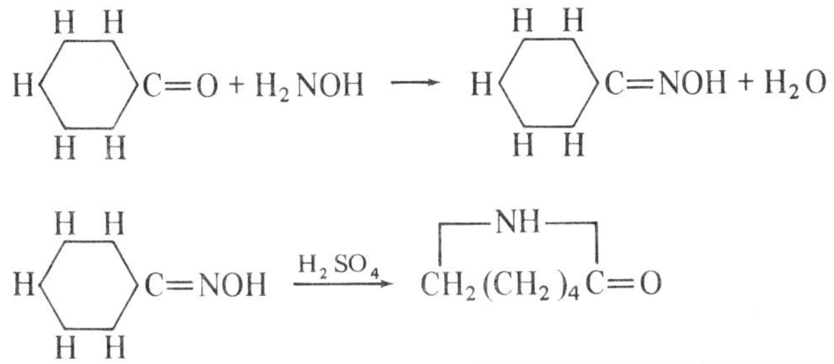

After purification, the lactam is polymerized by heating it at elevated temperatures in an inert atmosphere. During self-condensation the ring structure of the lactam is opened so that the monomer acts as an epsilon-aminocaproic acid radical. Unlike nylon 66, polymerization of caprolactam is reversible; the polymer remains in equilibrium with a small amount of monomer. As with nylon 66, nylon 6 is extruded in ribbon-like form, quenched, and broken into flakes for subsequent spinning, or the molten polymer is pumped directly to the spinning equipment.

Melt Spinning

Because of its extremely low solubility in low-boiling and inexpensive organic solvents, nylon 66 required a new technique for converting the solid polymer into fibers, hence the development of "melt" spinning, the third basic method for manufacturing man-made fibers. The following description refers essentially to nylon 66 since it was the first to use the method, but the process applies, in general, to all melt-spun man-made fibers.

In the original production of nylon fiber by melt spinning, the chips of predried polymer were fed from a chamber onto a melting grid whose holes were so small that only passage of molten polymer was possible. Both solid and liquid were prevented from contacting oxygen. The polymer melted in contact with the hot grid and dripped into a pool where it became the supply for the spinning itself. This melting operation has been almost entirely replaced by delivery of the molten polymer pumped directly from the polymerization stage or by "screw" melting. In the latter process the solid polymer in chip form is fed into an extrusion-type screw contained in a heated tube. The depth and helix angle of the grooves are engineered in such a way that melting takes place in the rear section, and the molten polymer is moved forward under increasing pressure to a uniformly heated chamber preceding the metering pump. An ingenious variation of the screw-melting operation, although not widely used as far as is known at this writing, uses the heat developed by the drag of a rotating/frictional member in the highly viscous molten polymer. In this device, at start up, the polymer is melted by application of external heat. Once this has been accomplished, the forward nongrooved head of the screw, rotating against a thin circumferential layer of molten polymer between it and the wall, develops sufficient frictional heat to melt the polymer which is fed forward to the area by the grooved section of the screw. When the desired pressure is reached in the chamber containing the liquid polymer the entire screw is pushed backward against a spring and in so doing the size of the opening through which the solid polymer is being fed is reduced.

Whatever means is used to secure the molten polymer, it is moved forward to a gear-type pump which provides both high pressure and a constant rate of flow to the final filter and spinnerette. The filter consists of several metal screens of increasing fineness or graded sand arranged in such a way that the finest sand is at the bottom. After being filtered, the molten polymer at a pressure of several thousand pounds per square inch is extruded through the small holes in the heavily constructed spinnerette. It is necessary to maintain the temperature of the pool, pump, filter, jet assembly, and jet at about 20–30°C above the melting point of the nylon, which is about 264°C for nylon 66 and 228°C for nylon 6.

When the extruded fibers emerge from the jet face into the relatively cool quench chamber where a cross current of air is provided, rapid solidification takes place. The solid filaments travel downward to cool, and an antistatic lubricant is applied before they make contact with the wind-up rolls to prevent static formation and to reduce friction in subsequent textile operations. Great care must be used in conveying the freshly spun yarn from the spinning chamber to the yarn package if it is to be "drawn" in a separate operation. (The combination of spinning and drawing, known as "spin-drawing" will be described later.)

Cold Drawing

It was learned early that the fibers made from nylon 66 could be extended to about four times their original length with very little effort, but that thereafter a marked resistance to extension took place. It was discovered that during this high extension the entire length of fiber under stress did not extend uniformly. Rather, a "necking down" occurred at one or more points, and when the entire length under tension had passed through this phenomenon, a high-strength fiber was obtained. It was also found that when more than one necking down was allowed to take place in a given length of fiber, a discontinuity occurred at the point where the two came together. Accordingly, the drawing operation is aimed at forcing the drawing to occur at a single point as the yarn advances from the supply to the take-up package.

Cold drawing consists essentially of removing the yarn from the package prepared in the melt-spinning operation and feeding it forward at a uniformly controlled rate under low tension. It is passed around a pulley or roller which determines the supply rate and prevents slippage; for nylon 66 it is then wrapped several times around a stationary snubbing pin. From there it goes to a second roller which rotates faster than the supply roller to produce the desired amount of stretch, usually about 400 per cent. The necking down occurs at the snubbing pin. In the case of nylon 6, drawing may be effected satisfactorily without passing the yarn around such a snubbing pin.

The long molecules of the nylon 66 or 6 polymer, which are randomly positioned in the molten polymer, when extruded from the spinnerette tend to form "crystalline" areas of molecular dimensions as the polymers solidify in the form of freshly spun fibers. In the drawing operation, both these more ordered portions as well as the amorphous areas tend to become oriented so that the lengthwise dimensions of the molecules become parallel to the long axis of the fiber and additional intermolecular hydrogen bonding is facilitated. It is this orientation which converts the fiber having low resistance to stress into one of high strength.

By controlling the amount of drawing as well as the conditions under which this operation takes place, it is possible to vary the amount of orientation. A minimal amount is preferable in the manufacture of yarns intended for textile applications wherein elongation of considerable magnitude and low modulus or stiffness is required rather than high strength. On the other hand, strength and high modulus are at a premium when fibers are to be used in products such as tire cords. High resistance to elongation is imperative if the tire is not to "grow" under conditions of use. In this connection it should be noted that nylon tire cord that has been produced by twisting the original tire yarn and plying the ends of these twisted yarns together is hot stretched just before use at the tire plant to increase strength and reduce even further the tendency to elongate under stress.

The separate operations of spinning and drawing nylon presented a challenge whose object was combination of these two into a single continuous step. But the problem was obvious, for the operating speeds of the two separate steps had already been pushed as high as was thought to be possible. How then would it be possible to combine them into a continuous spin-draw, wherein a stretching of about 400 per cent could take place? The answer appears to lie in the manner in which the cooling air is used and in the development of improved high-speed winding devices. By first cooling the emerging fibers by a cocurrent flow of air and then cooling it further by a countercurrent flow, the vertical length of the cooling columns can be kept within reason. In-line drawing may occur in one or two stages and relaxation may be induced if needed. The final yarn is said to be packaged at speeds of 4000 meters per minute, or even higher in the most recent installations.

POLYESTERS

Historical

Just as the original work of Carothers regarding polyamides inspired research which resulted in the development of nylon 6 in Germany, the same type of stimulus resulted in Whinfield and Dixon's invention of polyesters in England.[5] These men found that a synthetic linear polymer could be produced by condensing ethylene glycol with terephthalic acid or by an ester-exchange between the glycol and pure dimethyl terephthalate. A polymer could thereby be obtained which could be converted to fibers having valuable properties, including the absence of color. Unlike nylon, this material has not been popularized under its generic name, polyester. Those working with it commonly refer to it as PET. It first appeared under the trade name Terylene® (Imperial Chemical Industries, Ltd.) in England, and was first commercialized in the United States as Dacron® (E. I. du Pont de Nemours & Co.). In 1972 it was being sold under about 100 additional trade names throughout the world.

Manufacture

When the development of polyethlene terephthalate (PET) occurred, ethylene glycol was already being produced in large amounts from ethylene, a by-product of petroleum cracking, by oxidation of ethylene to ethylene oxide and subsequent hydration to ethylene glycol, which, in a noncatalytic process, uses high pressures and temperatures in the presence of excess water.

$$CH_2{=}CH_2 + O_2 \longrightarrow \underset{\text{(epoxide, O bridging)}}{CH_2{-}CH_2} \xrightarrow{H_2O} HOCH_2CH_2OH$$

On the other hand, although *o*-phthalic acid, or rather its anhydride, had long been produced in enormous amounts for use in the manufacture of alkyd resins, the *para* derivative was less well known and not available on a large scale. The synthesis is a straightforward one, however, from *p*-xylene which is oxidized to terephthalic acid either by means of nitric acid in the older process or by air (catalyzed) in the newer one. As already mentioned, in the early years this was then converted to the easily purified dimethyl ester in order to obtain a colorless polymer adequate for the manufacture of commercially acceptable fibers.

Several other methods were developed for producing the desired dimethyl terephthalate. The Witten (Hercules) process goes from *p*-xylene to toluic acid by oxidation of one methyl group on the ring, following which the carboxyl group is esterified with methanol. This process is then repeated with the second methyl group to secure the dimethyl ester of terephthalic acid.

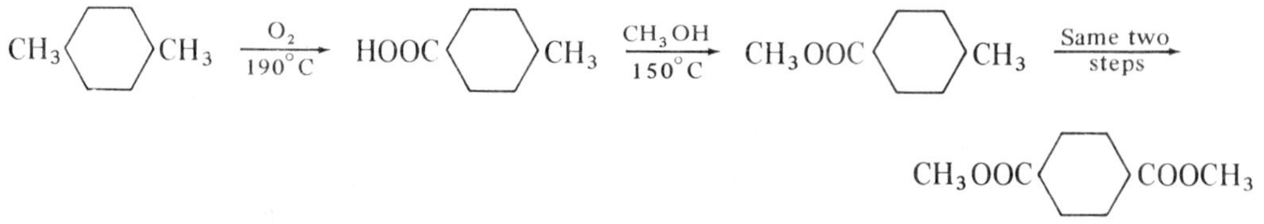

Either phthalic anhydride or toluene, both in ample supply as raw materials, may be used in the Henkel processes. Use of phthalic anhydride depends upon dry isomerization of the potassium salt of the *ortho* derivative to the *para* form at about 430°C and 20 atm pressure; or toluene is oxidized to benzoic acid, whose potassium salt can be converted to benzene and the potassium salt of terephthalic acid by disproportionation.

The first step in the reaction of dimethyl terephthalate and ethylene glycol is transesterification to form bis(*p*-hydroxyethyl) terephthalate (bis-HET) and eliminate methanol.

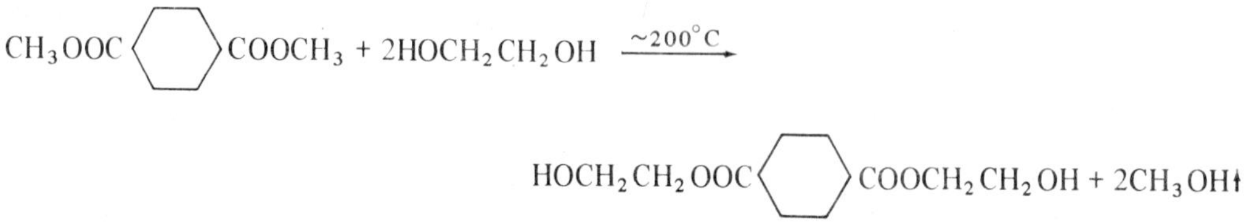

This product is then polymerized in the presence of a catalyst to a low molecular weight and the by-product glycol is eliminated. In a second stage, at a temperature of about 275°C and under a high vacuum, the molecular weight is raised to secure the melt viscosity desired for the particular material involved. Like nylon, this final material may be extruded, cooled, and cut into chips for storage and remelting, or it may be pumped directly to the spinning machines.

From the beginning it was obvious that it would be a considerable advancement in industrial chemistry, to say nothing of cost reduction, if the process could be simplified by making it unnecessary to go through the dimethyl derivative to secure a product of adequate purity. This was accomplished in the early 1960's when methods of purifying the crude terephthalic acid were developed, and conditions and catalysts were found which made possible the production of a color-free polymer. It is said that the selection of the catalyst is especially aimed at the prevention of other linkages in the polymer chain due to intracondensation of the glycol end groups.

It appears at this writing that another forward step is in progress. Two rather similar routes are reported. Both depend upon the reaction between ethylene oxide, rather than ethylene glycol, and terephthalic acid to form the bis-HET monomer already mentioned. The difference between the two methods lies in the point where purification is done; in one case the crude terephthalic acid is purified, in the other, the bis-HET monomer. In both cases this monomer is polymerized by known procedures to form a fiber-grade polyester. The titanium dioxide delustrant is added, as might be expected, early in the polymerizing process.

Only one other polyester has been produced in large volume up to this time. This is poly(1,4-cyclohexanedimethylene terephthalate) which is marketed under the trade name Kodel® by the Tennessee Eastman Co. in the

United States. The 1,4-cyclohexanedimethanol, which is used instead of ethylene glycol in this process, exists in two isomeric forms, one melting at 43°C and the other at 67°C. This makes their separation possible by crystallization so as to secure the desired ratio of the two forms for conversion to the polymer. This ratio determines the melting point of the polymer, this being a most important property for a material which is to be melt spun. The polymer from the 100%-*cis* form melts at 275°C and from the 100%-*trans* form at 318°C. Indications are that the commercial product is about 30/70 *cis-trans*.

Polyesters are melt spun in equipment essentially the same as that used for nylon, which has already been described. Wherever the volume is large and the stability of demand is adequate, the molten polymer is pumped directly from the final polymerization stage to the melt-spinning machine. The molten polymer is both metered and moved forward at high pressure by a gear-type pump through filters to the spinnerette which contains capillaries of about 9 mils (230 microns) diameter. Great care is used to eliminate moisture from the chips, if these are used, and from the spinning chamber. Since most polyester fibers are destined to become staple, the output of a number of spinnerettes are combined to form a tow which can be further processed as a unit. Continuous filament yarn is packaged for drawing. Spin-drawing is becoming more common.

Drawing

Unlike nylon, which contains a fairly high amount of crystalline component, polyethylene terephthalate fibers are essentially amorphous as spun. In order to secure a usable textile yarn or staple fiber this product must be drawn under conditions that will result in an increase in both orientation and crystallinity. This is done by drawing at a temperature above the glass transition point, tg, which is about 80°C. Conditions of rate and temperature must be selected so that the amorphous areas are oriented and crystallization will take place as the temperature of the drawn fibers drops to room temperature. An appropriate contact-type hot plate or other device is used and about 300 to 400 per cent extension is effected. As with nylon, the conditions of draw, especially the amount, determine the force-elongation properties of the product. Industrial-type yarns, such as those intended to be used as tire cord, are more highly drawn.

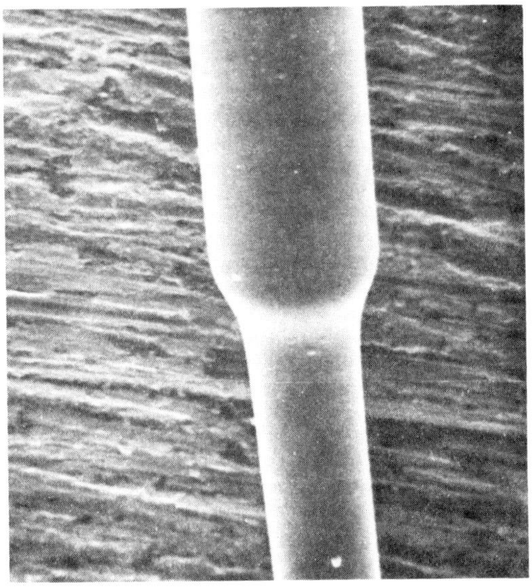

Fig. 11.17 Drawnecking-polyester single filament. (*Courtesy E. I. du Pont de Nemours and Co.*)

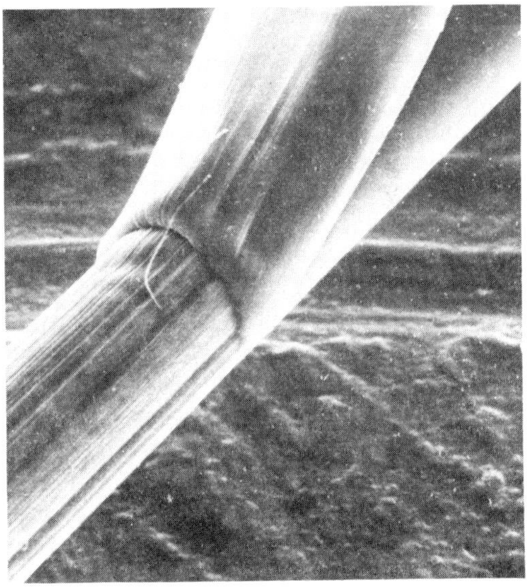

Fig. 11.18 Skin-peeling polyester filament showing fibrillar structure. (*Courtesy E. I. du Pont de Nemours and Co.*)

Heat Setting

The ability of textile fibers to be "set" is not characteristic of man-made fibers alone. Aided in many cases by the presence of starch, cotton fabrics have been ironed to a smooth and wrinkle-free condition; also, the sharp crease in wool trousers has been commonplace for generations. In other words, these fabrics were exposed to moisture at elevated temperatures while being held or pressed into desired geometrical configurations and then allowed to cool before being released from constraint. Such fabrics tend to remain unchanged while cool and dry, even though the fibers from which they are formed carry internal stresses. But reversion takes place upon washing or exposure to high relative humidity.

With the development of nylon, and especially polyesters, a durable kind of setting has become possible. When fabrics made from these fibers are shaped and then exposed to elevated temperatures either in the dry condition or, in the case of nylon particularly, in the presence of water vapor, thermoplastic relaxation of induced stresses in the fiber takes place and configurations at the molecular level adjust to a new and lower energy level. This depends on not only the temperature used but also the duration of the exposure. Thus a few seconds at 230°C will produce the same results as exposure for a considerably longer period at a temperature 50 to 75 degrees lower. The permanency of the setting, that is, the ability of a fabric or garment to return to its original configuration after temporary distortion even while exposed to moisture and raised temperatures, is a function of the severity of the heat setting.

It is this property of polyamides and polyesters that has been the main factor contributing to "ease of care" and the "wash and wear" characteristics of garments made from these polymers. In turn, these garments have revolutionized both the textile and apparel industries.

Textured Yarns

Fundamentally, the manufacture of "textured" yarns is closely related to the heat setting of fabrics, the difference being that the individual filaments or bundle of filaments in textured yarns are distorted from an essentially straight rod-like form and then heat set. In some instances the fibers are distorted in a more or less random way; at other times a regular pattern is desired.

The first commercially successful textured yarn was produced by highly twisting nylon-66, heat setting it as a full package of yarn, and then untwisting it through zero and applying a small amount of twist in the opposite direction. This process changed an individual filament from unitary close-packed structure to one which was voluminous due to mutual interference. The technique of heat setting the twisted yarn as a batch-unit operation has now been displaced by a continuous one, using what is known as a "false twisting" process. This is based upon the principle that if a length of yarn is prevented from rotating at both ends but is rotated on its axis at its center point, the resulting two sections will contain "Z" and "S" twists in equal amounts. When this occurs with a moving yarn, any element in it will first receive a twist in one direction but after passing the false twisting point must revert to zero twist. If now it is made to pass over a hot plate while in the twisted state and is heat set in that configuration, even after returning to the untwisted condition, the individual filaments will tend to remain distorted when lengthwise stress is released. Because of the low mass and diameters of textile yarns or monofilaments it is possible to false twist them at speeds as high as 300,000 rpm and thus make the operation a practical one, since twists of the order of 80 to the linear inch are required.

Toward the end of the 1960's machines were developed which combined in a single operation the drawing of yarn, already described, with texturing by the false twist technique. The undrawn yarn, or more probably a partially drawn product from a spin-drawing machine, passes over a hot plate, at which time it is both highly twisted and stretched in the amount required. Speeds as high as 800,000 rpm are attained either by spindles or frictional contact between yarn and driving source, thus making possible forward speeds of 400 meters per minute. The resulting yarn may be heat set as part of the same continuous operation by passing it through a heating chamber under conditions of overfeed or little or no tension in order to secure both thermally stable geometric configurations in the individual distorted filaments that comprise the yarn and the degree of "stretchiness" desired in the final product.

Since these yarns are being made in one less step and also within the plants spinning the parent product, this latest development may be said to constitute another advance in the industrial chemistry of man-made-textile products. This draw-texturing appears to be especially applicable to polyester yarns intended for fabrics known as "double knits," which are being manufactured in ever increasing amounts.

Yarn can be forced forward by means of "nip" rolls although this may seem to be quite contrary to the old adage that one cannot push on an end of string. When this is done so that the yarn is jammed into a receiver already full of the preceding material, it collapses with sharp bends between very short lengths of straight sections. In this condition heat is applied to set it. In practice, the mass of such yarn is pushed through a heated tube until it escapes at the exit past a spring loaded gate. During this passage it is heat set in a highly crimped configuration. It is cooled before being straightened and wound onto a package. In another continuous process, the yarn or monofilament is pulled under tension over a hot sharp edge so that it is bent beyond its elastic limit and heat set in that condition. The result is not unlike that produced by drawing a human hair over the thumb nail.

When such yarns are knitted or woven into fabric, the filaments tend to return to the configurations in which they were originally heat set. Contraction takes place in the direction of the yarn axis and this in turn converts the smooth flat fabric into a "stretch" fabric and gives the surface a textured appearance. These fabrics or the garments made from them, whatever the process used to produce the yarns, may be heat-treated to secure stability in a desired geometric configuration. A degree of stretch may be retained, or a flat and stable textured surface may be produced. There are a number of variations of the texturing process, these, combined with the many possibilities of heat-setting, impart considerable versatility to the final product. The growth in the use of these products in the 1960's is well known. Carpeting also provides a significant market since texture is one of the most important characteristics of soft floor coverings. Such products have been important to the successful use and expanded development of nylon and polyester yarns in recent years.

ACRYLICS

Polymer Manufacture

Acrylic (commonly not designated as polyacrylic) fibers are spun from polymers that are made from monomers of which a minimum is 85 per cent acrylonitrile. This compound may be made from hydrogen cyanide and ethylene oxide through the intermediate ethylene cyanohydrin,

$$\overset{\diagup\overset{O}{}\diagdown}{CH_2CH_2} + HCN \longrightarrow HOCH_2CH_2CN \xrightarrow[-H_2O]{catalyst} CH_2{=}CHCN$$

It may be made directly from acetylene and hydrogen cyanide,

$$CH{\equiv}CH + HCN \longrightarrow CH_2{=}CHCN$$

But the reaction which is currently preferred uses propylene, ammonia, and air,

$$3CH_2{=}CHCH_3 + 3NH_3 + 7O_2 \text{ (air)} \xrightarrow[<500^\circ C]{catalyst} CH_2{=}CHCN + 2CO + CO_2 + CH_3CN + HCN + 10H_2O$$

Pure acrylonitrile may polymerize at room temperature to polyacrylonitrile (PAN), a compound which, unlike polyamides and polyesters, does not melt at elevated temperatures but only softens and finally discolors and decomposes. Nor is it soluble in inexpensive low-boiling organic solvents. Since fibers made from it resist the dyeing operations commonly used in the textile industry, the usual practice is to modify it by copolymerization with other monomers, e.g., vinyl acetate, vinyl chloride, styrene, isobutylene, acrylic esters, acrylamide, or vinyl pyridine in amounts up to the 15 per cent of the total weight (beyond which the final product may not be termed an acrylic fiber). The choice of modifier de-

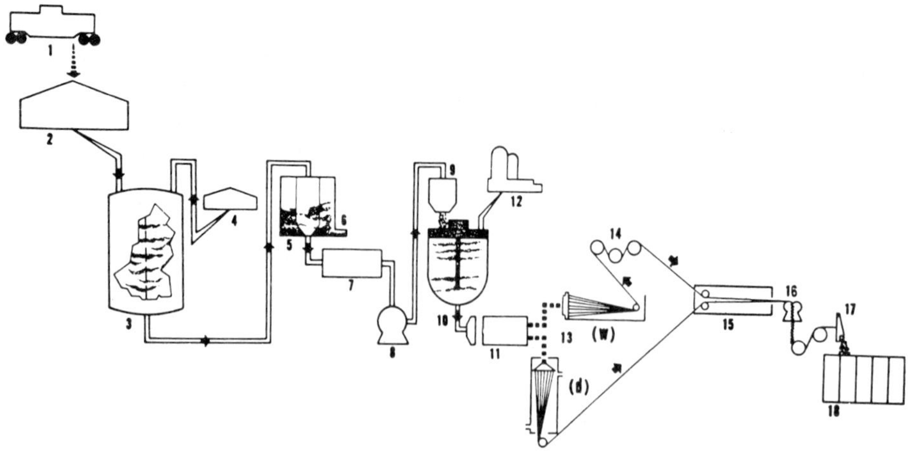

Fig. 11.19 Flow diagram for the manufacture of acrylic fiber: (1) acrylonitrile; (2) tank farm; (3) polymerizer; (4) co-monomer and catalyst; (5) centrifuge; (6) waste liquid; (7) dried polymer; (8) grinding; (9) polymer storage; (10) dissolver; (11) filter; (12) solvent plant; (13) spinnerette; (13w) wet spinning; (13d) dry spinning; (14) roller drier; (15) additional treatment; (16) crimper; (17) cutter; (18) acrylic fiber bale.

pends on the characteristics which a given manufacturer considers important in a fiber, the availability and cost of the raw materials in his particular area of production, and the patent situation.

In copolymerizating acrylonitrile with another monomer, conditions must be controlled in such a way that the reaction produces a polymer having the desired chain construction and length. Generally the reaction takes place in the presence of substances capable of producing free radicals. In addition, certain trace metals which have been found to increase reaction rates offer a means of controlling chain length. When polymerization is carried out in solution, after an induction period, the reaction is rapid and liberates a considerable amount of heat. Furthermore, since the polymer is not soluble in the monomer, a thick paste is formed. These facts limit the usefulness of such a process. Carrying out the polymerization in the presence of a large amount of water is a convenient method and the one most generally used. In this case the polymer forms a slurry and the water provides a means for removing the heat from the site of the reaction. Moreover, most of the common redox-catalyst systems are water-soluble. Polymerization may be carried out batchwise or by a continuous process.

In the standard batch method the monomers and catalyst solutions are fed slowly into an agitated vessel containing a quantity of water. The heat of reaction is removed either by circulating cooling water through the jacket surrounding the vessel or by operating the reaction mixture at reflux temperature and eliminating the heat through the condenser water. The monomer and catalyst feeds are stopped when the desired amounts have been added and polymerization is allowed to continue until there is only a small amount of monomer remaining in the reaction mixture. Then the slurry is dumped from the reaction vessel, filtered, washed, and dried.

If a continuous process is used, e.g., moving batch or pipeline, the raw materials are metered into one end and polymerization proceeds as the mixture moves forward. The resulting polymer slurry emerges from the other end. This method requires exceedingly careful control of feed ratios and temperatures to maintain consistent polymer quality.

In the continuous overflow method, rather than stopping the monomer and catalyst feed when the reaction vessel is full, the slurry is simply allowed to overflow; the solids are removed by filtration, washed, and dried. The filtrate contains a certain amount of unreacted monomer, and this is recovered by steam distillation after the catalyst has been destroyed to prevent further polymerization. The dried polymer is the raw material from which fibers are spun.

Spinning

As already mentioned, pure polyacrylonitrile softens at elevated temperatures and thermal decomposition starts before the molten state is reached. The same is true of the copolymers commonly used to produce fibers. Accordingly, melt spinning is impossible; spinning must be done from a solution of the polymer. Both dry and wet spinning are carried out in current commercial operations.

The operations which are used to either wet- or dry-spin acrylics are essentially the same as those already described for rayon and acetate, respectively. The polymer must be completely dissolved in solvent and the solution filtered to remove any impurities which would cause spinnerette blockage. Because acrylic polymers are not soluble in common nonpolar solvents, polar substances such as dimethyl formamide, dimethyl acetamide, or solutions of inorganic salts such as zinc chloride-calcium chloride mixtures are required. Only wet spinning is possible with the latter. Dimethyl formamide boils at 152.8°C and exerts a vapor pressure of 3.7 mm of Hg at 25°C compared with acetone (used in dry spinning of cellulose acetate) which has a vapor pressure of 228.2 mm of Hg at 25°C. It follows that, unlike acetone which requires an activated-carbon system for recovery, dimethyl formamide may be condensed directly from the air stream used to evaporate the solvent from the forming fiber.

Acrylics, like rayon, require stretching either during spinning or immediately after fiber formation, or both, followed by a relaxing treatment in order to obtain the desired characteristics of modulus, breaking tenacity, and breaking elongation. These same properties are influenced by spinning speeds, and the temperature of the drying air, if dry spun, or temperature and composition of the bath,

Fig. 11.20 Wet spinning of acrylic tow. (*Courtesy Monsanto Co.*)

if wet spun. The multitude of combinations made possible by the use of various comonomers and the flexibility of the fiber-forming operations furnish the several manufacturers with versatility and the users with a variety of acrylic fibers.

Acrylic fibers possess a property that made it possible for them in the late 1950's and early 1960's to find immediate, even spectacular, acceptance in the knitted sweater field, until then dominated by wool. When acrylic fibers, normally in the form of a heavy tow, are hot-stretched (for example, by being drawn over a hot plate and then cooled under tension) they are converted to a labile state. Upon immersion in hot water such fibers will contract considerably but not to their prior unstretched length. In practice this characteristic is used to produce a bulky yarn resembling the woolen yarns long accepted for use in sweaters. The process is described briefly below.

The stretched labile fibers are further cold-stretched to the breaking point so that non-uniform lengths are produced similar to the lengths found in wool. These are crimped and then mixed with thermally stable acrylic fibers which have been stretched and relaxed and have about the same length and degree of crimp. The blend is converted to a spun yarn by the same process used in making woolen yarns and in turn this yarn is knitted into sweaters and other similar products. When such garments are dyed in hot water, the labile fibers, intimately blended with stable ones, contract lengthwise individually. In the process segments of the stable units tend to be carried along physically by entrapment and friction. But since each such fiber does not change its overall length, only the yarn as a whole decreases in length. Lateral displacement of the large volume of stable fibers results in the formation of a more voluminous structure known as a "bulky" or "hi-bulk" yarn.

Bicomponent or Conjugate Spun Fibers

As will be shown, it should be theoretically possible to make any of the common man-made fibers in bicomponent forms. But acrylics have received the most attention for quite good reasons. Their general characteristics have tended to make them competitive with wool. This means that they should be processable upon machinery developed for handling wool, as well as being capable of being accepted into markets previously dominated by wool. It follows that since the natural fiber possesses crimp which produces the cohesion that determines its behavior in processing and in part its appearance and "hand" in usage, a similar crimp was desired for acrylics. Unlike the polyamides and polyesters, heat setting the crimp is not possible for very fundamental reasons.

The principle which is the basis for bicomponent fibers is usually likened to that which underlies the bicomponent metal strips often used in temperature controllers. With the latter, differential thermal expansion of the two joined components results in a bending of the thermal element. With fibers, moisture is usually the agent that acts upon the two side-by-side portions. Differential swelling or shrinkage causes the fiber to be brought into a crimped, or preferably, a spirally distorted condition. Incidently, the side-by-side structure occurs naturally in wool.

The combination of small size and large number of holes in a spinnerette might lead to the conclusion that it would be almost impossible to design a spinnerette assembly which would bring two streams of polymer or polymer solutions together at each such hole and extrude them side by side to form a single filament. Such designs have, in fact, been made. But solutions of fiber-forming polymers fortunately possess properties which encourage laminar flow and thus make other approaches possible. This phenomenon was remarked upon earlier in connection with dope dyeing; when a suspension of a colored pigment is injected into a dope stream, a considerable problem must be overcome in order to secure adequate mixing so as to secure "dope dyed" fibers of a uniform color. Thus, it was known that when two streams of essentially the same solution of a fiber-forming polymer are brought together, side-by-side, and moved forward down a pipe or channel by the same amount of pressure behind each, virtually no mixing takes place. By bringing these streams to each spinnerette hole in such an individual side-by-side arrangement and using appropriate mechanical separators, the extruded filament from each hole will have a bicomponent structure. In addition to producing fibers in which the two components form a bilateral symmetrical structure, ingenious arrangement of predividers of the two streams can produce from the full complement of holes in a single spinnerette a selected group of fibers wherein the amount and position of each of the two components is randomly distributed throughout their cross sections. It follows that curls of uniform or random geometry may be produced to meet required needs.

VINYL AND MODACRYLIC FIBERS

Vinyls

When nylon-66 was developed it was described as being "synthetic" or "fully-synthetic" in order to differentiate it from rayon and acetate. This was no small act of courage, since the word "synthetic," in that period just following the repeal of Prohibition in the United States, was often associated in the public mind with the least palatable kind of alcoholic beverages. In due time, what is known in the advertising business as "puffing" led it to be known as the "first fully synthetic fiber," which was an anachronism. It so happens that fibers based upon polyvinyl chloride predated it by several years.

About 1931, the production of fibers from polyvinyl chloride (PVC) was accomplished by dry spinning from a solution in cyclohexanone. But by chlorinating the polymer it was possible to secure solubility in acetone which has the advantage of possessing a boiling point about 100°C lower than cyclohexanone. Several million pounds per year of this fiber were produced in Germany during World War II to relieve the shortages of other materials. Unfortunately, PVC begins to soften at about 65°C and in the fibrous state shrinks disastrously upon heating. Because of its low softening point it cannot be dyed at the temperatures commonly used for this purpose and, furthermore, it resists dyeing.

Quite recently an improved PVC fiber has resulted from application of the newer techniques of industrial chemistry. A highly syndiotactic form can be produced by polymerization at −30°C, using a combination such as cumyl hydroperoxide and methyl sulfite as the catalyst. At that temperature the vapor pressure of vinyl chloride is below atmospheric and by operating in a partial vacuum the heat of polymerization is easily dissipated through refluxing of the boiling monomer. Continuous operation is possible with the polymer being removed from the slurry by centrifugation; unreacted monomer is distilled from the catalyst residues for re-use and the catalyst is replenished as needed.

This syndiotactic form has a considerably higher glass transition temperature and melting point than the previously available PVC. It can be converted to a higher level of crystallinity and has a greater resistance to solvents, including the chlorinated hydrocarbons used in dry cleaning. But this insolubility posed a problem in the conversion of bulk polymer to fibers. This problem was solved by wet spinning from a solution in cyclohexanone at 110°C into an aqueous bath at a lower temperature, the bath containing a mutual solvent for the cyclohexanone and the water. Afterward the fibers are stretched in boiling water to about seven times their original length, dried, heat set, and relaxed in steam.

As yet this fiber, produced in Italy under the trade name Leavil,®* has not been manufactured in the United States. Indications are that it can be processed in normal textile operations, including dyeing. It is nonflammable, and this characteristic alone may lead to its receiving increased attention since Federal regulations relative to fabric flammability are of considerable concern to the textile industry.

Modifications of PVC have been produced by copolymerization with other monomers. The first successful one consisted of 90 per cent vinyl chloride copolymerized with 10 per cent vinyl acetate. It was dry spun from acetone and given the trade name Vinyon by its producer, Union Carbide Corporation. (In 1960 vinyon was accepted as a generic name for fibers containing not less than 85 per cent vinyl chloride.) It has never been produced in

*Registered trademark, Chatillon Societa Anonima Italiana per le Fiber Tesili Artificiali S.p.A.

large volume and is used for heat-sealable compositions.

A copolymer of vinyl chloride with vinylidene chloride was used for a number of years to produce melt-spun, heavy monofilaments. These found use in heavy fabrics where the chemical inertness of the polymer was needed, in outdoor furniture and in upholstery for seats in public-transportation vehicles.

Another vinyl-based fiber which is not yet being produced in the United States is Kuralon®*. In 1970 some 180 million pounds of this material were manufactured, of which about 15 to 20 per cent was continuous-filament yarn and 80-85 per cent staple fiber. Acetylene made from calcium carbide is converted to vinyl acetate which, following polymerization, is saponified to polyvinyl alcohol.

$$CH \equiv CH + CH_3C(=O)-OH \rightarrow$$

$$CH_2 = CHOCCH_3 \xrightarrow[\text{catalyst}]{\text{heat}}$$

$$-(CH_2CH)_n- \xrightarrow{H_2O} -(CH_2CH)_n- + CH_3COOH$$

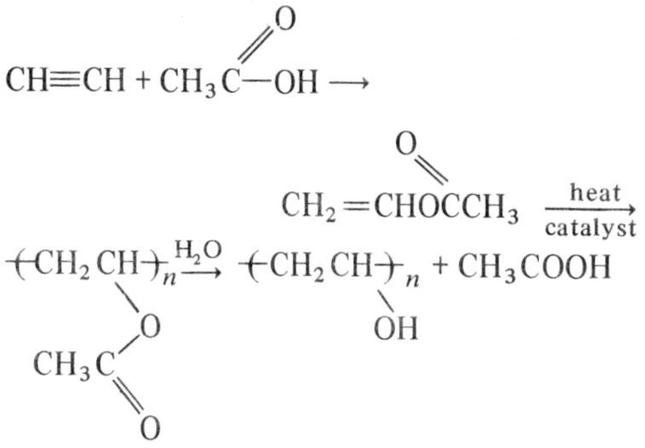

This is soluble in hot water, and the solution is wet spun into a coagulating bath consisting of a concentrated solution of sodium sulfate. The fibers are heat treated to provide temporary stabilization so that they may be converted to the formal derivative by treatment with an aqueous solution of formaldehyde and sulfuric acid. This final product resists hydrolysis up to the boiling point of water. It seems reasonable to assume that the final product contains hemiacetal groups and some unreacted hydroxyls on the polymer chain as well as crosslinking acetyl groups between the adjacent molecules.

The fiber has been produced in the form of continuous-filament yarn for fishing nets and industrial applications. As a staple fiber it has had considerable acceptance as the base material for fabrics used to produce uniforms for school children and industrial workers in Japan.

Modacrylics

In the United States the modification of PVC has moved in the direction of copolymerizing vinyl chloride with acrylonitrile, or perhaps it should be said that PAN has been modified by copolymerizing the acrylonitrile with chlorine-containing vinyl compounds. In any case, two modacrylic fibers are now produced in the United States, a modacrylic being defined as one containing at least 35 per cent but not over 85 per cent acrylonitrile. Dynel®* is 60 per cent vinyl chloride and 40 per cent acrylonitrile and Verel®** is said to be a 50-50 copolymer of vinylidene chloride and acrylonitrile with perhaps a third component graft-polymerized onto the primary material to secure dyeability.

It is not surprising that Dynel, which is available only in the form of staple fibers, is wet spun from a solution in acetone. The method used to produce Verel has not been revealed. Since it is being manufactured by a company experienced in and equipped for dry spinning cellulose acetate, it seems reasonable to suspect that this is the method used for this polymer.

Both of these modacrylic fibers require after-stretching and heat stabilization in order to develop the necessary properties. It seems probable that the stretching is of the order of 900–1300 per cent, and that, in a separate operation, shrinkage of about 15 to 25 per cent is allowed during the time the fibers are heat-stabilized.

The modacrylic fibers, like vinyon, and unlike the acrylic fibers, have not become general-purpose fibers. They can be dyed satisfactorily and, thus, are acceptable in many normal textile products. But their nonflammability tends to place them in uses where such a property is important, even vital. Blended with other fibers, they are used in carpets; but their largest market is in deep-pile products, such as "fake-furs," or in doll hair where a fire hazard cannot be tolerated.

---

*Registered trademark, Kuraray Co. Ltd.

*Registered trademark Union Carbide Co.
**Registered trademark Tennessee Eastman Co.

ELASTOMERIC FIBERS

The well-known elastic properties of natural rubber early led to processes for preparing it in forms which could be incorporated into fabrics used for garments. One such process used standard rubber technology. A raw rubber of high quality is compounded with sulfur and other necessary chemicals, calendered as a uniform thin sheet onto a large metal drum, and vulcanized under water. The resulting skin is then spirally cut into strips which may be as narrow as they are thick, for example, 0.010 in. by 0.010 in. square in cross section. These strips are desulfurized, washed, dried, and packaged. Larger cross sections are easier to make. This product, coming out of the rubber rather than the textile industry, is known as a thread.

Another method produces a monofilament known as a latex thread. As the name would indicate rubber latex is the raw material, and since extrusion through small holes is required, the purity of the material must be of a high order. With proper stabilization, the latex solution may be shipped from the rubber plantation to the plant where it is compounded with sulfur and other chemicals needed for curing and as well as with pigments, antioxidants, and similar additives. This is followed by "precuring" in order to convert the latex to a form which will coagulate upon extrusion into a precipitating bath of dilute acetic acid and form a filament having sufficient strength for subsequent operations. It passes out of the bath and is washed, dried, vulcanized in one or two stages, and packaged.

The rubber threads manufactured by either process can be used as such in combination with normal nonelastomeric yarns in fabrics made by weaving or knitting. But most of these, especially those made by the latex process, are first covered by a spiral winding of natural or man-made yarns. Often two layers are applied in opposite directions to minimize the effects of torque. Such coverings have two purposes. The first is to replace the less desirable "feel" of rubber on human skin by that of the more acceptable "hard" fiber. The second concerns engineering desired properties into the product to be woven or knitted into fabric. As an elastomeric material begins to recover from a state of high elongation it supplies a high stress, but as it approaches its original unstretched condition the stress drops to a very low order. When wound in an elongated state with a yarn having high initial modulus and strength, the elastomeric component cannot retract completely because its lateral expansion is limited and jamming of the winding yarn occurs. Thus the combination of such materials can be made to provide stretch and recovery characteristics needed for a broad spectrum of applications.

But the traditional elastomeric threads have been subject to certain inherent limitations. The presence of unreacted double bonds leads to a sensitivity to oxidation, especially when exposed to the ultraviolet radiation of direct sunlight. There is also low resistance to laundry and household bleaches and dry cleaning fluids.

During recent years elastomeric yarns or threads have been used in garments intended to restrain the female figure in a form approaching the ideal of youth and at the same time give the illusion of nonexistence. Such garments must be thin and highly effective per unit of weight. The materials from which they are composed must be compatible with these requirements.

Thus, it is not unexpected that the producers of man-made fibers, already eminently successful in meeting the needs of the market place, should look to the field of elastomeric fibers for new possibilities. Given the limitations of rubbers, both natural and synthetic, as well as the relationships between molecular structure and behavior of fiber-forming linear polymers, the scientists faced new challenges.

As an over-simplification it can be said that within limits a rubber-like material can be stretched relatively easily but reaches a state where crystallization tends to occur. The structure produced in this manner resists further extension, and the modulus rises sharply. In contrast with the conditions that occur when the man-made fibers discussed earlier in this chapter are drawn to form fibers of stable geometry in the crystalline and orientated states, the crystalline state of the elastomeric fibers is labile unless the temperature is lowered materially. This elastomeric aspect is reasonably straightforward;

synthetic rubbers have been around for a long time. But to improve on the chemical sensitivity of rubber, new approaches were necessary. The solution was found in linear polymers containing "soft" sections connected by "hard" components.

The soft, flexible, and low-melting part is commonly an aliphatic polyether or a polyester with hydroxyl end groups and a degree of polymerization of 10 to 15. The hard portion is derived from an aromatic diisocyanate supplied in an amount that will react with both end groups of the polyether or polyester to form urethane groups. The product, an intermediate known as a prepolymer, is a thick liquid composed essentially of molecules carrying active isocyanate groups at each end. For example:

$$HO\text{-}(RO)_n\text{-}H + 2\,O{=}CNR'NC{=}O \longrightarrow$$

$$O{=}CNR'N{=}\underset{\displaystyle OH}{C}O\text{-}(RO)_n\text{—}\underset{\displaystyle OH}{C}{=}NR'NC{=}O$$

where $\text{-}(RO)\text{-}$ is an aliphatic polyether chain, R' is one of several commonly available ring structures, and $n \sim 10$.

The elastomeric polymer is secured by "capping" or "extending" the prepolymer through its reaction with short-chain glycols or diamines, thus completing the formation of hard groups between soft, flexible chains. The conversion of these polymers into usable fibers may be accomplished by wet-, dry-, or melt-spinning operations, depending on the polymer. Additives to impart color or improve resistance to ultraviolet radiation and oxidation may be incorporated in the spinning solutions or in the melts.

The development of elastomeric fibers has resulted in a variant of wet spinning called "reaction" or "chemical" spinning. In point of fact rayon, the first wet-spun material, might properly be said to be produced by "reaction wet spinning" or "chemical wet spinning" since complex chemical reactions have always been involved in that operation. Be that as it may, it has been found that the prepolymer of an elastomeric fiber may be extruded into a bath containing a highly reactive diamine so that the chemical conversion from liquid to solid occurs there.

The elastomeric fibers produced in this fashion are based upon segmented polyurethanes and by definition are known generically as spandex yarns. Each manufacturer uses a trade name for the usual commercial reasons. Perhaps the most noteworthy aspect from the standpoint of industrial chemistry is the multitude of options available to the manufacturer through the ingenious use of various chemicals for soft segments, hard units, chain extenders, and conditions of chemical reaction, followed by numerous possibilities for extrusion and after-treatments. However, at this time most of the companies which entered the market with new products and enthusiasm in the middle 1960's have since dropped out of it.

POLYOLEFIN FIBERS

Although polyethylene was considered as a source of useful fibers at an early date, its low melting point (110-120°C) as well as other limitations precluded active development during the period when production of other fibers based upon the petrochemical industry expanded enormously. The higher melting point of high-density polyethylene gave some promise but this was overshadowed by the introduction of polypropylene around 1958-59. Great expectations were held out for the latter as a quick contender with the polyamides and polyesters, already successful, and the acrylics then coming into the fiber field in volume. Several facts were thought to be in its favor. The first was raw material costs—a few cents per pound; the second was the high level of sophistication in the spinning and processing of fibers and the presumption that this would readily lead to development of means of converting the polymer to fibers; and finally, there was the belief that the American consumer would be ready to accept, and perhaps even demand, something new and different. The limitations found in polypropylene fibers, such as extremely poor dyeability and low heat stability, combined with the lower prices and increased versatility offered by the already established fibers dashed the hopes for quick success. Nevertheless polypropylene has found an increasingly important place and its properties have led to new techniques of manufacture and specialized uses.

Early in the manufacture of polyolefins a concept was developed for dry spinning directly from the solution obtained in the polymerization operation. Had it been feasible it would have been the realization of a chemical engineer's dream; the gaseous olefin fed into one end of the equipment and the packaged fiber, ready for shipment to a textile mill, coming out the other end. But it did not turn out that way and today melt spinning is the accepted technique for the production of monofils and multifilament yarns. To this usual method there has been added a fibrillation or "split film" procedure, and, as might be expected from a film-forming polymer, splitting of thin sheets is also possible.

The polyolefins are completely resistant to bacterial attack, are chemically inert, and are unaffected by water. Monofilaments can be produced that possess high strength, low elongation under stress and dimensional stability at normal atmospheric temperatures. Polyolefin monofilaments have found broad application in cordage and fishing nets; when woven into fabrics they are used for outdoor furniture, tarpaulins, and similar applications.

Because of high melt viscosity, polyolefins must be extruded at 100–150°C above the melting point. The usual metering pumps follow the heated melt screws, and screen filtration units are placed above the spinnerette. The filaments are extruded into water for quenching and heat removal. The materials thus produced must be afterdrawn in a heated condition to several times their original length, this being determined by the molecular weight and the properties desired in the end product. Stabilization follows in a heat-setting operation at constant length or with a limited shrinkage being allowed.

The production of multifilament yarns follows along the lines commonly used in melt-spinning processes, with air quenching replacing the water quenching mentioned above. If anything, a greater drawdown, that is, a higher ratio of the velocity of solid fiber

Fig. 11.21 The melt spinning of polypropylene, showing a heavy yarn emerging from the cooling column, passing in contact with the lubricant applicator, and over the pair of draw-off godets that determine its speed, and finally to a tension-controlled packaging device. *(Courtesy Hercules Inc.)*

wound up to the average velocity of liquid polymer in the spinnerette capillary, is used in this spinning operation than is commonly used for other melt-spun fibers. This elongating operation does not preclude the need for an afterdrawing step to produce the orientation necessary to achieve the desired physical properties. As with monofilaments, hot drawing is required. Where the fibers are to be used as multifilament yarns a higher amount of stretch is used than when staple fibers are to be the end product. Higher strength and lower elongation are normally required of multifilament yarns as compared with staple fibers.

The production of "spunbonded" materials from a number of fiber-forming polymers has been especially applied to polypropylene (Typar®* to manufacture a sheeting which is used

---

*Registered trademark, E. I. du Pont de Nemours & Company.

Fig. 11.22 Spunbonded polypropylene shows embossing which helps bind the structure. (*Courtesy E. I. du Pont de Nemours and Co.*)

as the primary backing in tufted carpets. In this case the strength-contributing fibers are spun and drawn in a continuous operation. The oriented fibers are randomly laid as a web and bonded by fusing under heat and pressure at selected points.

As already noted, since polyolefins are converted in great volume to thin films, it is logical to slit these into narrow strips for uses where they can compete with the more conventional fibers. But the polyolefins have also made possible the "split-film" method of producing fibers. This is due to their ability to be cast into films which, when stretched, become highly crystalline and oriented in the direction of stretch. The very low strength in the direction perpendicular to the axis of orientation causes the film to split and fibrillate. The resulting web-like structure, comprising interattached fibrils of high longitudinal strength may be converted by twisting into yarn-like structures or cut into staple fibers.

POLYTETRAFLUOROETHYLENE

Teflon®* is unique in that it is not soluble in any known solvent and will not melt below about 400°C, at which point it is not stable enough to allow spinning. Such a combination would seem to pose an impossible problem. Research into the fundamental characteristics of the polymer, however, revealed that the submicroscopic particles precipitated from the polymerization reaction were about 100 times as long as they were thick. If a thin stream of polytetrafluoroethylene suspension is extruded, when the dispersion is broken the particles line up to form a discrete filament which has sufficient strength to allow its transportation to a sintering operation. This fiber has adequate tenacity (1.5 grams per denier) for ordinary textile handling, knitting, and weaving.

The chemical inertness and thermal stability of this material is so great that in spite of its extremely high price ($14–$27 per pound in 1973) it is used in chemical operations where drastic conditions exist and no other organic fiber is suitable.

GLASS AND CARBON FIBERS

Among the man-made inorganic fibers, glass is produced in by far the largest volume. There has been a rapid increase in the use of textile grades of glass fibers. Outside the textile field, enormous amounts of glass fibers are used in air filters, thermal insulation (glass wool), and for the reinforcement of plastics.

Glass possesses obvious and well-known characteristics which have largely determined the methods used to form it into large objects. It flows readily when molten and can be drawn into filaments, whose extreme fineness appears to be limited only by the drawing speed. The method used in producing textile-grade glass fibers follows this principle.

In the commercial operation the molten glass, produced either directly or by remelting

*Registered trademark, E. I. du Pont de Nemours & Co.

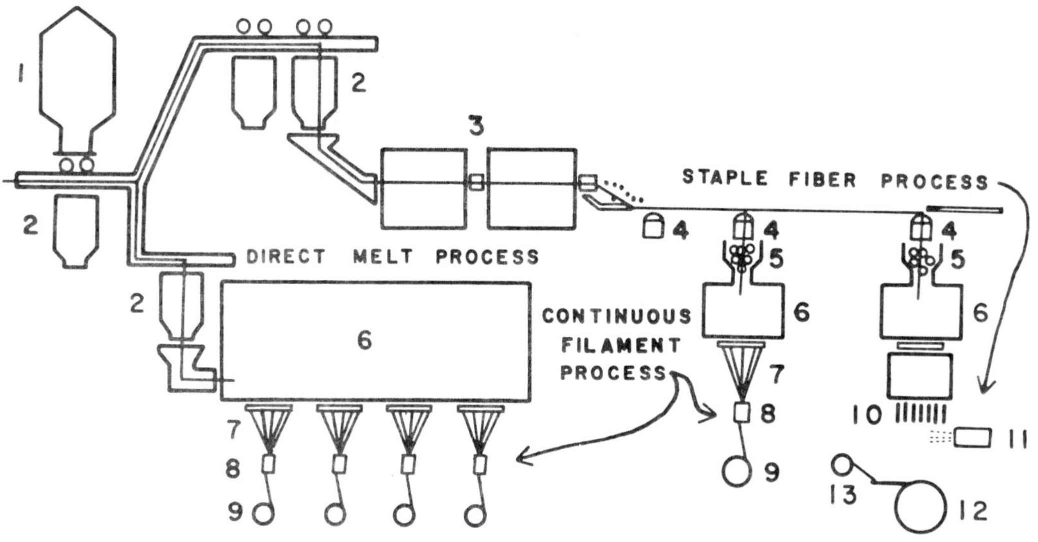

Fig. 11.23 Flow diagram for the manufacture of textile glass fiber: (1) glass batch; (2) batch cans; (3) marble-forming; (4) cullet cans; (5) marbles; (6) melting furnaces; (7) filament yarn formation; (8) gathering and sizing; (9) yarn packaging; (10) air jets; (11) lubricant spray; (12) collection for staple fibers; (13) staple fiber packaging. (*Courtesy Owens—Corning Fiberglas Corp.*)

of marbles, is held at a uniform temperature in a vessel, whose bottom carries a bushing containing small uniform holes. The molten glass flows through these as tiny streams which are attenuated into filaments at speeds of the order of 3,000 meters per minute, coated with a lubricant, gathered into groups to form yarns, and wound up. For a particular glass viscosity, the size of the individual filaments is determined by the combination of hole-size and speed of attenuation.

Because of the inherently high modulus of glass, very fine filaments are required in order to approach the required properties of textile materials. Thus, the diameter of glass filaments falls in the range of 3.8 to 7.6 microns whereas the average diameter of the finest organic fibers is about twice as large. The fiber- and yarn-numbering system is based on nomenclature used in the glass industry and differs from the traditional systems accepted in the textile and organic-fiber industries.

The method of manufacture of glass staple fibers differs from those used to produce the corresponding organic materials, all of which are based upon cutting the continuous-filament product. Air jets, directed in the same line of flow as the emerging streams of glass, attenuate them and break the solid glass into the lengths desired for further processing. These are gathered on an appropriate vacuum drum and delivered as slivers or a matte. To produce fibers which may be coarser and considerably less uniform to be used for the production of filters, paper, or thermal insulation, large streams of molten glass are cross blown by blasts of hot air, steam, or burning gas.

As might be expected from the nature of glass, the conversion of glass fibers into the final products has required the development of new lubricants, finishes, and processing techniques. For example, because glass fabrics cannot be dyed directly or printed with the colors demanded for acceptance as draperies, the colorant must be applied to a resin coating. But before applying the coating it is necessary to remove the lubricant which was placed on the fibers to permit their conversion into a fabric. This is done by burning. The elevated temperature resulting from this operation also relaxes the internal strains developed in the glass fibers during the steps of the textile operations and sets the yarns in the required geometry. The fabric is then resin treated, cured, and dyed or printed.

Another inherent property of glass is the tendency of unprotected fiber surfaces to abrade each other to destruction under the action of very little mechanical working. Also, it bonds poorly to the rubber of the intermediate adhesives commonly used between tire cords and rubber. For many years these limitations had frustrated attempts by tire manufacturers to take advantage of the very high tensile strength, completely elastic behavior, high modulus, and lack of moisture sensitity of glass fibers. Recently, it appears, it has been possible to modify the fiber surfaces so that satisfactory adhesion is achievable, and the impregnant can be applied in such a way that fiber to fiber contact is prevented. In any case, use of glass tire cord is increasing rapidly.

Following World War II the development of jet aircraft and rockets brought about demands for fibers having thermal resistance, strength, and modulus far beyond what could be obtained in existing organic fibers. Much of this need was for reinforcing materials which would be embedded in matrices of one type or another. As a result, techniques have been developed for preparing fibers from a good many metals and refractory inorganic compounds. Although these materials are essential for certain uses, the volume of production is still low and the prices correspondingly high (as much as $1,000 per pound).

Carbon and graphite fibers, although they have the economic limitations already noted, are produced from the organic fibers described in this chapter. The principle is essentially the same as that which brought about the formation of coal or, citing a more recent and dramatic example, the conversion of the original wooden beams of buildings in Herculaneum, buried by a flow of mud from Vesuvius in 67 A.D., to what appears to be charcoal. In the present commercial process an organic fiber, usually viscose rayon or an acrylic fiber, is prevented from oxidizing by excluding air, and at the same time it is subjected to conditions which cause the elements other than carbon to be driven off. High temperatures for very short time periods are substituted for the lower temperatures and millennia associated with the formation of the natural forms of carbon.

Use of organic fibers as the raw material for carbon or graphite fibers makes it possible to predetermine the morphology of the final fibers and their geometrical arrangement. Continuous fibers can be converted by passage through a furnace operating in an inert atmosphere or a vacuum. Woven fabrics, papers, and belt-like structures which cannot be made from the brittle carbon fibers may be obtained by prior weaving of the parent material and *in situ* carbonization.

HIGH-TEMPERATURE-RESISTANT FIBERS[6]

At present only one high-temperature-resistant fiber is produced in what may be termed commercial volume. The need for such fibers has arisen from both the general economy and the space program where the usual characteristics or organic-based fibers are desired but the high temperature resistance of inorganic fibers is required. Thus, they are expected to retain their structural integrity at temperatures of 300°C and above for considerable periods of time, but otherwise their properties should resemble those of the more common man-made textile fibers.

The one fiber referred to above was announced in early 1962 as HT-1 nylon, but it is presently sold as Nomex®*. It is generally accepted in the industry as being formed by solution polymerization of m-phenylenediamine and isophthalic acid chloride. Since it does not melt but rather decomposes at temperatures well above 300°C, it must be spun from solution. It appears that a dry-spinning technique is used. Polyamides of this type must be hot drawn and then relaxed in order for them to have the desired dimensional stability when later used at elevated temperatures.

Other fibers for use at elevated temperatures have been developed. One of these, Kynol®**, is based on a phenolic resin; the method of manufacture has not been revealed.

---

*Registered trademark, E. I. du Pont de Nemours & Co.

**Registered trademark, Carborundum Co.

Much is known, however, regarding the manufacture of yet another fiber developed for use at elevated temperatures. Poly-2,2′-(*m*-phenylene)-5,5′-bibenzimidazole, commonly designated as PBI, was developed under the aegis of the Air Force Materials Laboratory in recent years. It is a condensation polymer obtained from the reaction of 3,3′-diaminobenzidine and diphenylisophthalate in a nitrogen atmosphere at temperatures which may reach 450°C in the final stages. The polymer is dissolved at a high temperature under nitrogen pressure in dimethylacetamide to which a small amount of lithium chloride is added to increase the stability of the solution. It is then dry spun into an atmosphere of heated nitrogen (about 200°C), from which the solvent is recovered, stretched slightly in steam, washed to remove the lithium chloride and the last traces of solvent, dried, and packaged. Drawing and relaxing are done in an inert atmosphere, as might be expected, since temperatures as high as 250°C or higher are used.

The final yarn is golden yellow, and because this color appears to be an intrinsic property of the polymer, it may have a built-in limitation as far as the civilian market is concerned. But since it is capable of retaining about one half of its original strength (~5 g/den) upon exposure to air for 18 hours at 350°C or one hour at 425°C, it would appear to have considerable promise in a limited field.

REFERENCES

1. "Seven Centuries of Art," p. 12, New York, Time-Life Books, 1970.
2. Mark, H., and Whitby, G. S., "Collected Papers of W. H. Carothers," New York, John Wiley & Sons, 1940.
3. Markam, J. W., "Competition in the Rayon Industry," p. 16, Cambridge, Mass., Harvard University Press, 1952.
4. German Patent 748,253.
5. British Patent 578,079.
6. Preston, J., Ed., "Symposium on High Temperature Resistant Fibers from Organic Polymers," American Chemical Society Symposium, New York, John Wiley & Sons, 1969.

SUGGESTED FURTHER READING

The reader is referred to the two encyclopedias listed below for additional information. They contain enormous quantities of information on man-made fibers as well as comprehensive bibliographies.

1. Kirk-Othmer Encyclopedia of Chemical Technology, 2nd ed., Interscience Publishers, New York. (Twenty one volumes and a supplement.)
2 Encyclopedia of Polymer Science and Technology, Interscience Publishers, New York. (Sixteen volumes.)

The following books contain broad discussions of man-made textile fibers in their many ramifications.

1. Morton, W. E. and Hearle, J. W. S. "Physical Properties of Textile Fibres," Manchester, The Textile Institute, London, Butterworths, 1962.
2. Hearle, J. W. S. and Peters, R. H., Ed., "Fibre Structure," Manchester, The Textile Institute, London, Butterworths, 1963.
3. Moncrieff, R. W., "Man-Made Fibres," 5th Ed., New York, John Wiley and Sons, 1970.
4. Peters, R. H., "Textile Chemistry;" Vol. I, "The Chemistry of Fibers," and Vol. XI, "Impurities in Fibers; Purification of Fibers," New York, Elsevier, 1963 and 1967.
5. Mark, H. F., Atlas, S. M., and Cernia, E., Eds., "Man-Made Fibers; Science and Technology," Vol. I, Vol. II, and Vol. III, New York, John Wiley and Sons, 1967, 1968, and 1968.

12

Animal and Vegetable Oils, Fats, and Waxes

Glenn Fuller*

INTRODUCTION

Oils, fats, and waxes belong to the class of compounds called lipids, which are insoluble in water, but soluble in ether and other organic solvents. The lipids have long been considered important as one of the three major classes of food materials. They differ from the other two, proteins and carbohydrates, in that they provide more than twice the number of calories per unit weight when converted to carbon dioxide and water. Hence, they were thought to be primarily products for storage of energy in plants and animals. More recently, lipids have been shown to be extremely important in the organization of the living cell. Different metabolic functions are carried out in specific parts of the cell, and cell membranes which isolate the parts from one another are made up of lipids and proteins. Although not all of the functions of lipids in cell membranes are well understood, the selective permeability of lipid-membrane constituents certainly plays a role in the isolation of cell functions.

Oils and fats are usually distinguished from one another by their melting points and, to some extent, by their sources. Oils are generally liquid at ordinary temperatures while fats are solid or semisolid. The fats are usually of animal origin while oils are extracted from plant tissue, fish, or marine animals. Both fats and oils are composed largely of glycerol esters of fatty acids. The esters are formed by combination of carboxylic acids and alcohols with loss of water. Since glycerine (glycerol) is a trihydric alcohol, one molecule can combine with three molecules of fatty acid to form a triglyceride as illustrated:

$$\begin{matrix} CH_2OH \\ | \\ CHOH \\ | \\ CH_2OH \end{matrix} + 3C_{17}H_{35}COOH \longrightarrow$$

Glycerol — Stearic acid

$$\begin{matrix} CH_2OOCC_{17}H_{35} \\ | \\ CHOOCC_{17}H_{35} \\ | \\ CH_2OOCC_{17}H_{35} \end{matrix} + 3H_2O$$

Glycerol tristearate (tristearin)

*U.S. Department of Agriculture, Western Regional Research Laboratory, Berkeley, Ca.

Fats and oils obtained from animal or vegetable tissue are predominantly triglycerides with traces of mono- and diglycerides, free fatty acids, phospholipids, sterols, and other minor constituents. Procedures for refining and processing crude oils remove most of the trace constituents.

Waxes differ from fats and oils in that a mono- or dihydric alcohol has replaced glycerol in combination with the acid. Wax alcohols may be either aliphatic or alicyclic. In mammals, cholesterol esters of fatty acids are found, particularly in the adrenals, the liver, and blood plasma.[1] These cholesterol esters are of interest in nutrition because of the correlation of the incidence of atherosclerosis with their levels in blood plasma. Both the acids and alcohols in waxes show considerable variation in type.

In addition to cholesterol (illustrated), other sterols are found in minor amounts in fats either as esters or in an uncombined form. Some of the common sterols found in vegetable oils are stigmasterol, ergosterol, β-sitosterol, and lanosterol. Vegetable oils may also contain xanthophylls and carotenes which impart color to the oil, and which in many oils is considered desirable, from the standpoint of appearance.

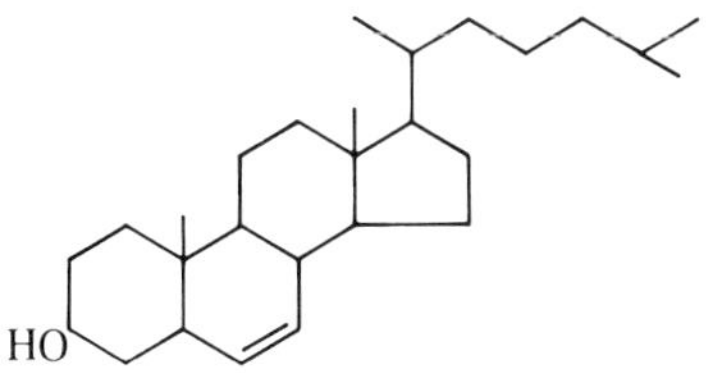

Cholesterol

One important group of compounds which are similar in structure to the triglycerides is the class of phospholipids. Three important phosphatidic esters are phosphatidylcholine, phosphatidylethanolamine, and phosphatidylserine. Phosphatidylinositol is also often present with sugar moieties attached. Phosphatides are separated by the addition of small amounts of water to crude oils. This degumming process produces "lecithin" from soybean oil, a product which has some commercial value because of its surface-active properties.

FATTY ACIDS

The physical and chemical properties of fats and oils are essentially determined by the fatty acid composition of their triglycerides. Most of the common fats and oils are principally mixed triglycerides of five common fatty acids; palmitic, stearic, oleic, linoleic, and linolenic acids. These five acids are unbranched carboxylic acids and except for palmitic acid, they all have eighteen carbon atoms in the chain. Their structures and properties, along with those of some of the many fatty acids appearing less commonly are illustrated in Table 12.1. The fatty acids of this table are only representative of the many hundreds of such acids which have been isolated and identified, but they illustrate some rules, which generally apply in their chemistry.

The common fatty acids have even numbers of carbon atoms. Odd-numbered fatty acids occur in many lipids, but almost always in very minor amounts. Unsaturated acids usually are in the *cis*-configuration, i.e., the hydrogens at each end of the double bond are on the same side. This preference for *cis*-fatty acids occurs in almost all living organisms, even though the *trans*-configuration is thermodynamically more stable. Also, in the common acids containing more than one double bond (e.g., linoleic and linolenic acids) the bonds are not conjugated, though the conjugated acids are also thermodynamically preferred. When double bonds are conjugated, however,

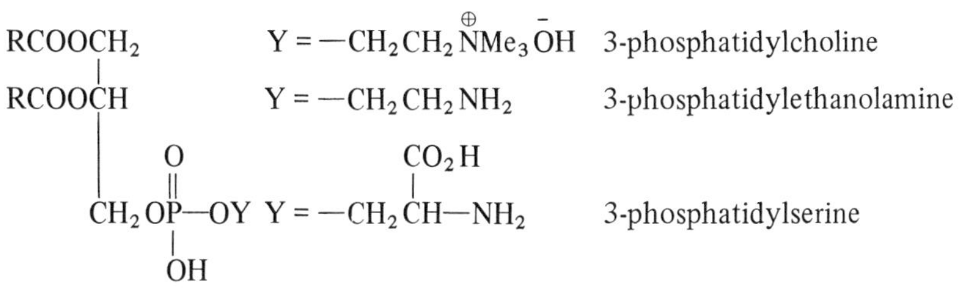

TABLE 12.1 Some Important Fatty Acids

No. of Carbon Atoms	*Systematic Name*	*Common Name*	*Melting Point, °C*	*Boiling Point, °C/mm*	*Methyl Ester (BP °C/mm)*	*Common Sources*
			Unbranched Saturated Fatty Acids			
4	Butanoic	Butyric	-5.3	164	103	Milk fats
6	Hexanoic	Caproic	-3.2	206	151	Milk fats and some seed oils
8	Octanoic	Caprylic	16.5	240	195	Milk fats and *Palmae* seed oils
10	Decanoic	Capric	31.6	271	228	Sheep and goats' milk, palm seed oils, sperm head oil
12	Dodecanoic	Lauric	44.8	130/1	262	Coconut oil
14	Tetradecanoic	Myristic	54.4	149/1	114/1	Palm and coconut oils
16	Hexadecanoic	Palmitic	62.9	167/1	136/1	Palm oil
18	Octadecanoic	Stearic	70.1	184/1	156/1	Animal fats
20	Eicosanoic	Arachidic	76.1	204/1	188/2	Some animal fats
22	Docosanoic	Behenic	80.0	-	206/2	Various seed oils incl. peanut oil
24	Tetracosanoic	Lignoceric	84.2	-	222/2	Minor amounts in some seed oils
26	Hexacosanoic	Cerotic	87.8	-	237/2	Plant waxes
28	Octacosanoic	Montanic	90.9	-	-	Beeswax and other waxes
30	Triacontanoic	Mellissic	93.6	-	-	Beeswax and other waxes
			Branched Saturated Fatty Acids			
5	3-Methylbutanoic	Isovaleric	-51.0	177	-	Dolphin and porpoise fats
19	10-Methylstearic	Tuberculostearic	11	-	-	Lipid of tubercle bacillus
			Monoenoic Fatty Acids			
10	9-Decenoic	Caproleic	-	142–8/15	115–6/12	Milk fats
14	9-Tetradecenoic	Myristoleic	-	-	108–109	Some seed fats–milk fats
16	9-Hexadecenoic	Palmitoleic	-0.5 to 0.5	-	134–5/1	Many fats and marine oils
18	6-Octadecenoic	Petroselinic	30–33	208–10/10	-	Parsley seed oil
18	9-Octadecenoic	Oleic	16.3	153/0.1	152.5/1	Almost all fats and oils
22	13-Docosenoic	Erucic	33.5	241–3/5	169–170/1	Rapeseed oil

			Dienoic Fatty Acids			
10	2,4-Decadienoic	Stillingic	–	–	–	Stillingia oil
18	9,12-Octadecadienoic	Linoleic	−5	202/1.4	149.5/1	Many vegetable oils
22	13,16-Docosadienoic	–	–	–	–	Rapeseed oil
			Trienoic Fatty Acids			
16	6,10,14-Hexadecatrienoic	Hiragonic	–	180–190/15	–	Sardine oil
18	6,9,12-Octadecatrienoic	–	–	–	–	Seeds of the evening primrose family
18	9,12,15-Octadecatrienoic	Linolenic	−11	157–158/0.001 0.002	184/4	Linseed oil
20	5,8,11-Eicosatrienoic	–	–	–	–	Brain phosphatides
20	8,11,14-Eicosatrienoic	–	–	–	–	Shark liver oil
18	9,11,13-Octadecatrienoic	Eleostearic	–	–	–	Tung oil
			Fatty Acids of More Unusual Structure			
18	12-Hydroxy-9-octadecenoic	Ricinoleic	–	–	–	Castor oil
20	14-Hydroxy-11-eicosenoic	Lesquerolic	–	–	–	*Lesquerella* seed oil
18	12,13-Epoxy-9-octadecenoic	Vernolic	–	–	–	Some *Compositae* seeds
18	8,9-Methylene-8-heptadecenoic	Malvalic	–	–	–	*Malvaceae* seeds
19	9,10-Methylene-9-octadecenoic	Sterculic	–	–	–	*Sterculiaceae* seeds
18	13-(2-Cyclopentenyl) tridecanoic	Chaulmoogric	–	–	–	*Chaulmoogra* oil

as in eleostearic acid, they may occur as a mixture of *cis-trans* or as all *trans* isomers.

Isolation and identification of fatty acids has been facilitated greatly by the development of gas-liquid chromatography. Triglyceride esters are not distillable, but they can be readily converted to methyl or ethyl esters by alcohol interchange, and it is as the methyl esters that acids are usually separated and identified. Even triglycerides may be partially separated by high-temperature gas chromatographic techniques according to the number of carbon atoms and the number of double bonds present.[2] Other techniques such as selective enzyme-catalyzed hydrolysis allow determination of the position of fatty acids on the glycerol moiety.

The physical properties of fatty acids are functions of their chemistry. Generally, melting points and boiling points rise with the number of carbon atoms. However, double bonds, particularly those of *cis*-configuration, lower the melting point so that saturated C_{18} stearic acid melts at 70.1°C, while monoenoic C_{18} oleic acid melts at 16.3°C and the diunsaturated C_{18} linoleic acid melts at −11°C. This profound effect on melting point comes about because double bonds break up the regular alternating structure of the hydrocarbon chain which allows the molecule to readily fit into a regular crystal lattice. The boiling points of the acids and their esters are not as greatly affected by double bonds, and separation by distillation is not ordinarily practicable. Branching of the carbon chain usually lowers the boiling point and melting point for acids with the same number of carbons, while polar groups, such as hydroxyl functions, attached to the chain increase the boiling point and the viscosity, though they may decrease the melting point.

TRIGLYCERIDES

Each molecule of triglyceride has three fatty acids and, in general, high-melting fatty acids produce high-melting glycerides. However, in natural fats properties are influenced to some extent by the distribution of isomers in the mixture. Consider a mixture of triglycerides with only two fatty acids, A and B. The following sequences are possible since the middle 2-position of glycerine differs from the 1 and 3 positions:

AAA, AAB, ABA, ABB, BAB, BBB

These may all occur; distribution may be random, but biosynthesis is usually directed so that a given triglyceride may appear in higher concentration than would be expected from strictly statistical considerations. Most oils have many more fatty acids than two, so that the distributional possibilities are numerous. A result of this multiplicity of molecular species is that fats do not possess sharp melting points; instead they freeze or melt over a wide temperature range. The first molecules to form solid crystals during cooling are the ones with the highest saturated fatty acid content. These may then be removed by filtration to leave an oil with higher unsaturate content. Such a process, called winterization, is used in making salad oils which can be kept under refrigeration. There are a few exceptions to the rule that fats behave as mixtures of many types of molecules. Castor oil contains 90 per cent ricinoleic acid and it can be considered under many circumstances to behave as triricinolein (glycerol triricinoleate). Cocoa butter which is used in confectionary coatings melts in the rather narrow range of 89–95°F. Its major ingredient is 2-oleopalmitostearin.

$$\begin{array}{l} CH_2OOCC_{17}H_{35} \\ | \\ CHOOC(CH_2)_7CH{=}CH(CH_2)_7CH_3 \\ | \\ CH_2OOCC_{15}H_{31} \end{array}$$

2-Oleopalmitostearin

Although a wide variety of molecular species exist in natural fats, the oils from many plant-seed sources show little variation in over-all fatty acid composition within a given species, especially if the plants are grown under similar climatic conditions. Interesting genetic and climatic variations do occur, however. Ordinary safflower oil usually contains about 75 per cent linoleic acid and 12–15 per cent oleic acid. In a new variety reported by Knowles[3] these ratios are essentially reversed. Sunflower oil exhibits considerable variation of fatty acid content with climate. Sunflower seeds grown in the Southern United States have a much lower linoleic to oleic acid ratio than seeds from the same variety grown in Minnesota or North Dakota.[4] The composition of depot fats of nonruminant animals is to a great extent a function of the fats ingested. If a high percentage of unsaturated oils is eaten, depot fats will contain more unsaturated fatty acids than if saturated fats are predominant in the diet. Ruminant animals show more consistency in fatty acid composition, since bacterial flora in the rumen modify diet fats.

EDIBLE VEGETABLE OILS

Soybean Oil

Soybean production in the United States amounted to more than one billion bushels in both 1968 and 1969, making soybean oil the most plentiful for domestic production and in world trade. During the year October 1969–September 1970, U.S. production of soybean oil amounted to 7.9 billion pounds. Of this amount, 1.4 billion pounds was exported.[5] A representative composition of soybean oil is: saturated acids, 14 per cent; oleic acid, 28 per cent; linoleic acid, 50 per cent; and linolenic acid, 8 per cent. Iodine value ranges from 120–141. Soybean oil which has not been hydrogenated tends to oxidize to give reverted flavors, often described as "painty" or "grassy." Consequently it is not commonly used for cooking. It can be partially hydrogenated to an oil of greater stability, and, in this form it is a major component of salad oils (after winterizing), shortenings, and margarines.

Cottonseed Oil

This is still a major oil in the United States, although it is a by-product with production dependent upon the amount of cotton grown. Most cottonseed oil is winterized and used for salad oil. Another important use of unhydrogenated cottonseed oil is in the frying of potato chips and other snack foods. A typical composition of cottonseed oil is about 21 per cent palmitic acid, 3 per cent other saturated acids, 27 per cent oleic acid, and linoleic acid comprising most of the other 49 per cent. Since it contains no linolenic acid it is not readily subject to flavor reversion. Hydrogenated cottonseed oil is very important in the manufacture of shortening. In 1969–70 production of U.S. cottonseed oil was almost 1.3 billion pounds.

Peanut Oil

Peanut oil is also a by-product oil, generally coming from low-grade nuts. It contains 13 per cent saturated fatty acids, 60 per cent oleic acid, 22 per cent linoleic acid, and minor amounts of fatty acids with more than eighteen carbon atoms. It is an excellent cooking oil, used by manufacturers of snack foods because of its good flavor. Domestic production of peanut oil has been close to 200 million pounds in recent years.

Sunflower Oil

Sunflowers grow well in cold climates; consequently sunflower seed is a major source of oil in northern countries such as the USSR and Canada. Only small acreages have been planted for oil in the United States, primarily in Minnesota and North Dakota. Southern farmers have recently been searching for an oilseed crop to use on land diverted from cotton; sunflowers may fill this need. As noted previously, the composition of sunflower oil changes when the seed is grown in warmer climates. Northern sunflower oils contains about 75 per cent linoleic and 14 per cent oleic acid, while southern sunflower oils may have over 50 per cent oleic acid and as little as 35 per cent linoleic.[6] Respective iodine values are 130 for northern sunflower oils and 106 for oil from plants grown in some southern states. While not used extensively in the U.S., sunflower oil is second only to soybean oil in world consumption.

Olive Oil

Olive oil is desired for its flavor; hence the best oil, called virgin oil, is pressed from the fruit pulp of the olive and not refined. Oil extracted with solvent, refined, and deodorized is called pure olive oil and is quite bland. Most olive oil sold in the United States is imported, though small amounts may be produced in California. Olive oil is expensive, and is therefore a specialty item in the United

States, used primarily for its flavor. It has been blended with other oils for use in salad dressings. A typical olive oil may contain 8 per cent palmitic acid, 2 per cent stearic acid, 82 per cent oleic acid, and 8 per cent linoleic acid. It is relatively low in polyunsaturation, with iodine values ranging from 80 to 90.

Safflower Oil

Safflower is an ancient crop, grown in the Middle East since prehistoric times, but only produced in quantity in the United States since the early 1950's. It has the highest linoleic acid content of any commercial oil, but it contains no linolenic acid. It is therefore not subject to the severe flavor reversion problems of soybean oil. With the increasing demand for polyunsaturated oils in the diet, safflower has found use in soft margarines, salad oils, dietetic imitation ice creams, etc. Because of its high polyunsaturate content, it is not a good frying oil, since it tends to produce off-flavors and polymeric products when heated in air.

A new safflower oil, a high-oleic acid variety, has recently been produced and sold in the United States. In composition it is quite similar to olive oil, with a low polyunsaturate content and high oxidative stability at frying temperatures. It can be used as a salad or cooking oil. Regular safflower oil contains 75-80 per cent linoleic acid and 13 per cent oleic acid, while in the new variety these values for the two fatty acids are reversed. Safflower grows well in hot, dry climates where soybeans and most other commercial oilseeds are not well adapted.

Coconut Oil

This oil differs from those previously described because it contains large percentages of saturated fatty acids with less than eighteen carbon atoms. Its composition gives it a high thermal and oxidative stability so that it is used in filled milks and imitation dairy products to replace butter fat and for other uses in which off-flavors develop quickly from oxidation. Coconut oil, however, is prone to hydrolytic rancidity, since it contains significant amounts of caprylic (C_8) and capric (C_{10}) fatty acids which, when converted from their glyceride esters to free acids have strong, unpleasant odors and flavors. The fatty acid composition of coconut oil is: caprylic acid, 8 per cent; capric acid, 7 per cent; lauric acid, 48 per cent; myristic acid, 18 per cent; palmitic acid, 8 per cent; palmitoleic acid, 1 per cent; stearic acid, 2 per cent; oleic acid, 6 per cent; linoleic acid, 2 per cent.

Rapeseed Oil

Rapeseed oil is unusual in that it contains a large percentage of the 22-carbon unsaturated acid, erucic acid. Rape plants are adapted to cold climates so that rapeseed oil has been used in Canada and northern Europe. Mustard seed and a new oilseed, Crambe, produce oils of similar composition. Before refining, these oils contain sulfur compounds which give a strong cabbage-like odor. Although rapeseed oil has been used in margarine for some time, current trends in Canada are toward the use of oils containing fatty acids of lower carbon number, in edible products, especially. A zero-erucic rape has been developed in which the erucic acid has essentially been replaced by oleic. A typical analysis for ordinary rapeseed oil is as follows: palmitic acid, 1 per cent; stearic acid, 1 per cent; oleic acid, 17 per cent; linoleic acid, 14 per cent; linolenic acid, 8 per cent; erucic acid, 51 per cent; other C_{20}-C_{22} unsaturated fatty acids, 4 per cent; and C_{20}-C_{24} saturated acids, 4 per cent.

INDUSTRIAL OILS

The oils described above are used primarily for edible purposes although all of them find some application in industry as sources of fatty acids, soaps, detergents, etc. Soybean and safflower oils, being semidrying oils, are used in paints and coatings. Some oils, however, are produced almost exclusively as industrial raw materials to produce a variety of products.

Linseed Oil

Linseed oil comes from varieties of the flax plant which are grown for oil production. It contains over 50 per cent linolenic acid, more than any other common oil. As already noted, the unique structure of linolenic acid makes it susceptible to oxidation, a reaction which is responsible for the drying action of polyunsaturated oils.[7] Initially, drying oils such as linseed oil react with the oxygen in the air to form hydroperoxides, but these, being thermally unstable, break down quite rapidly to undergo a series of complex reactions, including a free-radical polymerization process. Such polymerization takes place in oil-based paints and varnishes; it is often accelerated by adding to coating formulations catalysts such as cobalt naphthenate or other transition-metal compounds which increase the rate of peroxide decomposition. Linseed oil production has been trending downward in the United States, with the advent of newer synthetic coatings. Recent production has been about 300 million pounds per year. A representative composition for linseed oil is: saturated fatty acids, 10 per cent; oleic acid, 20 per cent; linoleic acid, 17 per cent; and linolenic acid, 53 per cent.

Tung Oil

This oil, produced from tung nuts, contains the conjugated trienoic acid eleostearic acid, which constitutes more than 80% of its component fatty acids. The susceptibility of eleostearic acid to oxidative polymerization makes it an excellent drying oil. Much of this oil used to be imported from China. When the supply was cut off, some tung production was begun along the Gulf Coast, particularly in Mississippi, Louisiana, and Florida. Recently, extensive freezes and the destruction of trees by tropical storms has cut tung production in the U.S. to an insignificant amount.

Castor Oil

Castor oil is an interesting raw material in that it has two chemical "handles," the double bond and the hydroxyl group of ricinoleic acid, which comprises over 90 per cent of the fatty acid in the oil. Castor oil is characterized by a high viscosity resulting from hydrogen bonding of its hydroxyl groups. It is a raw material for paints, coatings, lubricants, cosmetics, and a variety of other products. In France, a commercial nylon is produced from castor oil. Castor oil consumption in the United States is about 180 million pounds annually, but domestic production is only a small fraction of that amount. Most castor oil consumed in the U.S. comes from Brazil. The castor plant (*Ricinus communis*) produces an extremely toxic protein, ricin, several potent allergenic materials, and an alkaloid, ricinine, which all appear in castor pomace. Thus, the residual meal cannot be used as an animal feed without further treatment, a fact which has been a deterrent to economic production in this country.[8]

FATS OF ANIMAL ORIGIN

Lard

Lard is the fat rendered from the fatty tissues of the pig. For many years it was the major shortening product, but it has been supplanted to a great extent by hydrogenated vegetable oils. A great deal of lard is used in commercial frying operations. Although it has a relatively low content of unsaturated acids (iodine value 55-60) it is often hardened further by hydrogenation and blended into shortenings. Approximate composition of lard is: myristic acid, 1 per cent; palmitic acid, 27 per cent; palmitoleic acid, 3 per cent; stearic acid, 14 per cent; oleic acid, 43 per cent; linoleic acid, 10 per cent; and minor constituents, 2 per cent.

Tallow

Tallow is rendered from the fatty tissues of cattle in a manner similar to that use in the production of lard. After rendering, it is usually deodorized and used in shortenings and bakery margarines. It is also an industrial raw material, being used for the manufacture of soap and fatty acids, particularly industrial oleic acid. Tallow is somewhat higher in saturated fatty acids than lard. A typical composition is: myristic acid, 3 per cent; palmitic acid, 30 per cent; palmitoleic acid, 3 per cent; stearic acid, 19 per cent; oleic acid, 44 per cent; and minor constituents, 1 per cent.

Marine Oils

Marine oils are processed from herring, menhaden, pilchard, and sardines. They are not used extensively in the United States, but are

TABLE 12.2 Domestic Consumption of Some Important Fats and Oils, 1969-70

Oil Source	Millions of lbs Consumed[a] 1969	1970[b]
Edible oils		
Butter	1106	1075
Lard	1425	1649
Edible tallow	544	519
Corn oil	454	447
Cottonseed oil	1069	898
Soybean oil	6328	6267
Olive oil	63	67
Palm oil	133	191
Peanut oil	155	193
Safflower oil (est)	107	100
Coconut oil	401	343
Other uses		
Coconut oil	432	446
Inedible tallow and grease	2725	2593
Linseed oil	281	246
Tall oil	1174	1331
Castor oil	167	134
Tung oil	31	28

[a]Source: Fats and oils situation FOS 256-260 (1971).
[b]Preliminary.

hydrogenated and used for shortening in Canada and Europe. They are highly unsaturated, with fatty acids often containing five or six double bonds. Such polyunsaturation makes them highly susceptible to oxidation and formation of fishy flavors and odors.

Recent domestic consumption of some fats and oils is described in Table 12.2.

PROCESSING AND REFINING FATS AND OILS

The purpose of oil extraction is to obtain the maximum amount of good quality oil while getting maximum value from the residual meal. Oil is extracted from oilseeds by pressing, solvent extraction, or a combination of both. When brought in from the field, seeds are cleaned to remove stones, pieces of metal, and trash. They may then be stored for processing. Moisture content during storage is kept low to prevent microbial spoilage and spontaneous combustion. The seeds are cracked and, with some oilseeds, complete or partial decortication (removal of hull) is accomplished by a combination of air classification and screening or, in a few cases, flotation. The seeds are reduced by grinding or rolling, then steam cooked and flaked. The heat treatment coagulates the proteins and raises the moisture content so that extraction is accomplished most easily.

Cooked seeds may then be conveyed to screw presses for continuous pressing. The press cake emerging from a screw press will contain 8 to 14 per cent residual oil if solvent extraction is to follow, or 2 to 4 per cent if screw pressing is the only extractive process. Solvent extraction is done in a countercurrent procedure in which intimate contact is achieved between seeds and solvent. The apparatus may be the basket type arranged vertically or the rotary cell type in which containers rotate horizontally. The common solvent for edible oils is hexane, or actually a hexane-type naptha boiling in the range of 146-156°F. Extraction equipment must be enclosed and thoroughly protected because of solvent flammability. There has been interest for many years in extraction with nonflammable solvents, especially trichloroethylene, but the advantages of using a nonflammable solvent have been more than offset by the toxicity problems generated by chlorinated solvents.

After extraction, the solvent is recovered by distillation from the miscella (mixed solvent and oil). Solvent is also removed from the oilseed pomace by steam stripping. Recovered solvent is re-used for extraction while the extract oil is usually combined with the press oil for refining, except in cases such as olive oil in which the virgin press oil has special flavor qualities. While most oilseed meals are used for animal feed, especially soybeans, there is increasing use of the high-protein meal for human consumption. To produce protein isolates there must be as little denaturation as possible. Improvements are still needed in processing meal to keep temperatures low and protect the protein.

Marine oils are often solvent-extracted from fish meal; but animal fats such as lard or tallow, as well as many marine oils, are rendered by either dry or moist heat. Dry rendering in steam-jacketed tanks followed by pressing is satisfactory for inedible fats, but most edible

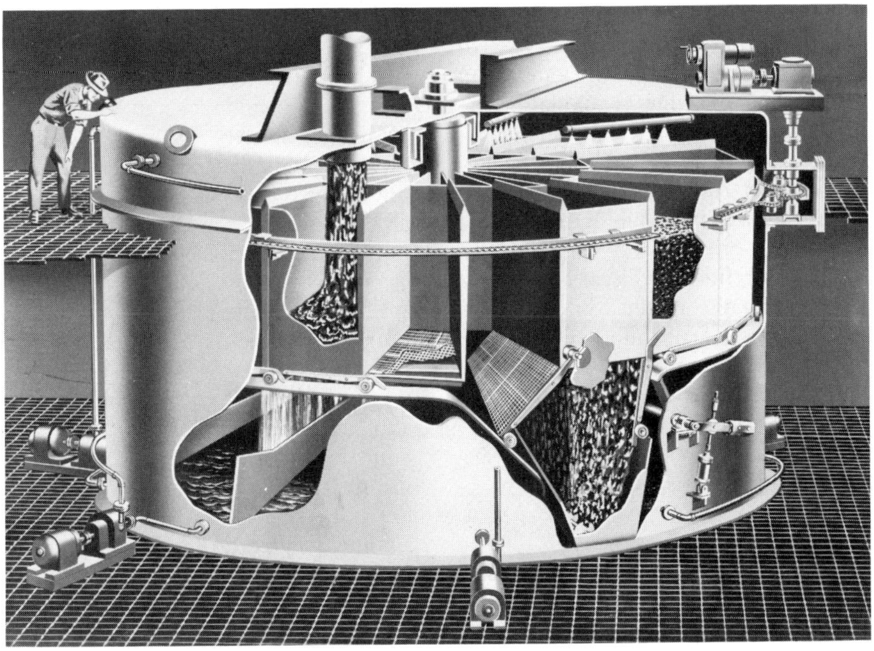

Fig. 12.1 "Rotocel" for extraction of oil from oilseeds or press cake. Compartments containing material to be extracted rotate slowly while solvent percolates through the solid mass. *(Courtesy Blaw-Knox Chemical Plants, Inc.)*

fats are produced by wet rendering, using steam at 40 to 60 psi. In the latter process 99.5 per cent or more of the fat is recovered from the raw material.

Edible oils have a number of trace impurities, most of which are undesirable. Refining with caustic soda is used to eliminate free fatty acids, phosphatides, and other impurities which may cause off-flavors and undesirable color. Usually the amount of free fatty acids is determined by analysis of the crude oil. The oil is then contacted with enough alkali solution to neutralize the free fatty acids plus an amount determined empirically to produce the desired effects with minimum loss. Reaction of caustic soda with the free fatty acids produces soapstock which can be continuously removed by centrifugation. Phosphatides are soluble only in the anhydrous oil, so these are taken out with the soapstock. If prior removal of lecithin is desired, it is accomplished before alkali treatment by degumming with water or steam. After separation of the soapstock, the oil may be water-washed and again centrifuged to bring soap concentration to a satisfactory low level. In the United States, refining of vegetable oils is usually a continuous process, though it may be done batchwise. Good alkali-refining reduces free fatty acid content to 0.01-0.03 per cent. In addition to free fatty acids, other undesirable components, such as gossypol (a somewhat toxic dark colored material of cottonseed oil) are removed by contact with the alkali solution.

Other refining methods, such as acid refining, are used for specialized purposes. Some solvent-extracted oils may also be miscella-refined with alkali before the solvent is removed.[9] Better contact with alkali solution and better separation of soapstock are among the claimed advantages.

Bleaching

Some pigmented materials such as chlorophyll, which are not eliminated by alkali refining, lend themselves to removal by absorption processes. Activated carbon has been used for bleaching, but acid-activated clay is less expensive and quite effective. The amounts of activated clay used depend upon the oil; for vegetable oils, 0.25-2.0 per cent by weight quantities are common. Atmospheric bleaching at 212-250°F is common, but changes can occur in bleached oils due to oxidation. If these changes are undesirable, vacuum bleaching at temperatures of 170-180°F may be employed. A contact time of clay with oil for 10-15 minutes is usually required. Continuous vacuum bleaching equipment is commercially available. Many refineries discard the clay without recovering the retained oil, although this results in a loss of oil almost equal to the weight of clay used. A number of processes are used to remove oil from spent bleaching clay. While the recovered oil is usually low in quality, in some circumstances it may still be worth enough to justify its recovery.

Hydrogenation

Edible oils are often hydrogenated to produce hard or plastic fats or to enhance the oxidative stability of the oils by lowering their content of polyunsaturated acids. In the hydrogenation process, one or more carbon-carbon double bonds in a triglyceride molecule are converted to single bonds by addition of hydrogen as illustrated below:

$$CH_3(CH_2)_7CH{=}CH(CH_2)_7COOR \xrightarrow[\text{Catalyst}]{H_2} CH_3(CH_2)_{16}COOR$$

The hydrogenation reaction requires hydrogen at moderate pressures and a catalyst, which is usually finely divided nickel. Such catalysts may be the type prepared by reducing nickel salts which have been precipitated onto an inert support of high surface area, or they may be Raney nickel-type prepared by dissolving aluminum from a nickel-aluminum alloy using strong alkali. Variables affecting hydrogenation include the catalyst, the temperature, hydrogen pressure, and the amount of agitation. If only partial hydrogenation is desired, the selectivity of the hydrogenation system is important. Consider, for example the molecule of oleolinoleolinolenin illustrated below:

If a light, partial hydrogenation is desired to increase oxidative stability, it would be preferable to first reduce all the bonds labeled *c* in the linolenic acid moiety. Because no catalyst is entirely selective, any of the other double bonds may be reduced. Reduction of one of the bonds labeled *a* or *b* would produce an "isolinoleic" acid from the linolenic acid portion of the molecule. Reduction of bonds *d* or *e* would produce isooleic or oleic acid from the linoleic acid and hydrogenation of bond *f* will form a stearic acid moiety from oleic acid. Not only is no catalyst completely selective, but hydrogenation catalysts may also cause isomerization of an unsaturated bond from a *cis* to a *trans* configuration, or the migration of double bonds up and down the fatty acid chain. There is a considerable amount of literature relating to the selectivity of hydrogenation systems. Some copper catalysts are reported to be considerably more selective for the hydrogenation of linolenic acid in the presence of other fatty acids than the commonly used nickel catalysts.[10]

Hydrogenation is a very versatile reaction allowing considerable variability in the properties of plastic fats formed an important factor in the preparation of the final products, such as margarines and shortenings. When very light hydrogenation is used to produce a liquid oil of relatively high stability, some of the fat molecules may become quite saturated. These can be removed by winterizing (chilling) the oil and removing the solidified portion, called stearine.

Deodorization

Bleached and hydrogenated oils often have undesirable off-odors which result from the presence of free fatty acids, aldehydes, or other slightly volatile materials. Such oils are deodorized by steam stripping under high vacuum at elevated temperatures (400-470°F). The oil must be carefully cooled before it is contacted with air, because undesirable oxidation can be quite rapid at the high tempera-

$$\begin{array}{l} CH_2OOC(CH_2)_7CH\overset{a}{=}CHCH_2CH\overset{b}{=}CHCH_2CH\overset{c}{=}CHCH_2CH_3 \quad \text{(linolenic)} \\ | \\ CHOOC(CH_2)_7CH\overset{d}{=}CHCH_2CH\overset{e}{=}CH(CH_2)_4CH_3 \quad \text{(linoleic)} \\ | \\ CH_2OOC(CH_2)_7CH\overset{f}{=}CH(CH_2)_7CH_3 \quad \text{(oleic)} \end{array}$$

tures in the deodorizer. Equipment for deodorization as well as most of the other processing steps is increasingly being built of stainless steel, since iron is a prooxidant at elevated temperatures. Copper is an especially strong oxidation catalyst, so that copper pipes or brass valves should not be used in oil processing. At the end of the deodorization step antioxidants and metal sequestrants suitable for food use are often added to oils.

ANALYTICAL PROCEDURES

Analytical methods for fats and oils have been published by the American Oil Chemists' Society.[11] There are a large number of procedures, though many of these have value only for testing properties which have special applications. It is useful, though, to describe a few of these tests because of their general applicability and wide use.

Iodine Value

Iodine value is defined as the number of centigrams of iodine absorbed by one gram of fat. Thus, it is a measure of unsaturation in fats and oils. In practice, it is determined by using iodine chloride or iodine bromide as a reagent, rather than free iodine. Even with these more reactive compounds absorption by double bonds is not usually complete, so that iodine value must be regarded as an empirical test which is of value to the processor. Since gas chromatography is now used widely, it is the practice in many research laboratories to first determine the composition of the fatty acid and then calculate the iodine value from that composition. Some iodine values of free fatty acids are presented in Table 12.3.

TABLE 12.3 Iodine Values of Unsaturated Fatty Acids

Fatty Acid	*No. of C Atoms*	*No. of Double Bonds*	*Iodine Value*
Decenoic	10	1	149.1
Dodecenoic	12	1	128.0
Tetradecenoic	14	1	112.1
Hexadecenoic	16	1	99.78
Oleic	18	1	89.87
Linoleic	18	2	181.04
Linolenic	18	3	273.52
Eicosenoic	20	1	81.75
Erucic	22	1	74.98

Source: Weiss.[6]

Saponification Value

The saponification value is the weight in milligrams of potassium hydroxide needed to completely saponify one gram of fat. It can give an indication of the average molecular weight of the fatty acids in a fat, but presents less useful information than the fatty acid composition determined by gas chromatography.

Melting Point

When applied to fats the term melting point has little significance, since melting usually occurs over a wide range of temperature. Some useful information may be gained, though, if melting points are determined by rigidly reproducible methods. A number of different methods are used, including the FAC melting point, the Wiley melting point, the congeal point, and the titer, all described in Reference.[11] In addition to these values, the Solid Fat Index (SFI) is often used for specifying oil properties. The SFI is related approximately to the percentage of solids in a fat at a given temperature. When determined at a number of specified temperatures it can be especially useful to margerine manufacturers or other processors who need to control the characteristics of their manufactured products by blending.

Peroxide Value

Oxidation of fats is desirable in drying oils, but undesirable in most other products. The first step in oxidation is the formation of hydroperoxides. These hydroperoxides have no odors or flavors, but they readily break down to produce aldehydes, hydrocarbons, ketones, and other volatile products which are characteristic of oxidative rancidity. Peroxides react with potassium iodide to liberate free iodine which may then be titrated. The peroxide value is expressed as milliequivalents of iodine formed per kilogram of fat. The peroxide value is used as an indicator of oxidative rancidity. Organoleptic evaluation by a trained odor panel can often detect rancidity at peroxide levels as low as 1 meq/kg.

Active Oxygen Method (AOM)

This method was developed to predict the shelf-stability of fats. It is useful with natural animal fats, but hydrogenated fats and oils which often are stabilized by the addition of antioxidants do not correlate well with AOM values. In this test, oils are heated at 97.8°C while air is blown through the sample. The AOM value is reported as the number of hours to reach a peroxide value of 100 meq/kg. A number other tests have been devised to predict oil stability, but none of them have yet shown accurate correlations for all types of oils.

Free Fatty Acid (FFA)

Free fatty acids result from hydrolysis of fats. Oilseeds contain lipase enzymes which catalyze such hydrolysis when seeds are damaged or become moist and begin to germinate. Lipases are also found in animal tissues. Consequently, if seeds are improperly stored, or if animal tissues are allowed to stand for a while before rendering, the FFA values will be high. Deodorization lowers the FFA to 0.05 per cent or less. To determine FFA, the sample is titrated with sodium hydroxide. It is calculated as free oleic acid and reported as a percentage.

Smoke Point

Smoke point is defined as the temperature at which a fat will produce continuous wisps of smoke. Impurities such as free fatty acids lower the smoke point drastically. A 1 per cent FFA content may lower the smoke point of a shortening from 425 to 320°F.[6]

Color

With a few exceptions, light color is desirable in oils. Color is best measured by recording the visible absorption spectrum. However, there are empirical methods which are quite satisfactory for quality control. For this purpose oils may be compared in a tube of standard depth with the color of glass standards, using the Lovibond Tintometer. Color is reported as yellow and red values, the former usually being about 10 times the latter. Reflectance values are measured for solid fat products.

TRENDS IN OIL PROCESSING

This chapter has dealt primarily with oil processing as it is done today. As with most manufacturing processes, innovation is almost constantly occurring. For example, new hydrogenation units have been built which use commercial liquid hydrogen rather than stored gaseous hydrogen; or hydrogen is manufactured at the site. Efforts are continuing to make processing more automated and more continuous. One factor, which is only beginning to have an impact now but will be increasingly important in the next few years, is the need to eliminate pollution in plant effluents. Such effluents may be both organic and inorganic. A large amount of pollutants result when soapstock is acidified (acidulated). Treatment of soap with sulfuric acid liberates free fatty acids, most of which are recovered, and large amounts of sodium sulfate. The aqueous phase from acidulation will have a high BOD*, due to the unrecovered acids, and a high concentration of inorganic salts, both of which place a burden on the local sewage system. Considerable effort is now being expended to reduce both the quantity and the concentration of effluents from oil refineries. This effort will no doubt have a large influence on oil processing in the future.

*Biochemical oxygen demand.

REFERENCES

1. Gunstone, F. D. "An Introduction to the Chemistry and Biochemistry of the Fatty Acids and their Glycerides," 2d Ed., pp. 2–4, London, Chapman & Hall, 1967.
2. Litchfield, C., Harlow, R. D., and Reiser, R., *J. Am. Oil Chemists' Soc.* **42**, 849–857 (1965).
3. Knowles, P. F., *Econ. Botany*, **19**, 53–62 (1965).
4. Kinman, M. L. and Earle, F. R., *Crop Sci.* **4**, 417 (1964).
5. *Fats and Oils Situation*, pp. 4–6, Economic Research Service, U.S. Dept. of Agriculture, June, 1971.
6. Weiss, T. J., "Food Oils and Their Uses," Westport, Conn., Avi Pub. Co., 1970.
7. Wexler, H., *Chem. Reviews* **64**, 591–611 (1964).

8. Fuller, G., Walker, H. G., Jr., Mottola, A. C., Kuzmicky, D. D., Kohler, G. O., and Vohra, P., *J. Am. Oil Chemists' Soc.*, **48**, 616–618 (1971).
9. Cavanagh, G. C., *J. Am. Oil Chemists' Soc.* **33**, 528–531 (1956).
10. Koritala, S., *J. Am. Oil Chemists' Soc.*, **47**, 269–272 (1970).
11. "Official and Tentative Methods," 2d Ed., Chicago, Ill., American Oil Chemists' Society.

SELECTED REFERENCES*

*A list of references on specific topics; published for the most part since 1962.

1. Allen, R. R., "Hydrogenation: Principles and Catalysts," *J. Am. Oil Chemists' Soc.*, **45**, 312A (1968).
2. Bates, R. W., "Edible Rendering," *J. Am. Oil Chemists' Soc.*, **45**, 420A (1968).
3. Boekenoogen, H. A., Ed., "Analysis and Characterization of Oils, Fats and Fat Products," Vol. 2, New York, John Wiley & Sons, 1968.
4. Crauer, Lois S., "Continuous Treatment of Refinery Waste Waters," *J. Am. Oil Chemists' Soc.*, **47**, 210A (1971).
5. Dugan, L. R. and Crauer, L. S., chairpersons, Symposium on Processing and Quality Control of Fats and Oils, 17th Annual Summer Program, American Oil Chemists' Society, Chicago, 1966.
6. Dutton, H. J., "Determination of Fat Composition," *J. Am. Oil Chemists' Soc.* **45**, 4A (1968).
7. Eckey, E. W., "Vegetable Fats and Oils," New York, Van Nostrand Reinhold 1954.
8. Eldridge, A. C., "A Bibliography on the Solvent Extraction of Soybeans and Soybean Products, 1944–1968," *J. Am. Oil Chemists' Soc.* **46**, 458A (1969).
9. *Fats and Oils Situation*, Economic Research Service, U.S. Dept. of Agriculture, Washington, D.C. (Published periodically.)
10. Gunstone, F. D., "An Introduction to the Chemistry and Biochemistry of the Fatty Acids and Their Glycerides," 2d Ed., London, Chapman and Hall, 1967.
11. Hilditch, T. P. and Williams, P. N., "The Chemical Constitution of Natural Fats," 4th Ed., New York, John Wiley & Sons, 1964.
12. Holman, R. P., Ed., "Progress in The Chemistry of Fats and Other Lipids," Vol. 8, Oxford, Pergamon Press, 1968.
13. Hutchins, R. P., "Processing Control of Crude Oil Production from Oilseeds," *J. Am. Oil Chemists' Soc.*, **45**, 624A (1968).
14. Kenyon, W. R., "Aerated Lagoons for the Treatment of Fats and Oil Industry Wastes," *J. Am. Oil Chemists' Soc.*, **47**, 402A (1970).
15. Kingsbaker, C. L., "Solvent Extraction Techniques for Soybeans and Other Seeds, Desolventizing and Toasting," *J. Am. Oil Chemists' Soc.*, **47**, 458A (1970).
16. Markley, Klare S., Ed., "Fatty Acids: Their Chemistry Properties, Reactions and Uses," 2nd Ed., Vol. 5, New York, John Wiley & Sons, 1968.
17. Masoro, Edward J., "Physiological Chemistry of Lipids in Mammals," Philadelphia, Saunders, 1968.
18. Pattison, E. Scott, Ed., "Fatty Acids and Their Industrial Applications," New York, Marcel Dekker, 1968.
19. Pierce, R. M., "Sunflower Processing Techniques," *J. Am. Oil Chemists' Soc.*, **47**, 248A (1970).
20. Simpson, C., "Gas Chromatography," New York, Barnes and Noble, 1970.
21. Swern, Daniel, Ed., "Bailey's Industrial Oil and Fat Products," 3d Ed., New York, John Wiley & Sons, 1964.
22. Swern, Daniel, Ed., "Organic Peroxides," Vol. I, New York, John Wiley and Sons, 1970.
23. Talen, H. W., "Recommended Methods for the Analysis of Drying Oils," London, Butterworths, 1965.
24. Thomas, R. and Ridlehuber, J., "Miscella Refining," *J. Am. Oil Chemists' Soc.*, **45**, 254A (1968).
25. Weiss, T. J., "Food Oils and Their Uses," Westport, Conn., Avi Pub. Co., 1970.

13

Soap and Synthetic Detergents

J. C. Harris*

It has been said that the amount of soap consumed in a country is a reliable measure of its civilization. There was a time when soap was a luxury; now it is a necessity. The current manufacture of vast amounts of soap in every civilized country is possible only because new raw materials have become available through chemical science; the tallows and animal greases of the old days have been supplemented by coconut, palm, cottonseed, and other oils. The "old" days are those when soap was practically the only detergent. Today syndets (synthetic detergents) account for more than 70 per cent of all detergents used.

SOAP

Soap is the sodium or potassium salt of stearic and other fatty acids; it is soluble in water and the solution has excellent cleansing properties. Other metals such as calcium, aluminum, and lead also form compounds with fatty acids, but these compounds are insoluble and serve several other purposes, e.g., lubricants, in paints; they are always designated as "calcium soap," "lead soap," etc.

Soap is made by the action of a hot caustic solution on tallows, greases, and fatty oils, with the simultaneous formation of glycerin which at one time was wasted or left in the soap, as it still is in certain cases; glycerin or glycerol is a valuable by-product. The reaction is as follows:

$$\underset{\text{Caustic soda}}{3NaOH} + \underset{\text{Glyceryl stearate (typical fat)}}{(C_{17}H_{35}COO)_3C_3H_5} \longrightarrow \underset{\text{Sodium stearate (soap)}}{3C_{17}H_{35}COONa} + \underset{\text{Glycerin}}{C_3H_5(OH)_3}$$

Soap may also be made by the action of caustic soda on fatty acid without producing glycerin:

$$\underset{\text{Caustic soda}}{NaOH} + \underset{\text{Stearic acid}}{C_{17}H_{35}COOH} \longrightarrow \underset{\text{Soap}}{C_{17}H_{35}COONa} + H_2O$$

*340 Colonial Lane, Dayton, O.

The glyceride used is never a single one, but a mixture of several, hence the soap produced partakes of the properties of each. Sodium stearate dissolves too slowly, while the sodium soap made from coconut oil dissolves quite rapidly; a mixture of the two has the right rate of solubility. Mixtures of these fats affect and control other physical and use characteristics. The reaction may be performed in steel vats or kettles, in which a solution of caustic soda is mixed with the fat or oil, and heated; the soap is partially in solution in the water and must be separated by the addition of salt (NaCl). For high-grade soap, a steel tank with the upper part stainless steel, is used. The glycerin remains in the alkaline water and is drawn off at the bottom. The raw materials in that case are caustic soda or caustic potash (KOH), fats, greases, fatty oils, and salt.

Raw Materials

The caustic soda is usually received in drums of 700 pounds of solid caustic; this is made into a strong solution by inverting the opened drum over a steam jet with provision to collect the resulting solution. The caustic may also be crushed and dissolved with occasional stirring. Caustic in flake form has advantages. Soap plants situated near an alkali factory receive the caustic in the form of 50 or 24 per cent solution; however, for long distances this more convenient form would mean a high freight bill. The caustic may also be made at the plant by causticizing soda ash with lime; this is rarely done. Caustic potash is usually received in drums as the solid. Bulk shipment of liquid caustic potash may be more economical.

Beef and mutton tallow of all grades are used, from the best No. 1 edible grade to the cheapest grade recovered from garbage; the grades chosen depend on the quality of the soap to be made. Tallow is not used without admixture with other fats or oils, for it gives a soap which is too hard and too insoluble; it is usually mixed with coconut oil.

Coconut oil is a solid (mp 20-25°C); it gives a soap which is fairly hard, but quite soluble. It is the basis of marine soaps, since it lathers even in salt water. According to the country of origin and to the manner of isolation, coconut oils differ in their content of fatty acids: the lower the fatty acids, the better the quality of the oil; this applies to all fatty oils. Of the coconut oils, Cochin is the highest grade.

Palm oil is usually colored orange to brown, and has 6.0 per cent free fatty acids; it is an important raw material, and is used for toilet soaps. Palm oil may be bleached by warming it and blowing air through it. Palm kernel oil is an oil of light color.

Castor oil is used for transparent soap.

Olive oil of the lower grades, no longer edible, is much favored by the soap maker; for fine toilet soaps, olive oil of the edible grade is used, but it is denatured by the addition of oil of rosemary, hence its import duty is low. Castile soap was originally a sodium-olive oil soap, as was the Savon de Marseille.

In the refining of cottonseed oil, treatment with a solution of caustic is the first step; the alkaline liquors contain the foots and are used in soap-making. Cottonseed oil itself is also used, usually combined with foots, or after being treated to form the free acids (Twitchell process).

The word "grease" to the buyer of soap stock means an animal fat, softer than tallow, obtained by rendering the carcasses of diseased animals, or from house and municipal garbage, from bones, and tankage. Rosin serves for laundry soaps.

Manufacture

Ordinary toilet soaps and laundry soaps may be made by the "boiled process," which is adapted to batches ranging from 1000 to 800,000 lbs; soap is also made by continuous saponification or neutralization of fatty acids. The "cold process" is used for special soaps.

Batch Kettle Soap. By the boiled process a batch of 300,000 lbs of soap, for example, is made in a steel kettle 28 ft in diameter and 33 ft deep, with a slightly conical bottom. A solution of caustic soda testing 18-20° Bé (12.6-14.4 per cent NaOH) is run into the kettle, and the melted fats, greases, or oils are then pumped in (a Taber pump is suitable). The amount of caustic is regulated so that there is just enough to combine with all the fatty acids liberated. Heat is supplied by direct steam

TABLE 13.1 United States Inedible Tallow and Grease Production and Utilization

(In Millions of Pounds)

		Utilization in Nonfood Products					
Year	*Apparent Production*	*Soap*	*Animal Feed*	*Fatty Acids*	*Lubricants and Similar Oils*	*Other*	*Total*
1956	3,232	813	296	286	34	276	1,686
1961	3,645	721	442	356	73	145	1,737
1966	4,480	667	893	583	107	214	2,264
1967[a]	4,753	643	972	546	92	151	2,401
1968[a]	4,745	639	990	576	89	288	2,575
1969	–	639	1,075	608	101	174	2.597

[a]Preliminary

Source: Agricultural Marketing Service, U.S. Dept. of Agriculture.

entering through a perforated coil laid on the bottom of the kettle. There is no stirrer, but agitation is provided by a direct steam jet entering at the base of a cental pipe (Fig. 13.1). The kettle is kept boiling until saponification is essentially complete; this requires about 4 hours. Salt (NaCl) is then shoveled in and allowed to dissolve, and the boiling is continued until the soap has separated, forming the upper layer. The lower layer contains glycerin (4 per cent) and salt, and is drawn off at the bottom of the kettle; its concentration is described under glycerin. The whole operation which was just described is termed the saponification change, and requires about 8 hours. The salt used is chiefly rock salt; most of it is recovered and used again.

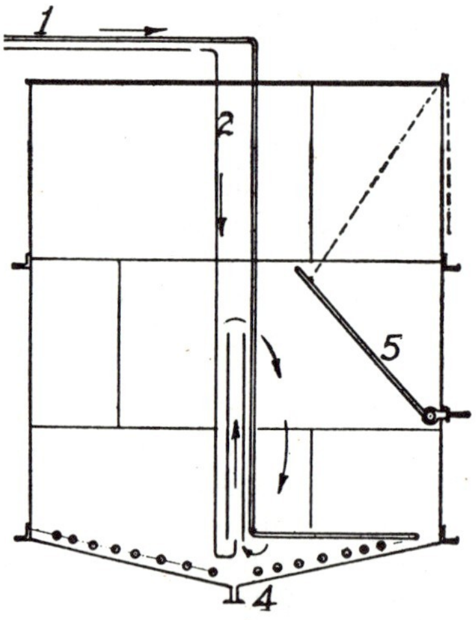

Fig. 13.1 Kettle for soap by the boiled process: (1) steam inlet for the perforated steam coil; (2) steam for agitation; (4) run-off for lye liquors and glycerin liquors; the soap is pumped out through swing pipe (5).

On the second day, water and some caustic is run in and boiled with the soap; any glycerin caught in the soap is dissolved, and the solution, a lower layer again, is run off at the bottom and combined with the first glycerin water.

On the third day, a 10° Bé fresh lye (6.5 per cent NaOH) is run into the kettle and boiled with the soap. Any glyceride which escaped the first treatment is saponified; any uncombined free acid is neutralized. The soap, which is not soluble in the alkaline liquor, acquires a grainy structure. This is called the strengthening change. After settling, the lye is run off and used in a new batch.

On the fourth day, the soap is boiled with water, which is chiefly incorporated in the soap. (Some salt is added at this point.) By this treatment the melted soap acquires a smooth, glossy appearance. On settling, three layers are formed: the upper layer, the melted soap; the middle layer, or nigre, dark in color, consisting of a mechanical mixture of soap in a soap solution and impurities; and a very small lower layer containing some alkali. The melted soap is pumped away by means of a swing pipe without removing the nigre; the latter may remain in the tank and be worked into the next batch; the smaller lower layer is wasted. This operation is the finishing change, and lasts several days because the settling must be very thorough. Approximately one week is customary for the entire cycle of operations.

The melted "neat" soap is pumped to dryers, crutchers, or storing frames; it contains 30–35 per cent water. One pound of fat or grease makes about 1.4 lbs of kettle soap; the factor varies with different raw materials.

Continuous Processing. Fig. 13.2 shows an up-to-date pilot plant for "neat" soap manufacture, either by the conventional kettle method or as a continuous neutralization of hydrolytic fatty acids. Processing the fats prior to kettle boiling is indicated, the kettle charge being adjustable as desired. Similar adjustments may be made in the stock delivered to the continuous fat splitter where the fatty acids are separated from the glycerin. The fatty acid is passed through the vaporizer where tars and decomposition products are removed; the water-white fatty acids pass either directly to the continuous Stratco contactor for continuous neutralization, or they may be vacuum fractionated to remove low molecular weight or special acids, thence to the continuous neutralizer.

While fatty acid-soap manufacture may be either batch or continuous, several continuous processes for manufacture from the fats have been developed.[5] The Monsavon process involves almost instantaneous saponification at 100°C of a previously colloid-milled emulsion of the fat in caustic solution. The soap flows across a column to a tank which acts as a buffer reservoir between saponification and washing. Saponification is completed in this tank. Countercurrent washing takes place in six chambers. Constant 85°C temperature is maintained by a jacket in which water is circulated and thermostatically controlled. The arrangement also allows for recovery of the heat of reaction to minimize heat input. The number of compartments may be increased for improved glycerin recovery.

A Sharples unit comprises each of the four steps comprising lye neutralization, completion of saponification, soap washing and glycerin recovery, and fitting to neat soap. Each of these sections consists of a pump-fed mixing unit and a supercentrifuge. The tank feeding the pump contains soap mass (called "reagent") from the preceding processing step; the mass is composed of the lyes returned from the following processing step to which have been added, if required, either caustic soda, saline solution, or water. The mass is conveyed by pump from the mixing unit to a supercentrifuge. The mixing vessel in the first

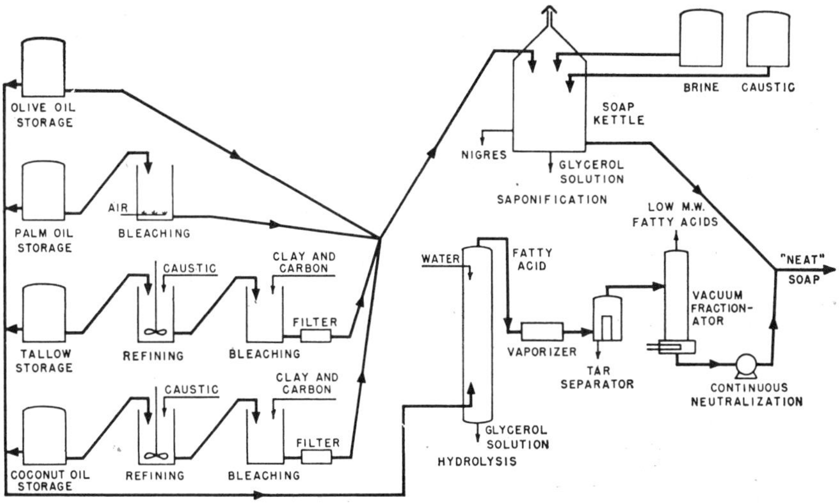

Fig. 13.2 Flow sheet for neat soap manufacture. [*Taken from Ind. Eng. Chem.*, **49**, *No. 3, 338 (March 1957); copyright 1957 by the American Chemical Society, and used by permission of the copyright owner.*]

section, which is slightly larger than the remaining ones in order to permit slightly longer contact time, is held at 100°C or slightly higher and under slight pressure. Into this section are charged the fats and oils, fresh caustic soda, and the lyes from the second section. Neutral lyes result from the first section since saponification is incomplete, completion being obtained in the second step. Settling is usually completed on lye and not nigre, though it is indicated that nigre can be separated. Neat soap is separated by two parallel centrifuges. Proportioning pumps and flow control systems are integral parts of the unit.

The DeLaval "Centripure" soap process[38] consists of two stages, viz., saponification and washing. In this process the fatty oils are fed into soap which is already present, this improves the degree of mixing; the soap acts as a "catalyst" in the saponification. A centrifugal pump provides intense mixing, saponification reaching 99.8 per cent in two-minutes passage time through one portion of the saponification column.

Correct soap phases are obtained by adding a definite excess of alkali to ensure complete saponification, and adequate control is provided in the system. Electrolyte content is carefully controlled to provide maximum mobility, and the mass is washed countercurrently with a carefully adjusted aqueous brine solution to provide a nongrainy or nigre form of soap; the washing occurs in three stages. The neat soap and spent lye are separated in a Hermetic centrifugal separator. The lye is then available for glycerin separation. The washing, concentration of brine, and viscosity measurement of the soap mass may be fully automated. Following the washing operations, deleterious substances, such as fat-soluble impurities comprising hydroxy compounds and low-molecular weight fatty acids, are still retained. The fitting operation is designed to remove them and to improve the milling properties of the soap. Investigation of viscosity relationships as functions of soap and sodium chloride percentages provides data by which relationship of viscosity to phase change may be determined. For a "hard" fitting, i.e., nigre with low soap-high electrolyte (1.2–1.4 per cent NaCl) content, viscosity measurements are used. After reaching the phase equilibrium for fitting, in the two columns, the fitted neat soap is separated by means of a Hermetic centrifuge. In the complete plant, the fitted soap then proceeds to an additive tower for mixing with builder or like substance. Table 13.2 gives a comparison of the glycerin lyes obtained by several different continuous processes.

Batch-Milled Toilet Soap. Neat soap is pumped from one of the kettles or from a soap storage tank to a paddle mixer or crutcher (Fig. 13.3). Two of these mixers (mounted on scales) are provided for alternate use. In these, soap is mixed with preservatives or special additives and then pumped to a chilling roll pass into a dryer where they make three passes through a hot-air zone on a wire mesh belt before they are discharged.

Chips from the dryer are collected in large four-wheeled carts having canvas walls and a cover, each holding about 1500 pounds of chips. These carts permit the chips to cool and reach a uniform humidity without any chance of sweating. The chips are then dropped through the floor into a scale hopper and discharged into a sigma-blade mixer called an amalgamator. Color, perfume, germicide, and other ingredients are added to the chips at this point and thoroughly mixed. The soap is made completely uniform by being put through a five-roll mill and a double-barreled vacuum plodder.

The extruded plodder bar is cut and pressed

TABLE 13.2 Characteristics of Glycerin Lyes Obtained by Different Processes[2]

Percentages	*Monsavon*	*Sharples*	*DeLaval*	*Kettle*
Total alkali as Na_2O	0.77	0.12	0.15–0.30	very variable
Glycerin	11	17–19	13 in two stages 18 in four stages	8–15
Fatty acids	0.2 + oxidized acids			

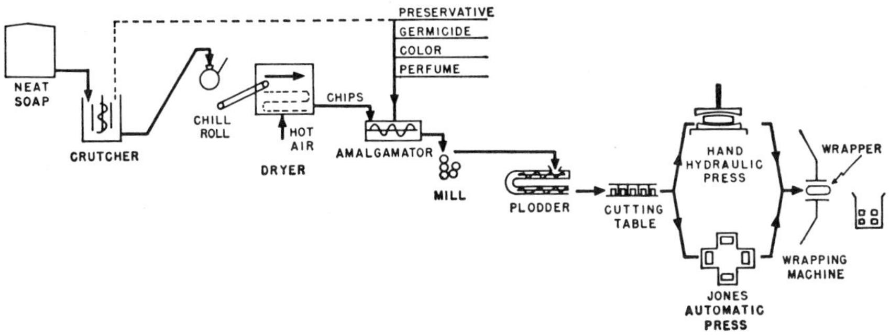

Fig. 13.3 Flow diagram for toilet soap manufacture. [*Taken from Ind. Eng. Chem.,* **49,** *No. 3, 338 (March 1957); copyright 1957 by the American Chemical Society, and used by permission of the copyright owner.*]

in dies. The bars are then automatically wrapped.

Continuous-Milled Toilet Soap. In addition to the continuous process of toilet soap manufacture described under fatty acid utilization, a European process has been used extensively. The Mazzoni process,[33] instead of using the continuous belt dryer, utilizes a vacuum chamber into which the neat soap is sprayed, reducing the water content from 30 to 21 per cent. A scraper arm removes the soap from the sides of the chamber and pushes it into a plodder which extrudes the mass through the bottom. A continuously operated amalgamator is followed by two additional plodders, the last forming the final bar which is fed to an automatic cutter. A vacuum system which is used between plodders, reduces the moisture content at each pass, and finally gives a desired 12–15-per cent water content. The cut bar soap passes through a skin-drying tunnel and then a cooling tunnel to the wrapper. The fully automatic and continuous synthetic toilet bar and soap-synthetic bar processing line is shown in Fig. 13.4. Since soap-synthetic mixtures are sticky, to eliminate bridging and to minimize surging or pulsation, twin-worm plodders with specially designed worms are used.

Laundry Soap. Though bar laundry soap is rapidly disappearing, it is still manufactured. The procedure for the boiled process is the same as that described above, but the raw materials are, for example, 4 parts tallow and greases, and 2–3 parts rosin. The latter is added, after the greases have been saponified, in the form of sodium resinate, which is made in a separate kettle by the action of soda ash (which costs less than caustic) on the rosin. Saponification does not take place with rosin, which consists chiefly of abietic acid rather than glycerides; the formation of sodium resinate is a neutralization reaction.

The kettle mixture for laundry soap is pumped from the kettle to a crutcher, which is a smaller tank (Fig. 13.5) fitted with a special agitator and with a steam jacket; one of the following materials is added: silicate of soda, 41° Bé (with 40 per cent solids), up to 30 per cent; soda ash, 2 to 5 per cent, either alone or with borax, 1 per cent; or trisodium phosphate, up to 5 per cent. The mixture is crutched until homogeneous, and then run off at the bottom of the crutcher into a "frame," a box 4 ft high, 5 ft long, and 15 in wide, with removable sides, two of metal and two of wood, and a base which is a small truck. The contents of such a frame, of which there may be hundreds, is 1000–1200 lbs of soap. A crutcher $4^1/_2$ ft high and $3^1/_2$ ft circular diameter has a capacity slightly greater than that of one frame. The contents of the frame hardens to a solid block in 3 days; the sides of the box may then be removed. The block is dried somewhat and then cut into slabs which are, in turn cut into bars and pieces. The water content of the finished soap is 8–10 per cent.

In addition to frame and bar methods there is the Mazzoni process for continuous produc-

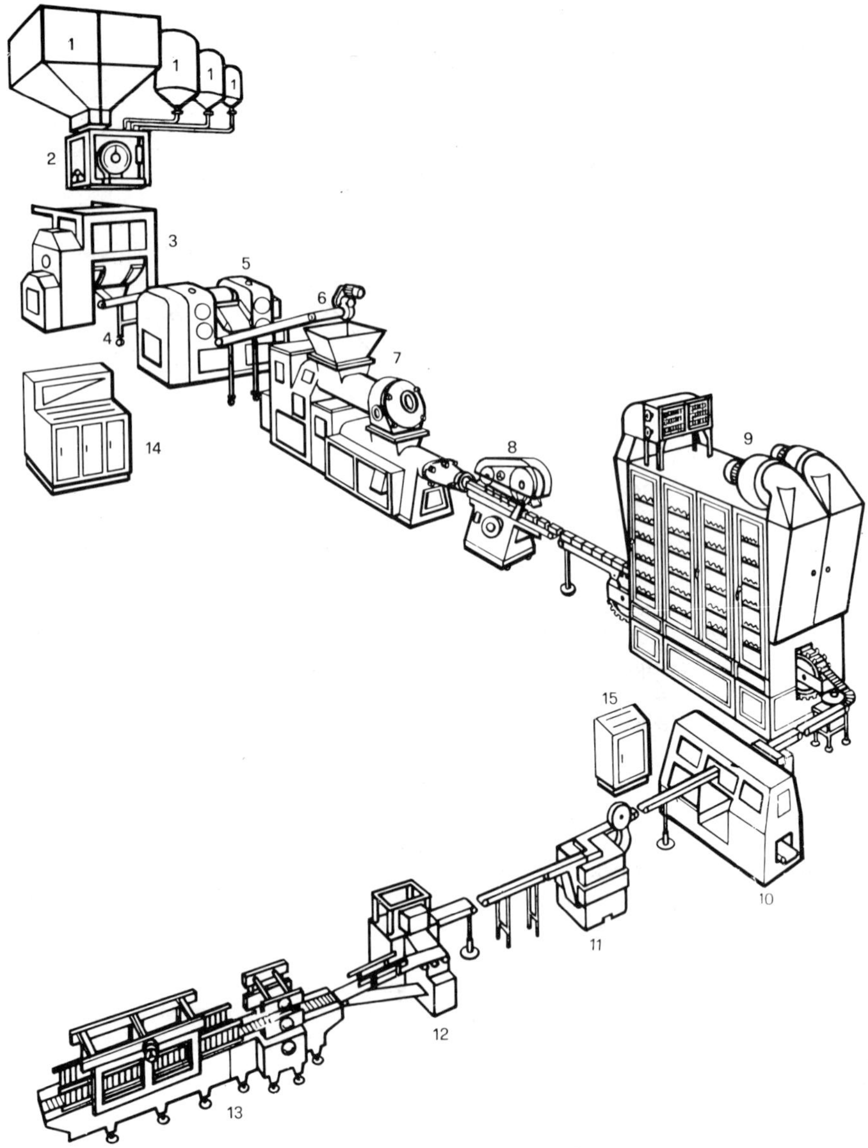

Fig. 13.4 Synthetic toilet bar and soap-synthetic bar processing line: (1) ingredient feeders; (2) weigh scale; (3) batch amalgamator; (4) belt conveyor; (5) three-roll milling machine or mixer refiners; (6) belt conveyor; (7) twin-worm duplex vacuum plodder; (8) cutter; (9) conditioning tunnel; (10) press; (11) wrapping machine; (12) case packer; (13) case sealer; (14) general switch board; (15) control board. *(Courtesy G. Mazzoni, S.p.A., Busto Arsizio, Italy.)*

tion of synthetic laundry bars which is somewhat similar to the continuous process for milled toilet bars cited above.

Cold Process Soap. In order to make soap by the cold process, either a vertical crutcher like the one described in the discussion of laundry soap (capacity 1200 pounds), or a horizontal one of greater capacity may be used. The fat is run in and heated to 130°F (54°C) by the steam in the jacket; then the lye is added and the mass is agitated. Since

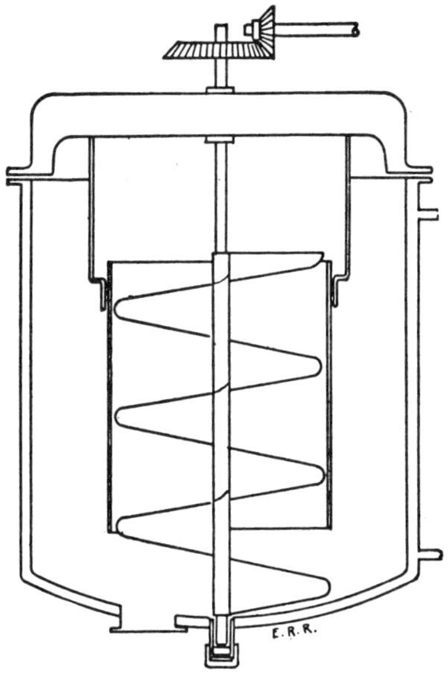

Fig. 13.5 The soap crutcher.

the reaction is exothermic, the heat may be turned off at this point. After crutching for about an hour, the mixture stands for $3^1/_2$ hrs; then it is agitated again, and again rested. The liberated glycerin remains in the soap, which is run off at the base of the crutcher into frames and cooled as in the process for laundry bars, and cut into smaller bars in the same way. Mixed sodium and potassium soaps are always made by the cold process. Potassium soaps are soft and cannot be salted out by potassium chloride, KCl. This process is used for shaving soaps, toilet soaps, and special soaps; certain laundry soaps are also made by this method. Shaving soap is a potassium-sodium soap containing free stearic acid to give the lather a lasting property.

Granulated Soap. Granulated soap may be prepared as suggested in the flow sheet of Fig. 13.6.

Miscellaneous Soaps. Transparent soap is made in a variety of ways. One method requires the best coconut oil, castor oil, and tallow, which are treated with caustic-soda lye by the cold process; the glycerin remains, and cane sugar and alcohol are added.

A "soap powder" is a mixture of soap and soda ash containing water in the form of crystal water, so that the powder is dry. The anhydrous soap content varies considerably (between 6 and 50 per cent) with appropriate variations in the amount of soda ash and water. A typical soap powder contains 20 per cent

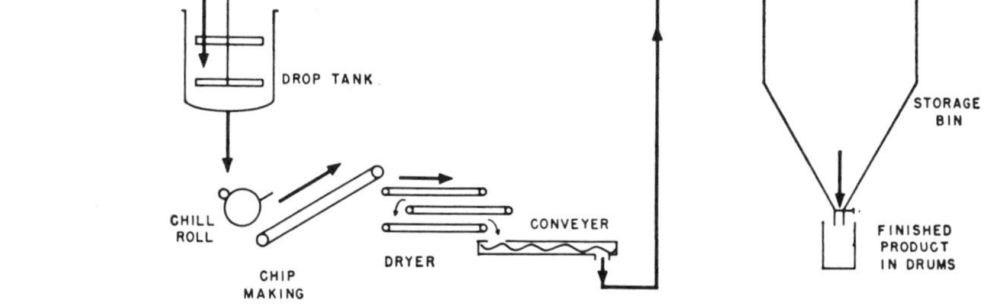

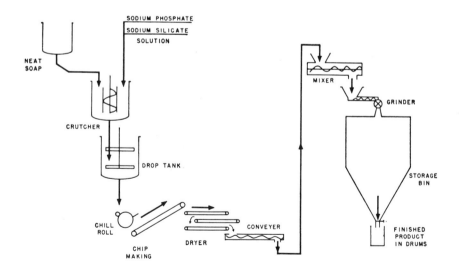

Fig. 13.6 Flow diagram for granulated soap manufacture. [*Taken from Ind. Eng., Chem., 49, No. 3, 338 (March 1957); copyright 1957 by the American Chemical Society, and used by permission of the copyright owner.*]

soap, 40 per cent soda ash, and 40 per cent water (as water of crystallization).

Spray Drying

The boxed detergents offered to the consumer market are almost universally dried in spray towers, and have loose bulk densities of from 0.20 to 0.50 g/cc. The beads may be regular or irregular globules, the walls of which are either continuous or broken; the particle size varies from very small to relatively large, representing a distribution of sizes, and is free-flowing in character.

Spray drying is not a new art, apparently having been used successfully since the last quarter of the 19th century. The initial controlling patent in the detergent industry was that of Lamont,[27] and was the basis for several long and expensive infringement cases.

Several advantages of spray drying have been summarized:[32] control of product density within a given range; control of particle shape and size; and preservation of product quality. Heat-sensitive materials may be dried, high tonnages may be produced at costs comparable with other procedures, equipment corrosion or product contamination can be minimized, and spray drying can simplify or eliminate other operations. Some disadvantages are the relative inflexibility of bulk density and particle-size variation (more fluid frequently, having to be evaporated before the soap can enter the slurry in pumpable form), higher initial investment than for other driers, and the existence of problems involving product recovery, and dust collection and utilization.

Atomization may be accomplished by liquid flow from a rotating smooth or vaned disk and by pressure atomization from single or two-fluid nozzles. The mathematics of atomization have been discussed rather extensively,[6,32] and factors such as tower design and operational procedures can be very important.

The application of chemical-engineering principles to the spray-drying art can prove valuable.[8] A spray-drying tower is essentially a thermally balanced heat-exchange operation in which atomization produces a large surface for heat and mass transfer. Though Fig. 13.7 is a simplified sketch of a countercurrent system, concurrent and mixed-flow systems have also been used.

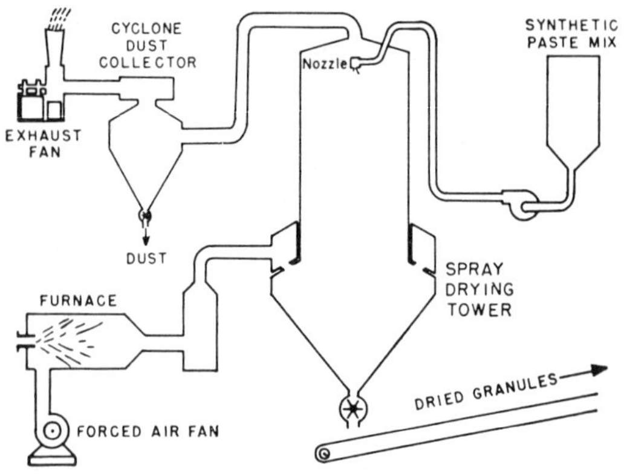

Fig. 13.7 Simplified sketch for countercurrent flow in a spray-drying tower. [*Adapted from Chem. Eng. Progr.*, **53**, *No. 12, 593 (1957); copyright 1957 by the American Institute of Chemical Engineers, and used by permission of the copyright owner.*]

The independent variables in the spray-drying process may be visualized in the diagram of Fig. 13.8, where feed moisture of the slurry affects the heat balance, dryability is the ease with which the slurry components lose water, and puffability is the characteristic of the mix to steam-puff; the other terms are self-explanatory.

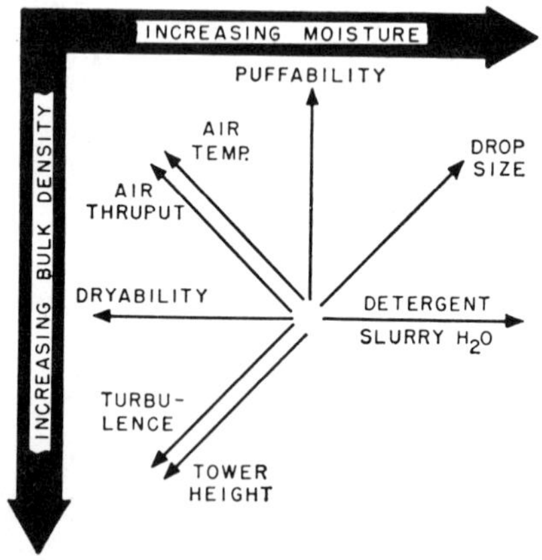

Fig. 13.8 Independent variables which affect density, moisture, and tower capacity. [*Taken from Chem. Eng. Prog.*, **53**, *No. 12, 593 (1957); copyright 1957 by the American Institute of Chemical Engineers, and used by permission of the copyright owner.*]

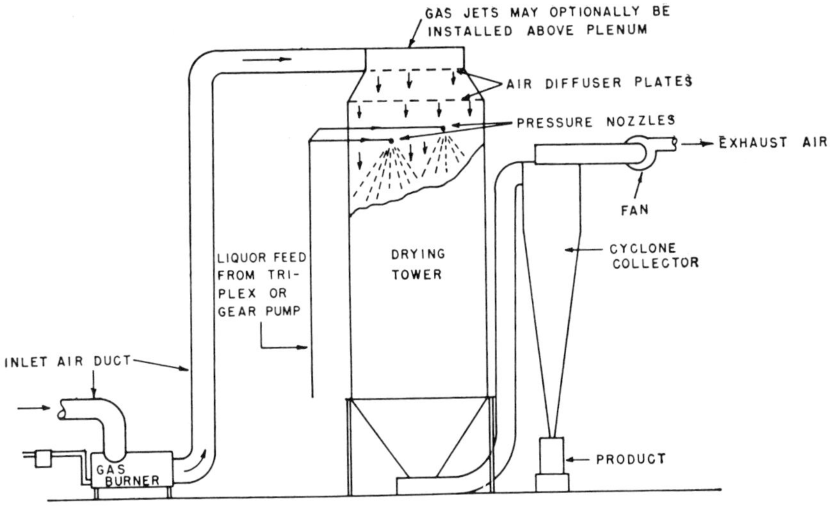

Fig. 13.9 Vertical downward concurrent spray drier (Zizinia type) with straight line air flow. [*Taken from Chem. Eng. Progr.,* **46,** *No. 10, 501 (1950); copyright 1950 by the American Institute of Chemical Engineers, and used by permission of the copyright owner.*]

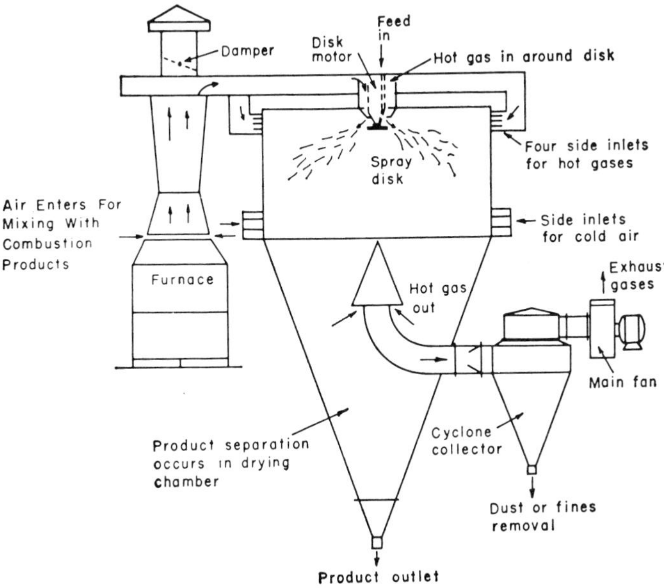

Fig. 13.10 Bowen spray drier for soaps, detergents, and other coarse-particle products. [*Taken from Chem. Eng. Progr.,* **46,** *No. 10, 501 (1950); copyright 1950 by the American Institute of Chemical Engineers, and used by permission of the copyright owner.*]

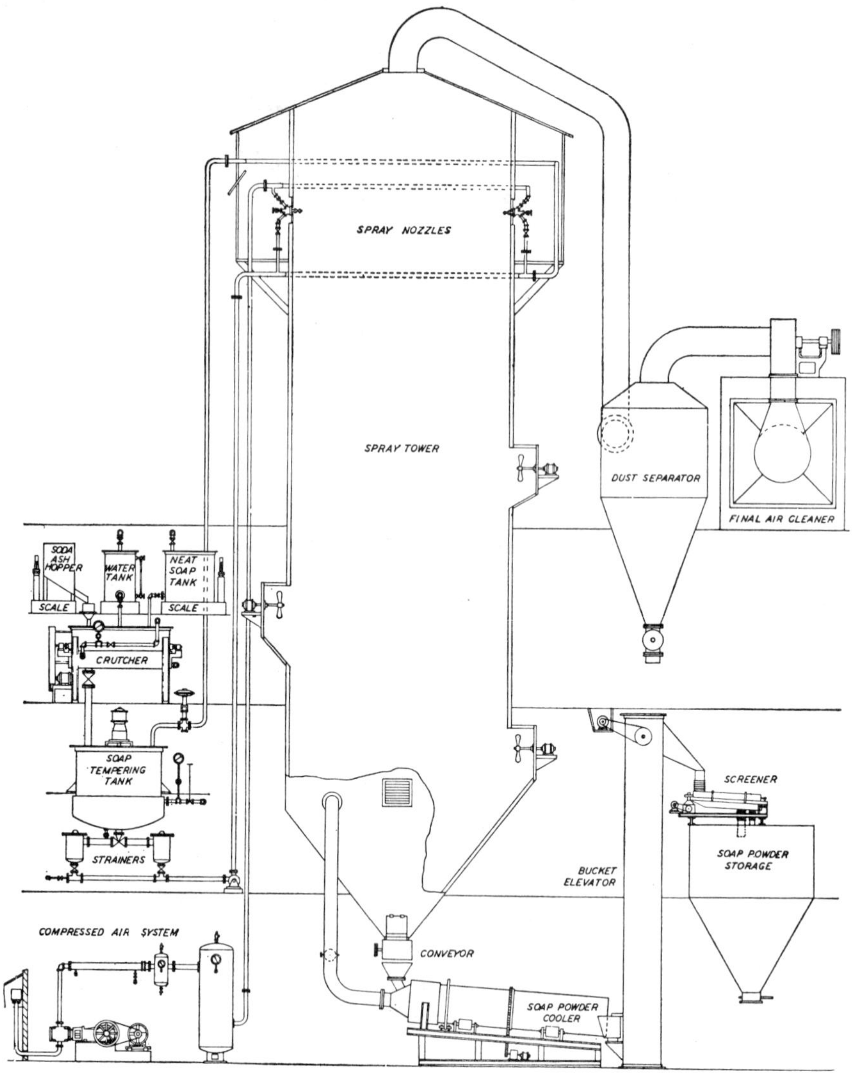

Fig. 13.11 Continuous spray-process soap powder plant. The spray tower is 100 feet high. (*Courtesy Wurster and Sanger, Inc., Chicago, Ill.*)

Some spray-drying units are shown in Figs. 13.9, 13.10, and 13.11.

The investment cost of spray-drying units is one of the principal considerations, if an alternative drying process is available. Figure 13.12 shows a chart of installed cost against pounds of water evaporated per hour, indicating that low-temperature, special-materials, or coarse-product costs remain high over the entire evaporation range. The second most important cost is that of operation, which includes fuel, labor, power, depreciation, and maintenance. Other costs may be involved if royalties are paid, and certainly overhead is an unavoidable item.

FATTY ACIDS

Manufacture from Glycerides

A glyceride may be hydrolyzed with the formation of glycerin and free fatty acids; the latter may then be changed into sodium soaps,

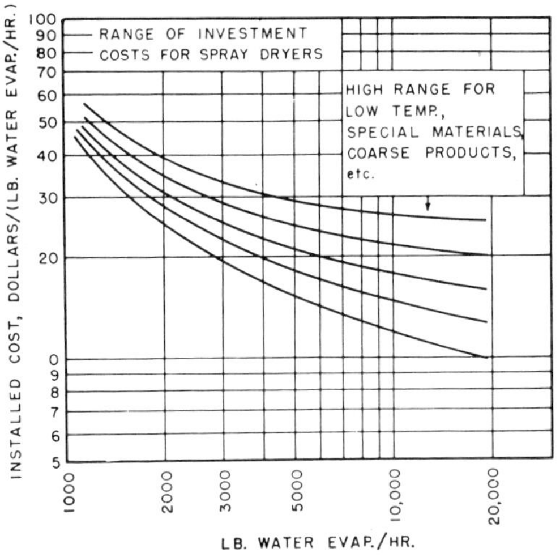

Fig. 13.12 Spray drier investment costs. [*Taken from Chem. Eng. Prog.*, **46**, *No. 10, 501 (1950); copyright 1950 by the American Institute of Chemical Engineers, and used by permission of the copyright owner.*]

or they may be used as such for different purposes.

Twitchell Process. In the Twitchell process, the foots or greases are heated with 30 per cent sulfuric acid in the presence of a small amount of catalyst (0.5 to 1 per cent) in an open lead-lined wooden tank. The catalyst is made by sulfonating a mixture of oleic or other fatty acid and naphthalene.[51] The melted fatty acids produced are washed free of acid with water and used as such, or are refined by distillation at very low pressures, supplemented by crystallization from solvents. The acid waters are treated with lime and filter-pressed to remove the calcium sulfate; the clear filtrate is distilled for its glycerin.

Hydrolysis. The hydrolysis of glycerides by means of water alone, at higher temperatures and higher pressures, has been carried out in batch processes[54] with metallic oxide catalysts; this is now a continuous process,[1] which does not require a catalyst. In the continuous process, the time involved is very short (two hours or less), compared with 6–10 hrs for the batch process, and 30–40 hrs for the Twitchell process. The temperature is held at about 485°F, and the pressure is maintained high enough to prevent the water from vaporizing (700 lb/sq in.), since the reaction is between liquid water and liquid glycerides.

The equipment is illustrated in Fig. 13.13; its essential part is the hydrolysis tower, in which preheated water travels down, and the

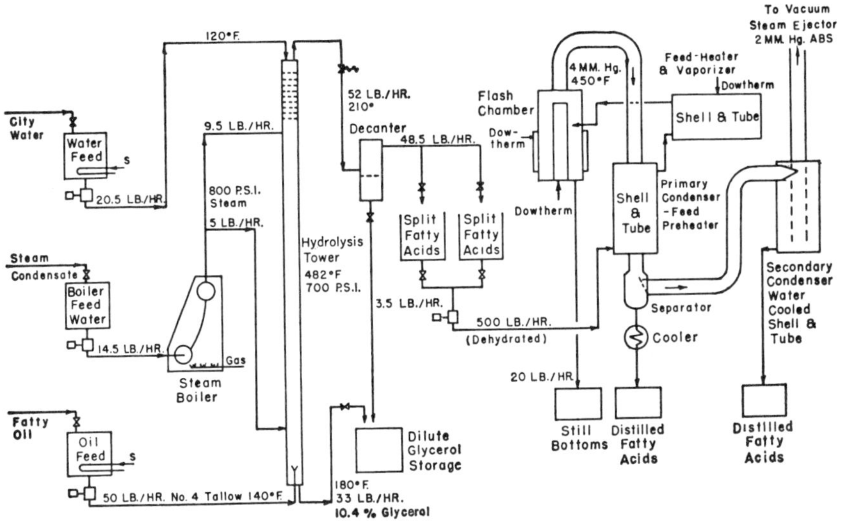

Fig. 13.13 Flow diagram for the continuous hydrolysis of glycerides by water alone. [*Taken from Chem. Eng. Progr.*, **43**, *459 (1947); copyright 1947 by the American Institute of Chemical Engineers, and used by permission of the copyright owner.*]

fatty acids formed by the hydrolysis rise. The tower in the pilot plant is 5 in. in diameter and 42 ft high, and is fitted with numerous trays. The trays in the upper part have one riser and one downspout; those in the lower part are perforated, and carry one riser for the fatty phase. Good contacting of the two layers is thus assured. The oil-water ratio is usually 1.6 to 1. The glyceride is fed in near the bottom of the tower; the fatty acids leave at the top, comparatively cool. They may be purified by heating to 445°F, for example, and flash-distilled at a pressure of 4 mm Hg. A water solution of glycerin containing 10–24 per cent of the latter leaves at the bottom of the still. Since it is comparatively pure, its concentra-

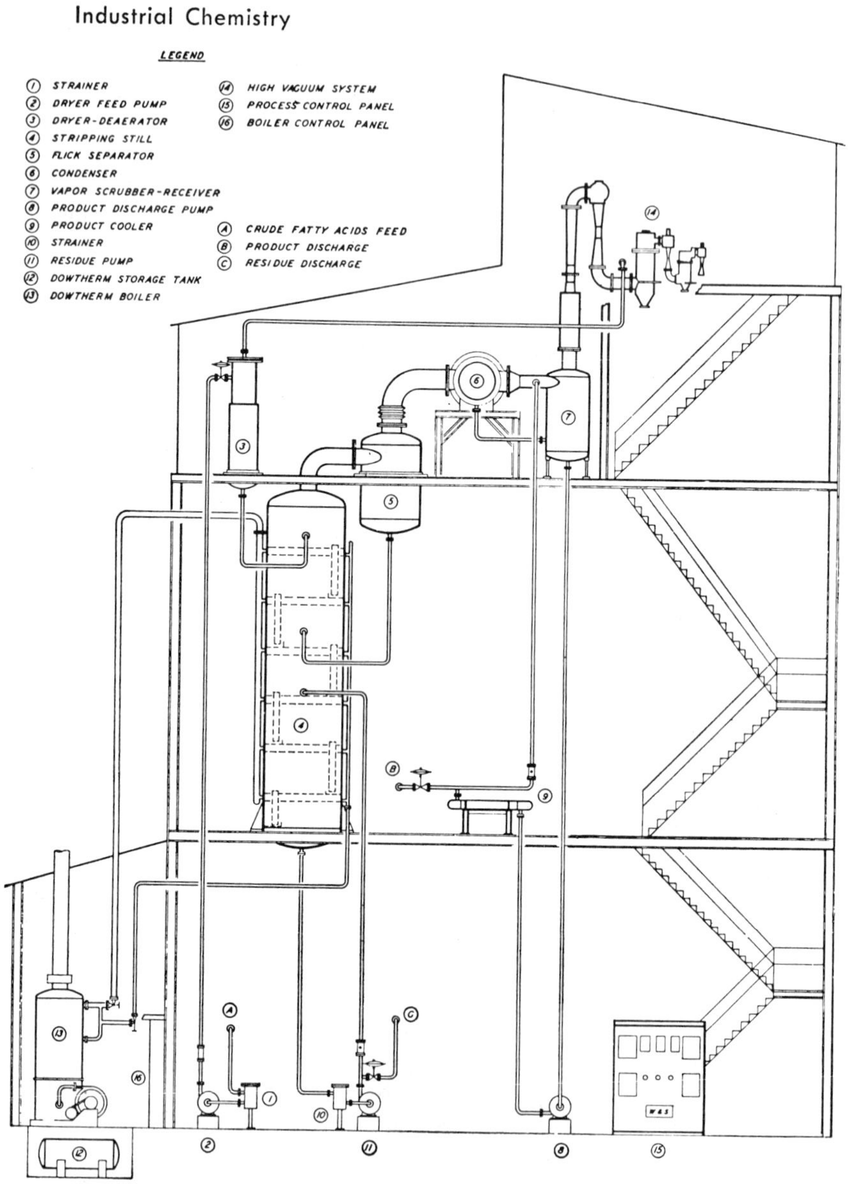

Fig. 13.14 A continuous fatty acid distillation plant. *(Courtesy Wurster and Sanger, Inc., Chicago, Ill.)*

tion offers no major difficulties. The over-all yield by this process is 96-99 per cent. The hydrolysis reaction may be represented as follows:

$$\underset{\text{A fat}}{(C_{17}H_{35}COO)_3C_3H_5} + 3H_2O \longrightarrow \underset{\text{Fatty acid}}{3C_{17}H_{35}COOH} + \underset{\text{Glycerin}}{C_3H_5(OH)_3}$$

Lime-Soap Process. Another process for liberating the glycerin is the lime-soap process, used to some extent in Europe. To the fatty acids from greases and glyceride oils there have been added, in the past, fatty acids obtained from tall oil, a by-product from the woodpulp industry. This ill-smelling low-grade raw material is separated by fractional distillation at low pressure into a fatty acid fraction, a rosin acid fraction, and a light-colored pitch residue. The first fraction may be used as such (88 per cent oleic-linoleic acids), or it may be redistilled to give a product containing 94 per cent oleic acid ("Neo-Fat D-142").

Emersol Process. Purification by fractional distillation under vacuum and redistillation is supplemented by separation using solvents; this may involve crystallization. In the Emersol process[22] the distilled fatty acid mixture is dissolved in methyl alcohol and the solution is chilled in a multitubular crystallizer, causing the formation of crystals such as stearic acid crystals, leaving an acid such as oleic acid in the solvent. The crystals are separated on a rotary vacuum filter, washed with cold alcohol while still on the filter, and then melted and passed to a still where the solvent is driven off; the pure fatty acid is left as the finished product. Similar treatment yields a finished oleic acid.

Henkel Process. A method for separating the fatty acids into liquid and solid fractions is the Henkel process, licensed by Lurgi. The solid-liquid fatty acid mixture is pumped through a scraper cooler to precipitate the stearin portion (stearic/palmitic acids). The cooled mixture is then passed through a hydrophilizer, a mixer where a concentrated solution of a surfactant (an alkylbenzene sulfonate or an alkyl sulfate) is used to wet and suspend the stearin crystals. Then these are separated from the olein or liquid fatty acids in a conventional solid-bowl centrifuge. The olein phase is drained, washed, and vacuum dried. The stearic acid portion is heated to the acid-melting point to separate the water, then the acid is washed and dried. The separated water phase is recycled for further use with the wetting agent.

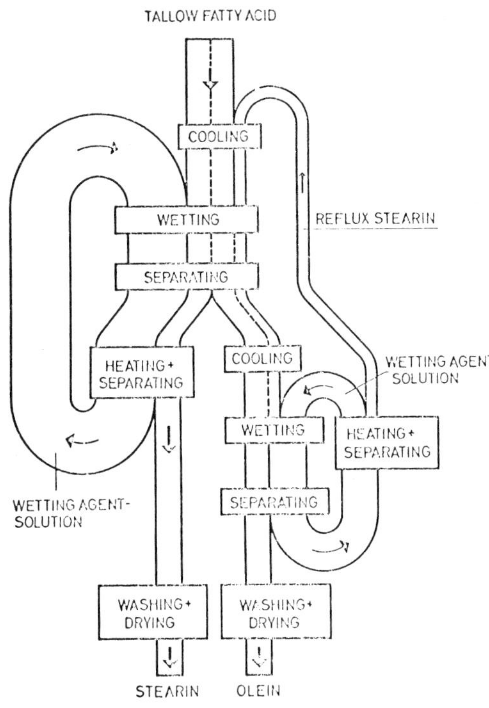

Fig. 13.15 Henkel stearin-olein separation process. *(Courtesy Lurgi Gesellschaft für Warme-und-Chemotechnik m.b.A.; process by Henkel & Cie., Düsseldorf.)*

Stearic acid may be cast into cakes or fed to a flaker. Powdered stearic acid may be made by atomizing the molten acid at the top of a spray tower.

The fatty acid in any glyceride, animal or vegetable,* may be obtained in the free state by boiling with a caustic solution, forming the

---

*It is interesting to note that all the naturally occurring fatty acids have an even number of carbon atoms which form a straight chain.

TABLE 13.3 Uses of Fatty Acids

Soap	Long-chain alcohols (by reduction) For detergents.	Blending agents For rendering certain synthetic resins oil-soluble (alkyds.)	Emulsifiers in insecticides upon addition of alkali.	Production of new chemicals.	Cosmetic Creams; Brushless Shaving Creams.	Surfactants

sodium soap, and treating the latter with sulfuric acid of medium strength; after good mixing, the warm fatty acid floats as an oil over the acidulated water and may be run off. The composition of the fatty acids varies as does their physical state,** their cost, and their applications. Their uses have increased rapidly, as indicated in the following chart (Table 13.3).

Soap-Making with Fatty Acids.

The fatty acids to be made into soap are neutralized with the proper equivalent of caustic in the form of a solution of definite water content. Salt is added (in a small amount), and several batches are blended at 200°F. The soap then goes to crutchers, to equipment for milled soap or flake soap, or to a spray-drying tower. Additions of builders are made in the proper places. In the modern plant,[34] a floating toilet soap is made by heating the soap to 400°F under a pressure of 700 psi, then flashing, and then reducing the temperature to 220°F; at this point the soap is a viscous pasty mass containing 20 per cent water. It is cooled further in special machines (Votators) where air is introduced. Then the soap is fed to an extruder which delivers a strip moving at a rapid rate. The strip is cut into a 3-bar length and cooled further to eliminate stickiness; then it is cut to size, embossed, and wrapped. The time from fat to finished soap is reduced to one day, as contrasted to the two weeks required in the traditional kettle process. (Refer also to continuous processing of soap.)

**After cooling, the free fatty acids in castor oil form a liquid; those in cocoa butter, a solid; those in tung oil, a soft solid; and those in coconut oil, a solid overlaid by a liquid.

TALL OIL

Sulfate paper-pulp mills can accumulate, as a result of digestion, an upper layer of black skimmings (soda salts of the resins, etc.) which can be further converted to an oil, called crude tall oil. The name tall oil was derived from Swedish, "tall" being pine; since a pine oil of commerce already existed in the United States, the Swedish term was adopted.

Whole Tall Oil

Skimmings are converted into crude or whole tall oil by treatment with sulfuric acid so that an upper dark brown oily layer separates. No further processing for the whole tall oil is necessary. Since they vary considerably in their proportions of fatty and rosin acids, depending on the kind of pine wood from which they originate,[40] tall oils are frequently sold under the brand names of the original producers.

A continuous process[60] for acidulation and recovery of crude tall oil from sulfate soap skimmings has been developed. This utilizes, in a single stage, a continuous centrifuge to separate the crude tall oil from the spent acid. Sludge of lignin and some tall oil is recycled. High yields and good product quality are claimed.

Tall Oil Refining

Solvent Refining. Whole tall oil is thinned with mineral spirits and treated with sulfuric acid. The separated acid sludge containing much of the color bodies is discarded, and the mineral spirits are removed by distillation. The acid treatment also causes some of the rosin acids to dimerize, making them more soluble in the fatty acids and reducing their tendency

TABLE 13.4 Fatty Acid Production, 1965-1969

(In Million Pounds)

	1965	*1966*	*1967*	*1968*	*1969*
Saturated Fatty Acids					
1. Stearic Acid–40-50% stearic content	67.1	76.0	75.8	79.2	77.8
2a. Hydrogenated fatty acids having a maximum titer of 60°C and a minimum I.V. of 5	89.3	97.4	92.6	101.5	102.9
2b. Hydrogenated fatty acids having a minimum titer of 57°C and a maximum I.V. under 5	89.4	90.5	102.1	109.9	122.9
3. High palmitic–over 60% palmitic, I.V. maximum 12	7.8	6.8	8.6	6.8	9.8
4. Hydrogenated fish and marine mammal fatty acids	10.8	14.1	13.3	12.7	13.3
5. Coconut-type acids, I.V. 5 or over, including palm kernel and babassu, hydrogenated coconut acid	23.7	26.2	30.8	43.4	41.4
6. Fractionated short-chain fatty acids, I.V. below 5, such as caprylic, capric, lauric, myristic	17.9	20.9	20.9	26.7	35.9
Unsaturated Fatty Acids					
7. Oleic Acid (red oil)	113.3	120.9	122.0	121.0	130.9
8. Animal fatty acids other than oleic–I.V. 36-80	41.4	39.0	29.2	36.4	32.5
9. Vegetable or marine fatty acids–I.V. maximum 115	9.8	9.6	8.7	9.0	8.9
10. Unsaturated fatty acids–I.V. 116-130	16.2	16.1	17.4	20.2	17.7
11. Unsaturated fatty acids–I.V. over 130	11.4	10.5	9.5	9.7	9.3
12. Tall oil fatty acids–containing less than 2% rosin acids and more than 95% fatty acids	147.2	173.0	176.2	185.4	196.5
13. Tall oil fatty acids–containing 2% or more rosin acids	151.0	164.2	156.9	153.4	174.2
TOTAL	791.8	865.2	864.0	915.5	974.1

Source: Fatty Acid Producers' Council: "Fatty Acid Production, Disposition & Stocks Census" (monthly); and the Pulp Chemicals Association for tall oil fatty acid statistics. Used by permission of the Soap and Detergent Assoc.

Note: Totals may not agree exactly with those shown due to independent rounding of figures.

to separate by crystallization. This causes a marked increase in the viscosity of the tall oil.

The resulting tall oil contains about 55 per cent fatty acids (oleic, linoleic, and palmitic), 35 per cent rosin acids (abietic, pyroabietic, and pimeric) and 10 per cent unsaponifiable matter (beta-sitosterol).

Distillation. Whole tall oil may be distilled into three or more fractions: a nonvolatile residue of tall oil pitch; a volatile fraction rich in fatty acids and lean in rosin acids; and a less volatile fraction rich in rosin acids and lean in fatty acids.

Figure 13.16 shows a flow sheet indicating the steps in both acid refining and refining by distillation.[16] Tall oil fatty acids analyzing 99 per cent fatty acids and 0.5 per cent unsaponifiables may be produced; and the rosin fraction may contain 95 per cent rosin and less than 3 per cent unsaponifiables.

In first-stage distillation, moisture is removed in the dehydrator (sodium sulfate). The crude tall oil is vaporized in a flash heater

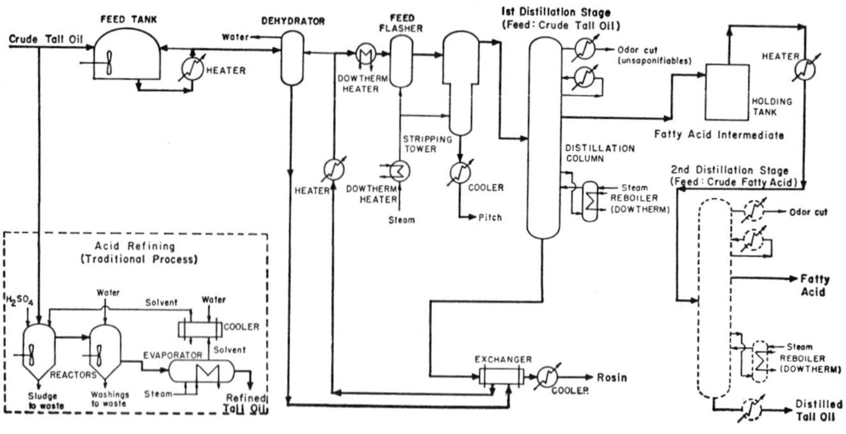

Fig. 13.16 Tall oil refining. [*Reprinted by special permission from Chem. Eng. (June 1957); copyright 1957 by McGraw-Hill, Inc.*]

which feeds a small stripping tower. There the crude is cracked into pitch bottoms containing much of the colorant materials and heavy unsaponifiables, while the overhead cut consists of rosin, fatty acids, and light unsaponifiables. The rapidity of the stripping operation minimizes damage to heat-sensitive components. The overhead of the first distillation stage consists of (1) a high intermediate cut of crude fatty acids containing rosin (this must be rerun); (2) an overhead cut containing some undesired light unsaponifiables and some low-molecular weight fatty acids; and (3) the bottom cut of rosin stream for market.

In the second-stage distillation the collected intermediate fatty acid fraction is preheated and pumped to the main distillation column (possibly a 125-foot high bubble-cap vacuum tower) where three fractions are separated: (a) fatty acid product; (b) bottom product of distilled tall oil, and (c) an overhead odor cut to remove unsaponifiables. Variation in operation can be made to control the compositions of these various cuts.

Tall Oil Soaps

About 12 million pounds of tall oil were used for soap production in 1958. This amount dropped to 4 million pounds in 1969. These soaps were frequently found in detergent compositions, or were used as liquid or paste products. (Table 13.5). The ease of neutralization of tall oil or of tall oil fatty acids makes production of soaps from them a relatively simple process, requiring a minimum of stirring and heat. The resulting products are highly soluble, even in cold water, and their neutrality is easily adjusted. Table 13.6 gives the proportions of various alkalies needed per unit of

TABLE 13.5 Tall Oil Disposition in Nonfood Products (Millions of Pounds)

Year	*Soap*	*Paint & Varnish*	*Linoleum & Oilcloth*	*Resins & Plastics*	*Other Drying Oils*	*Lubricants & Similar Oils*	*Fatty Acids*	*Other*
1956	17	48	31	28	39	–	–	356
1960	15	35	17	12	22	24	674	9
1963	9	47	18	9	35	14	779	12
1966	9	21	–	–	–	10	1171	87
1967	8	19	–	–	–	9	1167	86
1969	4	42	17	–	31	10	1154	12

Source: Agricultural Marketing Service, U.S. Dept. of Agriculture.

TABLE 13.6 Alkali Proportions for Tall Oil Soap Manufacture[a]

Alkali Grades	*Pounds per 1000 lb Tall Oil of Acid No. 175*
Caustic Soda	
Pure, 100% NaOH	125
Commercial, 76% Na_2O	
Solid, ground, flake, or powdered	127
Liquid, 50% NaOH	250
30%	417
20%	625
10%	1250
Caustic Potash	
Pure, 100% KOH	175
Commercial, 90% KOH	
Solid, ground, flake or powdered	194
Liquid, 50% KOH	350
30%	583
20%	875
10%	1750
Ammonia	
Anhydrous	53
Liquid, 26°Bé, 29.4% NH_3	181
Triethanolamine	443
Morpholine	278
Ethylamine, 70% solution	201

[a]*Tall Oil Assn. Bull.* No. 6 (1949).

tall oil for a neutral soap, and reflects the wide variety which can be manufactured.

GLYCERIN

Crude Glycerin

The spent lye drawn from the kettle in the boiled process contains the glycerin (5 per cent), the salt (10 per cent), some albuminous substances, free alkali, and soap in solution. The alkali content is normally 0.4 per cent or less; for purification, aluminum sulfate is added, and aluminum hydroxide and aluminum soaps, both insoluble, are formed. This precipitate settles well in the slightly acid solution; it is filter-pressed, and the filtrate is made slightly alkaline before concentration (Fig. 13.17). The concentration is performed in a closed, upright cylindrical steel vessel

TABLE 13.7 Tall Oil Utilization

Year	*Millions of Pounds*
1947	177
1952	270
1957	560
1962	859
1967	1309
1968[a]	1129
1969[b]	1173

[a]Preliminary.
[b]Estimated.
Source: Agricultural Marketing Service, U.S. Dept. of Agriculture.

with a conical bottom and fitted with a steam chest. The steam in the coils is at a low pressure, 5-25 lbs; this is sufficient because the pressure on the liquid is reduced to well below atmospheric by a vacuum-producing device, usually a steam jet ejector. Water vapor passes out, and when the specific gravity of the liquid has reached 29° Bé, the salt, which can no longer remain in solution, separates and drops into the cone. The entire contents of the evaporator is dropped into the salt box which has a false bottom covered with a filter screen; in this box, the salt settles on the screen. The clear liquor is sucked into the still. Glycerin water from the soap is run onto the salt, the suspension is settled again, and this liquor is also sucked into the still. The salt, now fairly free from glycerin, is blown with steam and then discharged by hand labor; it still retains approximately 0.5-per cent glycerin, but this is not lost since the salt is used over again in the kettle room.

The liquid drawn back into the still from the salt box is much lower in specific gravity, since it has lost the salt; its glycerin content is about 49 per cent. The evaporation is continued to 80 per cent, and the red liquid then obtained is the "soap lye-crude glycerin" of commerce. The test for this stage consists of heating a small sample in an open dish to a boiling point of 158-160°C.

When the drum is used, the salt is allowed to collect as it forms until the drum is filled with a sludge consisting of salt in the thickened glycerin. The sludge is then blown into a

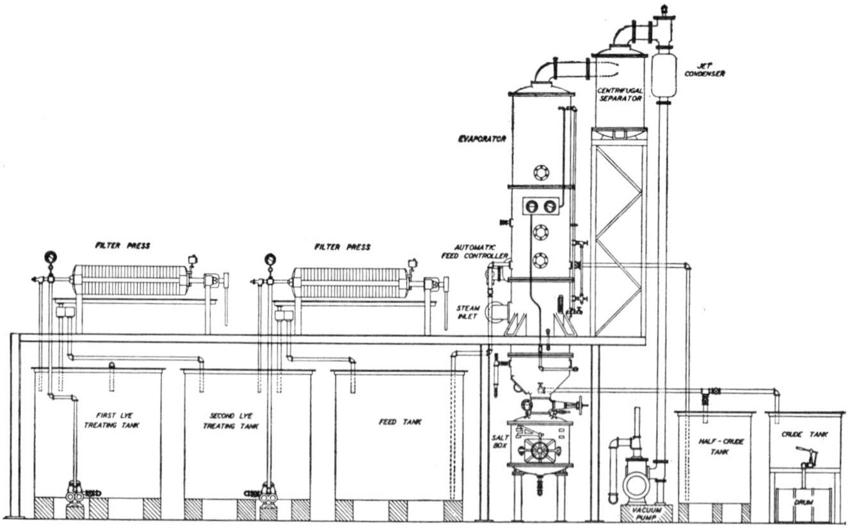

Fig. 13.17 Spent-soap lye-glycerin treating and evaporating plant. (*Courtesy Wurster and Sanger, Inc., Chicago, Ill.*)

hopper from which it drops to centrifugals; the glycerin is returned to the evaporator, and the salt is washed with a little water and discharged. In this operation the salt-drum functions as a blow case. A rotary drum-suction filter may replace the centrifugal.

There is a second kind of crude glycerin called "saponification crude," produced by concentrating the glycerin waters from the catalytic, batch-pressure (autoclave), or continuous-pressure process for fat hydrolysis. Soap lye crude contains 80 per cent glycerin, saponification crude, 88 per cent. Crude glycerin is the raw material of the glycerin distillers, a separate enterprise. Many large soapmakers, however, distill the glycerin in their own plants. Obviously, the growth of the synthetic detergent business has reduced the volume of glycerin obtained from soap manufacture.

Purification

The crude glycerin with a red color is refined to a purer straw-colored glycerin by distilling it under reduced pressure in a current of steam. The earlier batch stills have been replaced by stills with continuous operations. By reducing the pressure to 6–12 mm Hg (absolute), the distillation takes place with the liquid at a temperature of 315–320°F (157–160°C); at this comparatively low temperature no decomposition takes place. About 0.25 pound of steam is injected per pound of glycerin distilled. The steam economy is important; in recently constructed units,[61] the over-all steam consumption for the heating coil, vacuum producers, glycerin discharge pump, and injected steam is held to 2½–3½ pounds per pound of glycerin distilled. The steam and glycerin vapors pass to three heat exchangers in series. The first two are held at temperatures high enough to permit the steam to pass through, but cool enough to condense most of the glycerin. When the vapors reach the third cooler, the condenser, the remaining glycerin containing a little water is condensed to give a 90-per cent glycerin, which is concentrated further to dynamite glycerin. The rest of the steam is ejected. The condensates from the first two coolers enter a deodorizing unit; after leaving it they are high enough in grade and strength to serve as commercial distilled glycerin without additional operations. The glycerin-salt slurry which forms in the crude still is filtered and its salt is re-used, while the filtrate is distilled for its glycerin in the foots still at low pressure. The total recovery in the refining operation is 97-98 per cent.

The traditional procedure applies primarily to glycerin obtained in the kettle (or boiled) process. The procedure is modified for the purer product obtained by the continuous-pressure process.

A considerable amount of distilled glycerin is made into dynamite glycerin by concentrating it again at reduced pressure, and removing the water vapors until the glycerin content is 99.8-99.9 per cent and the specific gravity is 1.262 at 150°C. Dynamite glycerin is straw colored.

The distilled glycerin is made into **U.S.P.** grade by treating it, while lukewarm, with bone char, filtering, and redistilling the filtrate in a separate still used for nothing else. The distillate is treated with bone char a second time, and filtered; then it is as white as distilled water, has a strength of about 98 per cent, and a specific gravity of 1.258 at 150°C.

The yield of glycerin as dynamite glycerin or **U.S.P.** glycerin is about 90 per cent of the glycerin contained in the solution from the soap kettle.

If crude glycerin is purified by ion exchange prior to distillation, it has been demonstrated[49] that substantially all the material fed to the process is obtained as **C.P.** grade, the rest is dynamite grade or glycerin foots. The ion-exchanged glycerin generally contains less color, ash, and fatty acids and esters than the straight distilled product, and is more stable to sunlight.

The volume of glycerin from soap no longer leads that of the synthetic product; this is attributed to a reduction in kettle soap production and the increasing usage of syndets for laundry and toilet soap bars. In 1967 glycerin production reached 367 million pounds, the synthetic product increasing as soap production decreased. Table 13.8 shows the various usages for glycerin and the anticipated usage by 1975.

TABLE 13.8 Glycerin Usage

Use	*1955*	*1963*	*1975*
Alkyd resins	71	70	60
Cellophane	26	50	36
Tobacco	31	38	40
Drugs and cosmetics	24	45	62
Food and beverages	n.a.	25	30
Polyurethanes	0	9	27
Explosives	31	15	4
Cork and gaskets	n.a.	15	5
Paper	n.a.	10	2
Textiles	n.a.	5	5
Miscellaneous	52	3	4

Sources: J. W. Everson, Dow Chemical; Glycerine Producers' Association; Southern Research Institute. Taken from *Chem. & Eng. News*, 43, p. 27 (Aug. 23, 1965); copyright 1965 by the American Chemical Society and used by permission of the copyright owner.

Synthetic Glycerin

Propylene and Chlorine. Since mid-1948, a large-scale plant has been producing synthetic glycerin, an achievement preceded by some 10 years of intense study and experimentation.[14] Since it is made from propylene, a petroleum hydrocarbon, synthetic glycerin must be given a place on the list of petrochemicals. The success of the synthesis was practically assured when it was discovered that allyl chloride could be produced by the hot (500°C) chlorination of propylene. One hydrogen of the methyl group may be substituted by the chlorine without affecting the double bond:

$$CH_2{=}CH{-}CH_3 + Cl_2 \longrightarrow CH_2{=}CH{-}CH_2Cl + HCl$$

Propylene (a gas) — Allyl chloride (bp 44.9°C)

Allyl chloride may be reacted with chlorine at lower temperatures or with hypochlorous acid, and the product is readily hydrolyzed to give glycerin.

$$CH_2{=}CH{-}CH_2Cl + HOCl \longrightarrow CH_2OH{-}CHCl{-}CH_2Cl$$

Allyl chloride — chloroglycerolchlorhydrin

$$CH_2OH{-}CHCl{-}CH_2Cl + 2NaOH \longrightarrow CH_2OH{-}CHOH{-}CH_2OH + 2NaCl$$

Glycerin (a liquid) (bp 290°C)

In practice, the allyl chloride, purified of HCl and twice distilled, is reacted with chlorine, caustic soda, and water under controlled conditions to give a dilute solution of glycerin and salt. The solution is treated in a manner not unlike that used for the similar solution of natural glycerin, already described. To produce the allyl chloride, chlorine and preheated propylene are fed to a reactor where the reaction-heat added to the preheat produces a temperature of 500°C; at this temperature the substances react in seconds. Side reactions are few and limited. The vapors leaving the reactor are cooled and treated as follows: the hydrogen chloride is removed by water scrubbing; the allyl chloride and small amounts of other organic chlorides are absorbed in kerosene; the gas leaving the second absorber is unchanged propylene which is stored and recycled. The kerosene solution is fractionated, giving the allyl chloride fraction, among others.[45] Glycerin may be obtained from the dilute isomers 1,2-dichloro-3-hydroxypropane and 1,3-dichloro-2-hydroxypropane (called DCH) by any one of several methods.[58] In one instance a DCH solution (containing one mole HCl per mole DCH) can be reacted with 20–50 weight per cent aqueous caustic solution to yield 95–96 per cent glycerin. However, these solutions are dilute and contain impurities which are difficult to remove. Another method involves the conversion of DCH to epichlorohydrin, which is diluted with water and hydrolyzed with caustic to glycerin. The impurities are minimized, but the material loss in conversion is relatively high. A third process[55] combines the foregoing two methods. The processing details indicated in these patents are important to economical recovery.

TABLE 13.9 Synthetic Glycerin Processes

Company	*Process (Starting Materials)*
Shell	Propylene and chlorine via epichlorohydrin.
	Acrolein and hydrogen peroxide.
	Propylene and air via allyl alcohol.
Dow	Propylene and chlorine via epichlorohydrin.
FMC	Propylene oxide–allyl alcohol.
Olin Mathieson	Propylene oxide and chlorine.
Atlas	Hydrogenation of carbohydrates.

Competition for propylene-glycerin from a fermentation process is possible if blackstrap molasses prices remain sufficiently low. The glycerin in dynamite and explosives is being increasingly replaced by glycols and ammonium nitrate. More specialized polyols and modified alkyl formulations offer even greater competition in this large field. Some newer materials are also replacing glycerin in the toilet-goods field. However, glycerin (see Table 13.8) continues to find an expanding market as a humectant in cigarettes, and 50 million pounds were used in cellophane in 1958; but competition by other films should cause a reduction in glycerin usage by 1975. Increasing markets in cosmetic, tooth paste, and shaving and hand cream manufacture continues. Six million pounds of glycerin were estimated to be consumed in the edible emulsifier (margarine) field in 1958, and 25 million pounds in 1963.

The flow diagram of the Shell plant at Houston is shown in Figure 13.18 and the several possible routes from allyl chloride to glycerin are shown in the diagram in Figure 13.19. The description of the commercial unit[11,14] follows:

In the first step of the process the propylene feed is reacted with chlorine at elevated temperatures. Flow rates, temperature, and pressure are controlled to obtain optimum reaction conditions for the formation of allyl chloride. The reaction mixture is then purified through a series of distillation steps.

The pure allyl chloride reacts with caustic soda, chlorine, and water in the synthesis unit to form crude glycerin. The conversion proceeds under controlled conditions of pH, flow rate, temperature, and time. The effluent stream from the glycerin unit consists of a dilute solution of glycerin and salt.

The next steps consist of concentration, desalting, and purifying the crude glycerin. The raw solution is pumped to multiple effect evaporators which contain more than a quarter of an acre of heating surface. A concentrated salt-glycerin slurry is withdrawn from the last stage of the evaporators and pumped to a settling tank. From this vessel the glycerin solution is withdrawn and charged to another evaporator for further concentration.

After the concentrating operations the glycerin is desalted and purified. Desalting is per-

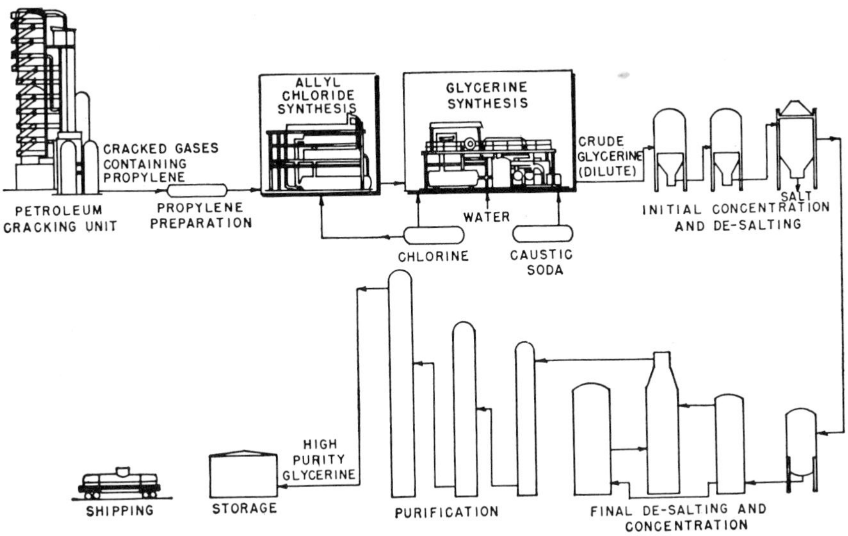

Fig. 13.18 Flow diagram of the manufacture of synthetic glycerin at the Houston plant of the Shell Chemical Corp. [*Taken from Chem. Eng. Progr.*, **44**, *No. 10, 16 (1948); copyright 1948 by the American Institute of Chemical Engineers, and used with permission of the copyright owner.*]

formed in a system of evaporators in which the glycerin is distilled under reduced pressure removing the last traces of salt. Following this the glycerin undergoes a series of purification treatments which result in a product of better than 99 per cent purity.

The purified glycerin is sent to specially lined storage equipment until ready for shipment. In order to prevent contamination which might result if ordinary materials were used, the product is transported in a fleet of made-to-order aluminum tank cars.

Formic Acid. Using allyl alcohol, 95 per cent formic acid, and methanol, glycerin of 99+ per cent purity is said to be produced[39] with yields of 94 per cent based on allyl chloride.

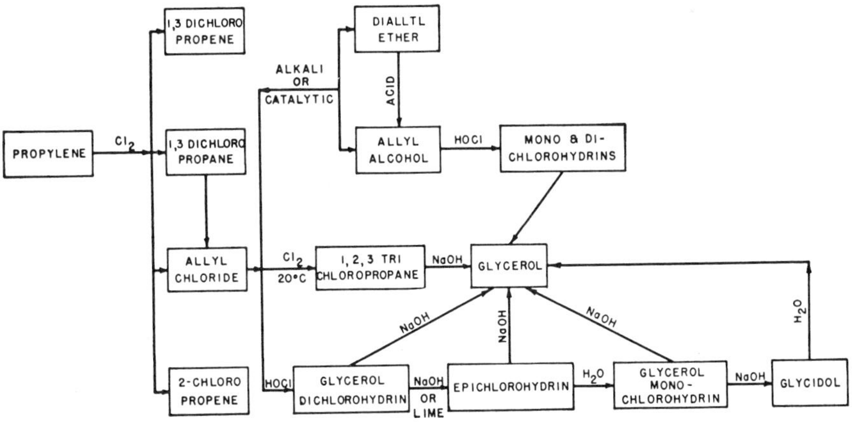

Fig. 13.19 Various possible processing steps for making glycerin from allyl chloride. [*Taken from Chem. Eng. Progr.*, **44**, *No. 10, 16 (1948); copyright 1948 by the American Institute of Chemical Engineers, and used with permission of the copyright owner.*]

The reactions are hydroxylation, ester interchange, and hydrolysis. Allyl alcohol is reacted with formic acid in the presence of hydrogen peroxide; the effective reactant is probably performic acid, and the reactions are:

$$\underset{H}{\overset{H}{C}}=\underset{H}{\overset{H}{C}}-\overset{H}{\underset{H}{C}}-OH + HCO_3H \rightarrow$$

$$HC\underset{OH}{\overset{H}{-}}\underset{OH}{\overset{H}{C}}-\overset{H}{\underset{H}{C}}-O-\overset{O}{\overset{\parallel}{C}}-H$$

$$HC\underset{OH}{\overset{H}{-}}\underset{OH}{\overset{H}{C}}-\overset{H}{\underset{H}{C}}-O-\overset{O}{\overset{\parallel}{C}}-H \xrightarrow{\text{MeOH}}{H^+}$$

$$HC\underset{OH}{\overset{H}{-}}\underset{OH}{\overset{H}{C}}-\overset{H}{\underset{OH}{C}}H + HCOMe$$

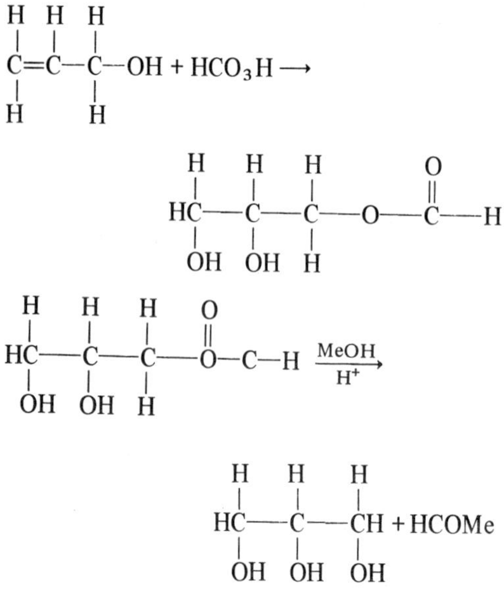

Acrolein and Hydrogen Peroxide. Figure 13.20 shows a flow sheet for this most recent process, based on a charge of propylene and oxygen coproducing glycerin and large amounts of acetone.[4]

Propylene is first hydrated to isopropanol using a standard process. The isopropanol is oxidized in the liquid phase by bubbling pure oxygen through a liquid mixture of isopropanol and hydrogen peroxide at a pressure of about 2.5 atmospheres and at temperatures between 90 and 140°C. The reaction mixture is diluted with water, stabilized, and fractionated to yield hydrogen peroxide solution, acetone, and unreacted isopropanol, which can be recycled.

A mixture of propylene with slightly more than an equal amount of steam is reacted with 25 per cent oxygen based on the weight of the propylene. The reaction is carried out over a fixed bed of catalyst based on cuprous oxide, supported on silicon carbide or some other thermal conductivity support. Close control of reaction temperature is necessary. The reaction pressure may be from 1 to 10 atmospheres and the temperature between 300 and 400°C. The reaction mixture is cooled and fractionated to give acrolein, unreacted propylene, and tarry by-products. The yield of acrolein is near 86 per cent of the propylene consumed.

The purified acrolein is mixed with pure isopropanol from the propylene hydration step and the mixture vaporized. The mixed vapors containing 2–3 moles of alcohol per mole of acrolein are passed through a catalyst bed containing both uncalcined magnesium oxide and zinc oxide, with the magnesia predominating. The reaction takes place at

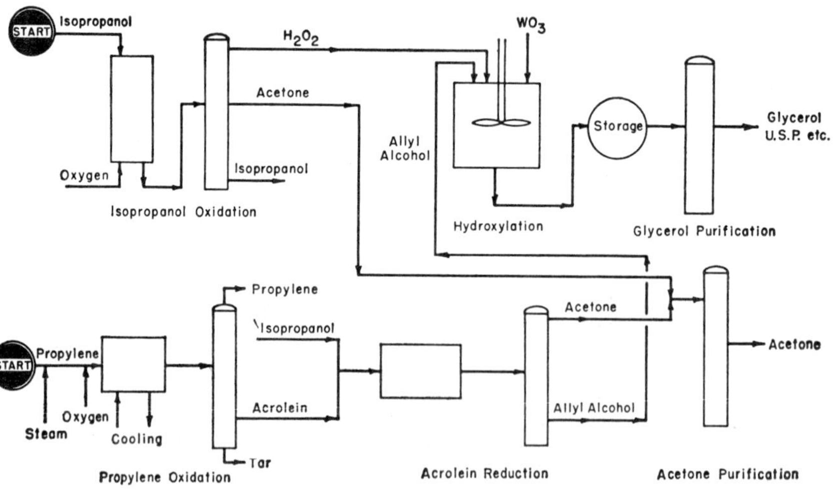

Fig. 13.20 Glycerin from acrolein and hydrogen peroxide. [*Taken from Petroleum Refiner, 36, No. 11, 250 (1957); copyright 1957 by the Gulf Publishing Co.*]

about 400°C and yields about 77 per cent allyl alcohol (based on acrolein charged) and an additional quantity of acetone which is added to that formed in the isopropanol oxidation step.

Purified allyl alcohol is agitated with a water solution of hydrogen peroxide containing a small amount of tungsten trioxide in solution. The effective catalyst is a 0.2 per cent pertungstic acid solution in 2 molar aqueous hydrogen peroxide. The reaction temperature is 60–70°C; reaction time is about two hours, producing glycerin in water solution.

The reaction mixture is pumped to receivers and distilled to yield high-purity glycerin. The recovered catalyst solution is recycled. The yield of the final step is 80–90 per cent based on the allyl alcohol charged.

The control of both the propylene oxidation step and the hydroxylation step must be exceptionally good for acceptable yields. The conditions quoted are approximate, and the practical operating limits are doubtless much less broad.

The over-all yield of useful products (glycerin and acetone) is not known. The over-all glycerin yield on propylene charged to the oxidation is in the range of 50–60 per cent.

SYNTHETIC DETERGENTS

Synthetic detergents were initially developed as soap substitutes in an economy which was running short of edible fats and oils. These compounds were made resistant to deleterious and insoluble hard-water salt formation, and were a marked improvement in wetting, cleansing, and surfactancy in general. The term "synthetic detergents" has been shortened to "syndets" to describe the detergent compositions comprising the synthetic active ingredient along with other detergent additives. "Surface active agent" has been shortened to "surfactant" to describe the surface active principal or active ingredient (AI). The rise of syndets in commerce in the United States has been meteoric: In 1940 only 1/100 as much syndet as soap was sold; in 1950 this had increased to 1/2; in 1953 slightly more syndet was sold, while in 1957 this had risen to a ratio of about 3 syndet to 1 soap. Since 1957 the ratio of syndet to soap has continued to rise, and in 1969 the ratio had increased to 6 to 1.

Even though surfactants have entered the toilet soap field, soap is still an excellent medium for combined usage, and fatty anionics are increasingly being used with it. While soap production in the United States has been reduced in proportion to syndets, increases in the population and general usage have resulted in an increase over the years, and in 1969 almost a billion pounds of soap were produced. Other advantages of soap are: (1) it is among the least toxic of all surfactants; (2) it is biodegradable and does not cause stream pollution; (3) it can be recovered where large amounts are used in an immediate area; (4) it makes antibacterial agents effective; and (5) it does not require added soil suspending agents as do the syndets.

The disadvantages of soap are pretty well recognized, not least among which is the use of fats and oils in competition with potential food uses.

Increase in liquid syndets has been a major continuing change. These liquid products were initially designed particularly for dishwashing and light-duty fabric cleansing or for medium-duty hard-surface cleaning. If the current trend continues, and in view of ecological considerations, liquid heavy-duty products may supplant the dry laundry detergents. Fully automatic washers and washer-drier combinations have triggered the demand for these easily dispensed products. At least half the total production of surfactants has applications in the household cleaning field, with smaller though appreciable tonnages used otherwise.

Surfactants may arbitrarily be subdivided into four categories, depending on ionic activity. Production data for 1968 are given in Table 13.10. Relative newcomers are the am-

TABLE 13.10 U.S. Soap and Syndet Sales (Figures in 1000's pounds)

Year	*Syndets*	*Soaps*	*Total*
1948	401,863	2,517,790	2,919,653
1953	1,866,883	1,645,431	3,512,314
1958	2,951,352	1,138,148	4,089,500
1963	3,863,352	1,026,164	4,889,516
1968	4,786,634	968,168	5,764,802
1969	5,011,748	942,364	5,954,112

Source: The Soap and Detergent Association.

pholytes and the nonionics; the latter have markedly increased in importance in the last several years, and it is forecast that nonionics will continue their growth from the 1968's 904 million lbs to 1,650 million lbs by 1980.[17] That this volume may be reached is suggested by the Department of Commerce whose "U.S. Industrial Outlook for 1970" predicts a 6 per cent per year volume increase during the 1970–1975 period.

World production of soaps and detergents for 1969, according to Henkel & Cie. GMBH, Düsseldorf, Germany, increased 4.3 per cent over 1968 to a total of 16.1 million metric tons (about 35.5 billion lbs). It was indicated that in the last decade the proportion of soaps decreased from more than 60 per cent to 40 per cent of the total washing agents while the proportion of syndets in the same period has increased from 30 to 50 per cent.

In number alone, in the U.S. syndets are continuing to increase; McCutcheon[35] listed 125 in 1947, while his latest compilation cites over 650 products.

Hydrophil-Hydrophobe Balance

Surface active agents are as the term indicates, active at surfaces by preferential orientation

TABLE 13.11 End-Use Applications[a]

	Consumption Million pounds[b]	*Per Cent of Total*
Household cleaning	548	51.5
Petroleum	182	17.0
Concrete	75	7.0
Formulated cleaners	46	4.0
Textiles	30	3.0
Food	25	2.0
Other aqueous cleaning	25	2.0
Metal cutting	21	2.0
Chemical intermediates	20	2.0
Cosmetics	18	2.0
Agriculture	16.5	1.5
Dry cleaning	10	1.0
Other	42.5	4.0
TOTAL	1,059	100.0

[a]Tariff Commission, *Chemical Week*, p. 90 (October 20, 1956).
[b]All figures are for 1955 on a 100 per cent active-agent basis; the total represents about 98 per cent of all United States production (1,071.5 million pounds in '55) as reported by the Tariff Commission.

TABLE 13.12 United States Production and Sales of Surfactants, 1968 (Millions of Pounds or Dollars)

	Production Pounds	*Sales Quantity*	*Sales Value*
Anionic	2,710	1,161	166
Amphoteric	8.4	8.2	4.8
Cationic	167	140	77
Nonionic	904	689	130
TOTAL 1958	1,335		
TOTAL 1968	3,789		

Source: U.S. Tariff Commission.

of the molecule. This suggests that some built-in characteristic appears to contribute to, or control, molecular activity. Because surfactants are effective in either aqueous or non-aqueous systems, depending on their solubility characteristics, the molecule may be tailored for either system. Starting with alkylbenzenes as an example, they are slightly surface active in nonaqueous media but insoluble in water, hence ineffective. By sulfonating an alkylbenzene, for example, dodecylbenzene (DDB), a single SO_3Na group provides high water solubility and excellent surfactant characteristics in water, but the compound is then essentially insoluble in petroleum solvents. If DDB is di- or trisulfonated, the compound becomes more water soluble and, in effect loses much of its surface activity, thereby approaching a simple electrolyte such as sodium sulfate or, by analogy, a simple benzenesulfonate. In neither case is there sufficient hydrophobe influence to increase preferential orientation, the hydrophil balance having been exceeded. However, the DDB monosulfonate, highly water soluble and an excellent surfactant for an aqueous system, may be rendered hydrophobic and useful in nonaqueous systems by neutralizing the SO_3H group with a long-chain amine to render the molecule water insoluble. It is also possible, in the case of a nonaqueous surfactant with a single SO_3Na group, to increase the C_{12} side chain to approach C_{18} or higher; the same effect can be obtained as that resulting from neutralization of the shorter alkylbenzene sulfonate with an amine.

For nonionic systems the same hydrophil-hydrophobe balance exists, except that in place of the SO_3 or SO_4 water solubilizing groups which form ionized aqueous solutions, nonionics depend on a multiplicity of oxygen groups or linkages which can unite with water by means of hydrogen bonds, thus inducing water solubility. Nonionics of the ethylene oxide-adduct type, therefore, introduce an extra dimension over ionics, since not only the hydrophobe may be varied but the hydrophil as well. This extra-dimensional feature may possibly account for some of the increasing usage of this class of compound.

In general, an optimum hydrophil-hydrophobe balance exists for a specific application, and for the class of compounds used for a given application. This optimum composition is generally arrived at by evaluating the hydrophil-hydrophobe characteristics of the specific purpose, and in at least one instance (emulsions), physico-chemical measurements have been used to predict the most effective surfactant or combination for the particular purpose.

In studying the literature one might be led to believe that the various compounds mentioned, such as lauryl sulfate, sodium stearate, dodecylbenzenesulfonate, octylphenyl nonaethylene glycol ether, and the like, are pure compounds. The compounds used commercially are mixtures, lauryl sulfate being a generic term for a mixture of sulfates whose largest fraction is derived from the C_{12} alcohol; the remainder comes from higher and lower alcohols, the amount depending on the sharpness of the original alcohol distillation cut. The same is true for alkylbenzene derivatives and octylphenol compounds, while soaps are mixtures of the various fatty acids natural to fats and oils. A further example of mixtures is that of ethylene oxide adducts. The nonaethylene glycol ether designation suggests that this compound is the main constituent, but ethylene oxide adducts are manufactured on the basis of weight addition; the ethylene oxide adds to individual hydrophobe molecules in a manner which can give a normal (Poisson) distribution of adducts, the largest proportion being represented by a 9-molar adduct. Both lower and higher adducts are also present. It might seem that a competitive edge could be gained by supplying a highly purified compound, but this is not necessarily true. Most surfactants are used for many different purposes having many varied requirements; mixtures frequently permit usage where the pure compound might be less effective. Long experience with soaps has shown that except for very specific purposes, pure soaps are not competitive with properly chosen soap mixtures. This experience carries over to other surfactants.

Anionic Surfactants

Alkylaryl Sulfonates. The surfactant currently used in the largest volume is represented by alkylbenzene sulfonates. The alkylbenzene portion is synthesized from petroleum tetrapropylene and benzene using either an aluminum chloride or hydrogen fluoride catalyst system. To the roughly 245 molecular weight of dodecylbenzene a further 103 is added for the SO_3Na group, the latter representing an appreciable portion of the molecular weight of the surfactant at a relatively low cost.

Early manufacture of *alkylbenzenes* was routed through chlorination of a suitable kerosene cut of alkanes in the C_{10-15} range obtained from high paraffin crudes. These were then used to alkylate benzene using aluminum chloride in the Friedel-Crafts reaction to provide an optimum amount of the monoalkyl product. The product of purification by distillation then represented an optimum molecular weight distribution for the detergent operation. More recently, the detergent alkylate has been generally manufactured by means of a tetrapropylene polymer in the C_{10-15} range. This is produced as a by-product of propylene polymerization for gasoline, using a phosphoric acid catalyst either in the vapor or liquid phase; or, to maximize the amount of desired product, the gasoline range may be recycled. The alkylation of benzene with the tetrapropylene polymer is catalyzed either with hydrogen fluoride or aluminum chloride. Purification again is by distillation, the final cut lying in the C_{11-15} range. A flow sheet[9] of this process is shown in Fig. 13.21.

The usual means for *sulfonation* utilizes an excess of strong sulfuric acid or oleum to approximate 100-per cent sulfonation of the detergent alkylate. The flow diagram[26] in Fig. 13.22 shows the major forms of the sulfonated alkylate. Excess spent sulfuric acid is generally separated, since some recovery through sale of

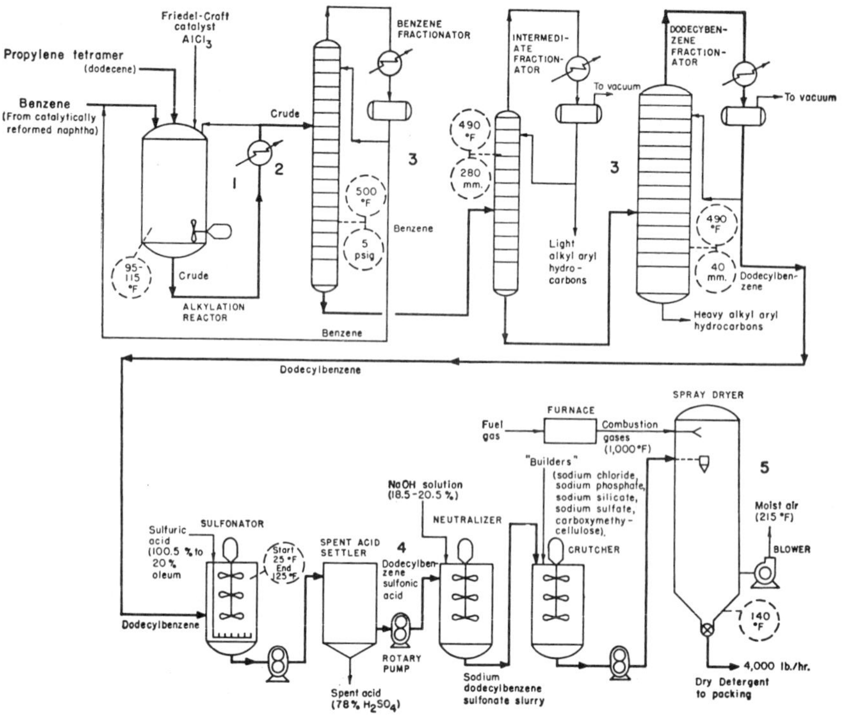

Fig. 13.21 Flow diagram for detergent manufacture: (1) Friedel-Crafts reactors; (2) heat exchangers for temperature control of benzene alkylation; (3) fractionating towers; (4) settling tank and neutralizers; (5) spray-drier equipment. [*Reprinted by special permission from Chem. Eng. (June 1954); copyright 1954 by McGraw-Hill.*]

the still useful acid is frequently of value. The neutralization may provide a highly concentrated active product which may be dried (generally drum dried) or used as a slurry for combination with other ingredients for a variety of products.

Sulfonation of the detergent alkylate using sulfur trioxide may be accomplished by passing the vaporized SO_3 (7 volume per cent) mixed with dry air (93 volume per cent) into the alkylate, stirring, and cooling.[18] The advantage of this process is that as little as 2.2 per cent sodium sulfate, basis 100 per cent sulfonate may be produced. The character of the alkylate will have a considerable effect on the amount of unsulfonateable residue, a *para* alkylated isomer being more difficult to sulfonate.

A package sulfonation plant occupying an area 20 feet square has been developed; it is capable of producing by continuous operation 20 million pounds of active ingredient per year, or 5 million pounds in an area 15 feet square. The flow diagram[47] is shown in Fig. 13.23. Using oleum, the unit includes the three steps of sulfonation, dilution, and separation of the excess acid, followed by concentration of the sulfonic acid. Correct adjustment of the variables of reaction time and temperature, acid strength, and ratio of acid to alkylate can control the sulfonate quality to minimize odor, double sulfonation, and prevent alteration of the alkyl side chain, thus assuring good color and odor. One time saving feature is the rapid separation of the sulfonation mass following dilution; this gives a final sulfonic acid containing about 7.5 per cent sulfuric acid (0.2 per cent sulfonic acid in the separated spent acid). The separation step takes place in 30 minutes as compared with 6-8 hours in the batch process.

It should be noted that this process is also

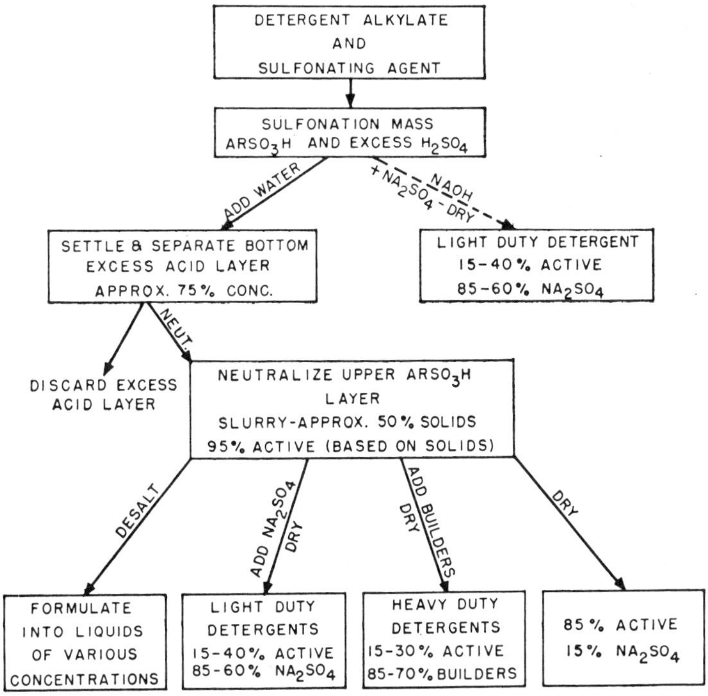

Fig. 13.22 Flow diagram for production of major forms of sulfonated detergent alkylate. [*Taken from Ind. Eng. Chem.*, **46**, *1925 (1954); copyright 1954 by the American Chemical Society and used by permission of the copyright owner.*]

useful for sulfating fatty alcohols, with either oleum or chlorosulfonic acid (for a high active ingredient content) as the sulfation medium. By tandem arrangement, both sulfonated alkylate and sulfated fatty alcohol can be produced (blending either before or after sulfonation) for use as the mixed active ingredients in some syndets (see section on Fatty Alcohol Sulfates).

Sulfur trioxide may be used in place of sulfuric acid with a reduction in the total amount of spent acid. However, the process requires close control to maintain satisfactory color and degree of sulfonation. A continuous SO_3 sulfonation unit has been developed which has been purchased by a considerable number of manufacturers. Figure 13.24 resembles the acid process of Fig. 13.23 except

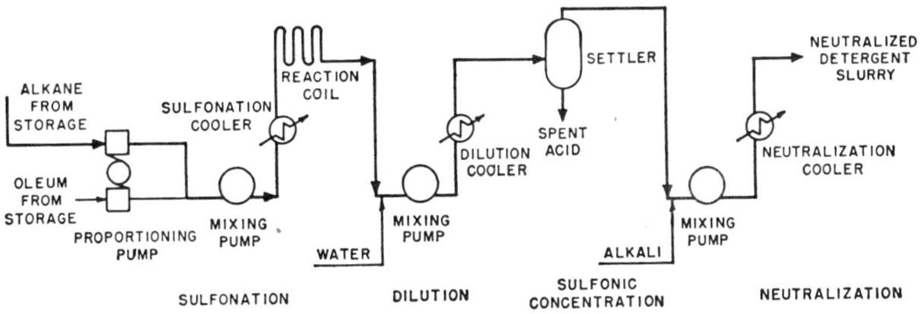

Fig. 13.23 Schematic diagram of continuous detergent slurry plant. [*Taken from Soap Chemical Specialties*, **35**, *No. 4, 131 (1959); by permission.*]

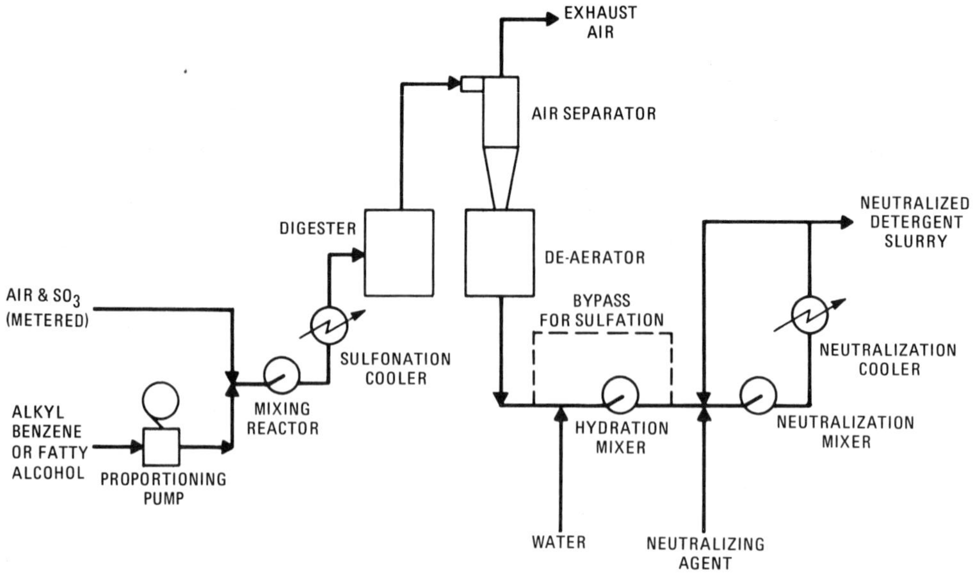

Fig. 13.24 Sulfur trioxide sulfonation. [*Excerpted by special permission from Chem. Eng.,* **70**, *(March 5, 1962); copyright 1962 by McGraw-Hill, Inc.*]

for the air separator-deaerator section, and the hydration mixer. It is claimed that the reaction from start to finish can be carried out in 30 mins. Other than the differences noted, the main variation lies in the SO_3 source. Three possible sources are cited: an air-SO_3 mixture produced by burning liquid sulfur; the vapor produced from liquefied stabilized SO_3; and the converter gas stream from any sulfuric acid plant.

Another approach to continuous sulfonation is one which relies upon nuclear radiation to perform the sulfonation. Figure 13.25 shows the flow sheet for the production of a sodium alkane sulfonate. The feed stock consists of a special kerosene which is fractionated to pass the straight chain hydrocarbon, and to reject branched or cyclic hydrocarbons. The straight chain product is then exposed to SO_2 and oxygen in a field of gamma radiation to perform the sulfonation. Since the sulfonation is incomplete, the unreacted oil must be separated and returned to the sulfonation unit, and the final product de-oiled prior to drying.

Other anionic hydrophobes can also be sulfonated with sulfur trioxide.[7]

The sodium alkylaryl sulfonate detergent solution or slurry (Fig. 13.26), is pumped to a mixer or crutcher where other ingredients are added; the mix is strained, deaerated in the Versator, and then forced by high pressure pump to the spray tower. Subsequently, adequate means for dust control and collection are maintained, and the powder is cooled and perfumed, passing from a hold silo or bin to the automatic packaging line.

Fatty Alcohol Sulfates. Two general methods of producing *fatty alcohols* from natural fats and oils are currently in use: (1) hydrogenolysis[21] by reduction of fatty acids, anhydrides, or their esters, with hydrogen at high temperatures (50–350°C) and pressures (10–200 atms) over a catalyst (copper chromite in the De Nora process):

$$C_{15}H_{31}COOR$$

$$C_{17}H_{35}COOR + 6H_2 \longrightarrow C_{15}H_{31}CH_2OH + 2C_{17}H_{35}CH_2OH + 3ROH$$

$$C_{17}H_{35}COOR$$

and (2) the sodium reduction process,[20] which requires a slightly smaller capital outlay, uses fatty glycerides, lower alcohols (for

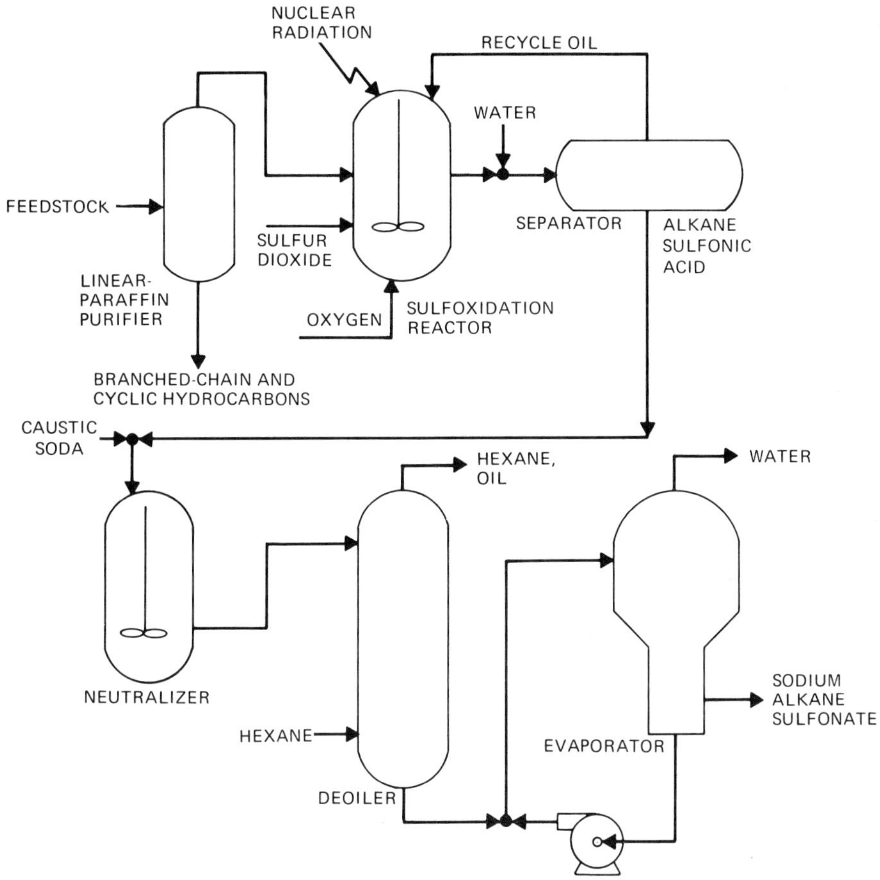

Fig. 13.25 Sulfonation with radiation catalysis. (*Excerpted by special permission from Chem. Eng.*, **100** *(Oct. 12, 1964); copyright 1964 by McGraw-Hill, Inc.*)

esterification), and metallic sodium:

$$\begin{array}{l} C_{15}H_{31}COO{-}CH_2 \\ \qquad\qquad\quad | \\ C_{17}H_{35}COO{-}CH \quad + 12ROH + 12Na \longrightarrow \\ \qquad\qquad\quad | \\ C_{17}H_{35}COO{-}CH_2 \end{array}$$

$$\begin{array}{l} C_{15}H_{31}CH_2OH \\ C_{17}H_{35}CH_2OH + \text{glycerin} + 12RONa \\ C_{17}H_{35}CH_2OH \end{array}$$

Glycerin is recovered as a solution containing caustic. The usual impurities, such as soap and unreacted ester, are removed through fractional distillation. The operating schemes are shown in Fig. 13.27

The advantages of hydrogenolysis are cheaper raw materials, and a wide choice of feed stocks and suitable locations. The advantages of sodium reduction are the production of both saturated and unsaturated alcohols, lower initial investment, simpler operation and maintenance, and superior quality products.

One large detergent manufacturer converted from the sodium process to catalytic hydrogenation, using essentially the scheme[15] shown in Fig. 13.27. It is claimed that the alcohols must be saturated for detergent use so that this process would be preferable to the sodium route. The alcohols are fractionated; the C_8 and C_{10} products are sold to plasticizer manufacturers, and the higher ones are retained for detergent manufacture.

Commercial production of oxo-process and

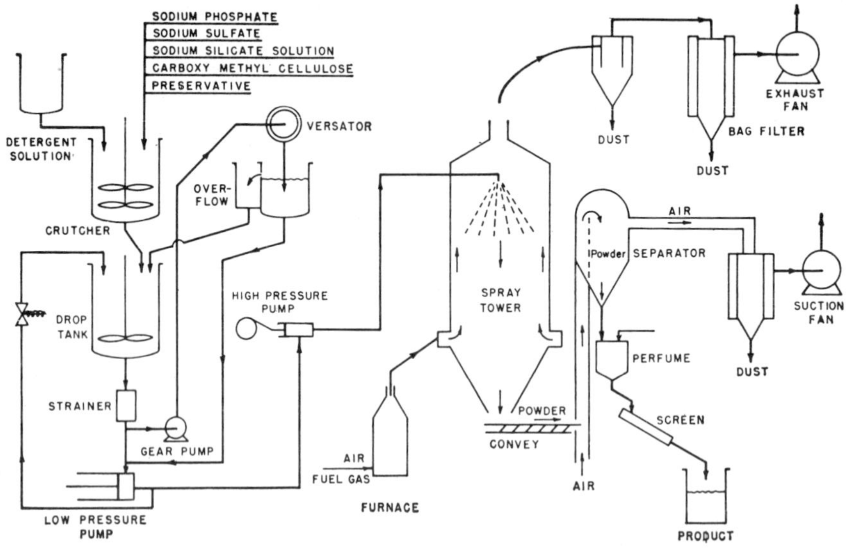

Fig. 13.26 Flow sheet for detergent spray drying. [*Taken from Ind. Eng. Chem.*, **49**, *No. 3, 338 (1957); copyright 1957 by the American Chemical Society, and used by permission of the copyright owner.*]

Ziegler-process long-chain alcohols has resulted in natural alcohols having to compete for both sulfation and use as alcohol-ethylene oxide adducts.

It was mentioned in the discussion of detergent alkylate sulfonation that a continuous process involving oleum or chlorosulfonic acid could be used for the sulfation of fatty alcohols. The preferred process is the one which

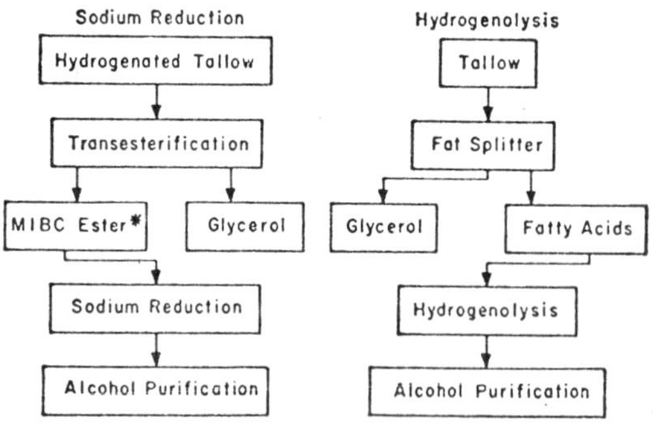

Fig. 13.27 Scheme of fatty-alcohol production. [*Taken from Ind. Eng. Chem.*, **46**, *1917 (1954); copyright 1954 by the American Chemical Society, and used by permission of the copyright owner.*]

utilizes oleum, although chlorosulfonic acid is used to produce a high active ingredient content. This latter process presents a corrosion problem; both the acid and the hydrogen chloride produced are difficult to handle and require corrosion-resistant equipment. Sulfamic acid and ammonium or amine sulfates are used where the end product is an ammonium or amine salt. Gaseous sulfur trioxide sulfation has also been employed successfully. In sulfating a fatty alcohol the desired product is the monoalkyl sulfuric acid ester.[15] Since the reaction is rapid, oversulfation may be a problem, causing olefin, ether, or aldehyde formation, although other reactions can occur. An optimum set of conditions exists for sulfation. Excess time or temperature reduces the completeness of sulfation and causes the product to deteriorate. When conditions are below optimum, equilibrium is not reached. This equilibrium point is determined by the free water concentration in the reaction mix; the SO_3 in oleum helps to maintain the water concentration at a low level, oleum being superior to sulfuric acid in this respect.

Alkylbenzenes are considerably simpler to sulfonate. The reaction, which is not reversible, follows a smooth time-temperature curve;

longer contact time or higher temperature alters the degree of sulfonation to a negligible degree. However, a proper excess of acid, to maintain the strength above 95 per cent, is required to prevent the product from darkening as a result of color formation.

Figure 13.28 shows a flow sheet of simultaneous alkyl sulfate-alkylbenzene sulfonate production. However, this plant is multipurpose and can produce alkyl sulfates, alkylaryl sulfonates and layered alkylaryl sulfonates. The reaction bath consists of a mixing pump and heat exchanger (only 8 inches in diameter and 5 feet long). The neutralization bath consists of a mixing pump, a heat exchanger (two units, each 2½ ft outside diameter and 20 ft long), a neutralizer, and a surge tank. Since the excess acid is ordinarily neutralized to form sodium sulfate, and is left in the product, a product with a higher active ingredient content requires spent-acid separation; this is accomplished by layering, or the separation of spent acid from the sulfonic acid. The process increases the active alkylaryl sulfonate to sodium sulfate ratio from 1:1 to 8:1.

Ethylene Oxide (EO) Adduct Sulfates. In the process of adding ethylene oxide to reactive products such as alcohols, fatty acids, etc., the effect is that of extending the fatty chain with EO groups such that the final product ends with a hydroxyl group. This group reacts to sulfonation in the same manner as a primary alcohol, to provide terminal $—O—SO_3H$ groups. This then can be neutralized with any alkaline compound to provide a variety of products of varying characteristics. But even more important than the characteristics of the sulfate group is the ability to add given numbers of ethylene oxide units to vary the characteristics of the compound. For example, for a hydrophobic character, if the sulfated compound has only one or two units of ethylene oxide, the product will be almost water-insoluble, but soluble in oil (this is further affected by choice of the acid-neutralizing compound). Conversely, sulfated products containing four or more ethylene oxide units are very water-soluble, with increasing surface activity in such a medium. A further virtue of the 4-5-EO sulfated products is their high water solubility, promoted by the residual EO units in the compound. Obviously, larger numbers of EO units may be used depending upon the hydrophobe unit, but for the approximate C_{12} hydrophobe, an increase

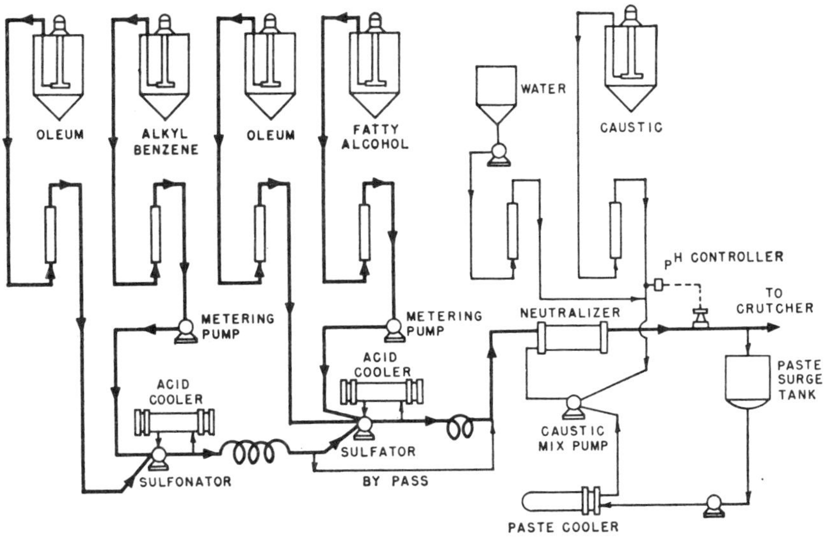

Fig. 13.28 Flow sheet of process to make sulfate and sulfonate detergents continuously. [*Taken from Ind. Eng. Chem.*, **51**, *13 (1959); copyright 1959 by the American Chemical Society, and used by permission of the copyright owner.*]

greater than four or five units fails to enhance activity. An extensive treatment of anionic surfactants based upon nonionic materials can be found elsewhere.[42]

Taurate Surfactants. In 1930 both H. T. Böhme, A.-G., and I.G. Farbenindustrie recognized that the weakness of soap was centered at the carboxyl linkage;[25] the former chose the alkyl sulfate route, the latter, esters of fatty acids. Subsequently, because the fatty esters were too unstable for many purposes, fatty amides which were taurine derivatives were developed. "Igepon T" (oleyl methyl taurate) was one of the first widely used surfactants which still has many applications.

The chemical reactions yielding Igepon T are:

$$\underset{\text{Oleic acid}}{3C_{17}H_{33}COOH} + PCl_3 \longrightarrow \underset{\text{Oleyl chloride}}{3C_{17}H_{33}COCl} + H_3PO_3$$

$$C_{17}H_{33}COCl + \underset{\text{N-Methyltaurate}}{HN(CH_3)\text{—}(CH_2)_2SO_3Na}$$

$$+ NaOH \longrightarrow \underset{\text{Oleyl N-methyl taurate}}{C_{17}H_{33}CO\text{—}N(CH_3)\text{—}(CH_2)_2SO_3Na} + H_2O + NaCl$$

It is of considerable interest that in the patent[52] covering compounds of this type

$$R^1\text{—}C(=O)\text{—}N(R^2)\text{—}R^3$$

in which R^1, R^2, and R^3 are branched or straight-chain aliphatic, cycloaliphatic, or aromatic hydrocarbon groups, or heterocyclic rings, and in which R^3 may be a sulfonic or sulfuric ester (these groups may have many forms), that over 1,620,000 variations are possible. Obviously, only relatively few of this number have been synthesized, and of these, fewer still have been merchandized.

Nonionic Surfactants

As the name indicates, these products are not ionic in nature, and in contrast to sulfates, sulfonates, or phosphates their solubility generally depends on hydrogen bonding through a multiplicity of oxygen groups in the molecule. The most widely manufactured products are ethylene oxide adducts, although mixed ethylene and propylene or butylene oxide compounds are, or can be, produced.

Ethylene Oxide Adducts. One main requirement of the nonionic hydrophobe used for ethylene oxide addition is that it contain a reactive hydrogen; the most important hydrophobes are given in Table 13.13.[23] As discussed under hydrophil-hydrophobe balance, these adducts are not single compounds, but represent mixtures approximating Poisson distribution. Production of ethylene oxide adducts is not difficult, but because of corrosion and explosion hazards the equipment is necessarily expensive, and handling and storing raw materials and products requires considerable capital. Operating costs are largely dependent on the volume produced; short-chain ethylene oxide adducts can be produced in much shorter cycles than the more generally used longer chain products.

Shick[42] has edited a very thorough volume concerning nonionic surfactants. Tall oil adducts probably comprised the largest single group of nonionics, largely because of their relatively low cost. However, they have been replaced by other nonionics for controlled low-sudsing detergents. The currently most used adducts are those derived from primary or secondary alcohols and from alkylphenols. Use of the fatty alcohol adducts as sulfates is discussed under Anionic Surfactants.

Polymeric Nonionics. Alkylene oxides such as ethylene, propylene, isobutylene, or 1,2-epoxybutane can be polymerized alone to form homopolymers, or alternatively, they may be copolymerized as "block-type" products:[44] $A_n\text{—}A\text{—}B\text{—}B_m$, where A represents propylene oxide and B ethylene oxide, the polyoxypropylene thus becoming the hydrophobe which is solubilized by addition of ethylene oxide.[56] Optimum polypropylene glycol molecular weights appear to lie between 800 and 2500 g. It is obvious that a large number of compounds varying markedly in characteristics can be produced; many of these have

TABLE 13.13 Preparation of Nonionic Surfactants from Ethylene Oxide[a]

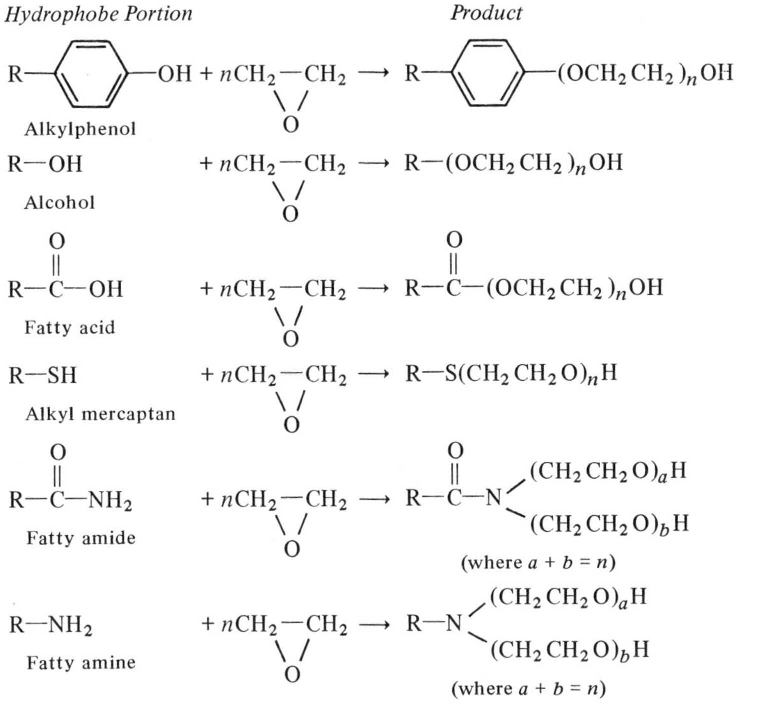

[a]Jelinek, C. F., Mayhew, R. L., *Ind. Eng. Chem.*, 46, 1930 (1954); copyright 1954 by American Chemical Society and used by permission of the copyright owner.

already been investigated. If, instead of polypropylene glycol, the central portion of the molecule becomes ethylenediamine, block polymers having four hydrophilic tails can be made by reaction with ethylene oxide[48] (Fig. 13.29). Where the molecular weights of the polyoxypropylene-polyoxyethylene compounds can approximate 10,000, molecular weights of the ethylenediamine products can approach 27,000.

Alkylolamides. These are reaction products of alkanolamines with fatty acids. In the 2:1 type[53] in which two moles of diethanolamine (DEA) are reacted with a single mole of coconut fatty acid, which is said[28] to be responsible for the surface-active properties of the mixture:

$$RCOOH + 2HN(C_2H_4OH)_2 \longrightarrow RCON(C_2H_4OH)_2$$

Excess DEA converts both amino ester and amido ester to the active 2:1 product N,N-bis(2-hydroxyethyl)lauramide.

New 1:1 types have been developed[27] having over 90 per cent crude amide content, achieved by an ester interchange of 1 mole DEA with 1 mole of fatty acid methyl ester under special synthesis conditions. The importance of these compounds lies in their detergent and foaming ability, and in the fact that they act as foam boosters and stabilizers for dodecylbenzene sulfonates. In addition they are compatible with both anoinic and cationic surfactants, are emollients, can affect the viscosity of liquid detergents, and are corrosion inhibitors.

Sugar Surfactants. Sugar is a desirably priced raw material on which a process for surfactant preparation has been based.[37] The product is a sucrose fatty acid monoester (the

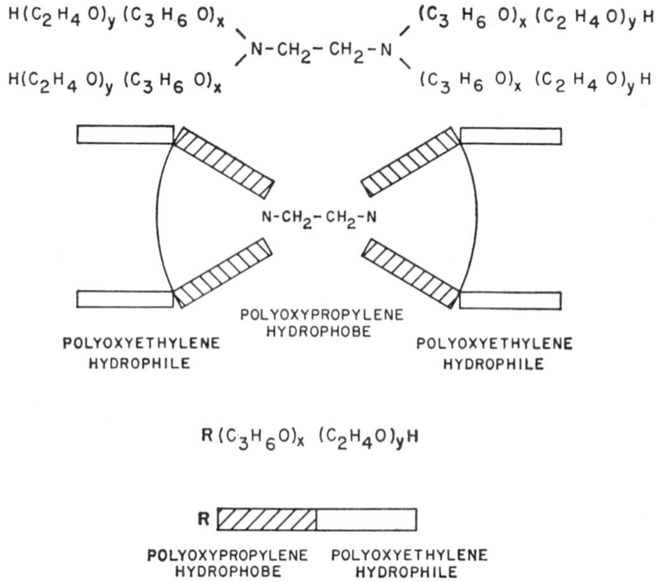

Fig. 13.29 Typical block polymer surfactants. [*Taken from Soap Chemical Specialties*, **33**, *No. 6, 47 (1957); by permission.*]

monostearate, for example), the eleven oxygen atoms of sucrose contributing about the same hydrophilic effect as a polyoxyethylene with the same number of oxygen atoms.

The process is typified by the following run:[13]

Three moles of sucrose, one mole of methyl stearate, and 0.1 mole of potassium carbonate (catalyst) are dissolved in dimethylformamide (or dimethyl sulfoxide). Potassium carbonate is a preferred catalyst because, unlike a more alkaline catalyst (e.g., sodium methoxide), it will not take part in undesirable side reactions at high temperatures.

The reaction mixture is agitated, heated at 90–95°C at 80–100 mm of mercury for 9–12 hours. The methyl stearate reacts with the sucrose to give a sucrose monostearate and methanol. The latter is stripped off.

After the solvent is distilled off and the product dried, it contains about 45 per cent monostearate, 1–2 per cent potassium carbonate, and about 54 per cent unconverted sugar (beacuse of the large excess used). The product can be used for many jobs, as is. More likely, however, the economics of the situation will dictate that the sugar be recovered by further purification of the product.

The sugar can be removed by adding toluene as a solvent.

Conversions of over 90 per cent are claimed. An excess of sucrose produces the monoester, best for detergent products, while an excess of nonsugar ester yields the diester product which is superior for food applications. Raw material costs are a controlling factor due to variation in world supplies. Purification is also difficult. The flow sheet is shown in Fig. 13.30.

In the equipment shown in Figure 13.31 the reactants, the solvent, and the catalyst are placed in the reaction vessel (5) in which the conversion from nonsugar ester to sucrose monoester is carried out. The product alcohol and part of the solvent are stripped from the system in a turbulent-film evaporator (1). The product alcohol and solvent are fractionated in the packed reflux tower (2). The solvent is returned to the system through (4) while the alcohol is condensed in (3) and collected in vessels (7).

Recovery of the sucrose ester from the slurry containing unreacted sugar can be accomplished using xylene as the ester solvent, since filtration is unsuccessful due to the sugar particle size. The xylene is then recovered by steam distillation and the sugar ester remains. Important to economic operation is minimization of diester formation (which is effected

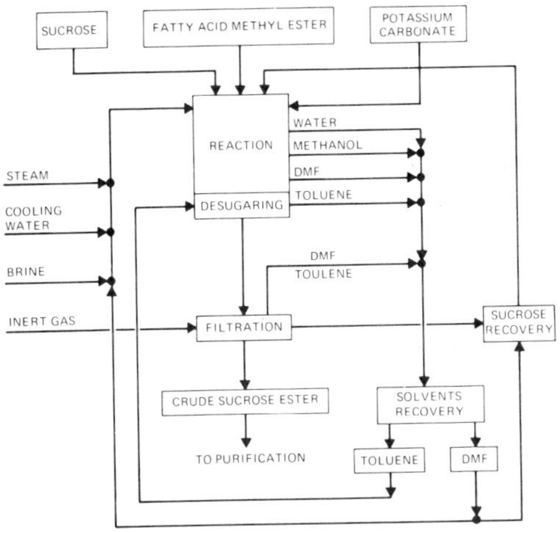

Fig. 13.30 Flow diagram for sucrose ester manufacture. (*Excerpted by special permission from Chem. Eng., p. 56 (Feb. 12, 1968); copyright 1968 by McGraw-Hill, Inc.*)

by controlling the water content during alcoholysis), recycling of unreacted sugar, and recovery of DMF and xylene. Several licenses have been granted for this process.

Sorbitol Compounds. Sorbitol may be produced by the hydrogenation of sugars such as glucose. Then this hexahydric alcohol may be reacted with ethylene glycol and the reaction product esterified to varying degrees with lipophilic fatty acids. Or sorbitol may be partially esterified with fatty acids and then these inner-ester sorbitol anhydrides may be further reacted with ethylene glycol.

Ampholytic Surfactants

These products are so named because they contain both cationic and anionic groups. A typical example is prepared by reacting a fatty primary amine and methyl acrylate to give the N-fatty-β-aminopropionic ester which is saponified to the water soluble salt.[2]

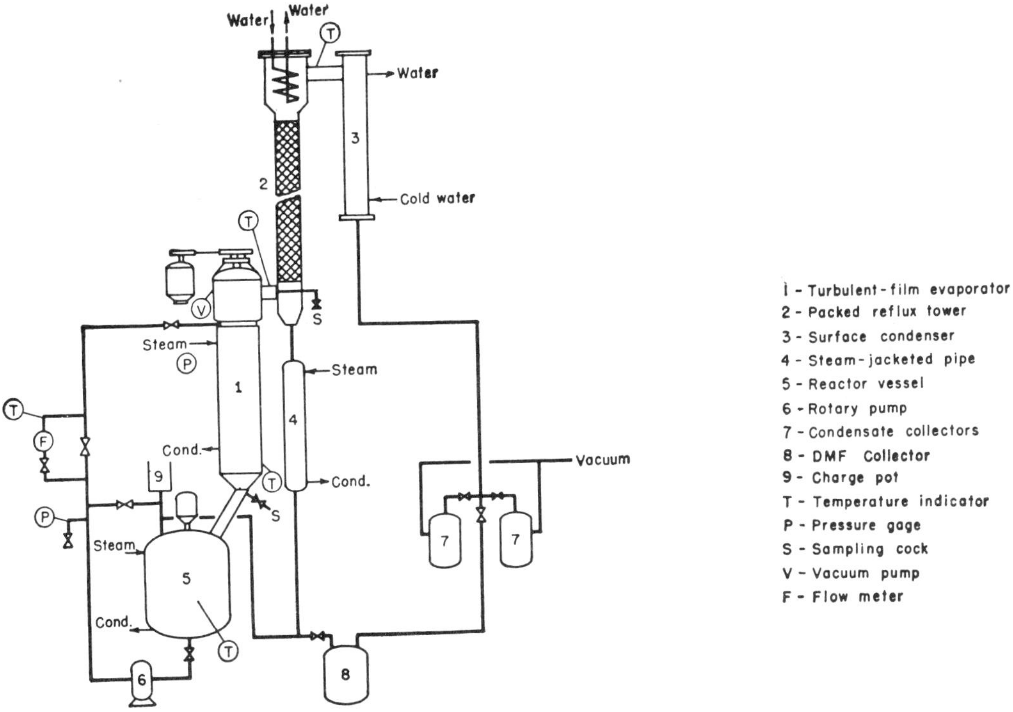

Fig. 13.31 Pilot-plant equipment for preparation of the sugar esters. [*Taken from J. Am. Oil Chem. Soc., 34, 185 (1957); by permission.*]

$$RNH_2 + CH_2{=}CHCOOCH_3 \longrightarrow RNHCH_2CH_2COOCH_3 \xrightarrow{NaOH} RNHCH_2CH_2COONa$$

A specific use for sodium lauroylsarcosinate is in toothpaste compositions[57] in which these enzyme inhibitive surfactants are said to reduce dental caries. These compounds are also used in shampoos, cosmetics, water-emulsion paints, corrosion inhibitors, detergent santitizers, etc.

Cationic Surfactants

Because these compounds derive their surface activity from the positively charged long-chain portion of the molecule, they are also known as invert soaps:

$$R_1NH^+{-}Br^-(\text{or } Cl^-) \qquad R_1COO^-{-}Na^+$$

cationic   anionic

These surfactants may be either nitriles, amines, amide-linked amines, or quaternary nitrogen bases such as

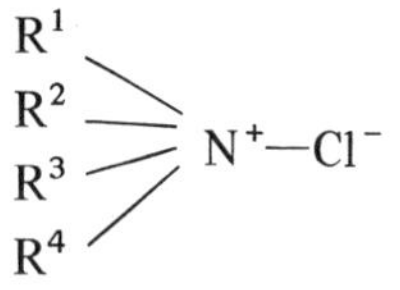

in which R^1 is a long-chain group, the other R groups may be alkyl chains or hydrogen, and the negative group may be a halogen or sulfate. Other quaternaries are exemplified by sulfonium and phosphonium compounds.[56]

Perhaps cationic surfactants were first used as softeners for the textile industry. Because these chemicals are substantive to, and exhaust onto, cotton fiber, imparting lubricity and softness, they are used to a considerable extent for this purpose. More recently they have been used as fabric softeners which the housewife may apply to her towels and garments following the usual laundering operation. Because they are antistatic agents they are used as finishes on synthetic fabrics which exhibit an undesirable static charge, and because of their bactericidal character they are also used as antibacterial and algicidal agents and detergent-sanitizers. Their manufacture has never been on the scale of fatty alcohol sulfates or alkylbenzene sulfonates; and while considerable tonnages are produced, the manufacturing procedures appear to be much less standardized than those for the anionic and nonionic surfactants. A factor which tends to minimize standardization is the apparent necessity for specialty compositions in which raw materials of rather diverse character are used.

Most of the information on the manufacture of these compounds is to be found in the patent literature, and excellent reviews of these have appeared.[43] Jungermann[24] has edited an extensive volume on cationic agents.

Detergent Builders and Additives

The dried, and even many of the liquid detergent formulations contain electrolyte salts which at first glance might be considered diluents. Examination of the voluminous literature shows, however, that these products have a definite function, and in fact make detergent products as effective as they are. For example, dilution of sodium dodecylbenzene sulfonate to a 40 per cent active ingredient (AI) level with sodium sulfate will yield somewhat superior wetting and detergent values over an equivalent weight of 100 per cent ingredient composition. Further, if the product containing 40 per cent AI and 60 per cent sodium sulfate is altered by substituting sodium tripolyphosphate for half the sodium sulfate, the detergent value of the composition is very markedly improved. Because of their effect, these inorganic salts are called detergent "builders."

Sodium Sulfate. Any surfactant prepared by sulfation or sulfonation can carry with it greater or lesser amounts of sodium sulfate, sodium sulfite, and/or sodium chloride; the latter two originate from sulfonation with either sulfur trioxide or chlorosulfonic acid. Many surfactant applications require a minimum of electrolyte, and for this purpose an alcohol purification may be used. A sulfur trioxide sulfonation may leave as little as 2–3 per cent electrolyte, while the other sulfonation procedures may leave as much as 60 per cent sodium sulfate. Oleum sulfonation may leave as little as 15 per cent sodium sulfate in the product by using proper spent acid layering and separation. A continuous sulfonation and neutralization procedure without spent acid separation may leave residual sodium sulfate of 20 to 40 per cent in the product. Since spent acid disposal may be a problem, and since sodium sulfate addition can improve surface activity when used in proper amounts, it frequently proves most efficacious simply to adjust the proportions to utilize all the sulfonation and neutralization products. In other instances it becomes desirable to add further quantities of sodium sulfate (salt cake), and in recent years the increased demand generated by this purpose has caused some scarcity and a price increase in this by-product raw material.

Sodium Silicate. Sodium and potassium silicates have long been used as soap and surfactant builders. Any one of several kinds may be used, varying from the semiliquid to liquid products of relatively low alkalinity to dry, highly alkaline powders. Several grades used in detergent products are the following:

	Ratio $Na_2O:SiO_2$
"N" Grade, liquid	1:3.3
1:2 Grade, liquid	1:2
Meta-, pentahydrate	1:1
Meta-, anhydrous	1:1
Sesqui	2:3
Ortho	2:1

The lower ratio Na_2O materials are used as buffering and deterging agents. Depending upon the composition for which they are used, they also exert corrosion inhibiting action. An example of their inhibiting action is found in their usage to prevent deleterious action on metals and the vitreous enamels of washing machines. Additionally, these materials exert functional effects in maintaining soil in suspension to ease removal and prevent redeposition.

Sodium Phosphates. The introduction of tetrasodium pyrophosphate ($Na_4P_2O_7$) in the mid 30's and early 40's improved the soap products then generally used, and made it possible for surfactants to compete quality-wise in soil removal for heavy-duty cleaning. The more recent product now almost universally used in this country is sodium tripolyphosphate (STP; $Na_5P_3O_{10}$). The polyphosphates sequester the water-hardening calcium and magnesium ions, soften water, help to adjust pH, and above all, sharply increase the soil removing and suspending characteristics of the surfactants with which they are used. In the spray-drying operation STP forms the hexahydrate, and this water of crystallization, because it cannot be reversibly removed, helps to stabilize the detergent moisture content. Spray-tower temperature-operating conditions are thereby imposed, however,[18] which demand that inversion to the ortho- and pyrophosphates does not occur.

It is estimated[30] that from 40 to 50 per cent of all phosphorus applications is accounted for by the sodium phosphates in detergents. A more precise estimate[41] states that the single largest use for phosphorus is in the production of STP (39.2 per cent) and tetrasodium pyrophosphate (6.6 per cent), totaling in excess of 1¾ billion pounds.

A full presentation of phosphates as detergents and syndet builders will be found in "Phosphorus and Its Compounds,"[59]

Brighteners

Another additive, whose importance is completely out of proportion to the amounts used in the formulation, is the group of colorless dyes known as brighteners; their function is to adsorb light on the washed garments and to reflect a visible blue color induced by the ultraviolet portion of the spectrum. The amounts in a formulation may vary from as little as a few hundredths of a per cent to 0.1 per cent. The visible improvement resulting from the use of the brighteners is so striking that all medium- to heavy-duty detergent products contain them.

Hydrotropes

While not necessarily builders in the usual sense, hydrotropes are almost a necessity for liquid detergents. Their function is to "drive" the surfactants and builders into solution, in effect a solubilizing action. Consequently, the built liquids remain clear and overcome freeze-thaw or high-temperature turbidity or sedimentation. The materials most widely used are short-chain compounds, i.e., methyl-, ethyl-, or propylbenzene sulfonates. These are produced by conventional sulfonation of toluene, xylene, or ethyl- or propylbenzenes.

Enzymes

A recent (1967) additive designed to reduce stains are enzymes. They were initially used in laundry products in 1913 in Germany, but the first really successful laundry product containing them was introduced in 1963 in Holland. A recent article[31] describes the use of proteolytic and amylolytic enzyme detergent additives. These are used in about 1000 to 1300 units (casein units per liter and alpha-amylase units per liter) in the wash. A typical alkaline protease may assay 330,000 units per gram.

Producers of enzymes for detergents number at least twelve, seven of which are foreign. An example is Mexatase, produced by the Royal Netherlands Fermentation Industries, Ltd., Delft, Netherlands, from a strain of *Bacillus subtilis.* The bacteria are cultivated in a protein-rich medium where the organisms are encouraged to attack protein and secrete proteolytic enzymes. The fermentation takes place in 21,000-gallon tanks until the enzyme reaches a certain maximum. The bacterial debris is filtered off and a clear solution containing enzymes and degradation products results. The liquor is concentrated in low-temperature evaporators to 1/10 its volume and the enzyme is then precipitated by adding a water-soluble organic solvent such as acetone or alcohol. After drying, the crude enzyme activity is standardized by dilution with sodium sulfate. Quality control is assured by decolorization with activated carbon, then homogenization in a conical solids blender; and sterility

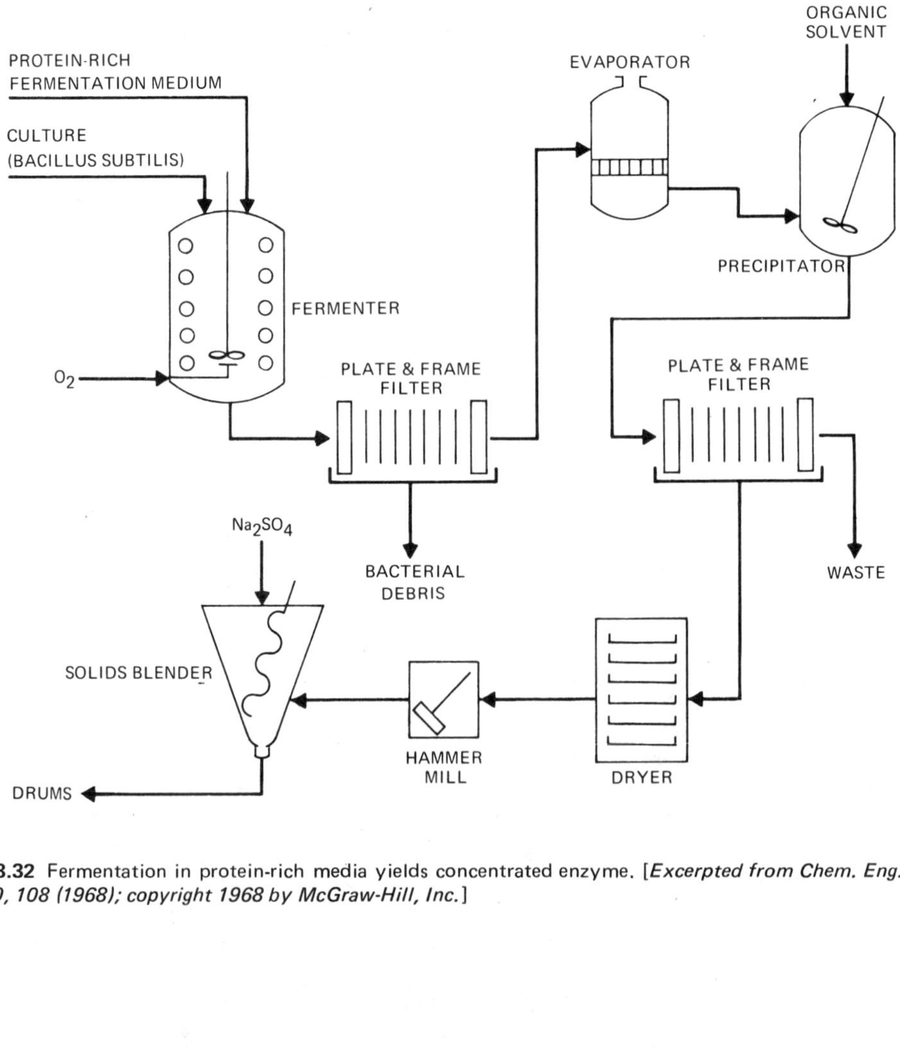

Fig. 13.32 Fermentation in protein-rich media yields concentrated enzyme. [*Excerpted from Chem. Eng.,* **75,** *No. 20, 108 (1968); copyright 1968 by McGraw-Hill, Inc.*]

is assured by using very fine filter cloths for the clear solution. Figure 13.32 outlines this process.

Opacifying Agents

While some liquid syndets can be formulated as clear, stable liquids, others may prove virtually impossible to clarify, or it may be desired that a cream-like liquid be prepared. In either case, it is possible to prepare shelf-stable creamy liquids by adding water-soluble polymeric compounds. An example of a compound which opacifies and increases viscosity is the ammonium or other water-soluble salt of the hydrolyzed copolymer of styrene and maleic anhydride. The literature is replete with other examples.

Sodium Carboxymethyl Cellulose (NaCMC)

In washing operations soil is disengaged from the surface being cleaned, but if either the soil load is too great or the rinsing operation dilutes the syndet too extensively, soil redeposition may occur. To prevent soil redeposition, NaCMC is the most extensively used of several candidates. It is most effective with cotton; but is less so with some synthetic fibers which have peculiar hydrophobic properties. Several possible modes of action for these agents have been advanced. The lock-and-key mechanism has been considered as a reasonable answer, while other investigators have shown that complexes with soil and dyes are formed which do not redeposit. Polymers used for preventing redeposition of soil on synthetic fibers are poly(vinyl alcohol) and poly(vinylpyrrolidone). NaCMC is used in formulations to the extent of fractions to about 2 per cent. It appears to help syndet formulations more than soap which seems to have a greater degree of built-in ability to reduce redeposition.

ECOLOGICAL EFFECTS

Very recently control of the environment and the effects of chemicals on the ecology have become prominent and well-publicized subjects. As far as the detergent industry is concerned, the control of foam and syndets in water systems as an indication of pollution and its effect on the ecology was the first step in adequate control.

Biodegradable Detergents

Soap use is historical and it is exceptional to find reference to water contaminated by it. The fact that water hardness reduces soap to a nonfoaming state is one reason why it has not been considered a major pollutant, and the further fact that it is almost without exception a straight-chain compound is another more pertinent reason. The latter characteristic is important to river water and sewage disposal as was evident when replacement of soap by syndets became more universal. Sewage plants began to display beds of foam at their effluent ends, some of them covering entire streams for several hundred feet. Then too, foaming of drinking water from ground-water supplies where septic fields were in use was a further indication of pollution, both by syndets and viable organisms. Presence of foam was a sure indication that biodegradation had not occurred.

It became evident in 1963 that something must be done about "hard" syndets. "Hard" because they were not straight chain compounds and resisted sewage treatment and river-water degradation. "Soft" or easily degraded surfactants were to become a necessity, particularly when Congress threatened legislation to ban "hard" syndets, a route for control that has been followed for some years in Germany. The industry in 1963 promised to develop "soft" syndets and had them on sale by the end of 1965. The surfactant chosen for study because of its wide usage was alkylbenzene sulfonate. The alkyl portion of the compound was the focus of attention: its source may be chlorinated kerosene or a suitable olefin. Extensive research[50] demonstrated that a tertiary carbon atom in the alkyl chain was the point at which biodegradation ceased. The organisms were unable to utilize the chain any further when such an atom was encountered, whereas with straight chain compounds such as soap no difficulty was found. The answer to this was a straight chain kerosene or olefin, and suitable synthesis control to prevent chain rearrangement, or more suitable use of a preferred straight-chain olefin in the synthesis.

The straight-chain product finally used was termed LAS, linear alkylbenzene sulfonate (as

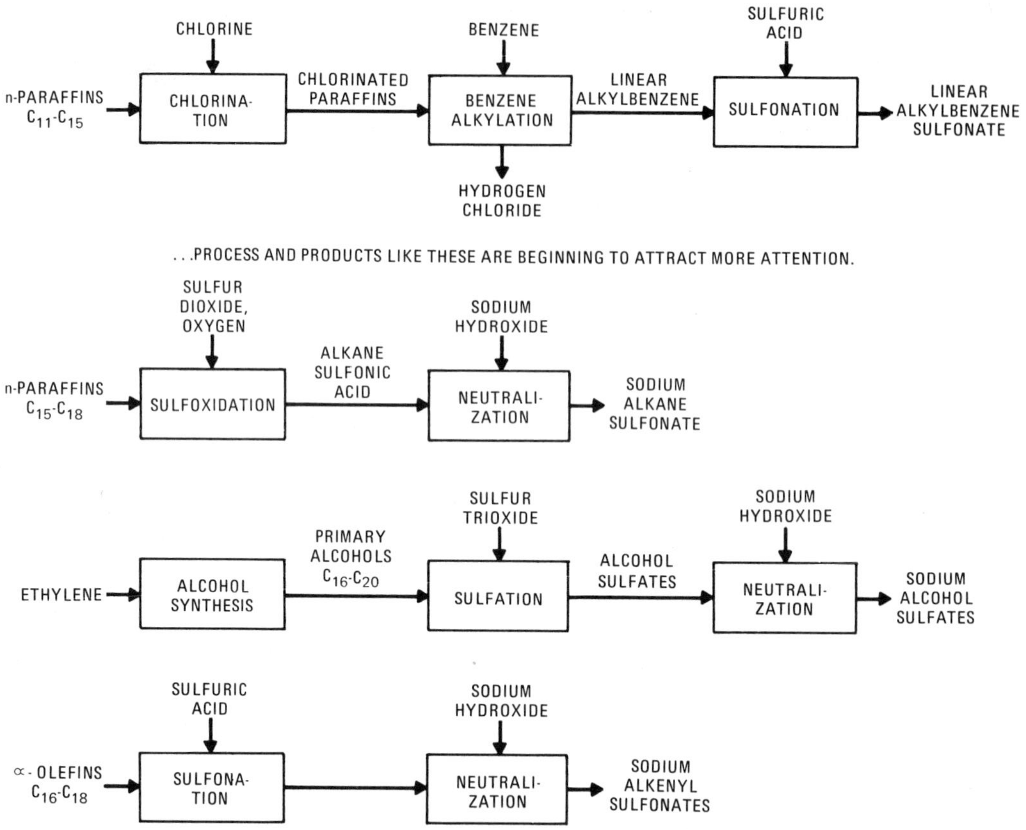

Fig. 13.33 Flow diagrams for manufacture of some biodegradable surface active agents. [*Excerpted by special permission from Chem. Eng.,* **74**, *No. 8, 108 (1967); copyright 1967 by McGraw-Hill, Inc.*]

contrasted with branched chain ABS). The manufacturers by a mass effort were successful, at considerable cost, in meeting the 1965 target date. Before LAS, sewage treatment plants were successful in degrading ABS only to 60 to 70 per cent; afterward a 90 per cent effectiveness was achieved. Today any surfactant to be used successfully must be biodegradable, as suggested by Fig. 13.33. A third generation of agents is beginning to draw attention, and includes nonionics and nonionic sulfates.

Detergent Additives

Phosphorus Compounds. Immediately following the development of biodegradable syndets to control pollution by surfactants, the overall pollution of rivers and lakes became the next focus of attention. Lakes in particular, as recipients of inflow from agricultural lands and either the effluent from inadequate sewage treatment plants or raw sewage were becoming so tainted that in effect it might be said the lakes were dying. The apparent effect was a change in the aquatic life with more frequent and extensive algal blooms. These blooms reduced the oxygen content of the water, killing much of the aquatic life, and in effect stagnating the water body by depositing organic debris which consumed still more oxygen. Reasons for this change were attributed principally to high phosphorus contributions from run-off and sewage effluent, and secondarily to nitrogen and carbon dioxide.

Ordinarily this process, called eutrophication, would take thousands of years, but it has been greatly accelerated by overnourishment.

Since the phosphorus content of sewage is not greatly reduced by treatment, syndets which contain from 5 to 50 per cent phosphates became the immediate target for control. The pros and cons continue,[62] while Congressional and state actions would ban phosphates in detergents. The problem of phosphate control is being attacked through more effective removal during sewage treatment and by substitution of agents capable of replacing phosphorus as a builder.

Phosphorus Substitutes. The usual builders which might partially substitute for phosphates are sodium or other water-soluble carbonates, and sodium or potassium silicates. Both are more effective in improving syndet soil removal than are essentially inert electrolytes such as sodium sulfate or sodium chloride. For liquid detergents perhaps a greater degree of dilution is possible, i.e., more water to replace phosphates, but this is not a good solution to the problem.

What is needed is a product which is completely biodegradable, is not a prime nutrient either as used or when degraded, and has the properties contributed by phosphates. That is, the replacement should soften water, preferably by sequestration, and be as effective in promoting detergency as are the polyphosphates. It should be noted here that both soap and surfactants require builders for effective detergency, and that reversion to soap with the aforementioned remaining builders will result in soil removal levels which are no longer considered satisfactory.

A variety of polymeric products have been tested[44] as sequestrants or chelating agents, but none has been found to be as effective as the original agents. Possibly the first compound to be fully investigated for detergency was ethylenediaminetetracetic acid, EDTA:

$$(HOOCCH_2)_2N-C_2H_4-N(CH_2COOH)_2$$

Each of the four carboxy groups is capable of reacting with metallic ions to form the tetrasodium salts, for example, or to sequester or chelate two calcium ions, forming soluble rather than insoluble salts. This compound and its variants have been extensively investigated as substitutes for STP. Although it is effective, its cost is excessive.

Chelating agents, organic phosphonates, polyelectrolytes, and sequestering agents were reviewed in some detail by McCutcheon.[35]

Another compound which received attention as a substitute for phosphates is also a chelating or sequestering agent. Nitrolotriacetate, NTA, when used in adequate amounts proved essentially as effective as EDTA but at lower

$$N(CH_2COOH)_3$$

cost. NTA was the compound which the detergent manufacturers started to use to replace STP. Its use was terminated, however, when it was disclosed that birth defects in rats and mice occurred when they ingested NTA complexed cadmium or methyl mercury, chemicals which are potentially present in ground waters. This left no suitable candidate for replacement, and for the time being the only replacements are the conventional alkaline builders and water.

REFERENCES

1. Allen, H. D., Kline, W. A., Lawrence, E. A., Arrowsmith, C. J., and Marsel, C., *Chem. Eng. Progr.*, **43**, 459 (1947).
2. Andersen, D. L., *J. Am. Oil Chemists' Soc.*, **34**, 188 (1957).
3. Association of American Soap and Glycerine Producers, Inc., May 1957.
4. Ballard, S. A., Finch, H. de V., and Peterson, E. A., British Patent 619,014 (Mar. 2, 1949; to N. V. Bataafsche Petroleum Maatschappij).
5. Berglron, J., *Rev. Franc. Corps Gras* (1958).
6. Buckham, J. A., and Moulton, R. W., *Chem. Eng. Progr.*, **51**, No. 3, 126 (1955).

7. Carlson, E. J., Flint, G., and Gilbert, E. E., *Ind. Eng. Chem.*, **50**, 276 (1958).
8. Chaland, J. H., Martin, J. B., and Baker, J. S., *Eng. Progr.*, **53**, No. 12, 593 (1957).
9. *Chem. Eng.*, **61**, 372 (1954).
10. *Chem. Eng. News*, **34**, 5752 (1956).
11. *Chem. Eng. Progr.*, **44**, No. 10, 16 (1948).
12. *Chem. Industries*, 247 (1949).
13. *Chem. Week*, p. 90 (Oct. 8, 1955).
14. Fairbairn, A. W., Cheney, H. A., and Cherniavsky, J. J., *Chem. Eng. Progr.*, **43**, No. 6, 280 (1947).
15. Fedor, W. S., Strain, B., Theoharous, L., and Whyte, D. D., *Ind. Eng. Chem.*, **51**, 13 (1959).
16. Forbath, T. P., *Chem. Eng.*, **64**, 226 (1959).
17. Garrison, L. J., *Detergents and Specialties*, p. 33 (May 1970).
18. Gilbert, E. E., and Veldhuis, B., *Ind. Eng. Chem.*, **50**, 997 (1958).
19. Hafford, B. C., *Ind. Eng. Chem.*, **46**, 1938 (1954).
20. Hatcher, D. B., *J. Am. Oil Chemists' Soc.*, **34**, 175 (1957).
21. Hill, E. F., Wilson, G. R., and Steinle, E. C., Jr., *Ind. Eng. Chem.*, **46**, 1917 (1954).
22. *Ind. Eng. Chem.*, **39**, 126 (1947).
23. Jelinek, C. F., and Mayhew, R. L., *Ind. Eng. Chem.*, **46**, 1930 (1954).
24. Jungermann, E., "Cationic Surfactants," in "Surfactant Science Series," Vol. 4, New York, Marcel Dekker, 1970.
25. Kastens, M. L., and Ayo, J. J., Jr., *Ind. Eng. Chem.*, **42**, 1626 (1950).
26. Kircheo, J. E., Miller, E. L., and Geiser, P. E., *Ind. Eng. Chem.*, **46**, 1925 (1954).
27. Kritchevsky, J., *J. Am. Oil Chemists' Soc.*, **34**, 178 (1957).
28. Kroll, H., and Nadeau, H., *J. Am. Oil Chemists' Soc.*, **34**, 323 (1957).
29. Lamont, D. R., U.S. Patent 1,652,900 (Dec. 13, 1927).
30. Latourette, W. I., *Barron's*, p. 11 (Sept. 8, 1958).
31. Liss, R. L., and Langguth, R. P., *J. Am. Oil Chemists' Soc.*, **46**, 507 (1969).
32. Marshall, W. B., Jr., and Seltzer, E., *Chem. Eng. Progr.*, **46**, No. 10, 501; No. 11, 575 (1950).
33. Mazzoni, G., S.p.A., Busto Arsizio, Italy, *Soap*, **34**, No. 3, 185 (1958).
34. McBride, G. W., *Chem. Eng.*, No. 8, 94 (Apr. 1947).
35. McCutcheon, J. W., "Detergents and Emulsifiers," 1969 and 1970 Annuals, Oak Park, Ill., Allured Pub. Co.
36. *Oil, Paint Drug Reptr.*, p. 3 (Oct. 5, 1950).
37. Osipow, L., Snell, F. D., and Finchler, A., *J. Am. Oil Chemists' Soc.*, **34**, 185 (1957).
38. Palmquist, F. T. E., and Sullivan, F. E., *J. Am. Oil Chemists' Soc.*, **36**, 173 (1959).
39. *Petrol. Refiner*, **36**, No. 11, 250 (1957).
40. Pollak, A., *Tappi*, **39**, No. 1, 60A (1956).
41. Riley, W. J., *Chem. Eng. News*, **34**, 5312 (Oct. 29, 1956).
42. Schick, M. J., "Nonionic Surfactants," in "Surfactant Science Series," Vol. 1, New York, Marcel Dekker, 1967.
43. Schwartz, A. M., and Perry, J. W., "Surface Active Agents," Vol. 1, New York, John Wiley & Sons, 1949.
44. Schwartz, A. M., Perry, J. W., and Berch, J., "Surface Active Agents and Detergents," Vol. II, New York, John Wiley & Sons, 1958.
45. Sherwood, P. W., *Can. Chem. Process Inds.*, **32**, 1102 (1948).
46. Sittenfield, M., *Chem. Eng.*, **55**, 120 (1948).
47. *Soap*, **35**, No. 4, 131 (1959).
48. Stanton, W. B., *Soap*, **33**, No. 6, 47 (1957).
49. Stromquist, D. M., and Reents, A. C., *Ind. Eng. Chem.*, **43**, 1065 (1951).
50. Swisher, R. D., "Surfactant Biodegradation," in "Surfactant Science Series," Vol. 3, New York, Marcel Dekker, 1970.
51. U.S. Patent 601,603 (1897).
52. U.S. Patent 1,932,180.
53. U.S. Patent 2,089,212.
54. U.S. Patent 2,154,835.
55. U.S. Patent 2,605,293.

56. U.S. Patents 2,674,619 and 2,677,700; and British Patent 722,746.
57. U.S. Patent 2,689,170.
58. U.S. Patent 2,858,345.
59. Van Wazer, "Phosphorus and its Compounds," Vol. II, New York, John Wiley & Sons, 1961.
60. Wetherhorn, D., *Tappi*, **40**, 879 (1957).
61. Wurster, O. C., "Glycerin Refining Plants," Chicago, Wurster and Sanger.
62. Yulish, J., *Chem. Eng.*, **77**, No. 12, 70 (1970).

14

Petroleum and Its Products: Petrochemicals

A. F. Galli*

PETROLEUM AND ITS PRODUCTS

When first obtained from the ground, before refining in any way, petroleum (rock oil) is called "crude oil." It occasionally appears at the surface of the earth through seepage; it usually occurs at moderate depths; and in some cases it must be sought by drill holes over a mile deep. When such a drill hole reaches an oil basin, the oil is frequently forced out under enormous pressures; gas, salty water, and sand usually accompany the oil. After a period which varies considerably, the flow becomes quieter; after some months it does not gush at all and the oil must be pumped out; finally, no oil is obtained even by pumping–the well is dry. New wells are therefore constantly being sought. The oil prospector chooses lands possessing a subsoil which has characteristics indicating petroliferous strata; these characteristics vary in different fields, and in no case is it beyond doubt that a drill hole will reach oil. The search for oil is supplemented by accidental discoveries, in the course of drilling for water, for example. Where natural gas occurs, it is reasonable to prospect for oil;‡ it is by no means certain that oil will be found, but since petroleum consists of a mixture of hydrocarbons, the lighter ones such as methane, CH_4, and ethane, C_2H_6, may have escaped, in part, leaving the main body of liquids and solids not very far away. The heaviest hydrocarbons, beginning, for example, with eicosane, $C_{20}H_{42}$, which melts at blood temperature, are solid; the intermediate ones are liquid.

Discovery of Petroleum

It was not the genius of man which discovered petroleum; the presence of petroleum was indicated by seepages which frequently coated small rivers, as in Pennsylvania, or contaminated brines, much to the disgust of the early

*College of Engineering, West Virginia University, Morgantown, W.Va.

†The Rodessa field, Caddo Parish, in Louisiana, was a gas field for many years; only in 1935 was it discovered to carry oil.

(1806) salt refiners, like the brothers Ruffner along the Kanawha River in West Virginia; or by escaping gas, rich in vapors and known for centuries, as in the Baku Peninsula, where the "eternal" fire gave powerful support to religious cults (Zoroasters).[32] Only in more recent times has a more intensive search taken place, partly because the evident clues had been exhausted, and partly because petroleum has become a necessity to national as well as to civil life. To the study of the soil for indication of petroliferous strata, scientific methods were added: the revival of the anticlinal theory, and the measurement of temperature in adjacent wells in order to locate the anticlinal axis; gravitometry, including seismic or sonic apparatus; the electric log; and the magnetometer.[33]

Petroleum is found only in sedimentary rocks, and has its origin in the petrified remains of plants and animals,† which accumulated with clays and muds along the seashore. At a later geologic period, the droplets of gas migrated from the source muds into a "reservoir" rock, chiefly sandstone. The strata were lifted and warped, forming arches (anticlines) and throughs (synclines), from one mile to several miles in width. Oil is found in the anticlines, having risen above the water; it may also be found in synclines, if the upper space (to the neighboring anticlines) is filled with the still lighter gas, also formed from these deposits, and tapped as natural gas. An oil or gas pool is not a subterranean pond, but merely an accumulation of petroleum or gas in rock pores. The container rock is of a highly porous type‡ in which the interstices are such as to permit the oil, gas, and water to move freely; the dense or nonporous retainer rock prevents the escape of the oil or gas from the container rock. The petroleum generally contains large amounts of gas in solution, under pressure.

It was in 1859, at Titusville, Pa., that Colonel Drake sank his now famous well; it was 59½ feet deep. Since then, many wells have been driven, and the depth has steadily increased. In 1968, a total of 30,621 wells were drilled in the United States, of which 14,342 were oil wells, 3,455 were gas wells, and 12,824 dry holes (Minerals Yearbook).

The producing oil wells in the United States on December 31, 1968, totaled 553,920, and the average production over the year per well per day was about 16.2 barrels. The cost of a 10,000-foot well is $350,000 or more. Depth drilling has been helped by the development of rotary drilling; several new chemical agents, the application of acids, and the use of special muds have also contributed to this development.

When first tapped, the well may be a gusher, with many thousands of barrels a day; this output begins to decrease at once, and gradually tapers off until a pump is needed to bring up the oil. A life of 20 years is a rough estimate for a newly tapped oil-bearing formation.

Petroleum Reserves

Taking petroleum from the ground means subtracting it from a finite store: less oil is left for the next generation. In past years the rate of discovery of new oil-bearing formations has been larger than the rate of oil removal so that the petroleum reserves in the United States have increased. Reserves are ample at present but small compared to the coal reserves. Thus there is concern over the diminishing rate of

†There have been many theories for the formation of petroleum deposits in the ground. Among the earlier ones was that of Engler who referred the formation of petroleum to a store of fatty remains of all kinds of life, but especially animal life; his opinion was based on laboratory experiments with menhaden oil, from which he produced a petroleum-like substance. C. Engler, *Ber. Deutschen Gesellschaft*, **21**, 1816 (1888). Among more recent workers, Alfred Treibs assigned to plants the dominant role in forming petroleum, to animal remains a secondary role. His statement is based on the spectroscopic study of the light absorption of petroleums in which he thus identifies and estimate quantitatively porphyrins of chlorophyll origin in a high percentage, while he also found porphyrins of hemoglobin origin, but in much smaller amounts, *Ann.*, **510**, 42 (1934). A brief summary of the older theories will be found in "The examination of hydrocarbon oils and saponifiable waxes," D. Holde, tr. by Edward Mueller, New York, John Wiley and Sons, 1922.

‡The porosity of a rock is defined as the property of containing interstices or openings, which may vary in size from caverns to subcapillary pores.

TABLE 14.1. Consumption, Production and Imports of Crude Oil in the U.S.[a] (Millions of Barrels Per Year)

	1964	*1966*	*1968*
Total consumption	3,959	4,325	4,788
Domestic production	2,787	3,028	3,329
Imports	1,172	1,297	1,459

[a]"Mineral Yearbook, 1968."

TABLE 14.2 Leading Crude Oil Producing States[a] (Millions of Barrels Per Year)

	1964	*1966*	*1968*	*1970*
Texas	989	1,060	1,135	1,151
Louisiana	549	677	819	845
California	301	346	373	375
Oklahoma	203	225	223	225
Wyoming	139	134	143	160
New Mexico	114	125	127	128
Kansas	106	103	97	85
Alaska	11	15	67	74

[a]"Mineral Yearbook, 1970."

new discoveries; the proved reserves—that is, known reserves in well established formations—may diminish, while on the other hand, the amount of oil removed continues to increase year by year. Another consideration is that of exports and imports. In the past, imports were less than exports of crude oil, but now they are much greater. It might be noted also that finding oil and drilling wells is far more costly than it was a decade ago. The military want to feel that there is a safe margin of oil supplies against a sudden emergency. The results are (1) the development and constant improvement of means for conserving oil, wasting none, and leaving the least possible amount unextracted in the formation; (2) supplementing these means to conserve by developing other sources of motor fuel, such as the distillation of oil shales; and (3) the manufacture of synthetic liquid and gaseous fuels from coal.

Means of conserving oil have commanded increased attention and study. One is to reduce or abolish waste caused by too rapid removal of oil from a formation which is accompanied by excessive lowering of the potential (pressure) of the well. Laws have been passed which regulate the rate of oil removal from wells and oil pools to conserve well pressure.

Another is the more complete extraction of oil from the oil sands. It should be explained that rarely is as much as 50 per cent of the total oil in a formation extracted; more often, it is 30 per cent or less. In order to recover the residual oil, the formation may be flooded with water, which drives the oil toward the well in which a pump may be working. For this method to be successful, the ground formations must be favorable. Much secondary oil is recovered also by pumping gas into the formation (repressuring). Flooding with water has given good results in some areas. Other methods are (1) fracturing oil bearing formations with explosives or hydraulic pressure, and (2) acid treatment to increase oil flow from wells (acidizing).

Another method of increasing the oil reserves is to utilize the oil bearing shales which occur in large amounts in Colorado, Utah, and Wyoming. High-yield processes have been developed but are not presently competitive with petroleum from conventional sources.

Synthetic petroleum very similar to the natural material can be produced from coal and other carbon containing materials. The cost of the synthetic product is considerably higher than it is for natural petroleum.

The United States continued to lead the world in production of crude oil in 1968 with a total of 3,329,042 barrels during that year. This represents about 24 per cent of the total world production. Table 14.4 shows the production of crude oil throughout the world.

TABLE 14.3 Proven Petroleum Reserves in the United States[a] (Millions of Barrels)

	1960	*1962*	*1964*	*1966*	*1968*	*1970*
Estimated reserves	31,613	31,389	30,991	31,452	30,707	39,001

[a]"Mineral Yearbook, 1970.

TABLE 14.4 World Production of Crude Petroleum[a] (Millions of Barrels)

	1964	*1966*	*1968*	*1970*
North America	3,227	3,526	3,917	4,187
South America	1,481	1,493	1,626	1,686
Europe	1,902	2,213	2,522	2,843
Asia	3,076	3,758	4,562	5,365
Africa	624	1,027	1,442	2,206
Oceania	1	3	15	65
TOTAL	10,311	12,020	14,084	16,689

[a]"Mineral Yearbook, 1970."

Nature of Petroleum

As is generally the case with naturally occurring materials, petroleum is complex and variable in chemical composition. It varies in color from light greenish brown to black, and it may be of low viscosity or so viscous as to be nearly immobile. The crude oil is usually dark colored, of low viscosity, and contains gases and solids, either dissolved or dispersed. Much of the gas separates when the oil reaches the surface, but it may also do so underground and be found as natural gas some distance from the oil.

The principal constituents of petroleum are the hydrocarbons; small amounts of combined sulfur, nitrogen, and oxygen are present as impurities. It is estimated that several thousand compounds are present in petroleum but relatively few have been positively identified. They vary in composition from one to more than 200 carbon atoms; those containing up to four are usually gases, five to sixteen, liquids, and those containing seventeen or more are usually solids.

The classes of compounds occurring in petroleum are as follows:

1. *Paraffin hydrocarbons*, general formula C_nH_{2n+2}. They have straight chains (normal paraffins) or branched chains (isoparaffins). These compounds may be gaseous, liquid, or solid (waxes), depending on their structure and molecular weight.
2. *Naphthene hydrocarbons*, general formula C_nH_{2n}. These are saturated hydrocarbons possessing a ring structure usually containing five to seven carbon atoms in the ring.
3. *Aromatic hydrocarbons*, formula C_nH_{2n-6}. The compounds are characterized by a six-carbon ring.
4. *Multiring cyclic hydrocarbons*. These are naphthene and aromatic compounds containing more than one ring in their structure.
5. *Olefin hydrocarbons*, formula C_nH_{2n} for monoolefins (can add one molecule of hydrogen), and C_nH_{2n-2} for di-olefins (can add two molecules of hydrogen). Since these compounds are very reactive they are found only in trace amounts, although large quantities may be formed during cracking.
6. *Sulfur compounds*. Sulfur is generally found in petroleum in the combined form in amounts up to 6 per cent. Hydrogen sulfide and thiophenes are the most usual forms but it may also be present as mercaptans, sulfides, and other compounds.
7. *Oxygen compounds*. Oxygen occurs in combined form in alcohols, phenols, resins, and organic acids.
8. *Nitrogen compounds*. These include pyridines, quinolines, indoles, pyrroles, and others.
9. *Inorganic compounds*. These include the salts (from salt water), and the clay, sand, and similar compounds which associate with the oil during its passage from the oil-bearing strata.

Since most of the compounds are hydrocarbons having a fairly constant hydrogen to carbon ratio, the ultimate analysis of all crudes falls within a narrow range.

	Per cent		*Per cent*
Carbon	83–87	Nitrogen	0.1–1.5
Hydrogen	11–15	Oxygen	0.3–1.2
Sulfur	0.1–6		

TABLE 14.5. Classification in 7 Classes of Petroleum Crudes, According to Their "Base," from the Distillation-analysis of 800 Samples from all Over the World[a]

	A *Paraffin Base Oil (Wax-Bearing)*	*B* *Paraffin Intermediate Base Oil (Wax-Bearing)*	*C* *Intermediate Paraffin Base Oil (Wax-Bearing)*	*D* *Intermediate Base Oil (Wax-Bearing)*	*E* *Intermediate Naphthene Base Oil (Wax-Bearing)*	*F* *Naphthene Intermediate Base Oil (Wax-Bearing)*	*G* *Naphthene Base Oil (Wax-Free)*
A.P.I. gravity	49.7°	39.2°	29.5°	39.6°	15.3°	29.5°	24.0°
Specific gravity	.781	.829	.879	.827	.964	.879	.910
Pour point	below 5°F	below 5°F	40°F	below 5°F	40°F	below 5°F	below 5°F
Per cent sulfur	0.1	0.28	0.32	0.33	3.84	0.16	0.14
Saybolt Universal viscosity 100°F	34 seconds	41	120	39	4000	47	55
Color	Green	Greenish black	Greenish black	Green	Brownish black	Greenish black	Green
Distillation 1st drop	93°F (34°C)	91°F (33°C)	176°F (80°C)	84°F (29°C)	280°F (138°C)	138°F (59°C)	315°F (157°C)
Distillates:							
gasoline and naphtha	45.2%	32.0	5.8	38.6	2.9	21.3	1.1
Kerosene	17.7%	17.2	nil	4.9	4.5	nil	nil
Gas oil	8.3%	10.6	27.8	17.3	10.6	34.6	55.5
Nonviscous lubricating	9.8%	10.9	20.4	9.4	8.6	10.4	14.2
Medium lubricating	3.4%	5.2	9.2	6.3	6.7	7.0	4.7
Viscous lubricating	nil	nil	nil	nil	1.020	4.7	11.6
Residuum	14.7%	23.5	36.4	22.1	58.4	21.4	12.7
Distillation loss	.9%	.6	.4	1.4	1.9	0.6	0.2
Carbon residue of residuum	1.1%	6.2	6.9	7.3	18.2	8.7	4.5
Carbon residue of crude	0.2%	1.5	2.5	1.6	10.6	1.9	0.6
Key fraction No. 1. (250–575°C) 482–527°F, 750 mm pressure							
Per cent cut	6.8%	6.5	7.1	5.8	5.1	10.1	19.6
A.P.I. of cut[b]	44.7°	40.6	36.4	37.0	37.0	30.2	27.9
Key fraction No. 2. (275–300°C) 527–572°F, 40 mm pressure							
Per cent cut	4.4	5.7	9.0	4.9	8.2	6.0	7.3
A.P.I. of cut[c]	34.4°	29.3	30.0	24.9	19.5	24.0	16.5
Viscosity at 100°F	110 seconds	120	120	165	240	230	over 400
Cloud test in °F	90	90	90	80	70	90	below 5

[a]R.I. 3279 U.S. Bureau of Mines, E. C. Lane and E. L. Garton (1935); condensed.

[b]If key fraction No. 1 reads 40.0° A.P.I. or lighter, the lower boiling fractions of the oil are paraffinic; if it reads 33.0° A.P.I. or heavier, they are naphthenic; if its gravity lies between 33.0° and 40.0° A.P.I., they are intermediate.

[c]If the gravity of key fraction No. 2 is 30.0° A.P.I. or lighter, the higher-boiling fractions of the oil are paraffinic; if it is 20.0° A.P.I. or heavier, the fractions are naphthenic; while if the gravity lies between 20° and 30° A.P.I., the fractions are intermediate.

Classification of Petroleum

An eminently satisfactory classification of crudes has been developed by the U.S. Bureau of Mines; the main features are given in Table 14.5. Class A comprises the paraffin-base crudes, which are wax bearing; these contain mainly paraffinic hydrocarbons in all their fractions; their residue in the still becomes the much sought after "cylinder stock." Straight-distilled gasolines from this crude would be paraffinic, and have "knocking" properties. Naphthene base oil, class G, contains mainly naphthenes, that is, cyclic compounds which are saturated, with both naphthenic and paraffinic side chains. Naphthene base and intermediate naphthene base, class E, may contain much black, brittle, almost infusible asphaltic material, although they often do not. The crudes differ also in the amounts of lower- and higher-boiling compounds present. This is shown in Table 14.5.

The petroleum refiner classifies crudes according to their "base" as follows: (1) paraffin-base crudes, high in wax and lube oil fractions, containing small amounts of naphthenes or asphalt and low in sulfur, nitrogen, and oxygen compounds; (2) asphalt-base crudes, which give high yields of pitch, asphalt, and heavy fuel oil; (3) mixed-base crudes, which have characteristics about midway between those of the paraffin- and asphalt-base crudes; and (4) aromatic-base crudes which contain large amounts of low-molecular weight aromatic compounds and naphthene, together with smaller amounts of asphalt and lube oils.

Petroleum is separated into fractions by distillation. The composition of each fraction is related to its boiling range, and no pure compounds are obtained. The fractions may be classified as shown in Table 14.6. The fractions require further treatment, including removal of impurities, blending, addition of certain chemicals (additives) to improve certain characteristics, and additional chemical processing (cracking, reforming) to increase the yield of desired products.

Transportation of Crude

Petroleum refineries received 75.8 per cent of their crude oil supply by pipeline, 23.0 by water, and 1.2 per cent by tank cars and tank trucks in 1968. Approximately 210,000 miles of pipelines transporting crude oil and refined products were in use in the United States in 1968. Although the greater part of crude oil is transported by pipeline, more than half of the refined petroleum products are transported by means other than pipeline.

TABLE 14.6 Fractions from Petroleum Distillation

Product	*Approximate Boiling Range, °F*
Natural gasoline	30–180
Light distillates	
Gasoline	80–380
Naphthas	200–450
Jet fuels	180–450
Kerosene	350–550
Light heating oils	400–600
Intermediate distillates	
Gas oils	480–750
Diesel oils	380–650
Heavy fuel oil	550–800
Heavy distillates	
Lubricating oils	600–1000
Waxes	Above 625
Residues	
Lubricating oils	Above 900
Asphalt	Above 900
Residium	Above 900
Petroleum coke	

PETROLEUM REFINING

Crude oil usually requires more than one operation for the production of finished products. Thus, a refinery consists of a number of individual processing units carefully designed and operated to produce competitively the products for a market which may vary from week to week.

Crude oil contains a number of inorganic impurities which are detrimental to the operation of the refinery units. For example, chlorides which may react with water to product corrosive hydrochloric acid, while sand and other suspended matter will cause plugging of trays in the distillation equipment. Water itself causes trouble during distillation and must be removed from the crude. Salt can cause fouling of heat exchangers,

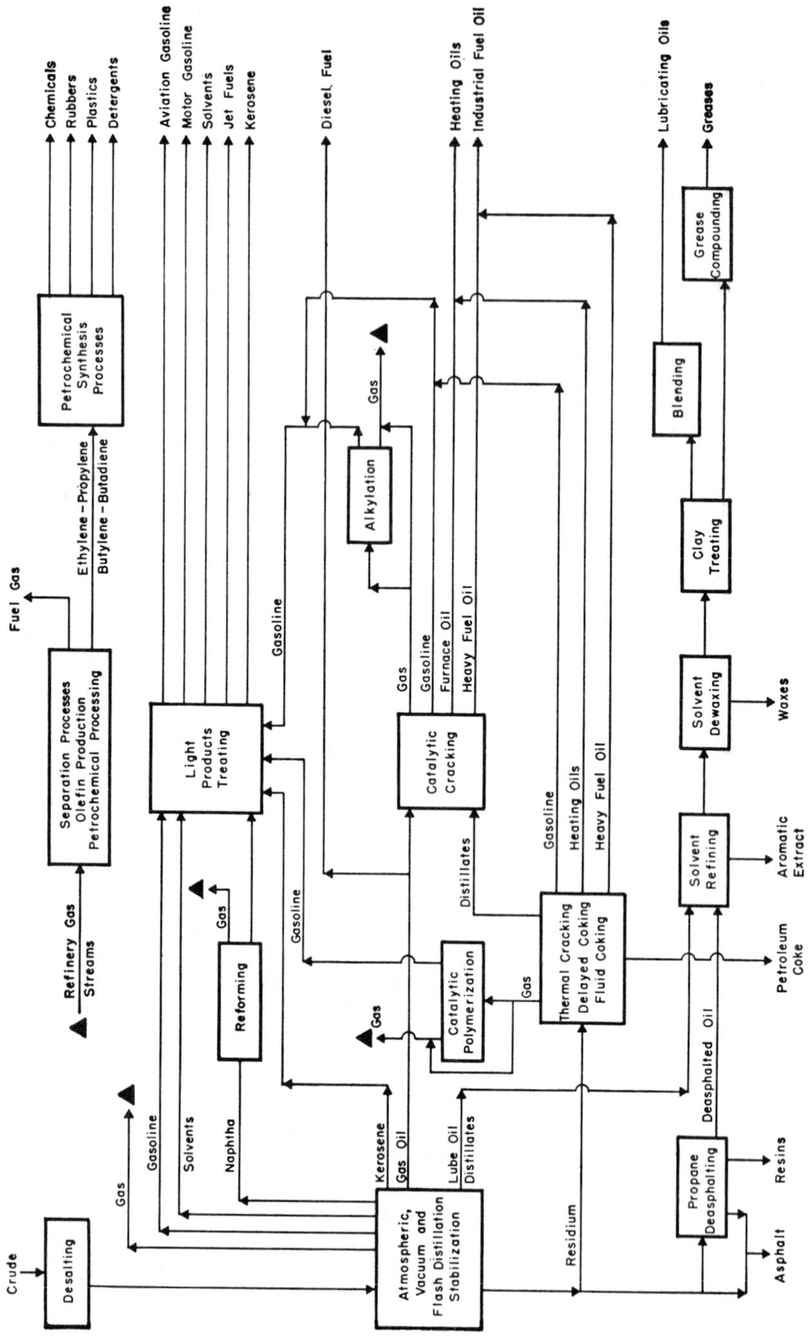

Fig. 14.1 Integrated flow sheet for petroleum refining. [*Adapted from "Petroleum Horizons," The Lummus Co. (1950).*]

while other impurities may poison the catalyst used in cracking or reforming operations.

Desalting

Chemical desalting of crude oil is accomplished by adding water in the amount of 6 to 15 per cent of the oil rate to the heated (200 to 300°F) oil under pressure sufficient to prevent vaporization. The mixture is emulsified, and the salt enters the water phase. Chemical additives may be used to break the emulsion, allowing the water phase to settle out. The water containing the salt is discharged from the system.

Electrical desalting[20] involves the addition of 4 to 10 per cent water under pressure at 160 to 300°F, emulsification in a mixture, and introduction of the emulsion into a high-potential electrostatic field. The field causes the impurities to associate with the water phase and at the same time causes the water phase to agglomerate so that it can be removed. The desalted crude continues to the distillation units.

Distillation

The various products obtained from distillation of the crude are given in Table 14.6. The number of fractions or cuts obtained depends on the crude base and operation conditions. Crude distillation systems can generally be classified into three types, viz., single stage, two-stage systems, and two stage with a vacuum tower.

Single Stage. A single-stage crude unit is shown in Fig. 14.2. The crude feed is preheated by the outgoing streams and then enters a direct-fired furnace-type heater (pipe still). The materials separate in the distillation column according to their boiling points, the lowest-boiling fraction leaving at the top of the column. Desired products may be withdrawn as side-streams at appropriate points on the column. The side-streams are further fractionated in small columns called strippers. Here, steam is used to free the cut from its more volatile components so that the initial boiling point of the product can be adjusted to its desired value.

Two-Stage Systems. The complexities of modern refinery operation frequently require the use of a two-stage system to provide sufficient cuts for the range of products desired. Such a system, shown in Fig. 14.3, includes a primary tower which operates at

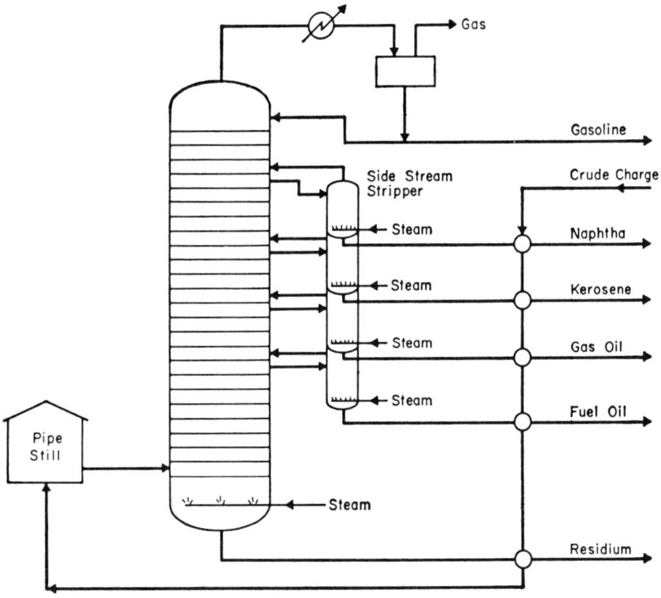

Fig. 14.2 Single-stage crude-distillation process.

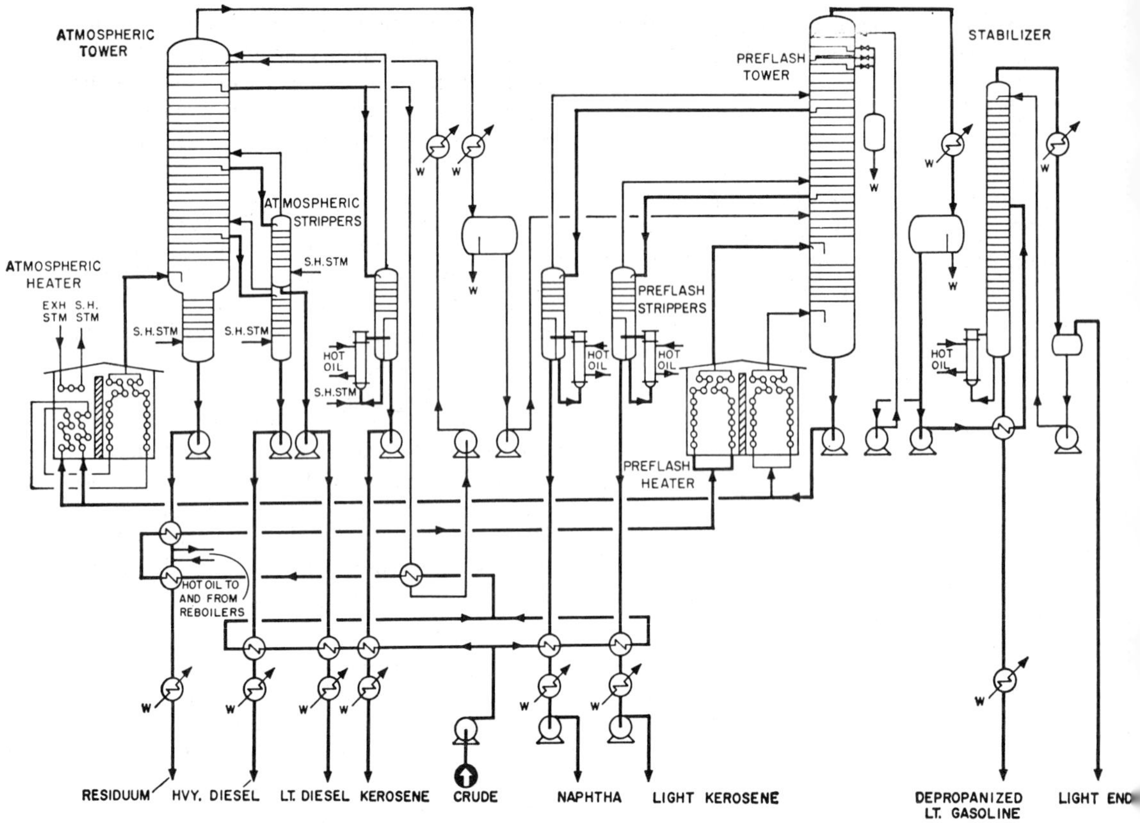

Fig. 14.3 Two-stage crude-distillation system. [*From "Petroleum Horizons," The Lummus Co. (1950).*]

about 50 psig, and a secondary tower which operates at atmospheric pressure, together with a stabilizing tower, and is used when the crude is to be separated into six to ten narrow cuts. Two or more side-streams may be withdrawn from the primary tower while the overhead becomes the feed to the stabilizer. The latter can be operated as a conventional stabilizer or as a debutanizer or depropanizer. The primary-tower bottoms become the feed to the secondary tower. The latter uses a circulating reflux stream, the overhead product returns to the primary tower, and side-streams are withdrawn at appropriate points. All the side-streams are run through strippers to remove the light ends.

Two-Stage with Vacuum Tower. Coking in heater tubes or on trays and thermal degradation may result from high temperature operation. It is necessary therefore, in certain cases, to operate under vacuum so that the operating temperatures may be reduced.

The petroleum refiner uses vacuum distillation to obtain lubricating oil fractions, catalytic-cracking-charge stock, and asphalt. A vacuum tower is needed for the production of lube oil fractions, and such a tower may be added to a two-stage distillation unit (see Fig. 14.5). The crude from the atmospheric tower is fed at about 800°F to the vacuum tower where the pressure is maintained at from 40 to 130 mm Hg absolute. The overhead gas-oil fraction is removed and the side-streams are withdrawn through strippers. The bottom product is residium or asphalt.

A vacuum unit for producing catalytic-cracking-charge stock is depicted in Fig. 14.6. Operating condition are about the same as for lube oil production.

Fig. 14.4 A distillation system for crude oil. (*Courtesy Esso Standard Division, Humble Oil and Refining.*)

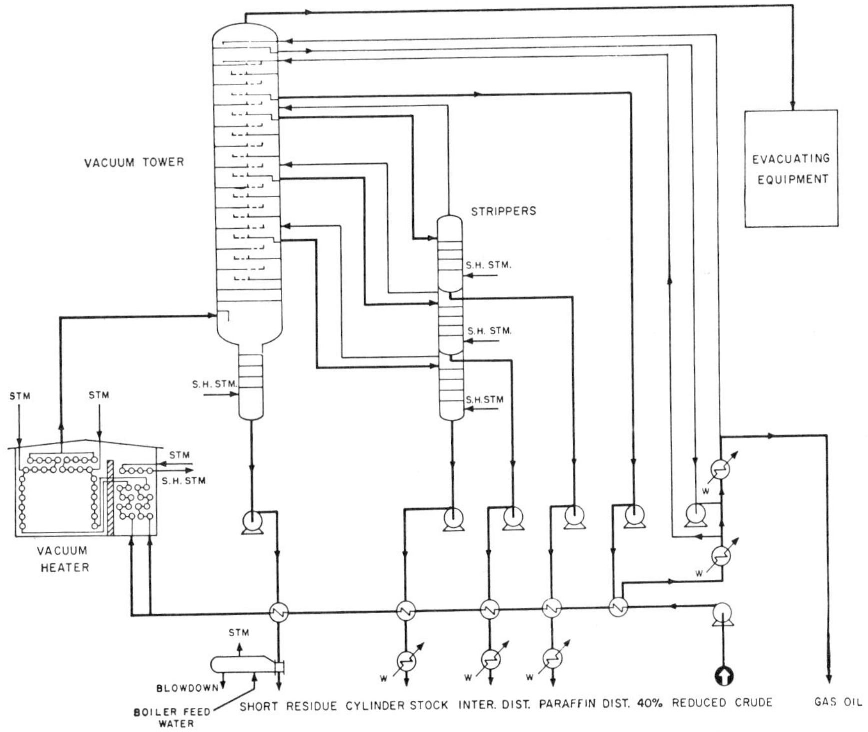

Fig. 14.5 Vacuum distillation for lube oils. [*From "Petroleum Horizons," The Lummus Co. (1950).*]

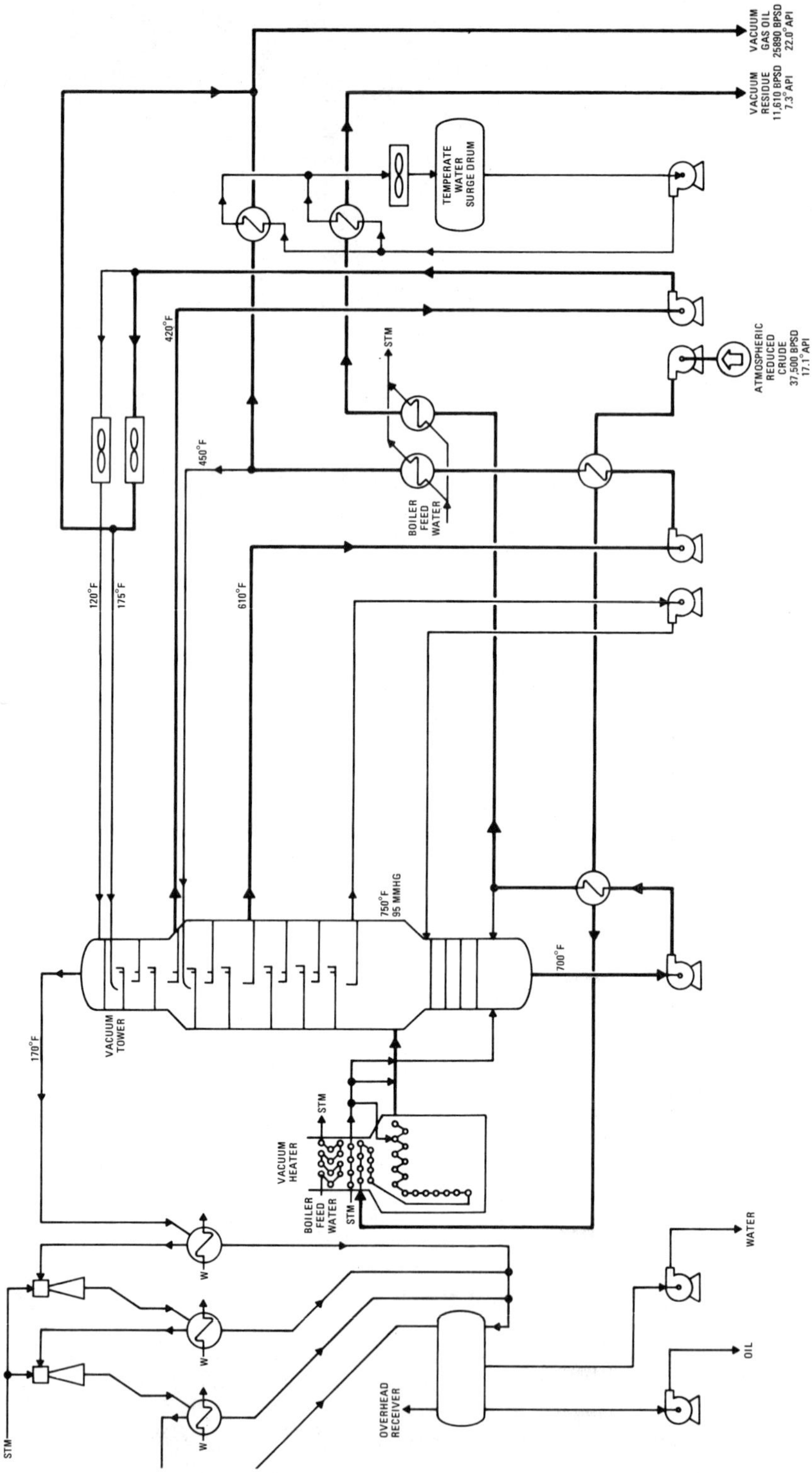

Fig. 14.6 Vacuum flash unit. *[Drawing provided by Lummus Co., Bloomfield, N.J. (1972).]*

Cracking Processes

Crude oil contains at most about 18 per cent gasoline. With the rapid growth of the automotive industry, it became necessary to devise methods whereby a larger portion of the petroleum could be utilized as gasoline. Thus, in 1959, the yield of gasoline from crude oil was 44.9 per cent. The difference is obtained by converting other fractions to gasoline. Motor gasoline may be obtained as shown below.

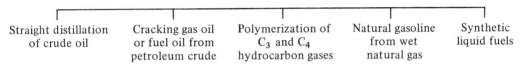

Motor Gasoline Including Aviation Gasoline are Obtained by

Straight distillation of crude oil	Cracking gas oil or fuel oil from petroleum crude	Polymerization of C_3 and C_4 hydrocarbon gases	Natural gasoline from wet natural gas	Synthetic liquid fuels

On distilling crude oil, there is obtained a large percentage of middle oil, including gas oil, light fuel oil, and heavy fuel oil, for which, until the last few years, there was very little demand. By heating gas oil in an oil heater and allowing it to expand in a fractionating tower, it was found that a high percentage of lower-boiling components were formed, which were in fact gasoline. The decomposition of an oil during its distillation is a common and generally unwelcome occurrence of the organic laboratory. The large molecule is sensitive to heat; it breaks down into smaller fragments, some of which reunite, or unite in new ways. The cracking art makes skillful use of this defect, and guides the breakdown to produce a maximum of the desired gasoline. Cracking produces a variety of products: gases, low- and medium-boiling naphthas suitable for gasoline and kerosene, gas oil and fuel oil fractions, residual oil and carbon, depending partly on the kind of stock fed to the unit, and partly upon temperature, pressure, and design of the equipment.

Thermal Cracking. In the early years of the cracking art, a flowing stream of gas oil, for example, was exposed for a period of time in steel tubes of high heat under a moderate pressure. It was then released to a large chamber at lower pressure, wherein gases, vapors, and liquid could separate. The condensed vapors and the liquid were then separated by fractionation in the usual way. The oil could also be vaporized first, and then subjected to the heat of cracking. A large number of processes were developed and operated successfully: the most common was the so-called "tube and tank" process. Thermal cracking is still practiced, especially in smaller refineries; in most of the larger refineries, as well as in a number of the smaller ones, thermal cracking has been displaced by catalytic cracking, which is really thermal cracking in the presence of a catalyst.

Thermal cracking of petroleum begins at slightly under 700°F, but the rate is too low at this temperature to be useful. Industrial practice employs temperatures in the range of 850–1050°F for gasoline production (see Table 14.7). The reactions occurring during thermal cracking include cleavage of C–C bonds, dehydrogenation, polymerization, and cyclization. Cleavage and polymerization are the most important, the others occurring only to a limited extent.

The operating conditions are fixed by the charge, the desired product, and undesirable reactions. It is usual to operate so as to obtain

TABLE 14.7. Reaction Conditions for Thermal Cracking

Feed	*Desired Product*	*Reaction Time, sec*	*Temp., °F*	*Press., psig*
Methane	Acetylene	0.01–0.1	Above 2000	Vacuum
Ethane	Ethylene	0.1–2.0	1350–1550	5–30
Propane	Ethylene	1–3	1250–1450	5–30
Gas oil	Gasoline	40–300	850–1050	200–900
Reduced crude	Gasoline	...	850–1000	10–70

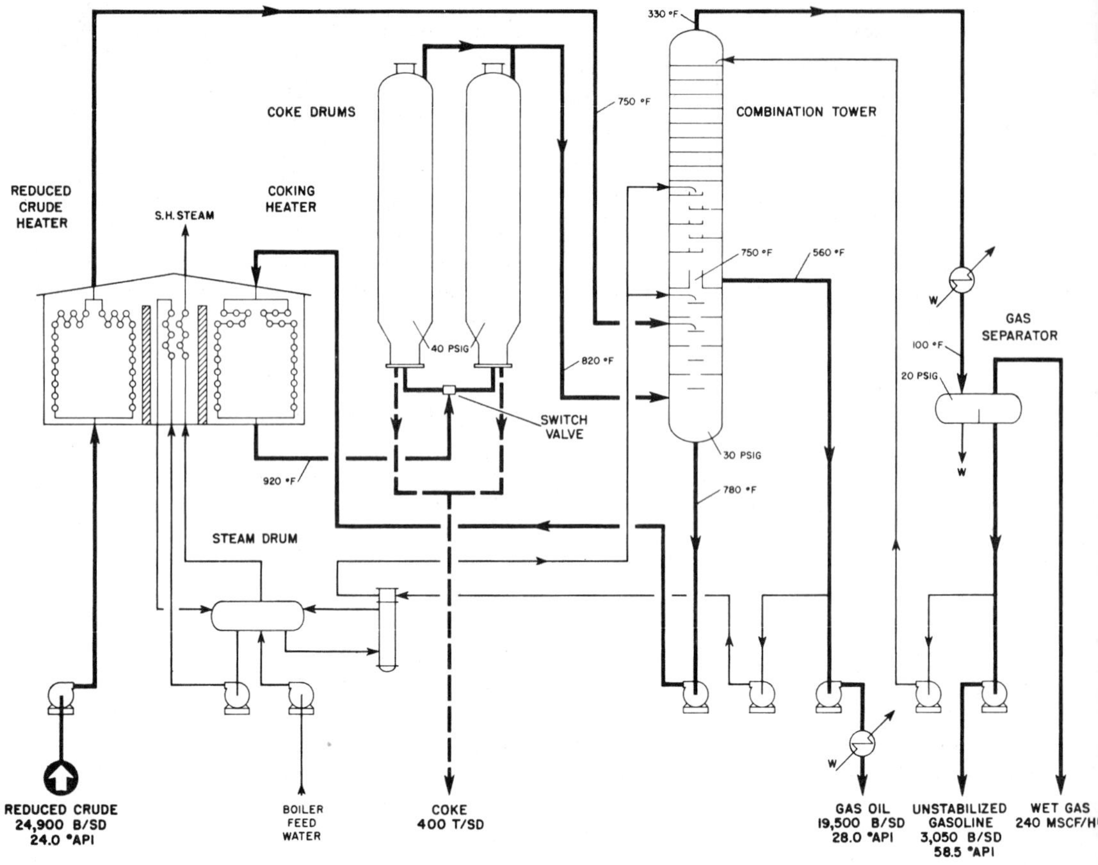

Fig. 14.7 Delayed coking unit for thermal cracking. [*From "Petroleum Horizons," The Lummus Co. (1954).*]

from 12 to 20 per cent conversion per pass and to recycle the uncracked stock. Increasing the conversion per pass decreases the ultimate yield of gasoline, increases coking, and affects the properties of the gasoline. It also results in shorter time on-stream between shutdown, due to the coking. The conditions chosen for a particular situation are, as usual, dictated by economic considerations.

It is necessary to maintain the charge in the liquid state during the cracking operation where gasoline is the desired product, and this requires operating under pressure.

Variations in charge stock lead to different performances at a given set of cracking conditions. Also, large differences in yields occur as operating conditions are changed. The reaction rate of petroleum hydrocarbons increases as its molecular weight increases[12] so that lighter-charge stock requires longer contact time to obtain a given conversion. Commercial practice is to raise the reaction temperature to increase the reaction rate rather than to use a longer contact time.

Prior to the introduction of catalytic processes, the two-coil Dubbs thermal cracking unit was the most widely used.[13] At present, however, coking processes predominate. These processes produce gas oil, gasoline, and coke from a charge of reduced crude, cracked tars, heavy catalytic-cycle oils and asphalts.[19] The feed is heated to about 900°F and charged into the reaction chamber (coking drum) where it remains until it is cracked to coke and volatile materials. The reaction chamber or drum is maintained at around 800–875°F. When the coke drum is filled it is taken out of service and steamed to remove volatile materials from the coke which is then sluiced out with water. The cleaned drum is returned to service. Two (or more) drums may be used in parallel and may be, typically, on-stream for 24 hrs and out for cleaning for 24 hrs.

In the fluid coking process[24] the charge is fed into the top of a fluidized bed of coke held at about 1000°F. The feed is cracked to lighter products and coke which deposits on the particles in the bed. A portion of the coke is continuously withdrawn from the reactor and sent to a coke burner where part of it is burned to provide heat for the reactor.

Catalytic Cracking. In the late thirties, the higher compression in gasoline motors demanded fuels with better antiknock properties, higher octane ratings, and higher lead susceptibility. It had been established that cracking in the presence of a catalyst produced a gasoline with a higher octane number than thermal cracking of the same stock. The requirements of the armed forces in the war years for high-octane motor fuel, especially aviation gasoline, stimulated intense activity in the invention and construction of catalytic processes. The first catalytic process was the Houdry, with fixed catalyst beds, which dominated the scene for several years. The two general types of catalytic cracking units now in use are the *fluid-bed* and *moving-bed* units. Catalytic cracking at present accounts for some 85 per cent of total cracking capacity.

The reactions occurring during catalytic cracking include cleavage, isomerization, alkylation, dehydrogenation, and aromatization, among others.

Two types of catalysts are used, natural and synthetic. The natural product is composed of silica and alumina with small amounts of other materials. Synthetic catalysts, on the other hand, are manufactured from pure materials to rigid specifications. The latest synthetic catalysts are manufactured from molecular sieves. These are synthetic crystalline zeolites consisting of the minerals mordenite and erionite, closely related to the aluminosilicates, in which the sodium cation has been replaced with a cation of a Group VIII or rare earth metal. These substituted molecular sieves are mixed with a binder and formed in the desired particle shape and size. Catalyst pellets vary in size from 3 to 4 mm for moving bed units and from 2 to 400 μ for fluid bed units.

In *the fluid catalyst cracking process*, the heated oil vapors pick up a fine powdered clay, or a synthetic catalyst in powder form,

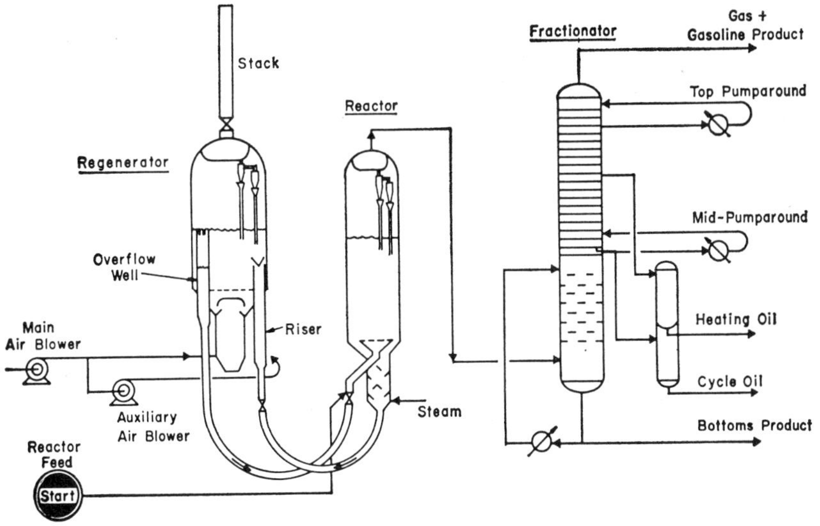

Fig. 14.8 Model IV fluid catalytic cracking process. [*Pet. Ref.*, **39**, *No. 9, 187 (1960) copyright 1960 by the Gulf Publishing Co.*]

Fig. 14.9 Fluid catalytic cracking unit. *(Courtesy Standard Oil Co. of Ohio)*

and the two whirl about in intimate contact in the reactor, where the cracking of the oil molecules takes place. The catalyst is said to be *fluidized*, for the mixture of powder and vapors behaves in many respects as if it were a single fluid. The density of the fluid may be varied by altering the ratio of catalyst powder to vapor. The cracked vapor enters an internal cyclone where most of its dust is released and returned to the reactor; then the vapor goes to a fractionator which condenses, in rotation, heavy gas oil, light gas oil, and raw gasoline or naphtha; and finally the gases pass from a cooler to a recovery system.

The catalyst becomes fouled and must be reactivated. To this end, a portion of the fluidized catalyst is constantly bled from the reactor, separated from its accompanying vapors (which return to the reactor) in the "spent-catalyst stripper," and travels downward to a stream of combustion air which carries it (upward) into the regenerator, where the carbon-rich deposit is burned off. The flue gases travel through an internal cyclone, where they deposit their dust, to a steam-producing heat exchanger, then to an electrical precipitator, where the last of the powder is caught, and thence into the atmosphere. The recovered catalyst, now clean, re-enters the regenerator, and is joined there by make-up catalyst, and the two together, at a temperature of 1100°F, are ready to enter the reactor. On the way there, they meet the incoming preheated oil feed, or charge, which they flash-vaporize. The catalyst and vapors in the reactor feed line are forced along and up by the velocity of the entering vapors and by the head of fluid in the regenerator standpipe, and enter the reactor at its lowest point, rising through the bed of fluidized catalyst. The temperature in the reactor is maintained at 900-950°F; the cracking reaction consumes heat, but the catalyst powders bring in the necessary heat from the regenerator. In this function, the catalyst may be thought of as a heat-transfer agent.

The full-scale unit may have a capacity of more than 100,000 barrels a day or more. In the full-sized unit, there are several hundred tons of catalyst in circulation. The cleaned catalyst from the regenerator is fed into the reactor at the rate of a box-car a minute, and let it be noted, without moving parts.

As already mentioned, the cracked vapors from the reactor enter the fractionator, in the bottom of which the slurry of heavier oils and residual catalyst is formed. If desired, the slurry oils may be recycled and returned to the reactor.

Operating conditions for the fluid process, and for the moving bed process described in the following section, are given in Table 14.8.

The *airlift thermofor catalytic cracking process* is depicted in Fig. 14.10.[18] Catalyst from the separator enters the reactor where it contacts the preheated feed. Catalyst and oil flow downward through the reactor. The product vapors from the reactor are sent to fractionators and the catalyst flows through a purge zone, out of the reactor and into regenerators. The carbon deposit is oxidized by air flowing countercurrent to the catalyst; the latter is

TABLE 14.8 Operating Conditions for Catalytic Cracking Units

	Fluid Units Fluid Units	*Moving Bed Units*
Reactor temp. °F	870–1000	830–975
Reactor press, psig	10–21	5–18
Space velocity[a]	0.5–3	1–4
Catalyst-oil weight ratio	5–20	1.5–7
Carbon on regenerated catalyst, wt %	0.4–16	0.1–0.6
Carbon on spent catalyst, wt %	0.5–2.6	1.2–3.1
Max. regenerator temp, °F	1275	1350

[a](Wt oil feed per hr)/(wt catalyst in reactor).

then sent to the cooling zone for temperature adjustment. From here it goes to the lift pot where a stream of low-pressure air picks it up and lifts it to the top of the reactor for another cycle.

Catalytic reforming is used to upgrade the octane rating of natural gasoline, naphtha, and similar compounds, because catalytic cracking units can no longer economically meet the demand for high octane gasoline. This has become the principal process for upgrading gasoline. It is a high temperature, catalytic process, and the action takes place in the presence of hydrogen; yet hydroforming is not hydrogenation; on the contrary it is dehydrogenation, and part of the hydrogen produced is recirculated merely to control the rate and extent of dehydrogenation. The catalyst is usually molbydena on alumina. The latest catalysts consist of rhenium and platinum on a support. The most important property of the catalyst is that which causes ring formation and permits ring preservation in molecules which have just undergone partial dehydrogenation (aromatization). The final product thus contains a high percentage of aromatic, and a small quantity of aliphatic hydrocarbons, compared with the feed. The product is unusually stable and may be blended directly with finished gasoline, after gas removal in a separator followed by distillation in a stabilization tower.

It is well to have two catalytic reactors so that while one is functioning, the other may be freed of the small amount of coke deposited on the surface of the catalyst granules. The coke is burned off with caution, to avoid harming the catalyst; this is accomplished by adding controlled amounts of air to an inert flue gas. For any given octane level, coke

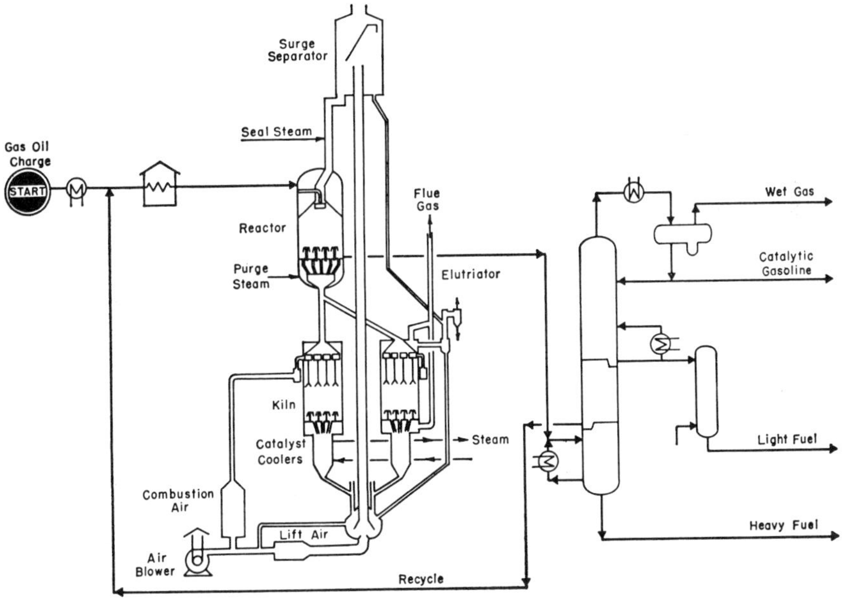

Fig. 14.10 Air lift Thermofor catalytic cracking process. [*Pet. Ref.,* **37**, *No. 9, 233 (1958); copyright 1958 by the Gulf Publishing Co.*]

deposition is a function of the amount of recycle gas and its hydrogen concentration. The removal of the coke restores the catalyst to its original activity.

Low-octane naphtha side-streams taken from the crude topping system are so enriched by the process of hydroforming that they contain 40 to 50 per cent aromatic hydrocarbons, of which 15 to 20 per cent is toluene. Removal of the toluene from accompanying hydrocarbons in the same boiling range (200–240°F) is performed by azeotropic distillation. Methyl ethyl ketone-water, the best azeotrope-former for this purpose[11] is added to the naphtha, and the mixture is introduced into a fractionating still. The azeotrope-former (MEK-water) passes out overhead, carrying with it the non-toluene hydrocarbons. The bottoms in the still consist of 99+ per cent toluene. The process is continuous. As would be expected, the number of reactions occurring during the process are numerous and complex. Several probable reactions[14] are given below:

1. Naphthene dehydrogenation

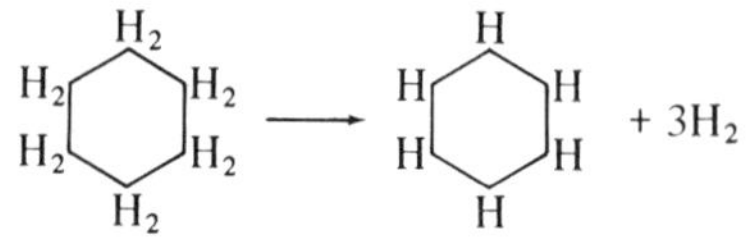

Cyclohexane — Benzene

2. Naphthene dehydroisomerization

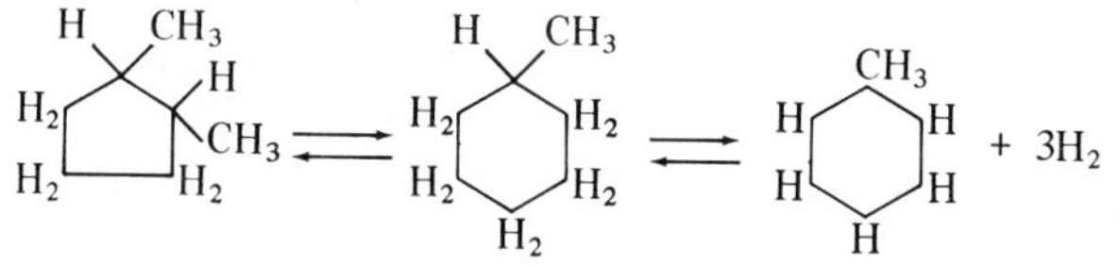

Dimethylcyclopentane — Methylcyclohexane — Toluene

3. Paraffin dehydrocyclization

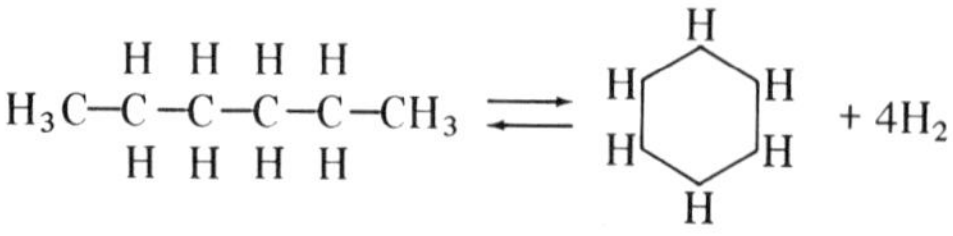

Hexane — Benzene

4. Paraffin isomerization

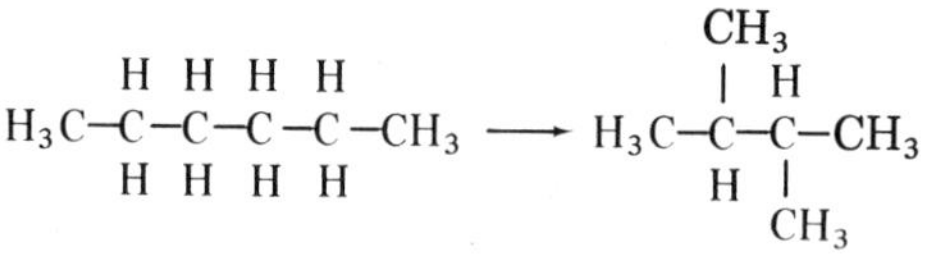

n-Hexane — Dimethyl butane

5. Paraffin hydrocracking

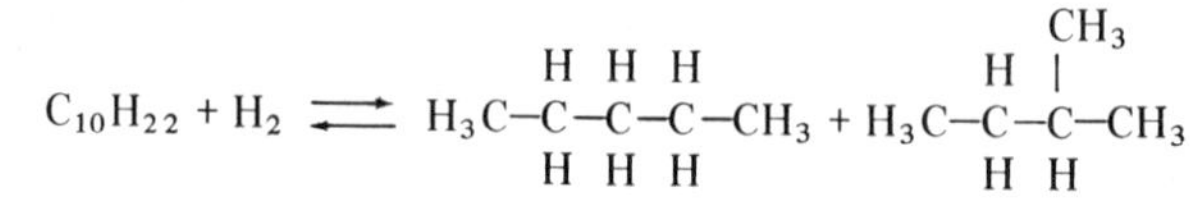

Decane — *n*-Pentane — Isopentane

6. Hydrodesulfurization

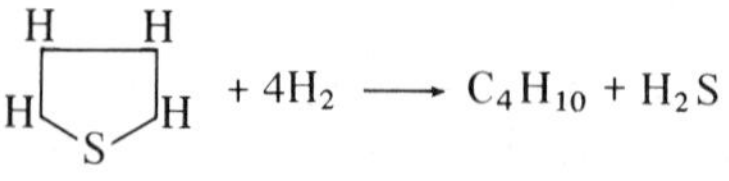

Thiophene

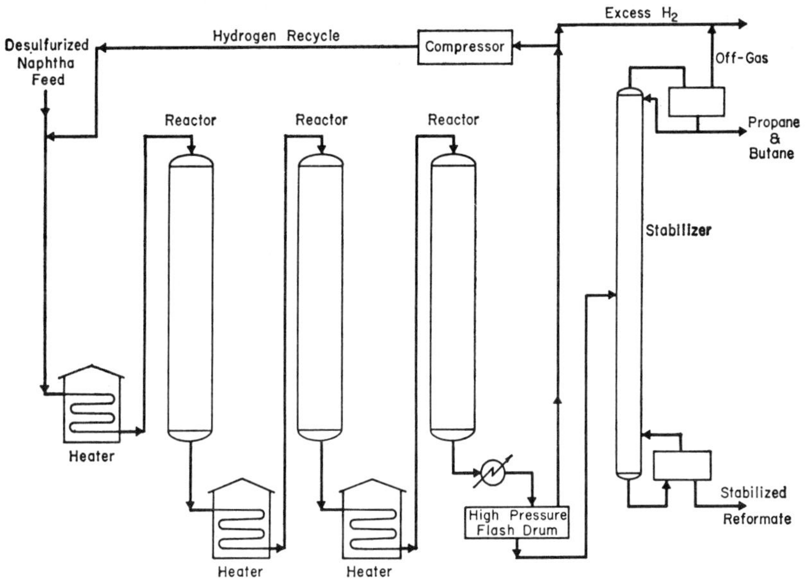

Fig. 14.11 Platforming catalytic reforming process.

Reactions 1, 2, and 4 predominate while the others may become significant at higher temperatures. Avoidance of hydrocracking is particularly important since it can lead to excessive coke deposition and reduced yields of liquid products.

Catalytic reforming units may be of the regenerative or nonregenerative type. A *platformer* (a nonregenerative-tyep reformer) is shown in Fig. 14.11.[17] Naphtha feed is mixed with hydrogen and passed through the catalytic reactors containing platinum catalysts. The charge is heated before entering each reactor to compensate for the endothermic reactions which occur. The final effluent is cooled, separated from the hydrogen which is recycled, and stabilized.

The *ultraforming* unit is of the regenerative type. Such a unit is shown in Fig. 14.12.[26] Operation is similar to that for the platformer

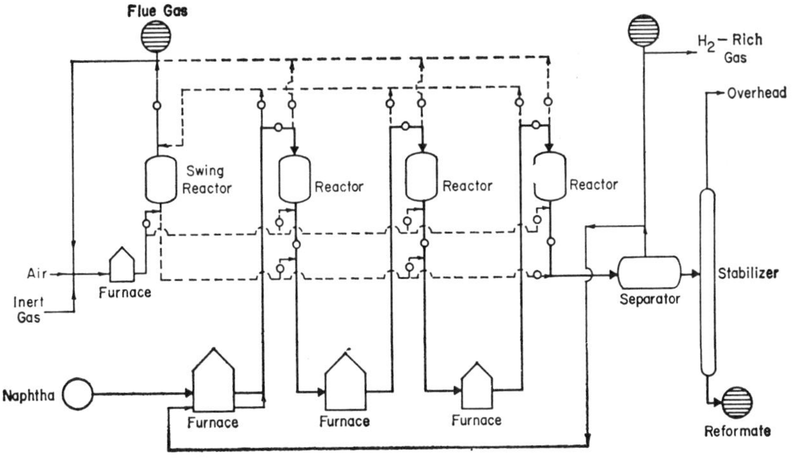

Fig. 14.12 Regenerative platinum reforming process. [*Pet. Ref.*, 34, *No. 9, 252 (1955); copyright 1955 by the Gulf Publishing Co.*]

except that a spare or "swing" reactor is substituted when the catalyst activity of an on-stream reactor declines. This permits regeneration of the catalyst without interrupting operations. One swing reactor may service several units.

Typical operating conditions are:

	Temp., °F	*Press., psig*	*Space Velocity (lb oil/hr)/(lb catalyst)*	*Hydrogen Rate (SCFH/bbl)*
Platformer	850–980	200–800	1–4	4,000–13,000
Ultraformer	900–950	200–350	1–5	3,500–8,000

Improving Gasolines by Additions and Reactions

Gasolines can be improved by the addition of foreign substances, by forming the hydrocarbon molecule in a definite manner, and by rearranging the molecular structure of the existing molecule, with or without simultaneous removal of hydrogen. The following processes alter the structure of the molecule, or add to its size, thus raising its octane rating:

stabilizing	alkylating	hydrogenating
reforming	hydrofining	polyforming
isomerizing	hydroforming	

Antiknock Compounds. In order to prevent the familiar knock in an internal combustion engine such as those in automobiles, when hot and laboring, and to prevent the loss of power which occurs as a result, a number of substances may be successfully added to the fuel. The effectiveness per unit weight of these substances varies greatly. The best known antiknock compound is tetraethyl lead, $Pb(C_2H_5)_4$, although tetramethyl lead is also used at present. The latter compound is more effective than tetraethyl lead in highly aromatic fuels.

The antiknock composition also contains ethylene dichloride or ethylene dibromide to prevent the build-up of lead in engines. The lead oxide which forms is changed to the volatile chloride or bromide which passes out with the exhaust gases.

Pollution legislation is forcing the automobile manufacturers to reduce the output of air pollutants from automobile exhausts, which will probably require the use of catalytic reactors in the exhaust systems. The catalysts are poisoned by lead compounds; thus the lead content of gasoline will be greatly reduced or eliminated from future gasolines. No adequate substitute for tetraethyl and tetramethyl lead has yet been found for octane improvement.

Other Additives. Gasoline properties which are controlled include boiling range, vapor pressure, octane rating, gum content, and sulfur content. The boiling range is adjusted by mixing straight-run gasoline, cracked gasoline, reformate, and others to provide the desired range and octane number before adding the lead compounds. Vapor pressure is controlled by additions of butane and natural gasoline. The vapor pressure is made high in colder climates or during the winter months to help with easy starting, and low during the summer to prevent vapor lock. Antioxidants are added to minimize gum formation, dyes for identification and eye appeal, and antifreeze to prevent icing of the carburetor.

Natural Gasoline. Natural gasoline is the liquid which accompanies "wet" natural gas in many regions; it is removed by compression, or by absorption, mainly in straw oil, or by a combination of both processes. The gasoline values are mainly low-boiling, and correspond to hexanes and heptanes, with some octanes and pentanes. Intermediate hydrocarbons may also be separated, although less completely, in the form of liquefied petroleum gases (L.P.G.) such as propanes and butanes. The relative amounts are reflected by the figures for the daily production in a plant treating 30 million cu ft of gas daily: propane 360 bbl; isobutane, 190; normal butane 210; isopentane 108, and debutanized gasoline 547 bbl. In this typical plant, the wet gas, compressed to 700 lbs/sq in gauge, enters two 40-plate towers operated in parallel, where it meets the absorber oil which enters at the top and

TABLE 14.9. Typical Properties of Motor Gasoline

	Regular	*Premium*	*Super Premium*
Distillation			
10%	116–147	116–147	116–147
50%	205–230	205–230	205–230
90%	325–370	315–360	315–360
E.P.	380–405	370–405	370–395
Reid vapor pressure	9.5–13.5	9.5–13.5	9.5–13.5
Gum, mg/100 cc max.	2.5	2.5	2.5
Octane number (research)	90–92	93–95	99–100
Sulfur, % max.	0.20	0.20	0.15

travels down, leaving the tower as saturated absorber oil. Unabsorbed gas leaves overhead; part serves as fuel gas in the plant; the balance is compressed and returned to the field. The "fat" oil is first flashed to a pressure of 385 lb/sq in., to release most of the absorbed methane and ethane with some propane and higher components. By means of a reabsorber, such values in propane and higher hydrocarbons are in large part recovered. The flashed oil passes to a primary or high-pressure still, where again the lower hydrocarbons are driven out and saved for other purposes. The oil then passes to the low-pressure or main still where the gasoline components, all of which may be compressed to a liquid at practically atmospheric pressure, are removed and collected. The stripped lean oil is cooled and returned to the absorbers.

The production of natural gasoline in 1968 was 145,200,000 barrels. Other products bring the total to 550,300,000 barrels for the natural gasoline industry.

Alkylation. The alkylation process for the production of high-octane gasoline results from the discovery that paraffin hydrocarbons unite with olefins in the presence of a catalyst. The "alkylate" produced from a properly selected stock, for example, the olefin-rich gas stream from cracking processes, can have an octane number as high as 95. The process may involve isobutane and olefins which produce high-octane dimers or trimers. The reaction is carried out at 60 to 95°F in the presence of sulfuric or hydrofluoric acid. The product is recovered from the acid in a settler and treated to remove propane, butane, etc.

Catalytic Polymerization. The polymerization of olefin-bearing refinery gases has as its objectives the production of high-octane motor fuel and petrochemicals. A process flow sheet for polymerization using a solid phosphoric acid catalyst is shown in Fig. 14.13. Preheated feed is passed over the catalyst in the reactor at 350 to 435°F and 400 to 1200 psi. The reactions are exothermic. The polymer gasoline thus produced is freed from butane and propane to complete the process. Compounds such as propylene dimer, trimer, and higher homologs can be produced by this method.

Lubricating Oil Refining

Lubricating oils make up only about two per cent of the petroleum products sold in the United States.[6] They are, however, high-profit items. Moreover, the incidental recovery of waxes, asphalt, and other by-product materials obtained in the refining processes serves to further enhance their importance in the over-all economics of petroleum refining.

The high-boiling paraffin hydrocarbons comprise the materials having the properties necessary for lubricating oils. These properties include stability at high temperature, fluidity at low temperatures, only a moderate change in viscosity over a broad temperature range, and sufficient adhesiveness to keep it in place under high shear forces.

A general flow sheet for the manufacture of lubricating oils is shown in Fig. 14.14. Since the desired fractions have high boiling points their separation into various boiling-range cuts must be accomplished under reduced pressures. This is done in a vacuum distillation

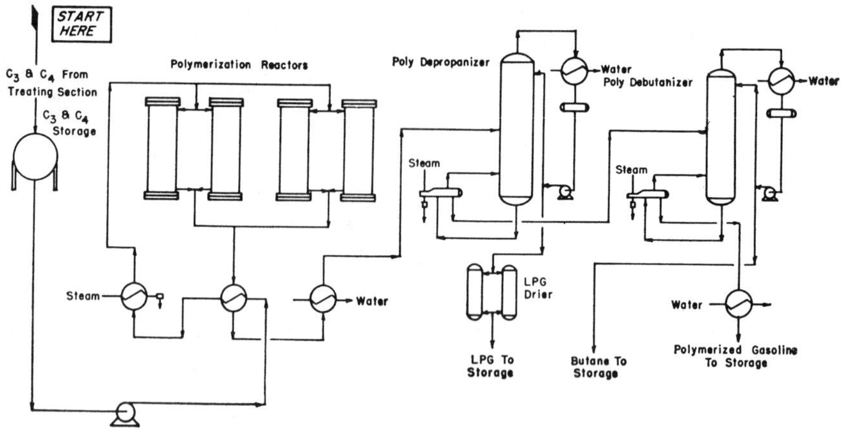

Fig. 14.13 Solid phosphoric acid polymerization process. [*From a booklet prepared by the Ashland Oil and Refining Co. (1960).*]

system (see Fig. 14.5). The cuts thus obtained are further treated to remove naphthenes, aromatics, nitrogen and oxygen compounds, sulfur compounds, and other impurities. Asphalt is found only in the residuum and therefore only this stream needs to be treated for asphalt removal.

Deasphalting. The residuum from the vacuum still is contacted in a tower with liquid propane which dissolves all the constituents except the asphalt which passes out the bottom of the column. The column is operated under pressures up to about 500 psi to maintain the propane in the liquid state at the operating temperatures. An interesting fact is that the column is kept about 50 degrees hotter near the top (140-200°F) than at the bottom. This reduces the solubility of the heavier hydrocarbons and helps to further narrow the

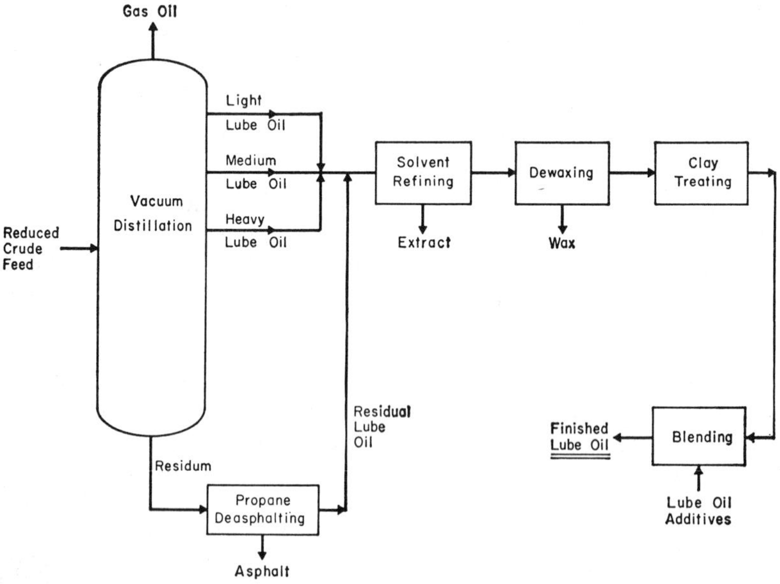

Fig. 14.14 General flowsheet for lube oil refining.

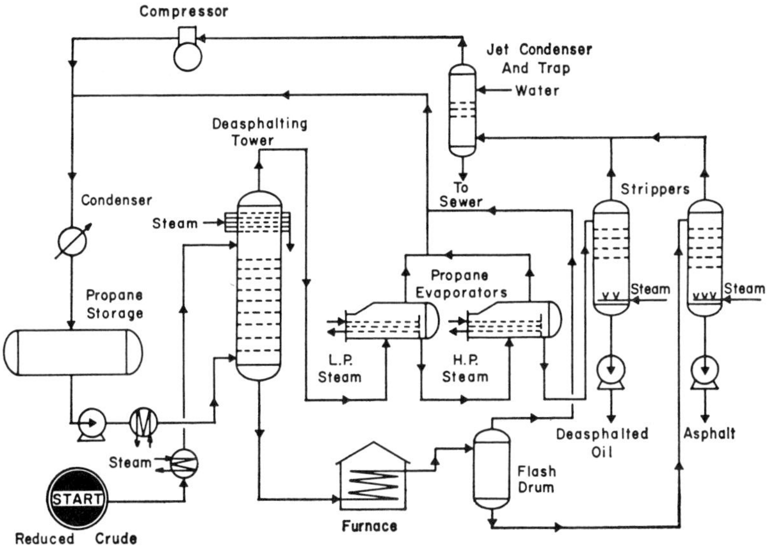

Fig. 14.15 Propane deasphalting process. [*Pet Ref.*, **39**, *No. 9, 239 (1960); copyright 1960 by the Gulf Publishing Co.*]

boiling range of the fraction thus treated. Propane is recovered from the oil and asphalt and recycled. This process has become important in decarbonizing and deasphalting the charge stock for catalytic cracking.[9]

Subsequent treatment of the oil fraction is similar to that used for the other fractions from the vacuum distillation. The steps are described in the following sections.

Solvent Extraction. The removal of nonparaffins by solvent extraction is usually the next step in lube oil refining. Two important solvents are furfural and phenol; other solvents may also be used.

The process using furfural is shown in Fig. 14.16. The lubricating oil fraction enters at the center of the extraction column, the solvent near the top. The two streams flow countercurrent to each other, the solvent leaving at the bottom and the oil at the top. Contained in the furfural solvent are nonlubricating compounds including aromatics, naphthenes, sulfur compounds, and nitrogen compounds. The solvent is recovered by flashing and steam stripping.

Dewaxing. Waxes present in lubricating oils may crystallize out at low temperatures and impair operations. They are, however, good lubricants and the amount permitted to remain in the oil is determined by the operating conditions envisioned for the oil.

Conventional dewaxing processes, which can yield a product with a pour point* as low as -15°F, involve contacting the oil fraction with a solvent, for example, methyl ethyl ketone, which will dissolve the oil and waxes. Chilling the solution causes the wax to crystallize, and it may then be removed by filtration. The wax may be fractioned to produce waxes of various melting ranges (from 90 to above 200°F). Solvent recovery systems recover the solvent from the oil and wax. A flow diagram for dewaxing is shown in Fig. 14.17.

Lubricating oils with pour points as low as -70°F can be obtained by a new process using urea.[30] Urea (three parts) and the oil (one part) are mixed and heated to about 100°F. There a wax-urea complex which is insoluble in the oil is formed and it can be filtered off. The urea and wax are recovered by treating the complex with water at about 170°F.

---

*The pour point is the temperature at which the oil ceases to flow.

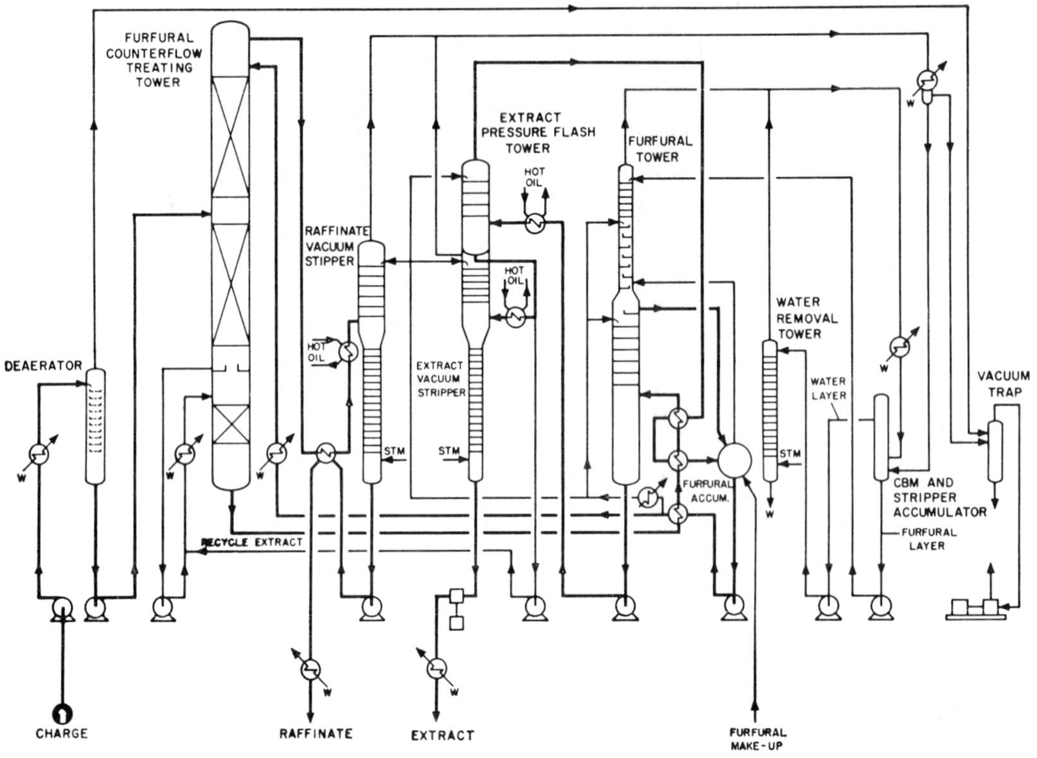

Fig. 14.16 Furfural refining process [*From "Petroleum Horizons." The Lummus Co. (1960).*]

Clay Treating. Oxygen, nitrogen, and sulfur compounds remaining in the oil following the refining steps described above are removed by adsorption on Attapulgus Clay, Fullers Earth, or other suitable adsorbents. Two methods in use are the percolation method and the contact-filtration process. In the former the oil is passed (percolated) through a bed of the adsorbing material. Depending on the adsorbent and the impurities to be removed, one ton of clay might treat well over a hundred barrels of oil before requiring regeneration. The latter operation consists of washing with naphtha and steaming, followed by heating at about 1000°F to remove the adsorbed impurities.

The contact-filtration process[16] is continuous. The adsorbent is mixed with the oil and the resulting slurry, heated to 225 to 550°F, is passed through a contact tower where the residence time may be from ½ to 1 hour. The effluent from the tower is cooled and filtered to remove the adsorbent. The adsorbent may or may not be regenerated, depending on the economics involved.

Blending of Lube Oils. Lubricating oils having particular properties are obtained by blending the various refined fractions and by the addition of certain nonpetroleum materials (additives). Modern lubricating oils contain a large number of additives including: oxidation inhibitors, to reduce susceptibility to oxidation, particularly at high temperatures; detergents to dissolve engine deposits; antifoam agents; and pour-point depressants, to name a few.

Greases. The semisolid consistency of greases is obtained by adding thickening agents to the oil. These may be metallic soaps, modified silicas, certain organic materials and special clays; they comprise from about 3 to 30 volume per cent of the grease.

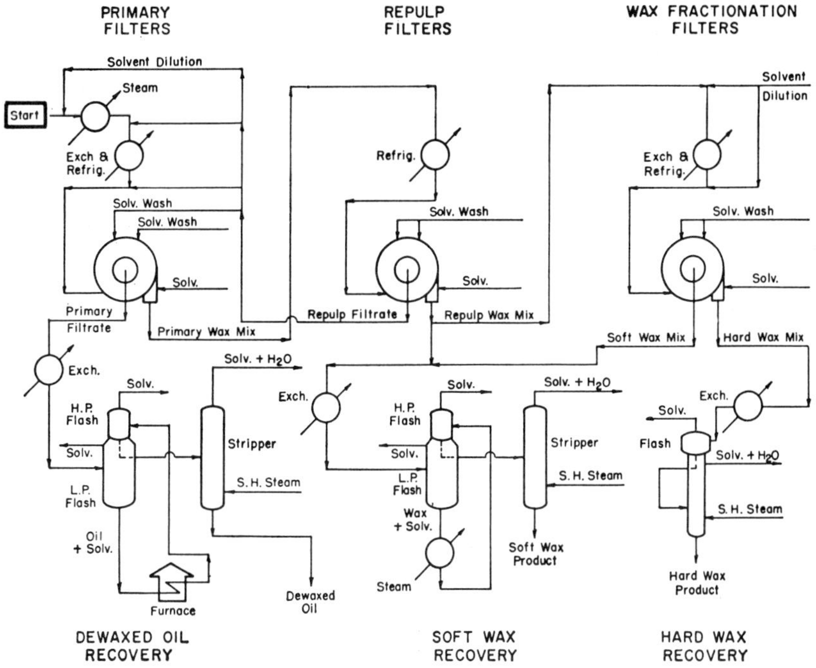

Fig. 14.17 Lube oil dewaxing. [*"The Story Behind One of the World's Largest Paraffin Wax Plants," Badger Manufacturing Co. (1959).*]

Hydrogen Processing of Petroleum

In the 1960's a large number of processes were developed for treating petroleum fractions with hydrogen. These processes are quite similar and usually consist of the following steps: (1) heating of liquid feed and hydrogen to reactor temperature; (2) contacting the feed with a catalyst, usually in a single- or two-stage fixed-bed reactor, (3) one- or two-stage separation of liquid and gases, (4) distillation of the liquid product to specification, and (5) purification of the hydrogen stream so that it can be recycled.

The operating conditions of the processes in general fall within the following ranges: (1) pressure, 200-6000 psig, (2) temperature, 350-950°F; (3) hydrogen feed rate, 400-10,000 SCF/bbl of oil; (4) liquid hourly space velocity (LHSV), 0.2-10.0 vol oil/vol cat; and (5) hydrogen consumption, 100-2,500 SCF/bbl oil.

These processes are used to improve some property of the petroleum being treated by using a catalyst specific to the desired reaction. In some cases several types of reactions are carried out simultaneously. Proprietary catalysts are used to promote the desired reactions which also determine the operating conditions. The catalysts in general do not require regeneration or reactivation, but in a few cases the catalyst may require reactivation every few months. Catalyst lives in general are very long and they may last from one to six years before being replaced.

Some of the hydrogen-treating systems for petroleum fractions are:

1. *Hydrodesulfurization.* Stringent new air pollution laws now require that fuels used for heating or power generation contain less than one per cent sulfur. Most heavy-oil fractions produced in refineries contain from 2.0 to 5.5 per cent sulfur and cannot be used as fuel unless the sulfur content is reduced. Hydrosulfurization converts the contained sulfur to hydrogen sulfide which can be easily recovered from the gas stream and converted to elemental sulfur, Figure 14.18 shows one of these processes.

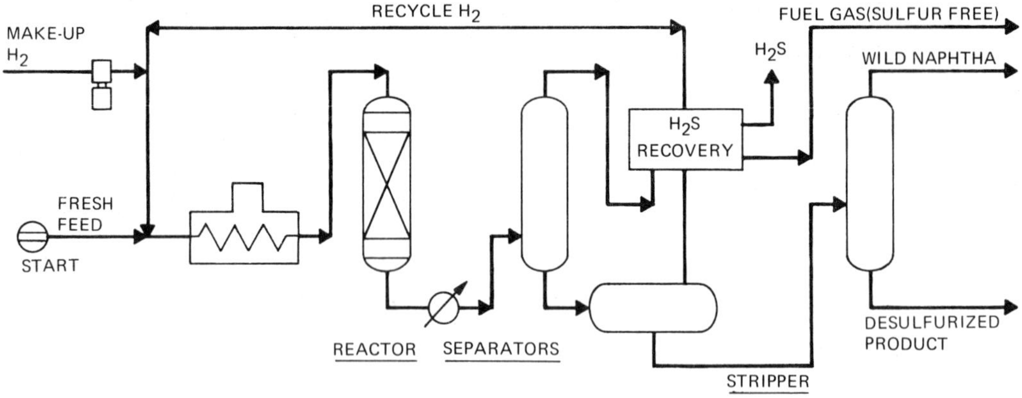

Fig. 14.18 Hydrodesulfurization process. [*Hydrocarbon Processing,* **48,** *No. 9, 133 (1969); copyright 1969 by Gulf Publishing Co.*]

2. *Hydrodenitrogenation.* Gas-oil fractions containing about 0.1 per cent or more nitrogen rapidly deactivate the catalytic cracking catalysts which must then be replaced. Hydrotreating processes can remove the nitrogen as well as oxygen, thus, improving the quality of the fractions for catalytic cracking.

Nitrogen and oxygen compounds in lubricating oil fractions are unstable and decompose or polymerize in service; thus they must be removed from the lubricating oil fraction. Hydrotreatment can remove these compounds and supplants the clay-treating step in lube oil refining.

3. *Hydrofining.* Hydrofining removes olefins from catalytic-cracking feed stock in order to reduce carbon deposition; it hydrogenates aromatics to naphthenes in kerosene and jet fuel in order to reduce smoking and improve the burning qualities of these fuels; it reduces vanadium and other metals to improve the charge for catalytic cracking; and it increases fuel oil storage stability.

4. *Hydrocracking.* Highly aromatic petroleum fractions and catalytic-cycle gas-oil are difficult to crack catalytically, but they can be hydrocracked quite readily to low-molecular weight fractions in high yield. Several hydrogen-treating processes have been developed to crack these fractions.

5. *Lube Oil Hydrorefining.* Catalysts have been developed for the removal of sulfur, nitrogen and oxygen; for the hydrogenation of aromatics to naphthenes, and for the reduction of the number of rings in multiring naphthenes by hydrogen treatment. For selected lube oil fractions this process can completely replace solvent refining and clay treating in the processing of lube oils.

THE PETROCHEMICAL INDUSTRY

The petrochemical industry may be generally defined as comprising those areas of chemical manufacturing which use raw materials extracted wholely or largely from petroleum or natural gas. Dominating the petrochemicals are organic compounds manufactured from such basic compounds as methane, acetylene, ethane, ethylene, propane, propylene, butane, butylene, butadiene, and benzene and other aromatic compounds. Additionally, many inorganic chemicals, including ammonia, urea, nitric acid, sulfur and sulfuric acid, carbon black, and others are being manufactured in ever-increasing amounts from petroleum sources.

One may logically divide the petrochemical industry into two sections, one comprising production and separation of the basic raw materials mentioned earlier, the other dealing with conversion of these raw materials into the numerous chemical compounds of commerce. The manufacture of several of these compounds are described in other chapters of this book; ammonia in Chapter 5, butadiene in Chapter 9, sulfur and sulfuric acid in Chapter 4, and many of the organics in Chapter 25. The remainder of this chapter will be devoted to discussion of the processes used to produce the basic raw materials from petroleum and natural gas.

Aliphatic materials are found in natural gas and refinery gas and are produced by cracking petroleum fractions. Aromatic compounds are produced by catalytic reforming and dealkylating processes; and they are purified by extraction and distillation.

Hydrocarbons from Gases

Natural gas, as taken from the earth, contains a large number of compounds which usually include hydrogen, methane, ethane, propane, butane, heavier hydrocarbons, and hydrogen sulfide. The methane may be recovered by an absorption process (the methane and hydrogen are not absorbed) and the remaining hydrocarbons fractionated to obtain pure components. Figure 14.19 shows a complete flow sheet for natural gas processing.[10] Since the natural gas leaving the well contains moisture,

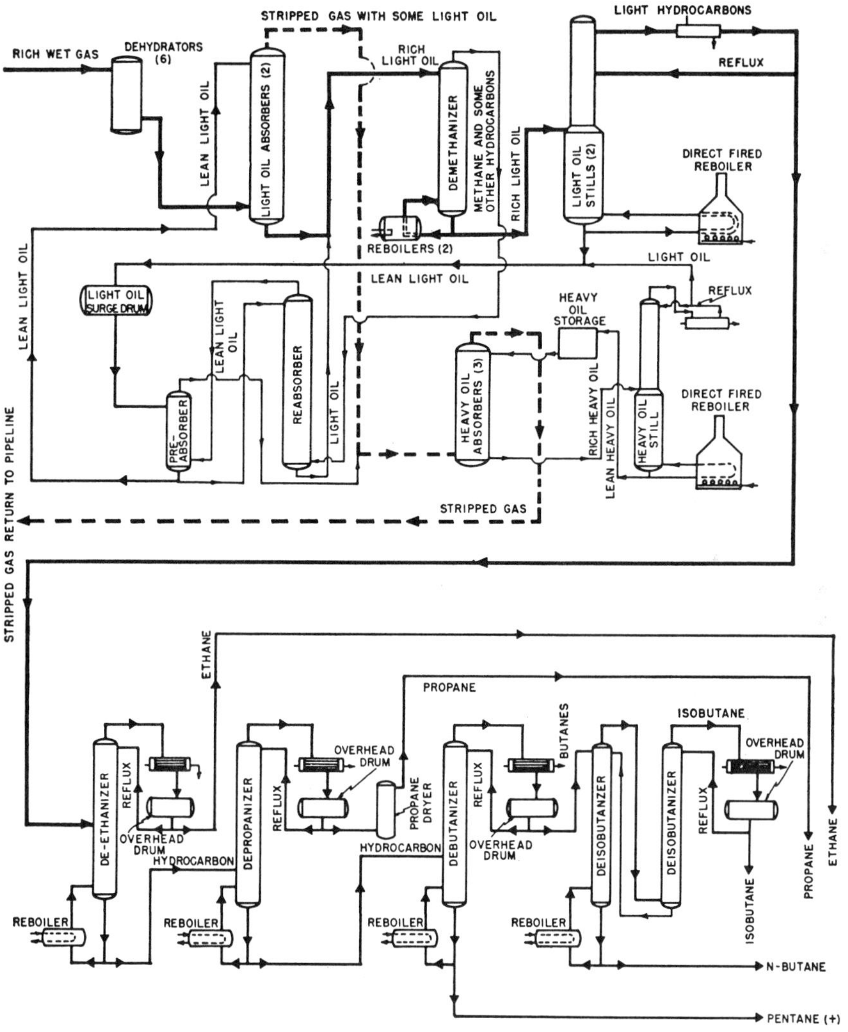

Fig. 14.19 Natural gas separation for petrochemicals. [*Ind. Eng. Chem.*, **48**, *No. 2, 170 (1956); copyright 1956 by the American Chemical Society.*]

it must be dehydrated to prevent freezing and hydration in the absorption column which operates at -20°C under a pressure of about 500 psi. The absorbing oil is normally a hexane fraction. The gas leaving the absorber is mainly methane and hydrogen although some methane is dissolved in the oil. The methane is recovered by passing the gas-laden absorption oil through a demethanizer; the methane leaving this apparatus contains some ethane and propane and these are removed in another absorber. The hydrocarbons dissolved in the oil from the two absorption columns are separated from the oil in a light-oil still. (The methane stream is passed through a heavy-oil absorber to recover the light oil it carries from the light-oil absorbers.) The hydrocarbon mixture from the light-oil still next goes to a distillation system where it is separated into its components: ethane, propane, *n*-butane, isobutane, and natural gasoline.

Refinery gases contain in addition to the materials present in natural gas, ethylene, propylene and butylenes. Therefore, if it is necessary to produce high purity ethane or propane, additional distillation equipment must be used to separate ethylene from the ethane and propylene from the propane. Extractive distillation with acetone as the entraining agent is commonly used to separate butylene from butane.

Acetylene Manufacture

In the past acetylene has been manufactured mainly from calcium carbide (Chapter 17). This picture has changed somewhat due to the development of economic petrochemical processes. One of these is shown in Fig. 14.20. Oxygen (98% pure) and natural gas are preheated to 950–1200°F and then mixed in the ratio 1.00 mole methane to 0.60 mole oxygen, less than enough to support combustion of all the methane. The reactor consists of three zones, viz., mixing, flame chamber, and quenching. After being rapidly and thoroughly mixed in the first zone, the mixture enters the flame chamber through the ports of a specially designed burner block and the methane is cracked according to the reaction:

$$2CH_4 \longrightarrow C_2H_2 + 3H_2$$

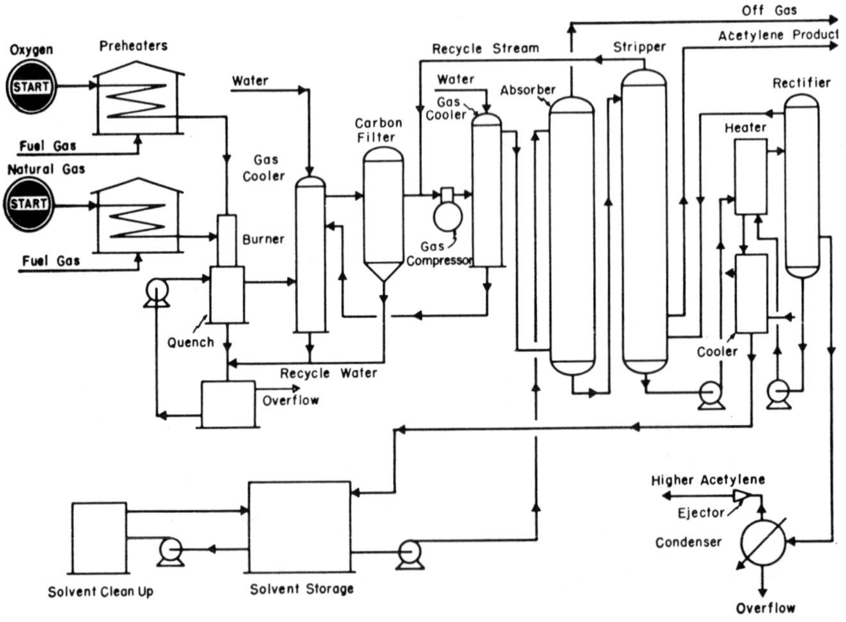

Fig. 14.20 BSAF acetylene process. [*Pet. Ref.*, **38**, *No. 11, 202 (1959); copyright 1959 by the Gulf Publishing Co.*]

Part of the methane is burned, the heat of combustion being used to raise the temperature of the reactants to reaction temperature and to supply the heat for the endothermic cracking reaction. About one third of the methane is cracked, most of the remainder being burned. The flame temperature is about 2700°F.

The gases from the flame chamber enter the quenching zone where they are cooled sufficiently to prevent further reactions, then further cooled in a spray chamber to around 100°F. The cooled gas is filtered to remove carbon, compressed and purified by absorption in selective solvents, such as dimethylformamide[28] and ammonia.[27]

Several compounds other than methane can be used to produce acetylene including heavy gas oil,[7] and naphtha. In some cases steam[3] may be used instead of oxygen, and in others[7] the hydrocarbons may be used alone. The trend is to use more of the heavier fractions as methane becomes more scarce and more costly.

Ethylene Manufacture

Based on the volume and number of derivatives, ethylene is one of the most important members of the petrochemical family. The most important uses of ethylene are for polyethylene plastics, ethylene oxide, ethyl alcohol, styrene, and halogenated ethylene compounds. Raw materials for ethylene production are ethane, propane, butane, refinery gas; and in recent years the use of naphtha, kerosene, and gas-oil from cracking has increased as the availability of the lighter materials has decreased and they have become more costly. Other raw materials for the production of ethylene include *n*-butane, which yields propylene and butylene as by-products, and certain crude oil fractions. The latter may be used to produce some aromatic compounds simultaneously with the production of ethylene. Figure 14.21 is a photograph of an ethylene purification unit.

In the production[21] of ethylene and propylene from ethane (see Fig. 14.22) and propane, the feed streams are first mixed with steam to promote higher yields of olefins and then cracked in separate furnaces at temperatures ranging from 1150 to 1500°F. The combined effluents are water-scrubbed and cooled to 100°F to condense polymers and aromatics. The gas streams are then compressed and, should an acetylene-free product be required, sent to a unit where the acetylene is hydrogenated. The gases are then chilled, dehydrated, and cooled before entering the separation process.

Fig. 14.21 Ethylene purification unit. (*Courtesy Esso Standard Division, Humble Oil and Refining.*)

Ethylene and lighter hydrocarbons are removed in the de-ethanizer, and leave in the overhead product stream. The bottom stream contains propylene and higher hydrocarbons. Recovery of ethylene is accomplished by chilling the overhead stream to about -195°F and treating the condensed liquid in a demethanizer. The bottom product from the latter apparatus next enters an ethylene fractionator for purification. Recovery of propylene involves treatment of the bottoms

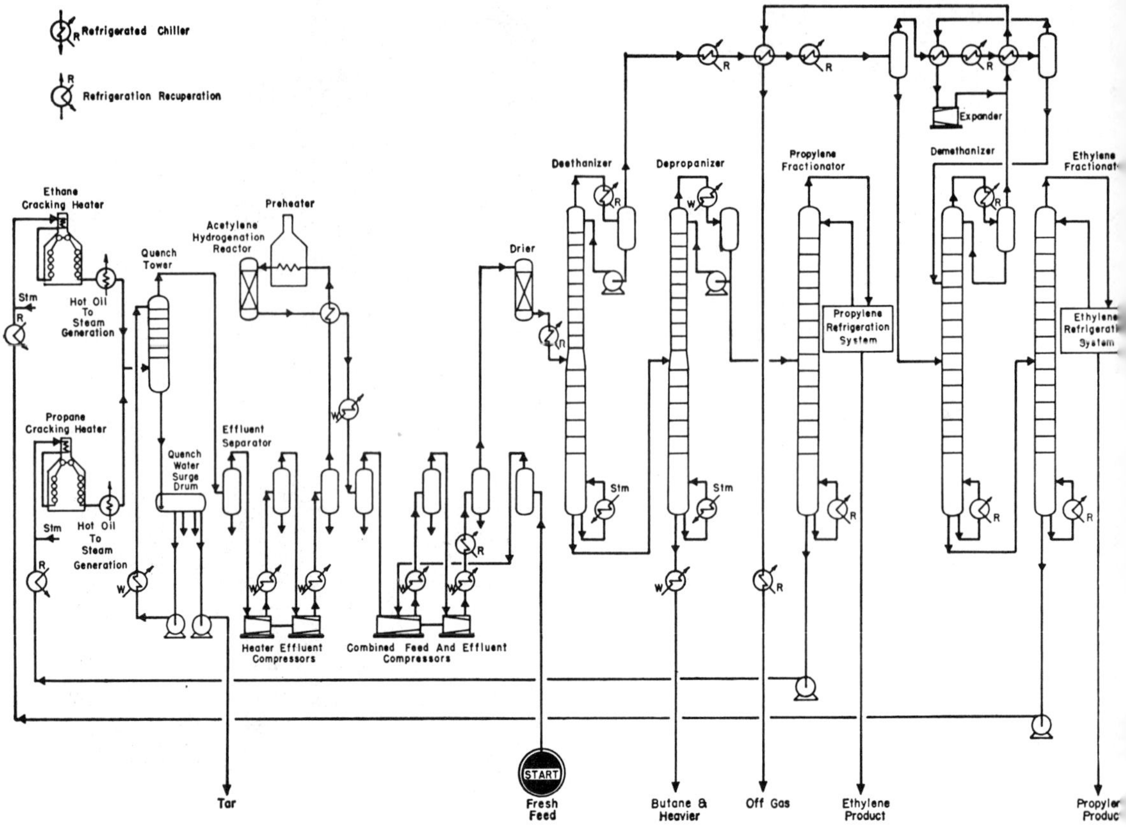

Fig. 14.22 Production of ethylene and propylene. [*Pet. Ref.*, **38**, *No. 11, 241 (1959); copyright 1959 by the Gulf Publishing Co.*]

from the de-ethanizer in a depropanizer and a propane splitter.

In addition to the low-temperature distillation process described above, ethylene may be purified in an absorption system operating as close to 0°F.

Propylene and Higher Olefins

The uses of propylene are in the production of polypropylene plastics, isopropyl alcohol, propylene trimer and tetramer for detergents, propylene oxide, cumene, glycerine, and gasoline. (About 80 per cent of the propylene produced is converted to gasoline.) Refinery off-gas contains large amounts of propylene and about 80 per cent of the present production comes from this source; the balance is manufactured as a by-product of cracking operations used for ethylene production.

Certain of the by-product streams in refinery operations contain large amounts of butane, isobutane, *n*-butenes and isobutylene. Isobutylene is used primarily for the manufacture of butyl rubber, and for polybutene production.[28] 1-Butene is produced mainly for use in butadiene production while *sec*-butanol, polybutene, and other compounds consume relativley small amounts.

The C_4 olefins can be produced by catalytic dehydrogenation of butane or butene in conjunction with selective solvent purification processes,[22] by recovery from certain refinery gases, and by cracking naphthas and other stock. The manufacture of butadiene is described in Chapter 9. With proper selection of operating conditions, the same processes can be made to yield butylenes as the main product.

The demand for olefins containing more than four carbon atoms is comparatively small at present. Some of these, together with the method of manufacture, are:

olefins

C_5 dehydrogenation of pentanes
C_6 dimerization of propane
C_8 dimerization of butane
C_9 trimerization of propylene
C_{12} tetramerization of propylene

Aromatic Chemicals

Benzene, Toluene, Xylene. During 1968 the chemical industry consumed about 1,700 million gallons of benzene, toluene and xylene; 900 million gallons of this amount was supplied by the petroleum industry which had a production capacity in reforming units of about 7 billion gallons. Most of this capacity is used to make aromatics for raising the octane rating of gasoline as described in an earlier section.

About 85 per cent of the benzene consumed by the chemical industry is produced by the petroleum industry. Benzene is used for the manufacture of styrene, phenol, decylbenzene, several synthetic fibers and plastics. Most of the toluene produced is used in gasoline manufacture but a substantial volume, 100 million gallons in 1968, is used by the chemical industry. Para-xylene finds use in the manufacture of polyester fibers and film while ortho-xylene is used to produce phthalic anhydride and other compounds.

Naphthalene. For many years naphthalene was derived entirely from the coking of coal and tar distillation. In 1959 all of the 485 million pounds consumed in the United States was produced from these sources. In 1960, however, three plants to produce naphthalene from petroleum were commissioned (see section on dealkylation of aromatics). In 1968 naphthalene production was greater than 1,500 million pounds, about 75 per cent coming from petroleum sources, and this can readily be expanded. With a plentiful supply of naphthalene assured one may expect the development of many new products and processes which will utilize this material.

Purification of Aromatics. Aromatic chemicals are found in the effluent streams from catalytic reformers and catalytic crackers, in amounts ranging from 10 to 65 per cent, together with paraffin and naphthene hydrocarbons. Thus, purification of the aromatics consists of (1) separating them from the non-aromatics, and (2) separating them into the pure compounds. Several methods are available for carrying out the first step. These are: (1) liquid-liquid extraction; (2) extractive or azeotropic distillation; and (3) adsorption.

The Udex[15,25] *extraction* process (Fig. 14.23) is widely used for separating aromatics from non-aromatic compounds previously mentioned. The solvent used in this process is aqueous diethylene glycol, which exhibits a selectivity directly proportional to the C/H ratio of the feed components and inversely proportional to their boiling points. The solu-

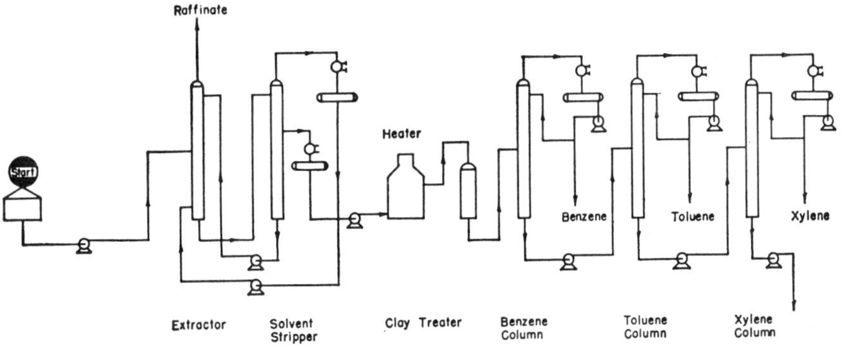

Fig. 14.23 Udex extraction process for purification of aromatics. [*Pet. Ref.*, **36**, *No. 11, 304 (1957); copyright 1957 by the Gulf Publishing Co.*]

bility of the hydrocarbons is dependent on the water content of the glycol.

Reformate from catalytic reformers is contacted with the glycol-water solvent in an extraction tower where the aromatic components dissolve in the solvent and leave the bottom of the column. The aromatics are recovered from the extract by stripping and then sent to the purification section for recovery of the pure components. The stripped solvent is recycled to the extraction tower.

Recovery of the individual aromatics is accomplished by heating and clay treating the aromatic mixture from the solvent extraction and then distilling it in a three-column distillation train. There are obtained relatively pure fractions of benzene and toluene, and a xylene fraction which is a mixture of the *ortho*, *meta*, and *para* isomers (20, 40, and 20 per cent, in the same order), together with about 20 per cent ethylbenzene. Some other C_8 compounds are present and must be removed. The process for separation and purification of the xylenes and ethylbenzene combines distillation with fractional crystallization.[8] Separation of *o*- from *p*-xylene is by distillation, *p*-xylene from ethylbenzene by fractionation (350 plates), and *m*- from *p*-xylene by fractional crystallization.

Dealkylation processes for producing benzene,[2,5] naphthalene, and phenol[29] were introduced during 1960. The raw material for the process can be either toluene or a C_8 fraction for benzene, alkylnaphthalenes for naphthalene, and cresols for phenol. The process is typified by the production of naphthalene (Fig. 14.24).

Alkylnaphthalene fractions from naphthenic crude oils, cycle oils (from catalytic cracking units), or other sources is combined with hydrogen and heated to about 1000 to 1200°F.[4] The heated mixture enters the reactor where dealkylation occurs:

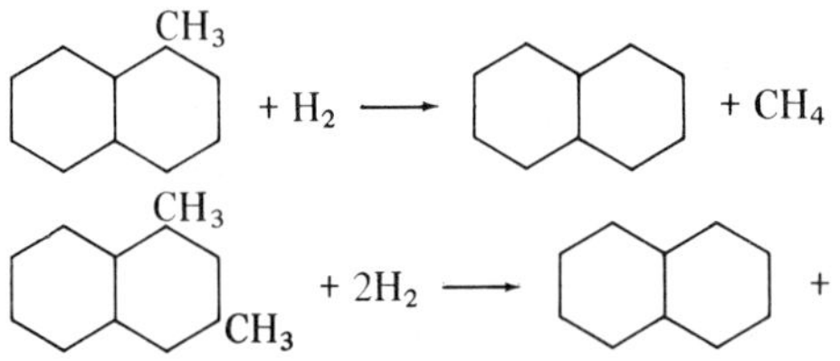

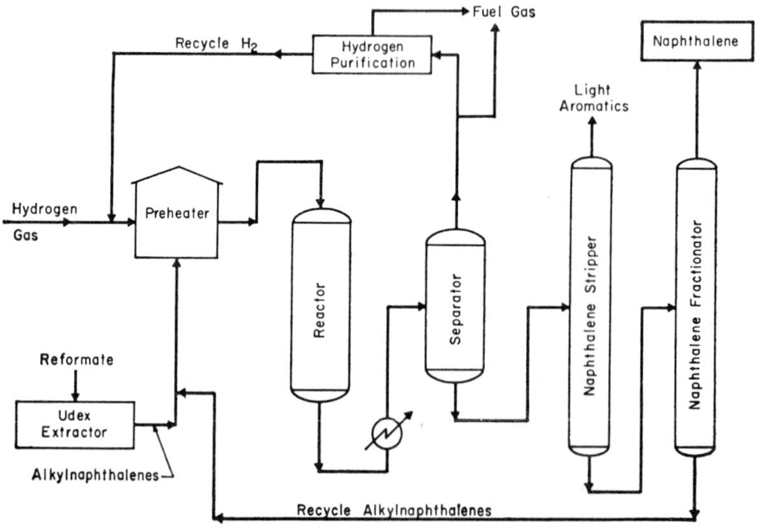

Fig. 14.24 Hydeal dealkylation for production of naphthalene. [*Adapted from Chem. Week*, **88**, *No. 11, March 18 (1961).*]

It is thought that the reactor contains an alkalized chromia-alumina catalyst and operates at about 800 psig.[4] The effluent from the reactor is cooled and hydrogen is separated from the liquid product which is stripped to remove light hydrocarbons and fractionated to remove the unreacted heavy materials. The product is naphthalene.

REFERENCES

1. Benedict, Q. E., *Petrol. Refiner*, **31**, No. 9, 103 (1952).
2. *Chem. Eng.*, **68**, No. 5, 76 (1961).
3. *Chem. Eng. Prog.*, **54**, No. 1, 41 (1958).
4. *Chem. Week*, **88**, No. 9, 47 (1961).
5. *Chem. Week*, **88**, No. 11, 32 (1961).
6. Guthrie, V. B., *Petrol. Engr.*, **30**, No. 1, C-7 (1958).
7. *Ind. Eng. Chem.*, **45**, No. 11, 2596 (1953).
8. *Ind. Eng. Chem.*, **47**, 250 and 1096 (1955).
9. *Ind. Eng. Chem.*, **47**, No. 8, 1578 (1955).
10. *Ind. Eng. Chem.*, **48**, No. 5, 168 (1956).
11. Lake, G. R., *Trans. Am. Inst. Chem. Engr.*, **41**, 327 (1945).
12. Nelson, W. L., "Petroleum Refiner Engineering," 4th Ed., p. 651, New York, McGraw-Hill, 1958.
13. *Ibid*, p. 682–686.
14. *Ibid*, p. 810.
15. *Petrol. Process*, **10**, No. 8, 1199 (1955).
16. *Petrol. Process*, **12**, No. 5, 97 (1957).
17. *Pet. Ref.*, **31**, No. 5, 97 (1952).
18. *Pet. Ref.*, **31**, No. 8, 71 (1952).
19. *Pet. Ref.*, **32**, No. 7, 102 (1953).
20. *Petrol. Refiner*, **32**, No. 9, 125 (1953).
21. *Petrol. Refiner*, **33**, No. 7, 135 (1954).
22. *Petrol. Refiner*, **33**, No. 12, 173 (1954).
23. *Petrol. Refiner*, **36**, No. 6, 179 (1957).
24. *Petrol. Refiner*, **36**, No. 9, 211 (1957).
25. *Petrol. Refiner*, **36**, No. 11 (1957).
26. *Petrol. Refiner*, **37**, No. 9, 223 (1958).
27. *Ibid*, p. 180.
28. *Petrol. Refiner*, **38**, No. 11, 204 (1959).
29. *Ibid*, p. 258.
30. Rogers, T. L., *Petrol. Refiner*, **36**, No. 7, 141 (1957).
31. *Texaco Star*, **47**, No. 4, (1960).
32. White, David, *Bull. Am. Assoc, Petrol. Geologists*, **19**, 469–502 (1935).
33. *Ibid*, 501.
34. Blume, J. H., *et al.*, *Hydrocarbon Process.*, **48**, No. 9, 133 (1969).

SELECTED REFERENCES

1. Sittig, M., "Catalytic Cracking Techniques in Review," *Petrol. Refiner*, **31**, No. 9, 263–316 (Sept. 1952).
2. Sittig, M., "How to Get Those Top Octanes," *Petrol. Refiner*, **34**, No. 9, 230–280 (Sept. 1955).
3. Kay, H., "What Hydrogen Treating Can Do," *Petrol. Refiner*, **35**, No. 9, 303–318 (Sept. 1956).
4. Process Handbook Issue, *Petrol. Refiner*, **37**, No. 9 (Sept. 1958).
5. Process Handbook Issue, *Petrol. Refiner*, **39**, No. 9 (Sept. 1960).
6. "Petroleum Horizons," New York, Lummus Company.
7. Hengstbeck, R. J., "Petroleum Processing," New York, McGraw-Hill, 1959.
8. Nelson, W. L. "Petroleum Refinery Engineering," 4th Ed., New York, McGraw-Hill, 1948.
9. Purdy, G. A., "Petroleum-Prehistoric to Petrochemicals," New York, McGraw-Hill, 1958.

10. Bell, H. S., "American Petroleum Refining," 4th Ed., New York, Van Nostrand Reinhold, 1959.
11. Kalichevsky, V. A. and Kobe, K. A., "Petroleum Refining with Chemicals," New York, Van Nostrand Reinhold, 1956.
12. Petrochemical Handbook Issue, *Petrol. Refiner*, **49**, No. 9 (Sept. 1970).
13. Petrochemical Handbook Issue, *Petrol. Refiner*, **50**, No. 11 (Nov. 1969).
14. Petrochemical Handbook Issue, *Petrol. Refiner*, **38**, No. 11 (Nov. 1960).
15. *Petrol. Process.*, **10**, No. 8, pp. 1157–1204 (Aug. 1955).
16. Astle, J. A., "The Chemistry of Petrochemicals," New York, Van Nostrand Reinhold, 1956.
17. Goldstein, R. F., "The Petroleum Chemicals Industry," New York, John Wiley & Sons, 1958.

15

Industrial Chemistry of Wood

Edwin C. Jahn* and Roger W. Strauss*

Wood was first used by man in his distant prehistoric antiquity. Certainly, the use of wood goes back to his most primitive culture. Wood has been employed by man for hundreds of millenia in two general ways, viz., (1) as found in nature and reshaped for specific uses, and (2) as a raw material to make something else, such as heat. Today, this ever-renewable resource from the forest provides the raw material for these same two basic classes of use, though on a greatly expanded level.

Present demands for wood are enormous and are continuously growing. Wood is almost as basic as food and is about as plentiful. Wood remains today the world's most widely used industrial raw material. It is the source not only of lumber and wood products but also of most of the world's packaging materails. Most of the world's rayon, cellophane, and cellulose lacquers and plastics are now made from chemical wood pulp.

Forest lands account for one-third of the total land area of the United States. These lands are the source of the raw material for the fifth largest industrial complex in the United States, namely, the forest-products industries which provide a wide range of wood and fiber commodities, from paper to houses. They are also a source of many other values, often harder to define economically, such as grazing, water supply, wildlife and fishes, recreation, and aesthetics.

This huge forest biome is rapidly being put under enormously increasing pressures due to a combination of causes, among which are: increase in per capita consumer demand for paper and wood products, accelerating population growth, and changes in patterns of use due to urbanization and increased leisure time.

A few examples[1,2] are sufficient to point up dramatically these increasing pressures on our forest resources. Per capita consumption of paper products has increased from 255 pounds in 1940 to about 570 pounds in 1970 and will be close to 800 pounds by 2000. Plywood and veneer consumption was 64 square feet per capita in 1962 and is expected to be about

*State University of New York College of Environmental Science and Forestry, Syracuse, N.Y.

100 square feet by 2000. By 2000 total demands for timber products are predicted to rise 80 per cent over the 1962 figure, the population by 74 per cent, and the gross national product to increase 3.5 times. At the same time there is a constant over-all reduction in timber-growing land due to the shifting use, such as for rights-of-way, urban and industrial expansion, parks and recreation areas, and protection of scenic and aesthetic values. A net nationwide reduction of timber lands on the order of 30 million acres or nearly 6 per cent of the 1968 total is considered possible over the next 30 years.[2] For these combined reasons it is expected that the demand for wood will exceed the supply by about the year 2000.

Besides these increasing human demands and pressures upon our forests there is the alarming sacrifice and tribute of trees to the relentless attack of diseases and insects upon our forests and ornamental trees. Wm. E. Waters and Robt. W. Brandt[3] of the U.S. Forest Service point out that 2.4 billion cu ft of timber are killed each year by insects and diseases. This, plus the resulting equal amount of growth-loss is about two-fifths of the annual harvest of timber. This is equivalent to 2.5 million low-cost homes. In dollars, the loss is $600 million, which with a 25-fold increase in value in finished products, amounts to an over-all loss of $15 billion per year. This does not include the value of losses to ornamental trees. For example, the Dutch elm disease is killing about 400,000 ornamental elms per year. The removal cost alone for these dead trees is about $80 million annually.

It is obvious that our present disease and insect pest-control methods by no means contain these depredations upon our forest and ornamental trees. A substantial cutting back on these losses would go a long way to extend our timber supplies.[4]

Furthermore, the manufacture of forest products utilizes only a portion of the available wood. There are great losses in terms of unsuitable or neglected species, in logging residues, in manufacturing residues, such as bark, slabs, edgings, trimmings, shavings, and sawdust left during the manufacture of wood products, and in the lignin and other organic materials dissolved in the pulping liquors. It is estimated that the combined wood residue not utilized is nearly 40 per cent of the forest drain.

Chemical utilization of wood has always been attractive to the chemist and the conservationist as a means of getting greater value from a given quantity of timber. Progress has been made in this direction through integrated forest operations and the use of wood residues from sawmilling for pulp.

Wood is not only one of our greatest resources, but, like agriculture, it is continually renewable. By most measures, forest products rank fifth in importance among industries in the United States. In terms of value of products produced, the pulp and paper industry by itself ranks fifth. Though the pulp and paper industry is a major industry, the other wood-chemical industries are small by comparison. Some have gone into decline, such as wood distillation, due to competition from synthetic processes, and others are comparatively new, such as chemically modified wood. The chemical wood industries also differ in other ways. Some, such as charcoal burning, which is centuries old, are still primitive in many areas. Others, such as paper, are modern and aggressive. The products which are treated in this chapter include pulp and paper, fiber boards, paper-base laminates, products derived from pyrolysis and chemical degradation, wood sugars and alcohol, tree exudates, extractives, modified wood, wood-plastic combinations, and composite boards.

CHEMICAL NATURE OF WOOD

Wood is a supporting and conducting tissue for the tree. To serve these functions, about 90 per cent of wood tissue is composed of strong, relatively thick-walled long cells. These, when separated from each other, are fine fibers very suitable for paper making.

Chemically, the cell wall tissue of wood is a complex mixture of polymers. These polymers fall into two groups, the polysaccharides and lignin. The polysaccharides of wood are collectively known as *holocellulose*, meaning total cellulosic carbohydrates. The holocellulose accounts for about 70 to 80 per cent of the extractive-free woody tissue, with lignin making up the remainder.

The holocellulose is composed of cellulose and a mixture of other polysaccharides, collectively termed *hemicelluloses.* Cellulose is a high-molecular weight linear polymer composed of glucose anhydride units. Hemicellulose is a mixture of shorter-chain polymers of the anhydrides of xylose, arabinose, glucose, mannose, and galactose, with xylan and galactoglucomannan as the most prevalent species. Cellulose makes up the main framework of the cell walls of the wood fibers. It is highly resistant chemically, whereas the hemicelluloses have relatively low resistance to acids and alkalis.

Lignin serves as the adhesive material of wood, cementing the fibers and other cells together to form the firm anatomical structure of wood. Lignin is a complex polymer of condensed phenylpropane units (see Fig. 15.5). It is susceptible to degradation and dissolution by strong alkalis at elevated temperatures, by acid sulphite solutions at elevated temperatures, and by oxidizing agents. Thus, lignin can be removed from the wood, leaving the separated cellulosic fibers in the form of a pulp.

In addition to the cell-wall substance, wood also contains extraneous materials present in the cavities of the cells. In some woods these are present in considerable amounts and are commercially important. The volatile oils and resins of the southern pines are an example. The extraneous materials are numerous and cover a wide range of chemically different materials. Most of these may be separated from the wood by steam distillation and solvent extraction as indicated below.

Steam distillation—terpene hydrocarbons, esters, acids, alcohols, aldehydes, aliphatic hydrocarbons.

Ether extraction—fats, fatty acids, resins, resin acids, phytosterols, waxes, nonvolatile hydrocarbons, and the above volatile compounds if not previously removed by steam distillation.

Alcohol-benzene extraction—most of the ether-soluble materials plus phlobaphenes, coloring matter and some tannin.

Alcohol extraction—tannin and most of the above organic material, except some resins.

Water extraction—sugars, cyclitols, starch, gums, mucilages, pectins, galactans, and some inorganic salts, tannins, and pigments.

The chemical composition of wood varies between species. The hardwoods and softwoods of the temperature zone show consistent differences. The hardwoods have less lignin and more hemicelluloses than the softwoods. Furthermore, the hemicelluloses of the hardwoods are high in xylan, whereas those of the softwoods are high in galactoglucoman-

TABLE 15.1 Chemical Analysis of Some North American Woods

	Hardwoods				*Softwoods*			
	Aspen	*White Birch*	*Beech*	*Basswood*	*White Spruce*	*Balsam Fir*	*Jack Pine*	*Eastern Hemlock*
Lignin[a]	20.9	19.0	22.1	19.8	28.6	30.0	28.6	32.5
Cellulose[a]	42.7	43.1	44.0	45.8	42.5	42.2	42.1	41.6
Acetyl-4-methyl-glucuronoxylan[a]	30.8	34.7	30.3	31.5				
Arabino-4-methyl-glucuronoxylan[a]					9.3	8.4	12.0	7.2
Glucomannan[a]	5.2	3.0	3.2	2.3				
Acetylgalacto-glucomannan[a]					19.2	19.1	17.1	18.5
Ash[a]	0.4	0.2	0.4	0.6	0.4	0.3	0.2	0.2
Solubility[b]								
Hot water	2.8	2.7	1.5	2.4	2.2	3.6	3.7	3.4
Ethyl ether	1.9	2.4	0.7	2.1	2.1	1.8	4.3	0.7

[a]Values based on oven-dry extractive-free (alcohol-benzene) wood. Analyses by Dr. Tore E. Timell.
[b]Values based on unextracted wood (oven-dry basis).

TABLE 15.2 Paper Consumption in the United States

Paper Consumption by Year		*Paper Consumption by Grades (1969)*	
Year	*Consumption (lb/capita)*	*Grade*	*Consumption (lb/capita)*
1810	1	Newsprint	95
1869	20	Book papers	70
1899	58	Writing papers	30
1919	119	Packaging	60
1929	220	Tissue and sanitary	35
1948	356	Paper board	250
1956	434	Construction	40
1969	576		

nan content and contain a smaller amount of xylan. The chemical composition of a few North American woods is shown in Table 15.1.

MANUFACTURE OF PULP AND PAPER

The importance of the pulp and paper industry to the American economy is exemplified by the growth rate in the use of paper and paper products. New uses are continuously being found for paper, and these developments together with the rising standard of living have resulted in a constant increase in the per capita consumption of paper. As shown in Table 15.2 the paper industry in the 19th century, dependent on rags as the source of fiber, was stagnant until the middle part of the century when processes were invented for the production of fiber from wood. Since that time the industry has shown a constant growth. Table 15.2 also shows the breakdown of paper consumption by various grades. The United States in 1969 produced a total of 53,490,518 tons of paper and paperboard in 825 separate mill locations. This paper and board was made from 41,647,000 tons of pulp, produced primarily from wood in 309 pulp mills, and 10,446,000 tons of recycled waste paper.

Table 15.3 lists the ten leading states for both pulp and paper production in 1968.

Since only a small amount of paper is made from rags or other fiber sources (agricultural residues), it is obvious that most wood pulp is produced from those areas of the country that are heavily forested. It is estimated that 63 million cords of pulpwood (a standard cord is 4 ft by 4 ft by 8 ft) were consumed in 1969 in producing the 42 million tons of pulp.

An examination of Table 15.3 shows that,

TABLE 15.3 Paper and Pulp Production by States

	Paper Production		*Pulp Production*	
State	*Tons*	*Ranking*	*Tons*	*Ranking*
Georgia	4,035,000	(1)	4,829,000	(1)
Wisconsin	3,027,000	(2)	1,444,000	(10)
Alabama	3,002,000	(3)	3,134,000	(4)
Louisiana	2,621,000	(4)	2,433,000	(6)
Washington	2,594,000	(5)	3,883,000	(2)
Oregon	2,532,000	(6)	2,503,000	(5)
New York	2,463,000	(7)	520,000	(22)
Michigan	2,227,000	(8)	575,000	(20)
Florida	2,123,000	(9)	3,165,000	(3)
Maine	2,106,363	(10)	2,173,000	(7)
South Carolina	1,826,000	(14)	2,013,000	(8)
Virginia	2,013,000	(12)	1,857,065	(9)

with the exception of Maine and Wisconsin, pulp production is concentrated in the southern and northwestern sections of the United States. While a high percentage of pulp is converted into paper or board at the same plant site, a significant portion (classified as "market pulp") is sold in bales to mills in other sections of the country for subsequent manufacture into paper. Thus, while New York and Michigan are relatively important in paper making they rank quite low as pulp producers. However, both New York and Michigan are large users of waste paper which is reprocessed into usable fiber. With the current emphasis upon recycling it is probable that the use of waste paper will be greatly accelerated in the near future.

Raw materials for the pulp and paper industry can be classified as fibrous and nonfibrous. Wood accounts for over 95 per cent of the fibrous raw material (other than waste paper) in the United States. Cotton and linen rags, cotton linters, cereal straws, esparto, hemp, jute, flax, bagasse, and bamboo are also used and in some countries are the major source of paper-making fiber.

Wood is converted into pulp by mechanical, chemical, or semichemical processes. Sulfite, sulfate (kraft), and soda are the common chemical processes while neutral sulfite is the principal semichemical process. Coniferous wood species (softwoods) are the most desirable, but the deciduous, broad-leaved species (hardwoods) have gained rapidly in their usage and constituted about 25 per cent of the pulpwood used in 1969. Table 15.4 lists the production of pulp by the various processes.

Nonfibrous raw materials include the chemicals used for the preparation of pulping liquors and bleaching solutions and the various additions to the fiber during the papermaking process. For pulping and bleaching, these raw materials include sulfur, lime, limestone, caustic soda, salt cake, soda ash, hydrogen peroxide, chlorine, sodium chlorate, and magnesium hydroxide. For papermaking they include rosin, starch, alum, kaolin clay, titanium dioxide, dyestuffs, and numerous other specialty chemicals.

TABLE 15.4 Woodpulp Production by Grade 1969

	Tons
Dissolving and special alpha grades	1,701,000
Sulfite paper grades	2,338,000
Sulfate paper grades	28,037,000
Soda	189,000
Groundwood	4,241,000
Semichemical	3,376,000
Other nonchemical	1,294,000

Wood Preparation

The bark of trees contains relatively little fiber, which is of poor quality, and much strongly colored nonfibrous material. It will usually appear as dark colored dirt specks in the finished paper. Therefore, for all but low-grade pulps, bark should be removed as much as possible, and this must be very thorough in the case of groundwood and sulfite pulps if the finished paper is to appear clean.

Debarking is usually done in a drum barker where bark is removed by the rubbing action of logs against each other in a large rotating drum (see Fig. 15.1). Hydraulic barkers using high-pressure water jets are excellent for large logs and are common on the West Coast. Mechanical knife barkers are becoming more common and are used extensively in smaller operations because of their lower capital cost. Also they have found widespread use in sawmills to debark logs prior to sawing so that the wood wastes can be used to produce pulp.

Wood cut in the spring of the year during the active growing season is very easy to peel. Much of the spruce and fir cut in the North is still hand peeled during this season and usually represents the optimum in bark removal.

The standard log length used in the Northeast is 48 inches while 63 inches is common in the South. Wood is generally measured by log volume, a standard cord being considered as containing 128 cubic feet. Large timber on the West Coast is generally measured in board feet of solid volume. Measurement and purchase of wood on a weight basis is practiced and has the advantage of being more directly related to fiber content.

The recent growth in the use of wood residues has been phenomenal. By barking the sawlogs, the slabs, edgings, and other trim-

Fig. 15.1 Wood being barked by two 12-ft by 45-ft barking drums. (*Courtesy Chicago Bridge and Iron Co.*)

mings that were formerly burned can now be used to make pulp. Table 15.5 shows the relative sources of wood in the United States.

TABLE 15.5 Sources of Wood for Pulp in the United States
(1970–Equivalent Cords of 128 Cubic Feet)

	Cords
Softwood pulpwood	34,063,000
Softwood residues	14,852,000
Hardwood pulpwood	13,054,000
Hardwood residues	3,241,000

Thus, almost 28 per cent of the wood used by the pulp industry in 1970 could be classified as waste wood. Several mills have been built that use no logs whatsoever but depend on residuals from satellite sawmill operations. Special sawmilling equipment has been developed to produce sawdust of a proper size so that it can be used also. These residuals are usually purchased in units of 2400 pounds of dry wood.

Wood used in producing groundwood or mechanical pulp requires no further preparation after debarking, but that used in the other chemical processes must first be chipped into small pieces averaging $^1/_2$-1 in. in length and about $^1/_8$-$^1/_4$ in. in thickness.

Chipping is accomplished with a machine consisting of a rotating disc with knives mounted radially in slots in the face of the disc (Fig. 15.2). Modern chippers have up to twelve knives, and the end of the logs are fed against the disc at about a 45° angle. Each knife cuts a layer of wood equal in thickness to the distance the knife protrudes. In chipping, it is necessary only to cut the wood across the grain since the layers of wood cut on the chipper immediately break up along the grain due to the forces exerted by the chipper. A typical chipper using pulpwood up to 20 in. in diameter would have a disc about eight feet in diameter rotating at 600 rpm and handle 15-25 cords per hour. Whole-log chippers on the West Coast have been built with discs up to 14 ft in diameter that will accept logs up to 36-40 in. in diameter and 30 ft long. Chip size is not uniform and screens are necessary to separate the oversize chips and sawdust from the acceptable chips.

Mechanical Pulping

Mechanical pulping, as the term implies, does not involve a chemical process. However, it is

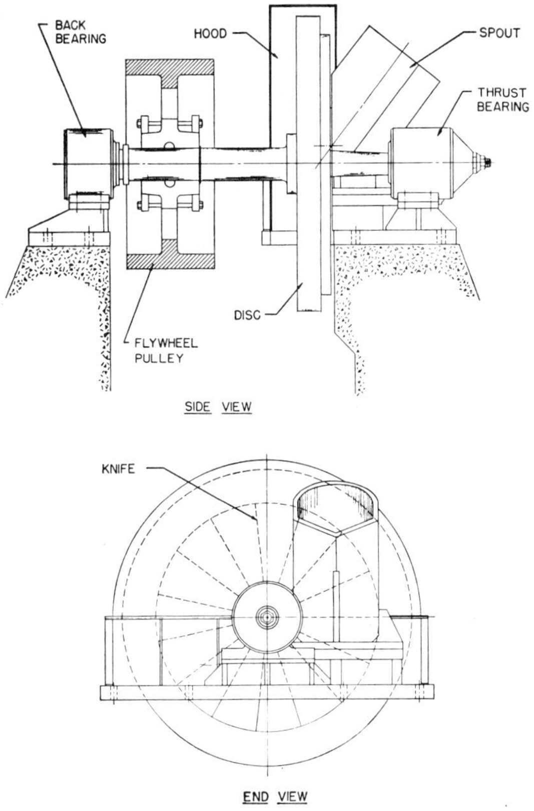

Fig. 15.2 Sketch of modern wood chipper. *(Courtesy Carthage Machine Co.)*

one of the more important methods of making pulp such as newsprint which consists of about 80 per cent mechanical (or groundwood) pulp.

Groundwood pulp is made by forcing the whole log against the face of a cylindrical abrasive stone rotating at relatively high speeds (Fig. 15.3). The logs are positioned so that their axes are parallel to the axis of the rotating stone. Sufficient water must be added to the stone to serve as a coolant and carry the pulp away.

At one time, natural sandstone was used for the grindstone but modern stones use either silicon carbide or aluminum oxide grits in a vitrified clay binder. Thus the characteristics of the stone can be varied to produce pulps "tailor-made" to fit their desired end use. Groundwood pulps for roofing or flooring felts must be extremely coarse and free draining and require a stone with large grits, whereas newsprint pulps are very fine and require the use of small grits.

Pulp characteristics can also be varied by changing the stone surface pattern, the stone speed, the pressure of the logs against the stone, and the temperature of the ground-pulp slurry. Generally a coarser and more freely draining pulp is obtained with a coarse surface pattern and high speed, pressure, and temperature. Type and condition of the wood are also factors, but groundwood pulps are usually made from the coniferous or long-fibered species since the deciduous or short-fibered species give very weak pulps.

Many designs of machines are used for grinding wood. The pocket grinder (Great Northern type) is usually equipped with two pockets for holding the wood and utilizing hydraulic pistons to force the wood against the stone. At the end of the cycle, the piston is retracted and a new charge of wood is placed in the pocket. Continuous or magazine type grinders use large chains to force the wood against the stone thus giving a more uniform grinding action. There are several advantages to each type and both are in common use. Modern grinders use up to 10,000 hp per stone and can produce up to 120 tons of pulp per day.

A recent development has been chip groundwood. Using the chips mentioned before it is possible to produce groundwood by using "refiners" (Fig. 15.4). The chips are fed between rotating discs containing steel plates with fine teeth. The rolling and cutting action of the plates as the chip moves from the center to the periphery of the discs reduces the chip to pulp. Although the process uses slightly more power it does produce a stronger groundwood and also has the added advantage of allowing some chemical pretreatment of the chips. Thus, a class of pulp called chemimechanical pulp is produced by treating the chips with a dilute solution of NaOH or Na_2SO_3 prior to refining. Hardwoods such as aspen have been used to produce acceptable pulps by this process.

Groundwood or mechanical pulp is low in strength compared to the chemical pulps. It is composed of a mixture of individual fibers, broken fibers, fines, and bundles of fibers.

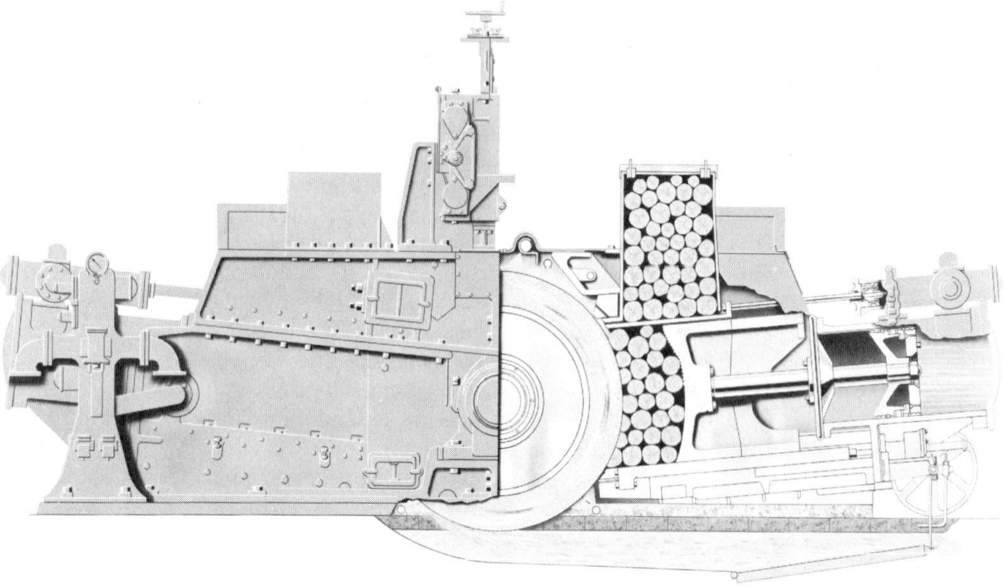

Fig. 15.3 Large two-pocket grinder for producing groundwood. (*Courtesy Koehring-Waterous Ltd.*)

Fig. 15.4 Single rotating disc refiner. (*Courtesy Sprout, Waldron and Co.*)

Papers made from groundwood also lose strength and turn yellow with time. Thus, groundwood pulps are used only in relatively impermanent papers such as newsprint, catalogue, magazine, and paperboard. Groundwood papers have excellent printing qualities because of high bulk, smoothness, resiliency, and good ink absorption. Newsprint contains about 80 per cent groundwood and the other papers mentioned about 30-70 per cent with the remainder being a chemical pulp for greater strength. Groundwood is the cheapest pulp made and also utilizes the entire wood giving essentially 100 per cent yield.

Chemical Pulping

Lignin is the noncarbohydrate portion of extractive-free wood and, as shown in Table 15.1, accounts for 20-30 per cent of the weight of wood. In manufacturing highly purified pulps of high whiteness, it is necessary to remove all of the lignin with minimum loss or degradation of the carbohydrate cell wall. Usually this is done in two steps involving pulping to liberate the individual fibers and then bleaching these fibers to the desired whiteness. Unbleached pulps are usually tan to dark brown in color and are used in that form for grocery bags, wrapping paper, and corrugated containers.

Lignin is not a specific chemical entity but is rather a very complex, heterogeneous, three-dimensional polymer consisting largely of phenylpropane units joined together by various ether and carbon-carbon linkages. It cannot be isolated from wood in its entirety without extensive alteration of its structure and requires rather drastic conditions for its removal, which in turn may cause it to undergo further condensation reactions. About 70 per cent of the lignin exists in the intercellular regions, and the rest is intimately associated with the carbohydrates in the cell wall. Thus lignin should be considered more as a *class* of related materials since its composition is not the same in all living plants. For example, significant differences exist between softwood and hardwood lignin and so it is probably correct only to refer to it in relation to its plant source and the method of isolation. Figure 15.5 shows a proposed structure that accounts for many of the reactions and chemical properties of lignin, but it should not be construed as a definite chemical formula.

The objective of chemical pulping is to solubilize and remove the lignin portion of wood leaving the industrial fiber composed essentially of pure carbohydrate material. While many variations are used throughout the world, the most convenient classification of pulping methods is by whether they are acidic or alkaline. Each has its own specific advantages and disadvantages but as can be seen in Table 15.4 the alkaline (sulfate) process accounts for over 90 per cent of all chemical pulp produced in the United States. All processes use aqueous systems under heat and pressure.

The sulfite process uses a cooking liquor of sulfurous acid and a salt of this acid. While calcium was the most widely used base at one time it has been supplanted by sodium, magnesium, and ammonia.

The sulfate process uses a mixture of sodium hydroxide and sodium sulfide as the active chemical. The term sulfate process is misleading, but it is called that because it uses sodium sulfate as the make-up chemical. The term kraft is also used to describe this process and is derived from the Swedish or German word for strength since it does produce the strongest pulp. Historically, sodium hydroxide alone (soda process) was first used as the alkaline pulping agent, but very few mills are still in operation since the pulp is weak and inferior to sulfate pulps.

Alkaline Processes. The pulping (cooking) process traditionally was performed on a batch basis in a large pressure vessel called a digester. Conditions used will vary depending upon the type of wood being pulped and the quality of end-product desired. Typical conditions for kraft cooking are listed in Table 15.6.

Digesters are cylindrical in shape with a dome at the top and a cone at the bottom. Ranging in size up to 40 ft high and 20 ft in diameter, the largest will hold about 7,000 cu ft of wood chips (about 35 tons) for each charge. The chips are admitted through a large valve at the top, and at the end of the cook they are blown from the bottom through a

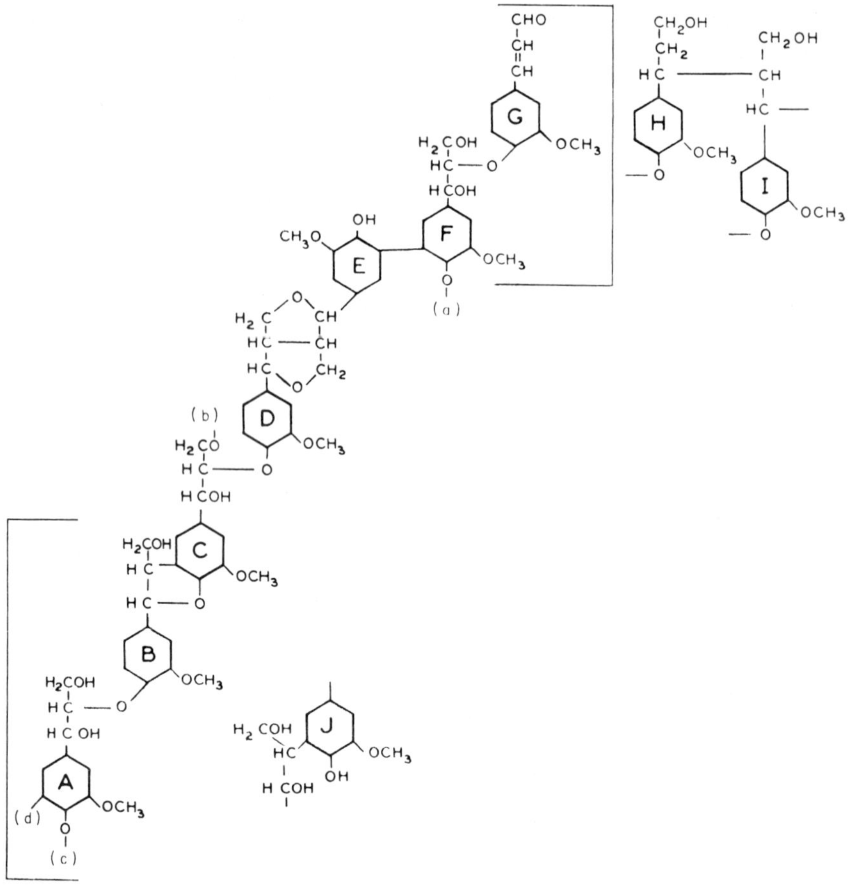

Fig. 15.5 Structure for protolignin from conifer or gymnosperm species. All rings are of the aromatic benzene type; bonds are "open" at positions *a*, *b*, *c*, and *d* to indicate other possible bonding points. (After Adler and co-workers.) (*From "Pulp and Paper Manufacture," Vol. I, R. G. Macdonald and Franklin, eds.; copyright 1969 by McGraw-Hill Book Co., and used by permission of the copyright owner.*)

valve to a large blow tank. During the cook the liquor is heated by circulation through a steam heat exchanger, which also avoids the dilution of the cooking liquor that would occur from heating by direct injection of steam.

TABLE 15.6 Typical Sulfate Pulping Conditions

Pressure	100–110 psig
Temperature	170–175°C
Time	2–3 hrs
Alkali charge	15–25% of weight of wood (calculated as Na_2O but consisting of approximately 5 NaOH + 2 Na_2S).

Liquor to wood ratio is 4 to 1 (by weight).

In recent years the development of the continuous digester (Fig. 15.6) has been a very important factor, especially in the kraft industry. Chips are admitted continuously at the top through a special high pressure feeder, and the cooked pulp is withdrawn continuously from the bottom through a special blow unit. Recent installations range in size up to 150 ft high and are capable of producing about 1000 tons of pulp per day in one unit. Cooking liquors and conditions are approximately the same as for the batch digesters. These units offer both good economics in the

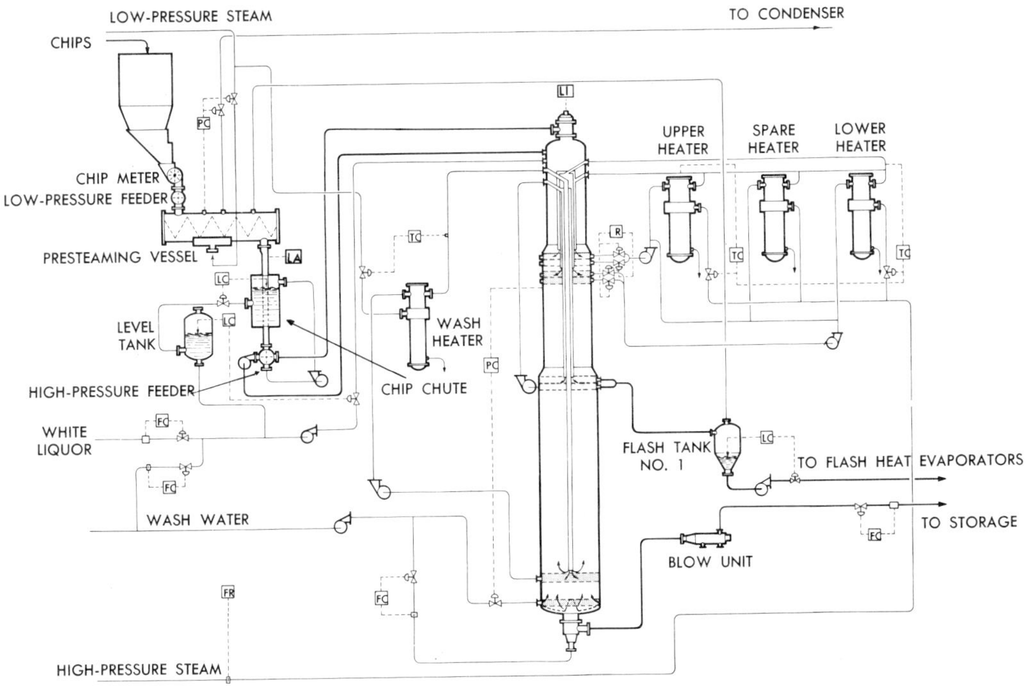

Fig. 15.6 Typical Kamyr continuous cooking system for sulfate pulp. (*Courtesy Kamyr, Inc.*)

production of pulp and a quality advantage compared to the batch digester. However, since capital investment is somewhat higher, both systems are being installed in new mills.

Due to the high alkali charge, the chemicals must be recovered and re-used. This also alleviates pollution problems since the yield of pulp is only about 45 per cent of the original wood weight and these organic residues must be eliminated. After being cooked in the digester, the pulp is washed in a countercurrent rotary vacuum washer system using three or four stages. The pulp is then ready for bleaching or for use in papers such as grocery bags where the brown color is not objectionable.

The separated liquor is very dark and is known as "black liquor." It is concentrated in multieffect evaporators to 60–65 per cent solids. At this concentration the quantity of dissolved organic compounds from the wood (lignin and carbohydrates) is sufficient to allow the liquor to be burned in the recovery furnace.

By controlling the amount of excess air admitted to the furnace and the temperatures, the organics in the liquor can be burned. The inorganics collect on the bottom of the furnace as a molten smelt of Na_2CO_3 and Na_2S. Sodium sulfate is added to the liquor as make-up and is reduced to Na_2S by carbon. After dissolving in water this mixture (called "green liquor") is reacted with slaked lime.

$$Na_2CO_3 + Ca(OH)_2 \longrightarrow 2NaOH + CaCO_3$$

Since the Na_2S does not react with the lime the resultant mixture of NaOH and Na_2S (called "white liquor") can be re-used to pulp more wood. The $CaCO_3$ sludge is filtered off, burned in a lime kiln and re-used. Thus, the chemical system is a closed one as shown in Fig. 15.7 and this minimizes costs and pollution.

The kraft process has had a serious problem with air pollution due to the production of hydrogen sulfide, mercaptans, and other vile-smelling sulfur compounds. In recent years the use of various techniques such as black-liquor oxidation, improved evaporators and furnaces, and control of emissions has greatly

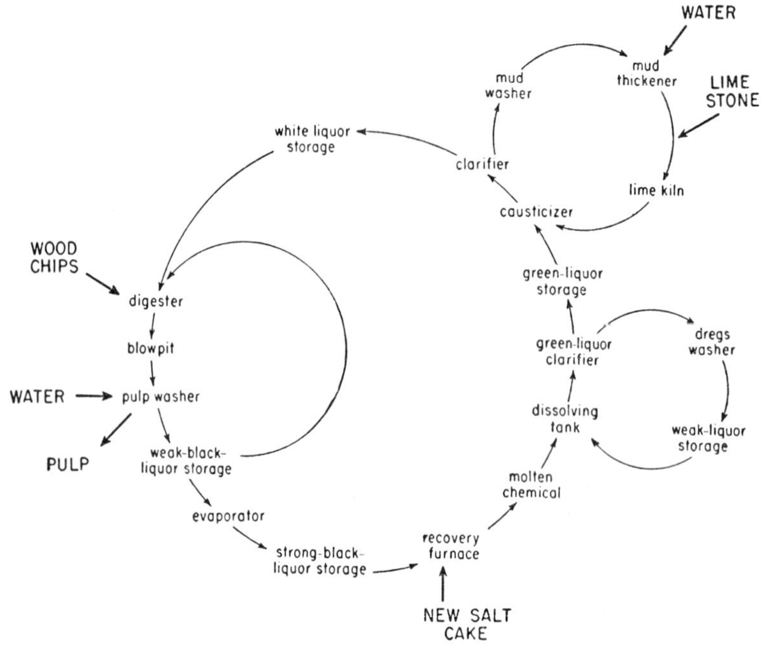

Fig. 15.7 Diagram showing cyclic nature of the kraft recovery process. (*"Pulp and Paper," 2nd Ed., Casey, J. P., Interscience Publishers, New York, 1967; by permission of John Wiley and Sons, Inc.*)

improved this situation. However, older mills are being forced to expend large amounts of money to bring their operations up to the new standards.

Practically any kind of wood can be pulped by the kraft process and, since it produces the strongest pulps with good economies, it has grown to be the dominant process in the world. While the odor problem is very severe it does not appear as though this process will be supplanted in the near future, instead it will be improved and modified. When pulping resin-rich woods such as southern pine the kraft process yields turpentine and tall oil as valuable by-products. The steam generated in the recovery furnace is almost enough to make the pulp mill self-sufficient.

Sulphite Process. Lignin will react with the bisulfite ion (HSO_3^-) under acidic conditions to form lignosulfonates that are soluble in water. For many years this was the preferred process since it produced pulps of light color that could easily be bleached, used cheap chemicals in fairly limited amounts so that no recovery was necessary, and was a relatively simple process to operate.

While production of sulfite pulp has remained relatively constant for the last thirty years the rapid growth of kraft pulping has reduced its share to less than 10 per cent of the chemical pulp produced. There are several reasons for this, but the primary ones are the inability to cook resinous woods such as pine, problems in producing strong pulps from hardwoods, and recently of the greatest importance is the lack of cheap and simple recovery systems such as in the kraft process to reduce water pollution problems. However, this process produces pulps with special qualities such as high-alpha-cellulose grades for rayon so that it will continue to be used.

Initially calcium was the preferred base since it was cheap and convenient to use. However, since no recovery system is available, most calcium-base mills have either ceased operation or have converted to sodium, magnesium, or ammonia for which recovery systems are available.

Regardless of the base used the initial step

is the burning of sulfur to produce sulfur dioxide (SO_2). The air supply to the burner must be carefully controlled, since too much air will enhance the formation of sulfur trioxide (SO_3) and subsequent production of sulfuric acid (H_2SO_4) which is very undesirable. The gas must also be cooled quickly from 1000°C leaving the burner to below 400°C also to minimize formation of SO_3. After cooling to 20–30°C the SO_2 gas must be absorbed in water and reacted with the proper base to form the cooking liquor.

For calcium-base liquor the gas was passed through towers packed with limestone with water flowing down through the tower. Because of the limited solubility of calcium bisulfite [$Ca(HSO_3)_2$] the pH of the liquor was very low (about 2) and free sulfurous acid was present. This is usually called the acid sulfite process. As mentioned before calcium-base mills have essentially disappeared in the United States.

The so-called soluble bases are now used with each having certain advantages. Since solutions of sodium, magnesium, and ammonium bisulfite are all soluble at pH 4.5, the current practice is to pulp at the higher pH and is usually called bisulfite pulping. Extremely long cooking times (7–10 hrs) are necessary with acid sulfite whereas 4–5 hrs are sufficient with bisulfite.

Sodium base is the easiest to prepare (Na_2CO_3 or NaOH is usually used as the make-up chemical) and gives the highest quality pulp. However, while recovery processes are available they are complicated and expensive. Magnesium base [from $Mg(OH)_2$] is somewhat more difficult to handle but good recovery systems are available and a majority of the sulfite pulp is now produced from this base. Ammonium base (from NH_4OH) has been used in the past. While the ammonia cannot be recovered, the liquor can be evaporated and burned without leaving any solid residue, thus reducing water pollution. As long as aqueous ammonia remains low in price this process will be attractive, since the SO_2 can be recovered from the waste gases by passing them through a wet scrubber flooded with fresh ammonium hydroxide.

Batch digesters are usually used in the sulfite process. Cooking temperatures are lower (140–150°C) and times are longer. Pulp yields are about the same as in the kraft process.

Spruce and fir are the preferred species for cooking by the sulfite process since they produce relatively strong, light-colored pulps. About 20 per cent of newsprint consists of this type of pulp that has not been bleached. Thus the sulfite industry is concentrated in Canada, northern United States, and the Pacific Coast where the supplies of spruce and fir are greatest and the largest quantities of newsprint are produced.

A large amount of research has been done on developing products from the waste sulfite liquor and some success has been achieved. Vanillin, alcohol, and torula yeast can be produced as by-products, while the lignosulfonates are used as viscosity modifiers in drilling muds and similar uses. However, the majority of waste liquor is burned to recover the cooking chemicals and the heat values.

Other Pulping Processes. Various combinations of chemical and mechanical treatment have been used to produce pulps with specific properties. Mild chemical treatments to give partial delignification and softening are followed by mechanical means to complete fiber separation.

The neutral sulfite semichemical (NSSC) process is one in which wood chips, usually from hardwoods, are cooked with Na_2SO_3 liquor buffered with either $NaHCO_3$, Na_2CO_3, or NaOH to maintain a slightly alkaline pH during the cook. Unbleached pulp from hardwoods cooked to a yield of about 75 per cent is widely used for corrugating medium. While bleachable pulps can be produced by this process, they require large quantities of bleaching chemicals and the waste liquors are difficult to recover. Currently many NSSC mills are located adjacent to kraft mills and the liquors can be treated in the same furnace. Thus the waste liquor from the NSSC mill becomes the make-up chemical for the kraft mill, solving the waste problem. NSSC hardwood pulp is the premier pulp for corrugating medium and cannot be matched by any other process.

Chemimechanical pulps are usually produced by soaking the chips in solutions of NaOH or Na_2SO_3, and then refining them in disc refiners to produce a groundwood-type pulp. Chemical consumption is very low and yields are usually 85–95 per cent. Currently much research is being done to improve the qualities of the pulp produced by this method.

Screening and Cleaning of Wood Pulp

The desired pulp fibers are usually between one and three mm in length with a diameter about one-hundredth as large. Any bundles of fibers or other impurities would show up as defects in the finished paper and must be screened out. Wood knots are usually difficult to pulp and must be removed.

Screening is usually a two-stage process with the coarse material being removed by screens with relatively large perforations ($^1/_4$ to $^3/_8$ in.). Additional fine screening is done with screens using very small (0.008 to 0.014 in.) slots to insure removal of oversized impurities. Screen size openings will depend on the species of wood being processed and the desired quality of end-product. Because of the tendency of the fibers to agglomerate when suspended in water it is customary to screen at the very low solids (consistencies) of about $^1/_2$ per cent fiber and $99^1/_2$ per cent water.

To meet the ever increasing demands for cleaner pulps the centrifugal cyclone cleaner (Fig. 15.8) has come into almost universal use. The screened pulp is pumped through these units at low consistencies and high velocities. The fiber slurry enters the cone tangentially at the top and a free vortex is formed with the velocity of the flow greatly increased as the diameter of the conical section is reduced. Heavier particles of sand, scale, or other dirt are forced to the outside of the cleaner and are discharged from the bottom tip through a small orifice. Due to the velocity gradients existing in the cone the longer fibers (75–95 per cent) are carried into the ascending center column and are discharged through the larger accept nozzle at the top. In a properly designed and operated unit a shape separation is also made, so that round particles, even though of the same specific gravity as the good fibers, will also be discharged as rejects through the bottom orifice. In this way, small pieces of bark are also removed. To reduce the quantity of rejects to an acceptable level they in turn are processed through a second, third, or even fourth stage of cleaners, thus holding the final loss of pulp to about $^1/_4$–1 per cent of the feed depending upon quality demands and the dirt level of the incoming pulp.

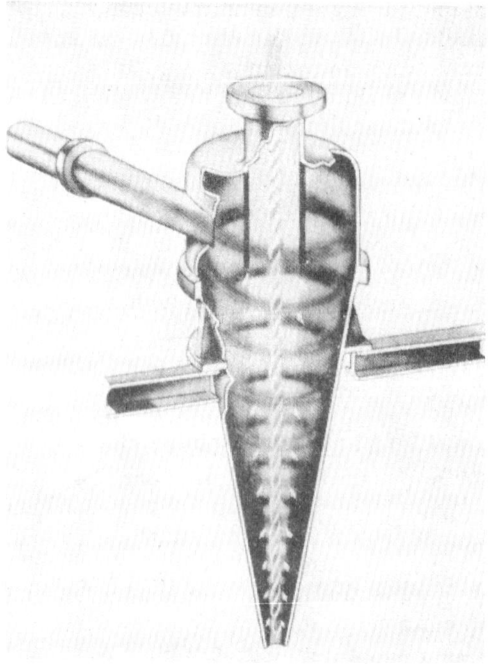

Fig. 15.8 Centrifugal cleaner for removal of dirt. (*Courtesy Bauer Brothers Co.*)

Bleaching of Wood Pulp

The color of unbleached pulp ranges from the cream or tan of the sulfite process to the dark brown of the sulfate (kraft) process. While about 75–90 per cent of the lignin has been removed by the pulping process, the remainder, along with other colored degradation products, must be removed by bleaching.

While it is possible to improve the brightness (whiteness) of the pulp in one stage, achievement of high brightness on an economic basis requires the use of several stages. Current practice uses combinations of chlorination with elemental chlorine (C), alkaline extraction with sodium hydroxide (E), and various oxidative stages using sodium or calcium hypochlorite (H), chlorine dioxide (D), or hydrogen peroxide (P). The pulp is washed between each stage to remove solubilized impurities. Many combinations are possible and each mill selects the sequence that fits their requirements the best. Those most commonly in use are:

CEH	CED
CEHD	CEDED
CEHDP	CEHDED

The greater the number of stages the higher the quality of the final pulp but at increased cost. Sulfite pulps are much easier to bleach and usually use only the 3- or 4-stage sequences, whereas kraft pulps require additional stages. Typical conditions for bleaching a kraft pulp would be as given in Table 15.7.

TABLE 15.7 Typical Conditions for Bleaching Kraft Pulp

Stage	*% Chemical on Pulp*	*Time (min)*	*Temperature (°F)*
Chlorination	5–6	30	70
Extraction	2½–3	60	140
Hypochlorite	1–2	90	105
Chlorine dioxide	½–1	240	160

Chlorine and caustic are purchased, but chlorine dioxide must be generated at the site using sodium chlorate as the basic chemical.

Effluents from bleach plants are a source of great concern in regard to pollution. Large quantities of water are discharged (typically 10–20 thousand gal per ton) which are high in color, especially the water coming from the chlorination and extraction stages. Two experimental bleach plants are now operating with a new sequence that should alleviate this problem, and as soon as these prove commercially feasible it is anticipated that a significant change in bleaching practice will take place.

These new developments involve the replacement of the chlorination and extraction stages with a single stage involving gaseous oxygen and sodium hydroxide. The pulp is dewatered to about 15-25 per cent solids, treated with about 4–6 per cent NaOH, and passed through a reactor in a fluffed condition using oxygen gas at about 150 psi. The pulp consumes about 1-2 per cent oxygen in about 15 mins and leaves in a semibleached condition. After washing, the use of conventional chlorine dioxide stages will produce the desired level of whiteness. The advantage of this process is that the effluent contains no chlorides and can be used as process water in the pulp mill. Any sodium and organics in the effluent will then enter the kraft recovery cycle and serve as make-up, thus reducing the pollution load. Currently the main disadvantage is the obvious one of introducing a new process involving rather unusual pieces of equipment, also there is some degradation of the cellulosic fiber since oxygen will attack cellulose as well as lignin. However, it is expected that these problems will not prove insurmountable and a rapid introduction of this process is anticipated.

These bleaching sequences are designed to remove lignin, yielding a highly purified fiber consisting only of carbohydrate material. When producing high-yield pulps such as groundwood where all of the lignin is retained in the pulp it is not possible to use these systems. However, extremely high brightness is not required, thus, some improvement is attained by using one stage with either peroxide or hydrosulfite (dithionite). No yield loss is encountered as the action of both of these is merely to decolorize the pulp rather than remove any impurities. Usually about ½ per cent of either of these chemicals will give a noticeable increase in brightness and are widely used to upgrade the quality of groundwood.

Pulping Other Fibrous Materials

Fibrous materials for papermaking other than wood are waste paper, cotton and linen rags, waste manila rope, cotton linters, and grasses such as bamboo, esparto, bagasse (sugar cane residue), cereal straw, flax, jute, and hemp. As improved technology is developed these other sources of fibers (particularly bagasse) will play a more important part in those countries where wood is in short supply. However, the United States, Canada, Scandinavia, and Soviet Russia will continue to depend on wood for all but a very small fraction of their needs. While the use of tropical hardwoods has been limited it is anticipated that countries such as Brazil will assume major importance as pulp suppliers in the near future.

Currently about 20 per cent of the paper produced in the United States is made from

TABLE 15.8 Consumption of Selected Paperstocks

Grade	*Paperstock (tons)*	*Consumption (%)*
Old newspapers	2,350,000	20.6
Old containers	3,500,000	30.7
Mixed waste	3,350,000	29.4
Special grades	2,200,000	19.3
TOTAL	11,400,000	100.0

Estimated Production of Selected Secondary Fiber Products

Product	*Production from Paperstock, Tons*	*Proportion of Total U.S. Consumption of the Grade Indicated, %*
Paper (from deinked waste)		
Newsprint	320,000	4.0
Book, uncoated	400,000	15.4
Printing, coated	300,000	9.1
Writing	300,000	10.7
TOTAL	1,320,000	
Paperboard (from waste used as is)		
Folding and Set-Up Boxboard	3,100,000	66.0
Other Boxboard	1,600,000	41.0
Linerboard	300,000	3.9
Corrugating Medium	800,000	18.2
Other Containerboard	1,500,000	55.6
TOTAL	7,300,000	
Building products	1,700,000	38.7

In the case of paper the amount made from waste paper is 4.3% of the total U.S. consumption of paper, whereas for paperboard it is 31% of the total.

waste paper. The majority of this recycled paper (about 80 per cent) is used "as is" without attempting to remove ink, dyes, or pigments from the paper. The resultant pulp is of rather poor quality and color and is used primarily as filler stock in paperboard. The higher quality waste paper is treated with sodium hydroxide and steam, and then bleached to produce a high quality pulp that is used to replace virgin pulp from wood. Due to the pressure of municipalities to reduce their solid waste disposal, it is anticipated that the amount of waste paper used will increase in the next few years. Table 15.8 shows the types and quantities of waste used as well as the quantities of paper and board made from that waste.

Stock Preparation

Stock preparation in a paper mill includes all intermediate operations between preparation of the pulp and the final papermaking process. It can be subdivided into (1) preparation of the "furnish" and (2) "beating" or "refining." Furnish is the term used to describe the water slurry of fiber and other chemicals which goes to the paper machine. Beating or refining refers to the mechanical treatment given to the furnish to develop the strength properties of the pulp and impart the proper characteristics to the finished paper.

Cellulosic fibers are unique in that when suspended in water they will bond to each other very strongly as the water is removed by filtration and drying, without the necessity of an additional adhesive. This is due to the large number of hydrogen bonds which form between the surfaces of fibers that are in close contact as the water is removed. This bonding is reversible and accounts for the well-known fact that paper loses most of its strength when wet. If paper is suspended in water and agitated it will separate into the individual fibers which allows the easy re-use of waste paper or the processing-waste from the paper mill itself.

In order to enhance the bonding capability of the fibers it is necessary to mechanically beat or refine them in equipment such as beaters, jordans, or disc refiners. This treatment of the pulp slurry at about 3-6 per cent consistency is done by passing the pulp between the two rotating surfaces of the refiner. These surfaces contain metal bars and operate at very close clearances. As the fibers pass between the bars they are made more flexible and a larger surface for bonding is developed by the mechanical action.

This refining brings about fundamental changes in the pulp fibers and increases the degree of interfiber bonding in the final sheet of paper. Thus the final properties of the paper can be significantly changed by varying the degree and type of refining. As additional refining is performed, properties such as tensile, fold, and density are increased while tear resistance, opacity, thickness, and dimensional stability are decreased. Thus, the proper refining conditions must be selected to bring out the desired properties without detracting too much from other properties.

The furnish of a paper machine varies widely depending on the grade of paper being made. Newsprint usually consists of about 80 per cent groundwood and 20 per cent chemical fiber (sulfite or semibleached sulfate). Bag papers and linerboard are usually 100 per cent unbleached softwood kraft, although obviously these could be made from bleached fibers if white containers are desired. Printing papers are made from bleached pulps and contain both hardwoods and softwoods. By selecting the proper pulps and refining conditions a wide variety of paper qualities can be achieved.

However, the paper industry is a large user of chemicals, as it has been found that relatively small quantities of additives can materially change the properties of paper. Use of 1-2 per cent rosin size and 2-3 per cent alum [$Al_2(SO_4)_3$] will greatly increase the resistance of paper to penetration by water or ink. Pigments such as kaolin clay, calcium carbonate, and titanium dioxide are added in amounts up to 15% to increase opacity and give a better printing surface. Organic dyes and colored pigments are added to produce the highly colored papers used in business and printing papers. Other additives such as wet-strength resins, retention aids, and starch can be used to give particular properties that are needed. Thus, in order to produce the wide variety of grades of paper now available the papermaker selects the proper pulps, refining conditions, and additives, and then combines the pulp and additives before sending them on to the paper machine for the final step in the process.

Papermaking Process

Some paper mills are not integrated with pulp mills and it is necessary for these mills to use dried, baled pulp manufactured at a separate location. Many mills making limited quantities of highly specialized papers fall into this category since it allows maximum flexibility in selecting the optimum pulps for a particular paper grade. However, the papermaking process is the same regardless of the source of pulp.

After the furnish has been prepared with the proper refining treatment and additives, it is stored in the machine chest and then fed continuously into the paper machine system. A refiner or jordan is placed in this line to give the paper-machine operators the opportunity to make small adjustments in the quality of the furnish as needed to give the desired paper properties. Screens and centrifugal cleaners are also included to insure high quality paper.

The papermaking process is essentially a system whereby the pulp is diluted to a very low consistency (about ½ per cent) and continuously formed into a sheet of paper at high speeds, and then the water is removed by filtration, pressing, and drying. The basic units of the Fourdrinier Paper Machine are diagramed in Fig. 15.9 and a picture shown in Fig. 15.10.

The section of the paper machine where the paper is formed is referred to as the "wet end." The fourdrinier machine is characterized by a headbox which allows the diluted stock to flow through an orifice (slice) onto the flat

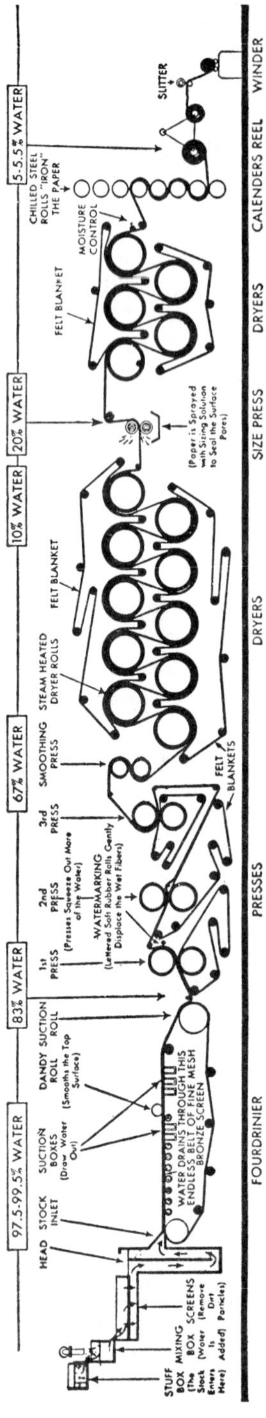

Fig. 15.9 Diagram of Fourdrinier paper machine. *(Courtesy Hammermill Paper Co.)*

moving wire. This is actually an endless wire belt which returns on the underside of the machine thus allowing the process to proceed continuously. As a low headbox consistency (about 1/2 per cent) is necessary for good formation the volumes of water handled are very large (about 10,000 gal/min for a machine producing 300 tons per day). Much of the water is removed through the wire by the action of the table rolls and foils that support the wires in the forming area immediately following the headbox. At this point the stock consistency has been raised to about 2 per cent and the stock no longer drains freely. By passing over suction boxes operating at fairly high vacuum (6-8 in. Hg) the consistency is raised to about 15-20 per cent. A dandy roll (covered with woven wire) rotating on top of the wet paper is used to improve the formation and can impart a watermark if it contains the proper pattern. The suction roll after the suction boxes serves several purposes as it removes additional water; serves as the driving roll for the wire, and serves as the point at which the paper separates from the wire and passes into the press section while the wire returns to pick up additional pulp at the headbox.

Modern fourdrinier machines are available up to a width of 360 inches and operate at speeds up to 3,000 ft/min. Newsprint machines are usually the widest and fastest while those making heavier grades such as linerboard run somewhat more slowly. In the United States, there are several machines capable of producing over 1,000 tons per day of linerboard. More specialized grades such as bond and printing papers are usually produced at a lower speed on narrower machines, and 150-250 tons per day would be considered a high output. Many specialized grades such as filter paper and tracing paper are produced on very small, slow machines producing only a few tons per day. Machines making tissue paper for sanitary purposes use modifications of the standard fourdrinier to produce tissue at speeds of up to 5,000 ft/min. Because of the lightweight of the paper it is necessary to make many modifications in the equipment shown in Fig. 15.9.

A second method used to form paper is the

Fig. 15.10 Modern high-speed paper machine. (*Courtesy Beloit Corp.*)

cylinder machine. Actually these machines are used primarily to form the multi-ply board used in packaging such as cereal boxes. The cylinder wet end consists of one or more (up to eight) cylinder vats, each of which forms a separate wet web of fibers. Each vat contains a woven wire-covered cylinder rotating in the diluted pulp slurry. The liquid head on the outside of the cylinder is greater than that on the inside resulting in a flow of water through the wire and out of the vat. The pulp mat that is formed on the face of the cylinder is removed by an endless woolen felt which moves in contact with the cylinder by means of a rubber roll riding on top of the felt. With machines having more than one cylinder, the same felt moves from one cylinder to the next and the wet sheets from each cylinder are laminated to each other on the bottom side of the felt. Thus, very heavy papers or boards can be fabricated by multivat machines. Since each vat can be supplied with a different type of pulp, it is possible to make boards with a white surface of high-quality pulp and a center of low-cost pulp from waste newspaper or other cheap grades. Due to hydrodynamic problems, cylinder-machine speeds are limited to about 250 feet per minute and widths of about 150 in. However, because of the heavyweight board produced and the widespread use of cheap waste paper for most of the furnish, the cylinder machine is widely used. New forming units looking like miniature fourdrinier units (Ultra-former, Inverformer) have been developed and are rapidly replacing the old-fashioned cylinder vat since their speed is not as limited.

From the wet end of the machine, the wet sheet is conveyed by woolen felts through a series of roll-type presses for further water removal, increasing the consistency to about 35 per cent. The sheet is then threaded through the dryer section consisting of a long series of steam-heated cast iron cylinders which reduce the moisture content to approximately 5 per cent which is about the equilibrium moisture content for cellulosic fibers at 40–50 per cent relative humidity. Tissue machines use one large dryer (called a Yankee dryer) ranging from 8 to 18 feet in diameter. Because of the light weight of the tissue paper, it can be dried at high speeds on a single dryer.

After drying, the paper is compacted and smoothed by passing through a calender stack consisting of a vertical row of highly polished cast iron rolls. The paper is then wound into rolls on the reel, and these rolls are then rewound on a winder into shipping rolls or

Fig. 15.11 Dry end of large Fourdrinier machine. (*Courtesy St. Regis Paper Co.*)

sheeted and trimmed to the desired size. Figure 15.11 shows the large reels at the dry end of a large fourdrinier machine.

The quality of many papers is improved by a surface treatment of the paper. A size press about two-thirds of the way along the dryer section can apply a solution of starch to improve surface bonding. More sophisticated coating equipment can apply a layer of pigments and binders to give the desirable properties associated with printing papers such as those used in high quality magazines and advertising.

Paper is converted into its end product by many different methods. Some papers are sheeted, shaped, or fastened into final forms while others require more elaborate processing. Corrugated boxboard is made by gluing sheets of linerboard to each side of a fluted sheet of corrugating medium. Papers for packaging may be laminated to polyethylene film, aluminum foil, or coated with waxes and hot-melt resins. The printing and bag- and box-making industries depend on the production of the many mills which produce the several hundred grades of paper used in the United States, and each user may require special paper characteristics to match his process.

BOARD AND STRUCTURAL MATERIALS

Board, sheets, panels and other structural materials are manufactured from wood fibers and various other vegetable fibers, from wood particles, and from paper. The industries making these products are not generally classified as chemical industries; nevertheless, they are closely related to chemical industry. Fiberboard manufacture is similar to papermaking; particle boards and paper laminates involve the use of synthetic resins and, therefore, chemical technology.

Fiberboard

Fiberboard is the term for rigid or semirigid sheet materials of widely varying densities and thicknesses manufactured from wood or other vegetable fibers. The board is formed by the felting of the fibers from a water slurry or an air suspension to produce a mat. Bonding agents may be incorporated to increase the strength, and other materials may be added to give special properties, such as resistance to moisture, fire, or decay.

Fiberboards are manufactured primarily for panels, insulation, and cover materials in buildings and other structures where flat sheets of moderate strength and/or insulating capacity are required. They are also used as components in doors, cupboards, cabinets, furniture, and millwork.

Classification of fiberboards is best done on the basis of density, since there is a great deal of overlap when classifying by use only. Table 15.9 shows the density classification of fiberboards as well as some of their major uses.

The production of fiberboards goes back to 1898 when the first plant was built in Great Britain. However, large-scale production, mainly of insulation board, developed in the United States between the two world wars. The United States is still the largest producing country and accounts for about a third of the world output. Fiberboard production figures for recent years are shown in Table 15.10. There has been a much more rapid increase in the production of compressed fiberboards (hardboards) than noncompressed fiberboards (insulation board) during recent years.

Wood is the principal raw material for the manufacture of fiberboards. The species used are numerous, including both softwoods (coniferous) and hardwoods (broad-leaved) and vary from region to region. The wood may be from the harvesting of commercial timber and pulp species, as well as from species not commonly used for lumber or pulp, from cull tim-

TABLE 15.9 Classification and Uses of Fiberboards*

Fiberboards	*Density Classification* g/cm^3	*Density Classification* $lb/cu\ ft$	*Major Uses*
Noncompressed (insulation board) semirigid insulation	0.02–0.15	1.25– 9.5	Heat insulation as blankets and batts; industrial cushioning
Rigid insulation board (includes wallboard and softboard)	0.15–0.40	9.5 –25.0	Heat and sound insulation as sheathing, interior panelling, base for plaster or siding, thick laminated sheets for structural decking, cores for doors and partitions, acoustical ceilings
Compressed			
Intermediate or medium density fiberboard (includes laminated paperboards and homogeneous boards)	0.40–0.80	25–50	Structural use and heat insulation as sheathing base for plaster and siding, interior panelling, containers, underflooring
Hardboard	0.80–1.20	50–75	Panelling, counter tops, components in doors, cabinets, cupboards, furniture, containers, and millwork, concrete forms, flooring
Densified hardboard (superhardboard)	1.20–1.45	75–90	Electrical instrument panels, templets, jigs, die stock

*From information in "Fibreboard and Particle Board," Food and Agriculture Organization of the United Nations, Rome, 1958.

TABLE 15.10 Fiberboard and Particle Board Production[a]

	Figures in 1000 metric tons 1966	1967	1968	1969
Hard pressed board				
(Compressed fiberboard)				
U.S.A.	980	874	1169	1245
World	4177	4241	5002	5384
Structural and insulating board				
(Noncompressed fiberboard)				
U.S.A.	1130	1090	1188	1143
World	2113	2119	2150	2132
	Figures in 1000 meters³			
Particle board				
U.S.A.	1155	1295	2522	3018
World	6237	7169	13690	16089

[a]Source: "Yearbook of Forest Products," Food and Agriculture Organization of the United Nations, Rome, Italy.

ber, from logging and forest management residues and from industrial wood residues. Other fiber raw materials for fiberboard manufacture are bagasse (sugar cane residue after sugar extraction) and waste paper. Only minor amounts of other plant fibers are used.

Wood handling and preparation for fiberboard manufacture is much the same as described for pulp and paper. Wood is debarked and chipped with the same type of equipment. If the chips are to be first extracted for resins or tannin, then cylinder or drum-type chippers may be used instead of disc chippers.

Fibers for fiberboard are coarser and less refined chemically than those used for paper. Processes are used which bring about fiber separation with minimum loss in chemical components and in maximum yield. The pulping processes used are generally the following: (1) mechanical, (2) thermal-mechanical, (3) semichemical, and (4) explosion methods.

Mechanical Pulping. The mechanical pulping process is the same as that described for making paper pulp. Stones of coarse grit are used to give somewhat coarser fiber of higher freeness than the usual groundwood for papermaking. Freeness is a measure of the ease of drainage of water from the pulp. A fast drainage rate is required since a thick mat is produced in forming the wet sheet and this must drain rapidly to maintain an economical rate of production.

A very coarse shredded wood fiber is made by a shredder, consisting of two cylinders to which are attached numerous small pointed hammers, which swing freely as the cylinders are rotated. These hammers "comb" or shred wet green wood, such as aspen, into stiff coarse bundles of fibers, producing a bulky pulp. This type of equipment is not widely used, and most mechanical pulp is made on conventional stone grinders.

Untreated green or water-soaked wood chips may also be directly refined in disc mills of either the single-rotary or double-rotary type similar to those used for making refiner groundwood. This gives a coarse pulp acceptable for insulation board.

Thermal-Mechanical Pulping. Generally, chips are given a steaming or other heat-treatment prior to or during defibering in a disc mill. Steaming or heating in hot water softens the wood so that, upon grinding, a pulp is produced with fewer broken fibers and with less coarse fiber bundles. Steaming is generally preferred and is carried out in a digester under a variety of conditions of time and temperature. A typical steaming period is about 30 min at 75 psi. If iron digesters are used, a small amount of alkali may be added to the chips to prevent corrosion by the organic acids produced by the hydrolytic action on the wood. The steamed chips are defibered in a disc-type attrition mill having two discs made of special alloys, one or both rotating, similar to those used for refiner groundwood.

The pulp may or may not be screened, as necessary. It is usually given some further refining to give maximum strength. Sizing and other additives are introduced and the pulp suspension is delivered to the wet-sheet-forming machine.

A special continuous thermal-mechanical process has been developed whereby the wood chips are steamed and ground while at elevated temperatures and pressures. The feature of the process is the combination of steaming and defibering in one unit in a continuous operation. The entire operation is carried out under pressure and no cooling takes place prior to defibering. Wood chips are continuously introduced by a plunger feed mechanism into a preheater where they are heated to 170–190°C by steam at 100–165 psi. Passage through the preheater takes 20–60 seconds, after which the hot chips are fed by a screw directly to the single rotating disc refiner, where they are ground while at the above temperature and pressure conditions. At these conditions the lignin, which is concentrated in the intercellular regions (middle lamella) of the wood, becomes somewhat thermoplastic, permitting easier separation of the fibers. The fibrous material is exhausted to the atmosphere through relief valves. Due to the very short steaming time, it is claimed that little hydrolysis takes place so that there is little loss in wood substance, the yields being 90–93 per cent. Additional refining is necessary for the preparation of insulation board stock, and slight refining may be desirable for hardboard stock, especially to break down slivers.

Semichemical Pulping. In some cases wood or other fibrous raw material may be given a mild chemical pretreatment prior to mechanical defibering. The processes are similar to those described for making paper pulps by semichemical methods. Generally, the conditions of treatment are somewhat milder than for paper pulp in order to get maximum yields. The chemical treatments usually involve cooking with neutral sulphite, caustic soda, or lime solutions. Yields from wood chips generally are above 80 per cent.

Explosion Process. A unique process for defibering wood was developed by W. H. Mason. Wood chips, about 3/4 in. long, prepared in conventional chippers and screened, are subjected to high pressure, in a cylinder, commonly called a gun, about 2 feet by 6 feet in size, and ejected through a quick-opening valve. The elevated temperature softens the chips and, upon ejection, they explode into a fluffy mass of fibers and fiber bundles. The process involves thermal plasticization of the lignin, partial hydrolysis, and disintegration by the sudden expansion of the steam within the chip.

About 260 lb of wood chips are fed into the cylinder and steamed to 600 psi for 30–60 secs. Then the pressure is quickly raised to 1000 psi (about 285°C) and held only about 5 seconds before suddenly releasing the charge into a cyclone. The time of treatment at this high pressure and temperature is critical and depends upon the species of wood, chip size, moisture content, and the quality of product desired. The steam is condensed in the cyclone and the exploded fiber falls into a stock chest where it is mixed with water and pumped through washers, refiners, and screens.

The high temperature to which the chips have been subjected causes appreciable hydrolysis of the hemicelluloses, resulting in a somewhat lower yield of pulp than obtained by mechanical or thermal-mechanical pulping. The hydrolysis results in a final board product with an enriched lignin content; about 38 per cent compared to about 26 per cent in the original softwood. The lignin content of the pulp can be varied by controlling the steaming process.

Board Forming. Pulp prepared by any of the above processes may be used for making insulation board. Mechanical or groundwood pulp was the first type of pulp used in large-scale production of insulation boards, and is still being used in many plants. Pulps from other sources, such as disc mills, many be admixed with it. Groundwood pulp is not considered satisfactory for hardboard, and most hardboard is made from pulp prepared by the explosion process or by defibering with disc refiners.

Board making is basically similar to papermaking and involves refining, screening, mixing of additives, sheet forming, and drying operations. Pressing is also required for hardboards. The pulp is refined and screened prior to sheet formation.

Sizing agents in amounts up to one per cent of the fiber are added to the pulp in mixing chests. Paraffin wax emulsion is commonly used for all types of board. For insulation boards rosin, cumarone resin, and asphalt are also used. Often a mixture of rosin and paraffin emulsion is used, with 10-25 per cent rosin in the mixture. For hardboards, paraffin wax is the most common sizing agent, though tall oil derivatives and phenol-formaldehyde resins are also used. The sizing agent is precipitated on the fibers by alum, with careful control of pH; the latter may be between 4.0 and 6.5, according to the conditions.

The strength properties of a fiberboard depend mainly upon the felting characteristics of the individual fibers and to a lesser degree upon their interfiber bonding. The felting or forming process is usually done from a water suspension of the fiber at a consistency of around one per cent. This is the *wet-felting* process. A relatively new *air-felting* or dry-forming process has been developed and is being used in a few American plants for hardboard manufacture.

The wet-felting process is generally carried out in a manner similar to papermaking, i.e., in a continuous operation on a fourdrinier machine or on a cylinder machine. The machines move more slowly than in the case of papermaking (5–45 ft/min on the fourdrinier) and a coarser mesh wire is employed. In the cylinder-machine method, a single large vacuum cylinder, 8–14 ft in diameter, or two cylinders counterrotating and forming a two-ply sheet are most commonly used. Further water removal is effected by suction boxes and press rolls. The wet sheet is cut to length and conveyed on rollers from the press sections to a tunnel type drier. When dried the sheets are cut into desired sizes.

A third type of wet-felting is a discontinuous method, known as the *deckle-box* method. The deckle box consists of a bottomless frame which can be raised or lowered onto a wire screen. A measured quantity of stock sufficient to form one sheet is pumped into the deckle box and vacuum applied to the lower side of the screen. After most of the water has drained off, pressure is applied from the top to express more water and compact the sheet, reducing its thickness. The deckle-frame is then raised and the sheet conveyed to the driers.

For the recent air-felting process, defibering is usually done in disc mills with control of the moisture content to give the minimum amount possible, consistent with good defibering conditions. Additional moisture may be removed by preheating the air that conveys the fiber from the refiners to a cyclone. The fiber may be further dried in a tunnel or other type of drier. Fines are removed either by air classification or screens after the drier. Wax for sizing is introduced either with the chips or added as a spray before or after passing the disc mills (about 2.5 per cent of the weight of the fiber). Sometimes 0.5–5 per cent of phenolic resin is added, depending upon the quality of board desired. The fiber-blend is fed to a moving screen by a metering unit through a combined air and mechanical action. The fibers felt as they fall on the screen and the fiber mat thus formed is precompressed between belts and/or rollers. If the board is to be wet pressed, water is added by spraying.

After the felting or sheet-forming operation, the subsequent operations differ for insulation board and hardboard (see Fig. 15.12). For insulation board, the sheets are dried without further compression and, for hardboard, the sheets are either pressed and dried simultaneously (wet-pressing) or are first dried and then pressed (dry-pressing). Air-felted sheets are pressed directly after forming.

Drying and Pressing. The wet-felted sheets for insulation board or for dry-pressed hardboard, containing 50–80 per cent water, may be dried by any of three methods, namely (1) tunnel kilns using racks or carts to support the sheets, (2) steam-platen dryers, and (3) continuous roller driers of single or multideck arrangement. Most widely used is the continuous roller-type multideck drier, which has an average length of 150 to 300 ft but may be more than 600 ft long. An average drier will have 8 decks and be 12 ft wide.

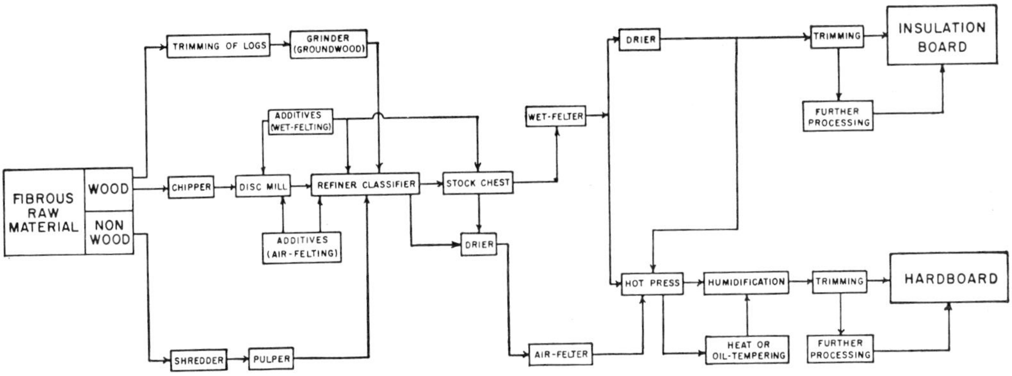

Fig. 15.12 Schematic outline for insulation board and hardboard manufacture. (*From "Fiberboard and Particle Board," Food and Agricultural Organization of the United Nations, Rome, 1958.*)

The pressing conditions greatly affect the board properties. The conditions of time, temperature, pressure, and moisture content will depend upon the fiber in the board and the product desired. In wet pressing, a typical cycle has 3 phases and lasts from 6–15 minutes. The first is a short high-pressure stage (up to 710 psi) to remove most of the free water and bring the board to the desired thickness; the second step serves to remove water vapor and requires most of the time; and the third stage is a final short period at high pressure to effect a final "cure" or bonding by plastic flow of the lignin. To secure this fiber-to-fiber bond a temperature of 185°C must be attained and temperatures up to 210°C may be used to increase production rate. In the dry pressing process the cycles are shorter (1½–3½ min) and the temperatures and pressures are usually higher.

Conditioning. After hardboard has been hot-pressed or has been heat-treated or oil-tempered, its moisture content is well below what it will reach in equilibrium with the atmosphere in normal use. Such very dry boards will change dimensions upon picking up moisture and may warp. It is important, therefore, to humidify the boards under controlled conditions before packaging. The desired equilibrium moisture content (E.M.C.) reached will vary from 5 to 12 per cent depending upon the nature of the board and the general humidity conditions in the region of use.

Most humidifying is done in chambers or tunnels, kept at 80–85 per cent R.H. and 38–50°C. A lesser-used system of conditioning is by water spraying and dipping followed by standing to allow uniform absorption.

Special Treatments. Hardboards are often given a special treatment to improve strength and resistance to moisture. *Heat treatment* is a method which has come into wide use. The boards, which are kept apart to permit hot air circulation, are heated in chambers by either batch, continuous, or progressive systems. Typical conditions are 5 hr at 155–160°C. The strength is increased (except impact), sometimes as much as 25 per cent, and the water resistance is improved. This operation may replace sizing wholly or in part. Some exothermic reaction takes place in the board and the heat developed must be removed by the hot circulating air to prevent burning. Probably some chemical condensations occur in the wood fiber, producing an internal resin system, and there is the possibility of some cross-linking of large molecules.

Some hardboard is *oil-tempered*. A drying oil, such as linseed, tung, perilla, soya, or tall oil, or an alkyd resin, is impregnated into the board, by passing the hot-pressed board through a hot oil bath. About 4–8 per cent of the oil is absorbed. The board is then heated in a kiln with circulating air at 160–170°C for 6–9 hr. This treatment hardens the oil as well as brings about chemical reactions in the fiber and results in greater strength and moisture resistance.

Various additives may be incorporated into insulation boards and hardboards, or they may be surface treated to bring about resistance to decay, insects, and fire. Pentachlorophenol, copper pentachlorophenate, and several arsenicals are commonly used for preservative treatments. The sodium salt of pentachlorophenol is added before sizing and is precipitated onto the fiber along with the size by the alum. Arsenic trioxide is usually added at the headbox of the board-forming machine. Special fire-retarding-paint coatings are sometimes used to give resistance to the spread of flame.

Particle Board

Particle boards are composed of discrete particles of wood bonded together by a synthetic resin adhesive, most commonly urea-formaldehyde or phenol-formaldehyde. The material is consolidated and the resin cured under heat and pressure. The strength of the product depends mainly upon the adhesive and not upon fiber felting as in the case of fiberboards, although the size and shape of the particles influence strength properties. They may be fine slivers, coarse slivers, planer shavings, shreds, or flakes. They are divided into two main groups, namely (1) hammer-mill produced particles (slivers and splinters from solid wood residues, feather-like wisps to block-shaped pieces from planer shavings) and (2) cutter-type particles sometimes called "engineered" particles (flakes and shreds). The various steps in particle-board manufacture are illustrated in Fig. 15.13.

Hammer-milled particles usually vary appreciably in size. Dry raw material produces greater amounts of fines than green wood. Cutting machines (either cylinder-type or rotating-disc-type) give more uniform particles, with the length dimension in the direction of the grain of the wood. The thickness, size, and shape of particles influence the strength of the board. Boards made from sawdust have the lowest strength properties, hammer-milled particles give boards of intermediate strengths and solid wood cut to flakes gives boards of highest strengths.

Particle boards may be made in a wide range of densities. Low density or insulating types are a comparatively recent development in Central Europe, whereas the high-density-hardboard types are an American development. Most particle-board production is in the middle-density range.

Particle boards are most commonly used as core stock for veneer in furniture and in doors, as interior panels for walls and ceilings, in sub-flooring, as sheathing and siding, and as components in interior millwork. The dense types are used in the same way as fiberboard hardboard, described in the previous section. Both dense particle boards and hardboards, after receiving a surface coating, may be printed with decorative designs.

Particle board production has increased rapidly, both in the United States and worldwide, in recent years. During the four-year period 1966–69, the increase was almost three-fold (see Table 15.10).

Paper-Base Laminates

Paper-base laminates are panels or other laminated assemblies composed of many plies of resin-treated paper molded together under high temperatures and pressure to produce

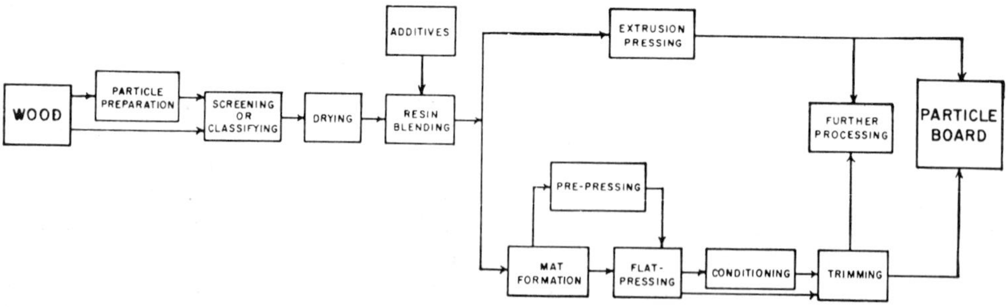

Fig. 15.13 Schematic outline for particle-board manufacture. (*From "Fiberboard and Particle Board," Food and Agriculture Organization of the United Nations, Rome, 1958.*)

rigid structures which no longer have the characteristics of paper. These products are widely used in the electrical and machine industries for insulators, gears, pulleys, and a multitude of machine parts. They possess high impact strength and toughness, good electrical insulation, high dimensional stability, and are not subject to corrosion and have a dampening effect on sound, eliminating rattle and drumming in steel cars and machinery. Furthermore, they can easily be manipulated into complex shapes and can be drilled, turned, and sawed. These properties make these products of great industrial value. They are also used in making trays, light flooring panels, table and counter tops and many other products employing panels.

Paper-base laminates fall into four major classes, namely (1) mechanical or structural, (2) electrical, (3) punching, and (4) decorative. Phenolic resins are especially suitable and mostly used where mechanical strength and resistance to heat, water, and electricity are required. For punching-grade laminates, phenolic resins are specially modified with plasticizers or drying oils to yield laminates having good plasticity and elasticity. For electrical grades the phenolics are generally catalyzed with ammonia, amines, or less conductive catalysts in place of the stronger alkaline catalysts otherwise used.

For decorative laminates, urea and melamine resins are the principal resins used. The melamines are used where translucent, light-colored products with good heat and water resistance are required. Polyesters and some melamines are used for low-pressure laminating, enabling the continuous production of counter and table tops by passing the assembly through a set of rolls and then through a heating chamber.

There are two broad general methods for introducing resins into papers for making paper-base laminates; namely (1) by the beater addition process and (2) by impregnation of the paper sheet. In the beater addition process the resin is added to the pulp in the beater and then precipitated on the fiber by alum or an acid. The resin adheres to the pulp fibers, which are then formed into a sheet. This type of paper-plastic combination is commonly termed *resin-filled paper.* These resin-filled papers may be made into flat or shaped preforms. The former are flat sheets, the latter are vacuum-felted to a shape closely conforming to that of the final molded product. Shaped preforms are used for deep forms requiring high strength contours. Little flow is required on molding, hence the paper sheet does not break and greater strength is thereby obtained. Stretchable cross-creped paper for postformable laminates is a new development.

The more common method of application of the resin is by impregnation of the wet or dry paper sheet (web) with a solution of resin. Such papers are termed *resin-impregnated papers.*

Water-soluble and alcohol-soluble types of phenolic resins are both used for paper-base laminates. The former tend to give more brittle but more dimensionally stable products than the latter. Phenolic resins are mostly applied by impregnation of the wet or dry paper sheet producing *resin-impregnated papers*, although some progress has been made in the slush stock addition procedures to produce phenolic *resin-filled papers.*

In the wet-web process for impregnating paper for laminates, the wet sheet of paper on the paper machine is carried through a resin bath while the sheet is supported by a wire. The sheet contains up to 65 per cent water and the amount of resin taken up depends upon the moisture content of the paper, solids content of the resin solution, viscosity of the resin solution, temperature of the bath, machine speed, and pressure of the squeeze-rolls. Only water-soluble or water-dispersible resins can be used in the wet-web process.

Generally, however, phenolic impregnations of paper for laminates and other purposes demanding a high resin content are done with dry paper on off-machine equipment. The dry paper sheet is continuously passed through a resin solution, generally an alcoholic solution, and then moves under and between two metering rolls, after which the paper is dried to remove solvent and to complete the condensation of the thermosetting phenolic resin.

The addition of phenolic resins in the form of emulsions to the slush stock by the beater-addition process has received considerable study in recent years. Papers containing 45 to 55 per cent resin and even as high as 65 per cent have been prepared. The resin is precipitated by adding alum to the beater. Papers prepared in this way are highly plastic and are suitable for low-pressure molding. The phenolic resins can also be used in combination with elastomers, such as GR-S types, neoprene, hycar and vinyl polymers, to produce laminates of high impact strength and greater elongation under tension.

Lignin is also commercially used as a resin for paper laminates. Its cheapness and availability as a by-product in pulp manufacture make it attractive. One of the most successful products is made with lignin recovered from the spent liquor of the soda pulping process. The lignin may be added to the pulp suspension in solution and precipitated directly among the fibers or added in a pre-precipitated form.

Properties of paper-base laminates depend both upon the resin and the paper. In general, the final product has the characteristics of the resin used, provided over 30 per cent is resin. The paper acts as a structural reinforcer, greatly enhancing toughness, and tensile, flexural, and impact strength.

Electrical properties of the laminate depend both upon the paper and the resin, though the amount of resin absorbed is the most important factor affecting electrical insulating properties. Papers of low-power factor, good dielectric strength, and high dielectric constant are necessary for electrical grades of laminates. Other desired paper properties for good paper-base laminates are uniformity, cleanliness (freedom from slime, dirt, and fiber bundles), low finish, neutral pH, freedom from chemicals (bleach residues, etc.), low and uniform moisture content (under 4 per cent), and uniform absorbency. High absorbency is desired when high electrical resistance and minimum water absorption are required, since greater amounts of resin are taken up by high absorbency papers. For high impact strength-punching grades of laminates, a low absorbency paper is necessary to reduce the amount of resin absorbed.

Substantial amounts of polymer resins are used in the production of paper laminates. During 1969 and 1970 an annual average of over 115 million pounds of phenolics and over 50 million pounds of urea and melamine resins were consumed in making paper laminates.

Polymer-Modified Papers, Overlays, and Honeycomb Cores

In addition to paper-base laminates, polymers are combined in many ways with paper to develop new properties or to modify or enhance certain properties. Paper is commonly combined with polymers, such as synthetic resins, elastomers, and plastics in general, to produce products which may be classified as follows: (1) wet strength papers, (2) special purpose papers containing large amounts of resins or elastomers (chemical resistant papers, sandpaper backings, gaskets, imitation leather, shoe parts, wood overlay and honeycomb core papers, air and oil filters, battery separators, etc.), (3) plastic coated papers, and (4) paper-base laminates. The use of plastics in combination with paper has grown enormously during the past two decades. For example, during 1969 and 1970 an annual average of 66 million lbs of urea and melamine resins were used for coating and treating paper and 326 million lbs of polyethylene were consumed in coating paper and paperboard. Many other polymers were also used in treating and coating papers; these included polyesters, styrene, polyvinyl chloride, their copolymers, and others. Polymer-modified papers are thus very important in our present-day economy.

The paper-base laminates have already been discussed. Two other polymer-treated papers used in construction and industrial materials will be briefly mentioned here, namely, overlay papers for lumber and plywood, and paper honeycomb cores. For the many other applications of polymer-modified papers the reader is referred elsewhere.

Overlaid lumber is a composite of lumber and phenolic resin-treated kraft paper. Similar paper overlays are applied to veneer, plywood, hard fiberboard and particle board. Paper honeycomb cores are also made from phenolic resin-treated kraft paper, which is formed into a honeycomb of different geometrical designs, such as figure eight or hexagonal, in special machines. Simpler types are made from resin impregnated corrugated sheets which can be assembled in several ways. This material permits construction of sandwich panels of light weight and high strength. Kraft papers for wood overlays or for honeycomb cores are treated with phenolic resin in the same ways as described above for paper-base laminates.

Paper overlays have three basic uses: masking, decorative, and structural. Masking overlays are used to cover minor defects and provide a more uniform paintable surface. Such overlays contain 20 to 25 per cent phenolic resin based on the weight of the paper. For structural purposes, one or more sheets of paper may be used in the laminate. High-density and medium-density types are produced for plywood. The high-density type contains not less than 40 per cent of a thermosetting resin, phenolic or melamine, and has a hard smooth surface not requiring further finishing and may be used for exterior service. The curing of the resin is completed at the same time the paper is bonded to the wood material in a hot press or, in the case of plywood, at the same time the veneer is assembled into plywood. The overlay sheet swells and shrinks much less than wood and thus exerts a resistance to the dimensional changes of the wood and may reduce lateral swelling by as much as 40 per cent. Overlay papers also upgrade the appearance of low grade lumber, increase strength properties, improve finishability, and increase resistance to weathering.

Decorative surfaces are obtained by applying a top sheet of white paper on which is a printed design. This is covered with a clear coating of melamine or vinyl resin. Or, a thin transparent (when cured) paper impregnated with melamine resin may be applied to a decorative veneer, providing a permanent protective finish.

For honeycomb cores, either water- or alcohol-soluble phenolic resins may be used. Many types of facings may be glued to the honeycomb cores, viz.: veneers, plywood, hardboards, asbestos board, aluminum, stainless steel, and paper-plastic laminates. Thin-sheet material may be used due to the almost continuous support of the core.

The honeycomb sandwich possesses great strength in relation to its weight. It may carry loads as much as 25 tons per sq ft. Strength and weight vary with weight of the paper, quantity of resin impregnated, and honeycomb design.

Honeycomb sandwich construction has many uses, such as in airplanes, cargo containers, truck and trailer bodies, railway passenger cars, cabins, barns, airplane hangars, house floors, walls, roofs, doors, and in a variety of other products. Besides combining strength with lightness, honeycomb sandwich material has high rigidity, good insulation properties, resistance to fungus and pests, and durability to temperature extremes.

NONWOVENS

Nonwovens are products made from fibers held together by a suitable adhesive. They are intermediate between true (woven) textiles and true (cellulose) papers and substitute for woven textiles in many uses, such as draperies.

The following processes are currently recognized:[5]

1. Carded (oriented), adhesive bonded.
2. Air-laid (random), adhesive bonded.
3. Water-laid (paper process), adhesive bonded.
4. Spun-bonded continuous filaments of synthetics, selfbonded (heat) or adhesive bonded.
5. Combinations of textile scrim and paper, adhesive bonded.

Currently, the textile oriented processes (carded and air-laid) account for over 80 per cent of production, whereas the paper oriented processes account for about 15 per cent of the total. The products made on the paper machine are stiffer, but they are less costly and can be made at relatively high speeds.

The typical water-laid furnish is composed of wood-pulp fibers and up to 25 per cent of synthetic or rayon fibers. After forming, the nonwovens are treated with an adhesive of suitable properties for the end use. The main purpose of the adhesive is to impart strength. It may be a latex, a water soluble polymer, or a monomer or low-molecular weight product that polymerizes to an insoluble resin upon drying or curing. Some latices may be added to the wet sheet (wet-end addition) before the sheet is dried. Other additives such as fire retardants may be added with the adhesive or applied by a separate treatment.

WOOD-PLASTICS COMBINATIONS AND MODIFIED WOOD

As in the case of paper, the use of synthetic resins in the wood industry has advanced with the developments in polymer chemistry. Polymers permit greater and better use of wood raw material, including waste and low-grade wood. They also improve strength and appearance, and decrease dimensional change, thereby improving the competitive position of wood in relation to other materials.

The combinations of polymer resins with wood in the manufacture of particle board, as well as the application of polymer-treated papers as overlays for lumber and plywood and as honeycomb cores for sandwich construction have been discussed above. Brief consideration will be given to the "modified" woods.

Modified wood refers to wood that has been subjected to chemical or heat treatment, with or without pressure, to bring about changes in its properties. Much research has been done, especially by the U. S. Forest Products Laboratory, on the modification of wood to overcome its dimensional instability and to impart other properties.

Reduction in hygroscopicity and improved dimensional stability of wood have been experimentally developed by the acetylation of wood and by heat stabilization. Heat stabilization is brought about by heating wood, preferably under the surface of molten metal or a fused salt to exclude oxygen; this procedure gives good temperature control and a minimum of loss in strength. Temperatures varying from 150 to 320°C have been used, with the used, with the time necessary to give a certain dimensional stability being reduced by half for each 10°C increase in temperature. Wood treated in this way has been termed *staybwood.*

When wood is heated, it exhibits a certain degree of thermoplasticity, as has been described in the discussion of fiberboards. Therefore, if heated wood is compressed (165–175°C at 100–140 kg/cm^2), the internal stresses due to compression are relieved by the plastic flow of the wood. Such a compressed wood, termed *staypak*, has improved resistance to moisture and to shrinking and swelling and improved strength properties. The mechanism of the stabilization of hot-compressed wood is probably due to the plastic flow of the lignin component of the wood.

The most effective means for modifying wood and developing improved properties is to produce a wood-plastics composite by forming a synthetic polymer within the interior of the wood. Two methods have been developed, one by impregnating the wood with a water solution of phenol-formaldehyde and the other by using liquid vinyl monomers.

To make a phenolic-wood composite, a water solution of low-molecular weight phenol-formaldehyde is impregnated into thin veneers or strips of wood under pressure. They are then cured slowly in a kiln to remove the water and set the resin, i.e., to polymerize it to a hard thermoset resin. This material, termed *impreg*, contains resin bonded to the internal capillary surfaces within the cell walls of the wood and, due to the volume of the resin, this keeps the wood in a partly swollen condition. Therefore, the wood retains a volume nearly corresponding to its water-swollen dimensions and goes through a much smaller volume change than untreated wood when it is immersed in water.

Wood impregnated with phenol-formaldehyde resins and dried, but not cured, has greater plasticity than untreated wood and, therefore, can be compressed at considerably lower pressures than untreated wood. For example, impregnated spruce, aspen, and cottonwood dried to 6 per cent moisture can be compressed at 149°C to one-half their original thickness at a pressure of 17.6 kg/cm.^{2} In contrast untreated woods of the same species and pressed under the same conditions will compress only 5–10 per cent.

There are two types of compressed resin-treated wood (*compreg*) being produced. One is a high resin-content (25–30 per cent) product of very high diemnsional stability and suitable for electrical insulators in high tension lines, knife handles, and most important of all, for tooling jigs, and forming dies.

Compreg dies are lower in cost, easier to repair, and harder to scratch than metal dies.

The other type of compreg contains 5-10 per cent resin and has higher shock resistance and greater toughness than the high-resin-content material.

More recently, composites of wood with vinyl polymers have been developed. Since the vinyl polymers are clear, colorless thermoplastic materials they do not significantly discolor the wood, so that its natural beauty is retained, whereas the phenolic resins darken the wood. Like the phenolic resins, the vinyl polymers also give improved mechanical properties and dimensional stability to the wood. The mechanisms of bringing about the property changes, however, are different. The phenolic resins are introduced in an aqueous solution, which swells the wood and at the same time penetrates within the cell walls. On drying and curing a high degree of permanent dimensional stability is thus achieved. The vinyl polymers, however, are introduced only into the cell lumens and other voids in the wood. They, therefore, act as a bulking agent, resisting and slowing down dimensional change when the material is subjected to water or high humidity.

A variety of vinyl monomers such as methyl methacrylate and styrene, may be used. Completely filling the cell lumens and other voids (the "full cell process") is easily accomplished by first subjecting the wood to a partial vacuum (about 0.3 in. of Hg), then covering the wood with the monomer and soaking for 2 to 6 hrs, depending upon the species of wood and its dimensions. Some penetration of the monomer into the cell walls may also be obtained by using a diffusion process, such as a solvent-exchange method.

Polymerization of the vinyl monomer in the wood may be done by either of two processes: (1) radiation and (2) free radical catalysts with heat. Some of the original research on the radiation process was done by Kenaga *et al*[6] of the Dow Chemical Co. Solvent-monomer mixtures were used to swell the wood before irradiation, thus holding the wood in its expanded state. If the wood can be anchored permanently in its maximum water-swollen state, then the antishrink efficiency should be 100 per cent. Kent and co-workers[7,8] at the University of West Virginia, during 1961-68, carried out an extensive study of radiation for the production of wood-plastic composites, Some of this technology is now in commercial practice. Gamma radiation is used for in-depth polymerization of vinyl monomers in wood, while beta radiation is used for polymerization of surface coatings on wood.

The catalyst-heat process was first mentioned[9] in a paper in 1936 describing the use of methacrylate resins by the Du Pont Company. This process has been intensively studied and further developed by Meyer and co-workers[10,11] at the State University of New York College of Forestry in Syracuse (1965-1971).

The polymerization of the vinyl monomers in both processes depends upon the same mechanism, namely initiation by free radicals. In the radiation process, the gamma rays passing through the monomer and the woody tissue creates a large number of excited and ionized molecules, many of which break into fragments, namely organic free radicals ($R^\bullet$). These act as the initiator for the polymerization of an unsaturated monomer:

$$R^\bullet + CH_2{=}CH_2 \longrightarrow R{-}CH_2{-}CH_2^\bullet$$

$$R{-}CH_2{-}CH_2^\bullet + {}_n\, CH_2{=}CH_2 \longrightarrow R(CH_2{-}CH_2)_n{-}CH_2{-}CH_2^\bullet$$

$$2R(CH_2{-}CH_2)_n{-}CH_2{-}CH_2^\bullet \longrightarrow R(CH_2{-}CH_2)_{2n+2}R$$

Alternatively, the free radicals may be formed by thermal decomposition of compounds involving a weak bond, such as benzoyl peroxide, employed by Meyer and co-workers, viz.:

$$C_6H_5{-}\overset{\overset{\displaystyle O}{\|}}{C}{-}O{-}O{-}\overset{\overset{\displaystyle O}{\|}}{C}{-}C_6H_5 \xrightarrow{\Delta} 2C_6H_5{-}\overset{\overset{\displaystyle O}{\|}}{C}{-}O{-} \longrightarrow 2C_6H_5^\bullet + 2CO_2$$

The free phenyl radicals will then initiate the polymerization of the vinyl monomer.

Commercially, the catalyst 2,2'-azobisisobutyronitrile is now most widely used, since it forms free radicals at a lower temperature than benzoyl peroxide.

$$CH_3-\underset{\underset{CN}{|}}{\overset{\overset{CH_3}{|}}{C}}-N=N-\underset{\underset{CN}{|}}{\overset{\overset{CH_3}{|}}{C}}-CH_3 \xrightarrow{\Delta}$$

$$2CH_3-\underset{\underset{CN}{|}}{\overset{\overset{CH_3}{|}}{C}}-N- \longrightarrow 2CH_3-\underset{\underset{CN}{|}}{\overset{\overset{CH_3}{|}}{C}}\cdot + N_2$$

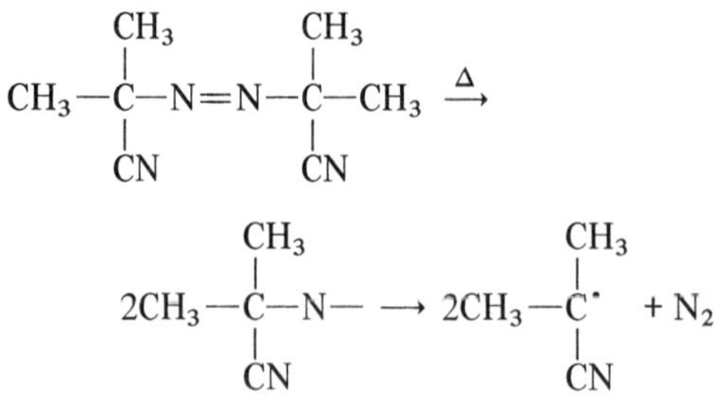

If the end-use of the wood-polymer composite requires an abrasive (sanding) or cutting process that brings about high temperatures, the thermoplastic polymer will melt, causing machining difficulties. To prevent such melting, a cross-linking substance such as diethylene glycol dimethacrylate is added to the monomer before impregnation into the wood (about 5 per cent of the volume of the monomer).[12] This brings about cross-linking of the long unbranched vinyl polymers, resulting in a substance which is not thermoplastic. The process is akin to the curing of rubber.

Both the radiation and heat-catalyst processes are used industrially. One of the major uses of the wood-vinyl polymer composite is for parquet flooring. It requires no finishing or waxing and is very easy to maintain. Other uses include bowling alley flooring, knife and brush handles, golf club heads, billiard cues, archery bows, shuttle cocks and kicker sticks, and tamping sticks for dynamite.

THE FORMING OF WOOD

In order to bend, shape, or otherwise form wood it is necessary to induce temporary plasticization. The steaming of wood has been an art for generations. As pointed out earlier, wood is somewhat plastic in the presence of heat and water, thus, by steaming it, this property can be taken advantage of in the bending of wood into various shapes for furniture and other uses.

A new process for the temporary plasticization of wood has been developed by Schuerch[13,14] using anhydrous ammonia, either in the liquid or gaseous phase. Anhydrous ammonia causes swelling of both of the major polymer systems of wood: the lignin and the polysaccharides. Ammonia wets lignin and causes a lower tack temperature than water. Ammonia also enters the crystal lattice of cellulose, causing it to swell and to form the crystal structure of an ammonia-cellulose compound. Upon removal of the ammonia the ammonia-cellulose reverts to cellulose in a more distorted crystalline form.

The forming of wood by temporary plasticization with anhydrous ammonia takes advantage of these principles. Wood that has been immersed in liquid ammonia or treated with gaseous ammonia under pressure until the cell walls have been penetrated, becomes pliable and flexible. In this plasticized condition it can be readily and easily shaped and formed by hand or mechanically. The ammonia readily vaporizes and evaporates from the wood, whereby the wood regains its normal stiffness, but retains the new form into which it has been shaped. The wood can be distorted into complex shapes without spring-back. For exact control of shape, restraint is needed for a short period, until the greater part of the ammonia has vaporized, but thereafter, there is no tendency for the wood to spring back to its original form.

The process is still developmental but particleboard, hardboard, and veneers, as well as wood strips may be formed by the ammonia process. Other possible uses, such as embossing, densification, and improved machining of wood may potentially be taken advantage of through the temporary modification of wood by anhydrous ammonia. Treating plants have been developed on a pilot plant scale, whereby temperatures can be controlled between −35 and +35°C.

PRESERVATIVE AND FIRE-RETARDANT TREATMENT OF WOOD

Wood Preservation

Wood, a natural plant tissue, is subject to attack by fungi, insects and marine borers. Some species of wood are more resistant to decay than others, as, for example, the heartwood of cedars, cypress, and redwood, due to the presence of natural toxic substances among the extractable components. Most woods, however, are rapidly attacked when used in contact with soil or water, or when exposed to high relative humidities without adequate air circulation. Wood for such service conditions requires chemical treatment with toxic chemicals, collectively termed *wood preservatives.* The length of service life of wood may be increased from 5–15-fold, depending upon the conditions of preservative treatment and the nature of the service.

The preservative treatment of wood is the second largest chemical wood-processing industry; pulp and paper manufacture is the most important. Since 1950, the average annual volume of wood treated has been 260 to 286 million cu ft (see Table 15.11). The more important types of wood products treated are also shown in Table 15.11.

Preservative Chemicals. Toxic chemicals used for the preservation of wood may be classified as follows:

1. Organic liquids of low volatility and limited water solubility
   - coal-tar creosote
   - creosote-coal tar solutions
   - creosote-petroleum solutions
   - other creosotes
2. Chemicals dissolved in organic solvents, usually hydrocarbons
   - chlorinated phenols (principally pentachlorophenol)
   - copper naphthenate
   - solubilized copper 8-quinolinolate
3. Water-soluble inorganic salts
   - acid copper chromate
   - ammoniacal copper arsenite
   - chromated copper arsenate
   - chromated zinc chloride
   - fluor chrome arsenate phenol

Creosote from coal tar is the most widely used wood preservative because (1) it is highly toxic to wood-destroying organisms; (2) it has a high degree of permanence due to its relative insolubility in water and its low volatility, (3) it is easy to apply and to obtain deep penetration; and (4) it is relatively cheap and widely available.

For general outdoor service in structural timbers, poles, posts, piling, mine props, and for marine uses, coal-tar creosote is the best and most important preservative. Because of its odor, dark color, and the fact that creosote-treated wood usually cannot be painted, creosote is unsuitable for finished lumber and for interior use.

Coal-tar creosote (see Chapter 8) is a mixture of aromatic hydrocarbons containing appreciable amounts of tar acids and bases (up to about 5 per cent of each) and has a boiling range between 200 and 355°C. The important hydrocarbons present include fluorene, anthracene, phenanthrene and some naphthalene. The tar acids are mainly phenols, cresols, xylenols, and naphthols; the tar bases consist of pyridines, quinolines, and acridines.

TABLE 15.11 Wood Products Treated with Preservatives, 1950–1970[a]

(Thousands of Cubic Feet)

Product	*1950*	*1957*	*1967*	*1970*
Crossties and switch ties	127,465	106,069	88,635	87,258
Poles	77,622	78,624	84,322	76,760
Lumber and timbers	32,821	37,402	62,241	55,699
Piling	12,523	15,746	16,627	15,128
Fence posts	12,269	12,706	21,018	15,196
Cross arms	2,493	4,209	4,605	3,454
All other	4,792	4,793	8,910	6,882
TOTAL	269,985	259,546	286,357	260,288

[a]Sources: "Statistical Abstract of the United States 1958," U.S. Dept. of Commerce, Bureau of the Census; and Wood Preservation Statistics 1970, Proceedings of the American Wood Preservers' Association, Vol. 67 (1971).

Often coal tar or petroleum oil is mixed with coal-tar creosote, in amounts up to 50 per cent, as a means of lowering preservative costs. Since coal tar and petroleum have low toxicity, their mixtures with creosote are less toxic than creosote alone.

A number of phenols, especially chlorinated phenols, and certain metal-organic compounds, such as copper naphthenate and phenyl mercury oleate, are effective preservatives. Pentachlorophenol and copper naphthenate are most commonly used, and they are carried into the wood in 1 to 5 per cent solutions in petroleum oil. Pentachlorophenol is colorless, and can be applied in clear volatile mineral oils to millwork and window-sash which require a clean, nonswelling and paintable treatment.

Inorganic salts are employed in preservative treatment where the wood will not be in contact with the ground or water, such as for indoor use or where the treated wood requires painting. They are also satisfactory for outdoor use in relatively dry regions.

Preservation Processes. The methods for applying preservatives to wood are classified as follows:

1. Nonpressure processes
   - Surface (superficial) applications by brushing, spraying, or dipping
   - Soaking, steeping, and diffusion processes
   - Thermal process
   - Vacuum processes
   - Miscellaneous processes
2. Pressure processes
   - Full-cell process (Bethell)
   - Empty-cell processes (Rueping and Lowry)

Brush and spray treatments usually give only limited protection because the penetration or depth of capillary absorption is slight. Dip treatments give slightly better protection. Organic chemicals dissolved in clear petroleum solvents are often applied to window sash and similar products by a dip treatment of 1 to 3 minutes.

Cold soaking of seasoned wood in low-viscosity preservative oils for several hours or days and the steeping of green or seasoned wood in water-borne preservatives for several days are methods sometimes employed for posts, lumber, and timbers on a limited basis. The diffusion process employs water-borne preservatives that will diffuse out of the treating solution into the water in green or wet wood.

The most effective of the nonpressure processes is the thermal method of applying coal-tar creosote or other oil soluble preservatives, such as pentachlorophenol solution. The wood is heated in the preservative liquid in an open tank for several hours, after which it is quickly submerged in cold preservative in which it is allowed to remain several hours. This is accomplished either by transferring the wood at the proper time from the hot tank to the cold tank, or by draining the hot preservative and quickly refilling the tank with cooler preservative. During the hot treatment the air in the wood expands and some is expelled. Heating also lowers the viscosity of the preservative so that there is better penetration. When the cooling takes place, the remaining air in the wood contracts, creating a partial vacuum which draws the preservative into the wood. For coal-tar creosote the hot bath is at 210–235°F and the cold bath at about 100°F. This temperature is required to keep the preservative fluid.

The hot- and cold-bath process is widely used for treating poles and, to a lesser extent, for fence posts, lumber, and timbers. The results obtained by this process are the most effective of the common nonpressure processes and most nearly approach those obtained by the pressure processes.

The vacuum processes involve subjecting the wood to a vacuum to draw out part of the air. The wood may be either subjected to a vacuum alone or to steaming and a vacuum before being submerged in a cold preservative. These methods are used to a limited extent in the treatment of lumber, timber, and millwork.

Commercial treatment of wood is most commonly done by one of the pressure processes, since they give deeper penetrations and more positive results than any of the nonpressure methods. The wood, on steel cars, is run into a long horizontal cylinder, which is closed and filled with preservative. Pressure is applied, forcing the preservative into the wood.

There are two types of pressure treatment,

the full-cell and the empty-cell. The full-cell process seeks to fill the cell lumens of the wood with the preservative liquid giving retention of a maximum quantity of preservative. The empty-cell process seeks deep penetration with a relatively low net retention of preservative, by forcing out the bulk liquid in the wood cells, leaving the internal capillary structure coated with preservative.

In the full-cell process the wood in the cylinder is first subjected to a vacuum of not less than 22 inches of mercury for 15-60 minutes, to remove as much air as possible from the wood. The cylinder is then filled with hot treating liquid without admitting air. The maximum temperature for creosote and its solutions is 210°F and for water-borne preservatives it is 120 to 150°F, depending upon the preservative. The liquid is then placed under a pressure of 125-200 psi and the temperature and pressure are maintained for the desired length of time, usually several hours. After drawing the liquid from the cylinder, a short vacuum is applied to free the charge of surface-dripping preservative.

In the empty-cell process the preservative liquid is forced under pressure into the wood containing either its normal air content (Lowry process), or an excess of air, by first subjecting the wood to air pressure before applying the preservative under pressure (Rueping process). In the former case the preservative is put in the cylinder containing the wood at atmospheric pressure, and, in the latter case, under air pressure of 25-100 psi. After the wood has been subjected to the hot preservative (about 190-200°F) under pressure (100-200 psi in Lowry process and 150-200 psi in the Rueping process) and the pressure has been released, the back pressure of the compressed air in the wood forces out the free liquid from the wood. As much as 20-60 per cent of the injected preservative may be recovered, yet good depth of penetration of preservative is achieved.

Preservative Retention. Retention of preservative is generally specified in terms of the weight of preservative per cubic foot of wood, based on the total weight of preservative retained and the total volume of wood treated in a charge. Penetration and retention vary widely between different species of wood, as well as with woods of the same species grown in different areas. In most species, heartwood is much more difficult to penetrate than sapwood. Also, within each annual growth ring there is variability in penetration, the latewood generally being more easily treated than the earlywood.

The American Wood-Preservers' Association Standards specify methods of analysis to determine penetration and retention. They also specify minimum retention amounts for different preservatives according to the commodity, the species, the pretreatment of the wood, such as kiln drying, and the end use of the commodity. Heavier retention is required for products in contact with the ground (poles, timbers, etc.) or with marine waters (piles, timbers, etc.). Unprotected wood in contact with the ground is subject to severe attack by fungi and insects and, in contact with sea water, it is quickly destroyed by marine borers. For wood products to be used in contact with the ground or marine waters, creosote is the major preservative employed, since it can be readily impregnated to give high retention and good protection and is not leached out by water.

Fire-Retardant Treatments

Protection of wood against fire may be accomplished by application of certain chemicals. Because of the cost, commercial treatments are limited to materials used in localities where fire building codes require fire-retardation treatment. Fire-retardant chemicals are water soluble, which limits the use of treated wood to places where it is not subjected to leaching.

Two types of treatment are used for improving the fire resistance of wood; namely (1) impregnation of the wood with fire-retardant chemicals and (2) coating the surface of the wood with an oxygen-excluding envelope. Among the most commonly used chemicals for impregnation treatments are diammonium phosphate, ammonium sulphate, borax, boric acid, and zinc chloride. These compounds have different characteristics with respect to fire resistance. Ammonium phosphate, for example, is effective in checking both flaming and glowing; borax is good in checking flaming but is not a satisfactory glow retardant. Boric acid is excellent in stopping glow but not so effective in retarding flaming. Because of these different characteristics, mixtures of chemicals are usually employed in treating formulations.

The American Wood-Preservers' Association Standards specify the following four types of fire-retardant formulations:

Type A	
Chromated zinc chloride—a mixture of sodium dichromate and zinc chloride having the composition: hexavalent chromium as CrO_3, 20% and zinc as ZnO, 80%.	
Type B	
Chromated zinc chloride (as above)	80%
Ammonium sulfate	10%
Boric acid	10%
Type C	
Diammonium phosphate	10%
Ammonium sulfate	60%
Sodium tetraborate, anhydrous	10%
Boric acid	20%
Type D	
Zinc chloride	35%
Ammonium sulfate	35%
Boric acid	25%
Sodium dichromate	5%

Minimum and maximum limits of variation in the percentage of each component in the above formulations are specified in the standards.

The impregnation methods are similar to those employed for the preservative treatment of wood by water-borne salts using pressure processes. The maximum temperature of the solution must not exceed 140°F for formulations Types A, B, and D, are not over 160°F for Type C. Subsequent to treatment, the wood must be dried to remove the water solvent to a moisture content of 19 per cent or less. For most uses the wood is kiln dried to a moisture content of under 10 per cent.

Larger amounts of chemical must be deposited in the wood for effective fire protection than is necessary for the water-borne chemicals used for decay prevention. Whereas retentions from 0.22 to 1.00 lb per cu ft of wood for the water-soluble toxic salts are specified according to commodity standards which give good protection against decay and insects, as much as 5 to 6 pounds of some fire-retardants may be required for a high degree of effectiveness against fire. Usually smaller amounts will give a good degree of protection. For example, formulation Type B when impregnated in amounts of 1.5 to 3 lb per cu ft of wood provides combined protection against fire, decay, and insects. The effectiveness of a fire-retardant treatment depends upon the performance rating of the treated material when tested in accordance with ASTM E84 (no greater flame spread than 25).

There are various explanations for the fire-retardant effect of the chemicals impregnated into wood for this purpose. There is probably no single mechanism but rather a combination of mechanisms which are operative, including (1) the fusing of the chemical within the wood at high temperatures to form thin noncombustible films which exclude oxygen; (2) the evolution of noncombustible gases in some cases; and (3) the catalytic promotion of charcoal formation, instead of volatile combustible gases, with the added benefit of increased thermal insulation.

Surface coatings for fire-retardation are less effective than impregnation of chemicals into the wood. Formulations are used containing either ammonium phosphate, borax, or sodium silicate, together with other constituents to provide good bonding to the wood. Paints containing substantial amounts (30 to 50 per cent) of these fire-resistant chemicals provide moderate protection against fire.

WOOD HYDROLYSIS

The availability of enormous quantities of wood residues from logging, sawmilling, and woodworking operations, estimated during the 1950's at 60 million tons annually, has stimulated much attention to the production of sugars and ethyl alcohol from this cheap raw material. Problems were encountered in developing commercial acid hydrolysis processes due to the serious corrosive action of the acid and the difficulty in removing the sugars from the scene of the reaction before they were subjected to too much degradation.

Cellulose, the major component of wood, gives about 90 per cent yield of pure glucose under laboratory conditions of hydrolysis, according to the following equation:

$$(C_6H_{10}O_5)_n + nH_2O \xrightarrow{\text{acid}} nC_6H_{12}O_6$$

where n is in the range of 10,000–15,000. The hemicellulose fraction gives a mixture of sugars, viz., xylose, arabinose, mannose, galactose, and glucose. Glucose, galactose, and mannose are yeast-fermentable sugars, whereas the pentoses (xylose and arabinose) are non-fermentable. The potential total reducing sugar yield from wood averages 65–70 per cent, whereas the fermentable sugar yield is abut 50 per cent for the hardwoods and 58 per cent for the softwoods. The lower quantity of fermentable sugar from the hardwoods is due to their higher content of pentosans, compared to the coniferous woods.

Hemicelluloses hydrolyze much more easily and rapidly than cellulose. Temperatures and acid concentrations that hydrolyze the cellulose to glucose in a matter of a few hours readily convert much of the hemicellulose into simple sugars in minutes or even seconds. Under industrial conditions of hydrolysis the sugars formed undergo decomposition, the pentoses decomposing more rapidly than the hexoses. Thus, the conditions of hydrolysis cause variations in the ratio and yields of the various sugars due to (1) their different rates of formation by hydrolysis and (2) their different rates of decomposition. Table 15.12 shows the relative rates of decomposition of the sugars obtained by the hydrolysis of wood.

The polysaccharides of wood (holocellulose) may be hydrolyzed by two general methods: (1) by strong acids, such as 70–72 per cent sulphuric acid or 40–45 per cent hydrochloric acid; or (2) by dilute acids, such as 0.5–2.0 per cent sulphuric acid. The hydrolysis by strong acids is constant, proceeds at a first-order reaction and is independent of the degree of polymerization. The reaction may be represented as follows:

$$\text{Holocellulose} \xrightarrow[\text{acid}]{\text{strong}} \left.\begin{matrix}\text{Swollen Cellulose} \\ \text{Soluble Pentosans}\end{matrix}\right\} \longrightarrow \text{Soluble Polysaccharides} \xrightarrow{\text{dilute acid}} \text{Simple Sugars}$$

In dilute-acid hydrolysis the reactions are heterogeneous and more complex because no swelling and solubilizing and the cellulose occurs. Cleavage of the insoluble cellulose takes place directly to low-molecular weight oligosaccharides (intermediate products), which are rapidly converted to simple sugars, as indicated below:

Holocellulose $\xrightarrow[\text{dilute acid}]{(1)}$

Insoluble "stable" cellulose
Soluble hemicellulose intermediates
Pentose sugars

$\xrightarrow{(2)}$ Oligosaccharides (cellulose intermediates) $\xrightarrow{(3)}$

Hexose sugars
Pentose sugars

Reaction (1) is rapid and occurs under mild conditions, hydrolyzing mainly the hemicelluloses. Reaction (2) is slow, proceeds as a first-order reaction and is the limiting reaction in this process. Reaction (3) is rapid.

Based on the above two methods, two industrial processes have been developed, namely: the Bergius-Rheinau process, based on the use of concentrated hydrochloric acid at ordinary temperatures, and the Scholler-Tornesch process, in which very dilute sulfuric acid is used at temperatures of 170 to 180°C (338 to 356°F). The latter method in an improved form is known as the Madison process, thanks to work done at the United States

TABLE 15.12 Decomposition of Wood Sugars in 0.8 Per Cent Sulphuric Acid at 180°C[a]

Sugar	*First-Order Reaction Constant K (min^{-1})*	*Half-life (min)*
Glucose	0.0242	28.6
Galactose	0.0263	26.4
Mannose	0.0358	19.4
Arabinose	0.0421	16.4
Xylose	0.0721	9.6

[a]Source: J. F. Saeman, *Ind. Eng. Chem.*, **37**, 43 (1945).

Forest Products Laboratory in Madison, Wis. A number of modifications have been developed, including four in Japan.

However, none of these processes offer sufficient economy in the United States to compete with agricultural sources for sugars and ethanol, or with petroleum for ethanol. In some countries, however, such as Soviet Union, the hydrolysis of wood remains economically competitive with other sources of sugar and ethanol. The Russians have a large industrial capacity for producing wood sugars which are used mainly for the manufacture of yeast. The Japanese have also been active in the development of wood-hydrolysis technology.

Bergius-Rheinau Process

Hydrochloric acid of about 40–45 per cent (by weight) is produced by reinforcing recovered, weaker acid with hydrogen chloride from salt-sulfuric acid retorts, or by burning chlorine with illuminating gas.

Wood chips are air dried, then charged into a tile-lined reactor and extracted countercurrently by the acid. The fresh strong acid enters that part of the battery of diffusers or reactors which contains the most nearly exhausted wood and is pumped through the next following containers until it is nearly saturated with the carbohydrates dissolved from the wood. Part of this solution is mixed, under slight cooling, with fresh wood and the mixture charged into the head container of the battery. After filling this and allowing a few hours for reaction time, an amount of solution is forced from this container through the pressure of the incoming acid, which corresponds to the yield from one charge.

The drawn-off solution contains hydrochloric acid and carbohydrates in about equal parts, and at a concentration of about 25 per cent (by weight) each. It is concentrated in stoneware tubes under vacuum. The distillate, containing about 80 per cent of the hydrochloric acid, with minor proportions of acetic acid and furfural is re-used after fortification. The concentrated sugar solution is dried to a powder in a spray-drier, where it also loses most of the remaining acid.

The dry, somewhat acid, powder contains the carbohydrates in the form of intermediate polymers (oligosaccharides). They are water-soluble and must undergo further hydrolysis to obtain simple sugars, either for fermentation or cystallization.

This is done by dissolving them, diluting the solution to approximately 20 per cent sugar concentration, and heating it for 2 hr in the presence of 2 per cent acid at 125°C. Part of the glucose can be crystallized from the neutralized and reconcentrated solution, while the mother liquors are fermented to alcohol or used for growing yeast. A diagrammatic flow chart for the Bergius process is shown in Fig. 15.14.

When hardwoods are to be used, it is necessary to remove a part of the hemicellulose first by prehydrolysis. This has been carried out on a large scale with straw, a substance chemically similar to hardwood, by heating it

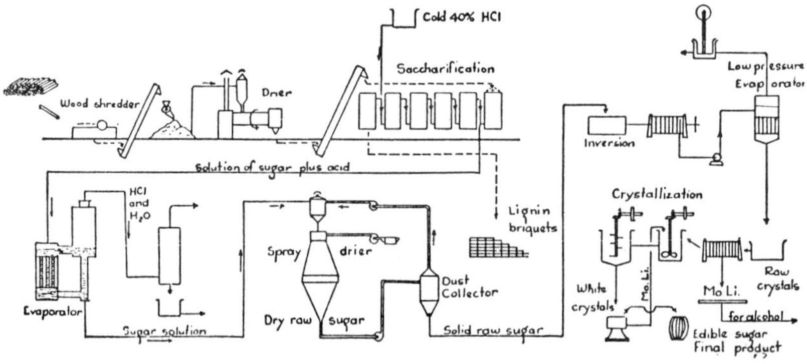

Fig. 15.14 Diagrammatic flow sheet for the Bergius wood-hydrolysis process.

in a 8 to 1 liquid to solid ratio, with 0.5 per cent sulfuric acid for 2 to 3 hr at 130°C (266°F). Without prehydrolysis, hardwoods and straws form slimy materials, probably because of their high hemicellulose content, which prevent the flow of the hydrolyzing acid.

In the Bergius-Rheinau process the concentrated hydrochloric acid employed requires dried wood, and recovery of the acid is essential. The process gives high yields of sugars (60 to 65 per cent) at high concentrations. The intermediate sugars first obtained, however, call for an extra processing step to reduce them to monomers, before fermentation or crystallization.

Madison Process with Continuous Percolation

In the Madison process, dilute sulfuric acid with an average concentration of 0.5 per cent is pressed through wood in the form of sawdust and shavings. Regular flow of the acid and of the resulting sugar solution is one of the two principal requirements; the other is a lignin residue which can be discharged from the pressure vessels without manual labor. Both depend upon careful charging of the wood, which should not contain too many very fine particles, and upon maintaining a pressure differential of not more than 5 to 6 lbs sq in. between top and bottom of the digester.

The digesters or percolators are pressure hydrolyzing vessels, commonly employing a pressure of 150 lb/sq in., and having a capacity of 2000 cu ft each. In the original Scholler plants in Germany, the digesters were lined with lead and acid-proof brick; in the Madison process a lining of "Everdur" metal was found to give sufficient protection.

The wood, about 15 tons, is pressed down with steam, and then heated by direct steam, after which the acid is introduced. The practice in Germany was to bring the dilute acid into the digester in several batches, with rest periods of about 30 min, heating the wood to temperatures of 130°C (266°F) at first, then to 180°C (356°F), while keeping the temperature of the entering acid 10 to 20°C lower. A total of about 14 hr was required to exhaust the wood, yielding about 50 lb of carbohydrates for 100 lb of dry wood substance. In the Madison process, continuous flow of the acid, and correspondingly, of the sugar solution, is provided–in other words, continuous percolation. The cycle is thereby reduced to 6 hr, and the yields are increased somewhat.

The lignin is blown out of the digester by opening the specially constructed bottom valves while the vessel is still under pressure.

The sugar solutions usually contain about 5 per cent carbohydrates and 0.5 per cent sulfuric acid. The solutions, still under pressure (150 lb/sq in.), are flash-evaporated to 35 lb/sq in., neutralized with lime at that pressure, and filtered. Calcium sulfate is much less soluble at the elevated temperature corresponding to the pressure than it is at 100°C. This is a fortunate circumstance, for it must be removed to an extent sufficient to avoid difficulties caused by the formation of incrustations in the subsequent alcohol distillation. The filtered solution is cooled by further flash evaporation and heat exchange with water to fermentation temperature.

Sugar yields from coniferous woods (softwoods) are about 50 per cent at an average concentration of 5 per cent. When fermented, the average ethyl alcohol yield per ton of dry wood is 50-60 gal and sometimes higher.

The dilute sulfuric acid employed in the Madison process gives lower yields (49 to 55 per cent) of sugars than the Bergius-Rheinau process, and only very dilute solutions are obtained directly. The difference in yield is not too important with a low-cost material such as wood waste. Recovery of heat is easier in the Madison process and the acid need not be recovered.

Japanese Developments

At an FAO meeting[15] on wood hydrolysis in Japan, 1960, four different modifications of the wood hydrolysis process were described, all of them in the pilot plant stage and with construction of one full-scale plant underway. All involve a prehydrolysis step to remove the hemicelluloses prior to hydrolysis of the cellulose. All operate at atmospheric pressure and at temperatures below 100°C. The main product in each case is crystallized glucose. The principal differences are in the acids used and in the acid recovery systems.

Two of the processes use concentrated sulfuric acid for the main hydrolysis step. In one of these, the *Nihon-Mokuzai-Kagaku* process, the acid is neutralized by lime, and the gypsum is used in gypsum-board manufacture and for other purposes. In the *Hokkaido* process the acid is recovered by dialysis.

The other two processes described use hydrochloric acid. The *Udic-Rheinau* process is a modification of the Bergius-Rheinau process. Strong (41 per cent) hydrochloric acid is used and recovered by distillation. In the *Noguchi-Chisso* process anhydrous hydrochloric acid is used.

Russian Developments

The Russians have energetically pursued the development of wood-hydrolysis technology and maintain a large Wood-Hydrolysis Research Institute. By 1934 the first industrial plant was built, and since then the number has grown to 30. Early production was directed to ethanol as a raw material for synthetic rubber. Today[16] the emphasis is on the production of yeast and furfural. Currently ten plants are producing 60,000 tons of dry yeast from wood sugar annually, using dilute sulfuric acid for hydrolysis. Crystalline glucose is produced by a modification of the Rheinau process, using supersaturated hydrochloric acid.

The present goal for yeast production is 250,000 tons annually. Furfural, which is obtained as a by-product during wood hydrolysis, is used as an intermediate in the production of organic chemicals and is converted mostly to tetrahydrofuran and furfuryl alcohol.

Prehydrolysis of Wood for Pulp Production

When wood is pulped by the sulfate process a large part of the hemicelluloses are converted to sugars and organic degradation products, using up pulping chemicals. The spent sulfate liquor is burned to recover pulping chemicals, and the sugars and their decomposition products contribute to the calorific value of the total liquor, though only to the extent of 18 per cent.

A process has been developed whereby the easily hydrolyzable sugars are removed from the wood by prehydrolysis and subsequently used for the production of fodder yeast or as a source of sugar for fermentation products. Hydrolysis of pine, fir, or spruce for 15 min at 165-175°C by 0.3 per cent sulfuric acid removes about 16 per cent of the wood. The sugars contain about 50 per cent hexoses (mainly glucose) and 50 per cent pentoses, and have been found to have a yeast yield of about 5 per cent of the weight of the wood.

The residual wood is pulped by the sulfate process and gives about 40 per cent yield of high alpha-content pulp for dissolving purposes (rayon, film, plastics).

Fermentation of Sulfite Waste Liquor

The sulfurous acid used in the sulfite pulping liquor causes hydrolysis of the more easily hydrolyzable components of wood, especially the pentosans in the hemicellulose. About 35 per cent of the potentially fermentable sugars in the wood are hydrolyzed. However, most of these are decomposed during the long pulping procedure, so that only one-fourth to one-third, including much of the more resistant hexoses, remains in the waste sulphite liquor. If these are fermented by yeast about 12.5 gal of 95 per cent alcohol per ton of wood may be produced.

A large number of plants in Europe and a few in North America have been constructed to utilize the sugar hydrolyzate in sulfite waste liquors. The procedure as carried out in one American operation is as follows:

The liquor is recovered from the digester by discharging it in such a manner that as much sugar as possible is removed with a minimum of dilution by washing. Free sulfur dioxide is removed and recovered by blowing steam through the solution, which decreases the acidity from a pH of 2.2 to about 3.9. The liquors are cooled by a vacuum flash and neutralized by lime to a pH of about 4.2. A small amount of inorganic nitrogen is added for yeast growth and about 1 per cent by volume of yeast is added continuously. Fermentation is carried out in a series of tanks, the solution flowing from one to the other with agitation to keep the yeast in suspension. The yeast is recovered by centrifuging and mixed with the new sugar solution entering the fermenter. About 30 hr is required for fermentation. The alcohol content of the fermented liquor is about 1 per cent by volume. The ethyl alcohol is recovered in stainless steel stills. Methanol and other alcohols are obtained in small amounts as by-products.

Vanillin

Vanillin is not a product of hydrolysis or of fermentation; in fact, it does not originate from the holocellulose, but rather from the lignin portion of the wood. The major organic material in sulfite waste liquor (spent liquor, after pulping wood by the sulfite process) is the lignin dissolved from the wood as lignosulphonic acid. Alkaline degradation of this lignin product produces vanillin, the same substance which occurs naturally in the vanilla

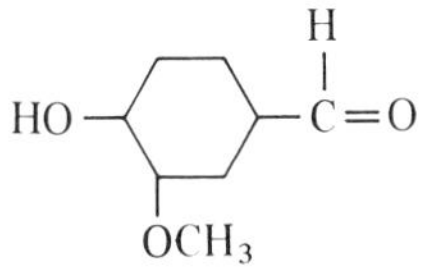

Vanillin

bean. By the Howard process 5-10 per cent of vanillin is produced, based on the lignin in the waste sulfite liquor. Some vanillin is produced from sulfite waste liquor in both the United States and Canada.

WOOD CARBONIZATION AND DISTILLATION

Wood has been carbonized from the beginning of recorded history to produce charcoal, which was valued as a light-weight and smoke-free fuel. Much later, the development of the iron smelting industry greatly increased the demand for charcoal. The charcoal was produced in simple earth kilns, often called "pit-kilns" which involved the slow burning of properly piled wood with a flue opening in the center and with the pile covered with turf and earth to reduce air circulation. Even today charcoal is still widely produced by this ancient method.

The yield of charcoal represents only about one-third of the weight of the wood, the remainder being accounted for as gases and vapors. The first commercial recovery of by-products from the gases and vapors was undertaken by James Ward at North Adams, Massachusetts, in 1830.[17] The gases and vapors were cooled and the condensible portions converted to liquors. From the aqueous portion, known as pyroligneous acid, acetic acid and its salts were recovered. This was the beginning of the hardwood distillation industry. Later, methanol was also recovered from the pyroligneous acid, and the tars were separated and used for fuel, or were fractionated into creosote oil, soluble tars, and pitch.

In time, improvements in equipment and chemical engineering technology resulted in better control, reduction in heat losses, and continuous operation, and led to increased yields, better quality products, and the separation and purification of additional products from the condensed aqueous and tar liquors. The noncondensible gases were in part recycled to bring fresh retorts to the desired temperature; the greater part was utilized by burning under the boilers to generate heat and power.

With the development of low-cost continuous processes for the synthetic production of methanol, acetic acid, and acetone, the hardwood-distillation industry went into a decline after World War II. However, the demand for charcoal has held firm and even increased. As a result, improved methods for charcoal production have been developed and the recent trend is toward continuous carbonization. Thus, the carbonization of wood has gone full cycle, from the production of only charcoal to the production of industrial organic chemicals, with charcoal playing a secondary role, and now back to the production mainly of charcoal.

The carbonization of wood is a process of thermal decomposition. When wood is subjected to temperatures above 100°C, thermal decomposition occurs, i.e., chemical decomposition brought about solely by elevating the temperature. The more active decompositions occur above 250°C and industrial carbonizations employ temperatures up to 500°C. Beginning at about 270°C exothermic reactions set in, which bring about complete carbonization without further application of external heat. A number of terms are used interchangeably for the thermal decomposition of wood, namely: carbonization, pyrolysis, wood distillation, destructive distillation, and dry distillation.

Thermal decompositions result in drastic changes in the wood. The large, complex polymeric molecules of the wood tissue are broken down to carbon and a wide variety of relatively simple molecules, producing a molecular debris. These products separate naturally into four groups; charcoal, pyroligneous acid liquor, tar, and noncondensible gases.

The products obtained by the distillation of hardwoods differ from those of the softwoods. This is due to the fact that only resinous softwoods could be profitably distilled, the resin giving rise to turpentine, pine oil, and rosin oil; products which are equally if not more important than the charcoal. Softwoods, however, yield only about one-half as much methanol and one-quarter as much acetic acid in the pyroligneous acid fraction as do the hardwoods. Generally, it was not profitable to isolate these chemicals from the pyroligneous acid of softwoods.

Hardwood Distillation

The once-flourishing hardwood-distillation industry has now ceased to exist in the United States due to the development of cheaper means of producing the chemical products from petroleum by synthetic methods. Therefore, this topic will be given only brief mention. The reader interested in further details is referred to the previous edition of "Riegel's Industrial Chemistry" (1962), Chapter 15.

Two methods of hardwood distillation were developed: (1) the externally heated process and (2) the internally heated process.

The Externally Heated-Oven Process. An outgrowth of the kiln method for producing charcoal was the hand-loaded iron retort developed in 1850. However, the large-scale development of the chemical-wood industry did not begin until 1875, when the large car-loaded ovens were employed.

The externally heated ovens (*retorts*) used steel cars or buggies. About 2.5 cords of wood in 4-foot lengths were piled on each car. The common 10-cord oven held 4 cars. The oven was made of steel, enclosed in a brick chamber with a space around the sides and top of the oven for circulation of the fire gases. Heat was applied by burning natural gas, oil, or coal. Two openings in the rear wall allowed the volatile products to pass out.

The cycle is 24 hrs. During the first few hours, the heating is rapid in order to reach the distillation temperature. Water comes over first. An exothermic reaction takes place next, after which the outside heat must be decreased. The vapors pass out to the condenser where they form the liquid condensate, the pyroligneous acid. The uncondensed gas is piped to the boiler house. After about 10 hrs, the flames in the burners are raised again, but not so high as at first. After 22 hrs the distillation is over; all burners are turned off, and the retort is allowed to cool for 2 hrs.

The buggies were placed in air-tight cooling chambers for two to three days, usually with a quenching spray of water after the first day. The charge shrinks considerably during the distillation, but the charcoal is obtained in the form of rather large pieces, with very little dust.

The Internally Heated Retort Process. Several processes were developed whereby the retort was heated internally by hot circulating gases. A vertical retort was employed and filled with wood blocks or chunks. Oxygen-free gases, resulting from carbonization of wood in neighboring retorts, was freed from condensible gases in condensers and scrubbers and part of it was used to heat the remaining gas in a furnace, usually to 500–600°C. These hot gases were then circulated through the wood charge. As carbonization occured, the vapors were passed to condensers and the charcoal dropped to a cooling chamber at the base of the retort.

Several variations of this process were developed, some of which are continuous instead of batch systems. In the continuous system, exothermic reaction was initiated by hot gases and then this reaction continued near the center of the retort as the wood in small sizes passed downward. The rising hot gases heated the incoming wood, and the charcoal dropped to a cooling chamber at the bottom (such as in Badger-Stafford, Lambiotte, and Pieters processes).

Chemical Recovery. The gases and vapors passing out of the oven or retort were cooled, the condensed pyroligneous acid was stored, and the gas was scrubbed and blown to the power house where it was burned under boilers, except the part used to bring a fresh retort to the proper temperature.

The pyroligneous acid was separated mechanically from the settled tar. Modern plants used a continuous refining process to fractionate the pyroligneous acid into water, tar, wood oils, and pure chemicals.

Destructive Distillation of Softwoods

Like the hardwood-distillation industry, the destructive distillation of softwoods is no longer practiced. Its major products, turpentine and pine oil, can be obtained in better quality and at lower cost by the extraction (so-called "steam distillation") process, or from the gum obtained by tapping southern pines, and as by-products of the sulfate pulp industry (see section in this chapter on Naval Stores).

Softwoods have less acetyl content than hardwoods and, therefore, yield less acetic acid upon distillation. The methanol yield is also much lower than from hardwoods. Recovery of these and their related chemicals was not practical in softwood distillation. Only resinous softwoods, especially longleaf and slash pine, were distilled because of the value of the products obtained from their resin content. Old stumpwood from logged-off areas and pitchy portions of fallen trees were the preferred material. This is because the sapwood, which is low in resin content had decayed away, leaving primarily the resin-rich heartwood. Some timber and sawmill wastes were also used, but generally only material which contained 20 per cent or more resin.

Both the older small, hand-filled retorts of 1- to 2-cord-capacity and the larger car-loaded ovens up to 10-cord-capacity, similar to those used in hardwood distillation, were used for the destructive distillation of resinous softwoods. Also some retorts were made of concrete and internally heated by flues of large iron pipe. In general the distillation procedure was similar to that for hardwoods.

Whether the crude oils were collected from the retort or oven as a whole, or in fractions, they were redistilled for separation into primary products. Copper stills, known as pine-tar stills, were commonly used, and were provided with both steam coils and steam jets for steam distillation. The products from the pine-tar still were usually light and heavy pine oils, pitch, and a composite of several light solvent oils. Further fractional distillation of the solvent oils yielded turpentine, dipentene, pine oil, and small amounts of other hydrocarbons. The yields vary greatly according to the resin content of the wood and the operating conditions. On the average, the yields per ton of southern pine stumpwood and "lightwood,"* are:

Total oils	35–40 gal
Wood turpentine	4–6 gal
Tar	20–30 gal
Charcoal	350–400 lb

Products of Wood Distillation

Charcoal is the major industrial product of hardwood distillation. It remains as a residue after most of the volatile decomposition products have been driven off. Charcoal consists mainly of carbon, together with incompletely decomposed organic material and adsorbed chemicals. The amount of these secondary materials (containing hydrogen and oxygen) associated with the carbon decreases rapidly with increase in the distillation temperature as shown in Table 15.13. Commercial charcoal corresponds to about the 400 to 500°C product and has a volatile content (organic residues) of 15–25 per cent. The average yield is 37–40 per cent.

Charcoal has a wide range of uses, which may be classified as fuel, metallurgical, chemical, and miscellaneous uses. Large tonnage outlets for fuel were for tobacco curing, restaurants, railway dining cars, and picnic fuels. The use of charcoal in metallurgy has largely

---

*Resinous portions of wood, mainly heartwood, after decay of the sapwood; also termed "fatwood."

TABLE 15.13 Composition and Amount of Charcoal Produced at Different Maximum Temperatures

Distillation Temperature	*Composition of Charcoal* Carbon	Hydrogen	Oxygen	*Charcoal Yield on Dry Wt. of Wood*
°C	%	%	%	%
250	70.6	5.2	24.2	65.2
300	73.2	4.9	21.9	51.4
400	77.7	4.5	18.1	40.6
500	89.2	3.1	6.7	31.0
600	92.2	2.6	5.2	29.1
1000	96.6	0.5	2.9	26.8

given way to coke. Charcoal is an important material for the chemical industry, for the manufacture of items such as calcium carbide, sodium and potassium cyanide, carbon disulphide, magnesium chloride, hydrochloric acid, carbon monoxide, electrodes, fireworks, black powder, catalysts, pharmaceuticals, glass, resin moldings, rubber, brake linings, gas cylinder absorbent, paint pigment, and, in the form of activated carbon, as an adsorptive agent for the purification of gases and liquids. Miscellaneous uses include nursery mulch, crayons, and poultry and stock feeds.

The pyroligneous acid is the dilute aqueous condensate obtained by cooling the vapors from the retort or oven. It contains acetic acid, methanol, acetone, and minor quantities of numerous other organic compounds, of which more than 30 have been identified. Formerly hardwood distillation was the exclusive source of acetic acid, methanol, and acetone, and these were considered the primary products of the process. These important industrial chemicals are now produced by other cheaper methods which led to the decline of the hardwood-distillation industry. The average yields based on the dry weight of the wood are: acetic acid, 4–4.5 per cent; methanol, 1–2 per cent; and acetone, 0.5 per cent.

The wood tars were largely used for fuel at the plant. There was a small demand for some of the tar components, however. The wood tars are of two types, namely (1) the soluble tars and (2) the settled tars. The soluble tars are those in the pyroligneous acid solution and they are separated as tars in the refining process. The settled tars are insoluble in, and heavier than, the aqueous pyroligneous acid and they are mechanically separated from it.

The settled tars can be fractionated into (1) light oils with boiling points up to 200°C, specific gravities less than 1.0, and contain aldehydes, ketones, acids, and esters, (2) heavy oils and boil over 200°C, have specific gravities greater than 1.0, and contain many phenolic components, and (3) pitch. Maple, beech, and birch give total tar yields of 10–12 per cent and oaks give 5–9 per cent tar.

The heavy-oil fraction contains phenols, especially cresols, and is known as *wood-tar cresote.* It was used as a preservative for timbers, as a disinfectant, and for staining. Another important product was medicinal beechwood creosote, which was used as a disinfectant. The light-oil fractions were used as solvents, and the pitch was used for waterproofing and insulating agents.

The noncondensible gases produced during the distillation of hardwoods vary widely in amount and composition with the distilling conditions. The average range of composition of the gas is: 50–60 per cent carbon dioxide; 28–33 per cent carbon monoxide; 3.5–18 per cent methane; 1–3 per cent higher hydrocarbons; and 1–3 per cent hydrogen. The hydrogen content of the gas increases with increasing distillation temperature. The gas mixture has a heating value of about 300 Btu per cu ft. The gas was normally used as a fuel in the distillation plant and as a heat-conveyor in the internal gas-heated processes.

The major products obtained by the destructive distillation of Southern pine wood were turpentine, dipentene, and pine oil. The wood turpentine obtained by destructive distillation differs chemically from gum turpentine, i.e., turpentine from the gum obtained by tapping living trees. Wood turpentine contains a large amount of alpha-pinene and also a large amount of dipentene, differing in these respects from gum turpentine. The general properties of the two turpentines and their uses, however, are similar. Likewise the properties and uses of destructively distilled dipentene and pine oil are similar to those of the extracted or steam-distilled products.

Some of the common uses of the less important products of the destructive distillation of resinous pines were:

Tar and tar oils	Cordage, rubber, oakum, fish nets, tarpaulins, paper, soaps, insecticides, roofing cements, and paints.
Pyroligneous acid	Limited use in meat smoking, leather tanning, and as weed killer.
Charcoal	Similar to those for hardwood charcoal.

Charcoal Production

As stated earlier, the manufacture of charcoal is the only significant industrial product now produced in North America by the carbonization of wood. Furthermore, the demand has been increasing in recent years. For example, from a low of 232,000 tons produced in 1958, production went up to 328,000 tons in 1961, and to 384,700 tons in 1967. Maximum production was 555,000 tons in 1909. The decline until 1958, was due (1) to losses in the heating and cooking markets, because of the advent of central heating, gas, and electricity, and (2) the substitution of coke for charcoal in the chemical and metallurgical industries.

The reversal of the downward trend in charcoal production since 1958 is due to the great demand for charcoal for outdoor cooking, a reflection of our growing affluence and increased leisure time. Today most charcoal is crushed or ground and compressed into briquettes, a form especially suitable for home and recreational use.

Charcoal production fits well into forest management and integrated forest-industry practices. It can use cull, logging waste, and low grade material from the forests and wastes from lumber and other forest-industry plants. No large capital investment or sophisticated personnel are required. Most plants are small. In 1961 there were 297 plants in the U.S.A. with nearly 2000 converting units.[18] The latter consisted of 262 brick kilns, 805 concrete and masonry kilns, 430 sheet-metal (bee-hive-type) kilns and 480 other units, such as retorts, ovens, and improvised chambers. One-third of the units had a capacity of less than a ton of stabilized charcoal per carbonization cycle and 95 per cent less than 25 tons.

The trend in the industry in recent years is in greater use of wood residues from other wood industry operations. This requires mechanized continuous methods of production. Although most charcoal is still produced by conventional kilns, the trend persists toward both vertical gas-recycle retorts operating batchwise and larger continuous conversion systems using sawmill wastes or prepared wood fines. Several new types of batch retorts, such as the Keil-Pfaudler, the Warner-Vega, and modified Cornell retorts have been developed.[19,20] These recycle the wood gases and volatile products back through the combustion system, thereby diminishing the need for outside sources of heat and decreasing the exhaustion of organic materials to the atmosphere. The yields of high grade charcoal are 30 to 38 per cent of the dry weight of the wood used. The labor requirements are low, and with good wood-handling facilities, including a mechanical conveyor and lift truck, charcoal can be manufactured in this type of retorts at the rate of 1 ton per 4 man-hours of labor, compared to 11 man-hours of labor to produce 1 ton of charcoal in a typical kiln.

NAVAL STORES

The naval stores industry in the United States began in the very early Colonial days, when wooden vessels used tar and pitch from the crude gum or oleoresin collected from the wounds of living pine trees. The demand for tar and pitch from crude gum is now of minor importance.

The industry is centered in the southeastern United States and is confined to the longleaf and slash pine areas. There is also a small, but locally important, naval stores-producing area in the Landes region of southwestern France, based on the maritime pine. Turpentine and rosin are the two important products obtained from the oleoresin gum of living pine trees.

The Federal Naval Stores Act, passed in 1923, recognized four types of turpentine, according to the methods of production, namely: (1) *gum turpentine*, made from the gum or oleoresin collected from living trees; (2) *steam-distilled* (*S.D.*) *wood turpentine*, obtained from the resin within the wood (scrapwood, knots, stumps or other wood residues commonly called "lightwood") formerly by steam distillation of the wood but today by solvent extraction of the wood, (3) *destructively distilled* (*D.D.*) *wood turpentine*, obtained by fractional distillation of the oily condensate recovered from the vapors formed during the thermal degradation of resinous pine wood, and (4) *sulfate turpentine*, recovered as a by-product during the sulfate (kraft process) pulping of resinous woods. Over half the world's supply of turpentine comes from the United States.

Turpentines differ in odor and composition according to the method of production. Gum turpentine is considered the highest quality and has a pleasant odor. Steam-distilled turpentine is but slightly inferior in odor and quality to the gum spirits. The destructively distilled turpentine has a pungent odor due to contamination with thermally decomposed substances. The gum turpentine differs chemically from the wood turpentines, though their properties are not greatly different. Gum turpentine is distinctive for its large beta-pinene fraction, which is virtually absent from the wood turpentines.

Turpentine is a volatile oil consisting primarily of terpene hydrocarbons, having the empirical formula $C_{10}H_{16}$. These 26 atoms can have many different arrangements. Only six are present in appreciable amounts in commercial turpentines, namely; alpha-pinene (b.p. 156°C), beta-pinene (b.p. 164°C), camphene (b.p. 159°C), Δ^3-carene (b.p. 170°C), dipentene (b.p. 176°C), and terpinoline (b.p. 188°C). The molecular configurations of most of these are shown in Fig. 15.15.

Rosin, the other major naval stores product, is a brittle solid which softens at 80°C. Chemically it is composed of about 90 per cent resin acids and 10 per cent neutral matter. The resin acids are mainly *l*-abietic acid and its isomers, $C_{20}H_{30}O_2$. These are tricyclic monocarboxylic acids.

Rosin is graded and sold on the basis of color, the color grades ranging from pale yellow to dark red (almost black). The color is due almost entirely to iron contamination and oxidation products. Fresh oleoresin, as it exudes from the tree, will yield a rosin that is nearly colorless. Color-bodies are removed by selective solvents and selective absorption from a 10–15 per cent gasoline solution passed through beds of diatomaceous earth. About 70 per cent of the world's rosin is produced in the United States.

There are three types of naval stores, according to method of production: gum, wood, and sulfate naval stores. Gum naval stores are obtained by tapping living pine trees. Wood naval stores are obtained by removing the resins from shredded wood, either by steam distillation and solvent extraction, or, as is the case today, by a solvent-extraction process using old pine stumpwood. Sulfate naval stores, commonly called tall oil, are obtained as a by-product of kraft pulp production from pines.

Gum Turpentine and Rosin Production—Gum Naval Stores

The crude gum or oleoresin is caused to flow from healthy trees by exposing the sapwood. The lower part of the tree is faced, i.e., a section of bark is removed, giving a flat wood surface for the gutters which are inserted into a slanting cut made by a special axe. The gutters conduct the gum to a container which can hold 1-2 qt of gum. At the top of the exposed face, a new V-shaped strip of bark is removed about every two weeks. Each time the bark is removed the wound is treated with sulfuric acid to prolong gum flow (up to 28 days). A 40–60 per cent solution of the acid is sprayed on the wound or a sulfuric acid paste formulation, introduced in 1966, is applied.[21]

The operations of inserting gutters, hanging cups, and cutting the first bark is preferably done in December or January, since early facing stimulates early season gum flow. The gum continues to flow until November, with the height of the season being from March to September. The average composition of crude

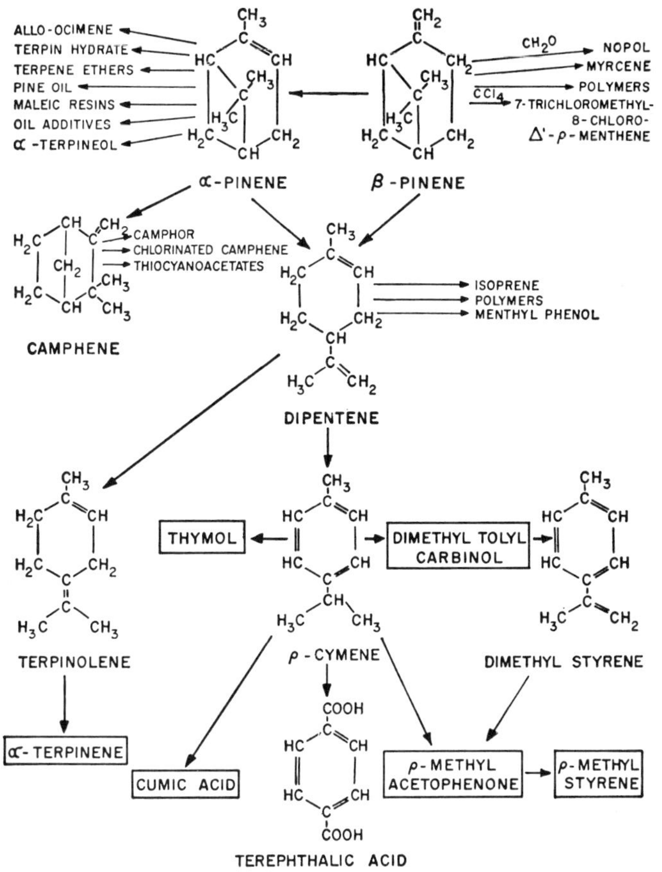

Fig. 15.15 Some reactions of alpha-pinene and beta-pinene. (*L. A. Goldblatt, "Yearbook of Agriculture," U.S. Department of Agriculture, 1950–51.*)

gum is 68 per cent rosin, 20 per cent turpentine, and 12 per cent water.

The collected gum is distilled from a copper still; turpentine and water pass over, and the rosin is left in the still. The remaining molten rosin, plus impurities, is passed through a series of strainers and cotton batting to remove dirt particles. The liquid rosin is then run into tank cars, drums, or multiwall paper bags for shipment. Annual production figures for gum turpentine and rosin are shown in Table 15.14.

Extraction Process for Wood Naval Stores

This process is applied to resinous wood chips from pine stumps and "lightwood." The latter refers to wood and stumps, mainly longleaf pine, left in or on the ground for 25 years or more after the tree is felled. The low-resin-content sapwood decays, leaving only the resin-rich heartwood stumps and wood chunks. These are removed with earth-moving equipment and shipped to plants where they are cleaned and shredded.

In modern practice all the resin products are removed by solvent extraction. The solvent retained by the extracted wood chips is recovered by steaming. Extraction is carried out with naphtha (b.p. 90–115°C fraction) or benzene. Multiple extraction is carried out in a series of vertical extractors in a countercurrent manner, whereby fresh solvent is used for the final extraction of a charge.

The solution from the extractors is vacuum

TABLE 15.14 Naval Stores Production for U.S.A.[a]

Rosin–1000 Drums of 520 lb net

Year	*Gum*	*Wood (Extracted)*	*Tall Oil*	*Total*
1955	528	1342	None	1870
1958	400	1196	269	1865
1960	370	1220	420	2010
1963	459	1098	528	2085
1966	273	986	699	1958
1969	119	830	793	1742

Turpentine–1000 Barrels of 50 gal

Year	*Gum*	*Wood (Extracted)*	*Wood (Destr. Dist.)*	*Sulfate*	*Total*
1955	176	208	2.41	232	618
1958	129	186	.84	312	628
1960	119	163	none	323	605
1963	141	157	none	376	674
1966	84	155	none	427	666
1970	35	103	none	480	618

Pine Oil and Other Terpenes–1000 Barrels of 50 gal

Year	*Pine Oil*	*Other Terpenes*
1955	189	108
1958	185	121
1960	197	94
1965	234	125
1970	278	100

[a]Source: Naval Stores Annual Report, U.S. Dept. Agr. (1958, 1955); Agricultural Statistics, U.S. Dept. Agr. (1970); Statistical Abstracts of U.S., U.S. Dept. Commerce, Bureau of Census, 91st Ed., Washington (1970).

distilled and the solvent recovered. The remaining terpene oils are fractionally redistilled under vacuum and recovered as turpentine, dipentene, and pine oil. The nonvolatile rosin is of dark color and is upgraded by clarification methods, such as selective absorption of its solution (bed-filtering). Annual production figures of the products are shown in Table 15.14.

Sulfate Turpentine and Tall Oil—Sulfate Naval Stores

Sulfate turpentine is obtained as a by-product during the kraft pulping of pine woods. Vapors are periodically released from the top of the digesters; these are condensed, and the oily turpentine layer is separated and purified by fractional distillation and treatment with chemicals to remove traces of sulphur compounds.

The spent black liquor from the sulfate or kraft pulping of pine woods contains the less volatile products of the wood resin in the form of sodium salts or soaps. These are removed as soap skimmings, called black liquor skimmings. When these salts are acidified by sulfuric or sulfurous acid, the organic acids collect on the surface of the aqueous phase and are separated mechanically as tall oil (tall is the Scandinavian word for pine), which is then fractionally distilled under vacuum.

The yield of tall oil varies from about 100 lb per ton of pulp from northern pines to 250 lb per ton of pulp from the southern pines. One ton of crude tall oil yields about 500 lb fatty acids, 800 lb rosin acids, and 700 lb of secondary products comprising distilled tall oil, tall oil heads, and tall oil pitch.[22]

Tall oil is particularly attractive as a raw material because of its availability. The gum-rosin industry, on the other hand, scars the trunks of living pine and involves manual tasks that are costly and time-consuming. And the wood-rosin industry depends on the decreasing availability of aged tree stumps whose removal and transportation are becoming increasingly costly. But crude tall oil is a "waste" product of pulp making. In fact, if the black liquor skimmings did not yield some useful products, they could be a disposal problem for pulp mills.

The growth of new capacity in kraft pulp mills has made available a large quantity of crude tall oil. Future growth of the paper industry—about 5 per cent a year—will provide increasing supplies of raw material for tall oil fractionators.

For these reasons there is a steady decline in naval stores production from gum and by the extraction of wood (so-called steam distillation process), and a corresponding production increase from tall oil and sulfate turpentine, as is shown in Table 15.14. Tall oil now (1971) accounts for about 50 per cent of the total rosin output and tall oil production is nearly 100 per cent more than in 1961, a remarkable growth.

Uses of Naval Stores Products

The naval stores products have a wide range of usefulness from ordinary household commodities to complex industrial uses, as outlined below.

Turpentine is used mainly as a raw material in the chemical and rubber industries. It is also widely used as a household solvent and paint thinner. The chemistry of the terpenes offers many possibilities for conversion to other substances, as illustrated by the chart in Fig. 15.15. The more common uses of turpentine are: paints, varnishes, enamels, waxes (shoe, floor, and furniture), polishes, disinfectants, soaps, pharmaceuticals, sealing wax, inks, wood stains, insecticides, crayons, synthetic camphor, and general solvent use.

Dipentene is present in large amounts in the higher boiling fractions of wood turpentine (not in gum turpentine). It is used in paints and varnishes and as a penetrating and softening solvent in rubber reclamation.

Pine oil has a wide range of industrial uses. A synthetic pine oil is also made from turpentine to meet the demands for the natural product. The more common uses of pine oil are: soaps, disinfectants, polishes, insecticides, deodorants, protective coatings, solvent and wetting agent in rubber and dye industries, and as a flotation agent in metallurgical operations.

Rosin is widely used in industry, with paper (38%), synthetic rubber (13%), adhesives (12%), and protective coatings (10%) being the major users. Inks (8%), wood resins (7%), and miscellaneous uses (12%, including chewing gum and rosin oil) account for the remainder of the consumer uses.[22] Because rosin has a carboxyl group and double bonds, it is reactive and can be used to produce salts, soaps, esters, amines, amides, nitriles, and Diels-Alder adducts; and it can be isomerized, disproportionated, hydrogenated, dimerized, and polymerized. When destructively distilled it produces a viscous liquid, termed rosin oil, which is used in lubricating greases.

The paper industry uses large amounts of the sodium salt of rosin as paper size, which accounts for the greatest single use of rosin. Synthetic rubber industry is the second most important user of rosin. In making styrene-butadiene rubber, disproportionated rosin soaps are used alone or in combination with fatty acid soaps as emulsifiers in the polymerization process. Disproportionation decreases the number of double bonds in the abietic acid of the rosin, making a more stable material.

The adhesives industry is the third most important market for rosin. Rosin, modified rosins, and rosin derivatives are used in several types of adhesives, including the pressure-sensitive, hot-melt, and elastomer-based latices, and solvent rubber cements.

Protective coatings are the fourth major user of rosin, either directly or in a modified or derivative form. Varnishes and alkyds are the most common types of protective coatings using rosin. Rosin is combined with a heat-reactive phenol-formaldehyde resin to produce a widely used varnish. Printing inks also use substantial amounts of rosin. There is potential for growth in the use of rosin, not only in its present uses, but in other ways. Biodegradable detergents and food packaging are two possibilities. The Food and Drug Administration has approved the use of rosin in food-packaging applications.

Tall oil fatty acids, obtained in about 25 per cent yield from crude tall oil, find a ready market in many applications. They are used in large amounts in protective coatings, soaps, detergents, disinfectants, intermediate chemicals, flotation aids, and tallate driers.

TANNIN AND OTHER EXTRACTIVES

The tissues of wood, bark, and the leaves of trees contain a great variety of chemical substances of considerable scientific interest and some of practical value. Turpentine, pine oil, and rosin from the resins of pines are the most important commercial extractives from American woods.

Tannin is a commercially important substance that can be extracted from the wood, bark, or leaves of certain trees and other plants. Tannins are complex dark-colored polyhydroxy phenolic compounds, related to catechol or pyrogallol, and vary in composition from species to species. They have the important property of combining with the proteins of skins to produce leather.

For many years most of the leather in the United States was tanned with domestic tannins from hemlock and oak bark and from chestnut wood. Today only a small amount of tannin comes from these and other domestic sources. The most important source of vegetable tannin today is the wood of the quebracho tree which grows mainly in Paraguay and Argentina. The tannin content of this tree and a few other sources of vegetable tannin are shown in Table 15.15

The wood or bark for tannin production is reduced to chips and shreds by passing the material through hoggers or hammer mills. It is then extracted with warm water in diffusion batteries. The dilute solutions are evaporated to the desired concentration. Loss of solubility of the tannin can be counteracted by treatment of the concentrate with sodium sulphite.

TABLE 15.15 Tannin Content of Some Plant Materials

Plant Material	*Per Cent Tannin*
Domestic sources	
Eastern hemlock bark	9–13
Western hemlock bark	10–20
Tanbark oak	15–16
Chestnut oak	10–14
Black oak	8–12
Chestnut wood	4–15
Sumac leaves	25–32
Foreign sources	
Quebracho heartwood	20–30
Mangrove bark	15–42
Wattle (*acacia* bark)	15–50
Myrobalan nuts	30–40
Sicilian sumac leaves	25–30

ESSENTIAL OILS

Various essential oils are obtained by steam distillation of wood chips, barks, or leaves of trees. Chemically they are related to turpentine, being mainly terpene hydrocarbons. In the United States the following oils are produced; cedar-wood oil from the wood of red cedar; conifer-leaf oils from the needles and twigs of spruce, hemlock, cedar, and pine; sassafras oil from roots, wood, and bark of the sassafras tree; and sweet-birch oil from the bark and twigs of birch (*Betula lenta*).

MEDICINALS

The bark of the cascara tree of the northwestern region of the United States yields cascara, a laxative used in medicine. Several hundred tons of bark are harvested annually.

The red gum tree of the southern United States exudes a yellowish balsamic liquid or gum from wounds. This is known as storax. It is produced by removing a section of bark and incising the wood much in the same manner as that used for the production of naval stores gum described above. Storax is used in medicinal and pharmaceutical preparations, such as adhesives and salves. It is also used as an incense, in perfuming powders and soaps, and for flavoring tobacco.

REFERENCES

1. U.S. Forest Service, "Timber Trends in the United States," Forest Service Report No. 17, U.S. Dept. of Agriculture, Washington, D.C. Feb. 1965.
2. "The United States Timber Supply," National Policy Conference, Arden House, Harriman, N.Y., Convened by State University of New York College of Forestry, Nov. 23–24, 1970.
3. Waters, W. E. and Brandt, R. W., *American Forests* (Sept. 1970).
4. "One Third of the Nation's Land," A Report to the President and to the Congress by the Public Land Law Review Commission, Washington, D.C., June, 1970.
5. Maxwell, C. S., *Tappi*, **54**, 1932–1934 (1971).
6. Kenaga, D. L., Fennessey, J. P., and Stannett, V. I., *Forest Prod. J.*, **12**, No. 4, 16–168 (1962).
7. Kent, J. A., Winston, A., and Boyle, W., "Preparation of Wood-Plastic Combinations Using Gamma Radiation," ORO-612, Office of Technical Services, Dept. of Commerce, Washington, D.C., Sept. 1, 1963.
8. Kent, J. A., Winston, A., Boyle, W., and Taylor, G. B., "Preparation of Wood-Plastics Combinations Using Gamma Radiation to Induce Polymerization (Copolymers, Additives and Kinetics)," Interim Report, U.S. Atomic Energy Comm., ORO-2945-7, 1967.
9. *Ind. Eng. Chem.*, **28**, 1160–1163 (1936).
10. Meyer, J. A., *Forest Prod. J.*, **15**, No. 9, 362–364 (1965).
11. Meyer, J. A. and Loos, W. E., *Forest Prod. J.*, **19**, No. 2, 32–38 (1969).
12. Meyer, J. A., *Forest Prod. J.*, **18**, No. 5, 89 (1968).
13. Schuerch, C., *Forest Prod. J.*, **14**, No. 9, 377–381 (1964).
14. Schuerch, C., U.S. Patent 3,282,313 (Nov. 1, 1966).
15. Locke, E. G. and Garnum, E., "FAO Technical Panel on Wood Chemistry: Working Party on Wood Hydrolysis," *Forest Prod. J.*, **11**, No. 8, 380–382 (1961).
16. Kratzl, K., *Holzforsch. Holzverwert.*, No. 1 (1971).
17. Brown, N. C., "Forest Products," 197, New York, John Wiley & Sons, 1950.
18. Hair, D., *Forest Prod. J.*, **14**, No. 2, 63–65 (1964).
19. Simmons, F. C., "Three New Charcoal Retorts," *Equipment Notes*, Forestry and Forest Products Div., Food and Agriculture Organization of the United Nations, Rome, Italy, Oct., 1960.
20. *Northern Logger*, **13**, No. 7, 12–13 (Jan. 1965).
21. Harrington, T. A., *Forest Prod. J.*, **19**, No. 6, 31–36 (1969).
22. *Chem. Engr. News*, **45**, 16–19 (Feb. 13, 1967).

SELECTED REFERENCES

Wood Chemistry

Bikales, N. M. and Segan, L., Eds., "Cellulose and Cellulose Derivatives," Parts IV and V, New York, John Wiley & Sons, 1971.

Browning, B. L., ed., "The Chemistry of Wood," New York, John Wiley & Sons, 1963.

Browning, B. L., "Methods of Wood Chemistry," Vols. I and II, New York, John Wiley & Sons, 1967.

Farmer, R. H., "Chemistry in the Utilization of Wood," Oxford, Pergamon Press, 1967.

Sarkanen, K. V., and Ludwig, C. H., eds., "Lignins," New York, John Wiley & Sons, 1971.

Stamm, A. J., "Wood and Cellulose Science," New York, Ronald Press, 1964.

Wenzel, H. F. J., "The Chemical Technology of Wood," New York, Academic Press, 1970.

Wise, L. E. and Jahn, E. C., "Wood Chemistry," Vols. I, II, New York, Van Nostrand Reinhold, 1952.

Pulp and Paper

Britt, K. W., "Handbook of Pulp and Paper Technology," 2nd Ed., New York, Van Nostrand Reinhold, 1970.

Casey, J. P., "Pulp and Paper," 2nd Ed., Vols. I, II, III, New York, John Wiley & Sons, 1960.

MacDonald, R. G. (Editor), "Pulp and Paper Manufacture," 2nd Ed., Vol. I, II, III, New York, McGraw-Hill, 1969.

Rydholm, S. A., "Pulping Processes," New York, John Wiley & Sons, 1965.

Fiber and Particle Boards

"Tomorrow's Technology Today Produces Medium Density Fiberboard," *Wood and Wood Products* 76, No. 7, 29–33 (1971).

Halligan, A. F., "Recent Glues and Gluing Research Applied to Particleboard," *Forest Prod. J.* **19**, No. 1, 44–51 (1969).

Report of an International Consultation on Insulation Board, Hardboard and Particle Board, Geneva, 1957, "Fibreboard and Particle Board," Food and Agriculture Organization of the United Nations, Rome, 1958.

Technical Papers submitted to the International Consultation on Insulation Board, Hardboard and Particle Board, Geneva, 1958, "Fibreboard and Particle Board," Food and Agriculture Organization of the United Nations, Rome, 1958.

Polymer Combinations with Paper and Wood

Jahn, Edwin C. and Stannett, V., "Polymer Modified Papers," in "Modern Materials," H. H. Hausner, Ed., Vol. II, New York, Academic Press, 1960.

Langwig, J. E., Meyer, J. A., and Davidson, R. W., "New Monomers Used in Making Wood-Plastics," *Forest Prod. J.* **19**, No. 11, 57–61 (1969).

Siau, J. F., Meyer, J. A., and Skaar, C., "A Review of Developments in Dimensional Stabilization of Wood Using Radiation Techniques," *Forest Prod. J.*, **15**, 162–166 (1965).

Siau, J. F., Meyer, J. A., and Skaar, C., "Wood-Polymer Combinations Using Radiation Techniques," *Forest Prod. J.*, **15**, 426–435 (1965).

U.S. Forest Products Laboratory, Report No. 2145, "Report of Dimensional Stabilization Seminar," Madison, Wis., 1959.

Plasticizing Wood

Davidson, R. W., "Plasticizing–A New Process for Wood Bending," *Furniture Methods and Materials*, pp. 26–29 (Feb. 1969).

Davidson, R. W. and Baumgardt, W. G., "Plasticizing Wood with Ammonia–A Progress Report," *Forest Prod. J.* **20**, No. 3, 19–24 (1970).

Wood Preservation

American Wood-Preservers' Association, "American Wood-Preservers' Association Standards," Washington, D.C., 1971.

Canadian Wood Council, "Wood: Fire Behavior and Fire-Retardant Treatment–A Review of the Literature," Ottawa, Nov., 1966.

Hunt, G. M. and Garratt, G. A., "Wood Preservation," New York, McGraw-Hill, 1953.

Siau, John F., "Flow in Wood," Syracuse, N.Y., Syracuse University Press, 1971.

U.S. Forest Products Laboratory, "Wood Handbook," Agriculture Handbook No. 72, U.S. Dept. of Agriculture, Washington, D.C., 1955.

Hydrolysis of Wood

Hägglund, Erik, "The Development of Wood Saccharification by the Rheinan Process," *The Organization* (Food and Agriculture Organization of the United Nations), pp. 8–12, 1954.

Harris, Elwin E., "Molasses and Yeast from Wood," in "Crops in Peace and War," The Yearbook of Agriculture, U.S. Dept. of Agriculture, Washington, D.C., 1950–51.

Mikhailov, M. T. and B. N. Tasinskii, "Outlook for the Development of Wood Chemistry and Hydrolysis Industry," Moscow, Goslesbumizdat., 1960.

Shlamin, V. N., "Wood Chemistry Industry in 1959–1965," Moscow, Gosplamizdat., 1965.

Working Party on Wood Hydrolysis, "Final Report of 2nd Meeting," Food and Agriculture Organization of the United Nations, Rome, 1960.

Carbonization of Wood

Beglinger, E., "Production and Uses of Charcoal," in "Crops in Peace and War," The Yearbook of Agriculture, U.S. Dept. of Agriculture, Washington, D.C., 1950–51.

Dargan, E. E. and Smith, W. R., "Continuous Carbonization and Briquetting of Wood Residue," *Forest Prod. J.*, **9**, No. 11, 395–397 (1959).

Lane, P. H., "Wood Carbonization in Kilns," *Forest Prod. J.*, **10**, No. 7, 344–348 (1960).

Northeast. Wood Utiliz. Council, Bull., No. 37 "Charcoal Production and Uses," (Jan. 1952).

Naval Stores

Durfer, J. M., "Flavor and Perfume Chemicals from Sulfate Turpentine," *Tappi*, **46**, 513 (1963); "Flavor Oils from Turpentine," *Tappi*, **49**, 117A (1966).

Goldblatt, Leo A., "Chemicals We Get from Turpentine," in "Crops in Peace and War," The Yearbook of Agriculture, U.S. Dept. Agriculture, Washington, D.C., 1950–51.

Lawrence, R. V., "The Industrial Utilization of Rosin," in "Crops in Peace and War," The Yearbook of Agriculture, U.S. Dept. of Agriculture, Washington, D.C., 1950–51.

Lawrence, R. V., "Naval Stores Products from Southern Pines," *Forest Prod. J.*, **19**, No. 9, 87–92 (1969).

Tannins and Wood Extractives

Hillis, W. E., Ed., "Wood Extractives," New York, Academic Press, 1962.

Northeast. Wood Utiliz. Council, Bull., No. 39, "Tannin from Waste Bark," 1952.

Rogers, J. S., "Native Sources of Tanning Materials," in "Crops in Peace and War," The Yearbook of Agriculture, U.S. Dept. of Agriculture, Washington, D.C., 1950–51.

16

Sugar and Starch

Charles B. Broeg* and Raymond D. Moroz*

INTRODUCTION

Sugar and starch are among those organic "chemicals" found so abundantly in nature that no serious efforts have been made to synthesize them commercially from coal, air, and water. Both are available at such concentrations in some plants that sizable industries have resulted from the production of these plants and the extraction of carbohydrates therefrom.

The primary use of sugar is in the manufacture of food or as a food in itself. Where used for such purposes, most of it is highly refined or purified, but considerable quantities are consumed in some areas as a crude product. Sugar is used to a limited extent in the production of other chemicals such as sucrose esters and in the form of by-product molasses as a substrate for fermentation processes.

Extensive use of starch is made in the textile and paper industries, but food is also a major outlet, especially if one considers flours that are primarily starch as a crude form of starch. However, a very important outlet for starch is its conversion to dextrose and glucose syrups for use in the food industry. The following explores briefly the sources of sugar and starch, and the processes by which they are converted into commercial products.

SUGAR

History

The ancestry of the sugar cane and its use as a food has been traced to the island of New Guinea. Around 8000 B.C., the plant started on its migration to the many areas of southeastern Asia, Indonesia, the Philippines, Malay, Indochina, and eastern India, with man probably as its main dispersal agent.[1] It is in Bengalese India that sugar cane was first cultivated as a field crop and that the juice was manufactured into various solid forms. A general knowledge of sugar was prevalent throughout India by 400 B.C. By the tenth century A.D., sugar cultivation and manufacturing had become important industries in Persia and Egypt.

*Sucrest Corp., New York, N.Y.

The early Islamic movement spread the knowledge of the sugar industry throughout the Mediterranean area. On the second voyage of Columbus to America in 1493, sugar cane was introduced to Santo Domingo. It spread rapidly through the West Indies and Central America. Cortez brought the cane to Mexico, and Pizarro introduced it in Peru. By 1600, the sugar industry was the largest in tropical America.[2]

The modern sugar industry dates from the end of the 18th century when steam replaced animal energy and made possible the development of larger and more efficient production units. The vacuum pan appeared in 1813, bag filters in 1824, multiple-effect evaporators in 1846, filter presses in 1850, centrifugals in 1867, driers in 1878, and packaging machines in 1891.[3]

Cultivation of the beet sugar plant and the manufacture of sugar from the beet developed in the industrial nations of Europe within the last two centuries. In 1747, a German chemist, Marggraf, established that sugar from beets was the same as sugar from cane. His pupil, Achard, in 1799, demonstrated that sugar can be commercially prepared from beets. During the Napoleonic wars, a short-lived sugar industry was established in France. It was slowly revived 20 years after Waterloo and spread to most of European countries.[4] Today, sugar from beets accounts for about 41 per cent of the world's supply.

Cane Sugar

Agriculture. The sugar cane is a large perennial tropical grass belonging to the genus *Saccharum.* There are three basic species, *S. officinarum, S. robustum,* and *S. spontaneum,* and a large number of varieties. Sugar cane is propagated commercially by cuttings, each cutting consisting of portions of the cane plant having two or more buds. The buds sprout into shoots from which several other shoots arise below the soil level to form a clump of stalks or a "stool." From 12 to 20 months are required for crops from new plantings and about 12 months for ratoon crops (that is, cane stalks arising from stools which have been previously harvested). Fields are replanted after 2 to 5 "cuttings" have been made from the original plantings. The cane stalk is round and from a few to more than ten feet long. It is covered with a hard rind which is light brown or green, yellowish green, or purple in color, depending on the variety. The stalk consists of a series of joints or internodes separated by nodes. The rind and the nodes are of a woody nature, while the internode is soft pith. The internode contains the greater part of the juice containing sucrose, and the nodes contain "eyes" or buds which "sprout" when planted.[5]

Harvesting. The sugar cane is still cut by hand with machete type knives in most of the producing areas. The canes are cut at ground level and at the same time leaves and tops of the stalk are removed. In areas where labor is scarce or expensive, machine harvesting of the sugar cane has come into widespread use. There are harvesters which will cut, chop, and load the cane into transporting vehicles in one operation at rates of 30 to 45 tons per hour. Other harvesters will cut, bundle, and dump cane in rows for pick-up by other machines. Transportation of field cane to the mill is accomplished mainly by railroad. The use of trucks, trailers, and cart tandems is increasing.

Preparation of Cane for Milling. Mechanically harvested cane must be washed before milling to eliminate soil, rocks, and field trash. In some instances, "dry cleaning" precedes washing. Washing systems can range from a simple wash with warm water on the carrier or table to an elaborate system consisting of conveyors with water jets, baths for removal of stones, and stripping rolls.

Juice Extraction. The cane is first prepared for grinding by one or more of the following operations: (1) chopping into smaller pieces with one or two sets of rotating knives (400–600 rpm); or (2) disintegrating the cane into finer sizes by a crusher and/or shredder. The crusher is usually a two roller mill of the Krajewski or Fulton types.

The crushed and shredded cane passes through a series of three horizontal rollers (mill) arranged in a triangular pattern with the top roll rotating counterclockwise and the bottom two rollers clockwise. A series of three-

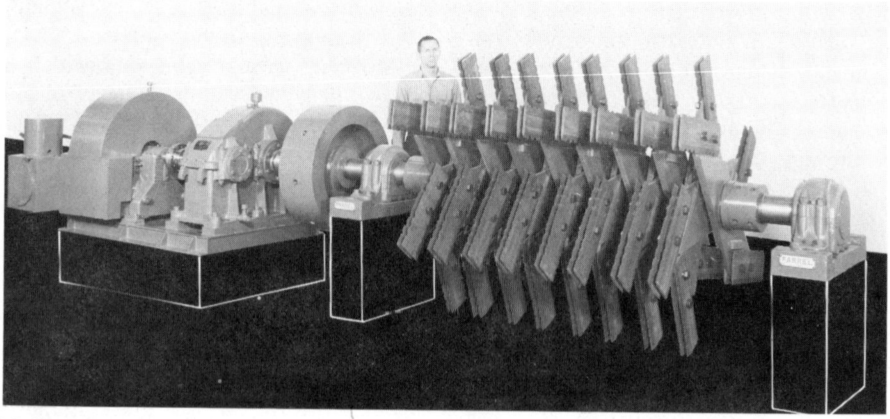

Fig. 16.1 Farrell type K-4 knife to handle 4500 tons of cane per 24-hr day. (*Courtesy Farrell Co., Division of USM Corp., Ansonia, Conn.*)

roller mills, numbering 3 to 7, is called a tandem. The pressure on the top roll is regulated by hydraulic rams and averages about 500 tons. Below each mill, there is a juice pan into which expressed juice flows. The crusher and first mill extract 60-70 per cent of the cane juice and the remaining mills take out 22-24 per cent.

When the fiber content of the bagasse reaches about 50 per cent, no more juice is extracted by pressure from the mills. The juice that remains with the fiber contains the same proportion of sucrose as the original juice of the cane. If this sucrose is allowed to remain with the bagasse as it leaves the mill, it would constitute a large loss to the factory. Consequently, a process called "compound imbibition" is used to reduce the sucrose in the fiber by repeated dilution. In the 5-mill tandem, water is added to the 4th mill and the expressed juice from that mill is brought back to the 2nd mill. The expressed juice from the 3rd mill is recirculated to the 1st mill and the 5th mill to the 3rd mill. In this way, the juice in the bagasse is always diluted before crushing. The amount of juice applied at each mill is approximately equal to the amount of water applied to the 4th mill or to the penultimate mill in a differently numbered tandum.[6]

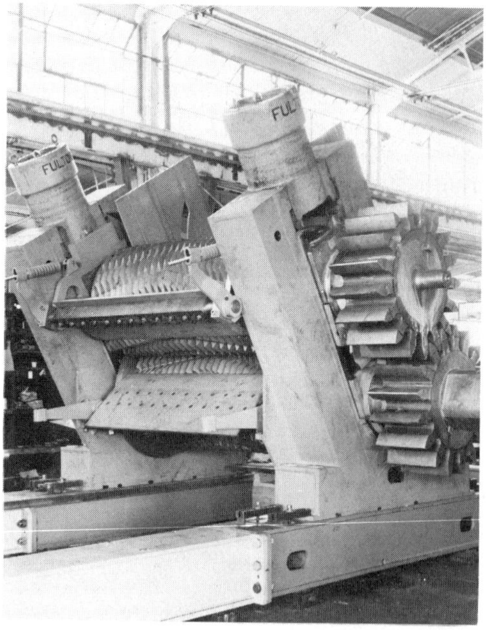

Fig. 16.2 Fulton crusher roll. (*Courtesy Fulton Iron Works Co., St. Louis, Mo.*)

Before the expressed juice goes to clarification, it is screened through perforated metallic sheets with openings of about 1 mm in diameter. There might also be further screening with stationary or vibrating metallic cloths.

Diffusion. Since 1962, there have been several diffusers installed in cane mills throughout the world. Cane diffusers are now manufactured by a number of companies including Braunschweigische Maschinenbauanstalt, CF&I Engineers Inc., De Danske Sukkerfabrikker, De Smet, S.A., Sucatlan Engineering, and Suchem Inc. of Puerto Rico.

Fig. 16.3 Farrell 4-unit milling train. (*Courtesy Farrell Co., Division of USM Corp., Ansonia, Conn.*)

Diffuser capacities range from 1000 to 6000 tons of cane per day. They are usually used in conjunction with some part of the mill, particularly with the crusher and the first mill. Preparation of the cane prior to its entry into the diffuser is essential for good extraction. In the diffuser, the crushed cane is countercurrently washed with imbibition water at temperatures ranging from 50 to 75°C. The last stage of a diffuser, the bagasse exit, receives water which becomes concentrated with sugar as it proceeds to the first stage. Most of the sugar is extracted in the first four or five stages by simple displacement of sucrose from the open ruptured cane cells. In later stages, diffusion appears to take place in those cane cells not originally ruptured. One or two mills are used to express water from the bagasse as it leaves the diffuser. The crusher and one mill extracts 70 per cent of the juice from the cane and the diffuser removes 30 per cent of the expected yield. Sucrose extraction from a mill-diffusion system averages 97 per cent compared to 90-94 per cent with a straight milling system.[7]

Juice Purification. The juice from the milling station is an acidic, opaque, greenish liquid containing soluble and insoluble impurities such as soil, protein, fats, waxes, gums, and coloring matter. The process designed to remove as much of these impurities as possible employs lime and heat and is called clarification or defecation. In simple line defecation, milk of lime is added to the cold juice in amounts (about one pound of calcium oxide per ton of cane) sufficient to bring the pH to the range of 7.5-8.5. The limed juice is pumped through heaters where the juice is heated to temperatures between 90 and 115°C. There are many modifications to this basic process involving different sequences of

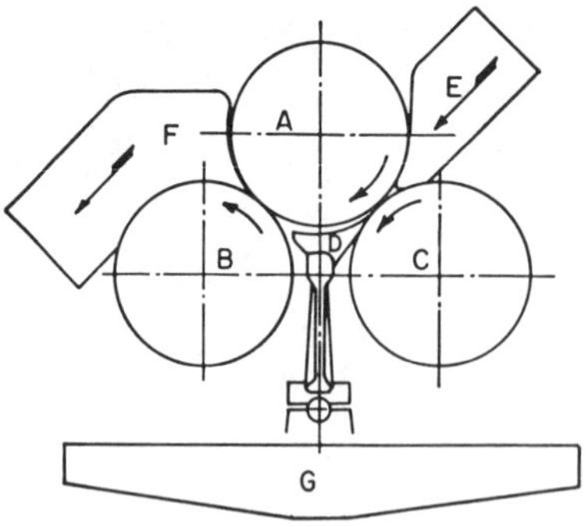

A - Top-roller.
B - Bagasse roller.
C - Feed roller.
D - Trash plate.
E - Feed of cane.
F - Discharge of crushed cane.
G - Juice-collecting trough.

Fig. 16.4 A cane mill with its three crushing rolls and with a typical turn-plate arrangement. An hydraulic piston maintains pressure on *A*, while *B* and *C* are fixed; the rolls are under pressure of 75–80 tons per foot of roll width.

lime and heat. For example, fractional liming and double heating involves liming the cold juice to pH 6.4, heating, liming to pH 7.6, and heating again before sending the juice to the clarifiers.

The combination of lime and heat forms a flocculent-type precipitate with various components in the juice. These consist mostly of insoluble lime salts, coagulated protein, and entrapped colloidal and suspended matter. The precipitate is removed by sedimentation or settling in continuous closed-tray clarifiers, i.e., Rapi-Dorr, Graver, BMA, and Bach Poly-Cell. The juice leaving the clarifier is a clear brownish liquid.

The flocculent precipitate or "muds" which settles on the clarifier trays contains about 5 per cent solid matter. Sugar is recovered from the muds by means of rotary vacuum filters equipped with perforated metallic screen cloth. The turbid filtrate is returned to the clarification system and the press cake is discarded or sent to the fields as fertilizer.

Recent innovations in the filtration of muds include the use of tighter filter media. A clear filtrate is obtained making it possible to send it on to the evaporators without reprocessing. The Eimcobelt rotary vacuum filter uses a tight filter cloth which is continuously removed from the drum and washed with hot water. Another type is the direct filtration of muds with horizontal leaf filters (Fas Flo Filters). One process uses polyelectrolytes to increase the filterability of the muds through cloth filters (Rapid-Floc system).

Good clarification depends upon the forma-

Fig. 16.5 Continuous ring-type (Saturne) total maceration sugarcane diffuser. *(Courtesy Sucatlan Engineering, Paris, France.)*

Fig. 16.6 Open top view of DDS double cylindrical tilted can diffuser. (*Courtesy De Danske Sukkerfabrikker, Copenhagen.*)

tion of a stable flocculent precipitate which settles rapidly. If the phosphate content of the preclarified juice contains less than 300 parts per million of phosphate, expressed as P_2O_5, it is customary to add a soluble phosphate. Polyelectrolytes are being used to improve the coagulation and the settling rate of the precipitate during clarification.[8]

Evaporation. The clarified juice (about 85 per cent water) is pumped to evaporators where it is concentrated to a clear heavy syrup containing about 65 per cent solids. Evaporation is carried out in "multiple effect" evaporators in order to achieve maximum steam economy. Each "effect" is arranged in series and operated so that each succeeding one operates under a higher vacuum. This arrangement allows the juice to be drawn from one vessel to the next and permits the juice to boil at a low temperature. The concentrated juice (syrup) is removed from the last effect by a pump. Three, four, and five effect evaporators are used but the four effect is most common. In a four (quadruple) effect, one pound of steam evaporates four pounds of water. Occasionally, surfactants are added to the juice to improve the rate of evaporation.[3]

Fig. 16.7 Oliver Campbell vacuum mud filter applied to cane-juice muds. (*Courtesy Dorr-Oliver, Inc., Stamford, Conn.*)

Crystallization. The syrup from the evaporator is pumped to a "vacuum pan" in which it is evaporated to supersaturation in order to cause sugar to crystallize. The vacuum pan is a single effect evaporator designed to handle viscous materials. It is a vertical cylinder with its bottom designed to allow easy removal of the crystallized mass. The heating elements used in vacuum pans are either short, large-diameter vertical tubes (calandria pans) or coils around the inner surface of the truncated cone of the pan (coil pans). A typical vacuum pan with a "catchall" or entrainment separator for separating syrup from vapors measures approximately 25 feet in height and 15 feet in diameter. The working capacity is about 1000 cubic feet of massecuite (mixture of crystals and syrup or mother liquor). The shape of the pan and the positioning of the heating elements within the pan are important design factors in maintaining good circulation of the massecuite. For example, floating calandrias (calandria not attached to shell of pan), horizontal pans, and pans having mechanical circulators[9] may be used.

Recently, a continuous vacuum-pan system was introduced (Fives Lille-Cail). The continuous vacuum pan is a horizontal cylinder with compartments in its lower part. The pan is provided with an additional evaporator called the concentrator, where the density of the syrup is raised to 78-80° Brix.* Seed is added in the first compartment and the resulting massecuite moves progressively through the compartments of the pan. Additional syrup is added to each compartment to control the fluidity of the massecuite. Approximately 20 per cent of the syrup is introduced at the concentrator and 80 per cent of the syrup is fed to the various compartments through special feed headers. The rate of massecuite discharge is controlled by the level in the pan.[10]

The crystallization of sucrose in vacuum pans is called "sugar boiling" and each boiling is termed a "strike." Since a single cyrstallization does not recover all the sucrose from the syrup, there are two to four boilings, the number depending upon the existing plant equipment. The three-boiling system is the most popular. In order to distinguish among the boilings, a letter is assigned to each boiling and its products, i.e., the first boiling (syrup massecuite) would be given the letter A and its products after centrifugation, A sugar and A molasses; the second and third boilings would be assigned the letters B and C, respectively.

The boiling systems are based on "apparent purity," the amount of sucrose remaining in solution, and is usually expressed as the ratio of the polarization value to the total solids as measured by a hydrometer. In a three-boiling system, the purity of the A massecuite is set between 80-85 by blending syrup with A molasses, the purity of the B massecuite is set between 70-75 by blending with syrup and A molasses, and the C massecuite is set between

Fig. 16.8 Top view of open Stork continuous crystallizer under construction. *(Courtesy Stork-Werkspoor Sugar Hengelo, The Netherlands.)*

*The Brix scale is a density scale for sugar (sucrose) solutions. The degrees Brix are numerically equal to the percentage of sucrose in the solution. The term "Brix solids" refers to the solid content as determined by the Brix hydrometer.

55-60 by blending with syrup and B molasses. Other variations are possible.[3]

The A and B massecuites, after being discharged from a vacuum pan, are sent to centrifugals for separation. The A and B sugars are combined to form the commercial raw sugar, the principal product of the cane sugar factory.

The C massecuite is a low-purity, highly viscous material which is not immediately sent to the centrifugals because of the large amount of recoverable sucrose remaining in solution. Instead, it goes to crystallizers, U-shaped or horizontal containers equipped with coils attached to a hollow rotating shaft through which water circulates. The massecuite remains in the crystallizer from one to four days to allow additional crystallization to take place. Centrifugation of the C massecuite yields a final molasses (blackstrap) and a C sugar which is used for seeding the A and B boilings.

Centrifugation. The discharge from the vacuum pans is sent to centrifugal machines wherein the crystals are separated from the mother liquor. A centrifugal consists of a cylindrical perforated basket lined with a screen of perforated sheet metal. The basket, enclosed in a metal casing, is mounted on a vertical shaft which rotates the basket.

In a batch centrifugal, hot massecuite is fed into the basket through a short chute from the holding vessel. As the basket rotates, the massecuite forms a vertical layer on the screen lining. When the machine reaches operating speed (1000-1800 rpm), the syrup flows through the perforated lining and basket and is removed through an outlet at the bottom of the casing. The sugar on the lining is washed

Fig. 16.9 Silver continuous centrifuge. *(Courtesy CF&I Engineers, Inc., Denver, Col.)*

with a spray of water to decrease the amount of molasses adhering to the crystals. The basket continues rotating until the sugar is fairly dry, at which time the machine is switched off and brakes are applied. The sugar is discharged manually or by a plough.

In continuous centrifugals, the machines do not stop but continue in motion while receiving fresh supplies of massecuite. The rate of feed must be carefully regulated to obtain optimum separation of molasses from crystals.[11]

Packaging and Warehousing. The packaging of raw sugar in jute bags has diminished to a very low level in most areas of the world. In fact, all of the raw sugar produced in Louisiana, Florida, Hawaii, Australia, and Puerto Rico is shipped in bulk. Where raw sugar is still bagged, bag weights vary from 140-150 lbs in the Philippines to 250 lbs in the West Indies. Raw sugar is brought from the factories to seaport terminals by means of dump trucks or railroad cars. These terminals with capacities of 75,000 tons or more are warehouses which were originally intended for bagged sugar, or they are specially designed structures. Unloading and loading of ships is done by bucket conveyors, cranes, or by gantries with traveling cranes.[6]

SUGAR REFINING

Affination and Melting

The first step in refining of raw sugar is called affination and consists of removing the film of molasses (in which a large portion of the impurities are contained) from the surface of the raw-sugar crystal. The process involves the mixing of raw sugar with heavy syrup (72-75° Brix) at about 120°F in a U-shaped trough called a mingler. The mingler has a rotating agitator to maintain maximum contact between the sugar and syrup. The mixture is centrifuged to separate the crystals from the syrup and the crystals are washed with hot water. The "washed" sugar is "melted" or dissolved in water to a density of 55-60° Brix. Potable water, steam condensate, and "sweet waters" are used for dissolution. The liquor from the melter is screened to remove non-sugar materials such as sand, stones, wood, cane fibers, and lint in a rectangular vibrating screen (Tyler-Hummer) or a round oscillating separator (Dorrclone).[6,8]

Purification

A number of processes are available for purifying the liquor from the melter prior to its crystallization into granulated sugar. The two combinations most commonly used are:

1. Clarification–filtration–decolorization.
2. Clarification–filtration–decolorization–ion exchange.

Clarification. One of three types of chemical treatment is used as the first step in the purification process. These are liming, phosphatation, and carbonation.

Liming is the simplest of the chemical treatments. Process liquor is "limed" to a neutral pH using milk of lime and heated. A slurry of diatomaceous earth is added and the treated liquor is pumped through a filter press, such as those discussed in the following Section.

When phosphatation is used, the screened liquor is heated to 140-160°F and mixed with phosphoric acid (0.005 to 0.025 per cent P_2O_5 on solids). The mixture is immediately limed to pH 7.0-8.0, aerated with compressed air, and sent to a clarifier (a rectangular tank equipped with heating coils). The liquor enters the clarifier at one end and is heated to 190°F while flowing to the outlet at the opposite end. The heated liquor releases air which rises, carrying the flocculent calcium phosphate precipitate to the surface. A blanket of this precipitate (scum) forms at the surface and is skimmed from the liquor surface by moving paddles. Clarifiers differ in shape and design; some of the best-known are Williamson, Jacobs, Buckley-Dunton, Sveen-Pederson, and SuCrest.

In carbonation, process liquor heated to 140-175°F is limed to about pH 10 (0.03 to 0.8 per cent CaO on solids), gassed with carbon dioxide, heated to 185°F, and regassed until the pH drops to between 8.4-9.0. Carbonation, because of the two-stage gassing, requires two clarifiers, primary and secondary. Washed flue gases are the source of carbon dioxide. The calcium carbonate precipitate that forms entraps colored matter, colloidals, and some inorganic compounds.[12]

Filtration. The liquor from a phosphatation or a carbonation clarifier contains small amounts of finely dispersed particulate matter which requires filtration for removal. The filtration process is similar for both types of clarified liquor.

A precoat of a filter aid (diatomaceous earth or perlite) is first placed on the filter surfaces of the press. Liquor at 160–185°F is fed to the press until a pressure of about 60 psig is reached or the flow rate drops below a predetermined level and ends the filtration cycle. Phosphate-clarified liquor requires more filter aid (0.3 to 0.7 per cent on solids) to maintain good flow for a long time and to improve the brilliancy of the filtrate. Carbonated liquor requires considerably less filter aid because calcium carbonate acts as a fairly good filter aid if the optimum particle size was formed during the carbonation process. A filter cycle may last from 2 to 12 hrs depending upon the quality of the feed liquor.

Three types of filter presses are generally used; all are leaf presses:

1. Horizontal body with vertical leaf filter, i.e., Sweetland, Vallez, and Auto-Filters.
2. Vertical tank with vertical leaf filters, i.e., Niagara, Angola, Pronto, and Industrial.
3. Vertical tank with horizontal plates, i.e., Sparkler and Fas-Flo.

The filter leaves are dressed with cloth made of cotton or synthetic fibers (nylon, dacron, etc.), or with a wire screen of 60–80 mesh size.

The sugar remaining in the filter cake is recovered by washing the cake in place (sweetening off) with hot water, or the cake is sluiced off the leaves with hot water and flushed to a holding tank for refiltration. The filtered sweet water is generally added back at the melter to dissolve washed raw sugar.[6,8]

Decolorization. Filtered clarified liquor is a clear dark brown liquid having a solids content between 55–65° Brix, a pH of 6.8–7.2, and temperatures between 150 and 185°F. If pulverized activated carbon is used as the decolorizing agent, it is usually added prior to filtration. Otherwise the liquor is subject to one or more decolorizing adsorbents, for example, bone char, granular carbon, and ion exchange resins. Bone char removes colorants, colloidal matter, and some inorganic substances while granular carbon is only a decolorizer. Ion exchange resins adsorb color and change the composition of the ash.

Bone char and granular carbon are generally used in fixed beds or in cylindrical columns, 20–25 ft high and about 10 ft in diameter. More recent systems percolate liquor upward or downward through a stationary bed of adsorbent or by countercurrent flow of liquor and adsorbent (CAP or continuous adsorption process).

Liquor flow through bone char is about 1500 gal/hr over a period of 30 to 60 hrs. For granular carbon the flow rate is 3000 gal/hr for 20 to 30 days.

After the decolorizing cycle is completed, the adsorbent is sweetened off by displacing the liquor with water and washed. It is transferred to regenerating equipment consisting of dryers, kilns, and coolers. Bone char is regenerated at 1000°F in a controlled amount of air. Granular carbon is revivified at 1750°F in a limited-oxygen atmosphere. The kilns can be either the retort type or multiple hearths in columns (Herreshoff). After regeneration the adsorbent is returned to the system for a new decolorizing cycle.

Ion exchange resins are used in columns 8–10 ft high and 6–10 ft in diameter holding between 100 and 300 cu ft of resin. The resin-bed depth is only 2–4 ft. Flow rates are rapid (3000 to 4500 gal/hr) and the cycle is short (8–12 hrs). Regeneration is accomplished in the column with a 10 per cent salt solution at 140–160°F. The chloride form of an anionic resin decolorizes the liquor, and the sodium form of a cation resin replaces calcium in the liquor.

The sweet waters from the various adsorbent columns are recovered by returning them to the melter.[6]

Crystallization

Decolorized liquor is a pale yellow liquid with a solids content of 55–65° Brix. In some refineries this liquor goes directly to a vacuum pan for crystallization. However, most refineries (both cane and beet) pre-evaporate the liquor in a multiple-effect evaporator to a solids content of about 72° Brix.

There are four stages in the process for crystallization of sucrose: (1) seeding or graining; (2) establishing the seed; (3) growth of crystals; and (4) finally concentration.

A sufficient quantity of evaporated liquor is drawn into the pan to cover the heating elements. Water is evaporated from the syrup until its supersaturation approaches 1.25. At this point, the steam pressure is lowered and seed crystals are added. The seed is finely pulverized sugar dispersed in isopropyl alcohol or sugar liquor. This method is called shock seeding because addition of the seed induces an immediate formation of crystal nuclei throughout the supersaturated syrup. The nuclei are "grown" to a predetermined size or grain. Once the grain is established, the crystals are "grown" to size by maintaining supersaturation between 1.25 and 1.40 through control of steam pressure, vacuum, and the feed rate of the evaporated liquor. Adequate circulation during crystal growth is important.

When the volume of the massecuite reaches the maximum working capacity of the pan, the syrup feed is shut off and evaporation is allowed to proceed until a thick massecuite is formed. When the massecuite concentration is considered "right," steam and vacuum are shut off and the contents dropped into a holding tank equipped with agitators where it is kept in motion until discharged into centrifuges.

Instrumentation is used extensively in sugar boiling to control the progress of crystallization. Some of the principles used to provide control of vacuum pans for boiling a strike of sugar are:

1. Boiling point rise (BPR)–thermometers are used to measure the temperature of the massecuite and its vapors (difference between these two temperatures is the observed BPR).

2. Electrical conductivity of the massecuite –based on the principle that conductivity is inversely proportional to the viscosity of the solution which in turn has a similar relationship to the water content and thus the degree of supersaturation.

3. Fluidity of the massecuite–an ammeter is used to measure the current used by the motor of a mechanical circulator; changes in current indicate changes in the fluidity of the massecuite.

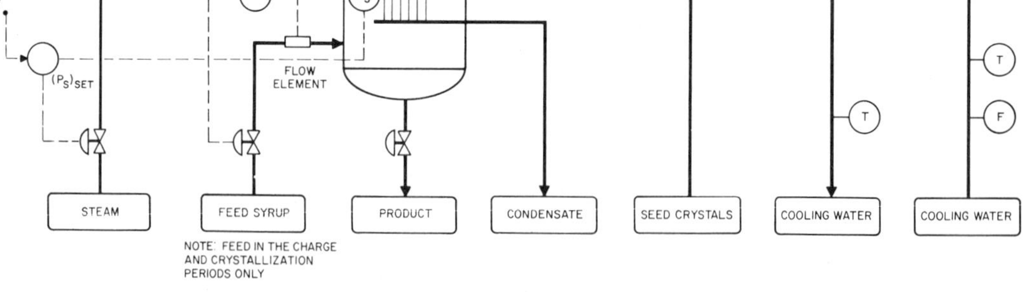

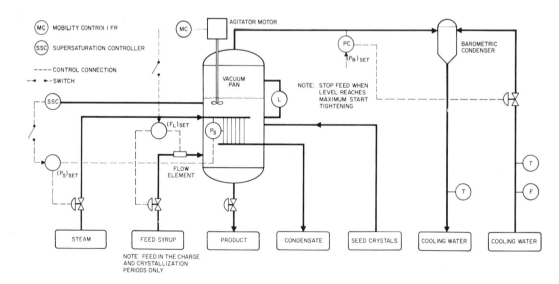

Fig. 16.10 Schematic drawing of vacuum pan and controls. [*Courtesy The Foxboro Co., Foxboro, Mass.; reproduced from Sugar y Azucar, p. 19 (Oct. 1970).*]

4. Soluble solids—measured by refractometer attached to the pan (pan refractometer).

Several systems employing instrumentation for automatic control of pan boiling by supersaturation have been used successfully. However, controls are empirical in nature and are based on the conversion of the various techniques and arts of a "sugar boiler" into mechanical operations which lend themselves to automation.

The boiling system of a refinery is straightforward; the first strike is boiled from evaporated liquor, the second strike is boiled from the runoff syrup of the first strike, continuing on to three or four strikes. The runoff syrup of the last strike can be used in a variety of ways: (1) as syrup for affination; (2) as syrup for remelt; (3) as syrup for producing a soft brown sugar; or (4) it can be reprocessed and reboiled.[6,9]

Remelt Sugars

The sugar contained in affination syrup must be recovered for economy. Recovery is accomplished by crystallizing the sugar in a vacuum pan. The resulting sugar is "raw" in composition and is returned to the refinery at the affination station. The residual syrup from the remelt station is known as refiners' blackstrap.

Centrifuging

Refined sugar crystals are recovered from the mother liquor by centrifuging the massecuite in equipment similar to that used for affining raw sugar. However, at this point the crystals are washed with a greater quantity of hot water. The washed crystals are discharged into a holding bin supplying a dryer.

Drying

Sugar from the centrifugals contains from 1 to 2 per cent moisture and is too wet to be placed in storage or packages. The wet sugar is fed to drying equipment called granulators, which are usually rotating horizontal drums, 15 to 35 feet long and 6–7 feet in diameter, inclined slightly so as to be discharged by gravity. Heated air is blown through the dryer cocurrent with the flow of the sugar. Two types commonly used are the Roto-Louvre granulator and the Standard-Hershey granulator. The Roto-Louvre is a single rotating drum in which hot air from louvres in the wall dries a moving bed of sugar. The Hershey granulator consists of a rotary dryer followed by a second unit in which the sugar is cooled to 110–130°F after leaving the dryer at 125–150°F.

Packaging and Storing of Refined Granulated Sugar

In the last three decades, there have been significant changes in methods of handling refined sugar as it leaves the dryer. Bulk deliveries of refined granulated sugar to customers has made it necessary for the refiner to store finished product in bulk rather than in bags. Granulated sugar remains free flowing for a longer period of time if it is "conditioned." "Conditioning" reduces handling, packaging, and storage problems resulting from caking. Conditioning involves several factors including control of the moisture content, the temperature, and the grain size, as well as good inventory management.[13]

As the sugar leaves the dryer, it is cooled in one of a variety of equipment types including the Holoflite screw conveyor, thermal disc processor, or a Buttner turbo tray dryer. The cooled refined sugar is classified according to crystal size by a screening or sieving operation. It is stored in large bins or silos provided with an atmosphere of controlled temperature and humidity. Circulation of dehumidified air through the sugar pile or above it and mechanical movement from silo to silo are also practiced to prevent caking.

Shipment of conditioned bulk sugar is made in specially designed rail cars (100,000 to 180,000 pounds capacity) or in hopper trucks (40,000 to 60,000 pounds capacity).

Conditioned sugar is also packaged in multiwall paper bags of 2-, 5-, and 10-pounds capacity. These sizes are overwrapped in 60 pound bales for shipment. Larger bags (100, 50, and 25 lbs) are also packed.

Granulated sugar is available in several sizes according to the needs of customers. These range from very large to very small crystals.

Direct—Consumption Sugar ("Plantation White")

Some cane sugar mills produce white sugar, usually by sulfitation or carbonation.

Sulfitation. The expressed juice from the mills is heated to 75°C, clarified with lime and sulfur dioxide, and in some cases filtered and then evaporated. After the syrup is treated with sulfur dioxide again, it goes through a three- or four-boiling system. The A and B sugars are mixed with a high purity syrup and centrifuged, dried, screened for size and distributed as white sugar.

Carbonation. The mixed juices from the mills are heated, clarified with lime, and evaporated to about 35° Brix. This syrup is relimed and treated with carbon dioxide, filtered, recarbonated, reheated, and refiltered. After carbonation, the syrup is given a double sulfur dioxide treatment and filtered. The resulting syrup is subjected to a three- or four-boiling system with the A and B sugars used as the refined product.

Both processes are subject to a number of modifications depending on the equipment of the mill and the quality of the sugar desired.[12]

Other Sugars

The cane and beet sugar industries manufacture several types of sugar in addition to granulated sugar. Among these are liquid sugars, brown or soft sugars, pulverized sugar, and special agglomerated sugars.

Liquid sugars (sugars dissolved in water). These are economically important largely because of the fact that a number of food manufacturers use sugar in the form of a syrup, and also because of the efficiency and ease of handling of a liquid product. Liquid sugars are produced by one of two methods: (1) dissolution or melting of refined sugar in water; or (2) further purification of in-process liquors by ion exchange treatment to remove minerals and additional decolorization using bone char, and pulverized or granular carbon. Liquid sugar has other advantages over granulated in that it can be delivered as sucrose, a mixture of sucrose and invert sugar, invert sugar, or as a blend of those sugars with various glucose syrups from starch hydrolysis and dextrose from the same source. Sucrose liquid sugars are usually distributed at a 67 per cent sugar concentration, while invert sugar and mixtures of invert sugar and sucrose are distributed at concentrations of 72–77 per cent sugar. Invert sugar is an equimolecular mixture of dextrose and levulose resulting from the hydrolisis of sucrose.

Brown or soft sugars. These are specialty "dry" products ranging in color from light to dark brown. In addition to sucrose, soft sugars contain varying amounts of water, invert sugar, and nonsugars. These sugars have a characteristic flavor. In some cases, soft sugars are "boiled" from a low-purity process liquor to obtain the desired color, flavor, and composition. In some instances, refined granulated sugar is "painted" with an impure syrup or molasses to produce a product of similar appearance and characteristics.

Pulverized sugars. These are manufactured by milling granulated sugar to the desired size. In most instances, pulverized sugars are mixed with small amounts of dried starch or tricalcium phosphate to prevent caking. A free-flowing agglomerated pulverized sugar without anticaking agents has been manufactured.

Agglomerated sugars. Several kinds are manufactured. Since soft sugars are subject to severe caking, one sugar refiner has resolved the problem by converting soft sugar into an agglomerated free-flowing dry product. Several companies manufacture "dry fondant" by agglomerating mixtures of invert sugar and very fine (less than 40 μ) sucrose particles.

BEET SUGAR

Agriculture

The sugar beet, *Beta vulgaris*, is a temperate zone plant grown largely in the Northern Hemisphere. A long growing season is required and in the United States it is one of the first crops in and the last out. The seed is planted in rows and after the plants have emerged, they are "thinned" to permit better development of a fewer number of beets. Unlike sugar cane, which cannot be cultivate after the cane has reached a certain height, sugar beets require a considerable amount of cultivation. Both cultivating and harvesting operations have been largely mechanized in the United States.

After plants have reached maturity in late fall (October and November), they are harvested by machines which remove the top growth of leaves, lift the root from the ground, and deliver them to a holding bin or a truck. Because the harvesting season is much shorter than the processing season, beets are stored in piles at the factory or at outlying points near transportation. Frequently, storage piles are ventilated to lower the temperature of the beets, thus reducing sugar-loss due to respiration during storage.[4,14]

Washing

Beets are transferred from storage into the factory in water flumes. These flumes lead directly into a rock-catcher which allows rocks to settle out and then on into a trash-catcher and a rotary washer.

Slicing

The washed beets are sliced into thin V-shaped cossettes by means of specially shaped knives set in frames mounted around the periphery of a rotating drum. Good removal of rocks and trash is essential to reasonable life for the knives.

Diffusion

The cossettes are weighed and transferred continuously into a diffuser where water passes countercurrently to the movement of the cossettes, and by dialysis, sugar and some of the nonsugars of the beet are extracted.

Both continuous and batch diffusers are used, but batch processes are largely being displaced. Continuous diffusers come in a variety of forms and shapes but all of them employ the same principle, i.e., movement of juice countercurrently to the movement of the cossettes. In batch diffusers the cossettes are held stationary and the diffusion juice is moved from cell to cell. A few of the well-known types of continuous diffusers are the Silver chain or scroll type, RT, BMA, DDS, DeSmet, Buckau-Wolf, and Olier.

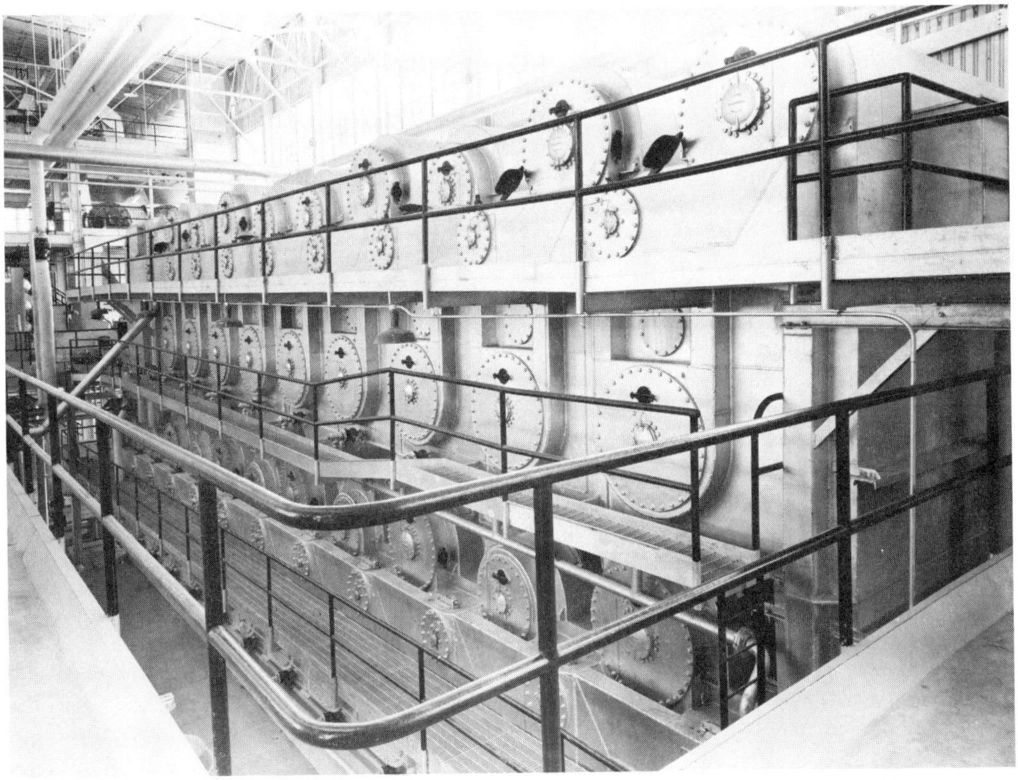

Fig. 16.11 Silver chain-type diffuser. *(Courtesy CF&I Engineers, Inc., Denver, Col.)*

The RT-type diffuser is a large revolving drum with an internal helix which separates the drum into moving compartments. As the drum revolves, the cossettes travel by the action of the moving helix from one end of the drum to the other end while the juice moves in the opposite direction.

In a Silver chain-type continuous diffuser, the cossettes move on a drag chain through a series of U-shaped cells about 2 ft wide and 12 ft deep.

The BMA diffuser is a cylindrical tower with a conveyor mechanism attached to a central rotating shaft. Guide plates on the shaft direct the cossettes upward, while at the bottom the juice exits through screens.

A well-known batch-type diffuser is the Robert battery. It usually consists of 12 to 14 identical columns connected in series so that water pumped into the end cell (tail) passes through each cell to the cossette entrance cell (head). Normal circulation of the juices within the column is from top to bottom. During operation of the Robert battery, three cells are not on stream, one is being washed out, another is being filled with cossettes, and the third cell is being emptied. The cossettes are distributed in the cell by hanging chains.

The temperature of the juices during difussion ranges from 60–85°C. Antifoaming agents are used to control foam in continuous diffusers and bactericides are added for microbiological control.[4]

Juice Purification

The diffusion juice is a dark turbid liquid containing 10 to 15 per cent sucrose and 1 to 3 per cent nonsugars including proteins, nitrogenous bases, amino acids, amides, inorganic material, and pectinous matter. These impurities are removed by a series of processes using lime and carbon dioxide (carbonation), sulfur dioxide (sulfitation), and crystallization. In recent years, there has been an increasing employment of granular carbon and ion exchange resins in those plants where refined granulated sugar is made.

Carbonation. The juice is heated to 80–90°C, limed, and sent to a carbonator for gassing with carbon dioxide. The resulting mixture containing insoluble lime salts, chiefly calcium carbonate, is pumped to subsiders (thickeners, clarifiers) to remove the insolubles by settling. From the subsider, the partially clarified filtrate is recarbonated and the residual lime is precipitated. This treatment is followed by press filtration. The sludge from the subsiders is filtered on rotary-drum filters, and the sugar-laden filtrate is returned to the first carbonation step. The carbonation process can be either continuous or batch.

One system of continuous carbonation widely used in the United States is the Dorr process. Heated juice is treated with milk of lime or lime saccharate in a cylindrical column (primary tank) where carbon dioxide gas is added. The contents of the secondary tank are continuously pumped into the bottom of the primary tank. The finished first-carbonated juice flows from an overflow line in the discharge of the recirculation pump to a subsider (Dorr multifeed thickener) in which the insoluble matter is separated from the juice. The thickener is a cylindrical tank consisting of a number of shallow compartments with trays and rotating paddles which continuously remove the settled sludge. The filtrate is reheated (90–95°C) and sent to another carbonation tank. The effluent from the second carbonation tank is pressure-filtered on a leaf press such as the Kelly, Sweetland, Vallez, or Angola. The filtered liquor is treated with a small amount of sulfur dioxide in a tower, heated, and refiltered.

Pre-liming. The European beet industry prefers a pre-liming system in which the difussion is sent through a series of tanks where lime is added to various levels of alkalinity. The limed juice is carbonated and generally follows a purification pattern similar to the Dorr process.

Ion exchange resins have been used in some plants to decolorize, deionize, and to delime the juices. However, their use is the exception rather than usual practice.

Evaporation and Standard Liquor

The thin juice (about 15 per cent solids) discharged from the filter presses after clarification is evaporated to thick juice (50–65 per cent solids) in multiple-effect evaporators. If decolorizing adsorbents are used, they are usually added to the thick juice. Granular carbon and pulverized carbon have been used for this purpose. Thick juice treated with pulverized carbon usually must be double filtered to remove all of the carbon.

Low raw sugars (sugars with purities below that of refined) are added to the thick juice in the melters to make standard liquor from which white sugar is crystallized. Standard liquor is usually filtered with diatomaceous earth before going to the vacuum pan.

Crystallization, Centrifuging, and Drying

Crystallization practices in United States beet sugar factories are similar to those in a cane-sugar refinery and result in a white granulated sugar comparable in quality to those of refined cane sugar. However, some European factories may make "raw sugar" as a separate product.

White-sugar centrifuge stations as well as drying operations are comparable to the same operations in a cane sugar refinery.

Sugar Recovery from Molasses

Beet sugar molasses from the normal operations is usually high in sucrose. Recovery of part of this sugar utilizes the "Steffan's Process" in which sucrose is separated by means of a lime salt known as calcium saccharate. In this process, molasses is diluted to 6-7 per cent sucrose and cooled to 18°C. Finely pulverized lime is added with agitation to form a precipitate known as cold saccharate cake. This precipitate (containing about 90 per cent of the sucrose) is filtered. The filtrate is heated, during which another precipitation occurs and this too is filtered. Cold saccharate precipitate is removed by rotary vacuum filters, and hot saccharate, by thickeners. Both precipitates are mixed with sweet water and returned to first carbonation where lime saccharate is decomposed into $CaCO_3$ and sucrose.

A second molasses-desugaring process is known as the "barium process." Sucrose is precipitated in the form of barium saccharate. Inasmuch as soluble barium solids tend to be toxic, a major process requirement is that all barium be removed before the recovered sucrose is returned to process. Furthermore, economics dictate that the barium be recovered for re-use.

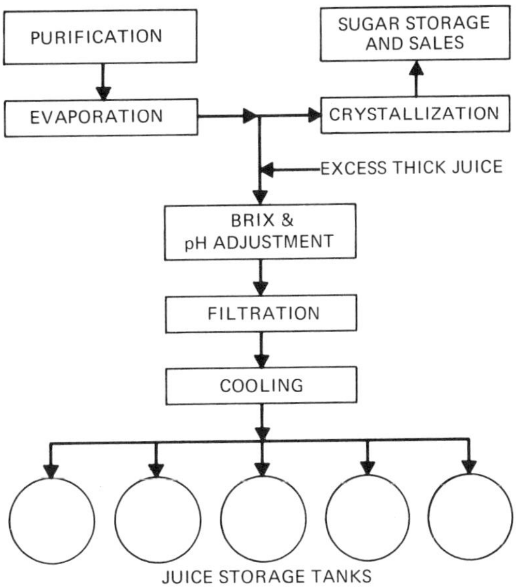

Fig. 16.12 Schematic flow diagram showing flow of thick juice to storage. (*Courtesy The Amalgamated Sugar Co., Ogden, Utah.*)

Sugar Recovery from Syrup

In 1962, a beet sugar company in the United States built a tank system for storing thick juice produced in excess of its crystallizing capacity. This storage system served two pur-

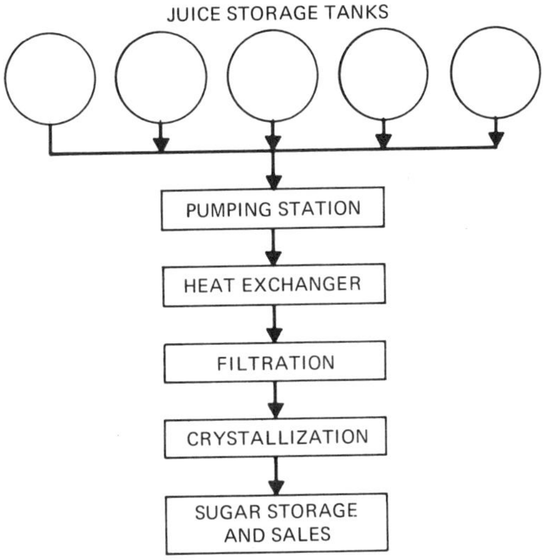

Fig. 16.13 Flow diagram showing flow of thick juice from storage through processing. (*Courtesy The Amalgamated Sugar Co., Ogden, Utah.*)

poses: (1) it enabled the "beet end" of the factory to operate at full capacity even though the "sugar end" could not handle the output; and (2) it provided raw material for the sugar end to operate at full capacity when the beet end was in trouble.

The system proved such a success that other beet factories have adopted the practice. Furthermore, the imporvements made in the storage systems have made it possible to extend the operating time of the sugar end far beyond the termination of the slicing campaign.[15]

Storage and Packaging

In the past 30 years, both beet sugar factories and cane sugar refineries have gone extensively to bulk storage of finished products. Beet sugar bulk storage is usually more extensive because of the need to hold sugar to meet customer needs from one processing season to the next. The daily capacities of cane sugar refineries are such, however, that it would be uneconomic to store more than a few day's output.

Prior to bulk storage, refined beet sugar was stored in bags (mostly 100 lbs). Such storage was acceptable as long as large customers purchased sugar in bags. However, a shift in the consumption of sugar from the home to food processors, and the development of bulk transportation facilities made bulk storage a natural consequence. Thus, beet sugar is packed in the usual customer-size packages and 100-pound bags, and it is also shipped in bulk containers ranging in size from "tote bins" (2,000 lbs) to rail cars holding several tons.

Cane and Beet Sugar Production

Cane sugar is primarily a product of tropical countries and beet sugar is produced in temperate climates. However, both are produced in a few countries, among which are the United States of America, China, Afghanistan, Spain, Iran and Pakistan.[16] Total annual production of centrifugal or crystalline sugar in 1969/70 and for estimated 1970/71 is shown in Table 16.1 together with average annual production for representative periods over the past 20 years.

Cane sugar production amounted to 59 per cent of the total in 1969/70. The largest producers of sugar in that year were the U.S.S.R. and Cuba with 9,500,000 and 9,400,000 short tons, raw value, respectively.[18]

The United States produced approximately 6.2 million short tons, raw value, of sugar in 1969/70.[18] About half of this amount was cane sugar produced in Hawaii, Louisiana, Florida, and Puerto Rico. The remainder was

TABLE 16.1 Centrifugal Sugar Production (Raw Value)[a] by Continent, 1950/51–1970/71[17]

(In Units of 1,000 Short Tons)

Continent	*1950/51 thru 1954/55 Average*	*1961/62 thru 1965/66 Average*	*1969/70*	*1970/71 (Est.)*
North America	13,655	16,061	21,448	19,120
South America	4,223	7,674	9,293	9,788
Europe (West)	6,677	9,114	11,706	11,540
Europe (East)	3,378	5,136	5,221	5,202
Africa	1,953	3,637	4,985	5,035
Asia	4,952	10,438	14,784	15,025
Oceania	1,290	2,244	2,734	2,954
U.S.S.R. (Europe and Asia)	3,010	8,443	9,500	10,500
WORLD TOTAL	39,138	62,746	79,671	79,164

[a]Raw value is a term defined mathematically to reduce sugars of differing sucrose contents (from raw sugar containing varying amounts of sucrose to refined sugar containing approximately 100 per cent sucrose) to an equivalent basis in pounds.

beet sugar produced in all the states west of the Mississippi River except Arkansas, Missouri, Oklahoma, and Nevada, and, in addition, in the states of Wisconsin, Michigan, and Ohio. (Small amounts of beet sugar have been produced recently in New York and Maine.) In spite of the fact that the United States is the second or third largest producer of sugar in the world, a substantial portion of its needs must be met by imports from other sugar-producing countries. Most of the imported sugar is produced from cane.

Noncentrifugal sugars are produced from sugar cane in substantial quantities (not shown in Table 16.1). In 1969/70, production amounted to 10.2 million short tons, raw value, with India's production accounting for more than 70 per cent of the total. These sugars are consumed almost exclusively in the area in which they are produced. They are known by such names as jaggery, gur, chancaca, muscavado, panela, piloncillo, and panocha. Many of them vary widely in composition and generally contain a larger amount of the nonsugars than do the centrifugal sugars.

Trade in Sugar

Many countries, especially the tropical cane sugar-producing ones, produce more sugar than is consumed locally. A number of beet sugar-producing countries, particularly in Europe, have become self-sufficient or nearly so in recent years. However, many of the European countries continue to import raw sugar to be refined and exported in accordance with historical trade patterns. Altogether, 21.3 million short tons, raw value, of centrifugal sugar were exported in 1968.[19] The United States, United Kingdom, Japan, and Canada are the largest importers, while Cuba, Australia, Brazil, Philippine Islands, Republic of South Africa, Dominican Republic, and Mexico are among the largest exporters.[19]

Trade in sugar throughout much of the world is influenced by agreements and special trading arrangements. For instance, the United States regulates sugar production and importation by means of the Sugar Act of 1948, as amended.[20] In addition to the quota system, an excise tax (in some instances an import tax) as well as a duty are applied to sugar. In some instances, trade is negotiated between governments and, in others, restrictions may be based on excessively high taxes and duties imposed on imported sugar. In still others, sugars from specified countries are given preferential treatment. The net result is that a relatively small proportion of the sugar produced in the world is traded freely in competitive markets.

Residual quantities of sugars that have no protected or preferential market are generally called "world sugars." The prices of such sugars are generally lower than the prices of protected sugar; however, on occasion, world sugar prices may exceed those of other sugars due to unusual supply-demand conditions. Efforts have been made to stabilize production and pricing through international agreements, the latest of which became effective on January 1, 1969.[21]

TABLE 16.2 Retail Price And Annual Per Capita Consumption in Selected Countries, 1969

Country	*Retail Price (U.S. Cents/lb)*	*Annual Consumption Kilograms, Raw Value*
United States	12.2	49.5
United Kingdom	8.8	52.9
Japan	16.4	25.6
Canada	8.8	50.9
Spain	10.8	27.3
Iran	14.0	25.2
Pakistan	17.3	4.8
Egypt	11.8	15.3
Switzerland	9.1	55.1
Chile	11.2	34.9
Sweden	13.0	41.8
Iraq	8.3	36.1
Ceylon	12.1	24.9
Algeria	15.2	18.3
Ireland	10.5	62.9
Norway	7.3	44.6
India	10.0	6.5
Mexico	5.6	38.3
South Africa	10.9	40.4
Australia	11.7	55.6
Philippines	9.6	17.8
Turkey	16.4	18.5
Cuba	6.9	77.1
Yugoslavia	9.7	27.8
Colombia	5.4	24.9
Venezuela	10.2	38.8
Peru	5.8	27.0
Austria	12.0	36.3
Denmark	14.0	50.8
Taiwan	13.0	14.9

Sugar Consumption and Uses

Per capita consumption of sugar varies widely as shown in Table 16.2.[22] Consumption is generally larger in highly industrialized nations than in the developing countries. Some countries deliberately discourage internal sugar consumption in order to make more sugar available for export, particularly when the sugar can be exported for "hard currencies."

Sugar prices in different nations vary just as widely as does consumption. The governments of most countries use sugar as a source of income in the form of tariffs, taxes, and monopolistic prices. Consequently, the cost of sugar to many consumers has little relationship to the economics of sugar production and distribution. The variation in sugar prices among nations is shown in Table 16.2.[23]

TABLE 16.3 Sugar Deliveries in the United States, 1969
(In Hundredweights)

Product or Business of Buyer	*Total All Sugar*
Bakery, cereal, and allied products	26,879,958
Confectionery and related products	20,744,252
Ice cream and dairy products	10,557,701
Beverages	41,981,078
Canned, bottled, and frozen foods; and jams, jellies, and preserves	18,324,024
Multiple and all other food uses	8,830,945
Nonfood products	1,442,451
Hotels, restaurants, and institutions	1,846,128
Wholesale grocers, jobbers, and sugar dealers	41,225,482
Retail grocers, chain stores, and super markets	23,864,073
All other deliveries, including deliveries to Government agencies	1,977,188
TOTAL	197,673,280

At one time, sugar was used primarily by families or households as a sweetening agent or in the preparation of food. Such is undoubtedly still the case in many countries, but in the United States household use of sugar accounts for only 20 or 25 per cent of the total. Food manufacturers use most of the sugar for the manufacture of all types of food products. Sugar consumption by type of user for 1969 in the United States is shown in Table 16.3.[24]

STARCH

Commercial Sources

Starch is an important class of carbohydrates found in plants, many of which store concentrations of starch in a relatively small part of the plant thus becoming important commercial sources for the extraction of starch. Starch is recovered primarily from three groups of plants—grains, tubers or roots, and the sago palm.

Grains are the most common raw materials for the commercial extraction of starch, though cassava starch may be consumed in larger quantities as a food staple. Among the grains, corn is the most important, followed by sorghum or milo maize, wheat, and rice. Starch from grain is obtained by complex "wet-milling" processes involving milling, wet gravity and sieving separations, and drying to remove excess water once the starch has been washed free of nonstarchy substances.

The main tubers and roots used for starch production are cassava, arrowroot, and potatoes. Cassava and arrowroot are tropical plants grown widely as a source of food in tropical countries. Arrowroot is grown less extensively, but in both cases the tubers are used as a carbohydrate food without completely separating the starch from the remainder of the tuber. Both white and sweet potatoes are sources of starch, but sweet potato starch because of its limited production and cost has few applications. White potato starch is extracted mainly in Europe and to a lesser extent in the United States. Starches from tubers or roots are extracted by a combination of grating or grinding operations followed by dispersion of the starch into a thin slurry. Fibrous fractions are separated from starch by wet screening through fine screens.

Sago starch is unusual in that it is found in the pith of sago palms, located in the trunk of the tree. When the tree is "ripe," it is cut down and the pith is taken from within the trunk. Starch extraction from the pith is similar to that from tubers.

The terms "starch" and "flour" are used synonymously when associated with cassava, arrowroot, and sago, but have different meanings when applied to grains and potatoes.

TABLE 16.4 Annual Production in Millions of Pounds[a]

Source	*1960*	*1961*	*1962*	*1963*	*1964*
Corn and sorghum	5,336	5,162	5,627	5,933	6,343
Potato	117	236	130	135	34
TOTAL	5,453	5,398	5,757	6,068	6,377

[a]Combined wheat and rice starch production amounted to 64 million pounds in 1963.

Production

Cassava flour and tapioca are produced commercially in Thailand, Brazil, Indonesia, and Madagascar. Arrowroot has a very limited production in the West Indies, while sago comes primarily from Indonesia. Corn starch is primarily a product of the United States but it is also produced in some European and South American countries. A few European countries manufacture potato starch, while rice and wheat starches are produced in small quantities in several places.

Starch production in the United States for the years 1960 through 1964 are shown in Table 16.4.[25]

In addition to the starch produced domestically, the United States imported approximately 325 million pounds of starches (cassava, tapioca, arrowroot, sago and potato) in 1964.[25]

Uses

Starch is widely used in the food-processing industry and in the manufacture of textiles, paper, adhesives, and laundry starches. However, the largest use in the United States is for conversion into dextrose and a number of glucose syrups.

Separation from Grain

Most of the starch converted into glucose products is obtained from corn,[25] and the starch conversion plant is most frequently a part of a larger plant which starts with whole grain as the raw material.

Cleaned shelled corn is soaked (steeped) in warm water (115–125°F) containing sulfur dioxide (0.15–0.2 per cent) for about 40–50 hrs. The softened grain is passed through degerminating mills which disintegrate the kernels. The grist (mixture of hull, germ, and endosperm) is carried by water to long horizontal V-shaped tanks in which the floating germ is skimmed off the top and the remaining mixture is discharged through screw conveyors at the bottom. The degermed mash is washed through a series of rotating metallic sieves (reels) and on vibrating sieves (shakers). These sieves are usually covered with bolting silk.

The screened (filtered) liquor is pumped to a battery of centrifugals for the separation of starch and gluten. The starch slurry from the centrifugals is washed with sulfurous acid and separated further from nonstarch materials in hydroclones and centrifugals. Starch is dewatered on rotary vacuum filters and is then ready for finishing or for conversion into other products:

1. Finished starch (pearl and powdered starch)–starch dried on belt, flash, or rotating-shelf dryers to a stable moisture content.
2. Dextrin and gums–starch is dry roasted at 390–480°F or heated at lower temperatures in the presence of acids to produce various grades of dextrins.
3. Thin boiling–starch is modified with acids to produce lower viscosities when heated.
4. Special starches–chemically modified starches and starch derivatives.[26]

Starch Hydrolisis

Undried starch is pumped as a slurry to the conversion plant where it undergoes one or more hydrolytic processes to yield mixtures of various sugars in the form of syrups or as crystalline dextrose. The kind and amount of the various sugars obtained depend upon the type of hydrolysis system used (acid, acid-enzyme, or enzyme-enzyme) and the extent to which the hydrolytic reaction is allowed to

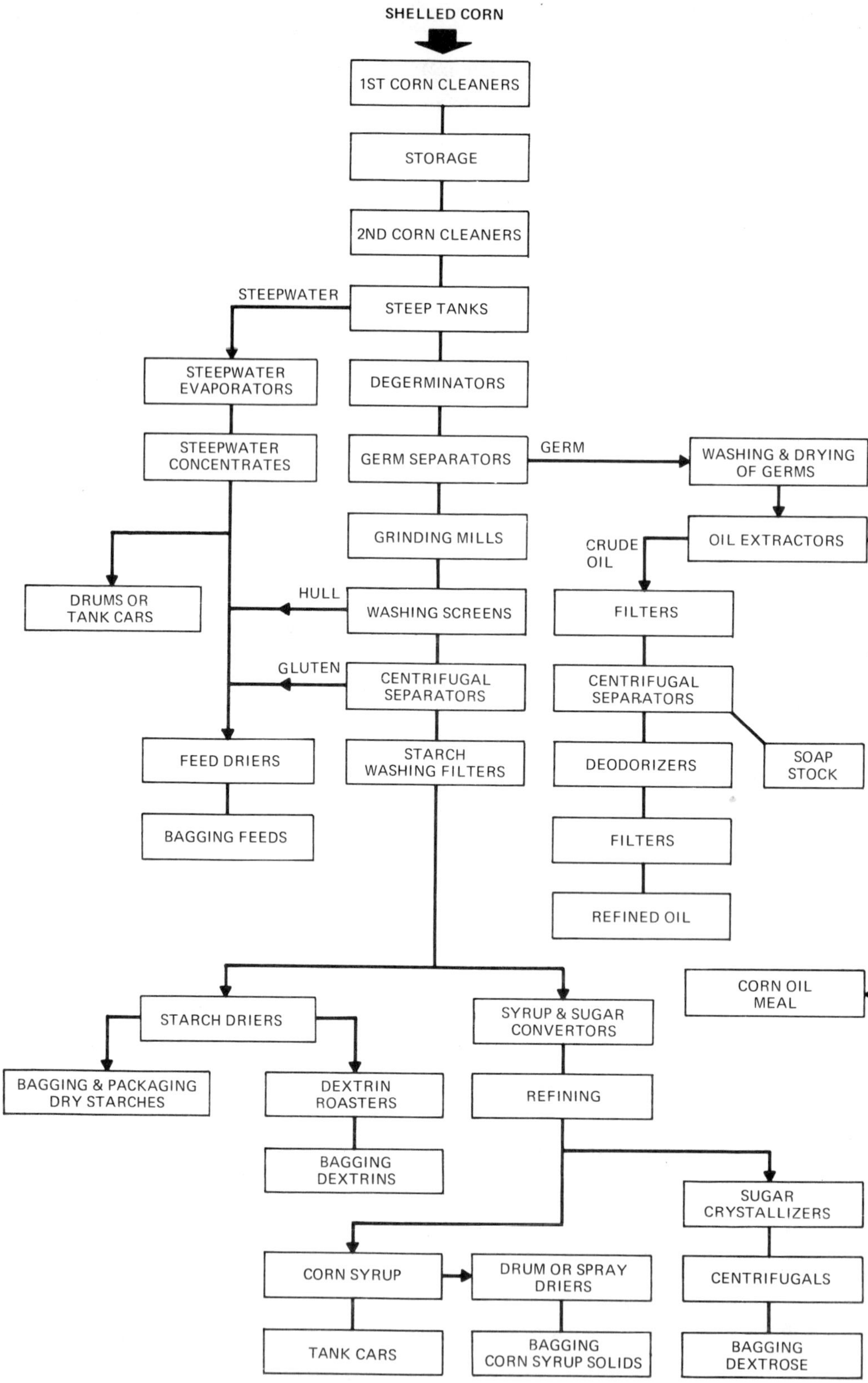

Fig. 16.14 The corn-refining process. (*Courtesy Corn Industries Research Foundation, Inc., Washington, D.C.*)

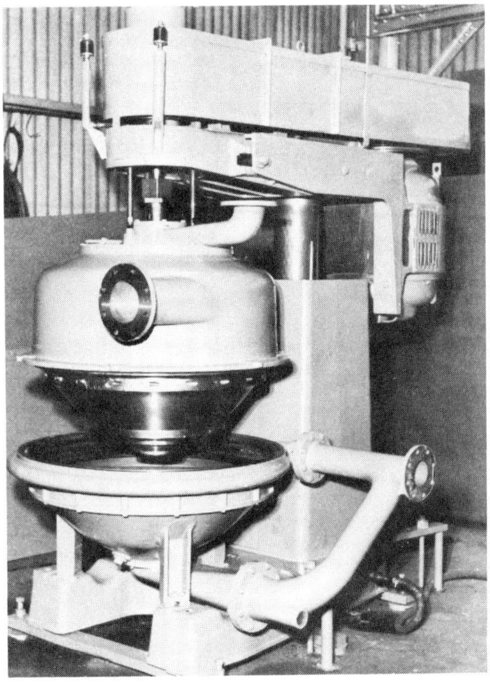

Fig. 16.15 Marco centrifuge used for starch separation in a wet-milling process. *(Courtesy Dorr-Oliver, Inc., Stamford, Conn.)*

proceed. The fact that most starches consist of two different kinds of polymers (amylose and amylopectin) also has an effect on the nature of the products obtained.

The extent to which starch is converted into sugars is indicated by the "dextrose equivalent" (D.E.), a measure of the reducing sugar content calculated as dextrose and expressed as a percentage of the dry substances.[27] Hydrolyzates having dextrose equivalents ranging from 5 to 100 are produced. Those having a very low dextrose equivalent are frequently referred to as "dextrines." They are produced by minimal acid hydrolysis or roasting. Starch-hydrolyzate syrups are commonly produced as "low," "regular," "intermediate," "high," or "extrahigh" conversion products, as more or less standard products. Table 16.5 shows the composition of some of the various syrups.[27] In addition to carbohydrates, the syrups contain minerals (largely sodium chloride) and nitrogenous substances.

Acid Hydrolysis. A starch slurry containing 35-40 per cent solids is acidified with hydrochloric acid to about pH 1.9. The suspension is pumped into an autoclave (convertor) where live steam is gradually admitted to a pressure of 30 psi. The conversion time largely determines the D.E. of the hydrolyzate, i.e., an 8-min conversion may produce 42 D.E. syrup, or 55 D.E. might require 10 mins.[28] Converted liquors are neutralized with sodium carbonate to a pH of 5-7, coagulating insoluble protein, fats, and colloidal matter. The scum is removed in skimming tanks. The dark-colored clarified liquor is pressure filtered, decolorized with bone char, and evaporated to about 60 per cent solids, after which it is further decolorized with bone char and powdered activated carbon. "Low ash" syrups are usually deionized with ion exchange resins. Finally, the decolorized liquor will be evaporated to a final solids content of 75 to 85 per cent solids.[26]

Acid-Enzyme Hydrolysis. Starch is first liquified and hydrolyzed to specific dextrose equivalents with hydrochloric acid. After evaporation to 60 per cent solids, a saccharifying enzyme (fungal amylases) is added to continue hydrolysis to the desired level. By choosing two or more types of enzymes (such as alpha-amylases, beta-amylases, or glucoamylases) and adjusting the initial acid

TABLE 16.5 Carbohydrate Composition of Glucose Syrups (Saccharide as a % of Total Carbohydrates)

DE	*Mono*	*Di*	*Tri*	*Tetra*	*Penta*	*Hexa*	*Hepta*	*Higher*
15	3.7	4.4	4.4	4.5	4.3	3.3	3.0	72.4
35	13.4	11.3	10.0	9.1	7.8	6.5	5.5	36.4
45	21.0	14.9	12.2	10.1	8.4	6.5	5.6	21.3
55	30.8	18.1	13.2	9.5	7.2	5.1	4.2	11.9
65	42.5	20.9	12.7	7.5	5.1	3.6	2.2	5.5

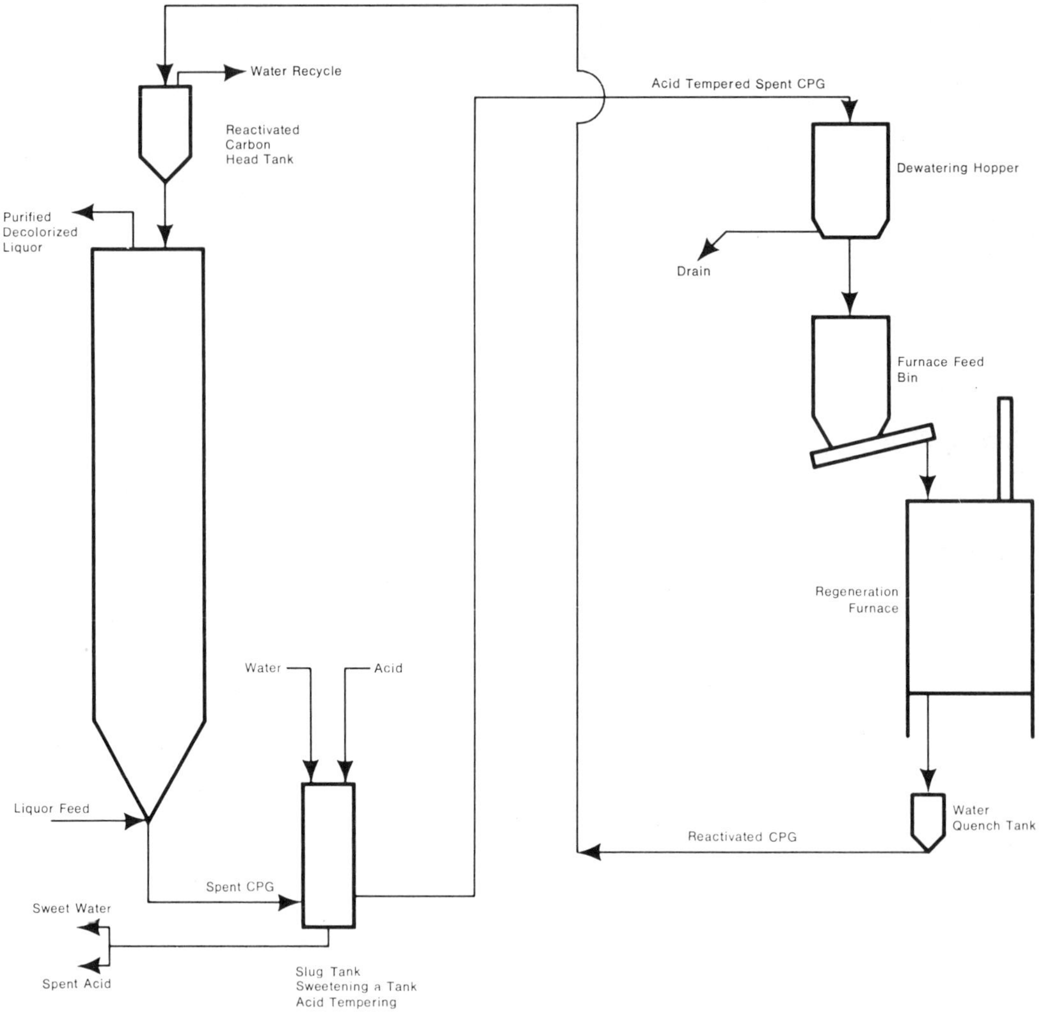

Fig. 16.16 Flow diagram of the Pittsburgh moving-bed system for glucose-dextrose. (*Courtesy Pittsburgh Activated Carbon Div., Calgon Corp., Pittsburgh, Pa.*)

hydrolysis, syrups with different ratios of dextrose, maltose, and higher saccharides can be obtained.[29]

Enzyme-Enzyme Hydrolysis. Enzyme-enzyme conversion employs an enzyme for starch liquefaction in place of acid. Subsequent hydrolysis is by enzymes as above. The choice of hydrolytic system depends upon economics and the kind of end product desired. Enzymes are usually inactivated by heating to 170°F.[29]

Conversion to Dextrose

The manufacture of dextrose requires conditions different from those used for the manufacture of syrups. When acid hydrolysis is used, the density of the starch slurry is lower (15 to 20 per cent). Hydrochloric acid is added to a concentration of 0.03 *N*, and higher steam pressures (40–45 psi) are used. The time of hydrolysis is extended to about 30 mins or until the purity reaches 90-91 D.E. The hydrolyzate is neutralized to a pH of 4-5 with sodium carbonate. Other purification steps are similar to those employed in syrup manufacture.

The high-purity liquor is pumped to insulated crystallizers fitted with slowly moving agitators for crystallization of dextrose monohydrate. A heavy seed base (about 25 per cent) from a previous batch is mingled with the syrup and cooled to about 38°C. The seeded liquor is held at this temperature for several days until about 60 per cent has crystallized. The mixture is centrifuged to separate crystals from the mother liquor. The wet sugar is dried in rotary dryers or recrystallized into anhydrous dextrose. The monohydrate may also be converted to anhydrous dextrose by drying the monohydrate in hot air. A second crop of crystals is taken from the mother liquor and the run-off syrup from this step is final molasses or "hydrol."[26,30]

Dextrose may also be made by acid-enzyme and enzyme-enzyme systems. In the acid-enzyme process, acid hydrolysis is stopped near 18 D.E. The liquefied slurry is evaporated to 60 per cent solids and treated with a fungal glucoamylase until the D.E. is above 90. With enzyme-enzyme processes, a 30 per cent starch slurry is liquefied with a heat-stable bacterial amylase followed by treatment with fungal glucoamylase. The liquor is purified in the same manner as others.[29]

In addition to syrups, several solid products are manufactured, including several kinds of dried sugars. "Crude sugar" is a solidified liquor containing large amounts of dextrose and substantial quantities of nondextrose solids as well. A solid product of recent development, "noncrystalline" or "enzyme-converted" dextrose, is highly refined, and contains about 92 per cent dextrose.

One of the most interesting syrup products developed within the past several years contains approximately 50 per cent dextrose, 42 per cent fructose, and 8 per cent higher saccharides on a solids basis. The development of this product commercially resulted from the use of glucose isomerase enzymes which isomerize dextrose into fructose.[31] Earlier isomerization of dextrose had been accomplished in a potentially commercial process by means of an ion-exchange resin.[32]

BY-PRODUCTS

Molasses

Molasses, a residual mother liquor from which little or no additional sugars can be recovered economically, is a by-product common to the cane and beet sugar industries and to the industry that converts starch into dextrose and glucose syrups. Each industry has designated this liquid by-product with a name peculiar to the industry: the molasses from cane sugar production is most commonly called "blackstrap," that from beet is simply "beet molasses;" while that from starch hydrolysis is known in the United States as "hydrol."

The molasses from each source differs considerably as shown in Table 16.6.

TABLE 16.6 Analysis of Molasses from Various Sources

Constituent	*Blackstrap[33] (Dry Basis)*	*Beet Molasses[34] (Dry Basis)*	*Hydrol[11] (Wet Basis)*
Sucrose	37.4	63.0	
Reducing sugars	32.7	0.7	55.0
Higher saccharides		1.5	
Ash	13.5	12.0	7.2
Nitrogen	0.4	2.0	0.07

Blackstrap contains a significant quantity of both sucrose and reducing sugars, some of which were present in the juice of the cane. A substantial amount of reducing sugar results from sugar-recovery processes which in the cane sugar industry tends to be slightly on the acid side of neutrality. Beet molasses, on the other hand, contains primarily sucrose and little or no reducing sugar because of highly alkaline processing conditions which destroy reducing sugars. Hydrol, of course, contains no sucrose. Higher saccharides result from incomplete hydrolysis or from the polymerization of "sugar" units during processing.

Recent annual world production of blackstrap and beet molasses combined is shown in Table 16.7.[18]

Published information on world production of hydrol is not available. However, production in the United States amounts to more than 20 million gallons annually.[35]

TABLE 16.7 World Production of Molasses

	Production Period			
	1966/67	*1967/68*	*1978/79*	*1969/70*
Production in 1,000 metric tons	18,634	19,140	20,141	21,138

TABLE 16.8 Uses of Molasses

(Millions of gallons)			
Year	*Industrial*	*Animal Feeding*	*Total*
1965	165.1	446.9	612.0
1966	159.3	490.7	650.0
1967	180.2	507.1	687.3
1968	170.3	566.3	736.6

The major outlet for all molasses at the present time is in feed for cattle or other animals.[35] Fermentation into rum, potable ethanol, citric acid, and vinegar continues to be a market for considerable quantities of molasses, and large quantities are used for growing yeast. These uses have diminished in relation to feed-use over the years. At one time, beet molasses, because of its high content of nitrogenous substances, was valued for the production of citric acid by fermentation. The use of molasses for various purposes are shown in Table 16.8[35]

Bagasse and Beet Pulp

An important by-product from a cane sugar mill is bagasse, the fibrous portion of cane from which juice is extracted. Bagasse, at the time it is discharged from the milling train, contains solid matter (short fibers and the spongy tissue of the pith) and 50 per cent by weight of water. Since the average fuel value of ash-free dry bagasse is 8300 Btu/lb[6] it is primarily used as fuel for the generation of steam in sugar factories. However, when there is an excess of bagasse available or alternate fuels are plentiful and reasonably priced, the bagasse is utilized as a raw material for the pulp, paper, paperboard, and wallboard industries, especially in wood-poor areas of the world.

Other successful uses of bagasses include its conversion into furfural and a filler for explosives. Since sugar cane contains about 12 per cent fibrous matter, approximately 110 million metric tons were produced in 1969/70.

Beet pulp is the structural portion of beet roots which remains after spent beets are discharged from a diffuser. Since it is highly hydrated at that time, the wet pulp is dewatered in pulp presses. It is then discharged into wet silos and may be sold as wet pulp for animal feed. However, it is more often dried or mixed with molasses and dried. In both cases, it is used in the manufacture of animal feeds.

Other By-Products

When grains such as corn, sorghum, and wheat are used as sources of starch, a number of important by-products are obtained. These include steep water, corn oil, gluten, and hulls. Most of these are used as ingredients in mixed feeds, but corn oil is widely used as a food product. By-products from the other commercial sources of starch are not as important as those from grains, but they usually find their way into feeds if they are salvaged.

REFERENCES

1. G. P. Meade, "The Sugar Molecule," **11**, No. 1, pp. 11–19 (1959–60).
2. N. Deerr, "The History of Sugar," Vol. 1, London, Chapman and Hall, 1949.
3. E. Hugot and G. H. Jenkins, "Hand Book of Cane Sugar Engineering," New York, American Elsevier, 1960.
4. R. A. McGinnis, "Beet-Sugar Technology," New York, Van Nostrand Reinhold, 1951.
5. Noel Deerr, "Cane Sugar," London, Norman Rodger, 1921.
6. G. P. Meade, "Cane Sugar Hand Book," 9th Ed., New York, John Wiley and Sons, 1964.
7. "BMA Cane Diffusion," *Sugar y Azucar*, p. 22 (July 1970).
8. V. Baikow, "Manufacture and Refining of Raw Cane Sugar," New York, American Elsevier, 1967.
9. P. Honig, "Principles of Sugar Technology," Vol. II, New York, American Elsevier, 1959.

10. B. Silver, "Continuous Vacuum System," *Sugar y Azucar*, p. 36 (June 1970).
11. P. Honig, "Principles of Sugar Technology," Vol. III, New York, American Elsevier, 1963.
12. P. Honig, "Principles of Sugar Technology," Vol. I, New York, American Elsevier, 1953.
13. T. Rodgers and C. Lewis, "Drying of Sugar and Its Effect on Bulk Handling," 15th Tech. Conf. Report, British Sugar Corp., *Intern. Sugar J.*, **64**, pp. 359–362 (1962).
14. W. G. Bickert, F. W. Bakker-Arkema, and S. T. Dexter, "Refrigerated Air Cooling of Sugar Beets," *J. Am. Soc. Beet Sugar Technologists*, **14**, No. 7, pp. 547–554.
15. "Thick Juice Storage for Off-Season Processing," *Beet Sugar*, **57**, pp. 46–49 (Jan. 1962).
16. F. O. Licht, *Intern. Sugar Reports*, **102**, No. 20, pp. 1–4 (1970).
17. *Foreign Agricultural Circulars*, FS2-59 (1959), FS2-69 (1969), and FS3-70 (1970), U.S. Dept. of Agriculture, Washington, D.C.
18. *Foreign Agricultural Circular*, FS3-70 (1970), U.S. Dept. of Agriculture, Washington, D.C.
19. *Foreign Agricultural Circular*, FS2-70 (1970), U.S. Dept. of Agriculture, Washington, D.C.
20. Public Law 331, 89th Congress, First Session.
21. International Sugar Agreement, 1968.
22. "Sugar Year Book—1969," pp. 352–360, International Sugar Organization, London.
23. "The United States Sugar Program," p. 9, Dec. 31, 1970, U.S. Government Printing Office, Washington, D.C.
24. "Sugar Reports," pp. 16–20, March 1970, U.S. Dept. of Agriculture, Washington, D.C.
25. "Summaries of Trade and Tariff Information," Schedule 1, "Animal and Vegetable Products," Vol. 6, "Cereal Grains, Malts, Starches and Animal Feeds," pp. 140–150, U.S. Tariff Commission, Washington, D.C., 1966.
26. R. W. Kerr, "Chemistry and Industry of Starch," New York, Academic Press, 1944.
27. *Corn Syrups and Sugars*, Corn Industries Research Foundation, Washington, D.C., 1965.
28. J. A. Kooreman, "Physical and Chemical Characteristics of Enzyme Converted Syrup," *Mfg Confectioner*, pp. 35–90 (June 1955).
29. G. Reed, *Enzymes in Food Processing*, New York, Academic Press, 1966.
30. G. R. Dean and J. B. Gottfried, "*Advances in Carbohydrate Chemistry*," Vol. 5, pp. 127–137, New York, Academic Press, 1950.
31. Technical Bulletin 1K1a -90402, "Isomerose 100 Corn Syrup," Clinton Corn Processing Co., Clinton, Iowa.
32. D. P. Langlois, U.S. Pat. 2,746,889 (May 22, 1956).
33. W. W. Binkley and M. L. Wolfram, "Composition of Cane Juice and Final Molasses," "Advances in Carbohydrate Chemistry," Vol. 8, New York, Academic Press, 1953.
34. J. B. Stark and R. M. McCready, "The Relation of Beet Molasses Composition to True Purity," *J. Am. Soc. Beet Sugar Technologists*, **15**, 61–72 (1968).
35. "Molasses Market News Annual Summary," U.S. Dept. of Agriculture, Washington, D.C., 1969.

17

Industrial Gases

R. M. Neary*

The industrial gases fall into two groups: hydrogen, helium, oxygen, nitrogen, argon, and carbon monoxide, the less easily liquefiable gases; and chlorine, sulfur dioxide, ammonia, nitrous oxide, carbon dioxide, propane, and fluorocarbon refrigerant gases, the more easily compressible gases. At ordinary temperatures, the former do not liquefy in spite of considerable pressure whereas the latter, at ordinary temperatures, form liquids under rather moderate pressures. Hence the content by weight of a standard cylinder for the gases in the first group is small, but for those in the second group, it is considerable. It follows that gases in the first group are generally used as free gases as soon as they are generated, or they will be shipped short distances, from many plants, each serving a small territory; gases in the latter group may be economically shipped long distances, from a few central plants.

The distinction between the two groups is far less sharp today, due to many recent advancements in low-temperature science. Handling liquefied hydrogen (-253°C), liquid helium (-269°C) and other gases that liquefy below -150°C have opened a new scientific field commonly called cryogenics. However, handling cryogenic gases is not new. By 1939, oxygen in liquid form was transported by tank cars and trucks. The other atmospheric gases, nitrogen and argon, have been shipped as liquids in the same type of containers since about 1940. Liquefied hydrogen, the coldest flammable gas, has been transported since 1960 across the United States in tank cars and trucks with super (vacuum) insulation so that no hydrogen is vented en route. Liquefied natural gas (LNG) and methane are transported by boat to the United States and other countries. Thus all of the above gases are being transported in commercial quantities in the liquid state. For miscellaneous industrial uses, however, the familiar steel cylinder with its charge of compressed gas remains standard.

Acetylene (C_2H_2) lies midway between the two groups. It is in a class by itself, partly because its potential explosive decomposition re-

*Union Carbide Corp., Linde Div., Tarrytown, N.Y.

TABLE 17.1. Properties of Industrial Gases[a]

Properties of Gases		Helium He	equilibrium-Hydrogen e-H_2	normal-Hydrogen n-H_2	Neon Ne	Nitrogen N_2	Air	Fluorine F_2	Argon Ar	Oxygen O_2	Methane CH_4	Krypton Kr	Xenon Xe	Acetylene C_2H_2	Carbon Dioxide CO_2
Atomic or molecular weight		4.0026	2.0159	2.0159	20.183	28.013	28.96	37.997	39.948	31.999	16.043	83.80	131.30	26.04	44.01
Normal boiling point (nbp)	degrees F	−452.1	−423.2	−423.0	−410.9	−320.4	−317.8 to −312.4	−306.7	−302.6	−297.3	−258.7	−244.0	−162.6	−118.5[b]	−109.4[b]
Triple point (tp)	degrees F		−434.8	−434.6	−415.5	−346.0		−363.3	−308.9	−361.8	−296.5	−251.0	−169.2	−116	−69.9
	psia		1.02	1.04	6.3	1.8		0.037	10.0	0.022	1.7	10.6	11.8	17.7	75.1
Critical point	degrees F	−450.3	−400.3	−399.9	−380.0	−232.5	−221.3	−200.2	−188.1	−181.1	−115.8	−82.8	+61.9	96.8	87.8
	psia	33.2	187.7	188.1	395.	493.	547.	808.	710.	737.	673.	796.	847.	905	1072.1
Density Gas, NTP	lb/ft^3	0.01034	0.005209	0.005209	0.05215	0.07245	0.07493	0.0983	0.1034	0.08281	0.0416	0.2172	0.3416	0.0678	
Density Gas, STP	lb/ft^3	0.01114	0.005611	0.005611	0.05618	0.07805	0.08072	0.106	0.1114	0.08921	0.0448	0.2340	0.3680		0.1234
Density Vapor, nbp	lb/ft^3	1.04	0.0835	0.0831	0.596	0.287	0.280	0.33	0.363	0.279	0.114	0.53	0.60		
Density Liquid, nbp	lb/ft^3	7.798	4.418	4.428	75.35	50.46	54.56	94.1	86.98	71.27	26.5	150.6	190.8		
Specific Heat, Cp, gas, NTP	Btu/lb, degree F	1.25	3.56	3.42	0.246	0.247	0.240	0.197	0.125	0.220	0.533	0.0593	0.0383	0.383	0.202
Specific heat ratio, Cp/Cv, Gas, NTP		1.66	1.38	1.41	1.66	1.41	1.40		1.67	1.40	1.31	1.68	1.66	1.26	1.307
Heat of vaporization, nbp	Btu/lb	8.72	193.	192.	37.0	85.7	88.2	74.0	70.2	91.7	219.	46.4	41.4		
Heat of fusion, T_p	Btu/lb		25.0	25.0	7.1	11.1		5.8	12.7	6.0	25.2	8.4	7.5	356[b]	246.3[c]

[a]Compiled from "Cryogenic Data Reference," Union Carbide Corporation, Linde Division.
[b]Sublimation temperature, because at atmospheric pressure acetylene and carbon dioxide go from gaseous to solid phase without entering the liquid state.
[c]Heat of sublimation.

quires that special precautions be taken in handling it.

In addition to a variety of technical uses, some members of the second group serve as ordinary refrigerants–ammonia, carbon dioxide, dichlorodifluoromethane (Refrigerant 12) and other fluorocarbons; fluorocarbons and other gases are also used as propellants in aerosols. Members of the first group are used as extraordinary refrigerants; for example, nitrogen is used in the liquefaction of liquefied hydrogen.

OXYGEN

The production and distribution of cryogenic liquefied gases, including liquid oxygen, is a major industry which has been undergoing very rapid growth in size and technology in recent years.

Refrigeration

The basic refrigeration systems used in the liquefaction cycle for most cryogens such as liquid oxygen, hydrogen, helium, and liquefied natural gas are essentially the same, and thus it is appropriate to discuss them first. The two basic systems employed include the Joule-Thompson expansion cycle and the expansion engine cycle. A third, the cascade system which involves the throttle-expansion of a liquid, is used in LNG plants.

All plants cool the incoming compressed gas by transferring heat to the cooler outgoing plant waste, recycle gas, and product in countercurrent heat exchangers. After preliminary cooling and purification, if necessary, the gas is refrigerated by expansion which may partially liquefy the gas.

Figure 17.1 illustrates the fundamental cycles used in liquefying cryogens today.[2] The left side of Fig. 17.1 shows the expansion of cold gas through a throttle valve or nozzle which produces a drop in pressure with a corresponding drop in temperature. This is a constant enthalpy process called a Joule-Thompson expansion. As it passes through

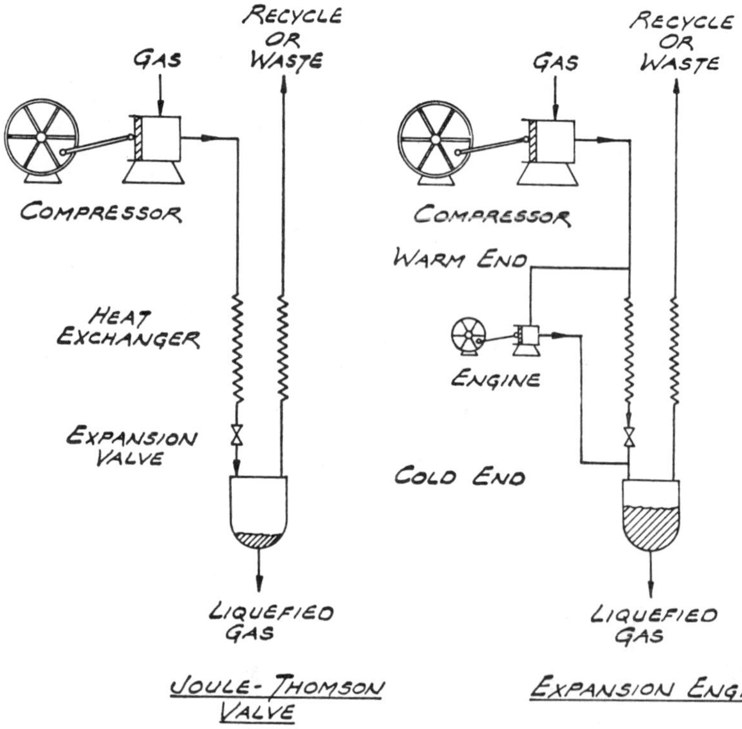

Fig. 17.1 Fundamental refrigeration cycles used in air-separation plants. (*Courtesy Linde Div., Union Carbide Corp.*)

the expansion valve some of the gas is liquefied. The liquid is then transferred to a separator, distillation column, or storage tank; the vapor passes through the heat exchanger. This simple liquefaction process was utilized in Dr. Karl von Linde's first cycle for the commercial separation of air.

A second method of producing refrigeration in a gas, as shown at right side of Fig. 17.1, is to expand it in an engine, doing useful work; the temperature is reduced because of the removal of energy. This cycle approaches a constant entropy expansion. The temperature reduction is greater than is possible by expanding gas through a valve, making the system more efficient than the Joule-Thompson cycle and better adapted to large-scale plants.

This expansion engine cycle, known as the Claude principle, does not produce liquid during the expansion phase, as liquid would damage the machine. This cycle is normally used in conjunction with condensers and expansion valves that produce liquid.

Uses

Oxygen is vitally important to industry, medicine, and to explorers of outer space because it is one of the primary tools of their trade. It is well established as a basic industrial chemical with a production of 12,200,000 tons in 1969, up from 3.3 million tons in 1959.[3] About 2 million tons were produced and distributed as liquid oxygen and the remainder as gaseous oxygen via pipeline to adjacent or nearby steel mills, chemical plants, etc. The latter is often referred to as "on-site," "tonnage" or "combustion" oxygen.

The primary use of oxygen stems from its ability to support combustion and to sustain animal life. Materials that burn in air burn faster in oxygen. Thus, the use of oxygen or oxygen-enriched air in place of ordinary air in many metallurgical and chemical processes increases the intensity and speed of reaction, resulting in shorter cycle time, greater yield per volume of equipment, and lower costs.

Fig. 17.2 Mechanized "hot" scarfing machine, utilizing many oxy-acetylene flames, removes surface scale and impurities prior to finish rolling. *(Courtesy Union Carbide Corp.)*

Over 7.6 million tons of oxygen are used in the production of ingot steel–in open hearth, blast furnaces, and the basic process. Another 1.5 million tons are used in the processing and fabrication of steel mostly in conjunction with the oxygen-acetylene flame, Fig. 17.2. Oxygen is used for the "cracking" of methane or natural gas by partial oxidation to produce acetylene, an important basic chemical. It serves as a raw material for synthesizing oxygen compounds (ethylene oxide, sodium peroxide).

Recent pollution-abatement pilot tests show that oxygen provides greater efficiency and lower costs than air as the oxidizing agent for biological treatment of a sludge waste-water stream to remove its organic matter and to lower its biological oxygen demand (BOD) rating. This applies to both commercial waste streams and community waste treatment.

The missile industry consumes large volumes of liquid oxygen in the testing and firing of rocket motors. Oxygen, stored as a liquid and converted to gas as used, provides a dry supply for aviator's breathing apparatus at high altitudes. Medicinal and breathing oxygen is piped from a central supply to rooms in most hospitals where it is administered to patients by tent, mask, and catheter. Most high-purity oxygen sold for industrial purposes is sufficiently pure to meet the requirements of United States Pharmacopoeia for breathing purposes.

Production

The oxygen industry is unique in that its raw material is available in abundance everywhere for over 99 per cent of it is obtained from atmospheric air, the remainder from the electrolysis of water. Generally, oxygen is produced by liquefying air and then separating it into its components (oxygen, nitrogen, and argon) by fractional distillation.

There are three fundamental steps in the production of oxygen: purification, refrigeration, and rectification. Purification is the removal of dust, water, vapor, carbon dioxide, and hydrocarbon contaminants. Refrigeration in an oxygen plant means cooling the compressed air until it becomes a liquid at about–190°C. Recitifcation is the separation of liquid air into its components, oxygen and nitrogen, by repeated distillation. In actual practice, these three steps overlap or are performed at the same time.

Purification. The atmosphere holds dirt, water vapor, and carbon dioxide which must be removed from the compressed air stream as these would "plug up" the rectification column. Purification can be accomplished in three ways: by a chemical method, a mechanical method, or a combination of both. Mechanical filters are usually employed ahead of the compressor to remove atmospheric dirt. An example of chemical purification is the removal of carbon dioxide by passing air up through towers filled with coke, down which a solution of caustic soda or caustic potash travels; also a sodium hydroxide solution will remove carbon dioxide from the air.

Removal of impurities from a large volume of air by chemical methods is expensive. Thus, mechanical methods are employed in modern plants. Water vapor is removed in traps. Ice and solid carbon dioxide are removed in cold heat exchangers, reversing heat exchangers and various types of filters.

The compressed air is cooled and refrigerated by the basic expansion cycles described earlier. In most liquid plants, the incoming compressed air is also cooled by a conventional refrigerant, such as Refrigerant 22 or ammonia. This is necessary to make up some of the refrigeration removed with the liquid products.

The cycles used in many air-separation plants today employ both of the refrigerating techniques shown in Fig. 17.1. About 40 per cent of the air is expanded in a turbo or reciprocating expander. Since some of the air is expanded in the engine a smaller portion is left to be cooled in the countercurrent heat exchanger. Consequently, the incoming air can be cooled to a lower temperature, with the net result that a much greater fraction of the air is liquefied by the expansion valve.

Rectification. For practical purposes air may be considered a binary mixture of oxygen and nitrogen. As illustrated in Table 17.2, nitrogen boils at -195.8°C and oxygen at -183°C, so the difference in boiling points is almost 13°C. It is upon this fact that fractionation is based.

If a body of liquid air is warmed, the initial

TABLE 17.2. Composition of Air

Atmospheric air contains:

	Per Cent by Volume	*Boiling Point °C*
Nitrogen	78.14	-195.8
Oxygen	20.93	-183.0
Argon	0.93	-186.0
Carbon dioxide	0.03	- 73.3 (Solidifies)
Water vapor (not included)	amounts variable	0° (Solidifies)

gas released is 93 per cent nitrogen and 7 per cent oxygen, leaving a liquid that becomes richer in oxygen. The last liquid to be evaporated would be about 45 per cent oxygen. Also when air is condensed, the first droplets in equilibrium with the air are about 45 per cent oxygen.

The distillation column consists essentially of a cylindrical shell that contains trays spaced at regular intervals. These trays are perforated metal, permitting the liquid to pass transversely over the trays while the gas rises through the perforations and bubbles through the liquid. This brings the liquid and gas into intimate contact. The lower-boiling constituents, particularly nitrogen, are boiled off at each tray so that the gas going up the column becomes richer in nitrogen. At the same time the higher-boiling constituents, particularly oxygen, are condensed at each tray so that the liquid becomes richer in oxygen. With a single column, 99.5 per cent oxygen accumulates at the bottom, but the waste gas at the top is 93 per cent nitrogen and 7 per cent oxygen, representing an efficiency of only 66 per cent.

All commercial plants employ a double rectification column separated by a boiler-condenser, commonly called a "reboiler." The lower or high-pressure column operates at 75 to 90 psi, while the upper column operates at 10 to 12 psi. The physical principle for operation of the double column is that increased pressure on the liquid increases its boiling point. The temperature of the 10-psi liquid oxygen on one side of the condenser is lower than the boiling (or condensing) point of nitrogen at 75 psi. Thus the oxygen boils producing vapor for the upper column while the nitrogen gas in the high pressure column that reaches the reboiler is condensed. It flows down over the trays in the lower column becoming increasingly rich in oxygen. The distillation in the lower column produces high-purity nitrogen at the condenser and oxygen-enriched liquid at the bottom. Each of these liquids is transferred to the upper column after being refrigerated through an expansion valve.

The high-purity-nitrogen liquid is introduced at the top of the low-pressure column where it acts as a reflux to strip out oxygen. As a result, relatively high-purity-nitrogen gas is vented from the top of the upper column so that the oxygen-recovery efficiency for a double rectification column is about 90 per cent.

Liquid Oxygen

All liquid oxygen is produced as high purity—99.5 per cent or higher. When the oxygen is produced as a liquid, a large quantity of refrigeration is removed from the cycle with the product and therefore, is not available to cool the incoming air. Thus, cycles for liquid oxygen must develop large amounts of refrigeration. A common way to accomplish this is by the Heylandt Cycle which includes compression of the air to a very high pressure and use of external refrigeration.

It will be noted that the Heylandt Cycle makes use of both the Linde principle of cooling by expansion from a high to a low pressure through a valve (Joule-Thompson effect), and the Claude principle of the conversion of energy (heat) into work by the expansion of a gas in an engine (or turbine). Figure 17.3 gives details of the cycle as used in most "high pressure" liquid plants.

The air is compressed to about 1500 psig in a 4-stage compressor and cooled in conventional water-cooled intercoolers and after-

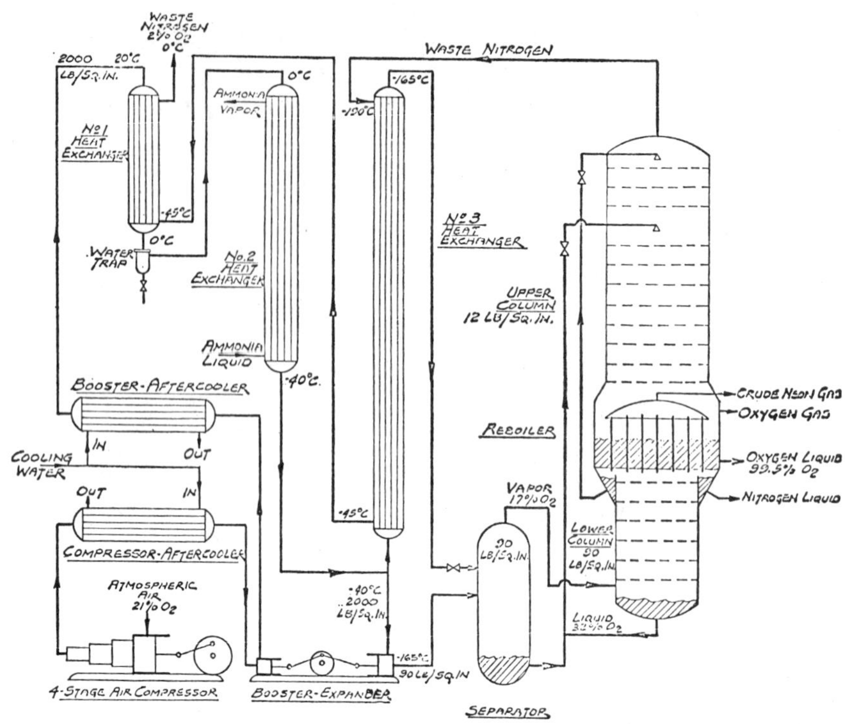

Fig. 17.3 Air-separation plant for liquid-oxygen production. (*Courtesy Linde Div., Union Carbide Corp.*)

coolers. The cooled air is compressed to 2000 psi in a fifth stage in a booster-expander, which will be explained later. The air is cooled to room temperature in an aftercooler and delivered to the heat exchanger section, consisting of three units.

In No. 1, the air is not permitted to cool below the freezing point of water. This permits most of the water to be condensed out and drained off through the water trap at the bottom of the unit. The cooling agent here is waste nitrogen. It enters the heat exchanger at about -45°C and leaves at about zero. This heat exchanger operates with a relatively large temperature difference.

The high-pressure air then enters the No. 2 heat exchanger, often called a forecooler, and is cooled by the evaporation of liquid ammonia or other refrigerant that is processed in an ordinary refrigeration system. This cools the air to about -40°C and freezes out any remaining water. Eventually this heat exchanger becomes plugged with ice and must be thawed out. Consequently, the No. 2 heat exchanger is provided in duplicate.

The air stream leaving the No. 2 heat exchanger is divided, 60 per cent passing through the No. 3 heat exchanger where it is cooled to an extremely low temperature, about -165°C. The air does not liquefy under these conditions because it is at a pressure of 2,000 psig, well in excess of the critical pressure of air (530 psig). This extremely cold air fluid is expanded through a valve to about 90 psig, which liquefies a sizable fraction of the air.

The 40 per cent portion of air that does not go through the No. 3 heat exchanger is expanded in an engine to about the same temperature and pressure as the vapor from the expansion valve. The air expanded in the engine is still gaseous. The work done by the engine is absorbed by direct coupling with the compression cylinder of the booster-compressor mentioned before. This expanded air stream enters the separator along with the throttled stream from the heat exchanger and the two are mixed together.

The vapor which leaves this separator is only about 17 per cent oxygen. Since the liquid formed is in equilibrium with the vapor in contact with it, it is about 23 per cent oxygen, much richer in oxygen than the entering air.

The liquid and vapor products of the separation are dealt with individually. The vapor stream enters at the bottom of the lower rectifying column at 90 psi. The liquid stream is combined with a similar liquid stream leaving the lower column, and the two are throttled to the pressure of the upper column (about 12 psig). The liquid stream then enters the middle of the upper column. The double rectification column previously described is used to separate the gases. The liquid oxygen accumulates in the main condenser and is piped through filters to an insulated storage tank.

Since liquid oxygen is continuously fed from the main condenser, the traces of hydrocarbon contaminants such as acetylene and methane are drained off with the oxygen. Thus, elaborate hydrocarbon removal systems are not needed in liquid-producing plants.

Gaseous Oxygen

Low-cost gaseous oxygen is produced in large quantities. This has made it economically feasible to use oxygen extensively in many steelmaking, copper-refining, and chemical processes. Oxygen-enriched air can be made economically by mixing the low-cost oxygen with appropriate volumes of air.

Gaseous oxygen is expensive to transport, therefore, these plants are built adjacent to the consuming plant, and are commonly referred to as "on site" plants. In the 1960's, the oxygen-producing facilities of the nation were more than tripled, and most of the expansions was by "on site" plants. These plants are equipped with modern instruments to improve their operating efficiency. Many facilities are completely automatic and operate unattended. In such cases, signal lights that indicate any operating difficulty are installed in the customers plant where they are continuously observed.

These new large-scale oxygen plants are being designed to produce a number of products to meet customer demands. These include high purity oxygen, 99.5 per cent; low purity oxygen, 95 per cent; high-purity nitrogen, 99.85 per cent for ammonia synthesis; ultrahigh-

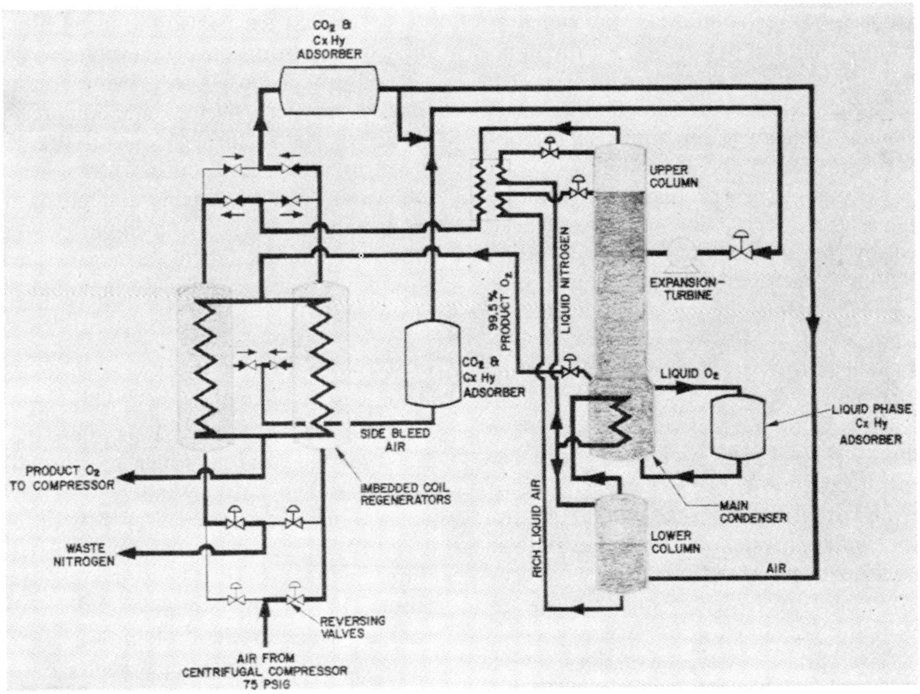

Fig. 17.4 Process diagram for an air-separation unit producing 80 million cu ft of oxygen per month. (*Courtesy Linde Div. Union Carbide Corp.*)

purity nitrogen, 99.99$^+$ per cent; and argon. Figure 17.4 illustrates the fundamental cycle used in many modern plants. Air is compressed to about 75 psig in a centrifugal compressor, cooled in an aftercooler and freed of liquid water. It then enters a regenerator (installed in pairs) where it is cooled to the saturation temperature. The air is then processed to remove small amounts of contaminating solids as well as the vapor-phase carbon dioxide and some hydrocarbons. The air then enters the lower column without any expansion.

To maintain self-cleaning conditions in the regenerators and to provide a preheat stream to the turbine, some air is withdrawn at the -100 to -120°C level. This air is cleaned of carbon dioxide and its temperature is adjusted so that the turbine discharge temperature remains just above the liquid region. The turbine discharge goes to the middle of the low pressure column. It should be noted that the refrigeration for this cycle employs primarily the Calude principle of conversion of energy (heat) into work by the expansion of gas in a turbine.

The conventional double column with condenser-reboiler is used to separate the gases. The oxygen-rich liquid air and the liquid nitrogen from the lower column are subcooled by waste nitrogen prior to transfer to the upper column. From this heat exchanger the waste nitrogen goes through the regenerators where it picks up the heat, water vapor, and carbon dioxide deposited by the incoming air.

The gaseous oxygen product in high purity plants passes through coils or heat exchangers heated by the incoming air. In low purity plants, the oxygen passes through a pair of reversing regenerators similar to those used for nitrogen.

The temperature of the waste nitrogen and product oxygen leaving the regenerators is a few degrees below that of the incoming air so that in this type of plant most of the refrigeraation is recovered. However, since no liquid oxygen is withdrawn from the reboiler, hydrocarbon contaminants tend to concentrate in the liquid oxygen. Thus, it is necessary to remove all traces of hydrocarbons from the reboiler liquid. One method of achieving this is by the appropriate placement of silica gel adsorption traps which remove most of the hydrocarbons from the process air before they reach the liquid oxygen in the reboiler. As a further safeguard, the liquid oxygen in the reboiler is continually circulated through another molecular sieve or silica gel adsorption trap to remove possible residual traces of those hydrocarbons that might concentrate to a hazardous level. In many plants, continuous circulation of "reboiler" liquid through an adsorption trap is the principal hydrocarbon removal device. Certain amounts of methane and ethane are normally found in the liquid oxygen in the reboiler and are nonhazardous if maintained below safe concentration levels in all parts of the reboiler.

In pressure-swing adsorption clean-up, hydrocarbons which are present in the feed air are safely adsorbed on the adsorbent bed and do not pass through the system with the oxygen product. The hydrocarbons are, in turn, desorbed into a waste stream which is depleted in oxygen to the extent that the oxygen concentration is about half that which exists in the atmosphere. In such an environment, one that probably would not support combustion, the dilute concentration of hydrocarbons in the stream are far below the lower flammable limit.

Gaseous oxygen is also produced in a pressure-swing adsorption plant using several parallel adsorbent vessels to provide continuous flow of oxygen. As feed air, compressed to 30–60 psi passes through one of the vessels, the absorbent (molecular sieve) traps (absorbs) the water, carbon dioxide and nitrogen gas, producing a relatively high-purity oxygen product. The absorbent is regenerated by blow down to low pressure during other phases of the cycle.

Distribution

Most people are familar with the steel cylinders in industry and hospitals where the oxygen is contained at very high pressure (1800–2200 psi). The large cylinders which have a water volume of about 1.5 cu ft hold 244 cu ft of oxygen measured at atmospheric temperature and pressure. These cylinders weigh about 215 lbs and contain about 15 lbs of product. Thus, the ratio of the weight of lading to weight of container is about 1 to 8, an uneconomical ratio. Oxygen is also transported at these high pressures in clusters of high pressure cylinders and long tubes permanently mounted on semitrailers. Although the production costs of gaseous oxygen are considerably less than for liquid oxygen, the transportation costs are usually higher because of the heavy containers employed.

The lowest-cost oxygen available today is produced in large-volume gaseous oxygen plants. As mentioned in the previous section, most of these are installed at the consumer's plant. However, recently, pipelines similar to natural gas pipelines have been employed to transport gaseous oxygen at moderate pressure (200 psi) from a central large-capacity plant to several nearby industrial users. This type of installation incorporates low-cost production with low-cost distribution.

Oxygen, nitrogen, and argon are transported commercially as pressurized liquids in (Department of Transportation) DOT 4L cylinders. The cylinders have a built-in or attached vaporizer that automatically converts liquid to gas as the products are withdrawn.

One cubic foot of liquid oxygen when evaporated or warmed to atmospheric temperature and pressure will produce 862 cubic feet of gaseous oxygen. Thus the Linde LC-3, shown in Fig. 17.5, weighs about 250 pounds and contains about 250 pounds of lading. It has a weight-of-lading to weight-of-container ratio of 1 to 1 as compared to 1 to 8 for the high pressure cylinder. This vessel contains the equivalent of 12 high pressure cylinders while occupying about 1/3 of the floor area, and weighing 1/4 as much (a weight saving of about 1/2 ton). It is about the same size and weight as a 55-gallon drum and can be moved with the same ease.

The vessel will stand several days with no withdrawal and no release of oxygen because of the unusually good properties of the fiberglass-aluminum-laminate insulation, the conductivity of which is about 5×10^{-4} Btu/hr/sq ft/°F. This is about 1/4 the value for 6 inches of powder-vacuum insulation or about 150 times better than 4 inches of cork.

Actually, liquid oxygen is pumped and handled in much the same manner as water except that the pumps and containers are well insulated (see Fig. 17.6). This liquid is normally transferred from the transport equipment to the storage tank by connecting a single pipe

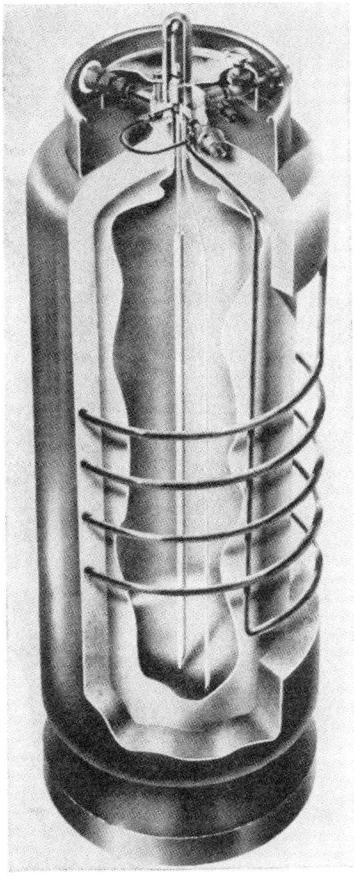

Fig. 17.5 LC-3 cutaway view. *(Courtesy Linde Div., Union Carbide Corp.)*

of hose to the liquid phase of the transport container. Pumps are usually used to transfer the liquid if the pressure in the receiving container is much above atmospheric pressure. If the storage tank operates at essentially atmospheric pressure, the transfer of liquid is often made by building a small amount of pressure in the transport vessel. Some pressure transfers are made up to 200 psi.

Actually liquid oxygen has been transported by tank truck since 1932 and by tank car since 1939. It is estimated that in 1970 there were about 1200 tank trucks and 525 tank cars in liquid oxygen-nitrogen-argon service.

Customer storage tanks range in capacity from 25,000 cu ft (220 gal) to 10 million cu ft (86,700 gal). (The capacity of a tank is usually expressed as equivalent to cubic feet

Fig. 17.6 Containers for liquid oxygen. (*Courtesy Linde Div., Union Carbide Corp.*)

of gaseous product at 70°F and one atmosphere.)

NITROGEN

The most economical way to utilize liquefied air is to make use of both the oxygen and the nitrogen. This is not always possible. In general, oxygen plants liquefy air, save the oxygen and waste the nitrogen, while the manufacturer of cyanamide, for example, uses the nitrogen and is willing to waste the oxygen. There are increasing demands for nitrogen, however, as well as for oxygen. Nitrogen production in 1969 was 4,530,000 tons, excluding the amount produced and consumed on-site in the production of ammonia. Over half of the above nitrogen production was delivered to users by pipeline for inerting and chemical operations.

Liquid nitrogen, which boils at −195.8°C at atmospheric pressure, is used as a coolant and purge in the space program, for the shrink fit of parts, and in the deflashing of molded rubber parts.

Liquid nitrogen has important applications in the field of health. Whole blood, which used to be limited to 21 days in storage to be suitable for transfusions, can now be frozen by a liquid-nitrogen process and stored for months and probably years. Bone marrow cells and other parts of the body can be preserved

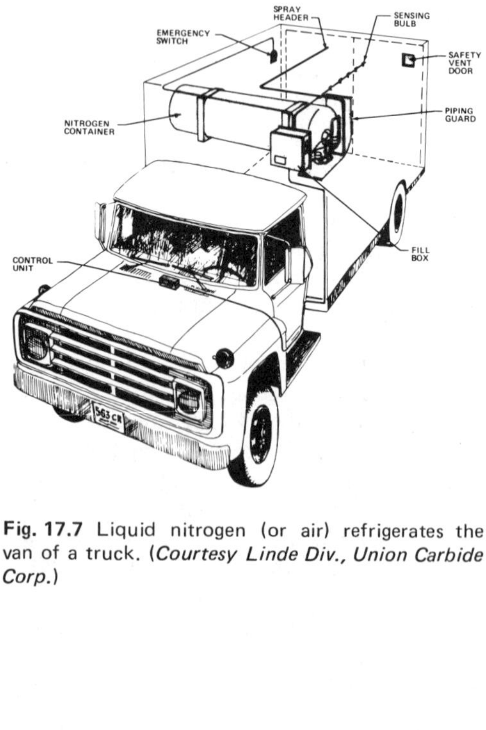

Fig. 17.7 Liquid nitrogen (or air) refrigerates the van of a truck. (*Courtesy Linde Div., Union Carbide Corp.*)

in liquid-nitrogen refrigerators. And cryosurgery has been used for Parkinson's disease and removal of other cell tissue by freezing.

Liquid nitrogen is also used for the refrigeration of frozen foods during transportation and storage. On trucks, the liquid nitrogen is held at 15–20 psig in an LC3-type container, which is not equipped with a vaporizer so that liquid is withdrawn. When the thermostat calls for refrigeration, a control valve is opened and liquid nitrogen is sprayed into the refrigerated van as shown in Figure 17.7 until the desired temperature is reached. Since the control valve is the only moving part, the reliability of this system is excellent. Another advantage in buildings is that the foods are held under an inert atmosphere, thereby preventing damage by rodents and fire.

Liquid nitrogen is transported in its own insulated containers, which are basically of the same construction as the liquid-oxygen containers.

ARGON

Mass markets for argon have developed, resulting chiefly from the development of gas-shielded arc-welding processes. These are used to join hard-to-weld metals such as aluminum, bronze, copper, Monel, and stainless steels. Once considered a "rare" gas, argon is available in tonnage quantities. Production in 1969 was about 1.47 million tons, a 275 per cent increase in five years. The metals titanium and zirconium, which have found wide application in our nuclear-space technology, depend heavily on argon or helium for their production. An inert gas envelops the manufacturing process for these metals from start to finish. These metals must also be welded under an inert atmosphere.

Argon is also used in incandescent lamp bulbs, in fluorescent luminous tubes, and in the manufacture of various semiconductor devices.

Production

Argon is relatively scarce; it represents only 0.93 per cent by volume of the earth's atmosphere. However, the advanced technology developed in air-separation plants plus the large air volumes handled in today's plants permits the recovery of this "rare" gas by the ton. The boiling point of argon, $-187^{\circ}C$, under conditions used in the fractional distillation of liquid air, lies between the boiling points of oxygen and nitrogen. Therefore, argon tends to accumulate in the middle of the low pressure (upper) column, readily building up to 10–15 per cent by volume. It is drawn off in a fraction that also contains oxygen and nitrogen and is further refined in other distillation columns that remove the nitrogen. The residual oxygen is removed by catalytic combination with hydrogen. Welding-grade argon is produced at a purity of about 99.995 per cent.

RARE GASES

The rare or "noble" gases include krypton, xenon, neon, helium, and argon. The availability of helium (other than atmospheric) and argon in large quantities makes the designation "rare" a misnomer insofar as these two are concerned. However, atmospheric helium is rare, as is argon specially purified for particular applications; the terminology for argon and helium appears to be a matter of philosophy. All these gases are characterized by extreme chemical inertness and some of them, primarily neon, krypton, and xeon, ionize or become electrically conducting at a substantially lower voltage than other gases. While passing current they also emit a brilliant colored light used to advantage in the tubular display signs so prominent in what has been termed the "neon jungle."

Present commercial applications for the rare gases rely principally on their inertness. They are used singly or in mixtures by the electronics industry in gas-filled electronic tubes. The lamp industry uses all the rare gases, including atmospheric helium and specially purified argon as fill gas in specialty lamps, neon and argon glow lamps, high output lamps, and others. The gases are also used as fill gas for ionization chambers, bubble chambers, and related devices.

Production

Partial separation of the rare gases is accomplished by the same liquefaction-fractional distillation process used to produce most other atmospheric gases, using liquid nitrogen as a refrigerant. Final purification requires the use of special processes and equipment.

HELIUM

Most of the helium produced for commercial use is obtained from certain natural gases. About 95 per cent of the known helium-bearing natural gases in the United States are contained in four helium-bearing gas fields: the Hugoton field of Kansas, Oklahoma, and Texas; the Panhandle field of Texas; the Greenwood field of Kansas; and the Keyes field of Oklahoma.

The rapidly increasing use of helium in the 1950's heavily taxed the production facilities which were operated almost exclusively by the Bureau of Mines under the Helium Act. The quantity of helium sold by the Bureau of Mines increased from about 50 million cu ft in 1949 to 477 million cu ft in 1959. Congress amended the Act in 1960 to provide for long-term purchase contracts for helium by the Bureau of Mines from private industry as part of an extensive helium conservation program. It is estimated that about 78 billion cu ft of helium will be collected under this program through 1985, about 36 billion cu ft will be purified and used, leaving 42 billion cu ft in storage at 1985.

In 1962-63, five large crude-helium plants were built by private industry in western Kansas and the Texas Panhandle to extract helium from the natural gas being piped to market. Crude (unrefined) helium is extracted by a cryogenic process that separates out a helium-nitrogen stream containing 50-85 per cent helium. Most of this crude is stored in an underground reservoir near Amarillo, Texas.

Refined helium is produced by five Bureau of Mines plants, Fig. 17.8,[4] having a capacity of about 800 million cu ft. Private industry added plants in 1961 and 1966 totaling about 360 million cu ft. The Bureau of the Census shows helium shipments to be 907 million cu ft in 1967.

Prior to 1946, about 99 per cent of the helium was used for dirigibles and balloons. Helium purity was gradually improved from 98.3 to 99.5 and then to 99.99$^+$, which greatly expanded its use. The biggest user is National Aeronautics and Space Administration and its contractors in missile development. Helium is also used for shielded-arc welding, nuclear energy development, and in breathing mixtures for deep underwater diving.

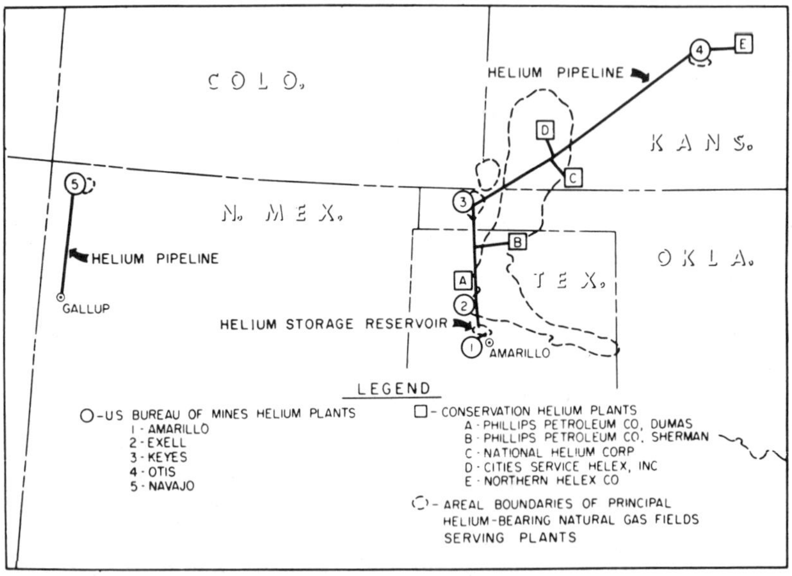

Fig. 17.8 Refined helium is produced at five Bureau of Mines plants.[4]

Production

Bureau of Mines and private plants take gas from natural gas pipelines, remove the helium and return the stripped gas to the pipeline. Plant capacities vary over a wide range, but are usually sufficient to handle the entire flow in a pipeline, except for a few cold winter days. Also the helium content in gas varies widely from plant to plant, ranging from 0.9 to 5.8 per cent. In general helium production involves two phases: (1) production of crude helium, and (2) purification to 99.99^+ purity.

An article[1] describes the process used at the Bureau of Mines Keyes plant, which is typical of all plants and includes the following: (1) scrubbing the natural gas to remove water and condensed hydrocarbons; (2) dust removal; (3) CO_2 removal in a scribbing tower; (4) drying; (5) chilling to liquefy and remove the natural gas; and (6) rectification of the gaseous materials from step 5 to produce the crude (75 per cent He, 25 per cent N_2, 0.1 per cent each of H_2 and CH_4) helium.

The pure helium is obtained from the crude by: (1) oxidizing the hydrogen with air; (2) removing the water formed by the oxidation; and (3) removing the last traces of hydrogen and nitrogen with activated charcoal. The final product is 99.99 plus per cent pure.

Many of the private plants and the Otis BM plant contain helium liquefiers that have sufficient capacity to liquefy most of the plant's output. The refrigeration techniques described earlier in Fig. 17.1 are used, plus precooling with liquid nitrogen as shown in Fig. 17.9.

Linde's helium liquefier at Amarillo, Texas has a capacity in excess of 100 l/hr. Pure helium gas is compressed 270 psig, passed through an aftercooler and into the cold box where it is cooled to 80°K against the exiting recycle helium and nitrogen. The helium is then refrigerated by liquid nitrogen, purified in a gel trap, and further cooled to 24.5°K by recycle helium gas. The stream then splits with part of the cold helium going through an expansion turbine, while the remainder goes through a heat exchanger and expansion valve.

Liquid helium is transported in vacuum-super-insulated tank trucks, portable tanks, and cylinders without nitrogen shielding. Many liquid-helium Dewars, 100 l and less, that op-

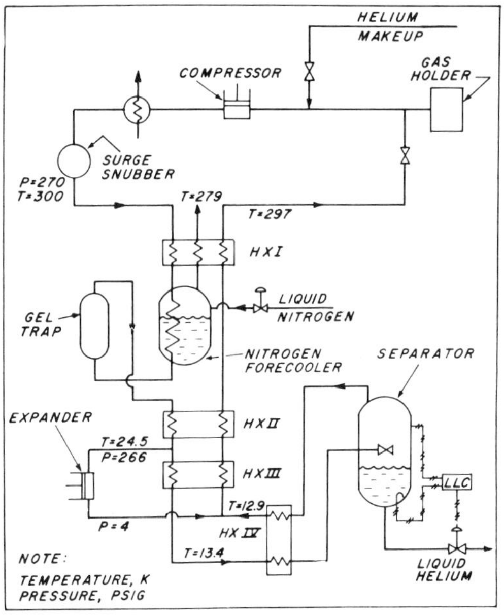

Fig. 17.9 Helium is liquefied by cryogenic refrigeration techniques, plus precooling with liquid nitrogen.[4]

erate at less than 15 psi, use nitrogen shielding. Since liquid helium at 4.25°K will readily solidify air, caution must be exercised to be sure that the Dewar neck tube does not become plugged with solid air, particularly during air transportation where it is subjected to wide changes in atmospheric pressure.[5]

Liquid helium boils at 4.25°K and this ultra-low temperature can liquefy any other gas. LHe is being used increasingly in low temperature research investigating the properties of matter, particularly electrical properties. LHe is also used in cryogenic chemistry, bubble chambers and for cryopumping.

HYDROGEN

Large quantities of hydrogen are used in the chemical syntheses of ammonia, hydrogen chloride, and methanol. Other alcohols are produced by the hydrogenation of the corresponding acids and aldehydes.

During recent years, hydrogen has become industrially important in the hydrogenation of the edible oil in corn, cotton seed, soy bean, and other vegetables for the production of shortening and other foods, many of which are low in cholesterol. Its other uses include the oxygen-hydrogen flame that has a temperature of about 2200°C (4000°F) and is well suited for low-temperature welding and brazing of thin metals, undersea cutting-welding where over 15 psi is involved, and in the fabrication of quartz and glass. Atomic-hydrogen welding, a form of shielded-arc welding, produces temperatures of about 6100°C (11,000°F). Hydrogen serves as de-oxidizing fuel in many applications such as annealing furnaces and fabrication of electronic components; it is used in large electric generators to reduce windage losses and heat.

Liquefied hydrogen is used as a rocket fuel by explorers of outer space; and in recent years it is the commercial source of ultrahigh purity hydrogen (99.999$^+$ per cent) required in parts of the electronic industry.

Hydrogen is obtained almost exclusively from water or hydrocarbons. From water gas, is it obtained by the removal of the nonhydrogen constituents. It is made by the catalytic action of steam on oil refinery gases and natural gas, by steam on heated iron, by the electrolysis of water, or by other miscellaneous processes. It is a by-product in the electrolytic cell for caustic,* in several fermentation processes, and in other types of processes. The choice of a process will be decided by the resources at hand, and by the degree of purity required. Should rapid generation with a minimum of apparatus in an isolated place be demanded, the steam-methanol or the ferrosilicon process would serve.

The Water Gas and Steam Process (Continuous Catalytic Process)

Water gas with steam in excess is passed over an iron oxide catalyst, just as is done in the Bosch process, except that since no producer gas is added, the amount of nitrogen is small. The converter has several trays, on which the catalyst rests. The reaction ($CO + H_2O \longrightarrow CO_2 + H_2$) is exothermic; as the temperature must be maintained at 450°C (842°F), the converters are insulated and the incoming gases heated in exchangers. Once the reaction has begun, no outside fuel is required. Three volumes of steam to one of gas are used; the great excess of steam drives the reaction to the right. After passing the exchangers the reacted gas is freed from its steam by water-cooling. The carbon dioxide formed, as well as the small amount which entered with the water gas (4 per cent), is removed by scrubbing with cold water while under pressures of 25 to 30 atmospheres, in tall steel towers; under such pressures carbon dioxide is freely soluble in water. The gas leaving the last scrubber has the composition:

	Per Cent
Hydrogen	92–94
Nitrogen	1–4
Methane	0.5
Carbon monoxide	2–4
Carbon dioxide	small
Moisture	small

The crude hydrogen may be further purified from carbon monoxide by scrubbing in ammoniacal cuprous chloride solution. The nitrogen impurity may be lowered by careful operation of the water-gas plant. The methane is not wanted, and may be almost avoided by using well-burned coke.

A similar process in which the catalytic agent is lime at the temperature of 450°C (842°F) instead of iron oxide has been proposed; its great advantage is that the carbon dioxide is simultaneously removed. Unfortunately this absorption is accompanied by a powdering of the lime granules, as the carbonate forms, and the powder tends to clog the lime towers.

Water Gas Process with Liquefaction of the Carbon Monoxide

There are two processes in which the carbon monoxide in water gas is liquefied by cold and pressure and removed in that state, leaving the hydrogen gas comparatively pure. The Linde-Fränkl-Caro process uses liquid air boiling under a few millimeters of pressure for the final cooling of the water gas already precooled by three steps, first by an ammonia refrigerating system (-35°C or -31°F), then in an exchanger wherein the uncondensed hydro-

*Hydrogen from mercury cells for caustic has a slight contamination of mercury, which for certain uses, must be removed.

gen takes up heat and finally by the liquid carbon monoxide separated in the final cooling. The carbon monoxide at the same time boils, and is used as a gaseous fuel. The other process is Claude's, which uses no liquid air, but obtains the necessary final cooling by the expansion of hydrogen from a pressure of 20 atmospheres to a lower pressure while doing work against a piston. The hydrogen is precooled in exchangers by outgoing gases, and by the evaporation of the carbon monoxide which has been previously liquefied.

Steam on Heated Iron

The interaction of steam and iron takes place at an elevated temperature, such as 650°C (1202°F) in a multiplicity of relatively small, upright steel cylindrical retorts. The iron packing selected should have a porous structure and little tendency to disintegrate; a calcined iron carbonate (spathic ore) has been found suitable. A plant for the production of 3500 cu ft (about 100 cu m) per hr would consist of three sets of 12 retorts each, each retort being 9 inches in diameter and 12 ft high. The action is intermittent. The steaming period (hydrogen production) lasts 10 min (upward travel of steam).

$$3Fe + 4H_2O \longrightarrow Fe_3O_4 + 4H_2$$

The iron oxide formed is reduced by water gas, for example, and the water period lasts 20 min (downward travel), because the reduction of the oxide is slower than the oxidation of the iron. A brief purging with steam sends the first hydrogen to the water gas holder. By the stepwise operation of such a plant a continuous flow of hydrogen is obtained. The spent water gas is cooled and burned (for it still contains combustible gases) around the retorts to maintain the reaction temperature. The steam reaction is exothermic; the over-all reduction by the water gas is endothermic.

The hydrogen passes out with the great excess of steam which is employed to drive the reaction to the right; it is cooled to remove the steam, and freed from carbon dioxide and hydrogen sulfide by lime purifiers. The gas obtained is 98.5 to 99 per cent pure; by careful purging, using closed condensers instead of scrubbing towers, and other modifications, the purity may be raised to 99.94 per cent. The steam-iron process is used chiefly in connection with the hydrogenation of fatty oils. The iron mass lasts six months, the retorts one year.

For the production of 1 volume of hydrogen, the continuous catalytic process requires 1.25 volumes of water gas, the liquefaction process (CO liquefied), 2.5 volumes, and the steam-on-heated-iron process, also 2.5 volumes.

Electrolysis of Water

In commercial cells, direct current is passed between iron electrodes, which may be nickel-plated, suspended in a bath consisting of a 10 to 25 per cent caustic soda or potassium hydroxide solution. Only distilled water is added to the cells, for the electrolyte is not consumed. Cells differ in the method of gathering the hydrogen (at cathode) and the oxygen (at anode), in size, and in details of construction. The efficiency is close to 7.5 cu ft of hydrogen per kw-h, and 3.8 cu ft of oxygen.

The decomposition voltage for the reaction $H_2O \longrightarrow H_2 + \frac{1}{2}O_2$ is 1.48; the operating voltage about 2. The amperage varies with the size and intensity of operation; with maximum load, it may be as much as 1000 amp for electrodes 40 in wide and 60 in high, with a number of electrodes in each cell. The electrodes are separated by an asbestos diaphragm, and suspended in cast-iron containers.

Steam Reformer Process

The most widely used process for producing hydrogen in large volume is by steam reformer; about 80 per cent of the hydrogen used for ammonia production is derived by that process.

Heated natural gas (or other hydrocarbon fuel) and steam are mixed and fed into a reformer furnace, consisting of large diameter, thick walled vertical tubes manifolded together and packed with a nickel catalyst. Heat input, 900°C (1600°F), for this step is from natural gas-fired burners. The reactions in the reformer are:

$$CH_4 + 2H_2O \longrightarrow 4H_2 + CO_2$$

$$CH_4 + H_2O \longrightarrow 3H_2 + CO$$

The crude hydrogen stream, containing 8.5 per cent CO and excess steam, is cooled in a steam generator to 400°C (750°F) and fed to a shift converter that reacts the CO to CO_2 while producing hydrogen. This reaction is slightly exothermic.

$$CO + H_2O \longrightarrow CO_2 + H_2$$

Liquefied Hydrogen

The above steam-reformer process is the source of hydrogen in a majority of liquefied hydrogen plants, which have four basic sections: steam reformer, purification, liquefaction, and storage as shown in Fig. 17.10.

Crude hydrogen from the reformer-shift converter contains about 20 per cent carbon dioxide and is saturated with water. The CO_2 is usually removed in a monoethanolamine (MEA) scrubber with the exiting crude being about 96 per cent hydrogen, 1.7 per cent CO, 2.2 per cent CH_4, and small amounts of nitrogen, CO_2, and moisture.

The moisture is removed to a very low dew point in a molecular sieve bed to prevent freeze-up of the low-temperature purification equipment shown in Fig. 17.11.[6] This equipment removes impurities in hydrogen to about 5 ppm, of which not more than 1 ppm is oxygen, in order to meet customer requirements and prevent freeze-up of the liquefaction section. Partial condensation is used to remove most of the methane and some of the nitrogen, and carbon monoxide is removed in propane and methane wash columns and other steps, using liquid nitrogen as the final cooling step. Then residual impurities are removed by low-temperature adsorption to obtain the desired hydrogen purity.

Pertinent properties of hydrogen are given here before discussing the liquefaction process. The boiling point of liquefied hydrogen is -252.9°C (-423.2°F), which is lower than the freezing temperature of all other gases except helium. Its density is only 4.418 lb/cu ft, about 1/14 that of water. Hydrogen can exist in two forms depending upon the rotation of the two atoms in the hydrogen molecule. In *ortho*-hydrogen, the atoms spin in the same direction; in *para*-hydrogen they spin in opposite directions. Also, hydrogen has a very low inversion temperature, -70°C, above which

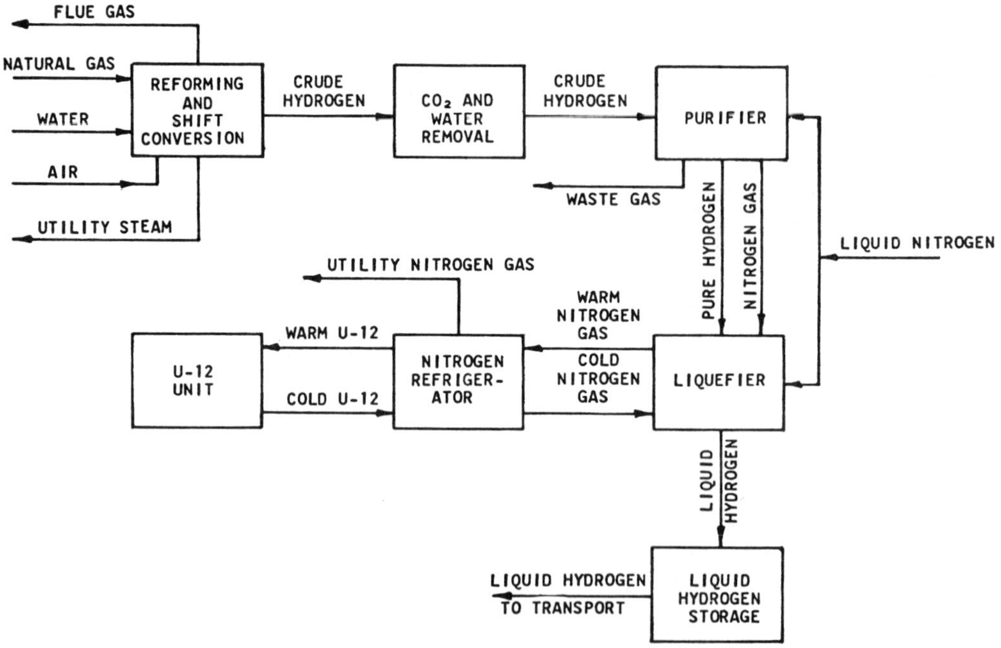

Fig. 17.10 Process block diagram of large liquid hydrogen plant.[6]

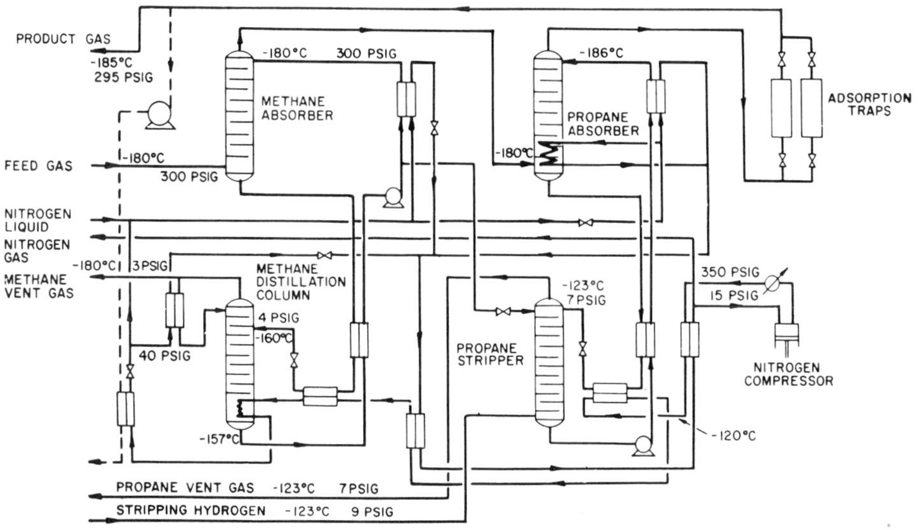

Fig. 17.11 Low-temperature purification equipment removes impurities from hydrogen. [*Chem. Eng. Progr.*, **59**, *No. 8, 61 (1963); copyright 1963 by The American Institute of Chemical Engineers, reprinted by permission.*]

Joule-Thompson expansion produces a rise in temperature, instead of lowering it.

The liquefier section produces liquefied *para*-hydrogen by removing heat from the purified hydrogen, and by converting the *ortho*-hydrogen to *para*-hydrogen. Conventional cryogenic refrigeration cycles described in Fig. 17.1 are used to liquefy the hydrogen. Liquid nitrogen is used to precool the hydrogen; then the hydrogen stream is split in two with portion one being cooled to near liquefaction temperature in a high speed expansion turbine. This cold, expanded hydrogen is used to cool portion two before it is partially liquefied through a (Joule-Thompson) expansion valve, Fig. 17.12.[7]

Ortho-to-*para* conversion is an important factor in the design of the liquefier cycle. Purified hydrogen, which is normal hydrogen, is only 25 per cent *para*-hydrogen at atmospheric temperature. Stable liquefied hydrogen is 99.7 per cent *para*-hydrogen, and the heat generated by conversion is about 11.5 per cent of the total heat removed during liquefaction. If normal (liquid) hydrogen were transferred to storage, up to two-thirds of the liquid would be vaporized as conversion took

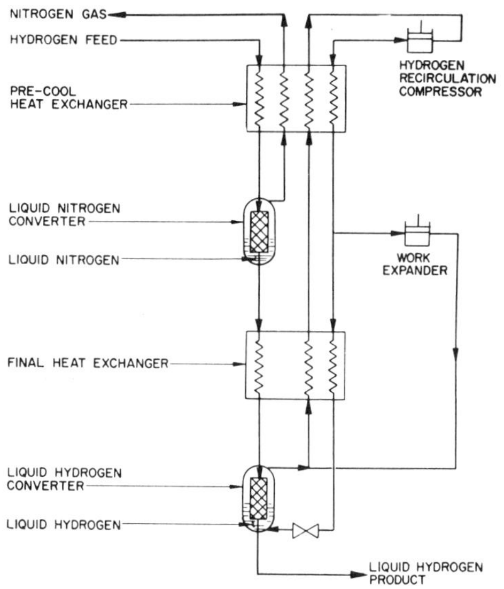

Fig. 17.12 Cycle for liquefaction and conversion of liquid hydrogen. (*Baker, C. R. and Matsch, L. S., "Advances in Petroleum Chemistry and Refining," Vol. X, Interscience Publishers; by permission of John Wiley and Sons, Inc.*)

place. So the liquefier cycle includes rapid conversion over ferric iron hydroxide catalyst, in two steps, at liquid nitrogen and at liquid hydrogen temperatures in the presence of liquid refrigeration to remove heat. Thus, liquefied hydrogen, which is over 95 per cent *para*, is produced by the liquefier and transferred to insulated storage tank. Purity exceeds 99.999 per cent.

Tank trucks and tanks cars with laminar and high-vacuum insulation are used to transport liquefied hydrogen from large production plants to any part of the country without loss of product enroute. Thus liquefied hydrogen is delivered commercially in containers and by procedures similar to those used for liquid oxygen, which were described earlier.

LIQUEFIED NATURAL GAS (LNG)

Natural gas as a fuel is discussed in Chapter 3. Since liquefied natural gas is a cryogen (boiling point about −162°C) and the refrigeration, handling, and transportation techniques discussed in this chapter for other cryogens is used for LNG, it seems appropriate to include a brief discussion of LNG.

The cryogenic phases used to produce LNG are purification and liquefaction, the same as for liquefied hydrogen. Natural gas contains some water vapor and carbon dioxide as it is withdrawn from the pipeline for liquefaction. Concentrations of these impurities are reduced to 1 ppm and 50 ppm respectively, to prevent plugging of the liquefying equipment. One purification method is to pass the gas through adsorption towers packed with molecular sieves, with an activated alumina bed at the inlet to remove oil vapor.

There are several liquefaction cycles used for LNG but the "cascade" concept in which several change-of-phase systems are combined is the most popular. Each system involves a fluid suitable for use at a given temperature level and is used to precool the fluid in the next change-of-phase system operating at a lower temperature level. Several levels are "cascaded" together to match the cooling curve of the natural gas, with throttling or Joule-Thomspon expansion of the refrigerant between each level. An advantage of this cycle for LNG is that the various temperature levels can be adjusted to provide the most efficient cooling with changes in composition of the natural gas to be liquefied.

The composition of the refrigerant is usually a mixture of hydrocarbons and nitrogen. The equipment consists of several heat exchangers and separators in series with throttle valves between them as shown in Fig. 17.13.[8] Thus, only one refrigeration compressor is needed and a typical cycle would include change-of-phase (system) of propane-, ethane- and methane-enriched streams in succeeding heat exchangers.

LNG is stored and handled in insulated containers, including seagoing vessels. Domestic transportation is by tank trucks which are similar to those used for liquid nitrogen. Most tank trucks built today for LNG are suitable for liquid nitrogen, and vice versa, to provide greater flexibility in a carrier's fleets and reduce fabrication costs.[9]

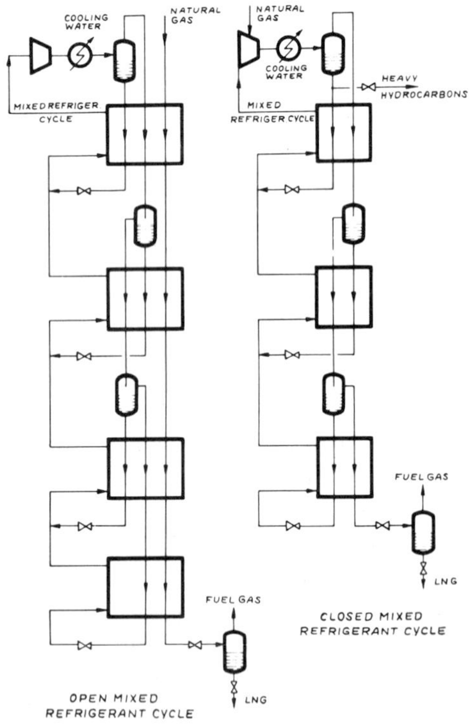

Fig. 17.13 Cycles commonly used in liquefying natural gas.[8]

ACETYLENE

Acetylene is an important industrial gas. The oxygen-acetylene flame produces the highest temperature of any combustible gas, hence its great value in welding and cutting of steel and other metals. A large volume of acetylene is being used at present as a basic raw material in the chemical industry for the production of synthetic rubber-like materials, flexible vinyl plastics, rigid plastics, paints, and textile finishes, to name a few. Acetylene is generally made by the action of water on calcium carbide, a product of the electric furnace.

$$\underset{}{CaC_2} + 2H_2O \longrightarrow \underset{\substack{\text{Acetylene}\\ \text{(gas)}}}{C_2H_2} + \underset{\substack{\text{Hydrated lime}\\ \text{(by-product)}}}{Ca(OH)_2}$$

The only method of distribution, other than pipeline, is by means of portable steel tanks, containing a porous solid filler saturated with acetone, or other suitable solvent, in which the acetylene is dissolved by pressure. Acetylene alone is not compressed to pressures higher than 2 atmospheres because of its tendency to decompose explosively into carbon and hydrogen; dissolved in acetone it may be under pressures of 10 to 15 atmospheres with safety. In order to fill the cylinders, the acetylene gas is dried over calcium chloride and compressed in a specially designed compressor. Cylinders are connected to the compressor and occasionally shut off to allow time for the dissolution in acetone.

In cases where the rate of consumption is high, it is sometimes feasible for the user to install and operate his own acetylene generator. This equipment reacts calcium carbide with water to produce acetylene gas. Calcium carbide is normally available in 100 lb drums, and in larger containers holding 250, 300, 500, and 600 lbs and 2½ and 5 tons.

In the low-pressure generator, the pressure is below 6 lbs; in the "medium-pressure" generators, it must not exceed 15 lbs per sq in. Carbide is fed a little at a time to a body of water. The volume of gas generated may be as low as 1 cubic foot an hour.

There are two kinds of welding torches, high pressure and low pressure, and they differ in important details in the internal mixer. Cutting torches differ from welding torches in the following way: in addition to the oxygen and acetylene conduits, which both torches have, the cutting torch receives a stream of oxygen around the flame; it is this oxygen which cuts the steel, by oxidizing it and forcing away the particles of oxide formed; the flame serves merely to attain the oxidizing temperature. The flame in each has an inner brilliant part, whose temperature is estimated at 3300°C; in the welding torch this is surrounded by a larger envelope into which the air penetrates. The inner portion, which does the welding, is sometimes called the neutral part. The use of acetylene is not without danger; the directions and cautions of the manufacturer must be observed.

The oxyacetylene torch has many uses besides the welding and cutting of steel. It serves for metal cladding, in certain special circumstances; and for steel conditioning, pressure welding, flame spinning, and flame hardening In shaping synthetic sapphire and ruby (hexagonal crystals of alumina) the torch is in constant application. Synthetic rubies, for example, are obtainable in the form of slim rods and boules (balls); the rods when heated in the torch may be bent, in the form of a thread guide, let us say, which is a complete loop. At the same time, the material acquires a flame finish of extreme smoothness. Ruby in rod form is made into precision gauges. It is also made into extrusion dies, phonograph needles, and knife edges on balances.

Economic petrochemical process have been developed for the manufacture of acetylene and now account for an important percentage of the total acetylene production. (See Chapters 14, 25.)

Acetylene undergoes a number of reactions which are the bases of several large industries. Acetylene as a chemical substance is presented in Chapter 25.

CARBON DIOXIDE

Sources

Carbon dioxide is used commercially as a gas (soda ash manufacture), compressed as a liquid in steel cylinders (for soda fountains, for refrigeration, and as convenient source of the gas), and as the solid. It is obtained from (1) the combustion of coke; (2) the calcination of limestone; (3) as a by-product in syntheses involving carbon monoxide; (4) as a by-product in fermentations; (5) by the action of sulfuric acid on dolomite; (6) and from wells. Gas from any one of these sources may be made into the gaseous, liquid, or solid form of carbon dioxide.

The utilization of the carbon dioxide in the combustion gases of coke involves the alternate formation and decomposition of alkali bicarbonates in solution. Hard coke is burned under boilers and the fuel gases are regulated in such a way that a maximum content of carbon dioxide, 16 to 17 per cent, is obtained. The gases enter a scrubber (tower) packed with limestone, to remove sulfur compounds, and fed with water, to cool the gas and arrest the dust. The cold gases enter the absorber, a tower packed with coke down which a solution of potassium carbonate passes; carbon dioxide is absorbed, and the saturated solution is run to a boiler where the absorbed gas is liberated by heat. It is under the boiler that the coke is burned. The operation is continuous; charged solution flows in constantly, the spent liquor is run off constantly. By means of an interchanger, the outgoing liquor heats the incoming liquor to some extent. The outgoing liquor, cold, returns to the absorber. The gas from the boiler is very pure. It is dried in a calcium chloride tower, and compressed to 100 atmospheres, at which pressure it liquefies at ordinary temperatures (see Figure 17.4).

In addition to the uses which have been mentioned, carbon dioxide serves as a chemical in the manufacture of salicylic acid, white lead, and other products, as well as in fire-fighting devices of various kinds. Fire extinguishers of the wall type, also called the soda-acid type, contain 2½ gal of saturated sodium bicarbonate solution and 4 oz of concentrated sulfuric acid in a small bottle. On inversion, the acid reaches the solution and liberates carbon dioxide; the pressure developed expels the liquid through a nozzle to a distance of 30 to 40 ft. The main extinguishing agent is the water. Liquid carbon dioxide under pressure in steel cylinders may be released so that a carbon dioxide snow forms which may be directed into the gaseous blanket over the fire. The "firefoam" extinguisher system relies upon the smothering action of a foam blanket produced by the interaction of a sodium bicarbonate solution with an alum solution, in the presence of a foam stabilizer.

Solid carbon dioxide is obtainable in commercial quantities. It is supplied in block form resembling the familiar artificial ice cake. Its uses are similar to the uses of ice, but it functions without melting, and without producing drips; it vaporizes, and leaves only a gas, which may be easily vented, so that it has received the rather apt name of "dry ice." Its manufacture will be described in terms of a particular plant.

Pure, liquid carbon dioxide under a pressure of 1000 lb and at a temperature of 70°F (21°C), is delivered to the plant by a pipe sys-

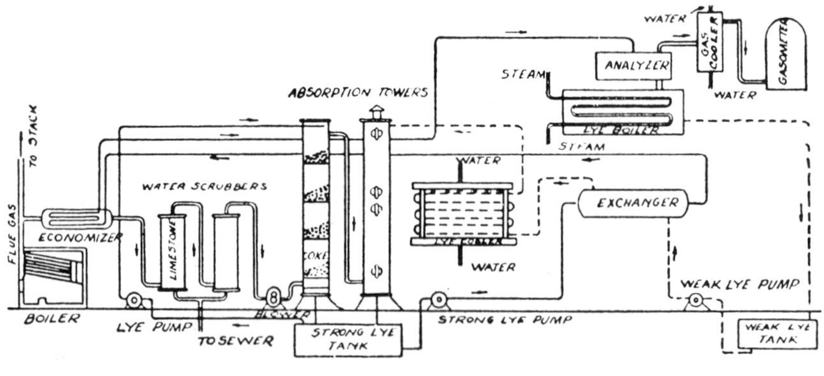

Fig. 17.14 Diagrammatic flow sheet for the absorption of carbon dioxide in the fire gases from burning coke. The gas dissolves in a strong lye solution from which it is driven out by heat, giving 100% CO_2 gas. *(Courtesy Frick Co., Waynesboro, Pa.)*

tem. It is sent to the "evaporator" (Figure 17.15), where its pressure is reduced to 500 lb with a simultaneous drop in temperature to 32°F (0°C). With the pressure set at 500 pounds, the liquid maintains itself at that temperature; as this is lower by several degrees than the room temperature, heat flows in and causes the liquid to simmer quietly. About 25 per cent of it boils away. The vaporized portion is sent to a special compressor which delivers it as gas to the main compressor gas line, at a pressure of 1000 lbs.

The 32°F liquid from the evaporator is admitted to the press chambers; these have movable tops and bottoms, worked by hydraulic pressure. The chamber is 20 by 20 in, and 24 to 30 in, deep. The liquid enters through an ordinary nozzle; part of it expands to gas and draws its heat largely from the incoming liquid, which is thus solidified to a fluffy snow. The gas formed is drawn off constantly by the suction line of the main compressors and recompressed. By operating the top and bottom walls, the snow is compacted to a solid block 20 by 20 by 10 in. Each press makes 6 to 8 cakes per hr. The density of the resulting cake is controlled by the amount of snow pressed into the 10-in space. After discharge to a conveyor, the block reaches band saws, which cut it into four smaller blocks, each a 10-in cube, weighing about 20 lbs. This is wrapped in brown paper and stacked in a specially insulated railway car for transportation to distant points, or into trucks for local delivery.

Of the liquid delivered to the press, 20 to 45 per cent is solidified; the rest turns to gas and must be reliquefied. The colder the temperature of the liquid CO_2 and the colder the press chest, the higher the percentage frozen. Based on heat content, it is found that it takes 3.75 pounds of liquid to produce 1 pound of solid. The expansion in the chest is due to atmospheric pressure.

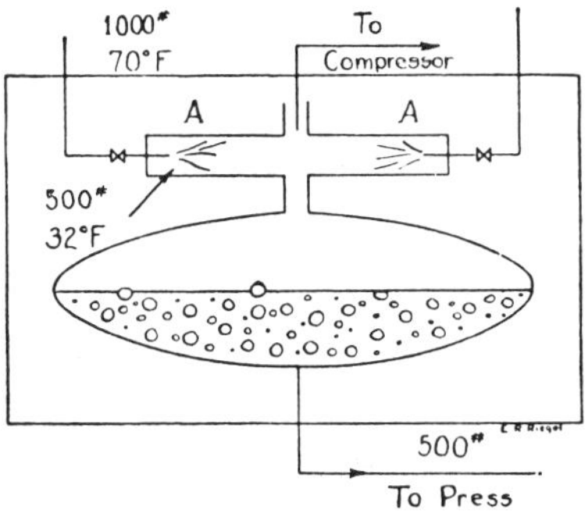

Fig. 17.15 The "evaporator," in which liquid carbon dioxide is formed and stored.

The critical temperature of carbon dioxide is 88°F (31.1°C), the critical pressure 1073 lb. At 70°F (21°C), it is considerably below the critical temperature, so that a pressure of 1000 to 1100 lb suffices to keep it in the liquid state.

It will be clear that much of the expense in the plant will be that for recirculating the carbon dioxide gasified at the presses. The compressors are four-stage machines: 0 to 5 lb, 65 to 70 lb, 300 to 325 lb, and 1000 to 1100 lb. From the last stage the gas enters oil-removing filters, then a condenser cooled with tap water, which reduces its temperature to about 70°F (12°C). In the condenser, the carbon dioxide liquefies, and enters the "evaporator" with the new liquid, at the same temperature and pressure. Carbon dioxide from any source may be made into the solid form.

The uses of carbon dioxide are as a refrigerant for the frozen food, dairy product, and meat packing industries; grinding of dyes and pigments; and in the manufacture of certain pharmaceuticals and chemicals.

SULFUR DIOXIDE

Of the more compressible gases, chlorine and ammonia are discussed elsewhere (Chapters 6 and 5, respectively). Sulfur dioxide, SO_2 anhydrous, liquefied under a moderate pressure (2 to 3 atm) at room temperature, is shipped in steel cylinders of 50- or 100-lb capacity, in 1-ton containers, and in single-unit 15-ton car tanks. It is used from such cylinders and tanks in preparing hydroxylamine sulfate, which in turn serves in making dimethyglyoxime, the nickel reagent; for refrigeration, for bleaching, and, increasingly, in petroleum refining. The boiling point is–10°C.

The burner gas from sulfur (or pyrite), freed from dust and cooled, is dissolved in water in two towers used in series; the solution from the second tower is elevated to the top of the first tower, where it meets the rich gas. Burner gas with 8 to 12 per cent sulfur dioxide yields a 1 per cent solution. In a third tower this solution is sprayed at the top, and flows down, while steam is injected at the base of the tower; previously the 1 per cent liquor was heated in a closed coil laid in the spent liquor from the base of the still. The packing in all the towers may be coke, or special earthenware cylinders. The gas issuing from the third tower is cooled to remove most of its moisture, and is passed up a fourth tower down which concentrated sulfuric acid flows. The dried gas is compressed in a bronze pump to 2½ atm, which suffices to liquefy it.

NITROUS OXIDE

Nitrous oxide (N_2O) is made by heating ammonium nitrate to 200°C in small lots (50 lbs) in aluminum retorts. The gas is cooled in a condenser, washed in a solution of sodium dichromate to remove nitric oxide, in caustic to absorb nitric acid, and in water. Under a pressure of 100 atm it liquefies, in small shipping cylinders, for instance; or it may be stored in a gas holder. The reaction is

$$NH_4NO_3 \longrightarrow N_2O + 2H_2O$$

Nitrous oxide is used as general anesthetic, usually mixed with oxygen, and sometimes with ether vapor.

REFERENCES

1. *Chem. Eng.*, **67**, No. 15, 96-99 (1960).
2. Shaner, R. L., "Production of Industrial Gases From Air," Linde Company.
3. "Industrial Gases–Series M28A," U.S. Department of Commerce, Bureau of Census, Washington, D.C., 1970.
4. Kropschat, R. H., Birmingham, B. W., and Mann, D. B., "Technology of Liquid Helium," *Nat. Bur. Stand. U.S., Monograph*, 111 (1968).
5. Neary, R. M., "Air Solidifying Cryogenics," American Society of Mechanical Engineers.
6. Baker, C. R. and Paul, R. S., "Purification of Liquefaction Grade Hydrogen," *Chem. Eng. Progr.*, **59**, No. 8, 61 (1963), and U.S. Patent 3,073,093 (to Union Carbide).
7. Baker, C. R. and Matsch, L. S., "Production and Distribution of Liquid Hydrogen," in "Advances in Petroleum Chemistry and Refining," Vol. X, New York, John Wiley & Sons.
8. "LNG Liquefaction Cycles," *Cryogenic and Industrial Gases*, **5**, No. 8 (1970).
9. Prater, P. G., "Design Considerations of an LNG Peak Shaving Facility," Cryogenic and Industrial Gases, **5**, No. 5 (1970).

SELECTED REFERENCES

"Handbook of Compresses Gases," Compressed Gas Association, New York, Van Nostrand Reinhold.

C. R. Baker, and L. C. Matsch, "Production and Distribution of Liquid Hydrogen."

"Standard Density Data–Atmospheric Gases and Hydrogen," Compressed Gas Association, New York.

Ruhemann, M., "Separation of Gases," International Series of Monographs of Physics, Oxford University Press, 1945.

R. M. Neary, "Handling Cryogenic Fluids," *Nat. Fire Protect. Assoc. Quart.* (Jul. 1960).

"Safety in Air and Ammonia Plants," Symposium, *Chem. Eng. Progr.*, **56**, No. 6, 73 (1960).

H. B. Sargent, "How to Design a Hazard-Free System to Handle Acetylene," *Chem. Eng.*, **64**, No. 2, 250 (1957).

18

Phosphates, Phosphorus; Fertilizers, Potassium Salts, Natural Organic Fertilizers, Urea

J. Q. Hardesty* and L. B. Hein**

PHOSPHORUS AND PHOSPHATES

Phosphorus

The principal source of phosphorus is from natural deposits of calcium phosphate containing fluorine and other impurities in lower concentrations. The basic chemical structure of the rock is that of apatite, which has resisted dissolution for thousands and probably millions of years because of its extremely low solubility. The apatite structure is also found in human teeth, which largely accounts for their durability.

Historically, the source of phosphate was from bones and guano. These sources are no longer important because the wide distribution of small quantities makes collection economically impractical. Also, early in the development of the domestic phosphate industry, phosphate nodules were gathered by hand in South Carolina. The increased cost of labor and the decrease in the availability of nodules has eliminated this source.

*Hyattsville, Md.
**Michigan Technological Univ., Houghton, Mich.

The value of phosphate rock is based on its P_2O_5 content. The trade uses the term Bone Phosphate of Lime (BPL), which means that phosphorus content is expressed as tricalcium phosphate $Ca_3(PO_4)_2$. To convert P_2O_5 to BPL, we use the molecular-weight ratios of the two materials, which is 2.185. The usual grades of commercial Florida rock vary from 68 per cent to 78 per cent BPL. The concentration is occasionally increased by calcining the rock before use to decrease the carbonate and organic contents.

Phosphate Rock Deposits

In 1970 there were four production areas of phosphate ores in the United States. The areas are considerably different in the type of ore present, the method of mining it, and its availability to the United States and world markets through channels of commerce.

Florida Deposits. The most important and greatest production is in an area of north central Florida known as "bone valley." The rock currently mined is of high quality and is easily separated into grades; most of it is white and of high P_2O_5 content. Large additional reserves of lower grade "Black Rock," containing a larger percentage of hydrocarbons, are located adjacent to the main deposits. During recent years this rock has received considerable testing in the wet process for production of phosphoric acid, which is discussed later. The ore is transported throughout the world through Port Tampa; and it is shipped within the United States by rail and by ship to ports on the East Coast, through the Gulf of Mexico to Southern ports, and up the Mississippi River to ports from Louisiana to Illinois.

The phosphate rock matrix, which is sedimentary, occurs in a horizontal layer from 15 to 20 feet deep and is usually located 20 to 30 feet below the surface of the ground. Immediately above the matrix layer is 6 to 8 feet of a material called leached zone, caused by the leaching of phosphate from the material over the ages. The phosphate in the leached zone is principally psuedo wavellite. The concentration of uranium in this material is severalfold that present in the deeper phosphate matrix. Successful processes for recovering the uranium from the leached zone material have been developed, but at present they are not economical compared with those processes which utilize uranium ores. Leached zone is a major reserve of uranium. The overburden which overlies the leached zone is of no value since it is principally quartz and clay.

The phosphates from Florida are used principally for the production of phosphoric acid. The lower grades of phosphate separated during beneficiation are used locally as feed to electric furnaces which produce elemental phosphorus. Small amounts of hard rock are also mined from a deposit which consists of rock rather than the usual pebble ore. The hard rock is used almost exclusively in one of the electric-furnace operations.

Western Deposits. The Western deposits in Idaho and Montana contain the largest-known deposits in the United States. They are in the form of shale and were probably originally laid down in horizontal layers. Subsequent volcanic action has altered the original flat beds to various other shapes, at least one of which resembles a horseshoe with its open end at the surface and its lower, rounded section 3,000 to 4,000 feet below the surface. Much of the phosphate shale is removed by underground mining in contrast to the open-pit mining in Florida. The deposits are thicker than those in Florida, which partially offsets the higher costs of underground mining. One deposit is mined by open-pit mining.

Due to the remote location of the deposits, transportation costs to markets become a major factor. This alone has caused processors to convert the phosphate to elemental phosphorus to obtain economies in transportation. Since the shale contains about 30 per cent P_2O_5 and elemental phosphorus is equivalent to 230 per cent P_2O_5, it is necessary to ship only one ton of phosphorus to market, compared to the 9 tons of shale which would be shipped to a furnace for subsequent processing. (Processing losses are included.) Some of the rock is processed locally by the wet phosphoric acid process. This is used to produce fertilizer for local consumption.

The shale varies in P_2O_5 content throughout its thickness from about 18 per cent P_2O_5 at one edge to 35 per cent at the other. All of the shale has a high organic content, which gives it an oily appearance. When the shale is calcined in preparation for an electric furnace charge, the heat generated by combustion of the hydrocarbons decreases the calciner fuel requirements by about 40 per cent.

Tennessee Deposits. Tennessee brown phosphate matrix at one time was the principal source of phosphate in the United States. The matrix is deposited in pockets, some of which were hundreds of acres in area. The larger beds have been depleted, and the remaining deposits are small and may contain only a few hundred tons or even less of matrix. Most of the mining is currently done by small contractors who supply the four major phosphorus producers in the area. Deposits containing low concentrations of phosphate are known to exist south and west of the principal deposits around Columbia, Tennessee. Currently these cannot be operated economically. The area is so depleted that the most concentrated phosphates are found in the old slime ponds formed by the process wastes deposited years ago.

Three types of phosphates are found in Tennessee. The only one of economic importance is the brown matrix. The brown matrix is bordered with brown silicate sand which contains insufficient phosphate to be processed economically. Although the inexperienced person cannot differentiate between worthless sand and the matrix, the operator of the recovery equipment can select the phosphate matrix and avoid sand without fail. A second type of phosphate is white rock. It also occurs in pockets which are so widely separated and so small that every business attempt to mine the rock has failed for economic reasons. There have been instances where white rock charged to an electric furnace has decrepitated to such an extent that the furnace exploded. The third type of phosphate in Tennessee is called blue rock. The deposits are usually very deep and very thin. They may be 200 to 400 feet below the surface and range in thickness from a few inches to a few feet. Testing of samples taken from outcroppings indicate that it is suitable for processing. Currently it is of no economic importance.

North Carolina Deposit. The North Carolina deposit is located near the Atlantic Coast in swampland. It is mined by creating a lake, floating a barge with a dragline on the lake, and bringing up the phosphate sand from under the water with a dragline. The North Carolina deposit is probably the third largest reserve in the United States, following the Western deposits and Florida, respectively. A processing plant is in operation at the site and the rock is shipped to other locations. Many problems were encountered in its development, including the inflow of salt solutions from underground streams.

Major World Suppliers. Morocco and Florida are major suppliers to European countries as well as other world markets. Uncalcined but beneficiated commercial phosphates have a P_2O_5 content of 28 to 35 per cent except for some lower grades used in the production of elemental phosphorus by the electric-furnace process. An unusual deposit is found in Uganda on the east coast of Africa, but is not a phosphate source at present. The deposit could easily be beneficiated to a P_2O_5 content of 45 per cent and could be processed by a small change in present techniques. The raw material contains enough niobium (columbium) to supply the world's requirement of this important metal if the phosphate ore were produced at about 1 million tons per year. Development of this source has been impractical because of its remote location.

Mining and Beneficiation of Phosphates

Mining. Phosphate mining is usually carried out by dragline. The magnitude of the operation varies tremendously from the huge 100-ton bucket used in Florida to the small 1-ton bucket used in Tennessee. The type of operation is also variable, from the usual dry line used in most fields to the underwater line used in North Carolina. In Florida the newly mined material is sluiced in water to a central point, in Tennessee it is hauled by trucks, and in the West it is either hauled by truck or conveyed by belt.

Beneficiation. Phosphates are usually beneficiated (concentrated) by washing and screening, or by washing and separating phosphates from gangue by use of liquid cyclones. Due to the wide variations in methods, this discussion will be limited to those in use in Florida, probably the most sophisticated methods used in this country.

The flow diagram for the flotation process, as used in Florida, is shown in Fig. 18.1. The process consists of washing and screening, classification, agglomeration, and flotation. The phosphate rock is slurried with water and pumped to a washing plant usually located several miles from the mining operation. The larger particles are screened from the slurry and crushed in a hammer mill. The combination of fines and crushed oversize are washed to remove mud, clay, and other foreign material. The material which passes through a ½-inch screen but is stopped on a 14-mesh screen is removed as product. The −14-

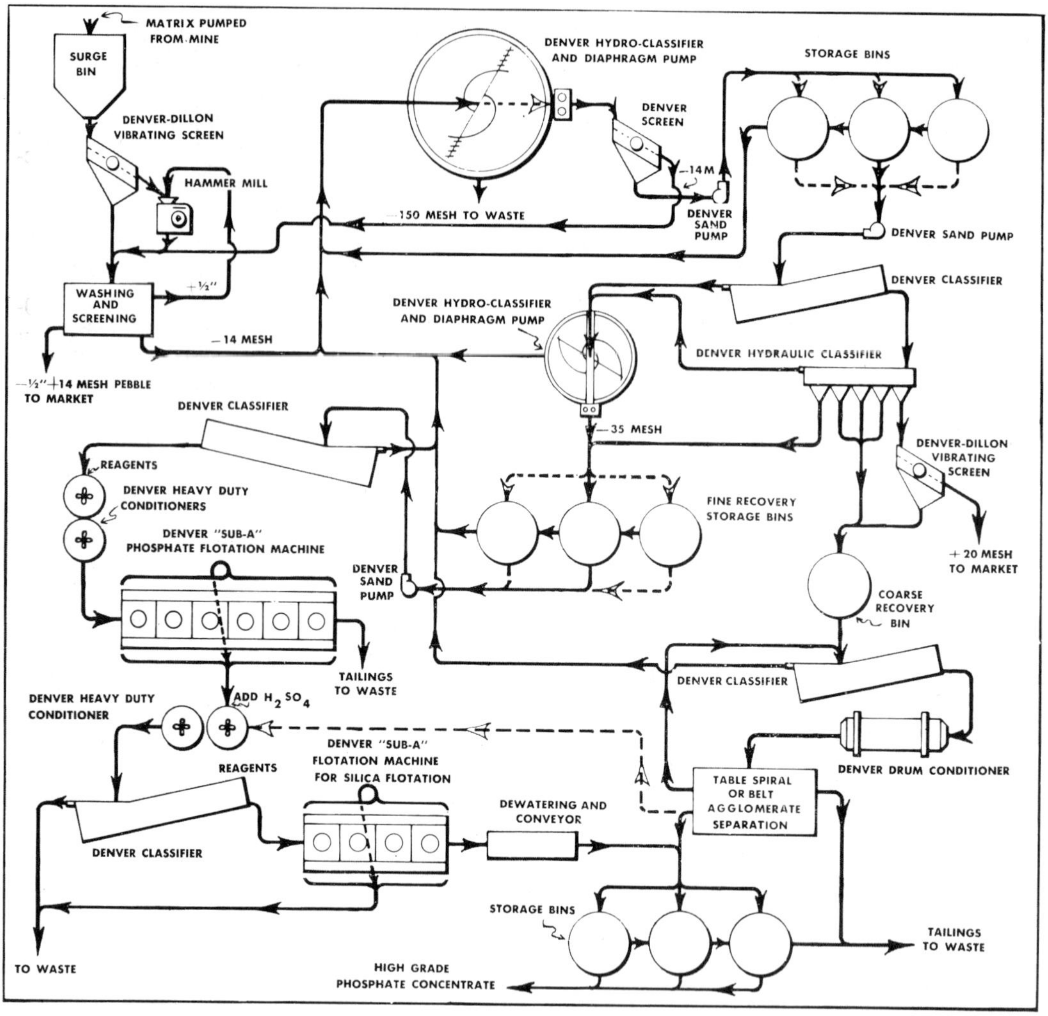

Fig. 18.1 Flow diagram for phosphate flotation process. (*Courtesy Denver Equipment Co.*)

mesh material is passed through a hydroclassifier from which the -150-mesh silica, colloidal clay, and phosphate is separated and moved to a settling pond. The +150, -14-mesh material is then treated in a classifier.

The overflow is passed to another smaller hydroclassifier and the larger-particle bottoms are passed to another larger classifier. The various-sized fractions are then introduced to either the agglomeration or flotation plant. The fraction consisting of -20, +35-mesh particles from the hydroclassifiers is agglomerated to separate the phosphates from unwanted materials. Sodium hydroxide, fuel oil, and fatty acids are added to this fraction. This treatment causes the phosphates to agglomerate or cluster. The separation of phosphate from unwanted material may be made by tabling, spiraling, screening, or a belt-agglomeration treatment. This processing produces a high-grade phosphate concentrate.*

In Florida the phosphate ore is stacked in large piles divided according to P_2O_5 content.

*For the principles involved in the separation equipment used see "Chemical Engineers' Handbook," by J. H. Perry, 4th Edition, 1963, chapter 19, p. 1, McGraw-Hill Publishing Company.

Underground tunnels under the piles lead to a central unit for blending the ores to the desired P_2O_5 content. The material is moved from under the piles to the mixing unit by belt conveyers. Mixing is necessary to make up the grades which are in demand.

In its statistical Abstract of 1970, the Bureau of Census, Department of Commerce reports the following production and values for phosphate rock produced:

States in Which Produced	*1960 Quantity (Short Tons)*	*Value*	*1967 Quantity (Short Tons)*	*Value*
Florida, Idaho Tennessee North Carolina	19,618,000	$117,041,000	39,770,000	$265,947,000

The table shows the growth of the industry and the effect of inflation.

Consumers require different grades of ore because of location and variations in processes. For example, a plant in the Midwest may find that using a 75 per cent BPL ore is more economical than using a 72 per cent BPL ore. Because of the long haul from the mine, transportation cost is greater than that of the ore so that the delivered unit-cost of P_2O_5 depends on concentration. Florida plants can use low grades because of proximity to the mine.

Production of Elemental Phosphorus and Phosphoric Acid

Phosphate rock is converted into usable chemicals by two major methods: (1) the wet acid process which produces an impure phosphoric acid, and (2) the burning of elemental phosphorus to give a pure phosphoric acid suitable for use in food products and for the preparation of pure phosphates. Production of the elemental phosphorus uses an electric furnace. The wet-process acid can be purified to produce pure phosphates.

Electric Furance Process. The electric furnace is used for the reduction of phosphate to elemental phosphorus because it is capable of producing the high temperature needed to reduce the ore and generate phosphorus vapor which is not diluted by other vapors. Figure 18.2 is a flow diagram for the furnace operation. In the past, the combustion of fossil fuels, rather than electricity, provided the heat required; but no such operation is now in domestic use.

The principal reaction in the furnace is the following:

$$12Ca_5F(PO_4)_3 + 43SiO_2 + 90C \longrightarrow 90CO + 20(3CaO \cdot 2SiO_2) + 3SiF_4 + 9P_4$$

The composition of phosphate rock is variable, but the ratio of phosphorus to fluorine is quite constant and approximately that shown in the equation above. Sand is the source of SiO_2 and metallurgical coke usually is the source of the carbon.

Furnaces usually are fed predried raw materials, although an undried feed is occasionally used. Moisture in a furnace can cause violent reactions, but the method of feeding in which hot off-gases pass through the charge in the top of the furnace dries it before it enters the reaction zone. The charge is sized to minimize segregation and some operations carefully drop it into the furnace with the objective of maintaining it in a well-mixed condition.

The furnace is fabricated from a carbon steel shell and cooled on the outside with water. The hearth and lower section of the interior walls are fabricated of monolithic carbon and the upper section and roof are of zircon cement. Power is introduced through three electrodes, and because 40,000 to 60,000 KVA of power is needed, the power source is a three-phase system. Electrodes 48 to 60 inches in diameter are needed to conduct the tremendous quantity of power required. The power flow through each phase is maintained about equal by a complicated system of raising or lowering electrodes to control their contact with the conducting fused mass. Control of power consumption from

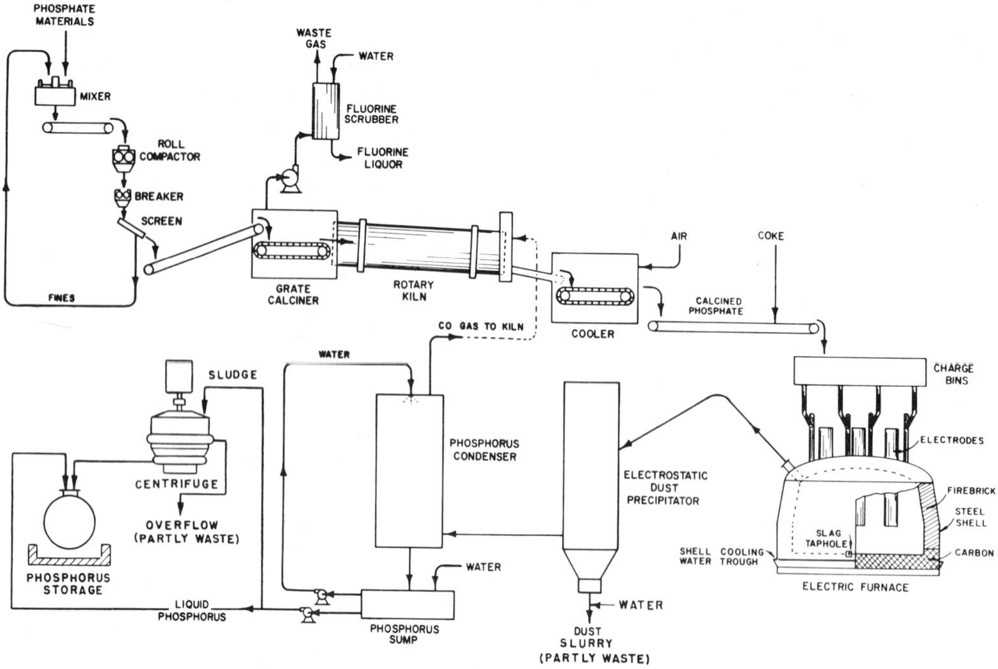

Fig. 18.2 Electric furnace process for production of elemental phosphorus. (*Courtesy Tennessee Valley Authority.*)

each phase is important to maintain a high power factor and thereby a minimum power cost.

Auxiliaries for the furnace include the equipment used to charge the furnace with raw materials, remove solids from the vapors from the furnace, condense phosphorus from the off-gas; and tap molten by-products from the furnace. These are also shown in Fig. 18.2. The charging equipment consists of weight belts, elevators, and charging chutes. The equipment is designed to insure a charge of constant composition to each chute. Solids are removed from the hot gases by use of an electrostatic precipitator consisting of a tube with a wire in the center to provide a positive and negative charge. Solids are electrically charged prior to entering the tubes and are removed by attraction to the opposite charge in the tubes, then emptied through the bottom of the tube. Phosphorus vapor and carbon monoxide pass from the top of the precipitator. The phosphorus is separated from other gases in a condenser equipped with water sprays directed at the gases to cool them. Carbon monoxide passes through the condenser. The liquid phosphorus and water flow by gravity to sumps and the phosphorus settles and is pumped to storage. Phosphorus-containing impurities are separated from pure phosphorus by differences in gravity and are subsequently handled in a separate system.

Phosphoric Acid from Phosphorus. Phosphoric acid is produced by burning elemental phosphorus with air and hydrating the resulting P_2O_5 with water. The chemical reaction is:

$$P_4 + \underset{\text{(air)}}{5O_2} + 6H_2O \longrightarrow 4H_3PO_4$$

Pure elemental phosphorus produces pure phosphoric acid except for the presence of a few metallic elements which originated in the phosphate rock.

Phosphorus is pumped from a storage tank under pressure to a burning nozzle in a vessel designed to atomize it and is mixed with predried air as shown in Fig. 18.3. Predried secondary air for combustion is supplied to the burner. The modern vessel is fabricated from stainless steel and the outside surface is

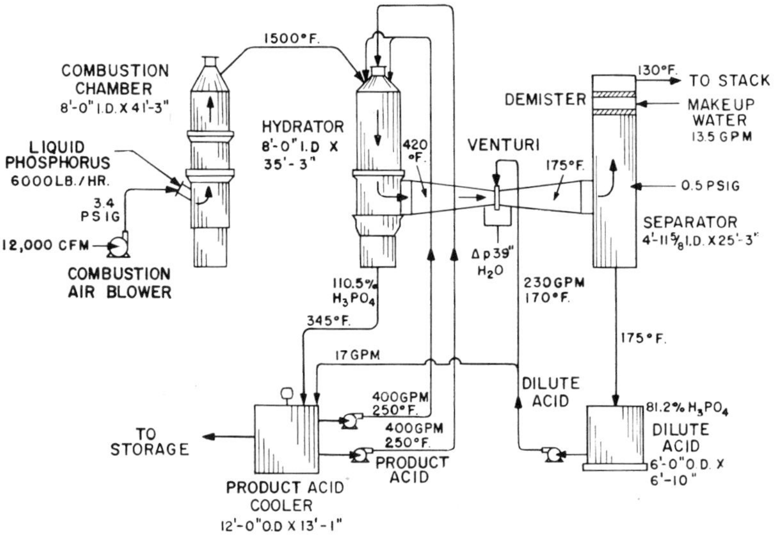

Fig. 18.3 Thermal process for the production of phosphoric acid from phosphorus. (*Courtesy Tennessee Valley Authority.*)

sprayed with water for cooling. Sufficient metaphosphoric acid of high viscosity condenses on the inside surface of the vessel to protect the stainless steel surface from corrosion. Otherwise, stainless steel would not be an adequate material of construction.

It is important that the air for combustion contain very little moisture to prevent the formation of acid mist which is difficult to liquify from the gas stream and which, when not liquified, passes out with the gases. The combustion gases are cooled and then sprayed with water to supply the ingredient needed to convert P_2O_5 to H_3PO_4 in a hydrator ($P_2O_5 + 3H_2O \longrightarrow 2H_3PO_4$). Cooling of the gas is essential to decrease the formation of acid mist. Phosphoric acid is drawn from the bottom of the hydrator. Gases from the hydrator are passed to an electrostatic precipitator or a venturi scrubber to separate the acid mist from the other gases. The combined product from the hydrator and mist collection system is 75 to 85 per cent H_3PO_4.

Superphosphoric Acid. In 1941 pilot plant studies showed that thermal phosphoric acid could be produced at concentrations of over 75 per cent H_3PO_4. In 1956 plant operation for the production of 76 per cent P_2O_5 or the equivalent of 105 per cent H_3PO_4 was demonstrated. The plant test was made in a unit constructed of graphite blocks which were cemented together. A rapid deterioration of the cement joining the blocks was caused by the higher temperatures required in the process. In 1962 the production of superphosphoric acid was demonstrated in a new plant constructed of stainless steel. The objective of the work in the stainless steel plant was to determine whether decreases in maintenance and in losses of P_2O_5 could be realized.

The plant layout was essentially that shown in Fig. 18.3. The changes in design consisted principally of making provisions for adding phosphoric acid to the hydrator and for cooling the recycled acid before it returned to the hydrator.

Phosphoric acid varying from 60 to 83 per cent P_2O_5 can be produced. Changes in concentration are made gradually by changing the flow rate of dilute acid to the hydrator to cause the concentration of product acid to change about 1 to 2 per cent P_2O_5 per hour until the desired concentration is obtained. Temperatures are also controlled.

A summary of heat removal at a phosphorus-burning rate of 3500 to 4300 pounds per hour follows:

Heat Removed in	*Area Cooled By Water (sq ft)*	*Heat Removal by Cooling Water*		
		Total Btu/hr	*Percentage of Input*	*Btu/hr (sq ft)*
Combustion chambers	1311	26,500,000	57.4	20,000
Hydrator	903	6,700,000	14.7	7,400
Product acid cooler	1300	9,100,000	19.7	7,000
Venturi scrubber	157	400,000	0.8	2,300
Separator tower bottom	396	900,000	2.0	2,400
Gas leaving separator	-	2,400,000	5.4	-

Corrosion of equipment used in processing did not present a major problem.

The production of acid higher in concentration than the usual 75 per cent-H_3PO_4 product offers the following advantages:

1. The higher concentration of P_2O_5 reduces transportation costs.
2. The product from ammoniation of superphosphoric acid gives some ammonium polyphosphate, which is a sequestrant. When wet process acid is ammoniated to manufacture liquid fertilizers, the sequestering action of the ammonium polyphosphate decreases the precipitation of solids. Solids in liquid fertilizers are undesirable because they settle in storage tanks and plug orifices in application equipment.
3. Superphosphoric acid is less corrosive than 75 per cent H_3PO_4.

Industrial Phosphates

Phosphorus is essential to all animal and plant life. Phosphate salts play a major role in industrial processing and they have been major additives to detergents for household and industrial use. (They currently face severe restrictions as detergent additives because of their role in water pollution.) The metallic phosphates are an important group of chemicals but the sodium salts are in greater use because they are low cost and are especially effective in many applications. This discussion will be limited to a brief review of those salts which are produced from phosphoric acid and sodium compounds such as sodium carbonate and sodium hydroxide.

Classification. The commercial sodium phosphates might be classified in a general way as (1) orthophosphates, (2) crystalline condensed phosphates, and (3) glassy condensed phosphates. Monosodium and disodium orthophosphates are produced according to the following reactions:

Monosodium phosphate: $$2H_3PO_4 + Na_2CO_3 \longrightarrow 2NaH_2PO_4 + H_2O + CO_2$$

Disodium phosphate: $$H_3PO_4 + Na_2CO_3 \longrightarrow Na_2HPO_4 + H_2O + CO_2$$

Trisodium phosphate (Na_3PO_4) cannot be produced by the addition of more sodium carbonate because the carbonate is not sufficiently basic. Sodium hydroxide must be used. The orthophosphates have a wide range of uses, but the production of condensed phosphates from them greatly extends their usefulness.

When some water is eliminated from the orthophosphates, compounds known as condensed phosphates are formed. Tripolyphosphate is the most important, based on tonnage sold. It is a crystalline material. The chemical reaction is:

$$2Na_2HPO_4 + NaH_2PO_4 \xrightarrow{-2H_2O} \underset{\text{Tripolyphosphate}}{Na_5P_3O_{10}}$$

Although the reaction is a simple one, careful control of the disodium to monosodium phosphate ratio is important, as well as control of temperature and rate of water removal. The chemical engineer must be most cognizant of such details as chemical analysis of raw materials, careful measurement of the amounts charged to a continuous process, and control of conditions. Unless great care of all these factors is taken, optimum results will not be obtained.

Sodium hexametaphosphate is a glassy material. The Na_2O to P_2O_5 ratio can be varied from 1:1 to 0.34:1. Ratios below 1 to 1 are acidic and are unstable in solution so they are not of value. The metaphosphates are produced at high temperature from mono- and disodium phosphates in ratios to give a product of the desired Na_2O to P_2O_5 ratio.

Another crystalline condensed phosphate is tetradosium pyrophosphate. The chemical equation for its preparation is:

$$2Na_2HPO_4 \xrightarrow{-H_2O} \underset{\text{Sodium pyrophosphate}}{Na_4P_2O_7}$$

Uses. Most of the sodium phosphates are used in the sequestration of metallic ions in water. The phosphates are capable of forming water-soluble complexes with these ions. Most water hardness in potable water is caused by the presence of soluble calcium and magnesium compounds. When soaps are used, they form insoluble compounds which precipitate causing grey deposits on clothes and rings in bathtubs and wash basins. Modern detergents are excellent cleaners and the phosphates prevent this precipitation.

One phosphate may sequester more of one cation, such as calcium, than a second. The second may sequester more magnesium than the first. Figures 18.4 and 18.5 show the capability for the various phosphates to sequester certain cations and their effect on pH at different concentrations.

At present the use of phosphates in detergents is under heavy attack because it is claimed that they cause the growth of algae in streams and lakes and thereby adversely affect our ecology. There are two schools of thought; those who believe that phosphates are harmful, and those who do not. Some experts believe that the limiting element for algae growth in our waters is phosphorus, and others believe that it is organic carbon. The solution to this problem was to be the development of nitrilotriacetic acid (NTA) as a replacement for phosphates in detergent formulations. Federal agencies have now shown that when NTA is given to animals in massive doses, it affects reproduction. Whether the massive doses given in order to accelerate effects are too large for realistic evaluation is unknown.

Large sums of money were invested in plants for producing NTA and production started prior to the ban on its use. Production equipment for NTA is being modified for

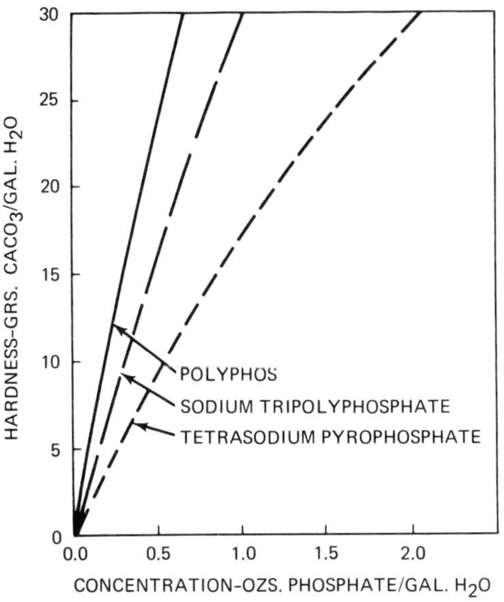

Fig. 18.4 Sequestration of calcium carbonate at 150°F and pH 10. (*Courtesy Olin Corp.*)

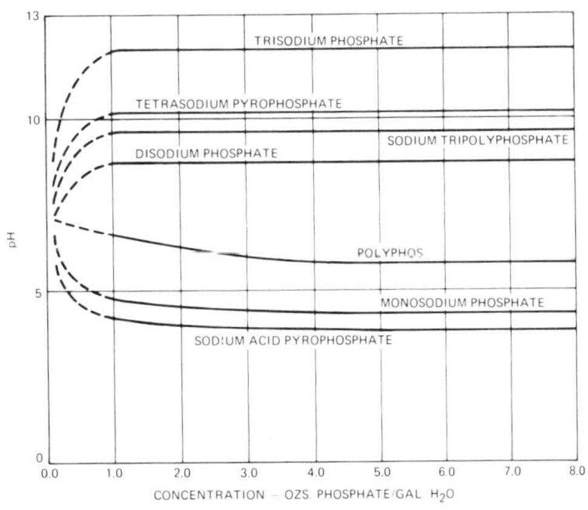

Fig. 18.5 pH of phosphate solutions at 75°F. (*Courtesy Olin Corp.*)

other uses. An eminent chemist said in early 1971 that pollution is of three catagories: (1) actual, (2) political, or (3) emotional. Because of this confusion it is not possible to make a factual presentation at this time.

Application to Water Softening. The principal use of condensed phosphates is in the sequestering of hardness in water. Figures 18.4 and 18.5 show the concentrations of polyphos (hexametaphosphate), sodium tripolyphosphate, and tetrasodium polyphosphate required to sequester calcium carbonate up to 30 grains per gallon of water at 150°F and a pH of 10. The data show that only 0.7 to 2.0 ounces of the various phosphates are required per gallon of water to sequester 30 grains of calcium carbonate.

Wet-Process Phosphoric Acid

Another method of preparing phosphoric acid is by the direct extraction of P_2O_5 from phosphate rock with sulfuric acid. The process consists of substituting the phosphate group in the rock with sulfate. The simplified reaction, ignoring the fluorine in the phosphate rock, is:

$$Ca_3(PO_4)_2 + 3H_2SO_4 + 6H_2O \longrightarrow 3(CaSO_4 \cdot 2H_2O) + 2H_3PO_4$$

The reaction is simple, but augmenting it is complicated. It is not sulfuric acid that extracts the phosphate but monocalcium phosphate $[CaH_4(PO_4)_2 \cdot H_2O]$. When sulfuric acid attacks phosphate rock, an impervious layer of calcium sulfate is formed over the surface of the rock. This layer prevents further reaction. It is, therefore, important to prevent high local concentrations of sulfuric acid in the slurry.

A flow diagram of the process is shown in Fig. 18.6. The first major step after taking the rock from storage is grinding. This may be done in various types of mills, a ball mill being commonly used. The fineness of grind desired depends on the type of rock but is usually controlled to 65 to 75 per cent through a 200-mesh sieve for 68 to 75 BPL* Florida rock. Florida black rock contains about 0.8

*BPL means Bone Phosphate of Lime ($Ca_3(PO_4)_2$), and 75 BPL means that 100 pounds of rock contains the equivalent of 75 pounds of calcium phosphate. The notation is a historical carry-over, but it is still used to designate the concentration of phosphorus by producers and consumers. The notation used by process engineers usually is based on the P_2O_5 content of the rock and intermediates during processing. A P_2O_5 content of 1 per cent is equivalent to 2.185 per cent BPL. Another notation is the elemental phosphorus content of the material. A P_2O_5 content of 1 per cent is equivalent to 0.0435 per cent P.

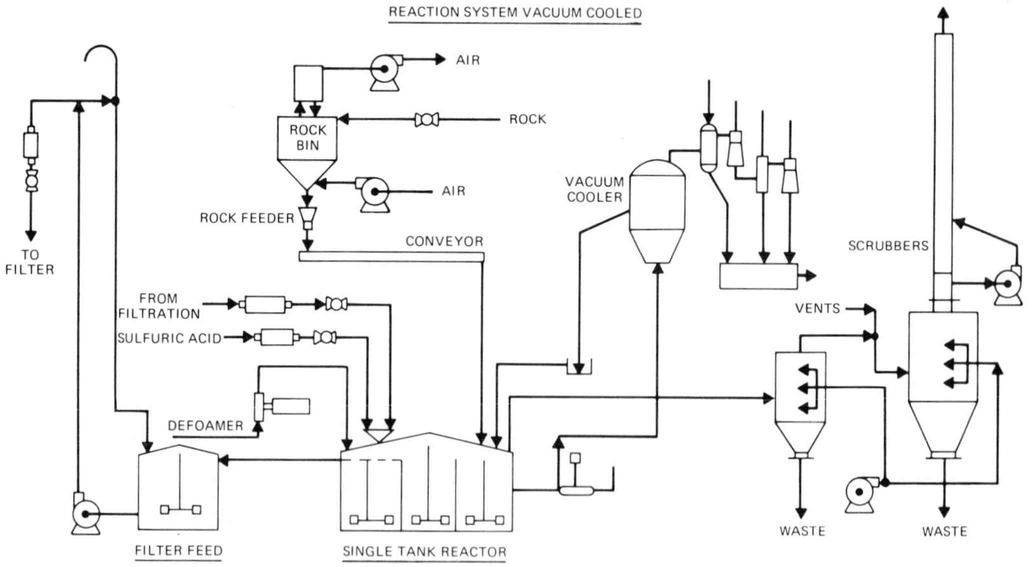

Fig. 18.6 Reaction system for wet-process phosphoric acid process. (*Courtesy Dorr-Oliver.*)

per cent hydrocarbons and is more dense and less reactive than white rock. It requires finer grinding and is beneficiated to 68 per cent BPL. Black rock is not extensively used at present but provides a large reserve for the future.

The ground rock is weighed continuously and introduced into the first-stage reactor. Sulfuric acid (often 93 per cent H_2SO_4) is usually introduced to the same reactor. The conditions of mixing are extremely important in order to prevent contact between the concentrated sulfuric acid and the raw rock. The method of control varies from plant to plant but often consists of mixing the rock with recycled weak phosphoric acid and recycled slurry, followed by the introduction of sulfuric acid to the three-component mixture.

The materials then pass successively by gravity through a series of reactors. The temperature in the reactors is controlled by blowing air into them or by cooling the recycled slurry under vacuum in a separate flash cooler. The amount of slurry recycled varies from zero to a very large ratio of tons of P_2O_5 per ton of P_2O_5 introduced as rock. Recently, extremely high recycle ratios well above those usually in use have been shown to greatly increase plant capacity without decreasing extraction efficiency or product-acid concentration.

The slurry, except for that which is recycled from the last reactor, passes to a filter-feed tank from where it is introduced to the filter at a controlled rate. Modern filters are specially designed for the separation of gypsum from acid, proper washing of the cake to remove a maximum amount of acid, dumping the cake, and washing the filter cloth with water from underneath to remove cake from the perforations in the cloth.

Figure 18.7 is a flow diagram for the filtration of wet-process phosphoric acid. The cake, practically depleted of soluble P_2O_5, is finally washed with water at the end of the filter. Then in successive steps to the left and toward the front of the filter, washes are made in increasing concentrations of phosphoric acid. Most modern filters are divided into separate sections which are containers with their own filter cloths. Each section is under vacuum to pull liquid through the cake. The liquid is either recycled back to the cake on the filter, or to the reaction system, or separated as product acid.

The product acid contains from 28 to 35 per cent P_2O_5. The product is impure and precipitates form for several months. If used within a short time for the production of fertilizers, the potential precipitates decrease the P_2O_5 content of the product only slightly.

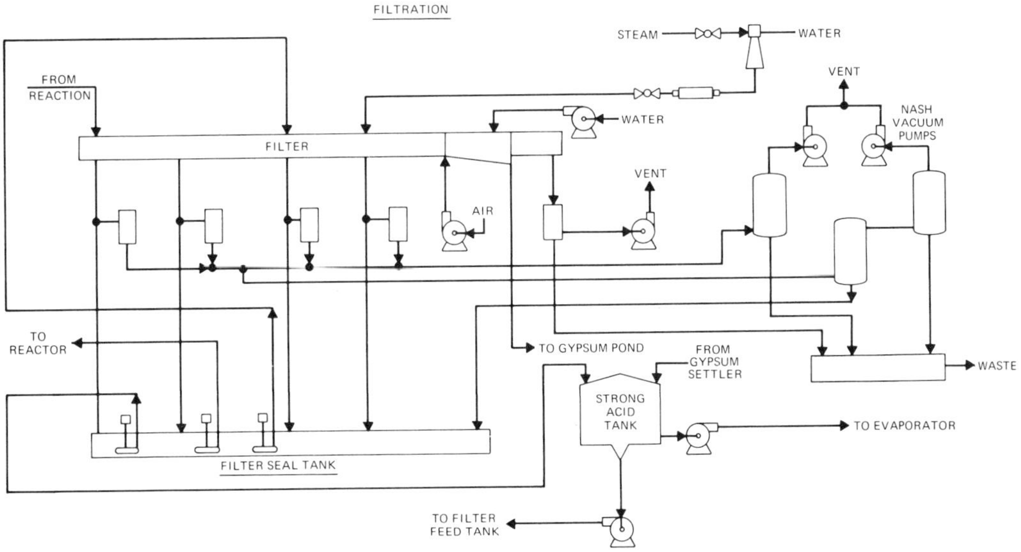

Fig. 18.7 Filtration section of wet-process phosphoric acid process. (*Courtesy Dorr-Oliver.*)

Some wet-process phosphoric acid is used for the production of ammonium phosphate or liquid fertilizers. In such uses it must be concentrated to about 40 per cent P_2O_5 for production of ammonium phosphates or 54 per cent P_2O_5 for liquid fertilizers. Concentration is usually made in a single-effect evaporator heated by steam with auxiliaries to control the formation of scales containing calcium sulfate. An evaporator may be in operation for about three days followed with a water wash of about one day. Unless some such schedule is followed, the tubes fill with scale to the extent that heat transfer is seriously decreased and cleaning becomes difficult. Accelerated cleaning is possible by washing the inside of the tubes with water at high pressure through specially designed nozzles. Sequestrants such as ethylenediaminetriacetic acid are also used as a wash, but this method is expensive.

POTASSIUM SALTS

Potassium salts are essential to plant growth and particularly needed for the formation of seeds.

Soluble Potassium Salts

European Deposits. The first deposit of soluble potassium salts was discovered in the area around Stassfurt, Germany. The deposit supplied nearly all of the world's requirements until 1914. After World War I deposits in Alsace were developed and became another important source.

The deposits near Stassfurt are located about 1,000 feet below ground level and consist of three distinct layers. The upper layer is principally carnallite ($MgCl_2 \cdot KCl \cdot 6H_2O$), while polyhalite ($2CaSO_4 \cdot MgSO_4 \cdot K_2SO_4 \cdot 2H_2O$) lies below the carnallite, and kainite ($MgSO_4 \cdot KCl \cdot 3H_2O$) lies below the polyhalite.

The Alsatian deposits lie 1,500 feet below the surface in two layers, both of which are principally sylvinite ($KCl \cdot NaCl$). Both the Stassfurt and Alsace deposits are crystallized from solution. Part of the material is purified to 98 per cent KCl (61 per cent K_2O) and some is sold at lower concentrations.

U.S. Deposits. Production was started in the Carlsbad, New Mexico area in 1961. This was the first production of sizable tonnages in the United States. The minerals are found in depths between 800 and 1,800 feet below ground level. Deeper deposits occur in nearby counties in Texas. Many layers exist with a total cumulative thickness of about 36 feet. The potassium minerals exist as sylvinite, kainite, and polyhalite. Potassium chloride is recovered by crystallization, and the production of some potassium sulfate provides a valuable by-product, because when applied to the soil, it supplies potassium without the introduction of chloride. The deposits near Carlsbad are the largest within the continental United States.

Potassium Salts from Brines. A unique approach to production of potassium from brines is exemplified by the processing of the solution from Searles Lake, located near the California-Nevada border in a dry area. The composition of the Searles Lake brine is given below.

Component	*% by Weight*
NaCl	16.35
Na_2SO_4	6.96
KCl	4.75
Na_2CO_3	4.74
$Na_2B_4O_7$	1.51
Na_3PO_4	0.155
NaBr	0.109
Miscellaneous	0.076
TOTAL solids	34.65

Potassium chloride, borax, sodium carbonate, sodium sulfate, lithium phosphate, sodium phosphate, potassium sulfate, boric acid, and bromine are produced from the brine. As should be expected, the process is complicated and consists principally of evaporations and filtrations of material in process. Development has continued since World War I when the first effort was made to recover potassium chloride from the brine.

According to the Bureau of Census, U.S. Department of Commerce, the following tonnage and evaluation was placed on the production and value of potassium salts in the United States for 1960 and 1967:

Source	*1960 Quantity Short Tons (K_2O Equivalent)*	*Value*	*1967 Quantity*	*Value*
New Mexico, Utah California and Michigan	2,638,000	$89,676,000	3,299,000	$105,313,000

Canadian Deposits. Work towards mining potash ore in Canada was begun during the mid-1950's.[2] The deposit, some 300 miles long and 10 miles across, is found largely at 3,000 to 3,500 feet below the surface, but in some cases it extends to 5,000 feet. Difficulties were anticipated partially because of the depth, but mostly because of water-logged sands existing below the 1,250-foot level. The sands, known as the Blairmore, are 200 to 300 feet in depth and presented a major obstacle because of this fluid consistency and the troublesome water that flowed into the mine below the Blairmore level. The first mining operation was carried out in 1956, but during the same year seepage of water into the shaft at the Blairmore layer made shutdown necessary for grouting and shaft repair. It was not until 1967 that the mine was back in production. In the meantime, another mine began to operate, and by 1967 four mines producing potassium chloride were in operation. In 1965 it was predicted that 25 per cent of all potash produced in the world by 1970 would come from Canada and 33 per cent would be produced by 1975. Actually, a world-wide oversupply of potash in 1968 caused a 40 per cent decrease in the selling price and destroyed any

Fig. 18.8 Surface structures at a Canadian potash mine. The shaft house is in the background, and processing and storage buildings are to its left. *(Courtesy Potash Company of America.)*

Fig. 18.9 Movement of ore in the mine to the shaft at a Canadian potash mine. (*Courtesy Potash Company of America.*)

incentive to expand production. Companies committed to new production cancelled plans and began efforts to merge with each other.

Figure 18.8 shows the surface installation at one of the mines. The shaft house is in the background in the photograph, processing and storage buildings are to its left. Figure 18.9 shows the movement of ore in the mine to the bottom of the shaft. The picture gives some idea of the size of the operation.

Another company started a mine in 1957; but because of problems in the Blairmore water sands, production was not achieved until 1962. Water was retained outside of the mine by installing a series of cast iron rings to seal off the Blairmore. Another operation in the area consists of solution mining. In this case a shaft is not needed because a solution is pumped down to the potash ore where it becomes more concentrated in potassium chloride and is returned to the surface for refining. The solution is recycled, but details of of the process have not been released.

Potassium Nitrate

Potassium nitrate has been produced historically either through the reaction of potassium chloride and sodium nitrate or reclaimed from natural deposits. The chemical reaction is:

$$NaNO_3 + KCl \longrightarrow KNO_3 + NaCl$$

The NaCl formed is crystallized and filtered from the hot solution. The KNO_3 remaining in the dilute solution is crystallized in standard processing equipment and dried. Since the raw-material costs are high, the process has limited commercial value. Production as practiced in Chile consists of leaching residues from the sodium nitrate operation to produce a weak brine of potassium nitrate which is concentrated by solar evaporation. When the proper concentration of potassium nitrate is achieved it is recovered by crystallization.

A process for production of potassium nitrate from potassium chloride and nitric acid after several years of development was placed in commercial production in the United States in 1963. The over-all process is represented by the following equation:

$$2KCl + 2HNO_3 + \tfrac{1}{2}O_2 \longrightarrow 2KNO_3 + Cl_2 + H_2O$$

It is rather complicated and details cannot be given here, but the key to its success is the oxidation of nitrosyl chloride according to the reaction:

$$NOCl + 2HNO_3 \longrightarrow 3NO_2 + \tfrac{1}{2}Cl_2 + H_2O$$

The nitrogen dioxide product is converted to nitric acid for recycle in the process. The corrosive nature of various intermediates requires the use of costly materials such as inconel, stainless steel, and titanium. The process has been described in detail by Spealman.[3]

The objective of the process is to produce potassium nitrate, which contains two of the three major plant-food elements (all except phosphorus), and eliminates the chloride ion, which is undesirable for some applications. Another product, chlorine, adds to the economics of the process. The premium price of potassium nitrate and the value of chlorine apparently compensate for the high capital and operating cost of the process.

THE FERTILIZER INDUSTRY

History

The fertilizer industry in the United States began during the middle years of the 19th century when Justus von Liebig (1803-1873) in Germany was teaching the dependence of vegetation on the mineral components of the soil. The first American edition of Liebig's book, entitled "Organic Chemistry in its Application to Agriculture and Physiology," appeared in 1841. Materials such as animal manures, wood ashes, bones, fish, guano, wool waste, chalk, and marl had been recognized as aids to plant growth in earlier times, but now they began to take on added significance because of the newly acquired knowledge that their agricultural value was due to their content of nitrogen, phosphorus, potassium, calcium, and other chemical elements.

It gradually became known that the phosphorus in bones and other natural phosphates (phosphate rock) could be made more available to crops by treatment with an equal weight of 60 per cent sulfuric acid, resulting in a product called "acid phosphate" or "superphosphate." James Murray (1788-1871) of Ireland and Sir John Bennet Lawes (1814-1900) of England each obtained English patents on the production of superphosphate on the same day, May 23, 1842.[4] In the United States, Philip S. and William H. Chappell of Baltimore, Maryland, obtained the first U.S. patent for mixed fertilizer on May 27, 1849. Baltimore rapidly became a manufacturing center for superphosphate fertilizers and the industry soon spread to other principal cities of the Atlantic seaboard.

In 1856, Samuel W. Johnson (1830-1909) returned to the United States from Europe, where he had studied under Liebig. Thereafter he contributed greatly to the technology and use of fertilizers in this country. He was one of the foremost leaders in the establishment of State Experimental Stations and the testing of the fertilizer industry's products. The search was on for more concentrated sources of the chemical elements essential to plant growth. Phosphate-rock deposits were opened in South Carolina in 1867, Florida in 1887, Tennessee in 1894, Idaho and Wyoming in 1906, Montana in 1921, and more recently, North Carolina in 1965. The first imports of potassium products, referred to in the trade as "potash," from Germany began in 1870. The last great barriers to our domestic sufficiency in the basic fertilizer elements were overcome with successful operation of the first U.S. plant for the direct synthesis of nitrogen from the atmosphere at Syracuse, N.Y., in August 1921 and the opening of vast deposits of potassium ores at Carlsbad, N.M., in 1931.

The average plant-food content (nitrogen, phosphoric oxide, and potassium oxide) of mixed fertilizers rose from 13 per cent in 1880, to 15 per cent in 1910, and to 20 per cent in 1940, but was in the neighborhood of

40 per cent in 1970. U.S. consumption of nitrogen, phosphoric oxide, and potassium oxide in all fertilizers has increased almost nine-fold in the last 30 years—from 1.8 million tons in 1940 to nearly 16.0 million tons in fiscal 1970. This increase in domestic fertilizer demand, as well as demand for more food by an increasing world population, has resulted in wide expansion of the fertilizer industry aided by a number of significant technological developments.

Following the advent of synthetic-nitrogen production in 1921, high-analysis synthetic nitrogen compounds largely replaced the low-analysis organic materials, such as cottonseed meal, fish scrap, etc., previously used as the chief sources of fertilizer nitrogen. By 1935, ammoniation of superphosphate fertilizers with low-pressure nitrogen solutions was well established, solid urea was introduced, and granulation of superphosphate to improve its physical condition became a commercial accomplishment.[5] During World War II, granular ammonium nitrate came on the market in substantial quantities. In 1945, the introduction of the cone mixer for the continuous acidulation of phosphate rock signaled a rapid rise in the production of concentrated superphosphate. Currently, there was initiated a rapid rise in the production of phosphoric acid and the ammonium phosphates.

Since 1950, there has been a phenomenal rise in the production of granular fertilizers, superphosphoric acid, and ammonium polyphosphates, and in liquid, slurry, and suspension fertilizers, bulk-blends, specialty grades, and slow-release nitrogen materials. The requirements of crops for small amounts of the micronutrient elements consisting of boron, copper, iron, manganese, molybdenum, and zinc, are widely recognized, as indicated by the domestic use of over 36,000 tons of these elements in fiscal 1970. There has been a steady rise in the production capacity of Canadian potash mines since the Saskatchewan field was opened in 1962. High-grade potassium nitrate has been in domestic production since 1963.

Early in the 1960's, supply was lagging behind demand for fertilizers in widely separated areas of the industry's operations. Mergers and acquisitions occurred between petroleum companies, mining companies, and fertilizer manufacturers to form large complex fertilizer establishments holding captive sources of the three primary plant-nutrient elements, nitrogen, phosphorus, and potassium. These combinations of previously separate operations apparently fortified the fertilizer industry with sufficient capital and know-how to meet any anticipated fertilizer demand that is surrounded by a favorable climate for investment.

The expansion in production, combined with a decrease in expected exports to the Emerging Nations and poor planting seasons in the U.S. owing to abnormally wet springs in 1966, '67, and '68, caused supplies to surge ahead of demand, and there was consequent erosion of market prices for fertilizer. Such conditions of oversupply tend to force smaller companies out of business and to compel the large complex companies to initiate either curtailment of production or general retrenchment and reduction in operating and production costs of large-scale operations to permit lower profit per unit of plant food. In the case of oversupply of Saskatchewan potash, the Canadian Government handled the situation by prorated restrictions on production and regulation of the minimum price per unit of potassium oxide. As of this year, 1973, the supply and demand for fertilizers is gradually being brought into balance, and the total world consumption of nitrogen, phosphoric oxide, and potassium oxide is estimated to increase from 59.5 million metric tons in 1969 to more than 180 million metric tons in the year 2000.

Plant Nutrient Elements

Sixteen chemical elements are known to be essential to the normal development and growth of plants. The percentage of these elements in the dry matter of the plant is known to vary widely with different species of plants, and with different soil and climate conditions under which plants are grown. However, the order of magnitude expressed as the average percentage of each element present in the dry weight of the whole plant[6] may be represented roughly as follows:

Element	*Amount in Whole Plant % (Dry Weight)*
Oxygen	45
Carbon	44
Hydrogen	6
Nitrogen	2
Phosphorus	0.5
Potassium	1.0
Calcium	0.6
Magnesium	0.3
Sulfur	0.4
Boron	0.005
Chlorine	0.015
Copper	0.001
Iron	0.020
Manganese	0.050
Molybdenum	0.0001
Zinc	0.0100
TOTAL	99.9011

Oxygen, carbon, and hydrogen make up the plant structure and comprise about 95 per cent of the elemental content of the plant. The plant obtains them from the atmosphere and water. The remaining 13 essential elements in plants are supplied in large part by soils and fertilizers. They comprise only about 4.9 per cent of the elemental content of the plant—3.5 per cent *primary nutrient elements*: nitrogen, phosphorus, and potassium—1.3 per cent *secondary nutrient elements*: calcium, magnesium, and sulfur—and about 0.1 per cent *micronutrient elements*, so-called because they are required in very small amounts; boron, chlorine, copper, iron, manganese, molybdenum, and zinc. Small amounts of aluminum, cobalt, fluorine, iodine, silicon, sodium, and many other elements are present in soil-grown plants, but they are not commonly considered essential to plant growth. Cobalt may be essential to some plants, and sodium is known to have a beneficial effect on plant growth, especially in potassium-deficient plants. The others are not known to be essential to plant growth, but cobalt, fluorine, and iodine are indispensable to animal nutrition.

Fertilizer Materials

The principal fertilizer materials (Table 18.1) vary widely in physical and chemical characteristics but most of them are suitable for either direct application to the soil or processing in various ways to produce mixed fertilizers, such as bulk or bagged blends of solids, chemically combined solids, clear liquids, and slurries or suspensions of solids in liquids.

Nearly 40 million short tons of fertilizer materials were consumed in the United States in fiscal 1970. About 47 per cent of the total tonnage was used for direct application to the soil, and about 53 per cent was used in the manufacture of mixed fertilizers. The primary nutrient content, calculated as nitrogen, phosphoric oxide, and potassium oxide, averaged 39.4 per cent of the total tonnage of all materials and mixtures used.[7]

The list in Table 18.1 by no means covers the vast number of materials that are suitable for use in promoting plant growth. The list might have been expanded to include a wide array of low-analysis seed and hull meals and ashes, and a variety of animal and vegetable wastes in the form of composts, manures, tankages, and industrial by-products that are low in nutrient content but may be used to make fertilizers near the source of supply, where transportation cost is not a large factor. These by-products include such materials as humates, dried blood, boneblack, coprolites, feather-, felt- and wool-wastes, leather scrap, cement flue dust, distillery waste, kainite, manure salts, kelp, and tobacco stems. Other materials of varying degrees of importance are ammonium, calcium, and potassium metaphosphates, calcium nitrate-urea, and potassium carbonate. High-analysis materials that have potential for future use in fertilizers are potassium hydroxide, potassium phosphates, and potassium ammonium phosphates.

Direct Application Materials. About 93 per cent of the fertilizer materials currently used for direct application is accounted for in eight types of chemical products: anhydrous and aqua ammonia, 24 per cent; nitrogen solutions, 18 per cent; ammonium nitrate, 17 per cent; potassium chloride, 13 per cent; superphosphates, 9 per cent; ammonium sulfate, 5 per cent; ammonium phosphates, 4 per cent; and urea, 3 per cent. The remaining 7 per cent consists of natural organic materials, phosphate rock, sodium nitrate, and a few specialty materials.[7]

TABLE 18.1 Principal Fertilizer Materials

Nitrogen

Material	*Nutrient, Per Cent in Typical Products*
Inorganic	
Anhydrous ammonia	N, 82
Aqua ammonia	N, 20–24
Ammonium chloride	N, 26
Ammonium nitrate	N, 33–35
Ammonium nitrate-limestone	N, 21; Ca, 8; Mg, 5
Ammonium nitrate-sulfate	N, 26; S, 13.5
Ammonium nitrate-phosphates	N, 20–30; P_2O_5, 10–25; S, 1–4
Ammonium sulfate	N, 20–21; S, 24
Calcium nitrate	N, 17; Ca, 24
Nitric acid, 60%	N, 13
Nitric phosphates	N, 20–22; P_2O_5, 20–22
Nitrogen solutions	N, 28–58
Sodium nitrate	N, 16
Synthetic Organic	
Calcium cyanamide	N, 21–22; Ca, 40
Urea	N, 45
Urea ammonium phosphates	N, 27–30; P_2O_5, 10–14
Urea-formaldehyde reaction products	N, 38
Urea-sulfur	N, 40; S, 10
Natural Organic	
Castor pomace	N, 4–7
Cottonseed meal	N, 5–7
Fish, scrap, meals, and emulsions	N, 6–10; P_2O_5, 5–8; Ca, 6
Guanos and manures	N, 1–12; P_2O_5, 10–25; K_2O, 0–4
Sewage sludge, processed	N, 5–7; P_2O_5, 2–4
Tankages, animal and garbage	N, 3–13; P_2O_5, 4–18; K_2O, 0–4; Ca, 1–11

Phosphorus

Material	*Nutrient, Per Cent in Typical Products*
Ammonium phosphates	N, 11–21; P_2O_5, 20–53; S, 16
Ammonium polyphosphates	N, 10–15; P_2O_5, 34–61
Basic slag	P_2O_5, 8–18; Ca, 32; Mg, 3
Bone meals, processed	N, 1–5; P_2O_5, 22–30; Ca, 24
Nitric phosphate-ammonium polyphosphate	N, 10–22; P_2O_5, 10–22; Ca, 5–10
Phosphoric acid	P_2O_5, 54; S, 2
Superphosphoric acid	P_2O_5, 68–80
Superphosphate, normal	P_2O_5, 19–22; Ca, 16–21; S, 12–13
Superphosphate, enriched	P_2O_5, 22–40; Ca, 11–19; S, 1–11
Superphosphate, concentrated	P_2O_5, 40–54; Ca, 10–14
Urea-ammonium polyphosphates	N, 21–28; P_2O_5, 28–42

Potassium

Material	*Nutrient, Per Cent in Typical Products*
Potassium chloride	K_2O, 50–62; Cl, 47
Potassium nitrate	K_2O, 44; N, 13
Potassium sulfate	K_2O, 50; S, 18
Sulfate of potash magnesia	K_2O, 21–27; Mg, 7–11; S, 18–22

In 1953, the domestic consumption of materials applied directly was only 33 per cent of the total fertilizer consumption, but the proportion has gradually increased to more than 47 per cent in fiscal 1970, largely owing to the very popular technical innovation of applying anhydrous ammonia directly to the soil. This practice, introduced in 1947, consists in placing the ammonia 3 to 6 inches beneath the soil surface through a delivery tube extending from the supply tank on the applicator to the bottom of the furrow behind a narrow plowshare. The ammonia gas is immediately covered as the soil falls back into place and is rapidly absorbed by the soil particles. Other liquids, such as nitrogen solutions, also are applied in this manner. Proper metering devices and pumps in the delivery line ensure uniform application. Nearly 3.5 million short tons of anhydrous ammonia and more than 3 million tons of nitrogen solutions were used for direct application in fiscal 1970.

Mixed Fertilizer Materials. About 53 per cent of all commercial fertilizer used in the United States in fiscal 1970 consisted of processed mixtures of individual fertilizer materials[7] to give products having balanced ratios of plant nutrients designed to meet the needs of various soils and crops. They have the advantage of fulfilling the crop's nutrient requirements in one application, rather than in multiple applications where individual materials are applied separately to provide specific amounts of more than one nutrient. Mixed fertilizers are frequently referred to as multiple, multinutrient, or compound fertilizers.

The first mixed fertilizers known to have been produced commercially in the United States in the early 1850's were simple mixtures of Peruvian guano, by-product ammonium sulfate, probably from the illuminating gas industry, and superphosphate made from the treatment of bones with sulfuric acid. By 1880, typical mixtures contained nitrogen as fish scrap or Chilean nitrate of soda, phosphoric oxide as superphosphate, usually produced from bones or from domestic phosphate rock first produced in South Carolina in 1867, and potassium oxide as wood ashes or a low-grade potash salt called "Kainit."

The total plant-nutrient content of mixtures in 1880 averaged 13.4 per cent, and continued to average below 16 per cent until 1925.[8] Early in this period, the waste nitrogenous materials, known in the trade as "natural organics" or "organic ammoniates" (Table 18.1), furnished much of the nitrogen in mixtures. The use of seed meals in mixtures doubled during the period between 1900 and 1909. Animal products, including animal tankage, dried blood, and fish scrap, generally among the most highly regarded of the natural organic materials, increased about 50 per cent in volume between the years 1900 and 1919.[9] Subsequently, the proportion of waste nitrogenous materials in mixed fertilizers declined, owing to utilization of the better ones in animal feeds where they commanded a higher price, and to increased supplies of cheaper chemical nitrogen materials.

Processed municipal sewage wastes and vegetable and animal tankages have been, for a long time, special sources of organic matter and some plant nutrients. With continuation of the present public interest in cleaning the environment, it is not altogether unlikely that a segment of the fertilizer industry in the future will develop economical methods of making and distributing greater quantities of organic manures produced from vegetable and animal wastes. The market for these products would not supplant the tremendous need for more chemical fertilizers as world populations increase. Organics are valuable as soil conditioners but they cannot supply the quantities of nutrients required by farm crops.

Nitrogen Sources. Production of by-product ammonium sulfate by coke producers was initiated about 1893 and the process was improved in 1910. During the rise and decline in the proportion of waste nitrogenous materials used in fertilizers, the production of by-product ammonium sulfate, mostly from the coke oven industry, increased from 6,500 short tons in 1901 to 744,200 tons in 1929.[8] In recent years, much by-product ammonium sulfate has been made from synthetic ammonia and waste sulfuric acid from other industries, such as the petroleum industry. By the early 1950's the production of the synthetic product had exceeded that of the by-product ammonium sulfate from the coke oven industry. Ammonium sulfate from either source remains an important material in the manufacture of mixed fertilizers. Total annual production presently (1970) is nearing 3 million short tons.

Calcium cyanamide, produced at Niagara Falls, Ontario, Canada, since 1910, was an important nitrogenous material for mixed fertilizers in the United States during the period 1910 to 1932, but the few thousand tons now used are mostly for direct application.

The first domestic production of synthetic ammonia directly from the atmosphere in 1921 stimulated intensive research aimed at the utilization of ammonia and synthetic nitrogen compounds, such as sodium nitrate, ammonium nitrate, and urea. Ammonia was, and still is, the most economical source of nitrogen in fertilizers.

Annual domestic production of anhydrous ammonia presently (1970) is over 13 million short tons as compared with production of about 5 million tons in 1960. Harre[10] estimates that, by 1972, North American gross nitrogen capacity will be nearly 16 million metric tons as compared with a world capacity of over 60 million metric tons. The world fleet of large ammonia tankers is growing rapidly and long-distance pipelines are transporting large tonnages of ammonia to the market areas. The world consumption of fertilizer nitrogen has tripled during the period 1958 to 1970. Fertilizer-nitrogen consumption is about 85 per cent of the total nitrogen consumption for all purposes.[11]

Domestic manufacture of synthetic sodium nitrate began at Hopewell, Virginia, in 1929, first by the reaction of sodium carbonate with nitric acid, and later by the reaction of sodium chloride with nitric acid. Chilean sodium nitrate of natural origin has been an important fertilizer in the United States for over 100 years, and, along with the domestic synthetic product, was formerly used in mixed fertilizers, but both are now used largely for direct application.

Ammonia reactions with superphosphate had been recognized for a long time, but it was not until 1928 that a commercial process was introduced in the United States for fixing anhydrous ammonia in superphosphate and fertilizer mixtures containing superphosphate. This was followed, in 1932, by the introduction of low-pressure nitrogen solutions for the same purpose.

A typical nitrogen solution is characterized by a code number, for example, 414(19-66-6) indicating that it contains 41.4 per cent total nitrogen, 19 per cent free ammonia, 66 per cent ammonium nitrate, 6 per cent urea, and, by difference, 9 per cent water. The solids dissolved in this particular solution reduce its vapor pressure to 11 psig at a temperature of 104°F as compared to 197 psig for anhydrous ammonia at the same temperature. Owing to the low vapor pressure, the nitrogen solution is more easily transported and handled than anhydrous ammonia. More than one hundred nitrogen solutions of different composition are marketed to fit different requirements of mixed fertilizer manufacture and direct application to the soil. Representative types are: 410(22-66-0) 12 per cent water, VP at 104°F, 10 psig; 454(37-0-33) 30 per cent water, VP at 104°F, 57 psig; 320(0-45-35); 210(0-60-0); 230(0-0-50); the latter three being aqueous solutions of urea and ammonium nitrate for direct application to the soil or for use in making other liquid mixed fertilizers.

Ammonium chloride has been produced and sold as a fertilizer in Europe and the Orient. Small amounts were imported by the United States in the late 1920's but the product did not gain favor with the farmer.

Ammonium nitrate-sulfate is a soluble salt produced by neutralization of a mixture of nitric and sulfuric acids with ammonia. It has excellent storage properties and is suitable for direct application or use in mixed fertilizers.

Calcium nitrate is manufactured and used in Europe but has not gained wide acceptance in the United States. Some solid material is imported and used in solution form on the West Coast. A more popular solution is a calcium-ammonium nitrate solution produced in California.

Nitric phosphates, made by the treatment of phosphate rock with nitric acid, are produced and used extensively in continental Europe but have not been widely accepted in the United States because of their hygroscopic tendencies and their low content of water-soluble phosphates. A few plants in this country use both nitric and phosphoric acids in the process to make nitric-ammonium phosphates having higher water-soluble P_2O_5 contents.

Urea-formaldehyde reaction products are nitrogen fertilizers belonging to the class of controlled-release nutrient products discussed later in this chapter in connection with specialty fertilizers.

Phosphate Sources. Normal superphosphate, made by the acidulation of natural phosphates with sulfuric acid, was the leading source of phosphorus in fertilizers for more than 75 years. In 1964, the domestic production of concentrated superphosphate, made by the treatment of phosphate rock with phosphoric acid, went ahead of normal superphosphate production on the basis of P_2O_5 content. Ammonium phosphate production reached over two million tons of P_2O_5 in 1967, surpassing the production of concentrated superphosphate, which had already surpassed normal superphosphate in 1965.[10]

The superphosphates have proven to be excellent sources of fertilizer phosphorus over many years of agronomic use. They also contain the important secondary element calcium, and normal superphosphate is a valuable source of the secondary element sulfur. The economics of production of these phosphates is favorable to their continued use in competition with other sources of phosphorus. The proportions of constituents in typical normal and concentrated superphosphates are given in Table 18.2.

Aside from minor domestic quantities of phosphate materials, such as bone products used in specialty fertilizers and basic slag from the steel industry used largely for direct application, the other phosphate materials (Table 18.1) are evidence of a veritable revolution in fertilizer production. Technical advances in production of superphosphoric acid and its reaction products with ammonia[12] have given a number of high-analysis fertilizers that promise to become increasingly important in lowering transportation and handling costs for an industry that moves millions of tons of products annually.

Potassium Sources. Most of the potassium chloride on the market contains 60 to 62 per cent K_2O, and is the chief source of potassium in fertilizers. The other sources of potassium (Table 18.1) are used on crops that will not tolerate large amounts of chloride, such as tobacco and potatoes. Potassium nitrate contains two of the primary nutrients. It is used largely in the production of quality crops fed with special types of soluble mixed fertilizers, either in solid or liquid form. Potassium sulfate and sulfate of potash magnesia have excellent handling and storage qualities. Both are valuable on sulfur-deficient soils and the sulfate of potash magnesia is also valuable for its content of the secondary nutrient element magnesium. Current annual U.S. consumption of K_2O as fertilizer is about 4 million tons as compared with world consumption of about 14.3 million tons.

TABLE 18.2 Composition of Typical Superphosphates

Constituent	*Normal Superphosphate*[a]	*Concentrated Superphosphate*[b]
Monocalcium phosphate	27%	68%
Dicalcium phosphate[c]	5%	15%
Calcium sulfate	50%	5%
Free phosphoric acid	3%	3%
Free water	5%	3%
Other[d]	10%	6%
TOTAL	100%	100%

[a]Containing 20% P_2O_5.
[b]Containing 48% P_2O_5.
[c]Including iron and aluminum phosphates.
[d]Including unreacted rock, trace elements, silicates, and combined water.

Conditioners. The term "conditioner" refers to any material added to a fertilizer for the purpose of improving its physical characteristics. Most conditioners are used to decrease caking of the product during storage and transportation and, thus, improve its flowing quality in farm distributing machinery. Caking is most frequently caused by crystalization of salts from a thin film of saturated solution existing at the surface of apparently dry particles in the product, thus cementing or binding the particles together. Consequently, caking occurs in products that have a high content of very soluble salts which contain moisture either absorbed from the atmosphere or already present because of insufficient drying during manufacture.

Examples of some 50 or 60 materials used as conditioning agents for nongranular fertilizers are cocoa shell meal, ground peanut hulls, rice hulls, vermiculite, and perlite. Those for granular products include finely divided clays, silicates, diatomaceous earth, Fuller's earth, fatty amine chemicals, and alkylaryl sulfonates, most of which are granule-coating agents. Enlargement of particles by the process of granulation itself reduces the number of contacts between particles per unit mass of product, and thereby decreases the caking tendency.

None of the conditioners give entirely satisfactory results unless the product is thoroughly dried in processing. Many plant operators prefer less than 1 per cent moisture in the granular product. Hygroscopic materials such as ammonium nitrate are dried to moisture contents of 0.2 per cent, or less, before coating with a conditioner. The proportion of conditioner or coating agent for granular products is 1 to 4 per cent, and that for nongranular products is 5 to 15 per cent.

Solid Mixed Fertilizer Manufacture

Formulation. Fertilizers in the United States are formulated on the basis of a 2000-pound ton, and a term called "the unit," representing 1 per cent or 20 pounds of nutrient, is used to signify the grade of the fertilizer. The primary fertilizer nutrients nitrogen (N), phosphoric oxide (P_2O_5), and potassium oxide (K_2O), are represented in that order (per cent N–per cent P_2O_5–per cent K_2O) as the grade of the fertilizer. Thus, a multinutrient fertilizer containing 6 units (6 per cent) of nitrogen, 24 units (24 per cent) of phosphoric oxide, and 12 units (12 per cent) of potassium oxide has the grade designation 6-24-12. A fertilizer formula for this grade appears as follows:

Mixed Fertilizer Formula
Grade 6-24-12

Material and Nutrient Content	*N*	*Units* P_2O_5	K_2O	*Pounds Per Ton*
Diammonium phosphate, 18% N, 46% P_2O_5	5.67	14.51		631
Ammonium nitrate, 33% N	0.33			20
Normal superphosphate, 20% P_2O_5		9.49		949
Potassium chloride, 60% K_2O			12.00	400
TOTALS	6.00	24.0	12.00	2000

When a fertilizer material or mixture does not contain all three of the primary plant nutrients, the missing nutrient is represented in the grade designation by a zero. For examples, the grades of the materials in the above formula may be expressed, in descending order, as 18-46-0, 33-0-0, 0-20-0, and 0-0-60. If a nonnutrient material had been used to replace potassium chloride in the above formula, the grade would have been 6-24-0.

With respect to fertilizer movement in the trade and in agronomic use, progress is being made toward the conversion of grade designations to the elemental basis. Several foreign countries report nutrient usage on the elemental basis. Fertilizer-control laws in several of the states in the United States permit reporting of guaranteed fertilizer analyses on the elemental basis and much of the scientific literature on fertilizer usage gives the data in terms of the element. However, grade designations for phosphorus and potassium in the form of their oxides, P_2O_5 and K_2O, respectively, have been used by the buyer and the seller for more than one hundred years. It is not surprising that the proposed change to the elemental basis is meeting great resistance.

Ammoniation. The term "ammoniation" is applied to the process in which free ammonia is caused to react with superphosphates, either alone or in admixture with other fertilizer materials, to form ammoniated superphosphates or ammoniated mixed fertilizers. The free ammonia may be in the form of anhydrous ammonia, aqua ammonia, or nitrogen solutions. Normal superphosphate (20% P_2O_5) has the capacity to absorb 9.6 pounds of ammonia per unit of P_2O_5 present, and concentrated superphosphate (46% P_2O_5), 6.4 pounds. However, ammoniation to maximum capacity may result in excessive loss of ammonia and reversion of the phosphorus to forms that are unavailable to crops. Therefore, commercial operation is in the range of 4.8 to 6 pounds of ammonia per unit of P_2O_5 in normal superphosphate, and 3.5 to 4 pounds per unit in concentrated superphosphate.

The principal chemical reactions in the ammoniation of typical normal superphosphate at the rate of 6 pounds of ammonia per unit of P_2O_5, and of concentrated superphosphate at the rate of 4 pounds per unit, may be represented by the following equations:

$$H_3PO_4 + NH_3 \longrightarrow NH_4H_2PO_4 \qquad (1)$$

$$Ca(H_2PO_4)_2 \cdot H_2O + NH_3 \longrightarrow CaHPO_4 + NH_4H_2PO_4 + H_2O \qquad (2)$$

$$NH_4H_2PO_4 + NH_3 \longrightarrow (NH_4)_2HPO_4 \qquad (3)$$

$$2CaHPO_4 + CaSO_4 + 2NH_3 \longrightarrow Ca_3(PO_4)_2 + (NH_4)_2SO_4 \qquad (4)$$

$$NH_4H_2PO_4 + CaSO_4 + NH_3 \longrightarrow CaHPO_4 + (NH_4)_2SO_4 \qquad (5)$$

Equations (1), (2), and (4) are common to both types of superphosphate. Equation (5) is important in normal superphosphate because of the large surplus of calcium sulfate, and equation (3) is important in concentrated superphosphate because of the lack of sufficient calcium sulfate to initiate any reaction by way of equation (5). Keenen[13,14] and White, *et al*[15] investigated the ammoniation reactions and showed that the stoichiometric quantities of reaction products according to these equations were largely in close agreement with analyses of products obtained in ammoniation practice.

The factors governing reaction efficiency are density, size, and structure of the superphosphate particles, moisture content of the superphosphate, concentration of P_2O_5 in the superphosphate, the reaction temperature, and the retention time of the material in the reactor.

The equipment used in the early practice of ammoniation with free ammonia was a closed rotary drum or batch mixer having a capacity of one or two tons and equipped with flights for mixing dry solids. A measured quantity of anhydrous ammonia or ammoniating solution piped from storage (Fig. 18.10) was the last ingredient to be added to the batch through the hollow axis of the mixer. Rotary mixing of the batch for 2 to 3 minutes was usually sufficient to ensure essentially complete reaction. The batch was then discharged from the drum and conveyed to storage. While a few plants still use this type of batch operation, most of the industry has converted to continuous operation.

Nielsson[16] describes a continuous rotary ammoniator into which ammonia or nitrogen solutions are introduced through a multiple-outlet distributor pipe extending beneath the surface of a rolling bed of superphosphate or mixtures held in the drum by retaining rings at each end. The dry charge and the ammoniating liquids are fed continuously to the drum at constant rates and the ammoniated product emerging from the outlet end of the drum is belt-conveyed to storage or to bagging machines. Modified versions of this ammoniator, widely known as the "TVA ammoniator," are in common use for ammoniation and granulation of present-day granulated fertilizers (Fig. 18.11).

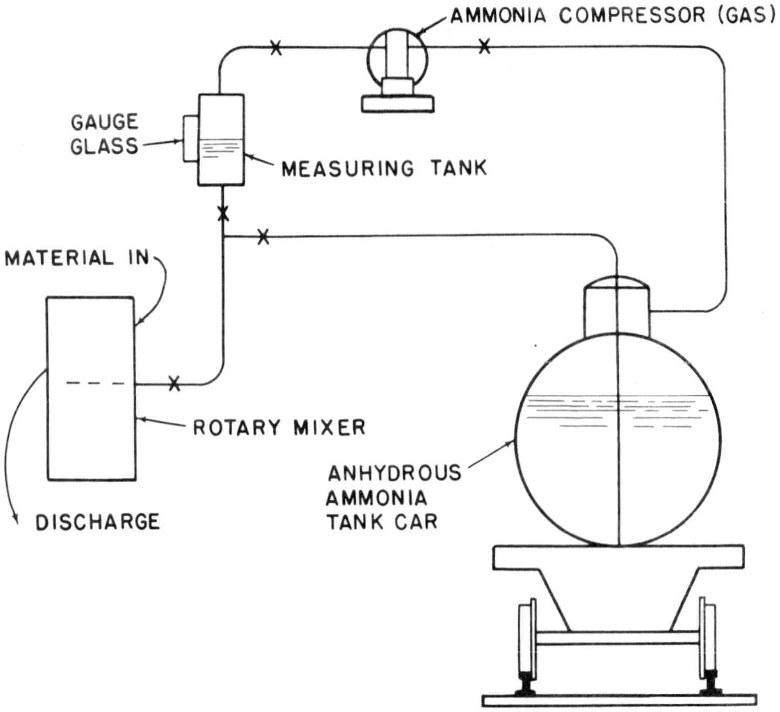

Fig. 18.10 Diagram of equipment used for batch ammoniation.

Granulation. Granular fertilizer is one in which 95 per cent or more of the product remains on any sieve within the range of 8 to 20 mesh (2.38 to 0.841 mm opening), and in which the largest particle passes through a sieve having an opening not larger than four times that of the sieve which retains 95 per cent or more of the product. Thus, as applied to fertilizers, 95 per cent of the particles in a truly granular fertilizer are of any size within the range of 3/8 in. to 20 mesh (9.51 to 0.841 mm), providing the largest particle present does not have a diameter more than four times that of the smallest particle in the product. This definition, with tentative status in the Official Publication, Association of American Fertilizer Control Officials, calls for uniformity of particle size within individual products (i.e., a mean ratio in diameter of 1:4) and, at the same time, allows for a wide range of particle sizes among different products according to the specific needs of various crops.

Nordengren[17] of Sweden has stated that undoubtedly the idea of making granular fertilizers came to Europe from the United States, and that the first fertilizers produced in granular form were possibly the Guggenheim nitrates (Chilean nitrate of soda). The first fertilizers in the United States that could be considered granular by today's standards were chemical nitrogen compounds that appeared on the market subsequent to the first domestic production of ammonia by direct synthesis from the atmosphere in 1921. Granulation decreased the sticky or caked condition inherent in many of these compounds. This improvement in physical condition helped to make these products popular for direct application to the soil, both in the United States and Europe.

Grained or granular multinutrient mixtures of chemical fertilizer salts under the trade name of "Nitrophoska"[18] were introduced in Germany in 1926. Research on the granulation of nitrogen compounds by the U.S. Department of Agriculture, beginning in 1922,

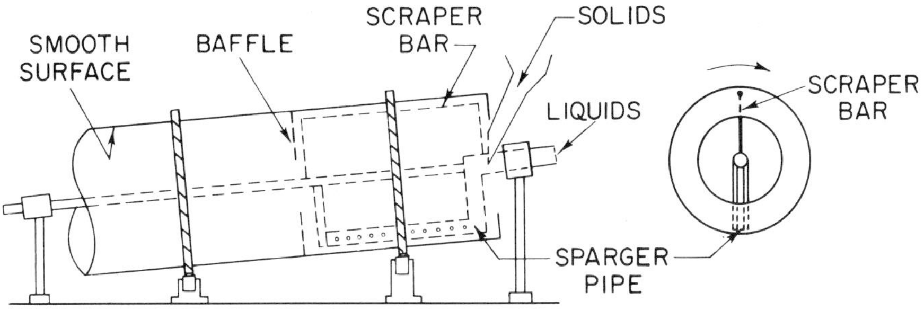

Fig. 18.11 Diagram of continuous rotary ammoniator-granulator.

was extended to the granulation of mixed fertilizers in 1930. More recently, engineering research and plant-scale demonstrations at the National Fertilizer Development Center, of the Tennessee Valley Authority, have stimulated modern commercial developments in the production of fertilizers.[4]

The exothermic nature of ammonia reactions with superphosphate had been known for many years prior to the accurate measurement of the heat developed by these reactions[17a] in the early 1930's. Also, it was well-known that heat was required in the agglomeration and drying steps of the granulation process. But it was not until 1950 that the first plant which incorporated ammoniation with the granulation of mixed fertilizer directly to grade was placed in operation. This was accomplished at Des Moines, Iowa, by the Iowa Plant Food Manufacturing Company, and marked the beginning of a far-reaching revolution in solid-mixed-fertilizer manufacture in the United States.

In 1950, the industry in the United Kingdom was granulating about 75 per cent of its compound (mixed) fertilizers. But the process usually consisted of formulating with nitrogen salts, superphosphate, and potassium salts; mixing and grinding; wetting and agglomerating; and then drying, cooling, and sizing the granular product. Ammoniation was not practiced in the United Kingdom at that time. But the industry was generous with technical literature on the subject of granulation[19,20] and contributed much to the practical application of granulation processes in the United States.[21]

The key steps in a continuous granulation process (Fig. 18.12) are the liquid-solids mixing or ammoniating operation and the agglomerating granule-compacting operation, whereby small particles are gathered into larger permanent masses. The remaining steps of formulating, mixing and screening of incoming solid materials, drying, cooling, and sizing of the granular product, the crushing and screening of oversize material, and the recycling of dust and fines in the process, while important to processing efficiency, are common unit operations in many industries.

The liquid-solids mixing operation is usually carried out in rotary cylinders, such as the continuous rotary ammoniator (Fig. 18.11), or in continuous twin-shaft pugmills or Blungers (Fig. 18.13). The thorough mixing of liquids and solids ensures rapid chemical reaction and allows sulfuric, phosphoric, or nitric acids to be used for neutralization of excess ammonia. The use of acids in this way provides products of exceptionally high analyses and induces additional heat of chemical reaction which frequently is sufficient to eliminate the need for any artificial drying of the product.

Efficiency in operation of the agglomeration step is the central objective of the entire granulation process. Methods of agglomerating small particles into larger permanent masses have been used in the manufacture of such products as brick, candy, briquetted coal, dishes, pelleted feeds, ore concentrates, powdered metal parts, tableted pharmaceuticals, plastic articles, sewer pipe, and tile,[4] as well

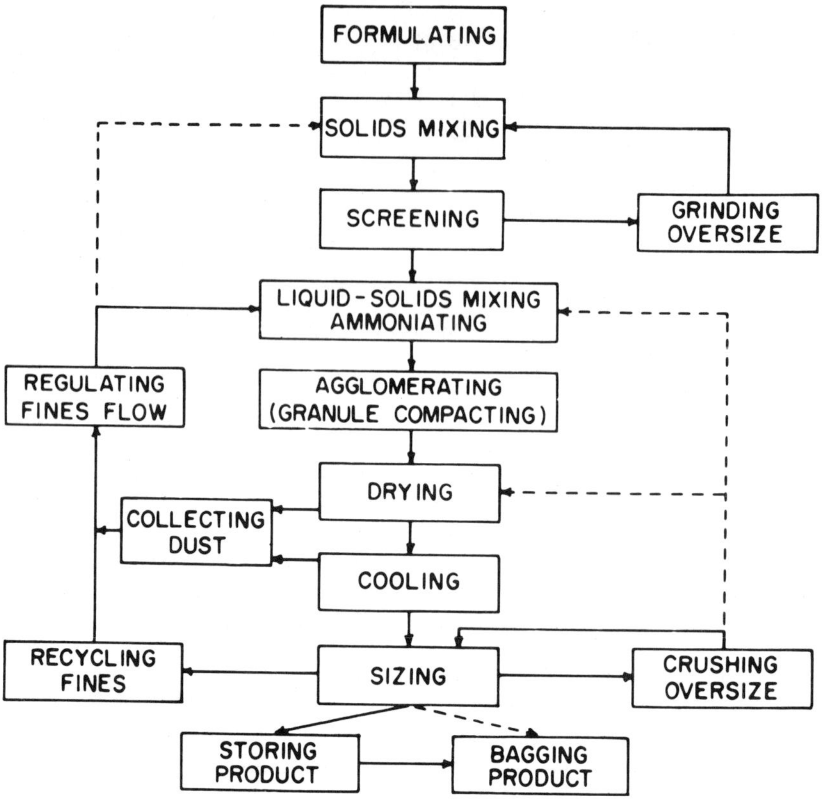

Fig. 18.12 Flow sheet of typical granulation process.

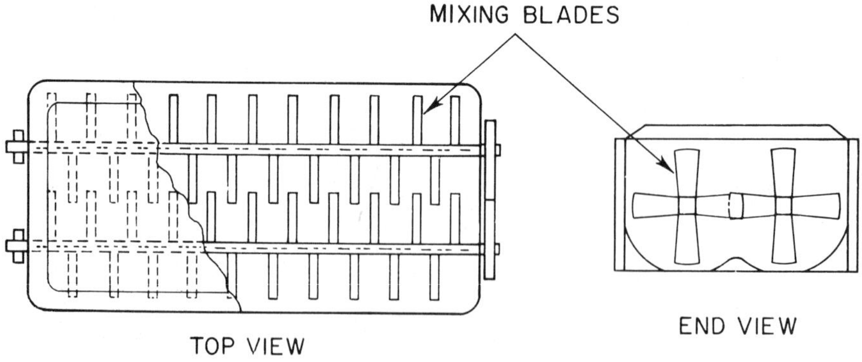

Fig. 18.13 Diagram of pugmill (Blunger); top and end views.

as fertilizers. Methods used in the agglomeration of such products include crystallization, flocculation, electrostatic precipitation, fusion or sintering, compression, congelation, and agitation. The latter three methods have gained varying degrees of use in the manufacture of fertilizers.

Compression techniques, such as extrusion through perforated plates, tableting, molding, and briquetting have been used to produce specialty fertilizers in the form of large pellets for tree culture or tablets for growing house plants. One compression method, used in granulating ammonium sulfate, diammonium phosphate, and potassium chloride[22] compacts the relatively dry powdered material under high pressure between smooth rolls to give a hard dense sheet which is then crushed and screened to give a product in the desired particle-size range.

Congelation techniques include spray drying or prilling, flaking, and cooling on moving belts. These methods are adapted to the granulation of fusible materials such as sodium, potassium, and ammonium nitrates, and urea. The typical spray-drying or prilling process consists in spraying a hot solution or melt in the top of a tower and allowing the droplets to congeal into solid granules as they fall through a countercurrent flow of air, or by spraying the melt into a vat of recirculated cooling oil. The product is then dried.

The flaking method consists of congealing a hot saturated solution or molten material on the outer surface of heated or cooled drums, depending on the state of the material. The thin layer of solid is removed from the surface of the drum by a scraper. The soft flakes are dried in a rotating cylinder that gives the final particles a spheroidal shape. The process has fitted into the economy of certain European producers of synthetic-nitrogen materials[23] but has not found favor in the United States.

Dorsey[24] describes the Stengel process for congealing a thin layer of molten ammonium nitrate on an endless, stainless steel, water-cooled belt. The sheet of solid material from the belt is crushed and screened to give the desired particle size of granular product.

Methods of agglomerating fertilizers by agitation, which consist of causing coalescense of the wetted particles of raw materials by bringing them into contact with one another, are by far the most used means of granulating compound, complex, and mixed fertilizers, both in this country and abroad.

The equipment generally used to provide continuous agitation in the agglomeration stage of the process is the horizontal rotary cylinder (Fig. 18.14) with a smooth interior surface, or with various types of internal structures such as lifting flights, rolling trays, or a concentric rotor.[4] Various types of horizontal pans, such as the Eirich mixer, have been used for batch granulation in Europe. Several modifications of the inclined pan (Fig. 18.15) have been used abroad, but they have not gained wide application in the United States. A typical unit is 12 feet in diameter and 14 inches deep. The raw materials enter the pan continuously at point X, and the plasticizing agent, generally water, is sprayed on the rolling material (point Y) at a sufficient rate to cause agglomeration of the fine particles which are deflected under the spray by the scraper. The large agglomerates migrate to the surface of the deep bed at the lower side of the pan and flow over the rim into a dryer while the fine particles at the bottom are carried to the shallow bed at the top of the pan for further agglomeration. In this way the pan tends to act as a particle-size classifier.

The continuous rotary ammoniator-granulator previously described (Fig. 18.11) which combines the operations of liquid-solid mixing, chemical reaction, and agglomeration[25] has been used widely in the granulation of a wide variety of fertilizers. Various modifications exist, but the typical size of the cylinder is 8 by 16 feet for a production rate of 20 tons per hour. The basic feature of this unit is the multiple-outlet distributor for introducing liquids including acids, gaseous ammonia, and steam beneath the surface of the rolling bed of solids. The after-section of the cylinder is for the purpose of rolling the agglomerates into firm, discrete granules. When the inherent operating variables are closely controlled, the process is capable of producing high-quality grades of high-analysis granular fertilizers.

In some high-analysis formulations, it is necessary to preneutralize the acids with am-

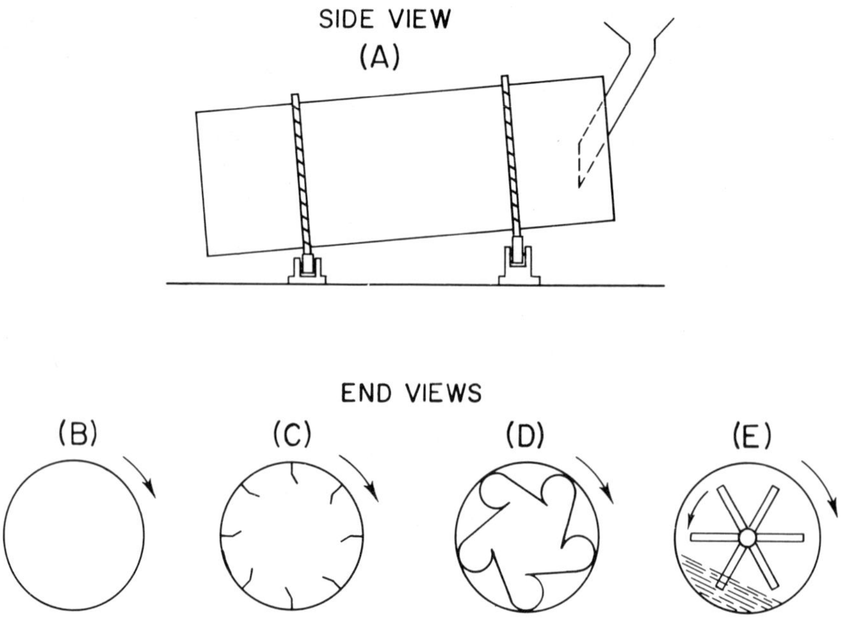

Fig. 18.14 The rotary cylinder: (A) side view; (B) end view of smooth shell; (C) end view of flights; (D) end view of rolling trays; and (E) end view of concentric rotor.

moniating solutions in a separate operation in order to avoid excessive temperatures in the ammoniator. These can lead to nitrogen loss and over-agglomeration. The liquid phase rather than moisture alone is the true plasticizer in the granulation process. The concentration of the liquid phase depends primarily on the quantity of moisture in the raw materials, the quantity and solubility of the salts present, the reaction temperature, and the degree of saturation of the solution as influenced by the contact time between solid and liquid.

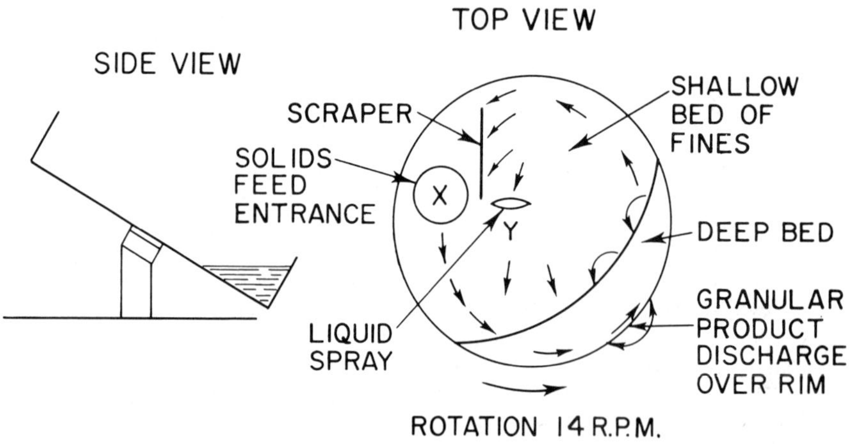

Fig. 18.15 Diagrams of inclined-pan granulators; top and side views.

Control of these variables determines the efficiency of the agglomeration stage of the process.

Slurry Granulation. The slurry, or "nucleation," process of granulation is one in which a hot slurry of the ingredients of the fertilizer is sprayed on undersize granular material and then dried at the surface of the granule so as to form successive "onion skin" coatings on the granule until it is large enough to pass over the product screen to storage. Distribution of slurry at the surface of the recycled undersize particles in this continuous process may be accomplished in continuous paddle mixers or pugmills, or the slurry may be sprayed on the solid at the inlet end of the rotary drier. The process is adapted to the production of ammonium phosphates,[26] concentrated superphosphate, nitric phosphates,[4] and NPK mixtures based on these materials, such as 15-15-15, 14-28-14, 12-36-12, and 12-12-18. This process has been used extensively in continental Europe for granulating nitric phosphates, and in Canada, the United Kingdom, and the United States for granulating ammonium phosphate fertilizers. The few plants in this country originally producing nitric phosphates have converted, at least in part, to production of granular ammonium phosphates by using some phosphoric acid in the process. This provides more water-soluble phosphate in the product than is normally present in conventional nitric phosphates. Davis[27] describes a process of this nature.

Bulk Blended Fertilizer. The practice of mixing dry, individual, granular materials or granulated base mixtures is referred to in the trade as "bulk blending."[28] The product is a mixture of granular materials rather than a granulated mixture. Bulk blending adheres to a marketing system intended to decrease the costs of transportation, chemical processing, and bagging of mixed fertilizers. It usually brings the fertilizer materials directly from the primary producer to a dry-mixing plant located in the market area. Bulk blenders often apply the fertilizer directly to the land according to specifications of the customer. This service to the farmer has been popular. About 47 per cent of all domestic solid mixed fertilizer currently is applied in bulk form.

Problems with stratification or segregation of materials having different particle sizes have been encountered. Bridger *et al*,[29] in testing various types of mixers for dry blends, found that rotary cylinders and ribbon mixers were satisfactory, but results with concrete mixers, cone mixers, and vertical mixers were poor. Gilbert *et al*[30] found that segregation of micronutrients added to bulk blends could be virtually prevented by coating the blends uniformly with the powdered micronutrient material. Poor mixing or segregation in bulk blends causes nonuniform distribution in the field.

Inadequate storage facilities at blending plants have frequently resulted in a short supply of fertilizer materials at the peak of the fertilizer season. These problems are not without parallel in the history of the industry.

Liquid Fertilizers

The term "liquid fertilizer" is commonly used in the trade to mean clear solutions, suspensions of solids in liquids, and slurries. The terms "suspension" and "slurry" are frequently used interchangeably, but suspensions contain a suspending agent, such as a gelling type of clay, to keep the solids dispersed in the solution for a longer period of time than is possible with slurries. Most liquid fertilizers, including some clear solutions, tend to separate into layers during long periods of storage. This is caused by changes in composition and specific gravity of the solution at different levels during storage or by settling of the solid ingredients. The degree of mixing required before or during use depends on the settling rate of the solids present.

Liquid fertilizers used in the continental United States[7] increased from about 27 thousand short tons in 1954 to nearly 9 million tons in 1969. They are as yet of limited importance in countries outside of North America. About three-fourths of the total domestic consumption of liquids is accounted for in the direct application of anhydrous ammonia, aqua ammonia, and nitrogen solutions. Some 2.25 million tons, or about 11 per cent of the domestic consumption of mixed fertilizers, are applied as liquid mixtures.

So-called "hot-mix" processes involve making base solutions of ammonium phosphates by neutralization of phosphoric acids with ammonia or nitrogen solutions. Although wet-process phosphoric acid forms some precipitates with iron and aluminum impurities, it is widely used in hot-mix processes to make 8-24-0 and 10-34-0 bases. In making 8-24-0, the precipitate is separated from the solution. The 10-34-0 is made with wet-process superphosphoric acid. The superphosphoric acid forms polyphosphates in the solution which act as sequestering agents and hold the impurities in solution. Furnace-grade superphosphoric is neutralized with nitrogen solutions to produce 11-37-0 which is a popular solution for use in making NPK grades of liquid fertilizer.

Use of superphosphoric acid in making high-analysis liquid fertilizers, as well as solid fertilizers, has considerable potential. The Tennessee Valley Authority presently is making a considerable tonnage of 12-44-0 polyphosphate base liquid by neutralizing hot, fertilizer-grade superphosphoric acid (80 per cent P_2O_5) in aqueous solution with gaseous ammonia to pH 5.5 in a stainless steel reaction tank, 6.5 ft in diameter and 9 ft high. This continuous process requires vigorous stirring in the reaction tank where the temperature is controlled between 150 and 180°F by recirculation of the liquid through a heat exchanger which removes about 800 Btu/lb of P_2O_5 throughput (600 Btu when anhydrous ammonia is used). Retention time in the reactor is about 30 mins and the production rate is about 15 tons per hr. The liquid product is passed through a second heat exchanger to storage, or to a mixing tank for incorporating Attapulgite clay when suspension fertilizers are being made. The operating procedure for making 12-44-0 is similar to that for 11-37-0 or 12-40-0, except that the latter two base liquids have a pH 6.0 to 6.2. Specific gravity measurements are used in controlling the concentration of the product.

Base liquids produced in hot-mix plants are frequently distributed to smaller plants in outlying areas where they are used in making the so-called "cold mix" by the addition of other materials to produce various grades of liquid fertilizer. In typical operations, a hot-mix plant makes 10-34-0 in 15-ton batches, each batch requiring about 45 minutes for completion. A 4-11-11 base mixture also is made from water, ammonia, phosphoric acid, and potassium chloride added to the reaction tank in that order. A nitrogen solution containing urea and ammonium nitrate (32 per cent N) is stored in advance of the peak fertilizer season. From these three base solutions, the cold-mix manufacturer can produce and deliver most grades of liquid fertilizer used within a distance of 50 miles from his plant.

Liquids appear to have an advantage over solid fertilizers in lower capital investment; less labor, handling, and conditioning costs; more uniform composition; and more uniform distribution on the land. Existing problems relate to crystallization of salts from solution in cold weather, corrosion of equipment, the economics of storage, and the restrictions that solubility imposes on the choice of nutrient materials, including the micronutrient carriers.

Hignett[31] reviews the technology and economics involved in making liquid fertilizers.

Special-Purpose Fertilizers and Soil Amendments

While the requirements of growing plants for the essential nutrient elements are not subject to change, there is an ever-widening market for special-purpose products designed to fit the needs of the homemaker in growing potted plants, or the farmer in growing a large acreage of a major crop.

Nonfarm Fertilizers. The annual domestic consumption of fertilizers for purposes other than growing major crops is nearly 1.6 million short tons.[32] These include fertilizers for house plants, flower and vegetable gardens, golf courses, parks, athletic fields, cemeteries, roadsides, and lawns around homes and public buildings.

Home Lawn and Garden Products. Probably the largest market in specialty fertilizers, at least dollar-wise, is in the home lawn and garden class of products. It is also a high-risk business. Several large fertilizer manufacturers have entered this market and have then withdrawn.[33] The older companies in the business cover the needs of the homeowner for lawn fertilizers, seed, pesticides, and lawn care equipment. A few companies operate on a national scale, and many others on a regional scale.

The lawn and garden segment of the industry features so-called "light-weight" products containing mixed fertilizer salts, with or without pesticides, absorbed on such low-density materials as vermiculite and peat. Many of the products also contain controlled-release nitrogen materials that become available very slowly and feed the lawn gradually over the entire growing season. This decreases the number of fertilizer applications during the growing season and eliminates the luxuriant growth that occurs when readily soluble nitrogen fertilizers are applied. Lawn fertilizers containing weed eradicators have special appeal to the homeowner.

Either dry or liquid fertilizer-pesticide mixtures are attractive to the manufacturer of farm fertilizers and the grower of major crops, but they present problems of compatibility, placement in the soil, timing of application, and labeling. Although the fertilizer-pesticide mixture is usually a custom blend for a special purpose, many factors may contribute to its misuse.

Controlled-Release Nitrogen Products. Urea-formaldehyde reaction products, referred to previously, are manufactured in considerable quantities and have gained wide use in specialty fertilizers. Economics of production has limited their use in farm fertilizers, but there is considerable evidence that the profits which can be derived from horticultural and farm use of controlled-release nitrogen products will eventually be recognized as offsetting the difference in price between these fertilizers and the more soluble forms of nitrogen.

There is a need for the controlled release of potassium as well as nitrogen in fertilizers. A promising method of delaying the release of both nitrogen and potassium is that of coating the fertilizer granules with sulfur.[34]

Nitrification inhibitors, such as sulfathiazole-formaldehyde reaction products[35] and 2-chloro-6-(trichloromethyl)pyridine,[36] are producing favorable results in delaying the release of nitrogen from soluble ammonium compounds.

The metal ammonium phosphates, such as magnesium ammonium phosphate, and oxamide have slow dissolution rates in soils and thus act as slow-release fertilizers.

Acid-Forming Fertilizers. The residual effects of fertilizers applied to the soil are either acidic, neutral, or basic depending on the balance between the potential neutralizing powers of acid- or base-forming elements in the fertilizer.[37] The acid-forming elements responsible for decreasing the soil pH are sulfur, chlorine, one-third of the phosphorus, and half of the nitrogen added in the fertilizer. The base-forming elements responsible for increasing the soil pH are calcium, magnesium, potassium, and sodium.

Materials having a balance in potential neutralizing power of acid- and base-forming elements are ammonium nitrate-limestone, superphosphate, potassium chloride, potassium sulfate, monopotassium phosphate, sulfate of potash magnesia, and gypsum. They do not influence the soil pH. Some materials that decrease soil pH, together with the number of pounds of calcium carbonate required to neutralize 100 lbs of each, respectively, are: ammonia, 148; ammonium sulfate, 110; urea, 84; ammonium nitrate, 59; diammonium phosphate (18-46-0), 64. Tables showing the potential acidity of most fertilizer materials appear in "Farm Chemicals Handbook" and other reference works on fertilizers.

Manufacturers of fertilizers to be used in regions where low pH of the soils is detrimental to acid-sensitive crops have made special nonacid-forming fertilizers by adding dolomitic limestone to correct the potential acidity of the product. However, this is not possible with such materials as anhydrous ammonia which is applied directly to the soil. One alternative in the latter case is to make a separate application of 184 pounds of calcium carbonate, dolomitic limestone, or equivalent base-forming material to neutralize the potential acidity of each 100 pounds of ammonia applied. This practice has not always been followed. The 3.5 million tons of anhydrous ammonia alone, which is currently used for direct application, would require 6.4 million tons of limestone to offset the potential acidity which the ammonia gives to the soil. Many soils require 2 tons and some require as much as 6 tons of limestone per acre to correct soil acidity.

Acid-forming fertilizers are beneficial on alkaline soils. Products such as sulfur, sulfuric acid, and special sulfur-containing compounds in liquid or dry form are used to correct alkalinity of soils usually occurring in low-rainfall regions. Highly soluble acid-forming fertilizers such as ammonia and phosphoric acid fit this purpose very well in irrigation farming.

Other Special-Purpose Products. Special soluble fertilizers for making "starter" or transplanting solutions to be placed in the root zone at planting time, or so-called "pop-up" fertilizers to be placed with or near the seed at seeding time, must be formulated with materials that have a low salt index in order to prevent injury to the tender roots of young plants or to seed sprouts during germination.[38] The salt index number of a material is a measure of its burning effect on growing plants. Tabulated salt index values for most fertilizer materials are available in "Farm Chemicals Handbook" and numerous texts on fertilizer technology and use.

Other special-purpose products include those for application by airplane and helicopter. Specifications frequently require uniformly sized, dust-free granular materials to avoid drift, or, for example, large pellets for application on forests to ensure that the fertilizer does not stick to, and damage, foliage.

Soluble fertilizer for making foliar sprays to correct or prevent nutrient deficiencies in a wide variety of horticultural and farm crops require a wide range of specifications relating to the concentration of primary, secondary, and micronutrient materials in the solution. Types of spreading agents in the solution and the selection of chelating agents to sequester specific micronutrients in handling and using foliar sprays are also special considerations.

State Fertilizer Control Laws

State fertilizer laws require registration and accurate labeling of all brands and grades of fertilizer sold in the respective state. They require guaranteed analyses of the manufacturers' products with penalties for failure to meet the guarantee. The various forms of tax, fee, or assessment on registrants are designed to make the enforcement agency self-supporting.

The Association of American Fertilizer Control Officials is a forum for the exchange of information and the sharing of experience with control problems in the various states, and for promoting the uniformity of inspection services and the enactment of uniform fertilizer legislation.

Looking Ahead

Engineering research on the production of high-analysis urea-ammonium phosphates[39] gives promise of future commercial production. Pilot-plant production of a 12-57-0 grade[40] of ammonium phosphate in conjunction with addition of solution melts of urea gives granular grades of 21-42-0 and 28-28-0. Potash addition produces a 19-19-19 grade. Such products afford savings in handling and transportation costs.

Several ultrahigh-analysis compounds of nitrogen and phosphorus, such as phosphonitrilic hexaamide (grade 55-92-0), under investigation at the National Fertilizer Development Center, TVA, may be the forerunners of major fertilizers of the future.

Success of present intensive investigation of the removal and recovery of sulfur oxides from stack gases will serve the double purpose of decreasing pollution of the atmosphere and providing valuable products for the manufacture of fertilizers.

Computerized techniques to aid in the formulation, manufacture, marketing, and use of fertilizers give promise of becoming commonplace in the industry.

Fertilizers containing radioactive nutrient elements continue to be valuable tools in tracing the course that nutrients take in soils and growing plants. Such agronomic research on how plants feed, coupled with advances in fertilizer manufacturing processes now on the horizon, will continue to shape the future of the fertilizer industry.

REFERENCES

1. Allgood, *et al.*, *Ind. Eng. Chem.*, **59**, No. 6, 19, (Jun. 1967).
2. Kapusta, E. C., "The Role of Potassium in Agriculture," Chapter 2, American Society of Agronomy, 1968.
3. Spealman, M. L., *Chem. Eng.*, **72**, No. 23, 198–200 (1965).
4. "Superphosphate–Its History, Chemistry and Manufacture," U.S. Department of Agriculture and TVA, U.S. Government Printing Office, 1964.
5. Mackall, J. N. and Shoeld, M., *Chem. and Met. Eng.*, **47**, 102–105 (1940).
6. "Farm Chemicals Handbook," 134, Meister Pub. Co., 1970.
7. "Consumption of Commercial Fertilizers in the United States," Preliminary Report, U.S. Department of Agriculture, Statistical Reporting Service, 1970.
8. Mehring, A. L. and Peterson, A. J., "Commercial Fertilizer Yearbook," 33–44, 1934.
9. Clark, K. G. and Bear, F. E., *Am. Fertilizer*, **108**, No. 3 7–10, 24, 26, 28, 30 (1948).
10. Harre, E. A., "Fertilizer Trends," TVA, Muscle Shoals, Ala.
11. Johnson, J. P., *Chem. Age*, **99** No. 2593, 17 (1969).
12. *Chem. Week*, **104**, No. 4, 26 (1969).
13. Keenan, F. G., *Am. Fertilizer*, **74**, No. 10, 19–22 (1931).
14. Keenan, F. G., *Ind. Eng. Chem.*, **24**, 44 (1932).
15. White, L. M., Hardesty, J. O., and Ross, W. H., *Ind. Eng. Chem.*, **27**, 562–7 (1935).
16. U.S. Patent 2,729,554.
17. Nordengren, S., *Fertiliser Soc. (Eng.) Proc.*, No. 2 (1947).

17a. Hardesty, J. O. and Ross, W. H., *Ind. Eng. Chem.*, **29**, 1283–90 (1937).

18. Merz, A. R., "New Fertilizer Materials," *U.S. Dep. Agric. Circ.* 185 (1931; revised 1940).
19. Angus, J. and Herdson, E. P., *Chem. Trade J.*, **121**, 423 (1947).
20. Proctor, J. T., *Fertiliser Soc. (Eng.) Proc.* 7 (1949).
21. Hardesty, J. O. and Clark, K. G., *Agr. Chemicals*, **6**, No. 1, 34–38, 95, 97 (1951).
22. *Chem. Eng.* **64**, No. 11, 154, 156 (1957).
23. Stanfield, Z. A., *J. Agr. and Food Chem.*, **1**, 1054–1059 (1953).
24. Dorsey, J. J., Jr., *Ind. Eng. Chem.*, **47**, 11–17 (1955).
25. Hignett, T. P., *Farm Chem.*, **126**, No. 3, 30–31, 56, 58 (1963).
26. Atwell, J., *Ind. Eng. Chem.*, **41**, 1318–24 (1949).
27. U.S. Patent 3,436,205.
28. Hignett, T. P., *Agri. Nitrogen Inst. Proc.* 53–8 (1969).
29. Bridger, G. L. and Bowen, I., *Fertilizer Industry Round Table, Proc.* **18**, 40–45 (1968).
30. Gilbert, R. L., Nau, H. H. and Cox, T. R., *Fertilizer Industry Round Table, Proc.* **18**, 66–68 (1968).
31. Hignett, T. P., *Phosphorus Agri.*, No. 51, 1–20 (1960).
32. Mehring, A. L., "Consumption of Fertilizers for Purposes Other than Farming," The Fertilizer Institute, Washington, D.C., 1967.
33. *Farm Chem.*, **130**, No. 4, 13–19 (1967).
34. *Chem. Eng. News*, **48**, No. 2, 49 (1970).
35. U.S. Patent 3,526,494.
36. Huber, D. M., Murray, G. A., and Crane, J. M., *Soil Sci. Soc. Amer., Proc.* **33**, No. 6, 975 (1969).
37. Pierre, W. H., *Am. Fertilizer*, **79**, No. 9, 5–8, 24, 26–27 (1933).
38. Hardesty, J. O., *Farm Chem.*, **130**, No. 10, 42–8 (1967).
39. U.S. Patent 3,425,819.
40. Lee, R. G., Meline, R. S. and Young, R. D. "Proceedings of Technical Conference of International Superphosphate and Compound Manufacturers Association," Sept. 8–11, 1970.

19

Chemical Explosives—Rocket Propellants

Dr. Melvin A. Cook* and Dr. Grant Thompson**

CHEMICAL EXPLOSIVES

Explosives serve two main purposes. First, they are utilized in industry to save billions of man-hours of work each year, e.g., in mining coal, and metallic and nonmetallic ores; in quarrying, clearing land, ditching, loosening formations in oil and gas wells, and in road building; for sporting ammunition; and for such important specialized applications as blind rivets and starter cartridges for aircraft and diesel engines, high-speed machining and metal forming, and in perforating oil-well casings. Second, they are of major importance in the field of rockets, missiles, space vehicles, and military and civilian weapons. The manufacture of commercial explosives is a growing industry; production has risen steadily in America from about 200 million pounds in 1920 to approximately two billion pounds in 1969. Bebie[1] lists some 135 chemicals and formulations which are useful as explosives; of these, 75 are used in the industry alone, 45 are primarily military explosives, and 15 are used for both purposes.

An explosive is a substance or mixture of substances which, when raised to a sufficiently high temperature, whether by direct heating, friction, impact, shock, spark, flame, or sympathetic reaction from a primary or donor explosive, suddenly undergoes a very rapid chemical transformation with the evolution of large quantities of heat and gas, thereby exerting high pressures on surrounding media. With some explosives the rate of this transformation (or burning rate) is so great that the explosive exerts a very great shattering action (or *brisance*), while with others the reaction may take place at a much slower, controlled, but still explosive rate to give pressure-time characteristics which make them suitable for use as propellants in guns, rockets, etc., where much lower rates of pressure development and peak pressures are required. Another characteristic property of explosives is *sensitivity*, or ease of initiating

*IRECO Chemicals (section on chemical explosives).
**Thiokol Chemical Corp. (section on rocket propellants).

the explosion, whether of the fast (shattering) type or the much slower, propellant type. *Strength*, or the maximum explosive energy available for useful work, is another important factor. It depends much less on the rate of reaction than does the brisance, peak pressure, or pressure-time curve of the explosive. The uses of explosives depend on all three of these characteristics (pressure or pressure-time curve, sensitivity, and strength); these are the bases for the groupings used in Table 19.1 and the classifications presented in the next paragraph.

The usual classification of explosives is into two general groups, *high* or *detonating* explosives, and *low* or *deflagrating*, sometimes also called *propellant* explosives. The latter have a low burning rate which permits them to have a relatively slow-rising pressure-time curve; the peak pressure seldom rises above 50,000 psi. The rate of burning of the low explosive directly into the grain never exceeds a few cm/sec, whereas in detonation the reaction rates are hundreds of times faster. The low explosives exert a powerful "heaving action" or push, and while they are used today primarily as propellants, they have a very desirable blasting action for lump coal. However, the most prominent of this type, from the historical viewpoint, namely black powder, has been used extensively in the past in borehole blasting. In high explosives the reaction takes place to a large extent in a peculiar type of shock wave known as the detonation wave which propagates in accord with well-known principles of hydrodynamics at velocities ranging from one to seven miles per second, depending on the density, heat of explosion; in some cases, the particle size and shape; and in gelatins, the air-bubble content and distribution. An important characteristic of detonation in condensed explosives is that the reaction zone comprises an ionized gas or plasma existing with high cohesion in a quasi-lattice structure, pictured resembling the metallic state.[2,3] This plasma causes the pressure rise in the detonation front to be much less steep than it was at first thought. That is, instead of being infinitely steep, the detonation rises to its characteristic pressure of 100,000-5,000,000 psi in a period ranging from a few tenths of a microsecond to several microseconds, depending on the explosive. Subclasses of high explosives are the primary explosives (used as detonators) and the secondary explosives. The former are characterized by the fact that even in very small quantities they develop (via the essential plasma formation) detonation waves in extremely short periods of time following simple ignition, e.g., by flames, sparks, hot wires, or friction. The secondary explosives, however, usually require detonators, and sometimes boosters also, to bring them to detonation, at least in practical applications. A booster may be one of the more sensitive secondary explosives, such as pressed tetryl, TNT, RDX, waxed RDX, or cast TNT and pentolites (TNT-PETN). The commercial detonators are the ordinary (or fuse) and electric (or EB–sometimes also called composition) caps which contain either mercury fulminate alone (fuse caps) or separate elements consisting of (1) an ignition element (e.g., lead styphnate), (2) a primary explosive (e.g., lead azide), and (3) a base charge comprising a secondary explosive (e.g., pressed tetryl, PETN, or RDX). These are the EB or composition caps. Commercial dynamites are all secondary explosives which may be detonated directly by commercial detonators or Primacord. The latter is a detonating fuse usually containing about 50 grams of PETN per foot in a special wax-impregnated cloth or plastic sheath which may or may not be reinforced by a binding wire. The least sensitive secondary explosives, of which the 94/6 prilled ammonium nitrate-fuel oil mixture is currently the most popular, require a relatively large booster. Other series of explosives of this type are the "slurry explosives" and "slurry blasting agents." The first *slurry* used commercially comprised essentially ammonium nitrate, coarse TNT, and water. Boosters are usually required for these relatively insensitive explosives. They comprise 0.1 to 1.0 pounds or more of cast 50/50 pentolite, cast TNT, or other water-insoluble material, usually cast, and water-insoluble explosives of very high brisance. Another early and still useful series of slurries substituted smokeless powder for TNT. Aluminum was added to these explosive-sensitized slurry types to increase strength. The most popular

TABLE 19.1 Characteristics and Uses of the More Important Explosives

PRIMARY EXPLOSIVES

Name	*Composition or Chemical Formula*	*Density (g/cc)*	*Detonation Velocity[a] (km/sec)*	*Detonation Pressure (kilobars)*	*Detonation Temperature[a] (°K)*
Mercury fulminate	$Hg(ONC)_2$	3.6	4.7	220	6900
Lead azide	PbN_6	4.0	5.1	250	5600
Lead styphnate	$C_6H(NO_2)_3O_2Pb$	2.5	4.8	150	–
Nitromannite (Mannitol hexanitrate)	$C_6H_8(ONO_2)_6$	1.73	8.3	300	6000
Dinitrodiazophenol (DDNP)	$C_6H_2N_4O_5$	1.5	6.6	160	–

SECONDARY HIGH EXPLOSIVES

Name	*Composition or Chemical Formula*		*Density (g/cc)*	*Detonation Velocity (km/sec)*	*Available Energy (kcal/g)*
Ammonia gelatin dynamites	30–90% grades same as straight gelatins except for some NG and $NaNO_3$ replacement by NH_4NO_3		1.2–1.5	4–6.5	0.75–1.15
Semigelatin dynamite	15–20% NG, 1–2% DNT oil, AN–SN dope		1.2	3.5–5 (depends on diameter)	0.9
Prilled AN-Fuel Oil	94/6 NH_4NO_3/oil		0.8–0.9	1.5–4	0.81–0.83
Slurry explosives TNT-SE	TNT	17–40	1.4–2.0	5–8	0.7–1.8
	Oxidizer*	30–65			
	H_2O	12–25			
	Al	0–20			
	Other	0.3–1.5			
Smokeless powder SE	SP	20–40	1.35–1.9	4–7	0.65–1.7
	Oxidizer*	30–60			
	H_2O	3–25			
	Al	0–20			
	Other	0.3–10			

[a]Most important properties of detonators.
*AN, SN, perchlorates, etc.

Sensitivity	*Major Characteristics*	*Uses*
Very high	Best primary explosive for single-component (fuse) detonators; easily detonated by flame, spark, heat, or friction; easily dead-pressed.	In fuse caps (mixed with $KClO_3$); propellant primer; in fuses for shells; small arms cartridge caps.
Very high (higher than NG; less than mercury fulminate)	Powerful detonator but requires strong igniters, e.g., lead styphnate.	Primary explosive in composition (EB) caps; military fuses.
Exceedingly high	Extremely sensitive to sparks, static electricity; explodes rapidly on ignition; good thermal stability.	Igniter in composition caps, military fuses; very satisfactory detonator explosive for fast ignition.
Very high (greater than NG; less than lead azide)	Stronger and more brisant than NG, RDX, PETN.	In composition caps and fuses.
Very high (less than lead azide)	Does not dead-press. About 3/4 as strong as TNT.	In composition caps and fuses.

Sensitivity	*Major Characteristics*	*Uses*
High	More economical; only slightly less brisant than straight gelatin; exhibits low-order detonation with threshold priming and high pressures.	General small and large diameter blasting in hard rock and under water.
High	Stringy, plastic; easily loaded in "uppers;" economical; high strength; moderate brisance.	Popular small diameter metal-mining explosive.
Low (requires booster)	One of the cheapest sources of explosive energy available today; flammable and will explode when ignited under strong confinement; no water resistance; adaptable to do-it-yourself operations.	Open-pit and underground blasting where dry conditions prevail; most adaptable to soft, easy shooting.
Low (requires boosters)	Gel or thick pea-soup consistency; capable of detonation at high pressures, excellent water resistance.	Large diameter, open-pit, small diameter underground, oil well, submarine, water-filled boreholes, deep-water bombs.
Low (requires boosters)	Generally similar to TNT slurry.	Large diameters, open-pit blasting.

TABLE 19.1 (cont'd.)

Name	*Composition or Chemical Formula*		*Density (g/cc)*	*Detonation Velocity (km/sec)*	*Available Energy (kcal/g)*
Al-SE	Al	1.0–10	1.1–1.5	2–5	0.7–1.3
Slurry blasting agents	Al Oxidizer* H_2O Other	0–35 50–80 4–18 0.2–10	1.1–1.6	2–6	0.7–2.0
Nitrostarch powders	Nitrostarch in place of NG		1.2	4–5	0.8–1.0
Composition B	40/59/1 TNT/RDX/wax		1.7	7.8	1.1
Composition B-3	40/60 TNT/RDX		1.73	7.9	1.15
Haleite or EDNA	$(CH_2NHNO_2)_2$		1.6 (pressed)	7.9	1.2
Ammonium picrate (Explosive D)	$(ONH_4)C_6H_2(NO_2)_3$		1.56 (pressed)	6.6	0.7
Nitrostarch	Mixtures of various nitro esters of starch		1.4 (pressed)	6.4	0.95
Tetryl	$(NO_2)_3C_6H_2CH_3N_2NO_2$		1.45 (pressed)	7.0	0.95
PETN (pentaerythritol tetranitrate)	$C(CH_2ONO_2)_4$		1.6 (pressed)	7.92	1.31
Pentolite	50/50 TNT/PETN		1.63 (cast)	7.7	1.1
Trinitrotoluene (TNT)	$CH_3C_6H_2(NO_2)_3$		1.59 (cast)	6.9	0.9
			1.45 (pressed)	6.9	–
			1.03 ("Pelletol")	5.1	–
			0.8 (grained)	4.2	0.8
Amatols	50/50 AN/TNT		1.55 (cast)	5–6.5 (depending on diameter)	0.95
	80/20 AN/TNT		1.0 (loose)	4 (large diameter)	0.93
			1.45 (pressed)	5.6 (large diameter)	

Sensitivity	*Major Characteristics*	*Uses*
From cap sensitive to very low—depends on aluminum fineness.	Gelatin; no explosive ingredients.	Small diameter, underground and general blasting.
Low (requires boosters)	Gelatin to thick or thin pea-soup consistency.	Large diameter, underwater, wet- and dry-hole blasting, large bombs.
Moderately high, but less than dynamites	Good "fumes;" fair water resistance; powerful; economical.	Small diameter blasting.
Average	Very high brisance.	Bursting charge and special weapons.
High	Very high brisance.	Experimental standard.
High	High brisance; less sensitive than RDX and PETN.	In Ednatols for bursting charges.
Very low	Insensitive to shock and friction; melts with decomposition; shells filled with high-pressure pressing.	Armor-piercing shells.
High	Highly inflammable white powder.	Demolition blocks and Trojan blasting explosives.
High	Very sensitive; rapidly reacting; easily pressed with 1–2% graphite; high brisance.	Booster; base charge in caps; in tetrytols for bursting charges.
High	Very powerful and sensitive (more sensitive than RDX, less than NG).	In Primacord fuse; base charge in caps.
Moderate	High pressure or brisance; primacord sensitive	Booster and special weapons; commercial booster for prilled AN-fuel oil and slurry explosives.
Low	Easily melted and cast; suitable liquid for slurrying with other explosives; easily pressed into blocks; completely waterproof.	Military; "Nitropel" TNT used in slurry explosives and in filling annulus between charge and borehole in water-filled holes; in amatols.
Low	Insensitive; hygroscopic, not waterproof; less brisant but stronger than TNT; 50/50 can be cast; 80/20 either pressed or granulated.	Military; oil well shooting; quarrying; dry-hole booster for very low-sensitive types.
Low		

TABLE 19.1 (cont'd.)

Name	*Composition of Chemical Formula*	*Density (g/cc)*	*Detonation Velocity (km/sec)*	*Available Energy (kcal/g)*
Dinitrotolune (DNT)	$CH_3C_6H_3(NO_2)_3$	1.28 (liquid) 0.8 (granular solid)	5 2–3.5 (depending on diameter)	0.7
Nitromethane (NM)	CH_3NO_2	1.12	6.2	–
Cyclonite (RDX)	$C_3H_6N_6O_6$	1.2 (loose) 1.6 (pressed)	6.8 8.0	1.32
HMX	$C_4H_8N_8O_8$	1.89	9.1	1.35
HBX	Mixtures of RDX, TNT, aluminum, and wax	1.78	7.5	1.5
Plastic Explosives (Compositions A, C, C-2, C-3, C-4)	Waxed RDX	1.45–1.6	8.0	1.1–1.3
PBX 9404	94/3/ HMX/binder/ nitrocellulose	1.84	8.8	1.3
Nitroglycerin (NG)	$C_3H_5(ONO_2)_3$	1.59	7.8	1.41
Ethylene glycol dinitrate (EGDN)	$C_2H_4(ONO_2)_2$	1.48	7.4	1.43
Straight dynamites[b]	20–60% NG, in balanced SN dope 20% grade ≡ 20% NG, etc.	1.3	4–6	0.55–0.85[b]
Ammonia dynamites (and permissibles)	As above except NH_4NO_3 replaces part of NG and $NaNO_3$	0.8–1.2[b]	1.5–5.5 Depends on AN particle size, NG content.	0.7–0.9[b]
Blasting gelatin	92/8/NG/nitrocotton ("Solidified" NG contains some wood pump to minimize low-order detonation)	1.55 (1.45)	7.5 (7.2)	1.45 (1.4)

[b]Depends on grade.

Sensitivity	*Major Characteristics*	*Uses*
Very low	Reddish brown or yellow liquid.	Sensitizer in "Nitramons"; 60/40 NG/DNT in oil well shooting; up to 20% in TNT bursting charges; in FNH (flashless) propellant; 6% in small-arms ammunition (with guncotton).
Moderate	Clear, watery liquid	Special demolition, experimental studies of liquid explosives.
High	High thermal stability in solid state; expressively sensitive in pure state; 1.65 times as strong as low density TNT; 1.45 times as strong as cast TNT.	Major ingredient in plastic explosives; one of most brisant explosives in cast TNT (composition B); base charge in caps.
High	Better than RDX in all respects.	Same as RDX.
Average	Very powerful.	Underwater explosive.
Moderate	Plastic, easily molded or pressed.	Specialized military demolition.
Moderate	Plastic bonded.	Specialized military demolition.
Very high (almost a primary explosive)	Oily, toxic liquid; volatile above 50°C; gelatinized by nitrocotton; exhibits low-order detonation with threshold priming.	Shooting oil wells; main explosive in dynamites; used in dynamites; used in double-base powders.
Very high	Closely resembles NG; more volatile, toxic, slightly stronger but less brisant (owing to lower density).	Used in solution with NG as freezing point depressant.
High	Cheesy, plastic substance; packed in paper cartridges; may be slit and tamped in borehole for greatest blasting effect; fired by detonator as are all dynamites; heat, friction, shock, and flame sensitive.	Ditching, stumping, other uses where high propagation-by-influence "sensitiveness" is required.
High	Cheaper than comparable grade straight dynamites; must be waterproofed by special additives.	General small and large dynamite blasting; permissible (some grades).
High	Strongest, most brisant dynamite; completely waterproof; exhibits low-order detonation with threshold priming and under high pressures.	Oil well and submarine blasting, tunnel drilling, demolition.

TABLE 19.1 (cont'd.)

Name	*Composition or Chemical Formula*	*Density (g/cc)*	*Detonation Velocity (km/sec)*	*Available Energy (kcal/g)*
Straight gelatin dynamite	20–90% grades	1.3–1.6[b]	4–7	0.75–1.15[b]

PROPELLANTS

Name	*Composition or Chemical Formula*	*Sensitivity*
Colloidal nitrocellulose (N.C.) powders	Pyrocotton: cellulose nitrate with 12.6% N.	Low
	Guncotton: cellulose nitrate with 13.3% N.	
Double-base powders	60–80% Nitrocellulose. 20–40% Nitroglycerin.	Moderate
Cordite	65% N.C., 30% NG, 5% vaseline	Low
FNH (Flashless nonhydroscopic powders)	Either straight N.C. or double-base powders with addition of coolants, etc., to prevent muzzle flash, and decrease water absorption.	Low
Albanite; DINA powder	Di(2-nitrooxyethyl)nitramine.	Low
Rocket powder (solventless powder)	Nitrocellulose plasticized with about 50% NG, plus stabilizers and potassium salts.	Low
Chemical propellants[c]	Hydrogen perioxide, 80–90% H_2O_2 plus Ca, Na, or K permanganate (solid or aqueous solution).	
	Hydrazine hydrate plus methyl alcohol.	
	Fuming nitric acid-aniline.	
	Mixed acid-monoethylaniline.	
	Liquid oxygen-kerosene.	
Black powder	75% KNO_3 (or $NaNO_3$), 15% charcoal, 10% sulfur [example].	High

[c]Many new rocket propellants have been described; the best ones are under security classifications.
[d]DDT = deflagration to detonation transition.

Sensitivity	*Major Characteristics*	*Uses*
High	Jelly-like substance; powerful, waterproof; exhibits low-order detonation under threshold primering and high pressure.	In hard rock; mudcapping, demolition; submarine blasting.

Major Characteristics	*Important Uses*
Burning rate controlled by graining, hygroscopic; smokeless flame, with intense flash; gelatinized with alcohol-ethel.	Combined with stabilizers and modifiers to make smokeless powders for artillery, small arms, and sporting ammunition.
Pyrocotton and guncotton are usually blended to secure an average of 13.15% N.	Dry guncotton in fiber form is used in primers fired by an electric current.
Very rapid burning rate, controllable by surface area; mor powerful and more readily ignitable than straight N.C. powders; causes erosion of gun bores; can be detonated and is subject to DDT.[d]	Propellant for mortars and sporting ammunition; not used by U.S. armed forces as cannon powder because of bore erosion.
Gelatinized with acetone.	Propellant for large caliber naval guns (English).
Like other smokeless powders, but can be rolled into sheets; flash reduced by DNT, potassium salts, etc.	Propellant for small armor-piercing rockets such as the "Bazooka" (NG base); for naval ammunition (NG base).
Better flashless powder than FNH powders.	Naval ammunition.
Very rapid, uniform burning rate; can be made with thick section since no solvent need be removed.	For rockets up to 4.5 inches.
Catalytic decomposition into water and O_2 releases about 1000 Btu per pound. Supplies oxygen to burn petroleum fuel.	For driving turbines on submarines: V-2 rocket-fuel pumps; jet motors; launching device for ram-jets.
Rapid combustion; fuel and oxidizer are both liquid.	For torpedo turbine drives.
Rapid rate of reaction generates heat and gases.	For launching device for ram-jets; jet motors.
As above.	As above.
	Rocket motors.
Cheap; excellent "heaving action," persistent smoky flame; very sensitive to friction, spark, and heat; hygroscopic.	Time (delay) fuses for blasting and shell; in igniter and primer assemblies for propellants; pyrotechnics; $NaNO_3$ powder, in commercial black powder and for practice bombs and saluting charges; (it is being discontinued as blasting charge).

and economical commercial slurries in use today are the so-called *blasting agents* in which no explosive per se is used. Foremost among these are the aluminum-sensitized slurries. The higher-grade dynamites are also sometimes used, though inefficiently, as boosters for the slurry explosives and slurry blasting agents. Most military explosives also require boosters; a 10-g tetryl or pentolite charge usually satisfies the booster requirements of the secondary military explosives, illustrating the level of sensitivity of the explosives used by the armed forces. Artillery rounds may contain igniters, primary explosives, boosters, and secondary explosives, all assembled so that each element serves a definite function.* Today cap-sensitive slurries, usually sensitized with fine aluminum, are rapidly coming into use for small-diameter, underground mining. In 1969 slurries comprised roughly 14 per cent of the total commercial market in the U.S.A. and 37 per cent exclusive of ANFO.

Detonation-wave velocities are measured by high-speed photography methods involving framing cameras capable of taking millions of frames per second, streak cameras with writing speeds of several miles per second, elaborate electronic timers called "pin-oscillographs," and centimeter-wave systems[5] which use wave guide and doppler principles. Figure 19.1 illustrates how the detonation velocity is measured by means of the streak camera. Since World War II, the above methods have largely replaced the older "Dautriche" and "Metegang" methods.[4]

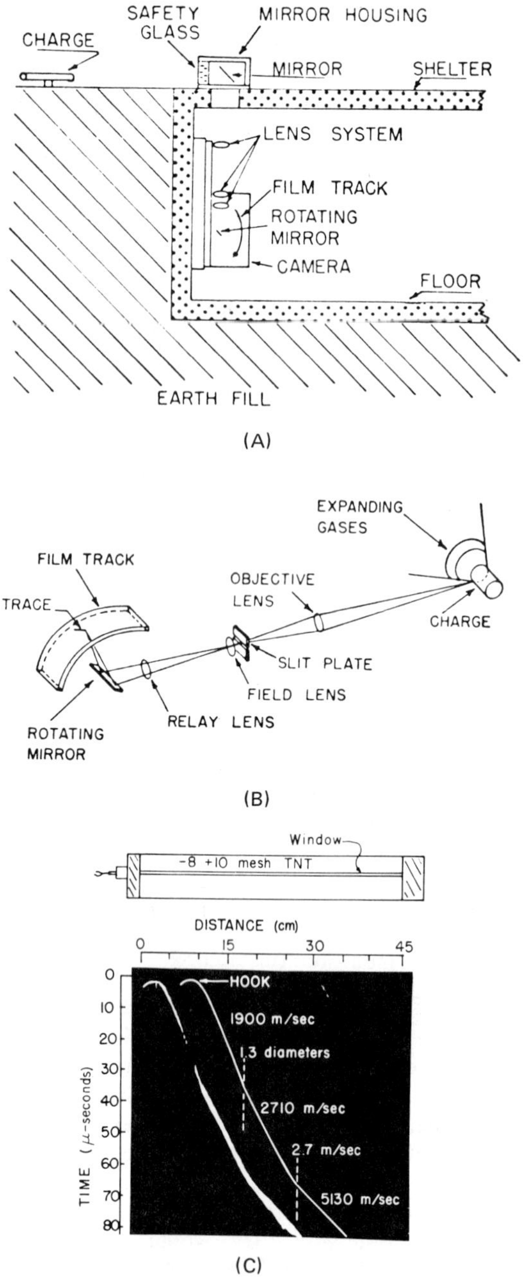

Fig. 19.1 Measurement of detonation velocity by means of the streak camera. (A) Arrangement in inside-out bombproof shelter. (B) Measurement of detonation velocity by means of the streak camera. (C) Streak camera "trace" illustrating detailed information made possible by following the detonation wave from the point of initiation until stable (steady state) detonation is established.

Some Principles of Explosives Technology

The most fundamental requirement of an explosive is that its characteristic chemical re-

*The design of ammunition is an interesting subject. A description of the function and requirements for cartridge primers and fuse primers (which are frequently different formulations) for propellant igniters and bursting charge detonators, etc., is available in a War Department publication, T19-4-205, "Coast Artillery Ammunition." This pamphlet also provides a description of mechanical base and nose fuses of various types, and explains how they work.

Strictly speaking, a "round of ammunition" includes "everything necessary to fire the gun once."

action be highly exothermic. The most prominent explosives exhibiting exothermic reactions are the ones which undergo oxidation-reduction reactions. These are of two general types: first, the *internal redox* compounds in which the oxidation-reduction involves oxidant and reducer radicals which are both *within* the same molecule; second, *redox mixtures*, i.e., simple mixtures in which the oxidant and reducer are separate and distinct molecules. Examples of the internal redox compounds are NG (nitroglycerin), EGDN (ethylene glycol dinitrate), PETN, RDX, TNT, tetryl, and Haleite (or EDNA). These internal redox compounds generally have a C_xH_y radical or trunk unit as the reducer part of the compound and a nitrate ($—ONO_2$), nitro ($—NO_2$), nitramine ($—NHNO_2$), or other similar oxygen-bearing radicals as the oxidant. For example, NG has as the internal reducer the $—CH_2\overset{|}{C}HCH_2—$ unit and as the oxidant, three nitrate groups; EGDN has $—CH_2CH_2—$ as the reducer unit, and two nitrate groups as the oxidant. PETN has $C(CH_2—)_4$ as the internal reducer and as the oxidant four nitrate groups; RDX has as the internal reducer three $—CH_2—$ units, and as the internal oxidant three $=N—NO_2$ units; TNT has as the internal reducer the unit

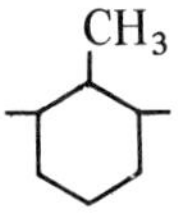

and as the internal oxidant three nitro groups; Haleite, or EDNA, has as the internal reducer the ethylene radical $—CH_2CH_2—$, and as the internal oxidants two nitramine groups; and finally, tetryl has two internal reducer units, the phenyl and the $—CH_2—$ radical, and two types of internal oxidant units, the nitro and the nitramine radicals. The practical internal redox compounds are necessarily stable compounds, otherwise they would be unduly hazardous. While nitrate, nitro, nitramine, chlorate, perchlorate, and similar groups are sometimes referred to as "explosophore" groups, this is a misnomer since they are actually entirely stable groups. For instance, there are many compounds containing nitrate and nitro groups that are nonexplosive, including many inorganic nitrates and nitro compounds. Nitric acid (HNO_3) and nitrous acid (HNO_2) are not explosive because their reducer unit (one hydrogen atom) can use up only a very small part of the oxygen of the oxidant radical. Also, ordinary metal nitrates and nitrites are nonexplosive. Ammonium nitrate is a very insensitive explosive even though its fuel radical $NH_4—$ reacts in a redox-type reaction with only two of the three oxygen atoms of the $—ONO_2$ radical. As evidence that the internal redox explosives are stable chemically, note that the heats of formation ΔH_f (positive for heat evolved and negative for heat absorbed, relative to the constituents in their standard states) are 83, 56, 123, and 13 kcal/mole for NG, EGDN, PETN, and TNT, respectively. Of the seven examples at the beginning of this discussion, only RDX and tetryl have negative heats of formation (-18.3 and -9.3 kcal/mole, respectively). The heats of explosion of RDX and tetryl are about 290 and 190 kcal/mole, respectively, showing that the "explosophore" character of the $—NHNO_2$ and $=N—NO_2$ groups is far less important than the internal redox character of these compounds.

Explosives of the redox-mixture type are exemplified by the popular 94/6 prilled ammonium nitrate-fuel oil, black powder, and dithekites (i.e., mixtures of nitric acid and nitrobenzene), and propellants based on ammonium perchlorate.

Many internal-redox explosive compounds may be made more powerful by combining their internal redox with the redox-mixture principle. For example, TNT has appreciably more reduction than oxidation potential, i.e., it is highly oxygen-deficient. Therefore its explosion potential is enhanced by mixing it with an oxidant. Likewise, an explosive compound that has an internal redox character but is rich in oxygen may be made more powerful by mixing it with almost any fuel or combustible substance. Ammonium nitrate and ammonium perchlorate are the outstanding examples of explosives having this character. An example of an oxygen-balanced mechanical mixture of both an oxygen-rich and an oxygen-deficient explosive substance is 80/20 amatol (AN/TNT). It is nearly always advantageous to oxygen-balance an explosive by mixing it with a suitable fuel or oxidant; although the highest strength frequently occurs in mixtures, especially the high temperature ones, having a somewhat negative oxygen balance. Aluminized and some other metallized explosives are exceptional, since the maximum strength occurs at a much more negative oxygen balance. For example, 80/20 tritonal (TNT/AL), although much more oxygen deficient, is appreciably more powerful than pure TNT.

Explosives based strictly on "explosophores," i.e., single groups or radicals whose rearrangement or decomposition generates large quantities of heat, are common among the primary explosives. For example, lead azide (PbN_6) contains only an "explosophore" group (—N=N=N or $-N\begin{smallmatrix}\diagup N\\ \| \\ \diagdown N\end{smallmatrix}$). All the heat of explosion, Q, of lead azide is therefore attributed to the rearrangement of this group to yield nitrogen, and in this case the heat of explosion is entirely the negative of the heat of formation ΔH_f. Another "explosophoric" group is the acetylide (—C=C—) group in which $Q = -\Delta H_f$ at low pressure. Under very high pressures the products from the detonation of acetylene consist of CH_4 and carbon instead of H_2 and carbon; then Q is somewhat greater than $-\Delta H_f$, i.e., about 73 instead of 55 kcal/mole. The fulminate group (—ONC) is another "explosophore" group, however $Q \doteq -2\Delta H_f$ because of the formation of CO and CO_2 in the products of detonation at high pressures.

The heat of explosion and the nT product, i.e., maximum temperature multiplied by the expanded volume of the gaseous products, are each roughly proportional to the total available energy. But the detonation pressure, P_2, is determined by the density and velocity. This is expressed approximately by the relation $P_2 = \frac{1}{4}\rho_1 D^2$, where ρ_1 is the density and D is the detonation velocity. Furthermore, D usually increases almost linearly with ρ_1; the pressure, therefore, involves terms in the first, second, and third power of the density. Hence in order to obtain high densities, casting a molten material into a shell is preferable to loading loose grains and tamping. However, if the melting point of the explosive is excessively high, the charge may undergo decomposition or even explode upon melting. TNT, with a melting point of 80.2°C, is easily and safely cast, whereas the casting of picric acid (m.p. 123°C) is hazardous; casting substances of still higher melting points is virtually impossible. For example, PETN and tetryl (m.p. 139 and 130°C, respectively) are extremely hazardous and may explode on melting. It is therefore common practice, in obtaining high density explosives, to make a slurry of the high-melting, more powerful explosive in TNT; or in some cases to make an eutectic solution of two explosives, e.g., tetryl and TNT, to permit casting at a lower temperature. High densities are also obtained by pressing the charges.

On the other hand, the burning rate of the low explosive decreases with increasing density; this is the basis for time-delay pellets used in fuses. The rate of burning of a propellant may be controlled by designing the grain so that the surface area available for burning increases as the weight of the grain decreases. This is called progressive burning and is typified by smokeless powders which are fluted on the outside and provided with one or more longitudinal perforations to control the burning rate. In the case of spherical propellant powders where the surface decreases with decreasing weight (regressive burning), the surface is coated with a chemical called a deterrent. This decreases the rate of burning while it is present, to match the slower rate prevailing when the diameter of the sphere is less. Control of the area of surface burning of the single-grained rocket propellants is tremendously important in rockets; it is accomplished by carefully shaped longitudinal perforations, each patterned to give a particular predetermined pressure-time curve.

Heat-absorbing compounds, such as mineral salts, powdered metals, and dinitrotoluene, may be added to smokeless powders to make them flashless. A graphite coating applied to grained propellants not only acts as a deterrent, but also decreases the hazard of handling them by removing static charges, acts as a lubricant to facilitate pressing them to a higher density, and increases the ignitability of the powder. Flame retardants and coolants are also used to some extent to cut down flame and increase safety in the detonating explosives ("permissibles") used in coal mining.*

The availability of raw materials frequently governs the choice of explosives in wartime. For example, during World War II the Italian forces used PETN rather than tetryl as a booster because the former is made from synthetic methanol, while the latter requires benzene, which was in short supply in Italy.

Many other factors are considered in evaluating an explosive. These include the toxicity of the gases produced on explosion,† hygroscopicity or water-resistance, stability in storage, and cost of manufacture.

The high explosives used for peacetime applications (industrial explosives) are chiefly the dynamites, slurries, and the ammonium nitrate mixtures. Coal mining utilized 41.2 per cent of all commercial explosives in 1950, metal mining 17.8, quarrying and nonmetal mining 19.6, railway and other construction 19.4, and all other uses, 2.0 per cent. In 1968 these figures were, respectively, 35.2, 20.7, 20.4, 21.1, and 2.6 per cent. (Seismic prospecting utilized 2.1 per cent). Prilled ammonium nitrate-fuel oil mixtures, which are being used extensively today, comprise roughly two-thirds of the total market in the U.S.A., thus effecting considerable savings in explosive costs in open-pit and underground blasting. Slurry explosives are also gaining rapidly in commercial blasting, not only in open-pit but also in underground mining.

Nitroglycerin

Nitroglycerin (glyceryl trinitrate) is a colorless (when pure) oily liquid which freezes at 13°C and is very sensitive to shock, especially when it contains air bubbles or when conditions are such that air might be trapped in the shocked liquid. It is made commercially in a steel nitrator which has steel cooling coils and a mechanical agitator. In the interests of safety and stability a very pure glycerin, above 99.9 per cent pure, is required. A typical process is described below.

About 6800 lbs of mixed acid having the composition 48 per cent nitric acid (HNO_3) and 52 per cent oleum (H_2SO_4)* is charged into the nitrator with agitation and is cooled by the circulation of calcium chloride brine at -20°C through the coils. A total of about 1300 lbs of glycerin is then added slowly in a steady stream. The temperature of nitration is 2-3°C,† and if there is any tendency for the temperature to rise, the flow of glycerin is stopped. The addition step takes about 50-60 minutes; agitation is continued a few minutes and then the spent acid is allowed to separate from the nitrated glycerin oil. The spent acid carries away in solution about 5 per cent of the total yield. The acid is given an additional settling treatment, then is denitrated for recovery of the HNO_3 and H_2SO_4. The glycerin trinitrate is washed with warm water, then with a 2-3 per cent Na_2CO_3 solution, and finally with a concentrated solution of NaCl which breaks down any emulsion present and partly dehydrates the nitroglycerin. The product, which contains about 0.5 per cent water, is stored in lead-lined tanks. A process developed in Italy for continuous nitration of glycerin is rapidly being adopted throughout the world.

The Dynamites

The high sensitivity of nitroglycerin to shock was a deterrent to its use until the chance discovery by Nobel in 1862 that large quantities of the liquid explosive could be absorbed in

---

*For use in gassy or dusty mines, explosives must pass tests conducted by the U.S. Bureau of Mines. If approved, they are placed on a "permissible list." Permissibles are usually ammonium nitrate explosives compounded to suit specific conditions. For a list of available U.S. brands, see Bebie, page 118.[1]

†Toxic gases produced in detonation are called *fumes*.

*A mixture of nitric and sulfuric acid is commonly used for organic nitrations because the sulfuric acid is capable of forming a hydrated molecule, thereby effectively removing one of the products of any nitration reaction, i.e., water. This shifts the equilibrium to the right, allowing more of the nitrated product to form and resulting in a higher yield than would otherwise be possible. The sulfuric acid does not otherwise take part in the reaction and normally is completely recovered for reuse, except for minor losses.

†This is below the freezing point of nitroglycerin but in acid solutions the oil will not freeze to the coils if there is sufficient agitation. Adapted from "Explosives Manual," Lefax, Inc., Philadelphia, Chap. XI, part 2.

kieselguhr (diatomaceous earth), forming a plastic cheesy substance which could then be transported and used with appreciably less hazard. This original formulation (kieselguhr dynamite) is now rarely used, however, since the presence of 25 per cent of inert "guhr" diminishes the blasting effect by absorbing energy. The absorbent used today is generally a mixture of wood pulps, meals, sawdust, flour, starch, cereal products, or the like, to which is added appropriate oxidizers, e.g., ammonium nitrate and/or sodium nitrate, and a small amount of an antacid ($CaCO_3$ or ZnO). The explosively active absorbents are called "balanced dope" because they are always carefully oxygen-balanced for maximum strength and minimum fumes. Various amounts of nitroglycerin may be used in the formula, depending on the strength desired. The commercial straight dynamites contain from 20 to 60 per cent, the grade being the same as the percentage of nitroglycerin present.* Ethylene glycol dinitrate ($O_2NO{\cdot}CH_2{\cdot}CH_2{\cdot}ONO_2$) or other explosive oils† are now added to all domestic dynamites to make them low-freezing. Nitroglycerin may be gelatinized with 7–8 per cent collodion cotton to make *blasting gelatin.* Gelatin dynamites are mixtures of gelatin and balanced dope ranging from 20 to 90 per cent grades, depending on the ratio of gelatin to balanced dope used. Gelatin dynamites comprise two types: the "straight" gelatins and the "ammonia" gelatins. The straight gelatins are based on sodium nitrate dopes, and the ammonia gelatins use NH_4NO_3 or mixtures of NH_4NO_3 and $NaNO_3$ dopes.

Straight dynamites have been replaced for many uses by the *ammonia dynamites*, which provide the same blasting strength but contain much less nitroglycerin for a given grade. Some of the high ammonium nitrate dynamites are really ammonium nitrate fuel or combustible explosives in which a small percentage of nitroglycerin is used to sensitize the NH_4NO_3 to detonation. Other ammonia dynamites (the highest grades) will contain appreciable quantities of nitroglycerin. Gelatin dynamites (ammonia and straight) have the greatest shattering action or brisance;* the straight dynamites are intermediate, and the ammonia dynamites have the least brisance for a given grade of dynamite, although there is some overlapping from one series to the other if grade is not also taken into consideration. The explosives with highest brisance are preferred for "mudcapping," hole "springing," and very hard rock shooting, although brisance must also be balanced against cost. Finally, mention is made of the *semigelatin dynamites* which contain sufficient gelation agent (nitrocellulose) to thicken the liquid NG appreciably, but not enough to make a stiff gel.

To make any of the above compositions, the carbonaceous material and the other "dope" ingredients are dried together in a steam-heated pan dryer, then mixed with ammonium nitrate and/or sodium nitrate in a spiral-blade mixer. For dynamite, the "balance dope" is placed in an edge runner (muller) and nitroglycerin is added carefully. For gelatin dynamite, the dope and nitrocellulose are blended with the nitroglycerin in a sigma-blade mixer. All equipment must be of non-sparking construction, and liquid nitroglycerin is handled in small lots in rubber-tired buggies and rubber pails. The blended products are packed in paper cartridges by automatic machinery and the cartridges are then paraffin-coated and boxed in sawdust.

TNT (2,4,6-trinitrotoluene)

One of the most important military explosives, TNT, is made either by continuous or by

*A "40% straight dynamite" contains, for example, 40% nitroglycerin, 44% $NaNO_3$, 14% carbonaceous material, 1% antacid, and 1% moisture. A "40% ammonia dynamite," with the same strength, may contain about 15% nitroglycerin, 32% NH_4NO_3, 38% $NaNO_3$, 4% sulfur, 9% carbonaceous material, 1% antacid, and 1% moisture.

†Dinitromonochlorhydrin and tetranitrodiglycerin are also used.

*A test for comparing the brisance of explosives is the sand bomb test: 80 g of standard Ottawa sand, sized -20 to +30 mesh, is placed in a thick-walled cylinder. A cap containing 0.400 g of the explosive is then inserted and the remainder of the cavity filled with 120 g of sand. The explosive is suitably detonated. The weight of sand which then passes the 30-mesh screen is a measure of the brisance of the explosive. For further details see Olsen and Greene, "Laboratory Manual of Explosive Chemistry," New York, John Wiley and Sons, Inc., page 80, 1943.

stepwise nitration of toluene, In the stepwise process, mixtures of nitric and sulfuric acid are used, and the reactions are carried on in well-agitated steel nitrators equipped with steel cooling (or heating) coils. In 1941, the accepted daily production of a TNT "line" was about 36,000 lbs. By the middle of 1945, however, the output of the same line, with but few modifications of equipment, was nearly 120,000 lbs per day. This increase was brought about by many factors, including adoption of the principle of adding the organic material to the mixed acid (indirect nitration) for all three nitrations. Previously, for mono- and trinitrations, the oil was first added to a large volume of cycle acid or oleum to "cushion" the reaction, and this was followed by slow addition of the nitrating acids. Since, in the newer method, only a small amount of nitratable material is present in the acid at any time, the hazards of the operation are appreciably reduced. Also, much less time is required for the reaction cycle.[5]

Temperature control and agitator speed are important. The acid charge is first preheated to near reaction temperature by passing hot water through the coils; then as the toluene, or mono- or dinitrotoluene is added, cooling water is turned on in the coils to maintain the proper temperature. In mononitration, 1,600 lbs of toluene are run into some 11,000 lbs of mixed acid, and the temperature is allowed to rise to 140°C. The batch is held at this temperature for a cooking period of about 25 min, or until the proper specific gravity is reached. In dinitration, the cooking is done for five minutes at 180°C. The temperature in trinitration is allowed to rise to 180°C, but after oil addition is completed the temperature is raised at a predetermined rate to 230°C where it is held for 30 mins. A strong upward circulation is maintained during addition of the oil, but the agitator speed is changed during the cooking period. When nitration is completed the charge is cooled somewhat, allowed to settle, and separated into waste acid and nitrated oil by visually examining the flow as the nitrator is emptied and switching valves as the low-gravity oil appears in the sightbox following the acid. Di- and trinitrations are carried on in the same nitrator; the mononitrator is located some distance away in another building. Nitrated oil is handled by pumps, while waste acid is moved by air pressure. The spent acids* are fortified in steel tanks similar in construction to the nitrators.

The trinitrotoluene is washed, while still in the liquid state, by agitation with a dilute ammoniated solution of sodium sulfite. The ammonia neutralizes the residual acid while the sulfite removes as soluble complexes any 3,5-dinitrotoluene and other undesirable by-products of nitration. The TNT is then crystallized by rapid addition of cold water, filtered on a Nutsch filter, melted with steam, dried by passing warm air through the melt, and then resolidified on a flaker drum. The flakes are packed in paper-lined boxes. Properly prepared product will have a setting point of 80.2°C, minimum. Two other types of TNT were also available, namely, the "shot tower" product sold by du Pont under the trade name "Pelletol" and a very fine-meshed fluffy graining-kettle product. The latter was the form of solid TNT common prior to World War II.

Cyclonite (RDX)

Cyclonite (RDX) (symmetrical trimethylenetrinitramine) is over 50 percent more powerful than low-density granular TNT and has good stability. Its excellent thermal stability makes it highly adaptable, especially at high temperatures. Its sensitivity and high melting point (about 200°C) were deterrents to its use until the discovery that it could be desensitized with beeswax for press-loading into shells, slurried with TNT for casting, and mixed with a special oil to make "plastic explosives"

---

*The strongest acid mix (concentrated HNO_3 plus oleum; no water) is required to introduce the third nitro group (trinitration). The diluted spent acid from the trinitration step is fortified with 60 percent HNO_3, and water is added to adjust the strength for dinitration. The waste acid from this operation is again fortified and diluted for use in the mononitration. The monowaste acid is denitrated with steam in a packed tower, the nitric oxides are recovered by absorption in water, and the 75 per cent sulfuric acid is concentrated to 66° Bé and remade into oleum. While some 220 lbs of 109 per cent oleum is required for each 100 lbs of TNT made, all but 2.5 lbs of this is recovered and recycled.

for demolition work.* The problem of its high cost was resolved first by a reduction in the cost of the raw material, hexamethylenetetramine, as it became more available, and second by the development of an ingenious method which combines two chemical reactions, one which does not require hexamethylenetetramine:

Reaction 1 (nitrolysis)

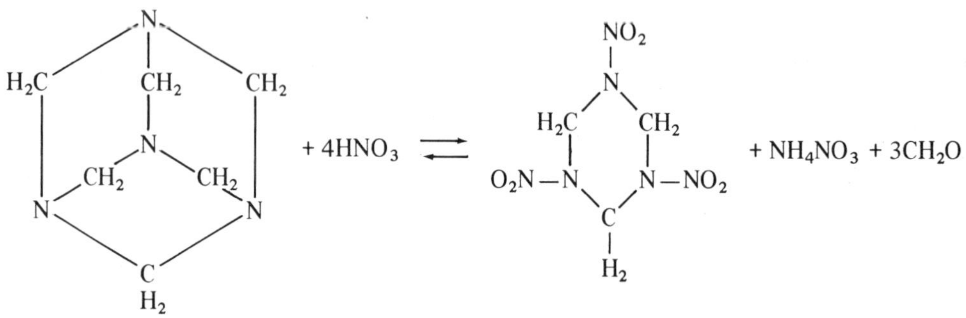

Hexamethylenetetramine
$(CH_2)_6N_4$

Cyclonite (RDX)
$(CH_2 \cdot N \cdot NO_2)_3$

Reaction 2

$$\underset{\text{Formaldehyde}}{3CH_2O} + \underset{\text{Ammonium nitrate}}{3NH_4NO_3} + \underset{\text{Acetic anhydride}}{6(CH_3CO)_2O} \rightleftharpoons \underset{\text{Cyclonite}}{(CH_2 \cdot N \cdot NO_2)_3} + \underset{\text{Acetic acid}}{12CH_3COOH}$$

Combined Reaction

$$(CH_2)_6N_4 + 4HNO_3 + 2NH_4NO_3 + 6(CH_3CO)_2O \rightleftharpoons 2(CH_2 \cdot N \cdot NO_2)_3 + 12CH_3COOH$$

The formaldehyde and part of the ammonium nitrate required by Reaction 2 are supplied as products of Reaction 1. Two moles of cyclonite are thereby theoretically produced from each mole of hexamine. The actual yield is about 1.25 moles per mole, due to the formation of some fifty undesired compounds, but this yield is better than the total obtained if the two reactions were carried on separately. The capacity reached at one point, during World War II, was over a million lbs per day of RDX-TNT and waxed RDX mixtures.*

EDNA

EDNA or Haleite† (ethylenedinitramine), while not quite as powerful nor as stable as cyclonite, is somewhat less sensitive, and is appreciably more powerful than TNT. Although several reactions have been developed for its production, the most economical appears to be the synthesis of ethylene urea, starting with formaldehyde and hydrogen cyanide,

*From a paper by Ralph Conner, formerly Chief, Division 8, NDRC, presented Sept. 9, 1946, at the Chicago meeting of the American Chemical Society. See also "Chemistry, Science in World War II," edited by W. A. Noyes, Jr., Chapters 2 to 11 by G. B. Kistiakowsky and Ralph Connor, Boston, Little, Brown and Co., 1948.

*RDX was used in World War II in mixtures: British and U.S. Composition "A": 91 per cent RDX, 9 per cent wax; Composition "B": 60 per cent RDX, 40 per cent TNT, 1 per cent wax; Composition C: 87 per cent RDX, 13 per cent wax mixture; Compositions C-2, etc.; Russian: 71.9 per cent RDX, 16.3 per cent TNT, 11.8 per cent tetryl. German compositions were similar to U.S. "A" and "B"; the Italians used an RDX-ammonium nitrate-wax mixture. Only the Japanese used RDX without a desensitizer, hence the shell loading had to be pressed to an extremely high density to avoid premature explosion.

†Named for Dr. G. C. Hale of Picatinny Arsenal.

and followed by nitration of the amine groups:

$$\underset{\text{Formaldehyde}}{CH_2O} + HCN \longrightarrow \underset{\text{Cyanohydrin}}{HO\cdot CH_2\cdot CN} \xrightarrow{NH_3}$$

$$H_2N\cdot CH_2\cdot CN \xrightarrow{H_2} \underset{\text{Ethylene urea}}{H_2N\cdot CH_2\cdot CH_2\cdot NH_2}$$

$$H_2N\cdot CH_2\cdot CH_2\cdot NH_2 + 2HNO_3 \longrightarrow \underset{\text{Haleite}}{O_2N\cdot NH\cdot CH_2\cdot CH_2\cdot NH\cdot NO_2}$$

Tetryl

Tetryl (2,4,6-trinitrophenylmethylnitramine), the standard booster used by the United States armed forces, may be made from dimethylaniline or from 2,4-dinitrochlorobenzene, both of which are made from benzene. Nitration of dimethylaniline is carried on by first dissolving the oil in 96 per cent sulfuric acid and adding the solution slowly to a mixed acid consisting of approximately 67 per cent HNO_3 and 16 per cent H_2SO_4. The temperature is controlled at about 70°C by cooling coils and agitation. In the nitrator, several reactions occur in sequence: (1) *ortho-para* nitration of the benzene ring, (2) oxidation of one of the N-methyl groups to carboxyl, (3) loss of CO_2, leaving an amine, (4) introduction of the third nitro group into the benzene ring, and finally (5), nitration of the amine to nitramine. The result is tetryl $[(NO_2)_3C_6H_2 \cdot N_2(CH_3)NO_2]$ The solid product is purified by boiling with water, followed by filtration. It may be recrystallized from benzene. In the alternate method the raw material is treated with methylamine to make monomethyldinitraniline, which is then nitrated with mixed acids to make tetryl directly. The yields are higher and the product does not require as many water washings. Also, the nitration is easier to control. On the other hand, the extreme toxicity of dinitrochlorobenzene is an unfavorable factor. Besides being used as a booster, where a small amount of graphite is added to facilitate pressing, tetryl is slurried with molten TNT and case for use as a bursting charge (Tetrytol).

PETN

PETN (pentaerythritol tetranitrate) is made by nitration of the four hydroxy groups of pentaerythritol $[C(CH_2OH)_4]$ by reaction with 94 per cent HNO_3 at about 50°C. Pentaerythritol may be made by the reduction of a mixture of formaldehyde and acetaldehyde, both of which are obtained from the corresponding alcohols. The pentolites made by casting slurries of PETN with TNT are used as boosters and in many specialized uses where relatively high detonator and primacord sensitivity is required. In 1956, cast 50/50 pentolite was introduced by M. A. Cook, H. E. Farnam, and the Canadian Industries Limited as a booster for slurry blasting agents and prilled ammonium nitrate-fuel oil mixtures.[6] It and related boosters are currently being used extensively for that purpose throughout the world, not only as pentolite but also as the core charge along with composition B or TNT as the main charge in the popular protected core or "Procore" boosters introduced in 1958.

Picric Acid

Picric acid (trinitrophenol) may be prepared* from benzene directly by simultaneous oxidation and nitration with nitric acid in the presence of mercuric nitrate (Russian), from monochlorobenzene, or from phenol. *Ammonium picrate*[9] is made rather easily by adding picric acid filter cake and aqua ammonia simultaneously and in small increments to a large amount of water in such a way that the batch is at all times slightly alkaline. These two explosives are of only limited importance today.

Initiators, Primers, and Igniters

The common primary explosives used in detonators and fuses (military detonators) are mercury fulminate, lead azide, lead styphnate, nitromannite, and diazodinitrophenol. These are highly sensitive to shock, heat or flame, spark, etc. Igniter compositions do not always include one of the highly sensitive initiators, but may be flame- or friction-sensitive mix-

---

*See the Lefax "Explosives Manual" previously mentioned, or refer to the 4th Edition, page 601.

tures of such compounds as potassium chlorate, antimony sulfide, and an abrasive. A black-powder charge is usually used in propellant and fuse primers. In the first instance it transmits the explosion to the igniter charge which may be black powder or a mixture of barium peroxide or nitrate with magnesium powder. In the second case it provides a delay action before detonating the booster charge.

Mercury Fulminate $[Hg(ONO_2)_2]$. This compound is made in small batches. One pound of mercury is added gradually to about 8 pounds of concentrated HNO_3 to make mercuric nitrate. This solution, which contains an excess of HNO_3, is refluxed with 10 lbs of 95 per cent ethyl alcohol. The reaction is vigorous, and is moderated in the later stages by addition of dilute alcohol. The solid fulminate crystallizes out as gray crystals which are screened, washed with cold water, drained, and stored in cloth bags under water. It is commonly used with other substances, such as potassium perchlorate, in a mixture which gives a larger flash. Powdered glass, TNT, $Pb(CNS)_2$, and Sb_2S_3 may also be added.

Lead Azide. This compound, used extensively in composition caps, is made by adding a 2 per cent aqueous solution of sodium azide* to a 5 per cent aqueous solution of lead acetate containing a small amount of dextrin which makes the product safer to handle. When lead azide is used in a percussion primer, it must be mixed with a suitable sensitizing explosive such as lead styphnate (lead trinitroresorcinate). While lead azide is the most common primary explosive or detonator element in blasting caps and fuses (a fuse is a military detonator), it is sometimes replaced by diazodinitrophenol, nitromannite, or other primary explosives.

Lead Styphnate. This compound is a common igniter for lead azide in the composition cap and is used in direct contact with the bridgewire to obtain fast ignition of the primary explosive. It is a primary explosive, but it serves better as the igniter because of its extremely high heat sensitivity and the great rapidity with which it ignites the detonator. It is used, for example, in bead form on the bridgewire as the igniter in the du Pont seismograph cap, which is one of the fastest of the EB caps when fired by small-current blasting machines. It is extremely sensitive to spark, static electricity, and flame, and therefore requires highly specialized equipment in processing and assembling in caps.

*Sodium azide, NaN_3, is made by heating a mixture of sodamide ($NaNH_2$) and lime or other water absorber in a stream of nitrous oxide, N_2O.

Black Powder

The many uses of this explosive are listed in Table 19.1. Since the 16th century, the formula for the standard fast-burning powder has been approximately 75 per cent potassium nitrate, 15 per cent charcoal, and 10 per cent sulfur. A slow-burning powder contains 59 per cent KNO_3. Powders suitable for time fuses are obtained by blending the two types. For blasting, a weaker powder, e.g., 40 per cent nitrate, 30 per cent charcoal, and 15 per cent sulfur, may be used. Its manufacture consists of a series of batch operations. The finely-ground components are moistened and kneaded together in a sigma-blade mixer, then milled in an edge-runner to make a more homogeneous mixture. The mass is pressed into dense cakes which are then granulated, dried, and screened to various sizes. The dust is used in pyrotechnics and to make fuses; the oversize is recrushed. The sized grains are "polished" and provided with a coating of graphite by tumbling them in rotating cylinders or coating pans. The blasting cartridge (coal mining) a single cylindrical piece with one perforation in the long axis, is formed by extruding a paste made of the proper sized grains. All the operations from granulation to screening are hazardous and are performed with remotely controlled machinery located in barricaded buildings. Black powder is rapidly becoming obsolete as a blasting agent, but it still remains the best fuse and igniter composition yet developed.

ROCKET PROPELLANTS

A rocket is a relatively simple energy-conversion device in which propellant chemical reactions take place and rapidly generate gases and heat. The acceleration and expansion of these gases through a nozzle results in the thrust that propels the rocket system. The uniqueness of the rocket is that it delivers very high power-to-weight ratios which make missile and space applications very feasible. In addition, since it carries along its own oxidizer and does not depend on air, it is the only internal combustion engine that can operate in a vacuum.

The use of rocket propellants has increased tremendously since the first Sputnik was orbited in 1957. This event triggered a large surge in the development of rockets for both space and military applications. While the Chinese used powder rockets as early as the thirteenth century, the modern era of rocketry began near the end of the nineteenth century. K. E. Tsiolkovskii in Russia, R. H. Goddard in the U.S.A. and H. Oberth in Germany were the pioneers. After World War II, the activity was intensified when the progress of German V-2 efforts became known.

The Principle of the Rocket

Figure 19.2 is a simplified illustration of how a rocket motor or engine* works. Figure 19.2a shows a chamber with a hole in one end in which gas is being generated rapidly enough to maintain pressure, P_c. Since the pressure is exerted equally in all directions, there will be a pressure imbalance forcing the chamber to the left because pressure is not exerted on the area where there is a hole on the right. The force, F, that will propel the rocket to the left is equal to the chamber pressure, P_c, times the area of the hole of throat, A_t.

$$F = A_t P_c$$

*A solid propellant rocket is usually referred to as a motor, while the liquid propellant type is designated as an engine.

In an actual rocket this accounts for only part of the thrust. As shown schematically in Figure 19.2b, an increase in thrust is obtained by accelerating the gas through the nozzle and expanding it to P_e, the pressure at the nozzle-exit plane. The new equation for force which is called thrust is

$$F = A_t P_c C_f$$

where C_f is the thrust coefficient whose value depends on the ratio of P_c to P_e and the area of the throat to the area at the exit plane.* Thus a nozzle-expansion cone is needed for high performance and is used on almost all rockets.

Since thrust is a function of motor design, it is not useful for comparing the performance of propellants. A much better criterion of rocket-propellant performance is specific impulse which can be defined as the thrust delivered per unit flow weight of propellant, which is the same as weight burned in unit time.

$$I_{sp} = \frac{1}{w}\int_{t_1}^{t_2} F dt$$

and the units are lb-sec (force)/lb of propellant. This is often abbreviated as sec. Measured I_{sp} values are determined by firing motors on elaborately instrumented thrust stands and converting the measured thrust according to the above equation.

It is time consuming and expensive to optimize propellant I_{sp} by firing numerous motors. Computer programs have been devised to calculate† the I_{sp} for various propellant

*A more detailed discussion can be found in Sutton.[7]
†A detailed discussion can be found in Siegel and Schieler.[8]

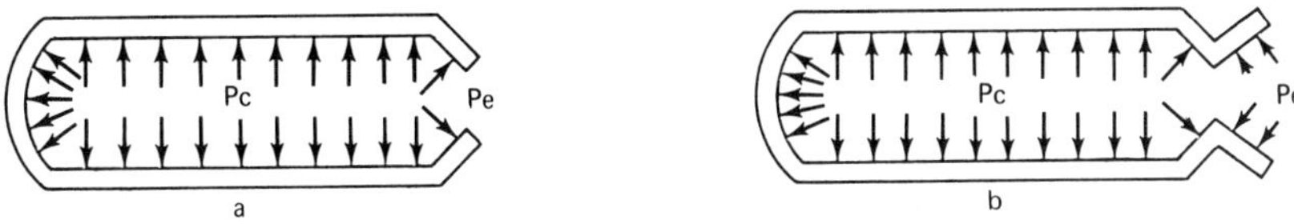

Fig. 19.2 Principle of rocket propulsion.

compositions. With a computer the calculations can be made rapidly, thus allowing many compositions to be investigated and the ratio of ingredients to be optimized. The results of the calculations are only as good as the thermodynamic data input to the computer program, but in most cases these data have been refined to the point that theoretical I_{sp}'s are dependable.

TYPES OF PROPELLANTS

There are two general types of propellants in widespread use; these are solid propellants and liquid propellants. Both of these require an oxidizer and a fuel. If a single compound is used then it is called a monopropellant. There is also a combination of these two types known as a hybrid propellant in which a solid propellant fuel is used with a liquid oxidizer.

Solid Propellants

A solid-propellant rocket motor is relatively simple in that it has very few moving parts (Fig. 19.3). The propellant in the motor is referred to as the grain. This is a carry-over from cannon powder terminology where each individual piece is called a grain. The grain is usually bonded to the metal- or fiberglass-reinforced plastic case and burns only on the interior surface called the perforation. Since the unburned propellant protects the case from heat this allows lighter weight cases to be used. The geometry of the perforation controls the shape of the pressure-time curve[9] after the motor is ignited. Since the propellant burns perpendicularly to the surface, the pressure-time curve can be neutral (level), progressive, or regressive. A progressive curve leads to increasing acceleration, while a regressive curve can yield constant acceleration because of weight loss as the propellant burns. Various perforation shapes can be used including cylindrical, star, and cruciform. In some cases, end-burning grains are used without a perforation. The latter produce rather low thrust but long burning times.

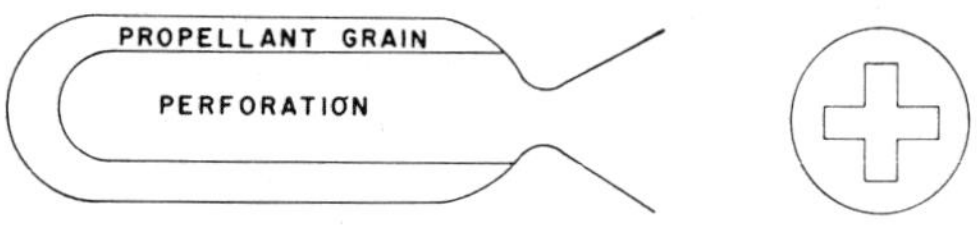

Fig. 19.3 Solid propellant rocket motor.

There are two general types of solid propellants, composite and double base. The composite propellants consist of a solid oxidizer and a metallic fuel intimately mixed in an elastomeric material which serves as both a binder and a fuel. Double base propellants use nitrocellulose plasticized with an energetic nitrato compound, and they may also contain a solid oxidizer and a metallic fuel. The double base propellants which contain a solid oxidizer and a metallic fuel are called composite-modified double base propellants.

Composite Propellants. Binders for composite propellants[9] are based on liquid polymers that contain functional groups which allow the polymer to be chain extended and crosslinked (cured) after they have been mixed with the solid oxidizer and fuel. When cured, the propellant exhibits rubbery properties. Table 19.2 gives information about the various common binders.

Table 19.2 Binders for Composite Propellants

Polymer Type	*Functional Group*	*Curing Agents*
Polybutadiene	COOH	Epoxides or aziridines
Polybutadiene	OH	Isocyanate
Polypropylene glycol	OH	Isocyanate
Polysulfide	SH	*p*-Quinone dioxime

The polybutadiene-type binders with the carboxyl functional group are most widely used in current solid-propellant rocket motors. Examples are first, second, and third stage Minuteman, first stage Poseidon, Genie, Sparrow, Phoenix, and Pershing. The polybutadiene polymers with hydroxyl functional group are newer and are still under development. The polyurethane binders based on polypropylene glycol have been used in motors such as the first stage Polaris. Polysulfide binders were used in some of the first successful composite propellants. They have been used in motors such as Sergeant, Falcon, and Bomarc.

The common oxidizers used in composite propellants are ammonoim perchlorate, ammonium nitrate, and potassium perchlorate. Some important properties of these oxidizers are shown in Table 19.3.

Table 19.3 Oxidizers for Composite Propellants

Compound	*Density (g/cc)*	*Available O_2%*	*H_f kcal/100 g*
NH_4ClO_4	1.95	34	-59
NH_4NO_3	1.73	20	-110
$KClO_4$	2.52	46	-32

Ammonium perchlorate is used in almost all composite propellants; ammonium nitrate and potassium perchlorate are used for specialized applications.

Metallic fuels are added to composite propellants primarily to increase the temperature of the gas which increases the specific impulse.[10] While the metal oxides produced are usually liquids or solids which reduce specific impulse, the amount of heat released more than offsets this loss. The common metallic fuels are aluminum and beryllium. The addition of metals also increases propellant density which allows more propellant to be put in a given volume. Since many missile or space rocket motors have only a certain volume available, the denser propellants greatly improve performance.

Theoretical performance values for several typical composite propellants are shown in Table 19.4. It is clear from these values that the amount of oxidizer in the propellant has a large effect on I_{sp} and density (ρ) as seen by comparing 80 per cent and 86 per cent NH_4ClO_4. An even larger effect is noted when a metal fuel like Al or Be is added. While Al yields a propellant with a higher density, Be propellants have higher theoretical I_{sp}'s because Be evolves more heat on oxidation. The propellants containing metal hydrides which produce considerable low molecular weight gas products yield an even higher specific impulse but lower propellant density.

Double-Base Propellants. Double-base propellants differ from composite propellants by having the oxidizing and reducing functions contained in the same molecule. Nitrocellulose is swollen by a nitrato plasticizer such as nitroglycerin (glyceryl trinitrate) to give a rubbery gelled structure which, although it is not crosslinked, does have viscoelastic properties. The nitrocellulose desensitizes the shock-sensitive nitroglycerin so that it can be used. Other plasticizers that are used for the nitrocellulose are trimethylol ethane trinitrate (TMETN), triethylene glycol dinitrate (TEGDN), and diethylene glycol dinitrate (DEGDN). Stabilizers such as *sym*-diethyldiphenylurea (ethyl centralite) prevent the decomposition of the nitrate compounds and are a very important part of the formulation. The compositions of JPN, an old extruded propellant, and OV, made by the casting-powder process are shown below:

	JPN	*OV*
Nitrocellulose (13.25% N)	51.50	–
Nitrocellulose (12.5% N)	–	58.6
Nitroglycerin	43.00	24.2
Dimethyl phthalate	–	9.6
Diethyl phthalate	3.25	–
Potassium sulfate	1.25	–
Ethyl centralite	1.00	1.0
Dinitrotoluene	–	6.6
Carbon black (added)	0.20	0.1
Wax (added)	0.008	–

Table 19.4 Performance Values for Composite Propellants

	T_c(°K)	*I°_{sp}(sec)*	*ρ(g/cc)*
80% NH_4ClO_4, 20% PBD	2345	229.1	1.57
84% NH_4ClO_4, 16% PBD	2750	241.8	1.64
86% NH_4ClO_4, 14% PBD	2940	246.5	1.68
70% NH_4ClO_4, 16% Al, 14% PBD	3398	262.7	1.77
73% NH_4ClO_4, 13% Be, 14% PBD	3302	282.1	1.67
55% NH_4ClO_4, 25% BeH_2, 20% PBD	2650	306	1.26

The addition of NH_4ClO_4 or HMX (cyclotetramethylene tetranitramine) or both, and Al increases both I_{sp} and density which can raise the theoretical I_{sp} to the 265 to 270 sec, and the density to 1.80 to 1.85 g/cc. The latter type of propellants are referred to as composite modified-double-base propellants.

Characterization of Solid Propellants. Solid propellants must be characterized so that a process can be designed which is able to handle the uncured and cured propellant. In addition, physical and mechanical properties must be defined so that a satisfactory motor can be designed, and the consequent changes in the propellant needed to meet design requirements must be determined. Safety tests such as impact, friction, static-spark sensitivity, and thermal stability are conducted to determine whether the propellant can be handled safely. Rheological properties, including viscosity and shear stress versus shear rate as a function of time, assist in deciding whether the propellant will process satisfactorily. To ascertain whether the propellant grain in the motor can withstand the thermal and gravitational loads, stress, strain, and modulus are evaluated at various temperatures and strain rates under uniaxial, biaxial, and even triaxial conditions; stress-relaxation and strain-endurance values are also obtained. The ballistic properties evaluated include I_{sp}, burning rate, and ratio of burning surface to nozzle throat area (K_n) as a function of pressure, temperature sensitivity of pressure (π_k) and burning rate (σ_p); these are used to design the propellant grain and nozzle throat size.

Manufacture. Composite solid propellants are relatively easy to manufacture. The first step is to grind part of the oxidizer (usually ammonium perchlorate) to a particle size of 1 to 30μ, depending on the burning rate required. The finer the oxidizer the higher the burning rate. A premix containing the liquid polymer, the metallic fuel, and the curing agent is often made. Either this premix is added slowly to the vertical or horizontal Baker-Perkins mixer containing the oxidizer, or the oxidizer is added slowly to the mixer containing the premix. After mixing a sufficient time to blend the ingredients, and raising the temperature to 60°C or so to increase fluidity, the propellant is cast (poured) through a slit plate into the evacuated motor. The vacuum removes entrapped air. A mandrel installed in the motor forms the perforation in the grain. After heating for several days at a temperature above room temperature (usually between 45 and 60°C) the propellant is a tough viscoelastic solid which is bonded directly to the walls of the motor case. A liner which is usually made from the same polymer as the propellant binder covers the case wall for better adhesion of the propellant.

There are two main methods of manufacturing double-base-propellant rocket motors, the casting-powder process and the slurry casting process.[12] In the casting-powder process, the solid ingredients, including the fibrous nitrocellulose, are mixed with a volatile solvent which yields a plastic mass. The plastic is consolidated in a blocking press, extruded, cut in small pieces and dried. The casting powder is placed in the motor and the casting solvent which includes the nitroglycerin or other nitroglycerin or other nitrato plasticizer is sucked into the evacuated motor to fill the voids. The motor is then heated to 40 to 60°C which causes the casting solvent to swell the casting powder and form a strong grain.

In the slurry casting process, a granular form of nitrocellulose is used. The plasticizer and solid ingredients including nitrocellulose, ammonium perchlorate, HMX, and aluminum can be added directly to the mixer. The mixture is stirred until it is homogeneous and then poured into the motor and cured. Since the plasticizer viscosity is low, it is sometimes desirable to predissolve some nitrocellulose in the plasticizer to increase its viscosity so that the solids do not settle.

Solid propellants have found wide application in rockets for both military and space applications because of their high reliability, low cost, and instant readiness. It is expected that they will continue to be used in many applications.

Liquid Propellants

The liquid bipropellant system is used today for the largest rockets. In this system the liquid fuel and the liquid oxidizer are stored in separate tanks and are fed separately to the combustion chamber. The propellants are fed either by means of pumps or by pressurization with an inert gas. In the case of the pump-fed rocket, the pumps are driven by gas turbines. The gas for these turbines is supplied by a generator which is actually a small rocket motor operating off a side stream of the main rocket propellants (Fig. 19.4) or, in some cases, a separate propellant system. In the largest rockets, thousands of pounds per second of propellants must be pumped to the engine, and the pump drives must develop thousands of horsepower. The propellants enter the thrust chamber through an injector which has a function similar to that of a carburetor in an internal combustion engine. The injector has to atomize and mix the propellant so that the oxidizer and fuel will be in the right proportions to vaporize and burn smoothly.

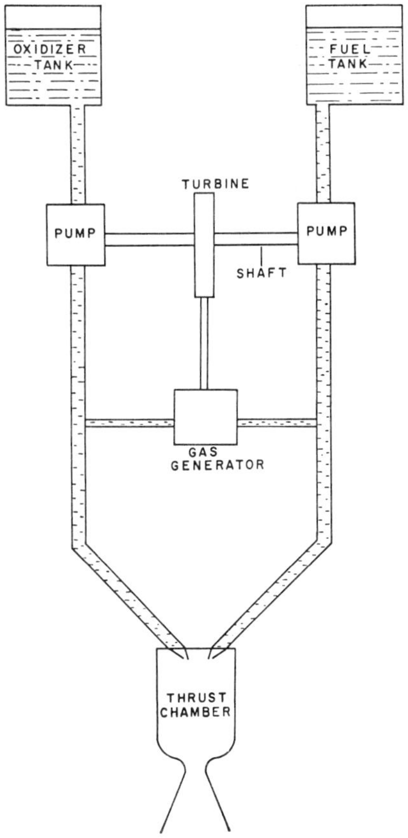

Fig. 19.4 Liquid bipropellant rocket engine.

Theoretical Performance. Some typical liquid oxidizers are liquid oxygen (LOX), liquid fluorine, oxygen difluoride (OF_2), chlorine fluoride (ClF_3) and nitrogen tetroxide (N_2O_4). Typical fuels are RP-1 (kerosene-like), liquid hydrogen, hydrazine (N_2H_4) monomethyl hydrazine (MMH), and unsymmetrical dimethylhydrazine (UDMH). Liquid propellant systems are usually classified as storable or cryogenic. Table 19.5 gives typical properites of several combinations of oxidizers and fuels. It is evident that the cryogenic propellants yield the highest I_{sp} values, however they also have the lowest densities, which is undesirable. Furthermore, they are more difficult to handle because of their low boiling points. They are, however, used in rockets, such as the Saturn, for space applications because of the heavy payload requirements, and because instant readiness is not a problem. The storable propellants, which have higher densities, have lower I_{sp}, but they can be used in systems that require readiness.

Properties and Manufacture. The liquid oxidizers and fuels that are in use today in operational and experimental rockets have been characterized rather thoroughly. The results of some of these studies and a brief description of the manufacturing methods follow.

*Cryogenic Propellants.** The ability to design and fabricate the large, high-performance rockets needed for space applications depends in large measure on advances in the engineering technology used in handling such intractable materials as liquid oxygen (b.p. −183°C), liquid fluorine (b.p. −188°C), and liquid hydrogen (b.p. −253°C).

As indicated in Table 19.5 liquid hydrogen allows the highest I_{sp} values to be obtained.

*A more detailed description can be found in the "Liquid Propellant Manual."[13]

TABLE 19.5 Properties of Several Common Liquid Bipropellants

Oxidizer	*Fuel*	$I_{sp}^{o\,a}$ *(sec)*	T_c *(°F)*	*Boiling Pt (°C)*	*Density (g/cc)*	*Type*
LOX				-183		Cryogenic
	H_2	389	4913	-253	0.28	Cryogenic
	H_2-Be (50/50)	457	4662	-253	0.23	Cryogenic
	N_2H_4	313	5453	113	1.07	Storable
	RP-1	300	6160	370–510	1.02	Storable
F_2				-188		Cryogenic
	H_2	411	6650	-253	0.40	Cryogenic
	H_2-Li (64.3/35.7)	434	3415	-253	0.20	Cryogenic
	MMH	346.2	7612	87	1.25	Storable
OF_2				-145		Cryogenic
	H_2	400	5864	-253	0.39	Cryogenic
	RP-1	340	7766	188–266	1.28	Storable
	MMH	353	7202	87	1.24	Storable
ClF_3				12		Storable
	MMH	283	6042	87	1.42	Storable
	N_2H_4-Be (81/19)	303	5855	113	1.43	Storable
N_2				21		Storable
	N_2H_4-UDMH (50/50)	289	5610	70	1.21	Storable
	MMH-Be (77/23)	314	5766	87	1.18	Storable
	N_2H_4	292	5406	113	1.22	Storable

[a]Chamber pressure of 1000-psi exhausting to 1 atm.

However, its low boiling point and low density (0.07) g/cc) make it difficult to handle and store. Also, its very low temperature can cause embrittlement of metals which results in the loss of impact strength. Storage tanks and transfer lines must be well insulated. Vacuum jackets, foamed plastics, and alternate layers of aluminum foil and glass-fiber mats have been successfully used.

Molecular hydrogen exists in two forms: *ortho*-hydrogen, wherein the nuclei of the two atoms are spinning in the same direction, and *para*-hydrogen, in which the nuclei spin in opposite directions. These two forms are in equilibrium with each other, and at room temperature the equilibrium mixture contains 75 per cent of the *ortho* form and 25 per cent of the *para* form. When cooled to the boiling point (-253°C), this mixture is unstable. Since the equilibrium concentration of *para*-hydrogen at this temperature is 99.8 per cent, the *ortho*-hydrogen will convert very slowly to the *para* form. The conversion from 25 to 90 per cent *para* takes about one month. This conversion is accompanied by the evolution of heat which is sufficient to cause severe evaporation losses of the liquid. In order to overcome this difficulty, a hydrous ferric oxide catalyst is used in connection with the liquefaction process to speed up the conversion so that a stable product may be obtained. Modern liquefiers can produce liquid hydrogen which is more than 99 per cent *para*-hydrogen.[14]

Hydrogen is recovered as a by-product from petroleum refining and from the partial oxidation of fuel oil.

Liquid monopropellants are "unstable" liquid chemicals which, in the presence of a catalyst or a small amount of initiating energy, will decompose to give heat and large volumes of gas. Examples of such materials are hydrogen peroxide and hydrazine:

$$H_2O_2 \longrightarrow 1/2\ O_2\ (g) + H_2O\ (g)$$

$$N_2H_4 \longrightarrow N_2\ (g) + 2H_2\ (g)$$

In general, monopropellants exhibit low performance (at 300-pounds chamber pressure, exhausting to 1 atm, hydrogen peroxide has a specific impulse of 133 secs; hydrazine, 174 secs); materials containing enough energy to give a high theoretical specific impulse are usually too shock-sensitive to be useful in rocket work. The monopropellants are very useful as gas generators where cool temperatures are required or high I_{sp} is not a requirement.

Since fluorine is the most electronegative of all the elements, there are handling problems in addition to those associated with its low boiling point. Fluorine will violently attack nearly all oxidizable substances including metals. Fortunately, metal surfaces can be passivated by forming thereon a fluoride film which protects the metal from further attack. Many metals have been found to be suitable for rocket applications; they include "Monel," stainless steel, copper, brass, and aluminum. It is extremely important that all equipment be clean and free from any organic material to prevent hazardous conditions. Fluorine is produced by the electrolysis of hydrogen fluoride in fused potassium fluoride.

Liquid oxygen is highly reactive with most organic materials, but it is not corrosive like liquid fluorine. However, the surfaces which will be in contact with this oxidizer must be kept extremely free of any contamination. The low boiling point can also cause problems due to low-temperature embrittlement. Metals that can be used are "Monel," copper, bronze, brass, aluminum, and stainless steel; plastics which are suitable include Teflon and Kel-F. Liquid oxygen is produced by the fractionation of liquid air.

Storable Propellants. Typical examples of storable liquid propellants are nitrogen tetroxide (N_2O_4), chlorine trifluoride (ClF_3), hydrazine (N_2H_4), and unsymmetrical dimethylhydrazine (UDMH). A brief description of the properties and production processes for these chemicals is given below.

Nitrogen tetroxide is very stable at room temperature, but at 150°C it begins to dissociate into nitric oxide and free oxygen. Upon cooling, nitrogen tetroxide is reformed. It is not corrosive to most common metals if the oxidizer is dry ($<0.1\%$ H_2O); carbon steels, aluminum, stainless steel, nickel, and Inconel can be used. Under wet conditions 300-series stainless steel should be used. As a result of its good properties, nitrogen tetroxide is currently the most important operational earth-storable liquid oxidizer.

The catalytic oxidation of ammonia is the most important commercial method of preparing N_2O_4. A mixture containing about 9 per cent ammonia in air is oxidized at pressures between 25 and 100 psig. The gaseous products containing nitric oxide (NO) and water are cooled and dried by scrubbing with cold nitric acid. The dry nitric oxide containing unreacted nitrogen from the air is further oxidized by the addition of air to form NO_2. Nitrogen dioxide forms an equilibrium mixture with its dimerized form, dinitrogen tetroxide, N_2O_4, and on cooling the equilibrium shifts to favor the formation of N_2O_4. The reaction sequence is illustrated by the following equations:

$$4NH_3 + 5O_2 \longrightarrow 4NO + 6H_2O$$

$$2NO + O_2 \longrightarrow 2NO_2$$

$$2NO_2 \longrightarrow N_2O_4$$

Chlorine trifluoride is highly corrosive but is stable to shock, heat and electrical spark. Because it forms an inert fluoride film on some metals, it can be used with copper, brass, steel, aluminum, nickel, and Monel.

Chlorine trifluoride has been prepared by the reaction of chlorine and fluorine at 280°C.

Hydrazine is stable to shock, heat, and cold. It begins to decompose at 160°C if no catalysts are present. It freezes at 1.4°C, but it contracts and does not decompose. It is compatible with stainless steel, nickel, aluminum, Teflon, and Kel-F.

Hydrazine is produced by the controlled oxidation of ammonia in the well-known Raschig synthesis. Sodium hypochlorite, formed by the chlorination of caustic soda, reacts with excess ammonia in a two-step process to form, successively, chloramine and aqueous hydrazine:

$$NH_3 + NaOCl \longrightarrow NH_2Cl + NaOH$$

$$NH_2Cl + NaOH + NH_3 \longrightarrow N_2H_4 + NaCl + H_2O$$

Hydrazine and water form an azeotrope which boils at 120.1°C. Anhydrous hydrazine is obtained by dehydration of the azeotrope with caustic, followed by a vacuum distillation. In another method, the azeotrope is broken by the addition of aniline as a third component, followed by a distillation in which the water-aniline azeotrope is taken overhead and anhydrous hydrazine is obtained as a bottoms product. Using this latter method hydrazine may be produced at 98+ per cent purity.

Unsymmetrical dimethyl hydrazine (UDMH) is neither shock- nor heat-sensitive, being stable to 249°C, its critical temperature. Furthermore, it is compatible with nickel, Monel, and stainless steel, but it does attack aluminum if water is present.

UDMH has been prepared by a modification of the Raschig process by reacting the chloramine intermediate with dimethylamine instead of ammonia. It can also be produced commercially by nitrosation of dimethylamine to N-nitrosodimethylamine followed by reduction and purification. UDMH is usually used as a 50/50 mixture with hydrazine; in fact, this is the most common earth-storable fuel.

Many applications have been found for liquid propellants because of their generally high performance. They have been widely used in space programs but have also found many military applications.

REFERENCES

1. Bebie, J., "Manual of Explosives, Military Pyrotechnics and Chemical Warfare Agents, New York, Macmillan, 1943.
2. Cook, M. A., "The Science of High Explosives," New York, Van Nostrand Reinhold, 1958.
3. Cook, M. A., and McEwan, W. S., *J. App. Phys.* **29**, 1612 (1958).
4. Davis, T. L., "The Chemistry of Powders and Explosives," New York, John Wiley and Sons, 1941.
5. Raifsnider, P. J., *Chem. Ind.* **57**, No. 7, 1054 (1945).
6. Farnam, H. E., Jr., Paper presented at Eighth Annual Drillers' and Blasters' Symposium, U. of Minn., Oct. 2, 1958; Cook, M. A., *Science*, **132**, 1105 (1960).
7. Sutton, G. P., "Rocket Propulsion Elements," p. 54, John Wiley and Sons, 1963.
8. Siegel, B., and Schieler, L., "Energetics of Propellant Chemistry," p. 6, John Wiley and Sons, 1964.
9. Sutton, G. P., *op. cit.* p. 360.
10. Klager, K. and Wrightson, J. M., "Recent Advances in Solid Propellant Binder Chemistry," in "Mechanics and Chemistry of Solid Propellants," Elmsford, N.Y., Pergamon Press, 1967.
11. Sarner, S. F., "Propellant Chemistry," p. 150, New York, Van Nostrand Reinhold, 1966.
12. Boynton, D. E. and Schowengerdt, J. W., *Chem. Eng. Progr.*, **59**, 81 (1963).
13. Cadwallader, A., "Liquid Propellant Manual," Chemical Propulsion Information Agency, 1969.
14. Von Doehern, J., "Propellant Handbook," p. 2.8, AFRPL-TR-66-4, Air Force Rocket Propulsion Laboratory, January, 1966.

20

The Pharmaceutical Industry

L. H. Werner*

INTRODUCTION

Economics

The pharmaceutical industry is one of the larger industries of the United States. Total output of prescription drugs for human use, the main product of the industry, amounted to 3.7 billion dollars in 1968 in domestic sales. The total for nonprescription or proprietary preparations was approximately 0.9 billion, and for drugs for veterinary use, approximately 0.2 billion dollars. By comparison, the GNP for that year was 739.6 billion, and the sales of General Motors, 22.7 billion. Government figures for 1963 show that 944 companies produced drugs.

The products of the pharmaceutical industry differ from those of other industries in that they serve the health of the nation in a rather direct way. In most instances the consumer does not select the product he purchases. Rather, it is prescribed by a member of the medical profession. It is the responsibility of the pharmaceutical industry to provide efficacious and safe drugs for the treatment of disease, and many government controls have been imposed on drug manufacturers to help ensure that these responsibilities are met.

Development of the Industry

Most of the drugs which were available from prehistoric times until the beginning of the sulfa drug and antibiotic age were, in the light of today's knowledge, crude nostrums. Few of them did the patient any good, and he was probably fortunate if they did him no harm. Nevertheless, a number of drugs of proven effectiveness were known before 1940, as shown in Table 20.1.

The discovery of the antibacterial activity of sulfanilamide in 1932 and of penicillin in 1940 opened a new era in drug research and manufacture which has contributed much to the practice of medicine as we know it today. During the period between 1940 and 1970 many new, potent, and effective drugs were introduced. Table 20.2 lists the year of intro-

*Research Department, Pharmaceutical Division, CIBA-GEIGY Corp., Summit, N.J.

TABLE 20.1 Some Early Drugs

Drug		*Year Used in Medicine*
Laxatives		used for centuries
Quinine	Antimalarial	1639
Smallpox vaccine		approx. 1800
Cocaine	Local anesthetic	1879
Digitalis	Cardiac stimulant	1890
Aspirin	Analgesic (mild)	1894
Phenobarbital	Hypnotic	1912
Morphine	Analgesic	1820
Insulin	Hypoglycemic	1921
Arsphenamine (Salvarsan)	Antisyphilitic	1925
Salyrgan	Diuretic	1928
Sulfanilamide	Antibacterial	1932

duction of each new drug which represented a major new development. In many cases, this new drug led to the synthesis of numerous related, and frequently more potent analogs.

Drug Nomenclature

United States law requires that all pure drug substances have a generic or nonproprietary

TABLE 20.2 Important Drugs Introduced Between 1940 and 1970

Therapeutic Class	*Name*	*Year of Introduction*
Antibiotic	Penicillin	1940
	Aureomycin	1948
Steroidal sex hormones	Testosterone, Progesterone, Estradiol	1936–1940
Analgesic	Meperidine	1944
	Methadone	1948
CNS-stimulant	Amphetamine	1944
Antihistamine	Benadryl	1946
Antispasmodic	Banthine	1950
	Probanthine	1953
Anti-inflammatory steroid	Cortisone	1950
Anti-inflammatory (nonsteroidal)	Phenylbutazone	1952
	Indomethacin	1965
Antihypertensive	Hydralazine	1952
	Reserpine	1953
Major tranquillizer	Chlorpromazine	1954
Minor tranquillizer	Meprobamate	1955
	Chlordiazepoxide	1960
Oral hypoglycemic	Tolbutamide	1957
Diuretic (nonmercurial)	Chlorothiazide	1958
	Hydrochlorothiazide	1959
Steroidal oral contraceptive	Enovid®	1957
Antidepressant	Imipramine	1959
Poliomyelitis vaccine		1961–1962
Hypocholesterolemic	Clofibrate	1962

name which is public property, apart from the trade-mark which is a protected name. The generic name of a drug is frequently derived from the chemical name and must be approved by the United States Adopted Names Council (USAN).

The trade-mark is a name selected by the manufacturer and registered with the U.S. Patent Office. It identifies the product with the manufacturer and his reputation for quality and integrity. Some drugs are manufactured by more than one company and sold under different names. Examples are given in Table 20.3.

TABLE 20.3 Trade-marks of Some Drugs

Trade-Mark	*Generic Name*
Equanil Miltown	Meprobamate
Esidrix Hydrodiuril	Hydrochlorothiazide
Erythrocin stearate Ilotycin	Erythromycin stearate Erythromycin
Meticortelone Hydeltra	Prednisolone

Regulation

Although fraud in food and medicine is an age old problem, it did not reach alarming proportions until the industrial 19th century. Already by 1892, Senator Paddock of Nebraska had sponsored a bill to control the adulteration of foods and drugs, but it died in the House. Finally, on February 1, 1906, the Pure Food and Drug Act was passed. However, from the start, inadequacies in this act were recognized. Sometime later, after a struggle lasting five years, the 1938 Food, Drug and Cosmetic Act was signed into law by President Roosevelt. The Act granted the authority needed to establish standards of identity and quality for most food products; hazardous cosmetics were outlawed, and for the first time, misleading cosmetic labeling was banned. Outlawed also were drugs which were dangerous to health when used according to directions (for example, radium water). Advances were made with respect to formula disclosure, and new drugs could not be marketed until the manufacturer had shown to the satisfaction of the Food and Drug Administration (FDA) that they were safe.

Intense drug research between 1940 and the present yielded many new drugs which revolutionized therapy. This accelerated development has led to closer scrutiny of the new products by the government. For example, greater authority was granted to the FDA by the Kefauver-Harris amendments which were adopted in 1962. These amendments required that the manufacturer of a new drug not only show that it is safe to use, but also that it is efficacious. All drugs introduced between 1938 and 1962 were submitted to various drug panels convened by the National Academy of Science–National Research Council in order to review their effectiveness.

Precise regulations govern both the study of new drugs in man and the preparation of documents which must be submitted to the FDA before such studies can be started. These studies are continuously monitored by the FDA, and when a new drug is finally approved after extensive clinical trials over a period of years, it is with the conviction that it is safe and effective. The Kefauver-Harris amendments also specifically require that the facilities, methods, and control procedures used by the manufacturer conform with "current good manufacturing practice" as established by the Food and Drug Administration to ensure the integrity of drug products. Plant inspections are carried out at prescribed intervals.

Products

The drugs manufactured today can be grouped into certain main classes as shown in Table 20.4. These classes do not cover all drugs but do include most of those which are widely used. The sales figures are a measure of their economic importance.

The pharmaceutical industry not only produces drug products that have wide application, but as a service to medicine also has made certain drugs available, often at considerable cost, for the treatment of infrequent special conditions such as iron poisoning (deferoxamine), botulism poisoning (botulism antitoxin USP), snake bite (polyvalent cro-

TABLE 20.4 Sales of Various Theraputic Classes of Drugs

Therapeutic Class	*Approximate Value at Manufacturers' level, 1969. (Millions of Dollars)*
Analgesics	340
Antacids	170
Antibiotics	490
Antihistamines	35
Antiarthritic (nonsteroidal)	80
Antiinflammatory (corticoids)	144
Antiobesity	85
Antinauseants	30
Cardiovascular agents	230
Contraceptive agents (oral)	100
Cough and cold preparations	194
Diuretics (nonmercurial)	115
Laxatives	111
Psychopharmaceuticals	475
Sedatives and hypnotics	80
Sulfa drugs	39
Vaccines	15
Vitamins	160

taline antivenin USP), and respiratory depression (e.g. doxapram).

Representative compounds in the categories listed in Table 20.4 are described below to illustrate the various types of chemical structures present in drug products.

Nonnarcotic Analgesics. This type of analgesic is used for the relief of mild pain. Examples are aspirin, acetaminophen, and phenacetin.

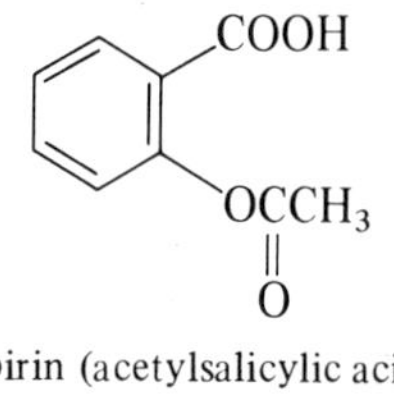

Aspirin (acetylsalicylic acid)

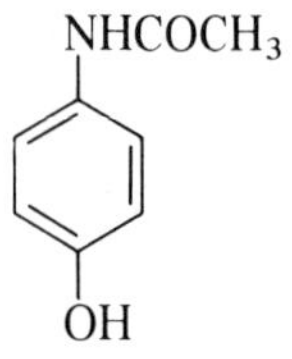

Acetaminophen

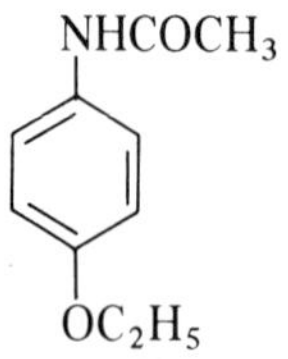

Phenacetin

Narcotic Analgesics. These substances are used primarily for the relief of severe pain; they are all addicting (i.e., producing psychological, and sometimes physical dependance) and they all produce side effects, e.g., respiratory depression.

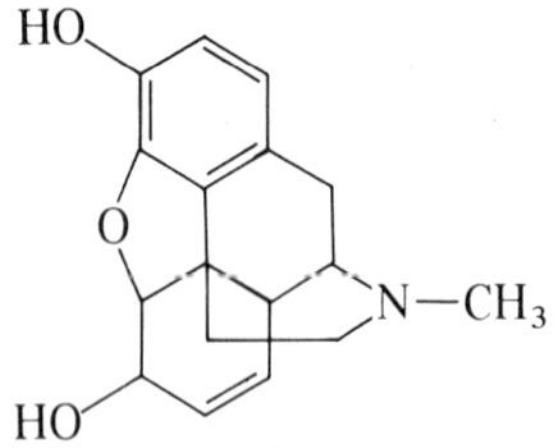

Morphine

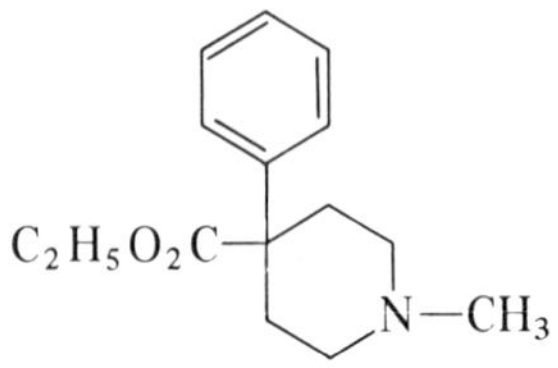

Meperidine

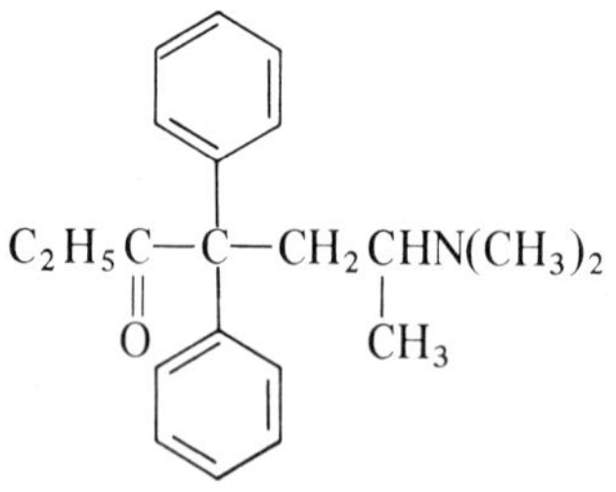

Methadone

Two more recent analgesics, propoxyphene and pentazocine, have a very low addiction potential and are more potent than aspirin, acetaminophen, and phenacetin.

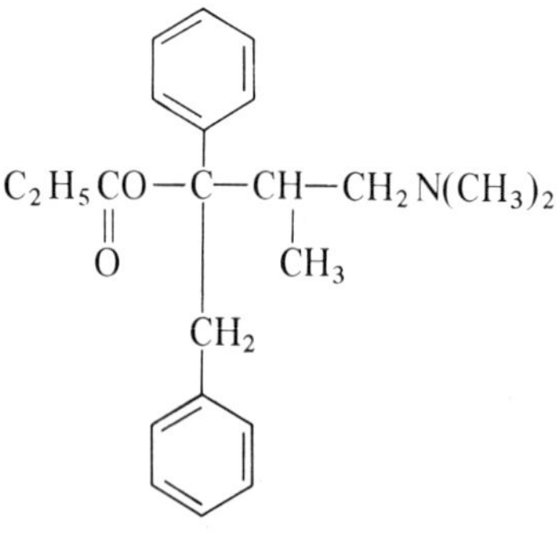

Propoxyphene

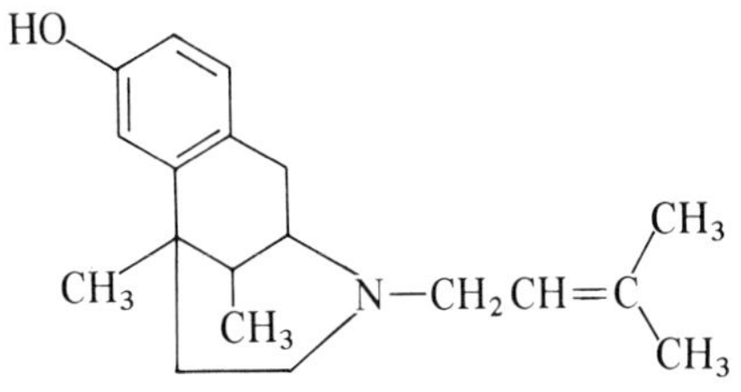

Pentazocine

Morphine is the principal opium alkaloid and is obtained from the seed capsules of *Papaver somniferum.* Meperidine, methadone, propoxyphene, and pentazocine are synthetic compounds.

Antacids. Gastric antacids are agents that neutralize or remove acid from the gastric contents. Generally they contain a weakly basic moiety. The following compounds are used:

$NaHCO_3$ Sodium bicarbonate	MgO or $Mg(OH)_2$ Magnesium oxide or magnesium hydroxide
$MgCO_3$ Magnesium carbonate	Aluminum hydroxide Magnesium trisilicate

Antibiotics. Antibiotics are widely used to combat bacterial infections. Numerous antibiotics have been developed since the introduction of penicillin in 1941. With the exception of chloramphenicol they are produced by growing the appropriate organism in a sterile medium and extracting the antibiotic from the broth. Chloramphenicol is now obtained by chemical synthesis. The structures of the most widely used antibiotics are given below.

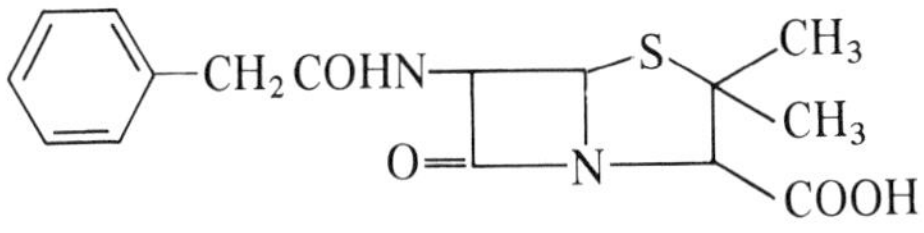

Penicillin G

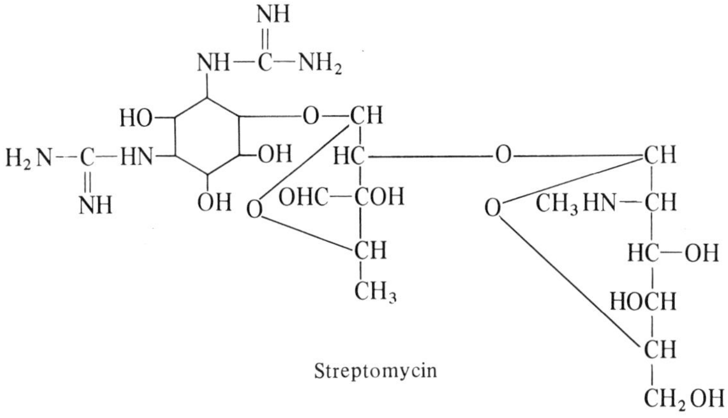

Streptomycin

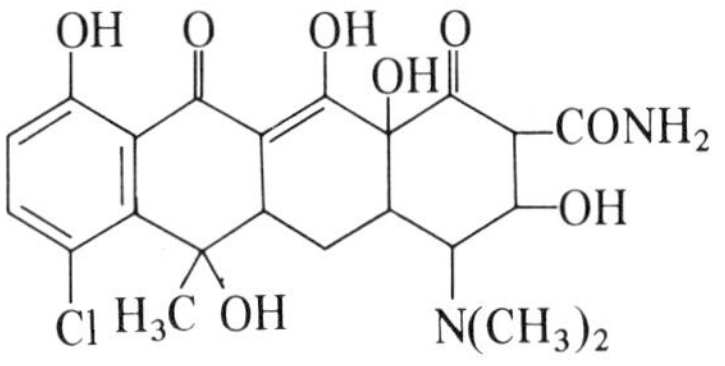

Chlortetracycline (Aureomycin®)

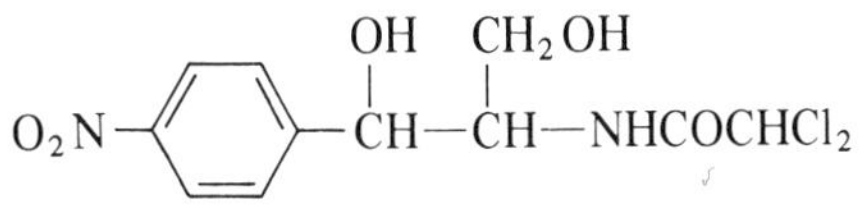

Chloramphenicol (Chloromycitin®)

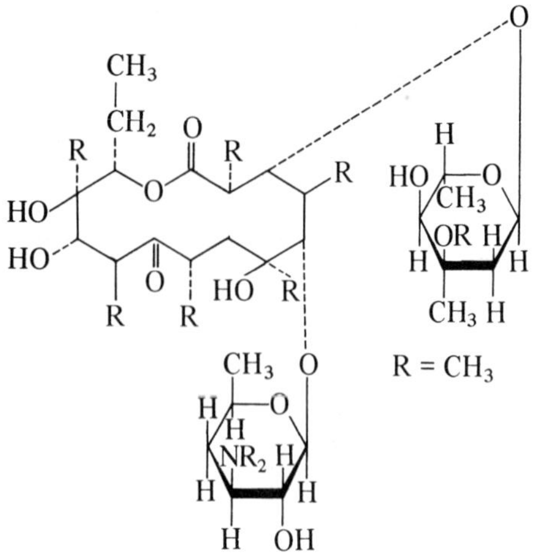

Erythromycin

Antihistamines. The antihistamines antagonize in varying degree most of the pharmacological actions of histamine and can also reduce the intensity of allergic and anaphylactic reactions. This is the basis for their major therapeutic applications. Structures for this type of compound are shown below.

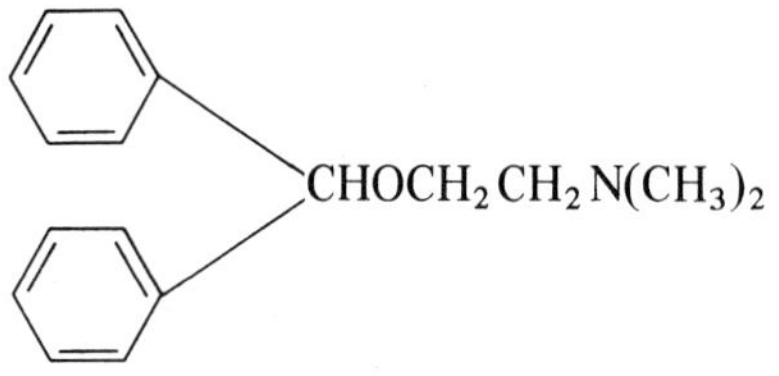

Diphenhydramine

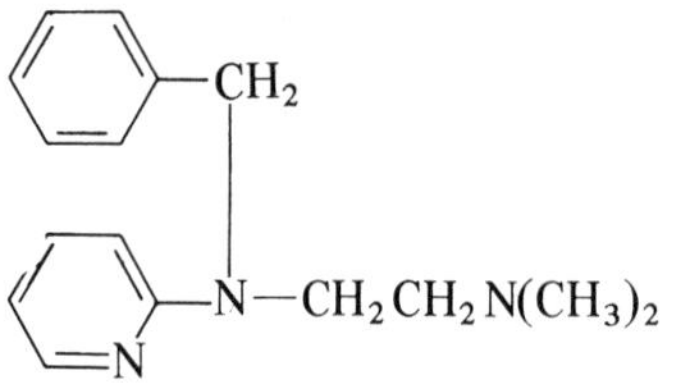

Tripelennamine

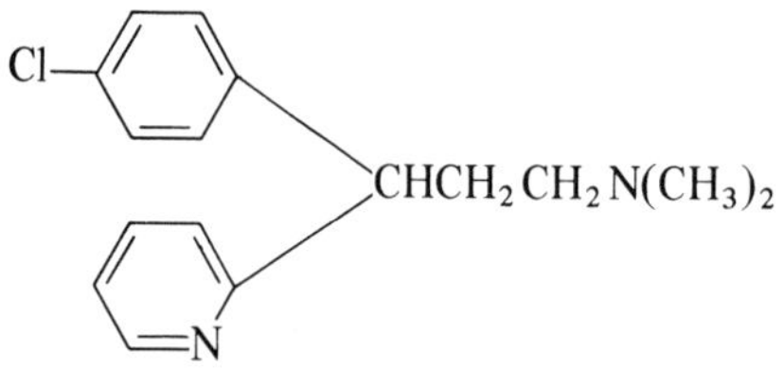

Chlorpheniramine

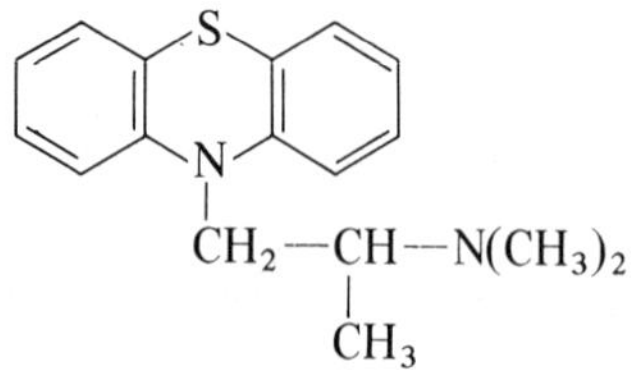

Promethazine

Steroidal Antiinflammatory Compounds. Corticosteroids include cortisol, a steroid hormone produced by the adrenal cortex, and its synthetic analogs which have the capacity to prevent or suppress the development of the redness, swelling, and tenderness by which inflammation is recognized. Clinically they will inhibit the inflammatory response regardless of whether the inciting agent is mechanical, chemical, or immunological. However, they only suppress the inflammatory response; the underlying cause is not affected.

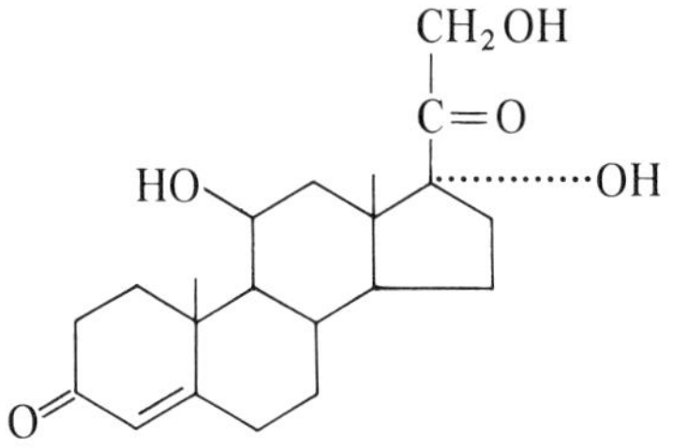

Cortisol

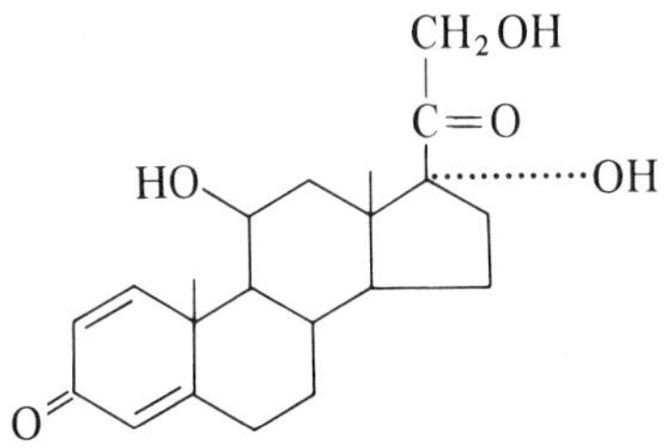

Prednisolone

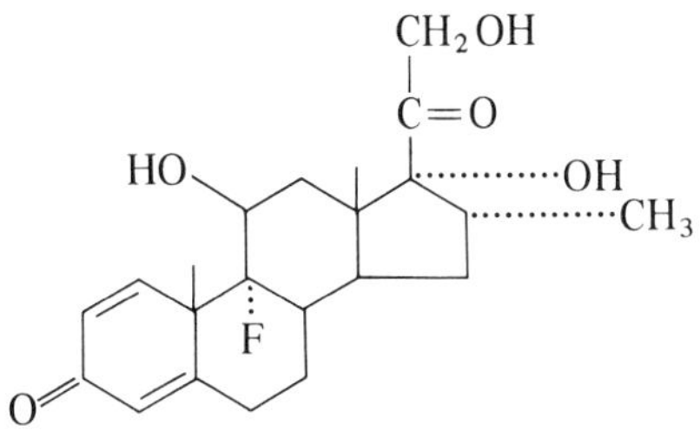

Dexamethasone

Nonsteroidal Antiinflammatory Compounds. Certain nonsteroidal compounds have been found to exert an analgetic and antiinflammatory effect. Examples of this type of compound are phenylbutazone and indomethacin.

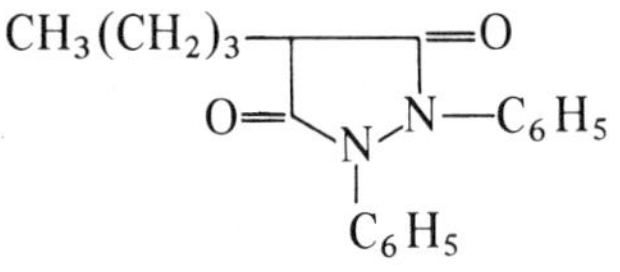

Phenylbutazone

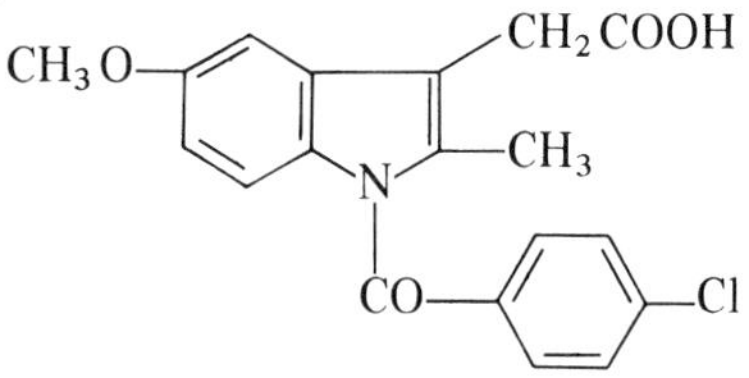

Indomethacin

Antiobesity Preparations. Antiobesity preparations are generally used as an adjunct to dieting as they tend to reduce the appetite. Amphetamine is used extensively for this purpose. It is, however, a powerful stimulant of the central nervous system.

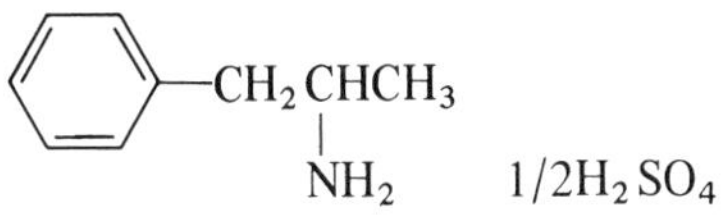

Amphetamine sulfate

Other compounds reported to have less stimulant action are also used, e.g. phenmetrazine and diethylpropion.

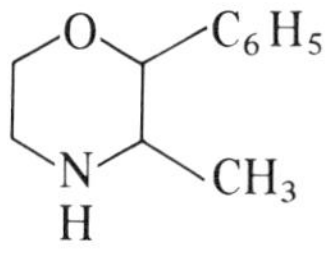

Phenmetrazine

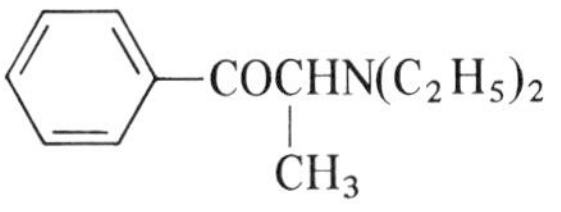

Diethylpropion

Antinauseants. Antinauseants are used mainly to control motion sickness; cyclizine and meclizine exemplify this type of compound.

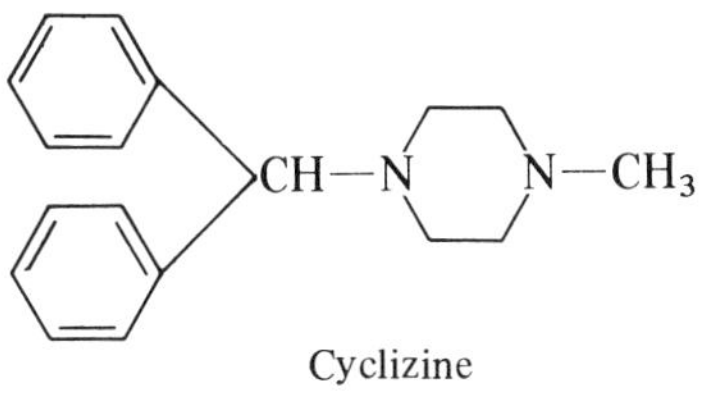

Cyclizine

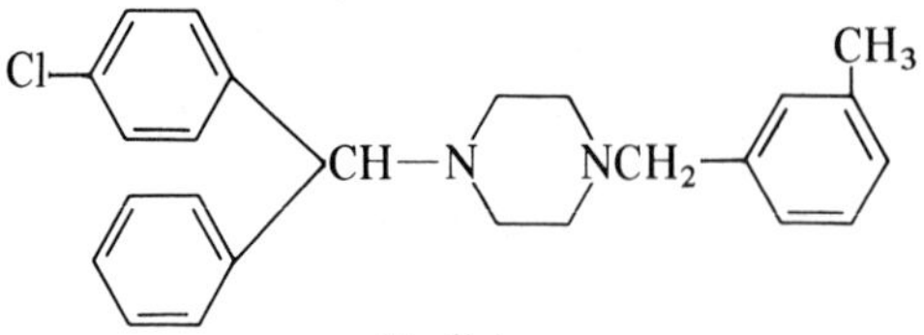

Meclizine

Cardiovascular Agents. A large number of drugs have as their major pharmacological action the ability to alter cardiovascular function.

Cardiac glycosides are obtained by extraction from plants, e.g. digitalis. Their main action is to increase the force of myocardial contraction. An important member of this group is digitoxin, a glycoside containing the aglycone digitoxigenin, three moles of digitoxose and one mole of glucose attached to the hydroxy group at the 3-position.

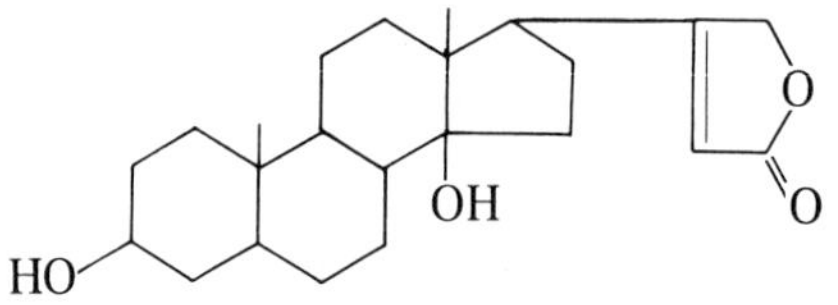

Digitoxigenin

Antihypertensive agents are found among a number of chemically and pharmacologically dissimilar compounds. The structures of a number of such compounds are given below.

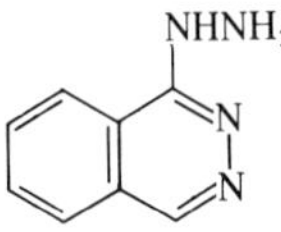

Hydralazine

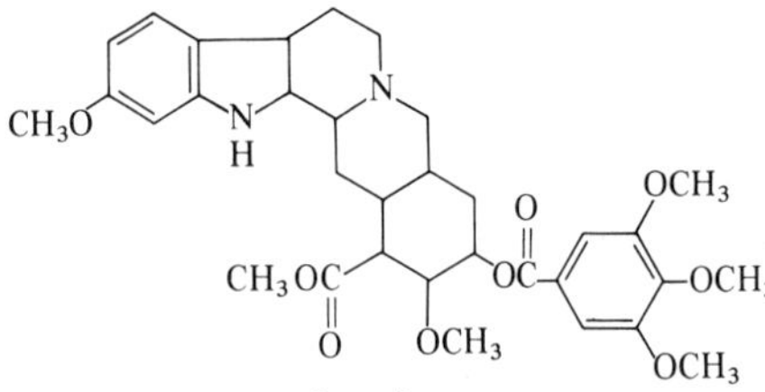

Reserpine

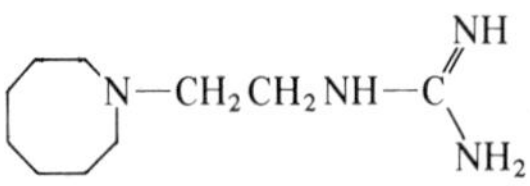

Guanethidine

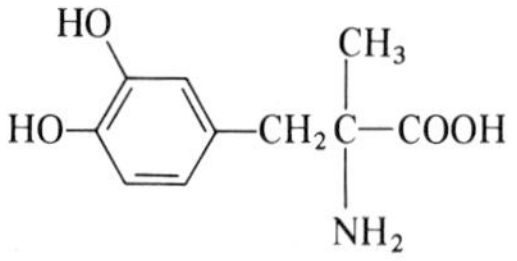

Methyldopa

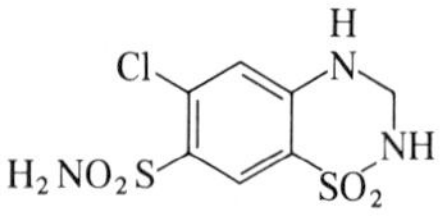

Hydrochlorothiazide

Vasodilator drugs are used to attempt to correct or improve an imbalance between the requirements of a tissue and delivery and removal of various materials by the bloodstream, particularly the delivery of oxygen. Examples are given below.

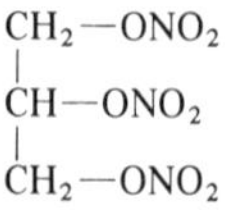

Glyceryl trinitrate

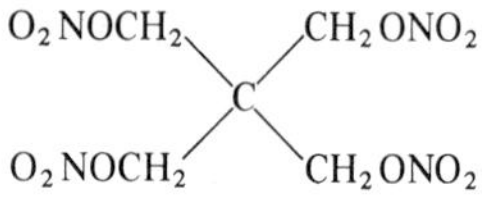

Pentaerythritol tetranitrate

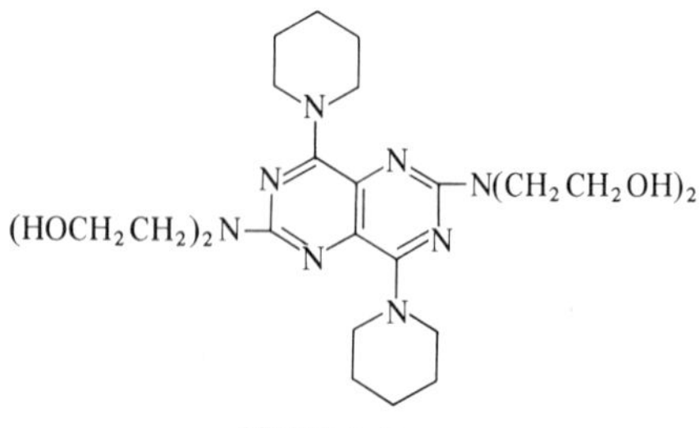

Dipyridamole

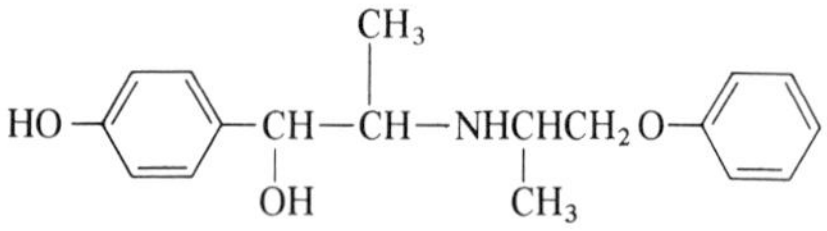

Isoxsuprine

Oral Contraceptive Agents. The combination of a progestin and an estrogen has been found to be highly effective as an oral contraceptive. Enovid®, introduced in 1957, was the first such preparation for this indication.

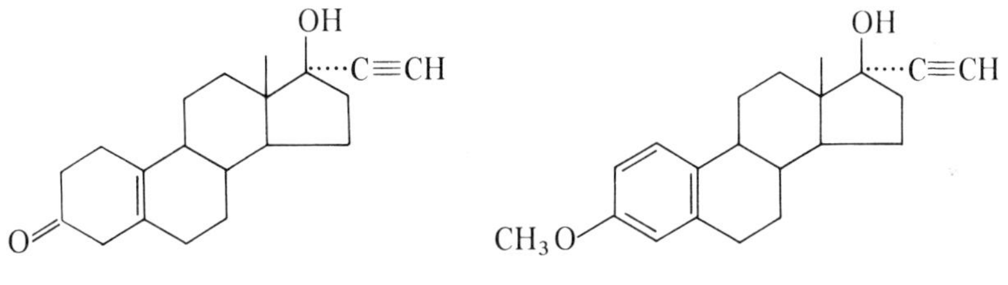

Norethynodrel Mestranol

Enovid®

Cold Preparations. Cold preparations treat the symptoms of a cold. They are generally a combination of a nasal decongestant, e.g. phenylpropanolamine, an antihistamine, and aspirin.

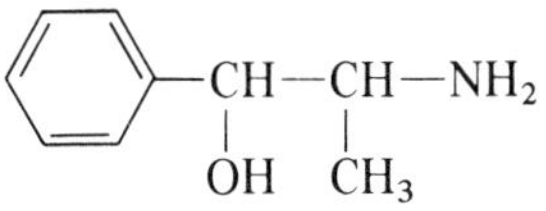

Phenylpropanolamine

Cough Preparations. Dextromethorphan is a relatively new product showing good antitussive properties. It is slightly more effective than codeine.

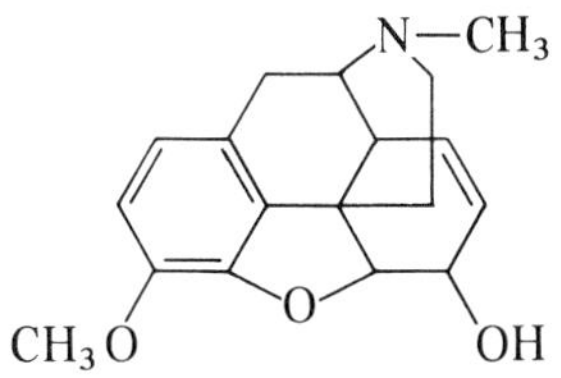

Codeine

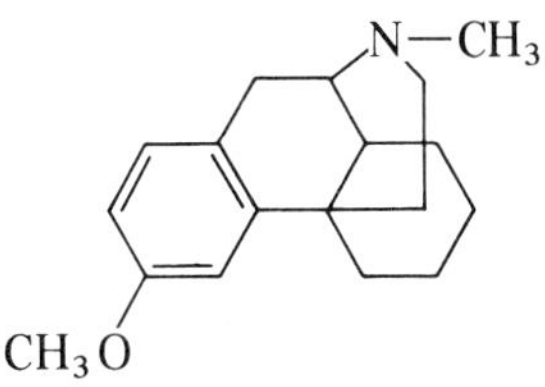

Dextromethorphan

Diuretics (nonmercurial). An important class of orally active diuretics is represented by hydrochlorothiazide and its derivatives.

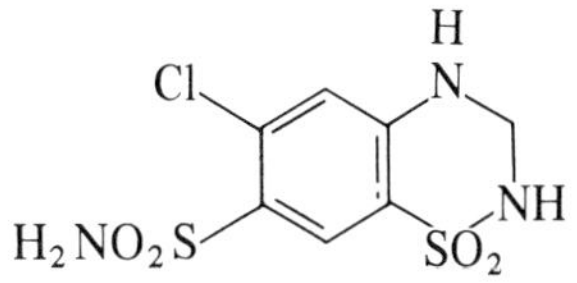

Hydrochlorothiazide

Chlorthalidone has similar activity but differs in structure.

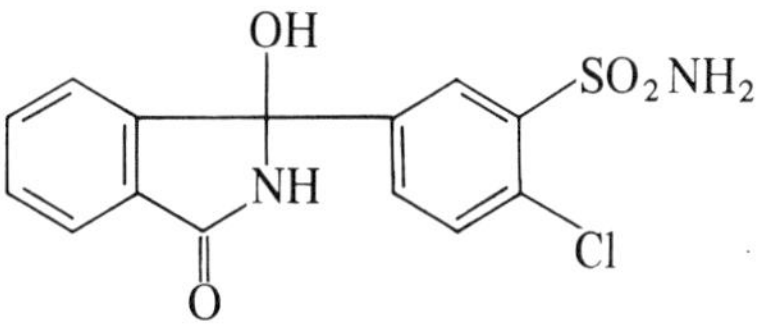

Chlorthalidone

An entirely different class of diuretics is represented by ethacrynic acid. It is orally active and has high potency.

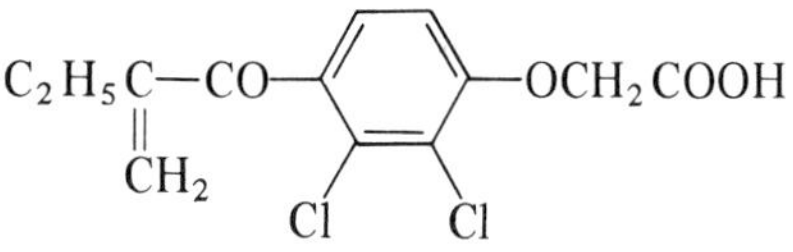

Ethacrynic acid

Laxatives. Laxatives or cathartics act by various mechanisms and can be classified accordingly. A large number of preparations are available.

Cathartic	
Stimulant	cascara, senna, castor oil phenolphthalein.
Saline	milk of magnesia, sodium sulfate, etc.
Bulk-forming	methylcellulose, plantago seed, bran, etc.
Lubricant	mineral oil.

Psychopharmaceutical Agents. These agents represent a very important group of drugs.

Major tranquilizers, neuroleptics, include more than twenty phenothiazine derivatives which are used in medicine at present. Chlorpromazine for one, has found widespread use. The first report on the effectiveness of chlorpromazine in mental illness appeared in 1952. Its chief usefulness lies in the treatment of psychoses.

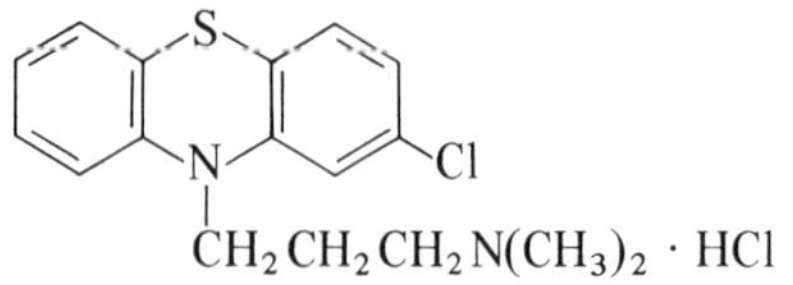

Chlorpromazine

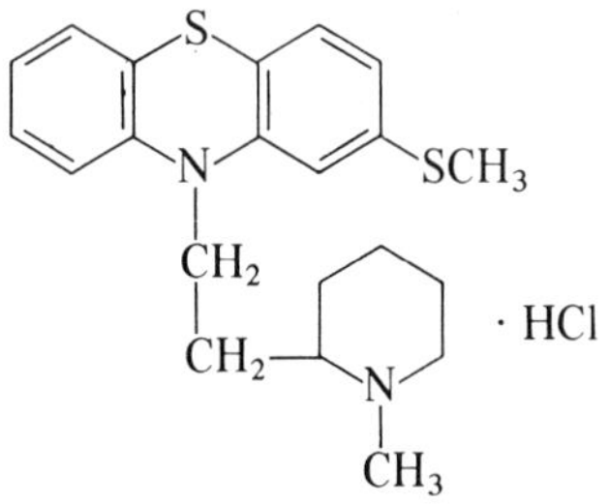

Thioridazine

Minor tranquilizers, drugs for anxiety, are widely used for the relief of mild anxiety and tension. They are of little or no value in the treatment of psychoses. The structures of three compounds of this group are given below.

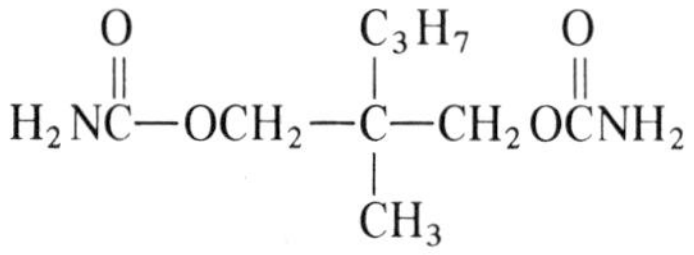

Meprobamate

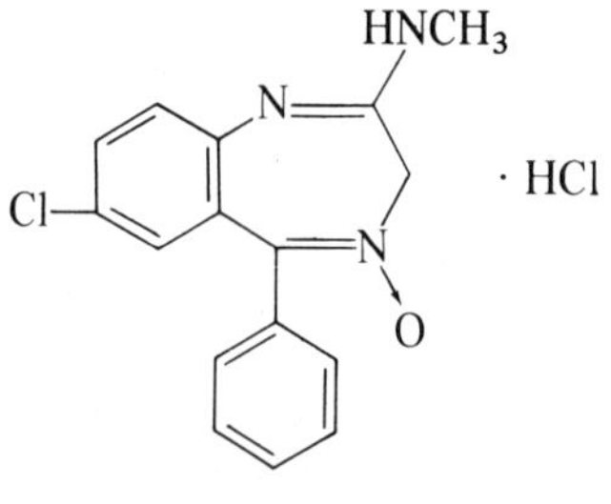

Chlordiazepoxide

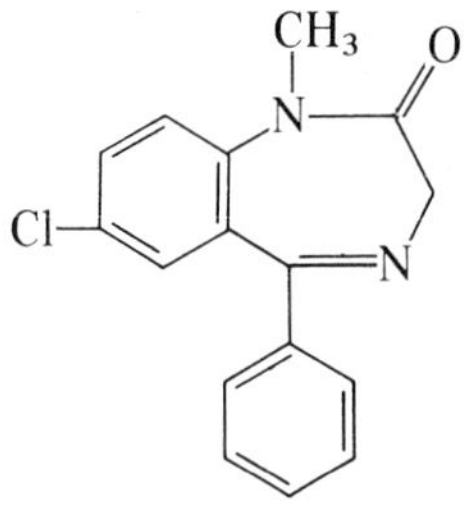

Diazepam

Antidepressive agents include monoamine oxidase (MAO) inhibitors and certain tricyclic compounds which have been found to be effective in the treatment of depression. Examples of such compounds are shown below.

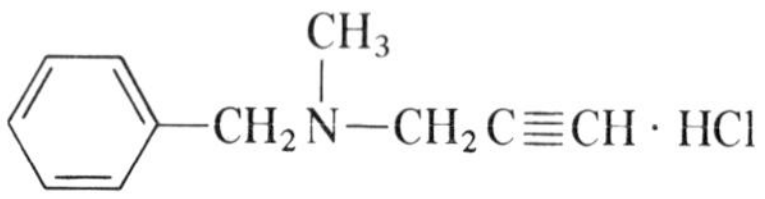

Pargyline
[MAO inhibitor]

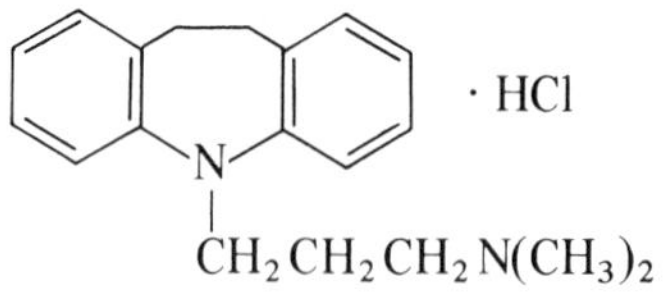

Imipramine

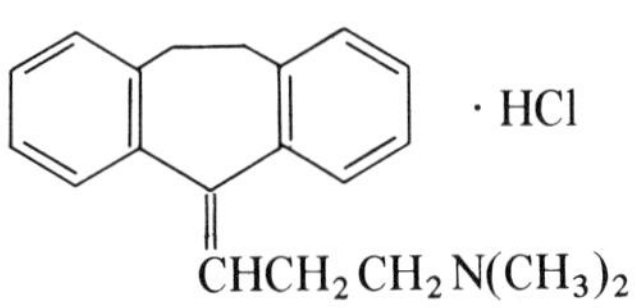

Amitriptyline

Sedatives and hypnotics are used mainly to produce drowsiness. They are general central nervous depressants. Derivatives of barbituric acid are widely used. Certain compounds not related to the barbiturates are also very useful hypnotics.

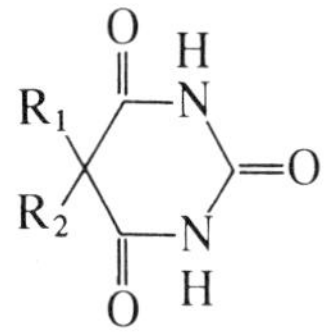

$R_1 = R_2 = C_2H_5$ Barbital
$R_1 = C_2H_5$; $R_2 = C_6H_5$ Phenobarbital
$R_1 = C_2H_5$; $R_2 = CH_3CH_2CH(CH_3)$— Butabarbital (as Na salt)

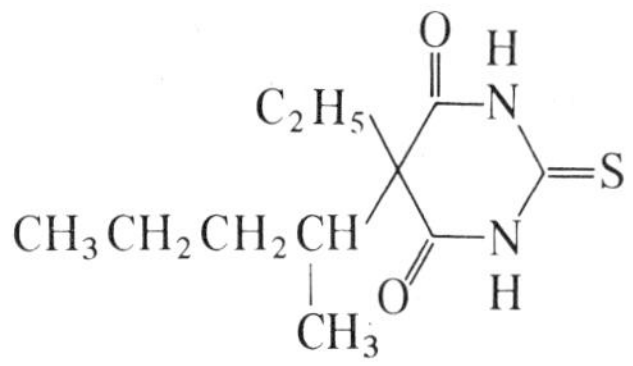

Thiopental

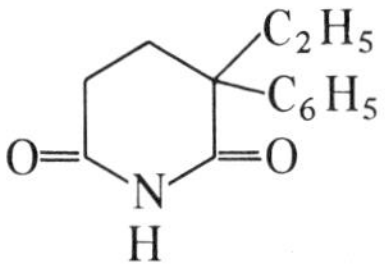

Glutethimide

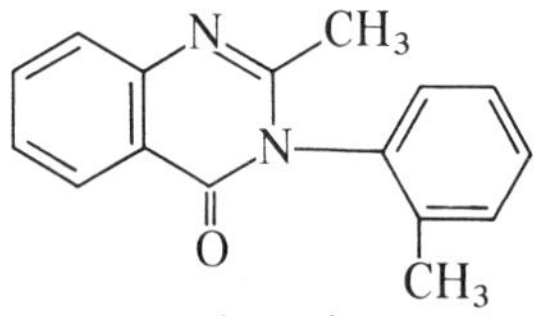

Methaqualone

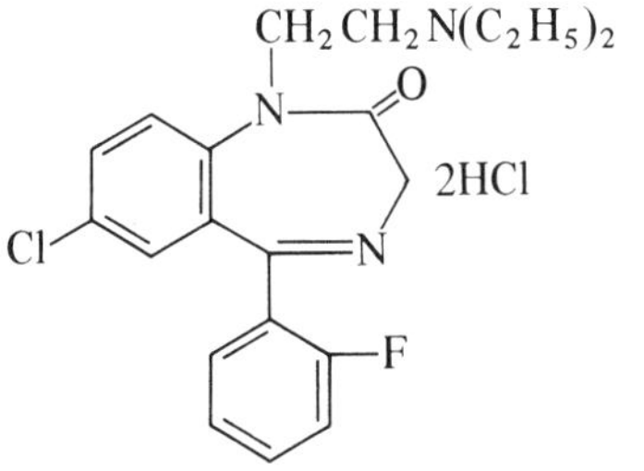

Flurazepam hydrochloride

Sulfonamides. The sulfonamide drugs were the first effective chemotherapeutic agents to be employed systematically for the cure of bacterial infections in man. Before penicillin and the other antibiotics became generally available, the sulfonamides were the mainstay of antibacterial chemotherapy. They still maintain an important but relatively small place in medicine. Structural formulas of some major sulfonamides are shown below.

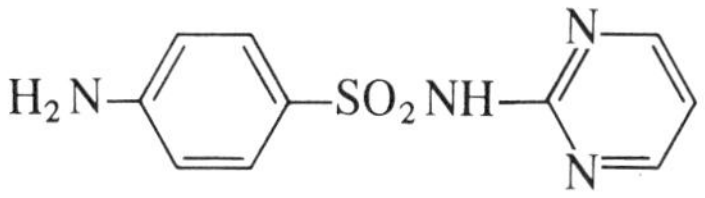

Sulfadiazine U.S.P.

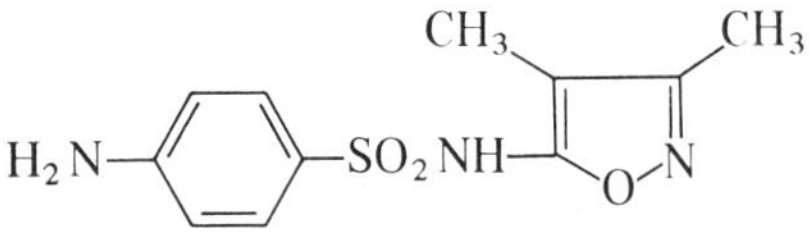

Sulfisoxazole U.S.P.

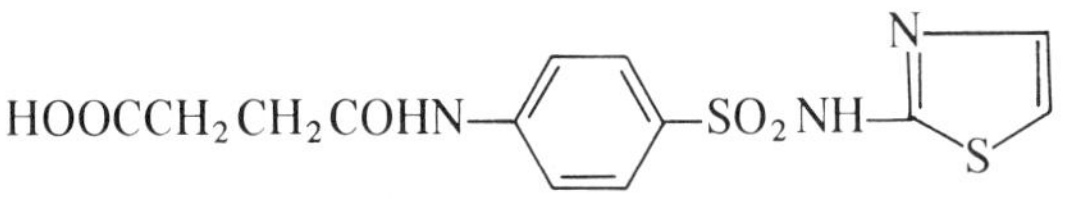

Succinylsulfathiazole U.S.P. (intestinal antibacterial agent)

Vaccines and other Immunizing Agents. Vaccines are available which will provide an immunity against certain bacterial and viral infections. Examples are given in Table 20.5.

TABLE 20.5 Some Important Vaccines

Vaccine	*State of Bacteria or Virus*
Pertussis U.S.P.	Killed bacteria
Typhoid vaccine U.S.P.	Killed bacteria
Smallpox vaccine U.S.P.	Live attenuated virus
Oral poliomyelitis vaccine U.S.P.	Live attenuated virus
Mumps vaccine	Live attenuated virus
Measles vaccine U.S.P.	Live attenuated virus
Influenza vaccine U.S.P.	Killed virus
Rubella vaccine	Live attenuated virus

The group of immunizing agents also include toxoids, e.g., diphtheria toxoid U.S.P., antitoxins, immune serums, e.g., antirabies serum U.S.P., and antivenins.

Vitamins. A vitamin may be broadly defined as a substance essential for natural metabolic functions, but not syntheiszed in the body. It must, therefore, be furnished from an exogenous source.

A well balanced diet should supply adequate amounts of vitamins. In certain cases, however, it is desirable to supplement the dietary intake with vitamins given in the pure chemical form. These may be administered either as individual compounds for certain specific indications, or as a multivitamin preparation. A large number of individual vitamin and multivitamin capsules and tablets are available. The only official multivitamin preparation is decavitamin capsules, U.S.P.

RESEARCH

Dedication to the search for useful new medicines is a characteristic of the drug industry. In 1968, 20,480 of the drug industry's employment of 134,700, or 13 per cent, worked in research. In excess of 11 per cent of the gross income of companies with research and development programs was required to carry out these programs.

From 1940 to 1967, 848 basic new drugs evolved through the efforts of the pharmaceutical industry; of these 525 were discovered and developed by American companies. Annual research expenditures grew from 50 million in 1951 to 500 million dollars in 1968. Of the many compounds prepared in the research laboratories, only one in 6000 tested compounds ever reaches the market place. It is quite possible that in the coming years this figure will approach 1 in 10,000.

The development of a research compound into a drug approved for use in patients may take from 6 to 9 years and require an investment of approximately 7 million dollars. At any time during this period an unfavorable side effect may be discovered which renders useless years of careful and patient work.

Figure 20.1 outlines the various steps in the development of a new drug. As can be seen, it requires the skills and experience of an organization of chemists, biochemists, biologists, microbiologists, doctors, pathologists, toxicologists, pharmacists, and engineers working together to discover and develop a new drug.

MANUFACTURING

Production

Most of the compounds used as drugs today are prepared by chemical synthesis, generally by a batch process. Antibiotics are an exception and are obtained by a fermentation and extraction process.

In designing a synthesis consideration must be given, as in all chemical processes, to the availability of starting materials and other reactants, and to the proper choice of equipment. Reactions are generally carried out in reactors varying in size from fifty to several thousand gallons. Depending on the reaction, either stainless steel or glass lined steel reactors are used. Figure 20.2 shows a drawing of a typical 100-gallon reactor, Fig. 20.3, a reactor installation, and Fig. 20.4, a complex installation of reactors required for a multistep chemical process.

Processes for acetylsalicylic acid (aspirin), phenobarbital, penicillin, and influenza vaccine are described below. They illustrate the chemical manufacturing procedures used in the pharmaceutical industry.

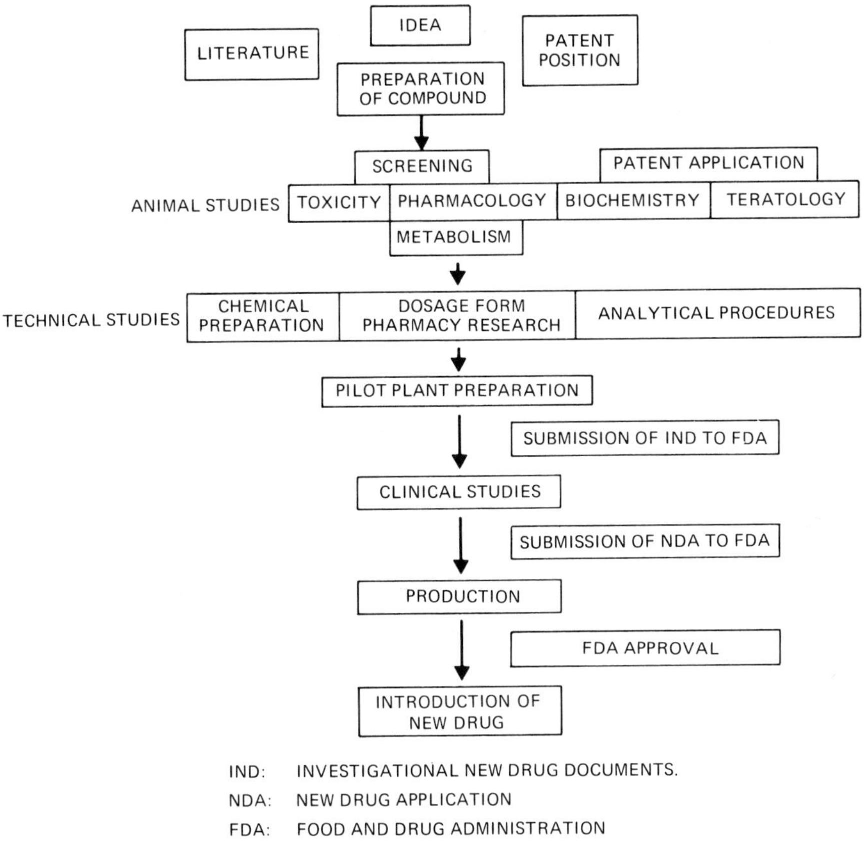

Fig. 20.1 Development of a new drug from idea to the market.

Acetylsalicylic Acid. A commercial process engineered for good yields, high purity, and low cost is shown in Fig. 20.5. Salicylic acid powder, acetic anhydride, and mother liquor are reacted in a 500-gallon glass-lined reactor for 2 to 3 hours. The mass is then pumped through a stainless steel filter to a crystallizing kettle which may be a 500-gallon glass-lined reactor. The temperature is reduced to 3°C, and after crystallization a portion of the mother liquor is drawn off for the next batch (step 1) in a modified nutsch-type slurry tank. The remaining slurry is centrifuged, and the crystals are dried in a rotary dryer to yield bulk aspirin. Excess mother liquor is stored and then distilled to acetic acid.

Phenobarbital. The reaction sequence and a schematic representation of a process for preparing phenobarbital are shown in Fig. 20.6. Dimethyl phenylmalonate and ethyl bromide are charged into a reactor, a methanolic solution of sodium methoxide is added at approximately 45°C, ethylation occurs and the reaction mixture is neutralized with sulfuric acid. Excess ethyl bromide and methanol are distilled off and water is added. The reaction mixture is then transferred to a separator and diluted with benzene. The aqueous sodium bromide solution is separated and the benzene solution of the product is washed with dilute sodium hydroxide at 5°C to remove unreacted dimethyl phenylmalonate, and the aqueous solution separated. Acidification of the sodium hydroxide extract yields phenylacetic acid after heating. The ethylation product is washed with dilute hydrochloric acid, and distilled to yield pure di-

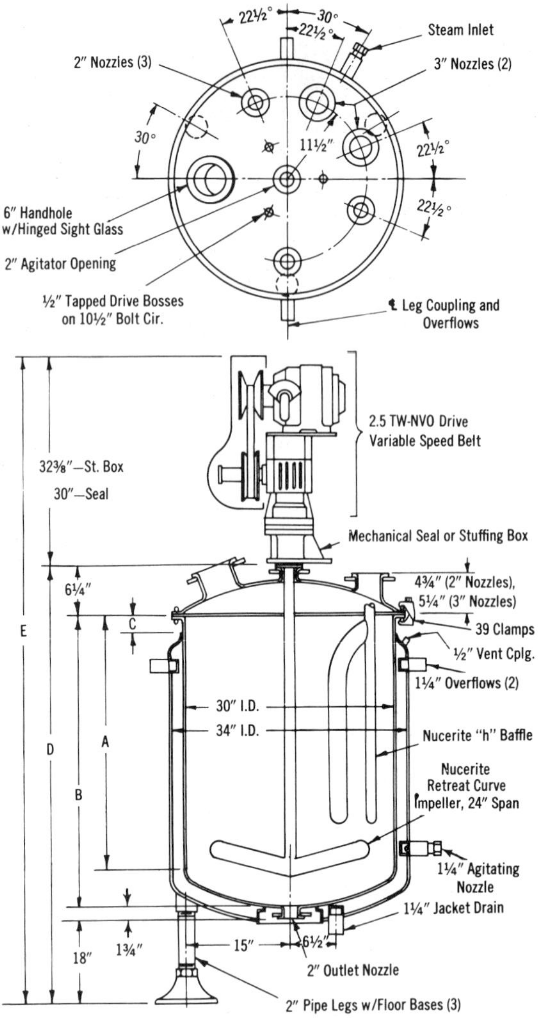

DIMENSIONS, IN.

25 PSI VESSELS

Capacity, Gallons	A	B	C	D	E Seal	E St. Box
100	36	41¼	3	67½	97½	99⅞

Fig. 20.2 Plan and elevation of a 100-gallon glass-lined reactor. (*Courtesy Pfaudler Co.*)

methyl ethylphenylmalonate. This is condensed with dicyandiamide and sodium methoxide in methanol, and the product is hydrolysed with sulfuric acid at reflux temperature. The crude phenobaıbitol is separated by centrifugation, washed, dried, recrystallized from an alcohol-water mixture, filtered, and dried.

Fig. 20.3 Chemical reactor installation in pilot plant. (*Courtesy CIBA-GEIGY Corp., Pharmaceuticals Division.*)

Penicillin. The production of an antibiotic is typified by the manufacture of penicillin. The mold used industrially is from the *Penicillum chrysogenum* group and is particularly effective against staphylococcus, streptococcus, and pneumococcus, mostly gram-positive microorganisms; generally it has little effect on gram-negative microorganisms.

The penicillin produced commercially is designated as penicillin G (benzyl penicillin) although there are several types produced by the mold. These compounds are carboxylic acids and since they are unstable as the free acids, the commercial products are the sodium, calcium, aluminum, potassium, or procaine salts. The formula for penicillin G is shown below. Other forms of penicillin contain different groups in the bracketed portion.

Fig. 20.4 Complex installation used in the chemical manufacture of drugs. (*Courtesy CIBA-GEIGY Corp., Pharmaceuticals Division.*)

In 1943 the means of production was the surface process in which the mold grows on the surface of a shallow layer of media in trays or bottles. With the development of the commercial method of submerged fermentation in 1944, reduction in space and labor requirements resulted in a tremendous reduction in cost.

In producing penicillin by submerged fermentation the inoculum or "seed" for the large fermentation tanks with a 5,000 to 30,000-gal capacity is prepared by growing a master stock culture of the mold from lyophylized spores on a nutrient agar substratum

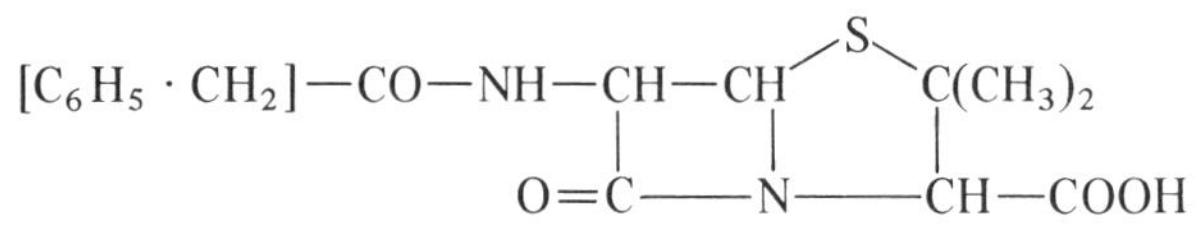

Penicillin G, also called penicillin II, or benzyl penicillin.

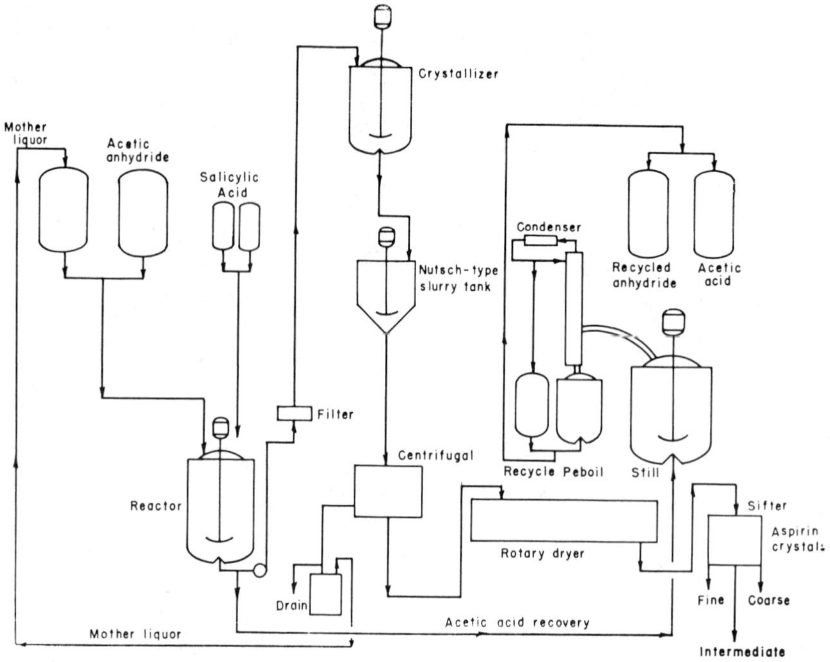

Fig. 20.5 Flow diagram for production of bulk aspirin. (*Excerpted by special permission from Chem. Eng.,* **60**, *No. 6, 116–120 (1953); copyright 1953 by McGraw-Hill, Inc.*)

with incubation. Several gallons of culture medium, generally constituting 5 to 10 per cent of the charge, are prepared in a series of seed tanks to "seed" a large tank.

The four main stages in the manufacture of bulk penicillin are:

1. Fermentation.
2. Removal of mycelium from the fermented broth and extraction of penicillin by solvents.
3. Solvent purification and formation of the penicillin sodium salts.
4. Testing, storage, and shipping.

The fermentation broth is made from corn steep liquor, to which 2 to 4 per cent lactose has been added, in addition to such inorganic materials as $CaCO_3$, KH_2PO_4, $MgSO_4$, and traces of iron, copper, and zinc salts. The addition of some compounds, although desirable for mold growth, must be omitted since they cannot be tolerated in the end use of the product nor economically removed. After adjusting the pH to 4.5 to 5.0, the culture medium is fed into the fermenter which is equipped with a vertical agitator, a means of introducing air which has been sterilized by filtration, and coils for controlling temperature. Following sterilization of the fermenter, the mold is introduced through sterile pipe lines by air pressure. During its growth the temperature is maintained at 23 to 25°C. Sterile air permits growth of the aerobic mold; agitation distributes it uniformly in the batch. One volume of air per minute is required for each volume of medium. Assays are made every 3 to 6 hours; after 50 to 90 hours, when the potency stops increasing, the mold is harvested. The batch is cooled to 5°C because of the instability of penicillin at room temperature, and the mycelia are removed by filtration on a rotary drum filter.

In the old process, penicillin was recovered from the filtrate by charcoal adsorption. It was eluted with amyl acetate, the eluate being concentrated, cooled to 0°C, and acidified with an organic acid to pH 2.0. In the solvent-extraction process the activated carbon step is omitted and the filtered liquor ["beer"] is adjusted to a pH of 2.5 with phosphoric acid in the line. Continuous countercurrent extrac-

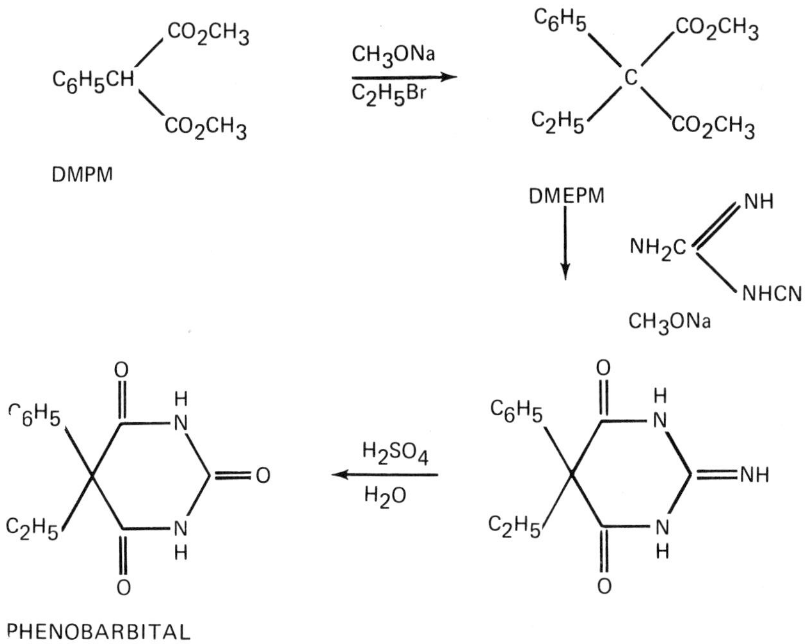

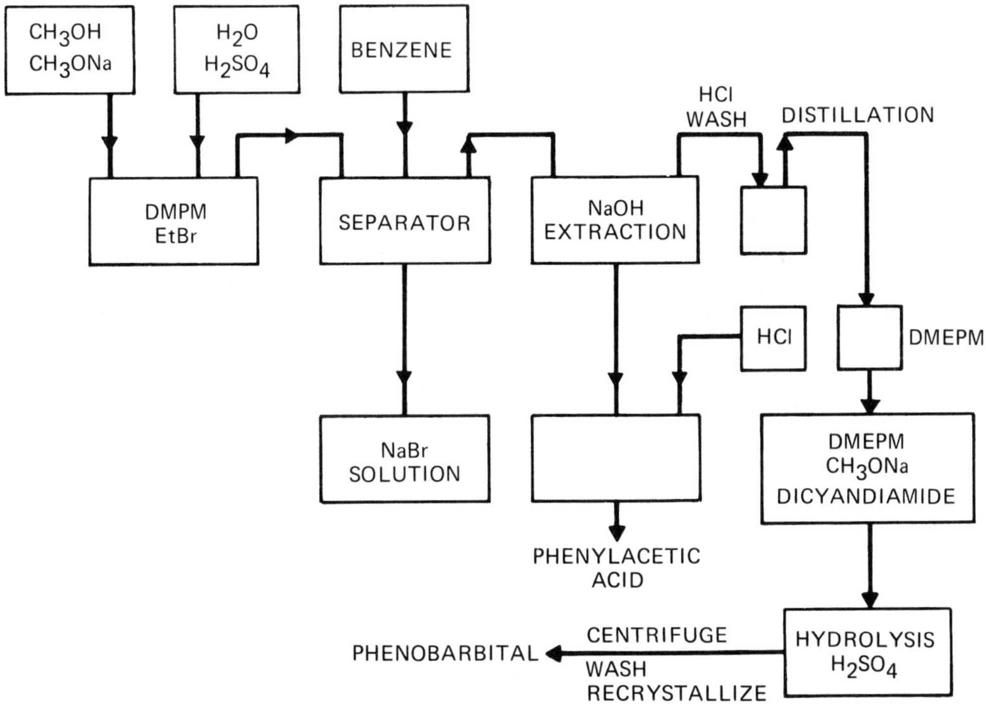

Fig. 20.6 Chemical reactions and flow diagram for manufacture of phenobarbital.

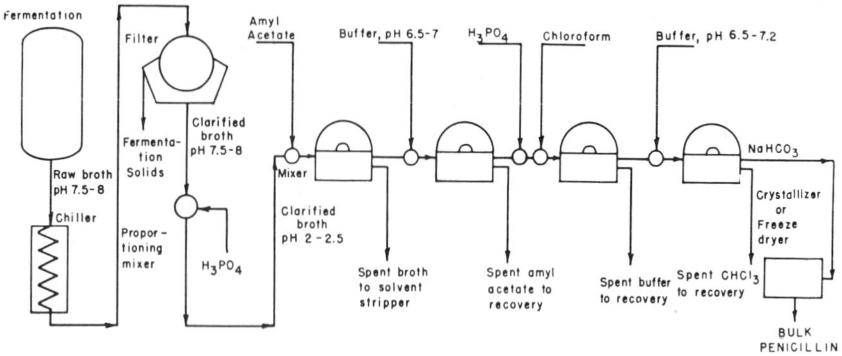

Fig. 20.7 Flow diagram for manufacture of penicillin. (*Excerpted by special permission from Chem. Eng.,* **64**, *No. 5, 247 (1957); copyright 1967 by McGraw-Hill, Inc.*)

tion is carried out with amyl acetate and then chloroform, with successive concentration in Podbielniak centrifugal extractors; the final liquor is treated with buffered phosphate and sodium bicarbonate to form the sodium salt. This material is made sterile by filtration and is freed aseptically of water and other solvents by crystallization; crystalline penicillin is thereby formed which, when dried, may be packed in bulk in polyethylene bags, or stainless steel containers.

Section 507 of the 1938 Food, Drug and Cosmetic Act requires that all forms of penicillin be tested prior to sale. Specifically, potency, toxicity, pyrogens, sterility, and moisture (since this affects stability) are extremely important. Potency is determined by comparison with a standard, to assure consistent clinical results; toxicity refers to a test made with mice that determines the toxic effect of the drug on humans when taken internally; pyrogen testing determines the pres-

Fig. 20.8 Antibiotic recovery plant for penicillin manufacture. (*Courtesy Pfizer, Inc.*)

ence of fever-inducing substances which may be undesirable in treatment; sterility, defined earlier, refers to the absence of micro-organisms.

A numerical lot number and code system is maintained for each lot from bulk to finished product so that in case of necessity (e.g., lapse of the "expiration" period) the material may be withdrawn from trade channels.

Penicillin is now produced in the United States, Canada, Mexico, Australia, Japan, Italy, England, and many other countries. Initial United States production was 250 billion units a year; in 1946 production was 25,000 billion units; in 1957, 525,738 billion units. The quality has improved with improvements in technology; the present product is practically pure, crystalline, and white.

Influenza Vaccine. Influenza vaccine is a sterile, aqueous suspension of suitably inactivated influenza virus. It usually contains two or more different virus strains. The strains used are those designated by the Division of Biologics Standards of the National Institutes of Health.

The actual production cycle starts with fertile eggs which are first candled, then disinfected by a coating of an iodine solution, and finally punctured with a small drill (Fig. 20.9). The virus is introduced through this hole which is then sealed with collodion gelatin and incubated at 99°F for 48 hours. A circular top section of the egg is removed and the allantoic egg fluid is extracted in the harvesting operation. The live virus is separated from the egg fluid by centrifuging at 50,000 rpm (Fig. 20.10); the heavier virus particles separate and cake out on the centrifuge cartridge. The virus is collected and reconstituted in a saline solution. The live virus is inactivated with formalin under prescribed conditions. The suspension is tested for inactivation of the virus, sterility and potency (antigenicity) and subsequently filled aseptically into vials.

Fig. 20.9 Seeding of eggs with influenza virus. *(Courtesy Pfizer, Inc.)*

Formulation

For medical practice it is rare for a drug substance to be administered as the pure chemical compound itself. Drugs are almost always administered in some kind of formulation.

Recently, great emphasis has been placed on the importance of formulations with the recognition that they can significantly influence the physiologic availability or "bioavailability" of drugs. The high degree of uniformity, the physiologic availability, and the therapeutic quality expected of modern medicinal products usually are the results of considerable effort and expertness on the part of the industrial pharmacist.

Drugs for specific applications are formulated in a large number of dosage forms. These are grouped into certain classes as follows:

Solutions and suspensions.
Parenteral preparations.
Ophthalmic solutions.
Extractives.
Medicated applications (e.g., ointments).
Powders.
Tablets, capsules, and pills.
Aerosols.

Tablets, Capsules, and Pills. Tablets and capsules are the most frequently used form for oral use. Advantages include relative simplicity in fabrication, stability, convenience in dispensing, accuracy, compactness, and relative blandness of taste.

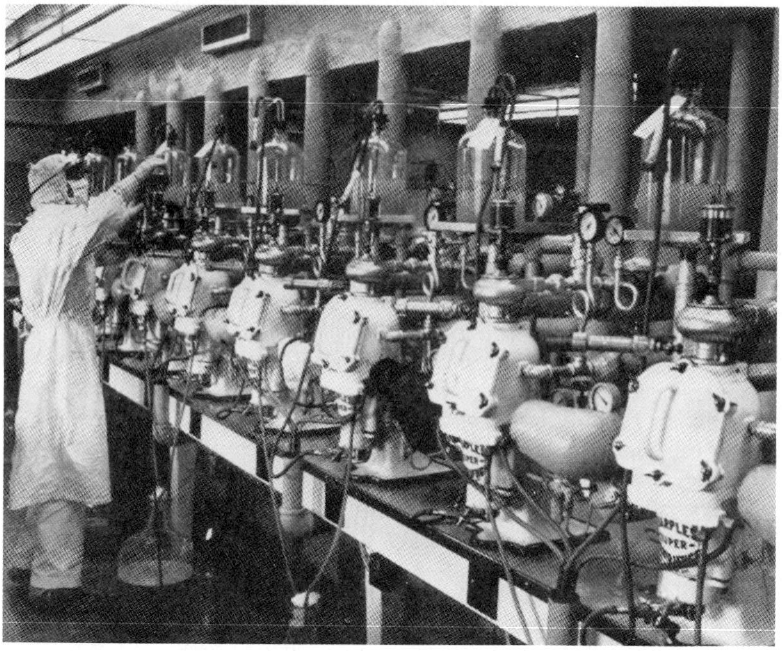

Fig. 20.10 Separation of virus by centrifugation. (*Courtesy Pfizer, Inc.*)

Tablets may be prepared without any special coating or they may be sugar coated, film coated, or enteric coated. Sugar and film coatings are used to enclose medicinals having an objectionable taste and to protect sensitive drugs from deterioration. Enteric-coated tablets resist solution in gastric fluid and do not release the medication until they have reached the intestine.

Tablets are compressed on a tableting machine as shown in Fig. 20.11. The formulation fed into the tableting machine contains the active drug combined with diluents (e.g., lactose), binders, lubricants (e.g., stearic acid), and frequently a disintegrator. A disintegrator is a substance or a mixture of substances added to a tablet to facilitate its breakup or disintegration after administration; corn starch is often used for this purpose. Coloring agents are added to assist in control of the product during manufacturing and to serve as a means of identification to the user.

Fig. 20.11 High speed Manesty tableting press. Output: 4000 tablets per minute. (*Courtesy CIBA-GEIGY Corp., Pharmaceuticals Div.*)

Capsules are a solid dosage form in which the drug, usually in a mixture with a diluent, is enclosed in a hard or soft soluble container of gelatin. The capsules are made tamper-proof by sealing the two parts of the capsule with a clear band of gelatin around the capsule joint.

In recent years, prolonged action formulations of drugs have been introduced. These eliminate the necessity for administration at frequent intervals. Federal regulations require the manufacturer to show that such a preparation is properly made and that the active components will be released at a safe and effective rate.

In one form of prolonged-action formulation the total dosage is divided into a number of small beads which are subdivided into a number (e.g., 3 to 9) of parts. One part is left uncoated while the others are given coatings of various thickness which should resist disintegration for 3 to 9 hours, thus providing medication for approximately 12 hours.

In another form, a tablet may contain a core of a wax and other materials in which the drug is imbedded. The outer layer contains the drug in a regular granulation which dissolves rapidly following administration. The drug in the core is then slowly leached out to provide sustained drug release. The core can also consist of a small plastic pellet containing thousands of small passages filled with the drug, which is slowly leaked out in the gastrointestinal tract. Numerous other modifications of these principles have been described.

Parenteral Preparations. Parenteral preparations are administered by injection and can provide immediate physiological action. They are of special usefulness in those cases where the patient is unconscious. The therapeutic response of a drug is also more readily controlled by parenteral administration since it avoids the irregularities of intestinal absorption. However, since the protective barriers of skin and mucous membranes are bypassed, it is imperative that parenteral solutions be as nearly perfect as possible with respect to purity and freedom from contamination by bacteria or viruses.

Single- and multiple-dose vials and ampuls are used in medical practice. The vehicle (solvent), which is of greatest importance for parenteral products, is specially distilled, pyrogen-free water; the container of choice is glass.

Prior to large scale production, painstaking studies must be carried out to determine the stability of the drug in its final dosage form. That is, its stability to heat, light, and humidity over an extended period of time.

Quality Control

In the industrial manufacture of drugs, the variety and complexity of operations make it necessary to assign to a separate and independent group of scientists within the company the responsibility for controlling the quality of the final product. This group is referred to as Quality Control.

To ensure the required high standards of safety, purity, and effectiveness of drug products, it is necessary to carefully control each step in the manufacturing procedure. This means control of all raw materials including packaging components and labels; control of the product during manufacturing by means of in-process analysis; and control during packaging of the final product. As an example, tablets are examined for the following properties:

Identity.
Content of active drug.
Size.
Physical appearance.
Disintegration time (time required for a tablet to disintegrate in water at 37°C).
Friability (mechanical stability of tablet).
Weight variation.

In addition, a representative sample of tablets is taken from each lot and each tablet is assayed individually for drug content. Automated equipment is frequently used for this test (Fig. 20.12). Injectable solutions are also tested for sterility and absence of pyrogens.

Spectroscopic, chromatographic, and other analytical methods are used to determine the purity and quantity of drug present in the various dosage forms.

Specifications for containers, closures, and other component parts of drug packages must be drawn up to ensure that they are suitable for their intended use. Also, packaging and

Fig. 20.12 Assayomat instrument used to determine drug content of industrial tablets. (*Courtesy CIBA-GEIGY Corp., Pharmaceuticals Div.*)

labeling operations must be adequately controlled to assure that correct labeling is employed for the drug. Finished products must be identified with lot or control numbers that permit determination of the history of the manufacture and control of the batch.

REFERENCES

1. L. S. Goodman and A. Gilman, Eds., "The Pharmacological Basis of Therapeutics," 4th Ed., New York, Macmillan, 1970.
2. J. E. Hoover, Ed., "Remington's Pharmaceutical Sciences," 14th Ed., Easton, Pa., Mack Pub., 1970.
3. L. C. Schroetor, "Ingredient X," Elmsford, N.Y., Pergamon Press, 1969.
4. Annual Survey Report, Pharmaceutical Industry Operations, Pharmaceutical Manufacturers Association, Washington, D.C., 1969.
5. The Pharmacopeia of the United States, 18th Rev. [USP XVIII], Sept. 1970.
6. National Formulary, 13th Ed. [NF XIII], 1970. (Prepared by the National Formulary Board by authority of the American Pharmaceutical Association.)
7. Pennsylvania Medicine, **73**, No. 2, 39–110 (Feb. 1970). (In Tribute to the Pharmaceutical Industry.)
8. R. E. Kirk and D. F. Othmer, "Encyclopedia of Chemical Technology," 2nd Ed., New York, John Wiley & Sons, 1963 to 1970.

21

The Pesticide Industry

Gustave K. Kohn*

INTRODUCTION

Scope of the Chapter

This chapter deals with the chemicals used in agriculture mainly to protect, preserve, and improve crop yields. Arbitrarily excluded are those substances that serve as fundamental nutrients, which are treated in Chapters 5 and 18 on nitrogen and phosphorus technology respectively. Nevertheless, it is the current practice of the farmer, particularly in advanced American agriculture, to integrate nutritional and plant-protection application schedules, and even provide single formulations that include both fertilizers and pesticides. Further, plant nutrition at this stage of scientific sophistication is far more complex than the older and conventional N—P—K applications. Many chemicals that accelerate plant growth act as hormonal agents modifying plant metabolic processes at some stage of development. Since these substances are manufactured and marketed by the pesticide industry, they are included as subject matter here.

Included in this chapter also, are chemicals which are significant in public health. Many organisms are vectors in the dissemination of human and animal disease. Since products of the pesticide industry control the insect or the rodent or the mollusk, etc. (the vectors), they are often the most effective means for controlling some of the most serious health problems of mankind, especially, but not exclusively, in the underdeveloped countries. Thus, the chemicals that control such vectors are also included in the subject matter of this chapter. In this area, and in the area of animal health products, Chapter 20 on the pharmaceutical industry overlaps to some extent.

Finally, the products of the subject industry are employed in urban, suburban, and rural areas in home and garden for a variety of useful and aesthetic purposes. Industry employs herbicides, algicides, fungicides, and bactericides, and the railroads utilize herbicides to keep right-of-ways free of vegetation.

*Ortho Div., Chevron Chemical Co.

The Lanauage of the Pesticide Industry

There is a need for a generic term to cover the diversity of functional applications. The term "insecticide" has been commonly applied to all control agents. The Federal Insecticide Act of 1910, provided an extension in meaning to include both insecticides and fungicides. The Federal Insecticide, Fungicide and Rodenticide Act (1947), uses "economic poison" for all these groups, and embraces herbicides as well. The term "pesticide" is now used officially to include all toxic chemicals, whether used against insects, fungi, weeds, or rodents, etc. Agricultural disinfectants and animal health products are in many instances also included under the term "pesticides."

"Plant protection agents" is another generic term. A common practice within the industry is to add the suffix "cide" to the group or biological unit under consideration. The term "insecticide" has both generic and specific meaning that depends upon context. Obvious in meaning, also, are such frequently used words as fungicide, bactericide, miticide or acaracide, ovicide, herbicide, rodenticide, muscicide, molluscicide, nematicide, and algicide. This practice of utilizing the "cide" suffix reflects also upon the state of agricultural technology as it was practiced during the first half of the twentieth century. Whereas the objective in most cases, until recently, was to kill or attempt to eradicate the offending pest, many of the newer programs reflect more complex approaches.

Hence the terms attractants, repellents, antifeeding compounds, hormones (plant- and insect-growth-control agents), defoliants, dessicants, etc., all describe functional chemical agents that are within the purview of the pesticide or plant-protection industry. There are many who believe that these chemicals will become more significant in the agriculture of the future.

History

It is probable that man's treatment of crops with foreign substances dates back into prehistory. The Bible abounds with references to insect depredations, plant diseases, and some basic agricultural principles such as periodic withholding of land in the fallow state. Homer speaks of "pest-averting sulfur."

More recently, in the nineteenth century, a great increase in the application of foreign chemicals to agriculture ensued. Discovered or rediscovered was the usefulness of sulfur, lime sulfur (calcium polysulfides), and Bordeaux mixture (basic copper sulfates). With the exception of formaldehyde, inorganic chemicals provided the farmer with his major weapons.

The earliest of the organic compounds were generally chemicals derived from natural products or crude mixtures of chemicals in states of very elementary refinement. Ground-up extracts of plant tissue were useful in the control of insects. Such extracts were employed in agriculture in many cases before the chemist had elucidated the structure or synthesized the molecule responsible for the biological activity. These extracts included the pyrethroids, rotenoids, and nicotinoids, which are still derived in large part from plant extracts. Crude petroleum fractions were recognized for their effectiveness in the control of mites, scale, and various fungi, as well as for their phytopathological properties.

Although a few synthetic organics were already known, the great revolution in the use of organic chemicals in agriculture roughly coincides with the period of the onset of World War II. The more important of these discoveries were DDT (Mueller–1939) 2,4-D (Jones patent–1945), benzenehexachloride (ICI and French development–circa 1940), and the organic phosphate esters (Schräder–begun in the late 30's, revealed in the forties). These new chemicals were so enormously more potent than their predecessors in their biological activity (frequently by orders of magnitude) that they very rapidly displaced almost all of the chemicals previously employed. The chemicals of today, largely discovered in the 1950's and 1960's, are predominantly extensions of this almost revolutionary transition from inorganics to synthetic organics that dates from the period of World War II.

These trends are vividly illustrated in Fig. 21.1. As late as 1945 inorganic chemicals accounted for almost 75 per cent of all pesticide sales, and oil sprays and natural products, most of the remainder. The rapid reduction in

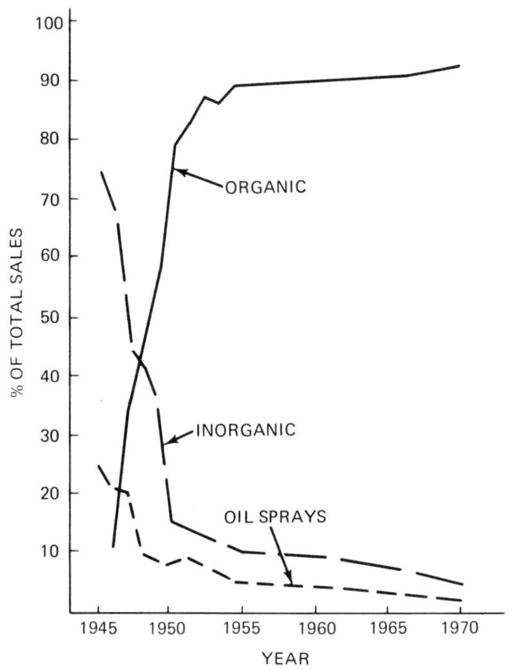

Fig. 21.1 Impact of synthetic organic chemicals on pesticide sales. [This chart is based on estimates for U.S. industry derived from data collected from one major chemical company (Chevron Chemical Co.) which is a manufacturer and marketer of organic and inorganic pesticides and oil sprays.]

the sales of inorganics was virtually complete by the early 1950's, and oil sprays and natural products reached a quantitatively low but constant level at approximately the same period.

Federal government statistics on the supply of pesticides in the United States emphasizes the dominance of the synthetic organics. Many of the botanicals are derived from imports to the U.S.

In contrast to the supply and sales of chemicals from the standpoint of their derivation, it is interesting to examine trends according to the biological function of the pesticide.

It is fair to say that within the United States the pesticide industry is *largely dominated by synthetic organic chemicals.* The inorganics, oil sprays, and natural products continue to share a small (less than 10 per cent) portion of the total pesticide market.

The data in Table 21.1 show the enormous growth of the pesticide industry (see also Fig.

TABLE 21.1 U.S. Production of Synthetic Pesticides

(Millions of lbs)

Year	*Insecticides*	*Fungicides*	*Herbicides*
1945	43	-	-
1950	177.3	28.5	1.1
1955	331.1	76.8	70.6
1960	351	98.2	102.5
1965	490	127	264
1968	569	154	469

21.3). Although in the United States fungicides are presently the smallest of the three main classifications of pesticides, it must be remembered that in tropical and high moisture areas

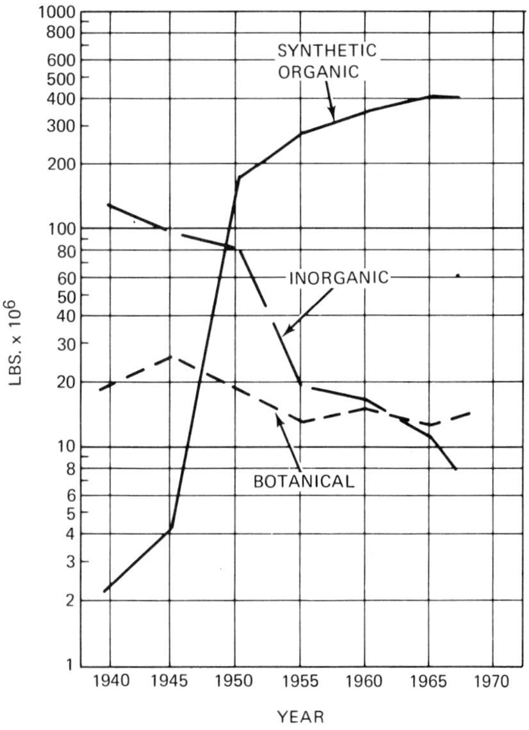

Fig. 21.2 Total U.S. supply of pesticides through 1967. (Note that the ordinate is on a logarithmic scale. The term *botanical* is used in the pesticide industry to describe compositions obtained from ground-up plant products, some of which are discussed elsewhere in this text. In essence, the figure demonstrates the logarithmic rise of the synthetic organics as compared with orders of magnitude lower, but fairly steady, supply or inorganics and botanicals.)

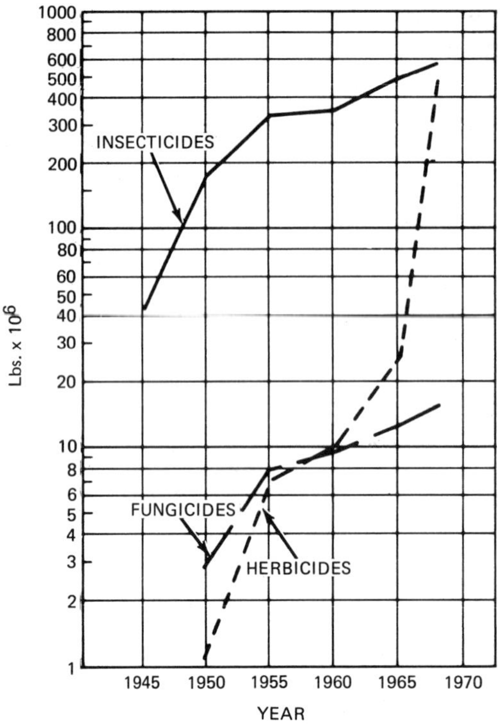

Fig. 21.3 U.S. production of synthetic pesticides.

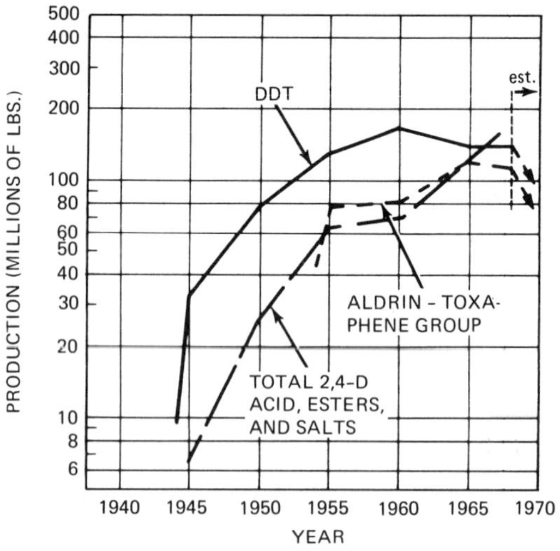

Fig. 21.4 U.S. production of certain halogenated pesticides, and projections for the future.

the problem of plant disease is much more serious. High labor costs are largely responsible for the extremely rapid growth rate of herbicide production. Herbicides, which diminish the need for cultivation (hoe and more modern mechanical weeding) in advanced agricultural countries, will probably find more use in the future. On the other hand, ecological pressures and the halting of herbicide use in war may lower the production figures of certain insecticides (DDT and chlorinated hydrocarbons) and certain herbicides (2,4,5-T, 2,4-D, and picloram) for 1970 and succeeding years.

CHARACTERISTICS OF THE PESTICIDE INDUSTRY

Among the distinguishing characteristics of the pesticide industry are (1) the multiplicity of chemical agents employed, (2) a limited price range (and hence limited chemical complexity governed primarily by the economics of agricultural production), (3) a fairly rapid obsolescence for the chemicals employed, and (4) a high degree of government regulation for the production, application, shipment, and use of pesticide chemicals.

Government Regulation

The first State laws on insecticides were enacted in 1900 to establish standards of purity for the arsenical, Paris green

$$Cu(CH_3{-}\overset{\overset{\displaystyle O}{\|}}{C}{-}O{-})_2 \cdot 3Cu(AsO_2)\ .$$

Gradually these laws were extended to cover a wide list of inorganic compounds and plant extracts, many of which are extremely toxic to man. Included in this group are such compounds as arsenic combined with copper, lead, and calcium; phosphorus pastes for ants and roaches; strychnine in rodent baits; thallium in ant and rodent baits; and selenium for plant-feeding mites. Mercury, both as corrosive sublimate and calomel, was used as an insect repellent and later as a seed disinfectant. Sodium fluoride was the common ant poison, and sodium cyanide, calcium cyanide, and HCN itself were the universal fumigants. Nicotine sulfate was used generally in the garden and on the farm. These compounds, among the most toxic of any known at that time, were widely marketed without supervision under any of the early state laws. There was no provision for public health, either in regulating the amounts applied, or the possible danger of minute amounts (residues) remaining on the marketed produce. The need to protect the operator and the public against the dangerous qualities of the insecticides, or their residues on crops, provided the motives for all the legislation that followed.

In response to reports from England of the poisoning of people eating apples grown in the United States, the Federal Food and Drug Administration in 1927 banned any fruit involved in interstate commerce if it possessed more than 3.57 ppm equivalent of arsenic trioxide.

Government concern was first related to standardization of the manufactured chemical and protection of the farmer in relation to the product that he purchased. This was then extended to the handling of the chemical in interstate commerce, and then to protection of the consumer of raw agricultural products (apples, corn, and lettuce, for example), and in other legislation to protection of the consumer of finished foods (canned juice, margarine, cereal food, meat, and milk, for example). Included in this legislation are provisions that protect the shipper of the chemical, the applicator of the chemical, and all personnel proximal to the application of the chemical.

TABLE 21.2. Development of Government Regulation of the Pesticide Industry in the United States

Date	*Legislation*	*Content*
1900	State laws on product purity (Paris green)	Quality control standard for protection of user of chemical.
1910	Federal Insecticide Act	First specific pesticide regulation standards for insecticides and fungicides moving in interstate commerce.
1938	Federal Food Drug and Cosmetic Act	Protection of public against contamination of foods, drugs, etc.
1947	Federal Insecticide Fungicide and Insecticide Act (FIFRA)	Regulates interstate shipment of pesticides safety standards for the handling of chemicals in interstate commerce. Required registration of labeling under fed. control.
1954	Miller Pesticide Amendment to FIFRA (Public Law 518)	Requires setting of residue tolerances for pesticides on *raw agric. products*; protects consumer.
1958	Food Additives Amendment (contains Delaney Amendment)	Protects public against contamination of processed agric. products—*foods*—from pesticide and other contamination; protects consumer; prohibits pesticide or food additive that causes cancer *at any dosage level* of test protocol.
1947–66	State enactments emboyding FIFRA and food additives legislation	By 1966, 49 States and Puerto Rico had added protection within state boundaries in varying degree for above fed. legislation.
1970–71	Environmental Prot. Agency by Executive fiat (EPA)	Regulates all manufacture and use of products from standpoint of *environmental considerations.* A super administrative and regulatory agency.

Legislation now regulates chemicals applied to crops or foods as protective agents—pesticides, emulsifiers, solvents, packaging materials (wax, container materials, plasticizers), antioxidants, etc. Table 21.2 summarizes government regulation of the pesticide industry.

Toward the end of the 1960's a new dimension of concern arose. The effect of the manufacture and application of the chemical to the environment was recognized. This has culminated in the establishment of the Environmental Protection Agency (EPA) which will supervise the total process involved with the manufacturing, application, registration, and labeling of agricultural chemicals.

At present, functions pertaining to pesticides, food, interstate shipment, consumer health, and the environment previously administered by the Department of Agriculture (PRD), the Department of Health, Education, and Welfare, the Food and Drug Administration, the Department of Interior, the Fish and Wild Life Service, Forestry, the Commerce Department, and other departments and agencies will be all directed by this new agency (EPA).

Many agricultural chemicals and their formulations manufactured in the United States are used throughout the world. Although the application of science and technology is high in the United States, and comprehensive legislation was developed first and most extensively in this country, all the technically developed nations regulate in some degree the manufacture, sale, and use of agricultural chemicals. Germany, Japan, Great Britain, France, and many smaller technically advanced nations have legislation and regulating agencies in this area. This is also true to lesser degree in most of the smaller nations of the world and for many of the so-called underdeveloped nations.

In the United States, enormous and complex questions of values and priorities—despite numerous studies and reports—remain unresolved. Many chemicals have been banned, or their use has been seriously restricted. Very few, if any, registrations of new chemicals to substitute for banned or restricted pesticides were issued in 1969 and 1970. Resolution of these matters relating to ecology, toxicology, and practical need is critical for the pesticide chemical industry, for agriculture, for the protection of the environment, and in the interests of public health.

Obsolescence of Pesticidal Compounds

Obsolescence of pesticidal chemicals derives from two main causes. The first of these is a consequence of the universal property of living things to adapt by selection to environmental changes, including foreign chemicals. From a practical standpoint, this is best exemplified by the tolerance of certain strains of the housefly to DDT. Certain resistant strains of the housefly require 2,000 times the topical application of this chemical to achieve an LD_{50} (the lethal dose required for a 50 per cent kill of a given population). In agriculture, resistance applies as a *practical* consideration almost exclusively to insecticides. The second cause of obsolescence is the development of more useful chemicals. The term 'useful' here covers a host of unrelated properties. The new chemical may be cheaper to produce, easier to apply, less hazardous (to the applicator, consumer, or the environment), possess a greater scope or more intense and extensive biological activity—or, in contrast, may be highly specific in its biological properties. The chemicals enumerated in the charts that follow were all largely the result of technology that contributed some advantageous property. In certain

TABLE 21.3 Obsolescence in the Pesticid. Industry, DDT and its History

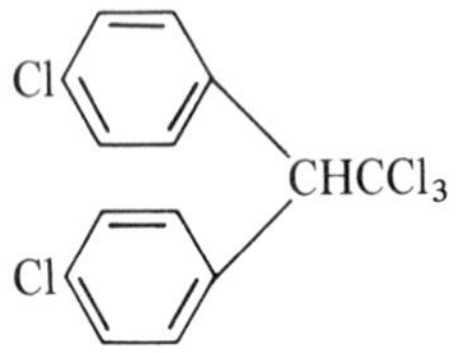

	Year
First synthesized, Zeidler, Germany.	1874
Laboratory curiosity until discovery of insecticidal properties, Paul Mueller, Switzerland.	1939
Nobel Prize (Medicine) Mueller.	1948
Mass introduction into agriculture and public health areas.	1949 Onward
Proliferation of modern analytical techniques. Recognition of insect resistance.	1950–1965
Recognition of residue, physiological, and ecological interrelations.	1960–1970
State and federal legislation prohibiting or severely restricting use.	1969–1970

cases, subsequent research and misuse may have demonstrated that these advantages were overshadowed by the discovery of undesirable consequences.

The case of DDT is an excellent example of the obsolescence of an important pesticidal chemical (Table 21.3). Many chemicals of the chlorinated hydrocarbon group were found which were more highly active—either broadly or narrowly effective—than DDT. Indeed this compound stimulated the research and development of toxaphene, the cyclodiene insecticides DDD, dicofol, etc., and even some of the fungicides and herbicides.

In the sixties, after the development of sensitive analytical techniques (mostly chromatography or spectroscopic methods) and concern for the proliferation of environmental hazards, DDT was virtually eliminated from agriculture. It is significant that just two decades after the discoverer of the insecticidal properties of DDT received the plaudits of the scientific world through the Nobel Prize, the chemical has been almost eliminated from use in the United States!

THE CHEMICALS OF THE PESTICIDE INDUSTRY

Complex chemicals such as streptomycin (which is employed against certain plant bacterial infections), rotenone (a nonpersistent insecticide), and gibberelic acid (a plant growth regulator) all have *limited* practical applicability in agriculture. By far the majority of pesticides are of simpler structure. This results, in part, from the realities of agricultural economics. Structures like those in Table 21.4 containing sequences of optically active sugars or peptides, *cis-trans*-isomeric multiple bonds,

TABLE 21.4 Typical Complex Structures Employed in Agriculture

Structure	*Name, Process, and Function*
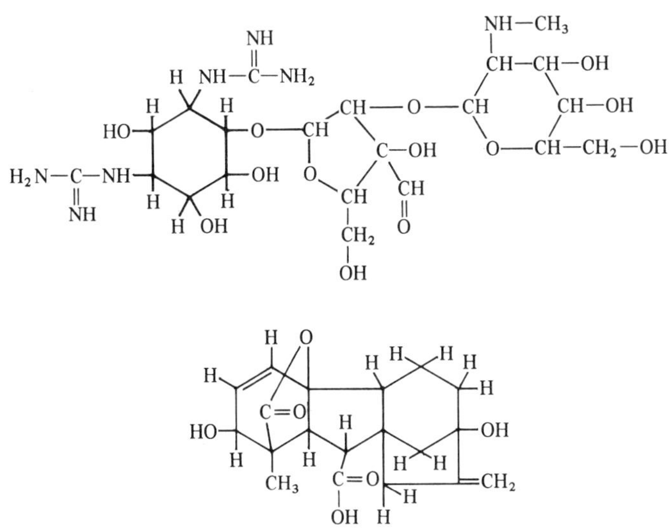	Streptomycin Produced by fermentation Plant bactericide
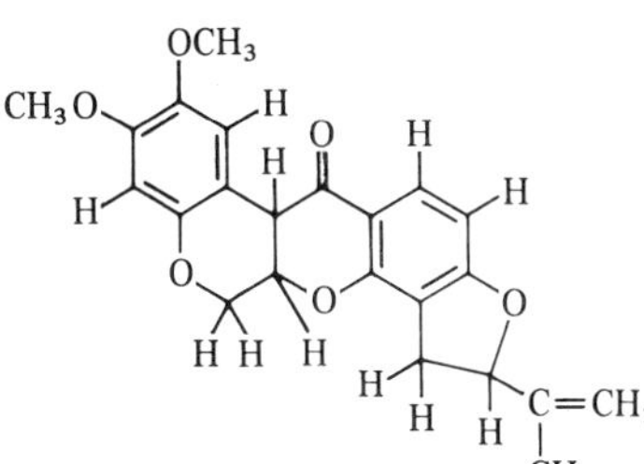	Gibberilic acid Produced by fermentation Plant-growth regulator
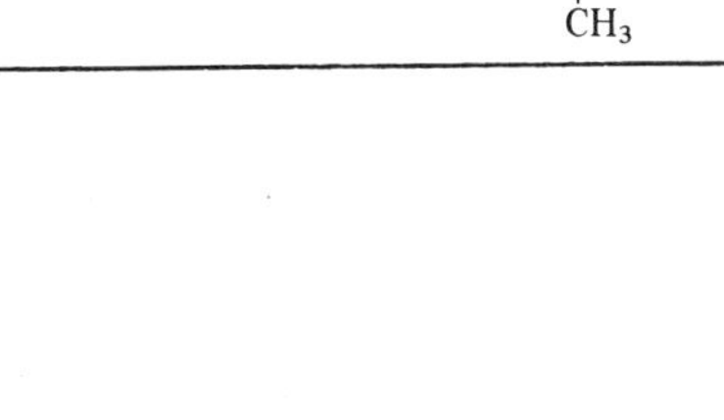	Rotenone Extracted from derris root Insecticide

numerous asymmetric centers, or multiple conjugated rings, present formidable obstacles to the synthetic organic chemist and well-nigh insuperable obstacles to engineers who would try to mass produce them at low to moderate cost.

Because pesticides frequently owe their activity to their "fit" on some biological surface e.g., an enzyme or cell membrane, configuration is frequently decisive for biological activity. The difference in activity resulting from isomeric variations is outlined in the following discussion. Structural isomerism, such as *ortho*, *meta*, or *para* orientation on a benzene ring can be reasonably controlled by the synthetic process. At times, the insecticide is defined by the process employed in its manufacture. Technical DDT is a pesticide which contains both isomers and impurities totaling as many as 14 distinct compounds. The technical product meets a standard defined by a minimum setting point at 89°C. The principal impurity is the *ortho-para* structure which has very little insecticidal potency (see Table 21.5).

The isomers of DDT illustrate the effects of ring substitution on biological activity. Large differences in activity also exist for *cis-trans* isomers. These are given for two organophosphate insecticides in Table 21.6. Optical isomers of pesticidal compounds also exhibit differences in biological activity. This difference is exhibited in the activity of pyrethrins, and in the isomers of benzenehexachloride, where lindane is the isomer that possesses almost all the useful insecticidal activity.

TABLE 21.5 Structural Isomerism of DDT

Structure	*LC_{50} Drosophila Melanogaster*	*% in Technical DDT*
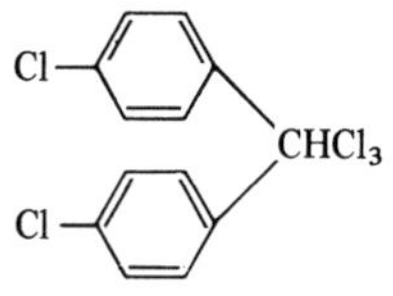 *p,p*-isomer	1	65–85
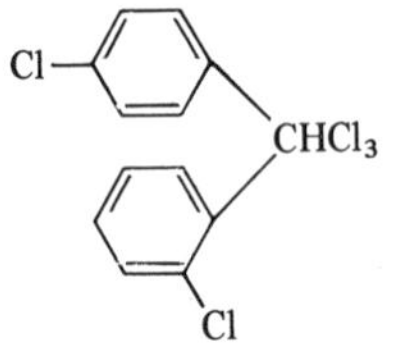 *o,p*-isomer	145	10–20

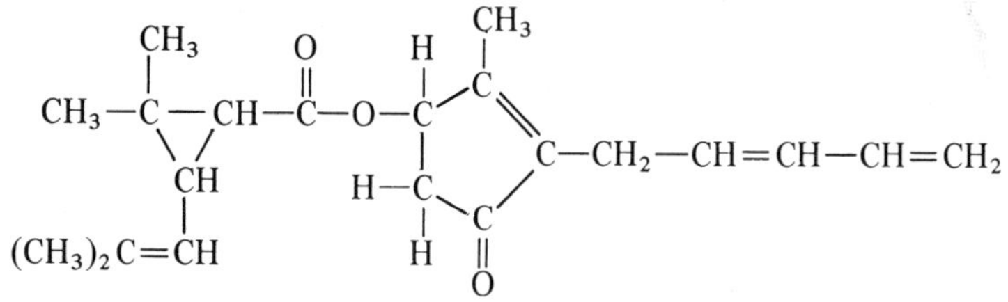

Structure of pyrethrin I (one of several isomers)

It can be concluded from these examples, as well as from the structures that follow, that relatively minor changes in chemical substitution or configuration can cause enormous differences in activity. These structural characteristics do provide some information about the requirements of the receptor site of the biological system affected.

Despite some very striking correlations found for the physical-chemical parameters of organophosphate and carbamate insecticides, the guiding principle for the development of new pesticides of the fifties and sixties was largely empirical.

The molecular weight of most pesticides provides a simple generalization of the size and complexity of the molecules that are commercially employed as pesticides. The common names of some extensively used pesticides are followed by their molecular weights in parentheses: DDT (354.5), 2,4-D (249), captan (300.6), parathion (291.3), atrazine (215.7), maneb (265.3), and carbaryl (201.2). These are small to medium-sized organic molecules frequently containing halogen, nitrogen, sulfur, and phosphorous as well as carbon, hydrogen, and oxygen.

TABLE 21.6 *Cis-Trans* Isomerism and Biological Activity[a]

Structure[b]	*Name and Isomer*	LD_{50} *Housefly (mg/kg)*
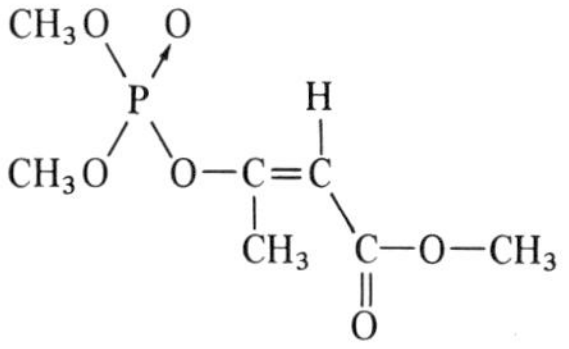	Mevinphos, *cis*.	0.27
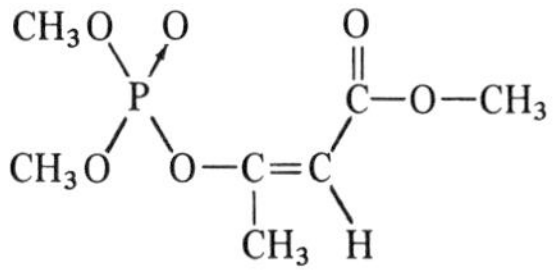	Mevinphos, *trans*.	23.0
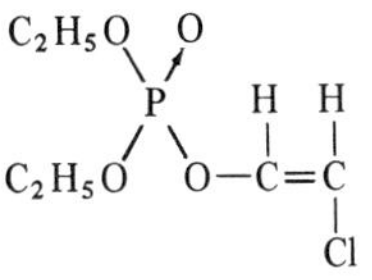	Diethyl β-chlorovinyl phosphate, *cis*.	1
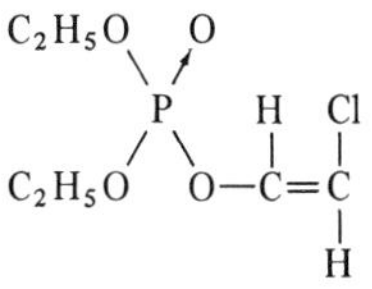	Diethyl β-chlorovinyl phosphate, *trans*.	80

[a]Based on Lichtenthaler, F. W., *Chem. Rev.*, **61,** 630 (1961).
[b]In both organophosphate insecticides, the *cis* isomer is about 80 times as active as the *trans* isomer as measured by topical application to the housefly.

Comprehensive descriptions of contemporary pesticides are provided in the bibliography. What follows here are groups of significant pesticides generally typical of a given class and commonly used at present.

Pesticide Nomenclature

Because no uniform practice exists for the naming of pesticides, and because commercial pesticides generally possess more than one designation, some clarification of the nomenclature adopted here is in order. In the charts that follow and in the text, the so-called generic or common name has been used.

Where a pesticide is a specific chemical composition, there is a systematic chemical name generally consistant with the rules of nomenclature in *Chemical Abstracts.* Such a name is often quite cumbersome. The systematic name for the structure of streptomycin given in the text occupies four or five lines of print and, from the standpoint of the chemist, is less informative than the two-dimensional structural formula that is provided.

Pesticides developed by private companies usually are initially given an arbitrary designation (generally containing a number) which is employed during the period of exploration and development. This name may prevail during the period that research teams, farm advisers and experiment stations work with the material. During the course of this period, a proprietary name is generally selected by the company responsible for the development of a

TABLE 21.7. Names of Some Pesticides

Generic Name	*Systematic Name*	*Proprietary Name*[a]	*Other Names*
dimethoate	O, O-Dimethyl S-(N-methylcarbamoyl-methyl) phosphorodithioate	Cygon®	Rogor®(Europe) Foslion®
carbaryl	1-Naphthyl N-methylcarbamate	Sevin®	
naled	1,2-Dibromo-2,2-dichloroethyl dimethyl phosphate	DIBROM®	RE-4355
captan	N-Trichloromethylmercapto-4-cyclo-hexene-*cis*-1,2-dicarboximide	Orthocide®	
endosulfan	6,7,8,9,10,10-Hexachloro-1,5,5a,6,9,9a-hexahydro-6,9-methano-2,4,3-benzo-dioxathiepin-3-oxide	Thiodan®	Malix®

[a]Proprietary names are generally capitalized for the initial letter and followed by the symbol ®, as in Sevin®; common names are written in small letters.

potentially successful pesticide. This proprietary name falls within the purview of the U.S. trade-mark laws, and the similar foreign trade-mark names which are selected are not always the same as the U.S. name. The product is usually sold under a variety of trade-mark designations which denote the company of its source. A name which is the generic or common name may then be selected for the commercialized chemical. Most scientific texts involved with the discussion of structure and biological activity, metabolism, toxicology, etc., of pesticides employ this generic or common name.

Table 21.7 provides some alternate names for pesticides whose structure is given in the text. For a given pesticide, one of the alternate names may be a more common designation than the generic name. It is obvious that for a relatively simple structure, such as endosulfan, the systematic chemical name is quite inconvenient.

Chlorinated Organic Group

The halogenated organics were among the earliest synthetic organics employed as pesticides. DDT exhibited a broad spectrum of insecticidal activity and its success stimulated the research that led to many of the other synthetic organic chlorine compounds. DDT has a relatively moderate acute mammalian toxicity—about 250 mg/kg for the technical product. It is oil-soluble and very insoluble in water—about one part per billion (1 ppb).

In the fifties and sixties there was on the one hand increasing evidence of resistance to DDT by various insect species, and on the other a proliferation of chemical analytical techniques enabling quantitative pesticide analyses in complex substrates down to the ppb level. These conclusively showed that DDT and its chief metabolite DDE (dichlorodiphenyl-dichloroethylene

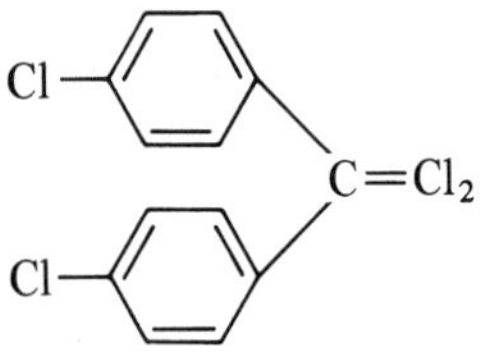

persisted in the environment and concentrated in fatty tissue, particularly in organisms at the top of any given ecosystem.

There is a considerable, but variable, cross resistance from DDT to other chlorinated hydrocarbons such as benzene hexachloride, and the cyclodiene insecticides. When one substitutes a *para*-methoxy, group for *para*-chlorine in the DDT molecule, the far less persistent—and somewhat less active—but useful insecticide methoxychlor is obtained. The substitution of OH for H in the trichloroethane group greatly reduces and almost eliminates the general insecticidal properties of the DDT configuration. Nevertheless, such molecules possess high acaracidal and ovicidal properties, e.g., dicofol.

The simplicity of the chlorination of benzene led to the discovery of the usefulness of benzene hexachloride and lindane as insecti-

TABLE 21.8 Group I—Miscellaneous Chlorinated Hydrocarbons

Name	*Structure*	*Significant Intermediates*	*Uses–Properties*
DDT	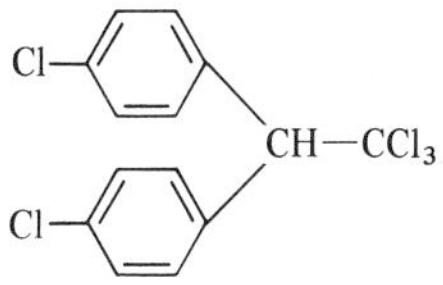	Chlorbenzene, chloral.	General insecticide (malaria control).
Toxaphene	Complex mixture	Chlorinated camphene (68-69% chlorine).	Insecticide (cotton).
BHC Lindane (99% pure γ-isomer)	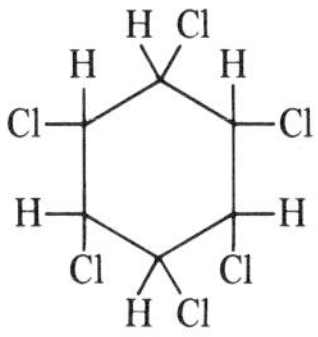	Benzene, chlorine ($h\nu$).	Insecticide used heavily in Japan (rice).
Hexachlorobenzene	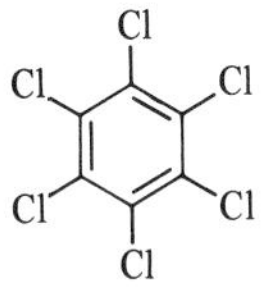	Benzene, chlorine.	Fungicide (cereal-seed treater).
dicofol	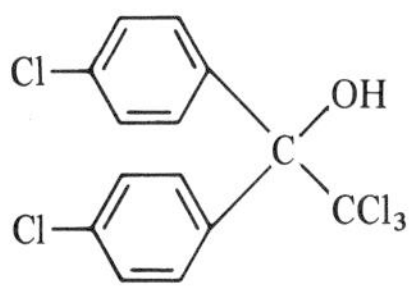	DDT, chlorine (hydrolysis).	Acaricide; miticide.

cides—and of hexachlorobenzene, the true hexachlorobenzene, as a grain fungicide. (Benzene hexachloride as a chemical is a misnomer. Benzene hexachloride, originally prepared by Michael Faraday in 1825, is hexachlorocyclohexane.) Lindane has the spatial configuration that possesses the highest activity—six hexachlorocyclohexanes are found in commercial preparations—although the stereochemistry of the cyclohexane ring does provide many more theoretical isomers.

Many economically important species of mites have a life cycle of two or three weeks. Mites have shown an enormous capacity for the development of resistance and can generally tolerate most chlorinated hydrocarbon insecticides as well as many organophosphorus insecticides.

The Cyclodiene Group

Hexachlorocyclopentadiene (C_5Cl_6)

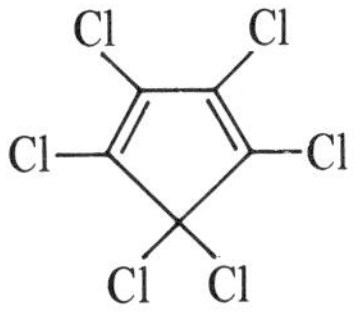

itself possesses herbicidal activity. As a diene combined with a variety of dienophiles and modified thereafter, it has given rise to a large number of very useful insecticides. The earliest of these was chlordane which was employed both in agriculture and for the control of home and garden pests. Chlordane is somewhat herbicidal as well and finds some use for crabgrass control.

Aldrin, dieldrin, heptachlor, and endrin are all very highly active toxicants with moderate to highly acute mammalian toxicity, great potency, and a wide spectrum of insecticidal activity. They and their metabolites tend to concentrate in fatty tissue and have considerable persistence in the environment. Recently, federal regulations as well as state legislation has severely limited their agricultural application. The names, uses and structural formulae of several of these chemicals are given in Table 21.9.

Despite the fact that this group of chemicals has had wide agricultural and other usage, very little has been unequivocally established concerning the essential biochemistry involved in their activity. Whereas the mechanism of resistance to DDT can be at least partially related to the enzymatic reaction

$$\text{DDT} \xrightarrow[\text{dehydrochlorinase enzyme}]{-\text{HCl}} \text{DDE}$$

no such simple mechanism can be associated

TABLE 21.9 Cyclodiene Group Toxicants

Name	*Structure*	*Significant Intermediates*	*Uses*
Chlordane	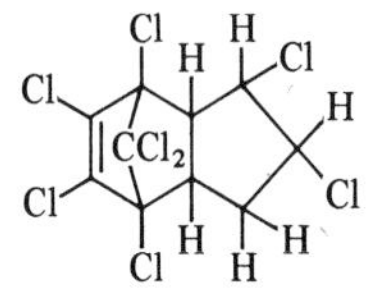	Hexachlorocyclopentadiene, cyclopentadiene, chlorine (isomeric mixture).	Insecticide; crabgrass control.
Aldrin	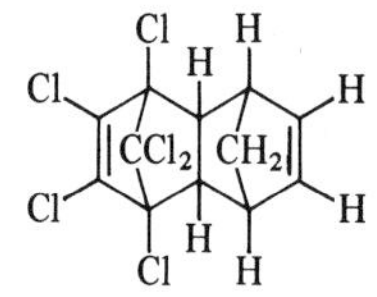	Hexachlorocyclopentadiene, acetylene, cyclopentadiene.	Broad spectrum insecticide (soil insects).
Dieldrin	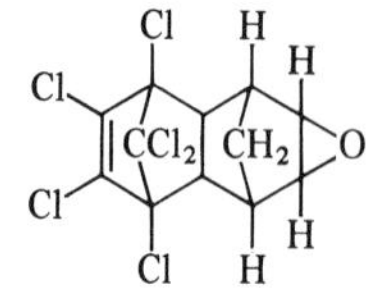	Hexachlorocyclopentadiene, acetylene, cyclopentadiene, peracetic acid.	Broad spectrum insecticide.
Mirex	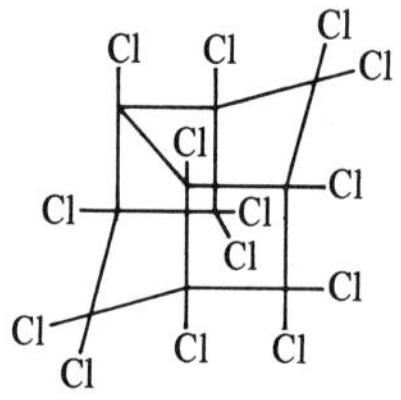	Hexachlorocyclopentadiene, chlorine.	Insecticide (fire ant control).
endosulfan	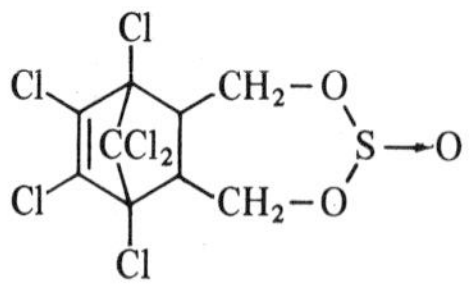	Hexachlorocyclopentadiene, 1, 4-butenediol, thionyl chloride.	Insecticide; aphicide.

TABLE 21.10 Pesticides Derived From Common Organic Acids

Name	*Structure*	*Significant Intermediates*	*Uses*
2, 4-D Acid salts Esters	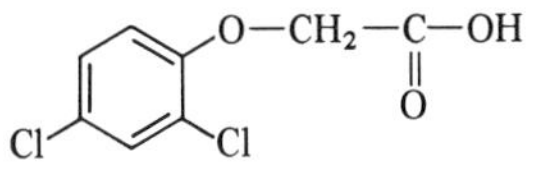	Phenol, chlorine, chloracetic acid.	Herbicide and growth regulator, broad-leaf control (grain and cereal).
TBA TBC	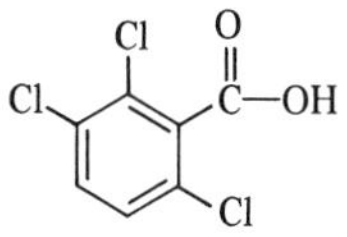	Toluene, chlorine (oxidation).	Preemergence broad leaf.
TCA	$Cl_3C-C(=O)-OH$	Acetic acid, chlorine.	Postemergence herbicide.
Captan	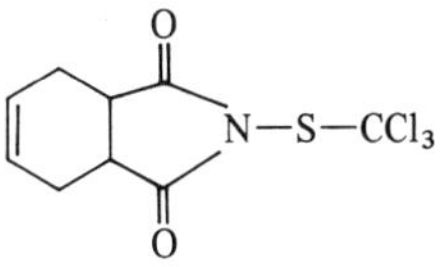	Maleic anhydride, butadiene, ammonia, carbon disulfide, chlorine.	General fungicide.
ANA	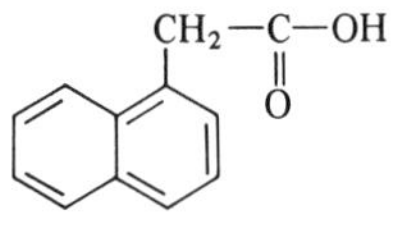	Naphthalene, chloracetic acid; or formaldehyde, hydrochloric acid, potassium cyanide (hydrolysis).	Plant-growth regulator.

with cyclodiene resistance mechanisms. Frequently, in the field, some insect varieties exhibit a generalized chlorinated hydrocarbon resistance.

Derivatives of Organic Acids

Many important pesticides may be considered chemically as derivatives of organic acids. (Table 21.10) For many years 2,4-D (2,4-dichlorophenoxyacetic acid) and 2,4,5-T (an additional chlorine in the 5-position on the phenoxy moiety—2,4,5-trichlorophenoxyacetic acid) dominated the herbicide market. They possess potent auxin-like behavior and have found their greatest usefulness in the control of dicotyledonous (broad leafed) weeds. 2,4,5-T has recently been severely restricted in sales, and 2,4,-D is being studied by several scientific and administrative bodies. In the manufacture of 2,4,5-T, particularly, an impurity, tetrachlorodioxin

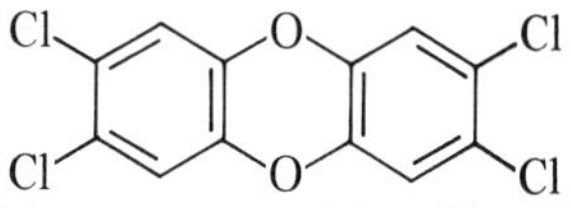

is formed in trace quantities. This dioxin is a highly toxic material and causes, among other effects, teratogenic (birth defect) manifestations on test animals. At present much scientific work is underway which will ultimately determine on the one hand the hazard, if any, relating to the parent chemical and/or a restriction on the allowed impurities.

Much chemistry has also been done (see formulation section) on altering the parent acids, e.g., to form esters (both volatile and nonvolatile), salts, and nonvolatile amides. Volatility is extremely important since, because of the potency of these chemicals, the dilute vapor as well as suspended droplets can injure broad leaf crops downwind from the point of application.

Trichlorobenzoic acid is one of a large number of related herbicides and growth regulants. (See Table 21.10.)

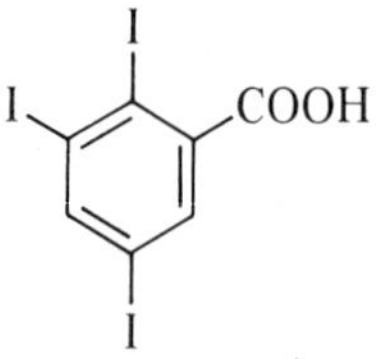

TIBA

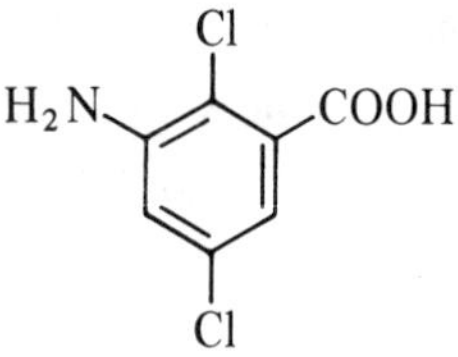

Amiben

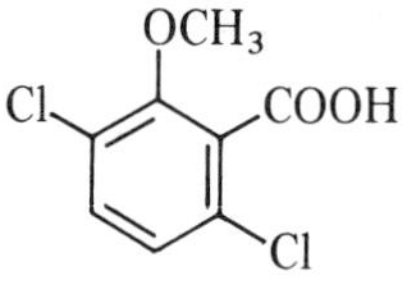

Dicamba

TIBA, amiben, and dicamba are among the more prominent. Halogenated terephthalics are also herbicidal. TCA (trichloroacetic acid) is one of the simplest herbicides, but is biologically weaker and less specific than many of those cited above. The related dichloropropionic acid ($CH_3CCl_2-\overset{O}{\overset{\|}{C}}-OH$ dalapon) is an important postemergent herbicide.

Folpet, a derivative of *o*-phthalic acid, and captan, from tetrahydrophthalic acid, are compounds of very low acute mammalian toxicity and low persistence in the environment, and they have a wide spectrum of useful fungitoxicity. They are used on fruit crops, grapes, potatoes, tomatoes, seed dressings, etc.

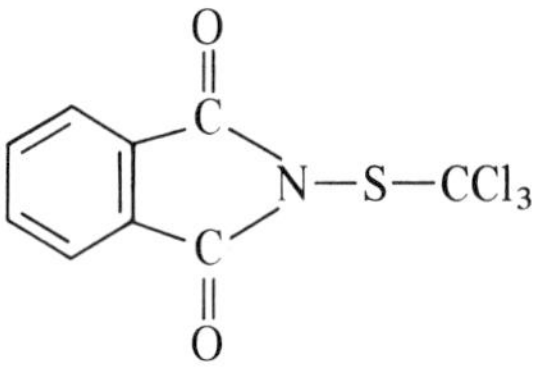

folpet

An important plant-growth regulator is alpha-naphthyl acetic acid (ANA). It stimulates root formation and affects the production of ethylene within the plant, and this compound, in turn, is implicated in many plant physiological processes. Many phenoxy, naphthyl, and naphthoxy acids, as well as more complex acids such as indoleacetic acid, influence plant physiological activity.

The Organic Phosphate Pesticides

These phosphorus esters, which have the generic structure:

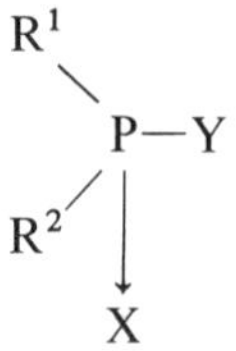

owe their insecticidal activity to their capacity to phosphorylate the enzyme cholinesterase, which moderates nerve-impulse transmission. The inactivated, phosphorylated enzyme reaction product possesses varying degrees of persistence. The effect of this deactivation is a build-up of the chemical acetylcholine ($(CH_3)_3\overset{+}{N}-CH_2CH_2O-\overset{O}{\overset{\|}{C}}-CH_3$). This results in energy transmission through the neuron and, among other effects, causes convulsive activity at some vital muscular center. R^1 and R^2 are generally lower alkyl, alkoxy, alkylthio, or alkylamino groups generally containing fewer than four carbons; X can be O or S, but if it is S, the activity is derived by an in situ oxidation to the oxo-derivative; Y is usually a hydrolytically unstable group which becomes detached from the molecules, thus providing a residue that can combine with the enzyme. The first six molecules in Table 21.11 all satisfy these generalizations, although

TABLE 21.11 Important Organophosphates

Name	*Structure*	*Significant Intermediates*	*Uses*
parathion	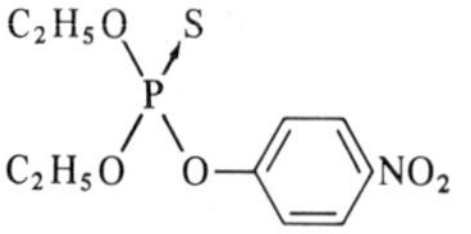	*p*-Nitrophenol, ethanol, phosphorus trichloride, sulfur.	General insecticide.
Methyl parathion	Methyl homolog of parathion.		Insecticide (cotton).

TABLE 21.11 (*Continued*)

Malathion	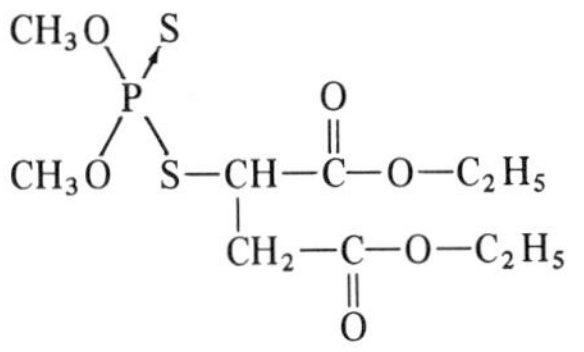	Methanol, phosphorus pentasulfide, ethyl maleate.	General insecticide.
Dimethoate	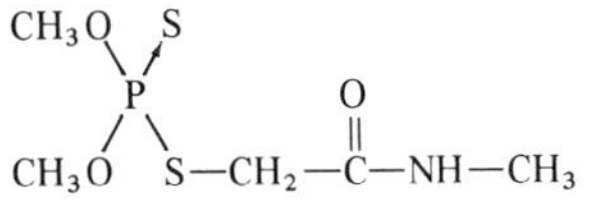	Methanol, phosphorus pentasulfide, chloroacetic acid, methylamine.	Systemic insecticide.
Naled	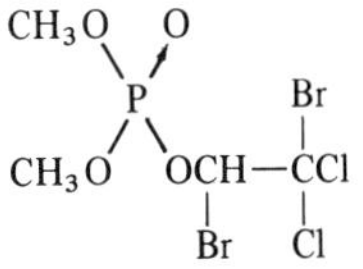	Chloral, bromine, trimethyl phosphite.	Nonpersistent insecticide.
Demeton (O) (Many variations of this structure are commercial insecticides.)	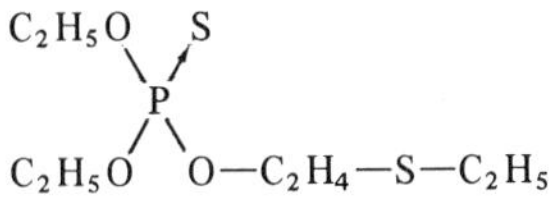	Ethanol, phosphorus thiochloride, ethyl thioethanol.	Systemic insecticide.
Folex®	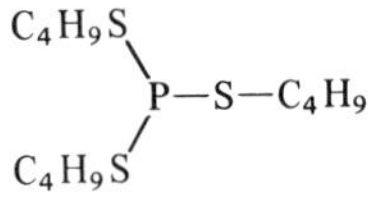	Phosphorus trichloride, butyl mercaptan.	Defoliant (cotton).
DEF®	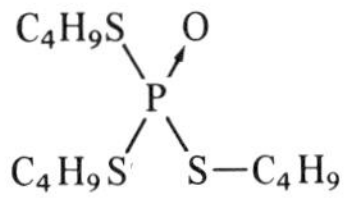	Phosphorus oxychloride, butyl mercaptan.	Defoliant (cotton).
ORTHENE	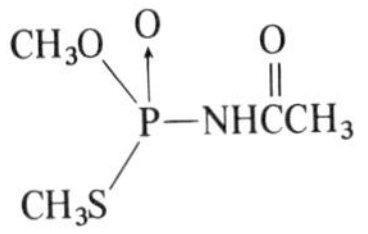	Dimethyl thionophosphorochloridate, NH_3, acetylchloride.	Cabbage looper, aphids.
Ethrel	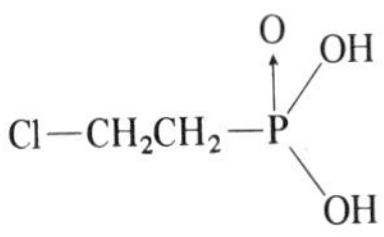	PCl_3 ethyleneoxide H_2O.	Plant growth regulator.
Glyphosate	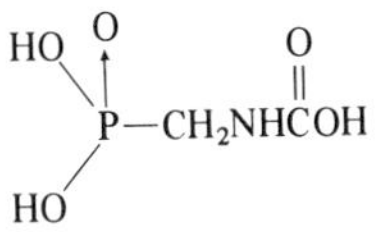	PCl_3, CH_2O, glycine.	Herbicide post emergent.

changes within the plant or soil or insect may be required before the biochemically active species is generated.

These organophosphates can act as contact, stomach, respiratory, or systemic toxicants. Many of them are highly toxic to mammals as well. In the case of malathion, mammals possess enzymes that can saponify the succinic ester before it arrives at vital centers, thereby providing relatively low mammalian toxicity. Insects appear to be less able to accomplish this detoxification. Many of the newer (and some of the earliest) organophosphorus derivatives possess amidate structures, that is X ↑ P—N bonds. Some of these are useful animal systemics, i.e., they possess a differential toxicity for the cow, sheep, fowl, horse, dog, etc., and the insects that attack them. They are tolerated by the host animals at concentrations that kill or suppress the particular parasitic species.

Two defoliants have been included in Table 21.11. There are other organophosphorus derivatives of the amidate types, as well as other classes which possess herbicidal and fungicidal properties. It is no wonder then that each year sees the development of many variations of organophosphorus structures for they have large capabilities for differential toxicity, variation in persistence, specificity, etc. Particularly with the interdiction of the chlorinated hydrocarbons, the organophosphate and carbamate groups provide opportunities for replacement of environmentally persistent pesticides. They are somewhat more expensive as practical pest-control agents.

Principles of physical organic chemistry can be successfully applied in relating the biological activity of the phosphate insecticides (also the carbamates). The Hammett σ and ρ functions for aromatic substitution provide useful correlations to anticholinesterase activity. Practical field-insect effectiveness involves more complex variables.

While the organo phosphorus insecticides are generally not highly persistent in the environment, and do not concentrate in fatty tissues, evidence of *resistance to this genus of chemicals exists and is mounting.*

The Carbamate Groups

Many of the more recently developed insecticides possess the carbamate structure.

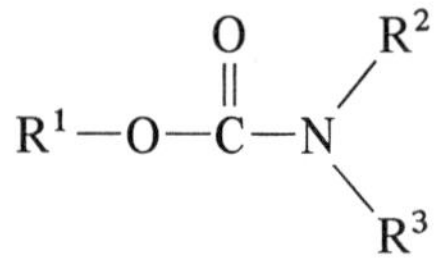

These compounds owe their activity to their interference with the enzyme cholinesterase which governs the concentration of acetylcholine at the synapse. Many of these carbamates compete with the normal substrate, acetylcholine, for positioning on the enzyme surface, thus preventing its saponification. This in turn governs the transmission of impulses through the synaptic junctions of the nervous system. In this way, their mode of action resembles, but is not quite identical to that of the organic phosphate esters previously discussed. For many of these carbamates R^1 is an aromatic moiety such as a substituted phenyl (BUX) or naphthyl group (carbaryl). Some of the more recent carbamate structures contain oximino groups, e.g., methomyl and the systemic and very potent temik

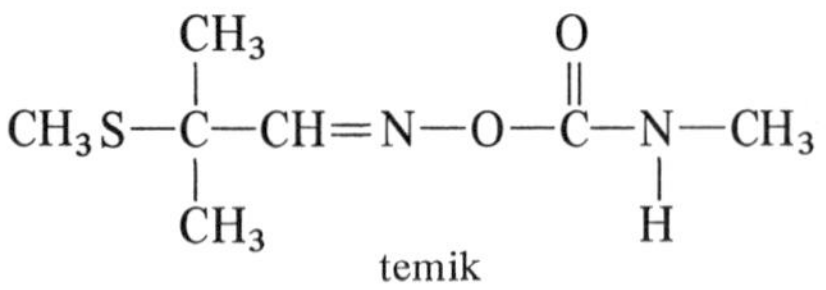

temik

as well as the heterocyclic variants such as carbofuran.

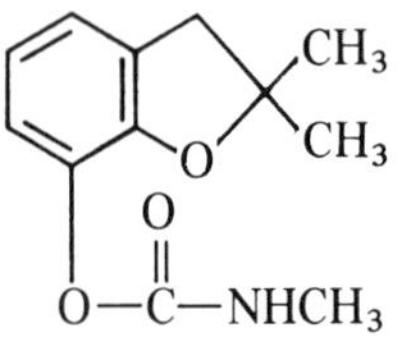

carbofuran

Usually R^2 is methyl and R^3 is hydrogen, although both R^2 and R^3 are sometimes methyl. In very recent pesticidal structures, the proton of R^3 is substituted by an acyl, sulfenyl, or phosphoryl moiety. The same fully oxygenated generic structure also provides important herbicidal compositions when R^2 is aryl or substituted aryl, R^3 is H and R^1 is aliphatic or substituted aliphatic. For example, there follows the substitutions in the order R^1, R^2, R^3 and the name of the herbicide, respectively:

Isopropyl, Phenyl, H, IPC,
Isopropyl, *m*-chlorophenyl, N. CILPC,
4-chloro-2-butynyl, *m*-chlorophenyl, H, barban

These carbamates are used in grain production and are involved with the inhibition of mitosis in the offending weed species.

While many of the derivatives of the fully oxygenated carbamates are useful insecticides, the *dithiocarbamate group*

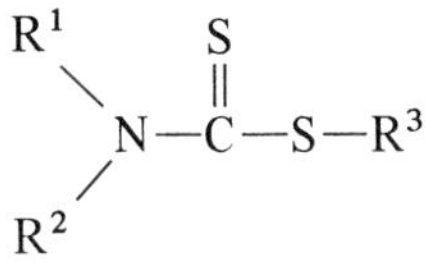

TABLE 21.12 Carbamate Pesticides

Name	*Structure*	*Chemical Intermediates*	*Uses*
Carbaryl	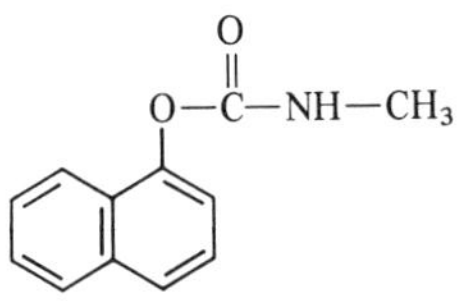	α-Naphthol, methyl isocyanate.	Broad spectrum insecticide (*Lepidoptera* larvae).
Methomyl	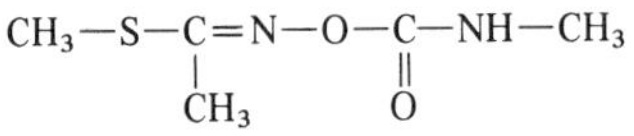	Acetyl chloride, methyl mercaptan, hydroxylamine, methyl isocyanate.	Contact insecticide.
BUX	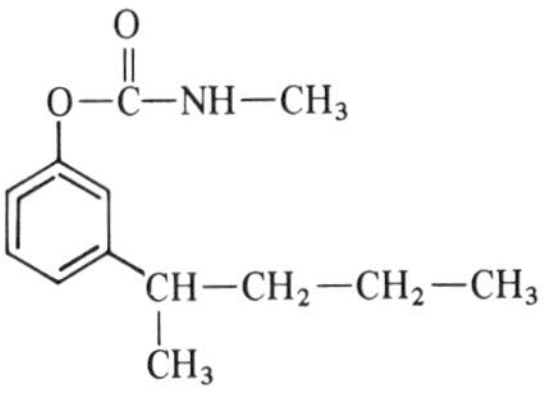	Phenol, pentene, methyl isocyanate.	Soil insecticide (corn).
Maneb	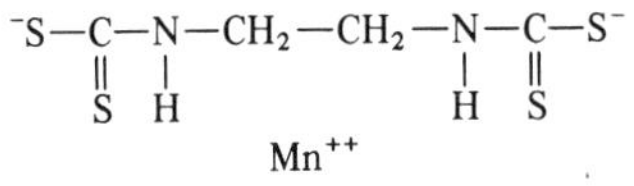	Ethylene diamine, carbon disulfide, caustic, Mn^{++}	Fungicide.
Vapam	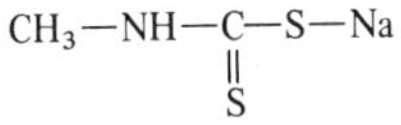	Methylamine, carbon disulfide, caustic.	Soil fungicide; nematicide.
Ferbam	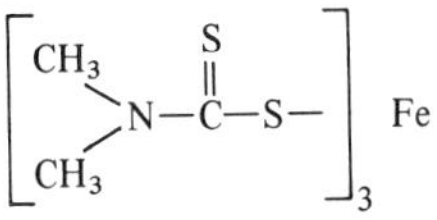	Dimethylamine, carbon disulfide, Fe^{++}	Fungicide (ornamental plants).
Eptam	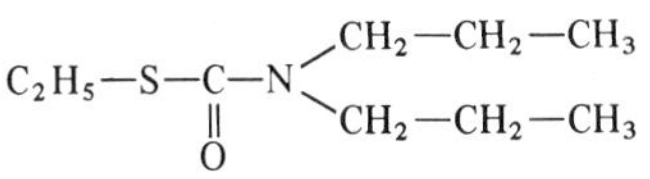	Dipropylamine, phosgene, ethylthiol.	Preemergence herbicide.
Sulfallate	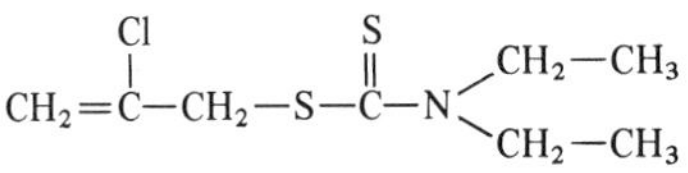	Diethylamine, phosgene, 2, 3-dichloropropene.	Herbicide.

provides many very significant fungicides. The sodium, ammonium, zinc, manganese, and iron dithiocarbamates derived from ethylene diamine (R^1 = $—CH_2CH_2—$; R^2 = H) and from dimethylamine (R^1 = R^2 = CH_3) are standard fungicides for a wide variety of fungal plant diseases. Two of these structures (maneb and ferbam) are given in Table 21.12.

Finally, the mixed oxygen-sulfur carbamates possessing the general structure

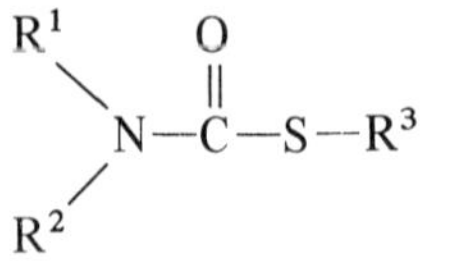

contribute important examples of herbicidal chemicals. The structure of eptam, an important pre-emergence herbicide, is given in Table 21.12. There are many homologues with variations of R^1, R^2, and R^3 of which pebulate (C_2H_5, n-C_4H_4, n-C_3H_7), vernolate (R^1 = R^2 = R^3, n-C_3H_7), and di-allate (i-C_3H_7, i-C_3H_7, $C_3Cl_2H_3$), are significant examples.

Although the above generalizations concerning the types of carbamates and their biological activity are generally sound, there are occasional exceptions. The biological activity of carbamates is quite broad and occasionally nematicidal, acaricidal, and other properties are exhibited. The major economic usefulness of the dithiocarbamate sulfallate is in the herbicidal area.

Nonsymmetrical Ureas

The nonsymmetrical ureas (Table 21.13)

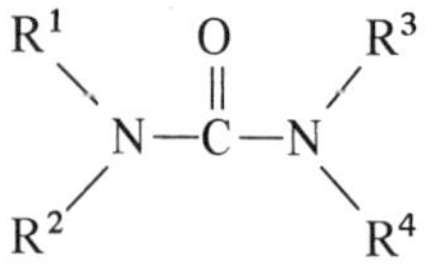

provide many valuable herbicidal compounds and are presently under active development.

Generally, one nitrogen is substituted with a cyclic group, such as an aryl, substituted aryl, heterocyclic, mono- or bicyclic group, and the other nitrogen with lower-alkyl or alkyoxy moieties. The compounds interfere with the photosynthetic process. The nature of the cyclic group provides interesting specificities, i.e., relative tolerance of certain crop species, and its effectiveness against groups of weed species. Many of the compounds possess de-

TABLE 21.13 Urea Herbicides

Name	*Structure*	*Chemical Intermediates*	*Uses*
Monuron	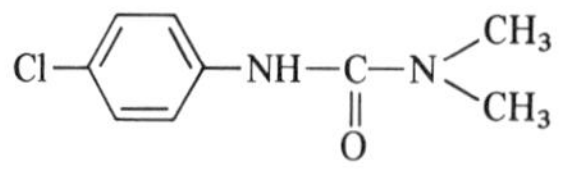	p-Chloroaniline, dimethylamine, phosgene.	Preemergence herbicide (cotton).
Linuron	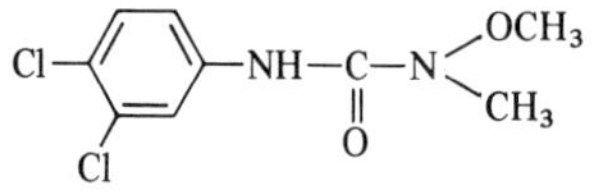	3, 4-Dichloroaniline, phosgene, hydroxylamine, dimethyl sulfate.	Herbicide.
Fluometuron	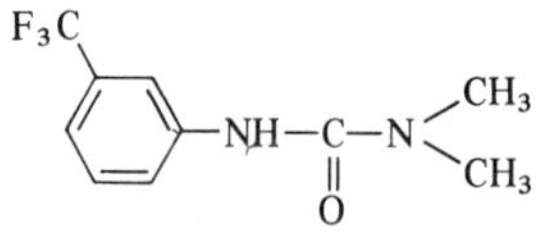	m-trifluoromethylaniline, dimethylamine, phosgene.	Herbicide (cotton).
Noruron	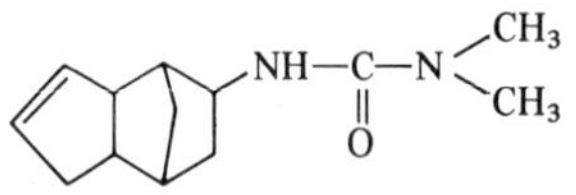	Dicyclopentadiene (reduction), ammonia, phosgene, dimethylamine.	Herbicide.
Cycluron	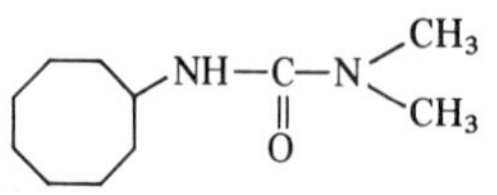	Butadiene, (dimerization; reduction), ammonia, phosgene, dimethylamine.	Herbicide.

grees of pre-emergence and postemergence activity.

A pre-emergence herbicide is one generally applied prior to the emergence of the weed seed from the ground. The chemical then may be applied at or near the time of crop seeding. Some pre-emergence herbicides must be worked into the soil for optimum activity; others are quite active when applied to the soil surface. A postemergence herbicide is one that exhibits its activity by killing or impeding the growth of the weed species from seedling stage onwards.

A sterilant is a chemical toxic to all forms of plant life, i.e., it exhibits very little useful *differential* plant toxicity.

Some Heterocyclic Pesticides

In recent years pesticides based upon heterocyclic structures (Table 21.14) have become

TABLE 21.14 Heterocyclic Pesticides

Name	*Structure*	*Significant Intermediates*	*Uses*
Atrazine	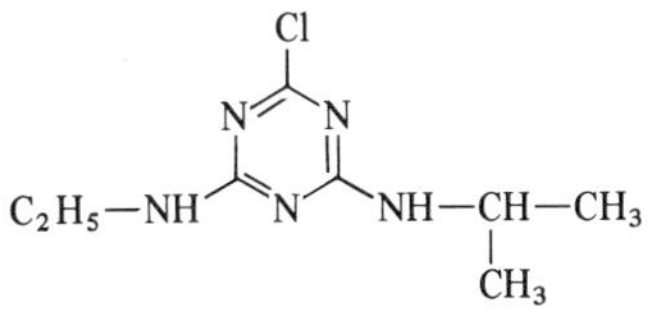	Cyanuric chloride, ethylamine, isopropylamine.	Preemergence herbicide (corn).
Paraquat dimethylsulfate	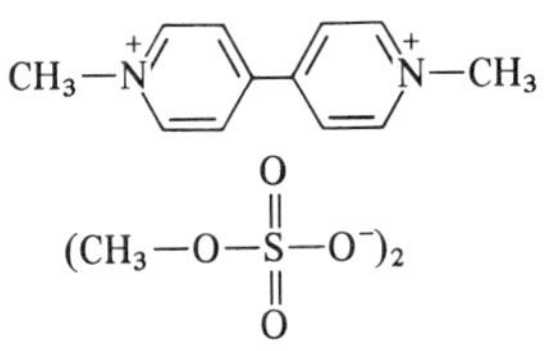	Pyridine, dimethyl sulfate.	Postemergence herbicide dessicant.
Picloram	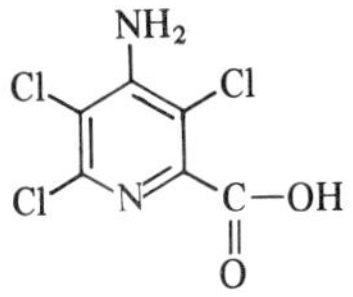	Chlorine, ammonia, α-picoline (oxidation).	General herbicide; plant sterilant.
Benomyl	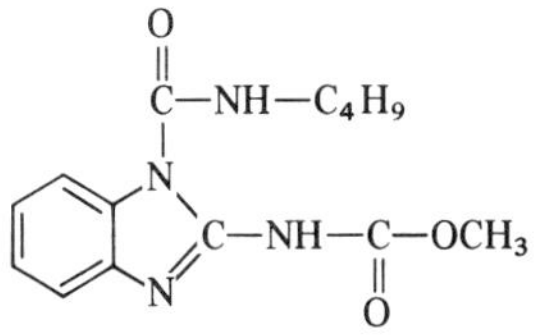	Phenylenediamine, phosgene, methanol, butylamine.	Systematic broad spectrum fungicide.
Bromacil	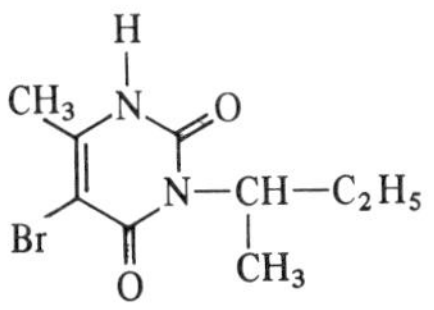	Acetoacetic ester, *sec*-butylamine, phosgene, ammonia, bromine.	General herbicide (garden and home).
Carboxin	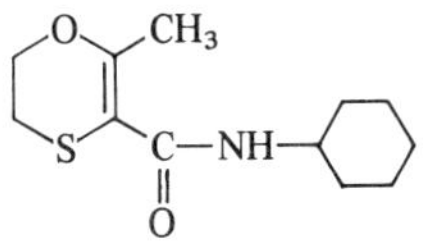	Aniline, ketene, mercaptoethanol.	Systematic fungicide (grain, soybeans–smut and rust).

more prominent. The corn, sorghum weed-control market has been dominated by the symmetrical triazine group, of which atrazine is a most important representative. The corn plant possesses an enzyme that degrades atrazine much more effectively than the weed species which compete with it. Many other triazine herbicides are based upon appropriate substitution of the chlorines in cyanuric chloride:

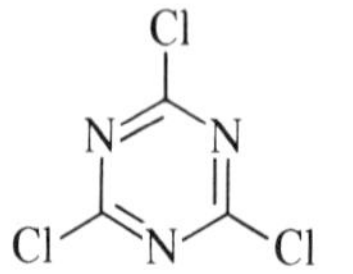

cyanuric chloride

e.g., simazine, prometrone, and propazine. Each gains an area of specificity by the substitution usually of methoxy or methylthio and lower alkylamino groups.

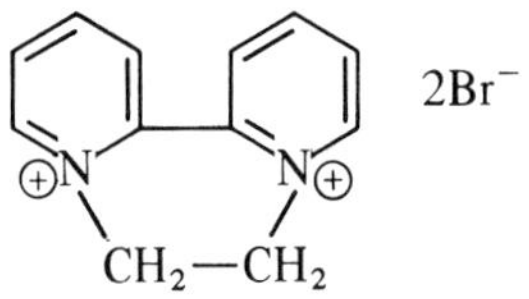

Diquat

Like paraquat, diquat is a member of a group of quaternary dipyridyls which owe their activity to their ability to produce inside the chloroplast a stable free radical that intercepts the transfer of energy in the early stages of the photosynthetic process. These compounds are irreversibly bound to montmorillonite clays and deactivated thereby, so that the surrounding clay soil becomes tolerant to plant growth. The trend to reduce labor in agriculture is exemplified by the use of paraquat as a "chemical hoe." Planting is performed with minimal cultivation in certain areas by the use of paraquat before the drilling of seeds.

Picloram is an extremely effective sterilant and, from a biological view, is perhaps one of the most powerful herbicides known. Its usefulness in the U.S. is presently limited because of the combination of the properties of persistence and almost universal phytotoxicity.

Two examples of interesting new heterocyclic systemic fungicides are provided in Table 21.14. These fungicides are absorbed by the plant and they, or their metabolites, inhibit the growth of the fungal organism. The sulfone homologue of carboxin, called oxycarboxin, is also a systemic fungicide.

There are many compounds, some of very recent origin, based upon cyclization of the phenylenediamine structure. One of these, benomyl, exhibits a broad scope of contact and systemic fungal activity.

Miscellaneous Chemicals

Dinitrophenols. One of the earliest groups of synthetic organic chemicals, predating World War II and still in use, are the alkyldinitrophenols. These chemicals have broad biological activity embracing the insecticide, fungicide, acaricide, and herbicide fields. This broad activity results from their capacity to block oxidative phosphorylation, hence, interfering with vital biochemical synthesis cycles found both in animal and plant life.

In addition to the example given in Table 21.15, there are many other structural variations with biological activity. Salts, esters, and free phenols are all utilized in commercial preparations.

Generic Formula	R^1	R^2	*Functions*	*Name*
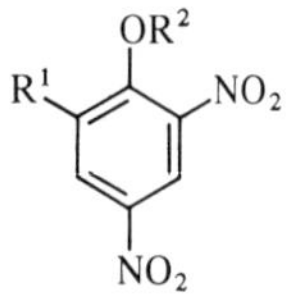	$-CH_3$	H	Insecticide Herbicide	DNOC
	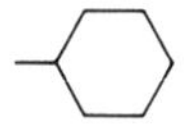	H	Insecticide Herbicide	DNOCHP
	$-CH(CH_3)-(CH_2)_5CH_3$	$-C(=O)-CH{=}CH-CH_3$	Acaricide Fungicide	Dinitrocaprylphenyl crotonate

TABLE 21.15 Miscellaneous Organic Pesticides

Name	*Structure*	*Significant Intermediates*	*Properties*
DNOBP	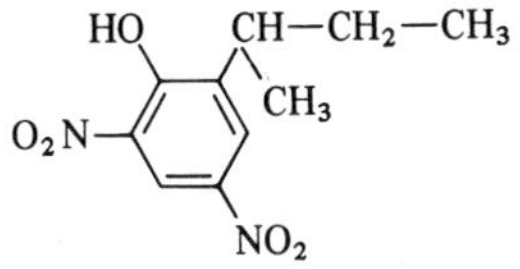	Phenol, nitric acid, butene-2.	Insecticide; herbicide.
Nemagon	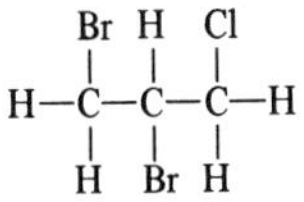	Propylene, bromine, chlorine.	Soil fumigant; nematocide.
EDB	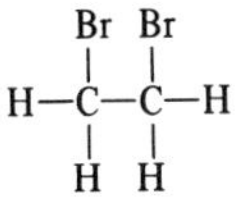	Ethylene, bromine.	Fumigant; nematocide.
PCP	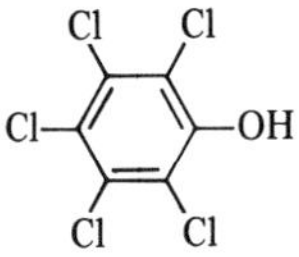	Phenol, chlorine.	Wood preservative; fungicide; herbicide.
Piperonyl butoxide	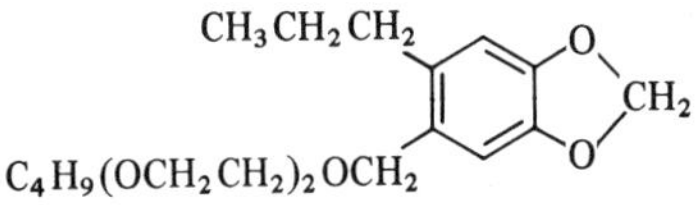	Safrole, hydrogenation, formaldehyde, hydrochloric acid, ethylene oxide, butanol.	Insecticide; synergist; aerosol spray.
Trifluralin	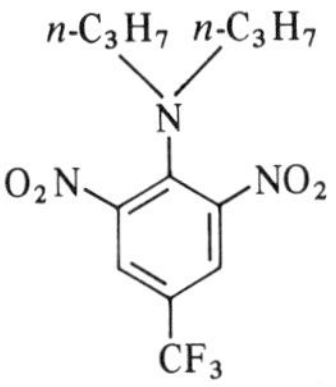	4-Difluorodinitrochlorobenzene, *n*-propylamine.	Preemergence herbicide (cotton).
Warfarin	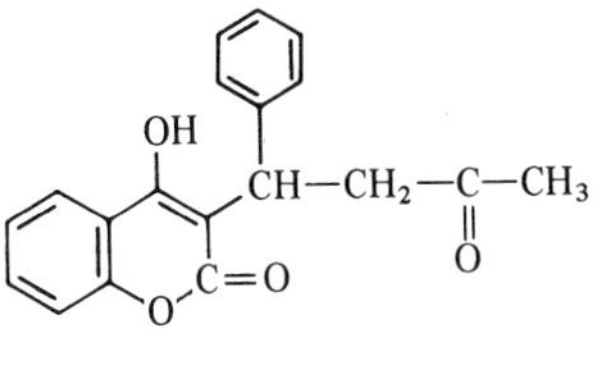	Benzalacetone, 4-hydroxycoumarin.	Rodenticide.
Plictran	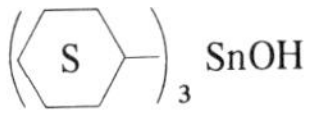	Chlorocyclohexane $SnCl_4$, Mg or Li.	Acaricide.

Soil Fumigants. In many areas the population of nematodes and pathogenic fungi is so high that seeds are destroyed, or the emerging seedling and plant are similarly attacked so that little or no crop can be harvested. Such soil is treated with a fumigant, usually a volatile general toxicant which is frequently injected into the soil prior to planting. These substances are generally, but not exclusively, low-molecular weight halogenated compounds. In addition to those chemicals given in Table 21.15, chloropicrin (CCl_3NO_2), methyl bromide (CH_3Br), and other halogenated C_1, C_2, and C_3 hydrocarbons find extensive application as soil fumigants.

Fumigants of various structures are also employed to preserve harvested crops, corn, small grain, etc., from insect and other pests.

Wood Preservatives. The pesticide industry also provides chemicals to prevent insect and fungal attack on lumber and wood products. In some cases, conventional pesticides are employed. Formulations of halogenated phenols are among the most common wood preservatives. Mercury and tin compounds have been also used to control wood rot organisms.

Synergists. It has long been known in medicine and in agriculture that the use of two chemicals can result in an additive, less than additive (antagonistic) or greater than additive biological effects. Where a greater-than-additive result occurs, the two chemicals are said to behave synergistically. In the pesticide industry there are a number of moieties which have little or no activity themselves, but when used with a toxicant improve the activity of the toxicant. These synergists are most advantageously employed when the toxicant is an expensive chemical, for the combination possesses a decisive, practical economic advantage.

The group

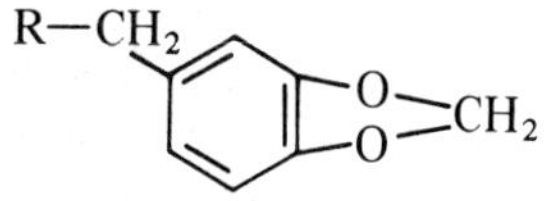

(piperonyl group) where R has a multiplicity of organic functions, synergizes many insecticides. Certain alkynes, heterocycles, imido, and thiocyanato linkages also possess degrees of synergy with insecticides. Synergists find considerable use in aerosol sprays with pyrethrins or synthetic pyrethroids such as allethrin.

Organic Mercurials. For many years, organic mercurials were important fungicides and bactericides. They were applied as foliage sprays, e.g., phenyl mercuric acetate, and especially as seed dressings. Cotton and grain seed, particularly, could be treated at the rate (for grain) of 1 to 4 oz per 100 lbs, or a 2 to 4 per cent solution. Enough volatility exists for many mercury compounds to sterilize not only the areas external to the seed coat, but to penetrate into the seed coat itself. In 1970, severe limitation of the use of mercurial pesticides was instituted, largely because of mercury contamination of the environment as well as the danger of mercury-treated seed entering (illegally) man's food supplies. The structures of some common mercurial pesticides are listed.

Ceresan L

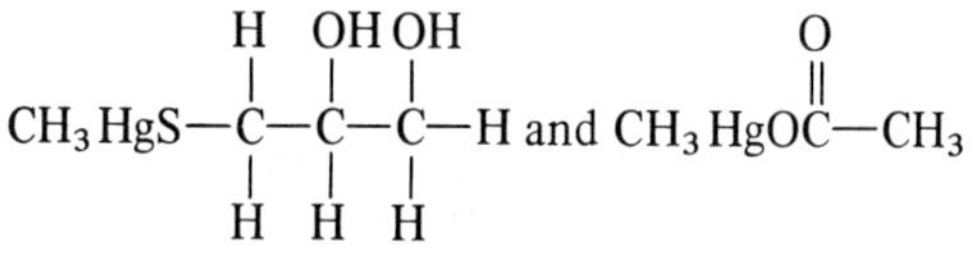

Phenyl mercuric acetate

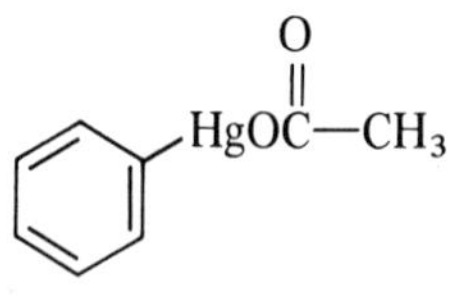

LM

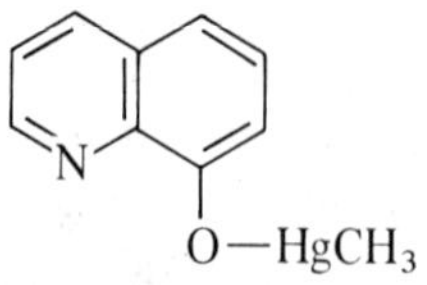

Fluorinated Herbicide. Included in Table 21.15 is an important pre-emergence cotton herbicide. It is noteworthy as one of the earliest fluorinated organic pesticides to satisfy the toxicological requirements for registration. Its effectiveness is markedly increased if it is "worked" into the soil.

Rodenticide. Damage to agricultural crops by rodents and the capacity of this group of mammals to harbor disease vectors has stimulated research for rodentidices. Effective control of various rodent species is achieved through impregnating grain with warfarin. This coumarin compound is related to the chemicals used in medicine to prevent blood clotting, particularly in cardiac disease. As a rodenticide, it causes internal hemmorrhage and eventual death.

Sulfur and Inorganic Sulfur Compounds. The oldest of the inorganics, sulfur and the water soluble polysulfides, continue to be employed, and their use may increase during the next decade in many agricultural areas. Sulfur, usually as a coarse powder or a small granule, is a soil additive. It is a required nutrient for all plant life, and the soils in many parts of this country and of the world are deficient in it. The oxidation of sulfur by soil microorganisms assists in the production of acid bodies that help solubilize other insoluble nutrients.

Sulfur is used as a wettable powder or dust, alone or in combination with other pesticides for foliar fungal control of mildews and other organisms. Calcium polysulfide (CaS_x, $x = 4.5$) is usually found in an about 30 per cent aqueous solution in many commercial preparations which are used for soil conditioning and fungal, mite, and insect control. The ammonium and sodium variants are also used.

Since sulfur is a low-cost chemical in abundant supply, is an essential nutrient, is relatively nontoxic to man, and is ecologically desirable (except in gaseous forms as an air pollutant), the agricultural usefulness of sulfur may expand in the seventies. It will be interesting to observe whether the removal of sulfur from fuels will increase the need for sulfur in agriculture.

Inorganic Pesticides. Though in countries with advanced agricultural technology synthetic organic pesticides dominate the market, the inorganics still find useful application in several areas. Copper salts, particularly the basic sulfate (Bordeaux mixture) and copper oxychloride, control many foliar fungal infections. In Europe, copper products continue to dominate in wine grape mildew-control programs, particularly for the "vin ordinaire" production in the Latin countries.

Basic sulfates of zinc and manganese are used as minor-element nutrient sprays as well as for fungicidal purposes, with or without copper. Basic phosphates are also employed. All of these salts are generally prepared under controlled pH conditions (sometimes *in situ* in the field) in such a way that a fairly insoluble salt is provided. Thus, its solubility limits the quantity of soluble metal ion available at any given time for absorption by the plant. The high quantities of Zn^{++}, Cu^{++}, Fe^{+++}, etc., which would be obtained from the normal sulfate, nitrate, or halide sprays would be too phytotoxic for practical use.

In this respect, modern chelation chemistry is also employed to provide metallic minor-element additives. The EDTA (ethylenediaminetetraacetic acid) group combined with metal salts is utilized both for soil and foliar application. Generally, however, less expensive chelants are employed. These include polyphosphates and partially refined products derived from the paper and lumber industry containing metal coordinating groups.

Calcium arsenate and lead arsenates (ortho-arsenates and more complex arsenates) are still employed for certain cotton insects and coddling moth control in apples, as well as for other insect-control purposes. These uses are dwindling as are the uses of arsenites for soil sterilization. At present, the residue and toxicological properties of the arsenicals in relation to human toxicity and ecological factors are under review by government, scientific, and administrative bodies. The above is true also for organic arsenicals such as cacodylic acid $[(CH_3)_2As(O)OH]$ and sodium methyl arsonate $[(NaO)_2As(O)CH_3]$ which are used as herbicides and defoliants.

CHEMICAL PESTICIDE MANUFACTURE

The essential raw materials and intermediates and, by implication, the routes employed in the manufacture of pesticidal chemicals are given in the foregoing tables. There follows in this section a discussion of some chemical, economic, and engineering aspects of a number of typical pesticidal processes.

Cyclodiene Group of Insecticides

The Diels-Alder addition of dienophiles to hexachlorocyclopentadiene (C_5Cl_6) is the first step in the preparation of all compounds in this group of insecticides. Chlorination of C_5H_6 (e.g., in the presence of caustic or other suitable catalyst) provides C_5Cl_6. One of the oldest and simplest examples is provided by the insecticide chlordane.

The equation illustrates the exothermic reaction which can be carried out neat or preferably with a low boiling solvent.

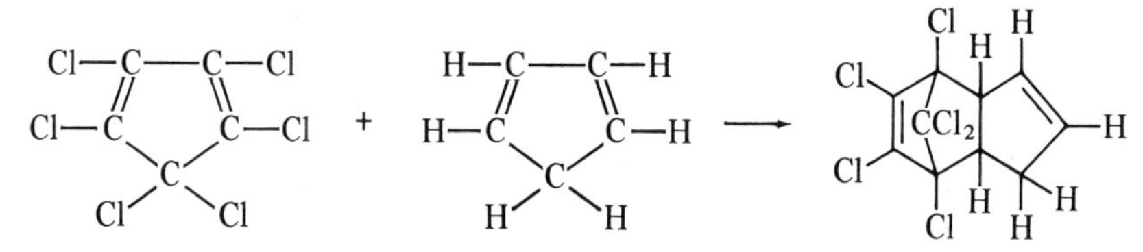

hexachlorocyclopentadiene cyclopentadiene chlordene (1)

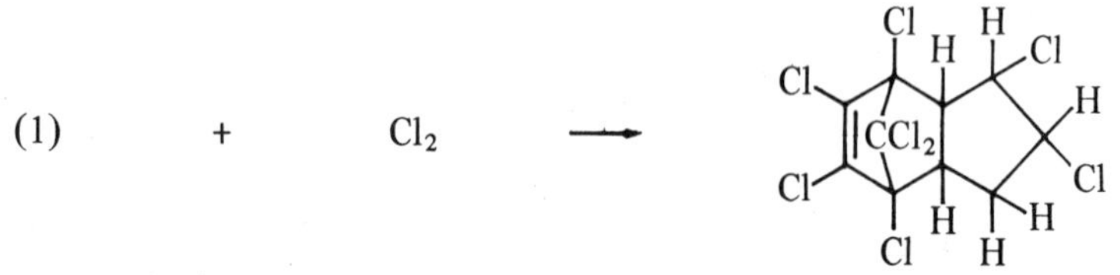

chlordene chlorine chlordane (a)

The second reaction is carried out with sufficient chlorine to provide a product containing 68-69 per cent chlorine. The technical product is a dark brown oil. It is obviously a mixture of a number of discrete chemicals (which can be separated chromatographically), and each component possesses its individual biological activity. In the Diels-Alder addition, *exo* and *endo* configurations can result. Some of each of these configurations are under- and overchlorinated, hence the crude final product with 68 to 69 per cent chlorine contains a large number of isomers and closely related compounds.

Chlorination by free radical mechanism (peroxides) yields the important insecticide heptachlor which is written below to portray spatial relationships. Such spatial differences as *exo* and *endo* attachments make for large differences in biological activity.

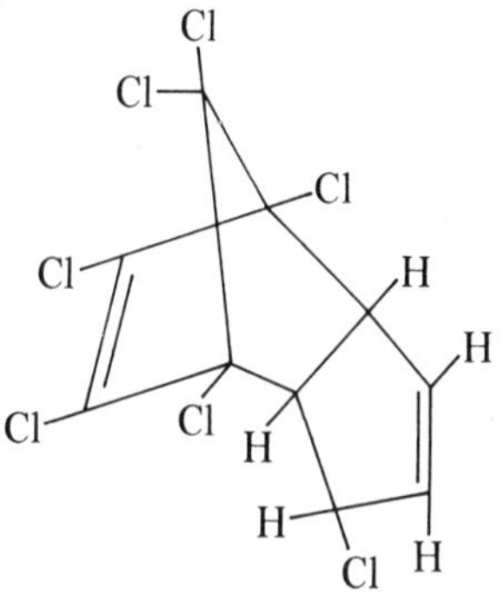

heptachlor

The preparative chemistry is very similar for aldrin, dieldrin, endrin, endosulfan, etc., the differences lying in the nature of the dienophiles and the subsequent chemistry and purifications.

Some Organophosphorous Pesticide Syntheses

A reaction distinctive to organophosphorous chemistry is the Arbusov reaction, which consists of the formation of a phosphonate by reaction of a phosphite with an organic halide. However, if a halogen is alpha to a carbonyl group a rearrangement, known as the Perkow rearrangement, takes place to produce a vinyl phosphate. This reaction is the basis for a whole series of commercially significant organophosphate insecticides including dichlorvos, naled, mevinphos, phosphamidon, monocrotophos,* dicrotophos,* etc.

The synthesis of naled is illustrative of the route followed to obtain these compounds.

$$\begin{array}{c} CH_3O \\ \diagdown \\ P-O-CH_3 + Cl_3C-CHO \rightarrow \\ \diagup \\ CH_3O \end{array}$$

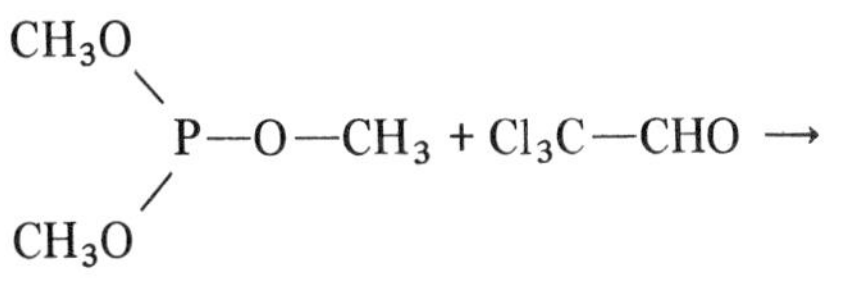

Trimethylphosphite   Chloral →

$$\begin{array}{c} CH_3O\ \ \ \ O \\ \diagdown\diagup \\ P \\ \diagup\diagdown \\ CH_3O\ \ \ \ O-CH=CCl_2 \end{array} + CH_3Cl$$

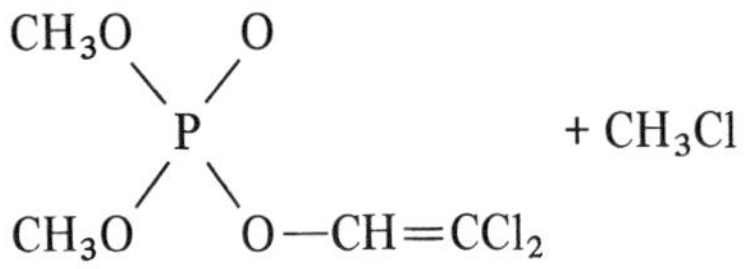

Dichlorvos (I)   Methyl chloride

$$(I) + Br_2 \xrightarrow[\text{or peroxides}]{h\nu}$$

Dichlorvos bromine

$$\begin{array}{c} CH_3O\ \ \ \ O \\ \diagdown\diagup \\ P\ \ \ \ \ \ \ \ H\ \ Br \\ \diagup\diagdown\ \ \ \ \ \ |\ \ | \\ CH_3O\ \ \ \ O-C-C-Cl \\ \ \ \ \ \ \ \ \ \ \ \ \ \ \ |\ \ | \\ \ \ \ \ \ \ \ \ \ \ \ \ \ \ Br\ Cl \end{array}$$

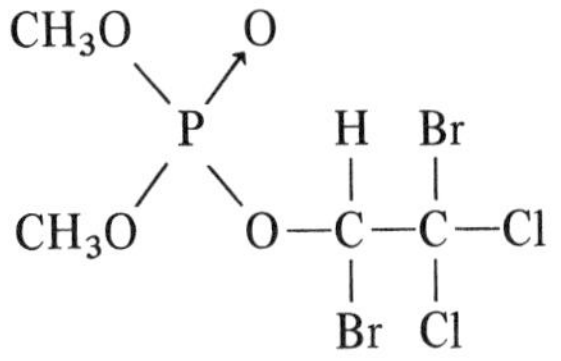

Naled

The Arbusov reaction combined with the Perkow rearrangement is generally on exothermic reaction, as in this case. Utilization of a low-boiling solvent is convenient. The bromination of the dichlorvos is carried out in the presence of light or peroxides to procure a fast rate of reaction with a minimum of destructive side reactions.

Another reaction frequently employed in organophosphate pesticide manufacture is the

*Space does not permit providing the structural formulae of all pesticides referred to in this section. The reader is referred to the references in the bibliography for the given structures.

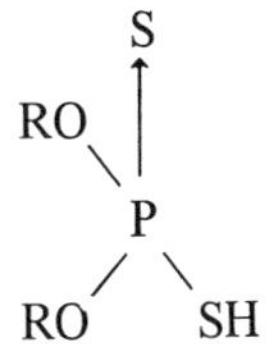

addition of $\begin{array}{c} S \\ \uparrow \\ RO\diagdown \\ P \\ \diagup\diagdown \\ RO\ \ \ SH \end{array}$ to an olefin or carbonyl, or its displacement of a reactive halide. $(CH_3O)_2$—PSSH is prepared by the reaction of methanol and P_4S_{10}.

An important commercial synthesis is the manufacture of malathion. By esterifying maleic anhydride with ethanol, the intermediate diethyl maleate may be prepared.

$$\begin{array}{c} S \\ \uparrow \\ CH_3O\diagdown \\ P \\ \diagup\diagdown \\ CH_3O\ \ \ SH \end{array} + \begin{array}{c} O \\ \| \\ CH-C-OC_2H_5 \\ \| \\ CH-C-OC_2H_5 \\ \| \\ O \end{array} \rightarrow$$

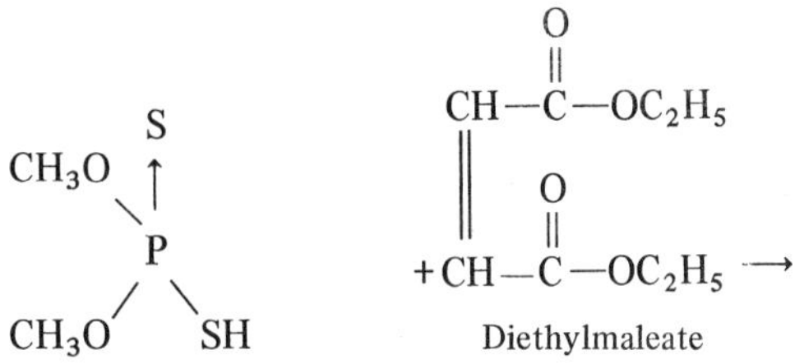

O, O-Dimethylphosphorodithioic acid   Diethylmaleate

$$\begin{array}{c} S \\ \uparrow \\ CH_3O\diagdown \\ P\ \ \ \ \ \ \ \ \ \ \ \ O \\ \diagup\diagdown\ \ \ \ \ \ \ \ \ \ \| \\ CH_3O\ \ \ S-CH-C-OC_2H_5 \\ \ \ \ \ \ \ \ \ \ \ \ \ \ \ | \\ \ \ \ \ \ \ \ \ \ \ \ \ \ \ CH_2-C-OC_2H_5 \\ \| \\ O \end{array}$$

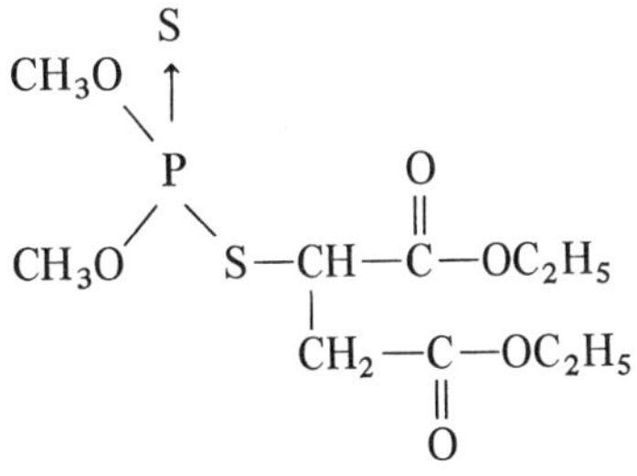

Malathion

Many commercial pesticides contain this dithioate group. Additional examples are: dimethoate, phosmet,* azinphos.*

Another frequently employed reaction is the displacement of a phosphorohalidate, usually a chloridate, by a neucleophilic reagent. A route to parathion is described. By reaction of P_4S_{10} with ethanol and subsequent chlorination the intermediate $(C_2H_5O)_2$—P—(S)—Cl is formed; this is then reacted with sodium p-nitrophenate in a solvent which is later removed.

$(C_2H_5O)_2P-(S)-Cl$ +

diethylthiophosphorochloridate

$Na^{\oplus\ominus}O$—⟨ ⟩—NO_2 $\xrightarrow{\text{low-boiling solvent}}$

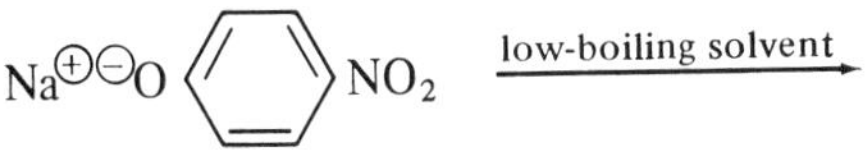

Sodium p-nitrophenate

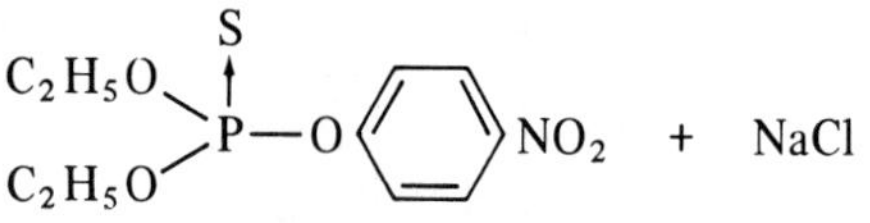

Parathion

Filtration and topping provide a crude insecticidal product that may be further purified.

Some Herbicide Manufacture

It was indicated that the triazine herbicides, including the very signficant atrazine and simazine, are prepared by the appropriate neucleophilic attack of amines and other neucleophiles on cyanuric chloride. The art centers upon the means of controlling the rates of displacement of the first, second, and third chlorine atoms (see Table 21.14).

The equally important group of urea herbicides depend directly or indirectly upon phosgene chemistry, as does the preparation of the insecticidal carbamates. The treatment of *p*-chloroaniline (as its hydrochloride) with phosgene proceeds as follows:

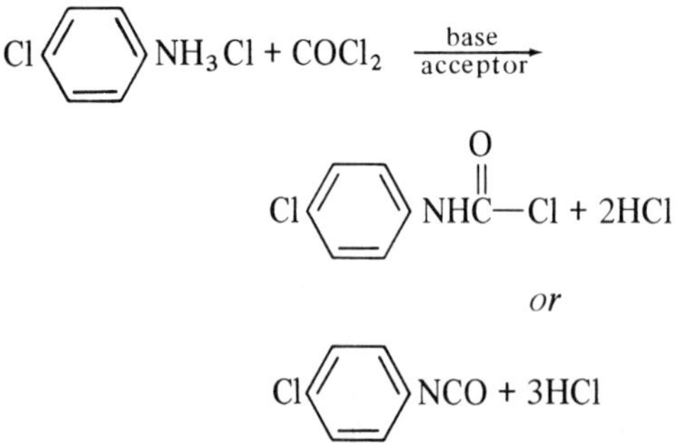

Either the above isocyanate or the carbamoyl chloride is then reacted with an amine to provide the urea derivative.

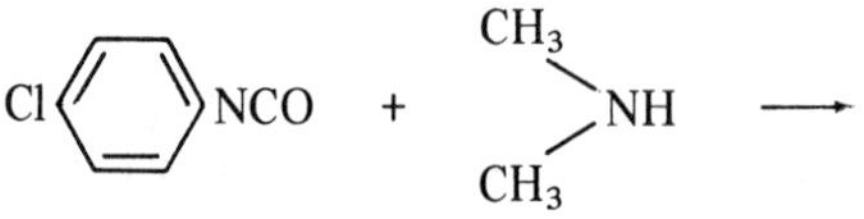

p-Chlorophenyl isocyanate Dimethylamine

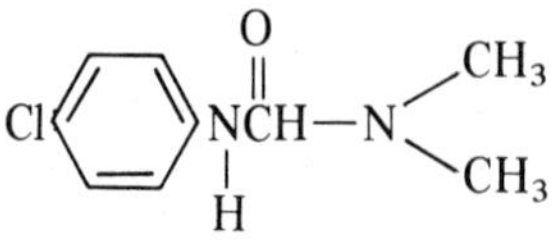

Monuron

These ureas are easily separated from the reaction mixtures as high-melting solids. The salt of the base-acceptor amine or alkali metal chloride can be removed by aqueous wash.

Insecticidal Carbamates

For the insecticidal carbamates, methyl isocyanate is the preferred intermediate. The new corn-root worm insecticide BUX is prepared by an interesting series of reactions.

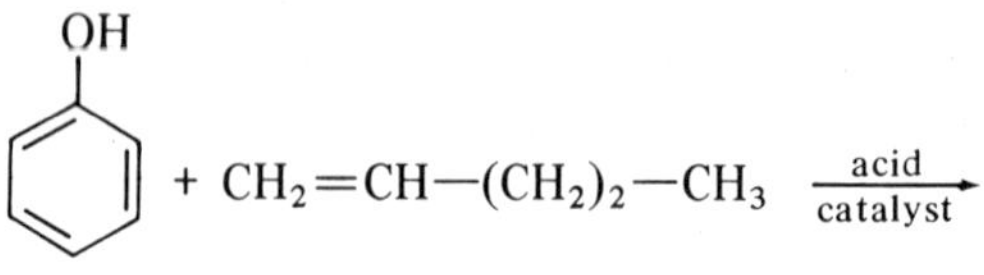

Phenol 1-Pentene

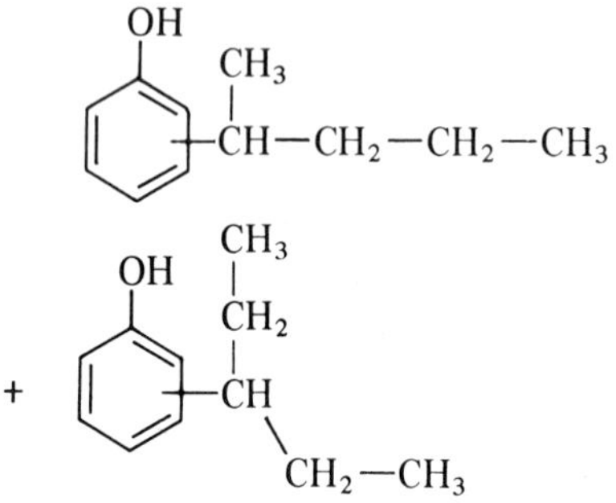

Mixed amylphenols, mostly ortho and para

This alkylation is performed at a low temperature using certain catalytically active clays. On elevating the temperature, an equilibrium mixture of amylphenols is formed in which *meta* isomers are predominant. This is a key step for the *o*- and *p*-amylphenols give rise to carbamates which are biologically relatively inactive.

The final reaction involves the condensation of the predominantly *m*-amylphenols with methyl isocyanate according to the equation:

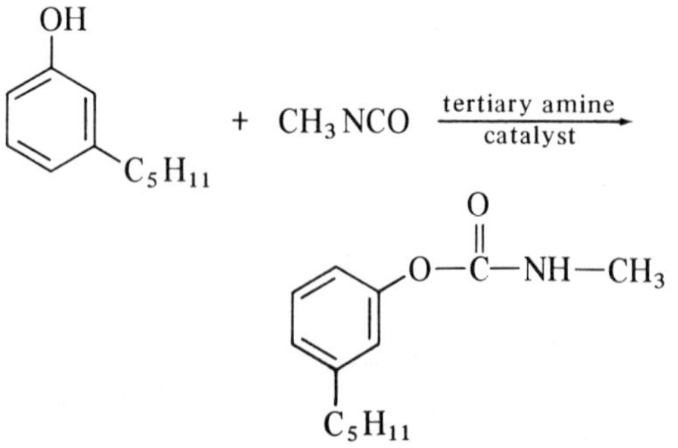

Because the alkylation step gives rise to both methylbutyl and the ethylpropyl isomers these species (both highly active carbamates) are found in the predominantly *meta* amylphenyl N-methylcarbamate which is BUX. Carbaryl is formed by the analogous reaction of alpha-naphthol with methyl isocyanate.

Fungicide Manufacture

The preparation of captan begins with the Diels-Alder addition of butadiene and maleic anhydride to form *cis*-Δ^4-tetrahydrophthalic anhydride.

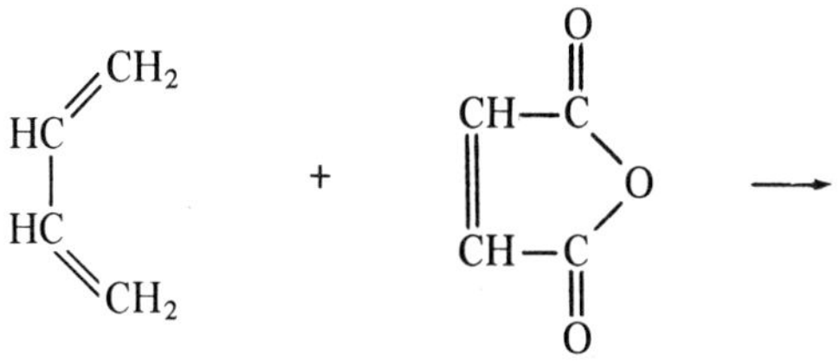

Butadiene   Maleic anhydride

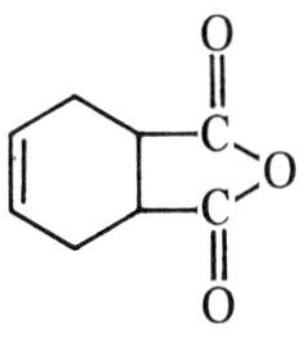

cis-Δ^4-Tetrahydrophthalic anhydride

The anhydride is then reacted with ammonia to produce the imide.

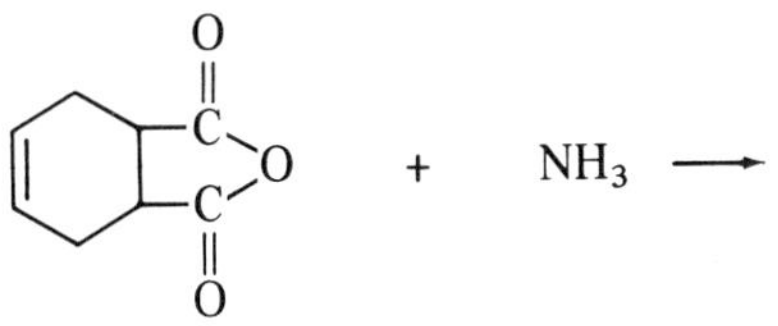

cis-Δ^4-Tetrahydrophthalic anhydride   Ammonia

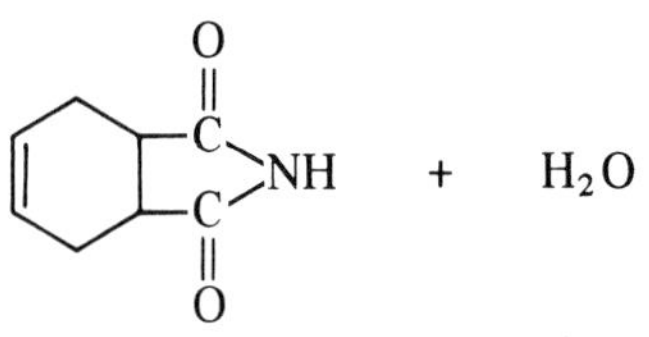

cis-Δ^4-Tetrahydro-phthalimide   +   Water

Carbon disulfide, which is derived from sulfur and hydrocarbons or carbon, is reacted with chlorine in the presence of catalytic quantities of iodine to provide, in an over-all reaction, trichloromethanesulfenyl chloride (which is known by the common name perchloromethyl mercaptan) and sulfur monochloride.

$$\underset{\text{Carbon disulfide}}{2CS_2} + \underset{\text{Chlorine}}{5Cl_2} \xrightarrow{I_2} \underset{\text{perchloromethyl mercaptan}}{2CCl_3SCl} + \underset{\text{sulfur monochloride}}{S_2Cl_2}$$

Finally, the tetrahydrophthalimide is dissolved in cold caustic and reacted with perchloromethyl mercaptan to give captan.

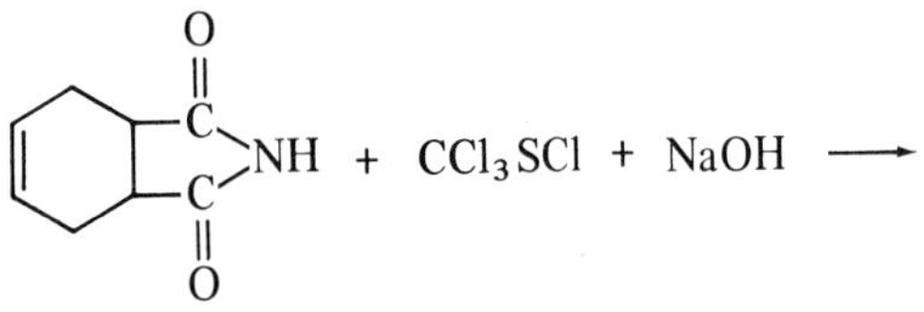

Tetrahydrophthalimide   Perchloro methyl mercaptan   Caustic

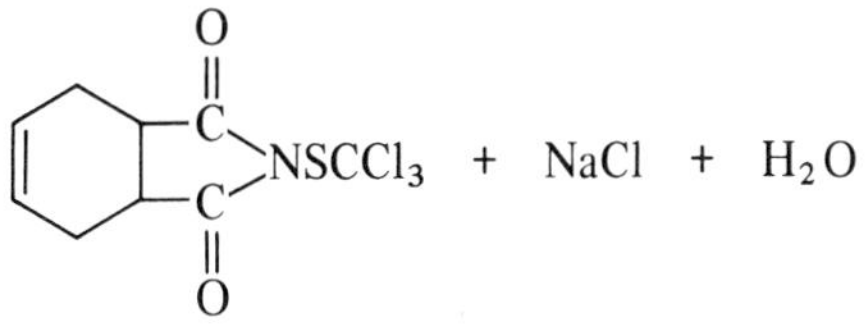

Captan   salt   water

Captan, a solid, is separated from the aqueous slurry by filtration, and air-dried. The low solubility of captan in water provides a finely divided powder which, after drying, is formulated most frequently as a wettable powder after some further grinding.

Some Special Considerations of Pesticide Manufacture

Because of the corrosive nature of many pesticides, their intermediates, and their by-products, exotic alloys, stainless steel, etc., are often employed as construction material for reactors, piping, columns, etc. Much glass or glass-lined equipment is also used for the same reason, or because of lability of the chemical to traces of iron or other metals.

Many of the pesticides and intermediates are highly toxic to humans and all of them, because of their biological activity, must be prevented from contaminating other manufacture and the environment. In the production of the organophosphates, carbamates, and certain chlorinated hydrocarbons both the intermediates and the final chemicals may be hazardous to man. Fume scrubbing, decontamination processes, and gas, liquid, and solid waste disposal require careful monitoring. This requires detection devices, automation of equipment, and reactors for the decontamination of by-products. Oxidation, hydrolysis, and reactions with neucleophilic reagents are all used for decontamination. The toxicity of pesticides to fish is often extremely high, and aqueous effluents are treated so that toxic components do not enter streams, rivers or bays. Experience has shown that 25 to 50 per cent of the total capital costs of the chemical plants must be expended to protect human and environmental resources. This proportion of total capital expenditure is increasing. There is also an increment of the operating cost that relates to the problems of human and environmental hazard. The choice of plant site is also related. There are an increasing number of examples in recent years where the combination of these factors (increased capital costs, increased operating costs, and plant-site location) has made a new pesticide-manufacturing project appear unfeasible. There is no doubt that the total recycling of by-products can be quite expensive, but that is the direction toward which pesticide manufacture is tending. The relatively rapid obsolescence of most pesticides and the increasing severity of regulations compounds the difficulties faced by the manufacturer.

FORMULATION OF PESTICIDES

The art and science employed in transformation of a manufactured chemical into a form that the farmer or the applicator can readily use in the field is of particular significance in the pesticide industry. This transformation process of chemical to agricultural product is the formulation of the pesticide.

The technical product, DDT, for example, is a waxy microcrystalline solid usually recovered in small lumps. The problem is to alter these lumps so that ¼ to 1 lb of product can effectively cover the foliage on an acre of farm land.

Formulations of DDT (and other pesticides) are usually of two major forms, solid or liquid. The solid formulations require aging, or hardening, of the microcrystals, followed by comminution and fracturing of the lumps into small particles—generally micron-sized (1-100 μ) particles. This requires a large expenditure of energy and complex engineering equipment. Among the techniques employed are air grinding, and rarely, wet grinding.

The formulation usually requires inert additives such as clay, celite, talc, etc., to absorb sticky particles and help prevent adhesion in the final product. If the micronized solid is to be used with water—if it is to be suspended in water and sprayed on the foliage as an aqueous suspension—then surfactants of various types must be added. Such a formulation of finely ground particles is called a wettable powder. Frequently a combination of surfactants are employed including those that wet the solid pesticide, maintain a stable suspension of particles in the water, and help in the adhesion of the suspended pesticide to the foliage surface. Hence the formulation art is replete with terms such as suspending agents, stickers, spreaders, and antiflocculants. Since pesticides are used on substances that will become either raw agricultural products or, ultimately, food products, the surfactants, additives, adjuvants, etc., must all be substances approved by the regulating agencies. In other words, the most efficient surfactant may or may not be acceptable for pesticide formulation. Toxicological and residue factors must be considered for the additives as well as for the pesticides themselves.

Liquid pesticides can also be transformed into wettable powders. This is generally done by dispersing the liquid onto fine particles of an absorbent clay, celite, etc. Again, surfactants and other adjuvants are added to provide the physical chemical properties necessary for optimum dispersibility in water, wetting of foliage and insect, etc.

Although used less in recent years, "dust" formulation are still employed in significant amounts. Solids can be ground into fine particles, mixed with inerts, such as talc and clay, and sometimes with adjuvants, and dusted on the plant surface or the soil.

Many pesticides are liquids and many of the solid ones are preferably employed as solutions. An emulsive concentrate is a formulation of a pesticide, which, when mixed with water, provides a dispersion or emulsion of a solvent containing the pesticide. Here again, choice of solvent and surfactant systems involve questions of art and knowledge of colloid and surface chemistry—as well as the toxicology and residue characteristics of the additives. In the formulation of petroleum oil sprays, heavy deposits with little runoff may be required. Formulation requires a compromise, then, between effective dispersion of the emulsive in the water during the spraying operation, and quick breaking of the emulsion upon the plant surface. In other cases, high surfactant activity is desired to perform the biological objective. Hence, for a single pesticide chemical the farmer may require many types of formulation. Each formulation requires a separate label approved by Federal and State regulations.

Nonionic, anionic, and cationic surfactants are employed in pesticide formulations. Though oil in water emulsions are most commonly employed, i.e., the pesticide system is dispersed in a continuous aqueous phase, the reverse (called an invert emulsion) is also, but less frequently, used.

One of the newer formulation techniques developed during the sixties is the use of low-volume and ultralow-volume sprays, particularly of liquid formulations. These formulations, developed to reduce application costs, contain concentrated pesticides with little or no solvent or additives. Mechanical equipment is used to form liquid pesticide droplets of fairly uniform size (40 to 100 μ). There is evidence to suggest that such formulations increase the biological effectiveness of insecticides, e.g., malthion and naled. The problem is one of providing uniform distribution; very fine, submicron droplets will drift past the objective, and large droplets are most inefficient, for many of them never reach the objective. Air is generally used for dispersion. Ground equipment, planes, and helicopters fitted with calibrated delivery systems are commonly used.

A variation of this technique which is applicable to ground equipment consists of the vaporization and rapid condensation of the pesticide solution to create a fog which is carried by air currents to its objective. In this case, the insect respires submicron particles of the toxicant.

Another expanding type of formulation is the pesticide granule. One area of its usefulness is the control of disease vectors in swampy areas. Such targets may be practically inaccessible except by air. However, conventional formulations employing droplets or fine solids are of little use since they will deposit on the foliage surrounding the watery areas in which the mosquitoes breed. Granules of a solid (compacted clay, organic materials, etc.) impregnated with the toxicant can be dispersed by air; they will penetrate through the foliage, and many of them will drop into the surface waters. These granules can be made to release the toxicant at various rates. Granules of various types are now employed in rice culture and in soil-insecticide formulations in the corn belt. They are increasing as new applications are found.

There are other types of formulations employed. Almost all cotton, corn, and wheat seed, and many other crop seeds are treated with special preparations which generally contain pesticides to combat insect and disease attacks, but they may also contain growth stimulants, nutrients, and other components. Not only must the additives employed be effective against the intended target, but they must also be tolerated by the seed and the emerging seedling under a wide variety of external conditions including variations of soil type, temperature, pH, and moisture. Such seed must be easily distinguished from untreated seed so that the toxicants employed in the treatment are prevented from entering into the animal and human food chain. Legislation requires and describes the coloring of such treated seed for ready identification. Such treated seed is proscribed from any but its intended use. Nevertheless, tragedies have occurred through human ingestion of seed containing organophosphates, mercurials, and other toxicants employed in the formulation of seed treaters.

Since in modern American agriculture, labor provides an increasing increment of the total cost, the farmer tries to reduce the number of applications of the formulations employed. To accomplish this, he sometimes combines fertilizer and pesticide applications. This leads to other special formulation problems. Most pesticides are covalent bonded—nonionic organic entities—and the surfactants used with them function best in aqueous solutions of low ionic strength. Fertilizers, on the other hand, are generally water-soluble ionic compounds. The problem of compatibility must be solved, and novel surfactants that will function in solutions of relatively high ionic strength must be found for combined fertilizer and pesticide formulation to be successful.

Aerosol formulations, dilute pesticide formulations containing a volatile component (usually a fluoromethane, carbon dioxide, or nitrous oxide) which propels the pesticide in fine droplets, are widely used in the home and garden.

No attempt is made here to cover the entire area of formulation, but rather to emphasize that formulation is frequently the key to successful application of the pesticide. The formulation process requires the application of physical-chemical principles, engineering skills, and familiarity with the art as a whole. Factors such as corrosion, solubility, surface activity, odor, vapor pressure, flammability, compatibility with additives and other formulations, stability, toxicity, and residue characteristics, are all part of this art and science.

RESEARCH IN THE PESTICIDE INDUSTRY; DIRECTIONS FOR THE FUTURE

During the past few years certain chemical companies, some quite large, have re-evaluated their research and development programs relating to agriculture in the light of social, political, and economic trends and have either abandoned pesticide research or have contracted it markedly. The industry most certainly is in a state of transition and it is interesting to consider some of the new directions being taken by research and development.

Hormone Research

One of the recent, most exciting, academic triumphs has been in the understanding of the physiology, the isolation, and the synthesis of the moulting hormone Ecdysone and the juvenile hormone. Figure 21.5 gives their structures.

Most insects proceed through an elaborate metamorphosis from egg to larvae to pupa to adult, including less dramatic intermediate stages as well. The juvenile hormone, at a given titre in the insect at each stage, tends to maintain the insect at any given stage. Ecdysone, on the other hand, tends to promote moulting. Through a complex interaction of these and other endocrine secretions the insect develops through its normal life cycle. By interfering in the normal equilibrium, one can prevent insects from reaching the sexually mature adult stage or from producing fertile mature eggs that will develop into larvae. This is the strategy then of the hormone approach to insect control. A hormone differs from a conventional insecticide in that its aim is not to kill the insect directly but to interfere with its life cycle.

Most current attention centers about compounds that will simulate the juvenile hormone or Ecdysone in biological activity. Both the juvenile hormone and Ecdysone appear to have little mammalian toxicity. At present, no successful large-scale field control of insects has been achieved, and there remain many biological, chemical, and economic problems that must be resolved.

Nevertheless, a more easily synthesized, smaller molecule within the confines of agricultural economics that possesses one or all of the above physiological properties would indeed provide a breakthrough in insect control.

Pheromone Research

Another area of active investigation is the use of insect signal agents (pheromones) combined with traps, biocides, or sterilants. The chemicals responsible for the sexual stimulation of

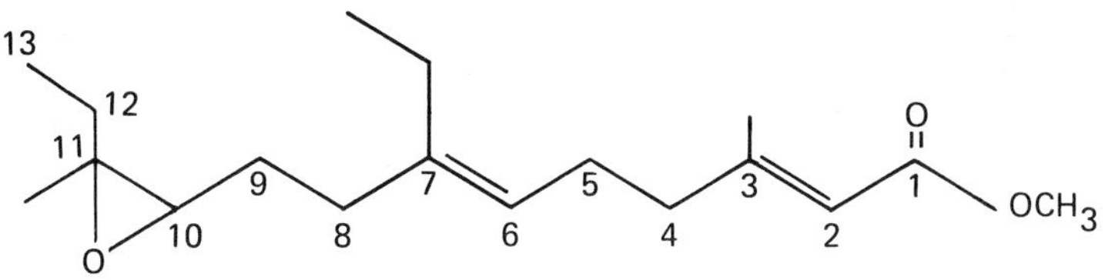

METHYL TRANS, TRANS, CIS-10-EPOXY-7-ETHYL-3,11-DIMETHYL-2,6-TRIDECADIENOATE

ECDYSONE - MOULTING HORMONE

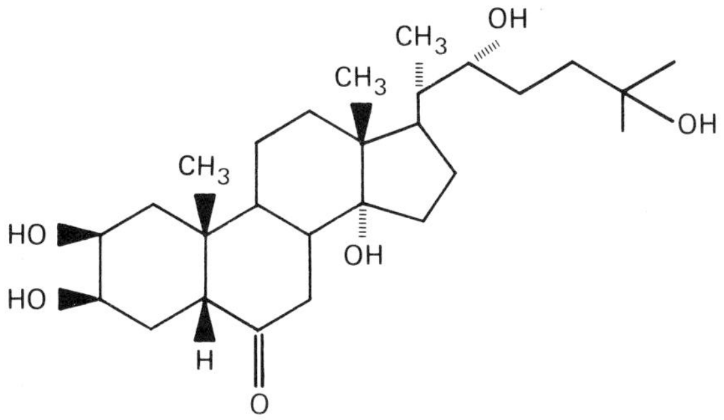

SEX PHEROMONE OF THE CABBAGE LOOPER
TRICHOPLUSIA NI

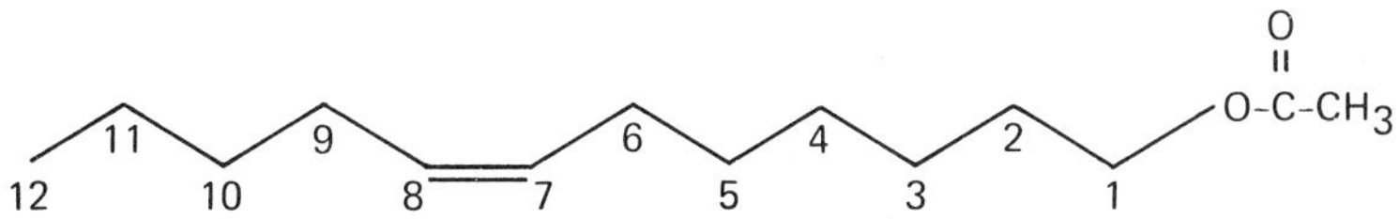

CIS-7-DODECENYLACETATE

Fig. 21.5 Juvenile Hormone, *J. H. Hyalophora Cecropia.*

certain male or female insects have been isolated in a number of cases. These are usually fairly long-chain, olefinic esters, alcohols, or acids (Fig. 21.5). In addition to sex attractants, there are other types of signal agents. There are chemicals that simulate attractive food entities, danger, and various social insect functions. As with the hormone approach, there appears to be less ecological hazard with the use of pheromones. The strategy here is to attract the insect and then either trap, or by the use of chemicals in confined locations, sterilize or kill the specific insect attracted to the given pheromone. Some success has been achieved, usually in locations where migration of insects is difficult. As an example, an infestation of the oriental fruit fly on Rota (a 3-sq mi island) was eliminated by the combination of an attractant and the insecticide naled.

The main attractiveness of the pheromone approach is the biocides need not be broadcast on the crop or soil. The problem of pheromones in present-day monoculture, i.e., where very large areas are mainly planted in one crop, such as the corn belt in the Midwest, remains to be solved. Although the sex pheromones of many economically important insects have been isolated and their structures elaborated, resent research points to somewhat greater complexity relating to the insects signalling systems than was originally indicated. Nevertheless, the area of pheromone research provides a broad field for fundamental chemical and field biological exploration.

Repellents and Antifeeding Compounds

Metadiethyltoluamide

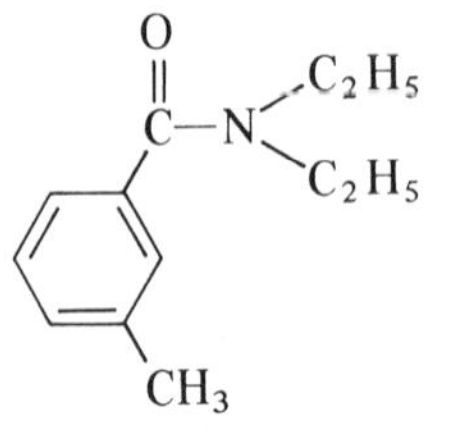

is used as a mosquito repellent, but no repellents have been successfully employed in large-scale agriculture to repel field insects from their normal food supply. Many chemicals exhibit varying degrees of laboratory repellent activity.

Certain chemicals exhibit the property of discouraging insects from feeding without killing the undesirable species. No large scale use of such chemicals exist.

Biological Control Methods

With the attention presently being given to ecological aspects of chemical pesticide control, increased exploration of biological methods for pest control appears timely and, indeed, various approaches are presently being investigated.

One biological method involves the culturing of a bacterium. *Bacillus thuringiensis* effectively infects the larval stage of certain lepidoptera. Economic control has at times been successful, with the use of preparations containing the spores of this bacillus for example, against the larval stage of the cabbage looper, an insect that has developed resistance to many pesticides including many potent organic phosphates. *Bacillus thuringiensis* preparations are being sold commercially. As a result of successful genetic experiments, highly active strains are presently available.

Experiments are also proceeding with the cultivation of the polyhedrosis virus for control of lepidopterous larvae. This approach is at an early stage of development. Some field work is presently being planned. More work will be necessary to determine whether, on the one hand, effective control can be achieved and, on the other, whether bacteria or virus species released to the environment produce effects other than those that were intended.

Another biological method that has been quite widely investigated and has been utilized with varying degrees of success, is the sterile-male technique. Male insects are treated with a sterilizing agent usually, but not exclusively, x-ray or ^{60}Co radiation. Then, large numbers of sterile males are introduced into the field. After mating, the females produce no offspring and the population of the next generation drops.

This method can be successful, but many problems arise, both mathematical and physiological, particularly where there are large populations of migrating adults. One as-yet-unsolved problem consists of the relative attractiveness and competitive vigor of the natural population of males as compared to the treated population. Under WHO auspices the sterile-male technique has been employed against the tsetse fly (sleeping sickness) in Africa. In North America, the campaign against the screw worm (a cattle parasite) using this technique was largely successful. At present, this technique is best suited for government or international agency supervision where logistical and political considerations can be resolved.

Another aspect of biological control consists of the release of parasitic and predator insects. Such predator and parasitic species attack crop-destroying insects. Certain wasps, lacewings, ladybugs, and predator mites, etc., suppress, but do not eliminate the undesirable species. The practical problems, including the economics of such control, are currently being investigated by government agencies and by certain pest-control companies.

Great enthusiasm existed a few years ago for the development of chemicals which caused sterility in insects. Unfortunately, most of the active chemicals, though quite effective, were powerful alkylating agents, and as such exhibited carcinogenic potential. Some very simple phosphorus amides containing the ethyleneimine $CH_2—CH_2$ group were among the

$$\begin{array}{c} \diagdown\ \diagup \\ N{=}A \\ | \end{array}$$

most active insect sterilants. $P(O)A_3$, $P(S)A_3$, and $[(PN)A_2]_3$, where A is the ethyleneimine group, are such agents. Even the solvent hexamethylphosphorictriamide $[[(CH_3)_2N]_3PO]$ possesses some sterilizing capacity.

At present, an integrated approach which will include some of the biochemical and biological methods as well as improved chemical agents in the more classic sense seems to be the direction in which current exploration is heading.

More Fundamental Investigation

Systemic agents for the control of plant disease are presently being studied and developed. These agents or their metabolic products provide a practical sort of plant immunity with little or no toxic residue on the plant surface. Progress has been achieved, and more is anticipated, in the genetic development of insect- and disease-resistant plant species.

The biochemistry of photosynthesis, the important energy and synthesis pathways, transpiration, respiration, and reproduction of plants is being studied both for fundamental knowledge and for new and better practical approaches to plant disease-control and growth-control agents, including selective herbicides. New chemicals for the control of insects are being developed which are more target specific, overcome prevailing resistance problems, and are nonhazardous to man and the environment.

Effort also is being expended to develop systems in agriculture that improve upon current practice. Examples include seed-treating methods which will permit fall planting, improvements in harvesting practices, and protection of harvested crops in storage. Many of these approaches employ chemicals in new and unforeseen ways.

Conclusion

The pesticide industry is an industry in transition. The interrelatedness of agricultural problems to environmental changes and to ecology in general has become recognized. With this recognition there is development of more restrictive legislation. Old and established methods are being challenged and new approaches are called for.

It is the conviction of the author that chemicals can be developed that are effective and that possess minimal ecological hazard, that are safe to man, and to animal species. These combined with some of the newer approaches will lead to integrated campaigns that can maintain an adequate food supply, and an improved public health. The optimum training for such a career must include specialization in chemistry and engineering, but must be broader so that biological and human consequences are fully appreciated. The methods and tools employed in 1980 will not be the same as those used in 1970. They will be better.

REFERENCES

General*

*General references relating to structure and/or mode of action, biological, chemical properties, etc.

1. Audus, L. J., Ed., The Physiology and Biochemistry of Herbicides, New York, Academic Press, 1964.
2. Crafts, A. S., "The Chemistry and Mode of Action of Herbicides," New York, John Wiley & Sons, 1961.
3. Frear, D. H., "Pesticide Index," State College, Pa., College Station Pub., 1968.
4. Horsfall, J. C., "Principles of Fungicidal Action," *Chronica Botanica* (Waltham, Mass., 1956).
5. Kohn, G. K., Ospenson, J. N., and Gardner, I. R., "Synthetic Pesticides from Petroleum," in "Advances in Petroleum Chemistry and Refining," Vol. VII, p. 323–363, New York, John Wiley & Sons, 1963.
6. Metcalf, R. L., "Organic Insecticides," New York, John Wiley & Sons, 1955.

7. Metcalf, R. L., Ed., "Advances in Pest Control Research," Volumes I–IX, New York, John Wiley & Sons, 1957–196
8. Silk, J. A. and Brown, S., "Crop Protection Chemicals Index," 5th Ed., Plant Protection, Limited, ICl, Jealott's Hill Research Station, Bracknell, Barkshire, 1969.
9. Spencer, E. Y., "Guide to Chemicals Used in Crop Protection," Publication 1083, Canada Dept. of Agriculture, Ottawa, 1968.
10. Torgeson, D. C., Ed., "Fungicides—An Advanced Treatise," Vols. I and II, New York, Academic Press, 1967 and 1969.

Specific

1. Hamner, C. L. and Tukey, H. B., *Science*, **100**, 154, 1944; Jones, F. D., U.S. Patent 2,390,941 (1945).
2. Mueller, P., U.S. Patent 2,329,074 (1943); Lauger, P., Martin, H., and Mueller, P., *Helv. Chim. Acta* **27**, 892 (1944).
3. Kittleson, A. R., *et al.*, U.S.P. 2,553,770-7.
4. Cupery, H. E., Searle, N. E., and Todd, C. W., U.S. Patent 2,705,195.
5. Lambrech, J. A., U.S. Patents 2,903,478 and 3,009,855.
6. Schrader, G., U.S. Patents 2,597,534 and 2,571,989.
7. Hyman, J., U.S. Patent 2,519,190.
8. Kohn, G. K., Ospenson, J. N., Moore, J. E., *J. Agri. Food Chem.* **13**, 232 (1965).

Statistical*

*Statistical data except for Fig. 1 was derived from references in this section.

1. U.S. Tariff Commission, "Synthetic Organic Chemicals, U.S. Production and Sales" (yearly through 1968). U.S. Government Printing Office, Washington, D.C.
2. Chemical Economics Handbook, Stanford Research Institute, Palo Alto, Cal.

Journals**

**Contents of these two U.S. journals devoted largely to matters pertinent to the pesticide industry were used in the preparation of this chapter.

1. *J. Food Agri. Chem.* **1–18** (ACS Publication).
2. Gunther, F. A. and Gunther, J. D., *Residue Reviews* **1–32.** (Springer-Verlag, New York.)

22

Pigments, Paints, Varnishes, Lacquers, and Printing Inks

Charles R. Martens*

PIGMENTS

Pigments are used in paints, plastics, rubber, textiles, inks, and other materials to impart color, opaqueness, and other desirable properties to the product. In paints, pigments may also adjust the gloss, impart anticorrosive properties, and reinforce the film. Extender pigments do not either have color or hiding power under most circumstances, but they impart other desirable properties to the product at low cost.

Pigments are insoluble powders of very fine particle size, i.e., as small as 0.01 micron and usually no larger than one micron. They are both natural and synthetic in origin; and organic and inorganic in composition.

Hiding power, or the ability of paint to obscure underlying color, varies with different pigments. In general, dark pigments, since they are more opaque, are more effective than light pigments in this respect. The difference between the index of refraction of the vehicle and that of the pigment largely determines the hiding power of paint, i.e., the greater the difference, the greater the hiding power. For example, white lead has a refractive index of 1.59, zinc oxide, 2.00, and titanium dioxide, 2.70. Most nonvolatile vehicles and resins have a refractive index of from 1.47 to 1.52 (see Table 22.1). Thus titanium dioxide is the most effective pigment for hiding power. The hiding power of pigments is also useful in minimizing the damaging effects of sunlight to the coating and the substrate.

Another factor affecting the hiding power is the particle size of the pigment. Within limits, the finer the pigments, the greater the hiding power. The optimum particle size for pigments to give maximum light scattering and hiding is approximately one-half the wave length of light in air or 0.2–0.4 μ. Below this size, the particle loses scattering power, above this size the number of interfaces in a given weight of pigment decreases.

Fading and color change in paints results largely from instability of the pigment. Pigment stability is particularily important for paints designed for exposure to sunlight and

*Sherwin Williams Co., Cleveland, O.

TABLE 22.1 Indices of Refraction of Some Common Paint Materials

Material	*Refractive Index*
Rutile titanium dioxide	2.76
Anatase titanium dioxide	2.55
Zinc sulfide	2.37
Antimony oxide	2.09
Zinc oxide	2.02
Basic lead carbonate	2.00
Basic lead sulfate	1.93
Barytes	1.64
Calcium sulfate (anhydrite)	1.59
Magnesium silicate	1.59
Calcium carbonate	1.57
China clay	1.56
Silica	1.55
Phenolic resins	1.55–1.68
Melamine resins	1.55–1.68
Urea-formaldehyde resins	1.55–1.60
Alkyd resins	1.50–1.60
Natural resins	1.50–1.55
China wood oil	1.52
Linseed oil	1.48
Soya bean oil	1.48

TABLE 22.2 Tinting Strength and Hiding Power of White Pigments

	**Tinting Strength*	*Hiding Power (Sq ft/lb)*
Rutile titanium dioxide (PSC)	1850	157
Rutile titanium dioxide (conventional)	1750	147
Anatase titanium dioxide	1250	115
50 per cent Rutile calcium-base	880	82
Zinc sulfide	640	58
30 per cent Rutile calcium-base	600	57
Lithopone	280	27
Antimony oxide	300	22
Dibasic lead phosphite	250	20
Zinc oxide	210	20
35 per cent Leaded zinc oxide	175	20
Basic carbonate	160	18
Basic sulfate white lead	120	14
Basic silicate white lead	80	12

[a]Tinting strength is the relative capacity of a pigment to impart color to a white base.

industrial fumes. Lead pigments should not be used in many industrial areas as they darken in the presence of sulfide fumes.

An ideal pigment should be chemically inert, free of soluble salts, insoluble in all media used, and unaffected by normal temperatures. It should be easily dispersed, nontoxic, and have low oil-absorption characteristics.

Pigments can be classified as follows:

Inorganic	*Organic*
White	Colors
Extender	Black
Colors	
Black	
Metallic	

Inorganic Pigments

All white pigments are inorganic compounds of titanium, zinc, lead, or antimony. They are classified as nonreactive and reactive, depending on whether they react with the vehicle, which may be acidic.

The most important of the nonreactive white pigments is titanium dioxide. Because of its high refractive index (2.76) it gives the greatest amount of hiding per dollar (see Table 22.2). Titanium dioxide is available in two forms, depending on the crystal structure—rutile and anatase. Anatase is considered to be a chalking-type pigment.

Two other nonreactive pigments are zinc sulfide and lithopone. Lithopone is a composite pigment of zinc sulfide coprecipitated upon calcium sulfate or barium sulfate crystals.

The reactive white pigments are basic carbonate white lead, basic sulfate white lead, basic silicate white lead, dibasic lead phosphite, zinc oxide, leaded zinc oxide, and antimony oxide. Reactive pigments serve several functions in coatings, for example, zinc oxide, being basic in nature, readily forms zinc soaps with the acid constituents of the paint vehicle. This reactivity is utilized by the paint chemist to increase the hardening of the paint film to increase the consistency of the fluid paint, to aid in mixing and grinding the pigments because of better wetting, to reduce after-yellowing, and to improve the self-cleaning and mildew-resistance of exterior house paints.

Antimony oxide is used because of its fire-retardant properties.

White extender pigments have several functions, such as controlling gloss, texture, suspension, and viscosity. Extenders have low refractive indices, i.e., 1.40 to 1.65. Common extenders are whiting (calcium carbonate), talc (magnesium silicate), clay (aluminum silicate), calcium sulfate, barytes (barium sulfate), silica, and mica. Particle size and shape are the most significant properties of extenders.

Colored inorganic pigments are both natural and synthetic in origin. These are:

Iron oxides–from both mineral and synthetic sources including Spanish oxide, Persian Gulf oxide, and domestic iron oxides; and siennas, ochers, umbers, and black iron oxide.

Lead chromate yellows and oranges–$PbCrO_4$ and modifications.

Molybdate oranges and reds–lead chromate modified with $PbMoO_4$.

Zinc chromates–approximate composition: $4(ZnO \cdot K_2O) \cdot 4(CrO_3) \cdot 3(H_2O)$.

Red lead–Pb_2O_3.

Cadmium colors–yellow, orange, and red–mixtures of cadmium and zinc sulfide, cadmium sulfoselenides, and cadmium selenides.

Iron blues–potassium, sodium or ammonia coordination compounds of ferriferrocyanide.

Ultramarine blues–formula unknown, made from china clay, sodium carbonate, silica, sulfur, and a reducing agent.

Chrome greens–blends of chrome yellows and iron blues.

Chrome oxide green–Cr_2O_3.

Chrome oxide green–$Cr_2O_3 \cdot 2H_2O$.

Organic Pigments

The difference between dyes and pigments is their relative solubility; dyes are soluble, while pigments are essentially insoluble in the liquid media in which they are dispersed. In the manufacture of organic pigments certain coloring materials become insoluble in the pure form, whereas others require a metal or an inorganic base to precipitate them. The coloring materials which are insoluble in the pure form are known as "toner pigments" and those which require a base are referred to as "lakes."

Organic pigments in general have lower hiding power but greater tinting strength then inorganic pigments. Although there are a great many organic pigments available, they may be classified into about six groups based on some general characteristic of their chemical composition. These are:

1. *Azo insoluble*–toluidine, *para*-chlorinated nitroanalines, naphthol reds, Hansa, benzidine, dinitroanaline orange; these are all members of the azo dyestuff family, which is insoluble in water.
2. *Acid-azo*–lithol, lithol rubine, BON colors, red lake C, Persian orange, tartrazine; these are all acid azo pigments obtained from dyestuffs which contain acid groups ($-SO_3H$, $-COOH$) in their structure, and they are insolubilized by reaction with sodium, barium, calcium, or strontium.
3. *Anthraquinone*–alizarin, madder lake, indathrene, vat colors.
4. *Indigoid*–indigo blue and maroons.
5. *Phthalocyanine*–phthalocyanine green and blue.
6. *Basic PMA, PTA*–PMA and PTA toners and lakes, rhodamine, malachite green, methyl violet, Victoria blue.

While normally not classified as such, carbon blacks are organic in nature. The carbon blacks are channel, furnace, and lamp blacks, and graphite.

Pigment Manufacture

Titanium Dioxide. Titanium dioxide pigments are manufactured by two processes, viz., the sulfate process and the chloride process.

The major portion of the world's production of titanium dioxide is by the sulfate process, the raw material for which is ilmenite ore. The ore, previously dried and ground, is digested with sulfuric acid (85-90 per cent). After digestion and solvation, the iron is reduced to the ferrous state and much of it is removed by crystallization as ferrous sulfate. The liquid is filtered, concentrated in vacuum evaporators, and boiled with sulfuric acid to precipitate the titanium dioxide pigment, which is washed, dried, and calcined at a temperature of about 1650°F. The calcining operation converts the titanium dioxide from the amorphous to the crystalline state, thereby raising the refractive index. Controlled grinding and bagging follow. After the purification steps, the process varies depending on the grade and type of product desired.

The chloride process produces pigment by the oxidation of titanium tetrachloride, itself obtained from mineral rutile by chlorination in the presence of carbon. The process is attractive because the titanium tetrachloride is a definite chemical compound which may be produced in a high degree of purity. All of the newer titanium dioxide plants use this process.

"Extended" titanium pigments are prepared either by mixing or coprecipitating TiO_2 with cheaper pigments of low hiding power. Titanium calcium may be prepared by two methods: (1) by precipitating hydrated TiO_2 in the presence of $CaSO_4$, the coprecipitate being filtered, washed, calcined, and dry ground; or (2) TiO_2 and $CaSO_4$ may be mixed as a wet slurry, filtered, dried, calcined, and dryground. The composite pigment contains 30 per cent TiO_2 and 70 per cent $CaSO_4$ and has much better hiding power than would be obtained from a simple dry mix of the two components in the same proportion.

Titanium dioxide paints which have controlled chalking are extensively used for house paints; they stay white longer because of the gradual erosion of the soiled surface. A large proportion of anatase, sometimes combined with a small amount of an oxide of antimony or aluminum, is used for this purpose.

Zinc Oxide. Zinc oxide is made in several ways. In one of these the ore (frankenite) is mixed with coal and burned on a grate. The natural oxide is first reduced, and then reformed by the air and carbon dioxide from combustion.

Lithopone. Lithopone is formed when a solution of zinc sulfate is mixed with one of barium sulfide. Barium sulfate and zinc sulfide are formed, both of which are white. The precipitate is not suitable for use as a pigment, however, until it has been dried, heated to a high temperature, and then plunged, while still hot, into cold water. Lithopone is 30 per cent zinc sulfide and 70 per cent barium sulfate, with slight variations.

Red Lead. Red lead is made by calcining litharge in a muffle furnace. A current of air is admitted into the muffle, the temperature is maintained within narrow limits near 640°F, and the time required is usually 48 hours.

Carbon Black. There are several kinds of carbon black: *thermal black* produced by the thermal decomposition of natural gas; *channel black* produced by the impingement of numerous small regulated flames against a relatively cold steel surface which is constantly scraped free of the soot deposit; *furnace black* produced by the partial combustion of the gas in a furnace with recovery of the carbon product in cyclones and electrical precipitators; and *lampblack*, used mainly as a tinctorial pigment.

Iron Oxide. Iron oxide (Fe_2O_3) is made on a large scale by roasting ferrous sulfate obtained from the vats used for pickling steel. Water and sulfur oxides are driven off and led through a stack to the atmosphere. The shade may be varied by altering the firing time, the temperature, and the atmosphere. It is a relatively cheap pigment and is usually used in red barn paint and metal primers. The use of selected grades for polishing glass and lenses is determined by their resistance to grit and the hardness of the glass; such grades of iron oxides are called *rouges.*

Metallic Pigments. Although gold, zinc, and copper bronze powders are used as pigments, powdered aluminum is the most important. In addition to its principal use in organic coatings it is also used in solid rocket propellants and as a filler for thermosetting resin systems. As a leafing pigment it reflects sunlight, thus preventing degradation of the organic film. As a nonleafing pigment it is used in automobile finishes to provide a metallic sparkle.

Aluminum powder is prepared in a stamping mill. Then aluminum sheets, usually mixed with a small amount of lubricant (e.g., stearic acid), are pounded into a powder, which is then screened and polished. Much of the powder is converted to a paste by grinding in a ball mill with mineral spirits.

PAINTS

Paint is a substance composed of solid coloring matter suspended in a liquid medium and applied as a coating to various types of surfaces.

Fig. 22.1 Bridge painted with aluminum paint. *(Courtesy Sherwin-Williams Co.)*

The purpose of the coating may be decorative, protective, or functional. Decorative effects may be produced by color, gloss, or texture. The protective coating may be the paint on a wooden boat which serves as a barrier against moisture and prevents rotting; the interior lining of metal cans or drums which prevents corrosion from foods or chemicals; the coating on electrical parts to exclude moisture; the fire-retardant paint which protects combustible surfaces; the coating on plaster or concrete which makes for ease of cleaning, etc. An example of the functional use of paint would be as a traffic paint which marks the center or edge of a road for safer driving.

There were between 600 and 700 million gallons of paint produced in the United States in the year 1970 with a value of over three billion dollars.

The paint industry serves two distinct types of markets, trade sales (shelf goods) and industrial sales (chemical coatings).

Trade sales is the large consumer-oriented portion of the business. Trade-sale paints are house paints and other products marketed through wholesale and retail channels to the general public and professional painters for use on new construction or for the maintenance of old buildings. Also included in this category are the paints sold to garages and repair shops for automobile refinishing, marine finishes for boats, paints for graphic arts such as signs, paints for refinishing machinery and equipment, and paints sold to government

Fig. 22.2 Florida exposure for paints. *(Courtesy Sherwin-Williams Co.)*

agencies, especially the traffic paints used for road-marking.

Industrial sales comprise the coatings sold directly to the manufacturer for factory application. The products on which paints are applied include durable goods such as automobiles, appliances, and house sidings, and nondurable goods such as cans for food and beverages.

All coatings contain a resinous or resin-forming constituent called the *binder*. This can be a liquid such as a drying oil, or a resin syrup that can be converted to a solid gel by chemical reaction. In some instances, where the binder is either a solid or is too viscous to be applied as a fluid film, a volatile solvent or *thinner* is also added. This evaporates after a film is deposited, causing solidification of the film. The binder plus solvent is known as the *vehicle*. Most paints also contain *pigments* which are described in the first part of this chapter.

In addition to pigment, binders, and thinners, a paint may contain many additives, such as defoamers, thickners, flow agents, and driers, to improve specific properties.

Paints can be classified according to the binder used, e.g., alkyds, vinyls, and epoxies. Paints are also classified according to their properties or end use. Alkyd enamels, for instance, are gloss paints with good abrasion resistance and good cleanability, while alkyd flat wall paints are characterized by very low sheen and good film build.

Paints applied directly to the surface are called undercoaters or primers. Primers are used to aid the adhesion of the top coat to a surface and to prevent absorption of the top coat into a porous surface. Primers can also be used to prevent corrosion of a metal surface. Fillers or surfacers are types of primers that fill scratches and surface imperfections to give a smooth surface.

Paints used as the final coat are referred to as finish coats or top coats. Some top coats are self-priming.

Binders

The protective properties of a coating are determined primarily by the binder. In the early days of paint technology, binders were limited to materials of natural origin, such as drying oils, congo resins, and asphalts. These still find some usage in the protective-coating industry. However, the chemical and plastic industries during the past sixty years have supplied a large number of synthetic binders which make possible paints with greatly improved protective and decorative properties.

Coatings in liquid form can be divided into two types, *solutions* and *dispersions*. In solution systems the resin binder is dissolved in a solvent. In dispersion systems the resin is in the form of tiny spheres (usually 10 microns or less in size) suspended in a volatile liquid carrier. If the liquid in a dispersion system is water, the system is called an emulsion; if it is an organic material, it is called an organosol. When the liquid evaporates, a mixture of soft resin and pigment is left behind and fuses into a continuous film.

The general types of resins used are listed below. Mixtures of two or more types may sometimes be used to improve certain properties.

Oils—easy application; soluble in aliphatic solvents.

Alkyds—all purpose; combined with other resins; most are soluble in aliphatic solvents.

Cellulosics—(nitrate and acetate)—used in lacquers; fast dry.

Acrylics—good color and durability.

Vinyls—good durability; abrasion resistant.

Phenolics—good chemical resistance; yellow color.

Epoxies—good chemical resistance.

Polyurethanes—good flexibility; abrasion resistance.

Silicones—good heat resistance.

Amino Resins (ureas and melamines)—blended with alkyds for baking finishes; tough, good color.

*Styrene-butadiene**—low cost, alkali resistant.

*Polyvinyl acetates**—low cost; good color retention.

*Acrylics**—good color and durability.

*Latex form

Solvents

The type of binder determines the type of solvent or thinner used in a paint formulation.

The preferred type of thinner is an odorless aliphatic hydrocarbon which can be used in all areas including the home. Unfortunately, aliphatic thinners do not dissolve all resins. In such cases, strong solvents such as aromatic hydrocarbons, esters, and ketones are used.

Solvents are generally classified as low boilers, medium boilers, and high boilers depending upon the evaporation rate. Various ranges are required depending on the specific application. Examples of solvents classified by their chemical composition are as follows:

Hydrocarbons, aliphatic–VM&P naphtha and mineral spirits; and aromatics–benzene, toluene, and xylene.

Alcohols–methyl alcohol, ethyl alcohol, and butyl alcohol.

Ethers–dimethyl ether and ethylene glycol monoethyl ether.

Ketones–acetone, methyl ethyl ketone, and methyl isobutyl ketone.

Esters–ethyl acetate, butyl acetate, and butyl lactate.

Chlorinated–tetrachlorethane.

Nitrated–nitromethane, nitroethane, and 1-nitropropane.

Latex or emulsion paints use water as the volatile component so that they may be used in the home.

General Properties of the Various Types of Paints

Acrylics. Acrylic resins are used in protective and decorative lacquers for paper, fabrics, leather, plastics, wood, and metal. White baking enamels having excellent resistance to chemicals and chemical fumes can be made from acrylic resins. Acrylic vehicles are used in luminescent paints. The present automobile finishes are based on acrylic resins.

Thermosetting types of acrylic resins have been developed which cross-link upon heating to form hard, insoluble, infusible coatings for appliance finishes that will withstand severe service on washing machines, stoves, and dishwashers.

Alkyds. Alkyds are oil-modified phthalic resins that dry by reacting with oxygen from the surrounding air. Alkyd finishes are usually of the general-purpose type and are available as clear or pigmented coatings. Paints are available in flat, semigloss, and high-gloss finishes in a wide range of colors. They are easy to apply and may be used on most surfaces with the exception of fresh concrete, masonry, and plaster which are alkaline. Alkyd finishes have good color and gloss, and retain these characteristics in normal interior and exterior environments except under corrosive conditions.

Fig. 22.3 Reactor for manufacturing alkyd resins. *(Courtesy Sherwin-Williams Co.)*

Alkyd finishes are available in odorless formulations for use in hospitals, kitchens, sleeping quarters, and other areas where odor during painting might be objectional.

In trade sales applications, alkyds are used for interior walls and woodwork in both flat and gloss, and on the exterior as trim paints.

In industrial finishes, alkyds are combined with amino resins to produce hard baking finishes for use on appliances, metal furniture, etc.

Cellulosics. The most widely used cellulosic derivatives are the esters, particularly cellulose nitrate and cellulose acetate. Cellulose nitrate has been combined with many different resins to produce useful coatings. Perhaps the most outstanding combinations have been those with alkyd and amino resins to produce tough, hard, durable lacquers capable of withstanding the severe service requirements of automotive, aircraft, and other industrial finishes.

Successful coatings have also been based on blends of nitrocellulose with natural resins and drying oils, resin derivatives, phenolic resins, acrylic resins, certain vinyl resins, and other materials. Applications include coatings for metal, wood, paper, fabrics, leather, and cellophane.

Epoxies. Epoxy binders are of two types: (1) the oil-modified compositions, which dry by oxidation; and (2) the two-component materials, which comprise epoxies and amine or polyamide hardeners and are mixed just prior to use. In the latter type, when the two ingredients are mixed they react to form the cured coating. These coatings have a limited pot-life, usually a working day. Anything left at the end of the day must be discarded.

Epoxy paints can be used on any surface and can be applied with high-solids content, thus producing high film build per coat. The cured film has outstanding hardness, adhesion, flexibility, and resistance to abrasion, alkali, and solvents, as well as being highly corrosion resistant. Their major uses are as tile-like glaze coatings for concrete and masonry and for the protection of structural steel in corrosive environments. Their cost per gallon is high, but this is offset by the higher solids content and the reduced number of coats required to provide adequate film thickness. When used on exterior surfaces, epoxy paints tend to chalk to low-gloss levels and fade. Apart from this their durability is excellent.

Epoxy-coal tar coatings are made by adding coal tar as an ingredient to epoxy paints, thereby reducing the cost. They have outstanding corrosion resistance and are used for interior and submerged surfaces. Color choice is limited to black.

Oils. Linseed oil is the major binder used in oil house paints. These paints are the oldest type of coatings in use. They are used primarily on exterior wood and metal since they dry too slowly for interior use. They are sensitive to alkaline masonry. Oil paints are easy to use and give high film build per coat. They also wet the surface very well so that surface preparation is less critical than with other types of paints for metal. Oil paints are not particularly hard or resistant to abrasion, chemicals, or strong solvents, but they are durable in normal environments.

Oleoresinous. These binders are made by processing drying oils with hard resins, such as natural resins, rosin esters, and hydrocarbon resins. They are generally used as spar varnishes or as mixing vehicles for aluminum paint.

Phenolics. Phenolic binders are made by processing a drying oil with a phenolic resin and, thus, are a type of oleoresinous binder. They may be used as flat (lusterless) or high gloss finishes which are clear or pigmented in a range of colors. The clear finishes may be used on exterior wood and as mixing vehicles for producing aluminum paints. The durability of the clear finish is very good for this class of material, i.e., one to two years; the durability of the aluminum paints is excellent.

Phenolic paints are used as topcoats on metal in extremely humid environments and as primers for fresh-water immersion. These paints require the same degree of surface preparation as alkyds, but they are slightly higher in cost than alkyds. Phenolic coatings have excellent resistance to abrasion, water, and mild chemical environments. They are not available in white or light tints because of the relatively dark color of the binder.

Phenolic and alkyd binders are often blended to combine the hardness and resistance of phenolics with the color and color-retention of the alkyds. This may be done either by blending phenolic varnish with the alkyd vehicle or by the addition of phenolic resins during the processing of the alkyd resin.

Silicones. Silicone resins are used for heat-resistant finishes. They have good water-resistance and outstanding gloss-retention. When pigmented with aluminum, heat-resistant organic finishes containing a high concentration of silicone resins have the ability to withstand temperatures up to 1200°F. A combination of silicone and alkyd resins provides some heat resistance at a lower cost.

Urethanes. Urethane binders are of three types: (1) oil-modified, which are cured by oxygen from the air; (2) moisture-curing, which are cured by moisture in the air; and (3) two-part systems which are mixed just prior to use.

Oil-modified urethanes are similar to phenolic varnishes. Although somewhat more expensive, they have better initial color and color-retention, dry more rapidly, are harder, and have better abrasion-resistance. They can be used as exterior spar varnishes or as tough floor finishes. Oil-modified urethanes can be used on all surfaces. In common with all clear finishes, they have limited durability when used on exterior surfaces.

Moisture-curing urethanes are used in a manner similar to other one-package coatings except that the containers must be full to exclude moisture during storage. They have outstanding abrasion resistance and chemical resistance.

Vinyls. Lacquers based on modified polyvinyl chloride resins are used on steel where the ultimate in durability under abnormal environments is desired. They are moderate in cost, but they have low solids and require the most extensive degree of surface preparation to secure a firm bond. Because of their low solids, vinyl finishes require numerous coats to achieve adequate dry film thickness; thus the total cost of painting is higher than with most other paints. Since vinyl coatings are lacquers, they are best applied by spray, and they dry quickly, even at low temperatures. Recoating must be done with care to avoid lifting by the strong solvents which are present. In addition, these solvents present an odor problem. Vinyls can be used on metal or masonry, but they are not recommended for use on wood. They have exceptional resistance to water, chemicals, and corrosive environments, but they are not resistant to solvents.

Vinyl-alkyd combinations offer a compromise between the excellent durability and resistance of vinyls with the lower cost, higher film build, ease of handling, and adhesion of alkyds. They can be applied by brush or spray, and they are widely used on structural steel in marine and moderately severe corrosive environments.

Rubber-Base. So-called rubber-base binders are solvent-thinned and should not be confused with latex binders which are often called rubber-base emulsions. Four types are available: chlorinated rubber, styrene-butadiene, vinyl-toluene-butadiene, and styrene-acrylic. They are lacquer-type products which dry rapidly to form finishes which are highly resistant to water and mild chemicals. Recoating must be done carefully to avoid lifting by the strong solvents used. Rubber-base paints are available in a wide range of colors and levels of gloss. They are used for exterior masonry, also for areas which are wet, humid, or subject to frequent washing, e.g., swimming pools, wash rooms, shower rooms, kitchens, and laundry rooms.

Latices. Latex paints are based on aqueous emulsions of three basic types of polymers: polyvinyl acetate, polyacrylics, and polystyrene-butadiene. They dry by evaporation of the water, followed by coalescense of the polymer particles to form tough insoluble films. They have little odor, are easy to apply, and dry very rapidly. Interior latex paints are generally used either as a primer or finish coat on interior walls and ceilings made of plaster or wall board. Exterior latex paints are used directly on exterior masonry or on primed wood. They are noninflammable, economical, and have excellent color and color-retention. Latex paint films are somewhat porous so that blistering due to moisture vapor is less of a problem than with solvent-thinned paints. They do not adhere readily to chalked or dirty surfaces nor to glossy surfaces under eaves. Therefore, careful surface preparation is required for their use.

Latex paints are very durable in normal environments, at least as durable as oil paints. The popularity of latex paints is due mainly to easy cleaning of brushes, etc. with water.

Inorganics. The major inorganic binders used in paints are sodium, potassium, lithium, and ethyl silicates. These binders are used in zinc-dust-pigmented primers in which they react with the fine zinc metal to form very hard films. These films are extremely resistant to corrosion in humid or marine environments. Many of these primers also contain substantial concentrations of lead oxides which react with the silicates in conjunction with the zinc to form an even more corrosion-resistant coating.

Another type of inorganic binder is Portland cement. The paint is supplied as a powder to which water is added before use. Cement paints are used on rough surfaces such as concrete, masonry, and stucco. They dry to form hard, flat, porous films which permit water vapor to pass through readily. Cement paints should not be used in arid regions. When properly cured, cement paints of good quality are quite durable; when improperly aired, they chalk excessively on exposure, and then they may present problems on repainting.

A typical paint formula for an outside house paint is shown in Table 22.3.

Manufacture of Paints

The manufacture of paint involves the following operations: mixing, grinding, thinning, adjusting, and filling.

One of the older methods consisted of mixing all the pigment and part of the vehicle to make a paste of suitable consistency in a tub with rotating blades. This paste was then fed by means of a trough into a slow-speed stone mill which had two circular stones, one stationary and the other rotating above it. The pressure developed between the two stones was sufficient to disperse the pigments and liquids intimately, eliminating unwetted agglomerates and air pockets. This paste was fed in a

Fig. 22.4 A three-roll mill. *(Courtesy Sherwin-Williams Co.)*

TABLE 22.3 Analysis of a White Outside House Paint

	Pounds	*Gallons*	*Per Cent (by wgt)*
Titanium dioxide (rutile)	105	3.05	7.4
White lead–basic carbonate	160	1.88	11.3
Zinc oxide (35 per cent leaded)	370	7.55	26.2
Talc	240	14.70	16.9
Linseed oil–bodied 3	85	10.54	6.0
Linseed oil alkali refined	350	45.40	24.8
Drier–manganese napthenate 6 per cent	2	00.26	–
Drier–lead napthenate 24 per cent	7	00.72	–
Mineral spirits	105	15.90	7.4
TOTAL	1419	100.00	100.0

Constants	
Viscosity	90 KU
PVC	30 per cent
Vehicle solids	80 per cent
Total solids	92.5 per cent
Wt/gallon	13.1 lbs

continuous stream into a third piece of equipment, the thinning tank, which had rotating blades. At this point the remainder of the liquids was added, and the completed product was tested for viscosity, color, and other physical properties pertinent to the formulation being prepared. The batch, having been approved, was then strained and filled into appropriate size containers.

The term "grinding" is a misnomer, as little or no breakdown of pigment particles takes place, but this term, nevertheless, is generally used in the trade. The word dispersion would be more descriptive.

Dispersion by old-fashioned low-speed stone mills is extremely slow and costly and is obsolete. Various types of more efficient equipment are now in use. The general principle, however, is the same, that is, to wet each individual particle thoroughly with the vehicle and to eliminate flocculated aggregates. Mills in use today include steel roller, ball, pebble, sand, high-speed impeller, Morehouse (a high speed stone type), and Cowles disperser.

Application of Paint

Paints can be applied in many different ways. Although most architectural paints are applied with a brush or roller, much paint is now applied by professional painters with air or airless spray equipment. With airless spray equipment, the paint is atomized by forcing it through a very small orifice under very high pressure. Other types of spray application are electrostatic spraying, hot spraying, steam spraying, two-component spraying, and aerosol spraying.

Fig. 22.5 Sandmills: one 16-gallon mill and one 8-gallon mill. *(Courtesy Glidden-Durkee Div. of SCM Corp., Carrollton, Texas.)*

Fig. 22.6 Phenolic-coated ducts. These exhaust-fume ducts are being coated with Bakelite® phenolic resin baking finishes. Resistance to most solvents and chemicals at elevated processing temperatures is an inherent property of this type of resin. Other uses include linings for lard pails and fish cans. *(Courtesy Sherwin-Williams Co.)*

In electrostatic spraying the atomized paint is attracted to the conductive object to be painted by an electrostatic potential between the two. Advantages of this process include efficient use of coating material, rapid application, and relative ease of coating irregular shapes uniformly. Two-component spray equipment consists of two lines leading to the spray gun so that two materials, for example, an epoxy and a catalyst, can be mixed in the gun just before application.

There are several other methods for industrial application of paints. Dip application of coatings is a simple method wherein objects to be coated are suspended and dipped into a large tank containing the paint. This method is often used for undercoating objects where the uniformity and appearance of the paint are not important. Electrodeposition consists of depositing paint on a conductive surface from a water bath containing the paint. The negatively charged paint-component particles are attracted to the object being coated which becomes the anode when an electrical potential is applied. Paint can be applied to very irregular surfaces with very uniform thicknesses and little loss. The system is limited to one coat of limited film thicknesses and the equipment-cost is high.

Yet, another method utilizes roller-coating machines to apply paint to one or both sides of flat objects, such as metal or fiberboard. The thickness of the coating can be controlled by the clearance between a doctor blade and the applicator rolls. Decorative effects such as wood-grain patterns can be applied with these machines.

Other methods include flow coating, where the paint is allowed to flow over the object being coated, and powder coating, in which paint in dry powder form is fused on the surface of the object being coated.

VARNISHES

The term "varnish" is applied to clear, transparent coating materials that dry by a process comprising evaporation of the solvent, followed by oxidation and polymerization of the drying oils and resins. Varnishes are a homogeneous mixture of resins, drying oils, driers, and solvents, and they contain no pigment.

Varnishes fall into three general classes, *spar varnishes* (exterior), *floor varnishes*, and *furniture finishes*.

The types of oils and resins and the ratio of oil to resin are the principal factors which determine the properties of a varnish. The selection depends on the compatibility of different oils and resins and on the intended use of the varnish. It is generally accepted that the oils in the finished coatings contribute to its elasticity, and the resin, to its hardness. In oleoresinous varnishes the ratio of oil to resin is expressed as the number of gallons of oil that are combined with 100 pounds of resin, and it is commonly referred to as the "length" of the varnish. Thus, where 50 gallons of oil are used with 100 pounds of resin, the varnish has a 50-gallon length. Varnishes containing less than 20 gallons of oil per 100 pounds of resin are usually classed as *short-oil* varnishes. A *medium-oil* varnish contains from 20 to 30 gallons of oil, and a *long-oil* varnish is one in which 30 gallons or more are used. Short-oil varnishes dry more rapidly than long-oil varnishes. They are used primarily where hardness, a high degree of impermeability, or resistance to alcohols, alkalies, or acids is desirable and where elasticity in the film is relatively unimportant. Short-oil varnishes are especially suitable where a "rubbed" finish is desired, e.g., on furniture, but they are too brittle for floors for which a medium-oil varnish is best suited. Varnishes for exterior exposure are usually of a long-oil formulation because of the beneficial effect of the drying oil in providing elasticity and resistance to weathering. Spar varnish, a high-grade exterior varnish formulated originally for use on the wooden spars of ships, is a varnish of this type.

Interior varnishes are often based on linseed oil in combination with ester gum or a maleic-modified-rosin resin. Rosin-modified phenolic resins form varnishes with fairly good water- and alkali-resistance, but for exterior durability a long-oil varnish is required. The most durable spar (wood) varnishes for exterior exposure are usually made from tung oil combined with 100 per cent phenolic resin. The phenolics as a group yield varnishes of relatively dark color. Terpene-phenolic resins are lighter in color and lower in cost than the comparable phenolic resins and have very good water- and alkali-resistance but only fair exterior durability. The terpene resins themselves are low in cost and make good light-colored interior varnishes with good color retention and good resistance to water and alkali.

Coumarone-indene resins are thermoplastic resins available in a wide range of hardness. Although poor in color and weathering, they contribute excellent water- and alkali-resistance and good dielectric properties to varnishes.

Petroleum hydrocarbon resins are soluble in drying oils and form varnishes with good water-, alkali-, and alcohol-resistance, but they have only fair color.

Natural resins, such as, congo, kauri, pontlianak, and batu are often used in varnishes because of their low cost and their ability to impart specific properties such as hardness, gloss, and moisture-resistance.

Alkyd varnishes differ from the conventional

type in that the resin is formed in the kettle and co-reacted with the oil.

The principal uses of varnishes are for interior woodwork, floors, and furniture, and outdoors for buildings, furniture, and boats.

A floor varnish should dry to a tack-free finish within four hours. It must be able to stand scuffing from shoes and moving of furniture, and have excellent adhesion, high gloss, good holdout, and good resistance to the water and alkali used in cleaning. Furniture varnishes should dry within about four hours and must not soften from body heat, otherwise clothing would tend to stick when people sit on furniture such as chairs. They should have good sanding and polishing properties, resistance to water, acids (food), alkalis, alcohol, etc.

Exterior varnishes should have good durability or weather-resistance. They must withstand the elements without failure by cracking, peeling, whitening, spotting, etc. and with minimum loss of gloss for a maximum period of time.

TABLE 22.4 Analysis of a Phenolic Marine Spar Varnish

Lbs		*Gals*
75	*p*-Phenylphenol pure phenolic resin	7.5
30	High m. p. modified phenolic resin	3.3
180	Tung oil	23.1
125	Alkali-refined linseed oil	16.5
328	Mineral spirits	48.0
2	Cobalt napthenate 6 per cent	.25
4.5	Lead napthenate 24 per cent	.5
2.0	Antiskinning agent	.25
743.5	TOTAL	100.4

Cooking directions: heat resins and linseed oil to 575°F; hold for one hour; add tung oil; reheat to 450°F; hold for body, then add drier and antiskinning agent.

Oil length	37.5 gallon
Viscosity	E-F Gardner Holdt
NVM	55 per cent
Wt/gal	7.40 lbs

LACQUERS

A lacquer is a protective coating which dries by evaporation of volatile components.

The film-forming constituent is usually a cellulosic ester, i.e., nitrate, acetate, acetate-butyrate, or other high molecular weight polymer. Other types of lacquers are based on acrylics, polyurethanes, vinyls, etc. There are several differences between varnishes and lacquers. As stated previously, lacquers dry essentially by evaporation, while varnishes dry by a combination of oxidation and polymerization. Lacquers are characterized by very rapid drying and distinctive odor. They are usually based on high molecular weight polymers which require low-boiling solvents of high solvency power such as alcohols, ketones, and esters. The film will redissolve in the original solvent. Lacquers are usually available in a solids range of 20–30 per cent, while varnishes have a solids range of 45–55 per cent.

Lacquers dry "tack-free" in 5 to 15 minutes and to a firm film in 30 minutes to 4 hours. They are usually applied by spraying, brush application often being impractical because of the very rapid drying. Pigmented lacquers predominate in use; clear lacquers are used where colorless, tough films are desired. Clear lacquers are not generally considered as durable as high grade varnishes on exposure to sunlight and moisture.

Nitrocellulose, which is one of the principal film-formers, will be considered in detail.

In addition to nitrated cotton, a nitrocellulose lacquer contains, for example, (1) a solvent mixture which usually contains a ketone, an alcohol, an ester, and frequently an ether-alcohol; (2) a resin such as an alkyd, a phenolic, or ester gum to increase solids and improve adhesion; (3) a plasticizer for flexibilizing the film; (4) an inexpensive volatile diluent such as toluene; and (5) a dye or pigment which is omitted if a clear lacquer is desired.

The main outlets for lacquers are automobile finishes, furniture finishes, metal finishes, and plastic, rubber, paper, and textile finishes.

PRINTING INKS

Printing inks are a mixture of coloring matter dispersed or dissolved in a vehicle or carrier, which forms a fluid or paste which can be printed on a substrate and then dried. The colorants used are generally pigments, toners, and dyes or a combination of these materials.

The vehicle acts as a carrier for the colorant during the printing operation and binds the colorant to the substrate.

Printing inks are applied to many different substrates such as, paper, paperboard, metal sheets, metallic foil, plastic film, molded plastic parts, textiles, and glass.

There are four major classes of printing inks, which vary considerably in physical appearance, composition, method of application, and drying mechanism. They are letterpress and lithographic (litho) inks, commonly called oil inks or paste inks, and flexographic (flexo) and rotogravure (gravure) inks, which are referred to as solvent inks.

Letterpress newspaper inks are based mainly on mineral oil which is sometimes combined with rosin. Drying is by penetration into the absorbent paper stock. Lamp black is used as the pigment for the black ink. Lithographic inks use heat-bodied oils and alkyds as the vehicles. They are heat-dried in ovens.

Flexographic and rotogravure inks are low viscosity materials and contain highly volatile solvents. Flexographic inks require solvents such as alcohols that do not attack the rubber rollers and printing plates. In the case of gravure ink no contact with rubber is involved, and it is permissable to use solvents such as ketones and aromatic hydrocarbons. Flexographic inks use nitrocellulose, polyamides, and acrylates as vehicles.

REFERENCES

Martens, C. R., "Technology of Paints, Varnishes and Lacquers," New York, Van Nostrand Reinhold, 1968.

Fuller, W., "Understanding Paint," *Am. Paint J.* (1965).

Larsen, L. M., "Industrial Printing Inks," New York, Van Nostrand Reinhold, 1962.

"Physical and Chemical Examination of Paints, Varnishes, Lacquers and Colors," 12th Ed., Gardner Laboratories Inc., Bethesda, Maryland, 1962.

Roberts, A. G., "Organic Coatings," NBS-BSS-7, U.S. Department of Commerce, Washington, D.C., 1968.

Roberts, A. G., "Paints and Protective Coatings," Army TM-5-6B, U.S. Government Printing Office, Washington, D.C., 1969.

Martens, C. R., "Emulsion and Water Soluble Paints and Coatings," New York, Van Nostrand Reinhold, 1964.

Horn, M. B., "Acrylic Resins," New York, Van Nostrand Reinhold, 1960.

Martens, C. R., "Alkyd Resins," New York, Van Nostrand Reinhold, 1961.

Blais, J. F., "Amino Resins," New York, Van Nostrand Reinhold, 1959.

Paist, W. P., "Cellulosics," New York, Van Nostrand Reinhold, 1958.

Skeist, I., "Epoxy Resins," New York, Van Nostrand Reinhold, 1958.

Gould, D., "Phenolic Resins," New York, Van Nostrand Reinhold, 1959.

Floyd, D. E., "Polyamid Resins," New York, Van Nostrand Reinhold, 1958.

Dombrow, B., "Polyurethanes," New York, Van Nostrand Reinhold, 1965.

Nuals, R. N., "Silicones," New York, Van Nostrand Reinhold, 1959.

Smith, W., "Vinyl Resins," New York, Van Nostrand Reinhold, 1958.

Marsden, C., and Mann, S. "Solvents Guide," New York, John Wiley & Sons, 1963.

23

Dye Application, Manufacture of Dye Intermediates and Dyes

D. R. Baer*

Dyes are intensely colored substances that can be used to produce a significant degree of coloration when dispersed in or reacted with other materials by a process which, at least temporarily, destroys the crystal structure of the substance. This latter point distinguishes dyes from pigments which are almost always applied in an aggregated or crystalline-insoluble form. Modern dyes are products of synthetic organic chemistry. To be of commercial interest, dyes must have high color intensity and produce dyeings of some permanence. The degree of permanence required varies with the end use of the dyed material.

All molecules absorb energy over various parts of the electromagnetic spectrum. The characteristic of dye molecules is that they absorb radiation strongly in the visible region, which extends from 4000-7000 angstroms. Only organic molecules of considerable complexity which contain extensive conjugation systems linked to electron withdrawing and attracting groups give sufficient absorption (tinctorial value) in the visible region to be useful as dyes. The shade and fastness of a given dye may vary depending on the substrate, due to different interactions of the molecular orbitals of the dye with the substrate, and the ease with which the dye may dissipate its absorbed energy to its environment without itself decomposing.

The primary use for dyes is textile coloration, although substantial quantities are consumed for coloring such diverse materials as leather, paper, plastics, petroleum products, and food.

The manufacture and use of dyes is an important part of modern technology. Because of the variety of materials that must be dyed in a complete spectrum of hues, manufacturers now offer many hundreds of distinctly different dyes. An understanding of the chemistry of these dyes requires that they be classified in some way. From the viewpoint of the dyer, they are best classified according to application method. The dye manufacturer, on the other hand, prefers to classify dyes according to chemical type.

Both the dyer and the dye manufacturer

*Jackson Laboratory, E. I. du Pont de Nemours & Co., Wilmington, Del.

must consider the properties of dyes with relation to the properties of the materials to be dyed. In general, dyes must be selected and applied so that, color excepted, a minimum of change is produced in the properties of the substrate. It is necessary, therefore, to consider the chemistry of textile fibers as a background for an understanding the chemistry of dyes.

The major uses of dyes are in coloration of textile fibers and paper. The substrates can be grouped into two major classes–hydrophobic and hydrophilic. Hydrophilic substances such as cotton, wool, silk, and paper are readily swollen by water making access of the dye to the substrate relatively easy. On the other hand, the ease of penetration also allows easy removal in aqueous systems and special techniques must be used where a high degree of wet-fastness is required.

On the other hand, hydrophobic fibers, such as the synthetic polyesters, acrylics, polyamides, and polyolefin fibers, are not readily swollen by water, hence, higher application temperatures and smaller molecules are generally required.

The polymer chemist has increased the versatility of the newer fibers by incorporating dye sites of a varying nature as needed to achieve dyeability with a predetermined class of dyes. It is now possible to have polyesters, acrylics, and polyamide fibers which can be dyed with positive (basic, cationic), negative (acid, anionic), or neutral (disperse) dyes. These recent developments have allowed the fabric designer to produce materials (textiles, carpets) fabricated in patterns which can be dyed three different colors from one dyebath containing three types of dyes. This concept is called cross dyeing and is becoming increasingly popular as a low-cost method of coloration.

TEXTILE FIBERS

Cotton and Rayon

Cotton and rayon (regenerated cellulose) fibers are composed of cellulose in quite pure form. Cellulose lacks significant acidic or basic properties but has a large number of alcoholic hydroxyl groups. It is hydrolyzed by hot acid and swollen by concentrated alkali. When cotton is swollen by concentrated alkali under tension, so that the fibers cannot shrink lengthwise, it develops a silk-like luster. This process is called mercerization. The affinity of mercerized cotton for dyes is greater than that of untreated cotton.

Cotton and rayon fibers are easily wetted by water and afford ready access to dye molecules. Dyeing may take place by adsorption, occlusion, or reaction with the hydroxyl groups. It is also possible to make cotton and rayon receptive to a variety of dyes by pretreatment or mordanting with a material capable of binding the dyes.

Wool and Silk

Wool and silk fibers are protein substances with both acidic and basic properties. They are destroyed by strong alkali. Strong acid causes hydrolysis, but the process may be controlled to permit dyeing from acidic solutions.

Wool and silk are wetted by water and are dyed with either acid or basic dyes through formation of salt linkages.* They may also be dyed with reactive dyes that form covalent bonds with available amino groups. Mordanting is sometimes used to alter the dyeability of wool and silk.

Cellulose Acetates

Acetylated cellulose fibers differ from cellulose fibers in that they are more hydrophobic and lack large numbers of free hydroxyl groups. The higher the degree of acetylation the more unlike cotton and rayon the acetates become. Strong acid and strong alkali degrade cellulose acetates, although the initial attack is slow under moderate conditions because of the difficulty of wetting the fiber. The triacetate is the most hydrophobic and the most stable.

Dyeing of cellulose acetates is effected with dyes of low water solubility which become dissolved in the fiber, or by occlusion of dyes formed *in situ*. Acid, basic, and reactive dyes cannot be used because of the lack of sites for attachment.

*This refers to ionic bonds, the forces acting between ions of opposite charges.

Polyamides

Polyamide fibers (nylon) are synthetic fibers possessing properties somewhat like those of wool and silk. They are more hydrophobic, however, with only a limited number of basic or acidic groups. Polyamides are degraded by strong acid but may be dyed from acidic dye baths under controlled conditions.

Polyamide fibers are dyeable near the boiling point of water with acid dyes that form salt linkages with basic sites. Dyeing by this means is limited by the availability of these sites. Dyes like those used on cellulose acetates (i.e., that dissolve in the fiber), or reactive dyes that bond to available amino groups may also be used.

Polyesters

Polyester fibers are synthetic fibers unlike any produced in nature. They are hydrophobic and possess good stability to acid and alkali as a result of this hydrophobicity. They are hydrolyzed under sufficiently drastic conditions, however. Some polyester fibers lack functional groups; others are provided with acidic groups or otherwise modified to make them more hydrophilic.

Unmodified polyester fibers are dyed by solution of dyes in the fiber or, to a limited extent, by occlusion of dyes formed *in situ*. Modified polyester fibers may be dyed in these ways or with dyes selected according to the nature of the sites introduced by the modification. Both unmodified and modified polyester fibers must be dyed under vigorous conditions, often with the assistance of a swelling agent to open up the fiber.

Acrylics

Acrylic fibers are hydrophobic synthetic fibers with excellent chemical stability. They do not resemble any natural product. The only functional groups present are those introduced for the purpose of providing sites for dyeing.

Acrylic fibers are dyed by solution of dyes in the fiber, by occlusion of dyes formed *in situ*, and by formation of salt linkages with dyes capable of attachment to sites provided for that purpose. Basic dyes are used on acrylic fibers bearing sulfonic acid groups, for example.

Vinyls

Vinyl polymers and copolymers make up a class of fiber-forming materials that varies greatly in properties, depending on constitution. Some vinyl fibers are very resistant to degradation by acids. Dyes are selected according to the nature of the specific polymer to be dyed.

Polyolefins

Polyolefin fibers are formed from the products of polymerization of unsaturated compounds of carbon and hydrogen, for example, propylene. They do not absorb water and are chemically quite inert. They can be dyed with special disperse dyes but are colored best by introducing a colorant into the polymer before the fibers are spun. Some types of polypropylene incorporate metal ions such as Ni^{++} to act as dye sites for chelatable dyes.

Glass Fibers

Glass fibers are used for special purposes, for example, where flammable materials cannot be tolerated. They are often colored during manufacture, but can be dyed by special techniques which involve the use of surface coatings that have affinity for dyes.

Paper

Paper is a nonwoven material made up primarily from cellulose of varying degrees of refining (see Chapter 15). Paper may be colored in the pulp as a watery fibrous slurry by either continuous or batch methods. The dyeing process takes place at ambient temperature and the dyes are adsorbed on the pulp by their affinity for the cellulose. Direct dyes are most commonly used. In continuous coloration the dye solutions are metered directly into a moving stream of pulp. In batch operations dye is added to a pulper, beater, or blending chest containing a given quantity of slurry.

Paper may also be colored on its surface after the initial sheet is formed, pressed, and partially dried. This can be done at the size press of the paper machine, or color can be carried by a calender roll for heavier sheets. A wide variety of low-cost dyes can be used for surface coloration.

THE PROPERTIES OF DYES

The properties of dyes may be classified as application properties and end-use properties. Application properties include solubility, affinity, and dyeing rate. End-use properties include hue, and fastness to degrading influences such as light, washing, heat (sublimation), and bleaching. Dyes are selected for acceptable end-use properties at minimum expense. Involved application procedures are used only when necessary to achieve unusually good results.

It has become common practice to treat dyed textiles with agents designed to improve resistance to shrinking, wrinkling, and the like. These agents frequently alter the appearance and fastness of dyes. Stability to after-treatments must therefore be considered as an important end-use property of dyes.

The amount of dye required to obtain a light shade is usually about 1 per cent of the weight of the fiber; heavier shades may require as much as 8 per cent. These values are very approximate, since dyes differ in color strength and are usually sold in diluted form. These amounts of dye are not sufficient, in most cases, to markedly affect the properties, other than color, of the fiber. Care must be exercised, however, to apply the dye under conditions that do not cause fiber degradation.

Color strength of a dye can be measured quantitatively as molar absorbitivity which falls within the general ranges given below.

Dye Type	*Molar Absorbtivity*
Anthraquinone	5,000–15,000
Azo	20,000–40,000
Basic	
cyanine	40,000–80,000
triarylmethane	40,000–160,000

It is obvious from the list above that many basic dyes have about 10–20 times the color value per molecule as the anthraquinone types. Unfortunately, light fastness is in the reverse order, the anthraquinones being used where maximum durability to light is needed. The challenge to the dye chemist or engineer is to increase the strength of the light-fast dyes or to increase the fastness of the strongest dyes.

CLASSIFICATION OF DYES

Dyes are classified according to application method, for the convenience of the dyer. The best classification method available is that used in the Colour Index, a publication sponsored by the Society of Dyers and Colourists (England) and the American Association of Textile Chemists and Colorists.

Acid Dyes

Acid dyes depend on the presence of one or more acidic groups for their attachment to textile fibers. These are usually sulfonic acid groups which serve to make the dye soluble in water. An example of this class is Acid Yellow 36* (Metanil Yellow).

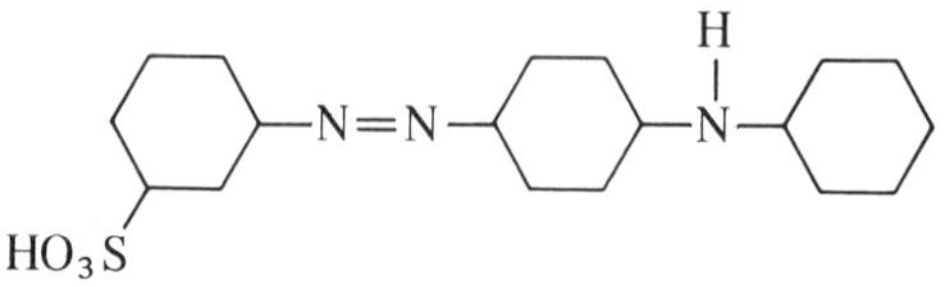

Acid Yellow 36

Acid dyes are used to dye fibers containing basic groups, such as wool, silk, and polyamides. Application is usually made under acidic conditions which cause protonation of the basic groups. The dyeing process may be described as follows:

$$Dye^- + H^+ + Fiber \rightleftarrows Dye^- H^+ \text{—} Fiber$$

It should be noted that this process is reversible. Generally, acid dyes can be removed from fibers by washing. The rate of removal depends on the rate at which the dye can diffuse through the fiber under the conditions of washing. For a given fiber, the diffusion rate is determined by temperature, size and shape of the dye molecules, and the number and kind of linkages formed with the fiber.

Chrome Dyes. A special kind of acid dye used mainly on wool, they possess improved fastness when converted to chromium complexes. A suitable chromium salt is applied to the fiber (1) before the dye, (2) at the same

*The dye names used in this chapter are those given in the "Colour Index."[10] Trivial names, when well established, are given in parentheses following the "Colour Index" name.

time as the dye, or (3) after the dye. All these methods are satisfactory, but more complicated than is desired. In recent years, manufacturers have made available dyes in which chromium is already a part of the molecule. These dyes are simpler to apply than the older types and, as a consequence, are increasing in importance.

Basic or Cationic Dyes

Cationic dyes become attached to fibers by formation of salt linkages with anionic or acidic groups in the fibers. Basic dyes are those which have a basic amino group which is protonated under the acid conditions of the dyebath. Cationic dyes can be divided into the three classes which are illustrated.

Basic Brown 1 (Bismark Brown) is an amino-containing dye which is readily protonated under the pH 2–5 conditions of dyeing.

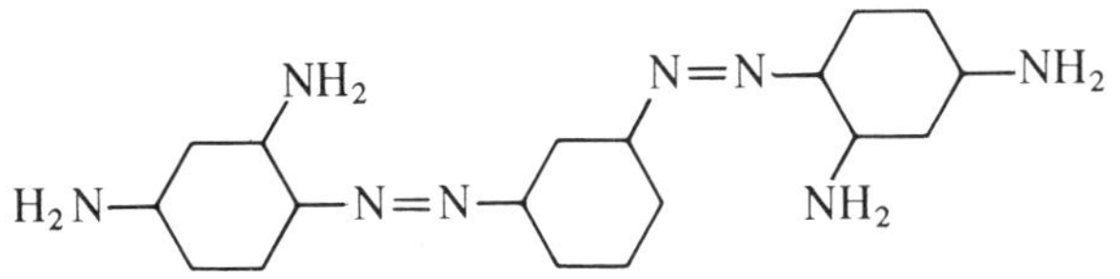

Basic Brown 1

Crystal Violet (Basic Violet 3) is an example of a cationic dye in which the cationic charge is delocalized by resonance and may be present at any one of the basic centers at any time. These resonance forms of almost equivalent energy are one of the reasons that Crystal Violet is among the strongest dyes known. This high color value (tinctorial strength) has important commercial interest in the hectograph copying system. In this system Crystal Violet in a wax base is transferred to the back of a typewritten copy sheet. By using paper moistened with alcohol more than 200 good copies may be made from the master.

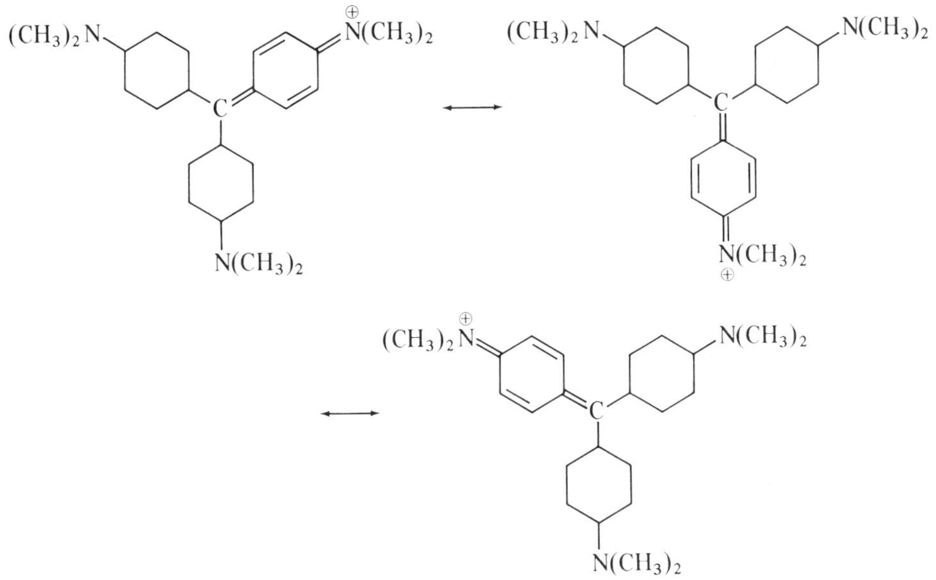

Basic Violet 3

The so-called "classical" basic dyes illustrated above are noted for their high color value, but on the average they also have very poor photostability. When the newer, more

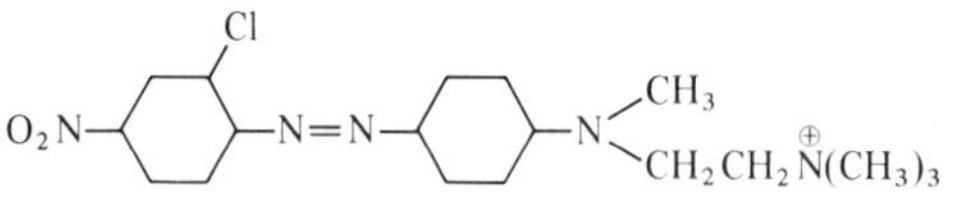

Basic Red 18

durable synthetic fibers emerged having affinity for cationic dyes, it was necessary to develop more light-fast types. These were formed by taking known light-fast azo or anthraquinone chromophores and attaching a cationic "tail" insulated from the chromophore. C.I. Basic Red 18 is a typical example of this pendant cationic dye type.

Further improvements in light-fast dyes for acrylic fibers are the diazo hemicyanines. These are unusual in that they contain a delocalized charged chromophore, which produces tinctorially strong bright shades, but they also have unusually good light fastness. The blue dye below is an example.

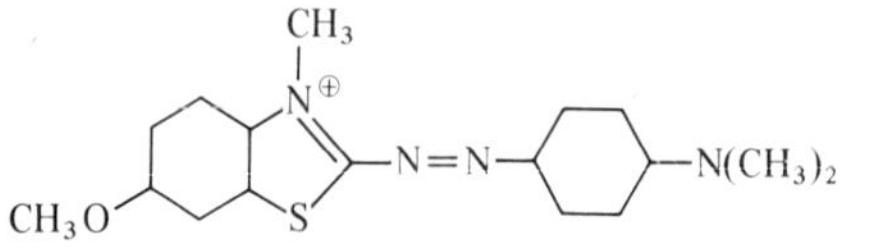

Basic Blue for acrylic fibers

Direct Dyes

Direct dyes are a class of dyes that become strongly adsorbed on cellulose. They usually bear sulfonic acid groups, but are not considered acid dyes since these groups are not used as a means of attachment to the fiber. Direct dyes are large, flat, linear molecules which can enter the water-swollen amorphous regions of cellulose and orient themselves along the crystalline regions. Common salt or Glauber's salt is often used to promote dyeing since the presence of excess sodium ions favors establishment of equilibrium with a minimum of dye remaining in the dye bath. Direct Red 28 (Congo Red) is a typical direct dye.

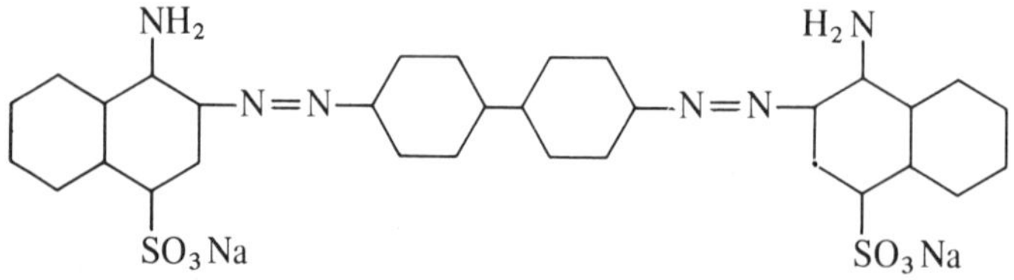

Direct Red 28

Since direct dyes are held in cellulosic fibers by adsorption, the dyeing process is reversible. Unless after-treated with resins and dye-fixing agents, direct dyes, as a class, have poor fastness to washing. They are used mainly because they are economical and easy to apply.

A special type of direct dye having free amino groups is designed to be diazotized and coupled (developed) in the fiber. This operation increases the size of the molecules and improves their fastness to washing. An example of this type is Direct Black 17 (Zambesi Black D).

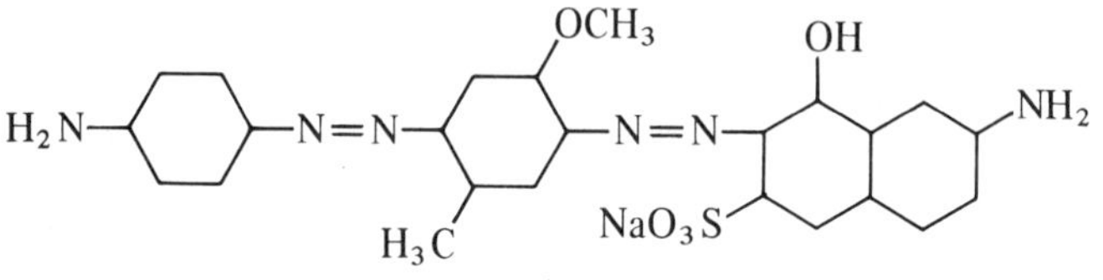

Direct Black 17

These dyes are used primarily to color plain grounds which are later to be printed in a pattern with vat dyes. The sodium hydrosulfite used to reduce the vat dyes destroys the ground color upon contact. Oxidation of the reduced vat dyes then produces two-color effects.

During the application of a "diazotized and developed" dye, the direct dye is first applied in the usual manner. The goods are then passed into a bath containing nitrous acid, where the amino groups are diazotized. Next, the goods are passed through an alkaline solution of beta-naphthol which couples with the diazonium groups to give the final dye.

Sulfur Dyes

Sulfur dyes are insoluble dyes which must be reduced with sodium sulfide before use. In the reduced form they are soluble and exhibit affinity for cellulose. They dye by adsorption, as do the direct dyes, but upon exposure to air they are oxidized to re-form the original insoluble dye inside the fiber. Thus, unlike the direct dyes, they become very resistant to removal by washing.

The exact constitutions of most sulfur dyes are unknown, although the conditions required to reproduce given types are well established. They are fairly cheap and give dyeings of good fastness to washing, as noted above. Their brightness and fastness to bleaching are often inferior, however.

Vat Dyes

Vat dyes, like sulfur dyes, are insoluble. They are reduced with sodium hydrosulfite in strongly alkaline medium to give a soluble form that has affinity for cellulose. This reducing operation was formerly carried out in wooden vats, giving rise to the name vat dye. After the reduced dye has been absorbed in the fiber, the original insoluble dye is re-formed by oxidation with air or chemicals. The dyeings produced in this way are very fast to washing, and, in most cases, the dyes are designed to be fast to light and bleaching as well. An example of a vat due is Vat Blue 4 (Indanthrone).

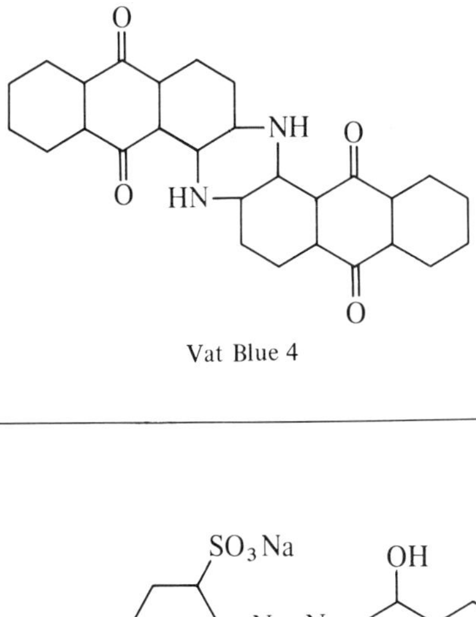

Vat Blue 4

Vat dyes are quite expensive and must be applied with care. They offer excellent fastness when properly selected and are the dyes most often used on cotton fabrics which are to be subjected to severe conditions of washing and bleaching.

It is sometimes impossible to tolerate the strongly alkaline conditions used to reduce vat dyes, for example, when dyeing fibers that are sensitive to alkali. For this reason, and for added convenience, some manufacturers offer soluble vat dyes. These are usually the sodium or potassium salts of the sulfuric esters of reduced vat dyes. When applied to the fiber and subjected to an acid treatment in the presence of an oxidizing agent, they hydrolyze, reverting to the original form of the dye.

Reactive Dyes

Reactive dyes are a relatively new class of dyes that form covalent bonds with fibers possessing hydroxyl or amino groups. One important type of reactive dye contains chlorine atoms that react with hydroxyl groups in cellulose when applied in the presence of alkali. It is believed that an ether linkage is established between the dye and the fiber. An example of this type is the orange azo dye shown below:

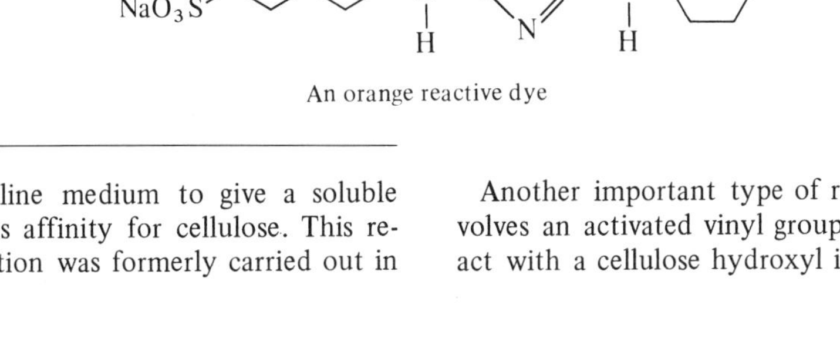

An orange reactive dye

Another important type of reactive dye involves an activated vinyl group which can react with a cellulose hydroxyl in the presence of a base acording to the following scheme:

$$\mathit{Dye}{-}SO_2CH_2CH_2OSO_3Na \xrightarrow{OH^-}$$

$$\mathit{Dye}{-}SO_2CH{=}CH_2 + HO{-}\mathit{Cellulose}$$

$$\downarrow$$

$$\mathit{Dye}{-}SO_2CH_2CH_2O{-}\mathit{Cellulose}$$

Reactive dyes offer excellent fastness to washing since the dye becomes a part of the fiber. The other properties depend on the structure of the colored part of the molecule and the means by which it is attached to the reactive part.

Disperse Dyes

Disperse dyes are nonionic dyes having low water solubility which are capable of dissolving in certain synthetic fibers. Their fiber attraction is the formation of a solid solution, since, being uncharged, there is no driving force to form salt linkages. Disperse dyes are now primarily used for polyester fibers although they were originally developed for cellulose acetate and polyamide fibers. In the decade 1960-1970 the production of disperse dyes increased by a factor of four, reflecting the tremendous growth in polyesters. This growth is expected to continue through the 1970's as polyesters achieve a larger share of the textile markets.

Disperse dyes like all other dyes go through a monomolecular form during the dyeing process. This means that when applied from an aqueous bath they must have a finite solubility under the dyeing conditions. This limits the size and polarity of the molecules used. The dyeing mechanisms involve the equilibrium processes shown below, with the net result being that the dye goes from solid state to solution in the fiber.

$$\underset{\text{(solid)}}{\mathit{Dye}} \leftrightarrows \underset{\text{(solution in dyebath)}}{\mathit{Dye}} \leftrightarrows \underset{\text{(adsorbed on fiber surface)}}{\mathit{Dye}} \leftrightarrows \underset{\text{(in fiber)}}{\mathit{Dye}}$$

Since the dye chemists design molecules which are in the lowest energy system where the dye is in the fiber, the dye will gradually move from solid phase to solution in the fiber.

This imposes a unique technological problem for the manufacture of disperse dyes. The rate of solution of the solid particle in the dyebath is a function of its particle size. In practice, commercial disperse dyes must be ground down or milled to particle sizes of $<5\ \mu$ and preferably <1 micron. The steps required to convert a disperse dye, after chemical synthesis, to a salable product may as much as double the cost of a commercial dye.

Disperse dyes may be applied by a dry heat (Thermosol) process to polyester fibers. In this case the dye achieves molecular form by sublimation (vaporization) from the solid dye to the fiber surface. Extremely small particle size is also important for this process.

Disperse dye molecules are generally small and have some hydroxyl or amino groups to give finite water solubility at dyeing temperatures. Examples of disperse dyes for polyester are shown below.

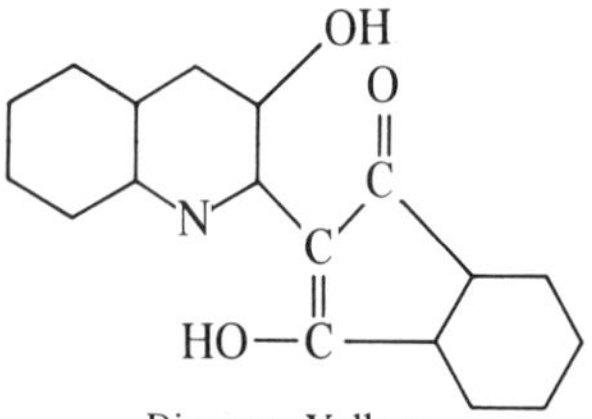

Disperse Yellow (Quinophthalone)

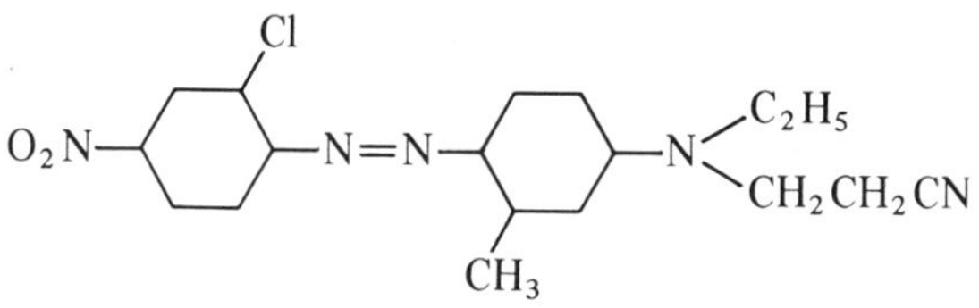

Disperse Red (Azo)

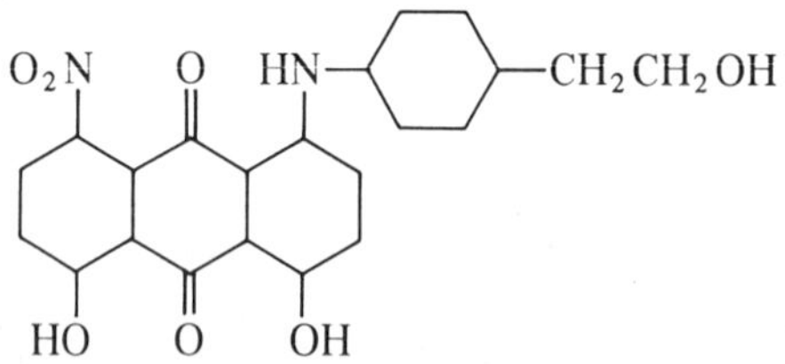

Disperse Blue (Anthraquinone)

Since disperse dyes become dissolved in textile fibers, the dyeing process is reversible. Fibers that are swollen easily by water can be dyed with disperse dyes under moderate conditions, but the dyeings have only modest fastness to washing. Fibers that are more difficult to swell must be dyed under more drastic conditions, but offer the advantage that these conditions are not duplicated during normal washing procedures. For example, most polyester fibers must be dyed under pressure or with the use of organic swelling agents. The washing fastness of disperse dyes on these fibers is excellent.

Mordant Dyes

Mordant dyes require a pretreatment of the fiber with a mordant material designed to bind the dye. The mordant becomes attached to the fiber and then combines with the dye to form an insoluble complex called a "lake." An example of a mordant is aluminum hydroxide that has been precipitated in cotton fiber. This mordant is capable of binding such dyes as Mordant Red 11 (Alizarin) by formation of an aluminum "lake."

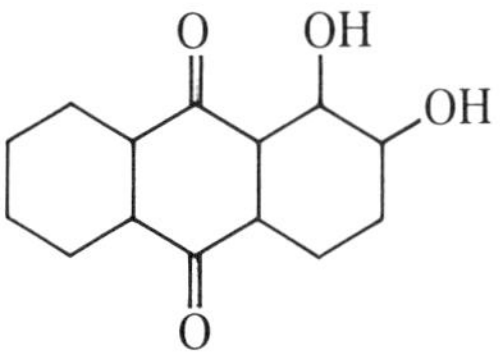

Mordant Red 11

Mordant dyes have declined in importance mainly because their use is no longer necessary. Equal or superior results can be obtained with other classes of dyes at less expense in time and labor.

Azoic Dyes

Azoic dyes are produced inside textile fibers, usually cotton, by azo coupling. The dye is firmly occluded and is fast to washing. A variety of hues can be obtained by proper choice of diazo and coupling components. For example, a bluish red is produced from diazotized Azoic Diazo Component 1 and Azoic Coupling Component 2 (Naphthol AS).

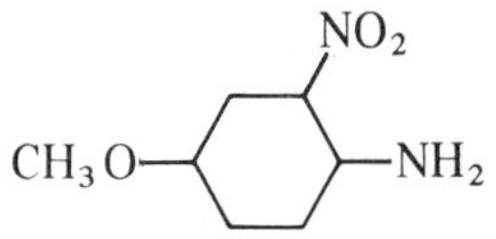

Azoic Diazo Component 1

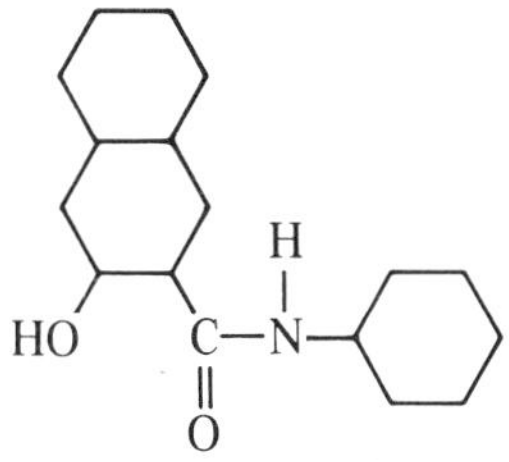

Azoic Coupling Component 2

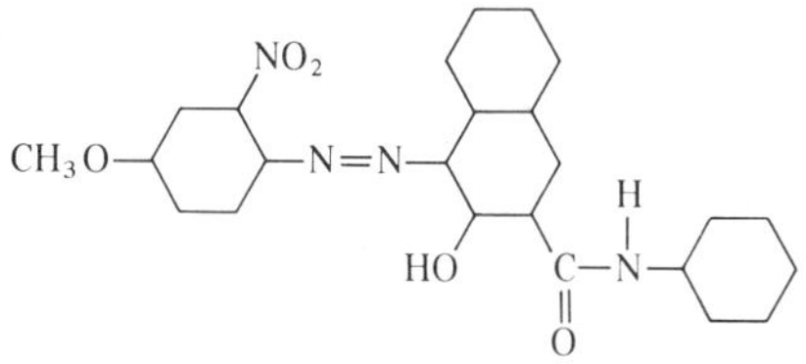

Bluish Red Azoic Dye

In the usual procedure for the development of azoic dyes, the fiber is first impregnated with an alkaline solution of the coupling component and then treated with a solution of the diazonium compound. Finally, the dyed goods are soaped and rinsed.

The diazonium compound may be produced in the dyehouse by diazotization of the azoic diazo component, or it may be purchased as a stabilized complex ready for use. Examples of stabilized diazonium complexes are the zinc double salts, nitrosamines and diazoamino compounds.

Special techniques have been developed for the use of azoic dyes on synthetic fibers. It is sometimes possible to apply both the diazo component and the coupling component simultaneously from aqueous dispersion and then to treat the goods with nitrous acid to produce the color.

Oxidation Dyes

Oxidation dyes are produced in textile fibers by oxidation of a colorless compound. For example, aniline may be oxidized in cotton with sodium bichromate in the presence of a metal catalyst to produce an aniline black. This is an economical way to produce full black shades. The appearance and fastness of the dyeings may be varied over a wide range by the choice of oxidant, conditions, and catalyst. The exact structures of the aniline blacks are not known.

Ingrain Dyes

Ingrain dyes are produced inside textile fibers. The azoic and oxidation dyes already discussed are examples. In addition, there is a small number of ingrain dyes on the market which do not fall into these classes. One such group, called precursors, is capable of generating the very bright blue dye copper phthalocyanine inside cotton fibers. The dyeings are extremely fast to light and washing. The structure of copper phthalocyanine is given elsewhere in this chapter.

THE APPLICATION OF DYES

The process of dyeing may be carried out in batches or on a continuous basis. The fiber may be dyed as stock, yarn, or fabric. No matter how the dyeing is done, the process is always fundamentally the same: dye must be transferred from a bath, usually aqueous, to the inside of the fiber. The basic operations of dyeing include: (1) preparation of the fiber; (2) preparation of the dye bath; (3) application of the dye; and (4) finishing. There are many variations of these operations, depending on the kind of dye. The dyeing process is complicated by the fact that single dyes are seldom used. The matching of a specified shade may require from two to a dozen dyes.

Fiber Preparation

Fiber preparation ordinarily involves scouring to remove foreign materials and ensure even access to dye liquor. Some natural fibers are contaminated with fatty materials and dirt. Synthetic fibers may have been treated with spinning lubricants or sizing which must be removed. Some fibers may also require bleaching before they are ready for use.

Dye Bath Preparation

Preparation of the dye bath may involve simply dissolving the dye in water, or it may be necessary to carry out more involved operations such as reducing the vat dyes. Wetting agents, salts, "carriers," retarders, and other dyeing assistants may also be added. "Carriers" are swelling agents which improve the dyeing rate of very hydrophobic fibers such as the polyesters. Examples are *o*-phenylphenol and biphenyl. Retarders are colorless substances that compete with dyes for dye sites or complex the dye in the bath and act to slow the dyeing rate. Their use is necessary when too rapid dyeing tends to cause unevenness in the dyeings.

Dye Application

During application, dye must be transferred from the bath to the fiber and allowed to penetrate. In the simplest cases this is done by immersing the fiber in the bath for a prescribed period of time at a suitable temperature. Unless the dye is unstable, the bath is usually heated to increase the rate of dyeing. To ensure an even uptake of dye, it is desirable to stir the bath. This is done with paddles, by pumping, or by moving the fiber. For continuous dyeing of fabrics, the dye liquor is picked up from a shallow pan, the excess is squeezed out by rollers, and penetration is assured by steaming or heating in air.

Finishing

The finishing steps for many dyes, such as the direct dyes, are very simple. The dyed material is merely rinsed and dried. Vat-dyed materials, on the other hand, must be rinsed to remove reducing agent, oxidized, rinsed again, and soaped before the final rinsing and drying steps are carried out. Generally, the finishing steps must fix the color (if this has not occurred during application) and remove any loose dye from the surface of the fiber. Residual dyeing assistants such as "carriers" must also be removed.

DYEING EQUIPMENT

The equipment used in modern dyehouses varies with the type of dyeing to be done. Stock dyeing is often carried out in large heated kettles made of stainless steel or other corrosion-resistant metal. These kettles can be sealed and used for dyeing at temperatures

Fig. 23.1 Jet dyeing machine. (*Courtesy Gaston County Dyeing Machine Co.*)

somewhat above the boiling point of water at atmospheric pressure. Yarns in packages are dyed in closed machines that circulate hot dye liquor through them. Fabrics are dyed in machines that move them through the dye liquor either under tension (jig) or relaxed (beck).

The newly developed pressure-jet dyeing machine is unique in that it has no moving parts. It is illustrated in the photograph Fig. 23.1, and it is illustrated schematically in Fig. 23.2. The cloth, in rope form, is introduced into a unidirectional liquid stream enclosed in a pipe. Liquor is pumped through a specially-designed Venturi jet imparting a driving force which moves the fabric. The two fabric ends are sewn together to form a continuous loop. The jet dyeing machine is becoming very popular for dyeing knit goods due to the absence of reels or drives which might chafe the fabric or tangle strands.

Fabrics can also be dyed in full width by

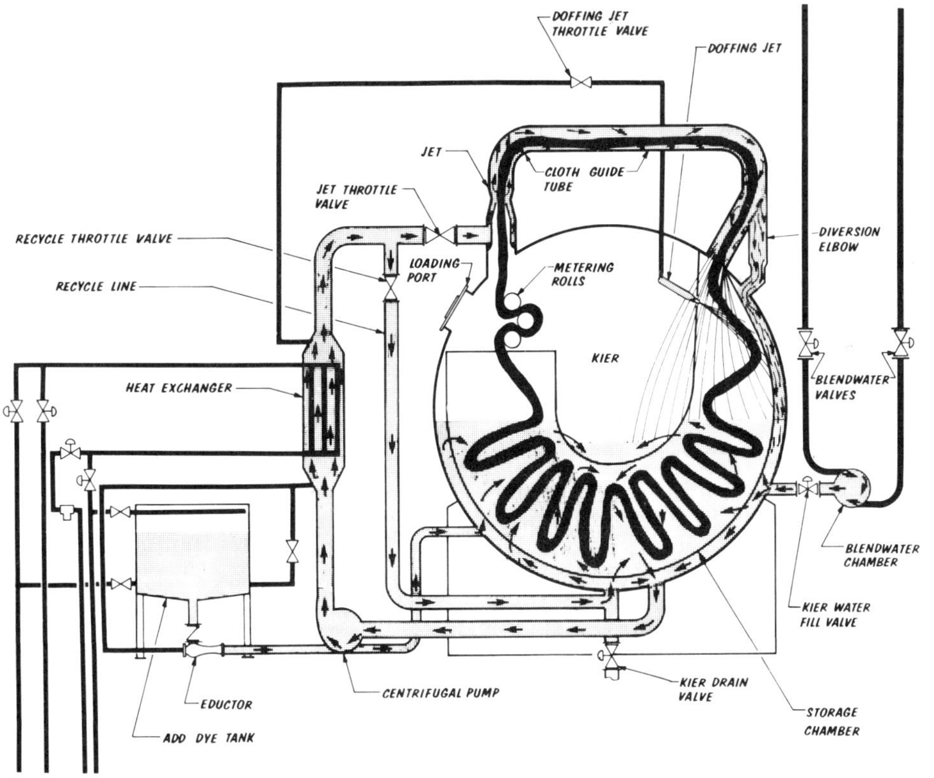

Fig. 23.2 Schematic diagram of jet dyeing machine. (*Courtesy of Gaston County Dyeing Machine Co.*)

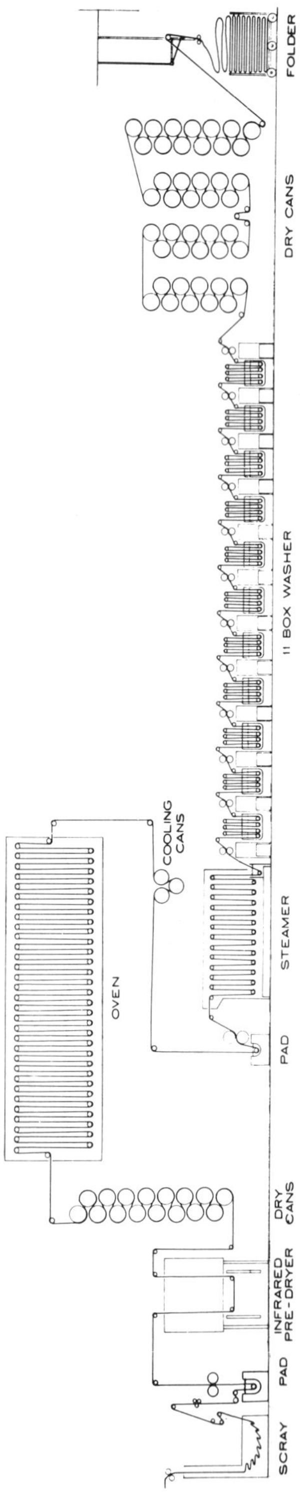

Fig. 23.3 Combination thermosol pad-steam dyeing system for continuous dyeing of polyester-cotton blends. *(Courtesy E.I. du Pont de Nemours and Co., Inc.)*

winding them on a perforated beam through which hot dye liquor is pumped. This is the principle of the Burlington beam dyeing machine.

Some of the most important developments over the past 35 years have been the evolution of continuous processes having high color reproducibility, high throughput, and relatively low labor costs. Batch dyeing of standard fabrics involves lengths of about 1,000 yards depending on weight, with a dyeing cycle of 4–8 hours. This contrasts with a modern continuous dyeing machine capable of dyeing at rates up to 100 yards/minute. One such machine is capable of producing 25 million yards of dyed cloth per year, while still leaving sufficient downtime for cleanup and repair.

The first volume-yardage continuous process was the continuous pad-steam process for vat dyes on cotton.[9] The vat-dye dispersion was was padded onto the cloth, dried, then passed through a reducing bath, steamed for 30 seconds, followed by an oxidizing bath, and then washing. When it was discovered that disperse dyes could be thermosoled into polyesters by dry heat for 60 seconds at 400°F, this procedure was readily adapted to continuous processing. The advent of large volumes of dyed polyester-cotton-blend fabrics in the late 1960's was made possible by combining these two processes into one thermosol pad-steam system. This is illustrated schematically in Fig. 23.3. Over one billion yards of cloth were dyed by this process in the United States in 1970.

Recently several machines have been developed in Europe which enable continuous dyeing of carpet at a rate of 10–15 yards/minute in 15-foot widths. The problems are much different than in the case of textiles due to the varying pile height, heterogeneity of the substrate, and the aesthetics of the final product. Nonetheless, continuous carpet dyeing was initiated on a commercial scale in the United States in 1969. Most of the mechanical engineering problems have been solved and a spectacular growth is forecast.[9]

In addition to the above, special machines are available to dye textile fibers at any stage during their conversion to finished goods. The principles involved are the same as those outlined in the preceding section "Application of Dyes."

PRINTING

Printing is a special kind of localized dyeing which produces patterns. Four kinds of printing are recognized: (1) direct, (2) dyed, (3) discharge, and (4) resist. In *direct* printing, a thickened paste of the dye is printed on the fabric to produce a pattern. The fabric is then steamed to fix the dye, and it is finished by washing and drying. *Dyed* printing requires that the pattern be printed on the fabric with a mordant. The entire piece is then placed in a dyebath containing a mordant dye, but only the mordanted areas are dyeable. Washing then clears the dye from the unmordanted areas, leaving the pattern in color.

In *discharge* printing, the cloth is dyed all over and then printed with a substance which can destroy the dye by oxidation or reduction, leaving the pattern in white. When a reducing agent such as sodium hydrosulfite is used to destroy the dye, the paste may contain a reduced vat dye. Finishing the goods by oxidation and soaping then produces the pattern in color. In *resist* printing, certain colorless substances are printed on the fabric. The whole piece is then dyed but the dye is repelled from the printed areas, thus producing a colored ground with the pattern in white.

Printing is most often done with copper rollers etched in the design to be printed. Printing paste is fed constantly to the roller from a trough. A scraper then clears the surface of the roller, leaving the dye paste only in the etched areas from which it is transferred to the fabrics.

PIGMENT DYEING AND PRINTING

Pigment dyeing and printing are processes that compete with the more conventional means of dyeing and printing described above. These processes use water-insoluble dyes or pigments which are bound to the surfaces of fabrics with resins. A paste or emulsion containing pigment and resin or resin-former is applied to the fabric. The goods are then dried and cured by heat to produce the finished dyeing or print. During the heating or cure, fabric, resin, and pigment become firmly bonded together. This method of color application is economical and produces good results. It should be noted that the pigment is confined to the surface of the fabric and can be selected without regard for fiber affinity.

NONTEXTILE USES OF DYES

Substantial quantities of dye are consumed by the paper industry. The most important paper dyes are selected from the direct, acid, and basic groups of textile dyes. The needs of the paper industry are such that it is frequently necessary to select and finish the dyes in a manner different from that employed for textiles.

Natural leather is a protein substance and can be dyed with acid dyes, among others. Finished leather also can be dyed or pigmented, the choice of color type depending on the nature of the finish.

Dyes for many materials such as gasoline, lacquers, inks, and varnishes are selected on the basis of solubility, as well as hue and fastness. It is necessary to modify some dyes to achieve the desired degree of solubility. For example, the solubility in alcohol of many acid dyes is greatly increased by forming the diphenylguanidine salts.

Important and growing markets for dyes are the plastic and metal industries. Dyed anodized aluminum, in particular, is finding increasing use in automotive and architectural applications.

Dyes for foods, drugs, and cosmetics must be chosen with great care to avoid toxic effects. In the United States, dyes are certified by the government when judged safe for such uses. Only the certified dyes may legally be used. This restriction does not apply to natural colors.

PRODUCTION AND USES

Table 23.1 shows the distribution of the free world's production of dyes by country.[11] In 1969 the United States and Germany were the two largest manufacturers, with Japan, Swit-

TABLE 23.1 Free World Production of Dyes by Country, 1968[a]

Country	*Metric Tons*	*Million Dollars*
Germany	98,912	361.2
Belgium	2,040	5.6
Spain	12,120	12.1
France	23,311	69.3
Italy	11,000	23.2
Netherlands	2,832	5.3
Portugal	251	0.3
United Kingdom	43,200	139.1
Switzerland	28,000	154.1
Japan	66,154	134.3
United States	113,000*	474.6

[a]"The Chemical Industry" 1969-1970, Organization for Economic Cooperation and Development, page 125. *1968.

zerland, and the United Kingdom about equal for third place in dollar value.

Table 23.2 gives the production of synthetic dyes in the United States according to application class. These figures are taken from the report of the U.S. Tariff Commission.

RAW MATERIALS FOR THE MANUFACTURE OF DYES

The raw materials for the manufacture of dyes are mainly aromatic hydrocarbons, such as benzene, toluene, naphthalene, anthracene, pyrene, phenol, pyridine, and carbazole. In the past, these aromatic hydrocarbons came almost exclusively from the distillation of coal tar, but in recent years increasing quantities, especially of benzene and toluene, have become available from petroleum and natural gas. The term "coal-tar dyes," still widely considered synonymous with synthetic dyes, is no longer an entirely correct description. A great variety of inorganic materials are required by the dye industry. These include sulfuric acid, oleum, nitric acid, chlorine, bromine, caustic soda, sodium nitrite, hydrochloric acid, sodium carbonate, sodium hydrosulfite, sodium sulfide, aluminum chloride, sodium bichromate, and manganese dioxide.

DYE INTERMEDIATES

The raw materials for dyes are almost never directly useful in dye synthesis. It is necessary to convert them to a variety of derivatives which are in turn made into dyes. These derivatives are called *intermediates.* They are produced by reactions such as nitration, reduction, sulfonation, halogenation, oxidation, and condensation. Most of these reactions lead to the formation of substituted hydrocarbons which are functional in nature, that is, they bear groups capable of undergoing further chemical reaction. The number of dye intermediates actually or potentially available is very large, and the technology of their manufacture is an important part of industrial

TABLE 23.2 Benzenoid Dyes: U.S. Sales, by Class of Application, 1971

Class of Application	*Quantity (million pounds)*	*Value (million dollars)*
Acid	24.1	60.5
Azoic dyes and components	7.8	12.3
Basic	15.5	44.2
Direct	34.1	58.0
Disperse	31.1	89.8
Fiber-reactive	3.5	12.5
Fluorescent brightening agents	27.2	41.6
Food, drug, and cosmetic colors	4.0	15.3
Mordant	1.3	2.2
Solvent	10.3	18.3
Vat	51.6	53.9
All other	18.9	13.4
TOTAL	229.5	422.6

organic chemistry. Intermediates are used not only for dye manufacture but also for the manufacture of other important products such as pharmaceuticals.

The substitution reactions of aromatic hydrocarbons, in which a hydrogen atom is replaced by another group, frequently lead to the formation of position isomers. Position isomers are compounds which are alike in the groups that they contain, but different in the relative positions of these groups, e.g., the two nitrophenols shown in the section on nitration. When position isomers are formed, it is almost always necessary to effect a separation since the isomers differ in properties and lead to different dyes. One of the most difficult problems in the manufacture of dye intermediates is the efficient separation of isomers. A further problem is the control of the relative quantities in which isomers are formed, or the discovery of uses for all of them in the event that control is impractical.

The many reactions employed in the manufacture of dye intermediates, and the delicate nature of many of the operations make it mandatory that all processes be well-planned and controlled. Some of the processes, nitration for example, are inherently dangerous if not properly run. For these reasons, close supervision by technically trained men is the rule in plants where intermediates are made.

In recent years there has been an ever increasing emphasis on high standards on uniformity and quality in chemical manufacturing. At the same time, the increasing cost of labor has discouraged the use of numerous manual controls to meet these standards. Consequently, many plants have turned to partly or fully automatic control. Such practice requires advanced instrumentation which can insure that all process variables are closely controlled.

Because of the large number of compounds that are required, often in limited amounts, most dye intermediates are manufactured in batches. Some of the more fundamental processes can be run continuously, however, with a decided economic advantage. Where continuous production is not justified, the largest practical batches are made to hold the costs of operation to a minimum.

A wide variety of batch-type equipment is used for the manufacture of dye intermediates. Reaction kettles are made from wood, cast iron, stainless steel, or steel, often lined with rubber, glass (enamel), brick, or other corrosion-resistant material. Usual production sizes are 500–10,000 gallons, and the kettles are equipped with mechanical agitators, thermometers, condensers, and cooling or heating coils or jackets, depending on the nature of the operation. Products are generally transferred by gravity flow, pumping, or blowing. Isolation is generally by plate and frame filter presses, filter nutsches, centrifugation, or continuous rotary filters. Both pressure and vacuum filters are used. Drying of dyes and intermediates is done in air or vacuum tray dryers or graining bowls. For larger volumes spray dryers are becoming increasingly important.

Nitration

The nitration of aromatic hydrocarbons is a fundamental operation in the manufacture of many dye intermediates. Nitration involves the replacement of one or more ring hydrogen atoms by the nitro ($—NO_2$) group. The nitration of benzene is an example.

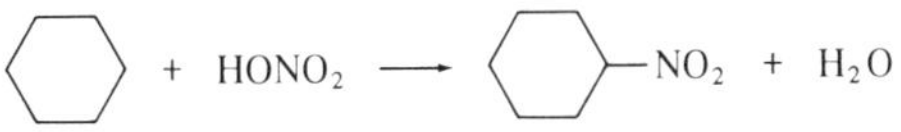

Nitrobenzene

Nitration of compounds of high reactivity is carried out with nitric acid in water or an organic solvent. Less reactive compounds are nitrated in a combination of nitric and sulfuric acids ("mixed acids"). The sulfuric acid serves as a solvent for the reaction and facilitates nitration by reaction with nitric acid to form the nitronium ion ($—NO_2^+$), generally believed to be the active nitrating agent. It also serves to maintain the strength of the nitrating mixture by combining with the water which is formed. A typical mixed acid consists of 33 per cent nitric acid, and 67 per cent sulfuric acid. Since nitric acid is a strong oxidizing agent, nitrations are carried out at low temperatures to avoid destructive side reactions. Some functional groups, such as the amino group, in compounds to be nitrated must be protected against oxidative destruction by acetylation. The acetyl group is removed by hydrolysis when nitration is complete.

When nitrating benzene to nitrobenzene, an excess of nitric acid must be avoided or dinitration will occur. The procedure is to run the "mixed acids" into the benzene, not the converse; this is a general rule when mononitration is desired. About 2,500 pounds of benzene are nitrated in one batch over a period of three to four hours. Heat is evolved during the reaction and is removed by means of cooling coils, or by means of cold water or brine circulated in a jacket surrounding the nitration vessel. Cast iron is used since it is not attacked by "mixed acids." The vessel is agitated to provide good contact between the two layers and to facilitate heat transfer. When the reaction is complete, the agitation is stopped and the nitrobenzene separates as an oil over the acid. This oil is drawn off and agitated with water or dilute alkali to remove residual acid. It may then be distilled if pure nitrobenzene is required. In recent years, continuous vapor-phase nitration methods have been devised. Aqueous nitric acid is used and the water resulting from the reaction is distilled off continuously to keep the nitric acid concentration high enough to be effective. Benzene can be nitrated continuously at about 80°C using 61 per cent aqueous nitric acid.

When benzene derivatives are nitrated, isomers of the desired product are obtained in addition to the product itself. For example, nitration of phenol by nitric acid gives *o*- and *p*-nitrophenol.

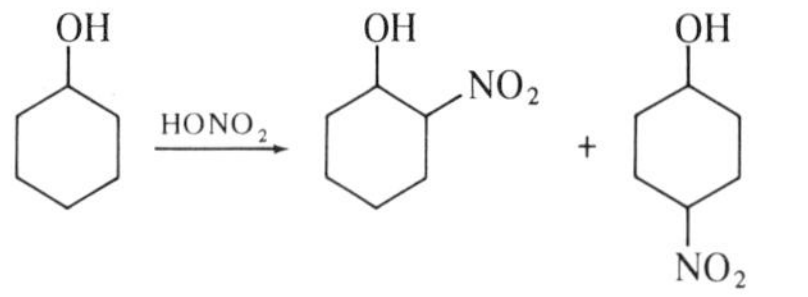

Phenol *o*-Nitrophenol *p*-Nitrophenol

The isomeric *o*- and *p*-nitrophenols can be separated by steam distillation. The *ortho* isomer is volatile in steam, while the *para* isomer is not and, therefore, remains in the distillation vessel. Commercially, *p*- and *o*-nitrophenols are made by hydrolysis of the respective chloronitrobenzenes.

Mononitration of anthraquinone at about 50°C gives mainly 1-nitroanthraquinone. At 80–95°C dinitration occurs to give a mixture of the 1,5- and 1,8-isomers. These isomers are important as starting materials for the preparation of other intermediates. In some cases the mixture is used; in others, isomer separation is necessary.

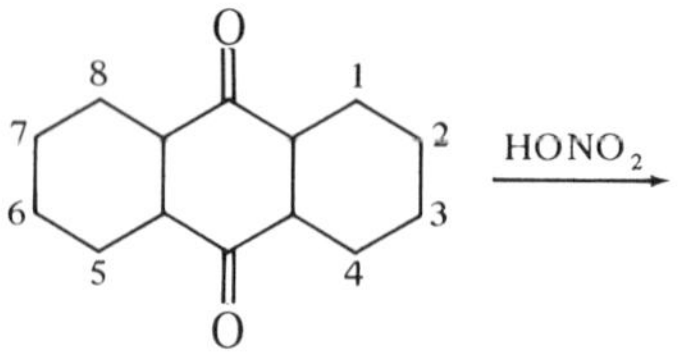

Anthraquinone

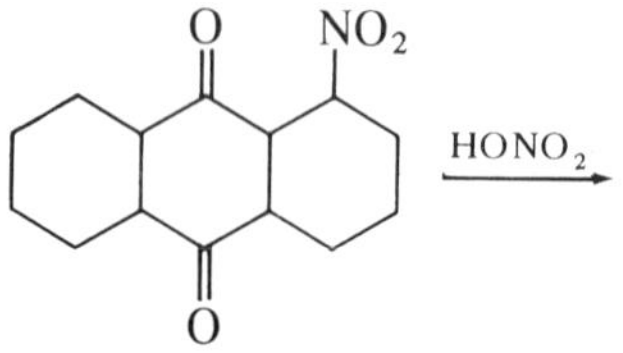

1-Nitroanthraquinone

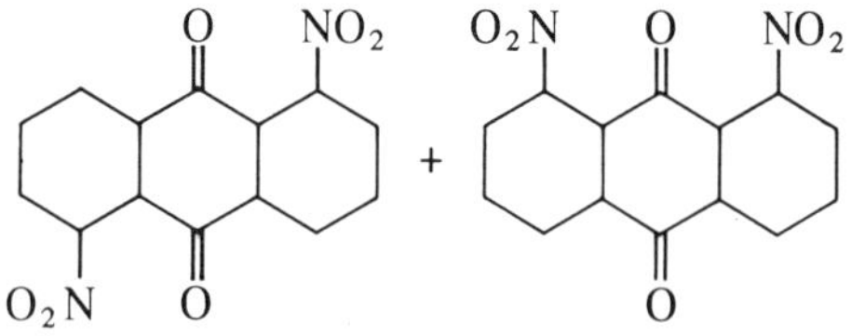

1,5-Dinitro-anthraquinone 1,8-Dinitro-anthraquinone

These nitrations are performed in cast-iron or steel vessels with steel agitators. Since the starting materials are solids they are first dissolved in sulfuric acid and then treated with "mixed acids;" the products are also solids.

Reduction

The most common reduction reaction in the manufacture of dye intermediates is the conversion of a nitro compound to the corresponding amine. This reaction is illustrated by the reduction of nitrobenzene to aniline.

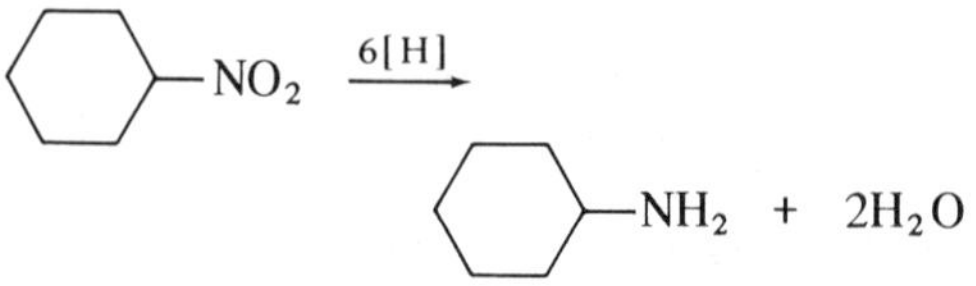

Reduction of nitro compounds is accomplished by: (1) catalytic hydrogenation, (2) iron reduction, (3) sulfide reduction, or (4) zinc reduction in alkaline medium. Generally, where the reaction is carried out on a large scale, the catalytic procedure is best. For small-scale batch operations, chemical reduction may be preferred.

Catalytic hydrogenation requires a catalyst such as nickel, copper, platinum, molybdenum, or tungsten. These catalysts are usually supported on other materials and are especially prepared for the type of reduction to be carried out. Reduction conditions vary widely, depending on the nature of the nitro compound and the catalyst. Reduction may be carried out in solvent, in the vapor phase, or in the liquid phase. Aniline can be made by continuous vapor-phase reduction of nitrobenzene at 350–460°C at nearly atmospheric pressure. Some reductions, on the other hand, are run at 1000 to 4000 psi.

Iron reduction is employed on a large scale because of its simplicity. Iron turnings are used in an agitated aqueous system containing a small amount of acid to promote reaction. The over-all reaction is illustrated for nitrobenzene as follows:

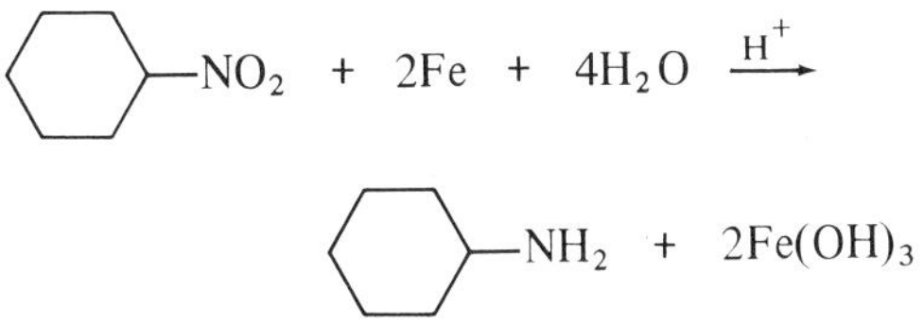

The nitrobenzene is placed in a reducer, a vertical cylindrical vessel provided with cover, steam jacket, and agitator. The iron turnings, or powder, and a small amount of hydrochloric acid are added in small portions. A brisk reaction is maintained by means of steam circulated in the jacket of the reducer or blown directly into the charge. A condenser returns to the reducer any vapors that escape. After the nitrobenzene is completely converted to aniline, a strong current of live steam is passed into the charge; a mixture of steam and aniline vapors passes to the condenser and is collected in storage tanks. The bulk of the aniline separates as a lower layer and is drawn off; the water over it still contains aniline, which must be recovered by distilling this "aniline water" again, or by extracting it with nitrobenzene. The iron sludge is washed out of the reducer through a side outlet by flushing. A reducer 6 feet in diameter and 10 feet high takes a charge of 5000 pounds of nitrobenzene in one batch and requires about 10 hours for reduction. The aniline may be redistilled, which renders it water white.

Iron reduction of nitro compounds is being de-emphasized in favor of catalytic reduction which is more efficient in a labor intensive industry such as dyes.

Sulfide reduction employs sodium sulfide, sodium polysulfide, or sodium hydrosulfide. An important feature of this type of reducing system is its adaptability to bring about stepwise reduction of dinitro compounds. Partial reduction is illustrated with *m*-dinitrobenzene which can be reduced to *m*-nitroaniline with sodium sulfide under controlled conditions.

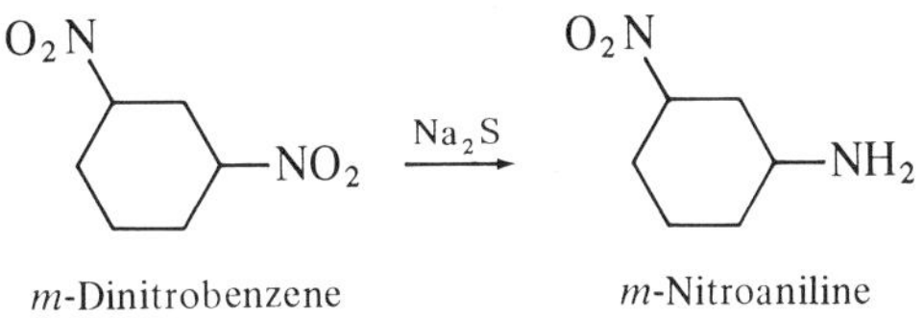

m-Dinitrobenzene *m*-Nitroaniline

The sodium sulfide is dissolved in alcohol and placed in a steam-jacketed reducer; the dinitrobenzene is added either in solid form or dissolved in alcohol. The mixture is boiled for two hours; then the alcohol is distilled off and collected for re-use. The *m*-nitroaniline mixed with inorganic salt remains in the reducer. The mass is agitated with water, which dissolves the salt, and is then pumped into a filter press. The press cake of *m*-nitroaniline is washed, and then discharged and dried in a vacuum drier.

Zinc reduction in alkaline aqueous or alcoholic medium is especially useful to bring about bimolecular reduction. This kind of reaction is illustrated by the conversion of nitrobenzene to hydrazobenzene. Rearrangement of hydrazobenzene with acid gives benzidine, a formerly important intermediate for azo dyes which is now banned due to its carcinogenic activity.

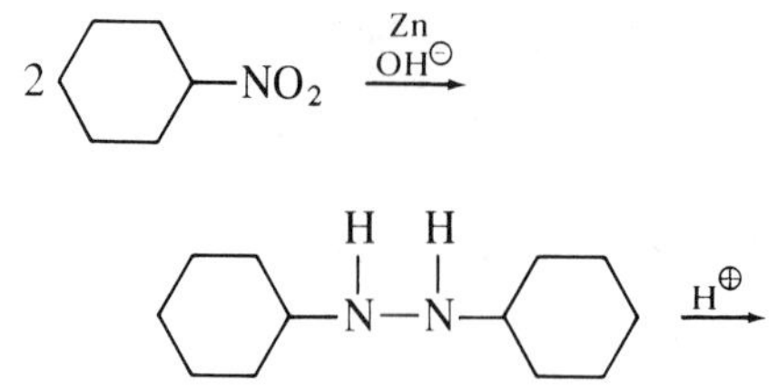

Hydrazobenzene

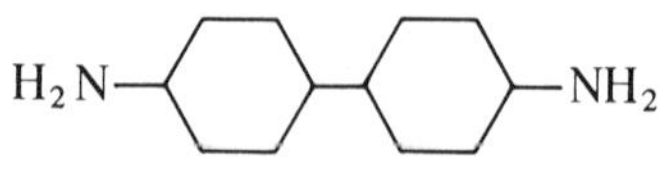

Benzidine

In a similar way, *o*-nitroanisole is converted to hydrazoanisole and then to *o*-dianisidine.

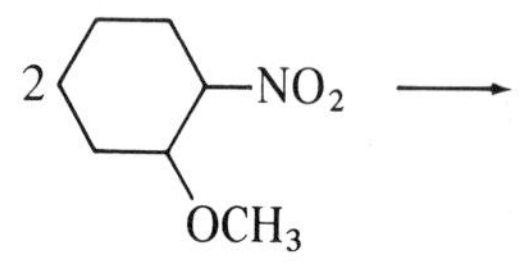

o-Nitroanisole

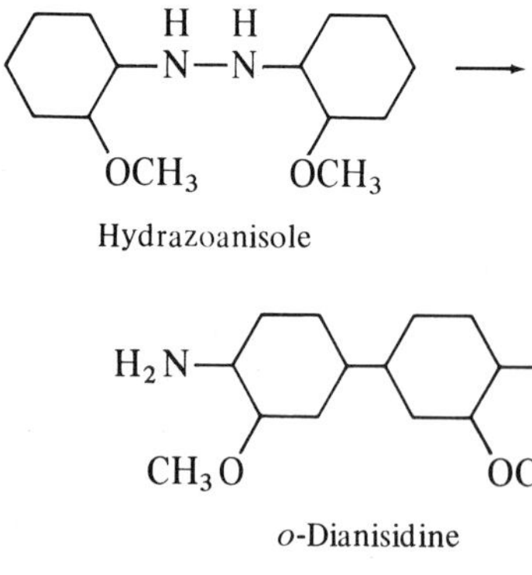

Hydrazoanisole

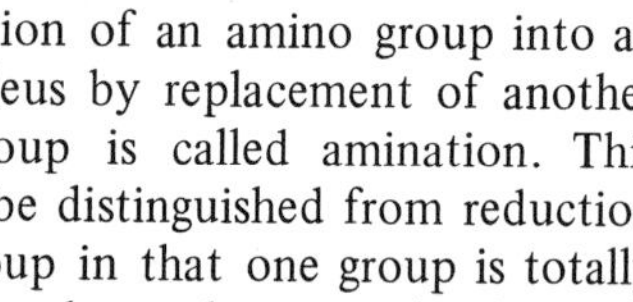

o-Dianisidine

Amination

The introduction of an amino group into an aromatic nucleus by replacement of another functional group is called amination. This process is to be distinguished from reduction of a nitro group in that one group is totally displaced by another and not simply altered in character.

An example of amination in the benzene series is the conversion of *p*-nitrochlorobenzene to *p*-nitroaniline with ammonia. This reaction may be carried out continuously with 40 per cent aqueous ammonia under 200 atmospheres pressure at 235–240°C.

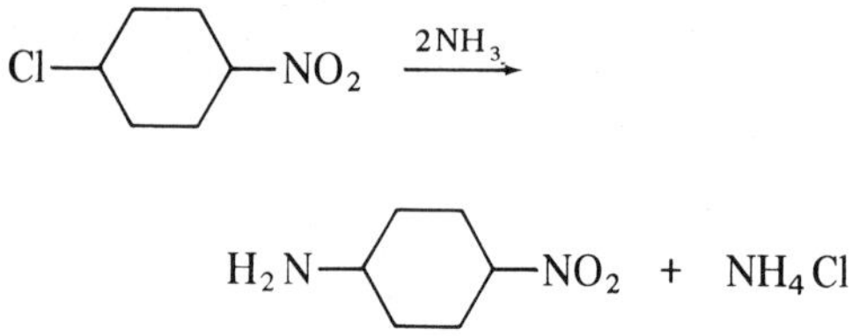

The Bucherer reaction illustrates amination in the naphthalene series. A naphthol is heated with ammonium sulfite or ammonia and alkali metal bisulfite. The result is replacement of the hydroxyl group by an amino group, probably by way of a bisulfite addition product of the keto form of the naphthol. 2-Naphthylamine is no longer used as a chemical of commerce due to its carcinogenic activity.

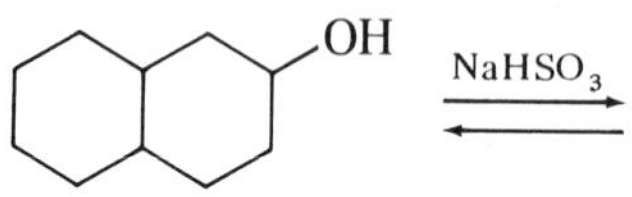

2-Naphthol

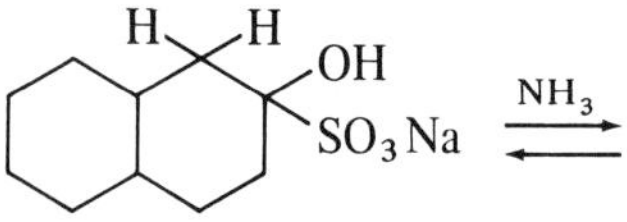

Addition product

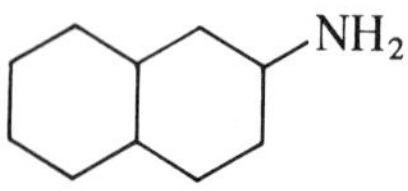

2-Naphthylamine

The reaction is reversible and may be used to convert naphthylamines to naphthols.

In the anthraquinone series, amination is frequently a convenient means of preparing amines. 1-Aminoanthraquinone-2-carboxylic acid can be made by reaction of 1-nitroanthraquinone-2-carboxylic acid with 15 per cent aqueous ammonia at 130°C. The nitro group is displaced, not reduced. In a similar manner 1-anthraquinonesulfonic acid can be aminated to give 1-aminoanthraquinone or, if desired, ammonia may be replaced by methylamine to give 1-N-methylaminoanthraquinone.

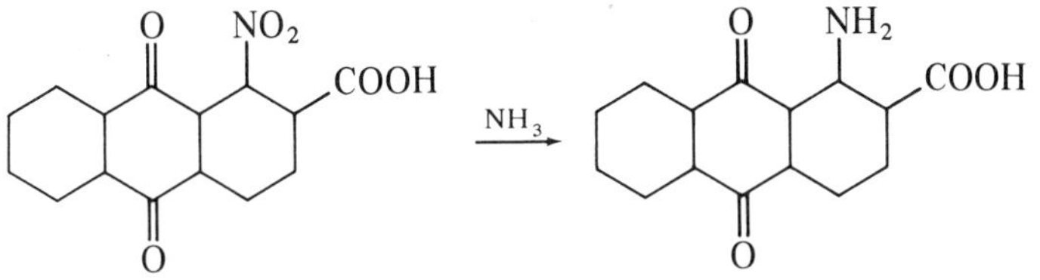

1-Nitroanthraquinone-2-carboxylic acid → 1-Aminoanthraquinone-2-carboxylic acid

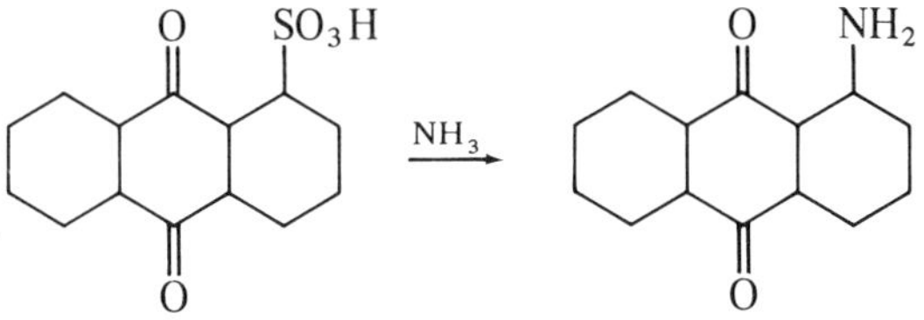

1-Anthraquinonesulfonic acid → 1-Aminoanthraquinone

Sulfonation

The sulfonic acid group (—SO_3H) is one of the more common substituents in dye intermediates. It is introduced to render intermediates soluble in water, or to provide a route to other substituents, such as the hydroxyl group which is obtained by subsequent alkaline fusion.

Direct sulfonation is achieved with: (1) strong sulfuric acid, (2) oleum (sulfuric acid plus sulfur trioxide), (3) sulfur trioxide in organic solvent or as a complex, or (4) chlorosulfonic acid. The sulfonation of benzene with sulfuric acid is illustrated by the following equation:

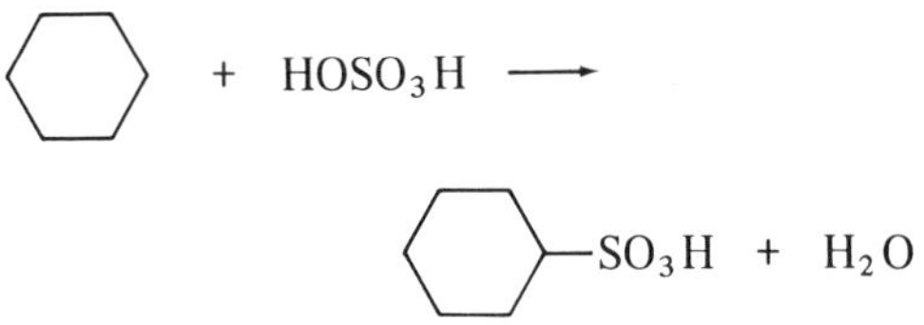

The actual sulfonating agent is believed to be the cation, $^+SO_3H$. In carrying out this reaction, oleum containing 8 per cent free sulfur trioxide is added slowly to offset the dilution caused by the water formed in the process. The temperature is maintained at 30°C until near the end, when it is raised to 50°C. When reaction is complete, the charge is diluted by running it into water, and the product is precipitated by adding salt. It is isolated by filtration as the sodium sulfonate.

Substitution rules for sulfonation are similar to those for nitration. The groups, —NO_2, —COOH, and —SO_3H direct the entering group to the *meta* position.

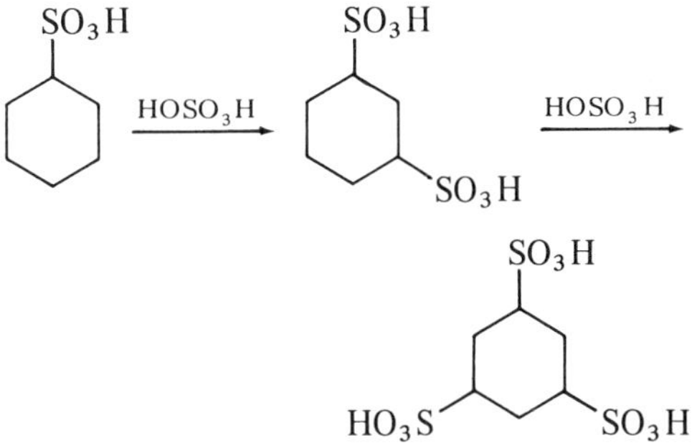

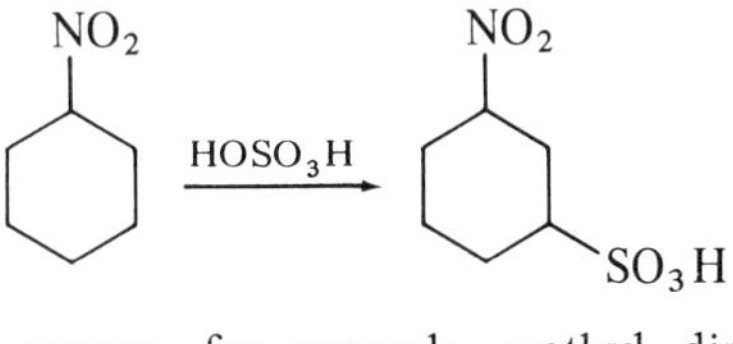

Alkyl groups, for example, methyl, direct the entering group predominantly to the

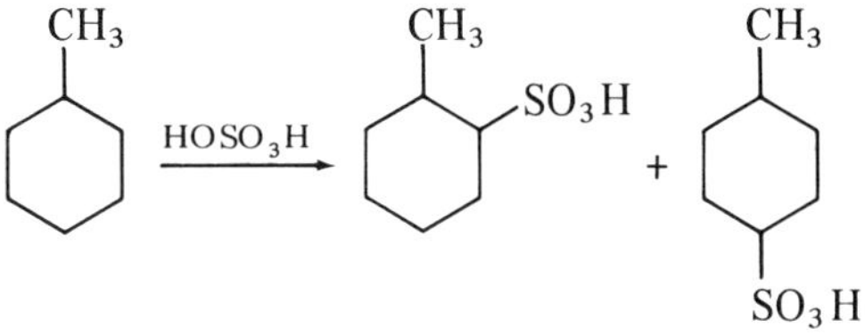

ortho and *para* positions. Usually, a mixture of isomers is formed.

Chlorine is similar to the methyl group in its effect on orientation but gives less of the *ortho* isomer.

The directing effect of amino groups depends on their basicity. The less basic amines are *ortho-para* directing. Aniline is an example of this type. More basic amines, for example, dimethylaniline, form *meta*-directing salts in acid.

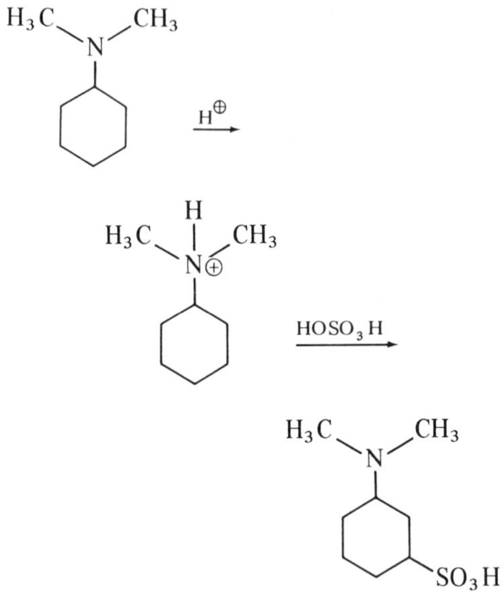

In addition to sulfonation with sulfuric acid or its equivalent, amines may be sulfonated by baking the sulfates at elevated temperatures. This procedure offers the advantage of giving fewer isomers. The baking of aniline sulfate at 260–280°C gives a high yield of the *para* sulfonic acid.

Indirect sulfonation may be achieved in a number of ways. Sodium bisulfite will often replace a labile functional group with the sulfonic acid group. An example is *o*-chlorobenzoic acid which is converted to *o*-sulfobenzoic acid by aqueous sodium bisulfite.

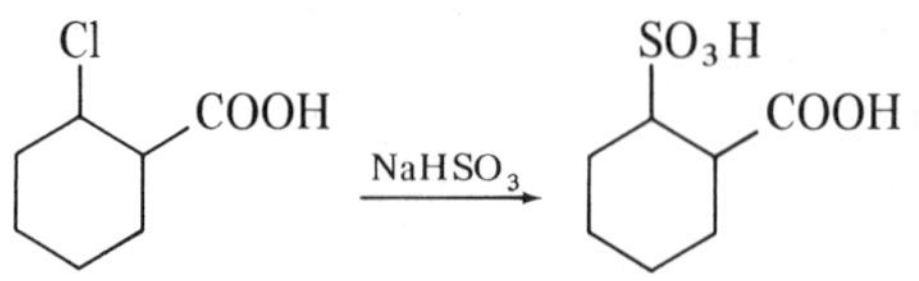

The sulfonation of naphthalene yields a number of isomers. The product obtained may be controlled to some extent by the choice of agent. With any one agent, temperature and time of reaction determine the result. It is rarely possible to obtain a single isomer; effort is directed toward forming a preponderant amount of one isomer. For example, for the monosulfonates, made by direct sulfonation with sulfuric acid, there is formed at 80°C in eight hours, 96 per cent 1-naphthalenesulfonic acid. As the temperature is raised, correspondingly less of this isomer is formed and more of the 2-sulfonic acid. At 150°C, for example, 18 per cent of the 1-isomer is formed and over 80 per cent of the 2-isomer.

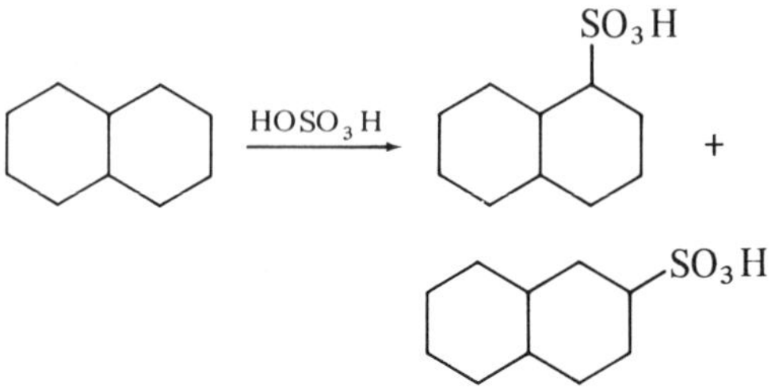

When carrying out this sulfonation, the acid is run into the melted naphthalene to avoid disulfonation. The amount of acid added is the calculated amount for one sulfonic acid group. The water formed during the reaction retards but does not prevent it. Oleum may be added toward the end to hasten the reaction.

The sulfonation of amino and hydroxy derivatives of naphthalene usually leads to a large number of isomers. To avoid isomer-formation as much as possible, naphthionic acid is often prepared by baking the sulfate of 1-naphthylamine.

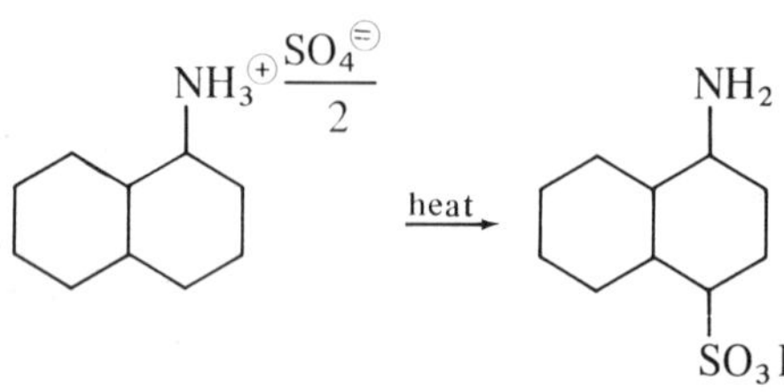

By direct sulfonation of 1-naphthylamine, four of the seven possible 1-naphthylamine-sulfonic acids may be formed. The main product under proper conditions in the 1,4-isomer. Direct sulfonation of 2-naphthylamine yields primarily a mixture of 2,5- and 2,8-isomers.

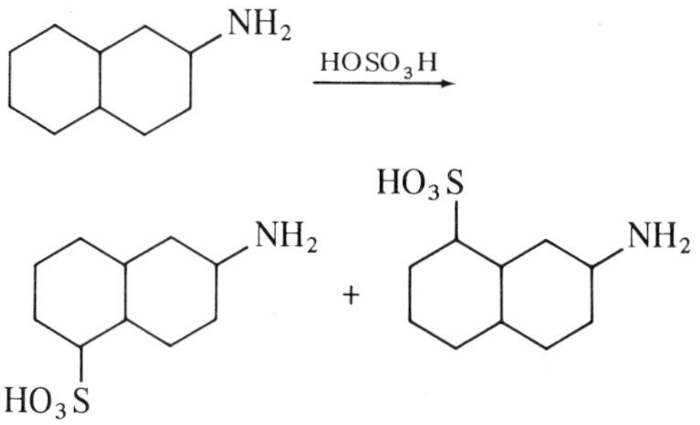

Two important monosulfonic acids resulting from the sulfonation of 2-naphthol are Schaeffer's acid and crocein acid. At 110°C, Schaeffer's acid is preponderant; at lower temperatures more crocein acid is formed.

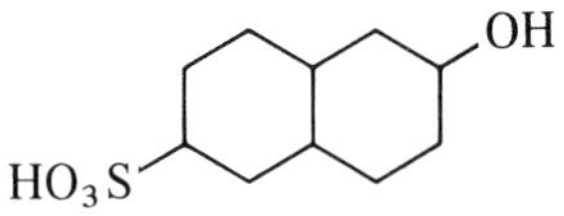

Schaeffer's acid

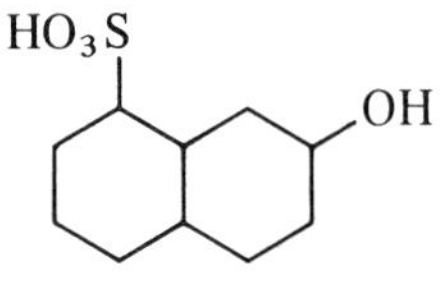

Crocein acid

By further sulfonation, two isomeric disulfonic acids are the main products. In the cold, G acid predominates, while at higher temperatures R acid is formed in greater amount. Both isomers are important intermediates for azo dyes.

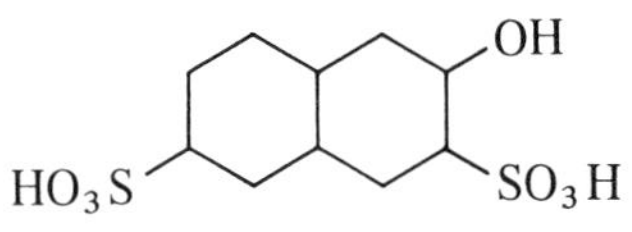

R Acid

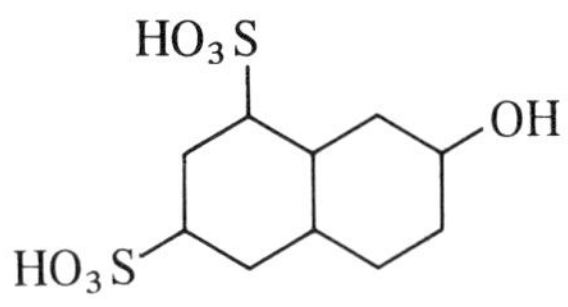

G Acid

Anthraquinone is sulfonated by suspending it in oleum containing 45 per cent free sulfur trioxide and heating at 150°C for one hour. The resulting melt is run into water and neutralized with sodium hydroxide while still hot. On cooling, the sodium salt of the 2-sulfonic acid separates. Further sulfonation produces a mixture of the 2,6- and 2,7-disulfonic acids.

When anthraquinone is sulfonated in the presence of mercury sulfate, the results differ from those just described. A single sulfonic acid groups enters at position 1. Two groups enter to form the 1,5- and 1,8-disulfonic acids. The 1,5-isomer is salted out from the more soluble 1,8-isomer after dilution of the sulfonation mass. The 1,5- and 1,8-disulfonic acids are of great importance for the manufacture of other derivatives which can be made by replacement of the sulfonic acid groups. Examples are the chloro- and hydroxyanthraquinones.

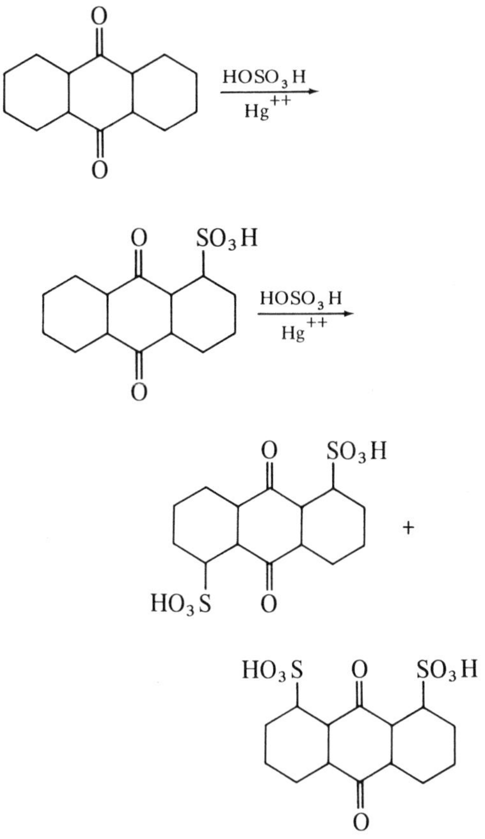

Halogenation

Chlorine is the most widely used of the halogens because it is comparatively economical. Most often, chlorinations are performed by dried chlorine gas, that is, by direct chlorination with or without a catalyst. An alternate procedure consists of generating nascent chlorine *in situ* by the oxidation of a chlorine-containing compound. In a few cases, chlorination may be achieved with reagents such as thionyl chloride, phosphorus oxychloride, phosphorus pentachloride, or sulfuryl chloride.

Chlorination of alkylated aromatic compounds, for example, toluene, can occur either in the aromatic ring or in the side chain. The use of an iron catalyst directs the chlorine to the aromatic ring, probably by inducing formation of the Cl^+ cation as the active agent. Without catalyst, chlorination takes place in the side chain, especially under ultraviolet light; the active agent in this case is believed to be the Cl· radical.

Chlorobenzene is made by passing a stream of dried chlorine into benzene in the presence of ferric chloride; some *p*-dichlorobenzene is formed at the same time. To produce monochlorobenzene as the sole product excess benzene is only partially converted and unreacted benzene is recycled. Even so, some polychlorobenzene is obtained.

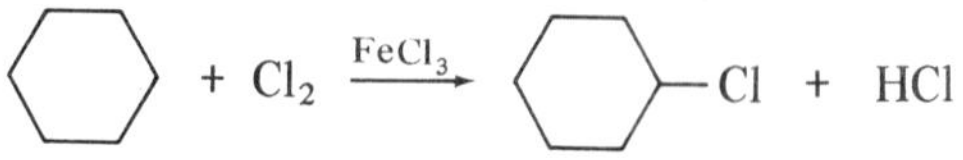

Chlorination of toluene in the complete absence of iron produces side-chain chlorination products; these are benzyl chloride, benzal chloride, and benzotrichloride. All these compounds are valuable, although the reaction is difficult to stop to produce pure intermediate products. Benzal chloride is converted to benzaldehyde with calcium carbonate in water, while benzotrichloride gives benzoic acid under the same conditions.

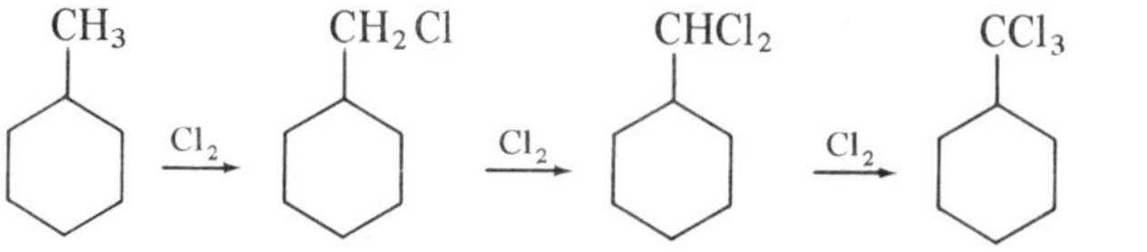

Toluene   Benzyl chloride   Benzal chloride   Benzotrichloride

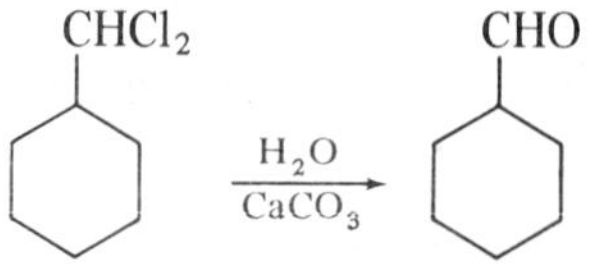

Benzaldehyde

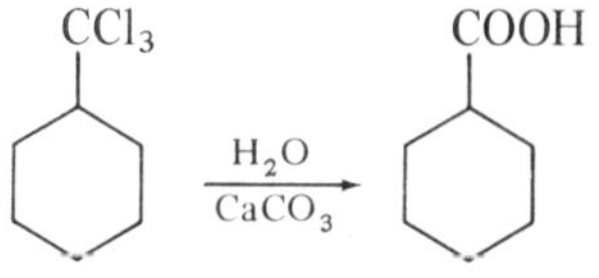

Benzoic acid

In the naphthalene series, direct chlorination is seldom used. The reaction takes place readily but leads to numerous isomers. In the anthraquinone series, both direct and indirect chlorination are employed. An example of indirect chlorination is the conversion of 1-anthraquinonesulfonic acid to the corresponding chloro compound. This reaction is carried out at approximately 100°C with sodium or potassium chlorate in hydrochloric acid.

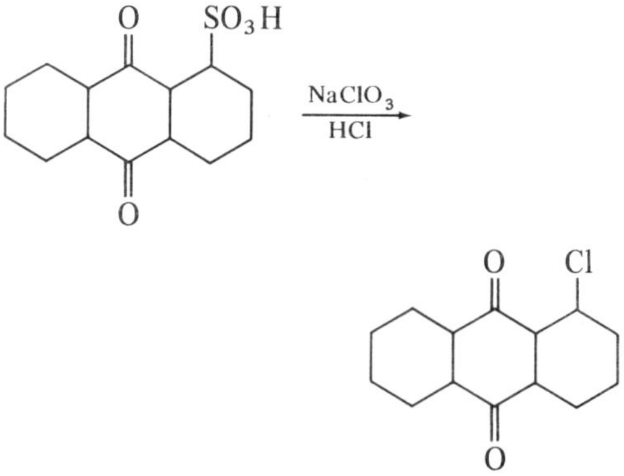

Chlorination of aliphatic compounds is illustrated by the chlorination of acetic acid. One, two, or three of the hydrogen atoms on carbon can be replaced by direct chlorination of the warm liquid in the presence of sulfur.

$$CH_3COOH \xrightarrow{Cl_2} ClCH_2COOH \xrightarrow{Cl_2} Cl_2CHCOOH \xrightarrow{Cl_2} Cl_3CCOOH$$

Monochloroacetic acid ($ClCH_2COOH$); Dichloroacetic acid ($Cl_2CHCOOH$); Trichloroacetic acid (Cl_3CCOOH)

Both fluorine and bromine find some use in the manufacture of dye intermediates, but high cost restricts them to applications where they offer some unique advantage over chlorine.

Alkaline Fusion

Alkaline fusion is an important procedure for the hydroxylation of aromatic compounds. In alkaline fusion, a sulfonic acid group is replaced by a hydroxyl group. This reaction cannot be used when nitro or chloro groups are present but is applicable to amino compounds.

Alkaline fusion is usually carried out with a concentrated solution of sodium hydroxide in a cast-iron pot which is equipped with a scraping agitator and is heated externally. The water is evaporated; then the mass fuses. The reaction temperature is between 190 and 350°C, depending on the reactivity of the sulfonic acid.

Phenol can be made by fusing benzenesulfonic acid according to the following equations:

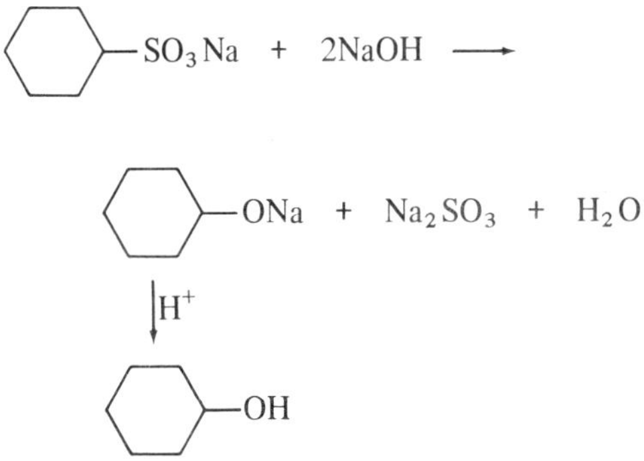

Phenol

In a similar way, resorcinol is made by fusion of *m*-benzenesulfonic acid.

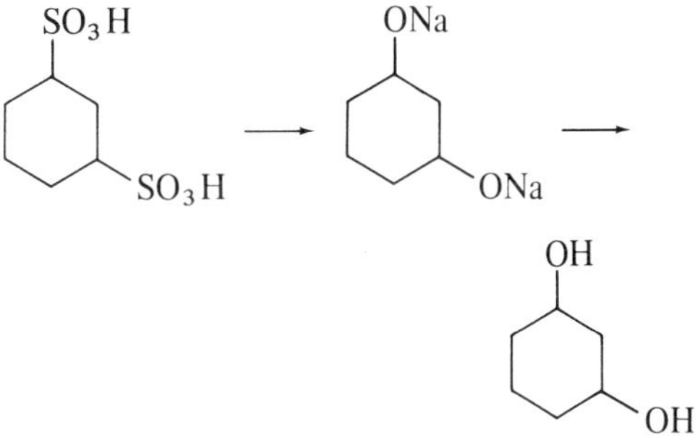

Resorcinol

2-Naphthol is made by fusion of the corresponding sulfonic acid with sodium hydroxide. The naphtholate is treated with carbon dioxide to precipitate the naphthol which is then purified by vacuum distillation.

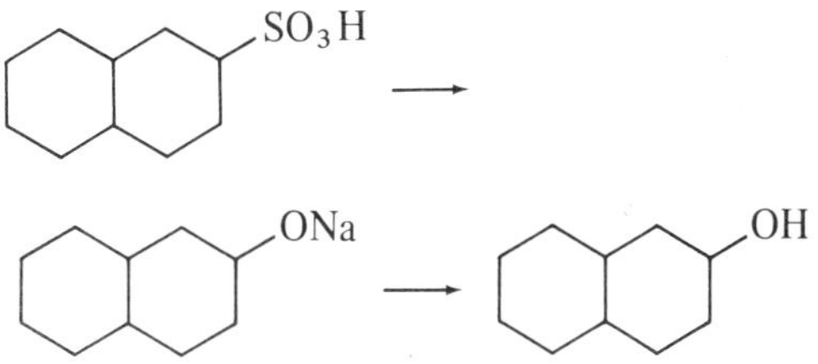

2-Naphthol

1-Naphthol is not made from the sulfonic acid. Instead, 1-naphthylamine sulfate is heated with water at 200°C in a closed vessel. On cooling, 1-naphthol crystallizes out.

For some compounds, the conditions of alkaline fusion are too severe. In such cases the reaction may often be effected in water solution under pressure. Chromotropic acid is prepared from 1-naphthol-3,6,8-trisulfonic acid in 60 per cent sodium hydroxide solution under pressure.

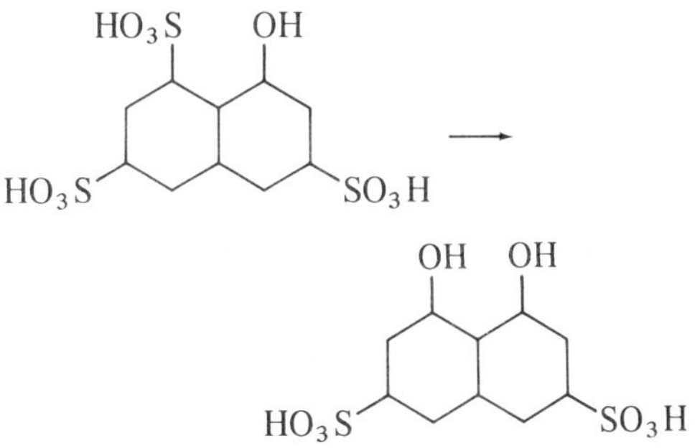

Chromotropic acid

Sulfonated naphthylamines may be fused without destruction of the amino group. The important azo dye intermediates, H acid and J acid, are made in this way.

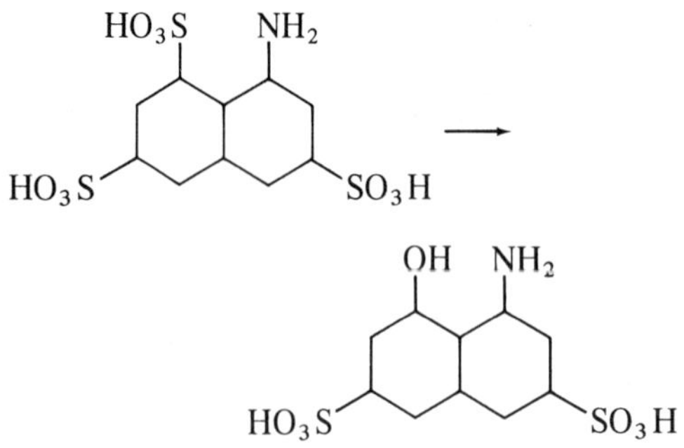

H Acid

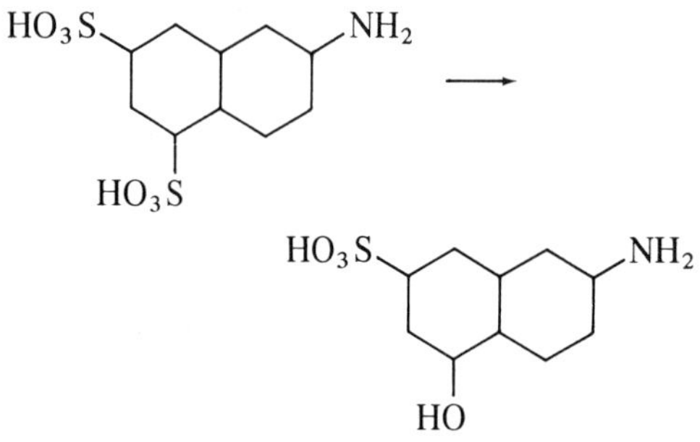

J Acid

In the anthraquinone series, 2-anthraquinonesulfonic acid can be converted to the hydroxy compound by heating with calcium hydroxide in water. Alkaline fusion of the same sulfonic acid gives alizarin. The latter reaction is discussed elsewhere in this chapter.

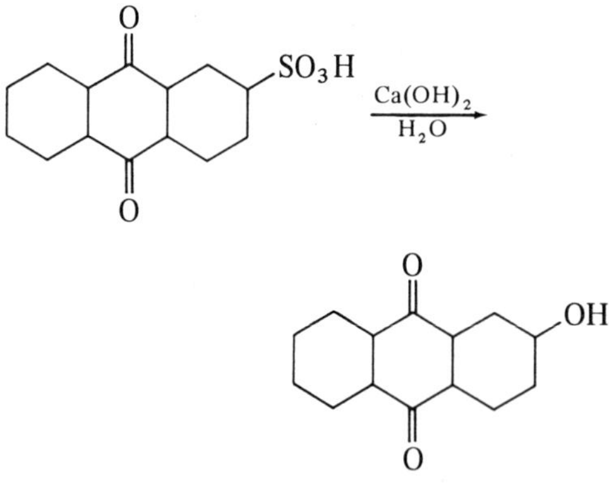

2-Hydroxyanthraquinone

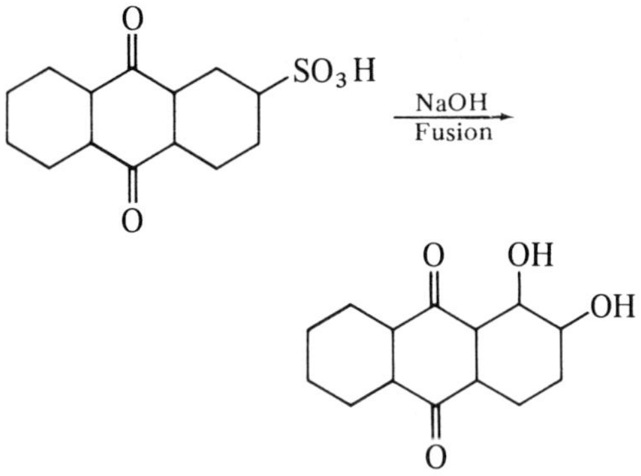

Alizarin

Oxidation

Oxidation may be effected by air in the presence of a catalyst or by a variety of chemical oxidants, such as manganese dioxide and potassium permanganate.

Catalytic vapor-phase oxidation is illustrated by the conversion of naphthalene to phthalic anhydride. This reaction is carried out over a vanadium pentoxide catalyst at 450°C.

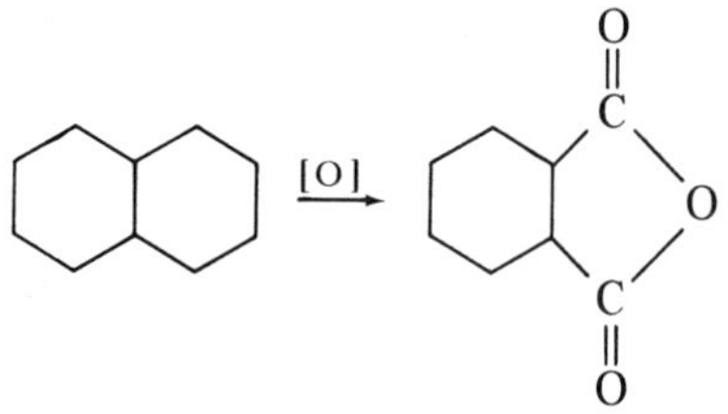

Another route to phthalic anhydride is oxidation of *o*-xylene, a product of the petroleum industry. The conditions for this reaction are similar to those for naphthalene oxidation, except that the temperature is higher (540°C).

Aniline sulfate can be chemically oxidized with manganese dioxide in sulfuric acid. The product is *p*-quinone.

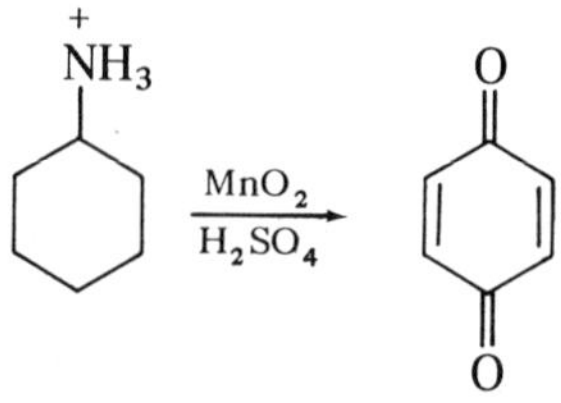

p-Quinone

The use of potassium bichromate in sulfuric acid effects the oxidation of 1-nitro-2-methylanthraquinone to the corresponding carboxylic acid. This reaction illustrates side-chain oxidation, an important route to carboxylic acids.

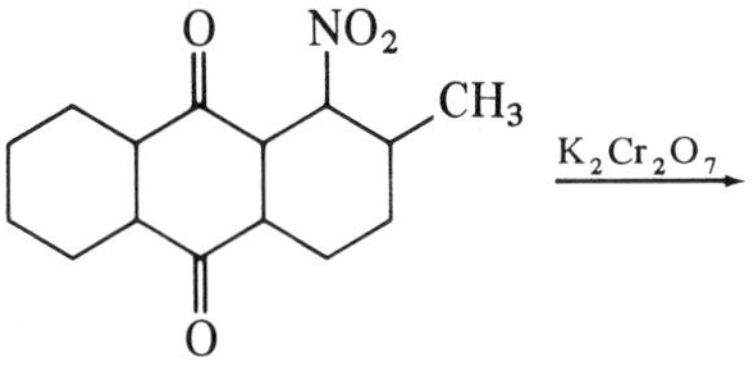

1-Nitro-2-methylanthraquinone

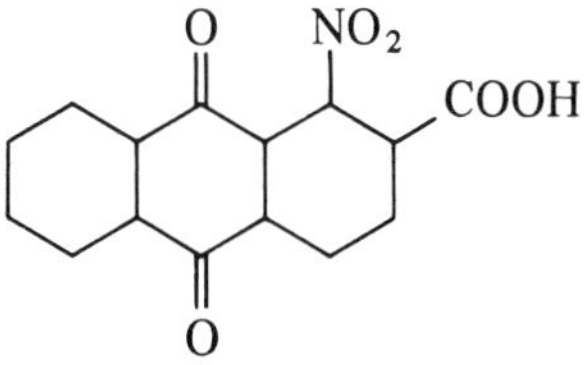

1-Nitro-2-anthraquinonecarboxylic acid

Other Important Reactions

Condensation. The term condensation describes a variety of reactions which join molecules or parts of the same molecule with elimination of a molecule of water or other low molecular weight substance. An example is the conversion of benzanthrone to dibenzanthronyl, described in detail elsewhere in this chapter.

Addition. An important intermediate, cyanuric chloride, is made by the addition of cyanogen chloride to itself. The reaction is catalyzed by a small amount of free chlorine.

3ClCN
Cyanogen chloride

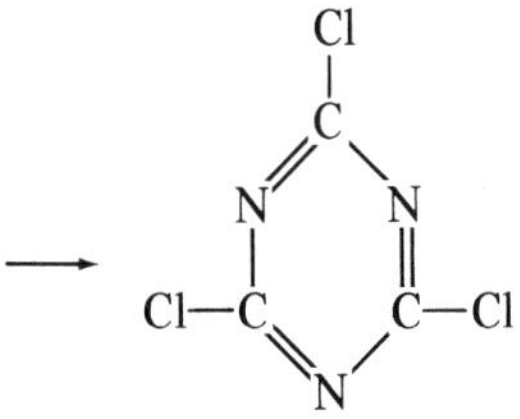

Cyanuric chloride

Alkylation. Alkylation refers to the introduction of an aliphatic group, such as methyl, into an organic molecule. Alkylation may occur on carbon, nitrogen, oxygen, or sulfur. Of these possibilities, alkylation on nitrogen and oxygen are most important in the manufacture of dye intermediates. A methyl group may be introduced into the amino group of aniline by heating with methyl alcohol under pressure in the presence of a mineral acid.

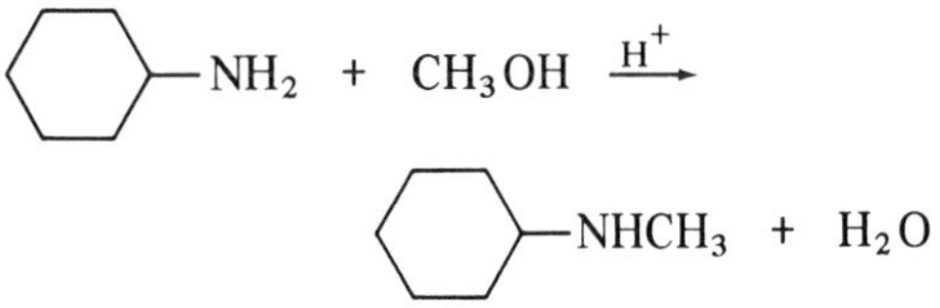

N-Methylaniline

Phenol may be methylated with methyl sulfate in cold alkaline medium to give anisole.

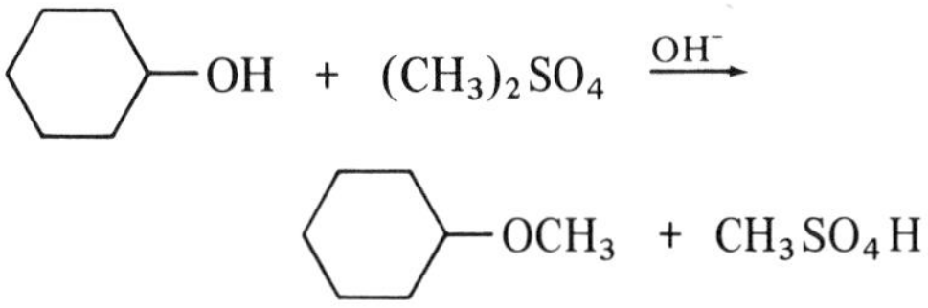

Anisole

Carboxylation. The carboxylic acid group may be introduced by side-chain oxidation as described above. In addition, it may be introduced by direct action of carbon dioxide on certain compounds. When sodium phenolate is treated with carbon dioxide under pressure at about 150°C, sodium salicylate is formed. Acidification of the salicylate gives the free acid which may be purified by vacuum distillation. 2-Hydroxy-3-naphthoic acid, an intermediate for developed azo dyes, is made from sodium 2-naphtholate.

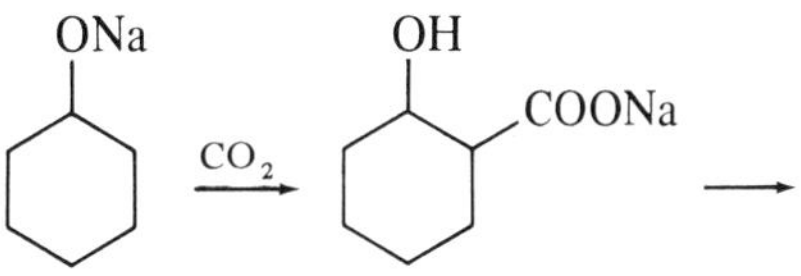

Sodium salicylate

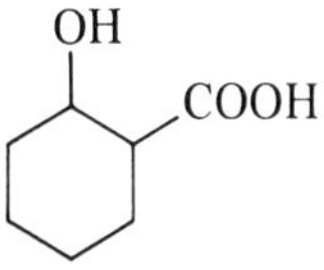

Salicylic acid

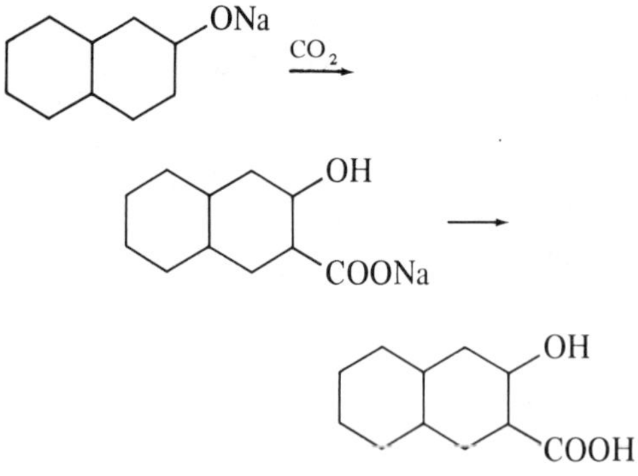

Hydroxynaphthoic acid

Sandmeyer Reaction. The replacement of a diazonium group by halogen, nitrile, nitro, sulfhydryl, and other groups in the presence of a cuprous salt, is known as the Sandmeyer reaction. An illustration of the use of this reaction is the preparation of 2-chloro-5-nitrophenol from the corresponding aminophenol.

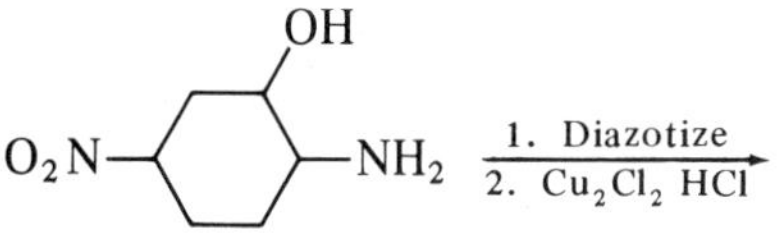

2-Amino-5-nitrophenol

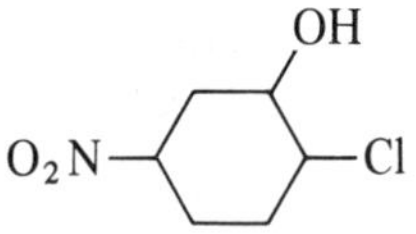

2-Chloro-5-nitrophenol

Cyanoethylation. Reaction of a primary or secondary aromatic amine with acrylonitrile results in N-cyanoethylation. The products are useful intermediates for azo and basic dyes. An example is the cyaroethylation of aniline with cupric sulfate catalyst.

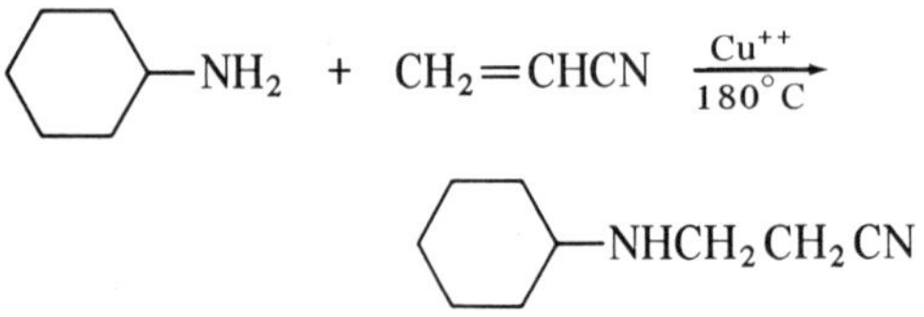

N-Cyanoethylaniline

PRODUCTION OF DYE INTERMEDIATES

Table 23.3 gives the production figures for some important raw materials and intermedi-

TABLE 23.3 United States Production of Some Raw Materials and Dye Intermediates, 1971

	In Millions of Pounds
Acetanilide	3.67
Aniline	385.99
o-Dichlorobenzene	53.64
p-Dichlorobenzene	70.41
N,N-Dimethylaniline	7.72
1,4-Dihydroxyanthraquinone	1.71
Dinitrostilbenedisulfonic acid	10.95
Nitrobenzene	444.87
Phthalic anhydride	794.42

ates in the United States for 1971. These figures are taken from the report of the U.S. Tariff Commission. It should be remembered that dye intermediates are often used for end-products other than dyes, and that the figures in Table 23.3 do not necessarily correlate with figures on dye production.

THE MANUFACTURE OF DYES

Dyes owe their color to their ability to absorb light in the visible region of the spectrum, between 4000 and 8000 angstrom units. Absorption is caused by electronic transitions in the molecules and can occur in the visible region only when the electrons are reasonably mobile. Mobility is encouraged by unsaturation and resonance. The main structural unit of a dye, which is always unsaturated, is called the *chromophore*, and a compound containing a chromophore is called a *chromogen*. Any substituent atom or group that increases the intensity of the color is called an *auxochrome*. An auxochrome may also serve to shift the absorption band of a chromophore to a longer wavelength, or may play a part in solubilizing the dye and attaching it to fibers.

Dyes are classified according to the type of chromophore that they contain. An example is the azo class, which is characterized by the presence of the —N=N— linkage. A typical chromogen is azobenzene.

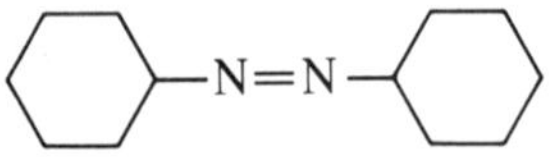

Common auxochromes are hydroxyl, amino, and carboxyl groups.

The hue, strength, and brightness of a dye depend on the entire light-absorbing system, consisting of chromophore and auxochromes, acting together. The nature of these groups and their relative positions in the molecule must be worked out correctly to produce a dye of desired appearance.

In general, for a given type of dye, extension of the unsaturated system and increased opportunities for resonance, shifts the absorption of light toward longer wave lengths. Assuming a single main absorption band, the color absorbed progresses across the visible spectrum from violet to purple. As this progression occurs, the light that is not absorbed is reflected and seen by the human eye as the color complementary to that absorbed. The wave lengths absorbed, the corresponding colors, and the observed complementary colors are given below for several major hues.

Wave Length Absorbed Å	*Corresponding Color Absorbed*	*Observed Color*
4300	Indigo	Yellow
4500	Blue	Orange
4900	Blue-green	Red
5300	Yellow-green	Violet
5900	Orange	Blue
7300	Purple	Green

In addition to securing the desired appearance, it is necessary in dye synthesis to provide the dye with any groups necessary to confer solubility and affinity for textile fibers. It is important also to use only color systems that have the required fastness to light and other degrading influences. Considering that all these properties must be exhibited by one compound, it can readily be understood that the development and manufacture of dyes require a high degree of technical competence in the laboratory and in the plant.

The "Colour Index"[10] lists 31 chemical classes of dyes. Many of these classes are closely related; others are of relatively minor importance. Only several of the more important classes will be discussed in the following sections of this chapter.

Azo Dyes

The azo dye class, one of the most important, includes many hundreds of commercial dyes of various application types. Azo dyes are characterized by the presence of one or more azo (—N=N—) groups.

The principal method of forming azo dyes involves *diazotization* of primary aromatic amines, followed by *coupling* with hydroxy or amino derivatives of aromatic hydrocarbons or with certain aliphatic keto compounds. Both the aromatic amine, which is diazotized, and the compound to which it is coupled may bear a variety of substituents, such as alkyl, alkoxyl, halogen, and sulfonic acid. Because of the large number of compounds that can be combined, often in more than one sequence, the number of possible azo dyes is almost infinite.

Diazotization takes place when nitrous acid (HNO_2) reacts with a primary aromatic amino group in acid medium. Usually sulfuric or hydrochloric acid is used and the nitrous acid is generated from sodium nitrite. The equation for diazotization, using aniline in hydrochloric acid, is as follows:

$$C_6H_5{-}NH_2 + HNO_2 + H^+Cl^- \quad C_6H_5{-}N_2^+Cl^- + 2H_2O$$

Diazotization is usually carried out with excess acid to prevent partial diazotization and to inhibit secondary reactions. If the reaction is to proceed easily, the amine must be in solution, or its hydrochloride, if insoluble, must be in a fine state of subdivision. Temperatures of from 0 to 5°C are usually employed for diazotization since diazonium salts are generally unstable. There are exceptions to this, and temperatures of 20°C or higher are occasionally preferred.

The ease of diazotization depends markedly on the basicity of the amine. Extremely weakly basic amines are diazotizable only by special methods.

The coupling reaction with aromatic hydroxy compounds is illustrated with benzene diazonium chloride and phenol.

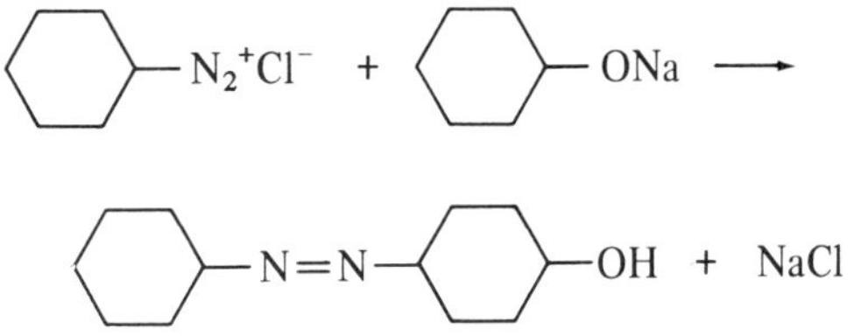

Coupling to phenols, naphthols, and related hydroxy compounds is carried out in alkaline solution. Under alkaline conditions the hydroxy compound is soluble, and coupling is usually rapid. Ordinarily the coupling must be carried out in the cold to prevent decomposition of the diazonium salt.

An example of coupling to an aromatic amine is the reaction of *p*-nitrobenzenediazonium chloride with *m*-toluidine.

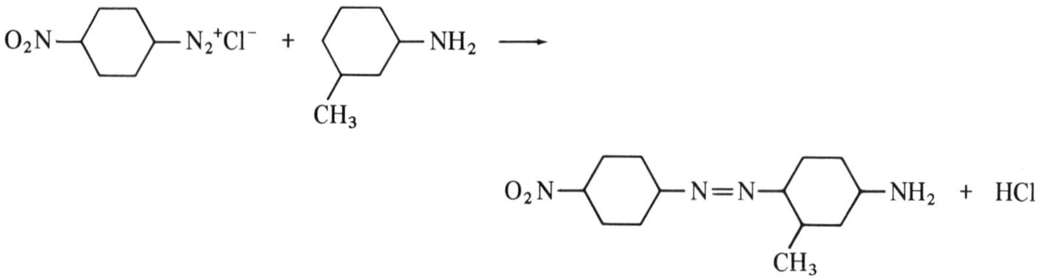

Couplings to amines are carried out in acid solution in which the amine is soluble. It has been shown, however, that the free amine couples, not its hydrochloride. For this reason coupling proceeds faster near the neutral point where the equilibrium concentration of free amine is highest. It is frequently best to start the reaction at a low pH and then to raise the alkalinity slowly with sodium acetate or soda ash as the reaction proceeds.

It should be noted that in the above examples of coupling the diazonium group is shown entering the position *para* to the hydroxyl and amino groups. This specificity is characteristic of azo coupling. The diazo group does not enter at random but in certain definite positions. For the benzene derivatives, the attack is always on the position *para* to the activating group. If this position is blocked, coupling will sometimes occur in one of the *ortho* positions, but at a much slower rate.

In naphthalene derivatives, orientation of the entering diazo group is somewhat different. In alpha-naphthol, the attack is at position 4. If 4 is blocked, the diazo group enters at 2. In beta-naphthol, coupling takes place at 1, never at 3 or 4. The same rules apply to the corresponding naphthylamines.

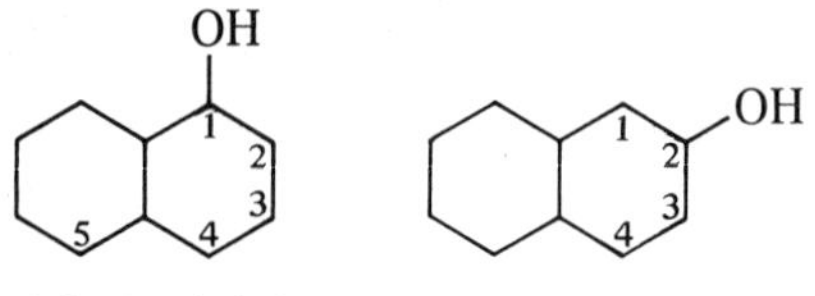

alpha-Naphthol beta-Naphthol

In the naphthalene series, it is found that the presence of certain substituents, especially the sulfonic acid group, can influence the position of coupling even when not in a directly blocking position. For example, a sulfonic acid group in position 5 of alpha-naphthol causes the coupling to occur predominantly at position 2, rather than 4.

Certain aminonaphthol sulfonic acids couple twice. In this case the place of entry depends on whether the coupling is performed in acid or alkaline solution. If acid, the coupling is *ortho* to the amino group; if alkaline, it is *ortho* to the hydroxyl. In the three examples following, the place of entry for acid coupling is marked X, for alkaline coupling, Z.

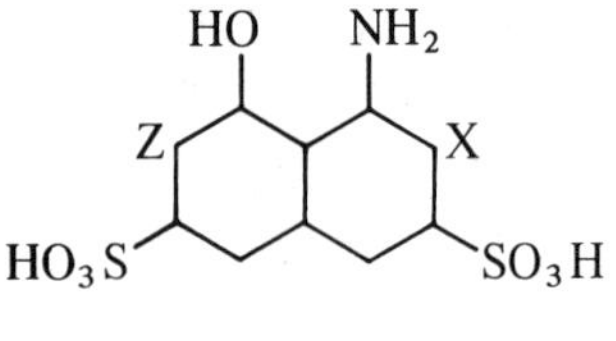

H Acid

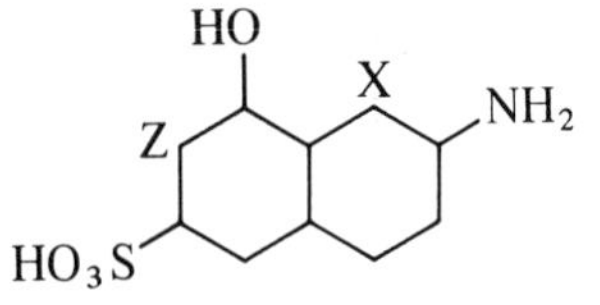

Gamma Acid

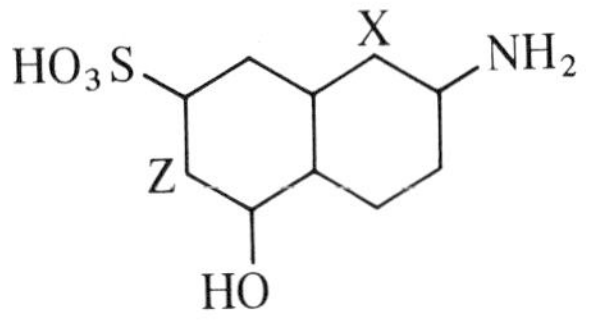

J Acid

Ortho and *para* diamines in both the benzene and naphthalene series do not couple at all; only the *meta* diamines do so.

A generalized procedure for diazotization and coupling as practiced in both laboratory and plant is as follows: Sodium nitrite is added slowly with stirring to an acid solution of the amine in a wooden or brick-lined tub. The tub is not externally cooled but ice is added directly to the amine solution to control the temperatures between 0 and 5°C. The progress of the reaction is followed by observing the disappearance of the nitrous acid generated from the sodium nitrite. This is done by spotting the reaction mass on a starch-potassium iodide test paper which turns black when nitrous acid is present. When no more nitrous acid is consumed, the diazotization is judged to be complete. The diazonium compound is not isolated, but is run at once at a slow rate into the alkaline (or acid) solution of the intermediate with which it is to couple. After addition of the diazonium compound, the batch is stirred for a period varying between a few minutes and three days, until coupling is complete. The solution of the dye is then warmed, and salt (NaCl) is added and allowed to dissolve. On cooling, the dye separates and is filtered. There are a number of variations in procedure. Some dyes are salted hot, some cold, some not at all. Filtration is carried out at a temperature best suited to isolate the dye without inclusion of impurities. As a result of salting, dyes with sulfonic acid groups are always isolated as the sodium sulfonates.

Diazotization and coupling are usually carried out as batch processes because of the limited production required and the necessity of testing the various batches of intermediates that are used. In a few cases, however, continuous diazotization and coupling are practical. When continuous processes are used it becomes necessary to devise a means of isolating the product as rapidly as it is made. Centrifugal filters, spray dryers, and other special apparatus are suitable for this use.

Monoazo Dyes. Monoazo dyes contain the azo (—N=N—) linkage once; the simplest example is aminoazobenzene. This compound is used as a dye for oils, lacquers, and stains under the name Solvent Yellow 1 (Aniline Yellow), and is also an intermediate for other dyes. It is made by coupling benzenediazonium chloride with aniline in acid medium:

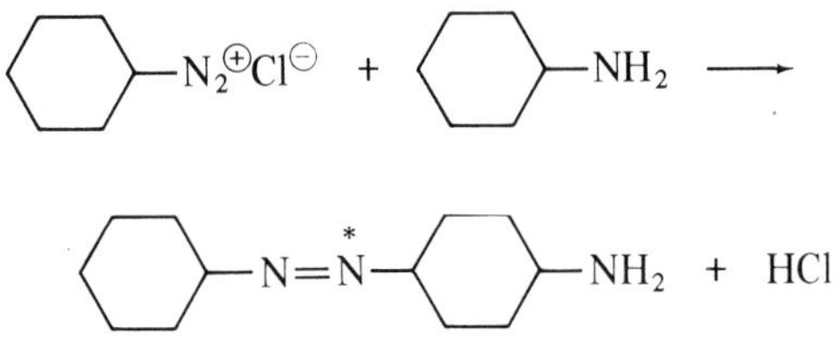

Solvent Yellow 1

To simplify the description of other azo dyes to follow, only the formulas of the dyes will be given, but the place of attack of the diazonium group will be indicated by an asterisk on the diazo nitrogen joined to the coupling component. It will be understood that the diazonium compound was prepared by diazotization of the corresponding amine and that coupling took place on the coupling component with the elimination of a hydrogen ion.

The coupling of benzenediazonium chloride with *m*-phenylenediamine in acid medium gives Basic Orange 2 (Chrysoidine), a dye of limited use on textiles but of importance for coloring paper, leather, and woodstains.

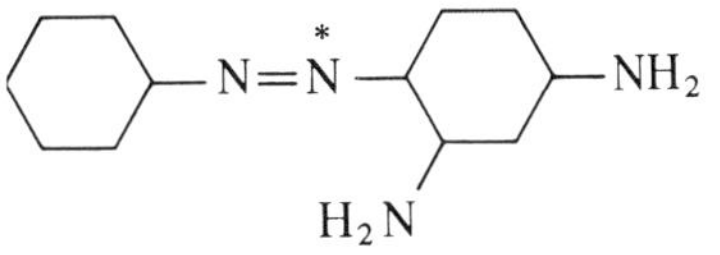

Basic Orange 2

Acid Orange 7 (Orange II) is prepared by coupling diazotized sulfanilic acid with beta-naphthol in basic medium. Acid Orange 7 is useful on textile fibers, such as wool, silk, and nylon, as well as on paper and leather.

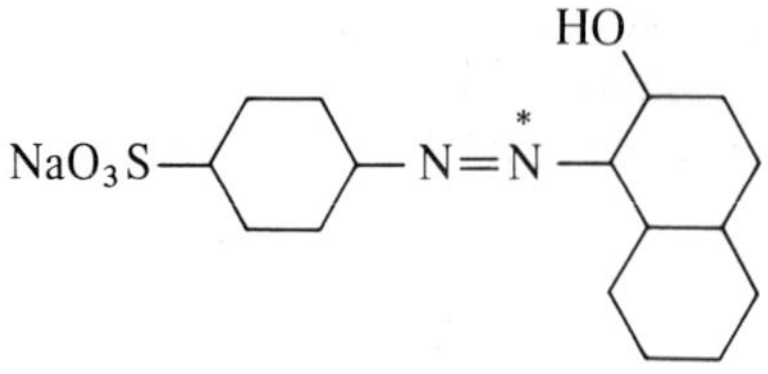

Acid Orange 7

Other acid monoazo dyes of importance are derived from a variety of sulfonated intermediates. Acid Red 26 (Ponceau R) is made from diazotized *m*-xylidine and R-Acid. Acid Red 14 (Azo Rubine) is made from diazotized naphthionic acid and Nevile and Winther's acid. Acid Violet 6 is *p*-aminoacetanilide coupled with chromotropic acid.

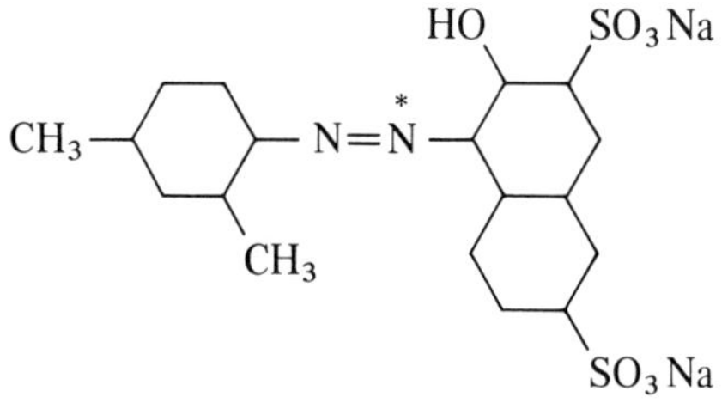

Acid Red 26

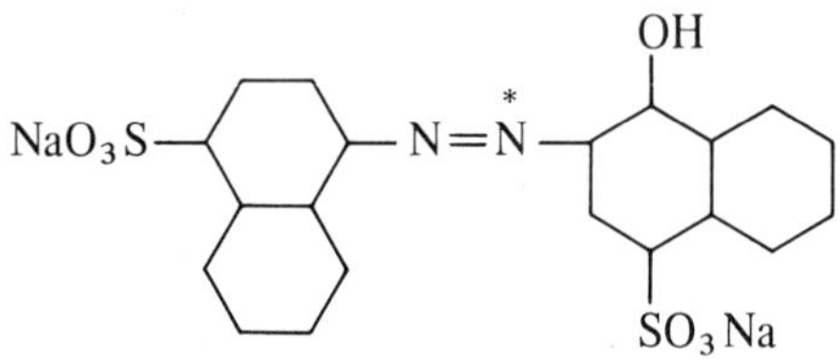

Acid Red 14

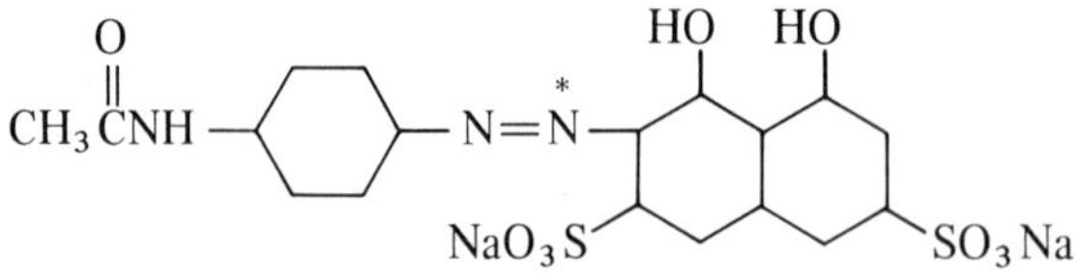

Acid Violet 6

Mordant Black 17 is an example of a type of dye that shows improved fastness on wool when treated on the fiber with a soluble chromium salt. Reaction occurs between the dye and the chromium ion to form a complex of enhanced fastness to washing and light. The component parts of Mordant Black 17 will be recognized as 1-amino-2-naphthol-4-sulfonic acid and beta-naphthol. Another dye of this type is Mordant Brown 4, which is made by diazotizing picramic acid and coupling it in acid medium with *m*-toluenediamine.

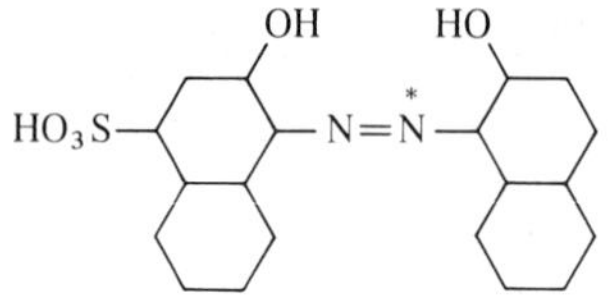

Mordant Black 17

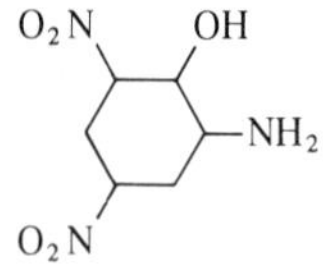

Picramic Acid

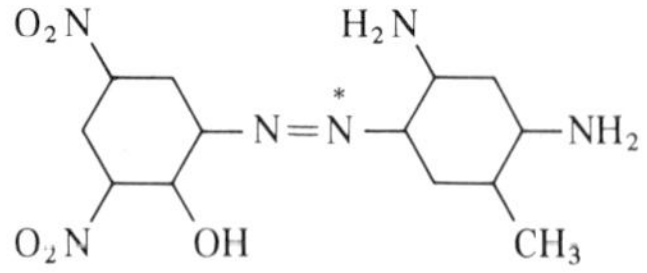

Mordant Brown 4

Acid Yellow 23 (Tartrazine) contains the pyrazolone nucleus. It is prepared by coupling diazotized sulfanilic acid with 1-(4'-sulfophenyl)-3-carboxypyrazolone-5, or by heating oxidized tartaric acid (1 mole) with phenylhydrazinesulfonic acid (2 moles). Acid Yellow 23 is useful as a dye for basic fibers, paper, and leather. It is also used as a filter dye in photography.

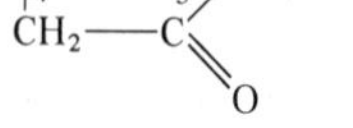

Pyrazolone

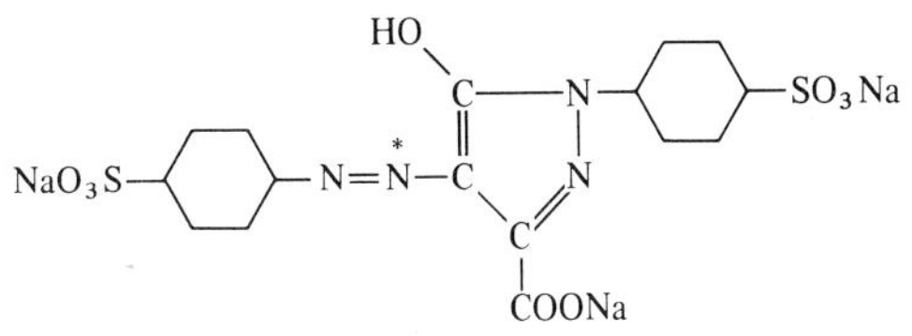

Acid Yellow 23

A further example of a pyrazolone dye is Acid Red 38. This dye is the chromium complex of the azo compound made by coupling diazotized 5-nitro-2-aminophenol with 1-(3′-sulfamyl)-3-methylpyrazolone-5. It is used as a dye for wool and nylon. Acid Red 38 has one advantage over dyes such as Mordant Black 17 in that the dyer does not have to form the chromium complex by a separate operation during the dyeing process.

Some insoluble azo dyes are used as pigments. Pigment Yellow 3 (Hansa Yellow 10G) illustrates this type. It is prepared by coupling diazotized 2-nitro-4-chloroaniline with *o*-chloroacetoacetanilide.

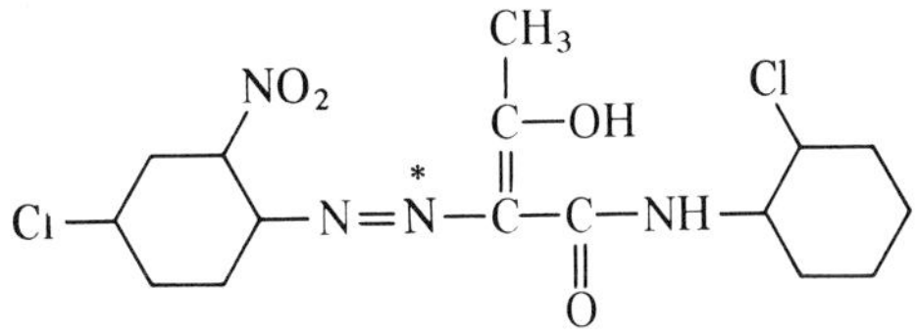

Pigment Yellow 3

A final example of a monoazo dye is a red fiber-reacting dye. This dye is prepared by condensing aniline-*m*-sulfonic acid and H acid with cyanuric chloride under carefully controlled conditions, followed by coupling with diazotized aniline-*o*-sulfonic acid. One chlorine atom remains for reaction with textile fibers bearing hydroxyl or amino groups.

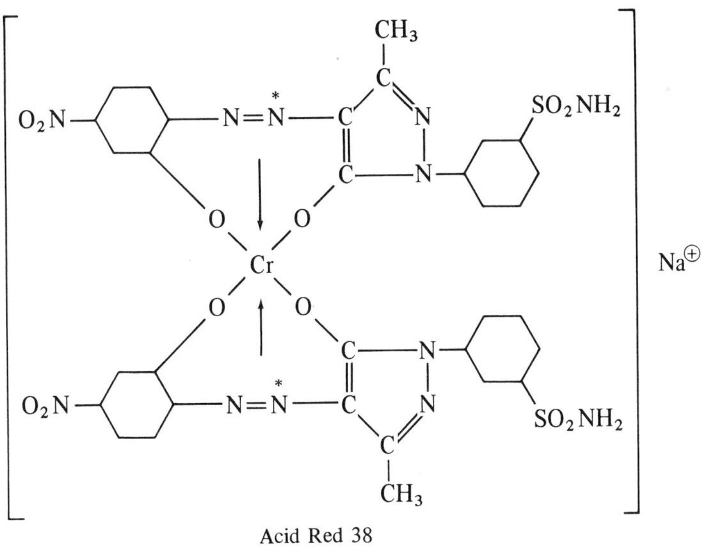

Acid Red 38

Disperse Red 1 is a dye for cellulose acetates, nylon, polyesters, and various plastics. It is prepared by coupling diazotized *p*-nitroaniline with N-ethyl-N-(beta-hydroxyethyl)-aniline in acid medium. It has low solubility in water and is supplied in finely divided form mixed with a dispersing agent.

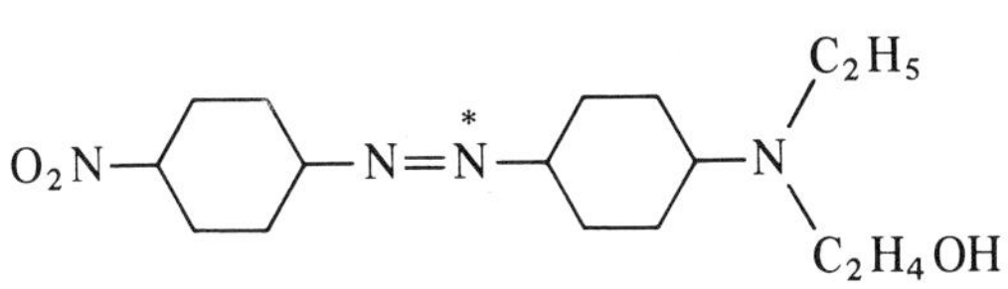

Disperse Red 1

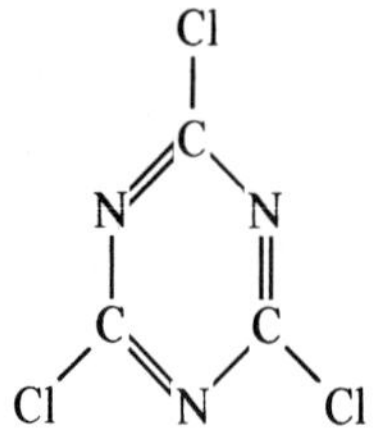

Cyanuric chloride

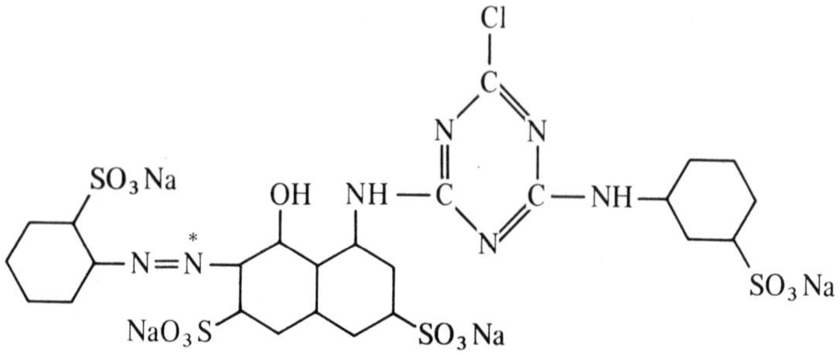

Red reactive dye

Disazo Dyes. Disazo dyes contain two azo linkages. They are formed: (1) by the further diazotization and coupling of monoazo dyes containing free amino groups; (2) by the coupling of two moles of diazonium compound to a coupling component that can couple twice; or (3) by the tetrazotization of a diamine followed by coupling to two moles of coupling component. When two moles of diazonium compound or coupling component are used, they may be the same or different in structure.

Acid Red 150 (Cloth Red 2R) is a disazo dye prepared by diazotizing aminoazobenzene and coupling to R Acid.

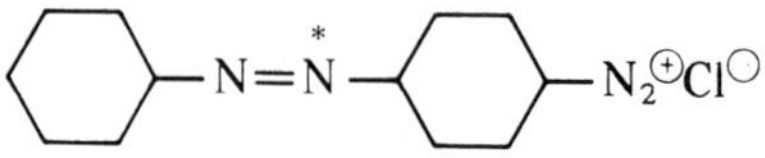

Diazotized aminoazobenzene

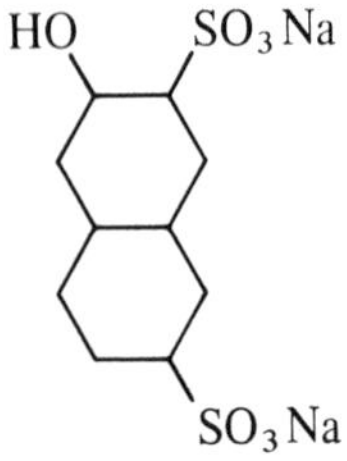

R Acid

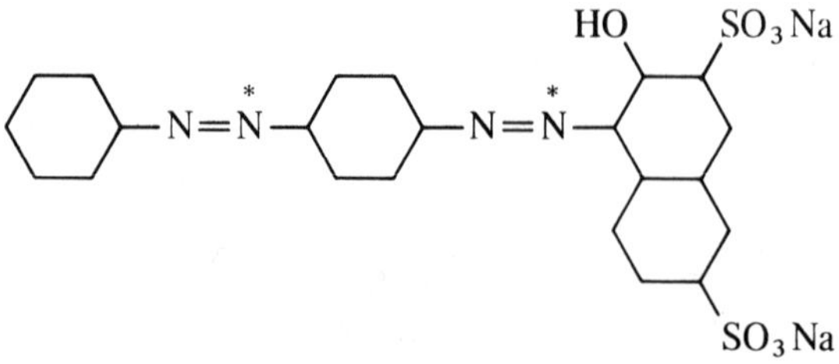

Acid Red 150

A more important disazo acid dye is Acid Red 73 (Brilliant Crocein M) which is used to color wool, leather, paper, and anodized aluminum. It is prepared from diazotized aminoazobenzene and G Acid.

Solvent Red 27 is a disazo dye used to color gasoline. Its formula indicates that it is made by diazotizing aminoazoxylene and coupling to beta-naphthol.

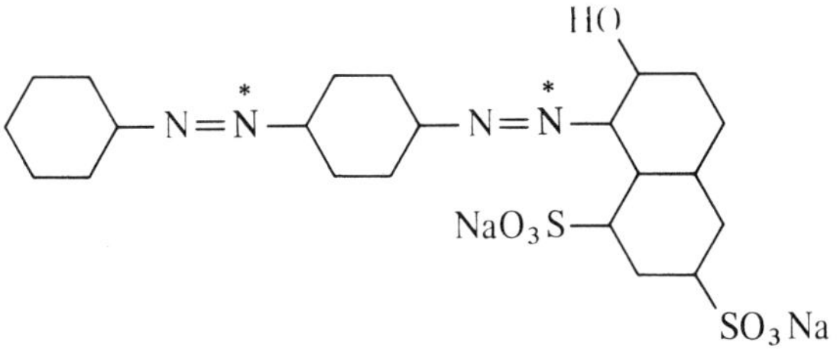

Acid Red 73

Other disazo dyes of this kind are Acid Red 115 (Cloth Red B), which is made by coupling diazotized aminoazotoluene with R Acid, and Acid Black 18. Acid Black 18 will be recognized as having been made by the combination, through diazotization and coupling, of sulfanilic acid, alpha-naphthylamine, and H Acid.

An example of a disazo dye prepared by coupling two different diazonium compounds to a single coupling component is Acid Black 1. This dye is made by coupling diazotized *p*-nitroaniline (acid) and aniline (alkaline) to H Acid.

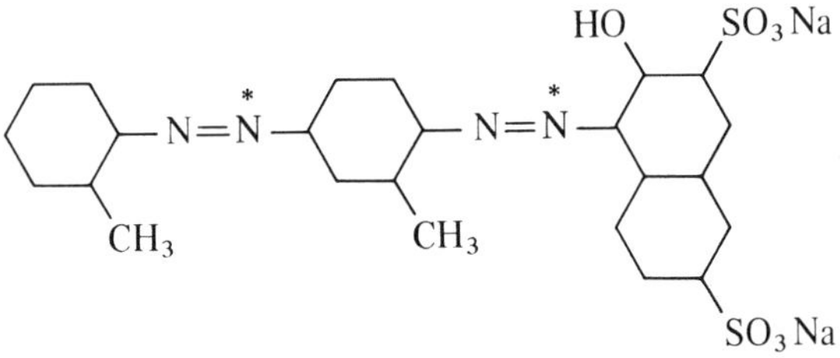

Acid Red 115

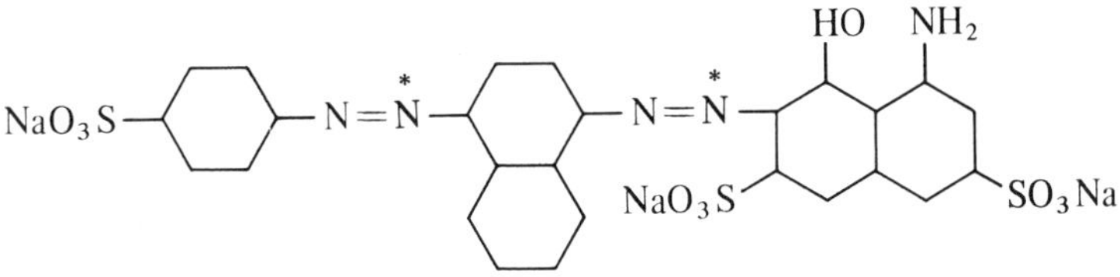

Acid Black 18

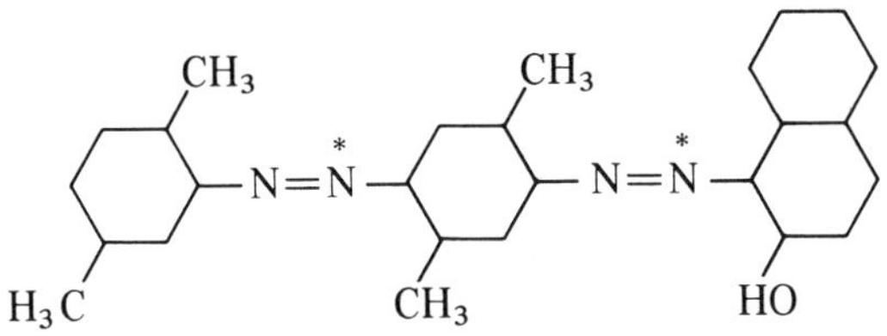

Solvent Red 27

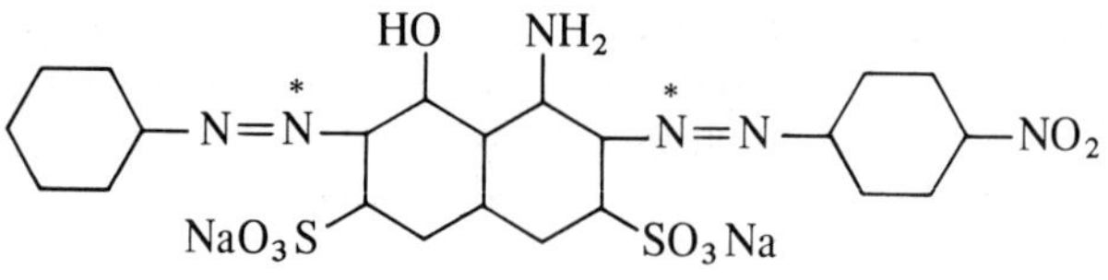

Acid Black 1

Direct Red 28 (Congo Red) is a disazo dye made by coupling tetrazotized benzidine twice with naphthionic acid. This dye is now of limited use on textiles and once was an important dye for paper. It is also useful as an indicator since it turns from red to blue in strong acid.

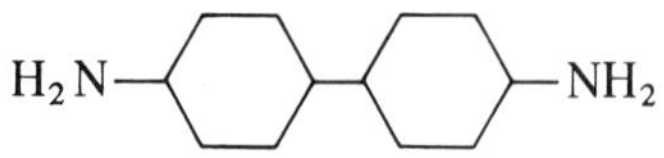

Benzidine

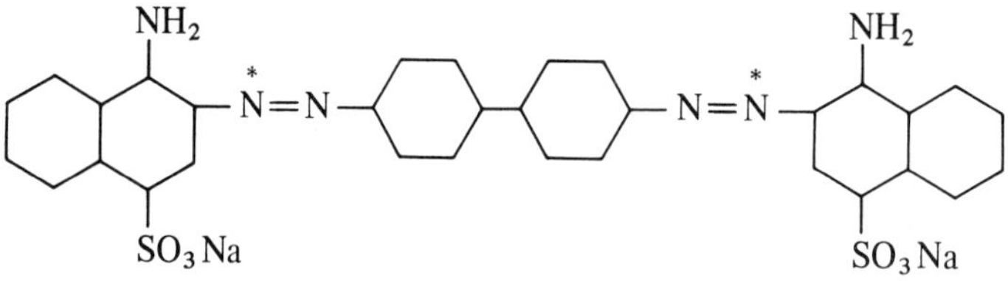

Direct Red 28

A direct dye of importance is Direct Blue 1 (Sky Blue FF). It is made by coupling tetrazotized *o*-dianisidine twice with Chicago acid. Direct Blue 1 is used as a dye for cellulosic textiles, for paper, and for nylon. It is applied to cellulosic textiles and paper under neutral conditions; it dyes them by adsorption. It is used as an acid dye on nylon and must be applied under acidic conditions.

As mentioned above, it is not necessary to couple both diazonium groups of a tetrazotized

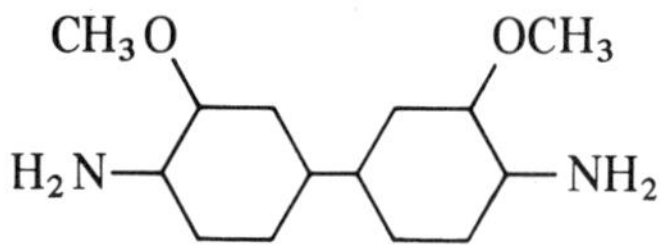

o-Dianisidine

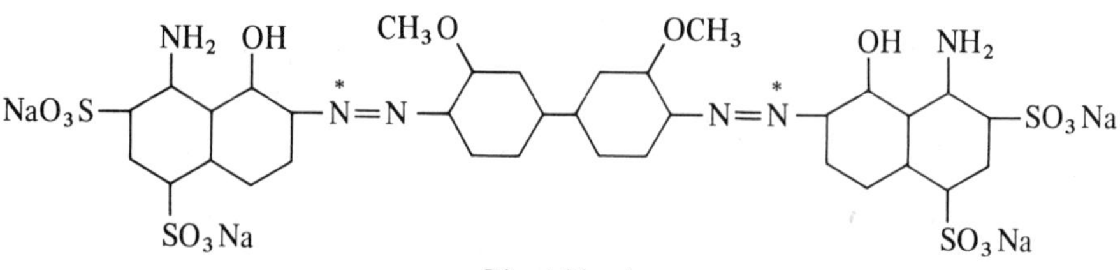

Direct Blue 1

diamine to the same coupling component. Acid Orange 49 is made by coupling tetrazotized *m*-tolidine with G Acid and with phenol, followed by formation of an ester of the phenol with *p*-toluenesulfonyl chloride. This dye is used on wool, silk, and leather.

Since dyes of this kind are usually made by reacting the coupling product with a cupric salt in aqueous ammonia, ammonia molecules are shown coordinated with the copper in the complex. This position may sometimes be filled by water or by various amines.

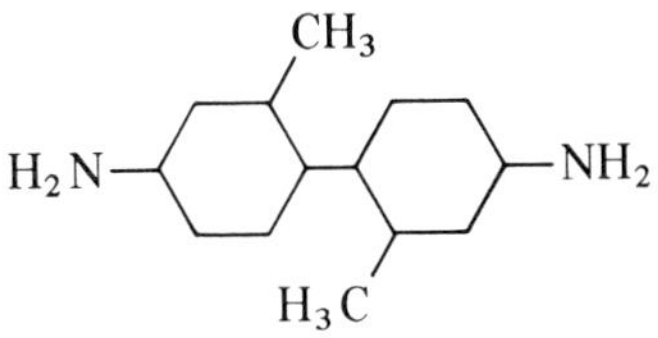

m-Tolidine

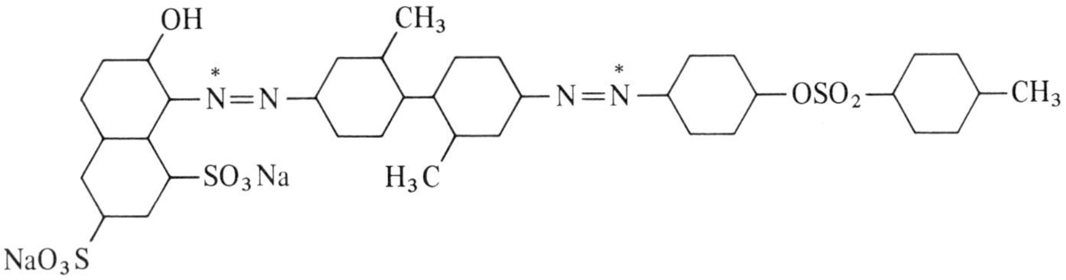

Acid Orange 49

The fastness to light of direct dyes is often improved by the formation of copper chelates. Direct Red 83 is the copper complex of the disazo dye prepared by coupling two moles of diazotized 2-aminophenol-4-sulfonic acid to one mole of J Acid urea.

A number of important dyes are based on diaminostilbenedisulfonic acid. An example is Direct Yellow 4 (Brilliant Yellow), which is made by coupling tetrazotized diaminostilbenedisulfonic acid twice to phenol. Direct Yellow 4 is an important dye for paper; it is

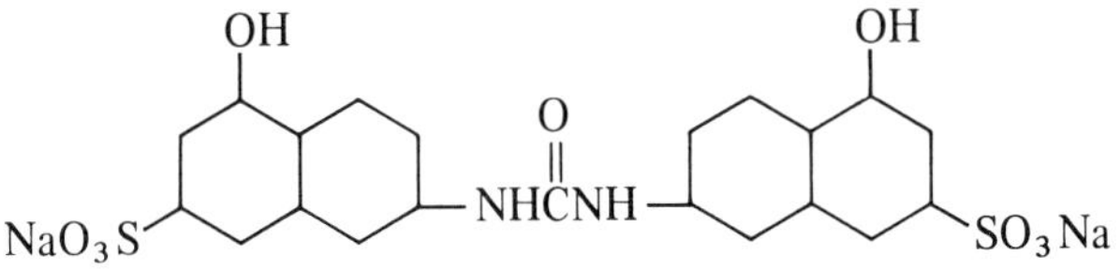

J Acid urea

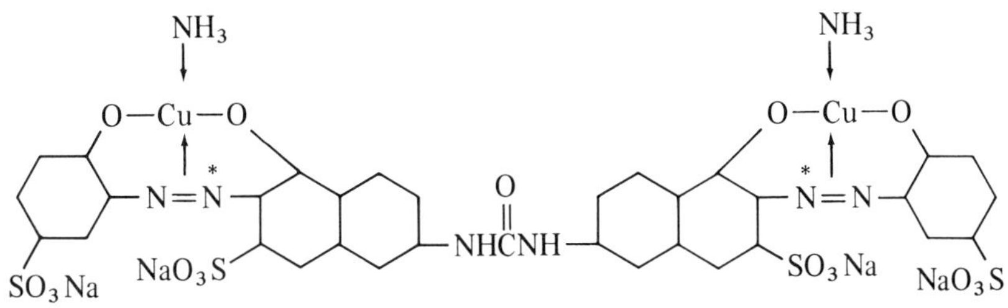

Direct Red 83

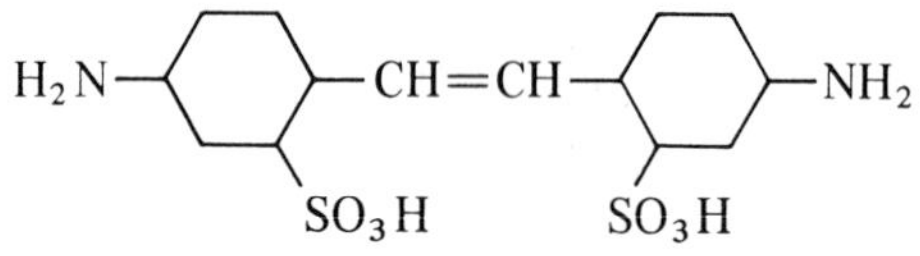

Diaminostilbenedisulfonic acid

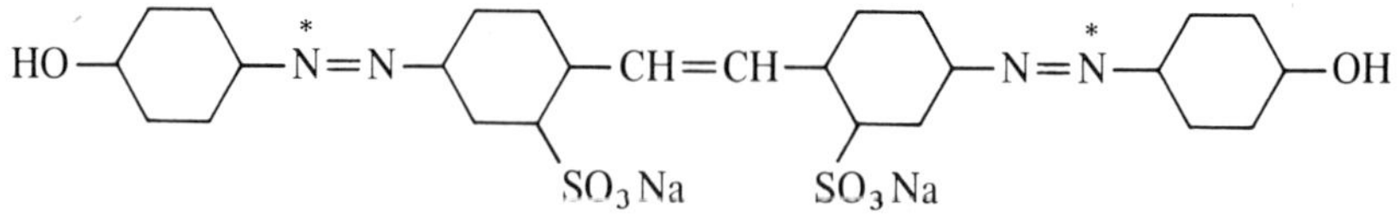

Direct Yellow 4

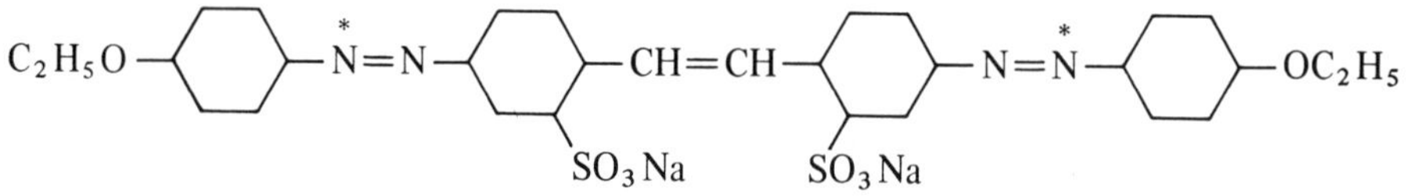

Direct Yellow 12

also used as an indicator since it turns red in strong alkali. Reaction of Direct Yellow 4 with ethyl chloride under pressure gives the ethyl ether, Direct Yellow 12 (Chrysophenine). This is an important dye for textiles and paper. Unlike Direct Yellow 4, it is not an indicator.

Some stilbene azo dyes are prepared by methods other than diazotization and coupling. Direct Orange 61, for example, is prepared by the condensation in alkaline solution of aminoazobenzenesulfonic acid and dinitrostilbenedisulfonic acid. The exact structure of Direct Orange 61 is not known, but the presence of an azo linkage formed by reaction of an amine and a nitro group can be demonstrated.

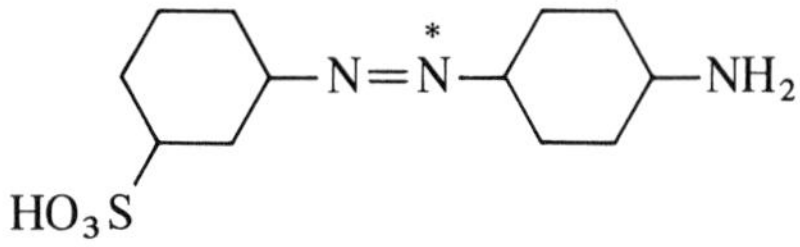

Aminoazobenzenesulfonic acid

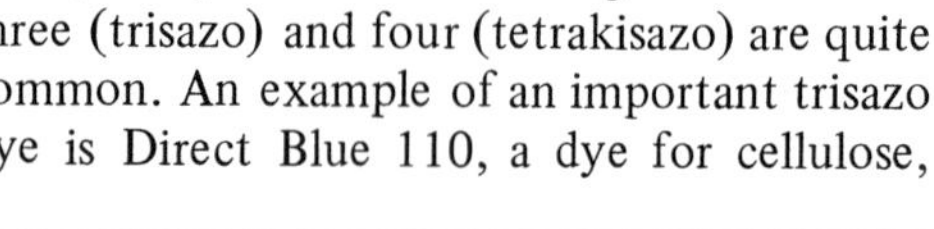

Dinitrostilbenedisulfonic acid

Polyazo Dyes. Azo dyes can be made with three, four, or more azo linkages. Those with three (trisazo) and four (tetrakisazo) are quite common. An example of an important trisazo dye is Direct Blue 110, a dye for cellulose,

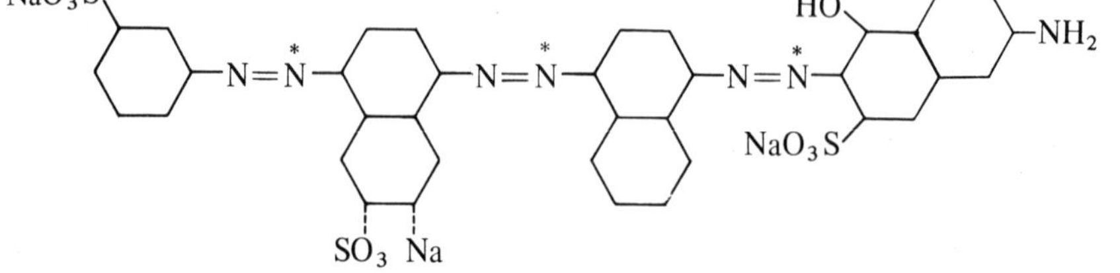

Direct Blue 110

wool, and silk. Examination of the formula shows that this dye is made from five intermediates: metanilic acid, 1,6 and 1,7 clevis acid, 1-naphthylamine, and J acid.

Manufacturing Processes for Azo Dyes

The manufacturing process for a typical azo dye may be illustrated with Direct Blue 6 (Direct Blue 2B), an important dye for cotton, viscose, silk, and paper.

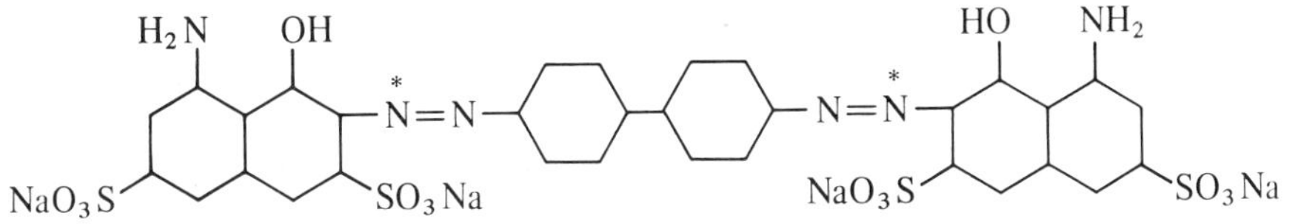

Direct Blue 6

The apparatus required for the manufacture of a typical azo dye is shown in Fig. 23.4.

p-Aminoacetanilide is suspended in water in a wooden or brick-lined tub (Tub 2) equipped with an agitator. Two moles of hydrochloric acid per mole of amine are added, and the whole is brought to a boil by passing in live steam. As soon as the *p*-aminoacetanilide has dissolved, ice is added. One mole of sodium nitrite, in solution in Tank 1, is run in slowly for about two hours. In the meantime, two moles of *p*-cresol are dissolved in an excess of soda ash in water (Tub 3), and to this the contents of Tub 2 are added with stirring. Enough ice is added to keep the temperature at about 5°C. The dye separates in part as the coupling proceeds. When coupling is complete, the contents of Tub 3 are warmed with steam.

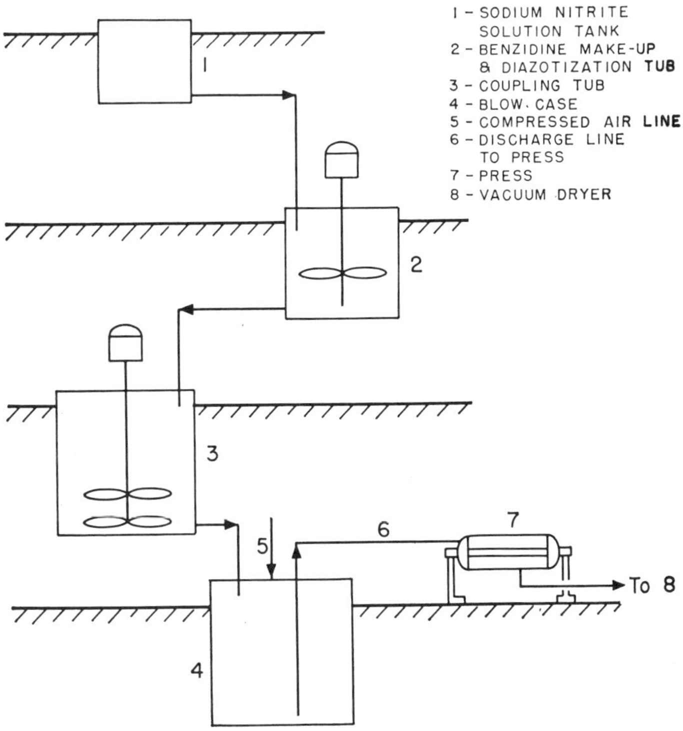

Fig. 23.4 Schematic diagram for manufacturing a benzidine dye.

The suspension of dye is run into a wooden plate-and-frame filter press. The filter cake is washed and freed from most of the adhering mother liquor by blowing with compressed air while still in the press. The moist cake is discharged into shallow trays which are placed in a circulating air drier, wherein the moisture is removed at temperatures between 50 and 120°C. Vacuum driers and drum driers may also be used. The dried dye is ground and mixed with a diluent and dispersant to make it equal in color strength to a predetermined standard. Dilution is necessary because batches differ in their content of pure dye; if sold "as is," the user would have to adjust his dyeing recipes for each batch. Uniformity is assured by dilution to a standard strength.

Triphenylmethane Dyes

Triphenylmethane dyes are characterized by the highly resonance-stabilized chromophore shown below. They are among the strongest and brightest of synthetic dyes, but usually do not exhibit good fastness to light. An exception is found with acrylic fibers, which are dyed in bright light-fast hues by selected triphenylmethane dyes.

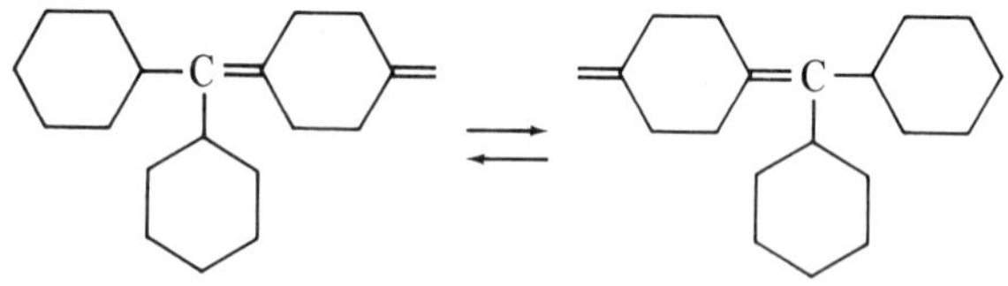

The triphenylmethane chromophore

Basic Red 9 (Pararosaniline) is the simplest triphenylmethane dye. It is prepared by the condensation of one mole of *p*-toluidine and two moles of aniline, usually by heating in nitrobenzene which serves as solvent and oxidant. The first product is a colorless carbinol which is converted to the dye with hydrochloric acid.

Closely related to Basic Red 9 is Basic Violet 14 (Fuchsine). This dye is made by condensing one mole each of aniline, *p*-toluidine, and *o*-toluidine in nitrobenzene. The dye is formed on neutralizing the carbinol with hydrochloric acid.

A second method for the synthesis of triphenylmethane dyes is illustrated with Basic Green 4 (Malachite Green). Benzaldehyde (one mole) is condensed with two moles of dimethylaniline to form a leuco base which is oxidized with lead peroxide and treated with hydrochloric acid to give the dye.

Basic Green 4 is used as a dye for bast fibers, acrylic fibers, leather, paper, and lacquers. When *o*-chlorobenzaldehyde is used in place of benzaldehyde, the product is Basic Blue 1, a more alkali-resistant dye which finds use on tannin-mordanted cotton, wool, leather, and paper.

Ketone condensation, a third method of manufacture of triphenylmethane dyes, is illustrated with Basic Violet 3 (Crystal Violet). Michler's Ketone, made by passing phosgene into dimethylaniline, is condensed with di-

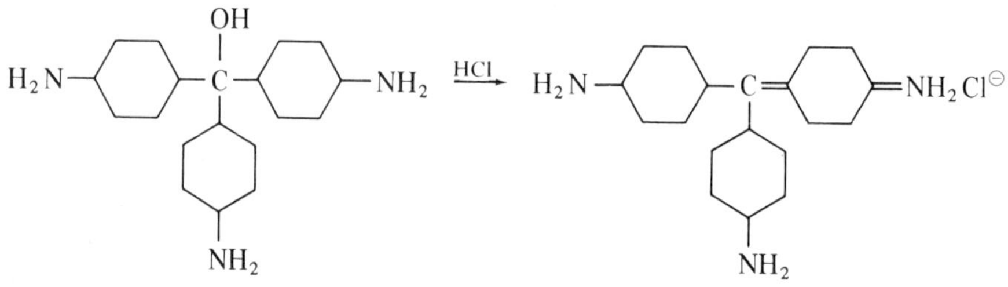

Colorless carbinol

Basic Red 9

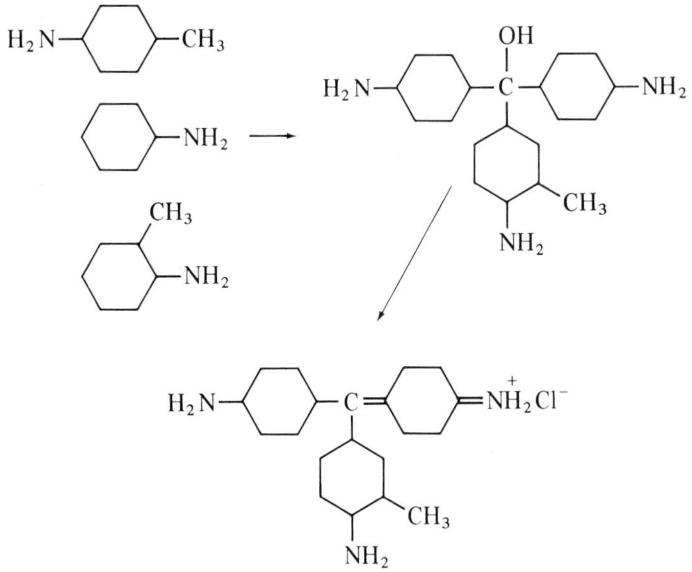

Basic Violet 14

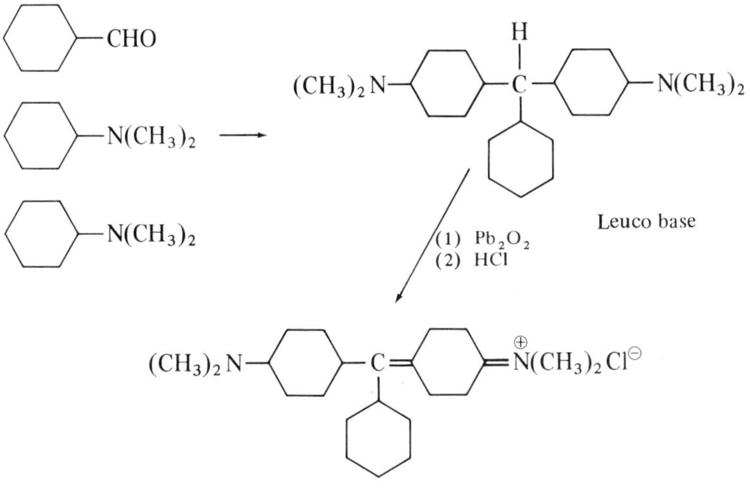

Basic Green 4

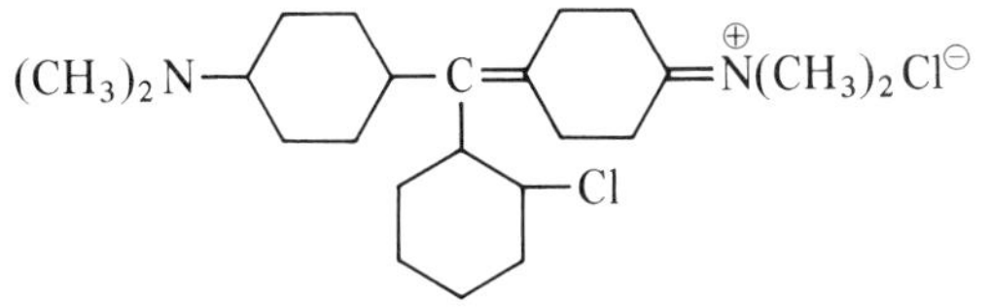

Basic Blue 1

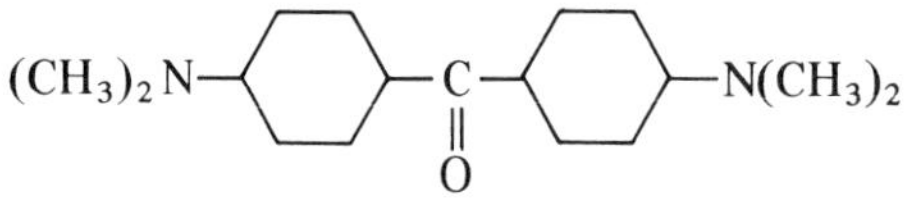

Michler's Ketone

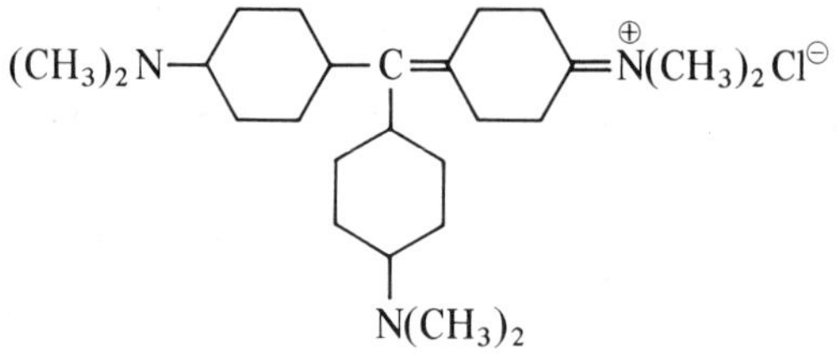

Basic Violet 3

methylaniline in a solvent using phosphorus oxychloride. A carbinol is formed which is converted to the dye with hydrochloric acid.

Not all triphenylmethane dyes are basic. Acid Green 3, for example, is a dye for wool, silk, and leather. There are two sulfonate groups in the molecule, one of which forms an inner salt with the positive charge on the chromophore.

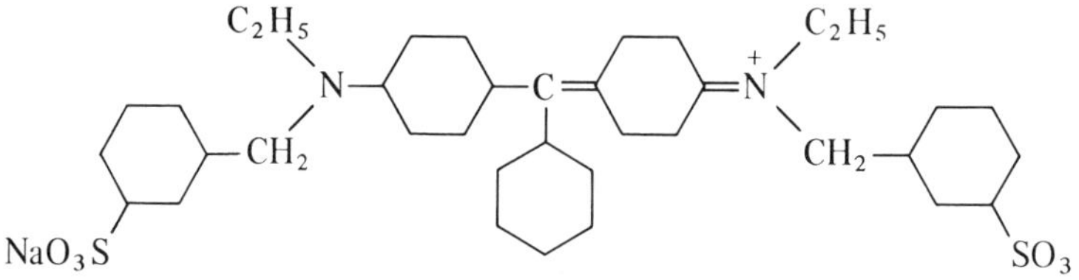

Acid Green 3

Aurin is a triphenylmethane dye containing no nitrogen. It is weakly colored, lacks solubility in water, and has no affinity for fibers. Some of its derivatives, however, with a carboxyl group *ortho* to one of the hydroxyl groups, are useful as mordant dyes.

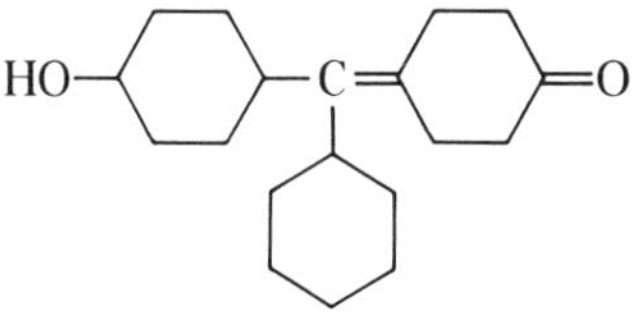

Aurin

Closely related to the triphenylmethane dyes are the diphenylmethane dyes. Only one of these, Basic Yellow 2 (Auramine), is important. It is useful on cotton, nylon, silk, wool, leather, and paper. It is sometimes classified as a ketonimine dye since it is the ketonimine hydrochloride of Michler's Ketone.

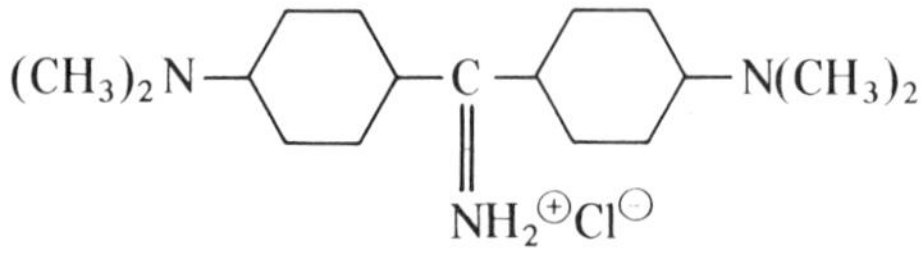

Basic Yellow 2

Xanthene Dyes

The chromophore of the xanthene dyes is the resonance-stabilized structure below. Acid Yellow 73 (Fluorescein), Acid Red 87 (Eosine), and Basic Violet 10 (Rhodamine B) each represent a series of xanthene dyes.

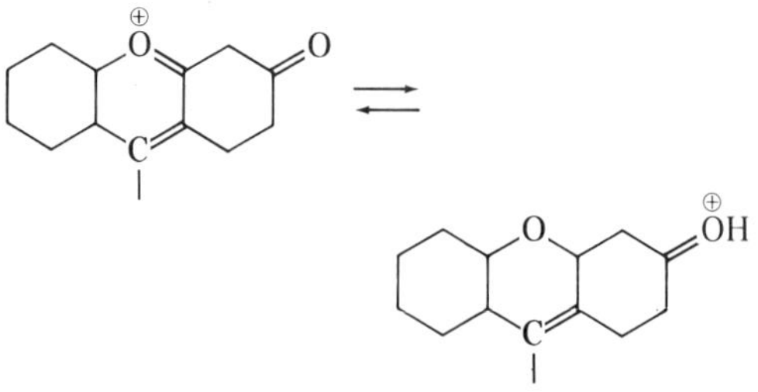

The xanthene chromophore

Acid Yellow 73 is made by heating phthalic anhydride (1 mole) and resorcinol (2 moles) in an iron kettle. The temperature is regulated by means of an oil or metal bath and kept at 220°C for seven hours. The melt is dissolved in caustic soda, and the product is precipitated by acidifying. It is a yellow-red powder that fluoresces green-yellow in alkaline solution. The sodium salt is called Uranine and is used to trace underground flow of water.

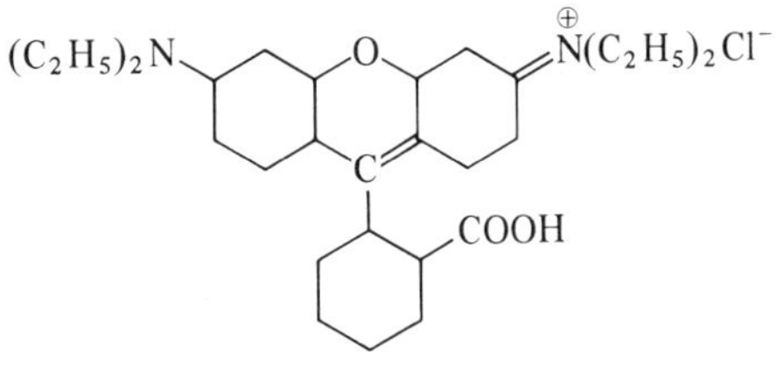

Basic Violet 10

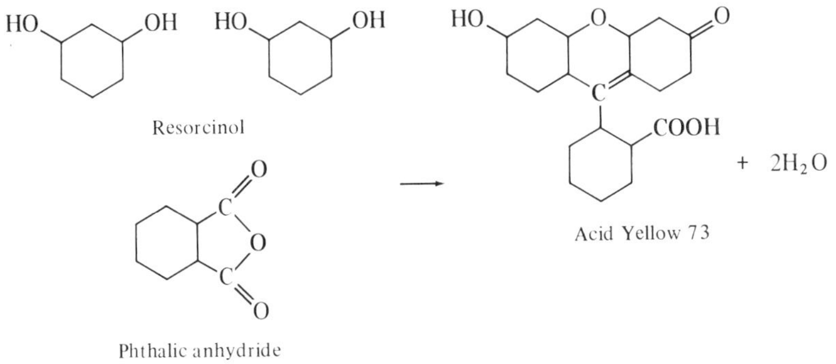

Acid Yellow 73

When Acid Yellow 73 is dissolved in alcohol and treated, while warm, with bromine, four equivalents are absorbed to from Acid Red 87 (Eosine). This dye is used to a limited extent on wool; its primary uses are for paper and inks.

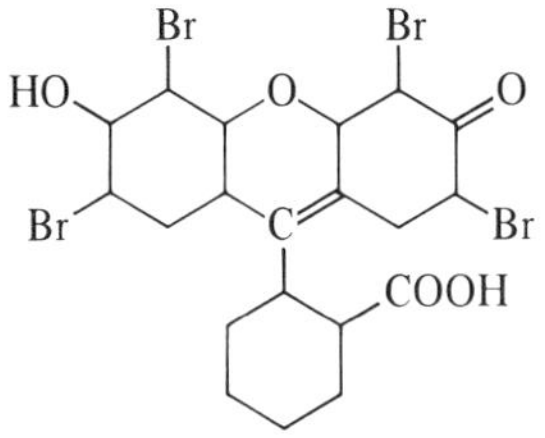

Acid Red 87

Other dyes related to Acid Red 87 are Acid Red 91 (dinitrodibromofluorescein), Acid Red 92 (Phloxine; tetrabromodichlorofluorescein), and Acid Red 95 (Erythrosine; tetraiodofluorescein).

Basic Violet 10 (Rhodamine B) is a dye for bast fibers, mordanted cotton, leather, and paper. It is prepared by condensing phthalic anhydride with *m*-diethylaminophenol and treating with hydrochloric acid. This is the general preparative route for rhodamines.

Related to the xanthene dyes are the *acridines*, *azines*, *oxazines*, and *thiazines*. Acridine Yellow results from the fusion of *m*-toluenediamine with glycerin and oxalic acid, followed by oxidation with ferric chloride. Basic Orange 15 (Phosphine) is a by-product of the manufacture of Fuchsine, and is an acridine dye.

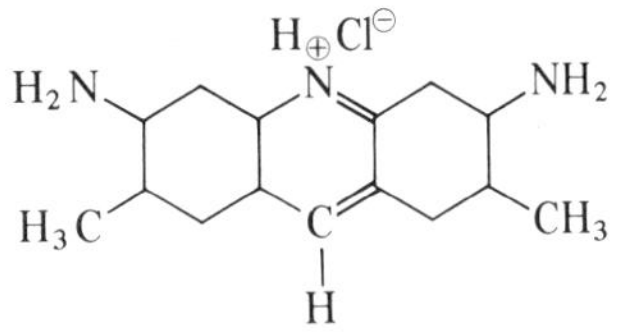

Acridine Yellow

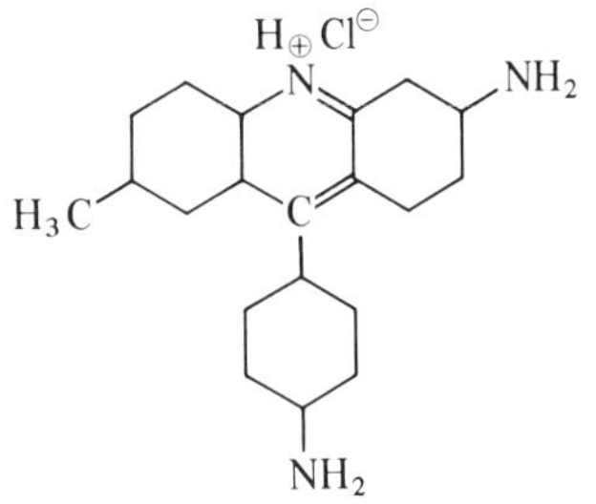

Basic Orange 15

Examples of azine dyes are Safranine B, Basic Red 2 (Safranine T), and Acid Blue 59. The preparation of azine dyes is illustrated with Safranine B. *p*-Phenylenediamine and aniline are reacted to form indamine. Further reaction of indamine with aniline under oxidizing conditions, followed by treatment with hydrochloric acid, gives Safranine B.

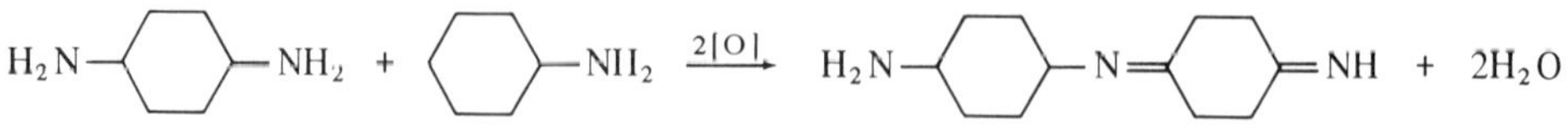

Phenylenediamine   Indamine

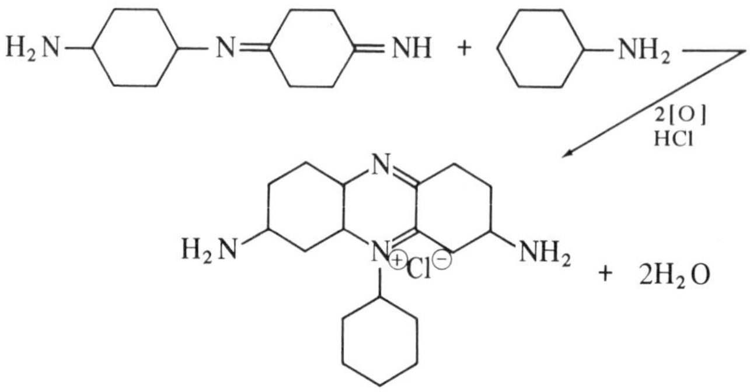

Safranine B

Indamine (above) is an example of a *quinone-imide* dye. Dyes of the indamine type are used in color photography; the dye is formed during development of the film.

Basic Blue 6 (Meldola's Blue) is an oxazine dye. Basic Blue 9 (Methylene Blue) illustrates the thiazine dye class.

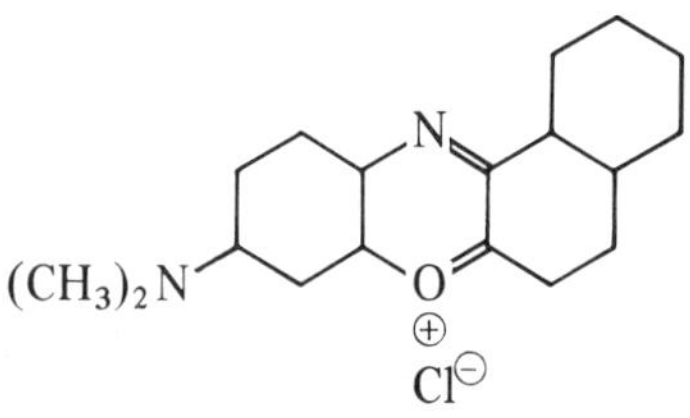

Basic Blue 6

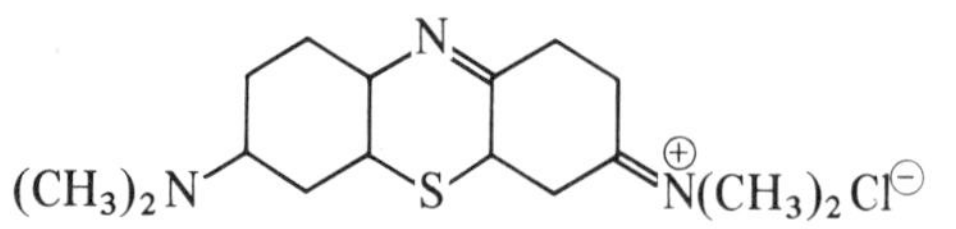

Basic Blue 9

Anthraquinone and Related Dyes

Dyes based on anthraquinone and related polycyclic aromatic quinones are of great importance. Many of the most light-fast acid, mordant, disperse, and vat dyes are of this kind. The chromophore is the quinoid group, $>C=O$.

Anthraquinone Acid Dyes. Anthraquinone acid dyes are illustrated by Acid Blue 25, Acid Blue 45, and Alizarin Cyanine Green G (Acid Green 25). These dyes are water-soluble anthraquinone derivatives which are used for dyeing wool, silk, nylon, leather, and paper. The chemistry of this type of dye is shown with Blue 25 as an example. The starting material is 1-aminoanthraquinone which is sulfonated and brominated to give Bromamine Acid. Condensation with aniline then yields Acid Blue 25 which is isolated as the sodium salt.

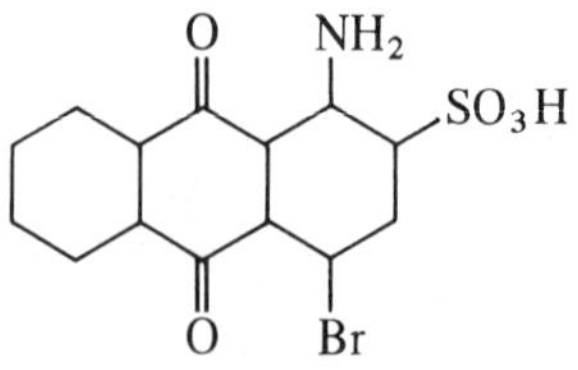

Bromamine Acid

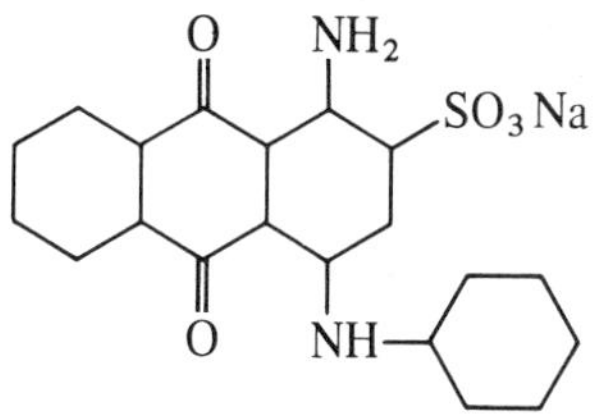

Acid Blue 25

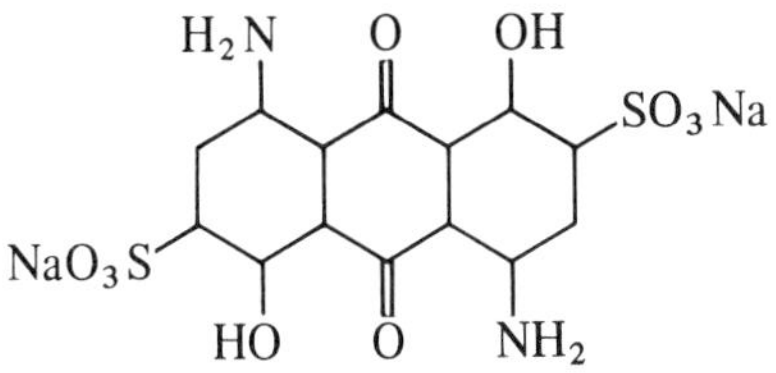

Acid Blue 45

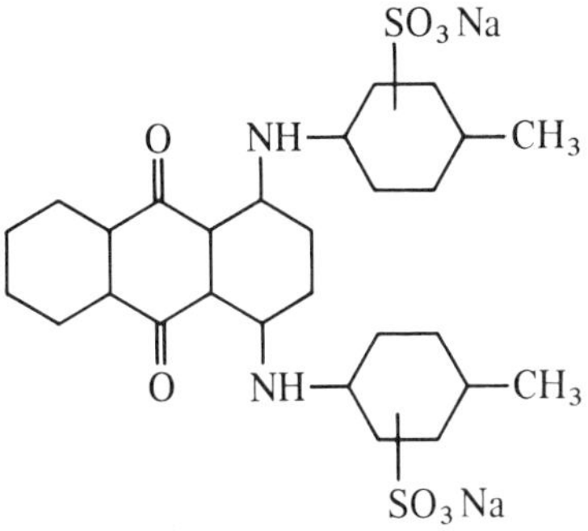

Acid Green 25

Anthraquinone Mordant Dyes. Anthraquinone mordant dyes contain groups, such as hydroxyl or carboxyl, which can combine with metal ions. The simplest member of this group, Mordant Red 11 (Alizarin), is 1,2-dihydroxyanthraquinone. When applied to wool with an aluminum mordant it gives the well-known Turkey Red, and when converted to its calcium salt forms a bluish red powder useful as a pigment.

Mordant Red 11 is made by heating, under pressure, Silver Salt (sodium anthraquinone-2-sulfonate, so called because of its silvery crystals), caustic soda, potassium chlorate, and water. A steel autoclave is ordinarily used and the temperature is maintained at about 180°C. The resulting melt is blown into water and acidified to decompose the sodium alizarate; the precipitated alizarin is filtered, washed, and standardized as a 20 per cent paste.

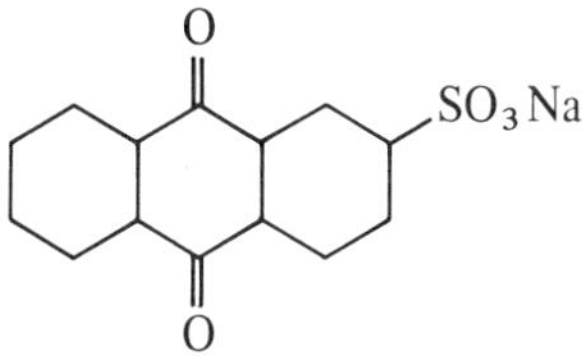

Silver Salt

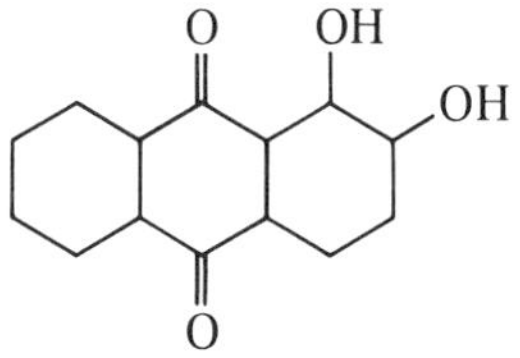

Mordant Red 11

While Mordant Red 11 is very sparingly soluble in water, the introduction of sulfonic acid groups gives soluble derivatives. The most important water-soluble dye of the alizarin mordant class is Mordant Black 13, a dye primarily for wool. It is made by condensing aniline with 1,2,4-trihydroxyanthraquinone and sulfonating the resulting base. It is applied to wool with a chronium mordant and is quite fast to light and washing.

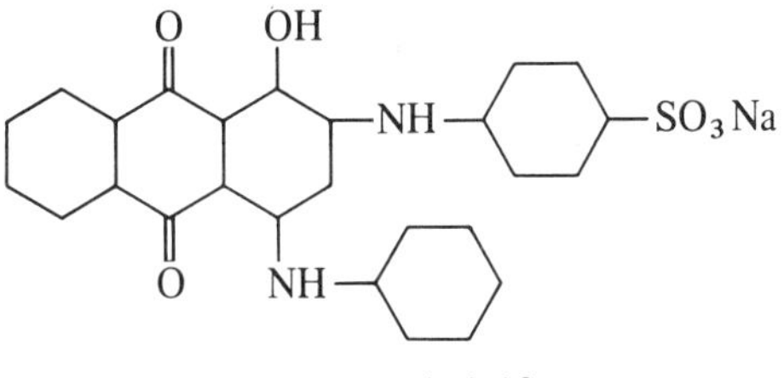

Mordant Black 13

Anthraquinone Vat Dyes. Many of the best vat dyes are derivatives of anthraquinone or related compounds. A relatively simple dye of this class is Vat Yellow 3, which is made by benzoylation of 1,5-diaminoanthraquinone. It is a yellow pigment that is used as such when properly ground and dried. As a vat dye, it is usually supplied as an aqueous paste. Reduction with sodium hydrosulfite in caustic soda

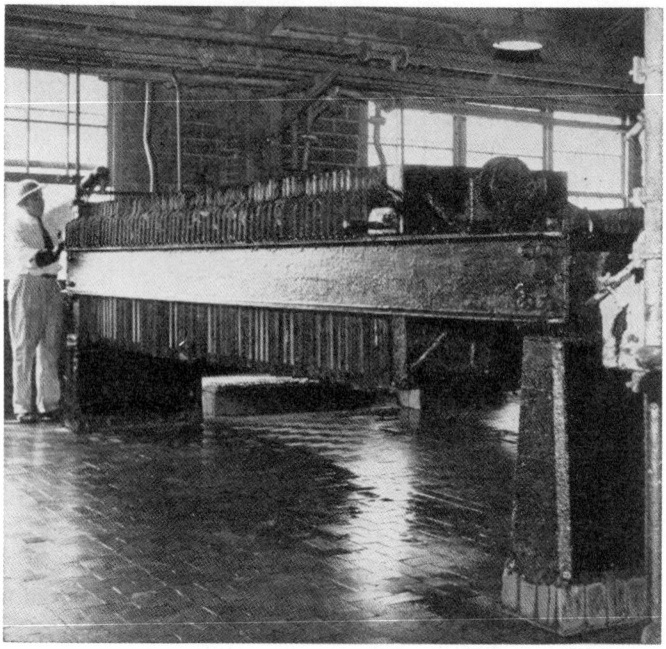

Fig. 23.5 A plate and frame filter press used for the isolation of intermediates and dyes. (*Courtesy Organic Chemicals Dept., E.I. du Pont de Nemours and Co., Inc.*)

solution gives the soluble salt of the hydroquinone, which has affinity for cellulosic fibers. After application to the fiber, the insoluble dye is re-formed by oxidation, as described earlier in this chapter.

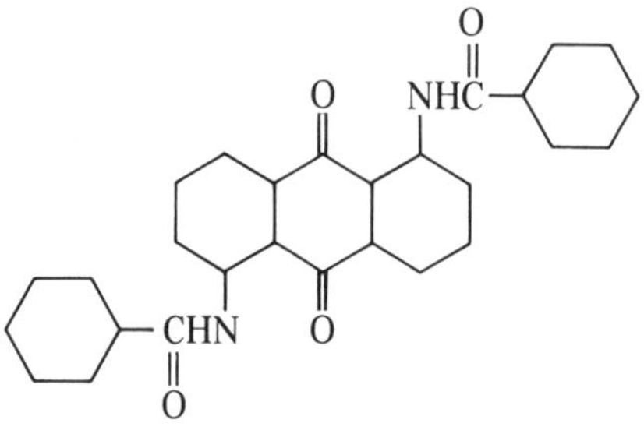

Vat Yellow 3

Perhaps the best known of the anthraquinone vat dyes is Vat Blue 4 (Indanthrone). This attractive dye is made by the fusion of 2-aminoanthraquinone with caustic potash. The fastness to chlorine bleaching of Vat Blue 4 is improved by chlorination. A number of chlorinated indanthrones are in commercial use and make up the most important group of fast vat blues. The preparative route for Vat Blue 4 is outlined below:

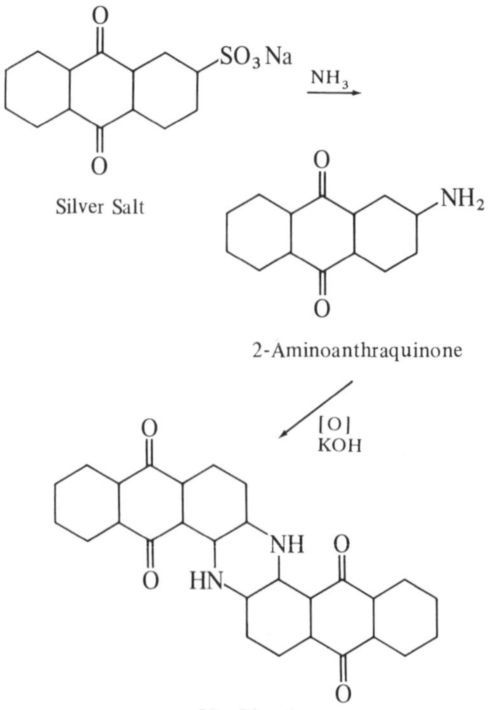

Vat Blue 4

Fig. 23.6 A low-pressure autoclave used in the manufacture of vat dyes. (*Courtesy Organic Chemicals Dept., E.I. du Pont de Nemours and Co., Inc.*)

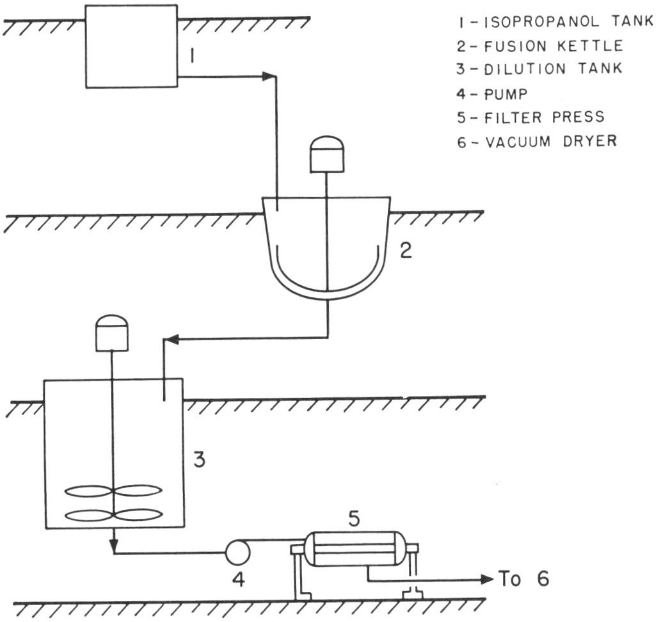

Fig. 23.7 Schematic diagram for manufacturing dibenzanthrone.

Related to Vat Blue 4 are Vat Yellow 1 (Flavanthrone) and Vat Orange 9 (Pyranthrone), as well as other important dyes.

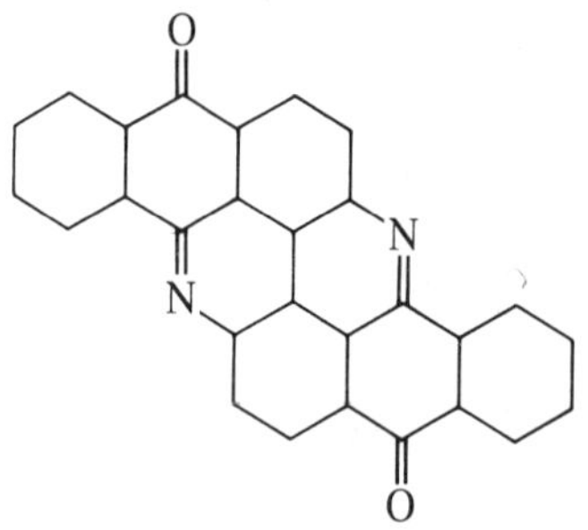

Vat Yellow 1

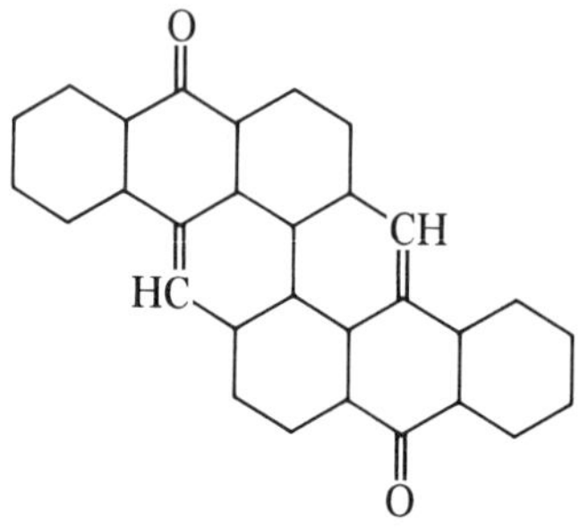

Vat Orange 9

One of the most attractive of all vat dyes is Vat Green 1 (Jade Green). This dye is a derivative of dibenzanthrone, which is itself a vat dye (Vat Blue 20 or Violanthrone).

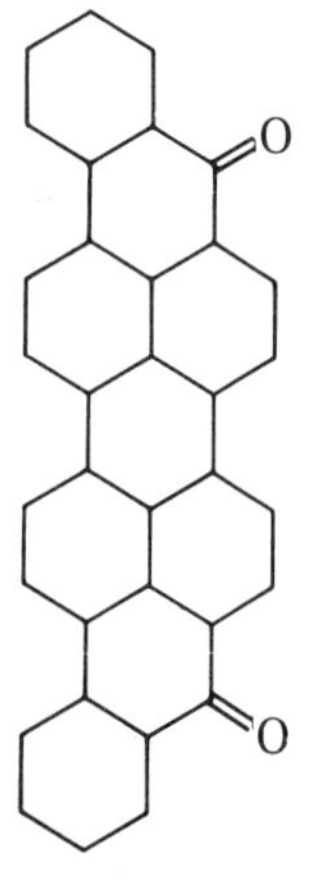

Vat Blue 20

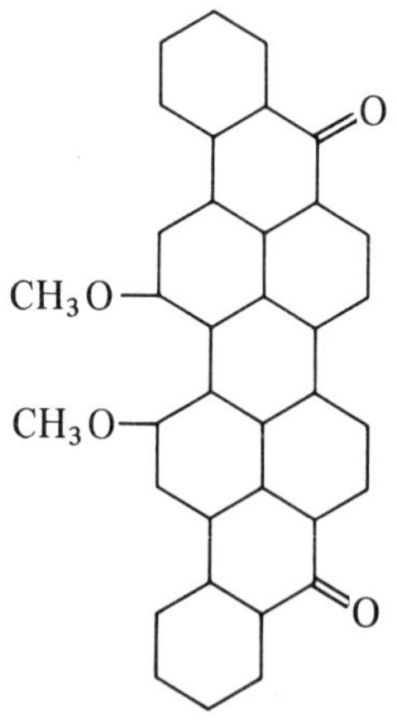

Vat Green 1

Anthraquinone-vat-dye manufacture is illustrated by the preparation of Vat Blue 20. Phthalic anhydride is condensed with benzene in aluminum chloride to give *o*-benzoylbenzoic acid which is ring-closed to anthraquinone with sulfuric acid. Treatment of anthraquinone in 82 per cent sulfuric acid with metal powder (iron, copper) and glycerin gives benzanthrone, probably by the series of steps shown below. Condensation of benzanthrone with caustic potash is isobutanol and sodium acetate gives dibenzanthronyl, which undergoes ring-closure to dibenzanthrone upon further heating. The dibenzanthrone usually is dissolved in sulfuric acid and reprecipitated as a paste of fine particle size by dilution with water. The paste is washed free of acid and standardized as Vat Blue 20. The equipment for this series of reactions must be able to resist the strongly corrosive conditions and must be externally heated and cooled. Cast iron, stainless steel, and enamel-lined closed vessels are used.

Indigoid and Thioindigoid Dyes

The parent compound of the indigoid dye class, indigo, has been in use as a vat dye since ancient times. Natural indigo is obtained from a species of plant, *Indigofera*. It can be reduced by fermentation as well as by the modern method using sodium hydrosulfite and caustic soda. Once reduced, it can be applied to cellulosic fibers and then oxidized by air to produce rich blue dyeings. Natural indigo is no longer important, but large quantities of the synthetic dye are used to produce low-cost blues on cotton. The oxidized form is also used as a pigment. The "Colour Index" designation for indigo is Vat Blue 1.[10]

A modern synthesis of Vat Blue 1 involves first the preparation of phenylglycine. To a water solution of sodium bisulfite are added formaldehyde, aniline, and sodium cyanide. The nitrile of phenylglycine which is formed is isolated and washed free of sulfite. Warming of the nitrile in an alkaline slurry then gives a solution of phenylglycine as its sodium salt. Ammonia is evolved during this step.

In the second step the phenylglycine salt is condensed by heating with a eutectic mixture of potassium and sodium hydroxides in the

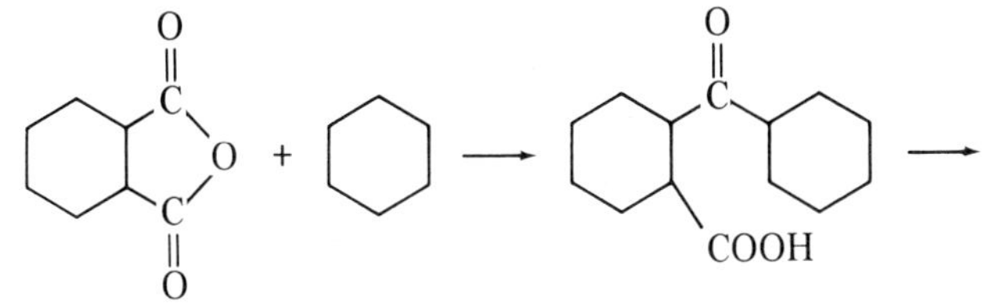

Phthalic anhydride o-Benzoylbenzoic acid

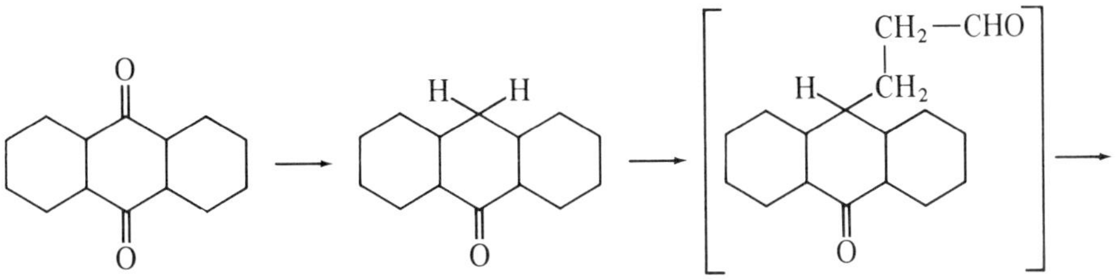

Anthraquinone Anthrone Intermediate

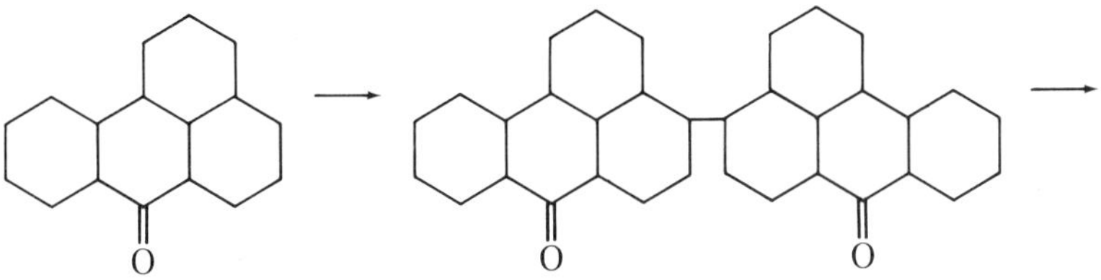

Benzanthrone Dibenzanthronyl

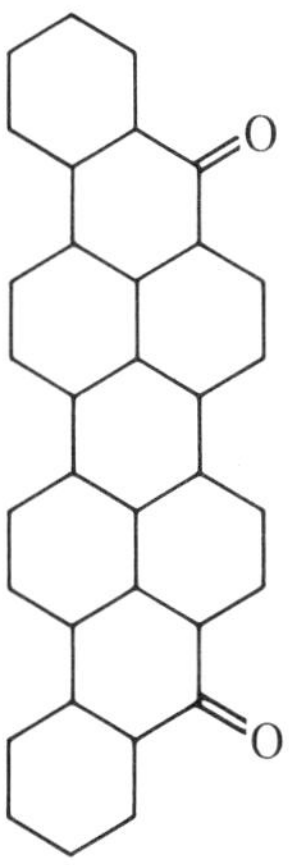

Dibenzanthrone

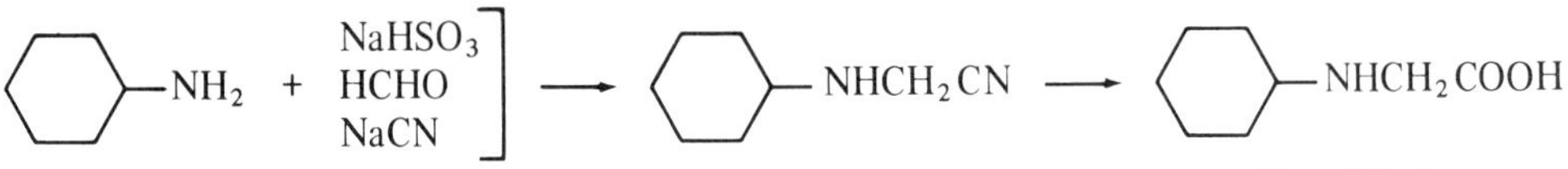

Nitrile of phenylglycine Phenylglycine

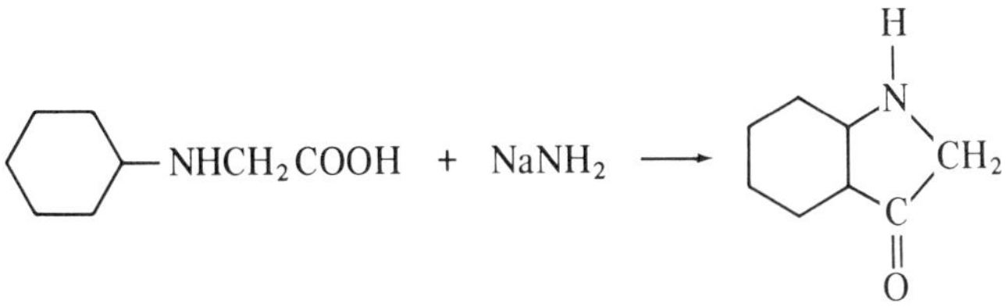

Sodamide Indoxyl

presence of sodamide, which removes the water formed. The reaction temperature is 200–220°C. The product, which is formed in high yield, is indoxyl.

After the fusion step, water is added and air is blown into the alkaline solution of indoxyl. Indigo forms and separates from solution. It is filtered, washed, and standardized as a paste of about 20 per cent solids.

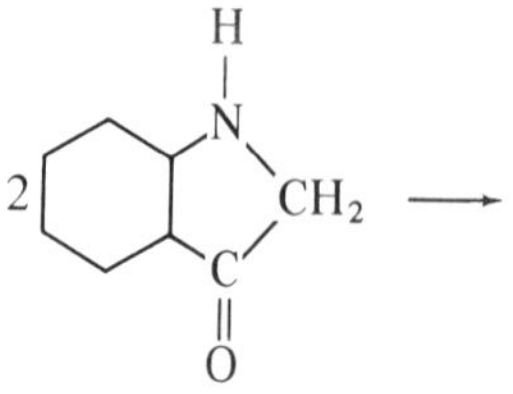

Indoxyl

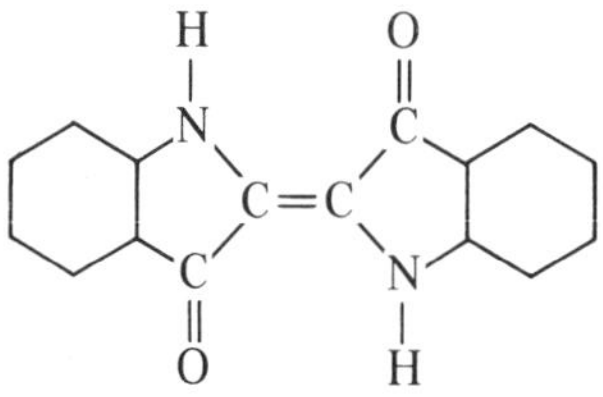

Vat Blue 1

Since very large volumes of indigo are produced, it has been worthwhile to devise continuous processes. Such processes were pioneered by the former I. G. Farbenindustrie in Germany.

Vat Blue 1 is available as a water-soluble salt of the reduced dye. When applied to the fiber and treated with a solution of sodium nitrite or sodium bichromate, it reverts to the insoluble dye.

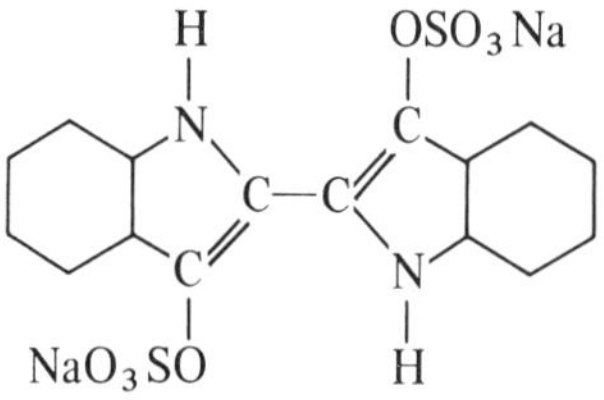

Solubilized Vat Blue 1

Important derivatives of indigo are the chloro and bromo derivatives. Vat Blue 5 is tetrabromoindigo. It is similar to Vat Blue 1 in shade but is considerably brighter. It is made by bromination of indigo in acetic acid.

Vat Red 41 (Thioindigo) is made by the reaction of thiosalicylic acid with chloroacetic acid, followed by fusion with caustic soda and oxidation by air. Thiosalicylic acid is made from anthranilic acid by diazotization and reaction with hydrogen sulfide.

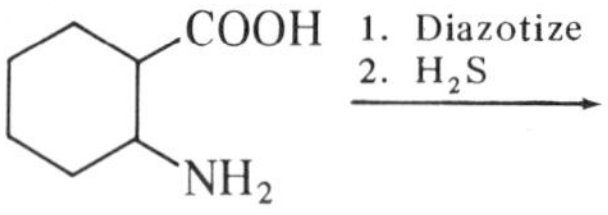

Anthranilic acid

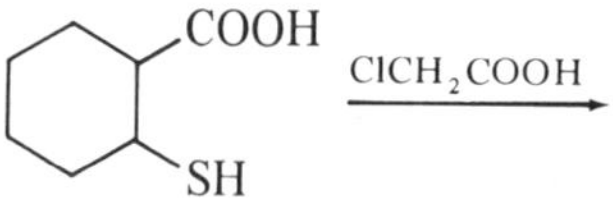

Thiosalicylic acid

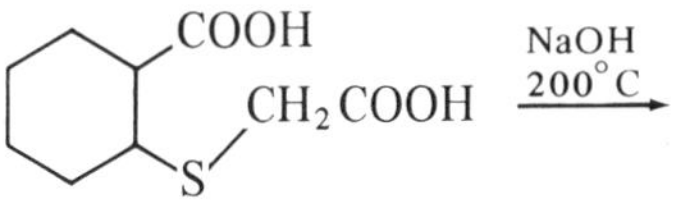

o-(Carboxymethylmercapto)benzoic acid

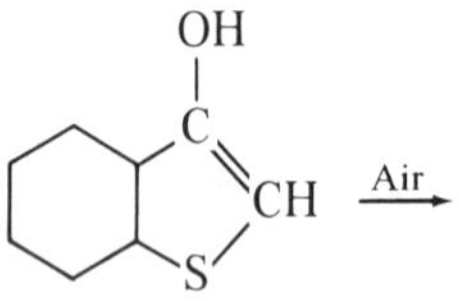

Thioindoxyl

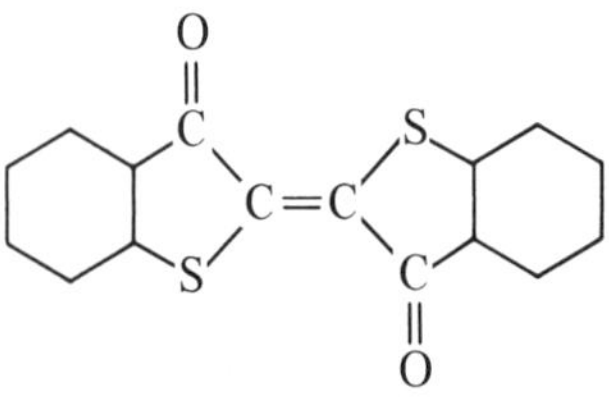

Vat Red 41

Vat Red 41 is a dull bluish red on cotton. On polyester fibers it produces bright pink shades. Many of its derivatives are of greater commercial importance on cotton. An example is Vat Orange 5.

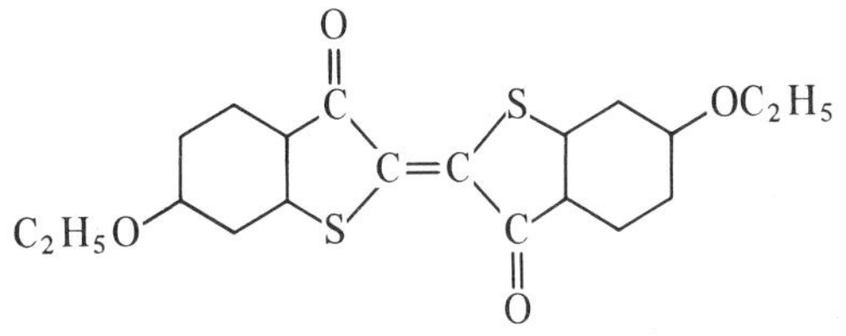

Vat Orange 5

Sulfur Dyes

Sulfur dyes are dyes of unknown constitution which can be applied to fibers when reduced with sodium sulfide. Most of them are insoluble in water before reduction. After reduction they are soluble and can be absorbed by fibers and then oxidized to an insoluble form with air. Although structures cannot be written for the sulfur dyes, the methods for reproducing individual types are well established.

m-Diamines (*m*-toluenediamine for example) fused with sulfur cause the evolution of hydrogen sulfide and produce a melt which yields brown sulfur dyes. Redder shades are obtained by using derivatives of azine, such as the compound below which produces Sulfur Red 6.

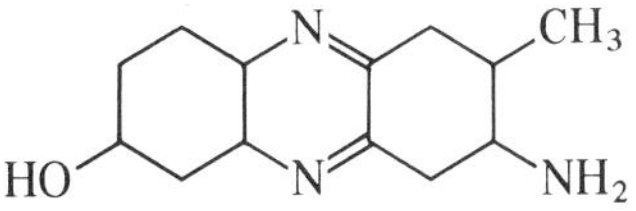

Intermediate for Sulfur Red 6

Blue sulfur dyes are obtained by treating diphenylamine derivatives with sodium polysulfide. Sulfur Blue 7, Sulfur Blue 9, and Sulfur Blue 13 are made from the intermediates below:

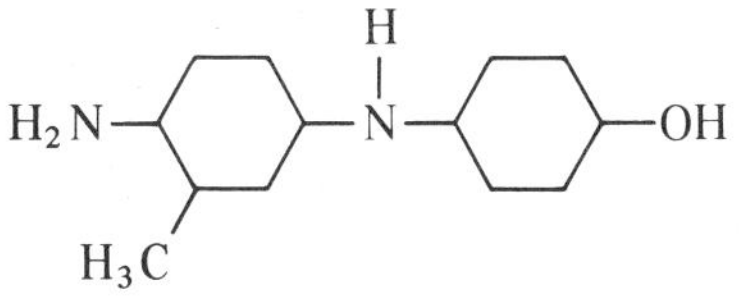

Intermediate for Sulfur Blue 7

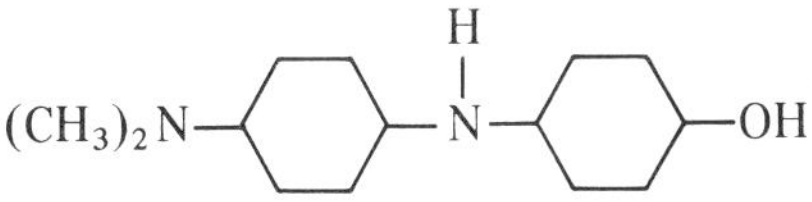

Intermediate for Sulfur Blue 9

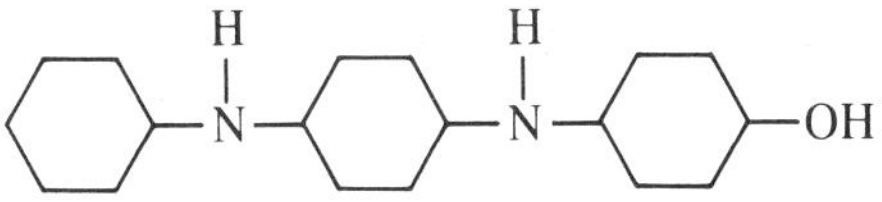

Intermediate for Sulfur Blue 13

Sulfur Black 1 is prepared by heating *m*-dinitrophenol with sodium polysulfide. The fusion mass is dissolved in water and blown with air until all the dye has separated. It is then filtered, washed, and dried.

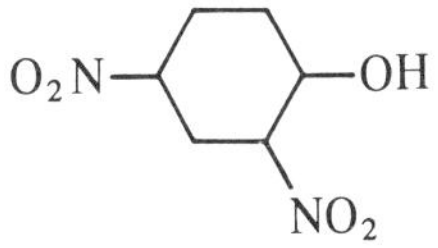

Intermediate for Sulfur Black 1

Vat Blue 43 (Hydron Blue R) is a kind of sulfur dye which, unlike typical sulfur dyes,

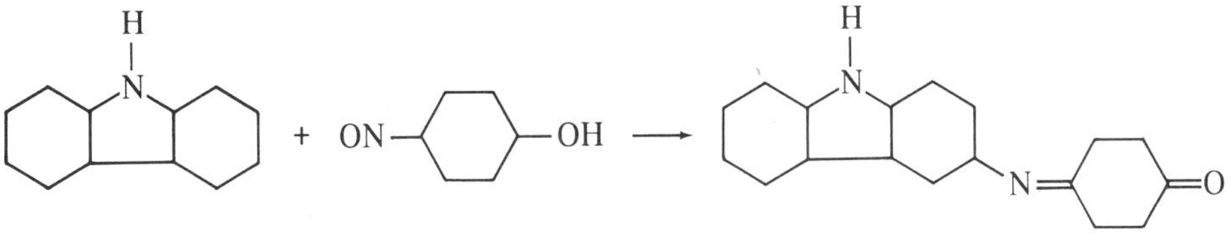

Carbazole *p*-Nitrosophenol Condensation product

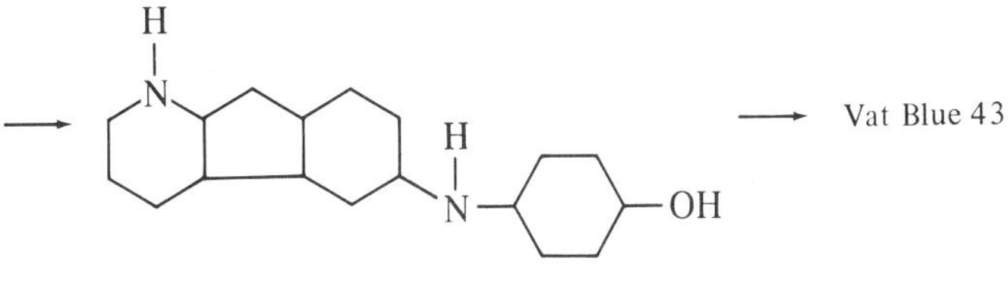

Reduced condensation product

may be reduced by sodium hydrosulfite without destruction. To prepare Vat Blue 43, carbazole is condensed with *p*-nitrosophenol in sulfuric acid to give an indophenol which is reduced and then fused with polysulfide.

Vat Blue 42 (Hydron Blue G) gives a greener shade than Vat Blue 43. It is prepared by using ethylcarbazole in place of carbazole.

Phthalocyanines

The phthalocyanines constitute an important class of synthetic pigments and dyes. The parent compound is Pigment Blue 16 (Phthalocyanine). One method of preparation involves the fusion of phthalonitrile with cyclohexylamine in an inert solvent. The two hydrogen atoms can be replaced by metals such as copper, nickel, iron, and cobalt. In actual practice, the metal derivatives are not made from the metal-free compound, but are synthesized directly. Pigment Blue 15 (Copper Phthalocyanine), for example, is made by the fusion of phthalonitrile with copper metal or a copper salt. An alternative method involves the fusion of phthalic anhydride with urea and a copper salt in the presence of a molybdenum catalyst. Pigment Blue 15 and its polychlorinated derivative, Pigment Green 7, are the most important of the phthalocyanines. They are distinguished by great brilliance, strength and stability, and are used in every field in which colored pigments are used.

A number of phthalocyanine derivatives can be made by substituting the hydrogen atoms on the benzene rings. Pigment Green 7, mentioned above, is an example. Vat Blue 29 is a partially sulfonated cobalt phthalocyanine. Reduction with sodium hydrosulfite in caustic soda solution converts it to a soluble form that can be applied to textiles and reoxidized to the pigment. The structure of the reduced form is not known.

Direct Blue 86 is copper phthalocyanine which has been sulfonated to the extent of between two and three sulfonic acid groups per molecule. It produces very bright blue-green shades on cotton, viscose, and paper. Solvent Blue 25 is a spirit-soluble derivative of copper phthalocyanine, useful for inks, lacquers, and stains. It is the reaction product of the tetrasulfonyl chloride of copper phthalocyanine with isohexylamine.

Fluorescent Brightening Agents

The fluorescent brightening agents, also known as fluorescent whiteners or optical bleaches, comprise a very valuable class of dyes. Although they were not commercialized until just prior to World War II, their use has grown at an exceedingly rapid rate. They are now one of the major dye classes produced in the United States, having a value of over \$51 million in 1971 (see Table 23.2).

Fluorescent brighteners derive their value

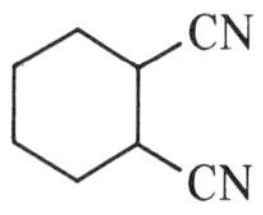

Phthalonitrile

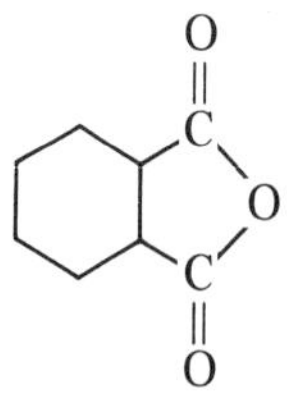

Phthalic anhydride

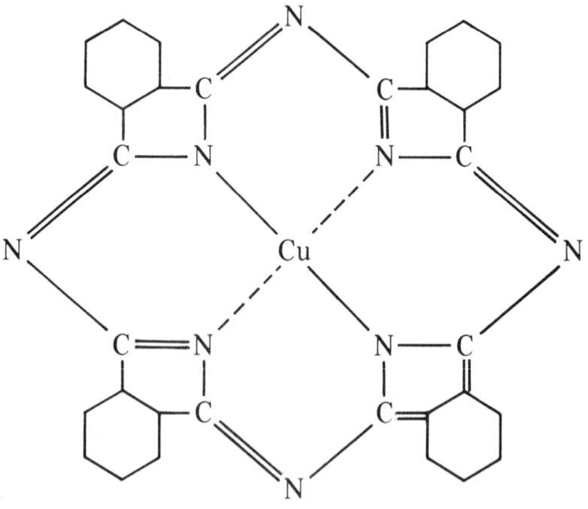

Pigment Blue 15

from the fact that although they are colorless, they strongly absorb light of shorter wave length in the ultraviolet region and fluoresce or re-emit light of longer wave length in the visible region of the spectrum. They, thus, produce a brightening or whitening effect which is useful in making yellowed products appear whiter and white materials brighter. The major use of optical brighteners is in detergents. However, considerable amounts are used in the paper and textile industries.

A number of molecules have the ability to fluoresce but relatively few have achieved commercial importance. The organic chemist can tailor fluorescent dyes to be useful on almost any substrate, that is, they can be made into direct, cationic, or disperse dyes. By far the major volume is in the direct-dye type for cotton and paper. The major fluorophore is the diaminostilbene nucleus which is then made substantive by incorporation of solubilizing and affinity-conferring groups. A typical commercial fluorescent brightener is made by condensing diaminostilbenedisulfonic acid with cyanuric chloride, and then, further substituting the cyanuric chloride residue with colorless amines to give a product such as the one shown below.

TABLE 23.4 Benzenoid Dyes: U.S. Production and Sales, by Chemical Class, 1971

Chemical Class	*Production (Thousand pounds)*
Anthraquinone	47,191
Azo	82,050
Azoic	10,385
Cyanine	871
Methine	3,889
Nitro	1,463
Oxazine	509
Phthalocyanine	1,811
Quinoline	2,237
Stilbene	32,162
Thiazole	369
Triarylmethane	7,718
Xanthene	1,060
All Other	49,646
TOTAL	243,729

NEW DEVELOPMENTS IN DYES

The most significant development influencing the dye industry during the past decade involved the rapid growth of synthetic fibers, which, in the late 1960's, surpassed natural fibers in over-all poundage. Commensurate with the rise of the synthetic fibers was the development of new dyes and dyeing methods which were needed to make these new fibers economically attractive. Higher labor costs and the need for greatly expanded output required the development of new low-cost continuous dyeing processes. Advanced instrumental techniques, with the aid of computers, are making exact color matching possible from lot to lot.

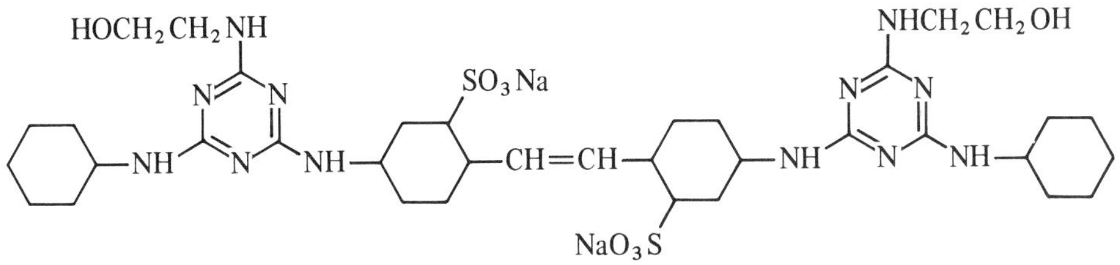

C.I. Fluorescent Brightening Agent

PRODUCTION STATISTICS

Table 23.4 gives the production of synthetic dyes in the United States for the year 1971, arranged according to chemical class. These figures are taken from the report of the United States Tariff Commission. As in Table 23.1, it will be noted that the classifications used are not exactly the same as those given in the "Colour Index."

The future will continue to see improvements and refinements in these techniques. Dye manufacture will be automated from batch to batch to produce spectrally reproducible products. New fiber variations will require continual development of new dyes to optimize fastness, application simplicity, and hue. Pollution abatement will be a major concern in the dye manufacturing as well as the dye application industries. Equipment and processes must be designed to eliminate noxious fumes and waste streams. This will require new designs for reactors and scrubbers, and efficient cleanup of effluents of both the colored and the not-so-obvious soluble colorless salts. Water, which is a major raw material for the manufacture of dyes and in dye application, must be conserved and left free of chemical or thermal contamination. There is much activity in the direction of solvent processing of textiles as a means of water conservation. Dyeing processes using recoverable organic solvents in place of water are attractive because of their reduced energy requirements for drying, and their reduced water requirements. As yet no large-scale applications of solvent dyeing have been commercialized. The period of the 1970–80's represents a real challenge to the engineer to demonstrate that man can freely use colored materials as a means of self-expression, motivation, and beautification, while maintaining a sound ecological environment.

REFERENCES

1. American Association of Textile Chemists and Colorists, *Technical Manual and Year Book*, an annual publication.
2. Fierz-David, H. E., and Blangey, L., "Fundamental Processes of Dye Chemistry," translation by P. W. Vittum, New York, John Wiley & Sons, 1949.
3. Friedlaender, Ed., Vols. 1–13, Fierz-David, H. E., Ed., Vols. 14–25, "Fortschritte der Teerfarbenfabriken," Berlin, Julius Springer, 1888–1942.
4. Groggins, P. H., Ed., "Unit Processes in Organic Synthesis," 5th Ed., New York, McGraw-Hill, 1958.
5. Houben, "Das Anthracene und die Anthrachinone," Leipzig, Thieme, 1929.
6. I.G. Farbenindustrie, FIAT 764 (PB 60946), "Dyestuffs Manufacturing Processes of I.G. Farbenindustrie A.G.," 1947.
7. I.G. Farbenindustrie, FIAT 1313 (PB 85172), "German Dyestuffs and Dyestuff Intermediates, including Manufacturing Processes, Plant Design, and Research Data," 1948.
8. Lubs, H. A., "The Chemistry of Synthetic Dyes and Pigments," ACS Monograph, New York, Van Nostrand Reinhold, 1955.
9. Meunier, *J. Am. Assn. Textile Chem. & Colorists*, **2** 386 (1970).
10. Society of Dyers and Colourists and American Association of Textile Chemists and Colorists, "Colour Index," 5 volumes, Third Edition, 1971.
11. "The Chemical Industry, 1969–1970," p. 125, Organization for Economic Cooperation and Development, Paris.
12. Thorpe, J. F., and Linstead, R., "The Synthetic Dyestuffs," 7th Ed., Cain and Thorpe, London, Charles Griffin and Co., 1933.
13. Venkataraman, K., "The Chemistry of Synthetic Dyes," New York, Academic Press, Vols. I, II (1952); Vol. III (1970); Vol. IV (1971); Vol. V (1971); Vol. VI (1972).
14. Vickerstaff, T., "The Physical Chemistry of Dyeing," 2nd Ed., London, Oliver and Boyd, 1954.
15. Zollinger, H., "Azo and Diazo Chemistry," New York, John Wiley & Sons, 1961.

24

The Nuclear Industry

Warren K. Eister* and Richard H. Kennedy**

INTRODUCTION

Thermal energy is the goal of the nuclear industry.[1] To achieve this goal it was necessary to develop a unique family of materials to meet nuclear specifications along with physical and chemical ones. Interest was centered in the atomic nuclides as well as in the chemical elements. Furthermore, because of the radiation properties of the essential nuclear fuel uranium-235, along with that of the actinide and fission-product materials produced, a tremendous effort had to be expended to evaluate their biological significance and to control their processing and use. Within ten years after the discovery that man could transmute one atomic nuclide into another, this knowledge was employed in the large-scale production of an essential nuclear fuel, plutonium-239. Also, for the first time there was the need for the large scale separation of a naturally occurring nuclide, the chemical isotope, uranium-235. Within a decade the research and production activities related to the nuclear industry led to the separation of all the naturally occurring nuclides and the preparation of a much greater number of unstable nuclides, the so-called radioisotopes. Both the stable and radioactive nuclides have found important uses in all fields of research, medicine, and industry, as well as in the nuclear field.

Thermal energy is used by man to improve his well being and it is projected that his requirement for this energy will double every ten years. The world now has a pretty good picture of the availability and requirements for thermal energy; and the first task in this chapter will be to review this situation to provide an appreciation of present and future roles of the nuclear industry. Next, the fission and fusion processes on which this industry is based will be considered. Then, attention will be turned to the established nuclear-fission power generation system and its future development. Finally, nuclear explosives, stable and

*Division of Isotopes Development, U.S. Atomic Energy Comm. Germantown, Md.

**Division of Raw Materials, U.S. Atomic Energy Comm. (section on raw materials). Germantown, Md.

radioactive nuclides, and the safety of this industry will be discussed.

Status and Outlook

The nuclear industry was created during World War II as a result of the successful development of the atomic bomb. As man came out of that war he found that he possessed a new source of energy to supplement those on which he had heretofore been dependent: wood, wind, water, coal, natural gas, and petroleum. At the time, few saw the need for nuclear power, but within two decades intensive development has made it economically competitive and perhaps essential for the generation of electric power. Nuclear energy can now be converted to electric power at a cost which compares favorably with that for coal and petroleum in many areas, 5 to 10 mills/kw-h (Fig. 24.1). The decision of whether the source of thermal energy in a new electric power station will be chemical or nuclear is based on local factors of fuel price and availability, transportation, plant costs, load factors, pollution, and public acceptance.

Nuclear energy is produced either by fission or fusion, that is, either by splitting the nucleus of an atom or by combining the nuclei of two atoms. In the early 1960's prototypes of various nuclear fission reactors were being built to demonstrate their economic capability to produce electric power. Today, the light-water-cooled converter reactor has been established as an economic device for this purpose. By 1980, 50 per cent of all new electric-power generating capacity may be nuclear, and the development of sodium-cooled breeder reactors may be nearing completion. On the other hand, production of thermal energy using the fusion reaction has thus far been limited to nuclear explosives, the so-called hydrogen bomb. The successes of recent experiments seem to assure the ultimate success of a controlled fusion reactor, but a tremendous engineering development effort spanning several decades may be required to produce commercial power.

Nuclear reactions depend on the properties of the nuclide and not those of the chemical element. For example, the nuclide uranium-235 is fissionable with thermal neutrons, but the nuclide uranium-238 is not. Nuclear reactions, in most cases, produce different nuclides than those found in nature. For example, the capture of a neutron by the nucleus of uranium-238 produces plutonium-239, a new

Fig. 24.1 Yankee Nuclear Power Station, Roe, Mass., 175 Mwe pressurized-water reactor, started operation in 1961, and has provided much of the data for nuclear-power cost evaluation. Designed by Westinghouse. (*Yankee Atomic Electric Co.*)

man-made nuclide that is fissionable in the same manner as uranium-235. Many other nuclides can be produced by the capture of nuclear particles to provide products of interest in various areas of human endeavor. These include iodine-131, technetium-99, cobalt-60, strontium-90, and plutonium-238. The nuclear industry has made both separated stable and man-made radioactive nuclides common products of commerce. These nuclides are particularly valuable for analytical and diagnostic purposes in research, medicine, and industry. In addition, the radioactive nuclides are finding use as long-lived thermal-energy sources, and for energizing certain types of chemical and biological processes. In less than a decade the world of chemistry went from less than 92 chemical elements to more than 2000 nuclides, most of them radioactive.

THE EARTH'S ENERGY SUPPLY AND USE

The Earth has both income and capital energy which is used by man (see Table 24.1 and Fig. 24.2). The income energy, equivalent to 178 million megawatts, is constantly replenished and includes solar, gravitational, and subterraneous sources. The capital energy, equivalent to 2.5×10^{11} million megawatt years (MMw-y) cannot be replenished and includes the chemical, fission, and fusion forms. The chemical and fission/fusion supplies are presently estimated to be 7.6×10^3 and 2.5×10^{11} MMw-y, respectively. Based on these figures and the present rate of energy-use of 4 MMw-y, these supplies should satisfy man's requirements for several hundred thousand years. The chemical fuels—coal, petroleum, natural gas, and tar sands and oil shale—resulted from the interaction of the sun with the earth's environment, a very slow process that no longer persists. This chemical deposit was equivalent to about 40 solar years of the five billion years the earth is believed to have existed. The fission and fusion fuels—uranium, thorium, lithium, and deuterium—were deposited on the earth at the time of its creation.

Little use was made of this stored energy until the last century, and now about 4 per cent of the chemical fuel has been consumed.

TABLE 24.1 Energy Status of the Earth, 1970

Income Supply	*10^6 Megawatts*	
Solar	178,000	
Gravitational	3	
Terrestrial core	32	
Stored Supply	*10^6 Megawatt Years*	*% Consumed*
Coal and Lignite	6400	1.8
Petroleum	370	13.6
Natural gas	340	
Tar sands and oil shale	90	
Uranium	5400[a]	
Thorium	3000[b]	
Lithium	30,000[c]	
Deuterium	2.5×10^{11}	
Electric Power Plant Capacity	*10^6 Megawatt*	*% Developed*
Coal	0.37	
Petroleum & gas	0.2	
Water	0.17	8
Nuclear	0.05	
Geothermal	0.001	2
Wind	0.001	nil
Tidal	0.0003	0.5

[a]Reasonably assured reserves at less than \$15/lb U_3O_8. There are similar terrestrial reserves of 10^{12} Mw-y at less than \$500/lb U_3O_8; and at this price, uranium may be recovered from the ocean, thus, expanding the supply by orders of magnitude.
[b]Reasonably assured reserves at low cost; similar reserves of 1.5×10^{12} mw-y at less than \$500/lb ThO_2.
[c]From known terrestrial deposits; the ocean would add another 6×10^{12} Mw-y.
Source: Hubbert[3] and U.S. Office of Science and Technology.

In spite of the fact that 96 per cent of the chemical fuel remains, there are increasingly serious problems today concerned with its extraction, distribution, and use. Petroleum production may reach its peak within the next few years in the United States and within the next 25 years on a world-wide basis. While the peak production rate for coal may not be reached for 150 years, the high-grade deposits in the industrialized regions of the world are being rapidly depleted.[3] Today, fission energy is competitive with coal in many parts of the United States and in other countries of the

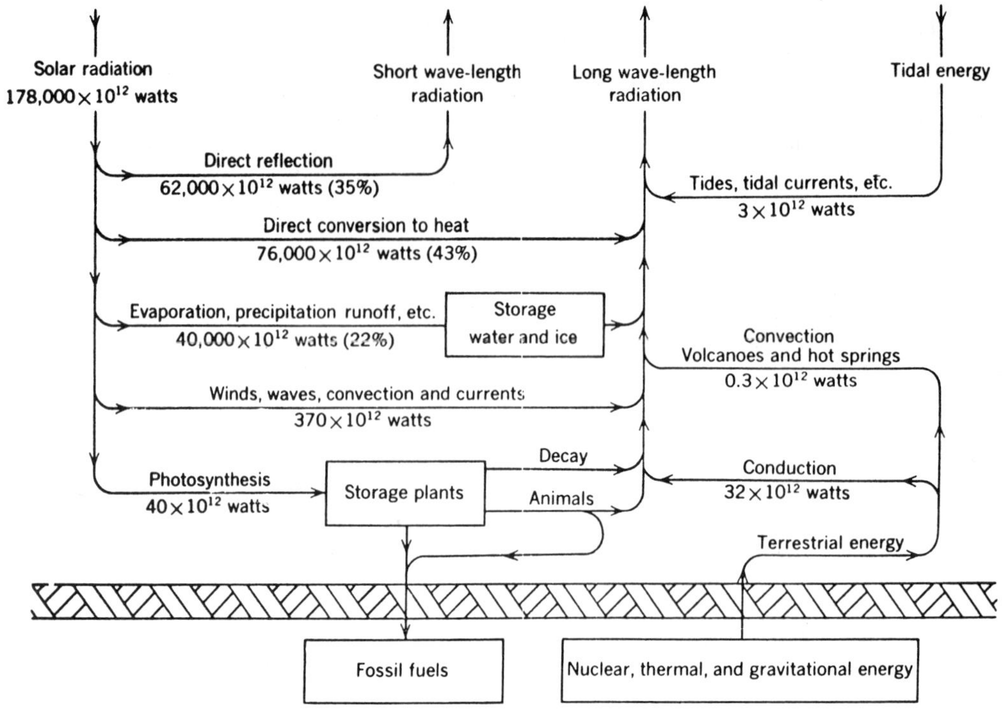

Fig. 24.2 World energy flow sheet. (*Hubbert, U.S. Geological Survey.*)

world. Power-generating systems using fusion energy have yet to be developed.

The demand for energy has increased as the world has continued to industrialize. The present total demand for all types of energy for all uses is about 1.2×10^{17} Btu (4 MMw-y) per year, with the annual per capita consumption ranging from 180 million Btu in the United States to 70, 10, and 5 million Btu, respectively, in Russia, Brazil, and India. Of course, only a fraction of this energy is consumed in the form of electricity—about 20 per cent in the United States in 1960 (Table 24.2).[4] Today in the United States the electricity-generating capacity is about 300,000 mwe, including 7,500 mwe from nuclear energy. In the next ten years more than 50 per cent of the new electric power stations are expected to be nuclear, and by the year 2020 there may be about 3,000,000 mwe of generating capacity operating in the United States, 90 per cent of this being nuclear (Table 24.3).[6]

TABLE 24.2 Energy Consumption in the United States, 1960 (10^{12} Btu)

Source	*Household*	*Commercial*	*Transportation*	*Industry*	*Other*	*Total*
Coal	1.81	1.35	0.1	6.25	1.6	11.1
Petroleum	3.00	1.00	8.7	2.9	3.6	19.3
Gas	3.71	1.22	0.4	6.0	1.9	13.3
Hydro	0.42	0.25	–	0.8	0.2	1.6
TOTAL	9.	3.84	9.2	16.0	7.4	45.4
% Electricity	26.4	36.5	nil	27.4	15.4	20.5

Source: Office of Science and Technology.

TABLE 24.3 Possible Construction Schedule for Electric Power Stations
(Number of 1000 Mwe Plants)

		Nuclear			
Year	*Fossil*	*LWR*	*HTGR*	*LMFBR*	*Total*
1970–79	91	121	–	–	212
1980–89	136	64	136	49	385
1990–99	165	–	–	453	618
2000–09	168	54	–	733	955
2010–19	–	–	25	1,369	1,394
TOTAL	560	239	161	2,604	3,564[a]

[a]About 3000 plants operating in year 2020, the others shut down at the end of their economic life.
Source: USAEC.

The increasing demand for energy can be largely attributed to the rapidly increasing population and the constant drive to raise the standard of living. More energy will be expended in the future than was required in the past to extract materials from nature. In addition, energy will also be required to treat the waste of all human activities in order to maintain a hospitable environment.

Nuclear energy is being considered as a means of providing water, fertilizer, and power for many arid coastal regions of the world in order to produce millions of tons of food for the world's expanding population. These arid regions have the necessary soil and climate, and the nuclear reactors, while providing several thousand megawatts of energy to purify sea water for irrigation, will also produce power to manufacture fertilizer, cultivate the fields, process the grain, and possibly extract minerals from the sea water. Such an agro-industrial complex would cost about a billion dollars and could economically feed millions of people.

NUCLEAR ENERGY ECONOMICS

The nuclear industry includes three major categories of activity (Fig. 24.3). The first category includes uranium and thorium mining, ore milling, feed materials processing, uranium-isotope separation, and reactor-fuel-element fabrication. The second includes nuclear reactor fabrication, electric power generation, coolant treatment, and waste disposal. The third category includes recovery of uranium, plutonium, thorium, actinide products, and fission products from the spent reactor fuel together with radioactive waste management.

Based on 1967 dollars and a 10-year doubling time for electric power requirements, the

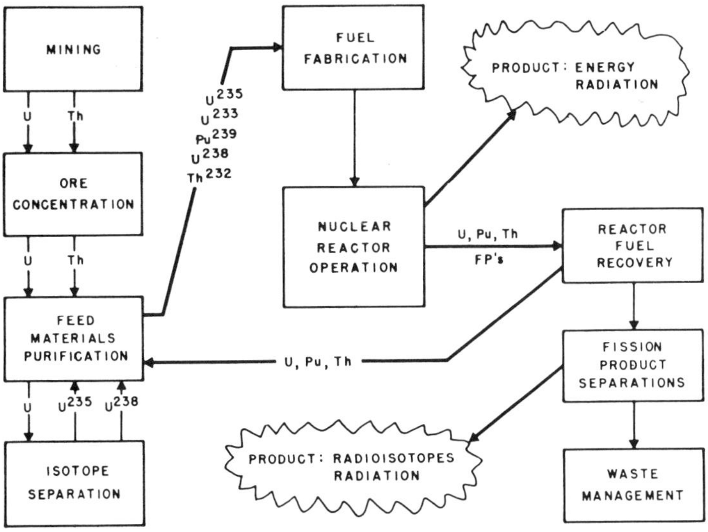

Fig. 24.3 Nuclear industry materials flow sheet; represented 115 organizations with 50,000 employees and product shipments valued at $562 million in 1969.

TABLE 24.4 Cost of a Nuclear vs a Coal-Fired Plant[a]

(Mills/kwh)

Capital Cost: 1100 Mwe Nuclear at $205.5/kw for completion in 1976
800 Mwe Coal at $170.6/kw for com- in 1975

	Nuclear	*Coal*
Fixed charges	4.95	4.16
Fuel charges	1.70	2.19[b]
Operating and maintenance	0.30	0.30
Insurances	0.11	–
TOTAL switchboard cost	7.06	6.65

Nuclear competitive with coal at 29.7 cents per million Btu

[a]Capital charge 16%, and capacity factor 75%.
[b]Coal at 25¢/million Btu, and 8750 Btu per kw-h.
Source: Nuclear News. (1970).

TABLE 24.5 Nuclear Industry Capital Costs

Basis: Installed Nuclear Capacity of 150,000 Mwe in 1980; 1967 dollars.

	$/kwe	*Billion $*
Light-water reactors	140 to 250	21 to 37
Reactor component fabricative plants		
Turbine generators	2	0.2
Pressure vessels	1	0.1
Steam generators	1	0.1
TOTAL	4	0.4
Fuel-cycle plants		
Uranium ore mills	10	1.5
Uranium refineries	1	0.1
Uranium enrichment	25	3.7
Fuel fabrication	5	0.8
Spent-fuel recovery	2	0.2
TOTAL	43	6.3

Source: USAEC

estimated cost of electricity over the next fifty years in the United States will be about $1600 billion. Nuclear fuels are expected to provide most of this energy at a cost of about 7 mills/kw-h (Table 24.4). The total capital and operating costs of a 1000 Mwe nuclear power station over the thirty years of its assumed economic life is estimated at about 1.5-billion dollars.[5] The fixed cost of the station is about 70 per cent of the total, operating costs 5 per cent, and the fuel cost 25 per cent. On this basis, by 1980 there will be a capital investment of about $28 billion in nuclear power plants, including the facilities for plant-component fabrication and the nuclear-fuel cycle (Table 24.5). This investment will be equivalent to about $187/kwe of the installed nuclear-power-plant capacity. Of this total the fuel-cycle investment represents $6.3 billions or $43/kwe.

The costs for LWRs (light water cooled reactors) when estimated in 1967 ranged from 91-million dollars for a 600-mwe unit to 134-million dollars for a 1000 mwe unit (Table 24.6). This is equivalent to about $150/kwe, but it is noted that a similar estimate made in 1969 (Table 24.4) showed the cost to be $205/kwe, while the actual contract prices ranged from $100 to $250/kwe during the late 1960's. This increase has been due to the cost of labor, materials, and money along with safeguard modifications in design. While Table 24.4 shows the cost of electricity from coal to be about 10 per cent cheaper than that from nuclear fuel, the competition between these two fuels will continue to keep the costs so close that local and general market conditions will be the determining factors.

The cost of the fuel cycle has been estimated to be 1.74 mills per kw-h (Table 24.7). This estimate includes working capital to cover the material in inventory, since it takes about 5 years from the time the uranium is mined until the spent fuel is processed. In spite of inflation the fuel-cycle costs have remained constant because of improving process efficiency and excess supply of uranium. In addition, with the assumed success of breeder reactors, continuing reduction in relative fuel costs is anticipated.

The success of nuclear power will be greatly enhanced if it is not completely dependent on the only naturally fissionable isotope, uranium-235. Fortunately, it is possible to produce fissionable material in a nuclear reactor. This is called conversion or breeding, depending on whether less or more fissionable material is produced than is consumed. Breeding increases the fuel supply by a factor of about fifty and significantly reduces the fuel-enrichment and fabrication requirements. Finally the higher operating temperature of the liquid metal type breeder promises a 30 per cent increase in energy-conversion efficiency, which not only further reduces the fuel re-

TABLE 24.6 Estimated Construction Costs—Light Water Reactors Single Unit Plants, March 1967.
(Thousands of Dollars)

Rating, Mwe	*600*	*800*	*1000*
Direct Costs	71,290	88,370	104,980
Indirect Costs			
Owner's G and A	2,340	2,590	2,870
A-E cost and constr. management	4,250	5,300	6,280
Misc. constr. cost	1,710	1,880	2,050
Start-up	410	440	470
Station crew training	350	360	370
Licensing and public relations	580	680	780
Spare parts	600	800	1,000
Land and land-rights	530	720	1,090
TOTAL Indirect Cost	10,770	12,770	14,910
SUBTOTAL	82,060	101,140	119,890
Contingency	2,130	2,600	3,010
SUBTOTAL	84,190	103,740	122,900
Interest during constr.	6,940	9,130	10,840
Capital cost	91,130	112,870	133,740
Capital cost per kwe[a]	0.152	0.141	0.134

[a]From 1967 to 1970 prices have ranged from $100 to $250/kwe for plants above 500 Mwe.
Source: USAEC

quirements but also reduces the waste heat that has to be dumped into the environment. The nuclear breeder reactor is one of the world's most important engineering-development goals today.

TABLE 24.7 Typical Fuel-Cycle Cost Breakdown for First 5 Years of Operation of a 1000 Mwe Single-Unit Light-Water-Reactor Plant[a]
(Mills/kwh, 1967)

Fuel-Cycle-Cost Component	*BWR*	*PWR*
Fuel element fabrication	0.54	0.52
Net uranium hexafluoride	0.40	0.42
Net separative work	0.39	0.48
Plutonium credit	−0.22	−0.22
Spent fuel recovery	0.20	0.17
Working capital	0.43	0.37
TOTAL	1.74	1.74

[a]Economic parameters and conditions assumed in developing these costs

Yellow cake (ore mill product)	$17.00
Feed preparation	$2.00
Enrichment (per separative work unit)	$26.00
Fuel-element fabrication	$90.00
Spent-fuel recovery	$36.00
Plutonium (239 + 241)	$7,700.00
Interest rate 11%	
Reactor-load factor 80%	

Source: USAEC

THE NUCLEAR REACTIONS

The reactions on which the nuclear industry is based involve the nucleus of the atom. These nuclear reactions result in the conversion of mass, m, to energy, E, as defined by Einstein's equation

$$E = mc^2$$

where c is the velocity of light. Based on this relationship, one atomic mass unit is equivalent to 931 Mev, or 4.2×10^{-17} kw-h. As noted earlier, there are two types of nuclear reactions that lead to the production of energy, fission of heavy nuclei and fusion of light nuclei. In addition, nuclear reactions involving the capture of neutrons are important in the production of additional fission and fusion fuels. The energy evolved from grams of nuclear fuel is equivalent to that evolved from tons of chemical fuels.[2]

Fission

Controlled nuclear fission was first achieved on December 2, 1942, at the University of

Fig. 24.4 Uranium metal fuel from the first atomic pile. This reactor contained 40 tons of uranium oxide along with 6.2 tons of uranium metal. (*ORNL Newa, 1-01-076.*)

Chicago by a group of scientists under the guidance of Enrico Fermi. The device in which it was accomplished, then called an "atomic pile," consisted of an ordered pile of graphite blocks along with natural uranium cubes and cylinders arranged in such a way that a sustained chain reaction could be obtained and controlled (Fig. 24.4). The fission reaction depended on uranium-235, the only naturally occurring nuclide that is fissionable with thermal neutrons. However, other fissionable nuclides are readily produced in the nuclear reactor, the most important being plutonium-239 and uranium-233.

There is a high probability of fission when a neutron is captured in the nucleus of a fissionable nuclide. A typical fission reaction is as follows:

$$^{235}_{92}\text{U} + {}^{1}_{0}\text{n} \longrightarrow {}^{236}_{92}\text{U} \xrightarrow{\text{fission}}$$

$$^{89}_{35}\text{Br} + {}^{145}_{57}\text{La} + 2.3\text{n} + 192\ \text{Mev} \quad (1)$$

At the time of fission 178 Mev is released, and additional energy, about 23 Mev, is subsequently released by the decay of the radioactive fission products. While the nuclear reaction is readily initiated at room temperature and pressure, nuclear-reactor operation at high temperatures is required for efficient electric power generation. The temperature, along with the associated pressure, is limited by the properties of the materials of construction as in the case of fossil-fuel systems.

Plutonium-239 and uranium-233 are produced by the capture of a neutron by the fertile nuclides uranium-238 and thorium-232, respectively.

$$^{238}_{92}\text{U} \xrightarrow{n} {}^{239}_{92}\text{U} \xrightarrow[\text{23.5 min}]{\beta}$$

$$^{239}_{93}\text{Np} \xrightarrow[\text{23.3 days}]{\beta} {}^{239}_{94}\text{Pu} \quad (2)$$

$$^{232}_{90}\text{Th} \xrightarrow{n} {}^{233}_{90}\text{Th} \xrightarrow[\text{23.3 min}]{\beta}$$

$$^{233}_{91}\text{Pa} \xrightarrow[\text{27.4 days}]{\beta} {}^{233}_{92}\text{U} \quad (3)$$

Neutrons are produced in fission (Equation 1), and the nuclear reaction can be made self-sustaining. In addition, since more than two neutrons are produced in the fission reaction, Reactions 2 and 3 may be used to produce more fissionable material than is consumed.

In the present generation of nuclear reactors, fission reactions are initiated by thermal neutrons, neutrons with an energy of 0.025 to 0.1 ev (electron volt), equivalent to temperatures from 22 to 870°C, which is the temperature of the environment in which the neutron is present. The 5-Mev kinetic energy of the fission neutron must therefore be reduced to thermal

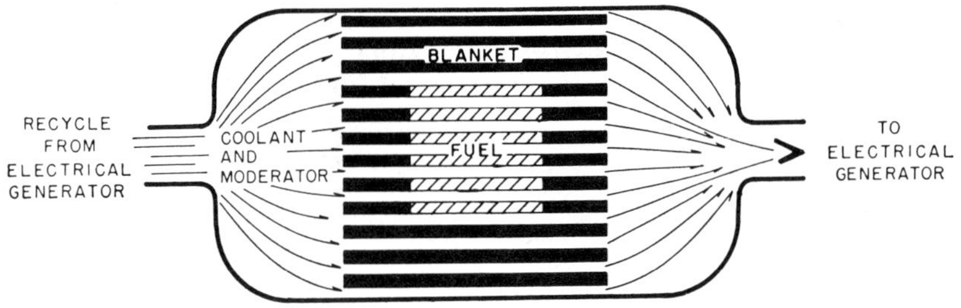

Fig. 24.5 Nuclear reactor core; fissionable material is principally in the fuel section with the fertile material in the blanket for greater neutron economy; but economics for converter-type power reactor require reversal of this pattern to achieve high fuel burn-up and uniform power distribution across the core.

energy before it escapes from the nuclear reaction zone or is captured by other materials in this zone (Fig. 24.5)[10]. Even when it is captured by fissionable material, fission does not always occur. The probability of a neutron reacting with materials is expressed in terms of their so-called cross sections (see Tables 24.8 and 24.9).

The fission reaction can also employ fast neutrons, as in the plutonium cycle which gives a high breeding ratio. The neutron cross sections of most materials are significantly less for fast neutrons, thus permitting greater freedom of choice in materials of construction and coolants. The neutron multiplication factor η for plutonium at 1 Mev is 2.99 compared with 2.28 and 2.07, respectively, for uranium-233 and -235 under thermal conditions, about 0.1 ev (Table 24.10). In addition, with fast neutrons uranium-238 has a significant fission cross section, thus adding to the interest in the fast-flux cycle. The thorium cycle provides the only possibility of breeding when using a thermal reactor.[7]

Fusion

Nuclear fusion, the stellar energy process, was first achieved in 1932 by the accelerator-produced collisions of deuterium nuclei. In 1952 the spontaneous release of fusion energy was demonstrated with the first test of a thermonuclear device, the so-called hydrogen bomb. Since that time research has been carried on to devise a method for carrying out controlled nuclear fusion. The fuels for the fusion reaction are hydrogen-2 and hydrogen-

TABLE 24.8 Nuclear Properties of Fissionable and Fertile Materials at Thermal Energy[a]

	σ_{abs}	σ_F	$1+\alpha$	η^b	υ^c
^{233}U	581 ± 7	527 ± 4	1.102 ± 0.005	2.28 ± 0.02	2.51 ± 0.03
^{235}U	694 ± 8	582 ± 6	1.19 ± 0.01	2.07 ± 0.02	2.47 ± 0.03
^{239}Pu	1026 ± 13	746 ± 8	1.38 ± 0.02	2.10 ± 0.02	2.90 ± 0.04
^{238}U	2.71 ± 0.02				
^{232}Th	7.56 ± 0.11				

σ_{abs} = cross section for neutron absorption in barns at a neutron velocity of 220 meters/sec; one barn = 10^{-24} cm^2.

σ_F = cross section for fission in barns at a neutron velocity of 220 meters/sec.

$$1+\alpha = \frac{\sigma_{abs}}{\sigma_F}$$

[a]Neutron cross sections, D. J. Hughes and R. B. Schwartz, BNL-325, July 1, 1958.
[b]η = Neutrons emitted/neutrons absorbed.
[c]υ = Neutrons per atom fissioned

TABLE 24.9 Thermal Neutron Cross Sections of Typical Nuclear Reactor Materials (Barns, σ)

Chemical Element	σ_{abs}	*Chemical Element*	σ_{abs}
C	0.0032	B	755
Be	0.010	Cd	3300
H	0.33	Hf	105
D	0.00057	Al	0.023
Na	0.53	Zr	0.18
K	2.0	Fe	2.5
He	0.007	Ni	4.6
N	1.9	Cr	3.1
O	0.0002	Xe	35

TABLE 24.10 Fission Cross Sections and Neutron Multiplication Factors for Fissile and Fertile Nuclei at High Energy

Neutron Energy (kev)	*10*	*100*	*1000*
Fission Cross Sections (Barns, σ)			
U-233	4.5	2.5	2.0
U-235	3.3	1.7	1.2
Pu-239	1.9	1.8	1.7
Th-232	–	–	0.076
U-238	–	–	0.411
Neutrons Produced per Neutron Absorbed			
U-233	2.24	2.26	2.40
U-235	1.77	1.90	2.32
Pu-239	2.00	2.41	2.99

3, more commonly called deuterium and tritium. Deuterium is a naturally occurring isotope of hydrogen with an abundance of 0.015 per cent. Tritium is the neutron-capture product of lithium-6, readily produced in a nuclear fission or fusion reactor. A third possible fuel for fusion is helium-3, the decay product of radioactive tritium. The fusion reactions of practical interest which have been identified so far are as follows:[11]

Reaction	Ignition Temperature in Millions of °C
${}^2_1H + {}^2_1H \longrightarrow {}^3_1H + {}^1_1H + 4$ Mev	100
$\longrightarrow {}^3_2He + {}^1_0n + 3.25$ Mev	
${}^2_1H + {}^3_1H \longrightarrow {}^4_2He + {}^1_0n + 17.6$ Mev	45
${}^2_1H + {}^3_2He \longrightarrow {}^4_2He + {}^1_1H + 18.3$ Mev	

In addition, the success of the fusion reaction depends on the capture of the fusion-produced neutrons in the lithium blanket of the reactor to produce both tritium and energy:

$${}^6_3Li + n \longrightarrow {}^3_1H + {}^4_2He + 4.6 \text{ Mev}$$

While the fission reaction, as noted earlier, can be initiated under conditions of room temperature and pressure, the fusion reaction requires very high temperatures and pressure.

The torodial type of fusion reactor is one of the most intensely studied systems today and gives promise of providing for continuous operation.[12] For this reactor the favored fuel is the deuterium-tritium system. In one version, an induction heater is used to bring the ionized reactants to temperature with the resulting plasma being magnetically retained in a torodial orbit. There are two products from this reaction, helium-4 and neutrons. The helium-4 is retained within the plasma and its associated energy, 3.5 Mev, maintains the plasma temperature. The neutrons escape and are captured in a lithium blanket. The energy of the neutrons, 14.1 Mev, and that from the neutron-lithium reaction, 4.6 Mev, are extracted and converted into electricity (Fig. 24.6). The conditions required for a productive reaction are about 10^{15} ions per cm^3 and a temperature of 10 kev for 700 milliseconds. In more familiar terms, this corresponds to a pressure of about 100 atmospheres at 10^8°C. The Russian "Tokamak" has achieved conditions about a hundredth of those required for the energy output to equal the energy input (Table 24.11). These results have been reproduced in the model ST Tokamak at the Princeton Plasma Physics Laboratory.

The timetable for the achievement of economic electric power from fusion is difficult

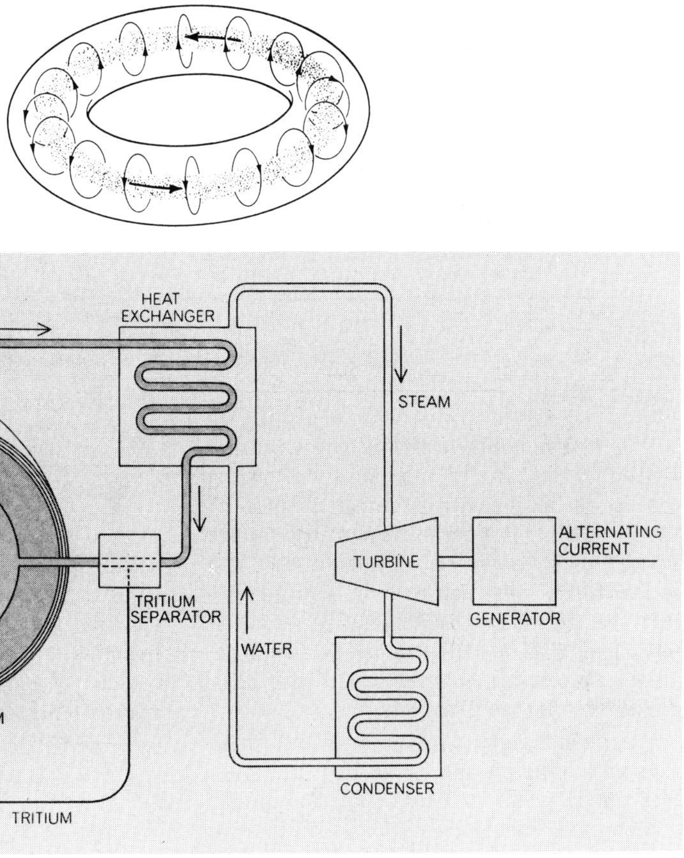

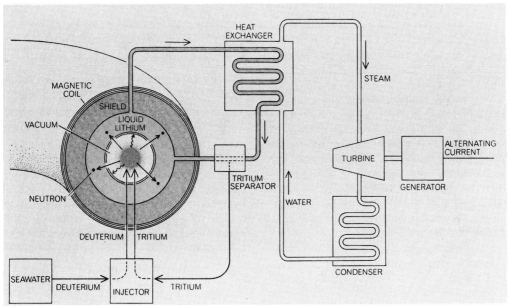

Fig. 24.6 Fusion-energy conversion system using the molten lithium blanket loop to extract the heat and the tritium; the plasma contained in the toroid is shown in the inset. (*USAEC*)

to predict, but it is clear, even in view of recent successes, that many more years and billions of dollars will be required for the engineering development.

TABLE 24.11 Tokamak Reactor, Status and Goal

	Achieved	*Required*
Fuel	T-D	T-D
Thermal power output		6000 Mw
Electric power requirement		2000 Mw
Plasma diameter		14 meters
Magnetic field		100 kG
Plasma density (particles/cm^3)	5×10^{13}	3×10^{14}
B[a]	0.03	0.3
Plasma temperature (kev)	0.5	10
Plasma volume (m^3)	1	200
Containment time (millisec)	25	700

[a]B = plasma pressure/magnetic field pressure.

EXPLOSIVES

When used as an explosive, a nuclear device creates cataclysmic physical, thermal, and nuclear effects of significance for both civilian and military applications. The fission of about 56 grams of uranium or the fusion of 6.8 grams of tritium is equivalent to the physical force of about one thousand tons of trinitrotoluene (TNT) (Table 24.12). The force of nuclear explosives ranges from fractions of kilotons to tens of megatons of TNT, whereas the cost is estimated to be about 1 per cent the cost of TNT (Table 24.13).

To generate a nuclear-fission explosion it is necessary to bring together a supercritical mass of fissile material, and then to hold it together long enough to produce the desired energy. In the fission-type explosive, the criticality event is initiated with chemical explo-

TABLE 24.12 Energy Equivalents of One Thousand Tons of TNT

Fission of 1.45×10^{23} atomic nuclei
Fission of 56 g Pu-239
Fusion of 16.5×10^{23} atomic nuclei
Fusion of 6.8 g TD
10^{12} calories
4.2×10^{19} ergs
1.2×10^{6} kilowatt hours
4.0×10^{9} British thermal units (Btu)

sives which either drive together two subcritical masses of fissile metal, or decrease the volume of a subcritical assembly. The duration of the fission reaction is less than one millionth of a second.

A fusion explosion using the deuterium-tritium reaction is initiated by the fission reaction which provides the necessary conditions of temperature and pressure. The fusion-type explosive produces a lesser quantity of radioactivity per unit of explosive force.

Plowshare

The Atomic Energy Commissions Plowshare program is studying the use of nuclear explosives for peaceful purposes.[14] These are both underground events carried out to create underground cavities and porosity, and surface craters. An underground explosion initially forms a bubble of vaporized material with a temperature of about one million degrees Celsius and a pressure of several million atmospheres. The high pressures cause the cavity, filled with hot gases, to expand rapidly and initiate an outward-moving spherical shock wave which crushes the surrounding medium.

TABLE 24.13 Comparative Energy Costs

Energy Source	*Cost per Million Btu*
2-Megaton thermonuclear explosive	$ 0.075
Lignite	0.14–0.17
Soft coal	0.15–0.20
Natural gas	0.20–0.15
Water power	0.89
Gasoline	1.50
Electricity ($0.006/kw-h)	1.78
Ammonium nitrate	4.50
10-Kiloton thermonuclear explosive	8.75
TNT	250.00

In deeply buried explosions in competent rock, excluding salt formations, a chimney of broken rock is formed when the cavity gases cool and the cavity ceiling collapses (Figure 24.7). The size of the chimney is a function of the magnitude of the explosion, the depth of burial, and the type of rock.

The resulting interstitial volume within the chimney and the surrounding fractures provides the basis for several potential applications classified as underground engineering. In the "Gasbuggy" shot a chimney with a volume of 220,000 cubic yards was formed; about 10^9 ft^3 of natural gas may be recovered from it, at least 5 times more than normally expected. In view of the anticipated shortage of natural gas within the next ten years, this method of gas

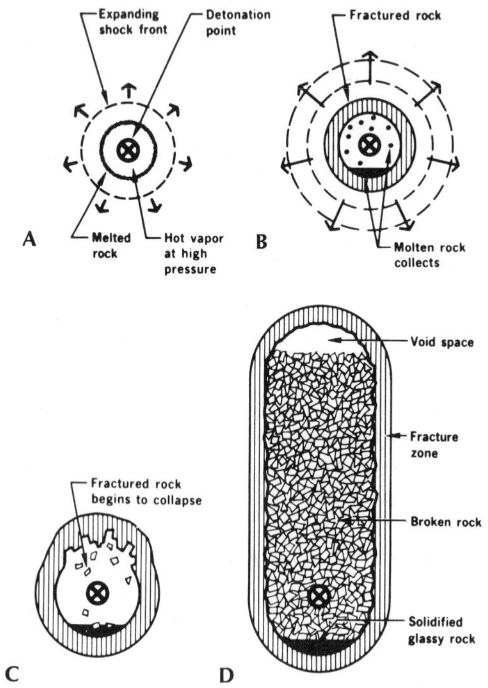

Fig. 24.7 An underground nuclear explosion (A) initially vaporizes, melts, and fractures the adjacent rock; (B) then sends out a shock wave; (C) the zone of thermal energy rises toward the surface continuing to fracture rock; (D) as the cavity cools the fractured rock is imploded into the cavity, resulting in a column of broken rock. If the cavity reaches the surface, the explosion lifts the rock and dirt and forms a crater. (*USAEC*)

Fig. 24.8 The first nuclear row-charge experiment created a channel 855 ft long, 255 ft wide, and 65 ft deep. The construction of a Panama sea-level canal has been estimated to cost $0.65 billion. (*USAEC*)

extraction from low-porosity deposits has considerable interest. Since the solid fission products remain deeply buried underground, the only radioactive species of consequence in natural gas from nuclear-stimulated wells are tritium and, to a lesser extent, Kr-85. Other applications under consideration include the *in situ* extraction of copper from low-grade ores by continuous leaching, and the extraction of geothermal energy.

The excavation of a crater 1280 feet in diameter by 320 feet deep has been demonstrated using a 100 kiloton thermonuclear explosive. Also, a ditch 850 feet long, 250 feet wide, and 65 feet deep was excavated by using a row of five simultaneously detonated 1.1 kiloton nuclear explosives (Fig. 24.8). The development and use of very low-fission thermonuclear explosives has reduced the production and release of radioactivity from cratering explosions.[15] This, in turn, has reduced the waiting time for reentry into the crater area to a few weeks. These and other nuclear cratering experiments have demonstrated the potential usefulness of nuclear explosives in such large excavation projects as canals, harbors, dams, and transits through mountainous terrain.

Military Applications

During World War II, nuclear energy for military applications was an important development. The subsequent nuclear-weapons development has been concerned with increasing their efficiency, evaluating their effects, and establishing protective measures. In military applications nuclear explosions may inflict severe damage from blast, heat and radiation effects at distances greater than 30 miles. Moreover, the radioactive debris from the explosion may extend the radiation effects from fall-out several hundred miles downwind. The radiation from stratospheric explosions may also affect radio and radar performance for several hours following the event.[16]

Studies of protective measures have been concerned with two aspects, the immediate effects of blast and neutron radiation, and the long term fall-out effects of the radioactive bomb debris. Blast-resistant structures, important in urban areas, generally provide adequate protection from all radiation effects, while fall-out shelters, adequate for rural areas, provide only radiation protection. In the design of certain new structures, blast or fall-out protection can be provided at nominal cost.

THE NUCLEAR REACTOR

The development of a commercial nuclear power industry has required about 30 years. While the first commercial electric power was produced using a pressurized water reactor (PWR) at Shippingport in 1957, only in the early 1970s did large numbers of power reactors come into operation. Nuclear-reactor development required the evaluation of many reactor concepts (Fig. 24.9) and involved two fuel cycles, the uranium-plutonium and the thorium-uranium. The light water reactors (LWRs) were the first to achieve wide-spread acceptance, with the high temperature gas cooled reactor (HTGR) being the only other thermal-converter type still being commercially developed in the U.S. In England and France a family of gas-cooled power reactors (GCR) were developed, and several were built in other countries. The current major development work is concerned with the liquid-metal fast breeder reactor (LMFBR), with more limited efforts being expended on the gas-cooled fast breeder (GCFBR), light water breeder (LWBR), and the molten salt breeder (MSBR) reactors (Table 24.14).[17]

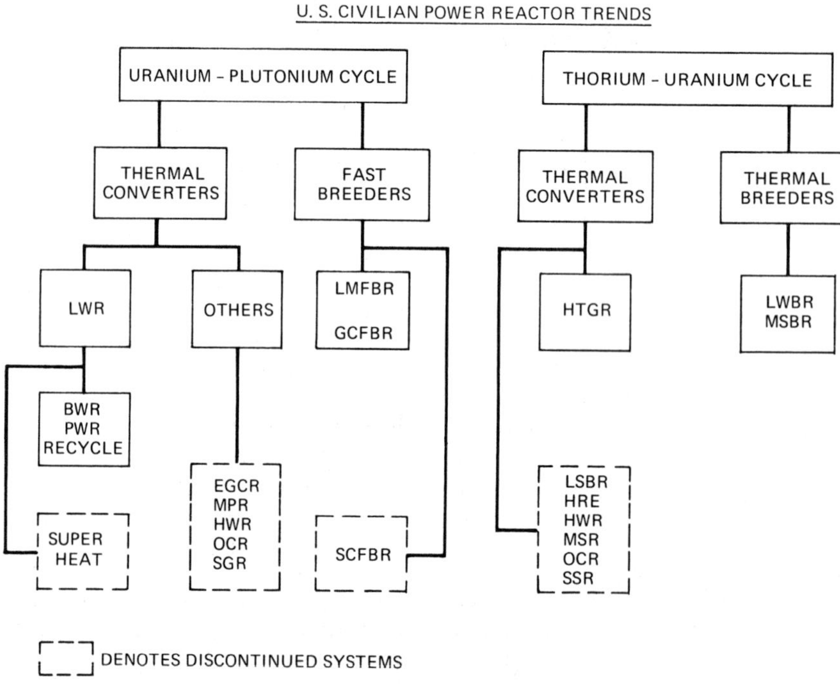

Fig. 24.9 U.S. civilian nuclear power reactors: *B*oiling *W*ater, *P*ressurized *W*ater, *L*iquid *M*etal, *G*as *C*ooled, *H*igh *T*emperature *G*raphite, *L*ight *W*ater, *M*olten *S*alt.—Discontinued systems: *E*nriched *G*as *C*ooled, *M*olten *P*lutonium, *H*eavy *W*ater, *O*rganic *C*ooled, *S*odium *G*raphite, *S*team *C*ooled, *L*arge *S*eed *B*lanket, *H*omogeneous *R*eactor *E*xperiment, *M*olten *S*alt, *S*pectral *S*hift.

TABLE 24.14 Characteristic of Nuclear Power Reactors

	PWR	*BWR*	*HTGR*	*LMFBR*	*MSBR*	*GCR*[a]
Thermal power (Mw)	3090	3293	2318	2417	2250	1500
Net station efficiency (%)	33.5	33.7	43.1	41.3	44	41.7
Coolant	H_2O	H_2O	Helium	Sodium	Molten salt	CO_2
Outlet temp (°F)	605	546	1525	1150	1300	1230
Pressure (psia)	2250	1050	695	30	65	600
Flow rate (10^6 lb/hr)	135	100	9.3		540	
Moderator	H_2O	H_2O	Graphite	Sodium	Graphite	Graphite
Fuel system	UO	UO	ThUO	PuUO	UThLiBeF	UO
Fissile, %	3.3	2.5	4.7	12	3	0.75
Fissile, kg	2900	4470	1470	1620	1468	300
kw/kg	35	22	60	157	2360	5.1
Lifetime at 0.8 power (yr)	3	4	4.0	2–3	2	
Burn-up (Mwd/MT)	30,000	27,500	65,000	100,000	140	
Conversion ratio	0.6	0.6	0.81	1.3	1.07	
Clad	Zr	Zr	Graphite	SS	N Hastelloy	Magnox
Costs capital ($/kwe)	133		125		135	
Fuel (mills/kw-h)	1.46		1.02		0.4	
O and M (mills/kw-h)	0.29		0.29		0.34	
Depreciation (mills/kw-h)	2.60		2.38		2.22	
TOTAL (mills/kw-h)	4.35		3.69		2.96	

[a]Hinkley, B., U.K. gas-cooled reactor.

There are two types of LWRs, the PWR and the boiling-water reactor (BWR).[5] In the PWR, the reactor water coolant transfers its energy through an intermediate heat exchanger to generate steam in a second cycle for electric power generation (Fig. 24.10). In the BWR, the steam is generated in the reactor (Fig. 24.11). Although the BWR eliminates the intermediate heat exchanger, it places the steam turbine in a radioactive environment. The two types are competitive, with costs ranging from $100 to $250/kwe over the last few years. The demonstrated cost of power in the larger plants has been in the 5 to 10 mills/kw-h range. Electric power from fossil fuels in the same geographic area averaged about 7½ mills/kw-h. The continuing interest in the HTGR is due to improved thermal efficiency, 43 per cent vs 33 per cent for the LWRs.

The first commercial breeders now scheduled for start-up about 1985 will follow several years of experience with demonstration units. Two breeder reactors now under development promise high thermal efficiency at reasonable operating pressures through the use of sodium (LMFBR), or molten salt (MSBR) as the coolant. The other breeder concepts, the LWBR and the GCFBR, will have thermal-pressure limitations. The LWBR development would, however, permit the conversion of LWR thermal converters to thermal breeders. In the AEC development program, primary emphasis has been placed on the LMFBR. Although the other breeder reactor concepts offer specific advantages including improved breeding ratio, lower capital cost, established technology, and reduced fuel cycle costs, there is broad international participation only in the LMFBR development.

Light Water Reactors

In the following sections three important aspects of the LWRs will be discussed: the fuel elements, the reactor vessels, and the reactor-containment system.

Fuel Elements. The fuel element is designed to provide primary containment of the radioactive fuel and fission products over the three year operating life of the PWR and BWR fuel (Fig. 24.12 and 24.13). This depends on the integrity of about 80 miles of 24 and 32 mil Zircaloy-4 tubing at temperatures up to 660°F. Zircaloy-4 has replaced stainless steel

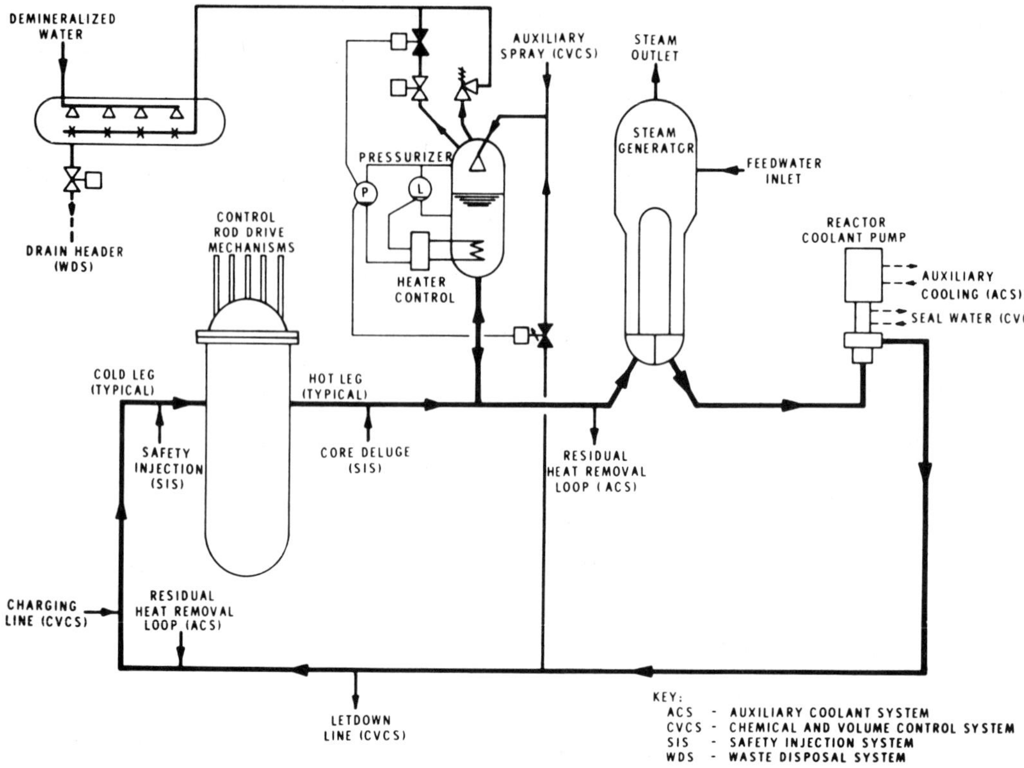

Fig. 24.10 Pressurized-water reactor flow sheet for power generation. (*USAEC*)

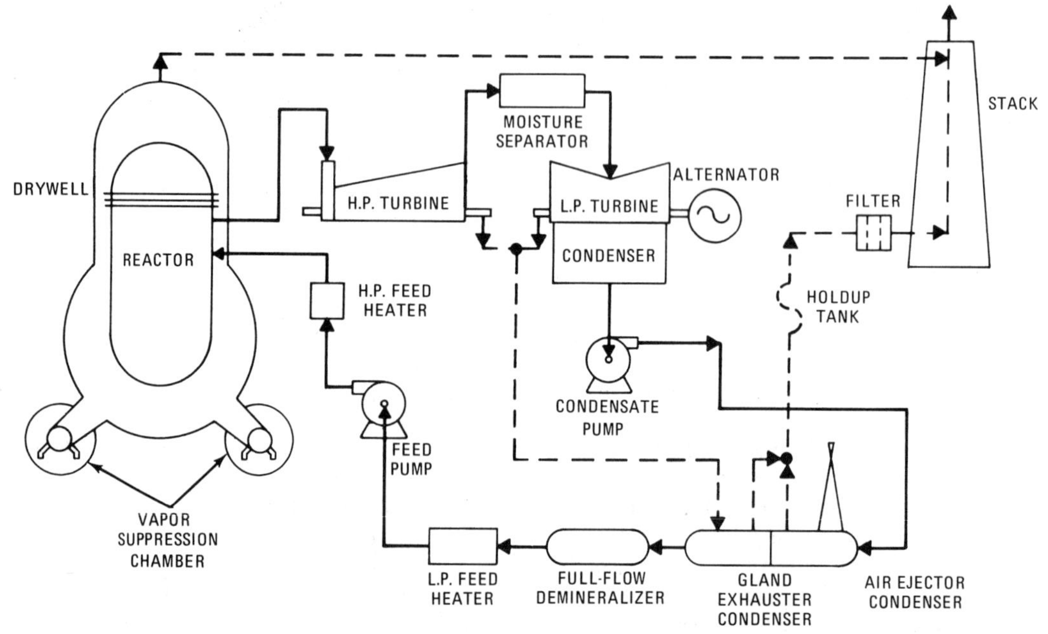

Fig. 24.11 Boiling-water reactor flow sheet for power generation. (*USAEC*)

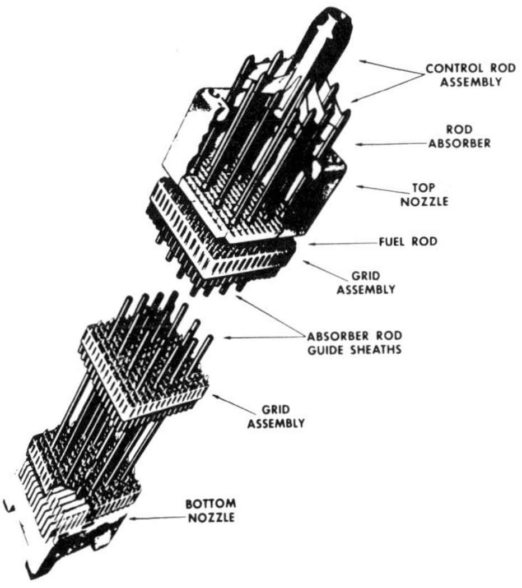

Fig. 24.12 Cutaway of PWR fuel element with the control-rod-cluster assembly. Element contains 1140 lbs UO_2 in 193 rods. (*USAEC*)

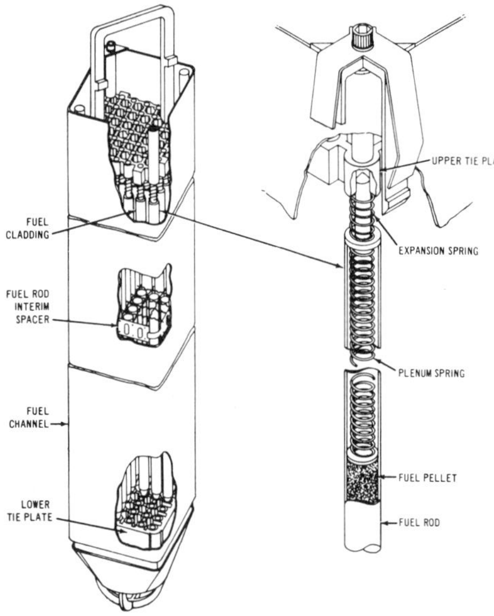

Fig. 24.13 Cutaway of BWR fuel element containing 488 lb UO_2 in 49 rods. (*USAEC*)

to increase neutron economy. The economics of the fuel cycle also favor low enrichment, long operating life and high burn-up. The 1000 mwe BRW core contains about 165 tons of UO_2 which is charged with 2.5 per cent uranium-235, the PWR, about 110 tons charged with 3.3 per cent uranium-235. Since these power stations represent a significant part of the power system capacity, frequent shut-downs are not desirable. Fuel is charged to the reactor about once a year with the new fuel being placed in the outer regions of the reactor. This serves to flatten the power distribution in the reactor while at the same time greater burn-up is achieved by placing the partially spent fuel towards the center of the reactor where the neutron flux is higher (Table 24.15). Significant improvements continue to be made in the design of the fuel elements.[19]

Two fuel-related projects are now under way to further extend the value of the LWRs: (1) recycle of the plutonium to reduce the enriched uranium-235 requirements, or (2) conversion to the thorium-uranium cycle to achieve thermal breeding. The demonstration phase of the plutonium recycle development is now in progress in several power reactors. Several LWRs were originally started up on the thorium-uranium-cycle but were changed over to the uranium-plutonium cycle to reduce fuel-cycle costs. A LWBR core will be tested in the Shippingport reactor to evaluate the feasibility of converting existing and future pressurized water reactors to self-sustaining breeders. In principle such reactors would make possible the conversion of 50 per cent of the mined thorium into energy sufficient

TABLE 24.15 PWR and BWR Fuel Burn-Up

	Fuel Burn-up (Mwd/MT)	*Fuel Composition (g/MT)* U-235	Pu-239	Pu-241
		PWR–Plutonium Withdrawal		
Charge	–	32,500	0	0
Discharge	32,000	8,359	5,327	1,213
		PWR–Plutonium Recycle		
Charge	–	6,869	19,849	4,082
Discharge	32,000	2,919	8,765	4,647
		BWR–Plutonium Withdrawal		
Charge	–	25,000	0	0
Discharge	27,000	6,403	4,808	1,034

for hundreds of years. This improvement could be realized without the major development required for other systems. The advantages of the other breeders are greater thermal and fuel cycle efficiency.

Reactor Vessels. While the nuclear industry has posed challenges on most frontiers of engineering, the fabrication of the reactor vessel is one of the most significant. The design criteria for optimum performance requires operating pressures of 1050 psia for BWRs and 2250 psia for PWRs. Typical dimensions for vessels serving 1000-mwe stations range from 21 ft diameter by 70 ft height for BWRs to 14.4 ft by 42.8 ft for PWRs (Fig. 24.14) with vessel weights of 782 and 459 tons, respectively. These are shop-fabricated (Figure 24.15). However, for larger plants both field fabrication and the use of prestressed concrete would be considered. For a 3000-mwe concrete vessel the cavity would be about 51 ft diameter by 50 ft height with side walls 25 ft thick. The use of a prestressed-concrete vessel combines the shielding with the pressure containment for the nuclear reactor. With a gas-cooled reactor, where prestressed concrete vessels are now used, the thermal insulation is placed inside the pressure vessel.

Containment. Under all conditions of normal operation and maximum credible accidents any release of radioactivity from the nuclear reactor to the environment must be below acceptable limits. The maximum credible accident in the LWR is postulated to be the loss of coolant. This could result in melt-down of the fuel with release of the radioactive fis-

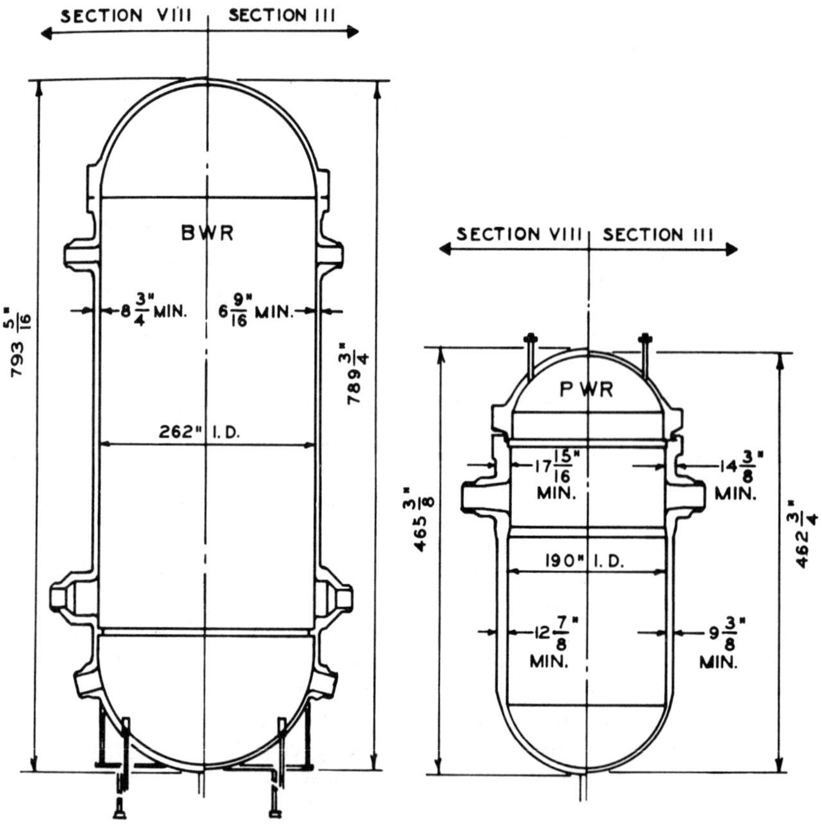

Fig. 24.14 BWR and PWR reactor vessels for 1000 Mwe plants; comparison of ASME Code Sections III and VIII. (*USAEC*)

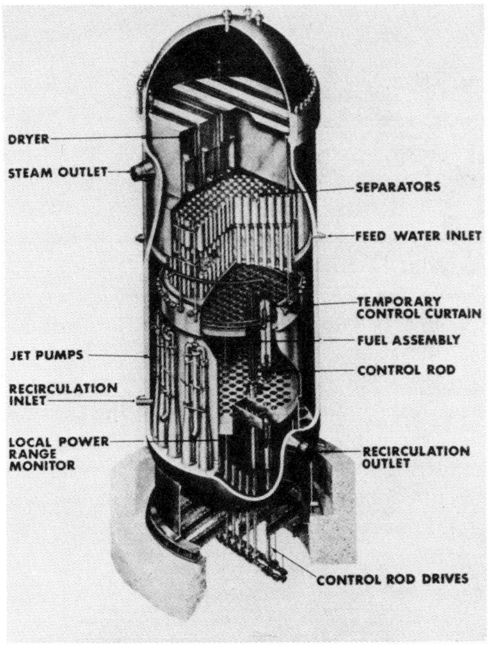

Fig. 24.15 BWR reactor vessel contains jet pumps to recirculate water to increase heat transfer; steam conditioner in the top of reactor requires bottom control-rod drives. (*USAEC*)

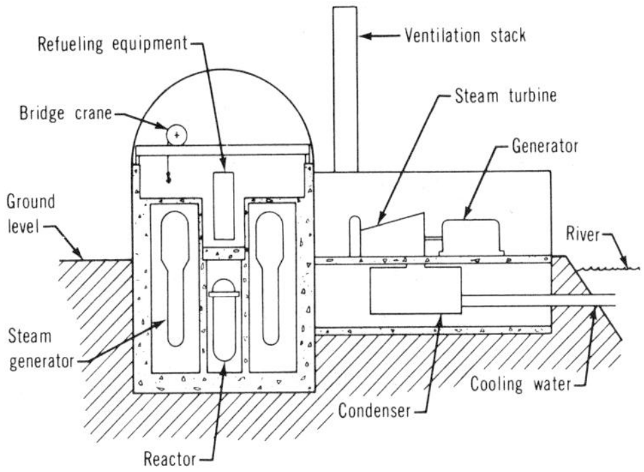

Fig. 24.16 Model of dual-cycle pressurized-water reactor, note man entering on right. Arrangement of components, shielding, and containment shown in insert. (*Combustion Engineering, Inc.*)

sion products. The fission products in turn could be transported out of the core in the escaping vaporized coolant. To prevent the escape of this radioactive coolant from the facility, a containment vessel is built around the reactor. Provision is made to rapidly shut down the nuclear reaction, condense the coolant, and scrub out and concentrate the radioactivity. The most recent PWR approach to accomplishing this uses a large-surface ice condenser. The ice is always present during reactor operation (Fig. 24.16). In the BWR the escaping steam is vented into a large water pool where it is condensed (Fig. 24.17). In both reactor systems there are redundant systems for cooling the reactor and treating the radioactive wastes, and an inert atmosphere is established to eliminate the possibility of metal combustion. Decontamination of the environment within the containment vessel is started within minutes.

As anticipated there have been occasional equipment failures involving reactors, but the safety systems have been sufficiently redundant that one or more have always worked. Two experiments are planned to better assess the margin of safety provided by the containment systems. The Power Burst Facility will study power transients and loss-of-flow at power capable of destroying fuel elements without damage to the facility. Various effects associated with the failure of present and planned fuel elements will be evaluated. In the second study, the Loss-of-Fluid Test (LOFT) project, integral tests with a 55-Mwt reactor will be carried out. Using the nuclear heat and fuel materials present in current power reactors, this will directly assess the margin of safety of the engineered safety systems.

High Temperature Graphite Reactor

High-temperature gas-cooled reactors are under development in the United States, the United Kingdom, and Germany.[8] The concepts for these reactors include: (1) only fissile, fertile, and modulator ceramic materials in the core for high temperature stability, (2) the good neutron economy of the thorium-uranium fuel cycle, and (3) high-temperature helium coolant coupled to a closed-cycle turbine for high-efficiency power conversion. The HTGR core has a large heat capacity that provides more time to provide alternate cooling if a loss-of-coolant accident occurred. Other fea-

Fig. 24.17 Oyster Creek Nuclear Power Station utilizing 640 Mwe boiling water reactor. (*Courtesy Jersey Central Power and Light Co.*)

tures of this reactor design include:

1. Field-constructed prestressed concrete reactor vessels that also serve as primary radiation shields, subject to in-service inspection to eliminate possibility of vessel failure.
2. Steam-driven helium circulators and control rods.
3. A fissile-fuel requirement about half that of the LWR cycle.

The core of the Gulf General Atomic HTGR consists of graphite-coated uranium and thorium particles embedded in a hexagonal graphite block (Fig. 24.18). The initial operation employs uranium-235 which must be rejected after one cycle to avoid neutron losses in the uranium-236 formed. Therefore, the size of the uranium and thorium particles differ so that they can be mechanically separated before processing the spent fuel. The uranium-233 is recycled with the thorium. The reprocessing of the spent fuel from this reactor could use the existing plants which treat LWR fuels, but a special head-end treatment facility would be required. The tentative process would involve crushing, burning in a fluidized bed, and leaching to prepare the uranium and thorium solutions for solvent extraction recovery of the uranium-235, uranium-233, and the thorium as separate products.

Liquid Metal Fast Breeder Reactor

The salient features of the liquid-metal fast-breeder reactor (LMFBR) include a fuel-doubling time of 7 to 10 years along with a high coolant temperature for more efficient energy conversion and low operating pressure through the use of metallic sodium as the coolant (Fig. 24.19). Operation of the reactor with

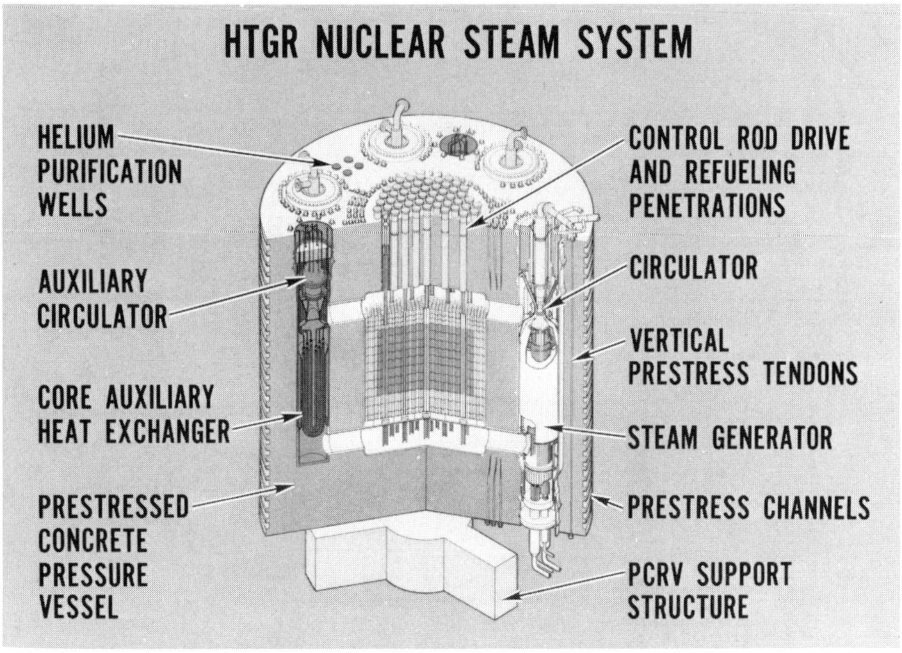

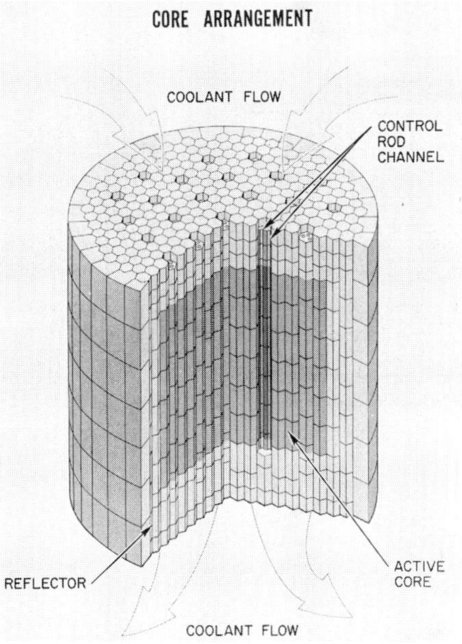

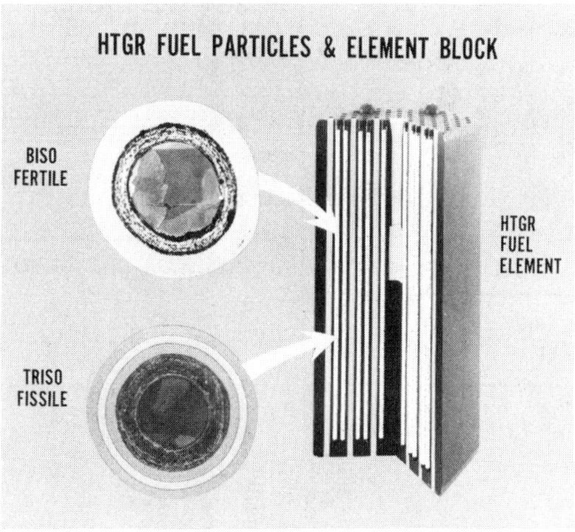

Fig. 24.18 High-temperature-graphite rector provides about 44 per cent energy conversion efficiency; to gain maximum fissile material conversion efficiency, the fissile uranium and fertile thorium oxides are separately fabricated into graphite-coated particles and then consolidated into the graphite fuel element block. Gulf Energy and Environmental, Inc. (*Gulf General Atomics.*)

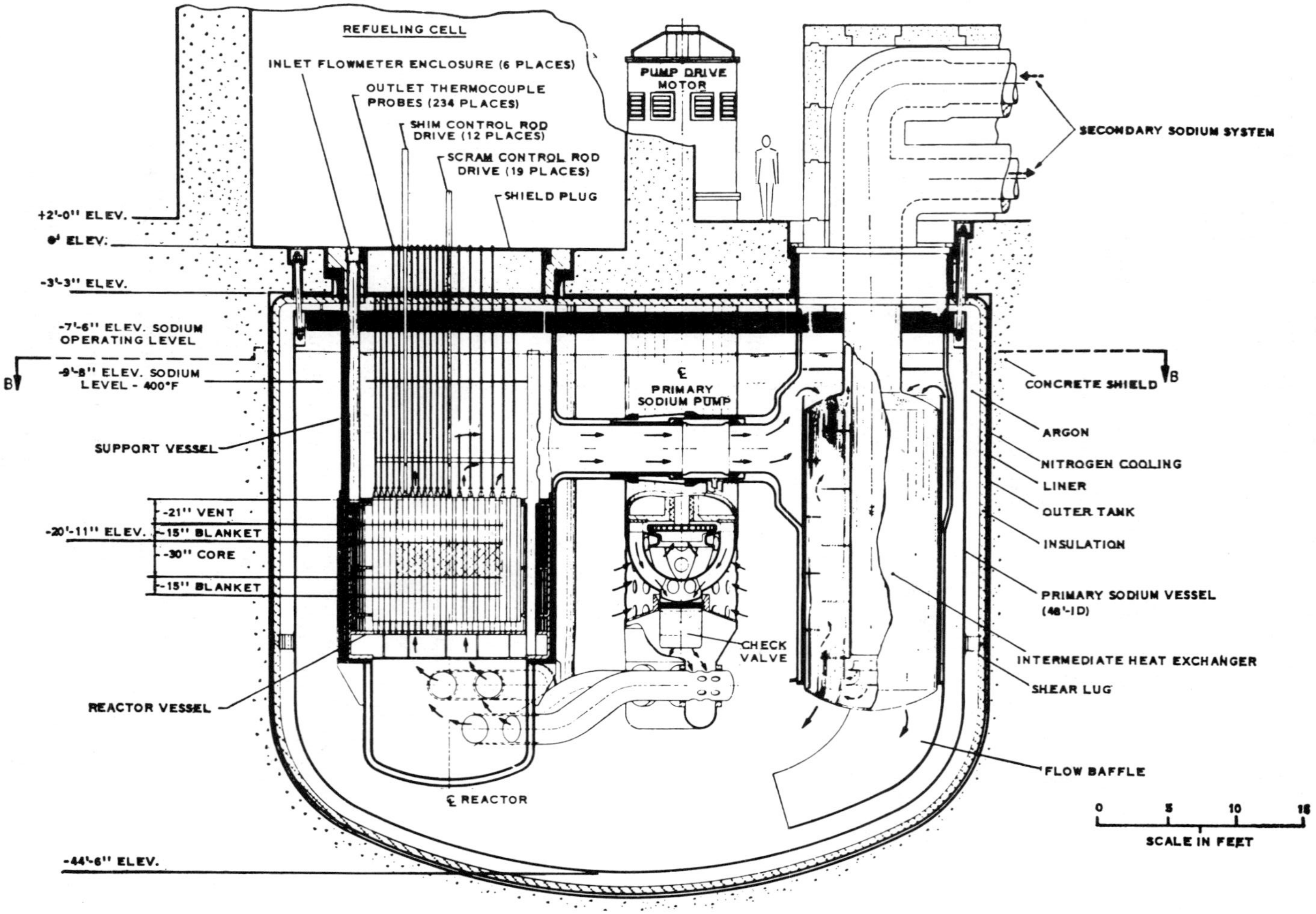

Fig. 24.19 Pool-type LMFBR for 1000 Mwe plant; both reactor and heat exchanger are submerged in sodium coolant to avoid loss-of-coolant accident with pre-stressed concrete reactor vessel and to eliminate vessel failure. (*General Electric*)

fast neutrons increases the neutron multiplication factor and reduces the parasitic capture of neutrons in the coolant and structural materials. This enhances the fuel-conversion ratio while permitting the use of high-temperature cladding material for the mixed uranium-plutonium oxide fuel. The power density in the LMFBR core is about 5 times greater than that in the LWR cores and therefore the former is much smaller in size. This core is surrounded with radial and axial blankets of fertile material. These blanket regions which absorb the leakage neutrons must be used to achieve breeding. Finally, the use of the uranium-plutonium fuel cycle makes optimum use of the plutonium produced in the LWRs.

All the major countries are participating in LMFBR demonstration plants ranging from about 200 to 500 mwe.[20] In 1971 the USSR was also building a second-generation plant incorporating advanced features in the 600-Mwe class. In the U.S., the Enriched Breeder Reactor-II (EBR-2) has operated at power up to 62.5 Mwt and demonstrated plutonium-uranium oxide fuel burn-ups in excess of the design goal of 100,000 megawatt-days per metric ton. In January 1971 the Southwest Experimental Fast Oxide Reactor (SEFOR) was operated at its licensed power of 20 Mwt, and tests were begun to investigate a key question regarding the inherent shutdown capabilities of the LMFBR type reactor. The 400 Mwt Fast Flux Test Facility is being constructed to test fuels and materials at a flux of 7×10^{15} and a power density of 300 to 700 kw per liter. EBR-1, the starting point of this program, went critical in 1951.

While the feasibility of this reactor system has been extensively demonstrated, the engineering development program aimed at the introduction of the first large scale LMFBR in 1984 will cost about two billion dollars. This cost may be better appreciated when considered in terms of the cost of the power to be generated over the next fifty years, about $1600 billion. The development cost-to-benefit ratio may be as great as 8, and it is principally achieved by the reduced cost for uranium ore and enrichment.

An alternate fast breeder reactor, the gas-cooled fast breeder, uses helium instead of sodium as the coolant. It incorporates important components of the HTGR including the prestressed-concrete reactor vessel and the turbine-driven helium circulators. While the fuel-doubling time for this reactor could be only half that of the LMFBR, a counter balancing factor is the higher capital costs attributable to the high-pressure-gas coolant system.

Molten Salt Breeder Reactor

The molten-salt breeder reactor (MSBR) is interesting from a chemical systems point of view, since it employs uranium and thorium fluoride in a solution of $^7LiF\text{-}BeF_2$ molten salt as the fuel.[21] This molten-salt homogenous reactor provides high thermal efficiency with low fissile material inventory and on-site fuel recycle (Fig. 24.20). A 7.3 mwt experimental reactor at Oak Ridge National Laboratory was operated for about five years with good results. During this period uranium-235, uranium-233, and plutonium-239 were used as the fuel. While the feasibility of this concept has been established, much chemical engineering development remains to be completed before commercial acceptance can be expected. These studies include the graphite moderator, the Hastelloy N for reactor vessel and plumbing, circulating pumps, heat exchangers, coolant salt, and separations processes.

Frequent processing will be required to achieve breeding. These processing operations include continuous removal of the xenon, protactinium-extraction every two or three days, and rare earth-fission-products removal about every 50 days. The xenon is readily removed from the recirculating molten-salt stream using a gas separator. The uranium may be recovered by fluoridation to the volatile UF, but the separation methods for protactinium and rare earth-fission products are not presently defined. As a breeder, only thorium would be added to the fuel cycle and the power level would be largely controlled by the uranium-233 concentration in the molten salt.

Other Nuclear Reactors

There are many other nuclear reactors that have been developed for research, engineering development, nuclide production, and mobile power. Most noteworthy are the nuclear propulsion systems for naval applications. The

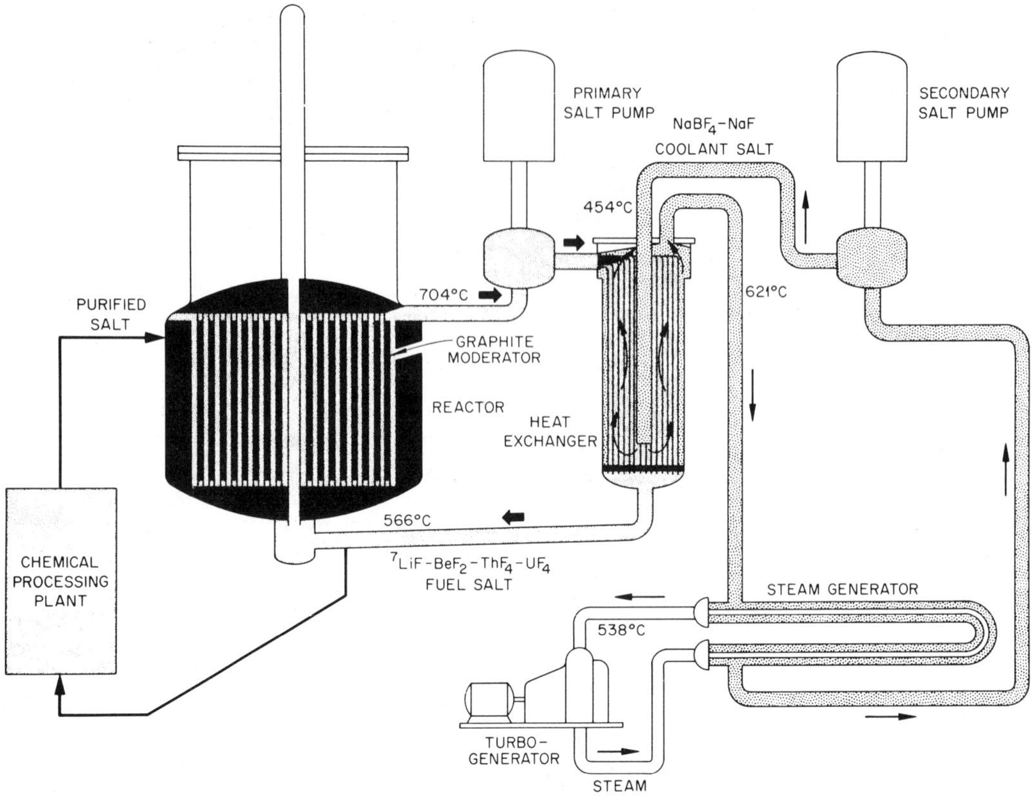

Fig. 24.20 Oak Ridge Molten Salt Breeder Reactor; most efficient thermal breeder using the thorium-uranium fuel cycle and frequent fuel processing, continuous xenon, 3-day protactinium, and 50-day rare earths. (*Oak Ridge National Laboratory*)

first nuclear submarine, the Nautilus, was commissioned in 1954. Today there are more than a hundred such submarines, along with a number of surface vessels. In the United States the PWR technology for electric-power generation was an outgrowth of the nuclear naval-development program for which Vice Admiral Hyman G. Rickover supplied the driving will.

Other nuclear propulsion systems have been studied for arctic tractor trains, aircraft, and interplanetary rockets. The nuclear rocket provides a significant extension of interplanetary transport capability, with final development dependent on a national commitment to space exploration. The others were terminated before completion due to insufficient need.

The remaining classes of nuclear reactors range from zero-power, subcritical neutron sources for university training to large-scale reactor systems for plutonium-239 production. Portable reactors provide heat, power, and water to U.S. bases in Alaska and the Arctic, and another provides electric power to the base in Panama. Private industry operates test reactors for development of reactor components and for commerical production of a wide variety of radioisotopes.

REACTOR MATERIALS PROCESSING

Reactor-materials processing is concerned with preparation of the special materials used for nuclear-reactor fuels, coolants, moderators, plumbing, heat exchangers, and other components. Attention in this chapter will be limited to uranium and thorium. However, complete coverage would require the consideration of structural materials including zirconium, stainless steel, and aluminum, coolants including water, sodium, and helium; moderators includ-

TABLE 24.16 Estimated Resources of Uranium, 1970 (10^3 Short Tons U_3O_8)

	Price Range < $10/lb		*Price Range $10–15/lb*	
Country	*Reasonably Assured*	*Additional Estimated*	*Reasonably Assured*	*Additional Estimated*
Argentina	12	18	10	30
Australia	92	102	38.3	38
Brazil	2.5	3.3	0.9	–
Canada	241	247	158	284
Central African Republic	10.5	10.5	–	–
Denmark (Greenland)	7.0	13	–	–
Finland	–	–	1.7	–
France	45	25	13.0	26
Gabon	26	6.5	–	6.5
India	–	–	3.0	1.0
Italy	1.6	–	–	–
Japan	3.6	–	5.4	–
Mexico	1.3	–	1.2	–
Niger	52	26	13	13
Portugal	9.6	7.7	–	30
South Africa and SW	263	10.4	80.6	33.8
Spain	11	–	10	–
Sweden	–	–	350	52
U.S.A.	337	700	183	300
Others	11.4	13.8	0.6	–
TOTAL (approximate)	1,126	1,185	869	814

Source: Nuclear Energy Agency

ing deuterium, graphite, and beryllium; and control rod materials including boron, cadmium, and hafnium.

Uranium Raw Materials

Uranium is the basic fuel of the nuclear power program. In 1970 it was estimated that the total reasonably assured resources of uranium outside the Communist bloc, in the price range under $15 per pound U_3O_8, amounted to 1,600,000 tons U_3O_8 (Table 24.16). Since that estimate was prepared, substantial new discoveries of uranium have been made in northern Australia and South West Africa; these may eventually increase the known resources by several hundred thousand tons. The world demand for uranium was projected (Table 24.17) to be 960,000 tons by 1985.[23] The reliability of the estimates for 1985-demand is significantly less than it was for earlier years, and in view of the fact that no allowance was made for the recycle of plutonium, the figure quoted for 1985 is more likely to represent the maximum than the midpoint of the range. If breeder reactors come into operation by 1985, the peak of uranium production may be reached by 1990; after that the uranium supply will be adequate to meet anticipated requirements for several centuries.[24]

TABLE 24.17 Estimated World Uranium Demand

Year	*Annual Demand Short Tons U_3O_8*	*Cumulative Demand from 1970 Short Tons U_3O_8*
1973	22,400	22,400
1974	30,900	53,300
1975	37,900	91,200
1976	44,600	135,800
1977	52,600	188,400
1978	62,300	250,700
1979	73,000	323,700
1980	81,300	405,000
1985	150,100	1,004,500

Source: WASH-1139 (72)

As of January 1, 1971, the U.S. reserves of uranium ore from which U_3O_8 was recoverable at \$8 per pound were 118,000,000 tons of ore at 0.21 per cent U_3O_8. This amounts to a total of 246,000 tons of U_3O_8 and represents about an 11-year supply. In anticipation of future requirements, a large exploration program is underway by private industry. Surface exploration and development drilling for the four years 1970-73 were projected at 78 million feet and a total cost of \$120 million, excluding land acquisition and exploration rights. Twenty-four million feet of drilling were performed in 1970, and a net 42,000 tons of U_3O_8 in ore recoverable at \$8 per pound were added to reserves.

In 1971 there were 17 urnaium-ore-processing mills in operation in the U.S., and three new plants were under construction and planned for 1972 start-up. The capacity of individual ore-processing plants varied from 450 to 7,000 tons of ore per day. The uranium-bearing ores processed commercially in the United States generally contain 0.1 to 0.5 per cent U_3O_8. The average concentration is about 0.2 per cent. In addition, some of the ores contain vanadium which can be recovered as a by-product (Fig. 24.21). Because the ores are quite low in uranium-content, the uranium-recovery plants or mills must be located near the mines to minimize transportation costs. The final product of ore treatment is a crude uranium concentrate containing 70 to 95 per cent U_3O_8. This material can be delivered to

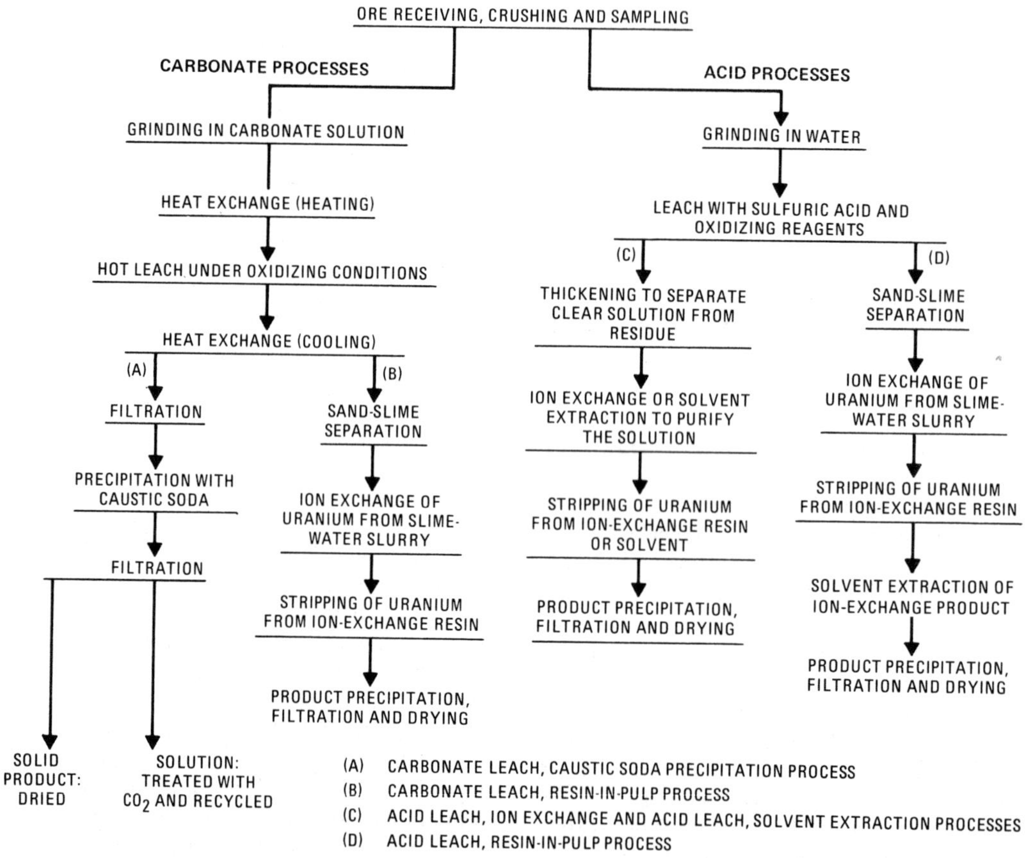

Fig. 24.21 Uranium ore mill processes; (A) carbonate leach, caustic soda precipitation process; (B) carbonate leach, resin-in-pulp process; (C) acid leach, ion exchange and acid leach, solvent extraction processes; and (D) acid leach, resin-in-pulp processes.

refineries located a long distance away at comparatively lost cost.[22]

Physical Methods of Concentration. Most uranium minerals are friable, and they become so finely pulverized and mixed with clay slimes in the grinding operation that they cannot be effectively separated from the gangue or waste constituents by physical methods. Flotation methods are successful only in the few cases in which the uranium is closely associated with another mineral that can be floated, such as pyrite.

Uranium Dissolution. Uranium can be dissolved from its ores by any mineral acid or by sodium carbonate solutions. If uranium is present in the ore in its chemically reduced state, as is generally the case, an oxidizing agent in conjunction with either acid or sodium carbonate must be used to dissolve it from the ore.

Carbonate Leaching. The ore is first crushed and ground to very fine particle sizes in an aqueous slurry, commonly called a pulp, containing sodium carbonate and bicarbonate. Then the pulp is heated to temperatures near boiling and agitated for approximately 12 to 24 hours with the continuous addition of air to supply oxygen, and in some cases with the addition of chemical oxidizing agents as well. Next the slurry is cooled in heat exchangers and filtered on rotary drum filters to separate the uranium-bearing solution from the residual ore solids. Sodium hydroxide is added to the filtrate to precipitate the uranium as impure sodium diuranate.

As the carbonate leach is quite selective for uranium, the product, after filtration and drying, may be sufficiently pure to meet specifications without further treatment. Additional steps are often required, however, to remove such impurities as vanadium or sodium, depending on the needs of the refinery to which the product is to be sent for subsequent treatment. The solution from which the uranium has been precipitated is regenerated with carbon dioxide, usually from boiler flue gas, and returned to the process.

Acid Leaching. After the ore is crushed and ground in water, the slurry is leached for about 8 to 12 hours in a dilute acid solution containing an oxidizing reagent. Because of its relatively low cost sulfuric acid is universally employed for leaching. Heat is often used to improve dissolution, but its effect is not as important as in the carbonate leach process. The two most commonly used oxidizing reagents are sodium chlorate and manganese dioxide. The manganese dioxide is often in the form of a crude pyrolusite ore containing roughly 30 to 40 per cent MnO_2. After the uranium is dissolved, the uranium-bearing solution is usually separated from the residual ore solids in a series of large settling tanks called thickeners. The solids are transferred from one thickener to the next countercurrent to the flow of wash water.

Many elements in the ore other than uranium are soluble in mineral acids. Thus, after separation from the ore pulp, the uranium-bearing solutions require further purification. Ion-exchange or solvent-extraction methods are employed for this purpose. If the uranium were to be precipitated directly from the leach solution with sodium hydroxide, as is done in the carbonate process, the resultant product would contain about 2 to 10 per cent U_3O_8, and would be contaminated with large quantities of iron, aluminum, silica, and many other impurities. The ion-exchange and solvent-extraction procedures make it possible to separate the uranium effectively from most of the impurities in the leach solution, and to obtain a product of satisfactory purity in a single unit operation.

Ion Exchange. The anion-exchange resins that are used for uranium recovery are small spherical beads ranging generally from $^1/_{10}$ to $^1/_{60}$ of an inch in diameter. In one procedure the resin is held in cylindrical tanks, and the clear uranium-bearing leach solution is passed through a bed of resin several feet deep. A uranium sulfate complex is preferentially retained by the resin, while the other soluble metals, which do not form such complexes, pass through the resin bed and are rejected. The uranium usually is removed from the resin by elution with an acidified chloride or nitrate solution. Partial neutralization precipitates the iron and aluminum, removing them from the solution. Then, the uranium is precipitated from the eluate with additional alkali, filtered, and dried.

Another ion exchange process known as resin-in-pulp is employed as an alternative to recovery of uranium from clear solutions. After the leach, the pulp is diluted so that the coarser sand particles can be washed and discarded. The remaining solution containing the very fine slime particles is then mixed with the ion-exchange resin. The presence of the slime does not interfere with uranium absorption from solution by the resin. Because the slime particles are very fine compared to the resin beads, screens can be used to separate the resin from the pulp after the absorption step. Uranium is then recovered from the resin in the manner previously described.

Solvent Extraction. In some plants, solvent extraction is the primary method for recovery of uranium from the acid leach liquors. However, several plants using resin-in-pulp as the primary separation method elute the resin with a strong sulfate solution and further purify the uranium-bearing eluate by a solvent-extraction step before precipitation. The final product of this process generally contains over 90 per cent U_3O_8.

The active solvents used for uranium recovery are either organic amines or phosphates in which each organic group generally contains 8 to 14 carbon atoms. The amines are more selective for uranium than the phosphates. However, the latter is sometimes preferred in plants which recover and purify vanadium as well as uranium. Due to excessive losses of expensive active solvent, the solvent-extraction process is not satisfactory for recovery of uranium from carbonate solutions, nor can it be employed to recover uranium from ore pulps. Thus, the procedure is limited to recovery of uranium from clear solutions produced by the acid-leach process.

The active solvent, which is a liquid ion-exchange material, is diluted to a concentration of about 5 per cent in kerosene. The diluted solvent is contacted with the clear leach solution in three or four countercurrent steps. Mixer-settler units are commonly used. In each stage the two immiscible solutions are agitated in a small mixing compartment from which they overflow into a settling tank. The organic solvent rises while the aqueous solution sinks to the bottom. After four contact stages, the organic stream containing the uranium is transferred to a separate circuit where it is stripped with any of several suitable aqueous reagents. The product is precipitated from the strip solution, then filtered and dried.

Uranium Product. The end product of all the milling procedures now being used is a crude concentrate, generally referred to as yellow cake. While some study is being given to the potential production of refined uranium compounds in the ore-processing mills, no mill is now producing such materials.

Thorium Raw Materials. Thorium, which is several times more abundant in the earth's crust than uranium, is a fertile material which can be converted to the fissionable isotope U-233 by bombardment with neutrons in a reactor.[25] Most of the presently known resources of low-cost thorium (500,000 tons ThO_2) are in placer deposits containing monazite sands in India and elsewhere. Thorium is also found in large quantities in vein deposits in the Lemhi Pass area of Idaho and Montana. It is present in significant concentrations in the uranium ores in the Elliot Lake district in Canada, and has been produced as a by-product of uranium in this area. These known reserves are adequate to support the present need. The future requirement for thorium depends on the acceptance of the HTGR and MSBR reactor programs.

FEED MATERIALS PROCESSING

Uranium Feed Materials

In uranium-feed-materials processing the crude uranium product from the ore mills—or the recycled product from the reactor-spent-fuel processing—is converted into the enriched uranium fuel required for the nuclear reactor (Fig. 24.22). For the light water reactor the fuel is slightly enriched uranium oxide, usually 2 per cent to 4 per cent uranium-235, contained in zirconium tubing. This zirconium-clad fuel rod is required to spend three to four years in a 600°F, 2000 psig water and steam

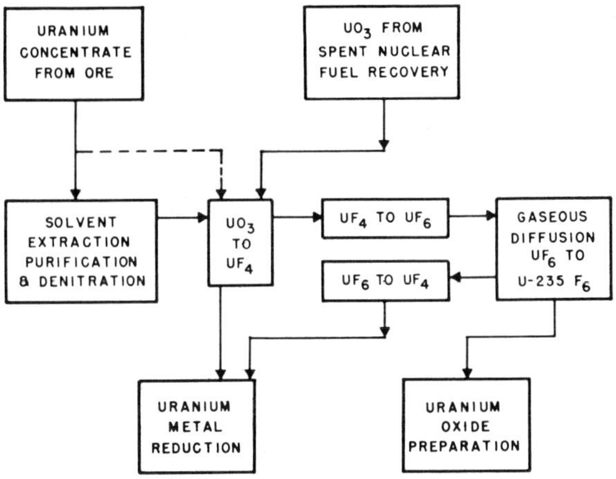

Fig. 24.22 Uranium-feed-materials flow sheet; by 1980 a $530 million business including fuel element fabrication.

environment. In the past, the standard LWR fuel was stainless steel-clad uranium oxide, but zirconium provides greater neutron economy.

Using high-grade ore concentrate it is possible to produce the uranium hexafluoride used for isotopic enrichment by hydrogen reduction followed by hydrofluorination and fluorination. Fluid-bed reactors are used for these conversion steps which are followed by fractional distillation of the UF_6. In another process, ore concentrates are first purified by tributyl phosphate solvent extraction to provide a product that requires no further purification and greatly reduces the operating problems in subsequent conversion to UF_6 in fluid-bed reactors. In addition, when uranium metal is to be the end-product, the intermediate uranium tetrafluoride product from the solvent-extraction route meets metal specifications.

The UF_6 product goes to the gaseous diffusion plant for enrichment followed by hydrogen reduction to UF_4 and, then, ammonia precipitation and calcining to UO_2. In the final step the UO_2 is pelleted, sintered in a hydrogen furnace, machined to size and loaded into zircaloy tubing on its way to becoming a fuel element for the nuclear reactor. As noted earlier, the fuel elements in the core of the LWRs contain about 80 miles of this zircaloy tubing filled with uranium oxide.

Uranium Solvent Extraction. The ore concentrate is dissolved in nitric acid and solvent-extracted with tributyl phosphate (TBP) in a kerosene or hexane diluent:

$$UO_2{}^{++}{}_{(aq)} + 2NO_3{}^{-}{}_{(aq)} + 2TBP_{org} \longrightarrow UO_2(NO_3)_2 \cdot 2TBP_{(org)}$$

This is an easily reversible reaction, directly dependent on the salting ion (NO_3^-) concentration in the aqueous phase. The salting agent is nitric acid. The extraction is very specific for uranium providing decontamination factors of 10^3 to 10^5. The uranium is extracted into the solvent, leaving most other ions in the aqueous phase; the extract is scrubbed with a small amount of water for further purification; and the uranium is then stripped from the solvent with water (Fig. 24.23). The product is an aqueous uranyl nitrate solution.[26]

Uranium Denitration. Denitration of the uranyl nitrate product from solvent extraction to UO_3 involves evaporation followed by calcination. The evaporation is carried out to a final boiling point of 120 to 143°C (9.8 to 12.5 lb U/gal) and the residue is transferred to a stirred-pot calciner where it is heated to 621°C (Fig. 24.24).

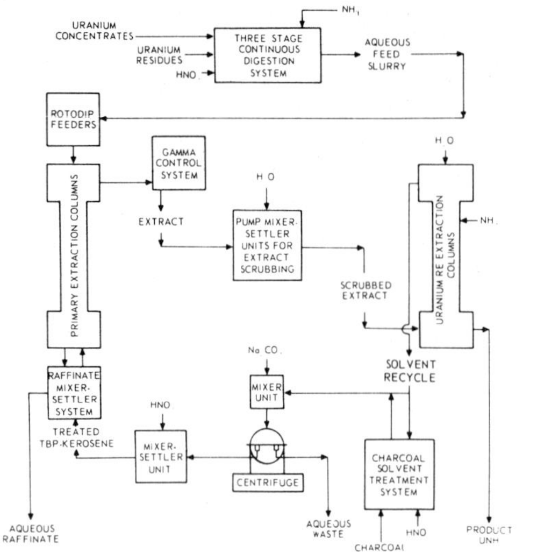

Fig. 24.23 Uranium solvent-extraction process for purification of ore concentrate and scrap; slurry feed eliminates clarification cost and losses. (*USAEC Feed Materials Plant, National Lead Co. of Ohio, Fernald, O.*)

$$UO_2(NO_3)_2 \cdot 6H_2O \xrightarrow[621^\circ C]{\Delta} UO_3 + NO_2 + NO + O_2 + 6H_2O$$

Other calciner types in use include the stirred-trough and the fluidized-bed types. The type of equipment and the processing conditions significantly affect the physical properties of the UO_3, and hence the kinetics of the subsequent reactions in the production of UO_2, UF_4, and the metal.

Uranium Oxide Conversion to Uranium Hexafluoride. At Allied's Metrolopis, Illinois plant using U_3O_8 as the feed, the conversion of uranium oxide to UF_6 is accomplished in a series of three fluid-bed reactors; first, for the hydrogen reduction of the uranium oxide to UO_2; second, for hydrofluorination to UF_4; and third, for fluorination to UF_6.[28]

$$U_3O_8 + 2H_2 \xrightarrow{650^\circ C} 3UO_2 + 2H_2O$$

$$UO_2 + 4HF \xrightarrow{400^\circ C} UF_4 + 2H_2O$$

$$UF_4 + F_2 \xrightarrow{500^\circ C} UF_6$$

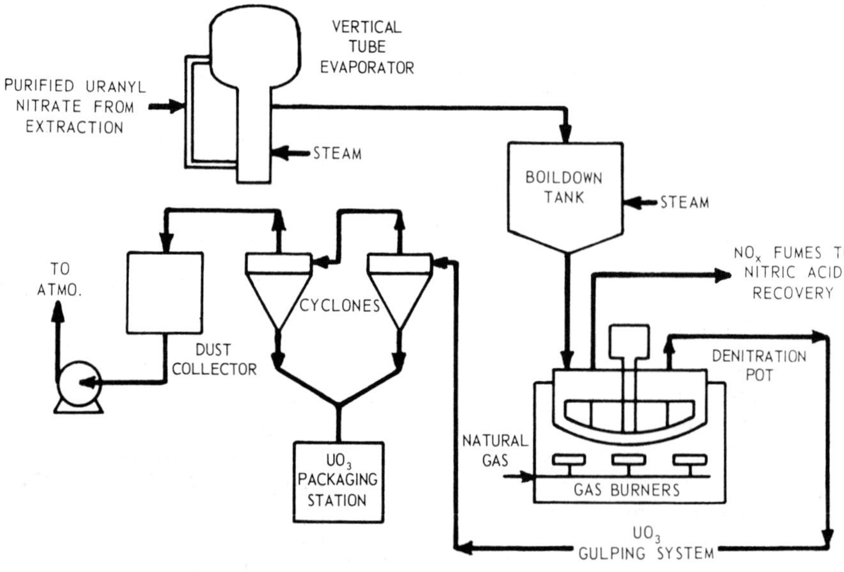

Fig. 24.24 Uranium denitration process prepares UO_3 for conversion to metal and UF_6. This batch-pot-type equipment usually used, however continuous fluid bed developed for large-scale use. (*USAEC*)

However, a second plant built by Kerr-McGee uses solvent extraction and denitration procedures, similar to those employed for metal production, providing chemically pure UO_3 before the reduction to UO_2 operation. In either case, the feed for UO_2 reduction is fluidized by the countercurrent flow of hydrogen at about 0.5 ft/sec with the temperature maintained at 600°C. The hydrofluorination step employs a two-stage fluid bed to achieve economic use of the hydrogen fluoride and maximum conversion (Fig. 24.25). The UF_4 product then goes either to metal production or on to UF_6 preparation.[27]

In the fluorination step complete use of the high-cost fluorine and complete recovery of the uranium as UF_6 are important. Here the fluorine feed first contacts the fluorinator ash for maximum uranium recovery, then goes to the fluorinator where most of the fluorine and uranium are reacted, and finally, the last traces of unreacted fluorine are stripped by passing through the UF_4 in a screw reactor that is feeding the UF_4 to the fluorinator.

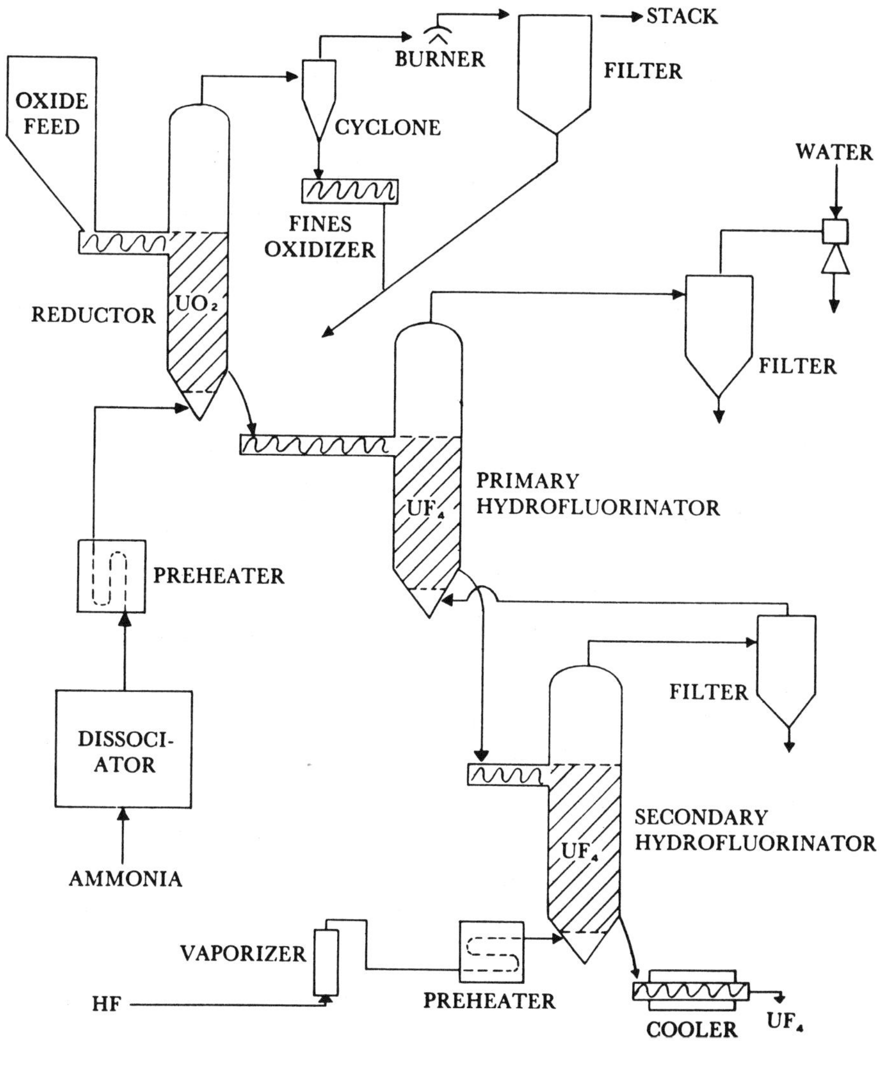

Fig. 24.25 Fluid-bed system for UO_2 conversion to UF_4. (Allied Chemical Company received award for this process-engineering achievement.)

Since there is no solvent extraction purification step in the Allied process, the UF_6 product still contains vanadium, molybdenum, and other impurities. These are removed by distillation, first in a 120-foot, 100-plate column at 200°F and 85 psia to remove the high-volatile impurities, then in a second 45-plate column operating at 240°F and 95 psia to remove the low-volatile impurities. The final UF_6 product is condensed and packaged in 10-ton cylinders, each holding about $200,000 worth of product for delivery to the gaseous diffusion plant for isotopic enrichment.

Reduction to Uranium Metal. Uranium metal is produced from UF_4 by reduction with magnesium.

$$UF_4 + 2Mg \longrightarrow U\text{ (metal)} + 2MgF_2 \quad (\Delta H_{298} = -83.5\text{ kcal})$$

This is a batch process, carried out in a steel reactor lined with the reaction by-product, magnesium fluoride (Fig. 24.26). The resulting 350-lb uranium regulus, called the "derby," is remelted, held at 1454°C in a vacuum furnace to vaporize and remove impurities, and then recast in a graphite mold to produce the ingot. The ingot is fabricated into reactor fuel by extrusion, rolling-mill operation, and machining. Uranium metal is primarily used for AEC production reactors. However, depleted metal also finds some limited use as radiation shielding for spent-fuel shipping casks, and in counter weights.

Uranium Enrichment. This phase of the nuclear industry represents about one third of the total capital funds invested to date in atomic energy.[29] Following early development work, gaseous diffusion was selected over electromagnetic separation and thermal diffusion to provide enriched uranium-235. While gaseous diffusion today is a well established large-scale method of isotopic separation, centrifugation is being investigated by several groups with the hope of finding a more economical method. This discussion will be limited to gaseous diffusion.

In gaseous diffusion the porous membrane or barrier, as it is usually called, and the temperature and pressure are adjusted to optimize the separation of gaseous molecules of differing mass (Fig. 24.27). The theoretical separa-

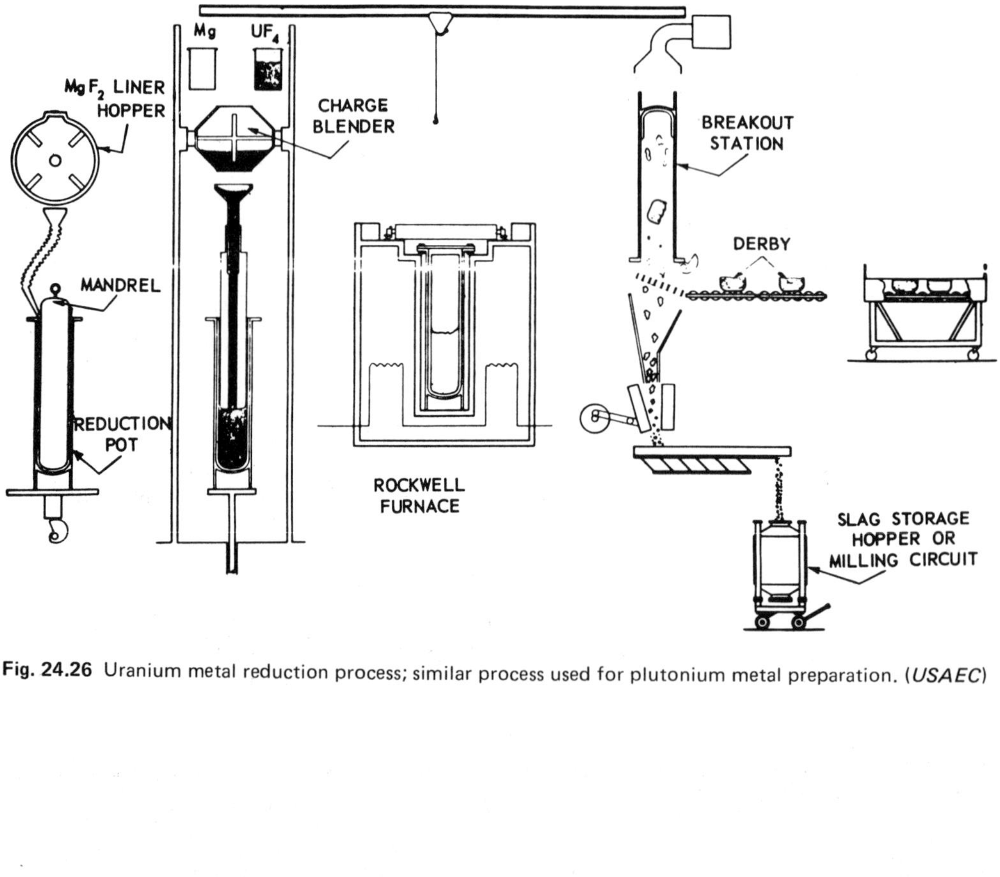

Fig. 24.26 Uranium metal reduction process; similar process used for plutonium metal preparation. (*USAEC*)

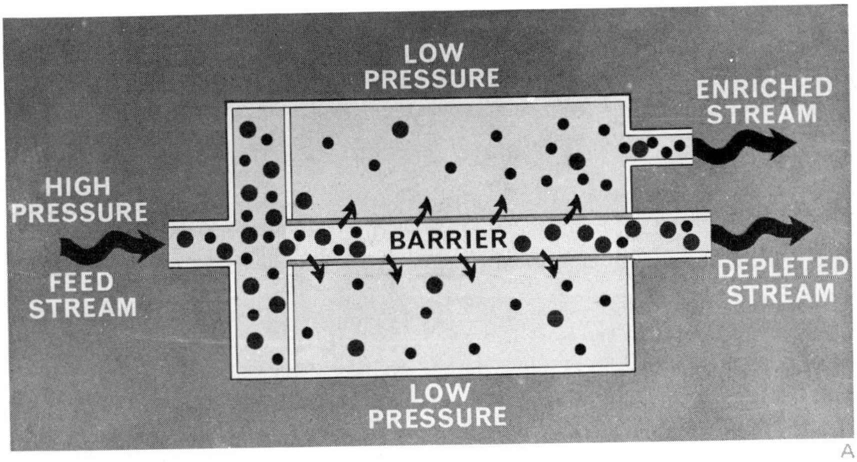

A

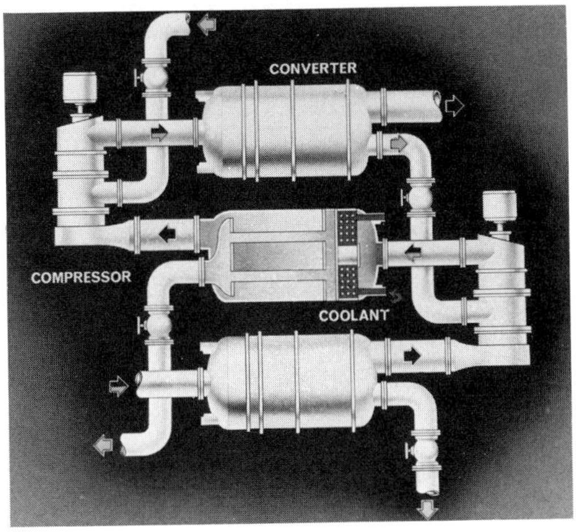

B

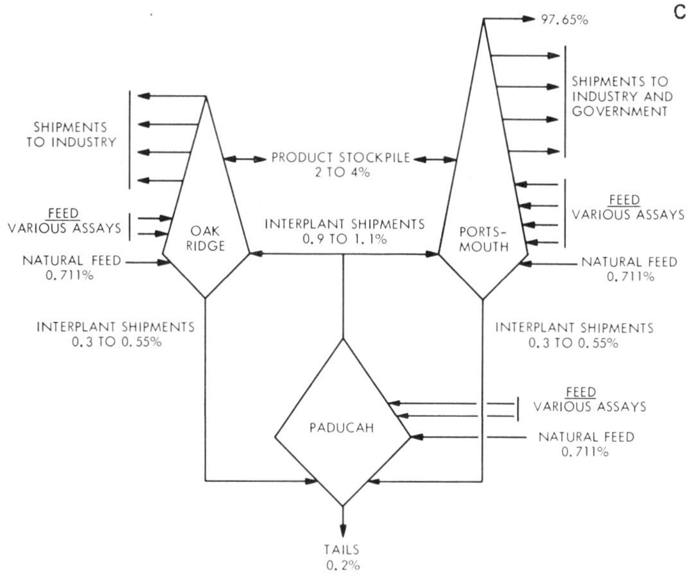

C

Fig. 24.27 U.S. AEC gaseous diffusion plant complex during 1967: A: Gaseous diffusion stage. B Stage arrangement. C. Mode of operation for gaseous diffusion complex during last half of 1970. (% values are weight % U-235.) (*USAEC*) tion factor for uranium-235 is:

$$\alpha = \sqrt{\frac{^{238}UF_6\ (352)}{^{235}UF_6\ (349)}} = 1.0043$$

The opimum conditions require the use of elevated temperature and reduced pressure with a pressure differential across the barrier sufficient to transport about half of the UF_6 while flowing through that stage. By varing this flow ratio it is possible to taper the stages in the plant. This is very important, since with the small separation factor a large number of stages are required to achieve the required enrichment. The large number of stages in turn require a large in-plant inventory and the large inventory results in a long operating time for the plant to reach equilibrium. By tapering the plant, the volume of the plant stages is decreased as the isotopic concentration of the uranium-235 increases. This volume reduction greatly reduces the time it takes to bring the plant to equilibrium, along with reducing the cost of the diffusion plant equipment, and plant operation.

The AEC gaseous diffusion plants contain 10,812 stages and consume up to 6000 mw of power and 1350 million gallons of water per day. The power and water requirements would meet the needs of a city of a million people. The buildings housing the system have a combined floor area of one square mile.

The separative capacity of these plants in full production is about 17,000 metric tons per year with the possibility that this can be increased to about 25,000. The 1972 enrichment cost is $32.00 per separation work unit per kg of enriched uranium product. The separation work unit is a calculated value which takes into account the uranium-235 concentration in the feed, product, and waste (Table 24.18). The uranium-235 concentration in the tails (waste) is generally assumed to be a constant 0.2 per cent U-235.

TABLE 24.18 Separative Work of Uranium Enrichment

Assay (wt.% U-235)	*Separative Work Unit*[a]
0.20	0
0.50	-0.173
0.60	-0.107
0.711 (natural)	0.000
1.000	0.380
2.00	2.194
2.60	3.441
3.00	4.306
3.60	5.638
4.00	6.544
5.00	8.851
10.00	20.863
20.00	45.747
50.00	122.344
98.00	269.982

[a]Negative units represent additional charges for depleted feed. Unit per kg uranium product.

Uranium Hexafluoride Conversion to Uranium (IV) Oxide. The production of uranium (IV) oxide (UO_2) for nuclear fuel is particularly important to power reactors which operate at high temperatures. The oxide has greater chemical and physical stability than uranium metal, and these factors override the disadvantage of lower neutron economy. To compensate for the loss of neutron economy, the usual practice is to use uranium with a slightly higher U-235 content. The uranium hexafluoride (UF_6) is hydrolyzed in an aqueous ammonia solution to precipitate ammonium diuranate, reduced with hydrogen, and then calcined to UO_2 (Fig. 24.28).

$$2UF_6 + 14NH_3 + 7H_2O \longrightarrow (NH_4)_2U_2O_7 + 12NH_4F$$

$$(NH_4)_2U_2O_7 + 2H_2 \longrightarrow 2UO_2 + 2NH_3 + 3H_2O$$

The oxide is ground to provide the optimum material for preparation of high-density oxide pellets. Close control of many empirical factors in this operation is necessary to produce a material having the desired properties.

Thorium Feed Material

Thorium Feed Material Processing. Thorium is at present essentially a by-product of rare earth recovery from monazite ore. The process involves grinding the ore to -200 to -325 mesh, followed by leaching with sulfuric acid. The thorium, rare earths, and uranium are recovered by precipitation (Fig. 24.29). The

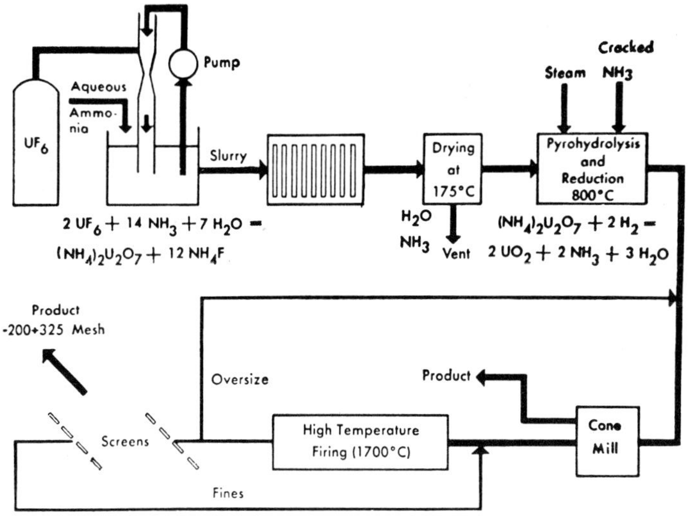

Fig. 24.28 UF_6 conversion to UO_2 by ammonium diuranate precipitation; about 923 metric tons of enriched uranium processed in 1969. (*USAEC*)

crude thorium product is purified by solvent extraction with tributyl phosphate from a nitric acid solution using a process quite similar to that used for crude uranium mill product.

SPENT FUEL RECOVERY

The end products from spent-fuel-recovery operations are purified uranium, plutonium, and thorium along with other actinides and fission products (Fig. 24.30). These products are generally in the form of nitrate solutions or oxides. The actinide products include neptunium, americium, curium, and californium, while the fission products of greatest interest include strontium, cesium, krypton, and promethium. The product specifications requires over-all reduction in radioactive impurties by a factor of 10^7 to 10^9 with a recovery of 99 per cent of the fissile materials.[30]

The steps in spent fuel reprocessing are dis-

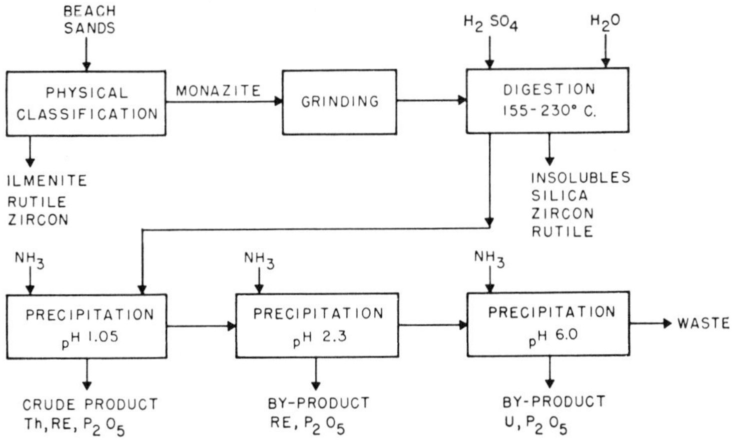

Fig. 24.29 Thorium recovery from monazite sand; estimated U.S. reserves are 100,000 tons with 1980 requirements of possibly 500 tons.

solution, separation and purification, and waste disposal. The unique feature that makes spent-fuel recovery different from other industrial operations is the radiation from the radioactive materials produced in the nuclear fission reaction (Table 24.19). Many of the fission products, a major part of the radioactive contaminants have sufficiently short half-lives that they may be avoided by allowing time for them to decay to their stable end-products.[31]

The most significant delay-and-decay contaminant is iodine-131 with a half-life of 8 days. The iodine properties of short half-life, high fission yield, biochemical action, and volatility makes it normal practice to delay the processing of spent power-reactor fuel for about 150 days. Even after 150 days, it is necessary to process and trap the off-gas in order to limit the iodine release to about 0.1 per cent and allow additional decay of the trapped iodine to provide another 10^3 reduction. When the LMFBRs come into operation, economics could require processing after only 30-days decay and under these conditions the release may be limited to only 10^{-6} per cent. Decay cooling time also reduces the irradiation and thermal effects on the process materials, and also eases the requirements for chemical purification. Radioactive-waste treatment is primarily concerned with reducing the volume of the radioactive material contained in the water and air associated with the spent-fuel processing operations. The guide lines recommended by international commissions limit the discharge of radioactivity to an amount that is equivalent to the natural radiation background; therefore, essentially all of the radioactivity from the spent fuel must be retained during reprocessing and placed in long-term storage.

Dissolution

While most spent nuclear fuels from LWRs are readily dissolved in nitric acid, the jacket material and associated fittings of the fuel element are not. Mechanical shearing of the fuel elements to expose the nuclear fuel material to permit efficient dissolution has been demonstrated in the West Valley Plant of the Nuclear Fuels Services Corp. The typical fuel element is a 12-ft long by 5- to 8-in square bundle of from 50 to 225 half-inch diameter fuel rods, and contains 500 to 1100 pounds of uranium oxide jacketed in zirconium or stainless steel. At West Valley, after the end fittings have been sawed off, the active tube bundle is placed in a 300-ton shear where the rods are cut in $\frac{1}{2}$- to 2-inch lengths; the cut pieces are dropped into a boron-stainless steel cannister. The cannister is transferred to the dissolver where the nuclear fuel material is dissolved in

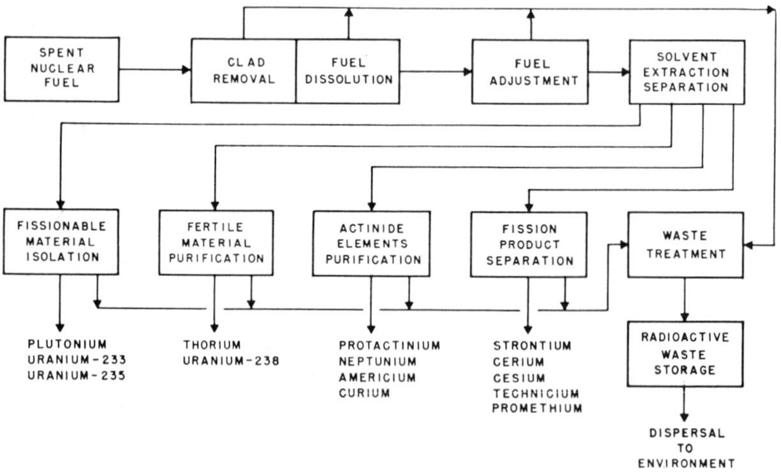

Fig. 24.30 General solvent-extraction process for spent nuclear fuel recovery; at \$33,000 per metric ton uranium; 4100 metric tons will be processed by 1980.

TABLE 24.19 Nuclear Reactor Fuel Composition[30]

	Half-Life	*LWR*	*LMFBR*
Burn-up (Mwd/MT)		33,000	80,000
Cooling time (days)		150	30
Specific power (Mw/MT)		30	148
		(Grams/Metric Ton)	
U-235	7.1×10^8 y	7950	6.44
U-236	2.4×10^7 y	4080	9.15
U-238	4.5×10^9 y	94,300	719,000
Np-237	2.2×10^6 y	744	179
Pu-238	87 y	160	1830
Pu-239	24,300 y	5300	117,000
Pu-240	6,600 y	2170	52,400
Pu-241	13 y	1030	14,500
Pu-242	3.8×10^5 y	349	9,020
Am-241	470 y	40	1,330
Am-243	8,000 y	90	711
Cm-244	18 y	31	42
TOTAL		116,000	915,000
		(Curies/Metric Ton)	
H-3	12.3 y	69	2,330
Kr-85	10.4 y	11,200	24,800
Sr-90	28 y	76,600	206,000
Zr-95	65 d	276,000	5,120,000
Nb-95	35 d	518,000	6,510,000
Ru-106	1.0 y	410,000	6,460,000
I-131	8 d	2	336,000
Xe-133	5.3 d		181,000
Cs-134	2.1 y	213,000	76,500
Cs-137	30 y	106,000	514,000
Ba-140	12 d	430	2,690,000
Ce-141	32 d	56,700	3,580,000
Ce-144	285 d	770,000	6,180,000
Pm-147	2.6 y	99,400	822,000
Sm-151	80 y	1150	10,900
TOTAL		4,390,000	48,800,000

nitric acid, leaving behind the jackets and associated fittings (Fig. 24.31). While there are chemical and electrochemical procedures capable of concurrently dissolving the zirconium and stainless steel components, the shearing procedure and preferential dissolution of only the nuclear fuel materials significantly reduces the volume of materials in solution to be processed and disposed of as radiochemical waste.

A critical aspect of the dissolution step is treatment of the off-gas. Nitrogen oxides are evolved from the reaction of nitric acid with the fuel material along with the release of volatile fission products: iodine, krypton, xenon, and tritium. Most of the tritium stays in the dissolver or is trapped in the down-draft condenser. Some of the iodine leaves the dissolver, and a significant fraction is scrubbed out in the condenser with the remainder being sorbed on a silver nitrate bed. The krypton and xenon could be removed by low-temperature charcoal adsorption or by freon scrubbing. When required for large-scale operations, the release of these radioactive gases from the dissolver may be greatly reduced by an additional head-end procedure.

Separation and Purification

Uranium, plutonium and the other transuranium products, and the fission products are separated and purified by solvent extraction followed by anion exchange and precipitation. The Purex Process employs tributyl phosphate as the solvent for this separation of products from the nitric acid solution of the spent fuel and is the only process now in commerical use for this purpose. About 99 per cent of the uranium and plutonium is recovered in separate product streams, and decontamination from fission products by a factor greater than 10^7 is effected. The solids content of the final radioactive waste is only about 150 pounds per ton of fuel processed, an important consideration. This is a great advantage over the earlier Bismuth Phosphate and Redox processes, since the radioactive waste must be stored for many years.

In the standard Purex Process the uranium is separated from the plutonium in a first cycle, and second cycles of solvent extraction are used for both the uranium and plutonium streams (Fig. 24.32). The wastes from the second cycles may be recycled back to the first cycle, both to reduce product losses and to minimize the quantity of radioactive waste.

There are two interesting modifications of this process. The first uses a co-extraction of the uranium and plutonium in the first cycle, with the partition being carried out in the second cycle. While this eliminates one set of solvent-extraction columns, a greater loss or recycle of plutonium may result. In the second modification—the Aquaflor Process developed

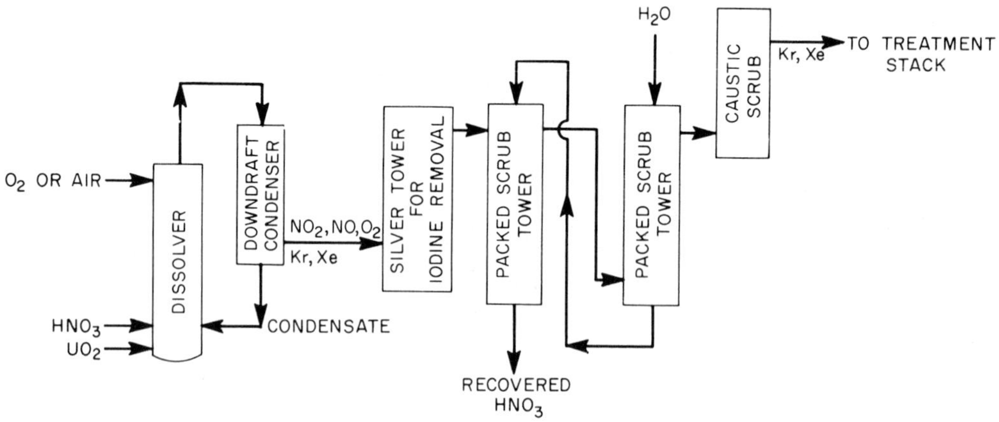

Fig. 24.31 Spent-fuel-dissolution flow sheet.

by General Electric for their Midwest Fuel Recovery Plant–following the normal Purex first cycle the uranium stream goes to a fluidized-bed operation for conversion to a purified UF_6. The plutonium stream goes to anion exchange as discussed next.

The uranium and plutonium product streams from the standard process go through sorption cycles for final purification and then they can be recycled to the feed materials operation. The uranium product passes through the bed with corrosion products being adsorbed on the silica gel along with final traces of plutonium and fission-product zirconium and niobium. The plutonium product is sorbed on an anion exchange bed, while corrosion products and traces of uranium pass through (Fig. 24.33). The plutonium is then eluted with dilute nitric acid, precipitated, and converted to the oxide.

The basic Purex process will be used to process the LMFBR fuel. But there are several conditions that will require several significant modifications to the flow sheet for LWR fuels as follows: the shearing and dissolution steps will have to accommodate the metallic sodium in failed fuel elements; the sodium, tritium, and possibly iodine will be removed by a high-temperature treatment before dissolution; the capability of the dissolver-off-gas treatment system must be increased for better iodine-removal efficiency (the iodine concentration will be much greater because fuel-inventory costs may justify rapid recovery and re-use of the plutonium); the solvent-extraction operation will have to be modified to minimize solvent exposure to the much higher fission-product radiation; and continuous equipment for plutonium purification will probably replace the present batch anion-exchange procedure.

RADIOACTIVE WASTE MANAGEMENT

Radioactive-waste management includes the treatment, storage, and disposal of liquid, gaseous, and solid effluents from nuclear reactor and fuel cycle operations. It employs three philosophies: delay-and-decay, concentrate-and-contain, and dilute-and-disperse. Combinations of all three are usually employed to solve each waste problem. Concentrate-and-contain is the major procedure, since far more radioactivity is produced by the nuclear industry than can be safely dispersed into the environment (Table 24.20). However, the maximum possible use is made of delay-and-decay.

Liquid Wastes

The most effective concentration procedure for liquid wastes is evaporation. The decontamination achieved by this approach is 10^3 to 10^5 per cycle for the water removed. Other more economical but less effective procedures

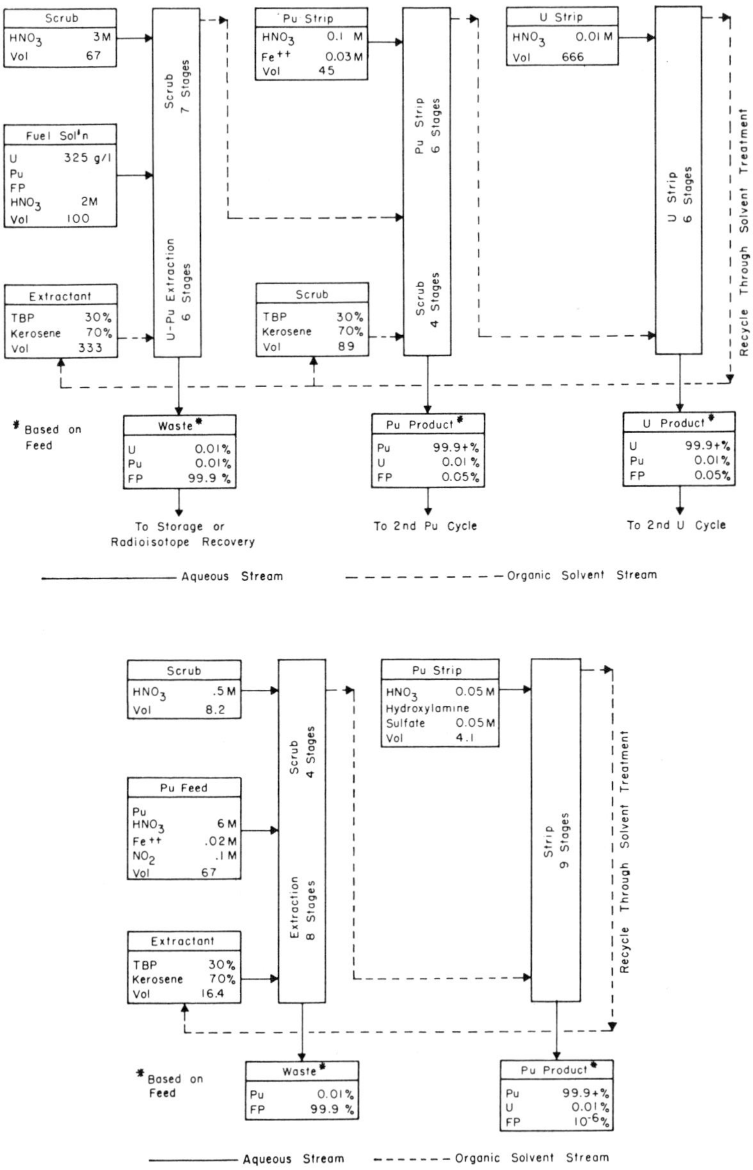

Fig. 24.32 Purex process showing (top) first solvent-extraction cycle, and (bottom) second plutonium solvent-extraction cycle; by 1980 the installed capacity will be 4000 metric tons per year.

include ion exchange and coprecipitation, with the choice of procedures depending on the degree of decontamination required. Tritium, the heaviest isotope of hydrogen, is the only radionuclide not removed.

Procedures have been developed to solidify the high-level liquid wastes from fuel reprocessing. The anticipated procedure includes evaporation, calcining in either fluidized bed or spray type reactors, fixation in silicate glass, and encapsulation in welded steel pots. The pot, approximately 1 foot in diameter and 10 feet long, provides the containment for handling, transport and interim storage.

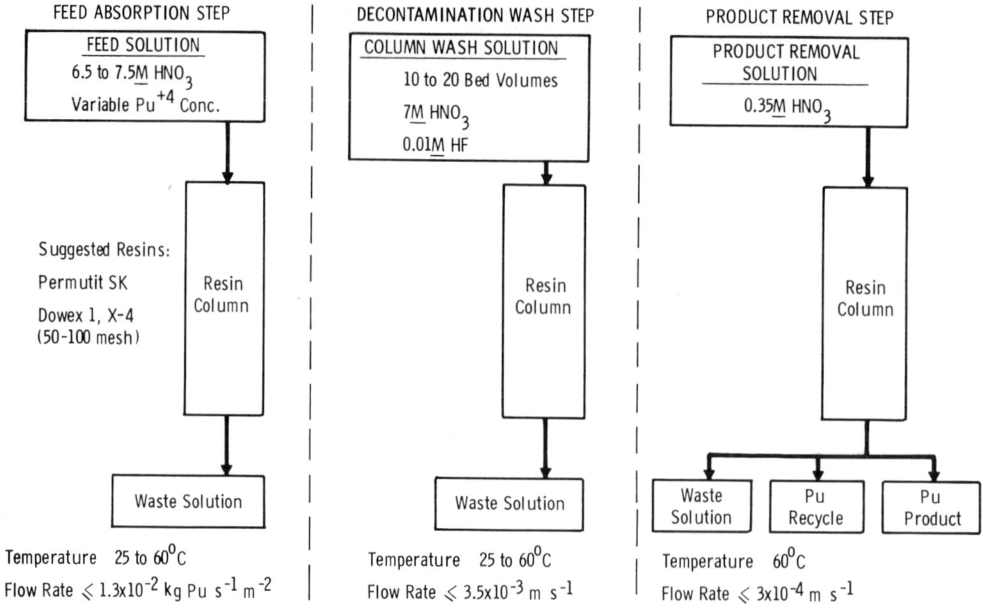

Fig. 24.33 Plutonium ion-exchange flow sheet.

Gaseous Wastes

Air contaminated with radioactive iodine, xenon, krypton, tritium, and particulates is the major problem. The iodine is removed by sorption on a silver catalyst bed followed by nitric acid scrubbers, and eventually it decays to stable xenon. Krypton can be recovered along with the xenon from the dissolver off-gas by low-temperature sorption or freon scrubbing, or it can be released to the environment. The krypton, when recovered, will require long-term storage because of the 10-year half-life of krypton-85. Since xenon represents about 8 times the volume of krypton and the half-life of the longest-lined xenon fission product is 12 days, it may be economically attractive after adequate decay to separate it from krypton-85 to reduce the volume of tankage required for the long-term krypton storage. Tritium may be removed from the spent fuel before dissolution by high temperature retorting.

Dry filtration with fibre filters is the standard final treatment for decontaminating radioactive-particulate-contaminated air. The paper filter developed for this purpose has a 99.97 per cent efficiency for removal of 0.3 micron particles. When necessary, these filters are preceded by scrubbing, sorption, and roughing filters.

Solid Waste Disposal

Encapsulated solid wastes represent the final storage form for high-level radioactive wastes, and federal repositories will be provided for disposal of these materials.

It is now planned to construct an engineered surface storage facility with a design life of a hundred years for the interim storage of these wastes. Bedded salt formations are being intensively studied for ultimate disposal of the high level radioactive wastes. A bedded salt pilot plant is part of this study. If necessary the waste can be retrieved from this pilot plant and sent to the engineered storage facility. Other geological formations are also being considered for this purpose. After the encapsulated waste is placed in the geological formation, the geological formation will provide the containment for the radioactive waste. The geological formation finally selected will provide the greatest isolation from the ground water system and from natural and man-made assults on its integrity.

TABLE 24.20 Projected Fuel Processing Wastes from Total U.S. Nuclear Power Economy[32] (Aqueous Processing of All Fuels)

	Calendar Year Ending				
	1970	*1980*	*1990*	*2000*	*2020*
Installed capacity, 10^3 mw (electrical)[a]	14	153	368	735	2210
Volume of waste generated, as liquid[b]					
Annually, 10^6 gal/year	0.017	0.97	2.69	4.60	13.7
Accumulated, 10^6 gal	0.017	4.40	23.8	60.1	238
Volume of waste generated, as solid[c]					
Annually, 10^3 ft^3/year	0.17	9.73	26.9	46.0	137
Accumulated, 10^3 ft^3	0.17	44.0	238	600	2380
Accumulated radioisotopes[d]					
Total weight, metric tons	1.75	451	2440	6200	24,600
Total activity, megacuries	210	18,900	84,500	209,000	666,000
Total heat-generation rate, megawatts	0.91	81.6	343	807	2520
^{90}Sr, megacuries	3.98	962	4640	9550	29,400
^{137}Cs, megacuries	5.27	1280	6540	15,600	57,500
^{129}I, megacuries	10^{-6}	10^{-4}	.003	.008	.03
^{85}Kr, megacuries	0.56	124	567	1190	3900
^{3}H, megacuries	0.033	7.29	36.2	89.5	332
^{238}Pu, megacuries[e]	0.002	1.20	8.28	30.7	166
^{239}Pu, megacuries	10^{-4}	0.022	0.235	1.31	8.45
^{240}Pu, megacuries	10^{-4}	0.0409	0.395	1.91	11.4
^{241}Pu, megacuries	0.0295	6.63	47.2	191	909
^{242}Pu, megacuries	10^{-7}	10^{-4}	.001	.004	.03
^{241}Am, megacuries	0.0089	2.31	22.7	121	763
^{243}Am, megacuries	0.0009	0.232	1.49	5.19	27.0
^{244}Cm, megacuries	0.128	29.9	137	255	700
^{242}Cm, megacuries	0.725	43.2	185	487	1490

[a]Data from Phase 3, Case 42, Systems Analysis Task Force (April 11, 1968).
[b]Assumes that wastes are concentrated to 100 gal per 10^4 mwd (thermal) and that there is a delay of 2 years between power generation and waste generation.
[c]Assumes 1 ft^3 of solidified waste per 10^4 mwd (thermal).
[d]Assumes that LWR fuel is continuously irradiated at a specific power of 30 mw/metric ton to a burn-up of 33,000 mwd/metric ton, and that the fuel is processed 90 days after discharge from reactor; LMFBR core continuously irradiated to 80,000 mwd/metric ton at 148 mwd/metric ton axial blanket to 2500 mwd/metric ton at 4.6 mw/metric ton, and radial blanket to 8100 mwd/metric ton at 8.4 mw/metric ton, and that fuel is processed 30 days after discharge.
[e]Assumes that 0.5% of the plutonium in the spent fuel is lost to waste.

Other, less toxic waste is buried in privately operated burial grounds, and requires about an acre of land for each 400,000 ft^3 of licensed-for-burial-type waste. At present about a million cubic feet of this waste is generated and buried each year.

ISOTOPE PRODUCTS

There are more than 2000 isotopes, or nuclides and isomers as they are more properly called. Isotopes are atomic species with the same atomic number but different mass numbers. Nuclides are characterized by a particular number of protons and neutrons in the nucleus of an atomic species, and isomers represent different energy states of the same nuclide. The stable nuclides number 276 with over 250 radioactive isomers, and there are approximately 1500 radioactive nuclides.

The nuclides are very significant by-products of the nuclear industry. Both the stable and the radioactive nuclides find considerable use as diagnostic agents in research, medicine, and industry, where they serve as tracers for physical, chemical, and biological processes. In addition the radiation from radioactive nuclides is used to detect changes in density and other characteristics of a material, to promote physical and chemical changes in materials, and to provide a source of thermal energy. Making these nuclides available has been a noteworthy contribution of the nuclear industry.[34,35,43,44]

Nuclide Conversion

The widely reported dream of the alchemist was to convert base metals into gold. In 1934 this dream was realized in principle when I. and F. Joliot-Curie announced that boron and aluminum could be made radioactive by bombardment with the alpha rays from polonium. This demonstrated that atomic particles could be captured by the nucleus to convert one nuclide into another. The Curies had produced phosphorus by the Al-27 (α, n) P-30 reaction. These reactions are a function of both the type and energy of the bombarding particles. In turn, the compound nucleus formed by particle-capture goes through an instanteous rearrangement emitting other particles and energy. The resulting nucleus may be stable or may continue to decay, emitting particles and energy. In some cases it decays through other unstable states before reaching a stable nuclide state. The time for half of the unstable nuclei to decay is a constant, the so-called half-life, which ranges from milliseconds to millions of years. Unfortunately for the modern alchemist, the cost of converting a base metal to gold would be several orders of magnitude greater than the present value of gold.[36]

Nuclide-conversion is primarily concerned with the preparation of the unstable (radioactive) nuclides called radioisotopes. Nuclear reactors and charged-particle accelerators are employed for this purpose. In the nuclear reactor, the most common reactions are neutron-capture and fission, while in the accelerators the reaction is charged-particle capture. The quantities produced are usually in the millicurie to curie range, this being sufficient to satisfy the tracer, diagnostic, and analytical applications. However, nuclear reactors have also been used to produce megacurie quantities of fissile and fusion nuclides—plutonium and tritium—for energy production as discussed in earlier sections of this chapter. A final point to be made is that most of the neutron-capture products are the same chemical species as the target material, while most of the charged-particle products are a different chemical atom.

Radioisotopes

Of the several hundred radioisotopes prepared and used in research, about twenty-five have been involved in extensive development efforts (Table 24.21). The fission products and transuranium (neutron-capture) products have received the most development attention since their production results directly from the operation of nuclear reactors. As the use of nuclear-power reactors increases so will the availability of these isotopes (Table 24.22). The processes to recover these products involve precipitation, solvent extraction, and ion exchange. Relatively large-scale production is required to achieve economic costs. However, the spent fuel is being processed for plutonium-239 recovery, and the rest of the products need only cover the incremental added cost.

In comparison with the fission and transuranium products, the production of the other reactor-neutron-capture and accelerator products is relatively simple. Most reactor products such as cobalt-60 are activation products in which no chemical separations are involved. For example, when cobalt-59 is irradiated with

TABLE 24.21 Major Radioisotopes and Applications

Am-241	A	I-131	M
Cf-252	AM	Ir-192	A
C-14	A	Kr-85	A
Cs-137	AMP	Mo-99	M
Cr-51	M	Ni-63	A
Co-57	AM	P-32	M
Co-60	AMP	Po-210	A
Cm-244	T	Pu-238	T
F-18	M	Pu-239	T
H-3	AT	Pm-147	AT
I-125	M	Sr-90	AT
Xe-133	M	Sr-85	M

A—Analysis and control.
M—Medical.
P—Process radiation.
T—Thermal.

TABLE 24.22 Long-Lived Radioisotope Availability From Civilian Nuclear Power

Product	*Half-Life (Yr)*	*Energy (Kilowatts/Year)* 1970	1980	1990
Ce-144	0.78	180	11,100	52,500
Ru-106	1.0	100	6,700	38,000
Sr-90	27.7	18	1,260	4,700
Cs-137	30	18	1,290	4,900
Pu-238	87.4	3.4	280	940
Cm-244	18.1	2.6	200	1,600
Pm-147	2.6	0.7	60	300
Kr-85	10.8	0.6	36	140
Am-241	458	1	1.8	3.2
Output of Projected (Gigawatts Electric) Reactor Systems				
Light-water reactors		12	148	390
Fast-breeder reactors		0	2	10
Fuel processed		(100)[a]	(4,270)[a]	(17,700)[a]

[a]Megawatt tons/year.

neutrons, chemical separation of the cobalt-60 product is not possible. Most of the accelerator products and the other reactor products are transmuted to different chemical elements, for example the reactor-neutron irradiation of sulfur to give phosphorous-32, or the cyclotron-proton irradiation of iron to produce cobalt-57. In this case chemical separations yield pure nuclide products, and the form of the target material may be selected to minimize the separation problem.

Fission Products. In the fission of uranium and plutonium the nuclides produced range in mass from about 72 to 162 with maximum yields occurring in two broad peaks in the regions of 95 and 138. In addition, some tritium results from triple fission. The maximum yields in the 95 and 138 regions are about 6 per cent. The major products recovered from spent nuclear fuel include strontium-90, cesium-137, promethium-147, and krypton-85. The strontium-90 is recovered from Purex waste by precipitation followed by solvent extraction, and the product is then converted to strontium titanate for heat-source applications. The promethium is separated by ion exchange and converted to the oxide for thermal and radiation applications. The cesium has been extracted by ion exchange and subsequently converted to the chloride for radiation-source applications. Krypton-recovery is part of the dissolver off-gas treatment system.[37]

Since about 1960, the Fission Product Development Laboratory at the Oak Ridge National Laboratory, using feed from Richland, Washington, has provided developmental quantities of strontium-90 and cesium-137. During this period, about 20 kilowatts of each product has been produced, strontium-90 (150 curies/watt) for thermal applications (Fig. 24.34) and cesium-137 (207 curies/watt) for process radiation. A promethium-147 capability was established in 1964 at the Pacific Northwest Laboratory and about 4 kilowatts (2750 curies/watt) processed for thermal and self-luminescent applications. The 5 per cent krypton-85 product from fission has been enriched by thermal diffusion to a 25 per cent product for luminescent applications.

The stable fission products are receiving increased attention. Rhodium and xenon are probably the most important materials because of their limited availability in nature and short-lived radioisotopic contamination. Developmental quantities of these materials along with technetium and palladium have been prepared.

The short-lived fission products represent a broad spectrum of nuclides. In this group, xenon-133 now has the greatest use, but the other products have unique features which justify continuing interest. These are produced

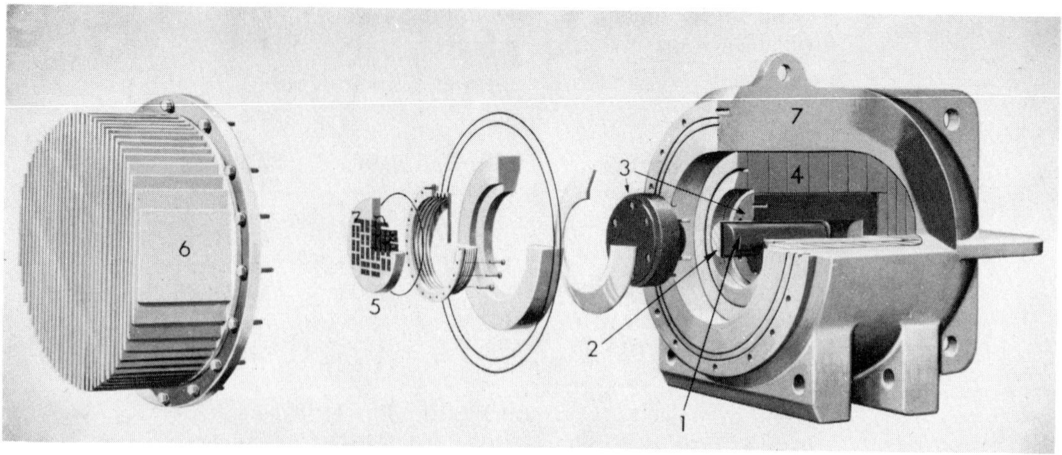

Fig. 24.34 Strontium-90-fueled radioisotope thermoelectric generator; provides electric power for 5 to 10 years at costs in the range of dry cell chemical batteries; (1) fuel; (2) capsule; (3) inner shield; (4) insulation; (5) thermoelectric module; (6) heat rejection head; and (7) outer shield and housing. (*Teledyne-Isotopes Corp.*)

by special irradiation of gram-size targets of uranium-235.

Reactor Products. The reactor products are made by neutron capture in target materials. There are three general groups of reactor products: (1) the transuranium actinide products resulting from the transmuting capture of neutrons in the reactor fuel and fertile materials; (2) the transmutation products made by irradiation of special target materials; and (3) the activation products also made by irradiation of special target materials. Most radioisotopes are produced in research and test reactors such as those described in Table 24.23. The cost of these products ranges from a few cents per curie for thulium-170, and 50 cents per curie for cobalt to ten dollars a microcurie for californium. Since most of the reactor-produced radioisotopes are the activation products of the same chemical element, very little chemical processing is required. In the case of cobalt-60, the naturally occurring cobalt-59 metal is irradiated and then encapsulated for use as a gamma-radiation or heat source (Fig. 24.35). In the few cases involving transmutation, the form of the target is selected to keep the processing to simple batch precipitation, distillation, and solvent extraction or ion exchange. The quantities involved are grams and millicuries. The irradiation is carried out in small quartz flame-sealed vials and the processing is performed in laboratory glassware (Table 24.24).

TABLE 24.23 Characteristics of Various Nuclear Reactors

		Neutron Flux (10^{14})	
Reactor	*Power (Mw)*	*Thermal*	*Fast*
ORG	3.5	0.01	nil
LITR	3	0.33	0.18
BSR	2	0.24	0.04
ORR			
(HT)	30	2.5	1.5
(S-1)		1.8	0.83
(S-2)		4.8	0.65
HFIR			
(T)	100	28	7.5
(RB)		15	2.6
SR[a]	700	60	20
PWR	3300	0.5	3.0
BWR	3300	0.4	1.9
LMFBR			
(core)	2600	nil	51
(blanket)		nil	9.7

[a]Savannah River C Reactor; flux demonstration.

The production of transuranium nuclides in the Savannah River reactor is a much more complex and challenging operation. The principle transuranium products include plutonium-238, curium-242 and -244, americium-241, and californium-252. The other nuclides

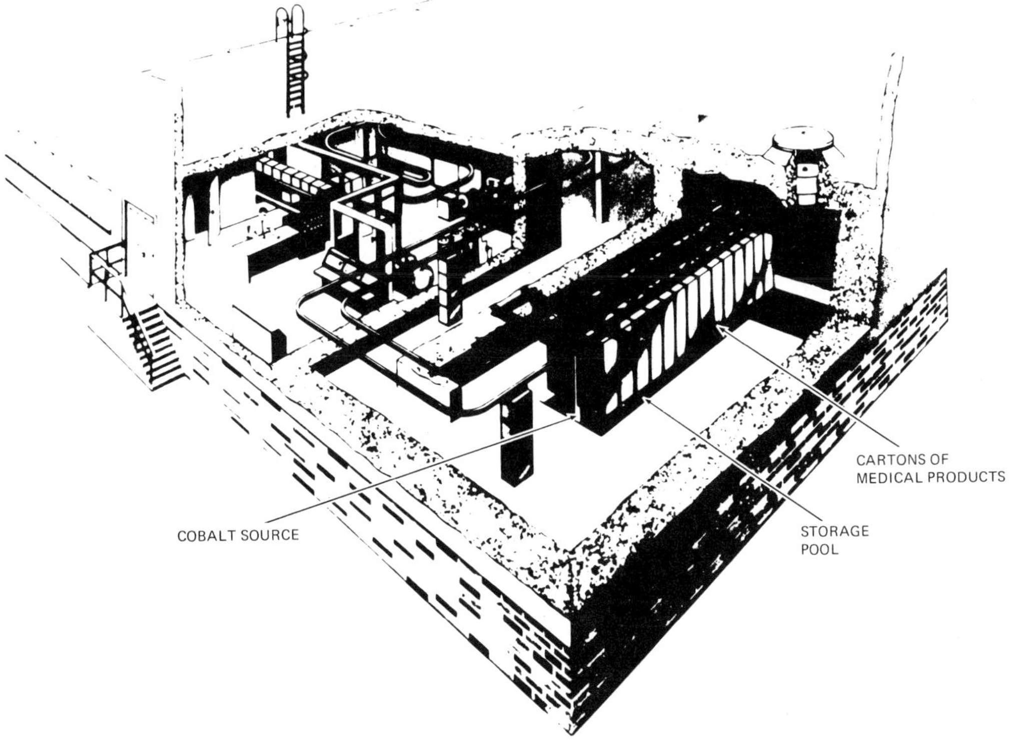

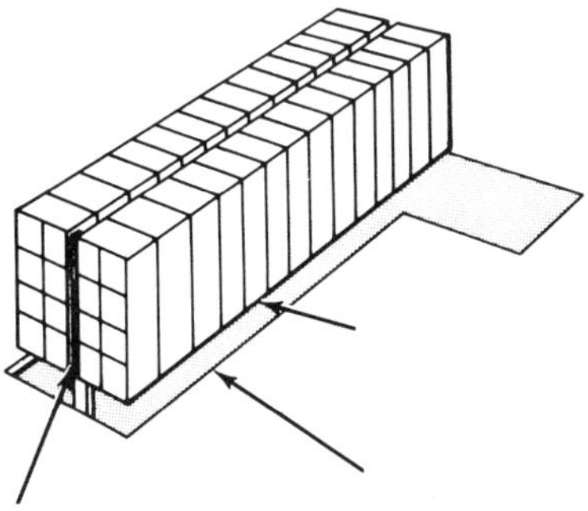

Fig. 24.35 Ethicon's medical-products irradiator, Somerville, N.J., sterilizes prepackaged sutures and syringes. Approximately 2.5 megarads is delivered to 140 cu ft of product per hour using 950,000 ci of cobalt-60. The cobalt-60 is installed in a slab geometry with the product cartons conveyed in two rows on either side. This slab is lowered into the water-filled pool below the conveyor when maintenance is required.

in this production chain have little usefulness because of half-life or other limiting nuclear characteristics. Although the production of plutonium-239 involves the nuclear capture of only one neutron in uranium-238, the production of either plutonium-238 or americium-241 involves the capture of 3 neutrons, while curium-242 requires 4, curium-244 requires 6, and californium-252 requires the capture of 14 neutrons. In each case there are intermediate products that require processing either to achieve more efficient conversion by recycling the target material or to recover a decay product from a long-lived nuclide parent.[39,40,41]

Solvent extraction and ion exchange compete as the principal techniques for use in these transuranium product separations. The chemical oxidation effects from the intense radiation (about 1 watt per liter) is one of the factors influencing the choice of technique. In solvent extraction these effects tend to neutralize the acid, while in ion exchange they cause gas-binding of the bed. The addition of methanol reduces the oxidation rate, and rapid

TABLE 24.24 Reactor Products

Activation Series		*Transmutation Series*	
Nuclide	*Half-Life*	*Nuclide*	*Half-Life*
Mo-99	2.75 d	F-18	110 m
Cr-51	27.8 d	K-43	22 h
Fe-59	45 d	Cu-67	61.6 h
Ir-192	74 d	I-131	8.06 d
Sn-113	118 d	P-32	14.3 d
Tm-170	130 d	P-33	25.2 d
Co-60	5.26 y	Po-210	138 d
Ni-63	125 y	Gd-153	242 d
Cl-36	3.1×10^5 y	H-3	12.5 y
		C-14	5730 y

high-pressure processing prevents gas-binding of the ion exchange columns. Other unique problems include the need for both neutron and gamma shielding, and the presence of considerable thermal energy, 0.5 w/g for plutonium-238 to 160 w/g for curium-242.

Accelerator Products. The accelerator products are prepared by capturing in special target materials the high-energy isotopes of hydrogen or helium: proton, deuteron, triton, alpha or helium-3 ions. The ion energy is usually in the range of 10 to 40 Mev with beam currents up to several milliamperes. For many years, the ORNL 86-inch cyclotron has been the principal source of accelerator products and this machine still has a greater isotope-production capability than any other accelerator employing low-energy protons. The BNL 60-inch and other cyclotrons provide the broader capability to use deuteron, alpha and helium-3 ions. In addition, BNL Van de Graaffs are being used to develop triton and deuteron-capture products. The characteristics of several machines are given in Table 24.25.

While the particle flux in an accelerator is very modest, the machine yield is adequate for most tracer-type needs. The irradiation cost is on the order of \$100 per hour, as compared with \$100 per week in a nuclear reactor, because accelerators are usually single-target machines. Accelerator products are somewhat more difficult to process than reactor products due to the differences in target-irradiation conditions. The charged particle beam in the accelerator is very hot, 10 to 1000 kw/cm^2 with the optimum target being a high-temperature metal plated on a water-cooled copper heat exchanger. When this situation is impossible, water-cooled capsules are used in many cases to contain the target material. In a few

TABLE 24.25 Characteristics of AEC Isotope-Producing Accelerators

Machine	*Particle*	*Energy (Mev)*	*Current (μAmp)*[a] *Internal*	*External*
		Cyclotrons		
ORNL 86-inch	p	17.5	3,000	–
		21	200–1,500	–
		22.4	–	15
BNL 60-inch	p	40	–	100
	d	20	–	
	α	40	–	50
	He-3	60	–	
		Van de Graaff		
BNL Physics		3.5	–	50
BNL Radiobiology	d	4.0	–	10
		Linear Accelerators		
BLIP	p	200	–	180
LAMPF	p	800	–	750

[a]μAmp (microamperes) = 2.3×10^{13} particles per hour.

TABLE 24.26 Accelerator Products

Nuclide	*Half-Life*	*Nuclide*	*Half-Life*
F-18	110 m	Rb-84	33 d
Fe-52	8.3 h	Be-7	53 d
I-123	13 h	Sr-85	65 d
Mg-28	21.3 h	Co-56	77 d
Cs-129	31 h	W-181	130 d
Au-194	39 h	Ce-139	137 d
Ga-67	78 h	Co-57	271 d
Y-87	80 h	Ge-68	287 d
V-48	16 d	Na-22	2.5 y
Cr-51	28 d	Fe-55	2.7 y

TABLE 24.27 Isotope Separation Processes

Process	*Applicable Isotopes*
Electromagnetic	All
Chemical exchange	H, Li, B, C, N
Thermal diffusion	H, He, N, O, Ne, A, Kr, Xe, Cl, C
Gaseous diffusion	U
Fractional distillation	H, O, N, C
Gas centrifuge	Ge, Xe, U
Ion migration	U, K, Cl, Cu, Mg, Li, Na
Molecular distillation	Li, K, Cl

cases, gas targets or gas-evolving targets can be successfully used. A few gaseous products for medical research, such as carbon-11 (dioxide), oxygen-15, and nitrogen-13, are produced by continuous methods in low-energy cyclotrons. These are very short-lived products (half-lives of from 2 to 20 minutes) for which simple, on-site continuous production systems were developed. The scale of accelerator production is usually in the millicurie to curie range where the processing can be carried out in lightly shielded facilities using laboratory glassware.

Several accelerator products are listed in Table 24.26.

Stable Isotopes

The separation techniques for stable isotopes are based on differences in their physical or chemical properties. For the lighter weight nuclides chemical exchange processes are employed because the differences in chemical equilibrium constants are sufficient to allow separation. For the heavier nuclides, the major techniques depend on differences in mass. They are electromagnetic, thermal diffusion, and gaseous diffusion (Table 24.27) Both the chemical-exchange and the mass-separation techniques are multiple-stage diffusion processes which, with the exception of the electromagnetic process, have very small separation coefficients, on the order of 1.1 to 1.001. Also characteristic of these processes are their relatively low yields and the long time required to reach equilibrium. For example, in the case of uranium enrichment by gaseous diffusion, the tails contain 0.2 mole per cent uranium-235 which represents 28 per cent of the uranium-235 entering the process in the natural uranium feed. This process requires months to reach equilibrium.[45]

In the electromagnetic process the separation factor may be on the order of 10 to 100. In this process the feed is ionized and then accelerated through a bending magnetic field. The differences in mass cause the heavier particles to pass through larger arcs with the products captured in slit-like pockets located 180° from the feed point (Fig. 24.36). The accelerating voltage and the magnetic field strength may be adjusted to give good separation although the yield per pass is only 1 to 10 per cent. The low yields result from inefficiencies in the ionizer, the accelerator, and the catchers. When the feed has value, the system is scrubbed and flushed to recover the material which is recycled. Although the cost of this process is high, it provides the most economical method for producing gram quantities, which are sufficient to meet most research and development needs.

Thermal diffusion is a second general purpose method for isotope separation and is applicable to all gases. The gas is circulated in a small column, about 1 inch in diameter. Chilling the wall and heating the center with a very hot wire, 700 to 800°C, causes the light isotopes to migrate to the center and move upward, while the heavier isotopes move to the wall and move downward. These columns are almost always operated on a batch basis with several months being required to reach equilibrium.

There are several types of chemical exchange systems involving gas-liquid, liquid-liquid, and liquid-solid equilibria operating in multistage fashion. Nitrogen-15 is usually enriched by stripping from nitrogen-14 using the gaseous-

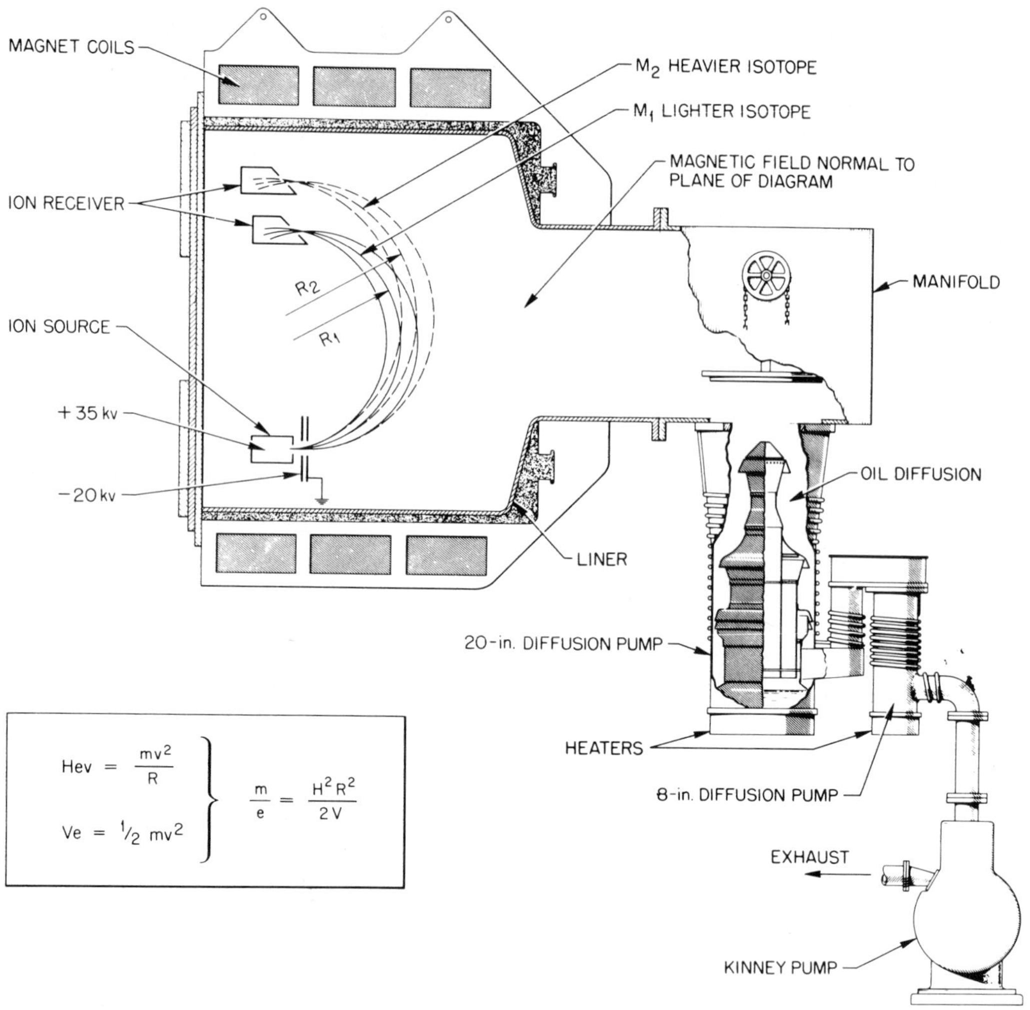

$$\left.\begin{aligned} Hev &= \frac{mv^2}{R} \\ Ve &= \tfrac{1}{2}mv^2 \end{aligned}\right\} \quad \frac{m}{e} = \frac{H^2R^2}{2V}$$

Fig. 24.36 Oak Ridge Calutron for electromagnetic separation of isotopes. First method to achieve large scale separation of U-235, and today the source of research quantities of most stable and a few radioactive nuclides. *(USAEC)*

nitrous-oxide/aqueous-nitric-acid system. The nitrogen-15 in this equilibrium favors the aqueous phase and therefore concentrates at the bottom of the plant. To achieve maximum concentration, the nitrogen is refluxed by decomposing the nitric acid in the bottom to nitrous oxide with sulfur dioxide. The nitrogen reflux at the other end of the plant reacts the nitrous oxide with oxygen and water to produce nitric acid.

The heavy water plants producing hundreds of tons per year represent the largest chemical exchange operations to date. Heavy water is a very efficient moderator for nuclear reactors and was used in the AEC's Savannah River reactors for nuclide conversion producing plutonium-239, plutonium-238, tritium, cobalt-60, and the transuranium elements through californium-252. It uses a dual-temperature process which is based on the fact that the separation factor for the exchange of deuterium between H_2S and water varies with temperature. Isotopic separation is effected by running a pair of columns, one at about 27°C and the other at about 220°C. Each column may have about two hundred plates. In the hot column, the deuterium is stripped from the water, enriching the H_2S in deuterium. The enriched H_2S then passes to the bottom of the cold column where the equilibrium now favors the water. The H_2S is on internal recycle, stripping the deuterium from the water. The fresh water feed, containing about 0.016 weight per cent of heavy water, passes through first the cold, then the hot columns. The deuterium concentration builds up at the bottom of the cold column and at the top of the hot column. After days of operation when equilibrium is reached, the gas and/or the liquid phase can be withdrawn and passed on to a subsequent cycle. In the AEC plant at Savannah River a 10 to 20 per cent deuterium product is achieved in two cycles. This is further enriched to 99.8 per cent by distillation.

Deuterium is produced for about $30.00/liter D_2O. The economics of the process depend on efficient exchange of heat from the hot to the cold stream to minimize the quantity of energy that must be added to the dual-temperature system. Other important factors include flow control, pumping energy, corrosion, product-extraction methods, and the number of cycles.

SAFETY

Safety has been and will continue to be the single most essential feature of the nuclear industry. Although radiation has always been part of our environment (Fig. 24.37), the nuclear industry has brought with it potential increases from man-made radioactive materials. Since the beginning of this industry there have been no accidents, with either reactors or radioisotopes, that have resulted in any significant exposure of the public. A few employee fatalities have resulted from nuclear criticality accidents. The excellent safety record is due in large part to development of both national and international agreements governing all aspects of the nuclear industry. Government and industry have cooperated with various standards and regulatory organizations to establish required performance criteria and to define the design specifications needed to meet these criteria.[46]

The International Commission for Radiation Protection (ICRP) has recommended that the radiation-exposure limit for workers be less than five roentgen per year after age 18. This limit is based on thorough and careful consideration of life-shortening and genetic damage that could be caused by radiation. The maximum allowable radiation exposure to the general public has been set at 0.17 roentgen per year. Moreover, the radioactivity discharged from nuclear facilities must be "as low as practicable" below these radiation guide values.

Through the cooperative efforts of industry and government systems have been developed and demonstrated to limit the release of radioactivity from the nuclear power reactors to a few per cent of natural background. Similar performance can be expected for the fuel cycle and waste disposal operations that support the operations of the nuclear power reactors.

The future well-being of our environment has been further assured by the National Environmental Protection Act of 1969. The National environmental goals as expressed by the National Environmental Protection Act are as follows:

> . . . it is the continuing responsibility of the Federal Government to use all practical means, consistent with other essential considerations of national policy, to improve and coordinate Federal plans, functions, programs, and resources to the end that the Nation may
>
> (1) fulfill the responsibilities of each generation as trustee of the environment for succeeding generations;
> (2) assure for all Americans safe, healthful, productive and esthetically and culturally pleasing surroundings;
> (3) attain the widest range of beneficial use of the environment without degradation, risk to health or safety, or other undesirable and unintended consequences;
> (4) preserve important historic, cultural, and natural aspects of our national heritage and maintain, wherever possible, an environment which supports diversity and variety of individual choice;
> (5) achieve a balance between population

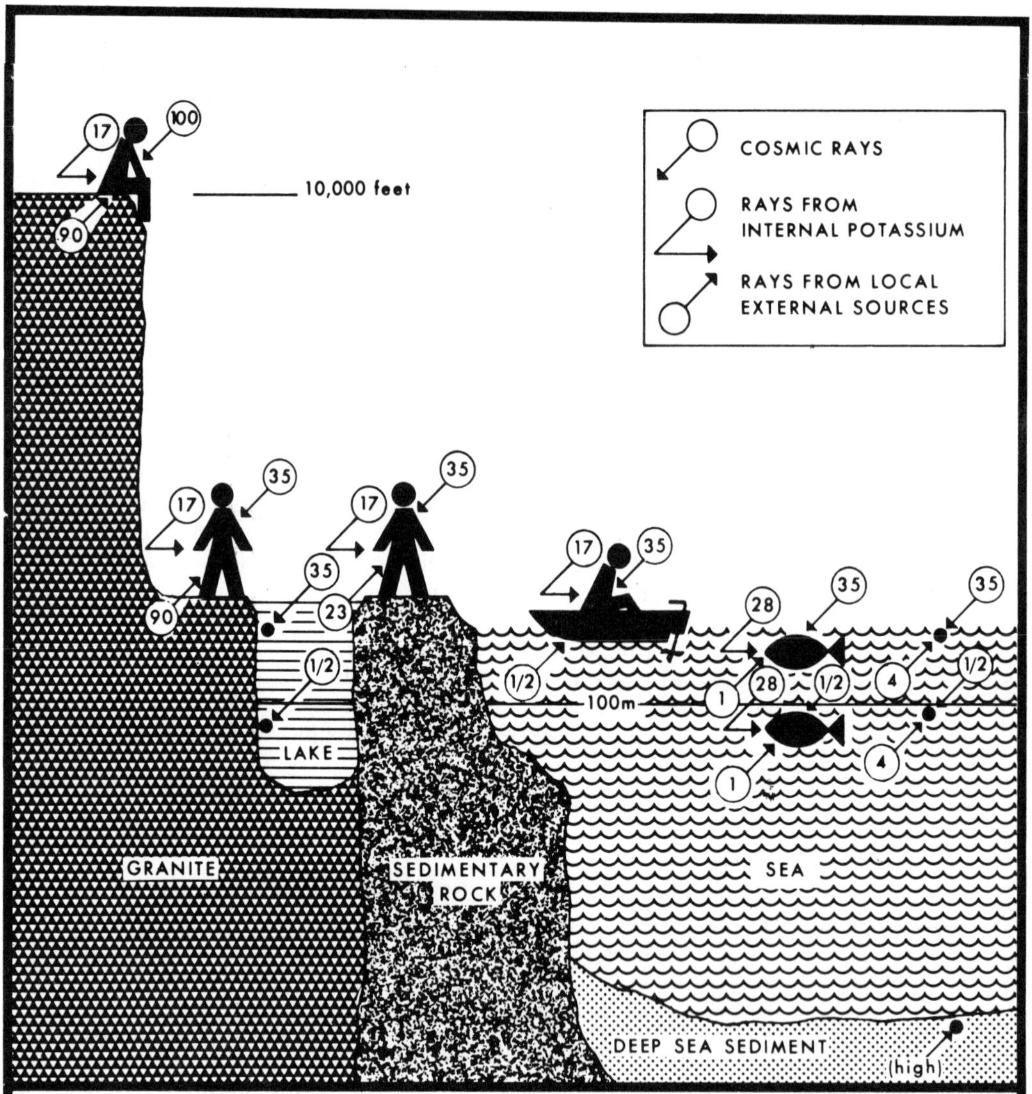

TOTAL NATURAL DOSES (mrad/year)✓

Man over granite		Man over sedimentary rock	Man over sea	Large fish in sea		Micro-organism in sea	
10.000	m.s.l.			at surf.	100m	at surf.	100m
207	142	75	52	64	30	39	5

✓ A mrad is one thousandth of a rad, the unit for indicating absorbed energy in tissues.

Fig. 24.37 Natural radiation. From "The Effects of Atomic Radiation on Oceanography and Fisheries" NAS-NRC Pub. 551/1957.

and resource use which will permit high standards of living and a wide sharing of life's amenities; and

(6) enhance the quality of renewable resources and approach the maximum attainable recycling of depletable resources.

These goals bring to focus interfaces between private and public needs, technology and quality-of-life concepts, industrial and ecological systems, and national and international economics. The nuclear industry is in the forefront of the studies in progress to assure these goals.

REFERENCES

1. "The Nuclear Industry," U.S. Atomic Energy Comm., Washington, D.C., 1970.
2. Glasstone, S., "Sourcebook on Atomic Energy," 2nd Ed., New York, Van Nostrand Reinhold, 1958.
3. Hubbert, M. K., "Energy Resources," NAS/NRC Report 1000-D, National Academy of Science, Washington, D.C., 1962.
4. Cambel, A. B., "Energy R & D and National Progress Interdepartmental Energy Study," U.S. Government Printing Office, Washington, D.C., 1964.
5. "Current Status and Future Technical and Economic Potential of Light Water Reactors," USAEC Report WASH 1082, U.S. Atomic Energy Comm., Washington, D.C., 1968.
6. "Cost-Benefit Analysis of the U.S. Breeder Reactor Program," USAEC Report WASH 1126, U.S. Atomic Energy Comm., Washington, D.C., 1969.
7. "The Use of Thorium in Nuclear Power Reactors," USAEC Report WASH 1097, U.S. Atomic Energy Comm., Washington, D.C., 1969.
8. "An Evaluation of Advanced Converter Reactors," USAEC Report WASH 1087, U.S. Atomic Energy Comm., Washington, D.C., 1969.
9. Glasstone, S., "Principles of Nuclear Engineering," New York, Van Nostrand Reinhold, 1955.
10. Glasstone, S. and Edlund, M. C., "The Elements of Nuclear Theory," New York, Van Nostrand Reinhold, 1952.
11. Glasstone, S. and Lovberg, R. H., "Controlled Thermonuclear Reactions," New York, Van Nostrand Reinhold, 1960.
12. Gough, W. C. and Eastlund, B. J., *Sci. Am.*, **224**, No. 2, 50–64 (Feb. 1971).
13. Glasstone, S., "Controlled Nuclear Fusion," in "Understanding the Atom Series," U.S. Atomic Energy Comm., Washington, D.C., 1968.
14. Greber, C. R., *et al.*, "Plowshare," in "Understanding the Atom Series," U.S. Atomic Energy Comm., Washington, D.C., 1966.
15. Knox, J. B., *Nucl. Appl. Techn.*, **7** (Sept. 1969).
16. Glasstone, S., "The Effects of Nuclear Weapons," U.S. Government Printing Office, Washington, D.C., 1962.
17. Seaborg, G. T., Shaw, M., *et al.*, *Nucl. Eng. Intern.*, **15**, No. 174 (Nov. 1970).
18. "The International Conference on the Constructive Uses of Atomic Energy," American Nuclear Society, Hinsdale, Ill., 1969.
19. Weber, C. E., *et al.*, Fuel Element Design section in "Reactor Handbook," New York, John Wiley and Sons, 1964.
20. "Proceedings of the International Conference on Sodium Technology and Large Fast Reactor Design," Nov. 1968, Clearinghouse for Federal Scientific and Technical Information, Springfield, Va.
21. Weinberg, A. M., *et al.*, *Nucl. Applications Techn.*, **18**, No. 2 (Feb. 1970).
22. Clegg, J. W., and Foley, D. D., "Uranium Ore Processing," Reading, Mass., Addison-Wesley, 1958.
23. "Uranium Resources, Production and Demand," A joint report by the European Nuclear Energy Agency and the International Atomic Energy Agency, E.C.D. Publications Center, Washington, D.C., Sept. 1970.
24. Singleton, A. L., Jr., "Sources of Nuclear Fuel," in "Understanding the Atom Series," U.S. Atomic Energy Comm., Washington, D.C.

25. Dukert, J. M., "Thorium and the Third Fuel," in "Understanding the Atom Series," U.S. Atomic Energy Commission, Washington, D.C.
26. Leist, N. R., "The Beneficiation and Refining of Uranium Concentrates," USAEC Report NLCO 1067, U.S. Atomic Energy Comm., Washington, D.C., 1970.
27. Mantz, E. W., "Production of Uranium Tetrafluoride and Uranium Metal," USAEC Report NLCO 1068, U.S. Atomic Energy Comm., Washington, D.C., 1970.
28. Ruch, W. C., "Production of Pure Uranium Hexafluoride from Ore Concentrates," Allied Chemical Corp., Engineers Joint Council, New York, April 1959.
29. "AEC Gaseous Diffusion Plant Operations," USAEC Report ORO 658, U.S. Atomic Energy Comm., Washington, D.C., 1968.
30. "Aqueous Processing for LMFBR Fuels–Technical Assessment and Experimental Program Definition," USAEC Report ORNL 4436, U.S. Atomic Energy Comm., Washington, D.C., 1970.
31. Long, J. T., "Engineering for Nuclear Fuel Reprocessing," New York, Gordon and Breach.
32. "Siting of Fuel Reprocessing Plants and Waste Management Facilities," USAEC Report ORNL 4451, U.S. Atomic Energy Comm., Washington, D.C., 1970.
33. Friedlander, G., Kennedy, J. W., and Miller, J. M., "Nuclear and Radiochemistry," New York, John Wiley and Sons, 1955.
34. Lederer, C. M., Hollander, J. M., and Perlman, I., "Table of Isotopes," 6th Ed., New York, John Wiley & Sons, 1967.
35. McKee, R. W., Nunn, S. E., and Haffner, D. R., "Nuclide Table," USAEC Report, U.S. Atomic Energy Comm., Washington, D.C.
36. Barbler, M., "Induced Radioactivity," CERN, Geneva, New York, John Wiley & Sons, 1969.
37. Beard, S. J. and Moore, R. L., "Progress in Nuclear Energy, Process Chemistry," Vol. 4, Oxford, Pergamon Press, 1970.
38. Mann, W. B. and Garfinkel, S. B., "Radioactivity and Its Measurement," New York, Van Nostrand Reinhold, 1966.
39. Groh, H. J. and Schlea, C. S., "Progress in Nuclear Energy, Process Chemistry," Vol. 4, Oxford, Pergamon Press, 1970.
40. Crandall, J. L., "The Savannah River High Flux Demonstration," USAEC Report DP-999, U.S. Atomic Energy Comm., Washington, D.C.
41. Leuze, R. E. and Lloyd, M. H., Progress in Nuclear Energy, Process Chemistry, Vol. 4, Oxford, Pergamon Press, 1970.
42. Gardner, R. and Ely, R., "Radioisotope Measurement Applications in Engineering," New York, Van Nostrand Reinhold, 1967.
43. "Understanding the Atom Series," U.S. Atomic Energy Comm., Washington, D.C.
    - Accelerators
    - Atomic Energy Basics
    - Atoms in Agriculture
    - Food Preservation
    - Nuclear Clocks
    - Our Atomic World
    - Power from Radioiostopes
    - Radioisotopes and Life Processes
    - Radioiostopes in Industry
    - Radioisotopes in Medicine
44. "Radioactive Pharmaceuticals," U.S. Atomic Energy Comm., Washington, D.C., 1966.
45. Baker, P. S., "Survey of Progress in Chemistry," New York, Academic Press, 1968.
46. Fitzgerald, J. J., "Applied Radiation Protection and Control," New York, Gordon and Breach, 1969.
47. "How to Get a License to Use Radioisotopes," U.S. Atomic Energy Comm., Washington, D.C., 1967.
48. "United States of America, Code of Federal Regulations," Title 10, Chapter 1, U.S. Atomic Energy Comm., Washington, D.C.
    - Part 20 Standards for Protection Against Radiation
    - Part 30 Radioisotope Licensing
    - Part 31 Radioisotope Licensing–General Licensed Products
    - Part 32 Radioisotope Licensing–Manufacture of General and Exempt Products
    - Part 33 Radioisotope Licensing–Broad License for Users
    - Part 34 Radioisotope Licensing–Radiography
    - Part 35 Radioisotope Licensing–Human Uses

Part 36 Radioisotope Licensing–Export and Import
Part 40 Natural Uranium and Thorium Licensing
Part 50 Nuclear Reactor, Fuel Reprocessing Plants and Other Large Scale Facility Licensing
Part 55 Operator Licensing
Part 70 Fissile Material Licensing
Part 71 Radioactive Material Transport Licensing
Part 100 Reactor Site Criteria

49. Wheelwright, E. J., "A. Generic Ion Exchange Process for the Recovery of Valuable Elements from the Nuclear Industry," Ion Exchange in the Process Industries Conference, July 1969, Society of the Chemical Industry, 1970.

25

Synthetic Organic Chemicals

Dr. William H. Haberstroh* and Dr. Daniel E. Collins*

Synthetic organic chemicals can be defined as derivative products of naturally occurring materials (petroleum, natural gas, and coal) which have undergone at least one chemical reaction, such as oxidation, hydrogenation, halogenation, sulfonation, and alkylation.

The volume of synthetic organic chemicals increased from 17 billion pounds in 1949 to more than 130 billion pounds in 1969. The production for the past two decades is shown in Fig. 25.1. Much of this phenomenal growth has been due to the replacement of "natural" organic chemicals. Since this replacement is now essentially complete, future growth for synthetic materials will be dictated by the expansion of present markets and development of new organic chemical end uses.

More than 2500 organic chemical products are derived principally from petrochemical sources. These are commercially produced from five logical starting points. Consequently, this chapter has been subdivided into five major raw material classifications: methane, ethylene, propylene, C_4 and higher aliphatics, and aromatics.

CHEMICALS DERIVED FROM METHANE

It has been stated that every synthetic organic chemical listed in Beilstein can be made in some way or other starting with methane. This section, however, deals only with the relatively small number which can be made economically and which are useful enough to warrant large volume production. A diagram of the principal materials covered is shown in Fig. 25.2.

Synthesis Gas

The most important route for the conversion of methane to petrochemicals is via either hydrogen, or a mixture of hydrogen and carbon monoxide. This latter material is known as "synthesis gas." The manufacture of carbon monoxide-hydrogen mixtures from coal was first established industrially by the well-known water-gas reaction

$$C + H_2O \longrightarrow CO + H_2$$

*Dow Chemical Company, Midland, Mich.

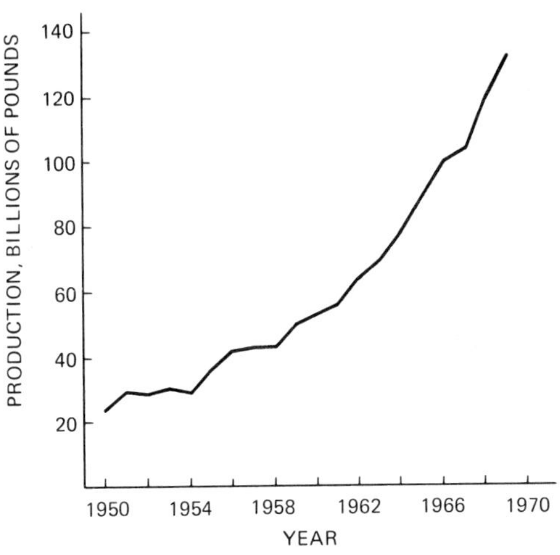

Fig. 25.1 Total production of synthetic organic chemicals. (*U.S. Tariff Commission.*)

Two important methods are presently used to produce the gas mixture from methane. The first is the methane-steam reaction, where methane and steam at about 900°C are passed through a tubular reactor packed with a promoted iron oxide catalyst. Two reactions are possible, depending on the conditions:

$$CH_4 + H_2O \longrightarrow CO + 3H_2$$

$$CO + H_2O \longrightarrow CO_2 + H_2$$

The second commercial method involves the partial combustion of methane to provide the heat and steam needed for the conversion. Thus the reaction can be considered to take place in at least two steps. The combustion step

$$CH_4 + 2O_2 \longrightarrow CO_2 + 2H_2O$$

followed by the reaction step

$$CH_4 + CO_2 \longrightarrow 2CO + 2H_2$$

$$CH_4 + H_2O \longrightarrow CO + 3H_2$$

The process is usually run with nickel catalysts in the temperature range 800–1000°C.

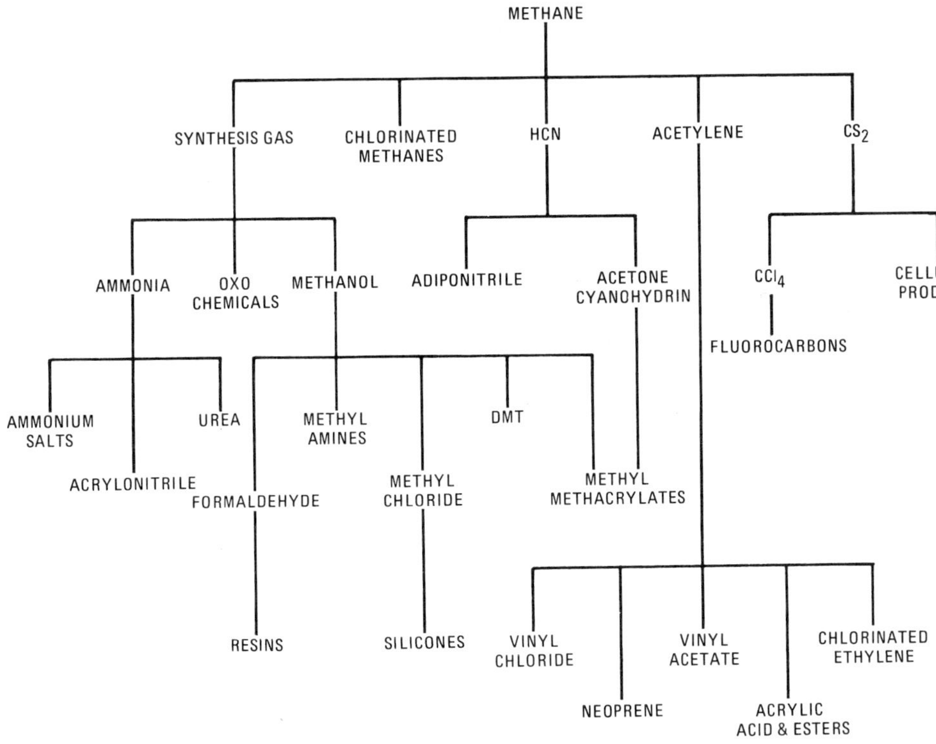

Fig. 25.2 Some important synthetic chemicals derived from methane.

The main outlets in the chemical industry for the gas mixtures obtained by the reforming of methane are in the manufacture of ammonia, the methyl alcohol synthesis, and in the Fischer-Tropsch and Oxo reactions.

Ammonia. Ammonia derived from petroleum and natural gas sources accounts for more than 97 per cent of the almost 26 billion pounds produced annually. Consequently, it can be termed the number one petrochemical in volume. About three-fourths of the ammonia produced goes directly into fertilizer uses, and the rest is used to produce such chemicals as ammonium nitrate, ammonium sulfate, caprolactam, nitric acid, urea, acrylonitrile, and organic amines.

Detailed description of the ammonia-manufacturing processes can be found in Chapter 5.

Methanol. Before 1926, all American methanol was obtained commercially as a by-product of the wood distillation process (wood alcohol). That year, however, marked the first appearance of German synthetic methanol. Presently only a negligible amount of the four-billion pounds produced comes from wood.

The methyl alcohol synthesis is well known. It resembles the synthesis of ammonia in that the catalysts operate only at high-temperature levels, and conversion and equilibrium are greatly assisted by high-pressure operation. The industrial reaction conditions are pressures of 250-350 atmospheres, and temperatures in the range of 300-400°C. The catalysts employed are based on zinc oxide, which is mixed with other oxides to provide temperature resistance. Variations between synthetic methanol plants are quite similar to those between synthetic ammonia plants. In fact, many ammonia operations are designed so that methanol could also be produced in them.

Methanol production has generally paralleled that of its largest end-use, formaldehyde. However, this is no longer completely true because fairly large quantities of formaldehyde now come from the hydrocarbon oxidation process. Other major uses of methanol are as solvents, inhibitors, and in the synthesis of methyl amines, methyl chloride, and methyl methacrylate.

Formaldehyde. Formaldehyde may be made from methanol either by catalytic vapor-phase oxidation,

$$CH_3OH + \tfrac{1}{2}O_2\ (\text{air}) \longrightarrow CH_2O + H_2O$$

or by a combination oxidation-dehydrogenation process

$$CH_3OH \longrightarrow CH_2O + H_2$$

It can also be produced directly from natural gas, methane, and other aliphatic hydrocarbons, but this process yields mixtures of various oxygenated materials.

Since both gaseous and liquid formaldehyde readily polymerize at room temperature, it is not available in the pure form. It is sold instead as a 37 per cent solution in water, or in the polymeric form as paraformaldehyde $(HO(CH_2O)_nH$, where n is between 8 and 50] or as trioxane $[(CH_2O)_3]$. The largest end use of formaldehyde is in the field of synthetic resins, either as a homopolymer or as a copolymer with phenol, urea, or melamine. It is also reacted with acetaldehyde to produce pentaerythritol $[C(CH_2OH)_4]$ which finds use in polyester resins. Two smaller-volume uses are in urea-formaldehyde fertilizers and hexamethylenetetramine, the latter being formed by condensation with ammonia.

Oxo Chemicals. The oxo chemicals are compounds—primarily C_4 and higher alcohols—made by the so-called "oxo" process. This process is a method of reacting olefins with carbon monoxide and hydrogen to produce aldehydes containing one more carbon atom than the olefin, these in turn are converted into alcohols. The earliest reaction studied used ethylene to produce both an aldehyde and a ketone. Thus the name "oxo," which was adapted from the German "oxierung," meaning ketonization. However, even though other names such as hydroformylation would much more accurately describe the process, the term oxo appears too deeply entrenched to be replaced.

A flow sheet of a typical process is shown in

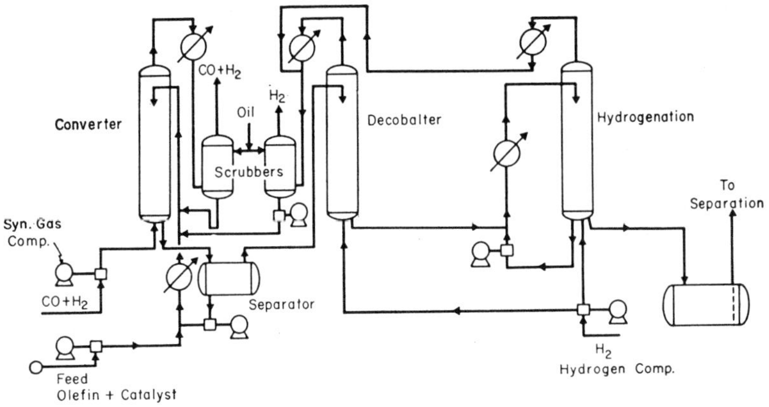

Fig. 25.3 The oxo process. [*Pet. Ref.*, **38,** *No. 11, 280 (1959), copyright 1959 by Gulf Publishing Co.*]

Figure 25.3. The steps involved in the reaction are

$$RCH{=}CH_2 + CO + H_2 \xrightarrow[250^\circ C]{Co \; 3000\ psi} \left\{\begin{matrix} RCH_2CH_2CHO \\ RCH(CHO)CH_3 \end{matrix}\right\}$$

$$\left\{\begin{matrix} RCH_2CH_2CHO \\ RCH(CHO)CH_3 \end{matrix}\right\} + H_2 \xrightarrow[200^\circ C]{3000\ psi} \left\{\begin{matrix} RCH_2CH_2CH_2OH \\ RCH(CH_2OH)CH_3 \end{matrix}\right\}$$

The cobalt catalyst used under these conditions is in the form of dicobalt octacarbonyl and cobalt hydrocarbonyl.

At the present time, the plant capacity in the United States amounts to more than 700 million pounds per year of oxo chemicals. These include such products as *n*-butanol, isobutanol, propionaldehyde, butyraldehydes, butyronitriles, isooctyl alcohol, decyl alcohol, and tridecyl alcohol.

Chlorinated Methanes

The chlorination of methane can be carried out either thermally or photochemically to produce methyl chloride (CH_3Cl), methylene chloride (CH_2Cl_2), chloroform ($CHCl_3$), and carbon tetrachloride (CCl_4). If only a particular chlorinated material is desired, other methods such as the chlorination of CS_2, or the reaction of methanol with HCl are generally used. These are fairly large volume chemicals, with the production of carbon tetrachloride alone estimated at almost a billion pounds.

Although largely replaced by other solvents in the dry-cleaning field, carbon tetrachloride has shown considerable growth as a raw material in the manufacture of chlorofluorohydrobons. Next in volume is methyl chloride, which is used to make silicones and tetramethyl lead. Methylene chloride finds use in paint removers, solvents, and aerosols. Chloroform is a raw material for fluorohydrocarbons.

Acetylene

Acetylene is made commercially in two ways: from calcium carbide, or from hydrocarbons. The choice of method is determined mainly by the fact that acetylene cannot be shipped easily, so large users must be at or near the point of origin. The carbide plant in turn must be near a cheap source of electric power, since each pound of carbide requires about 1.5 kw-h of electricity. The manufacture of acetylene is described in Chapters 14 and 17.

Acetylene has long been a valuable building block in the chemical industry. Major consumers of the more than a billion pounds produced are the manufacture of vinyl chloride, neoprene, vinyl acetate, acrylic acid and esters, and chlorinated ethylene. Growth in the demand for acetylene has been small in recent years, however, due to competition from cheaper raw materials. Ethylene is now preferred over acetylene as the starting material for vinyl chloride and vinyl acetate; propylene has completely supplanted it for acrylonitrile and is making inroads into the acrylates, and neoprene can now be made from butadiene.

Vinyl Chloride. Less than 20 per cent of the vinyl chloride (or about 700 million pounds) now comes from the addition of hydrogen chloride to acetylene. The process involves a mercuric chloride catalyst and temperatures around 200°C. All vinyl chloride is used to make plastics, the most important of which are the homopolymer (PVC), and copolymers with vinylidene chloride or vinyl acetate.

Vinyl Acetate. Vinyl acetate can be produced by combining acetylene and glacial acetic acid. This is a catalytic reaction (zinc or mercury compounds), and it may be carried out either in the liquid or vapor phase.

$$CH{\equiv}CH + CH_3COOH \longrightarrow CH_3COOCH{=}CH_2$$

Approximately 800 million pounds are produced annually, all of which is utilized in the polymeric form. Polyvinyl acetate (PVA) can be found in films and latex paints. It also can be used to produce polyvinyl alcohol (a water-soluble polymer), polyvinyl butyral (for safety-glass), polyvinyl formal, and various copolymers.

Acrylates and Methacrylates. The acrylates are esters of acrylic acid ($CH_2{=}CHCOOR$) with the R generally ranging from methyl to ethylhexyl. The main method of preparation involves reacting a mixture of acetylene, hydrogen chloride, nickel carbonyl, carbon monoxide, and the appropriate alcohol. About 80 per cent of the carbonyl group in the product ester is derived from the carbon monoxide, and the remainder from the nickel compound. Other methods involve ethylene cyanohydrin, ketene, the esterification of acrylic acid, or the oxidation of propylene to acrolein.

The most important products are ethyl acrylate and butyl acrylate. They are used in making emulsion polymers for latex paints and textiles.

Methyl Methacrylate. Methyl methacrylate is formed from acetone cyanohydrin in a two-step process.

$$\begin{array}{l} CH_3 \\ \,| \\ CH_3CCN \\ \,| \\ OH \end{array} + H_2SO_4 \longrightarrow$$

$$\begin{array}{l} CH_2{=}CCONH_2 \\ \qquad\;\, | \\ \qquad CH_3 \end{array} \cdot H_2SO_4 \xrightarrow{CH_3OH}$$

$$\begin{array}{l} CH_2{=}CCOOCH_3 \\ \qquad\;\, | \\ \qquad CH_3 \end{array} + NH_4HSO_4$$

While this is the major process in operation, there have been reports of a route involving the oxidation of isobutylene to methacrylic acid.

Production of methyl methacrylate now totals more than 450 million pounds. The largest part of this goes into cast sheet, where the clarity and resistance of poly(methyl methacrylate) are desirable. Other uses are in surface-coating resins and molding powders.

Hydrogen Cyanide. Hydrogen cyanide is prepared, as shown in Fig. 25.4, by passing a mixture of air, ammonia, and natural gas over a platinum catalyst. The converter is operated at a temperature of about 1800°F, and care must be taken to minimize the decomposition of the ammonia and methane, as well as the oxidation of methane to carbon monoxide and hydrogen. The effluent gases are cooled, washed with dilute sulfuric acid, and then passed through a column where the hydrogen cyanide is absorbed in water. This is concentrated by distillation, and an inhibitor is added to prevent polymerization. Although all new plants follow the methane-ammonia route, HCN can also be produced from coke-oven gas, from sodium and calcium cyanides, and by the decomposition of formamide.

Because of the safety problems, most production is captive to avoid the need for shipment. About one-third of the HCN goes into the production of acetone cyanohydrin, while almost another third is used to make adiponi-

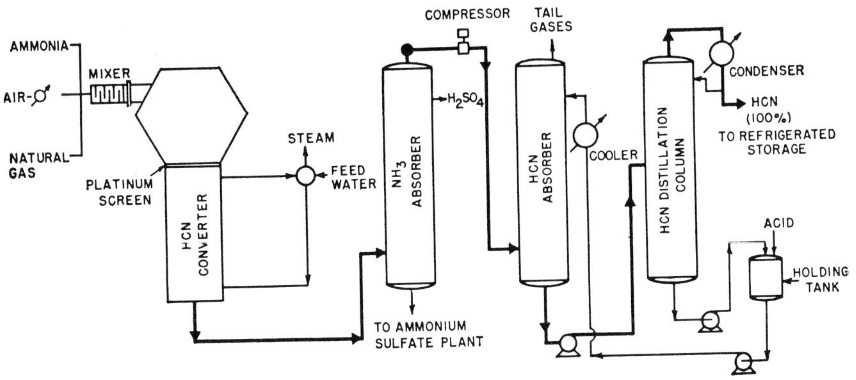

Fig. 25.4 The hydrogen cyanide process. [*Reprinted from Ind. Eng. Chem.*, **51**, *No. 10, 1235 (1959); copyright 1959 by the American Chemical Society, and reprinted by permission of the copyright owner.*]

trile. Other uses are for chelating agents and sodium cyanide.

The advent of the propylene-ammonia process for acrylonitrile has had an interesting effect on this material. Ten years ago acrylonitrile manufacture was a major consumer of HCN. Now almost 28 per cent of our HCN is produced as a by-product in acrylonitrile manufacture.

Carbon Disulfide

Carbon disulfide is made by the catalytic reaction of methane and sulfur vapor. Production is about 800 million pounds with the largest portion going to the manufacture of rayon and cellophane. The other major use is production of carbon tetrachloride.

CHEMICALS DERIVED FROM ETHYLENE

Ethylene far surpasses all other hydrocarbons both in volume and in diversity of commercial use. In the whole field of petrochemicals, it is exceeded in tonnage only by synthetic ammonia. Consumption of ethylene has grown remarkably in just the last 30 years. In 1940, 300 million pounds were produced, mostly for ethanol and ethylene oxide. The war time demand for styrene and the postwar impact of polyethylene aided in causing this figure to swell to almost 5 billion pounds in 1960. A boom in polyethylene use and strong growth in ethylene dichloride and ethylene oxide expanded ethylene production to over 16 billion pounds in 1968 and over 18 billion pounds in 1970. The major consumers of ethylene in 1969 are shown in Table 25.1.

Polyethylene

Polyethylene has shown a spectacular growth, accounting for only 4 per cent of total ethylene consumption in 1950, and almost 25 per cent ten years later. In 1961 polyethylene surpassed ethylene oxide as the principle ethylene consumer. In 1969, polyethylene and ethylene copolymer manufacture consumed nearly 40 per cent of all ethylene produced that year.

Three types of processes are used to produce polyethylene. The high-pressure process yields

TABLE 25.1 Ethylene Derivatives[a]

	Ethylene Consumed (Million lbs)	*(% of Total)*
Polyethylene	5,490	38.0
Ethylene oxide	3,253	21.5
Ethylene dichloride	1,979	13.0
Ethanol	1,480	9.8
Ethylbenzene	1,440	9.5
Acetaldehyde	485	3.2
Ethyl chloride	259	1.7]
Ethylene-propylene elastomers	98	0.6
Ethylene dibromide	52	0.3
Other	365	2.4
TOTAL	15,172	

[a]U.S. Tariff Commission Report, 1969.

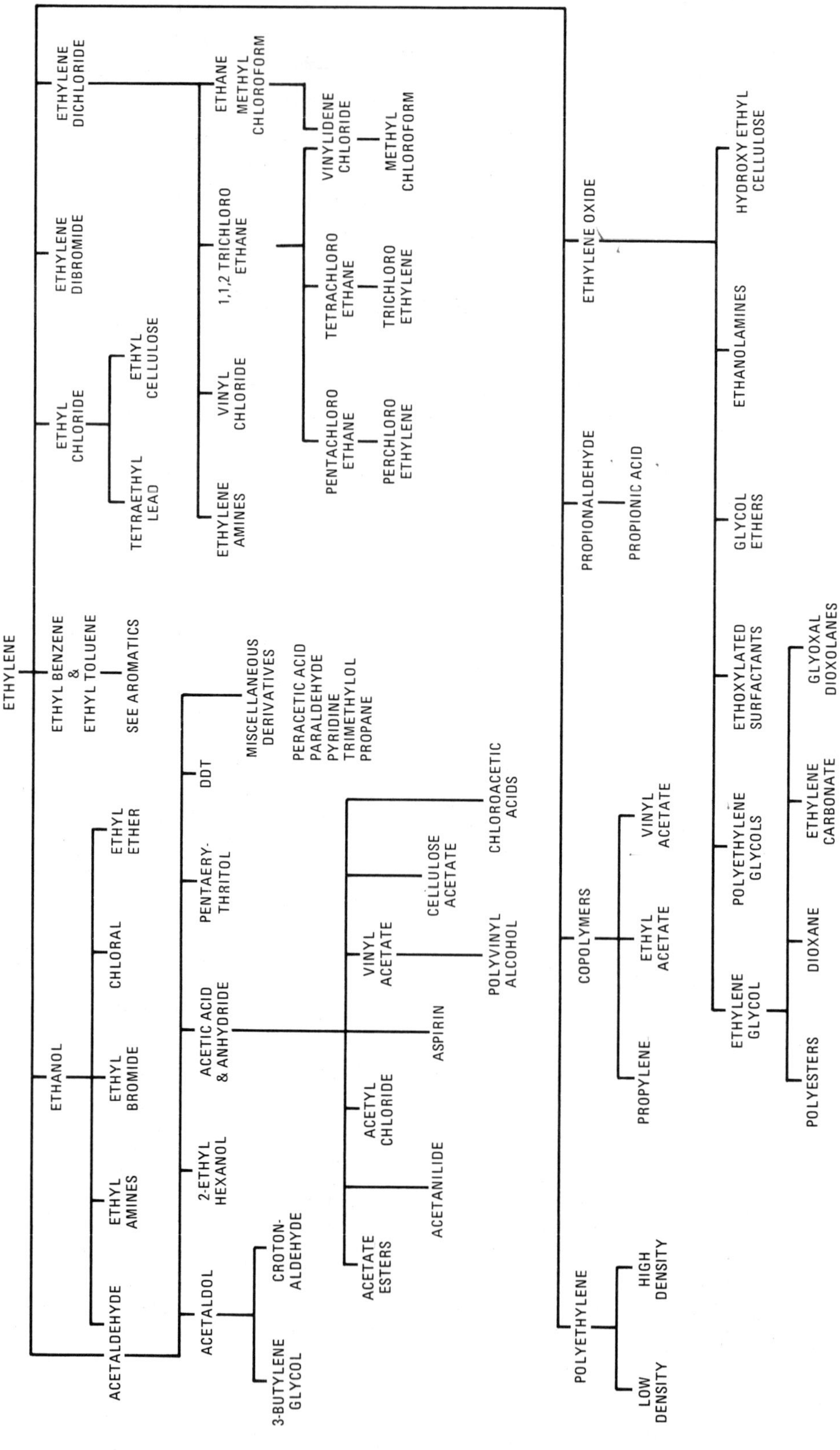

Fig. 25.5 Major ethylene derivatives.

a product of low density, and the low-pressure process yields either high density polymers via the Ziegler process or low or medium density polymers if the recent Phillips low-pressure process is used. Low density polyethylene and ethylene copolymers accounted for about 70 per cent of polyethylene produced in 1969. Polyethylene is covered in detail in Chapter 10.

Ethylene Oxide

Ethylene oxide was discovered in 1859 by Wurtz, who named it because of certain analogies with inorganic oxides. The method which he used was what is today known as the chlorohydrin process. He considered that direct oxidation was an impossibility, and stated flatly that "ethylene oxide cannot be made by the direct combination of ethylene and oxygen." It took almost eighty years to disprove this statement.

There are two basic processes presently used in the production of ethylene oxide from ethylene; the chlorohydrin process and the catalytic oxidation process. The chlorohydrin process is the older of the two and is based on the addition of hypochlorous acid to ethylene to produce ethylene chlorohydrin,

$$CH_2{=}CH_2 + HOCl \longrightarrow CH_2ClCH_2OH$$

which in turn is dehydrochlorinated to produce ethylene oxide.

$$2CH_2ClCH_2OH + Ca(OH)_2 \longrightarrow CaCl_2 + 2H_2O + 2\underset{\diagdown\ O\ \diagup}{CH_2{-}CH_2}$$

A flow sheet of this process is shown in the first part of Fig. 25.6. Ethylene, chlorine, and water are fed into the bottom of a large acid-proof brick-lined tower at somewhere below 50°C. The water and chlorine form hypochlorous acid, which then reacts rapidly with the ethylene. The dilute solution emerging from the tower contains about 5 per cent chlorohydrin. The major side reaction is the formation of ethylene dichloride. The solution

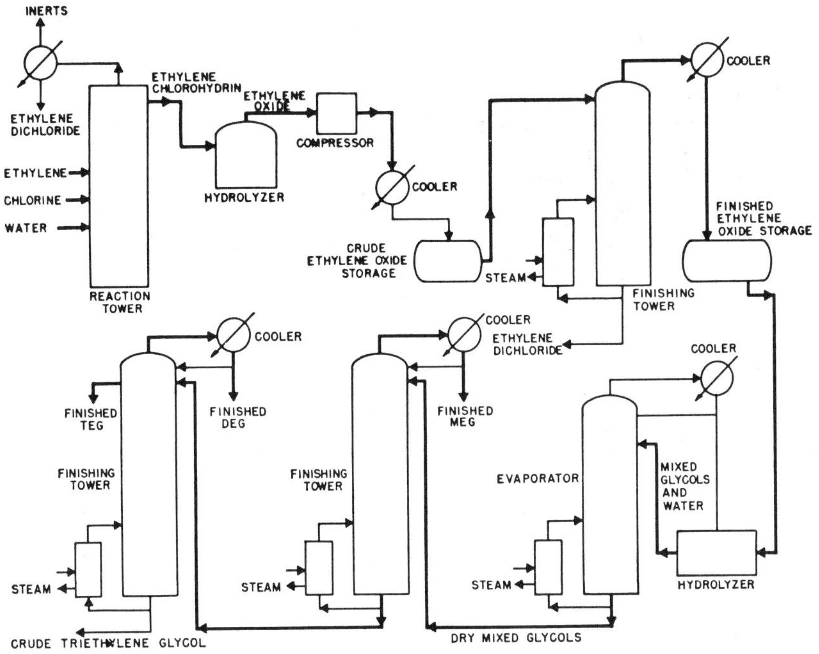

Fig. 25.6 Ethylene glycols and ethylene oxide by the chlorohydrin process. [*Ind. Eng. Chem.*, **51**, *No. 8, 896 (1959); copyright 1959 by the American Chemical Society, and reprinted by permission of the copyright owner.*]

next passes to a hydrolyzer where the chlorohydrin is treated with either slaked lime or caustic soda to produce the oxide. The crude ethylene oxide contains about 10 per cent ethylene dichloride which is removed by distillation. This process accounted for about 10 per cent of the ethylene oxide capacity in 1969, but many former ethylene oxide-chlorohydrin plants are now used for the production of propylene oxide.

The most important process involves the direct oxidation of ethylene with air in the presence of a silver catalyst.

$$2CH_2{=}CH_2 + O_2 \longrightarrow 2\,\underset{\diagdown\,O\,\diagup}{CH_2{-}CH_2}$$

A number of processes are available, and a typical one is shown in Figure 25.7. Ethylene, compressed air, and recycle gases are fed to a tubular reactor containing a silver catalyst. The oxygen and ethylene concentrations are maintained at a low level to avoid explosion hazards. The reaction temperature is 250-300°C with a pressure of 120-300 psi. Two competing side reactions which must be minimized are the total combustion of ethylene to carbon dioxide, and the isomerization of ethylene oxide to acetaldehyde. Some ethylene oxide direct oxidation plants use purified oxygen instead of air as the oxidizing agent.

Ethylene oxide is the most important of the olefin oxides. While it can be used directly as a fumigant for foodstuffs (usually mixed with carbon dioxide), it finds its chief outlet as a chemical intermediate. It owes its value to a combination of two types of reactivity: it can combine with chemicals containing a replaceable hydrogen, and it can polymerize to give a polyethenoxy chain. The first is typified in the formation of ethanolamines, while the second occurs in the synthesis of the polyglycols and higher glycol ethers.

TABLE 25.2 Production of Chemicals From Ethylene Oxide[a]

Product	*Output (Million lb)*	*Oxide Requirement (Million lb)*
Ethylene glycol	2,571	1928
Surfactants	750	409
Glycol ethers	474	245
Diethylene glycol	289	240
Ethanolamines	255	204
Polyglycols	—	118
Triethylene glycol	91	82
Others	—	182
TOTAL		3408

[a]Chemical Economics Handbook, Stanford Research Institute, 1970. U.S. Tariff Commission Report, 1969.

The estimated production for 1969 of ethylene oxide-derived materials, and the amount of ethylene oxide they consume is shown in Table 25.2.

Ethylene oxide is expected to maintain its present growth trend for at least the next few years. Ethylene glycol will remain the largest end use.

Ethylene Glycols. Ethylene glycol can be prepared directly by the hydrolysis of chlorohydrin, but the indirect hydrolysis via ethyl-

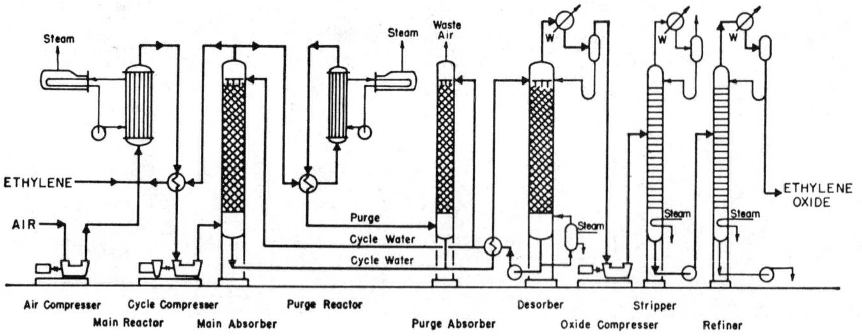

Fig. 25.7 Ethylene oxide by direct oxidation. [*Pet. Ref.*, **38**, *No. 11, 248 (1959), copyright 1959 by Gulf Publishing Co.*]

ene oxide is the preferred method. This is shown in the second part of Figure 25.6. The feed stream consists of ethylene oxide (either from the chlorohydrin or direct oxidation process) and water. This mixture is fed under pressure into the reactor vessel at about 100°C.

$$\overset{\displaystyle O}{\overbrace{CH_2\text{—}CH_2}} + H_2O \longrightarrow HOCH_2CH_2OH$$

By the end of the reaction, the temperature has risen to about 170°C. Some diethylene glycol and triethylene glycol are produced by the reaction of ethylene glycol with the oxide. The crude glycol solution is then concentrated in a multiple-effect evaporator. Final separation of the mono-, di-, and triethylene glycols is accomplished by distillation.

The main outlet, by far, for ethylene glycol is as the basic ingredient in permanent-type automotive antifreeze solutions. This accounts for more than 70 per cent of the total production. Other uses are in aircraft deicing fluids, solvents, hydraulic fluids, and as a plasticizer in cellophane. The chemical reactivity of ethylene glycol is utilized in a series of esters and polyesters. The most important polyester is that produced from ethylene glycol and terephthalic acid. This ester is sold both as a fiber and as a film, under trade-marks such as Dacron® and Mylar.® Another of the esters is ethylene glycol dinitrate, an essential component in low-temperature dynamites.

Diethylene glycol has properties similar to those of ethylene glycol. Its main uses are in unsaturated polyester resins and in polyester polyols for polyurethanes. Due to its hydroscopicity, triethylene glycol finds wide use as a tobacco humectant, and it is also used as a high-boiling solvent and plasticizer and as a drying agent for gases.

Polyethylene glycols are produced by passing ethylene oxide into a small amount of a low molecular weight glycol using a sodium or caustic soda catalyst. The molecular weight of liquid polyglycol products ranges from 200 to 1000. They are used as plasticizers, dispersants, lubricants, and humectants. Above a molecular weight of about 1000, the polyglycols are waxy solids, suitable for use as softening agents in ointments and cosmetics, and as lubricants.

Surfactants. In 1969, nonionics made up approximately 25 per cent of all synthetic detergents produced in the U.S., or nearly one billion pounds. In the years between 1960 and 1969, production of ethoxylated nonionics doubled. Behind this rise are several characteristics; most nonionics are liquid, they are low sudsing, and they can be built into readily biodegradable surfactants.

There are many nonionic surfactants, but

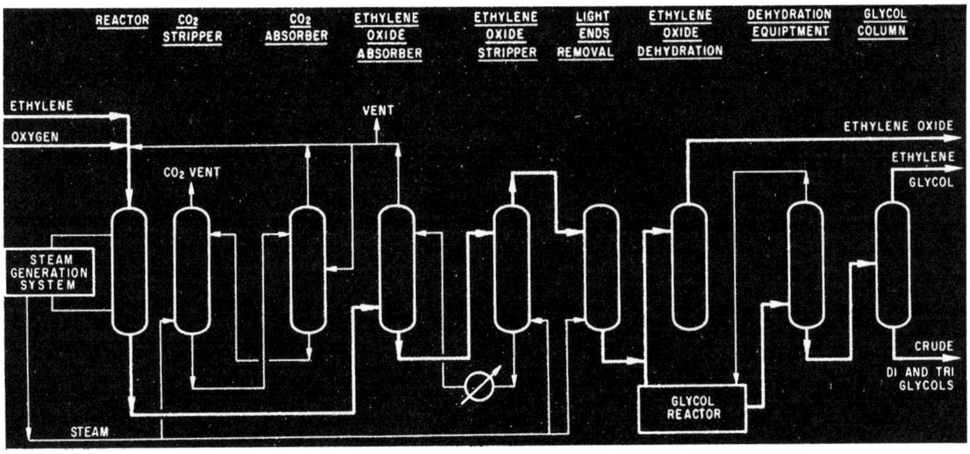

Fig. 25.8 Ethylene glycols and ethylene oxide by direct oxidation. [*Hydrocarbon Processing,* **50,** *No. 11, 158 (1971); copyright 1971 by Gulf Publishing Co.*]

four classes account for more than 80 per cent of the production. These are (1) the alkylphenol-ethylene oxide derivatives, (2) fatty acid–alkanolamine condensates, (3) tall oil–ethylene oxide adducts, and (4) the fatty acid–ethylene oxide adducts. The alkanolamine condensates are foam stabilizers in various detergent formulations. Tall oil adducts find use in household detergents, chiefly automatic washer products, because of their low-sudsing properties. The fatty alcohol adducts are used mainly as light-duty detergents, though some tridecyl adducts are utilized as foam stabilizers. The trend has been away from the benzenoid ethers because of their biodegradability limitations.

Ethanolamines. Ethanolamines are manufactured by reacting ethylene oxide and ammonia. The relative amounts of the three amines will depend primarily on the ammonia to oxide feed ratio.

$$NH_3 \xrightarrow{C_2H_4O} (HOC_2H_4)NH_2 \xrightarrow{C_2H_4O} (HOC_2H_4)_2NH \xrightarrow{C_2H_4O} (HOC_2H_4)_3N$$

The products from the reaction are separated by distillation. During the last few years, each of the amines has in turn been in the greatest demand, so processing flexibility must be maintained.

Monoethanolamine is used primarily in detergents and as an absorbent for acid-gas (H_2S, CO_2) removal. It is to a lesser extent, a chemical intermediate for compounds such as ethylene imine. Diethanolamine's major end use is in detergetns, but it is also utilized in textiles and as a gas-purification agent. Most of the triethanolamine goes into the production of cosmetics and textile specialities.

Isopropanolamines, derived from propylene oxide and ammonia, are competitive with the ethanolamines, and both are unique in that they are organic compounds and yet strongly alkaline.

Glycol Ethers. In the same way that water reacts with one or more molecules of ethylene oxide, alcohols react to give monoethers of ethylene glycol, producing monoethers of diethylene glycol, triethylene glycol, etc., as by-products.

$$ROH + CH_2\overset{O}{-}CH_2 \longrightarrow ROC_2H_4OH$$

$$+ CH_2\overset{O}{-}CH_2 \longrightarrow ROC_2H_4OC_2H_4OH$$

$$+ CH_2\overset{O}{-}CH_2 \longrightarrow ROC_2H_4OC_2H_4OC_2H_4OH$$

Since their commercial introduction in 1926, glycol ethers have become valuable as industrial solvents and chemical intermediates. Because glycol monoethers contain a $—OCH_2CH_2OH$ group, they resemble a combination of ether and ethyl alcohol in solvent properties. The most common alcohols used are methanol, ethanol, and butanol. Principal uses for the glycol ethers are as solvents for paints and lacquers, as intermediates in the production of plasticizers, and as ingredients in brake fluid formulations. The most common trade names are Dowanol®, Cellosolve®, and Polysolve®. Condensation of the monoethers produces glycol diethers which are also useful as solvents.

$$2ROCH_2CH_2OH \xrightarrow{H_2SO_4} ROC_2H_4OC_2H_4OR + H_2O$$

Alcohol-ethylene oxide products have also been developed in which the number of oxide units is considerably higher in order to improve water solubility. Long-chain fatty alcohols are condensed with 10-40 molecules of ethylene oxide to produce detergents for the textile industry. Also important are the water-soluble alkyl phenyl ethers of the higher polyethylene glycols. Phenols react in the same way as alcohols to give polyglycol ethers. The reaction is rapid, and essentially quantitative. These products are detergents of the same general class as the long-chain alcohol-ethylene oxide condensates.

Of recent origin is the interest in the extremely high molecular weight homopolymers of ethylene oxide. These resins, trade-marked

Polyox®, have good water and organic solvent solubility and thus are used for thickening agents and water-soluble films.

Chlorinated Hydrocarbons

The manufacture of chlorinated hydrocarbons forms an important part of industrial chemistry today. The products are useful as solvents, chemical intermediates, pesticides, monomers, and in many other ways. Table 25.3 shows some of the large-volume materials.

Chlorinated derivatives of aliphatic hydrocarbons are usually prepared by one of three general methods: (1) addition of hydrogen chloride to unsaturated hydrocarbons, (2) addition of chlorine to unsaturated hydrocarbons, or (3) substitution of chlorine for hydrogen in either saturated or unsaturated hydrocarbons. In the last case, hydrogen chloride is a by-product. Examples of the first method are the addition of HCl to ethylene to form ethyl chloride and the addition of HCl to acetylene to form vinyl chloride. Typical of the second method is the addition of chlorine to ethylene to form dichloroethane. The third method, the direct substitution of chlorine for hydrogen, usually involves a free-radical mechanism. The formation of chlorine free radicals occurs spontaneously at temperatures above 250°C and increases with temperature. The formation may also be brought about by the action of actinic light at lower temperatures. These light-activated, or photochemical, chlorinations may be carried out in either gas or liquid phase.

The principal uses for chlorinated hydrocarbons are as solvents, chemical intermediates, and insecticides. Their value as solvents derives from a combination of good solvent power, low flammability, and high vapor density. The latter property is particularly important in vapor degreasing of metal parts (a major use for trichloroethylene, perchloroethylene, and methyl chloroform). Certain of the compounds are of value as chemical intermediates. Thus, ethyl chloride is used to make tetraethyl lead; and ethylene dichloride is used to make vinyl chloride, which in turn is polymerized to polyvinyl chloride.

Ethylene Dichloride (EDC). While some ethylene dichloride (1,2-dichloroethane) occurs as a by-product of chlorohydrin and ethyl chloride processes, the bulk comes from the chlorination of ethylene. Almost 90 per cent of the EDC produced in 1969 went into the manufacture of vinyl chloride monomer. Fig-

TABLE 25.3 Production and Sales of Some Halogenated Aliphatic Hydrocarbons[a]

	Production (1000 lbs)	*Sales (1000 lbs)*	*Sales Value ($1000)*
Ethylene dichloride	6,037,428	1,226,845	36,345
Vinyl chloride	3,735,942	2,358,741	103,100
Carbon tetrachloride	882,735	785,884	42,698
Ethyl chloride	678,808	267,503	16,599
Perchloroethylene	635,251	611,625	42,529
Trichloroethylene	596,821	561,536	39,626
Methyl chloride	402,809	166,146	8,729
Dichlorodifluoromethane	367,658	335,409	85,203
Methylene chloride	366,005	338,107	25,891
Methyl chloroform	324,311	298,873	33,288
Ethylene dibromide	309,943	—	—
Trichlorofluoromethane	238,518	203,652	37,476
Chloroform	216,168	172,177	10,807
Chlorinated paraffins	61,935	59,124	7,584
All others	1,331,981	436,091	144,202
TOTAL halogenated hydrocarbons	16,186,313	7,899,651	677,396

[a]U.S. Tariff Commission Report, 1969.

ure 25.9 shows a vinyl chloride monomer and ethylene dichloride (EDC) process using ethylene, chlorine, and air, thus including oxychlorination.

In this process vinyl chloride is produced by thermal cracking of EDC. The feed, EDC, is supplied from two sources. In the first source, ethylene and chlorine are reacted in essentially stoichiometric proportions to produce EDC by direct addition. In the second source, ethylene is reacted with the HCl produced from the thermal cracking operation to produce EDC by oxychlorination. Chemical reactions are as follows:

Cracking EDC: $C_2H_4Cl_2 \xrightarrow{\Delta} HCl + C_2H_3Cl$

Direct chlorination: $Cl_2 + C_2H_4 \longrightarrow C_2H_4Cl_2$

Oxychlorination: $C_2H_4 + 2HCl + \frac{1}{2}O_2 \longrightarrow C_2H_4Cl_2 + H_2O$

The oxychlorination reaction is carried out in a fluid bed of copper chloride-impregnated catalyst. In direct chlorination, ethylene and chlorine gas are charged to a reactor containing catalyst and excess ethylene dichloride as solvent. The reaction is carried out at 50°C and 20 psig. Purified ethylene dichloride from both sources is cracked in a furnace at about 400°C and elevated pressure. The hot gases are quenched and distilled to remove HCl, then vinyl chloride. Unconverted EDC is returned to the EDC purification train.

Profitable disposal or utilization of HCl was once one of the major restrictions in the growth of ethylene dichloride as the source of vinyl chloride and other chlorinated hydrocarbon derivatives. The advent of oxychlorination which uses hydrochloric acid and molecular oxygen in contact with ethylene has opened the door for very rapid replacement of the acetylene process. The oxygen reacts with HCl in the oxychlorination step and generates chlorine *in situ* which reacts with the ethylene, forming the EDC.

Since oxychlorination has greatly improved the process economics of EDC, the trend now

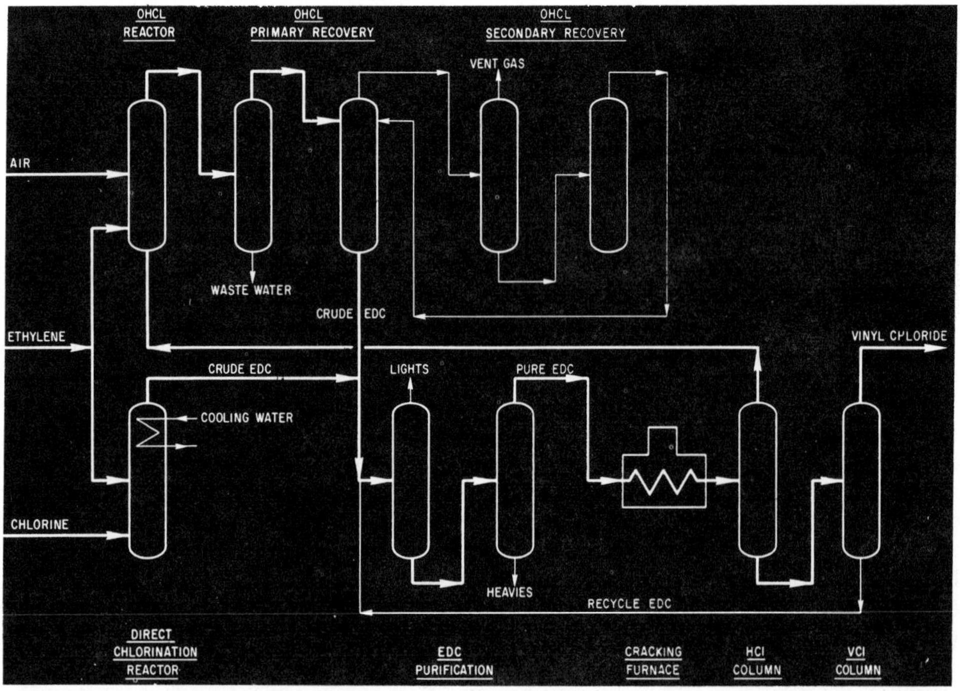

Fig. 25.9 Vinyl chloride and ethylene dichloride manufacture. [*Hydrocarbon Processing,* **50**, *No. 11, 220 (1971); copyright 1971 by Gulf Publishing Co.*]

is to make many of the chlorinated hydrocarbons previously derived from acetylene by this route. With proper conditions, EDC can be chlorinated to mainly tetrachloroethane, then catalytic dehydrochlorination of the tetra gives trichloroethylene. With different chlorination conditions, mainly pentachloroethane can be formed, and with dehydrochlorination, perchloroethylene is formed.

Another modification of EDC chlorination is to adjust the conditions to maximize 1,1,2-trichloroethane as the product. This product when dehydrochlorinated gives vinylidene chloride (1,1-dichloroethylene), a monomer used in a growing number of plastic polymers. Vinylidene chloride can, however, be hydrochlorinated to methyl chloroform, a very rapidly growing solvent, mainly by virtue of its low toxicity.

Almost 4 per cent of EDC produced in 1969 was used as a lead scavenger, while about 3 per cent was used as an intermediate to produce about 70 million pounds of ethylene amines.

Ethyl Chloride. About 90 per cent of the ethyl chloride is produced by the addition of hydrogen chloride to ethylene in the presence of an aluminum chloride catalyst.

$$CH_2{=}CH_2 + HCl \xrightarrow{AlCl_3} CH_3CH_2Cl$$

This is a liquid phase reaction, carried out at about 40°C. The remaining production comes from the catalytic reaction of hydrogen chloride with ethanol.

$$CH_3CH_2OH + HCl \xrightarrow{ZnCl_2} CH_3CH_2Cl + H_2O$$

The production of ethyl chloride is closely tied in with that of tetraethyl lead, which consumes over 90 per cent of the former. It is also used in the production of ethyl cellulose, and as a refrigerant and an anesthetic.

Tetraethyl lead is made by reacting ethyl chloride with a lead-sodium alloy (about 9 parts lead and 1 part sodium.)

$$4NaPb + 4C_2H_5Cl \longrightarrow Pb(C_2H_5)_4 + 3Pb + 4NaCl$$

The crude product is purified by vacuum distillation and water washing. It is then blended with ethylene dichloride and ethylene dibromide to make the gasoline antiknock compound. At present, the outlook for tetraethyl lead is dim due to serious concern for the environmental effects of lead compounds in automobile exhausts. This is leading to lead-free gasolines.

However, many possibilities similar to those of ethylene dichloride exist, ethyl chloride with properly controlled chlorination conditions yields methyl chloroform and 1,1,2-trichloroethane. The 1,1,2-trichloroethane can be chlorinated to tetrachloroethane, which can in turn be dehydrochlorinated to trichloroethylene. This use for ethyl chloride may utilize some of the ethyl chloride capacity idled by sharp cutbacks in the production of lead tetraethyl for gasoline.

Thus, starting with ethylene, a valuable large volume of chlorinated hydrocarbons can be produced with economics such that acetylene is rapidly being replaced. The new processes are also very flexible in the product mix they can produce. Prior to 1969, about 85 per cent of trichloroethylene was made from acetylene. In 1969, this proportion dropped to about 55 per cent, and the replacement of acetylene by ethylene is proceeding rapidly.

A similar shift has taken place with vinyl chloride where in 1962 about 50 per cent of vinyl chloride was based on acetylene and 50 per cent on ethylene via EDC. By 1969 ethylene-based vinyl chloride accounted for about 85 per cent of production.

Chlorinated Solvents. This group of compounds consists mainly of carbon tetrachloride, trichloroethylene, perchloroethylene, methyl chloroform, chloroform, and methylene chloride. Some of the intermediates to these products such as 1,1,2-trichloroethane, tetrachloroethane and pentachloroethane find special solvent uses. Since the manufacturing processes are difficult to untangle the sketches that follow are added for clarity, even though some start with methane, propane, or propylene.

The preparation of trichloroethylene from acetylene is a two-step process. In the first step, acetylene and chlorine are catalytically

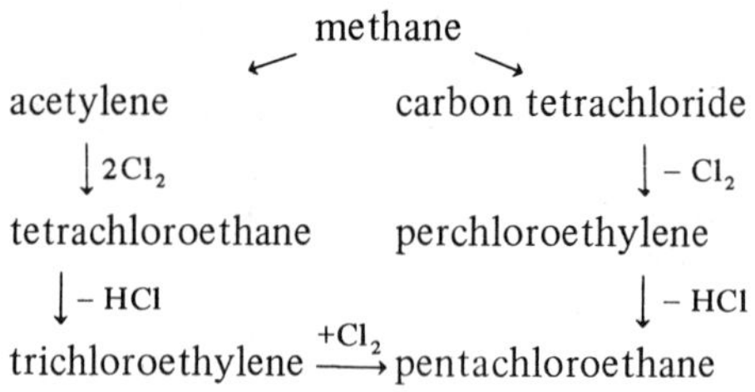

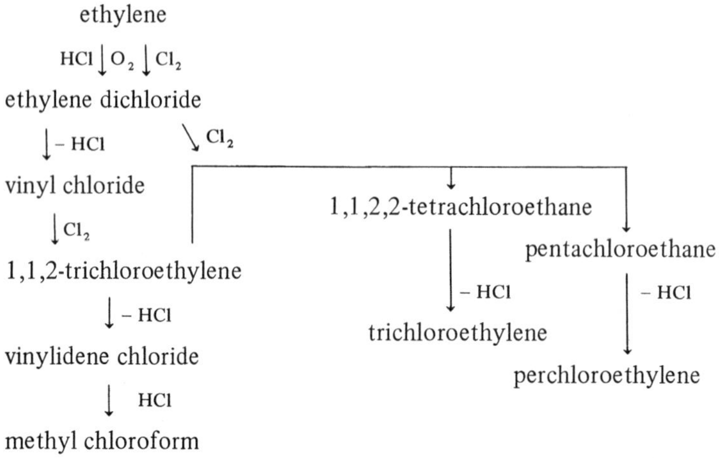

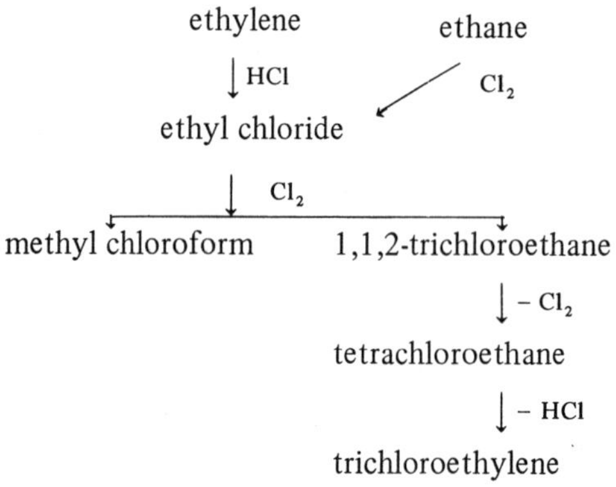

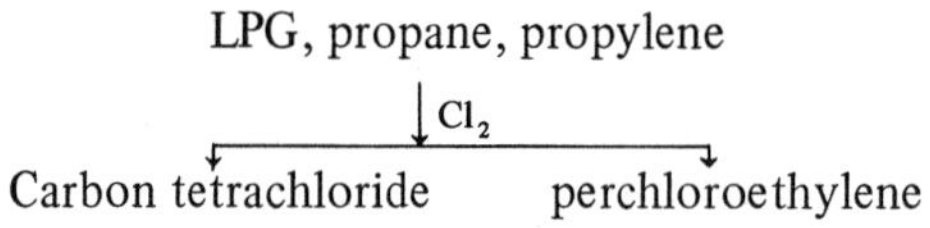

reacted in the liquid phase to produce tetrachloroethane.

$$CH{\equiv}CH + 2Cl_2 \xrightarrow[80^\circ]{SbCl_3 \text{ or } FeCl_3} CHCl_2CHCl_2$$

In the second step, the tetrachloroethane is dehydrochlorinated either thermally, catalytically, or by reaction with lime. The catalytic method is the most common commercially, and it involves a vapor-phase reaction over a bed of barium chloride at about 400–700°F.

$$CHCl_2CHCl_2 \longrightarrow CHCl{=}CCl_2 + HCl$$

More than 90 per cent of the trichloroethylene produced is for vapor degreasing of metals, a field in which it has almost completely supplanted carbon tetrachloride, but it is now meeting stiff competition from methyl chloroform.

Perchloroethylene is made by the pyrolysis of carbon tetrachloride, although it may also be produced by the dehydrochlorination of pentachloroethane. At a temperature of 800–900°C, carbon tetrachloride readily decomposes into perchloroethylene and hexachloroethane. The latter is recycled to produce more perchloroethylene. A modification of this process has been reported in which chlorine and a light hydrocarbon (such as natural gas or

LPG) are fed into a chlorination furnace at 900–1200°F. Chlorination takes place readily, producing carbon tetrachloride and perchloroethylene. Undoubtedly the latter is again formed by the pyrolysis of CCl_4. Most perchloroethylene is consumed by the dry cleaning industry. Some is used as a vapor degreasing solvent.

Other Halogenated Hydrocarbons. About 300 million pounds of ethylene dibromide is made annually by the addition of bromine to ethylene. It is used as a lead scavenger in antiknock fluids and as an agricultural fumigant. Smaller uses are as a solvent and in the synthesis of pharmaceuticals and dye intermediates.

Fluorocarbons made their first impact on the chemical industry in 1931 with the introduction of *Freon* refrigerants (Freon® and Genetron® are the trade-marks used by the two largest manufacturers). The next major advance came during World War II with the development of fluorocarbon polymers and nonfood aerosol propellents. By 1969, production was estimated at 700 million pounds, of which dichlorodifluoromethane accounted for more than half. The five main compounds today are as follows:

Fluorocarbon-12 (CCl_2F_2) is the most widely used fluorocarbon. It finds application in aerosol propellents, either alone or in combination with other gases, and as a refrigerant.

Fluorocarbon-11 (CCl_3F) is used with fluorocarbon-12 propellent to reduce the pressure in aerosols. It is also widely employed in air conditioning and process-water cooling.

Fluorocarbon-22 ($CHClF_2$) is used in small-scale refrigeration and air conditioning units. Tetrafluoroethylene can be produced from this compound.

Fluorocarbon-113 (CCl_2FCClF_2) is used to improve the solvent properties of fluorocarbon-12 propellents. It can be dechlorinated to produce chlorotrifluoroethylene.

Fluorocarbon-114 ($CClF_2CClF_2$) is also used with fluorocarbon-12 propellent, particularly where the aerosol product contains a large amount of water.

In addition to refrigerants and nonfood aerosol propellents, fluorocarbons are finding use in plastics, in particular the homopolymers of tetrafluoroethylene and chlorotrifluorethylene (Teflon® and Kel-F®, respectively). These polymers are specially noted for their temperature-resistance and chemical inertness. Also of importance is the elastomer (Viton®) produced by the copolymerization of vinylidene fluoride and hexafluoropropylene.

Fluorocarbons find smaller markets as fire extinguishers (bromotrifluoromethane), blowing agents for urethane foams, solvents, and specialty lubricants.

Ethanol

Less than twenty years ago, synthetic ethanol was the largest consumer of ethylene. Today it is in fourth place. Just prior to World War II, the fermentation of molasses accounted for about 72 per cent of the ethanol production. Today less than 10 per cent of the ethanol manufactured is made by this route; over 90 per cent is synthesized from ethylene by the esterification–hydrolysis process or by direct catalytic hydration. In 1969 over 2.36 billion pounds of synthetic ethanol was produced.

The esterification-hydrolysis process, such as that shown in Fig. 25.10, is the older of the two and still accounts for about 80 per cent of the production. This reaction takes place in two steps. The first occurs between sulfuric acid and ethylene at about 75°C in a plate column.

$$3CH_2{=}CH_2 + 2H_2SO_4 \longrightarrow CH_3CH_2HSO_4 + (CH_3CH_2)_2SO_4$$

The resulting mixture is then diluted in a hydrolyzer to produce the alcohol.

$$CH_3CH_2HSO_4 + H_2O \longrightarrow CH_3CH_2OH + H_2SO_4$$

The over-all yield is about 90 per cent ethanol and 5–10 per cent diethyl ether. If the emphasis is placed on ether production, the residence time in the hydrolyzer is increased and the alcohol is recycled. The reactions in the hydrolyzer will then include the following:

$$C_2H_5OH + C_2H_5HSO_4 \longrightarrow C_2H_5OC_2H_5 + H_2SO_4$$

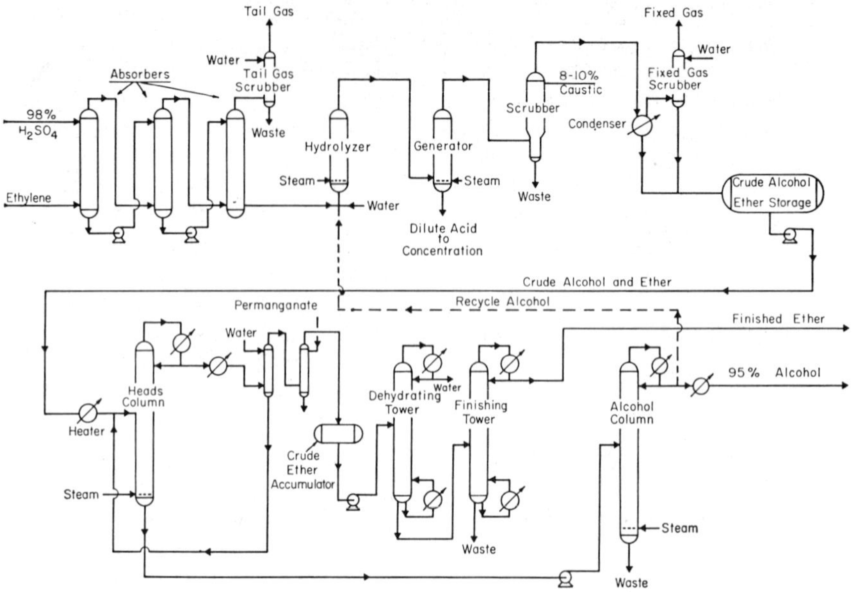

Fig. 25.10 Esterification-hydrolysis process for producing ethanol. [*Pet. Ref.*, **38**, *No. 11, 239 (1950); copyright Gulf Publishing Co.*]

$$C_2H_5OH + (C_2H_5)_2SO_4 \longrightarrow C_2H_5OC_2H_5 + C_2H_5HSO_4$$

The direct hydration process involves a water-ethylene reaction, over a phosphoric acid catalyst, at about 400°C and 1000 psi.

$$C_2H_4 + H_2O \xrightarrow{H_3PO_4} C_2H_5OH$$

This method has the advantage of producing less by-product diethyl ether. Over-all yield of ethanol is reported to be better than 97 per cent. A flow sheet of the direct hydration process is shown in Fig. 25.11. Forty-one per cent of the synthetic ethanol produced in 1969 was by direct hydration; in 1970 this had increased to 48 per cent and was expected to pass the 50 per cent mark in 1971.

The production of acetaldehyde was the principal use of industrial ethyl alcohol until 1968 when solvent uses surpassed it. Acetaldehyde-use is showing a steady decline while solvent-use continues to grow at a rate somewhat above 5 per cent per year. Other uses are in synthetic rubber, drugs, and in the synthesis of various chemicals such as acetic acid, ethyl chloride, and ethyl acetate.

Ethylbenzene

Ethylbenzene has but one end-use and that is the manufacture of styrene. In 1969 ethylbenzene was the fifth largest consumer of ethylene, accounting for about 1.44 billion pounds. In addition to being produced by the alkylation of benzene with ethylene, ethylbenzene is also isolated from mixed xylene streams or as a by-product of cumene manufacture. These latter sources account for less than 10 per cent of the ethylbenzene used today.

Ethylbenzene synthesis by reaction of ethylene with benzene is carried out either in the liquid phase using aluminum chloride as the catalyst, or in the vapor phase with a phosphoric acid or alumina-silica catalyst. Yields above 95 per cent are obtained in either case. Figure 25.12 shows a flow diagram of an ethylbenzene process.

The principal uses for styrene, made by catalytic cracking of ethylbenzene, are in polystyrene, styrene-butadiene latex and plastics, SBR rubber, and in the cross-linking of unsaturated polyester resins. For more details, see Chapter 10.

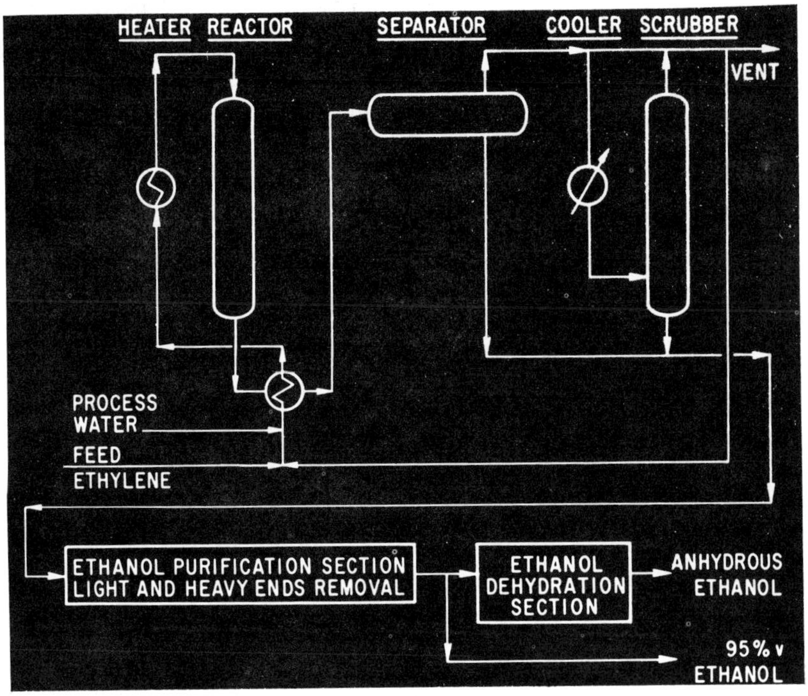

Fig. 25.11 Vapor-phase direct hydration of ethylene to ethanol. [*Hydrocarbon Processing,* **50,** *No. 11, 151 (1971); copyright 1971 by Gulf Publishing Co.*]

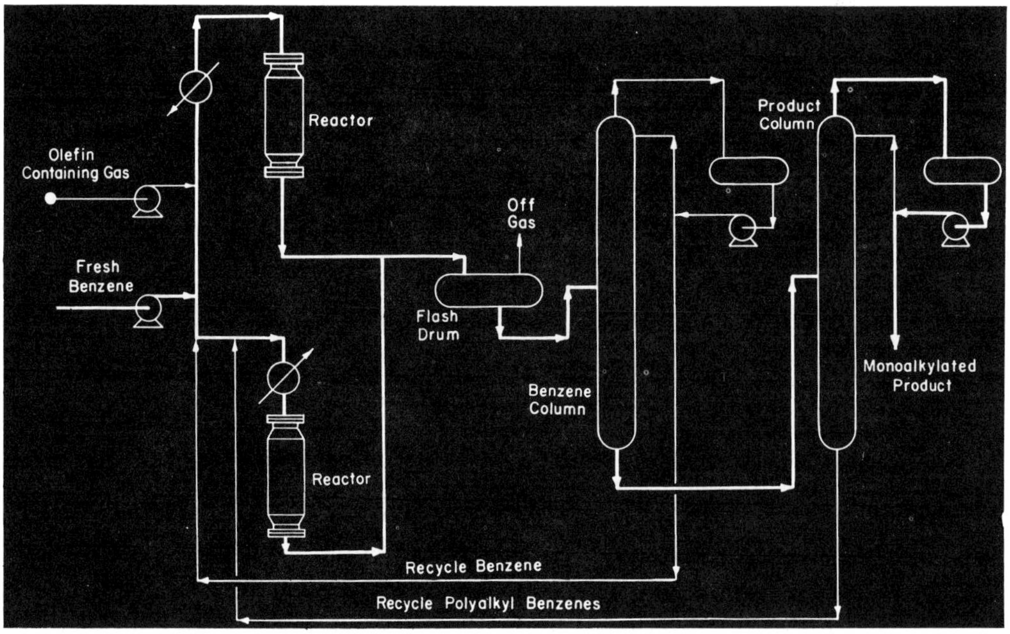

Fig. 25.12 Alkar ethylbenzene process. [*Hydrocarbon Processing,* **50,** *No. 11, 125 (1971), copyright 1971 by Gulf Publishing Co.*]

Acetaldehyde Acetic Acid, Vinyl Acetate

Acetaldehyde. Acetaldehyde production in 1969 was 1.65 billion pounds. Of this, 685 million or 42 per cent was made by direct oxidation of ethylene. The majority of this growth using direct oxidation has occurred since 1963. The direct oxidation process uses cupric chloride and a small amount of palladium chloride in aqueous solution as the catalyst. The reaction is exothermic and is controlled by evaporation. The gaseous reaction mixture is scrubbed to remove acetaldehyde and overhead gases are recycled. The liquid from the reaction mixture is regenerated with oxygen to return the copper to the cupric state and palladium metal back to palladium chloride.

The alternate processes consist of oxidation or dehydrogenation of ethyl alcohol. In one case air burns the liberated hydrogen to water, and in the other case, hydrogen is produced and can be recovered. A silver gauze catalyst and temperatures in the range of 375-550°C is used for the oxidation route; while in the dehydrogenation process, a chromium-activated copper catalyst and external heating to a temperature of 260 to 290°C is used. Unconverted alcohol and acetaldehyde are scrubbed with cold dilute alcohol, then run through a distillation recovery train.

Other smaller sources of acetaldehyde are from hydrocarbon-oxidation operations and as by-product from vinyl acetate operations.

Acetaldehyde is used mainly for the production of acetic acid and anhydride. *n*-Butanol, 2-ethylhexanol, and pentaerythritol combined probably consume less than 20 per cent of the acetaldehyde produced. Since the fate of acetaldehyde is determined by its use as an acetic acid–acetic anhydride intermediate, future acetic acid technology will have a strong impact on acetaldehyde use.

Acetic Acid. Acetic acid is made today primarily from acetaldehyde, although processes exist where the feed is ethyl alcohol and oxidation to acetaldehyde is carried out *in situ.* The acetaldehyde is recycled, and only acetic acid is taken as the product. Some acetic acid is synthesized by oxidation of *n*-butane and new processes have been announced for making acetic acid from methanol and carbon monoxide (see Fig. 25.13).

In the acetaldehyde process, acetaldehyde is fed to a reactor where air is passed through the liquid at 55 to 65°C and 70 to 75 psi. Manganous acetate, 0.1 to 0.5 per cent, is used in the liquid to control the formation of the explosive intermediate peracetic acid. Acetaldehyde is scrubbed from the off-gases with water and recovered for recycle, while the liquid reactor mass is distilled to 99 per cent glacial acetic acid. Continuous modern plants use cobalt acetate dissolved in acetic acid as the catalyst. Conversions of 20–30 per cent per pass are used and oxygen allows lower pressures and minimizes the off-gas.

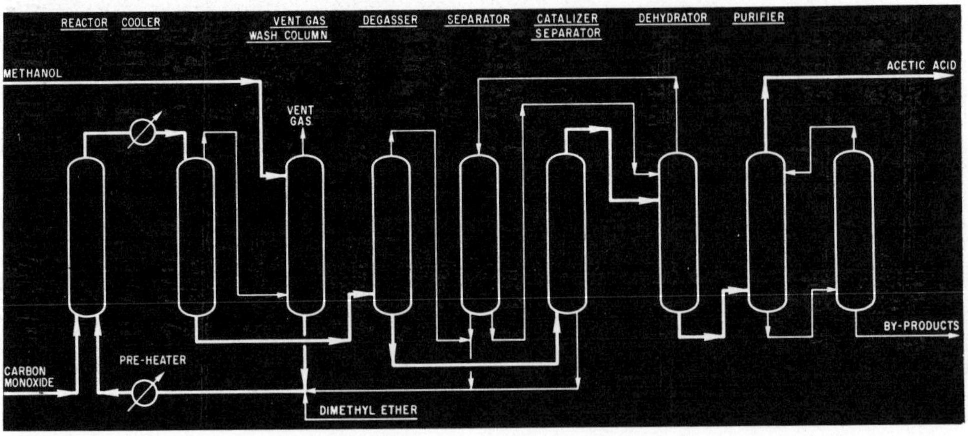

Fig. 25.13 High-pressure acetic acid from methanol and carbon monoxide. [*Hydrocarbon Processing,* **50,** *No. 11, 115 (1971); copyright 1971 by Gulf Publishing Co.*]

Acetic acid is used chiefly to produce acetic anhydride for cellulose acetate manufacture. In addition, sizeable quantities are consumed in the production of vinyl acetate, acetate esters and salts, and chloroacetic acid. Production of acetic acid was reported as 1.77 billion pounds in 1969.

Acetic Anhydride. Acetic anhydride is produced mainly by processes using acetaldehyde and acetic acid with copper and cobalt acetates with some manganese acetate. Concentrations of catalysts are generally 1 to 2 per cent. About 1.4 parts of acetic acid per part of acetaldehyde are fed to the reactor system. The reactor is run at 50–70 per cent and about 60 psi. Finishing is done in a vacuum distillation train.

In some newer processes oxygen replaces air as the oxidizing agent. The reaction is carried out at lower temperatures and pressures. Higher (95 per cent) conversions of acetaldehyde are obtained, and acetic anhydride and acetic acid are produced in a 50:50 weight ratio. Running the same process in the absence of diluent gives a higher oxidation rate but a lower (2:3) anhydride-acid ratio.

Of the 1.68 billion pounds of acetic anhydride reported produced in 1969, some 85 per cent probably was used to make cellulose acetates, about 10 per cent for vinyl acetate, and 5 per cent for aspirin and other esterification reactions.

Vinyl Acetate. Of some 729 million pounds produced in 1969, less than 100 million were produced by the new oxyacetylation process (using ethylene, acetic acid, and oxygen). The new process yields acetaldehyde as a by-product in adequate quantity to be converted to the acetic acid needed. This means that the only net feed to the complex is ethylene.

At present, some 600 million pounds of vinyl acetate capacity exist. It is apparent that this is another case where acetylene is rapidly being replaced. A flow diagram for the oxyacetylation process is shown in Fig. 25.14.

Ethylene Oligomers

Linear primary alcohols and α-olefins in the C_6 to C_{10} range and the C_{12} to C_{18} range have become important industrial chemicals in recent years. The linear alcohols in the C_6 to C_{10} range are used to make plasticizers for

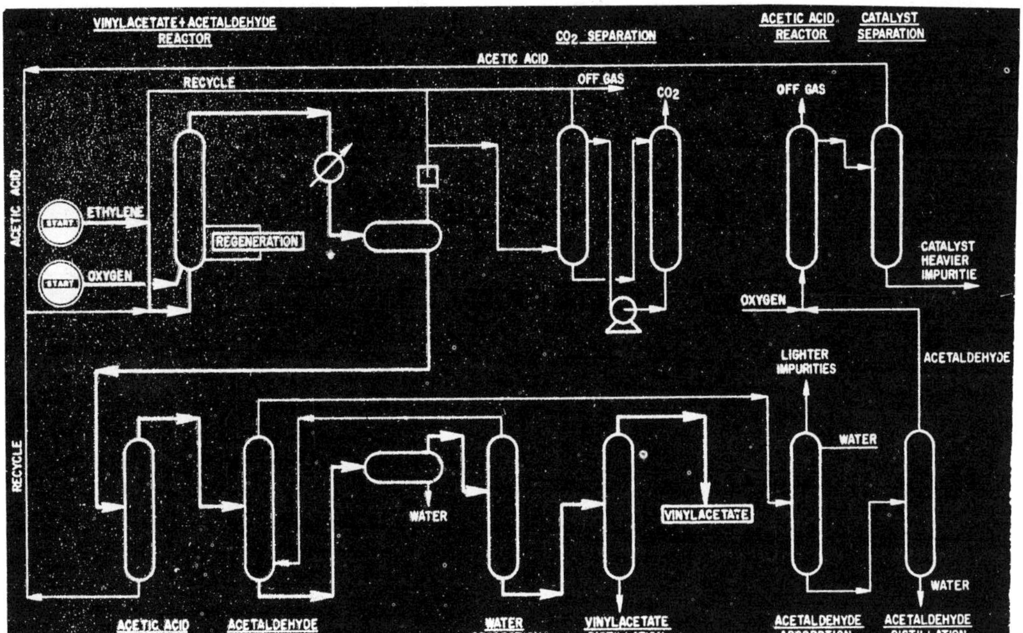

Fig. 25.14 Vinyl acetate by oxyacetylation process. [*Hydrocarbon Processing,* **50**, *No. 11, 219 (1971); copyright 1971 by Gulf Publishing Co.*]

flexible polyvinyl chloride. These plasticizers give more desirable properties than phthalates, adipates, or sebacates made from conventional alcohols, The C_{12} to C_{18} linear primary alcohols are used to produce highly biogradable surface active agents, which in the final forms are ethoxylates, alcohol sulfates, or sulfates of alcohol ethoxylates.

Production of linear primary alcohols approached 200 million pounds in 1969 and is expected to increase at a rapid rate. These compounds will replace many of the natural fatty alcohols now used for detergents. The production of α-olefin compounds as such is taking hold as uses develop for other than the alcohols. Production in 1969 was over 100 million pounds and projections for growth were very optimistic. Other low molecular weight oligomers appear to have considerable potential to become volume chemical intermediates.

Recently, a metallo-organic catalyst system was discovered which provides for the selective trimerization of ethylene to 3-methyl-2-pentene. Demethanation of this material at high temperature gives isoprene in good yields. Growth potential here is considerable in the synthetic rubber area.

Figure 25.15 is a flow diagram for the manufacture of Alfol® alpha-alcohols. In this process, aluminum triethyl is prepared by hydrogenation of aluminum powder in a solvent followed by ethylation with ethylene. The mixture must be kept scrupulously dry. Aluminum triethyl is then reacted with ethylene under pressure to form higher alkyls. A spectrum of alkyls from C_2 to C_{22} are obtained. The aluminum alkyls are then oxidized to aluminum alkoxides with very dry air. The alkoxides are hydrolyzed with sulfuric acid, after which the residual acid is neutralized with caustic and washed free of sodium sulfate with water. The alcohols are then dehydrated and fractionated into the desired cuts.

Figure 25.16 is a flow diagram of an alpha-olefin process. In this process, ethylene and a solvent containing an alkyl aluminum are fed to a reactor where the alkyl aluminum adds one or more ethylene molecules in sequence, forming linear alkyl groups averaging 10 to 20 carbon atoms. Eventually, ethylene displaces the alkyl groups to form α-olefins plus triethyl

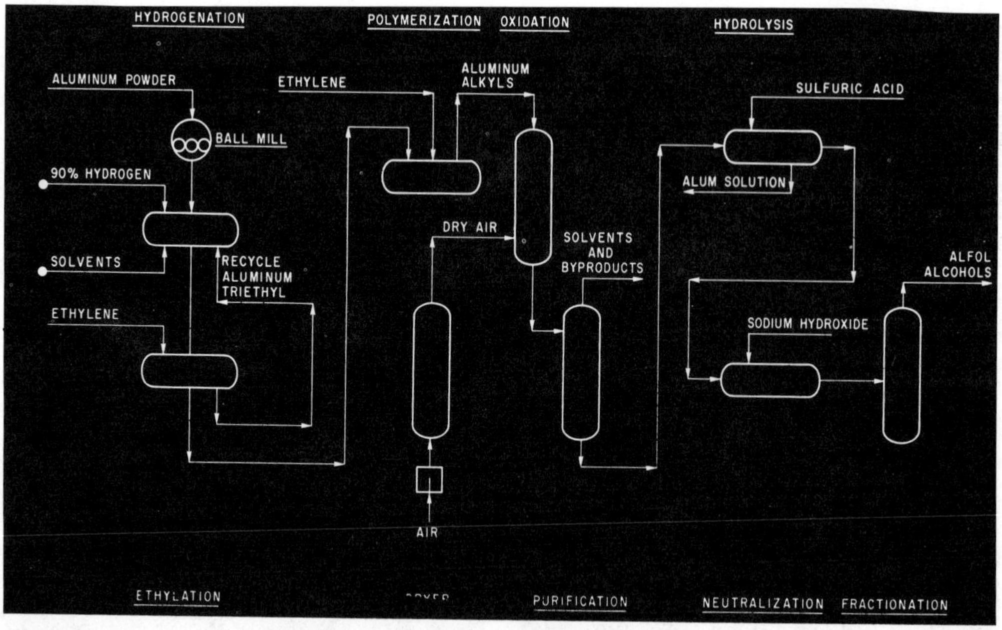

Fig. 25.15 Alpha alcohols by Alfol® process. [*Hydrocarbon Processing,* **50**, *No. 11, 126 (1971); copyright 1971 by Gulf Publishing Co.*]

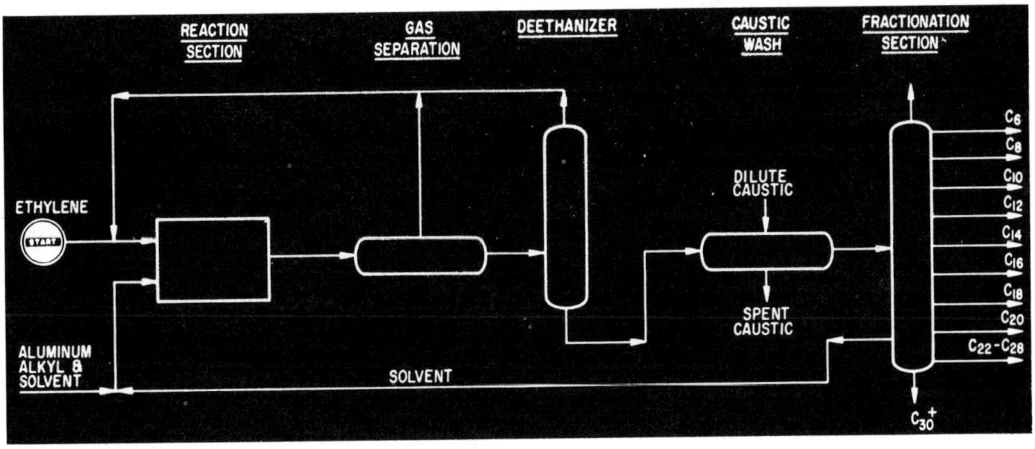

Fig. 25.16 Alpha-olefins process. [*Hydrocarbon Processing*, **50**, *No. 11, 127 (1971); copyright 1971 by Gulf Publishing Co.*]

aluminum that is re-used. The product is gas-liquid separated to recover ethylene for recycle, then washed and distilled into various fractions.

CHEMICALS DERIVED FROM PROPYLENE

The use of propylene as a chemical building block has been growing at a very rapid pace. Its current consumption in the chemical industry is about half as great as that of ethylene, but the total production of propylene is much larger because of nonchemical uses. In 1969, about 8.4 billion pounds was used to make chemicals, and about 10.8 billion pounds for the manufacture of gasoline alkylate and polymer gas. This chemical use is more than three times as great as that in 1959. The current sources of propylene are about 85 per cent from petroleum refineries and about 15 per cent as by-product from ethylene plants.

The most important chemicals derived from propylene are isopropyl alcohol, acrylonitrile, polypropylene, propylene oxide, and oxo chemicals. The approximate distribution of

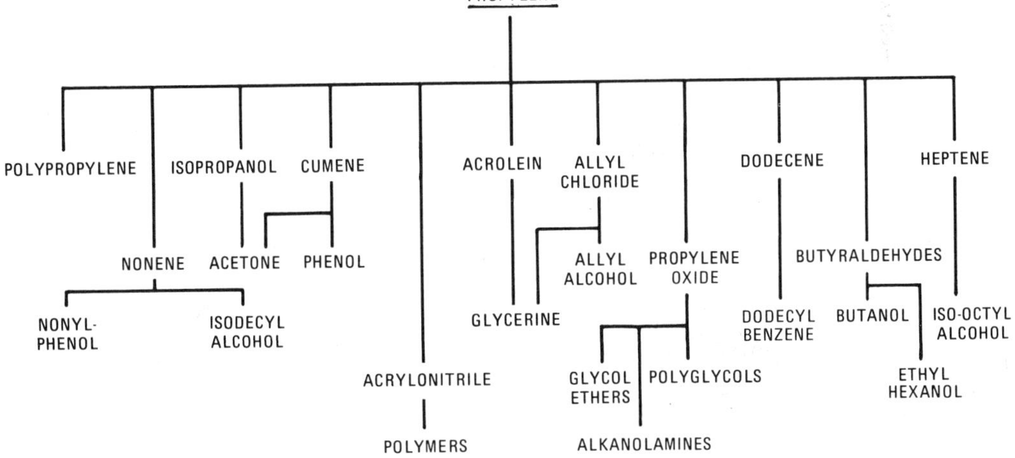

Fig. 25.17 Some chemicals derived from propylene.

propylene use is:

	%
Isopropyl alcohol	20
Acrylonitrile	17
Polypropylene	15
Propylene oxide	13
Dodecene	8
Cumene	7
Butyraldehydes	6
Others	14

It is expected that the chemical uses of propylene will grow at about 8–10 per cent per year, with the greatest growth rate being for polypropylene.

Isopropyl Alcohol

Isopropyl alcohol is claimed by some to be the first petrochemical. During the latter part of World War I the first isopropanol was manufactured by the Ellis process which was quite similar to the process used today. The flow sheet in Fig. 25.18 shows a typical isopropanol process scheme.

The dilute liquid-propylene feed stock is combined with recycled hydrocarbons and the mixture sulfated with H_2SO_4 at 300–400 psig. The sulfated mixture is then hydrolyzed to the alcohol, stripped from the acid, neutralized, degassed, and distilled to approach the binary azeotrope; then, it is finished in an azeotropic finishing column with an entrainer such as ethyl ether. The anhydrous product is taken from the bottom of the azeotropic still.

New plants have been installed in Europe utilizing a direct hydration process in which propylene and water are reacted in the presence of a catalyst such as phosphoric acid on bentonite. The elimination of sulfuric acid should decrease processing and maintenance costs. The drawback of this new route is that it requires a highly concentrated propylene feed rather than a dilute refinery stream.

During 1969, some 1900 million pounds of isopropanol were produced. The major use was in the manufacture of acetone, which accounted for about 55 per cent of the total. Other chemical uses were in the manufacture of acetates and xanthates.

Acetone. Acetone can be made from isopropanol by several routes, but catalytic dehydrogenation is the main one.

$$CH_3CHOHCH_3 \longrightarrow CH_3\overset{\overset{O}{\|}}{C}CH_3 + H_2$$

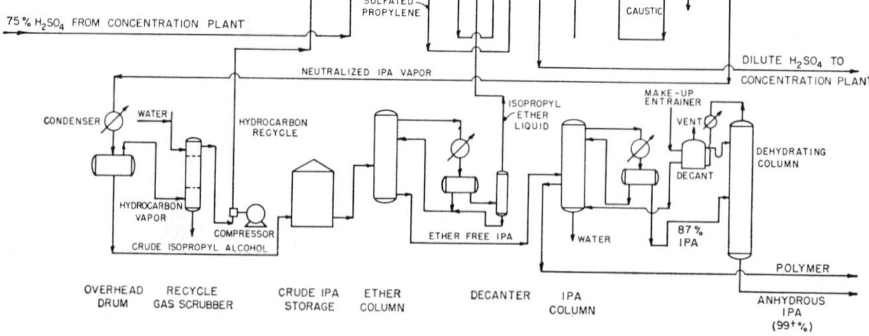

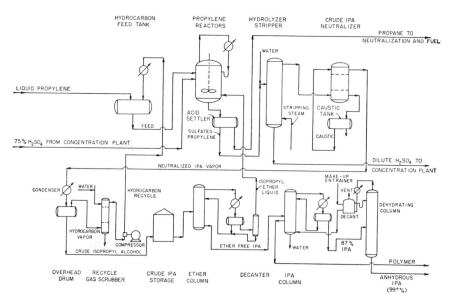

Fig. 25.18 Isopropyl alcohol manufacture. [*Pet. Ref.*, **38**, *No. 11, 264 (1959); copyright 1959 by Gulf Publishing Co.*]

Copper, brass, or supported zinc is used as the catalyst. Operation at a high temperature (400–500°C) and moderate pressure (45 psig) gives about a 90 per cent yield of acetone.

Less than 60 per cent of the current acetone production comes from isopropanol as opposed to more than 80 per cent in 1959. The decline in popularity of acetone from the isopropanol process can be attributed to the increasing number of phenol plants using the cumene route.

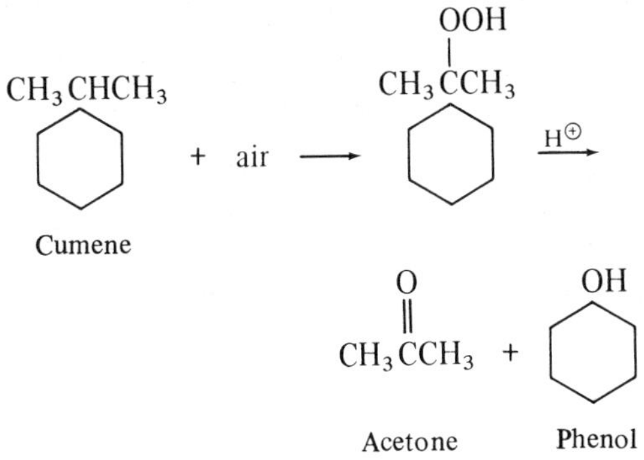

The oxidation of cumene leads to 0.6 pounds of acetone per pound of phenol produced. The importance of this source can be expected to grow as the demand for phenol increases.

The major chemical end-uses for acetone are in the production of methyl methacrylate and methyl isobutyl ketone (MIBK). Methyl methacrylate is made by the acetone-cyanohydrin process, and MIBK by condensing acetone to form mesityl oxide. Other uses are in making bisphenol A, methyl isobutyl carbinol pharmaceuticals, and as a solvent.

Isopropyl Alcohol Solvent. A significant amount of isopropanol is used as a solvent for essential and other oils, gums, shellac, rosins and synthetic resins. Isopropanol is an excellent solvent for these materials and thus finds extensive use for compounding or blending numerous incompatible substances. As a component of nitrocellulose lacquer solutions, isopropyl alcohol improves blush resistance and increases solvency in esters and ketones.

Polyproplene

Polypropylene is the fastest-growing end-use for propylene. According to U.S. Tariff Commission data, production exceeded 100 million pounds five years after its introduction and reached one billion pounds after only twelve years. It is produced by catalyzed polymerization of high-purity propylene in a slurry-type reactor to make a stereoregular polymer. The same process can also be used to produce high-density polyethylene.

The major uses are as an injection-molding resin and in fibers and filaments.

Acrylonitrile

One of the biggest "success stories" in the last decade in the field of synthetic chemicals resulted from the development of a process for making acrylonitrile directly from propylene, at the same time there was a huge growth in demand for the material. In 1960 almost all of the 260 million pounds of acrylonitrile produced came from acetylene. Ten years later, more than 1100 million pounds were produced and the industry had converted almost entirely to the propylene-ammonia process.

This process is shown in Fig. 25.19. Propylene, ammonia, and air are fed to a fluid-bed catalytic reactor operating at 800–900°F. The reactor effluent is scrubbed, and the organic materials are recovered by distillation. Hydrogen cyanide, water, and organic impurities are removed by fractionation. A high conversion to acrylonitrile is obtained with acetonitrile as a by-product.

The major growth in demand for acrylonitrile has been in the area of acrylic fibers. In 1969 more than 500 million pounds went into this use. Other areas are for ABS and SAN resins, and for nitrile rubber.

Propylene Oxide

Propylene oxide is manufactured primarily by the chlorohydrin process. This involves two steps:

1. Reaction of propylene with hypochlorous acid.

$$H_2C{=}\underset{\displaystyle CH_3}{\overset{|}{C}H}\;\text{(with } CH_3 \text{ on CH)} + HOCl \longrightarrow \underset{Cl}{CH_2}-\underset{OH}{CH}(CH_3)$$

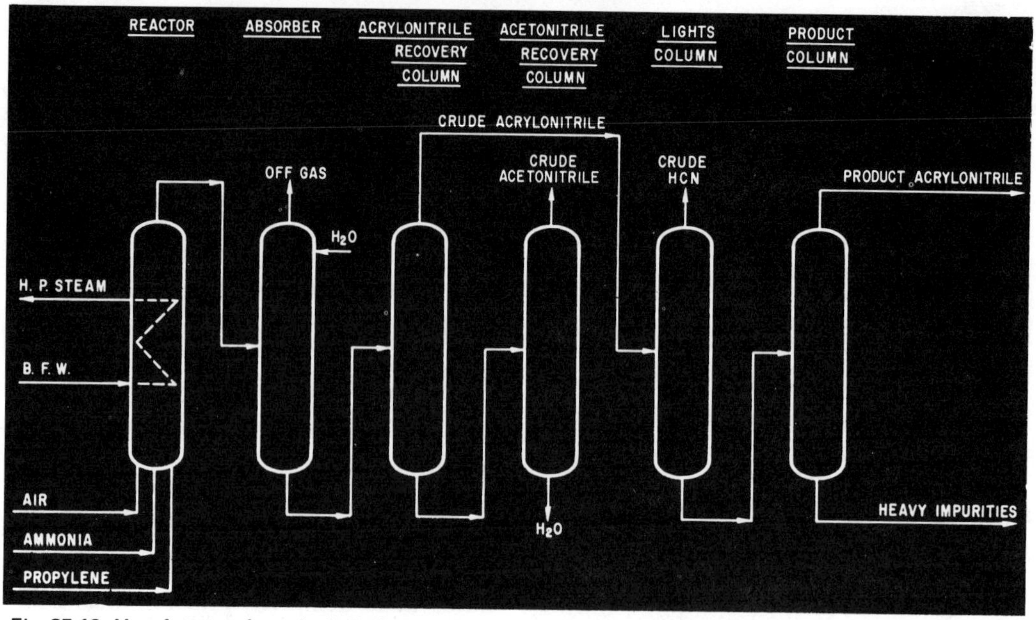

Fig. 25.19 Manufacture of acrylonitrile from propylene, ammonia, and air. [*Hydrocarbon Processing*, **50**, *No. 11, 122 (1971); copyright 1971 by Gulf Publishing Co.*]

2. Reaction of the propylene chlorohydrin with slaked lime.

$$2CH_2(Cl)-CH(CH_3)(OH) + Ca(OH)_2 \longrightarrow$$

$$2CH_2\!-\!CHCH_3\ (\text{epoxide, } \diagdown O \diagup) + CaCl_2 + 2H_2O$$

The process is basically similar to that which was used to make ethylene oxide. Depending on the reaction conditions, this process can generate appreciable amounts of propylene dichloride and propylene glycol as well as the oxide.

Before 1969, almost all propylene oxide was produced by the chlorohydrin process. In that year, however, the first plant was brought on stream using the direct oxidation route. In the direct oxidation route, isobutane is air-oxidized in the liquid phase to *t*-butyl hydroperoxide. After separation, the hydroperoxide is used to oxidize propylene to propylene oxide and is itself reduced to *t*-butyl alcohol. The alcohol can then be dehydrated to isobutylene. Depending on the starting material, several different coproducts are possible. For example, when ethylbenzene is used instead of isobutane, the coproduct is phenyl methyl carbinol which can be dehydrated to styrene. In late 1971, the start up of a second major facility was announced, and it is expected that any future new plants will use the direct oxidation process.

The major uses for propylene oxide are propylene glycol and polypropylene glycols. Other uses are in isopropanolamines, glycol ethers for hydraulic fluids, surfactants, and demulsifiers.

Glycols and Polyglycols. Propylene glycol manufacture is carried out with the same processes as those used for ethylene glycol, i.e., oxide hydrolysis, water removal, and product purification. The major uses are in resins and cellophane, as hydraulic fluids as a tobacco humectant, and in cosmetics. Production in 1968 was 33 million pounds.

Polypropylene glycols are prepared commercially by the base-catalyzed addition of propylene oxide to propylene glycol. Propylene oxide can also be added to such starting materials as glycerine, pentaerythritol, sucrose, and sorbitol, depending on the type of product desired.

Polypropylene glycols and the polyglycols made by cocondensing ethylene and propylene oxide are used as lubricants, hydraulic fluids, mold release agents, and as the polyether portion in polyurethane foam. Approximately 500 million pounds of propylene oxide is used to produce polypropylene glycols. Most of these polyols were used in the fast-growing polyurethane industry.

Dodecene, Nonene, Cumene

The manufacturing processes for these materials are very similar. The flow sheet in Fig. 25.20 specifically illustrates the manufacture of dodecene (propylene tetramer). When nonene is the desired product, additional fractionation is required, the extent of which is determined by product specifications.

In the reactor portion of this process, the olefin stock is mixed with benzene (for cumene) or recycle lights (for tetramer). The resulting charge is pumped to the reaction chamber. The catalyst, solid phosphoric acid, is maintained in separate beds in the reactor. Suitable propane quench is provided between beds for temperature control purposes since the reaction is exothermic.

Dodecene is an intermediate for surfactants, going mainly through two routes. One, the largest user, produces dodecylbenzene sulfonate for anionic detergents. The other goes through the oxo process to tridecyl alcohol, which is then converted into a nonionic detergent by the addition of alkylene oxides.

Nonene has two major outlets. The larger one is oxo production of decyl alcohol, which is used in the manufacture of esters, etc. for plasticizers. The other significant nonene use is in the manufacture of nonylphenol, an intermediate for the important series of ethoxylated nonylphenol nonionic surfactants.

Cumene is used mainly as an intermediate for phenol-acetone manufacture. A relatively small amount of cumene is used to manufacture α-methylstyrene.

Oxochemicals

A considerable amount of propylene finds its way either directly or indirectly into the manufacture of oxo chemicals. The estimated amount of oxo alcohols derived from butyraldehyde alone was some 500 million pounds. Description of the oxo process is shown under *Methane Chemicals.*

Butyl Alcohols and Aldehydes. Hydroformylation of propylene produces a mixture of *n*-butyraldehyde and isobutyraldehyde. The aldehyde ratio is approximately 2 to 1 in favor of *n*-butyraldehyde, but this ratio may be varied somewhat. The aldehydes may be used separately or the mixed aldehydes may be hydrogenated to the corresponding alcohols (*n*-butanol and isobutanol), which are then sep-

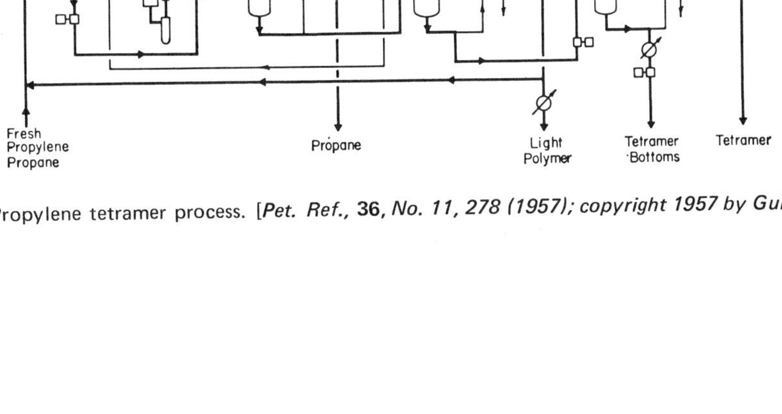

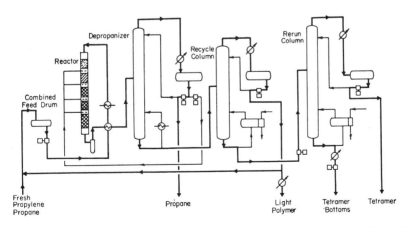

Fig. 25.20 Propylene tetramer process. [*Pet. Ref.*, **36**, *No. 11, 278 (1957); copyright 1957 by Gulf Publishing Co.*]

arated. The demand for *n*-butyraldehyde is much greater than that for the isobutyraldehyde.

The three major products obtained from the butyraldehydes are 2-ethylhexanol, *n*-butanol, and isobutanol. The ethylhexanol is used mainly to make phthalate plasticizers for polyvinyl chloride. The butyl alcohols are well-established solvents and intermediates for plasticizers and resins.

Isobutyraldehyde is used to produce higher molecular weight mono- and dihydroxy aldol condensation products such as 2-ethylisohexanol, neopentyl glycol, nonyl alcohol, and 2,2,4-trimethyl-1,3-pentanediol. Isobutyronitrile is made by reacting isobutyraldehyde with ammonia, then dehydrogenating to the nitrile.

An alternate route to butyraldehyde which is beginning to show promise involves the condensation of acetaldehyde.

Higher Oxo Alcohols. Oxo alcohols of major importance which are derived indirectly from propylene are decyl alcohol (from nonene) and tridecyl alcohol (from dodecene). Uses of these alcohols are described under *Dodecene*, *Nonene*, *Cumene*.

One other significant higher oxo alcohol is isooctyl alcohol, derived from a dimer of propylene and butylene. The isooctyl alcohol mixture derived from this oxo reaction contains 26 per cent 4,5-dimethyl-1-hexanol; 13 per cent 3,5-dimethyl-1-hexanol; 18 per cent 3,4-dimethyl-1-hexanol; 17 per cent 3-, and 5-methyl-1-heptanol; and 9 per cent other isomers. Esters of this mixture are used mainly as plasticizers for vinyl chloride resins.

The future of the oxo alcohols depends upon the expansion of their current uses in plasticizers and detergents, and the development of other new uses.

Glycerine

Synthetic glycerine (glycerol) is manufactured from propylene by two main routes as shown by the chemical reactions below:

1. Epichlorohydrin route.

$$CH_3CH=CH_2 \xrightarrow[\substack{15 \text{ psig} \\ 650-950°F}]{Cl_2} H_2C=CHCH_2Cl \xrightarrow[85-100°F]{HOCl} \underset{\underset{OH}{|}}{CH_2}-\underset{\underset{Cl}{|}}{CH}-\underset{\underset{Cl}{|}}{CH_2}$$

$$\underset{\underset{OH}{|}}{CH_2}-\underset{\underset{Cl}{|}}{CH}-\underset{\underset{Cl}{|}}{CH_2} \xrightarrow[<140°F]{Ca(OH)_2} \underset{\diagdown\diagup}{CH_2}\underset{O}{-}\underset{\underset{Cl}{|}}{CH}-CH_2 \xrightarrow{10\% \text{ NaOH}} \underset{\underset{OH}{|}}{CH_2}-\underset{\underset{OH}{|}}{CH}-\underset{\underset{OH}{|}}{CH_2}$$

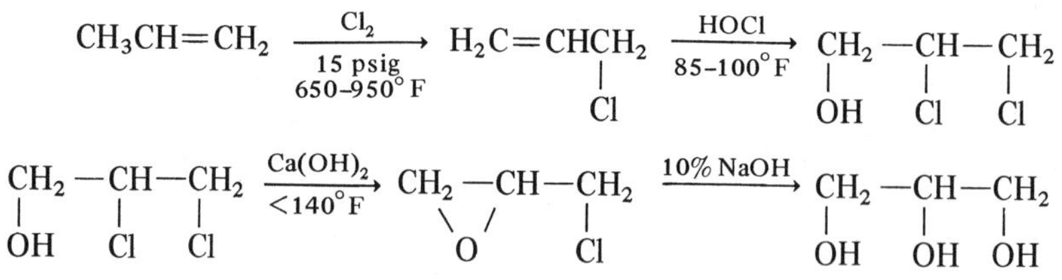

2. Acrolein and hydrogen peroxide.

$$CH_3CH=CH_2 \xrightarrow[\substack{H_2O \\ NaOH}]{H_2SO_4} \underset{\underset{OH}{|}}{CH_3CHCH_3} \xrightarrow[O_2]{H_2O} H_2O_2(\text{soln}) + CH_3\overset{O}{\overset{\|}{C}}CH_3$$

$$CH_3CH=CH_2 \xrightarrow[\substack{\text{Steam} \\ \text{Cat.}}]{O_2} CH_2=CH\overset{O}{\overset{\|}{C}}H$$

$$CH_2=CH\overset{O}{\overset{\|}{C}}H + \underset{\underset{OH}{|}}{CH_3CHCH_3} \xrightarrow[400°C]{\text{Cat.}} CH_2=CHCH_2OH + CH_3\overset{O}{\overset{\|}{C}}CH_3$$

$$CH_2=CHCH_2OH + H_2O_2(\text{soln}) \xrightarrow[60-70°C]{\text{Cat.}} \underset{\underset{OH}{|}}{CH_2}-\underset{\underset{OH}{|}}{CH}-\underset{\underset{OH}{|}}{CH_2}$$

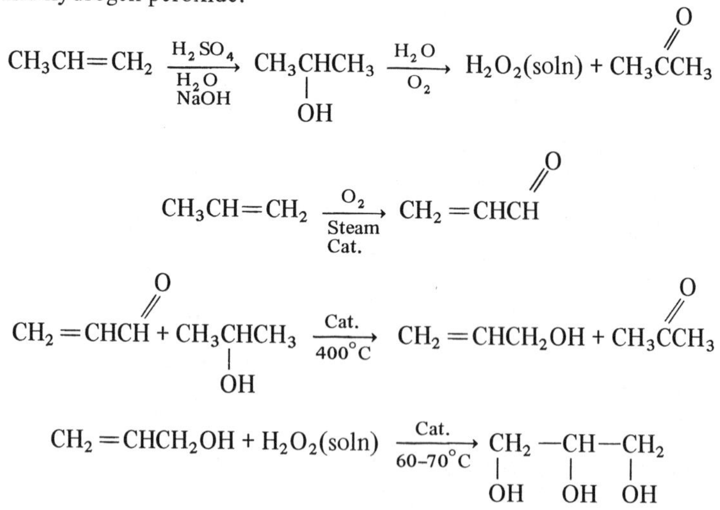

Approximately three times as much is made by the first route as by the second.

Glycerine by the Epichlorohydrin Process. In the epichlorohydrin process, synthetic glycerine is produced in three successive operations; the end products of these are allyl chloride, epichlorohydrin, and finished glycerine, respectively. A flow sheet for this process is shown in Fig. 25.21. A portion of the allyl chloride is used to manufacture allyl alcohol, and a portion of the epichlorohydrin is used in the manufacture of epoxy resins.

The key reaction in this process is the hot chlorination of propylene which fairly selectively gives substitution rather than the addition reactions. In this chlorination step fresh propylene is first mixed with recycle propylene. This mixture is dried over a desiccant, heated to 650–700°F, and then mixed rapidly with chlorine (C_3H_6 to Cl_2 ratio is 4:1) and fed to a simple steel-tube adiabatic reactor. The effluent gases (950°F) are cooled quickly to 120°F and fractionated. Yield of allyl chloride is 80–85 per cent.

Hypochlorous acid is then reacted with the allyl chloride at 85 to 100°F to form a mixture of dichlorohydrins. The reactor effluent is separated, the aqueous phase is returned to make-up hypochlorous acid, while the nonaqueous phase containing the dichlorohydrins is reacted with a lime slurry to form epichlorohydrin. Epichlorohydrin is steam distilled out and a given finishing distillation

Glycerine is formed by the hydrolysis of epichlorohydrin with 10 per cent caustic. Crude glycerine is separated from this reaction mass by multiple-effect evaporation to remove salt and most of the water. A final vacuum distillation yields a 99$^+$ per cent product.

Glycerine by Acrolein and Hydrogen Peroxide Process. This process is used to manufacture glycerine, but large amounts of acetone are obtained as a coproduct. Basic raw materials consumed are propylene and oxygen. Glycerine is synthesized in this process by hydroxylation of allyl alcohol with hydrogen peroxide.

An estimated 350 million pounds of glycer-

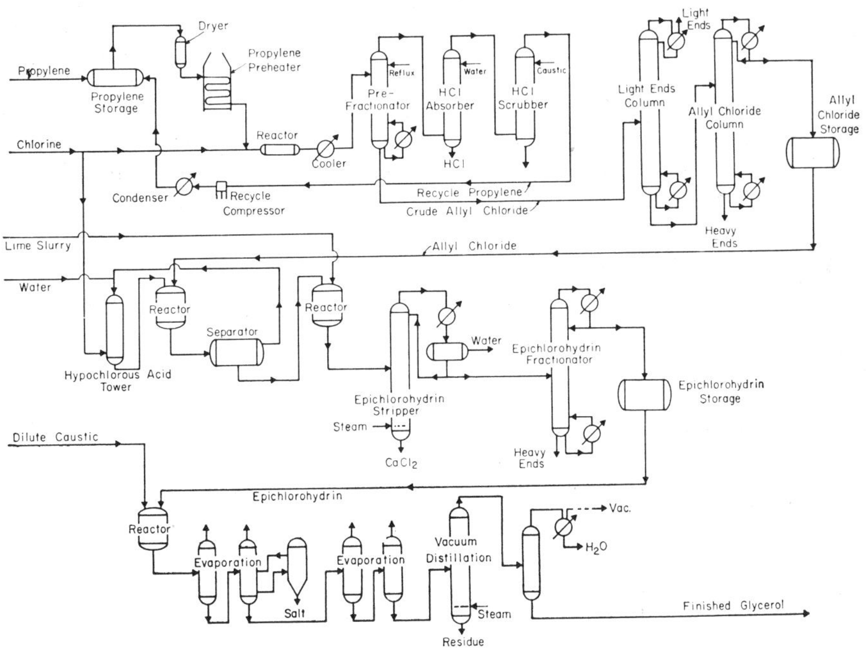

Fig. 25.21 Manufacture of glycerol. [*Pet. Ref.,* **38,** *No. 11, 252 (1959), copyright 1959 by Gulf Publishing Co.*]

ine were produced in 1969. Of this amount about 54 per cent was made synthetically, and the rest was obtained from natural sources, primarily as a by-product in the manufacture of soaps and fatty acids.

The major glycerine end-uses are: alkyd resins, 23 per cent; drugs and cosmetics, 19 per cent; tobacco, 15 per cent; cellophane, 14 per cent; and foods and beverages, 12 per cent. Other uses include explosives and polyether polyols.

BUTANES, BUTYLENES, LPG, AND HIGHER ALIPHATIC HYDROCARBONS

Saturated four-carbon hydrocarbons occur in natural petroleum products. They are found as heavy vapors in wet natural gas and in crude oil. The C_4s are also produced from other hydrocarbons during the various petroleum refining processes. The butylenes and butadiene–unsaturated C_4s–do not occur in nature, but are derived from saturated C_4s or from other hydrocarbons either as prime products or as by-products. The main sources of the C_4s are as shown in Table 25.4.

Routes 1 through 4 in Table 25.4 are primary sources of C_4 hydrocarbons. Processes 5 through 8 produce C_4 hydrocarbons from other C_4 hydrocarbons. A great majority of the C_4 hydrocarbons come from the first three sources. Over 20 per cent of the butadiene is obtained as a by-product of ethylene manufacture.

The complex interrelation between the C_4 hydrocarbons showing how they are produced and used is shown most descriptively in Fig. 25.22.

Chemical uses of C_4 hydrocarbons show butanes at about a 3-billion-pound-per-year level with *n*-butenes a close second at about 2.5 billion pounds and isobutylene at about one-half billion pounds. The chemical uses of the butanes, butenes, and butadiene, and some of the processes used in making the derivatives are described in succeeding pages.

To place the volume of C_4's used in chemical manufacture in perspective with refinery uses for gasoline, one finds that about 5 per cent of the butanes and under 15 per cent of the butylenes are used as chemical raw materials. Conversion of *n*-butane to gasoline in 1967 is estimated at about 6 billion gallons (30 billion pounds), while isobutane to gasoline was about 5 billion gallons (25 billion pounds). In the same year it is estimated that about 3.5 billion gallons (18 billion pounds) of butenes were used in gasoline.

As shown in the source chart Fig. 25.23, the trends that affect availability of C_4 hydrocarbons for chemical and energy end-uses are determined by the natural gas processors, petroleum refiners, and, to a growing extent, ethylene manufacture. Changes in technology and in the availability of optimum feed stocks have far-reaching effects on the entire product mix. For example, as the availability of LPG and ethane for ethylene manufacture wane,

TABLE 25.4 Main Sources of C_4 Hydrocarbons

	n-Butane	*Isobutane*	*Isobutylene*	*n-Butenes*	*Butadiene*
1. Wet natural gas	*	*			
2. Crude oil	*	*			
3. Petroleum refining					
Hydrocracking and reforming	*	*			
Thermal and catalytic cracking	*	*	*	*	
4. By-product of ethylene manufacture			*	*	*
5. Isomerization of *n*-butane		*			
6. Isobutane cracking			*		
7. Dehydrogenation of *n*-butane				*	*
8. Dehydrogenation of *n*-butenes					*

[a]"Chemical Economics Handbook," Aug. 1969, p. 620.5010 A, Stanford Research Institute, Menlo Park, Cal., 1969.

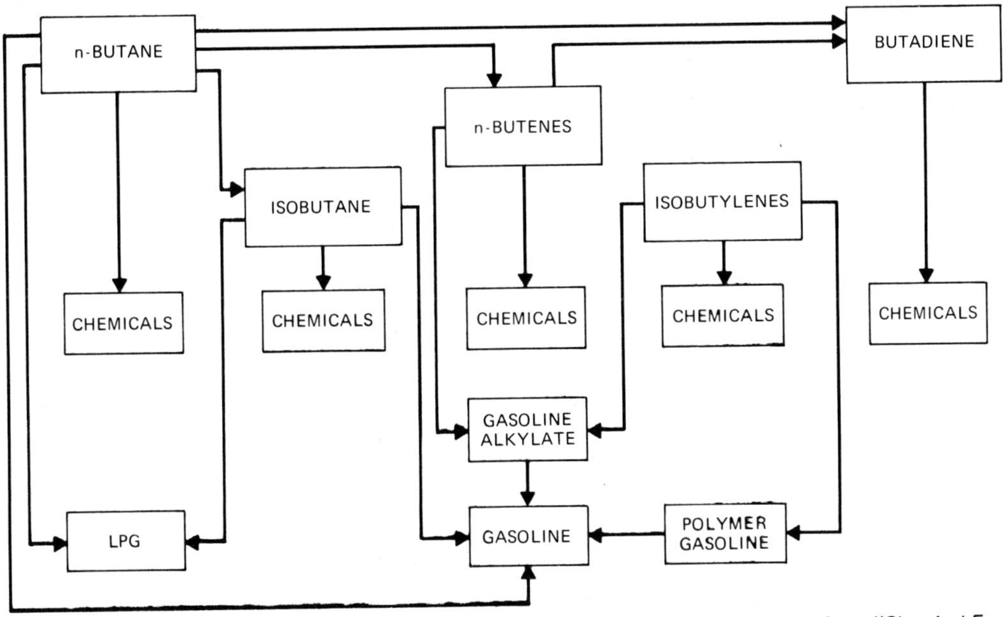

Fig. 25.22 Interrelation between production and use of C_4 hydrocarbons. [*Reproduced from "Chemical Economics Handbook," copyright 1969 by Stanford Research Institute, Menlo Park, Ca.*]

n-butane and higher cuts of crudes will be used and the proportion of by-product butadiene produced will increase since this butadiene will move into the market place first.

LPG and *n*-Butane

Oxidation of Propane-Butane (LPG) Mixtures. Liquified Petroleum Gas (LPG) is the name given to a mixture of hydrocarbons consisting primarily of ethane, propane, butane, C_5 (mixed 5-carbon materials) and naphtha. About 45 per cent of the 12.3 billion gallons produced in 1968 was sold to the chemical and rubber industry, an increase of more than 80 per cent over 1960. A majority of the LPG utilized by the chemical industry is consumed in the manufacture of ethylene, propylene, butadiene, and isoprene.

One of the smaller but very interesting chemical uses of LPG is the oxidation of butane and butane-propane mixtures.

In 1945, the first major plant for the oxidation of propane and butane was brought on stream. This is a noncatalytic, vapor-phase oxidation, which yields as its principal products formaldehyde, acetaldehyde, and methanol. In addition to these three chemicals, smaller amounts of propionaldehyde, acrolein, ethanol, glycols, and various alcohols and ketones are produced, depending on the nature of the hydrocarbon feed. The number of these products increases with the complexity of the hydrocarbon; propane gives a total of 16 oxygenated compounds, while the oxidation of *n*-hexane could produce about 60. A few of the possible propane-oxidation reactions are as follows.

$$C_3H_8 + O$$

$$\longrightarrow \text{acetaldehyde} + CO_2 + H_2O$$

$$\longrightarrow \text{methanol} + \text{formaldehyde} + CO + H_2O$$

$$\longrightarrow \text{formaldehyde} + \text{acetaldehyde} + CO + H_2O$$

$$\longrightarrow \text{propionaldehyde} + H_2O$$

$$\longrightarrow \text{ethylene} + \text{methanol} + CO + H_2O$$

The flow sheet in Figure 25.24 shows the process for the oxidation of a propane-butane mixture. This could be divided into four sections; primary oxidation, formaldehyde concentration, product separation and purification, and secondary oxidation.

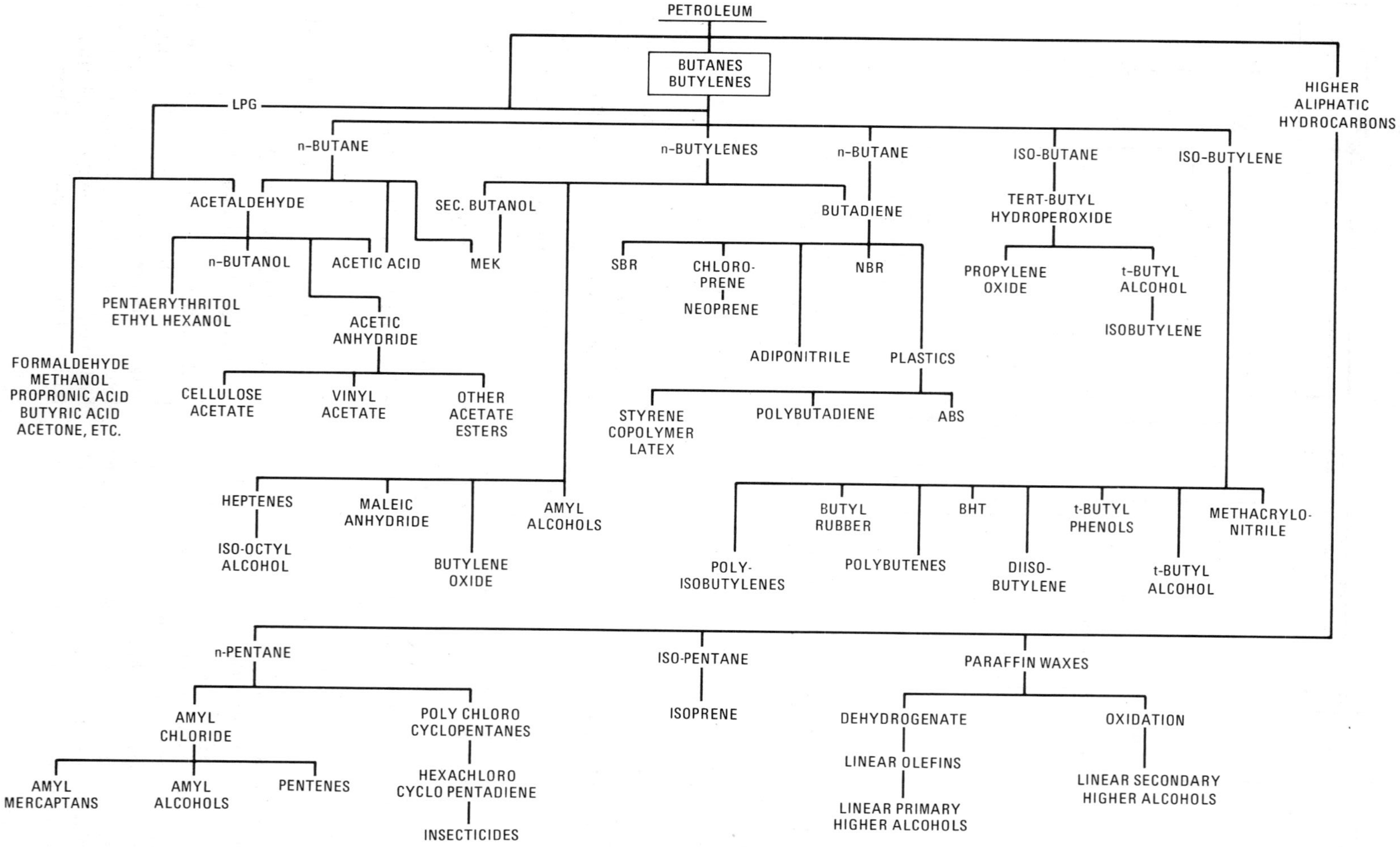

Fig. 25.23 Chemicals from butanes, butylenes, LPG, and higher aliphatic hydrocarbons.

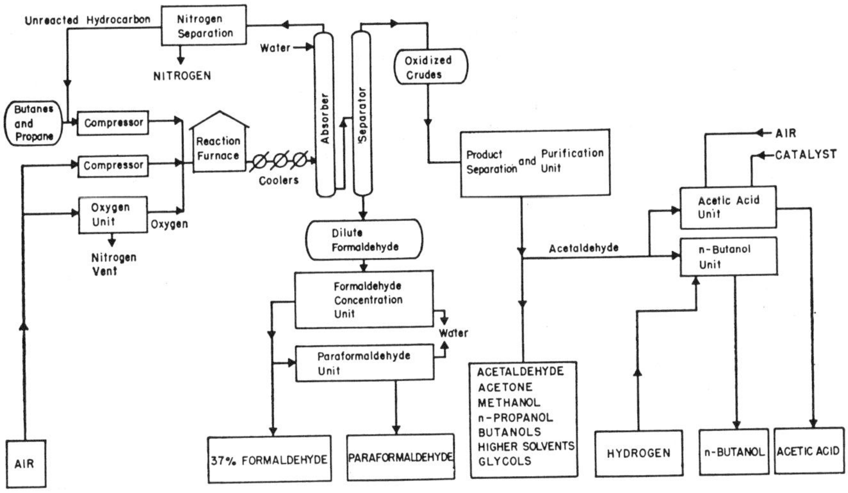

Fig. 25.24 Oxidation of propane and butane. [*Oil and Gas J., 115 (Sept. 2, 1957).*]

In the primary oxidation section the hydrocarbon mixture and compressed air (about a 1:2 ratio) are mixed with a recycle stream and fed into the reactor at about 100 psig and 700°F (recently, the operation was switched from air to 95 per cent oxygen). The reactor is a long empty steel tube in which the temperature rises to about 850°F due to the exothermic nature of the reaction. The reaction gases are cooled and passed through a water-scrubbing system. The unconverted material is recycled, while the water solution passes to a separator where dilute formaldehyde is drawn off.

The dilute formaldehyde solution contains about 20-25 per cent formaldehyde and 10-20 per cent volatile material. This stream is steam stripped and purified to produce a 37 per cent solution (formalin). Some of the formalin is further processed, essentially by dehydration, to yield paraformaldehyde and trioxane.

In the first step of the separation and purification section, acetaldehyde is removed from the crude mixture by fractionation and sent to storage, either for later sale or further upgrading. Next, the remaining water-crude mixture is distilled to separate the acetone and methanol. The remaining material is hydrogenated and then fractionated to yield propanol, isobutanol, glycols, and various special solvents. Some of the acetaldehyde undergoes aldoling to produce *n*-butanol and other materials.

A portion of the acetaldehyde is fed to the secondary oxidation section. Here it is readily converted by catalytic oxidation to acetic acid.

Oxidation of Butane. Of more recent origin is the catalytic oxidation of butane, which incidentally emphasizes the trend toward increased product selectivity in oxidation processes. This employs a liquid-phase, high-pressure (850 psi) oxidation using acetic acid as a diluent, and a metal acetate catalyst. Figure 25.25 shows the flow sheet for this process. From the reactor, the product mixture is passed through coolers, and then through separators where the dissolved gases are released. The major components of the oxidized crude are acetic acid, methyl ethyl ketone, ethyl alcohol, methyl alcohol, propionic and butyric acids, and acetone.

It should be noted that unlike the propane-butane oxidation, this process produces no formaldehyde. The separation and purification procedure is similar to that described previously. Part of the acetic acid, which is the major product, is converted in a separate unit at high efficiencies to acetic anhydride. Also connected with this plant are units for

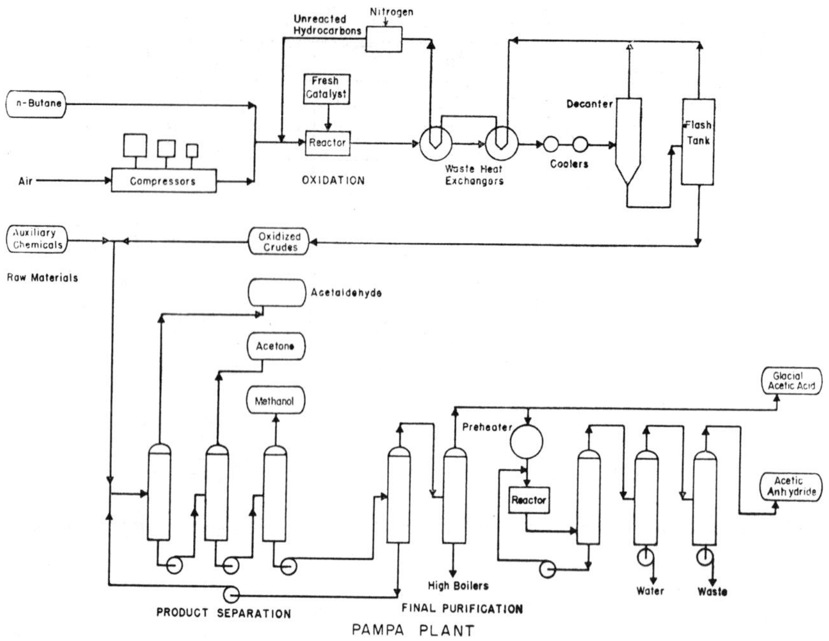

Fig. 25.25 Oxidation of butane. [*Pet. Ref.*, **38**, *No. 11, 234 (1959); copyright 1959 by Gulf Publishing Co.*]

producing vinyl acetate, methyl acrylate, and ethyl acrylate.

The two large butane-oxidation plants in the U.S. together produced approximately one billion pounds of acetic acid and about 125 million pounds of methyl ethyl ketone in 1967.

Reaction conditions in this process can be changed to produce more methyl ethyl ketone at the expense of some acetic acid.

Oxidation Products. Most of the chemicals produced by the oxidation of propane and butanes have been described in connection with other methods of preparation. Consequently, methanol, acetone, formaldehyde, propanol, acetaldehyde, and others will not be discussed here. Of interest, there remain methyl ethyl ketone, normal and secondary-butanols, and related materials.

Acetic anhydride can be produced by the catalytic oxidation of acetaldehyde. It can also be made from acetic acid (which is also made from acetaldehyde) through an intermediate called ketene.

$$CH_3COOH \longrightarrow CH_2{=}C{=}O + H_2O$$

$$CH_3COOH + CH_2{=}C{=}O \longrightarrow (CH_3CO)_2O$$

This reaction takes place at high temperatures, and uses a triethyl phosphate catalyst. It is also possible to produce ketene from acetone at high temperatures.

$$CH_3COCH_3 \longrightarrow CH_2{=}C{=}O$$

It is expected that the future of acetic anhydride will continue to be closely allied with that of its principal user, cellulose acetate.

In addition to being used to make acetic acid and anhydride, some of the acetaldehyde is up-graded to *n*-butanol by the aldol process. The steps in the reaction are: (1) the acetaldehyde is condensed to aldol; (2) the aldol is dehydrated to crotonaldehyde; and (3) the crotonaldehyde is hydrogenated (180°C, nickel-chrome catalyst) to yield *n*-butanol.

$$2CH_3CHO \longrightarrow CH_3CH(OH)CH_2CHO \longrightarrow$$

$$CH_3CH{=}CHCHO \longrightarrow CH_3CH_2CH_2CH_2OH$$

This route to *n*-butanol is currently losing out to the oxo process which uses propylene and carbon monoxide.

n-Butane Dehydrogenation. The Phillips and Houdry processes are used for dehydro-

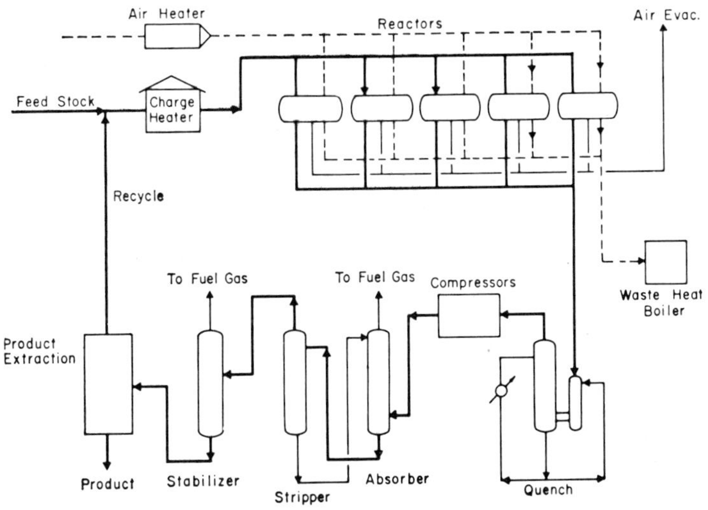

Fig. 25.26 Dehydrogenation. [*Pet. Ref.*, **38**, *No. 11, 233 (1959); copyright 1959 by Gulf Publishing Co.*]

genation of *n*-butane. In the Phillips process, high purity (98^+%) *n*-butane is first dehydrogenated to *n*-butenes, and then further dehydrogenated to butadiene. The Houdry or one-step process is fed 95^+% *n*-butane and conditions can be varied so that either butadiene or *n*-butenes are produced. Figure 25.26 shows a flow sheet of such a process. This is an adiabatic fixed-bed catalytic process. The catalyst is compressed cylindrical pellets consisting of activated alumina impregnated with chromic oxide. The reactor will normally operate in the range of 1000–1200°F, at a pressure of 1/6 atmosphere or higher. The feed is preheated to reaction temperature before contact with the catalyst in the reactors. The hot effluent from the reactors is cooled in a quench tower by direct contact with circulating oil, then goes to compression, cooling, and conventional absorption and stabilization. The stabilized product stream is then extracted for recovery of high-purity product. Monoolefins, when not taken as product, are returned with the parraffin stream for recycle.

This process is suitable not only for dehydration of *n*-butane to butadiene or *n*-butenes, but it can also be used for the production of propylene from propane; C_4 monoolefins from *n*- or isobutane; butadiene from *n*-butenes; C_5 monoolefins from *n*- or isopentane; isoprene from isopentane, isopentene, or mixtures thereof, or piperylene from *n*-pentane, *n*-pentenes, or mixtures thereof.

Typical ultimate yields of butadiene from an *n*-butane feed stream are shown in the table below.

Component Per Cent by Weight of Feed	*Feed*	*Net Products (Per Cent of Fresh Feed)*		
		Fuel Gas	*Butadiene*	*Un-recovered*
Dry Gas	–	16.3	–	–
Isobutane	1.0	–	–	–
Isobutylene	–	0.5	–	–
n-Butylenes	–	2.5	0.3	–
n-Butane	99.0	1.8	–	–
Butadiene	–	0.6	64.4	–
C_5^+, coke, C and H_2 to CO, CO_2 and H_2O	–	1.6	–	12.0
TOTAL	100.0	23.3	64.7	12.0

In 1967, one billion pounds of butadiene was produced by the dehydrogenation of *n*-butane.

Isobutane

It is believed that a new process for oxidizing propylene to propylene oxide uses isobutane-derived *t*-butyl hydroperoxide as the oxidizing agent. The products of this reaction are propylene oxide and *t*-butyl alcohol which is readily dehydrated thermally to isobutylene. This plant has recently been expanded to 500 million pounds of propylene oxide and has the capability of producing from 700 million to a billion pounds of by-product isobutylene annually.

n-Butylenes

Over 90 per cent of C_4 olefins come from refinery streams. This represents only about 10–15 per cent of the total available in these streams. The remainder comes from dehydrogenation of butane and as by-products of ethylene manufacture.

The basic problems of obtaining C_4 olefins is that of separation. Isobutylene is removed by absorption in 65 per cent sulfuric acid. With isobutylene removed, isobutane and 1-butene are separated from *n*-butane and the 2-butenes by fractionation. The olefins can then be separated from the paraffin hydrocarbons by extractive distillation (with furfural, acetone, etc.). Available today are commercial quantities of 1-butene and 2-butene, 95 per cent pure or higher. The 2-butene is a mixture of the *cis* and *trans* isomers.

The major derivatives of *n*-butenes are butadiene, *sec*-butyl alcohol, heptenes, butylene oxide, and crack-resistant high-density polyethylene.

Butadiene. Butadiene is produced by the dehydrogenation of *n*-butenes, or it may be produced as a coproduct with butenes in the dehydrogenation of butane. A flow sheet of the latter processing scheme is shown in Fig. 25.26. Both are fixed-bed catalytic dehydrogenation processes. Slightly higher yields of butadiene may be obtained from butene, but over-all economics, utilization of the resulting product mixture, and readily available feed streams determine the route.

Butadiene production in 1970 was approximately 3.1 billion pounds. Some 60 per cent of butadiene goes into styrene-butadiene rubber. Other uses are nitrile rubber, 3 per cent; adiponitrile, 9 per cent; polybutadiene, 17 per cent; high-impact polystyrene and paint latices, 7 per cent; export and miscellaneous uses, 4 per cent.

A rapidly growing new outlet for butadiene is the manufacture of chloroprene to be used in neoprene rubber. Butadiene is rapidly replacing acetylene in this use.

Cyclooctadiene and cyclododecatriene are very new butadiene derivatives. The former is expected to find uses in flame retardants, polymers, rubbers, perfumes, and metal chelates. The latter may find use as the intermediate for nylon 12 fiber and plastics.

For more details on the polymerization, manufacture, and uses of butadiene in rubber and plastics see Chapters 9 and 10.

sec-Butanol and Methyl Ethyl Ketone. The largest end-product chemical use for *n*-butenes is in the manufacture of *sec*-butanol. A mixed feed containing butane and *n*-butenes is contacted with 80 per cent sulfuric acid. Dilution and steam-stripping then produce the alcohol. The only important use of *sec*-butanol is the production of methyl ethyl ketone (MEK).

MEK is obtained by catalytic dehydrogenation of *sec*-butanol. A yield of approximately 75 per cent is obtained in this step. A flow sheet, Fig. 25.27, shows an integrated process for producing *sec*-butanol and MEK.

Approximately 85 per cent of MEK is manufactured by the *sec*-butanol route. MEK is also obtained as a by-product from the butane-oxidation process for acetic acid.

MEK is used almost entirely as a solvent in such applications as vinyl-resin lacquers, nitrocellulose, lacquers, lube oil dewaxing, paint removers, rubber cement, and adhesives.

U.S. consumption of MEK in 1969 was reported as 484 million pounds.

Heptenes and Isooctyl Alcohol. The heptenes comprise a C_7 olefin cut fractionated from polymer gasoline that is produced by polymerizing C_3-C_4 refinery gases. The C_4 used is generally *n*-butene. Heptenes are used mainly to feed oxo units for the manufacture

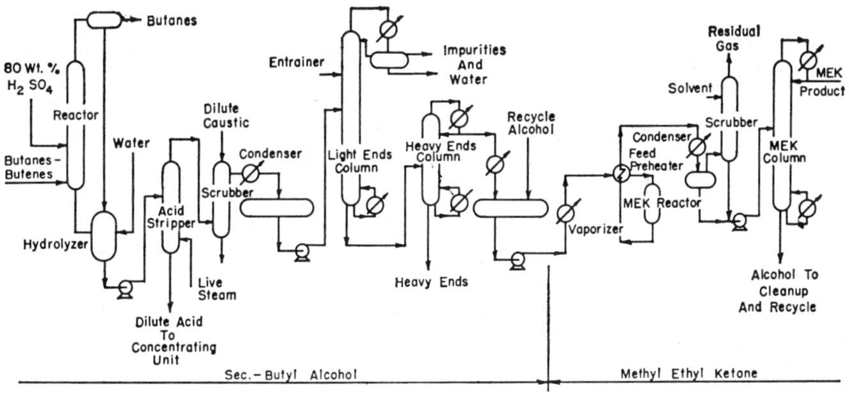

Fig. 25.27 Methyl ethyl ketone or acetone. [*Pet. Ref.,* **38**, *No. 11, 272 (1959); copyright 1959 by Gulf Publishing Co.*]

of isooctyl alcohol. Small quantities of heptenes are used to make isohexadecyl alcohol, heptylphenol, and heptylbenzene. In 1968, production of isooctyl alcohol was reported to be 132 million pounds.

Butylene Oxides. Approximately 20 million pounds of butylene oxides were produced in 1967. The feed used is either 1-butene to give 1,2-butylene oxide or a mixture of 1-butene and 2-butenes giving a mixture of 1,2- and 2,3-butylene oxide. Manufacture today is carried out by the chlorohydrin process. Butylene oxide is used as a corrosion inhibitor in chlorinated solvents such as methylchloroform and trichloroethylene.

Amyl Alcohols. Primary amyl alcohols are produced by the oxo reaction of *n*-butenes. The resulting alcohol mixture is about 60 per cent *n*-amyl alcohol, 35 per cent 2-methyl-1-butanol and 5 per cent 3-methyl-1-butanol. This alcohol may be fractionated or used as the mix for solvent uses, for acetate esters and esters of dithiophosphates which are used as lube-oil and hydraulic-fluid additives.

High-Density Polyethylene Copolymers. High-density polyethylene made with 3-5 per cent high-purity 1-butene ($98^{+}\%$) gives a plastic with improved crack-resistance under stress. This property has led to extensive use in blow-molding applications where this resistance is important. In 1967 some 800 million pounds of stress-resistant copolymer was produced.

Maleic Anhydride. There are no plants in the U.S. producing maleic anhydride from butenes; but plants are operating in Germany and Japan, and it has been noted that this technology has been offered for licensing in the U.S. A flow sheet for the German process is shown in Fig. 25.28. This process is based on catalytic air oxidation. The catalyst in a fixed-bed reactor is based on vanadium pentoxide, and the reactor is cooled with a liquid-salt cooling system. The gases from the reactor are cooled and scrubbed in aqueous maleic acid, and the maleic acid solution is evaporated to dryness, resulting in formation of the anhydride. Final purification is accomplished in fractionating columns.

Isobutylene

Isobutylene is more reactive than the *n*-butenes, but many of the compounds formed are quite readily reversible under less than extreme conditions.

Over 95 per cent of isobutylene used in the chemical industry goes into di- and trisobutylenes, butyl rubber, and other polymers. Total isobutylene consumption for chemicals was estimated at nearly 700 million pounds in 1967.

Dimers and Trimers of Isobutylene. A mixture of dimers and trimers of isobutylene is produced by absorbing isobutylene in 60-65 per cent sulfuric acid at 10-20°C, then reacting at 80-100°C for ½ hour.

The main chemical use of the dimer mixture

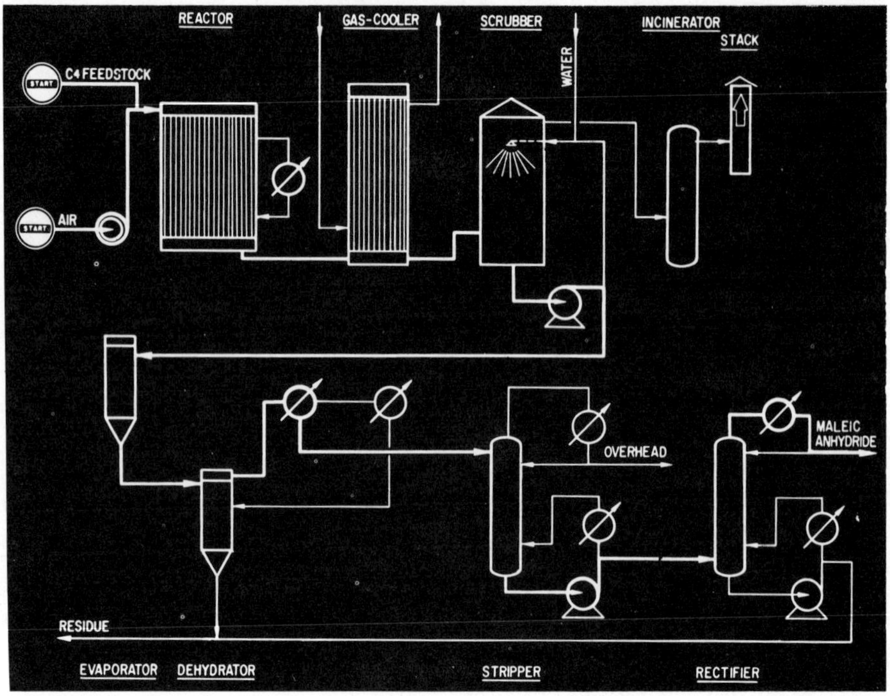

Fig. 25.28 Badische Anilin-und Soda-Fabrik A.G. process for maleic anhydride. [*Hydrocarbon Processing,* **50**, *No. 11, 175 (1971); copyright 1971 by Gulf Publishing Co.*]

is in the alkylation of phenols to octylphenols for detergent use. Nonyl alcohol is also produced from this dimer by the oxo process.

Use of diisobutylene is limited by its tendency to thermally depolymerize to isobutylene. Thus, only low-temperature reaction conditions can be utilized.

Approximately 40–50 million pounds of diisobutylene are produced annually for chemical uses.

Polybutenes and Polyisobutylenes. Polymerization of isobutylene can be carried out with such catalysts as boron trifluoride and aluminum chloride. This gives a wide range of polymers—from viscous liquids (called polybutenes) to semisolid and solid polymers (called polyisobutylenes).

Polybutenes are very stable materials with good oxygen or ozone resistance. These are highly saturated materials which do not set or "dry" on storage or use. Important industrial applications are in caulking and sealing compounds, adhesives, surgical tapes, vibration dampers, electrical insulation, and special lubricants. Output of polybutenes in 1967 was estimated at 230 million pounds.

Polyisobutylenes vary from soft, sticky gums to tough, elastic materials. Stability and resistance to chemical attack are excellent. Polyisobutylenes are used as tacky agents for oils to avoid dripping and splattering from bearings and rotating shafts, etc. These materials are also used as a viscosity-index-improver for oil and hydraulic fluids. Estimated 1967 output of polyisobutylenes was 34 million pounds.

A flow sheet showing a polybutene process is shown in Fig. 25.29.

Butyl Rubber. Low-temperature copolymerization of isobutylene (98 per cent) and isoprene (2 per cent) produces a solid, rubber-like, vulcanizable polymer. Butyl rubber has unusually low gas permeability and thus has found extensive use in tire inner liners and

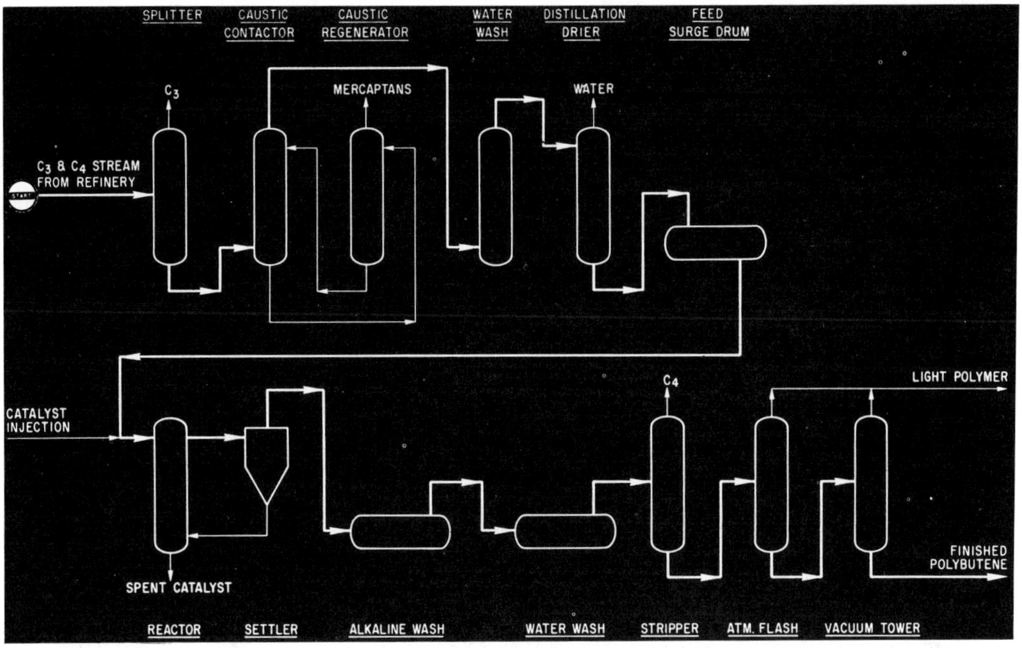

Fig. 25.29 Liquid polybutene from mixed C_4 hydrocarbons. [*Hydrocarbon Processing,* **50,** *No. 11, 194 (1971); copyright 1971 by Gulf Publishing Co.*]

tubes and pneumatic bags. Chemical variations such as chlorine-containing butyl rubbers have made butyl rubber compatible with SBR and natural rubber for blending purposes. See Chapter 9 for more details.

Output of some 260 million pounds of butyl rubber was estimated for 1970 and this volume is essentially static.

Butyl Hydroxy Toluene (BHT). The more proper chemical name for BHT is 4-methyl-2,6-di-*t*-butylphenol. The material is produced by alkylating *p*-cresol with high-purity isobutylene. BHT is an antioxidant and is sold in food grade for use in fats, oils, and fat-containing foods. The technical-grade product is used chiefly as a gum inhibitor in gasoline. Some 25 million pounds of BHT were produced in 1967.

t-Butylphenols. *t*-Butylphenols are formed by the reaction of phenolic compounds with isobutylene using sulfuric acid as catalyst.

These compounds are used as intermediates for bactericides, oil soluble phenol-formaldehyde resins, and antioxidants. Some 35 million pounds of compounds in this class are used annually, requiring about 20 million pounds of isobutylene.

t-Butyl Alcohol. The hydration of isobutylene to *t*-butyl alcohol goes readily under mild acid conditions. The alcohol is easily dehydrated under acidic conditions, thus it must be isolated from a dilute acid or neutral system. This property limits the use of this alcohol. Also limiting its use is its relatively high (25.6°C) freezing point. Production in 1967 was approximately 9 million pounds.

Isoprene. The Prins reaction between isobutylene and formaldehyde has been used somewhat to produce 4,4-dimethyl-1,3-dioxane as an intermediate. The second step catalytically converts this intermediate to isoprene and formaldehyde. Yields of approximately 76 per cent on formaldehyde and 83 per cent on isobutylene have been reported. Commercial use is still restricted by cost, since such materials as butadiene and styrene which are made on a mammoth scale make competition stiff. A more typical route to isoprene is shown in the C_5 section of this chapter.

Methacrylonitrile. This is produced by the ammoxidation of isobutylene in a process similar to that used to manufacture acrylonitrile from propylene, ammonia, and air.

n-Pentane and Cyclopentane

Derivatives of *n*-pentane are used for small specialty type products. Perhaps the most important are: (1) amyl chlorides, which in turn are converted to amyl mercaptan for use as a warning agent in natural gas; and (2) amyl alcohols, which are converted to sweet-smelling acetate esters for flavoring and photographic uses. Amyl chlorides can be dehydrohalogenated to specialty pentenes.

Cyclopentane finds its major outlet as an intermediate in the production of some Diels-Alder reactions with cyclopentadiene, in which the next step leads to such insecticide products as aldrin, chlordane, dieldrin, and heptachlor. Because these insecticides are very stable and difficult to biodegrade, they are losing their strength in the market place.

Isopentane

Of major importance is the use of isopentane as a base stock for dehydrogenation to isopentene and to isoprene. A flow sheet of a process for converting isopentane and tertiary amylenes to isoprene is shown in Fig. 25.30. This process exists essentially in two sections;

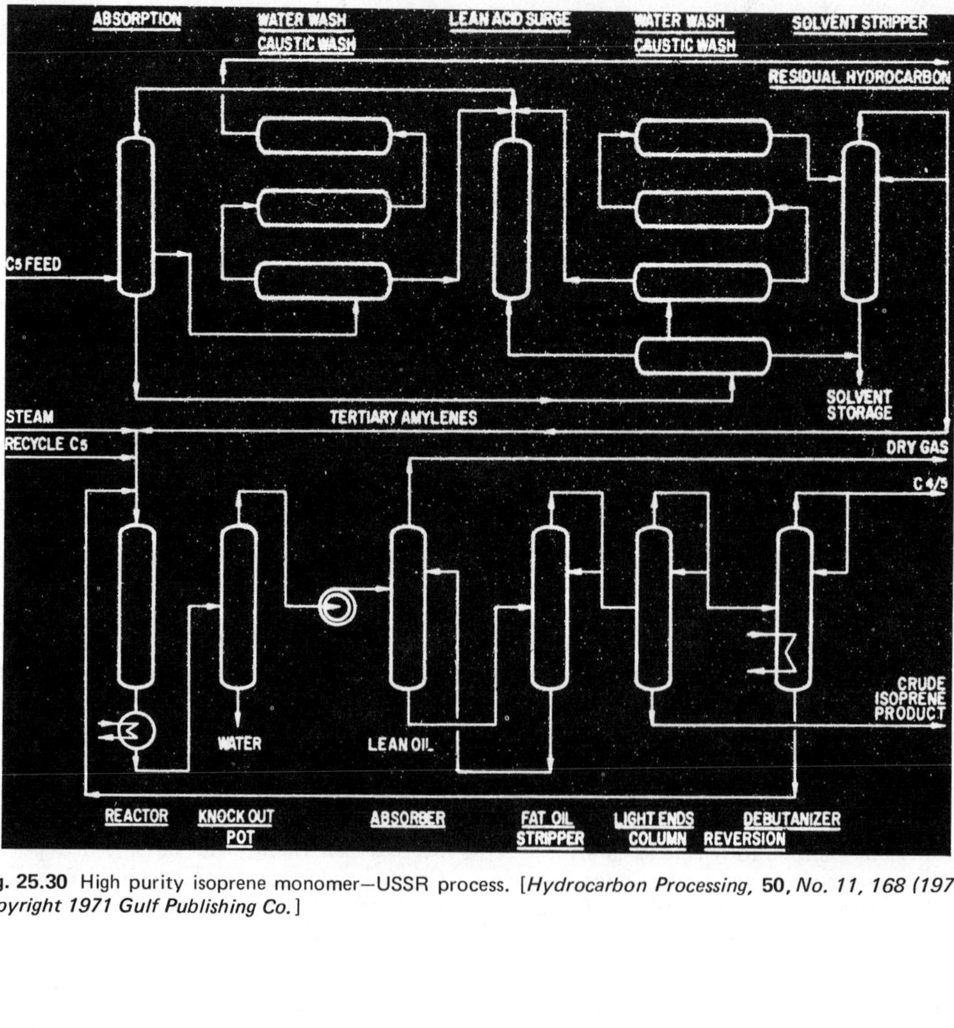

Fig. 25.30 High purity isoprene monomer—USSR process. [*Hydrocarbon Processing,* **50,** *No. 11, 168 (1971); copyright 1971 Gulf Publishing Co.*]

the first part is cleanup of cracked stock to obtain some tertiary amylenes of reasonable quality to be used in the final dehydrogenation to isoprene. The remainder of the process is cleanup and purification, where the final purification uses the Shell ACN (acetonitrile) extractive distillation to give high-purity isoprene.

Isoprene is used mainly in polyisoprene elastomers which are currently growing rapidly. A much smaller use is in the manufacture of butyl rubber, a copolymer of isobutylene and isoprene.

The two major uses indicate an isoprene consumption of 325-350 million pounds per year in 1970.

n-Paraffins, Monoolefins, Primary and Secondary Higher Alcohols

n-Paraffins. A process for separation of high purity normal paraffins in the C_{10} to C_{24} range from branched-chain and cyclic hydrocarbons is shown in Fig. 25.31. The principle employs the highly selective absorption action of molecular sieves in a multibed system. The beds are loaded with normal paraffins successively and desorption of normal paraffins is carried out with a purge stream. The purge material is then separated from the *n*-paraffin by distillation.

Monoolefins. Linear internal monoolefins are produced from the appropriate-carbon-number linear paraffins. Products are produced with a linear-monoolefin purity of about 94 wt per cent. A flow sheet of the processing scheme is shown in Fig. 25.32. In this process the linear paraffin is catalytically dehydrogenated to the monoolefin. The olefin and paraffin are then separated by using a fixed-bed adsorbent-extraction system. A desorbent is added to each stream, i.e., the olefin

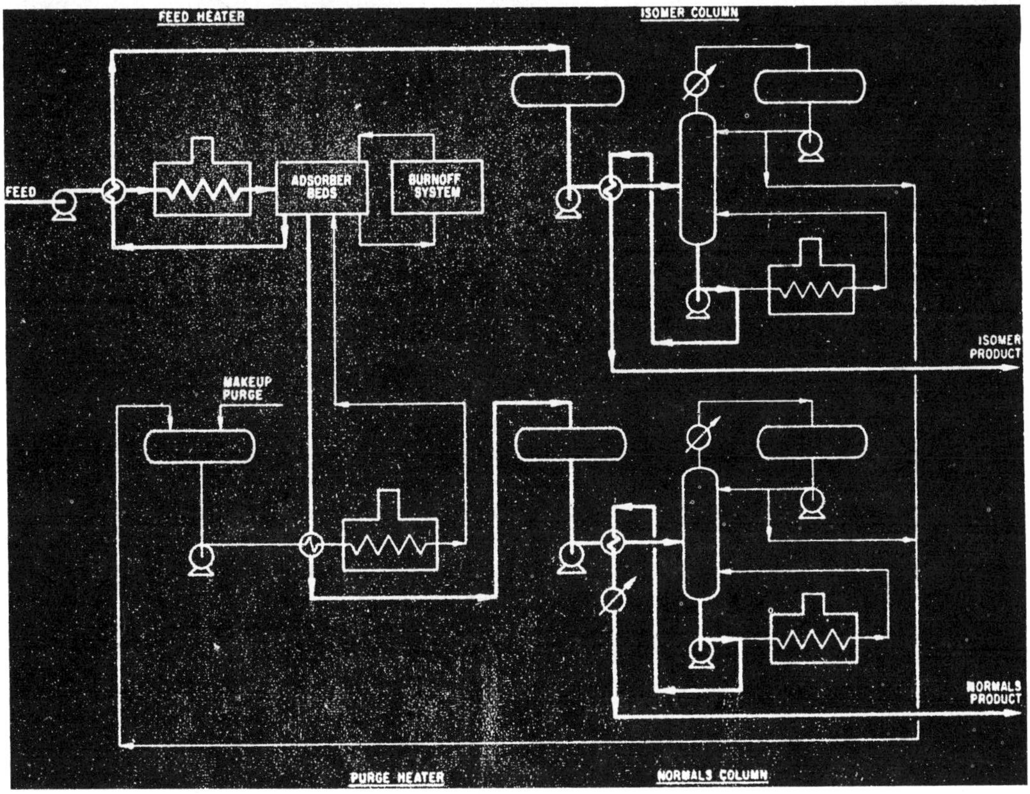

Fig. 25.31 Separation of high purity C_{10}-C_{24} normal paraffins from kerosene. [*Hydrocarbon Processing,* **50**, *No. 11, 184 (1971); copyright 1971 by Gulf Publishing Co.*]

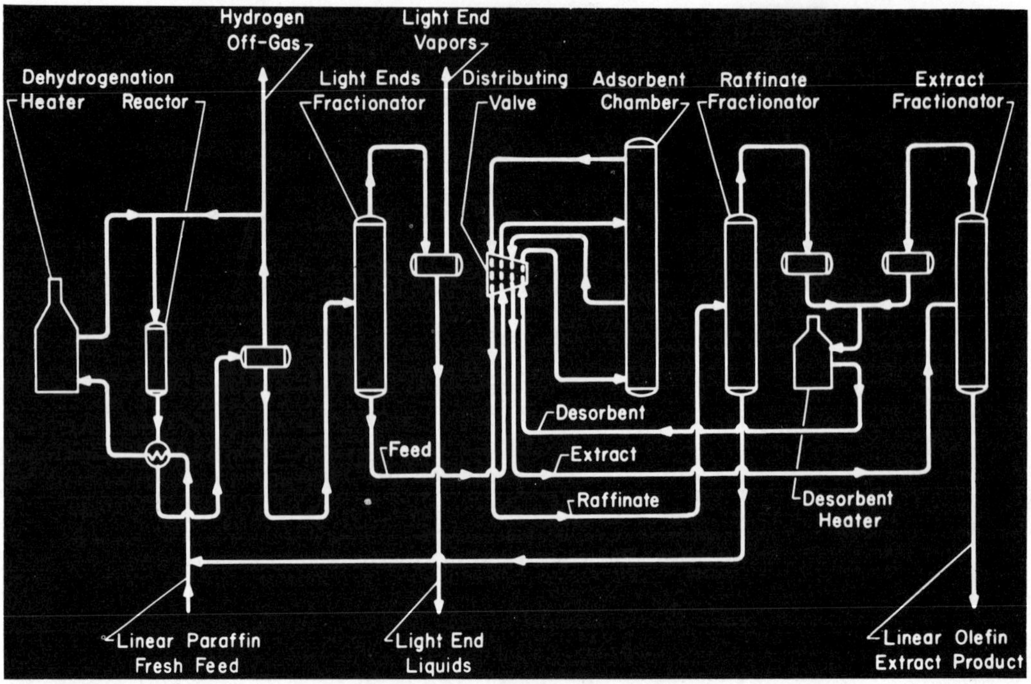

Fig. 25.32 Catalytic dehydrogenation process for linear internal monoolefins. [*Hydrocarbon Processing,* **50,** *No. 11, 174 (1971); copyright 1971 by Gulf Publishing Co.*]

extract and the paraffin raffinate, and is separated from each in a distillation column. The desorbent is a low-boiling hydrocarbon, chosen so that it will separate easily overhead in each of the final distillation columns.

Primary and Secondary Higher Alcohols. Essentially three methods are used to arrive at the higher alcohols. In one case *n*-dodecyl (lauryl) alcohol (C_{12}) is made by hydrogenation of coconut oil to give the linear primary alcohol.

In another case, the linear olefin is subjected to a modified oxo reaction to give a primary alcohol having some alpha-methyl branching.

In the third case, the linear paraffin is oxidized giving a linear secondary alcohol.

The major use for these higher alcohols, of course, is in the synthesis of biodegradable synthetic detergents. Depending on the final functionality desired the alcohols are ethoxylated, sulfated, phosphated, etc. to add the appropriate hydrophilic functional groups. These materials generally compete with the alpha-alcohols which are made directly or from alpha-olefins as described in the *Ethylene* section of this chapter.

AROMATIC CHEMICALS

Until World War II, coal tar was the only source of basic aromatic raw materials. Since then, petroleum as a source of aromatics has gained rapidly to the point where in 1970, 92 per cent of the benzene, 97 per cent of the toluene, and 99 per cent of the xylenes were derived from petroleum. The first petronaphthalene plant was put on stream as recently as early 1961.

Specific processing methods for obtaining these raw materials from petroleum are described in Chapter 14.

Benzene, toluene, and xylenes are also produced by fractional distillation of light oils obtained from high-temperature carbonization of coal, while naphthalene, anthracene, and other multiring compounds are recovered

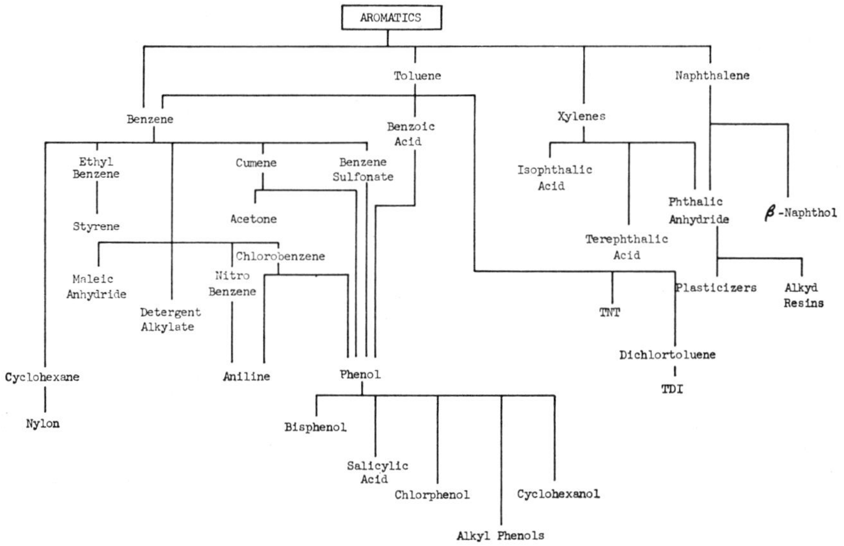

Fig. 25.33 Synthetic organic chemicals derived from aromatic compounds.

from tar oils, a high boiling fraction obtained from coal carbonization.

Benzene Products

Benzene is the most important aromatic chemical. During 1969, some 1.13 billion gallons of the approximately 1.23 billion gallons produced were consumed in the United States. Benzene is second only to ethylene as a synthetic organic chemical building block.

Benzene has a broad end-use pattern. Styrene monomer, cyclohexane (nylon), phenol, alkylbenzenes, aniline, and maleic anhydride are the major benzene derivatives. These materials appear to be heading for continued growth. Another important benzene outlet is the chlorobenzenes (other than for making phenol).

Styrene. Styrene manufacture is by far the largest use of benzene. Some 550 million gallons of benzene yielded about 4.65 billion pounds of styrene monomer in 1970. Styrene monomer is made by cracking (dehydrogenating) ethylbenzene (see Chapter 9 for details).

The major uses for styrene are in plastics, rubber latex paints and coatings, synthetic rubber, polyesters, and styrene-alkyd protective coatings. In these uses styrene is polymerized to a homopolymer or in copolymers with such materials as acrylonitrile, butadiene, maleic anhydride, and glycols. Further details may be found in the chapters on plastics, rubber, and paints.

Cyclohexane. Cyclohexane (mostly for nylon) is the second largest consumer of benzene. During 1969 some 257 million gallons of benzene were used in cyclohexane manufacture, yielding about 309 million gallons. Nylon 66, produced from adipic acid and hexamethylenediamine, is the major current domestic nylon. Nylon 6, derived from caprolactam, is growing more rapidly. Although these nylon intermediates are mainly derived from cyclohexane, there are various other routes. For example, adipic acid can be made by butylene oxidation as well as by cyclohexane oxidation; hexamethylenediamine can be made from butadiene as well as from cyclohexane; and caprolactam can be made by starting with phenol as well as with cyclohexane. See Chapter 11 for more details.

Phenol. Synthetic phenol is the third largest market for benzene. Approximately 230 million gallons of benzene were used to manufacture 1.64 billion pounds of phenol in 1969. Five different processes are currently used in the U.S. They are: cumene, sulfonation, chlorobenzene, Raschig, and benzoic acid. About 60 per cent of the phenol produced in 1969 was by the cumene process. All but the benzoic acid process require benzene as the raw material. The benzoic acid process uses toluene as the starting aromatic material. Details of the phenol manufacturing process are given in Chapter 10.

Some of the major uses for phenol are discussed below:

Phenolic resins are the major consumers of phenol and should maintain this position indefinitely. The major uses for phenolic resins are as plywood adhesives and as molding resins. See Chapter 10 for details on phenolic resins.

Bisphenol A is used in the production of polycarbonates and epoxy resins, which are growing plastics. Bisphenol A is in turn obtained from phenol by the following reaction:

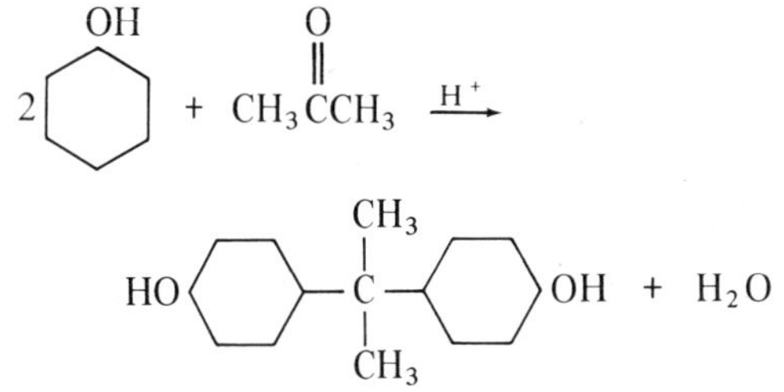

Bisphenol A

A flow sheet for bisphenol A manufacture is shown in Fig. 25.34.

Phenol and acetone in a molar ratio of 3:1 or 4:1 are charged to an acid-resistant stirred bisphenol A reactor. Glass-lined equipment is used ordinarily. A sulfur-containing catalyst is added, then dry HCl gas is bubbled into the reaction mass. The temperature is maintained at 30 to 40°C for 8–12 hours. The product crystallizes from the reaction mixture to form a slurry.

At the end of the reaction, the mixture is washed with water and treated with just enough lime to neutralize the free acid. Vacuum and heat are applied to the reaction kettle, and water and phenol are distilled separately from the mixture. The batch is finished by blowing the molten product with steam under vacuum at 150°C to remove the odor of the sulfur-containing catalyst.

The molten bisphenol A product is quenched in a large volume of water, filtered, and dried. It is a light tan powder, which may be further purified by recrystallization from solvents.

The patent literature indicates that ion-exchange resins promoted with various mercaptans or amines may be used in some of the newer plants in operation today.

U.S. consumption of bisphenol A during 1969 was 64 per cent in epoxy resins, 15 per

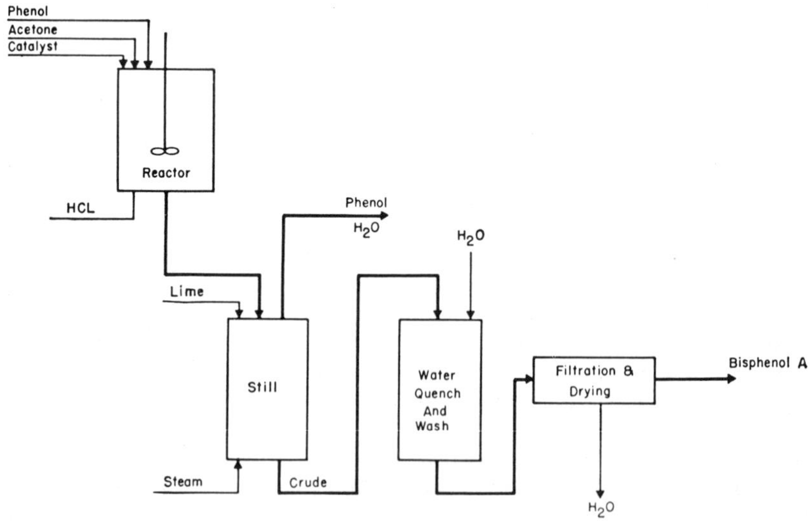

Fig. 25.34 Manufacture of bisphenol A. [*Pet. Ref.*, **38**, *No. 11, 225 (1959), copyright 1959 by Gulf Publishing Co.*]

cent in polycarbonate resins, and 21 per cent in other uses. Total consumption was over 182 million pounds.

Other products from phenol include aspirin, alkylated phenols, chlorinated phenols (to 2,4-D), and caprolactam. Chemical reactions for some of these materials are shown below:

Important by-products from the chlorobenzene process which are now substantial volume chemicals are phenylphenols, numerous chlorophenols, and Dowtherm® A (a eutectic mixture of diphenyl and diphenyl oxide sold as a heat transfer medium).

1.
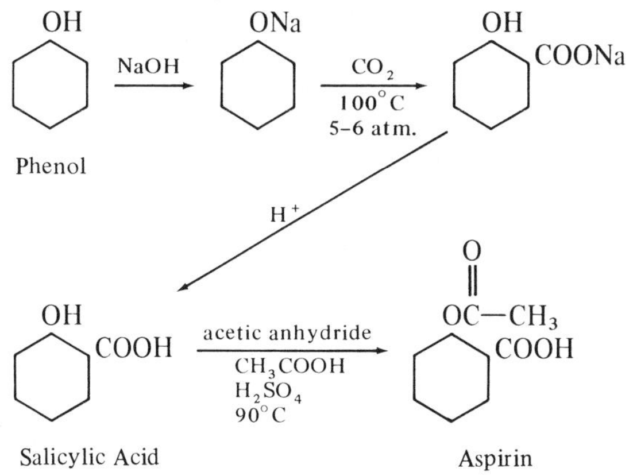

2.
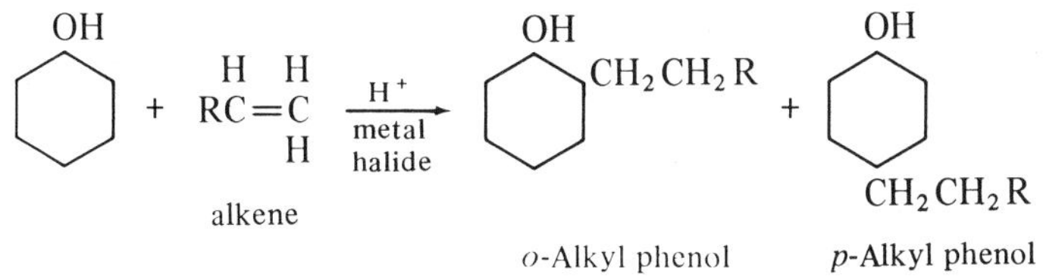

3.
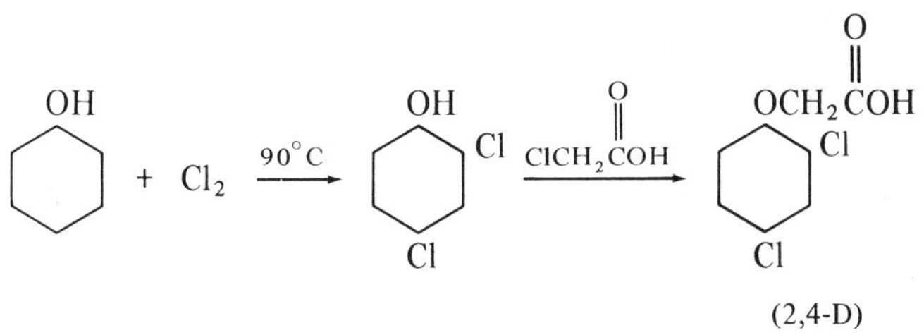

4.
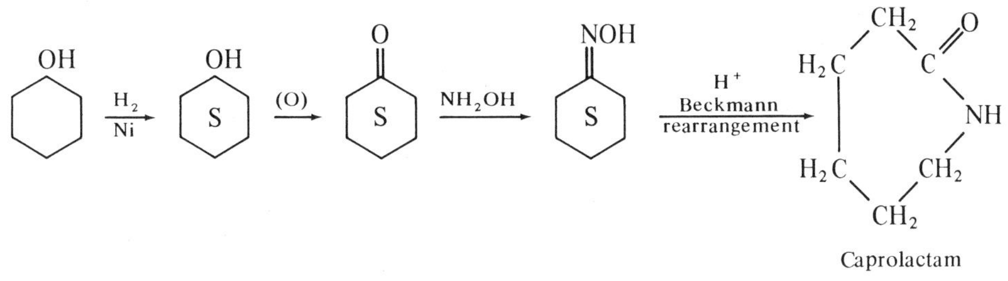

Detergent Alkylate. Alkylbenzene is an intermediate in the manufacture of synthetic detergents. Production during 1969 was approximately 530 million pounds from about 40 million gallons of benzene.

Alkylbenzenes are mostly dodecyl- and tridecylbenzene. To make synthetic detergents, α-olefins or internal linear olefins are reacted with benzene to form the alkylbenzene; this is then sulfonated, neutralized, blended with chemical "builders," and flake dried. The following chemical reactions are carried out in the processing;

The manufacture of detergent alkylate is described in Chapter 13, "Soaps and Detergents."

Maleic Anhydride. Maleic anhydride produced during 1969 reached approximately 200 million pounds. Some 92-95 per cent of this was produced from benzene, consuming approximately 250 million pounds of benzene. The flow sheet (Fig. 25.35) shows a manufacturing process starting with benzene. A butylene oxidation route to maleic anhydride is described in the C_4 section of this chapter.

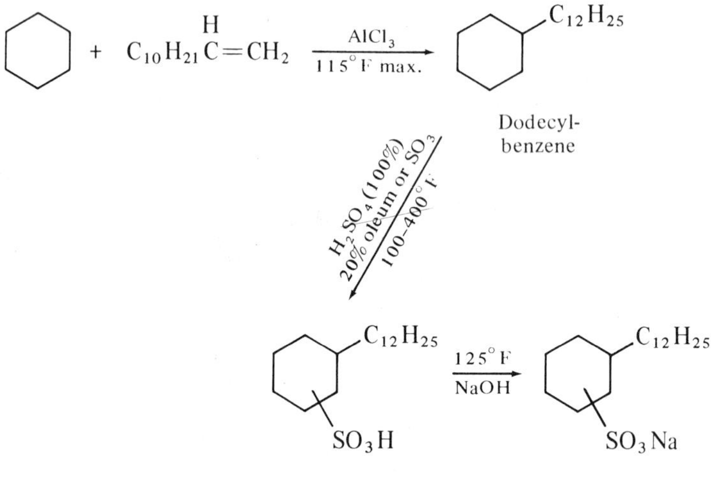

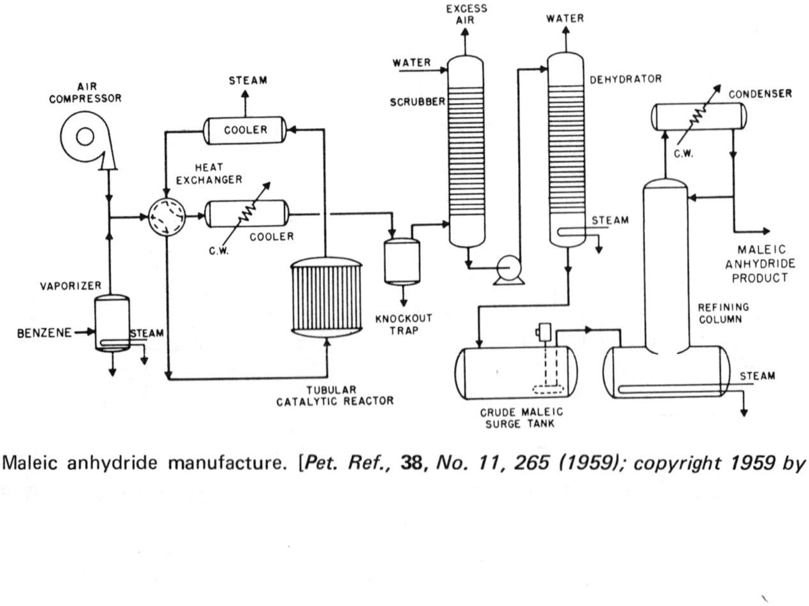

Fig. 25.35 Maleic anhydride manufacture. [*Pet. Ref.,* **38**, *No. 11, 265 (1959); copyright 1959 by Gulf Publishing Co.*]

Maleic anhydride is used for making polyester resins, alkyd resins, agricultural chemicals, paper-size drying oils, and styrene-maleic anhydride resins. Details on these products may be found in Chapter 10.

Aniline. Some 398 million pounds of aniline were produced from about 42 million gallons of benzene in the U.S. in 1970.

Aniline is produced in the U.S. by three processes: two start with nitrobenzene as an intermediate; the third starts with chlorobenzene. The chemistry of these processes is shown below and one is exemplified by the flow sheet, Fig. 25.36.

such as the thiazole derivatives. Use of aniline in isocyanates is booming because of the increased consumption of rigid polyurethanes, mainly for insulation.

Other important aniline uses are in dyes, drugs and veterinary medicines, and photographic chemicals (hydroquinone).

DDT (Dichlorodiphenyltrichloroethane). DDT is manufactured from chlorobenzene and chloral:

$$CCl_3CHO + 2C_6H_5Cl \xrightarrow{\text{oleum}} CCl_3CH(C_6H_4Cl)_2 + H_2O$$

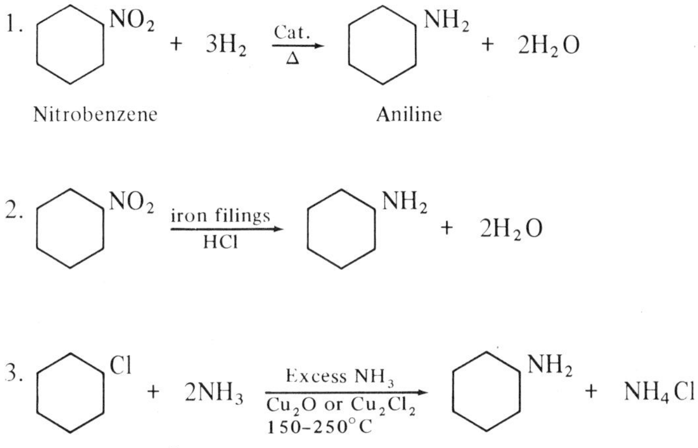

About 50 per cent of aniline currently goes into the manufacture of rubber chemicals

Approximately 123 million pounds were produced in 1969, but production had

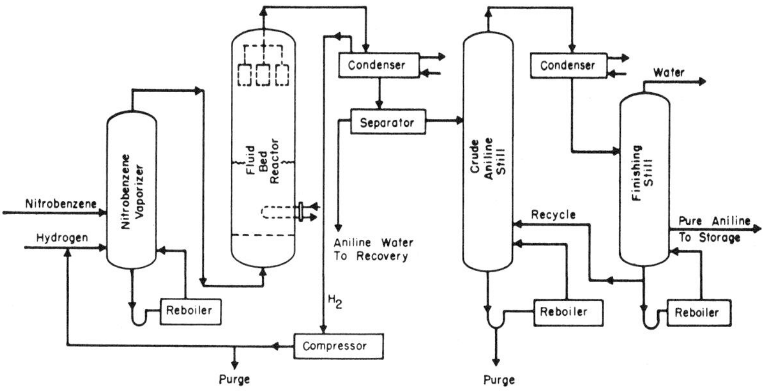

Fig. 25.36 Manufacture of aniline. [*Pet. Ref.,* **38,** *No. 11, 224 (1959); copyright 1959 by Gulf Publishing Co.*]

dropped to 59 million pounds in 1970. DDT is being replaced as rapidly as possible because of its extreme resistance to biodegradation.

Some Other Benzene-Derived Chemicals. Other significant chemicals derived from benzene are mono- and dichlorobenzene, nitrobenzene, and resorcinol.

Nitrobenzene is manufactured from benzene and mixed nitrating acids as shown below:

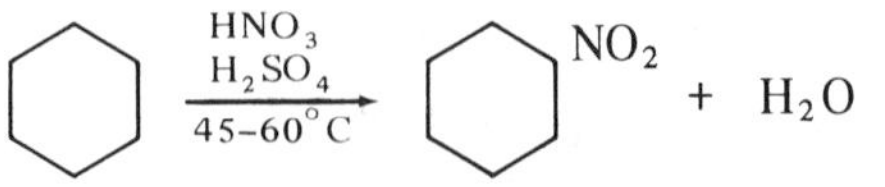

Nitrobenzene, except for aniline synthesis, is used mainly as a solvent. Of the 548 million pounds produced in 1970, over 90 per cent was used in aniline production.

Mono- and dichlorobenzene are made by two routes—direct chlorination or oxychlorination with HCl and air. Figure 25.37 shows a flow sheet of the Dow process for direct chlorination.

Monochlorobenzene is used for the manufacture of DDT, sulfur dyes such as sulfur black, drugs, perfumes, and numerous solvent applications. *o*-Dichlorobenzene is used mostly as a solvent or cleaner, particularly for metal degreasing. It has utility, when purified and stabilized, as a heat-transfer fluid in the temperature range of 150-260°C. *p*-Dichlorobenzene is used extensively in moth protection for wool. Small cakes of *p*-dichlorobenzene are used in the sanitary field. Its vapor pressure and pleasant odor make it highly suitable for this purpose.

Resorcinol (m-dihydroxybenzene) production for 1970 is reported at approximately 26 million pounds. Current production is reported to be entirely by the sulfonation process where the sodium salt of *m*-disulfonic acid is hydrolyzed with caustic, then neutralized with a strong acid. It is reported that a new plant will be starting up in late 1971 or early 1972 using a new process wherein *m*-diisopropylbenzene is oxidized to *m*-diisopropylbenzene dihydroxide which in turn is cleaved to yield resorcinol and acetone. The reactions involved are similar to those used in producing phenol from cumene.

Resorcinol is used by the tire industry in a

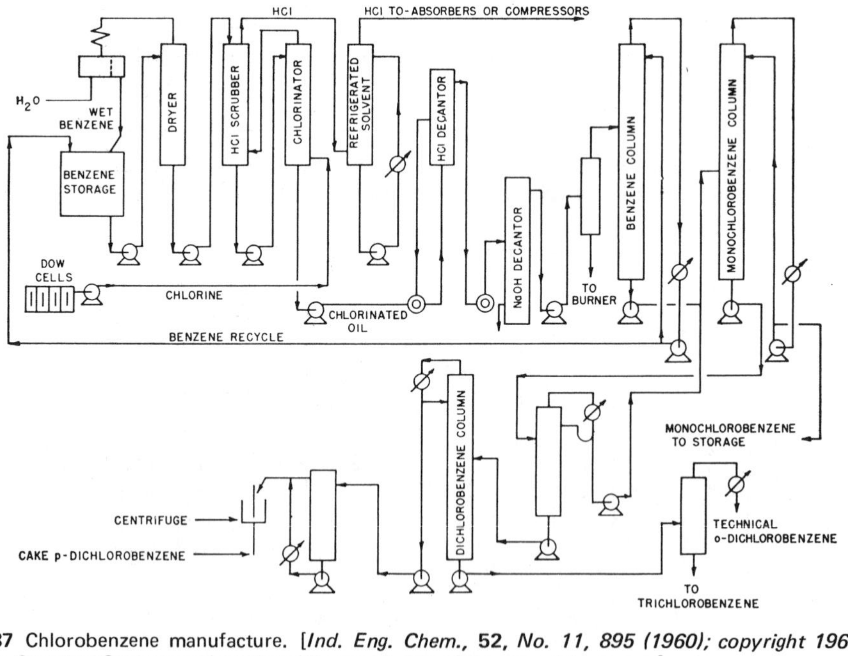

Fig. 25.37 Chlorobenzene manufacture. [*Ind. Eng. Chem.*, **52**, *No. 11, 895 (1960); copyright 1960 by the American Chemical Society, and reprinted by permission of the copyright owner.*]

resorcinol-formaldehyde resin form to bond the tire cord to rubber. This is a particularly effective adhesive for polyester and fiber glass cords.

Other uses are as a wood adhesive in the resorcinol-formaldehyde resin form–serving as a fast setting resin at low temperature–used mainly for laminated beam applications. Some resorcinol is used in the synthesis of ultraviolet absorbers, dyes, and for some pharmaceutical uses such as in skin ointments, medicated soaps, and antiseptic preparations for the mouth and throat. It is sometimes used as an anthelmintic.

Toluene Products

Approximately 3.3 billion pounds of toluene were consumed in the United States in 1966 for chemical manufacture and solvent uses. The breakdown of the major chemical usage is: benzene (70 per cent), toluene diisocyanate (4 per cent), TNT and other explosives (3 per cent), phenol (1.5 per cent), benzyl chloride (1.5 per cent), and benzoic acid (0.3 per cent).

Benzene is produced from toluene by the hydrodealkylation process. This use of toluene is very sensitive to benzene demand because it is more costly than the benzene isolated directly from reformate streams. About one-fourth of the benzene produced in 1966 was toluene-based.

Toluene Diisocyanate (TDI). Toluene diisocyanate can be manufactured from toluene by the route indicated in the following equations.

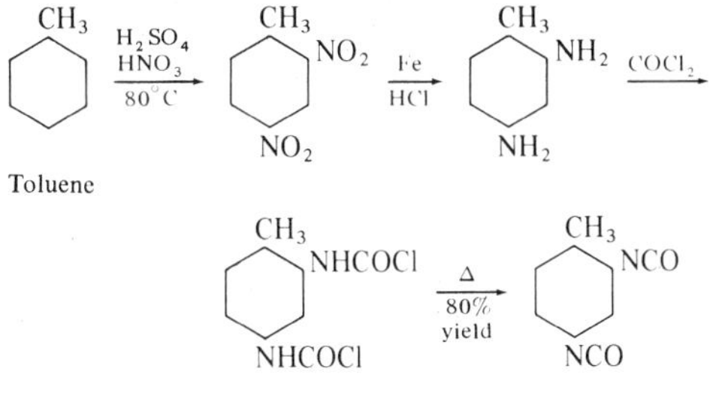

Toluenediamine is dissolved in a high boiling compatible solvent. The solvent mixture (10 per cent amine) is then mixed with phosgene (1.5 to 3.5 lb phosgene per lb diamine) in a reactor held at 20 to 50°C. In this temperature range, the first phosgene unit is reacting.

The resulting slurry is pumped to a second reactor where further reaction with gaseous phosgene is carried out at 185°C. Unreacted phosgene and HCl are vented from the top of the reactor. The final conversion to the diisocyanate occurs in a third vessel where an inert gas is blown through the solution at 110 to 115°C. The resulting solution of crude TDI is distilled at 2- to 5-mm Hg pressure. The diisocyanate product distills overhead.

Toluene diisocyanate is reacted with polyols or polyesters to make polyurethanes. Flexible polyurethane foams are used for cushioning and padding in automobiles, furniture, carpeting, etc. Semirigid urethane foams are used for padding such as crash pads on automobiles. Rigid urethane foams possess excellent thermal insulation properties, thus they are used for plastic panels for home construction and in insulation.

TNT (Trinitrotoluene). Manufacture of TNT is generally carried out by a three-stage batch nitration of toluene. Nitrating acids of various combinations of nitric and sulfuric acid are used to obtain maximum yield and efficiency. Strict temperature limitations and rates of heating are observed for safety and product quality. The crude "tri oil" TNT is treated by the Sellite process. This consists of hot and cold water washes, neutralization of free acids, sodium sulfite and sodium hydrogen sulfite treatments, "red water" separation, hot water wash, drying, and flaking.

Primarily a military explosive, consumption of TNT is directly influenced by government requirements.

Phenol. Relatively small quantities of phenol are made from the oxidation of toluene. It is not expected that this route will supplant the cumene process.

The process is a two-state oxidation. The first step oxidizes toluene to benzoic acid, part of the benzoic acid is oxidized to phenol in the second stage, and unconverted benzoic acid is recycled back to this second stage. Advantages claimed are cheaper raw material, very low by-product generation, and virtually no waste.

Benzyl Chloride. The principal method for producing benzyl chloride consists of chlorinating boiling toluene in darkness until there is a 37.5 per cent increase in weight. The reaction mixture is then treated with mild alkali and distilled.

Benzyl chloride serves as the starting material for many pharmaceuticals, insecticides, perfumes, and dyes. One of the most important uses of benzyl chloride is in the manufacture of benzyl alcohol by reacting it with alkalies.

Vinyltoluene. Vinyltoluene is produced by a process similar to the one used for styrene, i.e., alkylation to ethyltoluene by Friedel-Crafts reaction between toluene and ethylene, then catalytic dehydrogenation to vinyltoluene. Processing is complicated by the existence of 3 isomers; also, *o*-vinyltoluene undergoes side reactions in the dehydrogenation, and therefore, it is separated before dehydrogenation.

Vinyltoluene is in many uses a replacement for styrene, in some cases producing a superior product, as in certain types of coatings.

Chemicals from Xylene

Xylenes are obtained from petroleum reformate in the form of a "mixed xylenes" stream. A typical composition of this stream is 20 per cent ethylbenzene, 18 per cent paraxylene, 40 per cent *m*-xylene, and 22 per cent *o*-xylene. The major chemical uses of xylene require the pure isomers. *o*-Xylene can be separated by distillation, while most *p*-xylene is separated by low-temperature crystallization. About 400 million gallons of xylene go into the manufacture of chemicals. In terms of the specific isomers this amounts to 800 million pounds of *o*-xylene, 80 million pounds of *m*-xylene, and about 1600 million pounds of *p*-xylene. Their major uses are phthalic anhydride, isophthalic acid, and terephthalates, respectively.

Phthalic Anhydride from o-*Xylene.* The flow sheet, Fig. 25.38, shows the process for making phthalic anhydride from *o*-xylene.

The xylene is vaporized by injection into the hot air stream and then passes through a catalyst-filled 10,000-tube reactor. The crude phthalic anhydride desublimes in a switch condenser and any phthalic acid present is dehydrated in the predecomposer vessel. The crude is finally purified in two distillations. Although the fixed-bed process is currently the most important, there are a number of plants in operation which use a fluidized-bed reactor.

About 700 million pounds of phthalic anhydride are produced annually in the United States. More than 50 per cent of the present plant capacity is based on *o*-xylene and it is expected that the trend to this process instead of the naphthalene-based route will continue.

The major uses of phthalic anhydride are in plasticizers, 50 per cent; alkyd resins, 26 per cent; and unsaturated polyester resins, 13 per cent. The plasticizers are esters made by reacting two moles of an alcohol, such as 2-ethylhexanol, with one mole of phthalic anhydride. By far the largest use is in plasticizing vinyl chloride polymer and copolymers. Alkyd resins are a type of polyester resin used in surface coatings. The most rapidly growing end-use is in unsaturated polyester resins for reinforced plastics. Here the phthalic anhydride is used to modify the resin by replacing a portion of the unsaturated acid. Other outlets for phthalic anhydride are in dyes, agricultural chemicals, and pharmaceuticals.

Isophthalic Acid. Although *m*-xylene is the most abundant xylene isomer, it is in least demand as a chemical raw material. The only major outlet is in the manufacture of isophthalic acid, which is made mainly by liquid-phase oxidation, using heavy metal catalysts.

Terephthalic Acid. *p*-Xylene is in strong demand as the major raw material for the manufacture of terephthalic acid and dimethyl terephthalates, intermediates used in the production of polyester fibers and film. Processing techniques are very similar to those used for converting *m*-xylene. The crude terephthalic acid obtained after oxidation is either purified as the acid, or it is reacted with

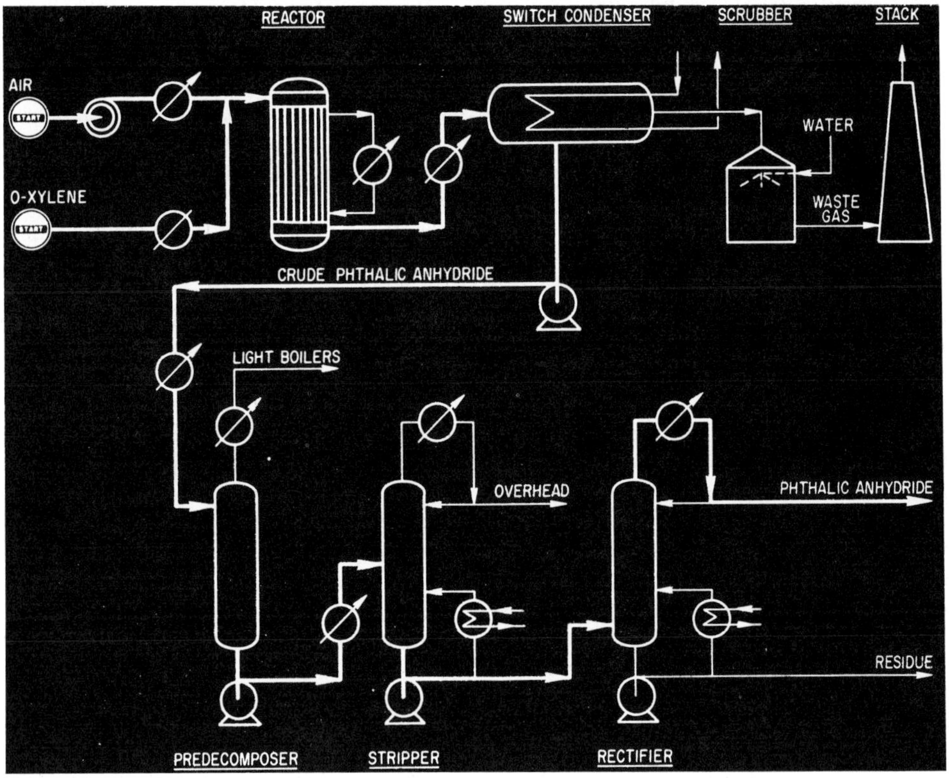

Fig. 25.38 Phthalic anhydride from *o*-xylene and air. [*Hydrocarbon Processing,* **50,** *No. 11, 188 (1971); copyright 1971 by Gulf Publishing Co.*]

methanol to give dimethyl terephthalate. Most of the present production of polyethylene terephthalate comes from the methyl ester, but there are indications that the acid will be the preferred route in the future.

More than a billion pounds of *p*-xylene are consumed in America alone with an almost equivalent amount in the rest of the world. Better than 90 per cent of this goes into fibers and the remainder is used to make films.

Naphthalene Chemicals

The U.S. supply of naphthalene was approximately 700 million pounds in 1970. Prior to 1960 all naphthalene came from coke, but now almost half comes from petroleum sources. About 75 per cent is consumed in the manufacture of phthalic anhydride. The remainder goes into insecticides, β-naphthol, mothballs, tanning agents, and surfactants.

Phthalic Anhydride. Phthalic anhydride is produced by catalytic oxidation of naphthalene. Two major types of processes are available today. One utilizes a fluid-bed oxidation reactor, while the other utilizes a fixed-bed reactor. Both processes utilize naphthalene and air as raw materials, carry out the oxidation under the influence of catalysts, and then finish the product by distillation. The fluid bed process is depicted in Fig. 25.39.

A process for phthalic anhydride derived from *o*-xylene is described in an earlier section.

Other Naphthalene Uses. The second largest use for naphthalene is in the manufacture of carbaryl (1-naphthyl N-methylcarbamate) for insecticides. About 65 million pounds went to this use in 1966. The naphthalene is first

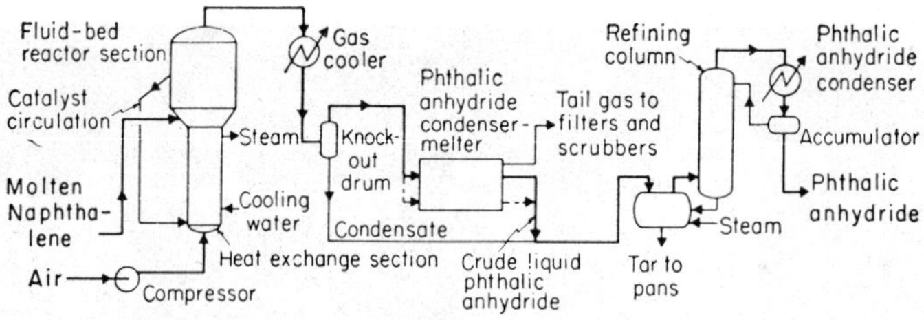

Fig. 25.39 Phthalic anhydride by Badger fluid-bed process. [*Reprinted by special permission from Chem. Eng. (Dec. 14, 1959); copyright 1959 by McGraw-Hill, Inc.*]

made into 1-naphthol and then converted to carbaryl by reaction with methyl isocyanate.

β-Naphthol is the most significant other use for naphthalene, taking some 30 million pounds in 1966. It is manufactured commercially by first making β-naphthalenesulfonic acid. The acid is then fused with caustic, acidified, washed, and vacuum distilled. β-Naphthol has numerous uses in dye, rubber, perfume, and pharmaceutical industries.

Sulfonated naphthalenes are available commercially as surface active agents of various sorts.

Other Polymethylbenzenes

Among the newer aromatic raw materials are the higher polymethylbenzenes. The most important of these is pseudocumene (1,2,4-trimethylbenzene) which is used in the manufacture of trimellitates for vinyl plasticizers. Other materials in this area are mesitylene and durene.

REFERENCES

1. R. Landau and G. S. Schaffel, "Recent Developments in Ethylene Chemistry," in "Origin and Refining of Petroleum," Part 8, Washington, American Chemical Society, 1971.
2. W. L. Faith, D. B. Keyes, and R. L. Clark, "Industrial Chemicals," 3rd Ed., New York, John Wiley & Sons, 1965.
3. H. Kehde, "C_4 Hydrocarbon Production and Distribution," New York, American Institute of Chemical Engineers, 1970.
4. "Petrochemical Handbook," *Hydrocarbon Process.* **50** No. 11, 113 (1971).
5. J. A. Kent, Ed., "Riegel's Industrial Chemistry," 6th Ed., Chapter 25, New York, Van Nostrand Reinhold, 1962.
6. "Synthetic Organic Chemicals," U.S. Tariff Commission Report, 1950–1970.
7. R. F. Fremed, "PMB's Begin Their Big Market Push," *Chem. Week* (Feb. 22, 34, 1969).
8. *Oil Paint and Drug Reporter* (Jan. 3, 10, 1972).
9. R. F. Goldstein, "The Petroleum Chemicals Industry," New York, John Wiley and Sons, 1958.
10. N. E. Ockerbloom, "Xylenes and Higher Aromatics," *Hydrocarbon Process.*, Series, beginning with **50**, No. 7, 112 (1971).
11. "Chemical Economics Handbook," Menlo Park, Ca. Stanford Research Institute.
12. "Chemical Technology," F. A. Henglein, London, Paragon Press Ltd., 1969.

26

Industrial Wastewater Technology

William J. Lacy*

INTRODUCTION

Industries use huge quantities of the nation's waters and are the major factor in the continuing rise in water pollution. They utilize over 15 trillion gallons of water but, prior to discharge, treat less than 5 trillion gallons. In terms of a single pollution parameter, BOD, the waste generated by industries is equivalent to that generated by a population of over 360 million people. Even more undesirable than the BOD loads of industrial effluents are the enormous quantities of mineral and chemical wastes from factories which steadily become more complex and varied. They include metals such as iron, chromium, nickel, and copper; salts such as compounds of sodium, calcium, and magnesium; acids such as sulfuric and hydrochloric; petroleum wastes and brines; phenols; cyanides; ammonia; toluene; blast furnace wastes; greases; many varieties of suspended and dissolved solids; and numerous other waste compounds. These wastes degrade the quality of receiving waters by imparting tastes, odors and color; and through excess mineralization, salinity, hardness, and corrosion. Some are toxic to plant and animal life.

The variety and complexity of inorganic and organic components contained in industrial effluents present a serious liquid waste-water treatment control problem in that the pollution and toxicity effects of these constituents are of greater significance than those found in domestic waste waters.

Conventional waste-water treatment technology which is often barely adequate for existing waste types offers even less promise of providing the type and degree of treatment which will be required in the near future. Therefore, industrial pollution-control technology must be developed to achieve effective and economical control of pollution from such varied industries as those producing metals and metal products, chemicals and allied products, paper and allied products,

*Director, Industrial Pollution Control, R&D Division, Environmental Protection Agency, Washington, D.C.

petroleum and coal products, food and kindred products, textiles and leather goods.

TERMINOLOGY

Several of the more common terms encountered in sewage-treatment technology are given below to assist the reader in developing his vocabulary in this field and in understanding what follows.

Activated sludge process removes organic matter from sewage by saturating it with air and biologically active sludge.

Adsorption is an advanced way of treating wastes in which carbon removes organic matter not responsive to clarification or biological treatment.

Aeration tank serves as a chamber for injecting air into water.

Algae are plants which grow in sunlit waters. They are a food for fish and small aquatic animals and, like all plants, put oxygen into the water.

Bacteria are the smallest living organisms which literally eat the organic parts of sewage.

BOD, or biochemical oxygen demand, is the amount of oxygen necessary in the water for bacteria which consume the organic sewage. It is used as a measure in telling how well a sewage-treatment plant is working.

Chlorinator is a device for adding chlorine gas to sewage to kill infectious germs.

Coagulation is the clumping together of solids to make them settle out of the sewage faster. Coagulation of solids is brought about with the use of certain chemicals such as lime, alum, or polyelectrolytes.

Combined sewer carries both sewage and storm water runoff.

Comminutor is a device for the catching and shredding of heavy solid matter in the primary stage of waste treatment.

Diffused air is a technique by which air under pressure is forced into sewage in an aeration tank. The air is pumped down into the sewage through a pipe and escapes through holes in the side of the pipe.

Digestion of sludge takes place in heated tanks where the material can decompose naturally and the odors can be controlled.

Effluent is the liquid that comes out of a treatment plant after completion of the treatment process.

Electrodialysis is a process by which electricity attracts or draws the mineral salts from sewage.

Floc is a clump of solids formed in sewage when certain chemicals are added.

Flocculation is the process by which certain chemicals form clumps of solids in sewage.

Incineration consists of burning the sludge to remove the water and reduce the remaining residues to a safe, nonburnable ash. The ash can then be disposed of safely on land, in some waters, or into caves or other underground locations.

Interceptor sewers in a combined system control the flow of the sewage to the treatment plant. In a storm, they allow some of the sewage to flow directly into a receiving stream. This protects the treatment plant from being overloaded in case of a sudden surge of water into the sewers. Interceptors are also used in separate sanitation systems to collect the flows from main and trunk sewers and carry them to the points of treatment.

Ion is an electrically charged atom or group of atoms which can be drawn from waste water during the electrodialysis process.

Lateral sewers are the pipes that run under the streets of a city and into which the sewers from homes or businesses empty.

Lagoons are scientifically constructed ponds in which sunlight, algae, and oxygen interact to restore water to a quality equal to effluent from a secondary treatment plant.

Mechanical aeration begins by forcing the sewage up through a pipe in a tank. Then it is sprayed over the surface of tank, causing the waste stream to absorb oxygen from the atmosphere.

Microbes are minute living things, either plant or animal. In sewage, microbes may be germs that cause disease.

Mixed liquor is the name given the effluent that comes from the aeration tank after the sewage has been mixed with activated sludge and air.

Organic matter is the waste from homes or industry which is of plant or animal origin.

Oxidation is the consuming or breaking down of organic wastes or chemicals in sewage by bacteria and chemical oxidants.

Oxidation pond is a man-made lake or body of water in which wastes are consumed by bacteria. It is used most frequently with other waste-treatment processes. An oxidation pond is basically the same as a sewage lagoon.

Primary treatment removes the material that floats or will settle in sewage. It is accomplished by using screens to catch the floating objects and tanks for the heavy matter to settle in.

Pollution results when something—animal, vegetable, or mineral—reaches water, making it more difficult or dangerous to use for drinking, recreation, agriculture, industry, or wildlife.

Polyelectrolytes are synthetic chemicals used to speed the removal of solids from sewage. The chemicals cause the solids to coagulate or clump together more rapidly than chemicals like alum or lime.

Receiving waters are rivers, lakes, oceans, or other water courses that receive treated or untreated waste waters.

Salts are the minerals that water picks up as it passes through the air, over and under the ground, and through household and industrial uses.

Sand filter removes the organic wastes from sewage. The waste water is trickled over the bed of sand. Air and bacteria decompose the wastes filtering through the sand. The clean water flows out through drains in the bottom of the bed. The sludge accumulating at the surface must be removed from the bed periodically.

Sanitary sewers, in a separate system, are pipes in a city that carry only domestic waste water. The storm water runoff is taken care of by a separate system of pipes.

Secondary treatment is the second step in most waste-treatment systems in which bacteria consume the organic parts of the wastes. It is accomplished by bringing the sewage and bacteria together in trickling filters or in the activated-sludge process.

Sedimentation tanks help remove solids from sewage. The waste water is pumped to the tanks where the solids settle to the bottom or float on top as scum. The scum is skimmed off the top, and solids on the bottom are pumped out to sludge-digestion tanks.

Septic tanks are used to treat domestic wastes. The underground tanks receive the waste water directly from the home. The bacteria in the sewage decomposes the organic waste and the sludge settles on the bottom of the tank. The effluent flows out of the tank into the ground through drains. The sludge is pumped out of the tanks, usually by commercial firms, at regular intervals.

Sewers are a system of pipes that collect and deliver waste water to treatment plants or receiving streams.

Sludge is the solid matter that settles to the bottom of sedimentation tanks and must be disposed of by digestion or other methods to complete waste treatment.

Storm sewers are a separate system of pipes that carry only runoffs from buildings and land during a storm.

Suspended solids are the wastes that will not sink or settle in sewage.

Trickling filter is a bed of rocks or stones. The sewage is trickled over the bed so the bacteria can break down the organic wastes. The bacteria collect on the stones through repeated use of the filter.

Waste treatment plant is a series of tanks, screens, filters, and other processes by which pollutants are removed from water.

TREATMENT LEVELS

There are at present two basic methods of treating wastes. They are called *primary* and *secondary*. In primary treatment, solids are allowed to settle and are removed from the water. Secondary treatment, a further step in purifying waste water, uses biological processes.

Primary Treatment

As sewage enters a plant for primary treatment it flows through a screen which removes large floating objects such as rags and sticks that may clog pumps and pipes. The screens vary from coarse to fine—from those consisting of parallel steel or iron bars with openings of about half an inch or more to screens with much smaller openings.

Screens are generally placed in a chamber or channel in a position inclined with respect to the flow of the sewage to make cleaning easier. The debris caught on the upstream surface of the screen can be raked off manually or mechanically. Some plants use a device known as a comminutor which combines the functions of a screen and a grinder. These devices catch and then cut or shred the heavy solid material. In this method the pulverized material remains in the sewage flow to be removed later in a settling tank.

After the sewage has been screened, it passes into what is called a grit chamber where sand, grit, cinders, and small stones are allowed to settle to the bottom. A grit chamber is especially important for cities with combined sewer systems because it will remove the grit or gravel that washes off streets or land during a storm and ends up at the treatment plants. The unwanted grit or gravel from this process is usually disposed of by filling land near a treatment plant.

In some plants, another screen is placed after the grit chamber to remove any further material than might damage equipment or interfere with later processes.

With the screening completed and the grit removed, the sewage still contains suspended solids. These are minute particles of matter that can be removed from the sewage by treatment in a sedimentation tank. When the speed of the flow of sewage through one of these tanks is reduced, the suspended solids will gradually sink to the bottom. This mass of solids is called raw sludge.

Various methods have been devised for removing sludge from the tanks. In older plants it was removed by hand. After a tank had had been in service for several days or weeks, the sewage flow was diverted to another tank. The sludge in the bottom of the out-of-service tank was pushed or flushed with water to a nearby pit and then removed for further treatment or disposal.

Almost all plants built within the past 30 years have included mechanical means for removing the sludge from sedimentation tanks. In some plants it is removed continuously and in others at intervals. To complete the primary treatment the sludge-free effluent is chlorinated to kill harmful bacteria and then discharged into a stream or river. The chlorination also helps to reduce odors.

Although 30 per cent of the municipalities in the United States give only primary treatment to their sewage, this process by itself is considered entirely inadequate for most needs. Municipalities and industry, faced with increased amounts of wastes and wastes that are more difficult to remove from water, have turned to secondary and even advanced waste treatment.

Secondary Treatment

Secondary treatment removes up to 90 per cent of the organic matter in sewage by making use of the bacteria it contains. The two principal processes for secondary treatment are *trickling filters* and the *activated-sludge* process. The effluent from the sedimentation tank in the primary stage of treatment flows or is pumped to a facility using one or the other of these processes.

Trickling Filter. A trickling filter is simply a bed of stones from three to ten feet deep through which the sewage passes. Bacteria gather and multiply on these stones until they can consume most of the organic matter in the sewage. The cleaner water trickles out through pipes in the bottom of the filter for further treatment. The sewage is applied to the bed of stones in two principal ways. One method consists of distributing the effluent intermittently through a network of pipes laid on or beneath the surface of the stones. Attached to these pipes are smaller, vertical pipes which spray the sewage over the stones. Another much-used method consists of a vertical pipe in the center of the filter connected to rotating horizontal pipes which spray the sewage continuously upon the stones.

Activated-Sludge Process. The trend today is toward the use of the activated-sludge process instead of trickling filters. The former process speeds up the work of the bacteria by bringing air and sludge heavily laden with bacteria into close contact with the sewage.

In the activated-sludge process the sewage from the settling tank in primary treatment is pumped to an aeration tank where it is mixed with air and sludge loaded with bacteria and allowed to remain for several hours. During this time, the bacteria break down the organic matter. From the aeration tank the sewage, now called mixed liquor, flows to another sedimentation tank to remove the solids. Chlorination of the effluent completes the basic secondary treatment. The sludge, now activated with additional millions of bacteria and other tiny organisms, can be used again by returning it to an aeration tank for mixing with new sewage and ample amounts of air.

The activated-sludge process, like most other techniques, has advantages and limitations. The size of the units needed is small so that they require comparatively little land space. Also, the process is free of flies and odors. But it is more costly to operate than the trickling filter, and it sometimes loses its effectiveness when faced with difficult industrial wastes.

An adequate supply of oxygen is necessary for the activated-sludge process to be effective. Air is mixed with sewage and biologically active sludge in the aeration tanks by three different methods. The first, mechanical aeration, is accomplished by drawing the sewage from the bottom of the tank and spraying it over the surface, thus causing the sewage to absorb large amounts of oxygen from the atmosphere. In the second method, large amounts of air under pressure are piped down into the sewage and forced out through openings in the pipe. The third method is a combination of mechanical aeration and the forced-air method.

The final phase of the secondary treatment consists of the addition of chlorine to the effluent coming from the trickling filter or the activated-sludge process. Chlorine is usually purchased in liquid form and injected into the effluent as a gas 15 to 30 minutes before it is discharged into a watercourse. If done prop-

Fig. 26.1 Comparison of three types of aeration techniques. Ditch on the left; lagoon, on the right; plastic media, trickling filter in the background. (*Courtesy EPA and Mead Paper Co.*)

erly, chlorination will kill more than 99 per cent of the harmful bacteria in an effluent.

Tertiary Treatment. Tertiary treatment is used when the waste stream must meet strict requirements governing recreational bodies of water, or must approach drinking-water standards. This may require one or several of the following processes: slow filtration; rapid filtration with activated carbon; adsorption by activated carbon; application of ozone; high-rate chlorination or use of other oxidizing chemical; or lagooning.

At each plant the question may arise as to what degree of treatment is actually required. Water quality criteria imposed by different waste-stream discharges may vary widely. Even within the same state, or for a particular river basin, different limits for each of the contaminants may be set for the section of the river under consideration.

LAGOONS AND SEPTIC TANKS

There are many well-populated areas in the United States that are not served by any sewer systems or waste-treatment plants. Lagoons and septic tanks are the usual alternatives in such situations.

A septic tank is simply a tank buried in the ground to treat the sewage from an individual home. Waste water from the home flows into the tank where bacteria in the sewage break down the organic matter and the cleaner water flows out of the tank into the ground through subsurface drains. Periodically the sludge or solid matter in the bottom of the tank must be removed and disposed of.

In a rural setting, with the right kind of soil and the proper location, the septic tank is a safe and effective means of disposing of strictly domestic wastes. Septic tanks should always be located so that none of the effluent can seep into wells used for drinking.

Lagoons or, as they are sometimes called, stabilization or oxidation ponds, also have several advantages when used correctly. They can give sewage primary and secondary treatment or they can be used to supplement other processes. A lagoon is a scientifically constructed pond, usually three to five feet deep, in which sunlight, algae, and oxygen interact to restore water to a quality equal to or better than effluent from secondary treatment. Changes in the weather affect how well a lagoon will break down the sewage.

When used with other waste-treatment processes lagoons can be very effective. A good example is the Santee, California, water reclamation project. After conventional primary and secondary treatment by activated sludge, the town's waste water is kept in a lagoon for 30 days. Then the effluent, after chlorination, is pumped to land located immediately above a series of lakes and allowed to trickle down through sandy soil into the lakes. The resulting water is of such good quality that the residents of the area can swim, boat, and fish in the lake water.

TYPES OF INDUSTRIAL WASTES

Industrial wastes exceed—if one includes the steam electric generating industry—the combined total of all other liquid wastes of human activities in terms of volume, probably averaging over 200 billion gallons per day; and they contain thousands of potentially polluting elements and compounds, often in high concentrations.

Because of the large gaps in the information base, it is not possible to hypothesize with any certainty that industrial wastes are more damaging on a national basis than the effects of unconstrained runoff; and in the Western States, damages from intense irrigation and water management practices may or may not exceed damages caused by industrial wastes. But there need be no hesitation in making the judgment that for the nation as a whole, routine discharges of industrial wastes exceed in polluting impact, sewered domestic wastes, urban runoff, mining, transportation, or accidental spills. Their sheer volume and the directly toxic influences of many kinds of industrial wastes (acids, heavy metals, some persistent organic compounds) are sufficient to justify the judgment, even though no comprehensive industrial waste inventory has been compiled.

The great variety of industrial pollutants argues against an attempt to catalog them, but for the purposes of general description we may recognize at least five distinct categories of wastes from industrial sources:

1. Oxygen-demanding materials wasted to water by industrial processes have been calculated to amount to 29.7 billion pounds of five-day chemical oxygen demand (BOD_5) per year (1968) having an aggregate discharged strength, after treatment, of 11 billion pounds of BOD_5 per year. The estimate, then, places the oxygen demand of industrial wastes before treatment at 3.5 times that of sewered domestic wastes, and the after-treatment strength at 5.5 times that of the after-treatment strength of domestic wastes.

2. Settleable and suspended solids wasted to water by industrial processes are calculated to amount to about 24 billion pounds per year before treatment, over 7 billion pounds after treatment, with the relationship between industrial and domestic wastes quite similar to that for BOD.

3. Many materials which impart acidity or alkalinity or which contain radioactivity are added to water in undeterminable amounts by manufacturing activities. Often these are directly toxic. Such materials cause a permanent change in water quality that is not reduced by natural assimilation. Dilution and neutralization—the latter usually adding to the concentration of total dissolved solids—are the only remedies in such cases.

4. Heat in the amount of about 9,152,000 million Btu's is currently discharged annually into water by industrial processes, most of it by the electrical generating industry. Although some control is exercised through recycling procedures, particularly in arid areas, the major part of the nation's industrial cooling water is discharged directly to streams. Waste heat is a pollutant in that it reduces the utility of water for additional cooling and may radically alter aquatic ecology. It also contributes to the polluting effect of water-borne materials by accelerating chemical reaction rates and by reducing the solubility of gases—including oxygen.

5. Toxic compounds occur in water as a result of industrial processes, either through direct discharge by establishments producing such compounds (e.g., factories engaged in the production of pesticides or pharmaceuticals), by the synergistic interaction of materials in water, or through the food chain. Again, no quantitative assessment is available, and for the second category of toxins, even determining probability of occurrence cannot be attempted.

Within these general categories of pollutants are included most of the possible sources of recognized water-quality problems. Only bacterial and viral presences fall outside of the group of polluting effects to which industry contributes materially. (Although these are not entirely out of the range of parameters to be included in the polluting activities of industry: meat packing plants, and other food processors in lesser measure, contribute to the presence and viability of water-borne bacteria.) Manufacturing must stand near the top, and it very probably leads, in any list of potential sources of water pollution.

And for any possible strategy of water-pollution control, industrial wastes are of critical importance. Over the last ten years, the amount and composition of industrial output has been such that for every incremental pound of BOD that has been generated directly by population increase, twenty more have been generated by increased industrial output. Increased per capita production is the essence of improvement in the standard of living, and the production of wastes is an inescapable concommitant of the production of goods. So that as population and per capita production continue to advance, we can anticipate a continuing and unavoidable advance in the volume of wastes to be managed.

Fortunately, industry has added rapidly to its inventory of waste-treatment facilities over the last decade and a half, and it appears that provision for waste treatment is routinely designed into new plants and plant additions. Estimated daily discharge of BOD from all sources in 1968 was little greater than in 1957, in spite of significant increases in population, the incidence of sewering, and industrial production. While additional treatment of municipal wastes had an unquestionable influence on the economy's ability to contain the level of waste discharges, the preponderance of industrial wastes and their more rapid rate of increase would have made containment impossible if incremental industrial-waste-treatment effectiveness had not occurred at more advanced rates than incremental industrial production.

Strategies for Management of Industrial Wastes

Unlike the public sector, where only variations on a single theme of waste treatment are possible, industrial pollutants can be managed through at least four distinct strategies. (1) The obvious procedure is the installation of industrial-waste-treatment plants. Treatment effectiveness through factory-operated plants is estimated to have been increasing at a 6.9 per cent annual rate (in terms of BOD reduction) since the early nineteen-sixties. (2) A second, and increasingly prevalent procedure is to discharge industrial wastes to public systems for treatment. The aggregare amount of BOD from industrial sources discharged to public sewers is estimated to have increased at an 8.4 per cent annual rate through the late nineteen-sixties; and industry is estimated to account for half of the current BOD loading to metropolitan area waste treatment plants. (3) Process modification and changed product formulations are probably the most effective—as well as the most efficient—means of reducing wastes. The outstanding example is the shift of the pulp and paper industry from the sulfite pulping process to the sulfate process, calculated to have been responsible for a greater reduction in the aggregate level of BOD than the combination of all of the waste-treatment plants in the U.S. More recent examples include the rising prevalence of cooling-water recycle, development of biodegradable detergents, and substitution of reclaimable hydrochloric acid for sulphuric acid in metal-pickling liquors. (4) In the absence of alternative control procedures, industry has on occasion abandoned a product line or production procedure. At the present time, phosphorus-based detergents, DDT, mercury-battery production of chlorine and alkali, and the Solvay process for production of soda ash all seem likely candidates for abandonment.

QUANTITY OF INDUSTRIAL WASTES

Table 26.1 shows reported quantities of industrial waste waters discharged in 1967. Waste-load estimates, based upon an estimate of the "average" quantity of pollutant per product unit, indicate that the chemical, paper, and food and kindred industrial groups generated about 90 per cent of the BOD_5 in industrial wastewater *before treatment.*

TABLE 26.1 Water Use in Manufacturing Industries[a]

Industry	*SIC Code Group No./Industry No.*	*Water Used (Billion Gals/yr)*	*Water Intake (Billion Gals/yr)*	*Water Discharged (Billion Gals/yr)*
Metals	331-336, 339, 341-349	7,841 (22.1%)	4,999	4,760
Chemicals	281, 283-289	7,820 (22.0%)	3,873	3,625
Pulp and paper	261-266	6,599 (18.6%)	2,264	2,089
Petroleum and coal	291, 295	7,220 (20.3%)	1,365	1,217
Food	201-209, 211	1,405 (4.0%)	818	760
Machinery and transportation	351-359, 371-374	1,225 (3.4%)	499	470
Stone, clay, glass	321-329	412 (1.2%)	251	219
Textiles	221-223, 225-229/ 2823, 2824	1,127 (3.2%)	498	461
Rubber and plastics	301, 306, 307/2821, 2822	1,106 (3.1%)	427	385
Other	242, 243, 249, 251, 311, 361-367, 369, 381, 382, 384, 386, 387, 391, 394, 396, 399	701 (2.0%)	361	327
TOTAL		35,556	15,355	14,313

[a]Data from 1968 Census of Manufactures.

Similar statistics on net waste-load discharges are not completely available. However, indications are that the extent of industrial waste-water treatment is not greater than that currently practiced for municipal wastewaters.

Industrial wastes differ markedly in chemical composition, physical characteristics, strength, and toxicity from wastes found in normal domestic sewage. Every conceivable toxicant and pollutant of organic and inorganic nature can be found in industrial wastewaters (see Table 26.2). Thus, the BOD_5 or solids content often are not adequate indicators of the quality of industrial effluents. For example, industrial wastes frequently contain persistent organics which resist the secondary treatment procedures applied normally to domestic sewage. In addition, some industrial effluents require that specific organic compounds be stabilized or that trace elements be removed as part of the treatment process.

It is therefore necessary to characterize each industrial waste water to permit comparative pollutional assessments to be made for individual industries as well as industry groups. Characterization will permit classifying the components of industrial waste waters into as few as four basic classes of pollutants to more readily collate pollution statistics and to evaluate economics of methods of treatment as well as to project least-cost methods. Proposed generalized basic classification parameters are BOD, COD, SS, and TDS into which all known pollutants can be classed. Also required is the establishment of a relative-pollution comparative index for all significant pollutants. This index, in combination with the known characteristics and volume of a waste water, will determine the relative gross-pollution severity of all industrial wastes and establish a basis for comparing the severity of pollution from different industries.

Table 26.3 presents permissible criteria for surface water for public supplies as obtained from the Report of the Committee on Water Quality Criteria, April 1, 1968. The addition of an assumed BOD_5 value of 5 mg/l to these criteria permits comparisons of the listed pollutants to be made against a unit of BOD. Under these circumstances it is relatively apparent that pollutants such as endrin and phenol (on a mg/l-concentration equivalent basis) are 5000 times more critical as pollutants than BOD. Further work in this area will permit establishment of more accurate priorities in terms of our nation's most critical needs.

Table 26.4 is a listing of all major industrial groups and industries suspected of making significant contributions to water pollution. These have been selected on the basis of a process-water intake of at least 1 billion gallons per year and with regard to the potential for pollution from the process-use of the water.

The industries listed in Table 26.4 number approximately 150 and potentially represent equally numerous waste waters of significantly different characteristics for which treatment technology must either be developed or upgraded. The interchangeability of treatment technology between similar types of wastewaters is anticipated.

A wide spectrum of technology is available for controlling and treating industrial water pollution. Some of the more important unit operations and unit processes are given in Table 26.5.

In spite of the complexity and magnitude of industrial pollution, initial estimates of the costs for clean waters from industrial sources have been made. Table 26.6 gives the types of treatments and costs for major industries, while Table 26.7 presents the estimated industrial capital requirements to abate pollution by 1973 to the extent of providing 85 per cent treatment effectiveness. It is seen that the capital requirements for this task are substantially less than the estimated capital requirements for municipal treatment or collection facilities for separating combined sewers, while the gross pollutional load contributed is substantially greater than either. This indicates that the average cost of industrial waste treatment is substantially less than for municipal waste treatment when based on treatment cost per lb BOD. If these estimates are reasonably accurate it would appear that for the most part industrial-pollution control to an equivalency with secondary treatment is within reach at a reasonable cost.

Alternatives in waste-water treatment are

TABLE 26.2 Wastewater Characteristics and Pollutants of Selected Industry

Liquid Waste Characteristic	*Domestic*	*Meat Products*	*Canned and Frozen Foods*	*Sugar*	*Textile Mill Products*	*Paper and Allied Products*
Unit volume	x	x	x	x	x	x
pH		x		x	x	x
Acidity				x	x	
Akalinity		x		x	x	x
Color					x	x
Odor					x	
Total solids		x		x	x	x
Suspended solids	x	x	x	x	x	x
Temperature		x				x
BOD_5/BOD ultimate	x	x	x	x	x	x
COD	x					
Oil and grease		x		x	x	x
Detergents						
(Surfacants)	x				x	x
Chloride		x				x
Heavy metals						
Cadmium						
Chromium					x	
Copper					x	
Iron						
Lead						
Manganese						
Nickel						
Zinc					x	
Nitrogen						
Ammonia		x		x	x	x
Nitrate					x	
Nitrite					x	
Organic		x		x		x
Total		x			x	x
Phosphorus		x				
Phenols					x	
Sulfide					x	x
Turbidity	x				x	
Sulfate					x	x
Thiosulfate					x	
Mercaptans						x
Lignins						x
Sulfur						x
Phosphates						x
Potassium						x
Calcium						x
Polysaccharides						x
Tannin						x
Sodium						x
Fluorides						
Silica						
Toxicity					x	
Magnesium						
Ammonia						
Cyanide						
Thiocyanate						
Ferrous iron						
Sulfite						x
Aluminum						
Mercury						x

[a]Source: *The Cost of Clean Water*, Volume II, FWPCA, U.S. Department of the Interior,

Groups[a]

Basic Chemicals	*Fibers Plastics and Rubbers*	*Fertilizer*	*Petroleum Refining*	*Leather Tanning and Finishing*	*Steel Rolling and Finishing*	*Primary Aluminum*	*Motor Vehicles and Parts*
X	X	X	X	X	X	X	X
X	X	X	X	X	X	X	X
X		X	X	X	X	X	X
X			X	X	X		X
X	X		X				X
X	X		X				X
X	X	X		X	X	X	X
X	X		X	X	X		X
X	X	X			X		X
X	X		X	X	X		X
X	X		X		X		X
X	X		X		X		X
X	X						X
X	X		X	X	X		X
X							X
X							X
X							X
X			X		X		X
X							X
X			X				X
X							X
X							X
X		X	X	X	X		
X							
X							
X			X	X			
X		X		X			
X		X					
X	X		X		X		X
X			X	X	X		
X			X				
X	X		X	X			
X							
X			X				
X							
X		X					X
X			X				
X			X				
X							
				X			
X			X	X			
X		X	X		X	X	
X		X					X
X	X						
			X				
X		X	X		X		
X				X	X		X
X					X		
X					X		
X							
X							
X		X	X				

U.S. Government Printing Office, Washington, D.C., April 1, 1968.

TABLE 26.3 Comparative Pollution Index Based on Surface Water Criteria for Public Water Supplies

Constituent or Characteristic	*Permissible Criteria*[b]	*Desirable Criteria*[c]
Physical		
color (color units)	75	<10
odor	Narrative	Virtually absent
temperature[a]	Narrative	Narrative
turbidity	Narrative	Virtually absent
Microbiological		
coliform organisms	10,000/100 ml[d]	<100/100 ml[d]
fecal coliforms	2,000/100 ml[d]	<20/100 ml[d]
Inorganic chemicals		
alkalinity	Narrative	Narrative
ammonia	0.5 mg/l (as N)	<0.01 mg/l
arsenic[a]	0.05 mg/l	Absent
barium[a]	1.0 mg/l	Absent
boron[a]	1.0 mg/l	Absent
cadmium[a]	0.01 mg/l	Absent
chloride[a]	250 mg/l	<25 mg/l
chromium[a] hexavalent	0.05 mg/l	Absent
copper[a]	1.0 mg/l	Virtually absent
dissolved oxygen	⩾4 mg/l (monthly mean) ⩾3 mg/l (individual sample)	Near saturation
fluoride[a]	Narrative	Narrative
hardness[a]	Narrative	Narrative
iron (filterable)	0.3 mg/l	Virtually absent
lead[a]	0.05 mg/l	Absent
manganese[a]	0.05 mg/l	Absent
nitrates plus nitrites[a]	10 mg/l	Virtually absent
pH (range)	6.0–8.5	Narrative
phosphorus[a]	Narrative	Narrative
selenium[a]	0.01 mg/l	Absent
silver[a]	0.05 mg/l	Absent
sulfate[a]	250 mg/l	<50 mg/l
total dissolved solids[a] (filterable residue)	500 mg/l	<200 mg/l
uranyl ion[a]	5 mg/l	Absent
zinc	5 mg/l	Virtually absent
Organic chemicals		
carbon chloroform extract[a] (CCE)	0.15 mg/l	<0.04 mg/l
cyanide[a]	0.20 mg/l	Absent
methylene blue active substances[a]	0.5 mg/l	Virtually absent
oil and grease[a]	Virtually absent	Absent
pesticides		
aldrin[a]	0.017 mg/l	Absent
chlordane[a]	0.003 mg/l	Absent
DDT[a]	0.042 mg/l	Absent
dieldrin[a]	0.017 mg/l	Absent
endrin[a]	0.001 mg/l	Absent
heptachlor[a]	0.018 mg/l	Absent
heptachlor epoxide[a]	0.018 mg/l	Absent
lindane[a]	0.056 mg/l	Absent
methoxychlor[a]	0.035 mg/l	Absent
organic phosphates plus carbamates[a]	0.1 mg/l[e]	Absent
toxaphene[a]	0.005 mg/l	Absent
herbicides		
2, 4-D plus 3, 4, 5-T, plus 2, 4, 5-TP[a]	0.1 mg/l	Absent
phenols[a]	0.001 mg/l	Absent
Radioactivity		
gross beta[a]	1,000 pc/l	<100 pc/l
radium-226[a]	3 pc/l	<1pc/l
strontium	10 pc/l[f]	<2 pc/l
BOD	5 mg/l	2 mg/l

[a]The defined treatment process has little effect on this constituent.

[b]Permissible criteria are defined as those characteristics and *concentrations of substances in raw surface waters* which will allow the production of a safe, clear, potable, aesthetically pleasing, and acceptable public water supply which meets the limits of drinking water standards *after treatment*. This treatment may include, but will not include more than, the processes described above.

TABLE 26.4 Standard Industrial Classification of Industries of Significance for Water Pollution

CODE	
20	FOOD AND KINDRED PRODUCTS
201	Meat products
2011	Meat slaughtering plants
2013	Meat processing plants
2015	Poultry dressing plants
202	Dairy products
2021	Creamery butter
2022	Natural and process cheese
2023	Condensed and evaporated milk
2026	Fluid milk
203	Canned and frozen foods
2033	Canned fruits and vegetables
2034	Dehydrated food products
2035	Pickled foods, sauces, salad dressings
2037	Frozen fruits and vegetables
204	Grain mill products
2041	Flour and other grain mill products
2043	Cereal preparations
2046	Wet corn milling
205	Bakery products
206	Sugar
207	Candy and related products
208	Beverage industries
2082	Malt liquors
2084	Wines and brandy
2085	Distilled liquors
2086	Soft drinks
209	Miscellaneous foods and kindred products
2091	Cottonseed oil mills
2092	Soybean oil mills
2094	Animal and marine fats and oils
2096	Shortening and cooking oils
22	TEXTILE MILL PRODUCTS
2211	Weaving mills, cotton
2221	Weaving mills, synthetic
2231	Weaving, finishing mills, wool
225	Knitting mills
226	Textile finishing, except wool
228	Yarn and thread mills
229	Miscellaneous textile goods
24	LUMBER AND WOOD PRODUCTS
2421	Sawmills and planning mills
2432	Veneer and plywood plants
2491	Wood preserving

CODE	
26	PAPER AND ALLIED PRODUCTS
2611	Pulp mills
2621	Paper mills, except building
2631	Paperboard mills
264	Paper and paperboard products
265	Paperboard containers and boxes
2661	Building paper and building board mills
28	CHEMICALS AND ALLIED PRODUCTS
281	Basic chemicals
2812	Alkalies and chlorine
2818	Organic chemicals, n.e.c.
2819	Inorganic chemicals, n.e.c.
282	Fibers, plastics, and rubbers
2821	Plastics materials and resins
2823	Cellulosic man-made fibers
2824	Organic fibers, noncellulosic
283	Drugs
284	Cleaning and toilet goods
2851	Paints and allied products
2861	Gum and wood chemicals
287	Agricultural chemicals
289	Miscellaneous chemical products
29	PETROLEUM AND COAL PRODUCTS
2911	Petroleum refining
295	Paving and roofing materials
30	RUBBER AND PLASTICS PRODUCTS, n.e.c.
3069	Rubber products, n.e.c.
3079	Plastics products, n.e.c.
31	LEATHER AND LEATHER PRODUCTS
3111	Leather tanning and finishing
32	STONE, CLAY, AND GLASS PRODUCTS
3211	Flat glass
3241	Cement, hydraulic
325	Structural clay products
326	Pottery and related products
327	Concrete and plaster products
3281	Cut stone and stone products
329	Nonmetallic mineral products
33	PRIMARY METAL INDUSTRIES
331	Steel rolling and finishing mills
332	Iron and steel foundries
333	Primary nonferrous metal
3341	Secondary nonferrous metals

[c]Desirable criteria are defined as those characteristics and concentrations of substances in the *raw surface waters* which represent high-quality water in all respects for use as public water supplies. Water meeting these criteria can be treated in the defined plants with greater factors of safety or at less cost than is possible with waters meeting permissible criteria.
[d]Microbiological limits are monthly arithmetic averages based upon an adequate number of samples. Total coliform limit may be relaxed if fecal coliform concentration does not exceed the specified limit.
[e]As parathion in cholinesterase inhibition. It may be necessary to resort to even lower concentrations for some compounds or mixtures.
[f]Maximum value found in tap water analysis of 20 communities.
Source: *Water Quality Criteria*, FWPCA, U.S. Department of the Interior, U.S. Government Printing Office, Washington, D.C., April 1, 1968.

TABLE 26.5 Unit Operations and Processes Applicable to Treatment and Control of Industrial Water Pollution

	Dissolved BOD Removed	*Suspended and Colloidal Solids Removed*	*Dissolved Refractory Organics Removal*	*Dissolved Inorganics Removal*	*Dissolved Nutrient Removal*	*Microorganisms Removal*	*Concentrate Removal*
Biological processes							
Activated sludge	x	x	–	–	–	x	–
Anaerobic digestion	x	x	–	–	–	–	x
Bio-filters	x	–	–	–	x	–	–
Biomass treatment (algae harvesting)	x	–	–	–	x	–	x
Biological PO_4 removal	x	–	–	–	x	–	–
Extended aeration							
Bio-denitrification	L[a]	–	L	–	x	–	–
Bio-nitrification	x	x	–	–	–	–	–
Pasveer oxidation ditch	x	x	–	–	–	x	–
Chemical processes							
Chemical oxidation							
Catalytic oxidation	x	x	x	–	–	x	–
Chlorination	x	x	x	–	–	x	–
Ozonation	L	–	x	–	–	x	–
Wet oxidation	x	x	x	–	–	x	–
Chemical precipitation	–	–	x	x	x	–	–
Chemical reduction	–	–	–	–	x	–	–
Coagulation							
Inorganic chemicals	x	x	–	–	x	x	–
Polyelectrotytes	x	x	–	–	–	x	–
Disinfection	–	–	–	–	–	x	–
Electrolytic processes							
Electrodialysis	–	–	–	x	x	x	–
Electrolysis	–	–	x	x	–	–	–
Extractions							
Ion exchange	–	–	–	x	x	–	–
Liquid-liquid (solvent)	–	–	x	x	–	–	–
Incineration							
Fluidized-bed	x	x	x	–	–	x	x
Physical processes							
Carbon absorption							
Granular activated	x	–	x	–	–	–	–
Powdered	x	x	x	–	–	x	–
Distillation	x	x	x	x	x	x	–

Filtration							
Coal filtration	L	X	—	—	—	X	—
Diatomaceous-earth filtration	—	X	—	—	—	X	—
Dual-media filtration	—	X	—	—	—	X	—
Micro-screening	—	X	—	—	—	X	—
Sand filtration	—	X	—	—	—	X	—
Flocculation-sedimentation	—	X	—	—	—	X	—
Foam separation	—	X	X	X	—	—	—
Freezing	X	—	X	X	X	—	—
Gas hydration	X	—	X	X	X	—	—
Reverse osmosis	X	X	X	X	X	X	—
Stripping (air or steam)	X	X	X	—	X	—	—

[a]Under specific conditions there will be limited effectiveness.

shown in the sketch below:

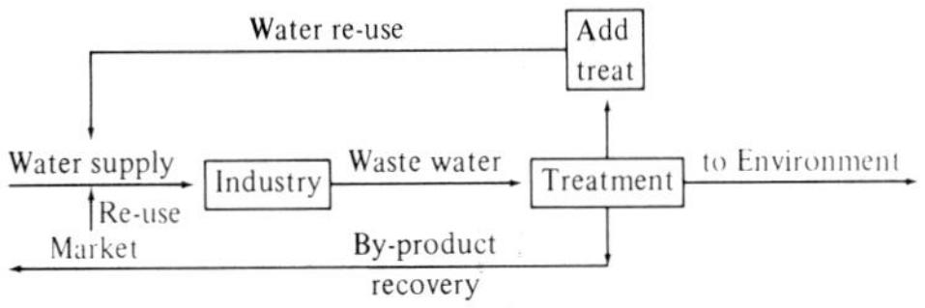

The alternatives shown primarily consist of:

a. Waste-water treatment (as required to abate pollution to meet water quality standards)
   (1) For discharge (to meet necessary water quality criteria).
   (2) For re-use (to meet industrial water-quality demands to conserve water and offset cost of treatment).
b. In-plant measures (to reduce pollutants and water discharge).
   (1) Operational (housekeeping techniques and manufacturing procedures).
   (2) Design (to permit re-use, to reduce waste-water generation).
c. Residue treatment.
   (1) By-product recovery (to reduce gross disposal, utilizes values).
d. Combined methods.
   (1) Joint treatment (to utilize scale factors, off-peak capacity, synergistic effects).
   (2) Other (combined a-b-c methods as appropriate).

The alternatives best suited for implementation will depend on many factors and local conditions. For nonprogressive industries, in-plant measures should be explored for potential application. For industries which have demonstrated effective treatment methods, lower cost alternatives of treatment stressing re-use and by-product recovery should be given consideration.

INDUSTRIAL WATER RE-USE

Waste products must be considered as an integral part of any manufacturing process and the cost of treatment of industrial wastes must be included in the pricing of the product. Waste-disposal operations normally result in a net cost to the industry producing the waste.

TABLE 26.6 Types of Treatment and Cost for Major Industries[c]

Industry	*Process*	*Primary*	*Secondary (Including Primary)*	*Tertiary (Including Primary and Secondary)*
		Screening Gravity Separation Chlorination Detention	Chemical Coagulation-clarification Flotation Filtration Bio-oxidation	Adsorption Ozonation Wet oxidation Ion exchange
Steel plants	Cold roll mills Roller grinding Wire drawing and coating			
Oil refineries	Coke plant by-product By-product extraction Recovery			
Chemical and Pharmaceutical	By-product extraction Evaporation Adsorption	Capital[a] 2–4%	4–6%	6–9%
Food processing	Washing, conveying Cooking, cooling	Operating[b] 4–8¢	8–16¢	16–24¢
Pulp and paper	Digestion, washing Bleaching, deinking			
Metal processing	Plating			

[a]Cost of treatment plant including required pumping, excluding land. Expressed in percentage of total capital cost.
[b]Total operating cost per 1,000 gallons of waste stream treated, including material, labor, supervision, maintenance, pumping, and capital charges.
[c]Lacy, W. J. and Cywin, A., "The Federal Water Pollution Control Administration's Research and Development Program: Industrial Pollution Control," *Plating*, 1968.

However by-product recovery and utilization techniques can reduce the cost of treatment and frequently prove to be less expensive than other methods of disposal.

Recycled water ultimately may be the most valuable product due to supply shortages, increasing water-supply costs, increasing water-treatment costs, and mounting municipal sewerage charges. The recovery of product fines, useable water, and thermal energy are key methods of reducing over-all waste treatment costs and must always be considered.

Frequently, waste streams can be eliminated or significantly reduced by process modifications. One notable example is the application of save-rinse and spray-rinse tanks in plating lines. This measure brings about a substantial reduction in waste volume as well as a net reduction in metal dragout.

Industries in general are becoming more aware of the need for over-all pollution control and product (or by-product) recovery. This is not only because pollution affects the environment but also because it affects the general public who are the customers. In addition, industries, too, depend upon our nation's rivers and streams for suitable water for their manufacturing processes.

METALS AND METAL PRODUCTS

Three broad categories of industrial activities are included in this area. These are the ferrous-metal industries, the nonferrous-metal industries, and metal-finishing operations.

The steel industry uses approximately 16 billion gallons of water per day or 19 per cent of the total industrial water usage. Most of the water is used for noncontact cooling purposes. Approximately 2.5 billion gallons per day are used in processing operations such as coke-oven-gas scrubbing, blast-furnace-gas washing, basic-oxygen-gas scrubbing, hot rolling, pick-

TABLE 26.7 Total Current Value of Waste Treatment Requirements of Major Industrial Establishments[a] and Total Investment Required to Meet 1/1/73 Air and Water Standards[b]

Industry	Value (Millions of 1968 Dollars) Fiscal 1969 (Estimate)	Fiscal 1973 (Census Projection)		Investment Required to Meet 1973 Standards (Millions of Dollars)
Food and kindred products	743.1	669.6		630
Textile mill products	165.2	170.9		140
Paper and allied products	321.8	917.6		1620
Chemical and allied products	379.7	1003.8		2250
Petroleum and coal	379.4	272.3		2740
Rubber and plastics	41.1	58.9		110
Primary metals	1473.8	1383.7		3980
Machinery	39.0	55.9		450
Electrical machinery	35.8	51.3		320
Transportation equipment	216.0	156.4		800
All other manufacturing	203.7	291.8		1160
TOTAL capital requirement	3998.6	5032.2	(Elec. Util.)	3880
			TOTAL	18080
Plant currently provided				
By Industry	2215.3	1752.3		
Through municipal facilities	731.4	635.9		
Current backlog	1051.9	2644.0		

[a]At least 85 per cent reduction of standard biochemical oxygen demand (determined according to the five-day test) and of settleable and suspended solids is assumed.
[b]Data Department, McGraw-Hill Economics Department.
Source: "The Cost of Clean Water," Vol. II, FWPCA, U.S. Department of the Interior, U.S. Government Printing Office, Washington, D.C., January 10, 1968.

ling and rinsing, cold rolling, and dipping and other finishing operations. These operations contaminate the water with large amounts of particulate matter, oil, acid, soluble salts, ammonia, cyanide, phenols, and other organic and mineral compounds. Most of the waste streams are better suited for chemical and/or physical treatment methods rather than biological methods.

The nonferrous-metal industries include aluminum, copper, zinc, lead, and nickel. These industries use approximately 1.3 billion gallons of water per day. As in the case of the steel industry, the wastes are high in particulate and mineral compounds and are best treated by chemical and/or physical means.

There are approximately 20,000 captive and independent metal-finishing operations within the United States. The principal operations include stripping, electroplating, anodizing, and etching (e.g., printed circuits). While waste volumes are not normally large compared to the steel, paper, food processing, or petroleum industries, their frequently corrosive and highly toxic nature make these wastes particularly hazardous. Waste treatment, usually by chemical processes, is necessary to protect sewer lines from corrosion, sewer maintenance crews from toxic gases and minerals, receiving municipal biological-treatment plants from deactivation by slugs or accumulations of toxic materials, and for protection of all species coming into contact with the receiving waters.

The suspended material from blast-furnace-gas-washer water and recovered mill scale are sintered and returned to the blast furnace or steel-making furnace for reprocessing. Oils recovered from cold mill rolling solutions are processed for re-use in the mill operation or for sale. Acid regenerated from spent pickle liquor is returned to the pickling tanks

Some other by-products are recovered in the course of gas-scrubbing cleaning operations

Fig. 26.2 Electrolytic recovery of copper from a brass acid-rinse waste stream. *(Courtesy EPA and Valco Brass Co., Kenilworth, N.J.)*

which reduce potential waste loads well ahead of the point of discharge of water effluents. Ammonium sulfate is produced in removing ammonia from coke oven gas with sulfuric acid; phenol is recovered from coke oven gas by vapor recirculation or solvent extraction. Copper is recovered from pickle liquor in a few brass plants, and one or two plants are able to dispose of pickle liquor to chemical or pigment manufacturers.

Chrome, nickel, and copper acid-type plating solutions may be reclaimed from the rinse tank by evaporation in glass-lined equipment and the concentrated solution returned to the plating systems. The water condensed from the steam is then re-used in the rinse tank following the plating tank to eliminate buildup of natural water salts. This process has proved effective for recovering valuable metal salts. The high initial cost for equipment is more than recovered, not only because of waste treatment, but also because of recovery of metals.

Geographical Aspects of the Industry

Most of the iron and steel industry in the United States is centered in the integrated facilities of large corporate enterprises; the eight largest producers are among the 100 largest industrial corporations in this country. The comparatively small integrated producer in this industry represents a very large industrial complex. Even the smallest producer in the industry may not be considered a very small industrial operation.

The primary metal industries are concentrated in the industrial Great Lakes–Northeast seacoast portion of the country, with additional manufacturing centered in California, Mississippi, Texas, and Washington, and minor participation in other parts of the country.

Technological Control Methods for Steel Mill Wastes

The primary method of treatment of by-product coke-plant wastes is to use recovery and removal units with high efficiencies, phenol being the main contaminant recovered. The BOD can be reduced by about one-third by recirculation and re-use of contaminated waters, and the recovery of by-products may be profitable in the case of ammonium sulfate, crude tar, naphthalene, coke dust, gas, benzene, toluene, and xylene. Quench water is usually settled to remove coke dust, and the supernatant liquor from the settling tanks is generally re-used for quenching. Gravity separators are used to remove free oil from the wastes from benzol stills, since the emulsified oils generally are not treated, and without separation the free portion of the oil would reach the sewers. Final-cooler water is also recirculated to reduce the amount of phenol being discharged to waste. Phenols may be removed by (1) conversion into nonodorous compounds, or (2) recovery as crude phenol or sodium phenolate, which has some commercial value. The conversion may be either biological (activated sludge or trickling filtration) or physical (ammonia-still wastes used to quench incandescent coke, a process which evaporates the NH_3). Although certain concentrations of phenol (0 to 25 ppm) may be handled by biological units, dilution with municipal sewage provides a buffering and diluting medium. If coke-plant wastes are added to sewer systems, careful control of the discharge is necessary so as not to upset the treatment-plant processes.

In treating flue dust, sedimentation, followed by thickening of the clarifier overflow with lime to encourage flocculation, has been found most effective for removing iron oxide and silica. Ninety to 95 per cent of the suspended matter settles within one hour, resulting in an effluent having less than 50 ppm suspended solids. Primary and secondary (lime-coagulated) thickened sludges are also obtained, these can then be lagooned without creating nuisances. The resulting waste water may be recycled.

The treatment of pickling liquor is a problem of considerable magnitude. The recovery of by-products from waste pickling liquor is not economically feasible for most small steel plants; therefore, they neutralize the liquor with lime. However, some companies do obtain by-products from this waste.

The recently developed Blaw-Knox-Ruthner process for the recovery of sulfuric acid involves the concentration, by evaporation, of waste pickling liquor before discharging it to a reactor where anhydrous hydrogen chloride gas is bubbled through it and reacts with the ferrous sulfate to produce H_2SO_4 and $FeCl_2$. The ferrous chloride is separated from the sulfuric acid (which is returned to the pickling line) and is converted to iron oxide in a direct-fired roaster. This liberates HCl, which is recovered by scrubbing and stripping and recycled to the reactor.

Neutralization of pickle-liquor waste with lime is costly because there is no salable end product and because there is a voluminous, slow-settling sludge which is difficult to dispose of. Neutralization takes place in four stages: (1) formation of ferric hydrate with a pH below 4; (2) formation of acid sulfate; (3) formation of the ferrous hydrate with a pH between 6 and 8; and (4) formation of the normal sulfate. Calcium and dolomitic lime are the least expensive neutralizing agents, caustic soda and soda ash being too expensive for this purpose. Even with the cheaper chemicals the overall cost of neutralization may range from $5 to $10 per 1,000 pounds of acid attributable to the pickling operation.

Hot-strip mills are becoming larger and faster, and they roll thinner gauges due to higher production requirements and to the market demand for thinner strip. Hot-scarfing is being increasingly used due to the market demand for higher quality products. Hydrochloric acid is substituted for sulfuric acid to obtain faster pickling and to produce a better surface on the product. Shot blasting is limited as a substitute technique for pickling due to the investment in pickling facilities and the limitations of available machines.

Spent pickling solutions and coke-plant wastes can be discharged to municipal sewers and treated with municipal wastes but this is not practiced widely in the industry. Acids can attack sewer materials and inhibit the biological processes in sewage-treatment plants; also, sulfates can cause corrosion in concrete sewers of certain types. Pickle liquor must be partially neutralized before discharge to municipal sewers and the flow rate must be regulated in relation to the relative volumes of sewage to which it is added.

Electrolytic tin-plating and galvanizing replace hot-dip processes due to the higher production rates possible with the newer processes and to the reduced plating metal used per unit of plated surface. The principal wastes—broadly suspended solids, lubricating oils, acids, soluble metals, emulsions, and coke-plant chemicals—appear in the effluents in various combinations, depending upon the particular manufacturing operation. The methods of treatment and the efficiencies of the treatment methods vary depending upon the sources of the wastes and the combinations in which they occur.

The general trend in the steel industry is toward the use of subprocesses which will produce products of lighter unit weight at increasingly higher speeds with minimum manual operations. Production units generally tend toward the largest sizes in order to realize the economies of scale. Thus, use of the Bessemer converter and open-hearth furnaces is declining, while the use of basic-oxygen furnaces and continuous-casting methods are increasing.

Particularly difficult waste problems arise from certain steel industry subprocesses, especially some of the newer production meth-

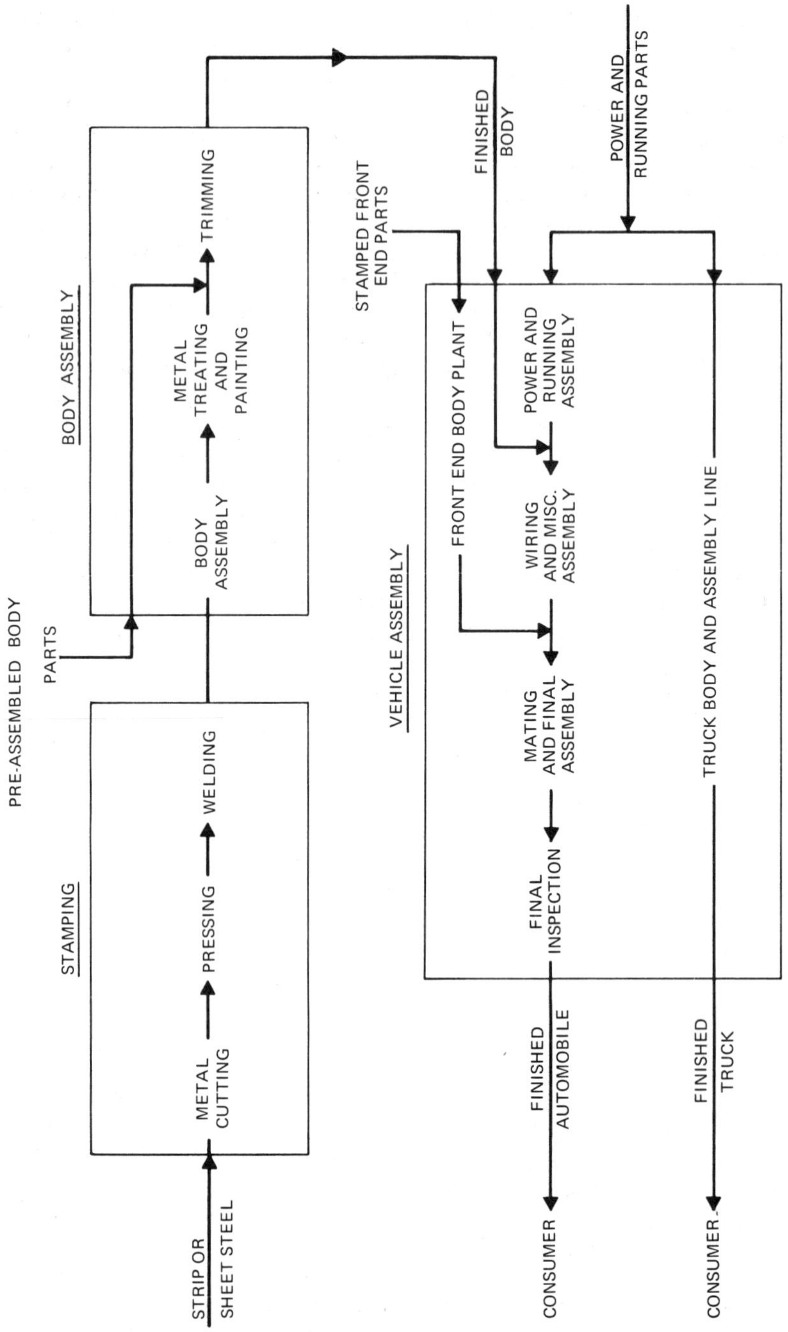

Fig. 26.3a Schematic flow diagram of automobile manufacturing.

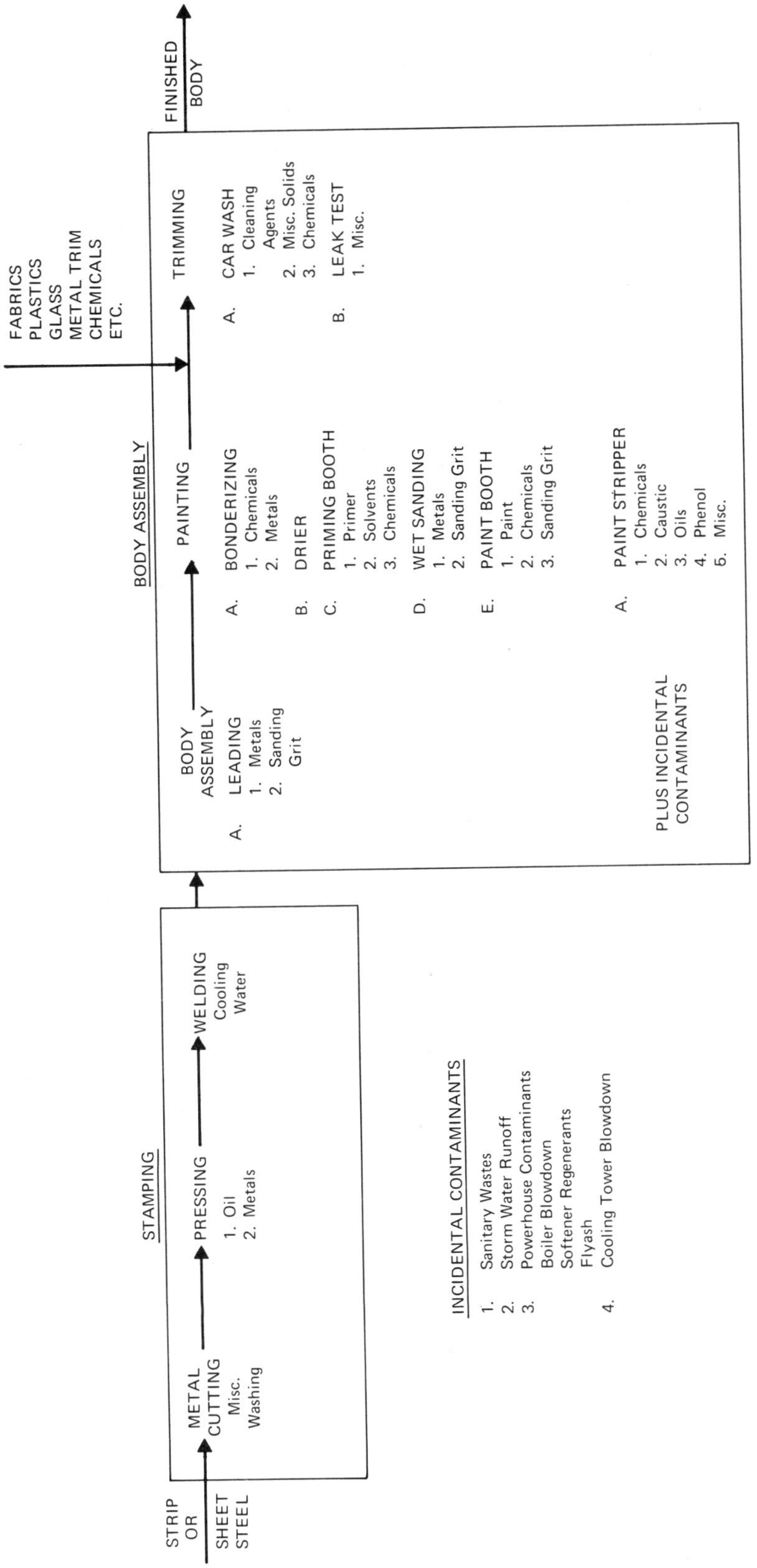

Fig. 26.3b Significant pollutants from body stamping and assembly operations.

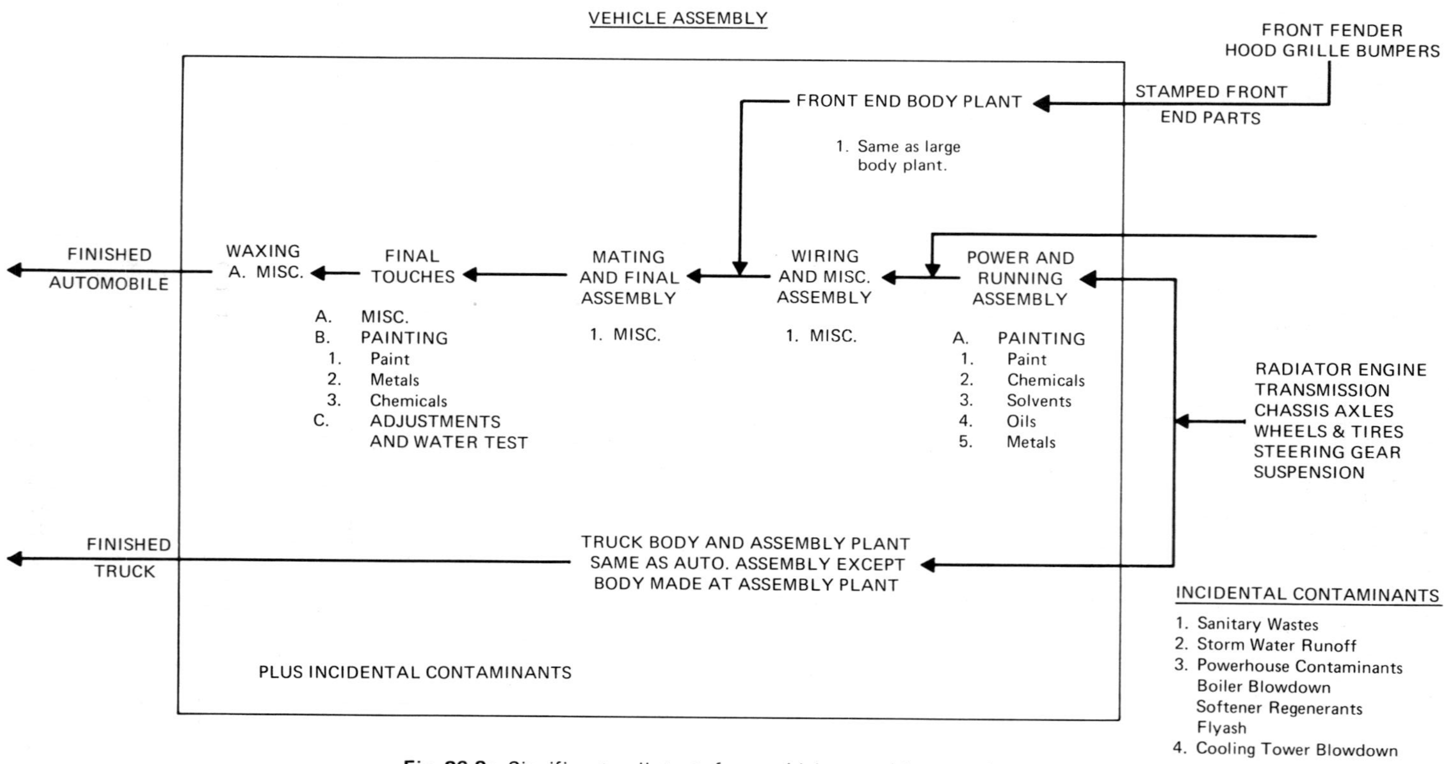

Fig. 26.3c Significant pollutants from vehicle assembly operations.

ods which are geared to high-speed, high-volume outputs. The gas-washer water from ferromanganese furnaces contains particles of submicron sizes which are extremely difficult to separate by sedimentation. Such furnaces also produce effluents which contain greater concentrations of cyanides* than those from iron-making furnaces. The waste waters from basic-oxygen furnaces contain very fine suspended particles which are difficult to treat; the waste waters from oxygen-lanced open-hearth furnaces pose similar problems.

The very fine particles in the waste waters from the newer hot-strip mills are not readily separated in ordinary sedimentation equipment and impart black or red colors to effluent streams, even in low concentrations. Effluents from cold mills, on the other hand, contain soluble oils which require extensive treatment for removal. Such emulsions are particularly stable when detergents are used in the rolling operation.

A characteristic of steel-industry effluents that makes for difficult waste-control problems is their large volume. Most industrial-waste discharges are not of the magnitude found in the steel industry. A waste stream flowing at the rate of 10,000 to 25,000 gallons per minute presents a difficult treatment problem.

Levels of technology in the steel industry may be described according to the relative prevalence of certain subprocesses in a particular plant.

In general, the newer technologies produce greater water-borne waste loads because lighter products are becoming more predominant in the industry and production rates are generally increasing. Wastes are generated in proportion to the surface area of steel exposed during rolling and finishing and to the relative gas-liquid interfacial areas in iron-making and steel-making; the newer technologies tend to maximize these areas. The newer installations, however, generally incorporate waste-treatment facilities, and actual discharges of newer plants are not necessarily greater. Continuous casting and shotblast cleaning as replacements for ingot molding and pickling, respectively, reduce waste loads.

One of the major users of steel is the auto industry where body manufacturing and assembly processes account for several pollutants from each major operation. These are shown in Figures 26.3a, 26.3b and 26.3c. The quantities of contaminants produced in an automobile body assembly plant are given in Table 26.8.

CHEMICAL AND ALLIED PRODUCTS

The dangerous and polluting effects of chemical industry discharges are felt in all areas of the country where this industry is represented. Principal concentrations are in the industrial Great Lakes and Northeast regions, but they spread into the Southeast, Louisiana, and Texas. Significant production also occurs in the West, particularly California and Washington. The regional growth rates reflect a continuing trend to move production facilities of inorganic chemicals closer to raw materials and markets. The industry, as a whole, is thus tending to concentrate in the Midwest and Southwest.

The production of organic chemicals results in many types of contaminated waste waters, and the treatment methods employed cover the range of known practical techniques. Typical waste-water characteristics associated with some chemical products can be seen in Table 26.9. In-plant control is the first step in instituting treatment practices. Such controls include the salvage of unreacted chemicals, recovery of by-products, multiple re-use of water, good housekeeping techniques to reduce leaks and spills, and changes in processing methods. These controls can result in reducing the concentrations of almost all potential pollutants and can, most importantly, reduce the volumes of waste waters requiring treatment. Physical treatment methods such as sedimentation or flotation are used primarily to remove coarse suspended matter and floating oils and scums. Filtration is used as a form of tertiary treatment for re-use or as a pretreatment prior to sedimentation, filtration, or biological treatment. Bio-

*The term "cyanide" as used here refers to a broad range of chemical compounds, all of which contain at least one cyanide (CN^-) ion.

TABLE 26.8 Contaminants Produced in an Automobile Body Assembly Plant

Process	Tank	Tank Contents	Capacity (Gallons)	Dump Schedule	Gallons/100 Bodies	pH	Pounds Per 100 Bodies: Hydroxide (OH)	Carbonate (CO_2)	Bicarbonate (HCO_3)	Total Acidity (as H_2SO_4)	Free Mineral Acidity (as H_2SO_4)	Orthophosphate (PO_4)	Total Chromium (CrO_4)	COD	Suspended Solids
Bonderite	1	Ridoline 76 Ridosol 250	5,465	1 week	131	8.6	–	0.053	0.865	–	–	0.465	0.008	19.706	1.916
	2	Rinse	2,430	1 day	292	7.1	–	–	0.413	–	–	0.207	0.004	1.850	0.451
	3	Rinse	2,430	1 day	292	6.2	–	–	0.059	–	–	0.159	0.002	1.266	0.401
	4	Granodine 18	4,056	6 mo	3.7	3.7	–	–	–	0.147	0.016	0.134	Neg	0.008	0.015
	5	Rinse	2,430	1 day	292	5.4	–	–	0.019	–	–	0.256	Neg	0.064	0.121
	6	Deoxylyte	2,430	1 day	1,446	4.7	–	–	–	2.772	–	1.506	2.952	0.314	3.010
	7	Deoxylyte	420	No dump	50	4.7	–	–	–	0.096	–	0.052	0.102	0.011	0.010
Undercoat, 1st prime	1	Metafloat 62	53,863	1 week	1,295	8.4	–	1.295	15.052	–	–	0.162	1.014	77.687	8.686
	Caustic	Caustic	4,309	1 mo	24	9.8	2.402	2.884	–	–	–	0.017	0.367	5.494	0.130
Undercoat, 2nd prime	2	Metafloat 62	53,863	1 week	1,295	8.4	–	1.295	15.052	–	–	0.162	1.014	77.687	8.686
	Caustic	Caustic	4,309	1 mo	24	9.8	2.402	2.884	–	–	–	0.017	0.367	5.494	0.130
Small parts bonderite	1	Durdine 210B	3,500	6 mo	3.2	5.4	–	–	0.003	–	–	0.145	Neg	0.045	0.007
	2	Rinse	1,500	1 day	180	6.7	–	–	0.067	–	–	0.037	Neg	0.029	0.014
	3	Deoxylyte	1,500	1 day	230	4.8	–	–	0.018	–	–	0.153	0.479	0.173	0.018
Lacquer	1	Metafloat 70	46,382	1 mo	253	6.4	–	–	1.935	–	–	0.002	0.007	19.009	1.352
	2	Metafloat 70	46,382	1 mo	253	6.4	–	–	1.935	–	–	0.002	0.007	19.009	1.352
	3	Metafloat 70	19,750	1 mo	108	6.4	–	–	0.824	–	–	0.001	0.003	8.094	0.576
	Caustic 1	Caustic	4,309	6 weeks	17	>10	0.424	0.173	–	–	–	0.002	0.006	3.453	0.069
	Caustic 2	Caustic	4,309	6 weeks	17	>10	0.424	0.173	–	–	–	0.002	0.006	3.453	0.069
	Caustic 3	Caustic	4,309	6 weeks	17	>10	0.424	0.173	–	–	–	0.002	0.006	3.453	0.069
Small parts booth	1	Enamel paint	9,636	2 days	579	8.7	–	0.502	7.055	–	–	0.034	0.084	25.094	6.9001
Deadner booth	1	Metafloat 70 Undercoat FS-1300	1,077	1 mo	6	8.8	–	0.023	0.044	–	–	Neg	Neg	0.020	0.003
Quarter panel	1	Metafloat 70	2,633	1 mo	14	8.9	–	0.042	0.068	–	–	0.003	0.002	0.036	0.005
Water-test booth	1	Sno-Flake No. 62	9,164	6 weeks	37	6.4	–	–	0.018	–	–	0.001	Neg	0.101	0.055
Grind and wire booth	1	Filings	17,954	1 mo	98	–	–	–	–	–	–	–	–	–	–

TABLE 26.9 Waste-water Characteristics Associated with some Chemical Products

	Flow (gal/ton)	*BOD (mg/l)*	*COD (mg/l)*	*Other Characteristics*
Primary petrochemicals				
ethylene	50–1,500	100–1,000	500–3,000	Phenol, pH, oil
propylene	100–2,000	100–1,000	500–3,000	Phenol, pH
Primary intermediates				
toluene	300–3,000	300–2,500	1,000–5,000	
xylene	200–3,000	500–4,000	1,000–8,000	
ammonia	300–3,000	25–100	50–250	Oil, nitrogen
methanol	300–3,000	300–1,000	500–2,000	Oil
ethanol	300–4,000	300–3,000	1,000–4,000	Oil, solids
buthanol	200–2,000	500–4,000	1,000–8,000	Heavy metals
ethylbenzene	300–3,000	500–3,000	1,000–7,000	Heavy metals
chlorinate hydrocarbons	50–1,000	50–150	100–500	pH, oil, solids
Secondary intermediates				
phenol, cumene	500–2,500	1,200–10,000	2,000–15,000	Phenol, solids
acetone	500–1,500	1,000–5,000	2,000–10,000	
glycerin, glycols	1,000–5,000	500–3,500	1,000–7,000	
urea	100–2,000	50–300	100–500	
acetic anhydride	1,000–8,000	300–5,000	500–8,000	pH
terephtalic acid	1,000–3,000	1,000–3,000	2,000–4,000	Heavy metals
acrylates	1,000–3,000	500–5,000	2,000–15,000	Solids, color, cyanide
acrolonitrile	1,000–10,000	200–700	500–1,500	Color, cyanide, pH
butadiene	100–2,000	25–200	100–400	Oil, solids
styrene	1,000–10,000	300–3,000	1,000–6,000	
vinyl chloride	10–200	200–2,000	500–5,000	
Primary polymers				
polyethylene	400–1,600		200–4,000	Solids
polypropylene	400–1,600		200–4,000	
polystyrene	500–1,000		1,000–3,000	Solids
polyvinyl chloride	1,500–3,000	50–500	1,000–2,000	
cellulose acetate	10–200	500–2,000	1,000–5,000	
butyl rubber	2,000–6,000	800–2,000	2,500–5,000	
Dyes and pigments	50,000–250,000	200–400	500–2,000	Heavy metals, color, solids, pH
Miscellaneous organics				
isocyanate	5,000–10,000	1,000–2,500	4,000–8,000	Nitrogen
phenyl glycine	5,000–10,000	1,000–2,500	4,000–8,000	Phenol
parathion	3,000–8,000	1,500–3,500	3,000–6,000	Solids, pH
tributyl phosphate	1,000–4,000	500–2,000	1,000–3,000	Phosphorus

Source: Roy F. Weston and REA.

logical treatment is most widely used in the industry due to the nature of the wastes, that is, their general susceptibility to biodegradation as evidenced by relatively high BOD values. The general waste-water-treatment-process/sequence-substitution diagram is given in Fig. 26.4.

The entire chemical industry has generally found that in-plant, separate treatment has economic advantages, particularly when significant quantities of waste water are involved. The composition of typical clean-water effluent streams such as cooling water and steam are shown in Table 26.10.

Many waste-treatment methods are available to the inorganic chemical industry, depending on the degree of treatment required. However, equalization, neutralization, sedimentation, and lagooning processes are most widely used. Biological treatment is not applicable since the contaminants are primarily dissolved or suspended inorganic materials. Plants with small discharges tend to employ only equalization and neutralization with total discharge to municipal sewer systems for joint treatment.

PAPER AND ALLIED PRODUCTS

Dun and Bradstreet lists 6,683 production establishments under "Standard Industrial

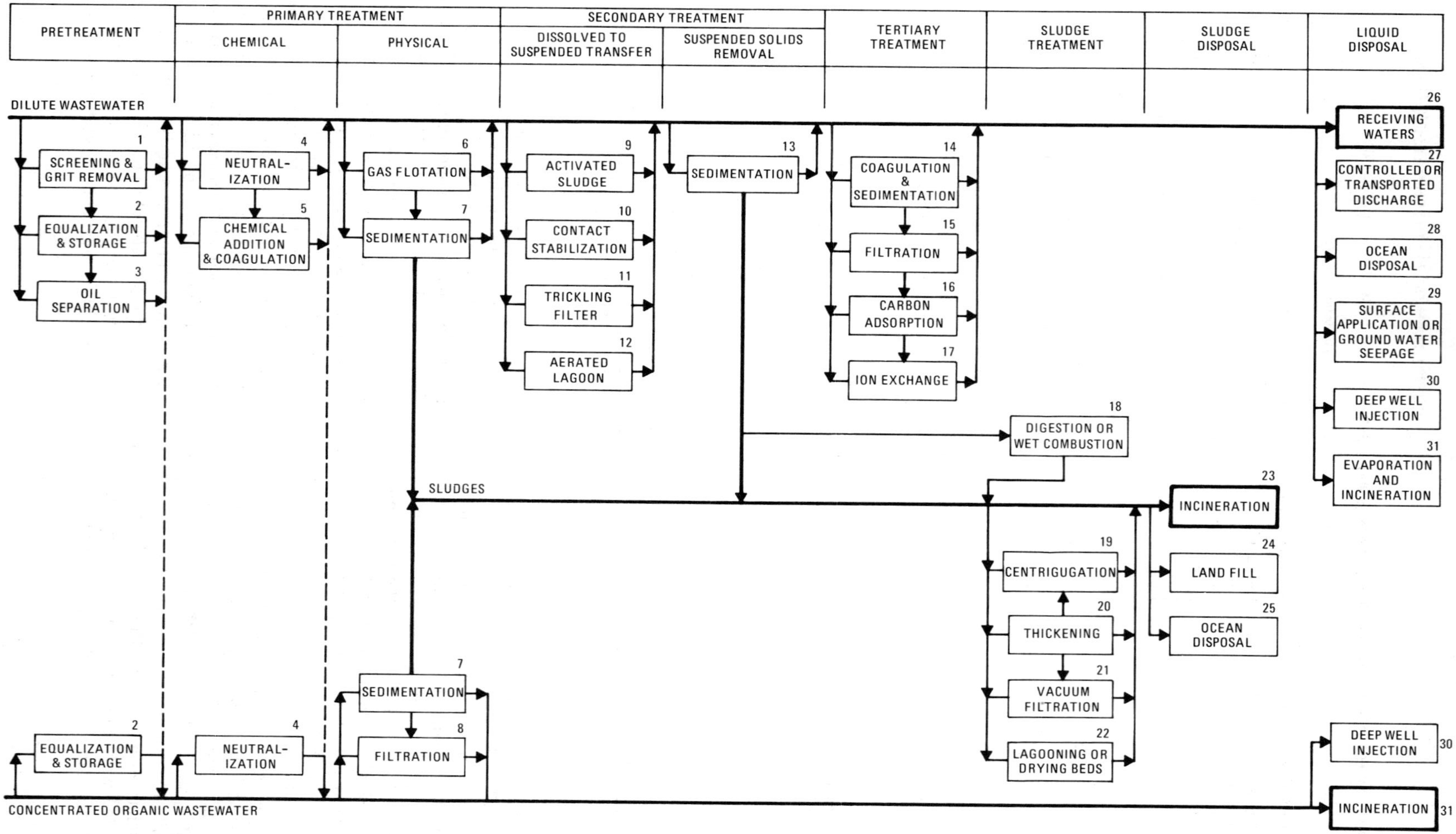

Fig. 26.4 Chemical industry waste-water treatment sequence/processes substitution diagram.

TABLE 26.10 Composition of Typical Clean-Water Effluent

			Potential Pollutants		
Water Sources	*% of Total Waste Water*	*Flow Range (GPM)*	*Sources*	*Type*	*Concentration Range (ppm)*
Cooling water (excluding sea water)	40–80	100–10,000 (500–200,000 gal water/ton product)	Process leaks Bearings, exchangers, etc.	Extractables	1–1000
				Mercaptans	
				Sulfides	
				Phenols	0–1000, but usually less than 1 ppm
				Cyanide	
				Misc. N-compounds	
				Acids	
			Water treatment	Chromate	0–60
				Phosphate	0–60
				Heavy metals	0–30
				Fluoride	0–30
				Sulfate	100–10,000
				Biocides, algacides	0–50
				Misc. organics	0–100
			Scrubbed from air through tower	Hydrogen sulfide	
				Sulfur dioxide	0–1000
				Oxides of nitrogen	
				Ammonia	
				Particulates	0–300
			Make-up water	Total dissolved solids	100–5000
				Particulates	0–100
				Phosphates	0–5
				Fluoride	0–2
Steam equipment	10	50–1,000	Boiler blowdown	Total dissolved solids	500–10,000
				Particulates	5–300
				Extractables	0–10
				Phosphate	1–50
				Sulfite	0–50
				Sulfide	0–5
				Misc. organic compounds	0–200
				Misc. N-compounds	1–100
				Heavy metals	0–10
				Alkalinity	50–400
			Waste condensate	Extractables	0–100
				Ammonia	0–10

Source: Freedman, A. J., *et. al.*, Natl. Petroleum Refiners Assoc., Tech. GC-67-19, 1967.

Classification" (SIC) number 26 titled "Paper and Allied Products," while *Lockwood's Directory* indicates that there are 542 independent paper mills, 278 paper mills with one or more associated pulp mills, and 32 independent pulp mills. It is toward these 852 production establishments that the effort of FWQA's research and development program is directed.

Pulp and paper making processes have changed little since the beginnings of this industry centuries ago. They generally cause considerable pollution, especially during pulp making. This industry uses large amounts of water, some 7 billion gallons per year in 1970. While this water is being used, and before it is discharged into the hydrographic system, it is loaded with pollutants. In particular, the pollutants are suspended and dissolved substances, both organic and inorganic. In 1970, the pulp and paper industry in the U.S. discharged a total of nearly 1.5 million tons of suspended matter and approximately 2.9 million tons of oxidizable matter (measured as BOD). Other water pollutants must also be taken into account for their possible toxic effects on fish and the biota. These pollutants may come from the raw material or from different stages in the production process. They include: resin acids, organic sulphides and traces of heavy metals. Mercury, which was commonly used in the past as a preservative agent or slimicide, is practically no longer used in the pulp and paper industry and will be completely abandoned. Finally, mention should be made of the color of the effluent, caused mainly by lignin from the wood, and of the eutrophication problems which could in special circumstances result from the careless use of high rate biological treatment in certain lake systems.

The problems of air pollution generated by the pulp and paper industry mainly concern the emission of particulate matter, sulphur dioxide, and odorous compounds of sulphur such as hydrogen sulphide and various mercaptans. Apart from emissions of sulphur dioxide and particulate matter from combustion plants, emissions which are found whenever solid or liquid fuels are burned, the particular emissions specific to the pulp and paper industry are the odorous gases emitted during manufacture of sulphate pulp and also sulphur dioxide from the recovery boilers used in the calcium-base sulphite process. These problems of air pollution are quickly detected by the general public, but they are usually restricted to the neighborhood of the works. They add, however, to the general problem of atmospheric pollution by sulphur, which might have adverse ecological effects in certain areas.

The disposal of solid wastes has so far received less attention than the more pressing water and air pollution problems. The more important of these wastes are: bark, sludges, knots and screenings. Bark is usually burned in steam-producing plants, which necessitates the installation of equipment to render it more suitable for burning. The increasing use of external methods of treating effluents will lead to the production of large amounts of sludges which are difficult to handle, especially those generated in secondary treatment processes. At present, there is no satisfactory solution to this problem, which, however, is not specific to the pulp and paper industry.

There is a growing interest in replacing virgin fiber with recycled wastepaper in the paper-making process. Environmental reasons as well as other pressures will most probably lead to a continuing increase in wastepaper recovery rates. However, when deinking is required, the resultant pollution load is similar to that of the chemical pulping process.

Pollution control techniques fall into two categories: *internal* preventive measures, applied to the manufacturing processes in the mill itself, reducing the pollutant load which must finally be handled; and *external* techniques for treating effluents before they are discharged into the environment. Implementation of anti-pollution measures now under way in most countries concerns the abatement of water pollution using external techniques, i.e. primary sedimentation with or without addition of chemical coagulants, and secondary biological treatment; in a few cases, treatment to remove color is also required. In specific cases, considerable reduction in pollution load can be achieved by merely applying chemical coagulation combined with efficient internal measures without using any biological secondary treatment. With respect to internal techniques of pollution control, the burning of waste liquor with or without recovery of chemicals and heat is the method most commonly used to reduce water pollution. Another important measure is to reduce the amount of fresh water used in the manufacturing process. This is achieved principally by recycling a portion of the used water which would otherwise be sewered. Such measures can substantially lower the volume of effluent, with consequent savings in the investment for external treatment plants, and also greatly reduce the amount of pollutants discharged.

With careful use of the various internal and external treatment processes that are available, the pollution load in effluents could be reduced to very low levels. Greater reduction of pollutants at less cost may be possible as a consequence of research and development work presently underway. It is expected that some of these new developments may radically alter the present concepts of effluent and emission treatment.

New manufacturing processes are being investigated; their advantage would lie mainly in reducing emissions of odorous gases and in allowing extensive recycling of the effluents; however, it is doubtful whether the new processes can be applied on an industrial scale before 1980, and a considerable amount of research and development remains to be done in this field. Finally, improvements to the qualities of mechanical pulp which are now being developed could have a significant effect by greatly reducing or even eliminating the chemical pulp needed for making newsprint.

Fig. 26.5 Sludge filter used to partially dewater sludges for cooperative EPA-Crown Zellerbach experimental study. (*Courtesy EPA and Crown-Zellerbach Co.*)

Substantial progress has recently been made in reducing emissions of particulate matter and of odorous gases in the pulping process. The techniques used at present are effective but still very costly in cases where regulatory criteria are very stringent or, especially in the case of sulphate process, for technical reasons. In some of these cases the cost of controlling air pollution could possibly equal or even exceed the cost of controlling water pollution. Here also a considerable amount of research and development remains to be done in order to develop cheaper methods.

The color of effluents is a problem for which effective technical solutions now exist; the cost of these techniques is still fairly high and this restricts their use to specific, local situations. A good deal of research, the results of which appear promising, is now going on in this field, and it will probably be possible in the near future to develop alternative methods.

It has been estimated that the total waste load originating from this industry represents 19 per cent of the total pollutional load attributed to manufacturing. These wastes are

Fig. 26.6 Rietz drier used in the pulp industry to dry and burn sludges. Notice the off-gas cleaner on the right. (*Courtesy EPA.*)

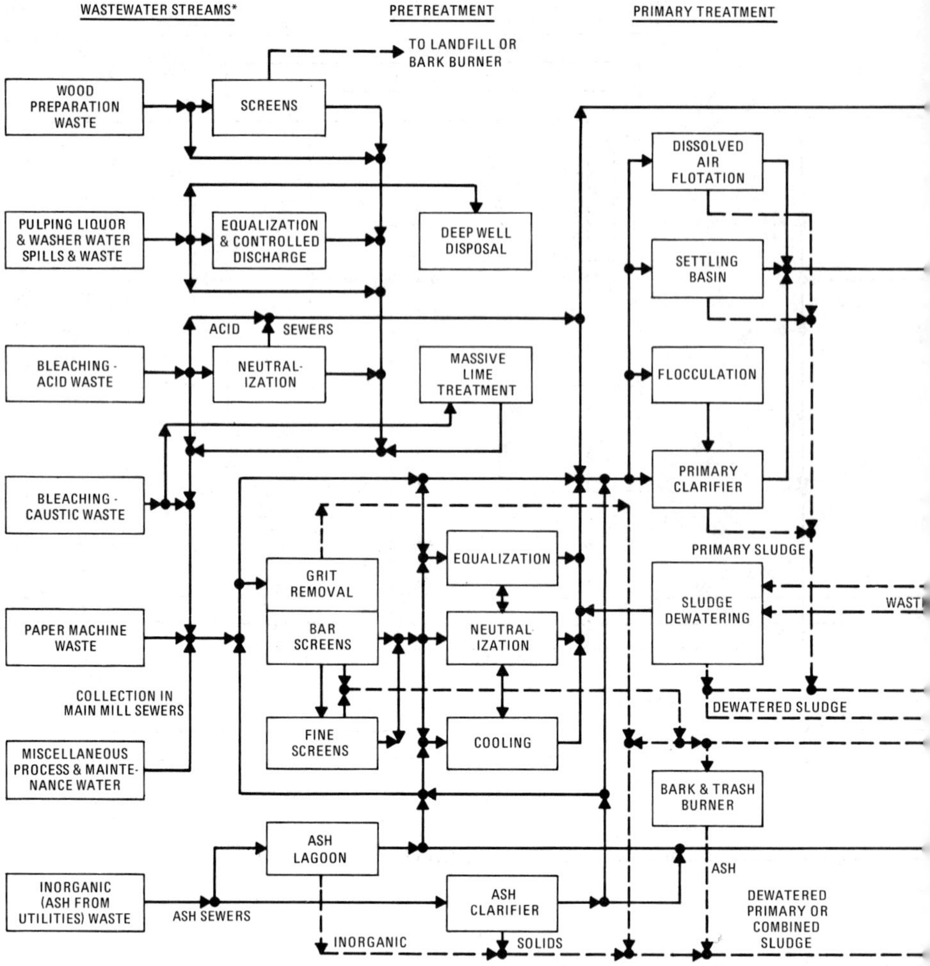

*EXCLUDING SANITARY AND COOLING WATERS

Fig. 26.7 F.P.C.A. paper industry profile sequence and alternatives in waste-water treatment practices.

characterized by extreme variations in BOD (depending on the pulp production process), high total organic carbon, TOC, and color levels; and variations in the organic (fiber), inorganic (fillers), and color (dyes) concentrations depending upon the paper production process.

Pulp and paper-mill wastes are treated by (1) recovery methods, (2) sedimentation and flotation to remove suspended matter, (3) chemical precipitation to remove color, (4) activated sludge to reduce the BOD, and (5) lagooning for purposes of storage, settling, equalization, and sometimes for biological degradation of organic matter. The cost of treatment is considered high in relation to the cost of the product; consequently, industry emphasis has been on recovery rather than treatment, and actual treatment equipment is installed only after exhaustive study of all other possibilities.

Facilities provided to treat process waste waters must either (1) remove contaminants so that waste water is suitable to discharge to

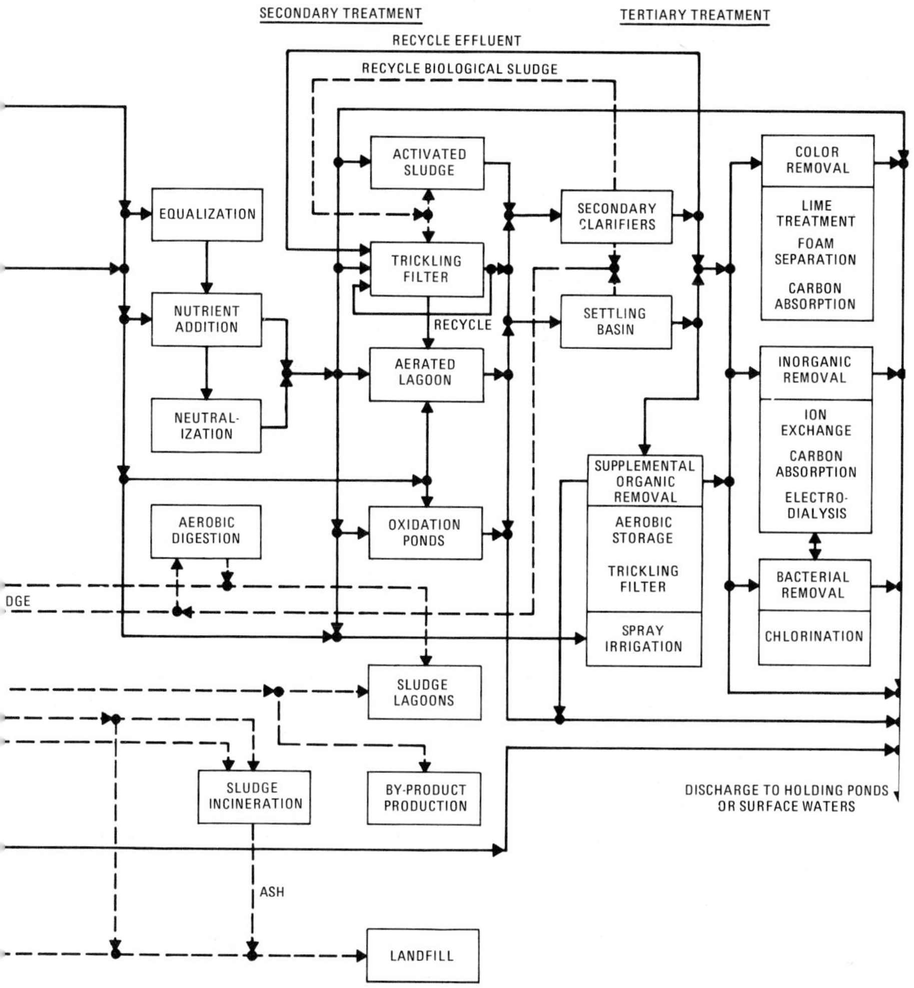

Fig. 26.7 Cont'd.

municipal sewers or receiving waters, or (2) improve the water quality so that it is satisfactory for re-use in the plant. Four functional groups that are included in waste-water treatment practices are: pretreatment and primary, secondary, and tertiary treatment.

Tertiary treatment is used to obtain additional removals or "polishing" of waste waters prior to discharge. To date, the industry has not applied tertiary treatment for removal of lignins in the form of dissolved color or bacteria, but laboratory studies are in progress. Processes being investigated include: holding ponds and filtration for the removal of BOD and COD; chlorination or ozonation for removal of bacteria; activated carbon adsorption, mass lime, and other chemical adsorption processes for color removal and foam reduction; and removal of inorganic solids using electrodialysis, reverse osmosis, and ion exchange.

Studies and operational data have shown that centrifugation is very effective in dewatering pulp and paper-mill sludges, both for

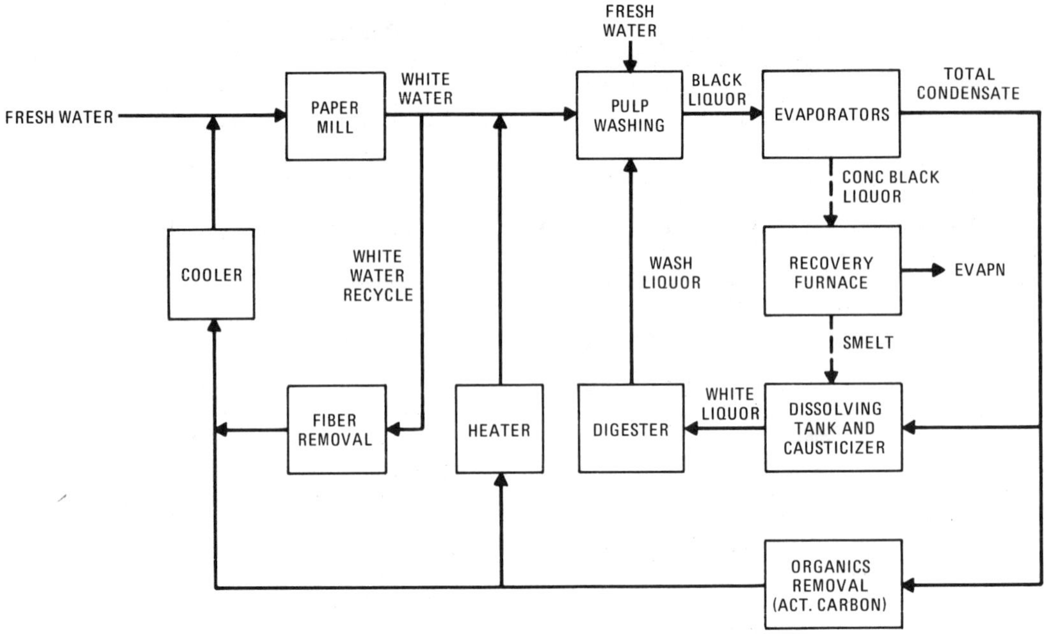

Fig. 26.8 Simplified flow diagram for potential total re-use of kraft mill water.

the primary sludge alone, and in combination with secondary treatment. However, secondary sludge usually requires chemical conditioning (polyelectrolytes) to be dewatered to 12 to 20 per cent solids.

Historically, sludges were disposed of by landfilling and lagooning together with other waste material such as bark, ash grit, and debris. Increased land costs and more stringent regulations have led to the incineration of primary and secondary sludges with solids concentrations of 30 to 40 per cent. Mechanical presses to further dewater sludges after vacuum filtration or centrifugation have been evaluated, but success has been limited by the low fiber content of the sludge.

The sequences and alternatives in wastewater-treatment practices in the paper industry are shown in Fig. 26.7. A flow diagram for potential total re-use of kraft mill water is given in Fig. 26.8.

PETROLEUM & COAL PRODUCTS

Petroleum

The petroleum refining industry is by far the larger water user in this industrial category. The industry uses 17 billion gallons of water per day or 20 per cent of total industrial water usage. Approximately 0.23 billion gallons per day is used in processing operations. Many distinct operations, such as crude oil distillation, reforming, catalytic cracking, thermal cracking, polymerization, alkylation, gasoline and middle distillate treating, and motor oil manufacturing, are utilized. Foul condensate is usually pretreated at the source, but by and large the refiners depend on central waste-treatment facilities for pollution control. Oil separation and recovery followed by biological conversion of phenols, sulfides, etc. are the treatment methods most frequently employed. Greater emphasis on more effective treatment, waste treatment at the source, product recovery, water re-use, and development of treatment methods requiring less land areas is needed.

The main remedial measures for reducing refinery wastes are (1) reduction of oil leakage by preventive maintenance of pipelines and equipment; (2) preventing formation of oil emulsions or, where they exist, isolating them and treating them separately; (3) removal of floating oil in separators located as near as

Fig. 26.9 Fluidizing-bed incinerator of 70 barrels per day capacity for the treatment of refinery chemical and oil-sludge wastes. (*Courtesy EPA and American Oil Co.*)

possible to the original source of waste; and (4) isolation and separate treatment of objectionable wastes.

Acidification of caustic wastes with sulfuric acid removes some of the objectionable compounds. Acid sludges may be used as a source of fuel or to produce by-products such as oils, tars, asphalts, resins, fatty acids, and other chemicals.

Evaluation of the effectiveness of in-plant processing changes in reducing waste-water pollution is somewhat qualitative and general, rather than quantitative and specific. In regard to the relative pollution effects of specific processes, the most significant developments have been in the area of hydrotreating. Each of these processes generates substantially lower pollution loadings than the processes they are replacing; available data on pollutant concentrations in the unit waste-water streams indicate that these processes have significantly reduced sulfide and spent-caustic waste loadings. A more general indication of pollution reduction by in-plant processing practices is the much lower pollutant loadings per unit of throughput for "newer" refineries as compared to "older" or "typical" refineries. This reduction is attributed in large measure to decreased losses to the sewers by sampling and water draw-off operations in the "newer" refineries, where facilities, controls, and general operation practices are likely to be superior.

Waste-treatment methods applicable to petroleum refineries can be divided into five types: physical, chemical, biological, tertiary, and special in-plant methods.

The subprocess services representative of a newer technology are shown in Fig. 26.10. The sequence/substitution diagram of waste-water treatment processes is shown in Fig. 26.11.

Coal

Over 500 million tons per year of coal is obtained by strip and deep mining operations in the U.S. and approximately one-half of this production is consumed by the utility and steel industries. Over 80 per cent of the coal is cleaned and classified by using water as the cleaning medium prior to marketing. Among the most troublesome wastes from coal processing and use are coal fines, sulfur, phenols, ammonia, and thermally polluted waters.

FOOD & KINDRED PRODUCTS

Activities under this industrial category encompass those industries dealing with the processing of products for human or animal consumption. Major categories included are (1) meat products, (2) dairy products, (3) canned and frozen foods, (4) grains, (5) bakery products, (6) sugar, (7) beverages, (8) candy and related products, and (9) miscellaneous products (coffee, edible oils, etc.). Further breakdown is given in the Standard Industrial Classification Manual.

Wastes generated by this industry, which comprises some 32,000 individual companies, represent 20 per cent of the total pollutional load attributed to manufacturing. The mainly organic wastes exhibit extreme variations in BOD, suspended and dissolved solids, pH, etc.,

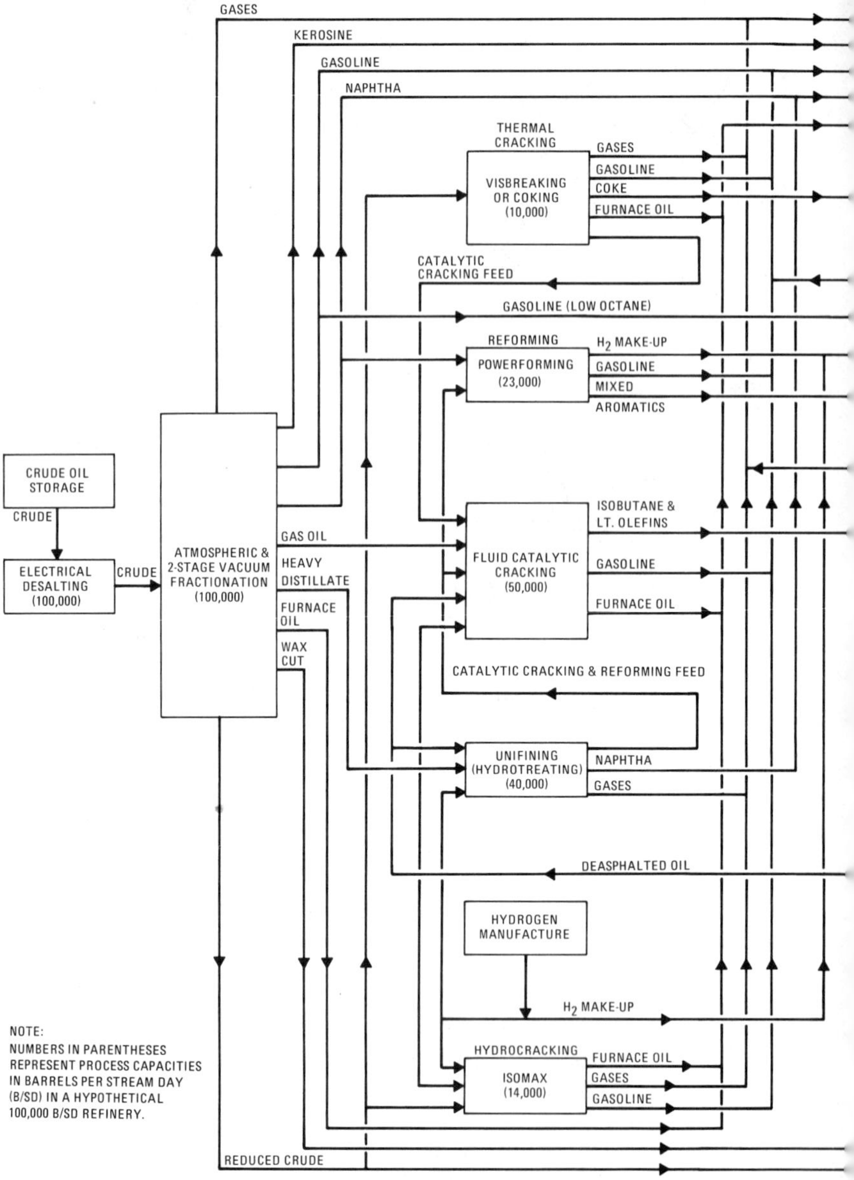

Fig. 26.10 Subprocess series representative of a newer technology.

as well as in volume and duration of processing operations.

Cannery Wastes

The treatment of cannery wastes can be divided into two categories: (1) partial or complete treatment by the industry, and (2) municipal treatment. Most canneries discharge into municipal sewers with minimal preliminary treatment—usually only vibrating screens to remove part of the suspended solids. In a recent survey of 113 canneries, 68 per cent

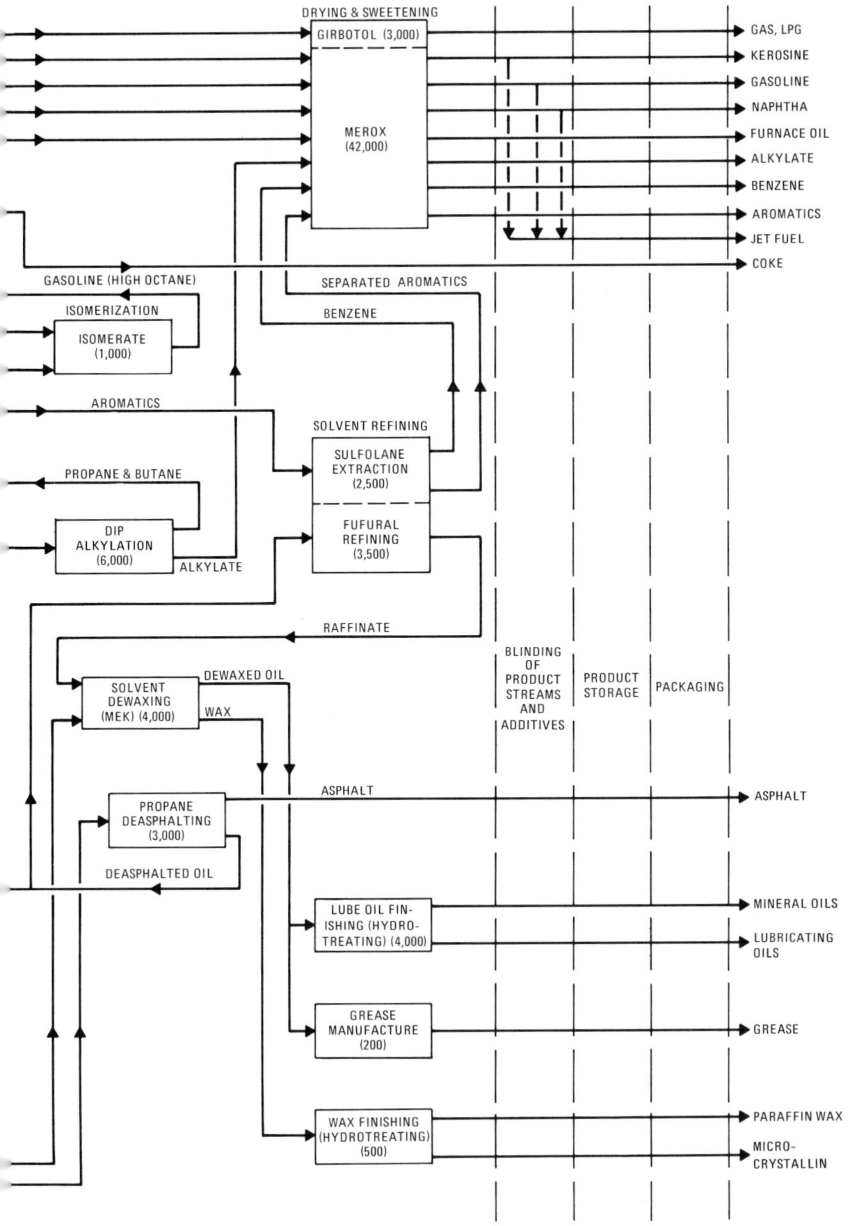

Fig. 26.10 Cont'd.

screened their wastes which were subsequently discharged to the municipal sewer system, while only 13 per cent used their own treatment plants; 2 per cent had no treatment; and 17 per cent either were no longer in operation or supplied no information. Of the 14 plants providing their own treatment, 3 used screens only; 6 used oxidation ponds after screening; 2 used spray irrigation after screening; 1 used screening plus neutralization; 1 used screening plus neutralization; 1 used screening, sedimentation, and ridge and furrow irrigation;

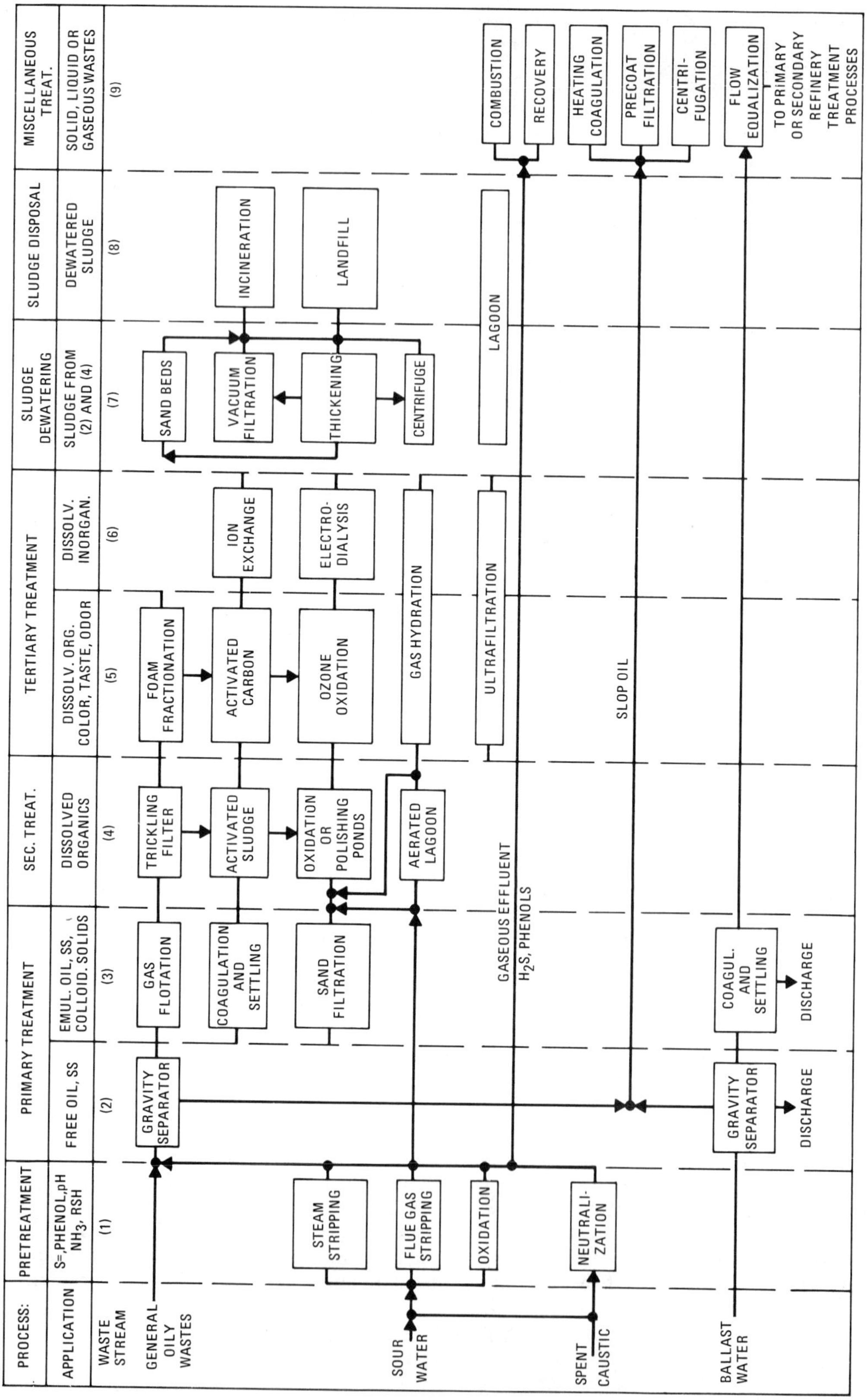

Fig. 26.11 Sequence/substitution diagram of waste-treatment process.

Fig. 26.12 A high-rate thermophilic trickling filter used to treat strong organic wastes encountered in vegetable and fruit processing. (*Courtesy EPA and National Canners Association.*)

and 1 used screening, sedimentation, neutralization, and nutrient addition followed by aeration.

Frozen Foods

The above information on types of treatment provided is also applicable to the frozen food industry. However, a higher percentage of frozen food plants have constructed their own lagoons or spray irrigation disposal facilities because a much lower percentage have access to municipal sewers.

Dairy Products

Milk and milk-processing-plant wastes are generally high in dissolved organic matter, they contain about 1,000 ppm BOD and are nearly neutral in pH. Since these wastes are mainly composed of soluble organic materials, they tend, if stored, to ferment and become anaerobic and odorous. Therefore, they respond ideally to treatment by biological methods. Aerobic processes are most suitable, but final selection of a treatment method hinges on the location and size of the plant. The six most effective conventional methods used are (1) aeration, (2) trickling filtration, (3) activated sludge, (4) irrigation, (5) lagooning, and (6) anaerobic digestion.

As there are several types of dairy processing operations, i.e., butter, cheese, condensed and evaporated milk, ice cream and frozen deserts, and fluid milk processing, with different quantities of wastes and several different treatment processes, the cost of waste treatment cannot be adequately summarized.

Meat

The most common methods used for treatment of meat-plant wastes are fine screening, sedimentation, chemical precipitation, trickling filters, and activated sludge.

Bakery Products

Among the nine basic food industries, baking ranks 6th in number of employees and 8th in value added by manufactured products. Despite the gigantic size of the food industry, the production of many bakery foods is accomplished in relatively small manufacturing plants scattered over the Nation. As most baked foods are perishable, distribution to the consumer is possible only within a localized area.

In general, the baking industry has a raw waste effluent that is high in soluble organic matter and greases. Fortunately, the wastes effluent from baking plants is not a health problem, from the standpoint of containing either pathogenic organisms or toxic contaminants. With proper design of facilities, these wastes are amenable to biological treatment for the reduction of the soluble organics.

Bakery waste are typically acid in nature. Lime can be used for pH control but ammonia can be used more effectively to treat this carbon-based waste since it contains nitrogen which is required for biological growth. The grease, fats and oils are easily emulsified and air flotation provides a very efficient treatment system.

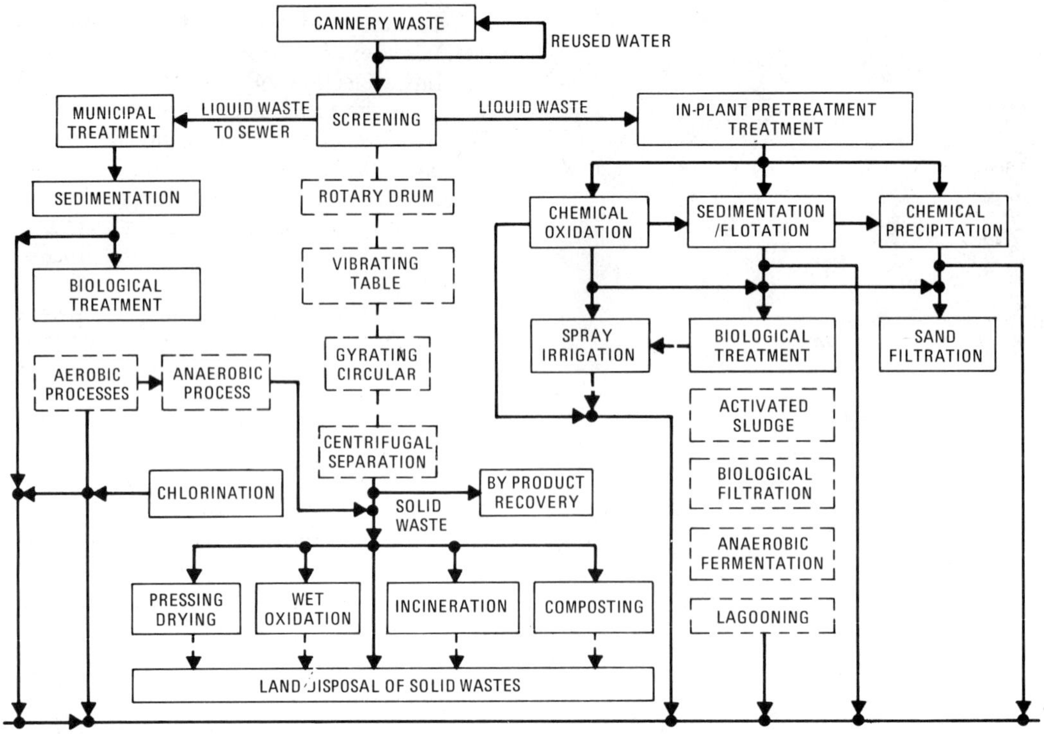

Fig. 26.13 Canned and frozen fruits and vegetables waste-treatment flow chart—SIC 2033, 2037.

Poultry

Poultry-plant wastes should and do respond readily to biological treatment, and if troublesome materials such as feathers, feet, heads, and so forth are removed prior to this treatment, satisfactory biological treatment is attainable. Treatment facilities for one poultry-dressing plant's wastes include stationary screens in pits, septic tanks, and lagoons; over-all removal of 93 per cent of the BOD was reported to result from the use of these measures.

The most typical waste-treatment flow chart is one from the canned and frozen fruits and vegetable processing area shown in Fig. 26.13.

MACHINERY AND TRANSPORTATION EQUIPMENT

Industrial activities in the Standard Industrial Classification Groups 34 (Fabricated Metal Products), 35 (Machinery), 36 (Electrical Machinery), and 37 (Transportation Equipment) are included in this section. Combined water usage is approximately 3.7 billion gallons per day or 4.4 per cent of industrial water use. Approximately 0.38 billion gallons per day is used in processing and operations. Oil, particulate matter, and cleaners constitute the principal contaminants in waste waters. The physical processes of sedimentation and flotation are the most frequently employed treatment methods but most of the installations provide no waste treatment facilities.

STONE, CLAY & GLASS PRODUCTS

Industrial wastes from the processing and manufacturing of flat glass and pressed and blown glassware, pottery, and similar products, as well as structural clay products are included in this category together with cement, concrete, and plaster products. Ready-mix and precasting operations as well as those activities involved with lime and gypsum products are also included. In addition, non-

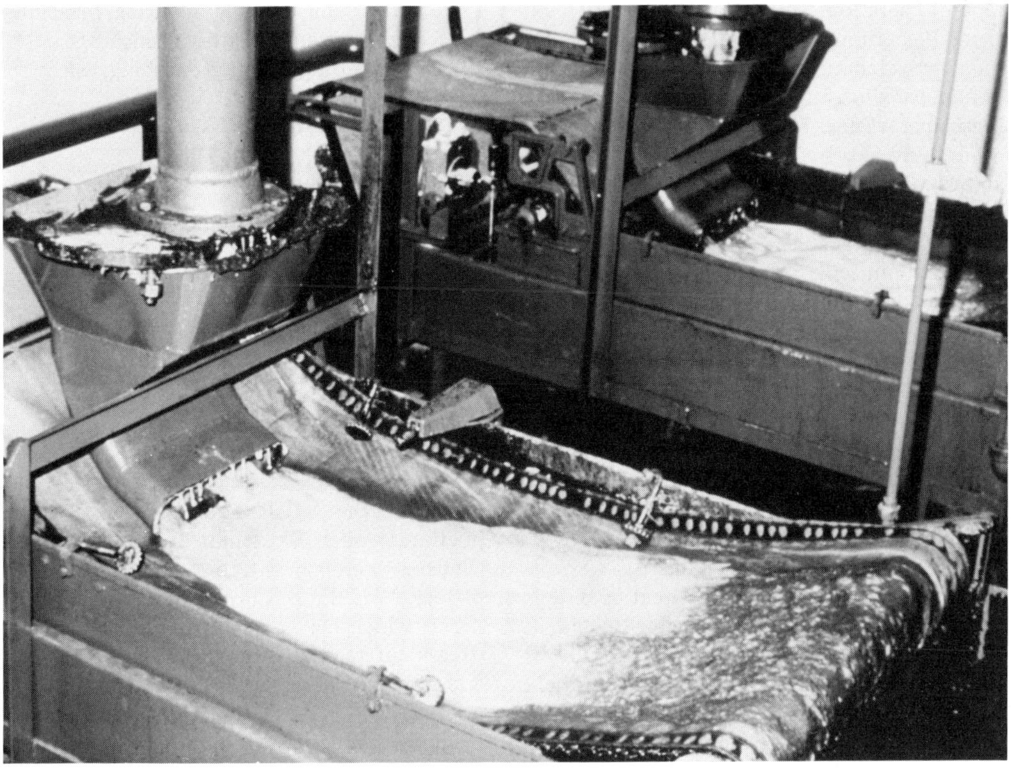

Fig. 26.14 Filtration of a glass-fiber-production waste stream to re-use the process water. (*Courtesy EPA and Johns-Manville Corp., Defiance, O.*)

metallic mineral items such as asbestos and abrasive products are included.

The wastes from these processes are composed in part of fine suspensions and sediments of the various nonmetallic minerals of the raw material, or the expended material used in the various manufacturing operations.

The wastes are varied and voluminous and their separation, transportation, and disposal or reprocessing constitute a significant part of the effort to alleviate the water pollution resulting from processing of the various materials in this category.

Under normal conditions the use of conventional treatment techniques usually results in effluent characteristics acceptable for discharge.

RUBBER AND PLASTICS

The most common current methods for treating rubber waste waters are aeration, chlorination, sulfonation, and biological methods. Coagulation, ozonation, and treatment with activated carbon are also employed. The phenolic constituents of rubber wastes may be reduced by chlorination, and sulfonation of styrene waste can render it almost odorless if sufficient time is given for the reaction.

While the above methods of treatment are valuable mainly for odor removal, biological treatment of rubber wastes affords the greatest reduction of BOD. For efficient biological treatment, either nitrogen and phosphorous must be added, or the wastes must be mixed with domestic sewage in a 1 : 3 ratio.

The principal requirement in treating plastics manufacturing wastes is BOD reduction.

Many renovation systems in current use were designed and developed to produce a water whose quality will not meet the standards of the next decade. Most fluidized culture systems for housing the biological oxidation process (conventional activated sludge, tapered aeration, contact stabilization, extended aeration, aeration ditch, completely mixed systems, oxidation ponds, aerated ponds, trickling filter) were developed to treat municipal wastes. This type of waste contains a relatively low fraction of readily biodegradable soluble organics with balanced nutrients. But wastes of the plastic industry are high in slowly biodegradable soluble organics without balanced nutrients. Therefore, attempting to upgrade the effluent quality from these marginally designed systems will reveal engineering inadequacies and subsequently result in waste-treatment quality and cost problems.

Increasing the per cent of waste removal in improperly or inadequately designed equipment will involve a disproportionate increase in the cost per unit of waste removed. The current cost of efficient primary plus secondary treatment for a given industrial waste is in some cases, five times as great, per unit volume of comparable waste water treated, as for token primary treatment system built 10 years ago. In general, recent plants reflect total treatment costs considerably higher for a comparable volume of water treated. The data available indicate that the initial cost of recently built equipment, which is designed to reduce effluent waste loads within anticipated legal limits, was considerably higher than that of older units built for less rigorous standards.

LUMBER AND WOOD PRODUCTS

There are 20,672 establishments under Standard Industrial Classification No. 24, "Lumber and Wood Products, Except Furniture." Wastes from the wood-preserving and veneer and plywood industries comprise perhaps the greatest problems. The use of such toxic compounds as pentachlorophenol and heavy metals ("CCA" copper, chromium, arsenic) in wood preserving, in addition to urea-formaldehyde and phenolic glues in the plywood industries provides a challenge to normal treatment techniques.

TEXTILE MILL PRODUCTS

The waste-water flows in the textile industry may be identified with the following textile fibers and processing operations:

Cotton—sizing, desizing, scouring, bleaching, mercerizing, dyeing, printing, finishing.

Wool—scouring, dyeing, washing, carbonizing, bleaching.

Noncellulose chemical fiber—scouring, dyeing, bleaching, special finishing.

Cellulose chemical fiber—chemical preparation, scouring, dyeing, bleaching, special finishes.

In 1968 the textile industry used 1,127 billion gallons of water for the manufacturing processes. The fresh water intake was approximately 498 billion gallons and consumption was 31 billion gallons.

Wool

The most common practices of waste treatment in the wool industry are biological methods, such as sedimentation, activated sludge, trickling filtration, and lagooning. Screening is almost universally used to remove fibers which may possibly damage subsequent treatment facilities. Equalization and holding are generally necessary due to batch dumping of many of the process wastes, an operation which produces shock loads and intermittent flows through the treatment system. Studies in recent years indicate clearly that the activated sludge process with modifications (primarily extended aeration time and influent pH adjustments) will consistently produce BOD reductions on the order of 90 per cent. As the discharge requirements imposed upon textile finishing plants are upgraded, it is probable that future waste treatment facilities will be predominately of the activated-sludge type.

The percentage of wool-finishing wastes treated by municipal plants is also increasing steadily as costs of building and maintaining in-plant treatment facilities increase. Newer finishing mills are being built close to municipalities to take advantage of the availability of municipal treatment. Industry wastes are generally pretreated prior to discharge to a municipal system since most municipal treatment plants are not equipped to readily handle the large amounts of grease produced by wool mills. Screening for removal of fibers is also necessary to prevent clogging of biological-treatment equipment and to reduce the quantity of suspended matter.

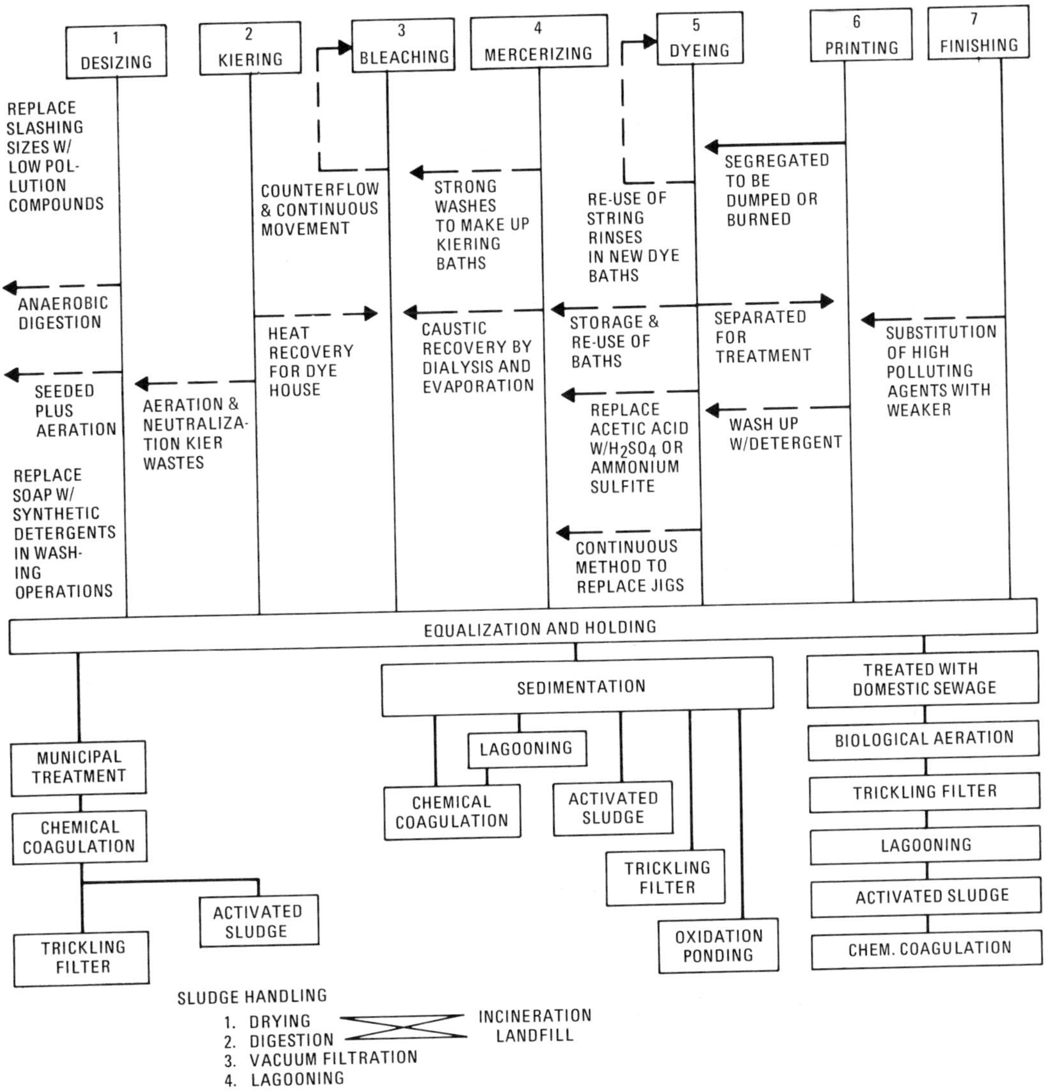

Fig. 26.15 Cotton-textile-finishing waste-treatment flow chart—SIC 2261.

Cotton

The activated-sludge process for cotton-textile waste is often modified by increasing the aeration time and carrying a higher concentration of mixed liquor-suspended solids in the aeration tank. With careful operation this process will produce excellent reduction of BOD and suspended solids. If some domestic sewage is available to mix with the textile waste, the efficiency of the plant is generally increased.

To reduce construction costs, an aerated lagoon is sometimes substituted for the activated-sludge process. Properly operated, it is capable of closely approaching the pollution-removal efficiency of the conventional activated-sludge process.

Many municipal waste-treatment methods will be susceptible to shock loads from the mills and, therefore, pretreatment should include flow regulation and equalization holding procedures to insure waste uniformity. In a large municipality, the mill waste would be diluted sufficiently before reaching the treatment facility and would not harm the opera-

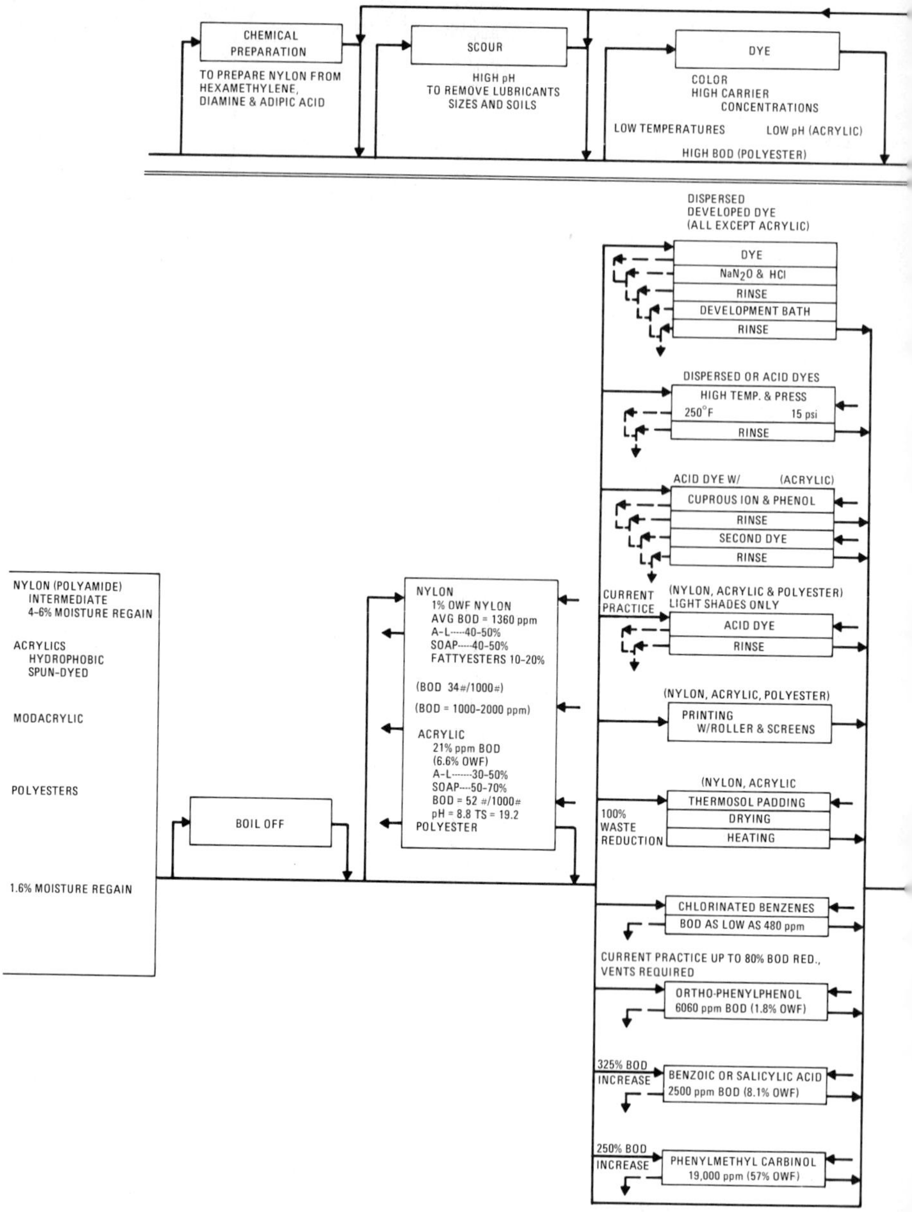

Fig. 26.16 Noncellulose-synthetic-textile finishing-process flow chart.

tion. Even so, most large municipalities require finishing plants to provide screening and constant-discharge holding basins. Normally, a cotton-finishing-plant waste is easily handled by conventional municipal treatment methods.

The waste-water treatment flow chart for the cotton textile finishing operation is shown in Fig. 26.15. Waste-water sources from a noncellulose type synthetic textile finishing process is shown in Fig. 26.16.

Synthetics

The tremendous growth rate of the synthetic (man-made) textile industry is expected to continue and even increase as various new materials (modifications of existing fibers) and

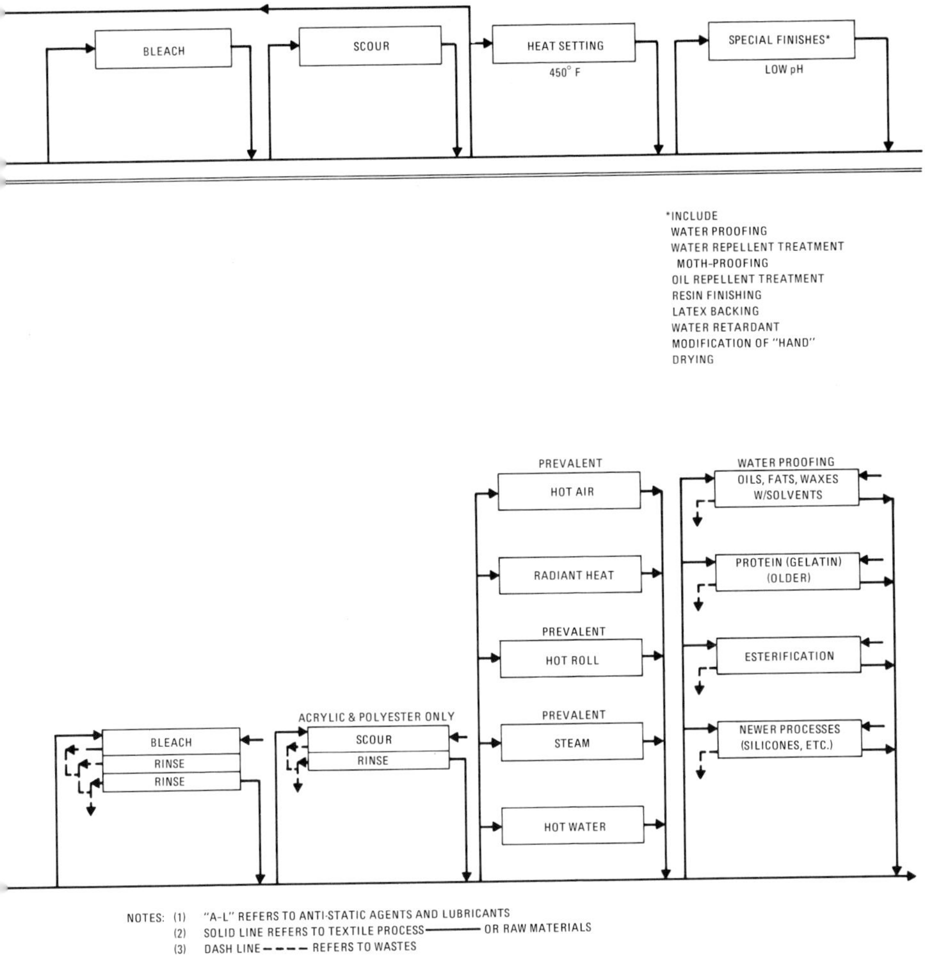

Fig. 26.16 Cont'd.

new fibers are introduced. Concurrent with this growth, a nearly equal increase in waste-load seems imminent. This, along with pressures by regulatory agencies regarding stream pollution, will lead to an increased rate of adoption of waste-treatment practices in the future.

Synthetic-textile wastes have generally been treated by biological methods with good removal efficiency at reasonable cost. In the future it is expected that through process changes, water requirements per unit production will be reduced, resulting in a plant effluent which will be higher in pollution concentration and lower in volume. Therefore, the adoption of more elaborate waste-treatment facilities utilizing pretreatment and tertiary polishing can be expected.

Toxic metallic ions in dye wastes can retard biological oxidation when present in high concentrations. Chemical pretreatment may therefore become a requirement, or the industry may choose to adopt treatment by chemical coagulation as the principal method.

Problems associated with treating textile wastes in municipal facilities are dependent primarily on the volume ratio of the domestic sewage to the industrial waste. If the textile waste is only a small percentage of the total volume entering the municipal plant, no problems are encountered. If, however, the textile waste constitutes a significant percentage of the total volume, it may be necessary to make special provisions. These would normally include holding and equalization at the textile finishing plant, pH control, chemical precipitation of any toxic constituents, and, possibly, chemical treatment for color removal.

COMBINED INDUSTRIAL-MUNICIPAL WASTES

Combined industrial-municipal wastes are those wastes, treatable at a municipal waste-treatment plant, which contain an appreciable amount of waste originating from industrial sources. Industries involved in the processing of minerals, carbohydrates, hydrocarbons, refractory material, and protein material are of greatest concern. The wastes are voluminous and of great variety.

Use of municipal facilities is preferred by a majority of the smaller industries and accounts for the largest number of establishments whose wastes are treated. On a volume basis about three-fourths of industrial wastes are currently treated in industrial waste-treatment facilities and one-fourth are treated in municipal facilities.

Although only 7.5 per cent of the waste waters of major industrial establishments are being disposed of to municipal sewers, sewering presently provides the principal waste-disposal method for seven of the eleven industrial sectors. The seven industries include food processing, textiles, rubber and plastics, machinery, electrical machinery, transportation equipment, and miscellaneous manufacturing. The wastes of these seven industries are more amenable to treatment at municipal treatment plants than are the wastes of the four other industries: (1) paper and allied products, (2) chemicals, (3) petroleum and coal, and (4) primary metals.

In connection with the trend toward increased use of municipal facilities by many industries, it is important to note the rapid increase in municipal treatment capabilities. Both the number of treatment plants and the average level of treatment have risen steadily, the growth being most marked since the institution of Federal grants for construction of waste-treatment plants. As recently as 1960, almost 30 per cent of the nation's sewered communities did not have waste treatment provided to them. By 1962, less than 20 per cent of the total number of sewered communities were without waste treatment. In 1970, less than 10 per cent were without some degree of waste treatment. Moreover, well over two-thirds of the sewered communities now have secondary waste-treatment facilities. Thus, municipal facilities have an increasing potential capacity for handling many industrial wastes.

Joint systems for treating both municipal and industrial wastes in many cases are likely to provide the means of attaining adequate water-pollution control most effectively and least expensively. The extent to which joint handling systems will increase over the next five years depends largely upon the managerial ability of municipal and industrial officials and their willingness to enter into such cooperative arrangements. This, in turn, will depend upon the costs which industrial establishments are required to pay to use municipally-operated facilities. To the extent that appropriate charges and pretreatment requirements are fixed and that joint treatment facilities are designed and operated effectively, increased use of such facilities by industry may well lower over-all pollution control costs significantly in the short-term future.

Reduction of many industrial wastes is often accomplished most efficiently and economically by process modifications. While the rate and effects of technological change are difficult to evaluate, quantities of water used per unit of production have been decreasing in most industries, while recycling to make more efficient use of water is increasing. Moreover, modern operational practices and engineering design increasingly stress waste control.

Industrial waste-treatment costs are affected significantly by the methods industry employs to reduce its wastes. In general, waste reduction may be accomplished through treatment by municipal facilities, by on-site treatment, through process changes which lessen the amount or strength of wastes generated, by ground disposal, or by combinations of these alternatives.

There are potentially great savings through the "economy of scale" when the treatment facilities are designed to serve joint municipal-industrial needs, and this possibility needs to be explored in detail. Included here is the possibility of encouraging industries to utilize the municipal waste-handling systems on a special fee basis.

Although the treatment of industrial wastes by public agencies has only begun to receive attention, the practice is well established. A very substantial portion of the hydraulic loading of the total national system of public treatment plants is of industrial origin—the precise relationship varies according to the definition of "industry" that is employed. Because of the higher materials concentrations of industrial wastes, it is probable that well over half of all wastes removed or stabilized by public treatment are of industrial origin. Table 26.11 shows the relative loadings of municipal waste-treatment plants in 1968.

COST ASPECTS OF POLLUTION CONTROL IN THE PULP AND PAPER INDUSTRY

In Table 26.12 is given detailed information regarding expenditures already committed (reference year: 1970) for controlling water pollution in the pulp and paper industry, and the cost of applying further measures planned for the periods 1971-1975. The figures given take into account the cost of capital and the net operating costs (giving credit for recovery of heat, chemicals, fibers, and so on). These figures are expressed in dollars per ton of production and, as far as possible, they have been separated into the cost of measures for controlling water pollution (external measures only). It should be stressed that the cost figures presented are estimates, and that they refer only to those mills which were operating in 1970.

The data show clearly that the expenditures to be committed for controlling pollution in the pulp and paper industry are likely to rise sharply in the next few years. These expenditures are a new and increasingly significant cost factor in the economic viability of many pulp and paper mills. By 1975, averaged pollution control costs are expected to rise from a 1970 level of about 0.6 per cent of product

TABLE 26.11 Relative Domestic and Industrial Loading of Municipal Waste Treatment Plants, 1968

Community, Population Category	*Number of Plants*	*Gross Indicated Loading*	*Domestic Component at 100 G/C/D*	*Domestic Component at 65 G/C/D*	*Industrial Remainder*	*Per Cent Industrial*
		Million Gallons Per Day				
under-500	1400	64.0	49.0	32.0	5.0-2.0	23-50
500-999	1600	156.0	120.0	78.0	36.0-78.0	23-50
1,000-2499	2400	588.0	420.0	273.0	168.0-315.0	29-54
2,500-4999	1300	682.5	487.5	317.0	195.0-366.0	29-54
5,000-9999	1000	1050.0	750.0	487.5	300.0-562.5	29-54
10,000-24,999	800	2010.0	1400.0	910.0	610.0-1100.0	30-55
25,000-49,999	300	1687.5	1125.0	731.0	562.5-956.2	33-57
50,000-99,999	160	2040.0	1200.0	780.0	840.0-1260.0	41-62
100,000-249,999	85	2677.5	1487.5	967.0	1190.0-1710.0	44-64
250,000-500,000	28	2100.0	1050.0	682.5	1050.0-1417.5	50-68
over 500,000	24	2700.0	1800.0	1170.0	900.0-1530.0	33-57
TOTAL	9100	15,756.0	9890.0	6430.0	5870.0-9325.0	37-59

TABLE 26.12 Pollution Control Cost by Pulp and Paper Industry from NCASI (National Council on Air and Stream Improvement) in $\$10^3$

Item	*Baseline Pollution Control Costs, 1970*	*Semichemical Pulp (Including Integrated Paper Production)*	*Sulphite Pulp (Including Dissolving)*	*Integrated Sulphate Pulp and Paper*	*Newsprint (Including Mechanical Pulp)*	*Other Paper and Board (Including Mechanical Pulp & Building Board)*	*Summary*
1	Production in 1970, '000 tonnes	3035	2968	26,798	3,000	19,000	53,069
2	Annual Capital Costs in 1970	3490	5,780	17,884	1,252	6,846	34,000
3	Net annual operating costs in 1970	2000	3,000	21,145	1,480	19,055	40,200
4	Total annual pollution control costs in 1970	5490	8,780	39,029	2,732	15,901	74,200
5	Pollution control costs per tonne in 1970, $/tonne	1.81	2.96	1.45	0.91	1.20	1.40
	Estimated Pollution Control Costs, 1975						
6	Production in 1975, '000 tonnes	3035	2968	26,798	3,000	19,000	53,069
7	Value of total investments in 1971–75	97,290	161,000	498,648	35,000	194,062	948,000
8	Annual interest rate	10%	10%	10%	10%	10%	10%
9	Total capital costs in 1975 for 1971–75 investments	15,800	26,200	81,109	5,678	31,091	154,200
10	1970 capital costs in 1975	3,490	5,780	21,145	1,252	3,585	34,000
11	Net annual operating costs in 1975	6,000	10,000	61,542	4,308	39,458	117,000
12	Total annual pollution control costs in 1975	25,290	41,980	163,796	11,238	74,134	305,200
13	Pollution control costs per tonne in 1975, $/tonne	8.33	14.14	6.11	3.75	3.90	5.74

Note: Capital Costs for 1975 calculated on 10 year Capital Recovery Factor (0.16275) and 10% interest.

price to a level of 2.0 per cent in the paper industry and 2.0 per cent in newsprint. The corresponding figures for sulphate pulp range from approximately 1 per cent in 1970 to a level of 1.3 to 6.8 per cent in 1975. On the basis of data available it can be assumed that these costs will further increase by 1980.

The cost of pollution control is especially high in the case of semichemical pulp and in the case of sulphite pulp. For sulphite pulp, pollution control costs are expected to rise from a level of about 2 per cent of product price in 1970 to a level of nearly 8 per cent in 1975. The application of anti-pollution measures, and the ceiling placed on price by the possible substitution by sulphate pulp, will further aggravate the already difficult economic position of sulphite pulp and hasten the current trend in sulphite mills, i.e. the closure of less efficient mills and the conversion of others. Social problems may result, locally, from such structural changes.

Until recently, it has been usual for the pulp and paper industry in the U.S. to fund the total cost of controlling pollution. Since adoption of stricter regulations in other countries, many governments now intervene in some way to subsidize directly or indirectly this expenditure. The nature of government intervention varies from one country to another; it may take the form of direct grants, tax reductions or low interest loans. The extent of such subsidization also varies among countries, and this may strengthen the possible trade effects of differences in pollution control costs.

In many countries, government financial support for pollution control is considered as a temporary measure aimed at encouraging industry to accelerate the pollution abatement programs. Some countries justify such subsidization on the basis that the balance between old and new mills must be restored from an anti-pollution cost standpoint, that pollution control costs will account for a substantial part of total capital investments for old mills in the short term, and that, due to its record of low profitability, the industry in many countries may find it difficult to obtain the necessary capital even for productive investments.

Most of the countries are engaged in research and development into methods of controlling water and air pollution. The cost of this work is usually divided between industry and the government, in proportions which differ considerably from one country to another, as do the procedures employed. However, the expenditures under this heading cannot be neglected (depending on the country it varies between \$0.09 and \$0.34 per ton of production) and is in addition to the cost of applying control measures. It must be stressed that the amounts which will have to be spent on development and full scale testing of new pollution control methods are considerable when compared to the cost of research itself.

SELECTED REFERENCES

General

1. "Liquid Waste of Industry," N. Nemerow, Ed., Addison-Wesley, Reading, Mass., 1971.
2. "The Chemical Process Industries," 3rd Ed., R. N. Shreve, Ed., McGraw-Hill, New York, 1970.
3. "Water Pollution Control, Experiment Procedures for Process Design," W. W. Eckenfelder and D. L. Ford, Eds., Jenkins Pub. Co., Austin, Tex., 1970.
4. Proceedings, First National Symposium of Food Processing Wastes, Portland, Oregon, April 6–8, 1970, 12060–04/70, EPA, Water Pollution Control Research Series.
5. Nation Symposium on Food Processing Wastes, Second Food Wastes Symposium Proceedings, Denver, Colorado, March 23–26, 1971, 12060–03/71, EPA, Water Pollution Control Research Series.
6. "Surfactant Adsorption onto Porous Media" (Ph.D. dissertation), G. Jackson, University of Maryland, College Park, Md., 1971.
7. "Correlation of Gravitational Forces for Absorption in Packed Columns," G. S. Jackson and J. M. Marcello, *Ind. Eng. Chem. Process Design Develop.*, **7** (1968).
8. "Effective and Wetted Areas for Absorption in Packed Columns," G. S. Jackson and J. M. Marcello, *J. Chem. Eng.* (Japan), **3**, 359 (1970).

9. "Removal of Viruses from Water and Wastewater," G. Berg, EPA, Cincinnati, Feb. 1971.
10. "Single-Cell Protein," R. I. Mateles and S. R. Tannenbaum, Eds., M.I.T. Press, Cambridge, Mass., 1968.
11. "Coliphages as Virus Indicators in Water and Wastewater," 16030 DQN, EPA Second Annual Report, Haifa, Israel, Jan. 1971.
12. Public Health Series, *A WTR Series*, 1964 to 1971.
13. "By-Products from Milk," 2nd Ed., B. H. Webb and E. D. Whitter, Eds., AVI Pub. Co., Westport, Conn., 1970.
14. "Industrial Water Reuse and Future Pollution Solution," G. Rey, W. J. Lacy, and A. Cywin, *Environ. Sci. Techn.*, p. 760 (September 1971).

Chemical and Allied Products

1. "Chemical Process Industries," 3rd Ed., R. N. Shreve, McGraw-Hill, New York, 1967.
2. "The Petrochemical Industry–Market and Economics," A. V. G. Hahn, McGraw-Hill, New York, 1970.
3. "Aqueous Wastes from Petroleum and Petrochemical Plants," M. R. Beychok, John Wiley & Sons, New York, 1967.
4. "Liquid Waste of Industry–Theories, Practices and Treatment," N. L. Nemerow, Addison-Wesley, Reading, Mass., 1971.
5. "Advanced Wastewater Treatment," R. L. Culp and G. L. Culp, Van Nostrand Reinhold, New York, 1971.
6. "Water Pollution Control–Experimental Procedures for Process Design," W. W. Eckenfelder and D. L. Ford, Jenkins Pub. Co., Austin, Tex., 1970.
7. "Microorganic Matter in Water," ASTM Special Technical Publication No. 448, Philadelphia, 1969.
8. "Water Treatment Plant Design," AWWA, New York, 1969.
9. "Organic Chemical Pollution of Freshwater," in "Water Quality Criteria Data Book," Vol. I, D. Little, Inc., Cambridge, Mass., EPA Project 18010 DPV, Dec. 1970.
10. "Water Supply and Treatment," 10th Ed., M. L. Riehl, Bull. 211, National Lime Assoc., Washington, D.C., 1970.
11. "Inorganic Chemicals Industry Profile," Datagraphics, Inc., Pittsburgh, Pa., EPA Project 12020 EJO, July 1971.
12. "Organic Chemistry of Synthetic High Polymers," R. W. Lenz, John Wiley & Sons, New York, 1967.
13. "Water Demineralization and Decontamination Studies Utilizing Electrodialysis," P. E. Des Rosiers, USAERDL, Ft. Belvair, Va., 1961.
14. "Analytical Parameters of Petrochemical and Refinery Wastewaters" (paper presented at 159th A.C.S. meeting), D. L. Ford, J. M. Eller, and E. F. Gloyna, Houston, Texas, Feb. 22, 1970.
15. "Minimizing Waste in the Petrochemical Industry," S. K. Mencher, *Chem. Eng. Progr.*, **63**, No. 10 (Oct. 1967).
16. "Petrochemical Effluents Treatment Practices-Summary," Engineering-Science, Inc. Texas, Austin, Tex., NTIS PB 192 310.
17. "Polymeric Materials for Treatment and Recovery of Petrochemical Wastes," Gulf South Research Institute, New Orleans, La., Government Printing Office, Washington, D.C.
18. "Preliminary Investigational Requirements-Petrochemical and Refinery Waste Treatment," Engineering-Science, Inc. Texas, Austin, Tex., Government Printing Office, Washington, D.C.
19. "Water Pollution and Its Control in the Inorganic Fertilizer and Phosphate Mining Industries," Battelle-Northwest, Richland, Wash. (under review).
20. "Inorganic Chemicals Industry Profile," Datagraphics, Inc., Pittsburgh, Pa. (in press).
21. "Projected Wastewater Treatment Costs in the Organic Chemicals Industry," Datagraphics, Inc., Pittsburgh (in press).
22. "The EPA R&D Program for Environmental Control in the Agricultural Chemical Industries," Environmental Control and Fertilizer Production Conference, Washington, D.C., May 3, 4, 1972.

Food and Kindred Products

1. *Proceedings: First National Symposium on Food Processing Wastes*, FWQA, USDA, National Canners Association, and Northwest Food Processors Association, available from Government Printing Office, Washington, D.C., 04/70 $3.00.
2. *Current Practice in Seafoods Processing Waste Treatment*, Oregon State University, Corvallis, Oregon, available from Government Printing Office, Washington, D.C., April, 1970.
3. *Waste Reduction in Food Canning Operations*, National Canners Association, Berkeley, California, available from Government Printing Office, Washington, D.C., 08/70 $1.00.
4. *Treatment of Citrus Processing Wastes*, The Coca-Cola Company–Foods Division, Orlando, Florida, available from Government Printing Office, Washington, D.C., 10/70 $2.75.
5. *Dry Caustic Peeling of Tree Fruit for Liquid Waste Reduction*, National Canners Association, Berkeley, California, available from Government Printing Office, Washington, D.C., 12/70 $.60.
6. *Proceedings: Second National Symposium on Food Processing Wastes*, EPA, Pacific Northwest Water Laboratory and National Canners Association, available from Government Printing Office, Washington, D.C., 03/71 $4.50.
7. *State-of-the-Art, Sugarbeet Processing Treatment*; by Beet Sugar Development Foundation, Ft. Collins, Colorado, available from Government Printing Office, Washington, D.C., 07/71 $1.25.
8. *Membrane Processing of Cottage Cheese Whey for Pollution Abatement*, Crowley's Milk Company, Binghamton, New York, available from Government Printing Office, Washington, D.C., 07/71 $1.25.
9. *Complete Mix Activated Sludge Treatment of Citrus Process Wastes*, Winter Garden Citrus Products Cooperative, Winter Garden, Florida, available from Government Printing Office, Washington, D.C., 08/71 $1.25.
10. *Elimination of Water Pollution By Packinghouse Animal Paunch and Blood*, Beefland International Inc., Council Bluffs, Iowa, available from Government Printing Office, Washington, D.C., 11/71 $.50.

Pulp and Paper

1. "Pulping Processes," S. A. Rydholm, John Wiley & Sons, New York, 1965.
2. "The Chemical Process Industries," R. N. Shreve, McGraw-Hill, New York, 1956.
3. "Pulp and Paper, Chemistry, and Chemical Technology," J. D. Casey, McGraw-Hill, New York, 1960.
4. "Handbook of Pulp and Paper Technology," K. W. Britt, Van Nostrand Reinhold, New York, 1965.
5. "Bergstrom Makes Waves With '100%' Concept," *Chem. 26 Paper Process.*, **7**, 56–59 (July 1971).
6. "Feasibility Test II," *Chem. 26 Paper Process.*, **7**, 26–27 (Aug. 1971).
7. "The Polysolv Process," *Chem. 26 Paper Process.*, **7**, 24–30 (Sept. 1971).
8. "The Paper in Your Hand," *Chem. 26 Paper Process.*, **7**, 28–30 (Oct. 1971).
9. "Development of the Carbonating Stage of the Tampella Recovery Process," H. Romantschur and J. Mattila, *Tech. Assoc. Pulp Paper Ind.*, **54**, 1495–1499 (Sept. 1971).
10. "An Evaluation of Four Chemicals for Preserving Wood Chips Stored Outside," E. L. Springer, *et al.*, *Tappi*, **54**, 555–560, April, 1971.
11. "Upgrading Waste Treatment Plants," *Chem. Eng.*, G. L. Shell, *et al.*, **78**, 97–102 (June 21, 1971).
12. "Advanced Pollution Abatement Technology in the Pulp and Paper Industry," Organization for Economic Cooperation and Development, Paris, France, 1972.

Petroleum and Coal Products

1. "'Torrey Canyon' Pollution and Marine Life," J. E. Smith, Ed., Cambridge University Press, London, 1968.
2. "The Impact of Oily Materials on Activated Sludge System," 12050 DSH 03/71, Hydroscience, Inc., Westwood, New Jersey, Government Printing Office, Washington, D.C.

3. "Fluid-Bed Incineration of Petroleum Refinery Wastes," 12050 EKT 03/71, American Oil Company, Mandon, N.D., Government Printing Office, Washington, D.C.

Metal and Metal Products

1. "Wastewater Treatment Technology," State of Illinois Institute for Environmental Quality, IIEO Document No. 71–4, Aug. 1971.
2. "A State-of-the-Art Review of Metal Finishing Waste Treatment," 12010 EIE 11/68, Battelle Memorial Institute, Columbus, O. Government Printing Office, Washington, D.C.
3. "Limestone Treatment of Rinse Waters from Hydrochloric Acid Pickling of Steel," 12010 DUL 02/71, Armco Steel Corporation, Middletown, O., Government Printing Office, Washington, D.C.
4. "Treatment of Waste Water-Waste Oil Mixtures," 12010 EZV 02/70, Armco Steel Corporation, Middletown, O.
5. "Brass Wire Mill Process Changes and Waste Abatement, Recovery and Reuse," 12010 DPF 12/71, Volco Brass and Copper Company, Kenilworth, N.J., Government Printing Office, Washington, D.C. 20402.

Textile Mill Products

1. "The EPA R&D Program for Water Quality Control in the Textile Industry" (presented at the American Association of Textile Colorist and Chemist Symposium), C. H. Ris and W. J. Lacy, Atlanta, Georgia, Mar. 31, 1971.
2. "Reuse of Chemical Fiber Plant Wastewater and Cooling Water Blowdown," 12090 EUX 10/70, Fiber Industries, Inc., Charlotte, N.C., and Davis and Floyd Engineers Inc., Greenwood, S.C., Government Printing Office, Washington, D.C.
3. "Bio-Regenerated Activated Carbon Treatment of Textile Dye Wastewater," 12090 DWM 01/71, C. H. Mosland and Sons, Wakefield, R.I., Government Printing Office, Washington, D.C.
4. "Fine Precipitation and Recovery from Viscose Rayon Wastewater," 12090 ESG, 01/71, American Enka Company, Enka, N.C., Government Printing Office, Washington, D.C.
5. "State-of-the-Art of Textile Waste Treatment," 12090 ECS 02/71, Clemson, S.C., Government Printing Office, Washington, D.C.

Joint Industrial-Municipal Treatment

1. "Joint Municipal and Semichemical Pulping Wastes," City of Erie, Pa., and Hammermill Paper Co., Government Printing Office, Washington, D.C.
2. "Feasibility of Joint Treatment in a Lake Watershed," Onondaga County, N.Y., NTIS PB 201 698.
3. "Onondaga Lake Study," Onondaga County, Syracuse, N.Y., Government Printing Office, Washington, D.C.
4. "Joint Treatment of Municipal Sewage and Pulp Mill Effluents," Green Bay Metro Sewage District, Green Bay, Wisc. (under review).
5. "Combined Treatment of Domestic and Industrial Wastes by Activated Sludge," City of Dallas, Ore., Government Printing Office, Washington, D.C.
6. "Biological Treatment of Chlorophenolic Wastes," City of Jacksonville, Ark., Government Printing Office, Washington, D.C.
7. Rey, G. and Lacy, W. J., "Industrial Water Closed Cycle Research Programs and Needs," Delaware River Basin Conference, Trenton, N.J., September 12, 1972.

Miscellaneous

1. "Water Treatment for Industrial Uses," E. Nordell, Van Nostrand Reinhold, New York.
2. "Water Pollution Control," W. W. Eckenfelder and D. L. Ford, Jenkins Pub. Co., Austin, Tex.
3. "Handbook of Industrial Water Conditioning," Betz, Betz Labs., Trevose, Pa.
4. "Industrial Water Pollution," W. W. Eckenfelder, McGraw-Hill, New York.

5. "Disposal of Wastes from Water Treatment Plants," American Water Works Association Research Foundation, New York, NTIS PB 186 157.
6. "Activated Sludge Treatment of Chrome Tannery Wastes," A. C. Lawrence Co., Peabody, Mass., Government Printing Office, Washington, D.C.
7. "Treatment of Sole Leather Vegetable Tannery Wastes, J. D. Eye, University of Cincinnati, Cincinnati, O., Government Printing Office, Washington, D.C.
8. "Anaerobic-Aerobic Lagoon Treatment for Vegetable Tanning Wastes," University of Virginia, Charlottesville, Va., Government Printing Office, Washington, D.C.
9. "Treatment of Wastes from a Sole Leather Tannery," J. D. Eye and L. Liu, *J. Water Pollution Control Federation*, **43**, 2291–2303 (Nov. 1971).
10. Rimer, A. E. *et al.*, "Activated Carbon System for Treatment of Combined Municipal and Paper Mill Waste Waters in Fitchburg, Mass.," *Technical Association of Pulp and Paper Industry*, **54**, 1477–1483 (Sept. 1971).
11. "A Comparison of Macroscopic and Microscopic Indicators of Pollution," J. D. Gallup, *et al.*, *Okla. Acad. Sci.*, **50**, 49–56, 1970.
12. "The Evaluation of Water and Related Land Resource Projects: A Procedural Test," J. D. Gallup, *et al.*, Bureau Water Resources Research, University of Oklahoma, 1970.
13. "Industrial Waste Survey and Treatability Study," J. D. Gallup, Oklahoma Vegetable Oils Products Co., Durant, Okla. 1970.
14. "Investigation of Filamentous Bulking in the Activated Sludge Process" (Ph.D. Thesis), J. D. Gallup, University of Oklahoma, 1971.

27

Air Pollution Control

William R. King*

INTRODUCTION

Air pollution control (APC) is in a state of flux both because of rapidly changing technology and a shift from state rule-making and enforcement to joint state-federal efforts. Because of this rapidly changing climate, a presentation of control techniques, allowable emission levels, and design rules would soon be obsolete. Instead, a case study approach is used in this chapter. These studies illustrate control trends, some special considerations that must be given to pollution-control-equipment design, and the application of conventional engineering principles in nonprofit situations. The goal of this chapter is to develop a series of concepts that the reader can apply to problems of immediate concern. A great deal of space is devoted to control costs since the problem is usually not "can a process be controlled" but "how much does control cost." In addition to forming the basis of a meaningful comparison of alternatives, costs are important inputs to the yet unresolved questions on the effects air pollution control will have on our economy. For instance, will control cause the disappearance of plants, companies, or dislocation of entire industries due to competition, imports, or substitute products that are less burdened by A.P.C. costs? What relative advantages or disadvantages will the laws give competing geographic regions? Are the social benefits purchased—reduced material damage, more attractive environment, better health—worth the cost of control?

Since air pollution control is not an end in itself, as the manufacture of polyethylene is, but a response to legal and moral pressures, reasonable extrapolation of the case studies cannot be made without some knowledge of the law.

This chapter covers four general topics:

1. The law and the thrust of its application.
2. Case studies of three chemical manufacturing processes—sulfuric acid, nitric acid, and wet process phosphoric acid.
3. An examination of two cases of pollution

*Division of Applied Technology, Environmental Protection Agency, Research Triangle Park, N.C.

control's effects on chemical markets—trichloroethylene and lead alkyls.

4. An examination of the capital and operating cost of air pollution control equipment.

THE LAW

With the passage of the Clean Air Amendments of 1970[1] the movement toward effective nationwide control quickened, and the Federal Government assumed a more active role. Air Quality Control Regions (AQCR) were expanded from the 92 covering the large population centers to 247 covering the entire country.[2] (Figure 27.1 shows these regions.) Designation of these additional AQCR's put the entire country under the same rule-making obligations, which eliminated to some extent the incentive for industry to avoid control by relocation and some competitive advantages due to geographical location.

The 1970 Clean Air Amendments have provided three rule-setting mechanisms that will directly affect the chemical industry:

1. National primary and secondary ambient air quality standards.
2. Standards of performance for new stationary sources.
3. National emission standards for hazardous air pollutants.

National Primary Air Quality Standards are time-based, maximum-allowable ambient air pollutant concentrations that can be tolerated without adversely affecting the *public health*. *Secondary Standards* are those concentrations of a pollutant that can be tolerated on a time basis without affecting the *public welfare*. The law requires that these standards be justified or supported with documents on air quality criteria and control techniques. The former examines the kind and extent of identifiable effects of a pollutant on people and property, and its interactions with other pollutants; the latter examines control methods and costs.

Setting and enforcing emission-control regulations to insure that ambient air standards are met is a state function in keeping with historical precedent. However, the trend will probably be toward much more uniformity. EPA is implicitly encouraging this uniformity through the publication of typical regulations,[3] and in the required Federal review of state implementation plans.

Both primary and secondary standards have been published for six classifications of pollutants—sulfur oxides, particulate matter, carbon monoxide, petrochemical oxidants, hydrocarbons, and nitrogen dioxide.[4] These are summarized in Table 27.1. The allowable emission levels are coupled to specific sampling and analytical methods, described in references 4 and 22 because different scnemes give different results. Correlations between methods have not been developed for the most part.

Ambient air standards seem to be most applicable when applied against pollutants that are emitted by a large number of different sources. The sources of other pollutants—fluorides, odors, and hydrogen chloride, for instance, are less widely distributed. This, coupled with the meager existing data base from which the necessary criteria documents can be developed, leads one to believe that this enforcement tool will not be extensively used in the near future.

New *Stationary Source Performance Standards*, as their name implies, are allowable emissions for new or modified* emission sources. This mechanism regulates a specific industry or segment of an industry, rather than a pollutant. Thus far, standards have been set for five industries; the industries and allowable emissions are listed in Table 27.2.

Three of these might be broadly classified as "chemical industry sources." Other areas for which new source-emission standards are tentatively scheduled include segments of the petrochemical, thermal phosphoric acid, phosphate fertilizer, elemental phosphorus, aluminum reduction, and nonferrous smelting industries. Primary responsibility for enforcement of new source standards is the Federal

*Modified is defined in the 1970 law as any change that "increases the amount of any air pollutant emitted" or results in the emission of a new pollutant. The rule-making procedure[5] has specifically exempted equipment replacement, changes in production rates less than design rates, changes in on-stream time, and use of alternate fuels and raw materials.

UNITED STATES
DEPARTMENT OF COMMERCE

BUREAU OF THE CENSUS

Boundaries of counties and county equivalents as of January 1, 1970

For sale by the Superintendent of Documents, U.S. Government Printing Office
Washington, D.C. - Price 30 cents

Fig. 27.1 Federal Air Quality Control Regions

TABLE 27.1 National Primary and Secondary Standards

Pollutant	Time Basis	Primary Standard ($\mu g/m^3$)	Secondary Standard ($\mu g/m^3$)
Sulfur oxides (measured as SO_2)	Annual arithmetic mean.	80	60
	Maximum 24-hr concentration (not to be exceeded more than once/yr).	365	260[a]
	Maximum 3-hr concentration (not to be exceeded more than once/yr).	–	1300
Particulate matter	Annual geometric mean.	75	60
	Maximum 24-hr concentration (not to be exceeded more than once/yr).	260	150
Carbon monoxide	Maximum 8-hr concentration (not to be exceeded more than once/yr).	10,000	10,000
	Maximum 1-hr concentration (not to be exceeded more than once/yr).	40,000	40,000
Photochemical oxidants (measured as ozone)	Maximum 1-hr concentration (not to be exceeded more than once/yr).	160	160
Hydrocarbons, except methane (measured as methane)	Maximum 3-hr concentration, 6–9 a.m. (not to be exceeded more than once/yr).	160	160
Nitrogen dioxide	Annual arithmetic mean.	100	100

[a]A proposal to drop this standard is being considered. (See Reference 21.)

Government's, however, it can be delegated to qualified state control agencies.

As presently* set up, the method for insuring new source compliance is shown in the flow chart, Fig. 27.2.[5] When Federal new-source performance standards are set, the law requires that the individual state governments establish emission standards for the same industries or industry segments unless the pollutants emitted by these existing sources are controlled under previously announced national ambient-air or hazardous-pollutant standards.* Although this rule-making is a state function, the variation of the standards across the nation may be relatively small.

This section of the law will be applied frequently since the criteria that must be met in setting new-source standards is that the proposed regulations be consistent with the degree of control attainable with the best "adequately demonstrated" control system. The admonition in the law that the standards set shall take "into account the cost of achieving

*Procedures for delegating authority for control of new sources to the states were established in late 1971 (Ref. 23). This will probably yield a period in the mid 1970's where new plant controls in some sections of the country are state responsibility, while in others they will be federal responsibility.

*For example, SO_2 emissions from existing sulfuric acid plants would be controlled by states under the previously announced National Ambient Air Standards for Sulfur Oxides; to control acid mists, states would set rules under this section.

TABLE 27.2 New-source Emission Standards

New Source	*Source*	*Pollutant*	*Allowable Emission*[a]
Fossil-fuel-fired steam generators of more than 250 × 10^6 Btu/hr heat input (200,000 lb/hr steam rate).	Boiler stack[c]	Particulate matter	a. 0.10 lb/10^6 Btu. b. 20% opacity (40% opacity 2 min/hr; except where high opacities are due to water vapor).
		Sulfur dioxide	a. 0.8 lb/10^6 Btu (liquid fuel fired). b. 1.2 lb/10 Btu (solid fuel fired).
		Nitrogen oxides	a. 0.2 lb/10^6 Btu (gaseous fuel fired). b. 0.3 lb/10^6 Btu (liquid fuel fired). c. 0.7 lb/10^6 Btu (solid fuel fired).
Incinerators with changing rate greater than 50T/24-hr day. (Solid wastes is defined as material containing more than 50% municipal-type waste).	Incinerator stack	Particulate matter	0.14 g/nm^3 (stack gas is corrected to 12% CO_2).
Sulfuric acid plants (plants using sulfur, hydrogen sulfide, organic sulfur compounds, or oil refinery-spent acid as their sulfur source).	Tail-gas stack[b]	Sulfur dioxide	4 lb/ton of 100% sulfuric acid produced
		Acid mist and SO_3	0.15 lb/ton of 100% sulfuric acid produced (expressed as H_2SO_4).
Nitric acid plants producing 50–70% nitric acid solutions.	Tail-gas stack[b]	Nitrogen oxides (except nitrous oxide)	3 lb/ton of acid of 100% to HNO_3.
Portland cement plants	Kiln	Particle	a. 0.30-lb/ton kiln feed (dry basis). b. 10% opacity (except where higher opacities are due to water in the stream).
	Clinker	Particle	a. 0.10-lb/ton kiln feed (dry basis). b. 10% opacity
	Raw material preparation Product sizing Raw material storage Clinker storage Product storage Converter transfer points Bulk loading and unloading	Particle	10% opacity

[a]All process emission rates are based on 2-hour average rates.
[b]The rule-making procedure requires that automatic monitoring devices be installed on these sources to record the emissions of all controlled pollutants.

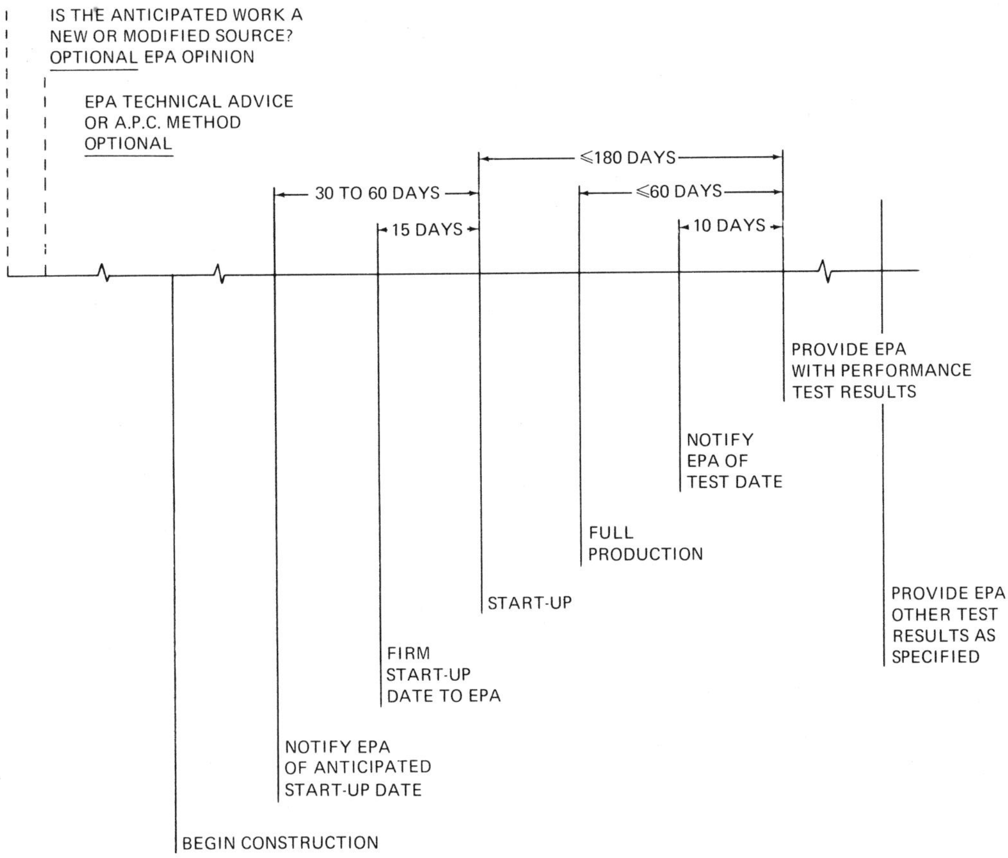

Fig. 27.2 Approval procedure for plants for which new-source standards apply.

such reduction" is presently interpreted within EPA to mean that the standards shall be cost effective; that is, the regulations will measurably improve the atmosphere.

National Emission Standards for Hazardous Air Pollutants provide the third and final standard-setting tool in the law. A hazardous air pollutant is defined as one that in the administrator's judgment will "cause or contribute to an increase in mortality or an increase in serious irreversible or incapacitating reversible illness."[2] As for new-source standards, the Federal Government has the primary enforcement responsibility but it can be delegated to qualified state organizations.

Asbestos, beryllium, and mercury have been designated hazardous pollutants. Table 27.3 lists the proposed allowable emissions standards.[6] The standards, as the table shows, have been applied on a source-by-source basis. The number of additional sources of these pollutants regulated will depend upon the extent and depth of governmental exploratory source-sampling programs, since for many potentially significant sources, these pollutants appear as trace, undetected, or not as presently unmonitored pollutants.

Modification to or construction of new units in the regulated categories require prior EPA approval; registration of all controlled units are required.[7] In addition, emission tests of controlled sources of mercury and beryllium are required every three months. At present, quantitative chemical tests for mercury and beryllium exist, therefore definite emission limits could be set. In contrast, no good way exists to identity and separate asbestos from other particulate matter, so instead of the

TABLE 27.3 Hazardous Pollutants Emission Standards

Source	*Acceptable Control Method*	*Emission Rate*
Asbestos		
Asbestos conversion operations	Fabric filters or wet scrubbers	No visible emissions.
Ore processing		
Cloth, cord, paper manufacture		
Insulation manufacture		
Coating and adhesive manufacture		
Paint formulation		
Plastic processing		
Chlorine manufacture (Cell diaphragm fabrication)	Fabric filters or wet scrubbers	No visible emissions.
Construction		
Application (i.e., fireproofing)	Fabric filter or wet scrubber	No visible emissions.
Demolitions	Prescribed procedures	
Roads	Use of tailings forbidden except in mines	
Beryllium		
(All sources in the following units handling beryllium-containing materials).		
Machine shops		10 g/day or 0.010 g/m^3 average 30-day atmospheric conc.
Ceramic plants		
Propellant plants		
Foundries (alloys with more than 0.1% beryllium).		
Extraction plants		
Incinerators		
Rocket-motor test sites		2 g/hr to a maximum of 10 g/day.
Mercury		
All sources within ore-processing plants		
Mercury cell chlor-alkali plants–hydrogen product stream, process vents, and room vents		2300 g/day total.

more traditional emission limits, minimum equipment-performance standards will be required.

The extent to which this rule-setting tool will be used to set standards for other sources of asbestos, beryllium, mercury, and other hazardous pollutants is unclear at present.

Contact-Process Sulfuric Acid Plants*

Higher efficiency sulfuric acid-manufacturing technology has evolved over the last decades. Table 27.4 shows the effect of these efficiency improvements on sulfur oxide pollution from sulfur burning, contact-process acid plants. The emission rates in the table are based on efficiencies at design production rates. However, like most processes, sulfuric acid plants are built with a little "fat." This design safety margin is usually excess combustion-air capacity and over-designed heat-exchange surface. If the extra acid can be sold, there are good economic reasons to run the plant at the maximum possible capacity. To illustrate, the profit potential of a 600 T/day, 4-contact stage, single-absorption unit designed to op-

*Refer to Chapter 4 for a complete description of sulfuric acid manufacture.

TABLE 27.4 Approximate Design Emissions for Sulfur Burning Plants (Producing 93–98% H_2SO_4)

Plant Description	*Period During Which Design was Popular for New Plants*	*Typical Design Conversion Efficiency* (SO_2–SO_3)	*Emission Rates: lbs of Sulfur/ton of 100%* H_2SO_4 *(ppm in stack gas)* SO_2	SO_3
Three contact stages Single-absorber Pad-type mist eliminators.	pre-1960	96	28 (4000 ppm)	4
Four contact stages Single-absorber Pad-type mist eliminators.	1960–1970+	98	13 (1900 ppm)	4
Four contact stages Double-absorber Panel-type mist eliminators.	after 1970	99.5	2 (270 ppm)	~0.1

erate 330 days/year on a burner effulent of 8 per cent SO_2 will be considered. The assumptions will be made that (1) the plant can burn enough extra sulfur to produce a 9 per cent burner effluent; (2) all acid produced can be sold at the same price; and (3) at design capacity the profits are sufficient to yield a 5-year capital payback** on the entire investment. This return is equivalent to a profit of \$2.50/T before 50 per cent taxes. Operating the plant with 9 per cent-SO_2 burner discharge rather than 8 per cent increases its production rate to 660 T/day, 10 per cent over-design. However, sulfur-to-acid efficiency drops to about 96.8 per cent. Table 27.5 compares the economics for the plant operating at 600 T/day and 660 T/day. As the similar dollar-per-ton manufacturing costs at the two production levels show, the unit-cost savings made by spreading the capital, maintenance, and labor charges over the higher production rate are nearly offset by the lower sulfur efficiency. The improved profit picture comes mainly from the higher sales. As the table shows, the extra 10 per cent production increases the after-tax cash flow by nearly 15 per cent and drops the capital payback period 0.4 year. Therefore, barring the complicating factor of air pollution control, it makes economic sense to operate at as high a rate as sales and plant bottlenecks allow. However, even this small production increase nearly doubles the stack-gas sulfur concentration from 1700 ppm to 3000 ppm (22.5 lb/T of 100 per cent acid).

During the 1960's air pollution control regulations were applied to sulfuric acid plants. Comparison of Figure 27.3 with Table 27.5 shows that regulations roughly paralleled the control attainable in plants utilizing the most modern technology. For example, the double-absorption, double-contact process just recently introduced into the U.S. will be able to meet the current Federal new-source performance standards.*

However, existing plants which utilize older technology were usually faced with modification. The federal suggested standard for existing sources[3]—3.4 pounds of sulfur in the stack gases per ton of 100 per cent acid—is below that attainable by good maintenance and careful operation in single-absorption units. For these plants conversion to dual-absorption units is possible as is the addition of a large variety of tail-gas purification units. The available add-on processes dispose of the removed sulfur oxides in one of four ways:

1. By recycling sulfur oxides to the main unit.

**Capital payback = $\dfrac{\text{Capital investment}}{\text{After-tax profits + Depreciation}}$

*Reference 8 reports that stack measurements on a double contact double absorption unit showed emission rates of 1.7 lb/T of 100 per cent acid.

TABLE 27.5 Sulfuric Acid Plant Profits at Design and Higher Rates

Capital Investment [Nominal 600 T/Day (198,000 T/Yr) Unit]

Battery limits	$\$1.90 \times 10^6$ [a]
Off sites	$\$0.57 \times 10^6$
TOTAL	$\$2.47 \times 10^6$

Production Costs	*Unit Operating at 600 T/Day ($/Yr)*	*Unit Operating at 660 T/Day ($/Yr)*
Variable costs		
Sulfur ($20/short ton)		
.333 T sulfur/ton 100% H_2SO_4 at 98% conversion	1,300,000	
.337 T sulfur/ton 100% H_2SO_4 at 96.8% conversion		1,470,000
Cooling water[b] 84 × 10^3 gal/T at $.03/$10^3$ gal	50,000	55,000
Power 48 kw-h/T at $.01/kw-h	95,000	
45 kw-h/T at $.01/kw-h		98,000
Boiler and process water 400 gal/T at $.20/$10^3$ gal	16,000	18,000
Steam 2300 lb/T at $1.20/T	(274,000)	(300,000)
Semifixed costs		
Labor 1 man/shift at $4.25/hr	37,000	37,000
Supervision 1/3 time at $15,000/yr	5,000	5,000
Maintenance 4% capital	99,000	99,000
Payroll added costs 30% of labor and supervision	13,000	13,000
Fixed costs		
General overheads (50% semifixed)	77,000	77,000
Taxes (1%), insurance (.5%), depreciation (10%)	283,000	283,000
TOTAL	1,701,000	1,855,000
Total manufacturing cost $/T	8.60	8.56
After tax profit $/yr	247,000	285,000
Capital payback–years	5.0	4.6

[a]4th Quarter 1971 Costs. All electric drives; steam produced for export.
[b]The effects of efficiency difference on utility usages are ignored since their magnitude is 2% in all of the tabled figures in all cases.

2. By recycling to the main unit weak sulfuric acid manufactured in the recovery unit.

3. By producing a by-product with some marketable value.

4. By transforming the waste gases into a more socially acceptable waste.

A discussion of the many processes available and dual-absorption process is available in References 7 and 8.

In order to illustrate the effect of control on existing sulfuric acid plants' costs, lime scrubbing of the effluent gases, a minimum-capital process, will be added to the previously described 600 T/day, single-stage absorption plant.

Lime scrubbing, as shown in Fig. 27.4, was chosen for this study over other process tail-gas purification schemes for the following reasons:

1. The process is a simple, demonstrated sulfur-removal system.

2. While the capital requirements are minimal, the total impact on acid costs are reasonably representative of most recovery processes.

3. The economics are not complicated by by-product sales.

4. Except for the highly reliable scrubber, the acid plant can be operated independently of the air pollution control system.

5. SO_2, SO_3, and acid mist can be controlled.

Slurry and $CaSO_4$-handling difficulties are the major detractions of the process. The basic

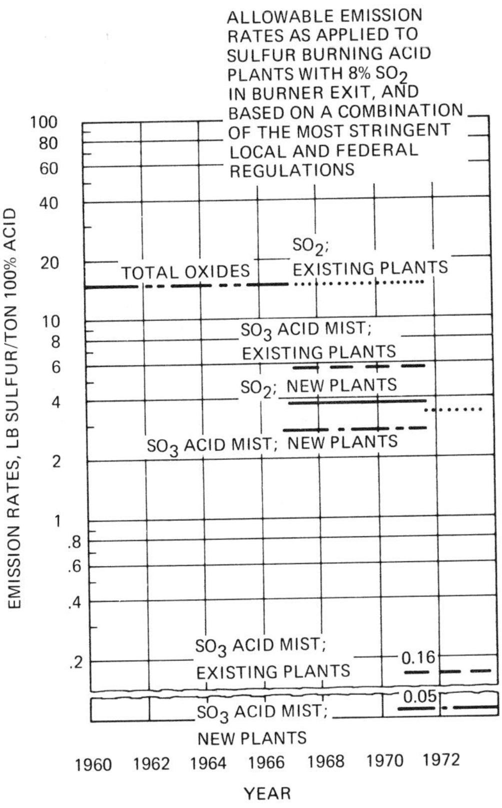

Fig. 27.3

chemical reactions involved are:

$$CaO + SO_2 \longrightarrow CaSO_3 \cdot 2H_2O \downarrow$$

$$+ \tfrac{1}{2} O_2 \longrightarrow CaSO_4 \cdot 2H_2O \downarrow$$

$$CaO + 2H_2O + SO_3 \longrightarrow CaSO_4 \cdot 2H_2O \downarrow$$

The capital cost shown in Table 27.6 is based on indoor installation for cold-weather protection, engineering and construction by an outside contractor, and 4th-quarter 1971 costs. Two cost studies are presented—the plant operating at design and at 10 per cent over design. In both cases, air pollution control drives the manufacturing cost up by about a dollar a ton. (Compare Table 27.6 with Table 27.5). The small per-ton-manufacturing-cost advantage gained by running the uncontrolled plant at over-design rates ($0.10/T) is eliminated by the higher lime-use rates. The effect of control on profits depends upon the competitive situation—supply and demand, shipping radius, competitors' plant size, and whether competition must meet the same rules.* Two extremes are illustrated in the table: (1) the net selling price remains the same; (2) the net selling price rises to yield an acceptable profit on investment. If the market does not allow a price increase, capital-payback periods slip from 5 to 6.7 years for the plant running at design rate. An increase in the net selling price of $1.40/T (about 13 per cent) is necessary to bring the payout of the plant to 5 years. In both cases, running the plant at higher than design rates becomes less attractive than it did without control.

Cost estimates for a slightly larger—700 T/D—double-absorption, double-contact unit have shown a capital cost difference between it and a single-absorption unit of about $950,000 (1971). Total manufacturing cost, when put on the same basis as the lime-scrubbing case, is about $0.95/ton of 100 per cent acid.

In summary, examination of the control of atmospheric emissions from sulfuric acid plants shows:

1. Changes in control regulations tend to parallel improvements in manufacturing efficiency and tend to become more stringent with time.
2. Maximizing profits and minimizing emissions may be incompatible goals.
3. The cost impact of air pollution may be significant.

Nitric Acid

In contrast to sulfuric acid, nitric acid plants were seldom specifically controlled before new-source standards were set. Usually, nuisance ordinances and opacity regulations were the main regulations applied to these plants.

Modern nitric acid plants* pass the absorber off-gases through expansion turbines to furnish 75 per cent or more of the feed-air-compressor energy requirements. In order to maxi-

---

*The ability of sulfuric acid manufacturers to pass costs on to consumers is discussed in "The Economics of Clean Air," published about March 1972, U.S. Government Printing Office, Washington, D.C. 20402.

*See Chapter 5 of this book for a complete process description.

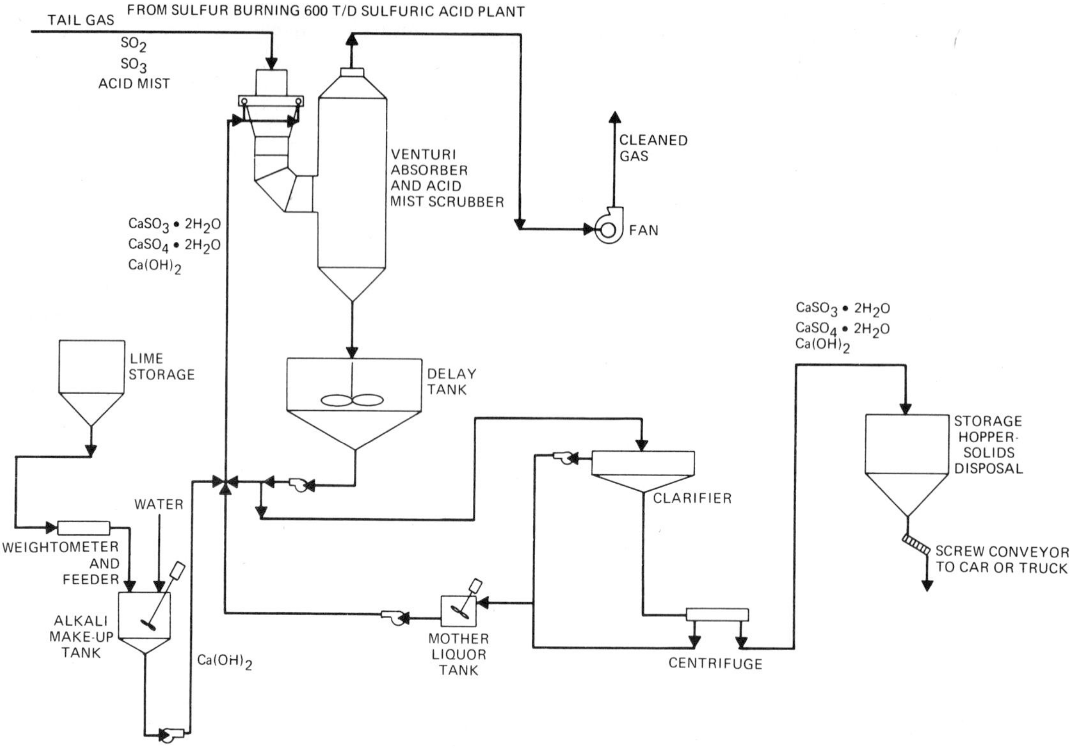

Fig. 27.4 Lime-solution scrubbing system to remove sulfur oxides from the absorber off-gases of a 600-T/D sulfuric acid plant.

mize the power recovery, plants built in the 1960's add fuel—methane or hydrogen—to the tail-gas stream and pass it through a catalytic combustor to raise its temperature. Expansion turbine limitations usually hold the maximum combustor temperature to about 1250°F. This operating innovation produced a side benefit—the reduction of NO_2 to NO—which eliminated the color from the tail-gas stream.[9] However, the total NO_x concentration was not changed since NO was not reduced in the oxidizing atmosphere of the reactor.

By adding excess fuel, about 10 per cent more than the amount necessary to consume the free oxygen and the oxygen bound on the NO_x, the system can be made to reduce the NO_x concentrations to acceptable levels. However, the process runs somewhat on the ragged edge of the equipment capabilities. With methane used as fuel, the reaction mass must be heated to about 900°F before the combustor will operate. For each 1 per cent free oxygen in the gas entering the combustor, the stream temperature increases about 230°F across the unit. Equipment-design and catalyst-deactivation problems limit the maximum reactor temperature to about 1500°F. Apparently a minimum reactor-effluent temperature of about 1050°F must be reached before the NO_x-reduction reaction proceeds at a satisfactory rate.[10] Therefore, the free-oxygen content of the combustor inlet stream can range only between 1 and 2.5 per cent.

Figure 27.5 is a process sketch of the tail-gas-processing and power-recovery section of a nitric acid plant. The heavily outlined equipment items are those which must be modified or added to achieve NO_x burnout instead of simple power recovery. Hydrogen, if available, can be used as fuel[10] in the catalytic reduction unit. Ammonia can be used to reduce NO_x, however, no power recovery is possible.[11]

In order to develop comparative economics

TABLE 27.6 Air Pollution Control Cost for Sulfuric Acid Plant Using Lime Scrubbing

Capital Investment (Lime-Scrubbing System for Nominal 600 T/Day Unit)
Total plant 0.42×10^6

Production Costs (Lime-Scrubbing Only)	*Unit Operating at 600 T/D ($/Yr)*	*Unit Operating at 660 T/D ($/Yr)*
Variable costs		
Power 15 Kw-h/T[a] at $0.01/Kw-h	30,000	
14 Kw-h/T at $0.01/Kw-h		31,000
Process water 20 gal/T at $.20/$10^3$ gal	8,000	9,000
CaO .0106 T/T (90% utilization) at $16/T	34,000	
.0197 T/T		68,000
Waste disposal at $3/T		
(30% water) .0387 T/T	23,000	
.0565 T/T		37,000
Semifixed costs		
Labor 1 man–daylight at $4.25/hr	12,000	12,000
Supervision (no additional)	0	0
Maintenance (4% capital)	17,000	17,000
Payroll added costs (30% labor and supervision)	4,000	4,000
Fixed costs		
General overheads (50% semifixed)	16,000	16,000
Taxes (1%), insurance (.5%), depreciation (10%)[b]	48,000	48,000
TOTAL air pollution control costs	192,000	242,000
Base plant-manufacturing costs (from Table 27.5)	1,701,000	1,855,000
TOTAL manufacturing costs	1,893,000	2,097,000
TOTAL manufacturing cost $/T	9.60	9.60
Case 1–no change in net selling price		
After tax profit $/yr (net selling price of $11.10/T)	148,000	164,000
Capital payback on 2.89 $\times 10^6$ in years	6.7	6.5
Case 2–five-year capital payback at design rates		
Net selling price to yield a 5-yr payback on base case $/T	12.50	–
After tax profit at net selling price of $12.50/T in $/yr	289,000	320,000
Capital payback–years	5.0	4.8

[a]Variable-use rates are per ton of 100% H_2SO_4.
[b]For plants existing before 1/1/69 accelerated depreciation allowances are allowed for pollution-control equipment put into operation between 12/31/69 and 1/1/75. (Federal Register Vol. 36, Number 189, pp. 19132–19134 contains guide lines.) Such accelerated allowances were not used in these studies because they tend to distort the economics.

showing the impact of air pollution control, the question "Where does the process end and the APC system begin?" must be answered. Catalytic combustors, as previously mentioned, were originally installed to improve energy recovery and lower operating costs. The fact that the tail-gas stream was decolorized is incidental. Therefore, the original conbustor system cannot be considered an APC device. However, the additional capital necessary to modify the basic combustor design to enable NO_x burnout, and the additional costs of operating the unit to achieve burnout should certainly be charged to APC. Table 27.7, developed on this basis, shows that new stationary-source-performance standards add about a dollar per ton to the transfer price (net selling price) of nitric acid. Increased capital charges are responsible for half the increase, reduced catalyst life and higher natural gas use account for the rest.

Reduction of NO_x is accomplished by op-

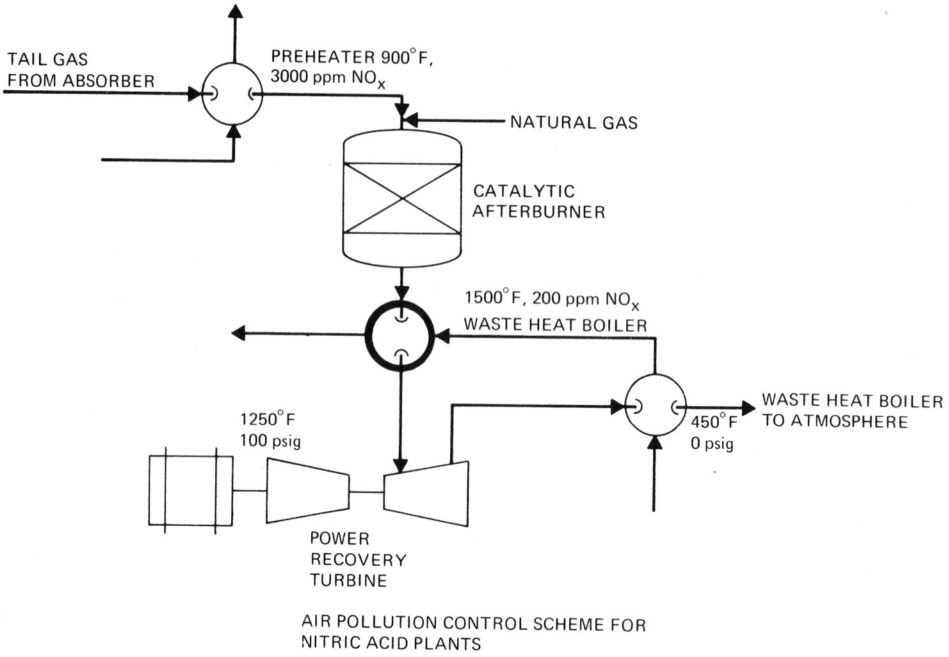

Fig. 27.5 Air pollution control scheme for nitric acid plants.

erating the catalytic reactor in a reducing atmosphere. Measurements show that when NO_x is reduced to 200 ppm in the tail-gas stream (3 lb/T of acid) about 500 ppm of CO and 5 ppm of HCN can be expected. In addition, unreacted fuel (CH_4) and partially reacted fuel (CH_4O, for example) may be present. Currently the Environmental Protection Agency apparently feels that the emission levels of these pollutants are acceptable. However, as operating experience is gained, emission limits may be set on these reducing compounds.

This section, as does the previous section covering sulfuric acid, shows the relatively large cost impact of control and the tendency of A.P.C. regulations to require the application of the most modern-available manufacturing processes. The possibility of integrating A.P.C. into the basic process, and the problem of dividing costs between A.P.C. and product manufacture are illustrated. Finally, this section points out the fact that with available technology, control may require trade-off decisions. These decisions would recognize that a high level of control for a specific pollutant may create additional A.P.C. problems. (Reduced NO_x concentrations result in finite reductant emissions.)

Wet-Process Phosphoric Acid Manufacture

In the previous two sections, the major pollutant has been essentially unrecovered product; here the major pollutant—water-soluble fluorides—is usually wasted by-product. Also in contrast to the previous case studies, A.P.C. requirements are not pushing available technology in this industry, thus, projecting the evolution of A.P.C. will be a useful exercise.

A typical phosphate ore contains 30 to 35 weight per cent P_2O_5 and 3 to 4 weight per cent fluorine. Assuming 54 per cent* acid is being produced, the fluorides are typically distributed as follows:[12]

	lb F/ton P_2O_5
With feed	210
With product acid	33
With waste gypsum	98
Evolved as gases	79

*Refers to the P_2O_5 content of the product phosphoric acid.

TABLE 27.7 Nitric Acid Manufacturing Costs with and Without Control

	300 T/D (100,000 T/Yr) Plant	
	With Power Recovery	*With Single Stage* NO_x *Burn-Out*
Capital Costs[a], *4th Quarter 1971*	*\$2.6* × *10*^{6}	*\$2.7* × *10*^{6}
Production costs \$/yr		
Variable costs		
Ammonia 0.29 T/T 100% of HNO_3 at \$35/T	1,010,000	1,010,000
Power 115 Kw-h/T at \$0.01/Kw-h	115,000	115,000
Main reactor catalyst 0.14 g active material/T at \$3.50/g	49,000	49,000
Cooling water 30 × 10^3 gal at \$.03/10^3 gal	90,000	90,000
Boiler & process water		
550 gal/T at \$0.20/10^3 gal		11,000
370 gal/T	7,000	
Power recovery–NO_x		
Reduction-reactor catalyst .60 × 10^3 ft at \$700/ft^3 1 year life		42,000
4 year life	10,000	
fuel at \$.50/10^6 Btu		
1.8 × 10^6/T		90,000
0.9 × 10^6/T	45,000	
Steam at \$1.20/T		
4400 lb/T		266,000
3900 lb/T	234,000	
Semifixed costs		
Labor 1 man/shift at \$4.25/hr	37,000	37,000
Supervisor (1/3 time at \$15,000/yr)	5,000	5,000
Maintenance (5% capital investment)	130,000	135,000
Payroll added costs (30% labor and supervision)	13,000	13,000
Fixed costs		
General overhead 50% of semifixed costs	92,000	95,000
Taxes, insurance, depreciation	300,000	311,000
TOTAL manufacturing cost \$/yr	1,669,000	1,737,000
TOTAL manufacturing cost \$/T	16.70	17.40
Transfer cost–assuming a 5 year capital payback \$/T	21.90	22.80

[a]Electrical supplemental drive on main compressor–all steam produced for export.

Most of the available evidence indicates that the greater portion of the gaseous fluorides evolved in wet-process plants is SiF_4. A small portion is evolved in the reaction and filtration steps, and in product and in-process storage, but most come off during concentration of the acid from 32 to 54 per cent. A typical A.P.C. scheme is shown in the process sketch, Fig. 27.6. Water recycled from cooling ponds is utilized as the major source of process water. While passing through the cooling pond, the process water loses much of the absorbed fluorides through reaction with other ionic species in the water, such as calcium, followed by precipitation. The long-term, steady-state fluoride-ion concentration is in the range from 0.5 to 1.0 wt. per cent. The equilibrium concentration of gaseous fluoride in the air in contact with this liquid is 1–2 ppm depending on temperature.[13] The theoretical minimum emission rate, assuming complete fluoride collection and about 60,000 SCF of fluoride-containing gases to be scrubbed per ton of P_2O_5 processed, is on the order of 0.025 pounds of fluoride per ton of P_2O_5. Existing state regulations* allow 0.3 to 0.4 lb/ton of P_2O_5 processed. As noted previously, pond water has a finite concentration

*Florida, Montana.

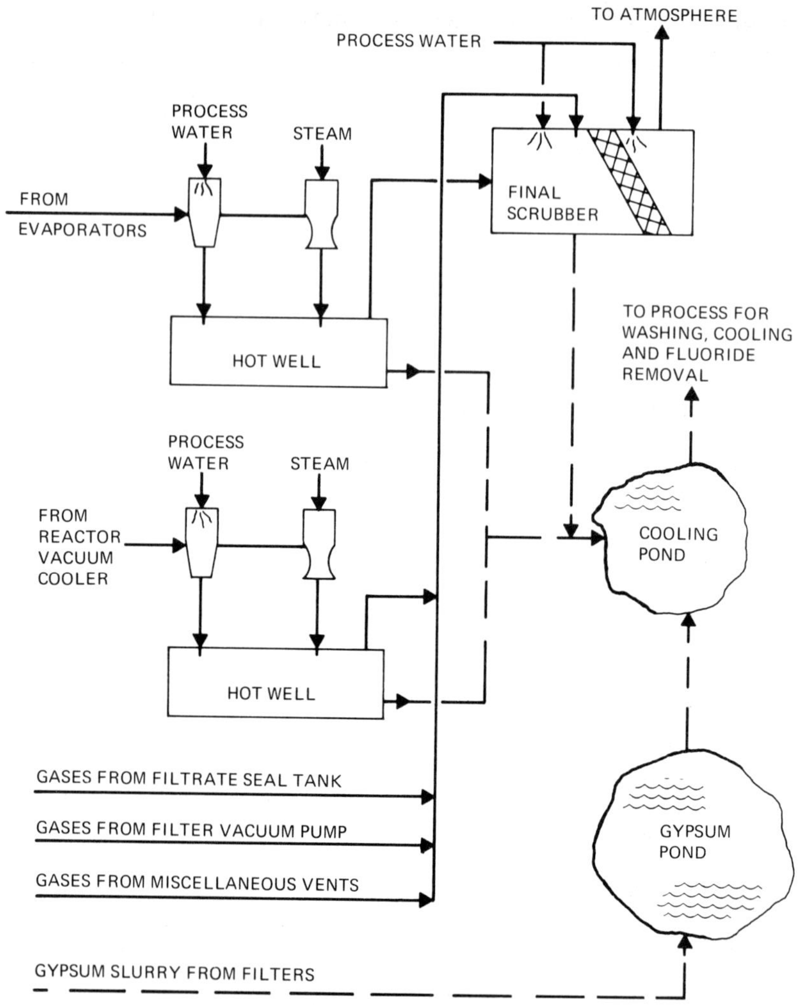

Fig. 27.6 Air pollution control scheme for wet-process phosphoric acid plants.

of fluoride vapor in equilibrium with it. An analysis of the available data[14] disclosed that wet-acid cooling and gypsum ponds might emit on the average 5 lb of fluorides/acre-day (3 lb/ton P_2O_5 processed to 54 per cent acid). No regulation covers these ponds at present. Both pond emission and reduced plant emission are areas in which future regulations can be expected.* The cost impact of such regulations is discussed in a later paragraph.

*The state of Florida proposed emission limits from wet-process phosphoric acid plants of 0.02/lb F/T P_2O_5 about December 1971. Pond emissions were not covered.

As in the nitric acid case study, before control costs can be defined, process and A.P.C. equipment must be separated. Neither the reactor or the evaporator can be operated without the barometric condensers associated with them; therefore, the condensers are process equipment even though a large portion of the gaseous fluorides evolved are reabsorbed in them. Fluoride and gypsum ponds are necessary to prevent fluoride water pollution. Since these ponds would be necessary to contain process water such as the barometric-condenser discharges, the ponds cannot be considered primarily A.P.C. devices. The final scrubber, along with a portion of the process-

TABLE 27.8 Phosphoric Acid Manufacturing Costs with and Without Control

Capital Costs 4th Quarter 1971	*Without Control*	*With Control*
Battery limits	$\$6.72 \times 10^6$	$\$6.72 \times 10^6$
Offsites	2.01	2.01
Pollution control	—	.27
TOTAL	8.73	9.00
Production Costs	*$/T* P_2O_5	*$/T* P_2O_5
Variable costs		
Phosphate rock (3.48 T of 30% P_2O_5 rock)		
at $5.52/T	19.20	
at $5.60/T		19.50
Sulfuric acid (2.56 T of 100% acid/T P_2O_5)		
at $11.10/T	28.50	
at $12.50/T		32.10
Power ($0.01/Kw-h)		
250 Kw-h/T	2.50	
260 Kw-h/T		2.60
Steam at $1.20/T		
1.15 T/T	1.36	1.36
Water-cooling process at $\$0.03/10^3$ gal/T 35×10^3 gal/T	1.05	1.05
Semifixed costs		
Labor 5 men/shift at $4.25/hr	0.73	0.73
Supervision 1 time at $15,000/yr	0.11	0.11
Maintenance 5% capital investment	1.89	1.96
Payroll added costs 30% labor and supervision	0.25	0.25
Fixed costs		
General overheads (50% semifixed)	1.49	1.52
Taxes, insurance, depreciation	4.40	4.52
TOTAL manufacturing cost $/T	61.48	65.70
Transfer cost—assuming a 5-year capital payback	68.90	73.40

water circulating system, then, is the A.P.C. system needed to meet present requirements. The cost impact of this control is shown in Table 27.8. In this table, sulfuric acid and phosphate rock were charged to the uncontrolled and controlled processes at different prices to illustrate the pyramiding effect of pollution control on costs. Direct pollution control, based on the discussion in the previous paragraph, adds about $0.60/T to the manufacturing cost; pollution control on the rock grinding and drying and the sulfuric acid plants add $3.60/T to the cost.

Table 27.9 illustrates the potential cost impact of the more stringent air pollution control which may be applied to the industry. The estimated control cost, which is a factor of 5 greater than present costs, is based on reducing plant emission to 0.1 lb F/T P_2O_5 via more efficient scrubbing and pond emissions to 0.15 lb F/T P_2O_5 by liming the pond to a pH of 4 to 5. Lime accounts for about 70 per cent of the costs as presented here. Unfortunately, essentially no data are available in the open literature as to the necessary lime-use rates so this analysis must be considered speculative.

The foregoing case study has illustrated the following points:

1. Long range profit evaluations should be concerned with what might be done in A.P.C. as well as what is being done.

2. A.P.C. costs incurred by raw materials must be recognized in any economic study.

3. Proposed solutions to A.P.C. problems must be evaluated to make sure that the problem is solved, not shifted to some other source.

TABLE 27.9 Potential A.P.C. Costs for Wet Process Phosphoric Acid Manufacturing Subject to More Stringent Regulations

Capital Investment (A.P.C. for 700 T/D Unit)
Total Plant $0.56 × 10^6

Operating Costs (A.P.C. Control Equipment Only)	*$/Yr*
Variable costs	
Power 25 Kw-h/T[a] at $0.01/Kw-h	58,000
CaO 0.116 T/T[b] at $16/T	416,000
Waste disposal[c]	0
Semifixed costs	
Labor 1 man-daylight at $4.25/hr	12,000
Supervision (no additional)	0
Maintenance (5% capital)	28,000
Payroll added costs (30% labor and supervision)	4,000
Fixed costs	
General overheads (50% semifixed)	22,000
Taxes, insurance, depreciation	64,000
TOTAL A.P.C. costs	604,000
TOTAL A.P.C. costs $/T	2.60

[a]All uses are per T P_2O_5.
[b]A lime-use rate twice the stoichiometric amount necessary to form CaF_2 with the evolved gaseous fluorides was assumed.
[c]Solid wastes are slurried to gypsum pond; disposal costs included in power costs.

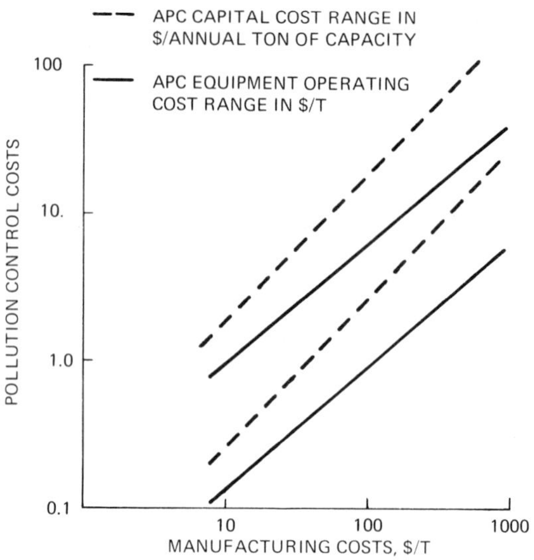

Fig. 27.7 Pollution control costs versus production cost.

Generalized Control Costs

Figure 27.7 shows the capital cost of air pollution control equipment plotted against the uncontrolled-process manufacturing cost; control-equipment operating costs are plotted against the same independent variable. The data on which the plots were based are taken from a number of sources.[12,15,16,17] The wide correlation range is due in part to different design philosophies, different by-product credits, different methods of handling the cost of money, and different manufacturing rates. These correlations exist because the same things that drive the manufacturing costs up—process complexity, corrosiveness, low production rates—tends to drive the control costs up. If the graph is used for preliminary or scope cost estimates, values should be chosen from the higher portion of the ranges. This recommendation is made for two reasons:

1. The cost estimates that this writer has examined in detail and found that he can agree with tend to lie in the higher-cost end of the ranges.

2. The cost of control is escalating faster than other costs, either because early optimism is being worked out of estimates, or because earlier estimates incompletely considered the problem.

A striking example of this escalation is brought out by comparing the 1969 and 1970 EPA reports to Congress.[15,16] The estimated average-annual-control cost for 15 industries jumped 145 per cent; only four decreases were recorded.

PRODUCT MIX CHANGES

While previous sections have discussed the changes that A.P.C. may effect in chemical manufacturing processes, this one is concerned with the change APC may make in the chemicals manufactured.

1,1,2-Trichloroethylene (tri) and 1,1,1-trichloroethane (1,1,1) are both used primarily for metal degreasing; ninety per cent of all tri and 95 per cent of all 1,1,1 manufactured are used for this purpose. Ninety per cent or more of the solvent-degreasing market is held by these two chemicals. Traditionally, tri is the favored material because it is cheaper, has superior properties (e.g., higher boiling point), and its corrosion-inhibitor system is better developed.

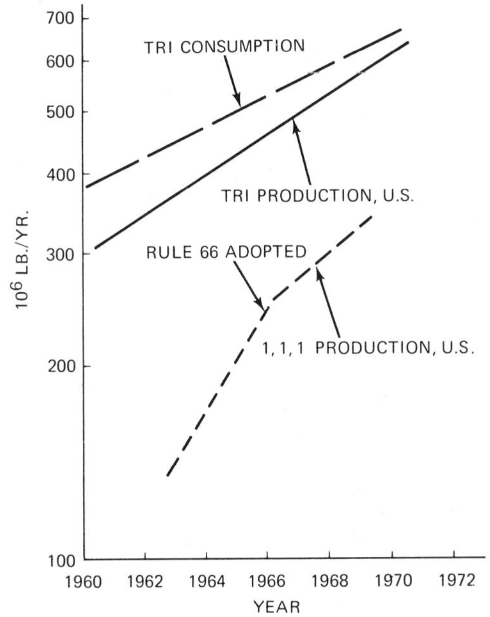

Fig. 27.8 Growth trends for trichloroethylene (tri) and 1,1,1-trichloroethane (1,1,1).[19]

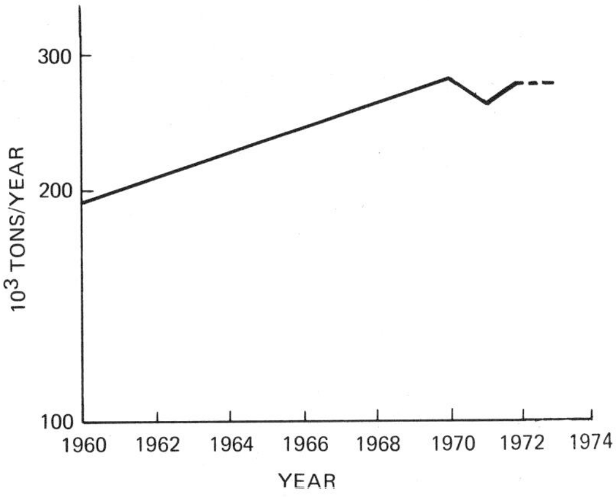

Fig. 27.9 Use of lead in lead alkyl manufacture.[17,20]

In 1966, Los Angeles[18] listed tri among the photochemically active solvents, and therefore, it was subject to emissions limitations; 1,1,1 was considered unreactive. The A.P.C. regulations required tri users to add vapor-recovery equipment or change to 1,1,1, an exempt solvent. Los Angeles' action and the adoption of similar rules by other localities would, it was feared, force tri out of its traditional market. While local markets may in truth have suffered from restraints on tri emissions, the national sales picture as shown in Fig. 27.7 has not been affected. Tri production and sales (the difference being imports) do not show a decrease in the growth rate after 1969. Neither has 1,1,1-trichloroethane production shown the growth rate increase that would be expected if it were increasing its penetration into metal degreasing. Therefore, the available evidence does not show that solvent-emission restrictions have affected the balance between 1,1,1 and tri.

Concern over lead emissions from automobiles began to grow in the late 1960's. The effects of this concern and impending legislation* on the market for lead-based anti-knock compounds–tetramethyl lead and tetraethyl lead–is not clear since there has not been enough time to develop long term trends. However, the available information, Fig. 27.9, shows a leveling and perhaps downward trend in the amount of lead consumed in the manufacture of these compounds. EPA projections of the use of these compounds confirm the trend.

In summary, then, it seems likely that some chemicals will be driven from the market place in the future due to air pollution control regulations, although demonstrating that air pollution has affected markets in the past is difficult. A second tentative conclusion may be drawn from the two case studies–chemicals that are part of widely dispersed emissions (or perhaps consumer products) are more likely to be affected by pollution regulations than chemicals that are part of less widely dispersed emissions.

---

*At this writing legislation controlling the lead content of gasoline is being considered because lead tends to render ineffective certain types of pollution reducing afterburners on automobile exhausts, not because of its toxic properties.

REFERENCES

1. Clean Air Amendments of 1970, PL 91-604.
2. "Title 40–Protection of Environment," *Federal Register*, **36**, No. 228, 22369-22480 (Nov. 25, 1971).

3. "Requirements for Preparation, Adoption, and Submittal of Implementation Plans," *Federal Register*, **36,** No. 158, 15486–15506 (Aug. 14, 1971).
4. "National Primary and Secondary Ambient Air Quality Standards," *Federal Register*, **36,** No. 84, 8186–8201 (Apr. 30, 1971).
5. "Standards of performance for New Stationary Sources," *Federal Register*, **36,** No. 247, 24875–24876 (Dec. 23, 1971).
6. "National Emission Standards for Hazardous Air Pollutants," *Federal Register*, **30,** No. 234, 23239–23256 (Dec. 7, 1971).
7. "Engineering Analysis of Emissions Control Technology for the Sulfuric Acid Manufacturing Industry," PB Number 190–39 (Bibliography 190–471), National Technical Information Service, U.S. Dept. of Commerce, Washington, D.C.
8. "Background Information for Proposed New Source Standards," Environmental Protection Agency, Office of Air Programs, Report Number APTD-0711, Aug. 19, 1971.
9. Newman, D. J., "Nitric Acid Pollutants," *Chem. Eng. Progr.*, **67,** No. 2, 79–84 (Feb. 1971).
10. Adlhart, O. J., Hindin, S. G., and Kenson, R. E., "Processing Nitric Acid Tail Gas," *Chem. Eng. Progr.*, **67,** No. 2, 73–78 (Feb. 1971).
11. Anderson, H. C., Green, W. J. and Steele, D. R., "Catalytic Treatment of Nitric Acid Plant Tail Gas," *Ind. Eng. Chem.*, **53,** No. 3, 197f (1961).
12. "Engineering and Cost Effectiveness Study of Fluoride Emissions Control," Vol. I, PB 207–506, Biography APTD0944, National Tech. Info. Service, U.S. Dept. of Commerce, Washington, D.C.
13. Tatera, B. S., "Parameters Which Influence Fluoride Emissions from Gypsum Ponds" (Univ. of Florida, PhD, 1970), University Microfilms Number 71-275, Ann Arbor, Mich.
14. King, W. R., Robinson, J. M., "Fluoride Emissions from Wet Phosphoric Acid Plant Gypsum and Cooling Ponds," paper presented at 164th Annual Meeting of the American Society, Division of Fertilizer and Soil Chemistry, August 1972.
15. "The cost of Clean Air," Second Report of the Secretary of Health, Education and Welfare to the Congress of the United States, March 1970, U.S. Government Printing Office. Washington, D.C.
16. "The Economics of Clean Air," Report of the Administrator of the Environmental Protection Agency to the Congress of the United States, March 1971, U.S. Gov. Printing Office, Washington, D.C.
17. Various private communications.
18. "Rules and Regulations, Rule 66," Air Pollution Control District, County of Los Angeles (adopted July 28, 1966).
19. "Stanford Research Institute, Chemical Economics Handbook."
20. Minerals Yearbook, U.S. Bureau of Mines, Washington, D.C., 1971, and 1972, 1973 "Lead Bulletins."
21. Proposed Sulfur Oxides Secondary Standards, *Federal Register*, Vol. 38, No. 87, 11355–11356, May 7, 1973.
22. Proposed Reference Method for Determination of Nitrogen Dioxide, *Federal Register*, Vol. 38, No. 110, 15174–15180, June 8, 1973.
23. Title 40, EPA Rules and Regulations, *Federal Register*, Vol. 36, No. 228, para. 51.18, 22404, Nov. 25, 1971.

Index